Principles of Field Crop Production

FOURTH EDITION

John H. Martin
Late, of Oregon State University
& United States Department of Agriculture

Richard P. Waldren
University of Nebraska–Lincoln

David L. Stamp
Late, of Texas Technological University

PEARSON
Prentice
Hall

Upper Saddle River, New Jersey
Columbus, Ohio

Library of Congress Cataloging-in-Publication Data

Martin, John H. (John Holmes)
  Principles of field crop production / John H. Martin, Richard P. Waldren, David L. Stamp.—
4th ed.
    p. cm.
  Includes bibliographical references and index.
  ISBN 0-13-025967-5
1. Field crops—United States. I. Waldren, Richard P. II. Stamp, David Lee, 1936- III.
Title.
  SB187.U6M3 2006
  633'.0973—dc22

                                                                         2004028707

Executive Editor: Debbie Yarnell
Assistant Editor: Maria Rego
Production Editor: Alexandrina Benedicto Wolf
Production Coordination: Carlisle Publishers Services
Design Coordinator: Diane Ernsberger
Cover Designer: Ali Hohrman
Cover art: Index Stock
Production Manager: Matt Ottenweller
Marketing Manager: Jimmy Stephens

Earlier editions, by John H. Martin and Warren H. Leonard, © 1949 and 1967 by Macmillan Publishing Co., Inc.
Previous edition, by John H. Martin, Warren H. Leonard, and David L. Stamp, © 1976 by Macmillan
Publishing Co., Inc.

Pearson Education Ltd.   Pearson Education Australia Pty. Limited
Pearson Education Singapore Pte. Ltd.   Pearson Education North Asia Ltd.
Pearson Education Canada, Ltd.   Pearson Educación de Mexico, S.A. de C.V.
Pearson Education—Japan   Pearson Education Malaysia Pte. Ltd.

ISBN 0-13-025967-5
29 2021

Contents

Preface

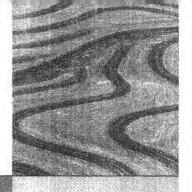

Extensive progress has been made in the science and practice of field crop production since the third edition was published in 1976. Advances in the knowledge of how plants grow and respond to their environment, a better understanding of plant processes at the molecular level, and the ability to directly manipulate chromosomal material using biotechnology are but a few examples. Additionally, improvements in the production equipment have allowed producers to manage soil resources more efficiently, provide accurate supplies of soil nutrients, and manage harvest crops more efficiently. Although more of the information has been extensively updated, some of the older, more traditional information has been retained to show how crop production has progressed over the past six decades.

All U.S. maps and production data are courtesy of the USDA National Agricultural Statistics Service unless otherwise noted. World production data are from the United Nations Food and Agriculture Organization FAOSTAT database unless otherwise noted. Special thanks to the agricultural businesses and manufacturers who provided many of the photographs in the book.

English units have been converted to metric units throughout the book, and both are provided to assist readers outside the United States. Metric values have been rounded for readability.

ACKNOWLEDGMENTS

I offer special thanks to Susan Waldren for her support and her extraordinary mastery of English grammar. Also, my thanks goes out to the following colleagues who provided insight, information, and initial review of selected chapters: Bruce Anderson, Max Clegg, Bill Fagala, Steven Mason, Dennis McCallister, Lowell Moser, Lenis Nelson, Alex Pavlista, and Leland Tripp. Thanks to the following reviewers for thier valuable feedback: E. James Dunphy, North Carolina State University; Adam Khan, Morrisville State College, NY; and Jeffery C. Wong, California Polytechnic State University.

Richard P. Waldren

Online Supplements Accompanying the Text

An online Instructor's Manual and TestGen are also available to instructors through the Martin catalog page at www.prenhall.com. Instructors can search for a text by author, title, ISBN, or by selecting the appropriate discipline from the pull down menu at the top of the catalog home page. To access supplementary materials online, instructors need to request an instructor access code. Go to www.prenhall.com, click the **Instructor Resource Center** link and then click **Register Today** for an instructor access code. Within 48 hours after registering you will receive a confirming e-mail including an instructor access code. Once you have received your code, go the site and log on for full instructions on downloading the materials you wish to use.

AGRICULTURE SUPERSITE

This site is a free on-line resource center for both students and instructors in the Agricultural field. Located at http://www.prenhall.com/agsite, students will find additional study questions, job search links, photo galleries, PowerPoints, *The New York Times* eThemes archive, and other agricultural-related links.

Instructors will find a complete listing of Prentice Hall's agriculture titles, as well as instructor supplements supporting Prentice Hall Agriculture textbooks available for immediate download. Please contact your Prentice Hall sales representative for password information.

THE NEW YORK TIMES THEMES OF THE TIMES FOR AGRICULTURE

Taken directly from the pages of *The New York Times*, these carefully edited collections of articles offer students insight into the hottest issues facing the industry today. These free supplements can be packaged along with the text.

AGRIBOOKS: A CUSTOM PUBLISHING PROGRAM FOR AGRICULTURE

Just can't find the textbook that fits *your* class? Here is your chance to create your own ideal book by mixing and matching chapters from Prentice Hall's agriculture textbooks. Up to 20% of your custom book can be your own writing or come from outside sources. Visit us at http://www.prenhall.com/agribooks.

SAFARIX: TEXTBOOKS ONLINE

SafariX Textbooks Online™ is an exciting new choice for students looking to save money. As an alternative to purchasing the print textbook, students can subscribe to the same content online and save up to 50% off the suggested list price of the print text. With a SafariX WebBook, students can search the text, make notes online, print out reading assignments that incorporate lecture notes, and bookmark important passages for later reviews. For more information, visit www.safarix.com.

PART 1 GENERAL PRINCIPLES OF CROP PRODUCTION

The Art and Science of Crop Production

▇ 1.1 CROP PRODUCTION AS AN ART

Primitive people lived on wild game, leaves, roots, seeds, berries, and fruits.[1] As the population increased, the food supply was not always sufficiently plentiful or stable to supply their needs. Crop production began at least 9,000 years ago when domestication of plants became essential to supplement natural supplies in certain localities. The art of crop production is older than civilization, and its essential features have remained almost unchanged since the dawn of history. First, people gathered and preserved the seeds of the desired crop plants. In preparing the land, they destroyed other kinds of vegetation growing on the land and stirred the soil to form a seedbed. They planted when the season and weather were right as shown by past experience, destroyed weeds, and protected the crop from natural enemies during the growing season. Finally, they gathered, processed, and stored the products.

The early farmer cultivated a limited number of crops, the cereals being among the first to be grown in most parts of the world.[2] The same crop often was produced continuously on a field until low yields necessitated a shift to new land. This temporary abandonment of seemingly partly worn-out land has been almost universal in the history of agriculture. It is still practiced in some less-developed regions of the world but is rapidly decreasing. Pressure from population growth and dwindling land resources do not allow the abandonment of land today.

The primitive farmer removed by hand the destructive insects in fields and appeased the gods or practiced mystic rites to drive away the evil spirits believed to be the cause of plant diseases. With advancing civilization, materials such as sulfur, brine, ashes, white-wash, soap, and vinegar were applied to plants to suppress diseases or insects.

Cultivated plants are a product of human achievement and discovery that has enabled people to provide food and fiber needs with progressively less labor. The first successful domestication of plants is thought[3] to have occurred in Thailand in Neolithic times. Remnants of rice and broadbeans or soybeans from 10,000 years ago have been discovered. Emmer and barley specimens dating about 6750 BC were recovered at the Jarma site in Iran. Harlan[4] lists both the Middle East and Near East as centers and noncenters of agricultural origin.

Much of the record of early agriculture comes from the writings of Greek and Roman scholars such as Herodotus about 500 BC and Pliny about 50 AD. Hieroglyphs of harvest scenes and remains of plants and seeds in ancient tombs show an Egyptian agriculture as early as 5000 to 3400 BC with the cereal grains emmer and barley of major significance. This successful agricultural economy enabled rulers to construct pyramids and beautiful tombs and develop fine arts.

Romans of the first century AD intertilled crops with iron hand knives. In 1639, Wood wrote (New England Prospect) in great detail about the skillful use of "clamme shell hooes" used in maize fields to control weeds. Intertillage with animal power was advocated in England in the seventeenth century.

The value of lime, marl, manures, and green manures for the maintenance of soil productivity was recognized 2,000 years ago. Books on agriculture written by the Romans (Pliny, Varro, and Columella) around the first century AD describe the growing of common crops including wheat, barley, clover, and alfalfa by procedures very similar to those in use today. Much of the work was done with hand labor and the farm implements used were crude.[5] However, in the experimental nursery plots of present-day agronomists, as well as in thousands of home gardens and on the small farms of many lands, one sees crops being grown and harvested using hand methods almost identical with those followed by the slaves in the Nile Valley in the time of the pharaohs 6,000 years ago.

The old art of crop production still predominates in farm practices throughout the world. Plant pathologists and entomologists have found ways to control plant diseases and insect pests more effectively. Chemists and agronomists have found supplements for the manure and ashes formerly used for fertilizers. Rotations perhaps are slightly improved. Many new crop hybrids and varieties (cultivars) have been developed. The control of weeds with herbicides began in the twentieth century.

Improved cultural methods followed observations made by primitive farmers. They found better crops in spots where manure, ashes, or broken limestone had been dropped; where weeds were not allowed to grow; where the soil was dark, deep, or well watered; or where one crop followed certain other crops. Observations or empirical trials quickly revealed roughly the most favorable time, place, and manner of planting and cultivating various crops. These ideas were handed down through the generations. Observation, the only means of acquiring new knowledge until the nineteenth century, continued to enrich the fund of crop lore. Eventually, the exchange of ideas, observations, and experiences, through agricultural societies and rural papers and magazines, spread the knowledge of crop production.

■ 1.2 CROP PRODUCTION AS A SCIENCE

Agronomy is the branch of agriculture that studies the principles and practice of crop production and field management. The term was derived from two Greek words *agros* (field) and *nomos* (to manage). Scientific research in agronomy may be said to have begun with the establishment of the first experiment station by J. B. Boussingault in Alsace in 1834 and was given further impetus by Gilbert and Lawes who established the famous research facility at Rothamsted, England, in 1843 to study fertilizer use. It was 1870 before such tests were undertaken in the United States at the land grant agricultural colleges. However, long before these landmark efforts were launched, many empirical tests had established numerous facts about crops and soils.

Agronomy has been a distinct and recognized branch of agricultural science only since about 1900. The American Society of Agronomy was organized in 1908. Agronomy had its origins largely in the sciences of botany, chemistry, and physics. Botanical writings describing crop production began with the ancient Greeks.

Theophrastus of Eresus, who lived in about 300 BC, was given the name "Father of Botany." This student of Aristotle listed plant differences that distinguish between monocots and dicots, gymnosperms and angiosperms, and described germination, development, and annual growth rings. He also mentioned "dusting" (pollination) of date palms, which first demonstrated sexuality in plants. Botanical writings continued by herbalists in monasteries and medical practitioners were concerned chiefly with the use of plants for medicinal purposes. It was not until the twelfth century that interest in plants evolved to modern systematic botany and later to other plant sciences. Chemistry had its origin in ancient mystic alchemy and in the work of people who compounded medicines. Lavoisier, often called the father of chemistry, did his work in the second half of the eighteenth century. The application of chemistry to agriculture dates from the publication of the book by Sir Humphry Davy entitled *Essentials of Agricultural Chemistry* in 1813. Physics arose from ancient philosophies. Agricultural engineering is largely applied physics.

The science of agronomy was developed by coordination of knowledge derived from the natural and biological sciences with written records of observations and empirical trials. Later, controlled experiments that dealt with crop production were conducted.

Better crop production results from the use of new knowledge, improved machines, use of pesticides, and the development of superior cultivars. Informed farmers in advanced communities quickly adopt these improvements. This is evident from the worldwide acceptance of hybrid cultivars of corn, sorghum, and sugarbeet, of dwarfed types of rice, wheat, grain sorghum, and sunflower, and of farm mechanization, fertilizers, and chemical pesticides. This results in sharp decreases in the labor expended to produce crops (Figure 1.1). Improvements in crop production from research and invention often arrive after long and painstaking trials.

Crop production today is rapidly changing from historic times. Farm machines speed up the process or enable the producer to better accomplish tasks. Electronic technology enables the producer to manage fields on a site-specific basis (Figure 1.2), and biotechnology is changing the very genetic makeup of the plants grown. There has been more advancement in crop science since the start of the twentieth century than in all of recorded history preceding it.

■ 1.3 POPULATION AND FOOD SUPPLY

Crop culture will always be an important industry because crop products are essential to the existence of humans. It has been stated that a person who goes without food for 24 hours will quarrel; one who is denied food for 48 hours will steal; and one who is without food for 72 hours will fight. Thus, the difference between peace and anarchy in most countries is a matter of only a few days without food.

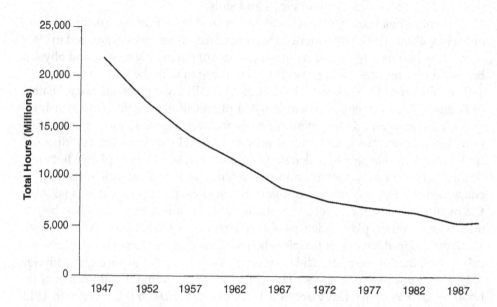

Farm Labor Hours In United States

FIGURE 1.1
Labor used to produce crops has steadily declined as farms become more efficient. [United Nations FAO Statistical Database]

FIGURE 1.2
Using Global Position System equipment on the farm for precise application of fertilizer, pesticides and herbicides. [Courtesy USDA NRCS]

1.3.1 Malthusian Theory

Aristotle and Plato agreed that the population of a civilized community should be kept within bounds. Thomas R. Malthus raised the problem of sufficient food for a population that continues to increase in a world of limited land area again in 1798.[6]

He argued that people could increase subsistence only in arithmetic progression, whereas human population tended to increase in geometric progression (i.e.,

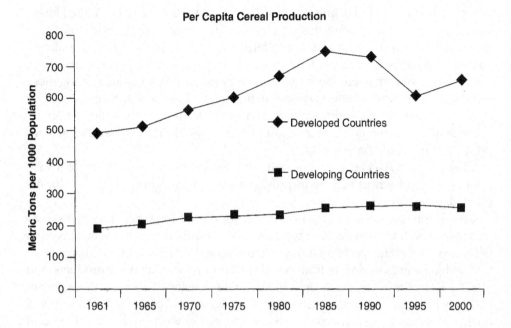

Per Capita Cereal Production

◆ Developed Countries

■ Developing Countries

Metric Tons per 1000 Population

FIGURE 1.3
Cereal production has increased faster than the population in more developed countries but has barely kept up in lesser developed countries. [United Nations FAO Statistical Database]

by the compound-interest law). So far, Malthus's predictions have not been realized because modern technology has increased production beyond levels predicted even 100 years ago. The basic check on population increase is the maximum limit set by the food supply. Population was formerly contained within that limit by war, famine, pestilence, and premature mortality. This will again occur unless the less-developed populations reduce their birth rate and increase their farm productivity.

The human population at the beginning of the Christian Era has been estimated at roughly 250 million. The aggregate world population doubled to about 500 million by 1650. It doubled again to 1 billion in only 200 years, or by 1850. The United Nations estimate of the world population only 100 years later, in 1950, was 2.4 billion people. It reached 4 billion in 1973 and 6 billion by 2000.

The production of cereal grains that comprise the base of the food supply has exceeded population growth in the United States and other advanced countries (Figure 1.3). This resulted from increased production and lower birth rates. In recent years, per capita production has declined slightly but developed countries still produce a surplus that is exported to other countries.

The Malthusian Theory has suffered setbacks since 1798.[7] Sometimes food production has outrun population growth in a few countries, following political schemes to inflate farm commodity prices. Despite this, one-third to one-half of the world population suffers at times from malnutrition, hunger, or both and one-sixth is chronically malnourished (United Nations Population Fund, 1999). Areas of the world experiencing the most rapid population increases are Latin America, Asia, Africa, and Oceania. Oceania is the only area of the four capable of excess food production on a relatively predictable basis. FAO reported that in 1998, 62 percent of the world population was in developing countries, where economic growth is slow and large majorities of the people are illiterate and poverty stricken.

Yields of grain crops have increased significantly in the United States since 1900, but there is a limit beyond which increases are impossible. Additional land for increasing production is limited. While the area of potentially arable land is three times the area harvested in any one year (The World Food Problem, White House, May 1967), more than half of this untilled land lies in the tropics. Such land usually requires clearing, drainage, heavy fertilization, new plant varieties, and pest control for successful crop production. Much of it is tropical rain forest and there is increasing concern about the long-term ecological ramifications of clearing those forests, particularly as it affects species extinction and global warming.

In Asia, there is little additional potentially arable land except by irrigation development. It is estimated that in southern Asia over 200 million additional acres (81 million ha) could be irrigated.

Crop yields in Asia are currently being increased by improved short-stem varieties of rice and wheat and by improved maize and sorghum hybrids with expanded geographic adaptation.

Cereal grains are a suitable common denominator for food production, consumption, and trade. On a calorie basis, cereals account for an average of 53 percent of human food in the world by direct consumption. Probably another 20 percent of human food comes indirectly from cereals in the form of meat, dairy products, and eggs.[8] Diets of most of the people in the underdeveloped countries remain appallingly inadequate, primarily due to shortages of animal proteins. The nutritional quality of the diet is usually considered unsatisfactory when more than 80 percent of the calories are derived from cereals, starchy roots, and sugar. For the world as a whole, food supplies for adequate nutrition will need to be greatly expanded.[9, 10]

1.3.2 Means for Increased Food Production

Agronomists, as well as all other agriculturists, are confronted with the problem of providing food for a world population that continues to grow at an accelerated rate. Underdeveloped countries of Asia, Africa, and Latin America are now in a deficit position with respect to food production for their own populations. World food production can be augmented by expansion of the cultivated land area or by increased yields on present agricultural land (Figure 1.4). Aside from conventional

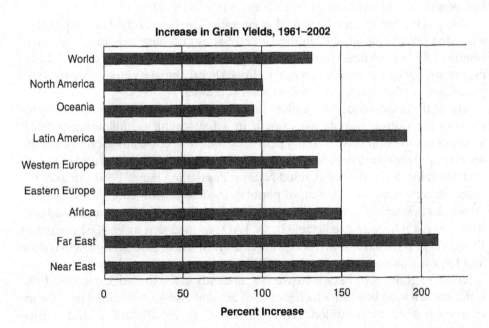

FIGURE 1.4

Grain yields have more than doubled in most parts of the world since 1961. [United Nations FAO Statistical Database]

agricultural means, food supplies may also be increased by the synthesis of foods and by the culture of certain lower plant forms such as yeast or chorella.

In traditional agriculture, new land was plowed by the peasant or farmer for increased food production. This simple means of expansion of arable land is seldom possible today. Many countries of the world are now essentially fixed-land economies.[5] Roughly 3,400 million acres (1,379 million ha) or about 10 percent of the total world land area is classified as arable land, fallow, and orchards. Another 7,400 million acres (3,000 million ha), or 19 percent of the total, is used for grazing or permanent vegetation. The remainder of the world land area, or about 70 percent, produces little or no food.[8, 11] Potential total arable land has been estimated at about 6,600 million acres (2,700 million ha).

Increased world grain yields (Figures 1.5 and 1.6) were achieved by growing more productive varieties and hybrids, greater fertilizer use (Table 1.1 and Figure 1.6), and other improvement practices. Increases in Asia, Oceania, and Latin America (Figure 1.4) are largely a result of new wheat and rice varieties and increased fertilizer use. Increased agricultural production has kept pace with population growth in many less-developed countries (Figure 1.3). However, these countries are not increasing production enough to provide the increases in food supply necessary to improve their diets.

Higher yields involve technology (applied science) plus capital. Probably 90 percent of the increase in world food production since 1950 has come from higher yields on present agricultural land. Improved crops, fertilizers, irrigation, drainage, pesticides, more effective farm implements, multiple cropping, fallow, improved cultural practices, or some combination of these can increase yields. One of the preconditions for a yield increase usually is literacy of the rural population.[5,12,13] World cereal-grain yields increased about 140 percent between 1961 and 2002 (Figure 1.4). The area of land irrigated on farms in the United States increased from about 20 million acres (8 million ha) in 1944 to 53 million acres (21 million ha) in 1997, with resulting increases in crop yields.

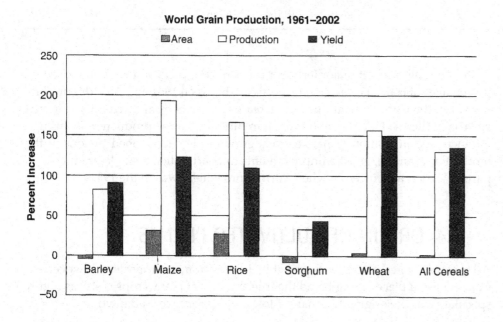

FIGURE 1.5

Changes in world production and yield of cereal crops from 1961 to 2002. [United Nations FAO Statistical Database]

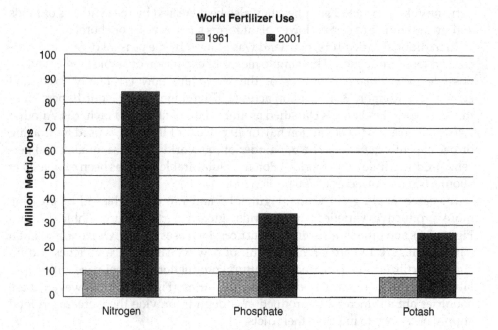

FIGURE 1.6

Comparison of world consumption of N, P_2O_5 and K_2O from 1961 to 2001. [United Nations FAO Statistical Database]

TABLE 1.1 Consumption of N, P_2O_5, and K_2O by Geographic Location (100 Metric Tons)

	N			P_2O_5			K_2O		
	1948–49 1952–53	1970–71	2001	1948–49 1952–53	1970–71	2001	1948–49 1952–53	1970–71	2001
World	43,086	316,077	819,696	60,514	206,423	330,495	45,027	165,380	227,106
Europe	18,999	96,748	136,347	25,527	81,456	40,831	25,334	74,846	48,412
North America	12,075	74,765	137,861	21,343	59,100	50,807	13,032	39,929	50,345
Latin America	1,160	14,073	51,960	616	3,464	38,941	549	6,905	35,659
Near East	938	8,003	39,665	178	1,838	13,044	51	371	3,310
Far East	6,172	40,187	428,367	2,500	12,829	159,288	1,626	12,382	78,898
Africa	325	4,752	25,057	1,344	6,856	9,248	277	2,342	5,055
Oceania	174	1,629	13,167	4,691	10,160	16,746	154	1,954	3,897

Source: FAO Statistical Database

World application of major fertilizer nutrients (N, P_2O_5, and K_2O) increased by percentages of 600, 206, and 161, respectively, between 1961 and 2001 (Figure 1.6). However, there are limits to further increases in yields from conventional inputs such as fertilizers. One possible avenue to further increase production is through the increased utilization of native crops grown in less-developed regions of the world. For example, genetic improvements to pearl millet, a staple grain crop in parts of Africa, are currently where corn improvement was in the 1930s.[14]

1.4   ORIGIN OF CULTIVATED PLANTS

All basic cultivated plants were probably derived from wild species. However, the exact time and place of origin and the true ancestry of many crops is still as highly speculative as the origin of humans. Most crop species were adapted to the needs of people before the dawn of recorded history. Evidence from archeology and ancient writings support the early origin of most present-day crop plants.[3] The centers of origin of both agriculture and culture were in populated areas favored by a more or less equable climate.[1] Vavilov[15] predicted the center of origin of a crop by finding the region where the greatest diversity of type occurred in each crop plant. However, some plants appear to have originated in two or more centers or over wide zones.[16] DeCandolle[17] concluded that 199 crops originated in the Old World and 45 in the Americas. The crop plants peculiar to the Western Hemisphere include maize, potato, sweet potato, field bean, peanut, sunflower, Jerusalem artichoke, and tobacco. Eurasia yielded wheat, barley, rye, rice, pea, certain millets, soybean, sugarbeet, sugarcane, and most of the cultivated forage grasses and legumes. Sorghum, cowpea, yam, pearl millet, finger millet, and teff were domesticated in Africa. Cotton originated in both American hemispheres.

▦ 1.5 VARIATION IN CULTIVATED PLANTS

Cultivated plants have undergone extensive modifications from their wild prototypes as a result of the continuous efforts of people to improve them. The differences between cultivated and wild forms are largely in their increased usefulness to humans, due to such factors as yield, quality, and reduced shattering of seed. Through the centuries, people selected from many thousands of plant species the few that were most satisfactory to their needs and which, at the same time, were amenable to culture. Primitive people were masters in making these selections, and modern times have added little of basic importance.

All cultivated plants were divided by Vavilov into two groups: (1) those such as rye, oats, and vetch that originated from weeds, and (2) fundamental crops known only in cultivation. Cultivated rye is believed to have originated from wild rye that even today is a troublesome weed in wheat and winter barley fields in certain parts of Asia. Oat is said to have come into culture as a weed found among ancient crops such as emmer and barley. Maize is known only in cultivation.

▦ 1.6 SPREAD OF CULTIVATED PLANTS

In their migrations, people invariably have taken their basic cultivated plants with them to insure a permanent food supply and to support their culture. This occurred in prehistoric as well as in recorded times. People also transported weeds, diseases, and insect pests along with the crops. Pre-Columbian American agriculture was based strictly on plants and animals native to America. None of the many plants involved were known in Europe or Asia prior to 1492, nor were cultivated plants native to Eurasia known in America before that time.[1]

▦ 1.7 CLASSIFICATION OF CROP PLANTS

Crop plants may be classified on the basis of a morphological similarity of plant parts.[16, 18, 19] (See Chapter 3.) From an agronomic standpoint, they may be partly classified on the basis of use, but some crops have several different uses.

1.7.1 Agronomic Classification

Cereal or Grain Crops: Cereals are grasses grown for their edible seeds. The term *cereal* can be applied either to the grain or to the plant itself. They include wheat, oat, barley, rye, rice, maize, grain sorghum, millets, teff, and Job's tears. *Grain* is a collective term applied to cereals. Buckwheat is used like a grain, but it is not a cereal. Quinoa *(Chenopodium quinoa)* is grown for its edible seeds in the Andes highlands of South America.

Legumes for Seed (Pulses): These include peanut, field bean, field pea, cowpea, soybean, lima bean, mung bean, chickpea, pigeonpea, broadbean, and lentil.

Forage Crops: Forage refers to vegetable matter, fresh or preserved, utilized as feed for animals. Forage crops include grasses, legumes, crucifers, and other crops cultivated and used for hay, pasture, fodder, silage, or soilage.

Root Crops: Crops designated in this manner are grown for their enlarged roots. The root crops include sugarbeet, mangel, carrot, turnip, rutabaga, sweet potato, cassava, and yam.

Fiber Crops: The fiber crops include cotton, flax, hemp, ramie, phormium, kenaf, and sunn hemp. Broomcorn is grown for its brush fiber.

Tuber Crops: Tuber crops include the potato and the Jerusalem artichoke. A tuber is not a root; it is a short, thickened, underground stem.

Sugar Crops: Sugarbeet and sugarcane are grown for their sweet juice from which sucrose is extracted and crystallized. Sorghum as well as sugarcane is grown for syrup production. Dextrose (corn sugar) is made from corn and sorghum grain.

Drug Crops: The drug crops include tobacco, mint, wormseed, and pyrethrum.

Oil Crops: The oil crops include flax, soybean, peanut, sunflower, safflower, sesame, castor, mustard, rape and perilla, the seeds which contain useful oils. Cottonseed is an important source of oil, and corn and grain sorghum furnish edible oils.

Rubber Crops: The only field crop grown in the United States that has been used for rubber is guayule, but other plants such as koksagyz (Russian dandelion) have been tested.

Vegetable Crops: Potato, sweet potato, carrot, turnip, rutabaga, cassava, Jerusalem artichoke, field pumpkin, and many of the pulses are utilized chiefly as vegetable crops.

1.7.2 Special-Purpose Classification

Cover Crops: Cover crops are those seeded to provide a cover for the soil. Such a crop turned under while still green would be a *green manure* crop. Important green manure crops are the clovers, alfalfa, the vetches, soybean, cowpea, rye, and buckwheat.

Catch Crops: Catch crops are substitute crops planted too late for regular crops or after the regular crop has failed. Short-season crops such as millet, sunflower, and buckwheat are often used for this purpose.

Soiling Crops: Crops cut and fed green, such as legumes, grasses, kale, and maize, are soiling crops.

Silage Crops: Silage crops are those preserved in a succulent condition by partial fermentation in a tight receptacle. They include corn, sorghum, forage grasses, and legumes.

Companion Crops: Sometimes called nurse crops, companion crops are grown with a crop such as alfalfa or red clover in order to secure a return from the land in the first year of a new seeding. Grain crops and flax are often used for this purpose.

Trap Crops: Planted to attract certain insect or phanerogamous parasites, trap crops are plowed under or destroyed once they have served their purpose.

1.8 BOTANICAL CLASSIFICATION OF CROP PLANTS

1.8.1 Method of Botanical Classification

Botanical classification is based upon similarity of plant parts. Field crops belong to the Spermatophyte division of the plant kingdom, in which seeds carry on reproduction. Within this division, the common crop plants belong to the subdivision of angiosperms, which are characterized by having their ovules enclosed in an ovary wall. The angiosperms are divided into two classes, the monocotyledons and the dicotyledons. All the grasses, which include the cereals and sugarcane, are monocotyledonous plants. The legumes and other crop plants except the grasses are classified as dicotyledonous plants because the seeds have two cotyledons. These classes are subdivided into orders, families, genera, species, subspecies, and varieties (cultivars).

1.8.2 Families of Crop Plants

Most field crops belong to two botanical families, the grasses *(Poaceae)*, and the legumes *(Fabaceae)*.

THE GRASS FAMILY The grass family includes about three-fourths of the cultivated forage crops and all the cereal crops. They may be annuals, winter annuals, or perennials. Grasses are almost all herbaceous (small, nonwoody) plants, usually with hollow cylindrical stems closed at the nodes.[20,21] The stems are made up of nodes and internodes. The leaves are two-ranked (alternate) and parallel-veined. They consist of two parts—the sheath, which envelops the stem, and the blade. The roots are fibrous. The small, greenish flowers are collected in a compact or open inflorescence, which is terminal on the stem. The flowers are usually perfect, small, and with no distinct perianth. The grain or caryopsis may be free, as in wheat, or permanently enclosed in the floral bracts (lemma and palea), as in oat.

THE LEGUME FAMILY Legumes may be annuals, biennials, or perennials. The leaves are alternate on the stems, pedicillate, with netted veins, and mostly compound. The flowers are almost always arranged in racemes as in the pea, in heads as in the clovers, or in a spike-like raceme as in alfalfa. The flowers of field-crop species of legumes are papilionaceous or butterfly-like. The irregular flowers consist of five petals: a standard, two wings, and a keel that consists of two petals more or less united. The calyx is normally four or five toothed. The fruit is a pod that contains one to several seeds. The seeds are usually without an endosperm, the two cotyledons being thick and full of stored food. Legumes have taproots. Often, the roots have abnormal growths called nodules caused by the activities of a bacterium, *Rhizobium*, which has the ability to fix atmospheric nitrogen and supply it to the host plant. Eventually, surplus nitrogen is left in the plant residues.

The principal genera of legume field crops, all of which belong to the suborder Papilionaceae, are: *Trifolium* (clovers), *Medicago* (alfalfa, burclovers, and black medic), *Glycine* (soybean), *Lespedeza*, *Phaseolus* (field bean), *Pisum* (field pea), *Melilotus* (sweetclovers), *Vigna* (cowpea), *Vicia* (vetches), *Stizolobium* (velvetbean), *Lupinus* (lupines), *Crotalaria*, *Lotus* (trefoils), and *Pueraria* (kudzu).

OTHER CROP FAMILIES Among the other botanical families that contain crop plants are: *Cannabaceae* (hops and hemp), *Polygonaceae* (buckwheat), *Chenopodiaceae* (sugarbeet, mangel, and wormseed), *Cruciferae* (mustard, rape, canola, and kale), *Linaceae* (flax), *Malvaceae* (cotton), *Solanaceae* (potato and tobacco), and *Compositae* (sunflower, Jerusalem artichoke, safflower, and pyrethrum).

1.8.3 Binomial System of Nomenclature

In a botanical classification, each plant species is given a binomial name. Binomial names are sometimes called botanical names or scientific names. This provides two names for a plant: the genus and species. The binomial system of nomenclature is founded upon the 1753 publication of *Species Plantarum* by Swedish botanist Carl Linneaus. A letter or abbreviation indicates the name of the botanist who first proposed the accepted name. For example, the letter *L* following the botanical name of corn or maize (*Zea mays* L.) means that Linneaus named it (see Appendix Table A-1). Letters and abbreviations associated with binomial names have been omitted in this book to improve readability.

The binomial system provides a practically universal international designation for a plant species, which avoids much confusion. Some crops (e.g., proso millet and roughpea) are known by several different common names in the United States but are immediately identifiable by their botanic names. Corn is called maize in most of the world outside North America.

A species is a group of plants that bear a close resemblance to each other and usually produce fertile progeny when intercrossed within the group. Nearly every crop plant comprises a distinct species or, in some cases, several closely related species of the same genus. Within a species, the plants usually are closely enough related to be interfertile. Interspecies crosses are infrequent in nature, but many of them have been made artificially.

Varietal (cultivar) names are sometimes added to the species name to make them trinomial, but ordinarily crop varieties are given a common name or serial number to designate them. Classifications of agricultural varieties have done much to standardize variety names. The American Society of Agronomy has adopted a rule to use a single short word for a variety (cultivar) name, and the variety is not to be named after a living person. That society also registers properly named improved varieties of several field crops.

1.8.4 Life Cycles

The life cycles of plants provide us with a simple yet universal means of plant classification. All higher plants can be classified as summer annuals, winter annuals, biennials, or perennials. A summer annual is a short-lived plant that completes its entire life cycle from seed to seed in a single growing season and then dies. A winter annual utilizes parts of two growing seasons in completing its life cycle. Winter annuals are planted in the fall and are vernalized during the fall and winter after which they produce seed and die the following summer. A biennial on the other hand normally utilizes two complete growing seasons to complete its life cycle. Vegetative growth occurs during the first season resulting in a rosette form of growth. This is then followed by flowering and fruiting, a process known as bolting, which occurs in a second growing season. A stalk emerges from the center of the rosette and a flowering inflorescence forms on the terminal end of the stalk. Perennials have an indefinite life period. They do not die after reproduction but continue to grow indefinitely from year to year.

▓ 1.9 THE LEADING FIELD CROPS

World production of the principle field crops (exclusive of forage crops) is shown in Table 1.2 and Figure 1.7. Wheat, rice, corn (maize), soybean, barley, sorghum, and millet occupy the largest areas. These seven crops account for approximately 75 percent of the total. The acreage, yield, and production of the different field crops in the United States and Canada are shown in Tables 1.2, 1.3, and 1.4. Corn, wheat, soybean, alfalfa, and sorghum occupy the largest acreages of specific crops. Soybean, grain sorghum, corn, wheat, and alfalfa production has increased markedly since about 1940. The production of buckwheat, syrup sorghum, cowpea, and broomcorn continues to decline. Former domestic crops that apparently are no longer grown include fiber flax, teasel, hemp, and chicory.

The total land area in the United States is about 2,264 million acres (917 million ha). Less than half of this area, 932 million acres (377 million ha), was in farms in 1997. Of this, 431 million acres (174 million ha) were classed as cropland (Table 1.5). The total area for grazing was 492 million acres (199 million ha), not counting the harvested cropland in grains and forages that were grazed for part of the season or after harvest. There has been a steady decline in farmland over the years from expansion of urban areas, roadways, and other uses (Table 1.5). According to the USDA's 1997 National Resources Inventory, an average of 3.2 millions acres (1.3 million ha) of forest, cropland, and open space were converted annually to urban and other uses from 1992 to 1997. An average of 1.4 million acres (567,000 ha) was converted from 1982 to 1992. Texas and Pennsylvania had the biggest development rate increases, followed by Georgia, Florida, North Carolina, California, Tennessee, and Michigan.

TABLE 1.2 Area, Production, and Yield of the Principle Crops, Except Forages, in the World and the United States in 1999 (FAO Statistical Database)

Crop	Area Hectares (Millions)		Production (Million Metric Tons)		Yield (Metric Tons per Hectare)	
	World	USA	World	USA	World	USA
Wheat	213.8	21.9	583.9	62.8	2.7	2.9
Rice	153.1	1.5	588.8	9.6	3.8	6.6
Maize	139.9	28.7	604.4	242.3	4.3	8.4
Barley	57.6	1.9	133.6	6.1	2.3	3.2
Sorghum grain	45.9	3.4	68.1	15.2	1.5	4.4
Millets	37.0	0.1	28.6	0.2	0.8	1.5
Oats	13.8	1.0	26.1	2.1	1.76	2.1
Rye	10.0	0.2	20.3	0.28	2.0	1.8
Buckwheat	2.7	0.04	2.8	0.04	1.0	1.1
Soybeans	71.9	29.5	154.9	72.8	2.1	2.4
Peanuts (in shell)	24.8	0.6	32.8	1.7	0.9	2.9
Dry beans	26.9	0.8	19.4	1.4	0.7	1.9
Dry peas	6.1	0.1	11.3	0.3	1.9	2.1
Cowpeas	7.5	0.0	3.2	0.0	0.4	0.9
Lentils	3.4	0.07	3.0	0.1	0.8	1.5
Vetch	1.1	0.0	1.0	0.0	0.9	
Lupins	1.5	0.0	1.5	0.0	1.1	
Cottonseed	33.9	5.4	54.7	9.3	1.6	1.7
Rapeseed (canola)	27.1	0.4	42.2	0.6	1.5	1.4
Sunflower seed	24.5	1.5	28.8	2.3	1.2	1.6
Linseed	3.5	0.5	2.7	0.2	0.8	1.1
Sesame seed	6.2	0.0	2.4	0.0	0.4	
Castor	1.3	0.0	1.2	0.0	0.9	0.0
Tobacco	4.6	0.3	7.0	0.57	1.5	2.1
Potatoes	17.6	0.6	288.5	21.8	13.3	37.3
Sweet potatoes/yams	9.0	0.03	156.4	0.6	22.9	17.1
Sugarcane	19.6	0.4	1276.9	34.1	65.3	85.3
Sugarbeets	6.8	0.6	259.8	30.3	38.1	49.1

The labor required in 1971 to produce the major field crops, except tobacco, was only ⅙ to ⅑ of that required in the 1910–1914 period. Labor has continued to decrease as farms get larger and more mechanized (Figure 1.1). Yields of crops have increased tremendously since 1900 as a result of increased fertilizer and pesticide use, more irrigation, more productive cultivars, improved cultural practices, timelier field operations, and shifts of many of the crops to more productive areas or to areas where they could be produced more economically with less labor.

The use of chemical pesticides is not the only method for controlling diseases and insects. Breeding crops for resistance to diseases began after 1890 and is a much expanded and necessarily continuous enterprise today. Selection and breeding for resistance to insects has been in progress for about 80 years. Sex

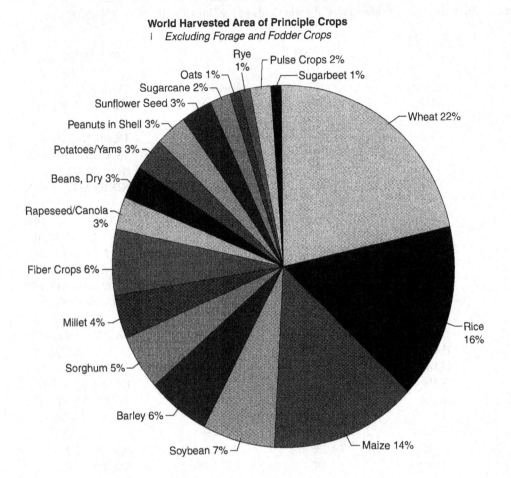

World Harvested Area of Principle Crops
| *Excluding Forage and Fodder Crops*

Rye 1%
Oats 1%
Sugarcane 2%
Sunflower Seed 3%
Peanuts in Shell 3%
Potatoes/Yams 3%
Beans, Dry 3%
Rapeseed/Canola 3%
Fiber Crops 6%
Millet 4%
Sorghum 5%
Barley 6%
Soybean 7%
Pulse Crops 2%
Sugarbeet 1%
Wheat 22%
Rice 16%
Maize 14%

FIGURE 1.7
Wheat, rice, and maize are grown on over one-half of the cropland in the world. [United Nations FAO Statistical Database]

attractants are being developed to permit sterilization of male insects by radiation. Other attractants are used as baits for trapping insects. Natural enemies of insect pests and disease organisms and viruses may help to control some pests. Natural defenses are being incorporated into existing hybrids and cultivars using biotechnology. It will be many years, if ever, before pesticides can be replaced for protecting all farm crops. Chemicals are not the only hazard to mankind. Sunshine, in excess, can cause sunburn, skin cancer, and sunstroke, but we cannot live without it.

Destructive insects were a problem before the beginning of agriculture in an area. The following item, dated July 19, 1806, appears in the diary of Captain Wm. Clark while in Montana en route east from the Pacific Coast: ". . . emence sworms of grass hoppers have destroyed every sprig of grass for many miles on this side of the [Gallatin] river."[22]

Even as late as 1951 to 1960, losses in various crops ranged from 3 to 28 percent from plant diseases, 3 to 20 percent from insects, and 3 to 17 percent from weed competition. This justified greater pesticide use. Current fertilizer use has not restored the fertility levels of most of the virgin soils. Water pollution has been present since the earth was created, as is evident from the considerable amounts of mercury present deep in the Greenland icecap. Many other countries are modernizing their agriculture along lines similar to those in North America.[9, 10]

TABLE 1.3 Acreage, Production, Yield, and Principle States of Farm Crops in the United States in 2000–2003

Crop	Harvested Acreage (1,000)	Acre Yield	Production (1,000)	Leading States
Corn				
Grain (bushel)	70,419	137	9,624,576	IA, IL, NE, MN, OH
Silage (ton)	6,469	16	103,000	WI, NY, CA, PA, MN
Sorghum				
Grain (bushel)	807	56	439,129	TX, KS, NE, AR, MO
Silage (ton)	345	10	3,564	TX, KS, SD, NE, AZ
Oat (bushel)	2,130	62	131,855	ND, MN, SD, WI, IA
Barley (bushel)	4,571	58	267,281	ND, ID, MT, WA, MN
Wheat (bushel)				
Winter	33,113	43	1,139,728	KS, OK, WA, TX, OH
Spring	13,953	36	497,000	ND, MN, MT, SD, ID
Durum	2,984	31	92,490	ND, MT, AZ, CA, SD
Rye (bushel)	287	27	7,756	OK, GA, ND, SD
Rice (pound)	3,139	6,500	20,406,475	AR, CA, LA, MS, TX
Soybean (bushel)	72,550	37	2,705,551	IL, IA, MN, IN, NE
Flaxseed (bushel)	595	19	11,119	ND, MT, MN, SD
Peanut (pound)	1,338	2,801	3,751,850	GA, TX, AL, FL, NC
Cotton (pound)	13,058	725	8,747,520	TX, GA, MS, CA, AR
Alfalfa (ton)	23,578	3.2	79,307	CA, SD, IA, MN, WI
Other hay (ton)	39,310	1.9	76,354	TX, MO, KY, TN, OK
Dry bean (pound)	1,489	1,657	2,474,500	ND, NE, MI, MN, CA
Sugarbeet (ton)	1,348	23	30,605	MN, ID, ND, MI, CA
Sugarcane (ton)	998	35	34,503	FL, LA, HI, TX
Sunflower (pound)	2,392	1,255	3,019,915	ND, SD, KS, MN, CO
Canola (pound)	1,324	1,340	1,773,399	ND, MN
Buckwheat (bushel)	161	19	2,985	ND, WA, MN, NY, PA
Potato (100 wt)	1,250	367	459,045	ID, WA, WI, ND, CO
Proso millet (bushel)	463	21	10,460	CO, NE, SD

Source: USDA National Agricultural Stastistics Service

■ 1.10 FUTURE INCREASES IN PRODUCTIVITY

"Give a man a fish and he can live for a day; teach him to fish and he can live forever!" This philosophy, from an ancient Chinese proverb, is reflected in the growing realization that less-developed countries of the world must have the desire to succeed in modernizing agriculture. Leaders of these nations must make an effort to provide the multiple factors involved in modernizing agriculture with help, where available and necessary, from outside.

Several promising research developments are pointing toward continued future increases in productivity:

1. Collection of plant species and broadening of germplasm bases
2. Improved breeding techniques including the use of biotechnology

TABLE 1.4 Acreage, Production, and Yield of Crops in Canada in 2002 (FAO Statistical Database)

Crop	Harvested (1,000 Ha)	Production (1,000 Mt)	Yield (Kg/Ha)
Wheat	8,836	16,198	18,332
Barley	3,348	7,489	22,369
Oats	1,379	2,911	21,110
Rye	77	134	17,442
Mixed grains	132	359	27,231
Corn for grain	1,283	8,995	70,111
Corn for forage	220	6,356	289,163
Soybeans	1,024	2,335	22,811
Buckwheat	12	12	10,252
Rapeseed	3,262	4,178	12,810
Mustard seed	255	154	6,056
Sunflower seed	95	157	16,638
Sugarbeets	10	345	341,287
Potatoes	171	4,697	274,816
Linseed	633	679	10,726
Canary seed	214	164	7,645
Beans, dry	215	225	18,930
Chickpeas	154	157	10,182
Lentils	561	354	9,144
Broad beans, dry	5	9	17,500

TABLE 1.5 Land Utilization on Farms in the United States in 1969 and 1997 (USDA NASS)

Number of Farms	1969 2,858,051	1997 1,911,859
	Million Acres	
Cropland		
Harvested	300	309
Fallow or crop failure	36	21
Crop failure	NA	3
Cover crops or idle	51	33
Pastured	88	65
Total	475	431
Woodland—not grazed	50	42
Farmsteads, roads, waste	25	32
Grazed area		
Grassland pasture	452	397
Woodland pasture	62	30
Conservation or wetlands reserve	NA	30
Total land in farms	1,176	932

3. Crop physiology advances in understanding plant efficiency in utilization of light and nutrients

4. Interactions between plant breeders, geneticists, soil chemists, engineers, etc., are increasingly bringing together the best of many disciplines

5. Extension of areas of adaptation of previously restricted varieties

6. Improved disease and insect control with advances in chemical and biological methods

7. Improved protein quality factors such as high lysine corn

8. Improved medication and breeding principles in livestock

9. Reduced production costs of nitrogen fertilizer

10. Improved winter hardiness of varieties to extend crop production

These are but a few of the areas in which progress is being made. A combination of enhanced agricultural production and population control to achieve subsistence must be a primary goal of all peoples of the world.

REFERENCES

1. Merrill, E. D. "Domesticated plants in relation to the diffusion of culture," *Bot. Rev.* 4 (1938): 1–20.

2. Heiser, C. B., Jr. *Seed to Civilization*, San Francisco: Freeman, 1981.

3. Flannery, K. V. "The Origins of Agriculture." *An. Rev. of Anthropology.* 2: 271–310. An. Reviews Inc., Palo Alto, Cal. 1973.

4. Hitchcock, A. S., and Chase, A. "Manual of the grasses of the United States," *USDA Misc. Pub.* 200, 1950, pp. 1–1051.

5. Brown, L. R. "Increasing world food output," *USDA Foreign Agricultural Economic Rpt.* 25, 1965, pp. 1–140.

6. Malthus, T. R. *An Essay on the Principle of Population, 1798–1803.* Reprint New York: Macmillan, Inc., 1929, pp. 1–134.

7. Borlaug, N. E. *Mankind and Civilization at Another Crossroad.* Madison, Wis.: Wise. Agri-Business Council, Inc., 1971, pp. 1–48.

8. Brown, L. R. "Man, land, and food," *USDA Foreign Agricultural Economic Rpt.* 11, 1963, pp. 1–153.

9. Aldrich, D. C. *Research for the World Food Crisis.* Washington, D.C.: AAAS, 1971, pp. 1–320.

10. *Agricultural Research: Impact on Environment.* IA State University Spec. Rpt. 69, 1972, pp. 1–84.

11. Cook, R. C. "Population and food supply," *Freedom from Hunger Campaign, FAO Basic Study,* 7 (1962): 1–49.

12. Army, Thomas J., Frances A. Greer, and Anthony San Pietro. *Harvesting the Sun; Photosynthesis in Plant Life.* New York: Academic Press, 1967.

13. Columella, L. Junius Moderatus. *Of Husbandry* (translated into English with several illustrations from Pliny, Cato, Varro, Palladius, and other ancient or modern authors), London, 1797, pp. 1–600.

14. National Research Council. *Lost Crops of Africa, Vol. I: Grains.* Washington, D.C.: National Academy Press, 1996.

15. Vavilov, N. I., and D. Love. *Origin and Geography of Cultivated Plants.* London: Cambridge University Press, 1992.

16. Harlan, J. R. "Agricultural origins: Centers and non-centers," *Science* 174, 4008 (1971): 468–74.

17. De Candolle, A. *Origin of Cultivated Plants*, 2nd ed. New York: Hafner, (reprint), 1959, pp. 1–468.

18. Hitchcock, A. S. *A Text Book of Grasses*, New York: Macmillan, Inc., 1914, pp. 1–276.

19. Pawley, W. H. "Possibilities of increasing world food production," *Freedom from Hunger Campaign, FAO Basic Study* (1963): 1–231.

20. Schooley, J. *Introduction to Botany*, Stamford, CT: Delmar, 1997.

21. Theophrastus. *Enquiry into Plants.* Translated by Sir Arthur Hort. New York: Putnam, 1916.

22. Lewis, Merriwether, and William Clark. Edited by Elliott Coues. *The Journals of Lewis and Clark.* New York: F. P. Harper, 1893.

Crop Plants in Relation to the Environment

2.1 FACTORS IN CROP DISTRIBUTION

Staple agricultural crops show a marked tendency to geographic segregation despite the fact that they may grow well over wide areas. Thus, corn and oat, although concentrated in the Corn Belt, are both grown successfully in the 48 contiguous states in the United States. The principal factors that influence localization are climate, topography, character of the soil, insect pests, plant diseases, and economic conditions.[1]

Crops are generally profitable only when grown in regions where they are well adapted. The best evidence of adaptation of a crop is a normal growth and uniformly high yields.[2] Adapted crops usually produce satisfactory yields even on the poorer soils in a region.[3] The farther a crop is removed from its area of good adaptation, the more care is necessary for satisfactory production.

2.2 CLIMATE

Climate is the dominant factor in determining the suitability of a crop for a given area. Knowledge of the crops and crop varieties grown in a given region is a better measure of climate, as applied to crop production in that region, than are complete climatic records. Thus, it is known that the climate in the Puget Sound region of Washington is milder than that of eastern Maryland, and that the latter region is milder than northern Texas because winter oat survives the winter more regularly. It is also known that conditions in the winter wheat region of Sweden are not as severe as in many winter wheat regions in the United States because the Swedish varieties are less resistant to cold. Climatic counterparts of American agricultural regions are found in various parts of the world. The Great Plains region is similar to the Ukraine because the same crop varieties thrive in both places. Many crop varieties are equally adapted to Australia and California. The irrigated regions of southern California, southern Arizona, and southern Texas are comparable to irrigated sections of the Mediterranean region. The wheats of Great Britain and Holland fail miserably in the United States except in the mild, cool, humid Pacific Northwest where they thrive. The Corn Belt of the United States has a counterpart in the Danube Valley of Europe.

2.2.1 General Types of Climate

Climatic differences are due chiefly to differences in latitude, altitude, and distance from large bodies of water, ocean currents, and the direction and intensity of winds.

Continental climates, which occur in interior regions, are characterized by great extremes of temperature between day and night and between winter and summer. The temperature ranges increase, in general, with the distance from the ocean. Some of these regions, such as the steppes of Russia and the Great Plains of North America, are further characterized by an irregular approach of seasons, deficient rainfall, low humidity, and generally unobstructed winds.[4] The limited rainfall that occurs is usually sporadic and often torrential. The great wheat areas of the world are found in such climates. Other hardy crops also are grown there.

Oceanic or marine climates are more equable. Moderate temperature changes occur between day and night and between winter and summer. The sea or other large bodies of water influence the temperature changes predominantly. This influence results from differences in the specific heat of water and land. Water takes up and gives off heat only one-fourth as rapidly as does land.

Three distinct climatic regions are recognized in the United States.[3] The first is a narrow strip of territory from the Pacific Coast to the Cascade and Sierra Nevada mountains, a purely oceanic climate in which the rainfall ranges from less than 10 inches (250 mm) in southern California to over 100 inches (2,500 mm) per year in the Northwest. In this region, the winters are mild, while the summers in the northern part and along the southern coastline are cool. The second region is the upland plateau from these mountains eastward to the one-hundredth meridian. The climate is continental over most of this area. The third region is from the one-hundredth meridian, where a continental climate prevails, to the Atlantic, where conditions again are modified by the ocean. The change from one type to the other is gradual in these areas.

2.2.2 Precipitation

Rainfall has a dominant influence in crop production. In semiarid regions, such as the Great Plains and Great Basin, the conservation and utilization of the scanty rainfall is so important that it relegates all other factors, including soil fertility, to minor positions.

In regions of low rainfall, the deficiency may be partly overcome by moisture-conserving tillage and rotation practices or by irrigation. Under dry farming conditions, red clover and sugarbeet are failures and alfalfa is unprofitable except on bottomlands. When such lands are irrigated, these crops may be highly successful in the same region.

CROP AREAS BASED UPON RAINFALL Crop regions are frequently classified on the basis of average annual rainfall. It is obvious that these regions are arbitrary since the actual boundaries may fluctuate from year to year. (1) The arid region has an average annual rainfall of 10 inches (250 mm) or less. Irrigation is necessary for successful crop production in most of such areas (Figure 2.1). (2) The semiarid region is arbitrarily considered to be where the annual rainfall varies from 10 to

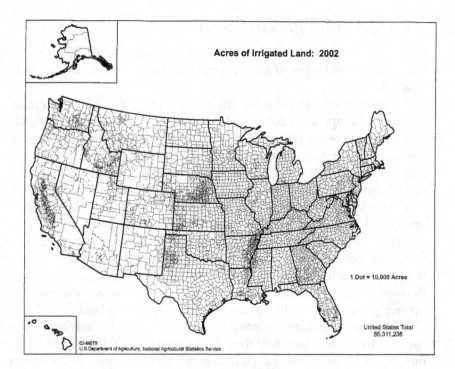

FIGURE 2.1
Irrigated land in farms. About 40 million acres of land in the United States and one-eighth of the world's cropland is irrigated. [Source: 2002 Census of U.S. Department of Agriculture]

20 inches (250 to 500 mm). Successful crop production in this region will need tillage methods that conserve moisture, crop varieties adapted to dry farming regions, or irrigation. (3) Annual rainfall in most subhumid areas varies from 20 to 30 inches (500 to 760 mm) (Figure 2.2). This amount of rainfall often is inadequate for satisfactory crop yields unless methods that utilize the rainfall to best advantage are followed in regions such as the southern Great Plains where seasonal evaporation is high. (4) The humid region is regarded as that in which the annual precipitation is more than 30 inches (760 mm). Conservation of moisture is not the dominant factor in crop production in this region.

EFFECTIVENESS OF RAINFALL The effectiveness in crop production of a given quantity of rainfall depends upon the time of year that it falls and the rapidity and intensity of individual rains. Seasonal evaporation is even more important.

The total rainfall fluctuates widely from year to year. Over a ten-year period, the annual precipitation at North Platte, Nebraska, in the Great Plains area, varied front 10 to more than 40 inches (250 to 1000 mm).[5] In general, the lines of equal rainfall from the Rocky Mountains eastward lie north and south but shift eastward as they extend to the north. In the southern states east of Texas, equal rainfall lines follow the outline of the Gulf coast.

Rainfall has its greatest value when it falls during the growing season, ordinarily between April 1 and September 30, for summer crops. The critical period for water uptake for most crops occurs just before or after flowering.[6] In corn, the ten days after tasseling and silking have an almost dominant effect on yield. In regions of winter rainfall, a greater portion of the water can be conserved by fallowing compared to regions of summer rain because it falls during a period of low evaporation. Winter rainfall regions are especially suited to the growing of winter grains,

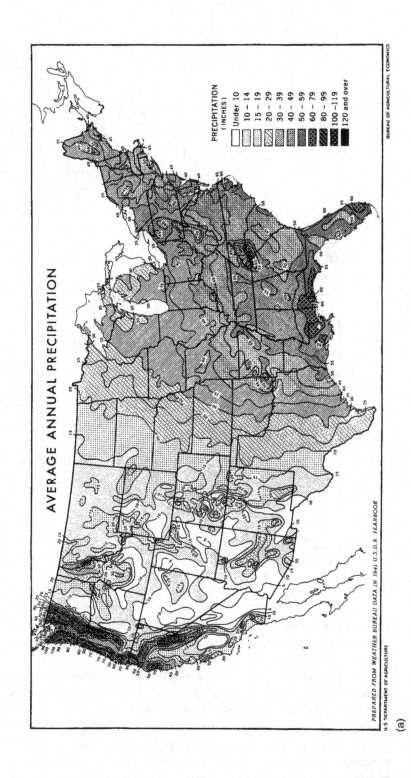

FIGURE 2.2
Precipitation in the United States. (a) Average annual precipitation. (Continued)

25

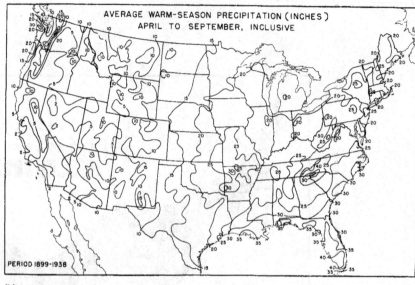

(b)

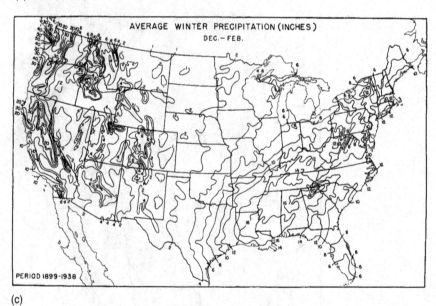

(c)

FIGURE 2.2

(Continued) Precipitation in the United States. *(b)* Average warm-season precipitation. *(c)* Average winter precipitation.

although these crops thrive in summer rainfall regions also. Corn and sorghum are poorly adapted to winter rainfall regions because their greatest demands for moisture occur during the dry, hot season.

East of the Mississippi River, the monthly precipitation distribution is rather uniform throughout the year, especially north of the Cotton Belt.[7] However, in Florida, the six months from November to May are comparatively dry and the rainfall is heavier in the summer months. On the Great Plains, approximately 70 to 80 percent of the annual precipitation falls from April to September, inclusive. Most of the precipitation in the Pacific Coast and intermountain regions comes in the winter months. In Arizona and adjacent areas, rainy periods frequently occur in both summer and winter.

Ordinarily, summer showers of less than ½ inch (13 mm) are of very little value in storing soil moisture for the use of crops unless they immediately precede or follow other rains. Light showers are largely lost by evaporation from plant foliage and soil surface. However, such showers may benefit crops by reducing evaporation and cooling the atmosphere. Frequently, rain comes in torrential showers in the Great Plains and states to the east, with the result that much of it is lost through runoff, especially on steep lands with heavy soils.

Water is the chief component of living plant tissues. It is involved in the transfer of nutrient elements from the soil into the roots and upward throughout the plants. It supplies hydrogen as a component of organic materials formed in photosynthesis. The transpiration of water cools the plants under high-temperature conditions.

2.2.3 Temperature

Temperature also is an important factor in limiting the growing of certain crops (Figure 2.3). The 50°F (10°C) isotherm of monthly mean temperature during each summer month marks the growth limits for most plants. The great crop areas have four to twelve months in which the mean monthly temperature is between 50 and 68°F (10 and 20°C). This is in the temperate belt. Temperature is influenced by latitude and altitude as well as by exposure. A difference of 400 feet (120 m) in altitude or 1 degree in latitude causes a difference of approximately 2°F (1°C) in mean annual and mean July temperatures and 3°F (1.5°C) in mean January temperatures in the United States.[8] Slopes that face south and west receive the most sunshine and are regularly warmer than slopes facing north and east. Crops can be grown at much higher mountain elevations on the warm slopes.[9]

EFFECT OF TEMPERATURE ON PLANTS Each crop plant has its own approximate temperature range (i.e., its minimum, optimum, and maximum for growth). Although they are subjected to a rather wide range of temperature, most crop plants make their best development between 60 and 90°F (16 and 32°C). They either cease growth or die when the temperature becomes either too low or too high. Temperatures of 110 to 130°F (43 to 54°C) will kill many plants. Most cool-season crop plants cease growth at temperatures of 90 to 100°F (32 to 38°C), while annual crops are killed by low temperatures that range from about 32°F (0°C) down to –40°F (–40°C). The minimum temperature at which any plant can maintain life during the growing season is the approximate freezing point of water. The threshold value, or point at which appreciable growth can be detected, is usually taken at 40 or 43°F (4 or 6°C). This value varies for different crops. Winter rye in the sun may show a sign of growth even when the temperature recorded in a shaded shelter is below freezing. Sorghum practically ceases growth at a temperature of 60°F (16°C),[10] sugarcane at 70°F (21°C), and tomatoes at 65°F (18°C). Table 2.1 lists the optimum root and shoot temperatures of several crops.[11]

Crop areas have more or less definite boundaries that extend in an east-west direction in conformity with the isothermal trend.[6] The time to plant crops is defined by the temperature conditions of the locality, but the planting time is generally later as one proceeds northward. According to the Hopkins bioclimatic law,[8] such events as seeding time vary four days for 1 degree of latitude, 5 degrees of longitude, and 400 feet (120 m) of altitude, being later northward, eastward,

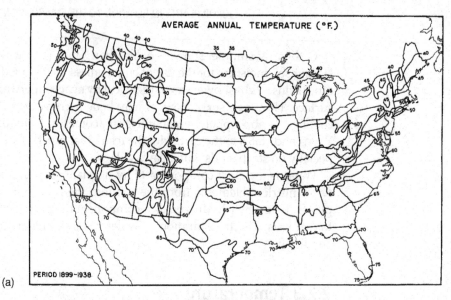

(a)

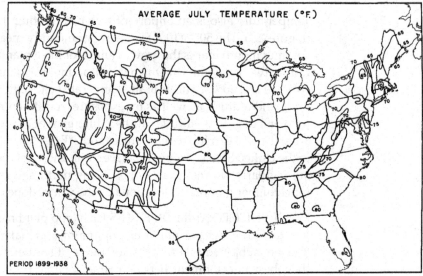

(b)

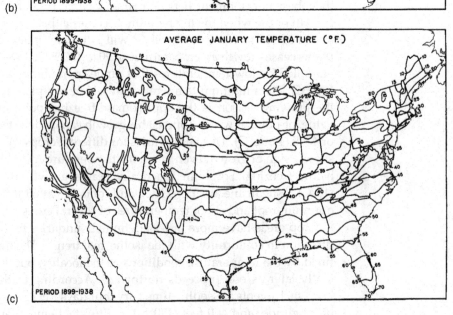

(c)

FIGURE 2.3
Temperatures in the Unted
States. *(a)* Average annual
temperature. *(b)* Average
July temperature.
(c) Average January
temperature.

TABLE 2.1 Optimum Root and Shoot Temperatures of Various Crops

Crop	Roots		Shoots	
	°C	°F	°C	°F
Alfalfa	20–28	68–82	20–30	68–86
Barley	13–16	43–55	15–20	60–69
Corn	25–30	77–86	25–30	77–86
Cotton	28–30	82–86	28–30	82–86
Oats	15–20	60–69	15–25	60–77
Rice	26–29	80–85	26–29	80–85
Wheat	18–20	64–68	18–22	64–72

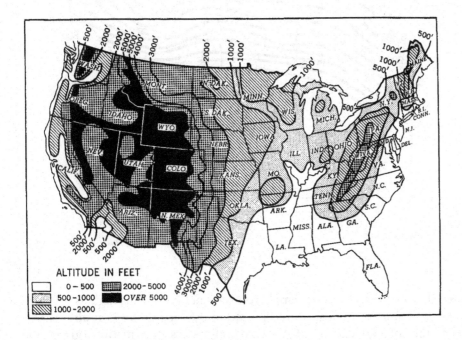

FIGURE 2.4
Approximate general altitudes in the United States.

and upward in the United States (Figure 2.4). The normal daily temperature for spring planting of a given crop is rather uniform throughout the country. The seeding of spring wheat begins when the normal daily temperature rises to about 37°F (3°C), spring oats at 43°F (6°C), corn at 55°F (13°C), and cotton at 62°F (17°C).

Within suitable temperature ranges, growth increases as the temperature increases. This increase loosely corresponds to the Van't Hoff-Arrhenius law for monomolecular chemical reactions (i.e., a doubling in activity (or growth) for each rise in temperature of 18°F [10°C]).

Crops differ in the number of heat units required to mature. Heat units, sometimes called growing degree days, are usually computed by a summation of degree days for a growing season. A degree day is the number of degrees Fahrenheit above an established minimum growing temperature such as 50°F (10°C) for corn. Thus, where the sum of maximum daily temperatures above that minimum is 3,000 during a 100-day growing season there are 3,000 total heat units (Figure 2.5). The number of heat units necessary to mature a given crop differs with the region. Increased growth is not directly proportional to increased temperatures. Also, fewer units are

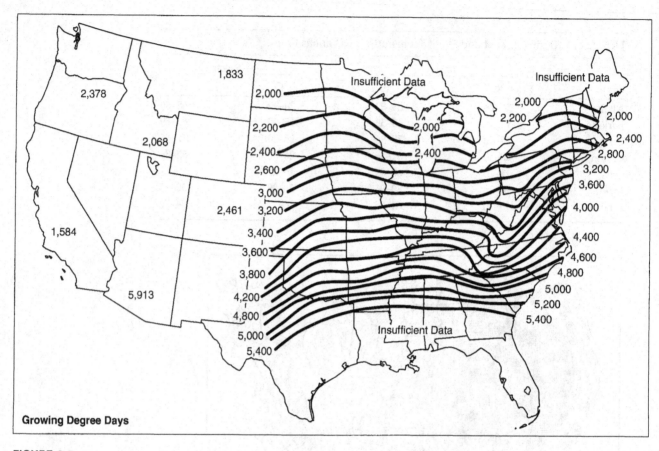

FIGURE 2.5
Growing degree day accumulation during the growing season.

required in a cool region where temperatures do not become as high as to retard plant growth. This relationship is recognized in the Linsser law—in different regions, the ratio of total heat units to the heat units necessary to mature a given crop is nearly constant.[12]

COOL VERSUS WARM SEASON CROPS Crops may be classified broadly as warm-season or cool-season crops (Appendix Table A–1). The crops that make their best growth under relatively cool conditions, but are damaged by hot weather, include wheat, oat, barley, rye, potato, flax, sugarbeet, red clover, vetch, field pea, and many grasses. Field pea, the small grains just mentioned, flax, and northern grasses may withstand freezing temperatures at any time up to about the flowering stage. The 10°F (–12°C) isotherm for daily mean temperatures for January and February coincides remarkably well with the northern boundary of the Winter Wheat Belt, beyond which spring wheat is more important.[13] The isotherms for 20 and 30°F (–7 and –1°C) coincide closely with the northern limits of winter barley and winter oat, respectively, up to about 1950 when more hardy varieties were available.

The crop plants that prefer a warm season include corn, cotton, sorghum, rice, sugarcane, peanut, cowpea, soybean, velvetbean, dallisgrass, and bermudagrass.

Warm-season crop plants, after they reach appreciable size, are killed by temperatures slightly below freezing, and sometimes at several degrees above freezing when exposed for prolonged periods.

2.2.4 Length of Growing Season

The number of days between the average date of the last spring frost and first fall frost usually is considered as the length of the growing season (Figure 2.6). A frost-free season of less than 125 days is an effective limitation to the production of most crops. Wheat, oat, and barley mature under a shorter frost-free period than corn or sorghum because wheat, oat, and barley can be sown safely two months before the average last spring frost. Cotton requires 200 days of frost-free weather, a fact that limits its growth to the South. The length of the period for growth of a crop may be limited also by drought.

2.2.5 Humidity

Humidity is water vapor in the air. Relative humidity is the vapor pressure in the air in terms of the percentage necessary to saturate the atmosphere at the particular temperature. A saturated atmosphere that causes fog, dew, or rain has a theoretical relative humidity of 100 percent. The lower the relative humidity at a given temperature the more rapidly the air will take up water transpired by the leaf or evaporated from a moist soil surface. At the same relative humidity, the vapor pressure and drying capacity at 95°F (35°C) is 9.2 times that at 32°F (0°C).

Evaporation and transpiration increase with higher temperatures and lower relative humidity. A high seasonal evaporation from a free-water surface is generally reflected in a high water requirement of crops. Evaporation determines the efficiency of rainfall, particularly under a rainfall of less than 30 inches (760 mm) per year. There is an increase in the water requirement of crops and in the seasonal evaporation from north to south in the Great Plains. In North Dakota, a ton (907 kg) of alfalfa can be produced with 500 tons (454 MT) of water, while in the warmer Texas Panhandle 1,000 tons of water (907 MT) would be required.[14]

Each additional inch (25 mm) of seasonal evaporation increases the water requirement to produce a ton (907 kg) of dry matter about 14 to 18 tons (13 to 16 MT).[15] The maximum transpiration for young sorghum plants occurs in soils well above field capacity (a soil moisture potential of 0.04 to 0.2 atmospheres) with a rapid decline to the permanent wilting point at a soil moisture potential of about 14 atmospheres.[16] The consumptive use of water (transpiration plus evaporation from the soil) by different irrigated crops varied from 13 to 74 inches (330 to 1,880 mm) in Arizona.[17] This varied with transpiration efficiency, total plant yield, and the length and time of year of the growing season for the particular crop. Short-season crops and those grown during the cool winter period required less water.

The rainfall-evaporation ratio is sometimes used as an index of the external moisture relations of plants, with a ratio of 100 percent meaning that rainfall and evaporation are equal.[3] This ratio ranges from 110 to 130 percent along the Atlantic Coast, but westward it drops gradually to 20 percent at the base of the Rocky Mountains. Only xerophytic shrubs and grasses can thrive where rainfall is so far below evaporation. True forests occur mostly where rainfall exceeds evaporation.

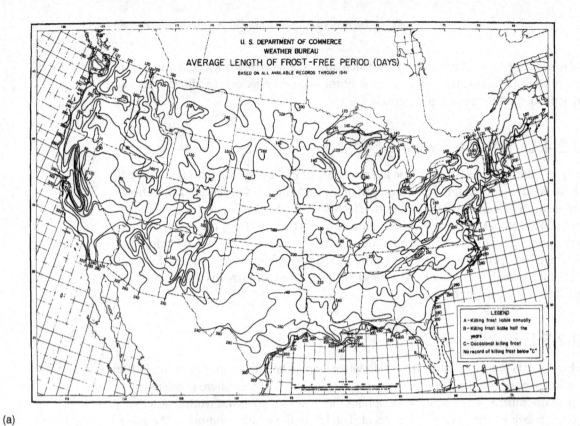

(a)

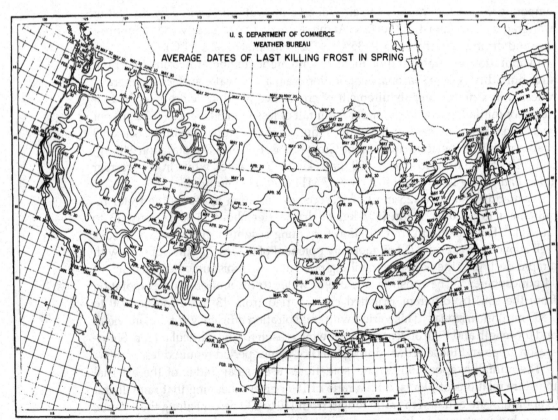

(b)

FIGURE 2.6

Frost-free periods with frost dates. *(a)* Average length of frost-free period. *(b)* Average dates of last killing frost in spring. *(Continued)*

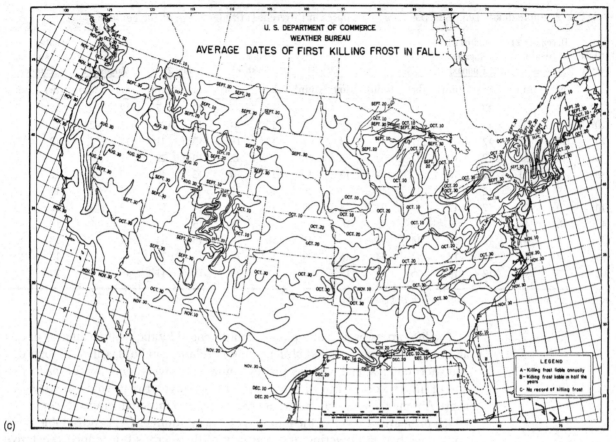

(c)

FIGURE 2.6
(Continued) (c) Average dates of first killing frost in fall.

2.2.6 Light

Light is necessary for the formation of chlorophyll in green plants and for photosynthesis, the process of manufacturing food essential to their growth. Plant growth may increase with greater light intensities of 1,800 foot-candles (19,375 luxes or meter-candles) or more, an intensity of about one-fourth to one-half of normal summer sunshine. Maximum solar radiation may exceed 10,000 foot-candles. The hours of sunlight each day vary from a nearly uniform 12-hour day at the equator to continuous light or darkness throughout the 24 hours for a part of the year at the poles (Table 2.2). The day length is approximately 12 hours everywhere at the time of the vernal and the autumnal equinoxes. At the Canadian border of the United States (latitude 49° N), the day length (sunrise to sunset during the year) ranges from about 3 hours, 12 minutes to about 16 hours, 6 minutes. At the southern tip of Florida (latitude 25° N), the seasonal range is only from 10 hours, 35 minutes to 13 hours, 42 minutes. The long summer days at high latitudes enable certain plants to develop and mature in a relatively short time. When timothy was grown from Georgia to Alaska, the vegetative period was progressively shortened from south to north.[8] Flowering was hastened by the longer days.

TABLE 2.2 Approximate Length of Day on Various Dates at Different Latitudes North of the Equator

Latitude (North)	December 21 (Winter Solstice)		March 21 (Spring Equinox)		April 21		May 21		June 21 (Summer Solstice)		July 21		August 21		September 21 (Autumn Equinox)	
	(hr)	(min)	(hr)	(min)	(hr)	(min)	(hr)	(min)	(hr)	(min)	(hr)	(min)	(hr)	(min)	(hr)	(min)
0°	12	7	12	7	12	7	12	7	12	7	12	7	12	7	12	7
10°	11	33	12	7	12	24	12	37	12	43	12	38	12	24	12	8
20°	10	56	12	9	12	42	13	9	13	19	13	11	12	43	12	10
25°	10	35	12	10	12	53	13	28	13	42	13	29	12	54	12	10
30°	10	11	12	10	13	4	13	47	14	4	13	48	13	6	12	12
35°	9	48	12	12	13	17	14	9	14	32	14	12	13	18	12	12
40°	9	20	12	12	13	31	14	35	15	2	14	36	13	32	12	13
45°	8	46	12	14	13	48	15	4	15	38	15	6	13	49	12	15
50°	8	4	12	15	14	9	15	41	16	24	15	44	14	9	12	17
55°	7	10	12	17	14	34	16	29	17	24	16	33	14	35	12	19
60°	5	52	12	18	15	8	17	38	18	54	17	41	15	9	12	21

Source: Data from Allard and Zaumeyer, *USDA Tech. Bull*, 867(1944)

The life processes of many plants are influenced by the relative lengths of day and night in a process called *photoperiodism*.[18] Plants can be classified as long-day and short-day plants. Long-day plants require a relatively long day for the formation of inflorescences, but they increase in vegetative growth when the days are short. Among the long-day plants are red clover and the small grains (except rice), which ordinarily flower in the long days of early summer (Appendix Table A–1). Long days hasten flowering and maturity of these crops but reduce vegetative growth. Short-day plants are stimulated to vegetative growth, with delayed flowering and maturity, when the days are long, and produce flowers and fruits when the days are relatively short. Corn, sorghum, rice, millet, and soybean are short-day plants. In one experiment, Biloxi soybean plants grew 9 inches (23 cm) tall and flowered in twenty-three days under a 10½ hour photoperiod but were 30 inches (76 cm) tall and flowered in sixty days under a 14½ hour photoperiod.

The length of the night (dark period), rather than the length of day, is the critical determinant although light is also essential for the complete photoperiodic reaction. Short-day plants require daily prolonged darkness to induce flowering, and a short interval of light during the night may prevent flowering. Long-day plants may flower under continuous light. Flowering in long-day plants is hastened by even short light exposures during the night.

Other crop plants and certain varieties of some of the previously mentioned crops tend to be rather indifferent in their responses to photoperiodic influences. Cotton, cultivated sunflower, and buckwheat are in the day-neutral, indeterminate, or intermediate group.

The distribution and time of maturity of different crops and crop varieties are influenced by photoperiodic responses. Corn or sorghum varieties from the tropics seldom mature when planted in regions where the days are long. Corn hybrids and soybean cultivars are adapted to particular zones of latitude (i.e., different day length). When planted outside these zones, they usually are too early or too late for best performance. Native grass species that extend from the northern to the southern portions of the United States are differentiated into numerous local strains (ecotypes), each adapted to its particular photoperiodic range. The northern strains are small and early when grown in the South, whereas southern strains are large and late when moved to the North. Wild rice and gramagrass exhibit this variation to a marked degree.

Day length and temperature compensate for each other to some extent.[19] Photoperiodic responses of a number of plants may be modified by temperature.[1] Quick-maturing sorghum varieties may grow larger in the North where days are long than they do in the South where temperatures are more favorable. The time a crop is planted affects the periods required to reach maturity due to temperature and photoperiodic influences. The growing period for spring-sown crops decreases with later seeding, except in the case of very late seeding, which forces the crop to complete its growth at a slower rate induced by cool autumn conditions. Most cool-season crops have a long-day response, whereas most warm-season crops have a short-day response.

Supplementary light to shorten or to interrupt the dark period usually is effective in the control of flowering at intensities of 30 foot-candles or even less. Red light at wavelengths of about 660 nanometers is most effective in inducing flowering in long-day plants or in preventing flowering of short-day plants. Far red light (wavelength about 730 nanometers) nullifies the effect when applied immediately after a short exposure to red light. Red or far red rays are absorbed by *phytochrome*, a photo-reversible blue pigment of protein nature. Phytochrome is present in minute amounts in all flowering plants, even in seedlings without chlorophyll that were grown in the dark. It is present in some seeds that require exposure to light to trigger germination. The pigment behaves like an enzyme when it triggers flowering reactions.[5]

■ 2.3 AIR

Air not only supplies carbon dioxide for plant growth but also furnishes nitrogen indirectly. It supplies oxygen for respiration in the plant as well as for chemical and biological processes in the soil. Air sometimes contains gaseous substances in concentrations harmful to plant growth; these gases usually come from fuel combustion or industrial fumes. Of these, ozone and PAN (peroxyacetyl nitrate) are photochemical oxidants formed by sunlight acting on products of fuel combustion. Excess ozone causes spots on the leaves of tobacco, maize, and other plants; and PAN causes young leaf tissue to collapse. Sulfur dioxide causes blotches between leaf veins. Nitrogen dioxide suppresses plant growth. Fluorides cause chlorotic patterns or the death of the margins and tips of the leaves. Industrial smoke and auto exhaust contribute an average of 50 pounds per acre (56 kg/ha) of nitrogen over the whole state of Connecticut.[20] A concentration of sulfur dioxide of one part or more in 2 million parts of air is injurious to plants.

Carbon monoxide levels in the air over city streets are no higher with the present heavy motorized traffic than they were in the earlier years when coal stoves and furnaces, artificial gas fuel, and kerosene lamps were used. The pretty blue flame in a wood fireplace displays the release of carbon monoxide. Artificial (coal) gas is a mixture of carbon monoxide and methane. Lead levels in human tissues are lower than they were three generations previously. Then, coal burning, lead paints, lead plumbing, pewter utensils, and thickly soldered tin cans prevailed in the homes, and lead arsenate was a popular pesticide.

All air pollutants, including dust, soot, and ashes, return to the earth eventually in rain, snow, or by gravity or downward air currents. The solid particles become a part of the mineral or organic soil constituents. Ashes and dust contain essential plant nutrients such as potassium, phosphorus, and calcium. Gaseous compounds of nitrogen and sulfur are converted into solid plant nutrients, but some atmospheric nitrogen is released by soil organisms that use the oxygen in nitrogen oxides during anaerobic conditions. Carbon dioxide, the chief component of consumed fuels, is absorbed directly by plants in photosynthesis. Some of the carbon monoxide is converted to carbon dioxide or carbonates by lightning or solar energy in the air.

Phosphorus and its compounds from air, soil, and sewage, when emptied into still waters, promote the growth of algae. When algae die, microorganisms use the oxygen from the water to promote decay. This process, called eutrophication, reduces the oxygen supply, which suffocates the fish. Eutrophication also occurs in ponds remote from farmlands.[21] Eutrophication has been operating since the beginning of time to produce peat. Heavy pressure in later geological ages converted some of the peat into coal and petroleum.[22] Most of the nitrogen and phosphorus that reach the waters are in industrial and sewage wastes rather than from farms.[23]

Wind may cause mechanical lacerations and bruises to the tissues of crop plants. High winds may bruise and tear the leaves or, as often happens with small grains grown on fertile soils, the plants may be blown over or lodged. The rate of transpiration or water loss from plants increases in proportion to the square root of the wind velocity. In high winds, crop plants may be cut off by moving sand particles, completely uprooted, or buried under soil drifts.

█ 2.4 SOIL REQUIREMENTS

Although less important than climate, soil texture and soil reaction (acidity) play a major role in determining which crops are grown. Despite popular belief, soil texture and reaction have only a minor effect on determining the variety of a particular crop that is grown. Surveys of crop varieties show that their distribution is determined largely by climatic and soil-moisture factors.

Soils provide naturally, or after treatment, certain favorable conditions for plant growth. Soil furnishes a stratum for germination of the seed as well as spread and anchorage of the plant roots. In addition, soil maintains a reasonably satisfactory balance of moisture and essential mineral elements for nutrition of the plants.

Soil is not essential for plant growth, however, as evidenced by the frequent growing of many plant species in sand and water cultures, to which nutrients are supplied artificially. Spearmint was grown in water cultures by John Woodward[24] as early as 1699. Some 240 years later, the process of growing plants in water cultures was named hydroponics.[25, 26]

2.4.1 Texture

Soil texture has an important influence upon crop adaptation. Medium or fine-textured soils are best for fine-rooted grasses, wheat, and oat, whereas coarser crops including rye, corn, and sorghum plants can thrive, and are commonly grown, on the light, sandy soils. Rice demands a reasonably heavy soil with a nearly impervious subsoil that prevents excessive water losses from leaching.

Water percolates more quickly and more deeply into coarse-textured soils, which help check runoff, although fine-textured soils have the greater water storage capacity by volume. Partly for this reason, crops in semiarid regions are less subject to drought on sandy soils than on fine-textured soils, whereas in regions of high rainfall where much of the water may percolate downward below the root zone, crops on sandy soils are more subject to drought.

Texture of a soil refers to the size and distribution of its individual grains or particles, grouped on the basis of diameter as sand (2 to 0.05 mm), silt (0.05 to 0.002 mm), and clay (0.002 mm and less) separates. Soil textural class is recognized on the basis of the relative percentages of these separates. The principal textural classes are: sand, loamy sand, sandy loam, silt loam, clay loam, silty clay loam, and clay, in increasing order of their content of clay separates. The farmer commonly calls a soil with a large proportion of clay particles a heavy soil, and one with a large proportion of sand a light soil, due to the ease of cultivation. In weight per unit of volume, "light" soil is heavier than "heavy" soil.

The finest particles, having a diameter of 0.001 millimeter or less (usually less than 0.0001), constitute the colloids. They retain many available plant nutrients and water by adsorption. The degree of adsorption depends upon the amount of surface. Fine particles have an enormous surface area per unit weight of soil. One pound of average soil colloidal particles has about 5 acres of surface (4.5 ha/kg).

In most soils, the fine clay and colloidal fractions contain most of the available nutrients. They play a major role in determining its properties, which include cation exchange. The percentage of particles of clay size and smaller approximates the percentage of moisture at the field capacity of many soils. At field capacity, where the water is retained against the force of gravity, the soil moisture potential is –⅓ bar or less. At the permanent wilting point for plants, the potential is –15 bars. Plant roots take up water readily from loam soils with potentials of up to about 8 atmospheres (50 percent of field capacity), but water stress, or even wilting, may develop above this potential in hot, dry weather. Irrigation is advisable at moisture potentials of about 6 atmospheres, the stage at which about one-third of the available water capacity of the soil remains.

2.4.2 Structure

Soil structure refers to the manner in which the individual particles are arranged. An aggregated or compound structure, in which the particles are grouped into crumbs or granules from 1 to 5 millimeters in diameter, provides irregular spaces with larger pores through which water and air can circulate. This also favors water infiltration, good seedbed preparation, ease of cultivation, and protection from wind or water erosion. Soil aggregation normally occurs when ample organic matter is present. It increases sharply with increases in the soil carbon content from 0 to 2 percent or more. A moist soil of good structure crumbles in the hand. This granular condition or tilth may be destroyed by flooding or by packing or tilling the soil when it is too wet. It is then puddled; that is the soil is made up of single unaggregated grains.

2.4.3 Relation of Soil Groups to Crop Production

There is a systematic classification for soils similar to the botanical classification system for plants.[27] There are six levels of classification in *Soil Taxonomy*: (1) *order*, the broadest category, (2) *suborder*, (3) *great group*, (4) *subgroup*, (5) *family*, and (6) *series*, the most specific category. The system is hierarchical; that is, each order has several suborders, each suborder has several subgroups, and so forth (Table 2.3). There are twelve orders of soils as shown in Table 2.4. The orders that offer the fewest constraints to crop production are the Mollisols, Alfisols, and Ultisols, although crop production is possible with the others with the possible exception of the Gellisols that are found only in arctic environments.

The Mollisols, which develop most commonly under prairie vegetation, are the most adaptable to crop production. These soils are usually deep with well-developed structure. Since prairie grasses die back each winter, much organic matter accumulates in the topsoil. In the United States, there is one contiguous region of Mollisols that reaches from Canada to Texas and from western Minnesota and Iowa to Colorado. There is another area of Mollisols in northern Illinois. A third region stretches across northern and western Montana, southern Idaho, and eastern Washington and Oregon. Practically all the land is under cultivation with a wide variety of crops.

TABLE 2.3 Comparison of the Classifications of a Plant, Wheat *(Triticum aestivum)*, and Holdrege Series Soil

Plant Classification		Soil Classification	
Division	Spermatophyta	Order	Mollisol
Class	Angiospermae	Suborder	Ustall
Subclass	Monocotyledonae	Great group	Argiustoll
Order	Graminales	Subgroup	Typic Argiustall
Family	Poacae	Family	Fine-silty, mixed, mesic
Genus	Triticum	Series	Holdrege
Species	aestivum	Phase[1]	Holdrege silt loam

[1]Not a category in Soil Taxonomy but used in field surveys to identify the texture (silt loam).

TABLE 2.4 Descriptions of Soil Orders

Soil Orders	Descriptions
Gellisols	Soils with permafrost or churning from permafrost
Histosols	Soils with high organic soil material extending down to an impermeable layer or more than 40 cm thick
Entisols	Soils with little or no profile development, very recently deposited
Inceptisols	Soils with the beginning or inception of profile development
Andisols	Soils formed from volcanic ash or cinders
Aridisols	Soils formed in arid regions
Vertisols	Soils with high concentrations of swelling and cracking clays
Mollisols	Soils formed under prairie grasses
Alfisols	Soils formed under native deciduous forests or savanna
Ultisols	Soils with a well-developed clay horizon
Spodosols	Soils formed from coarse-textured, acid parent material
Oxisols	Soils that are highly weathered, formed in hot tropical regions

Alfisols, which developed under native deciduous forests of the United States, are found east of the Mollisols across Illinois, Indiana, and western Ohio and from eastern Minnesota and southern Wisconsin and Michigan to northwest Missouri. There is another area of Alfisols from southern Oklahoma through Central Texas. A third area is found in western Colorado and a fourth in California. Many of these soils are cultivated where permitted by topography and produce a wide variety of crops. They usually have less soil organic matter than the Mollisols (Figure 2.7). Deciduous trees put their roots deep into the soil and recycle much less organic matter than prairie grasses.[27]

Ultisols are similar to Alfisols in their development. They have a well-developed clay horizon and are found mostly south of the Alfisols and east of the Mollisols in

 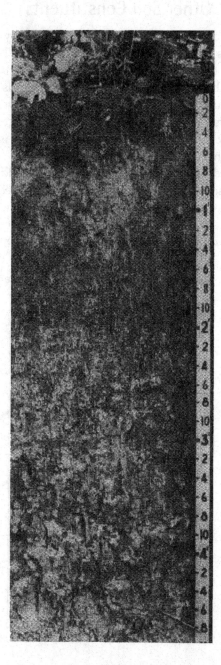

FIGURE 2.7

Soil profiles formed from loess deposited in Tama County, Iowa *(Left)* Black praire soil with the top upper 18 inches high in organic matter. *(Right)* Gray-brown podzolic soil formed under oak-hickory forest.

the southeast United States and east of the Appalachian Mountains north to New Jersey. Ultisols differ from Alfisols in that Ultisols are generally more highly weathered and of lower native fertility. These soils are mostly cultivated where permitted by topography, producing a wide variety of crops.

An area of Inceptisols under cultivation in the United States is the Mississippi Delta from the southeast corner of Missouri through Louisiana. These soils are very productive when water is plentiful. Some Aridisols in New Mexico, Arizona, Nevada, eastern California, Colorado, and Wyoming are under cultivation where irrigation water is available.

2.4.4 Other Soil Constituents

Water occurs in three liquid forms. Capillary water, held in the soil by surface tension, is the water mainly used by plants. When plants wilt, the soil may still contain 2 to 17 percent moisture, depending upon its texture and humus content. Water below this permanent wilting point is largely unavailable to plants. Gravitational water is that which moves downward by gravity and may percolate beyond reach of the roots of crop plants. Hygroscopic water, which is retained by an air-dry soil kept in a saturated atmosphere, is adsorbed on soil particles with such force that it is not available to plants.

Field capacity is the amount of water held by a soil against the force of gravity. Most soils at field capacity will hold between 1 and 2 inches (25 and 50 mm) of plant available water in each foot of soil. The finer the soil particles, the more water the soil holds. Extremely coarse, sandy soils are unable to store moisture in sufficient quantities for crop growth (Figure 2.8).

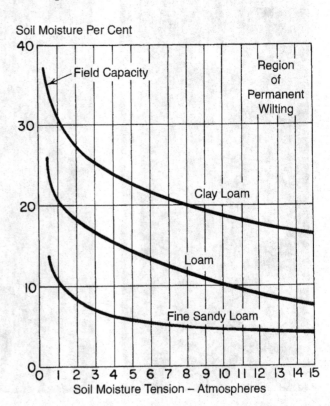

FIGURE 2.8

Field capacity of three soil types, with soil moisture contents, different tensions, and wilting points.

Air, which constitutes from 20 to 25 percent by volume of an ordinary moist soil, supplies the oxygen necessary for root growth and for oxidation of organic matter and other soil constituents. Aeration may be too poor for plant growth in a soil with the water table too near the surface or in one with a heavy, impervious subsoil that cannot be drained satisfactorily. Water-logged soils are often relegated to native vegetation or low-grade grazing.

An adequate supply of plant nutrients is essential for the growth of crop plants. About 25 or 30 chemical elements are found in plants. Nutrients used in the greatest quantities by plants are called macronutrients. Carbon, oxygen, and hydrogen that come from air and water are the most abundant macronutrients. All other nutrients come from the soil. The macronutrients that are most frequently deficient in soils are nitrogen, phosphorus, and potassium. Calcium, magnesium, and sulfur are other macronutrients that are absorbed by crops in considerable quantities.

Nutrients absorbed in lesser amounts are called micronutrients. The micronutrients are manganese, iron, boron, chlorine, copper, zinc, molybdenum, and cobalt. Zinc is essential for plant growth and development, including cell division, starch synthesis, and seed formation. Manganese and iron serve as catalysts in chlorophyll synthesis, while chlorine is involved in electron transport during photosynthesis. Molybdenum is essential for nitrogen fixation, functions in nitrate reduction, and is a component of enzymes. Cobalt is essential for nitrogen fixation. Copper functions in nitrate reductase and oxidizing enzymes, and boron in cell division and growth processes.

The plant can take up other elements that are not considered to be essential. Traces of otherwise toxic elements, such as selenium, nickel, and arsenic, may increase plant growth. Silicon, a compound abundant in the soil, is an important constituent of rice hulls and other plant parts. Sodium is required by sugarbeet and mangel, while chlorine is needed to produce a suitable burning quality in flue-cured tobacco. Usually sodium and chlorine are abundant in soils, except locally in the humid Southeast.

Two soil factors that greatly influence nutrient availability and retention are soil pH and cation exchange capacity. Soil pH affects nutrient availability by affecting the degree of solubility in the soil solution, and each nutrient has a pH range when it is most available. Below and above that range, a nutrient will become less soluble and hence less available. Cation exchange capacity (CEC) is a measure of the number of anionic sites available on soil clay and humus particles. These negatively charged sites will attract and hold cations (positively charged) in the soil solution. Some nutrients that are found primarily as cations will be held on the cation exchange sites until needed by the plant. Representative CEC values for common soil orders are found in Table 2.5.[28]

TABLE 2.5 Cation Exchange Capacity of Surface Soils by Soil Order

Soil Order	CEC (molc/kg)	Soil Order	CEC (molc/kg)
Alfisols	0.12	Mollisols	0.22
Aridisols	0.16	Oxisols	0.05
Entisols	0.43	Spodosols	0.11
Histisols	1.40	Ultisols	0.06
Inceptisols	0.19	Vertisols	0.37

Soils deficient in cobalt, iodine, selenium, or fluorine may produce crops that are inadequate in these elements from the standpoint of animal or human nutrition. Crops may take up so much selenium, molybdenum, and arsenic from soils with excessive quantities of these elements that they are toxic to livestock and humans.[29] Aluminum and manganese may be absorbed in quantities harmful to the plant when the soil is too acid.

Mineral deficiencies can often be detected or at least suspected from symptoms seen in crop plants.[5, 14, 30] The symptoms, which differ in various crops, have been summarized in detail and are illustrated in the book *Hunger Signs in Crops*.[14]

Some of the general characteristic symptoms indicating mineral deficiencies, which apply to some but not all crops, are included in the following list.

Diagnosis of Mineral Deficiencies

Plants stunted, leaves pale green: nitrogen, sulfur
Plants stunted, leaves dark green or tinted with purple: phosphorus
Plants stunted, leaves bluish or dark green, margins and areas between veins scorched or brown: potassium
Old leaves chlorotic between the veins: magnesium
Leaf margins rolled upward: calcium
Chlorotic mottling or striping (especially on calcareous soils): iron
Chlorotic mottling and small, discolored spots or lesions: manganese
Root crops with decayed crown, distorted leaves, and hollow or decayed root: boron
Leaves chlorotic, crinkled, and drooping at tips: molybdenum

Good soils contain adequate amounts of available nutrients to meet the needs for normal growth of the crop plant and do not contain harmful concentrations of any constituent. They are neither excessively acid nor excessively alkaline. In general, relatively permeable and fertile soils with a good water-holding capacity are favorable for the growth of the most important crop plants. Some soils seem to possess an especially wide range of crop adaptation. Crops, on even the best soils, usually respond to the addition of suitable fertilizers when growing conditions are favorable.

Arable soils require a topography suitable for culture operations and must be free from stones that would interfere with tillage. A soil depth of 1½ to 3 feet (0.5 to 1 m) or more is usually essential to successful cropping. Clay soils that are difficult to till may be an effective barrier to profitable utilization for some crops. The peculiar tendency of certain silt loams and very fine, sandy soils to erode reduces their long-time use without protective measures. The crop grower has numerous methods for improvement of the soil environment.

2.4.5 Crops for Special Soil Conditions

CROPS TOLERANT TO SALINITY Crops such as barley, bermudagrass, wheatgrass, sugarbeet, rape, cotton, rhodesgrass, rescuegrass, and tall fescue may be grown where the soil solution from the root zone contains as much as 0.8 to 1.6 percent salts (8 to

16 millimhos* per centimeter). White, red, and alsike clover, as well as meadow foxtail, burnet, and field bean have a low tolerance, (i.e., the salt content should not exceed 0.3 to 0.4 percent). Most of the other field crops have a medium tolerance (i.e., the salt content should not exceed 0.8 percent).[31]

CROPS TOLERANT TO ACID SOILS A soil with a pH of 7 is neutral, one higher than pH 7 is alkaline, while a soil with a pH below 7 is acid. These pH values for hydrogen-ion activity are reciprocal logarithmic expressions to the base or power of 10. Therefore, a soil of pH 5 is 10 times as acid as one of pH 6, while one of pH 4 is 100 times as measured by hydrogen-ion activity. Usual groupings for soil reaction are as follows:

Group	pH	Group	pH
Extremely acid	Below 4.5	Neutral	6.6–7.3
Very strongly acid	4.5–5.0	Mildly alkaline	7.4–7.8
Strongly acid	5.1–5.5	Moderately alkaline	7.9–8.4
Medium acid	5.6–6.0	Strongly alkaline	8.5–9.0
Slightly acid	6.1–6.5	Very strongly alkaline	9.1+

Many factors other than direct acidity may complicate the growth of crop plants on soils with high acidity. High acidity may increase absorption of aluminum and manganese by plants and retard absorption of calcium or phosphorus as a result of ion exchange. Excessive alkalinity may also interfere with iron or phosphorus absorption. Microbial activities in the soil are greatly influenced by the soil reaction. Fortunately, most economic plants will grow well on acid soils. Among the plants that will respond well on highly acid soils (pH 4 to 5) when a moderate amount of active calcium is present are lespedeza, redtop, millet, buckwheat, rye, soybean, cowpea, and alsike clover. Those that will grow on strongly to medium acid soil (pH 5 to 6) with a moderate amount of active calcium present are corn, wheat, oat, barley, timothy, potato, field bean, and vetch (Figure 2.9). Certain plants require a neutral or only slightly acid soil with an abundant supply of active calcium. These are alfalfa, red clover, sweetclover, and sugarbeet. Abundant calcium, rather than low acidity, appears to be the critical factor in these crops. A good quality of tobacco cannot be grown on an alkaline soil.

▓ 2.5 INDICATOR SIGNIFICANCE OF NATIVE VEGETATION

Native plant cover was used by primitive people, as well as by early settlers, as a clue to the lands most productive for crops (Figure 2.10).[32] The type of vegetative cover often serves as an indication of overgrazing.

Farmers in the Ohio Valley who settled on lands that were covered with sugar maple and beech were more prosperous than those on oak and pine lands. The lands occupied by southern hardwoods, river-bottom forests, southeastern pine

*Millimho is a unit of electrical conductance that is the reciprocal of a milliohm. A soil is classified as saline by the USDA Salinity Laboratory when the soil saturation extract is 4 millimhos or more per centimeter at 25°C.

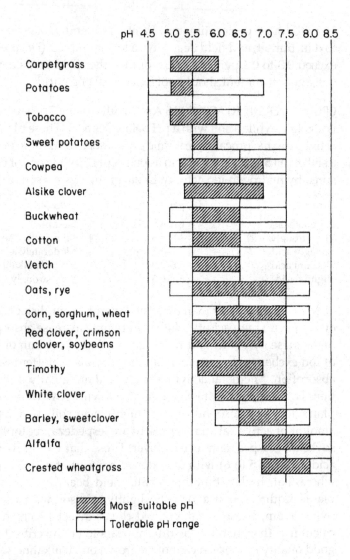

FIGURE 2.9
The range of pH
adaptability for major crops.

forests, and northeastern hardwoods are regarded as valuable for agricultural production. In the West, yellow pine forests often occupy slopes that are too steep or soils that are too stony and shallow for cultivated crops.[9] Small grains are grown for forage in the more favored spots. Lodgepole pine forests occupy areas too cool for cultivated crops, where the frost-free season is seldom longer than 75 days.

The true prairies, for the most part, were originally made up of tall, coarse grasses such as big bluestem (*Andropogon gerardii*), little bluestem (*Schizachyrium scoparium*), and Indian grass (*Sorghastrum nutans*). These lands are valuable for production of corn and most field crops. The short-grass or plains grassland association, largely consisting of blue grama (*Bouteloua gracilis*) and buffalograss (*Buchloë dactyloides*), extends through the Great Plains from Montana southward.[33] Buffalograss grows on more clayey soils where the soil moisture, while the land is in grass, rarely penetrates below the second foot (60 cm). The moisture often penetrates deeper when the land is kept cultivated between crops. Crop failures are to be expected there in years of much less than average moisture. Corn and wheat are the chief crops in the northern part of this region and wheat and grain sorghum in the southern part. The sandier soils in the Great Plains are occupied by the taller

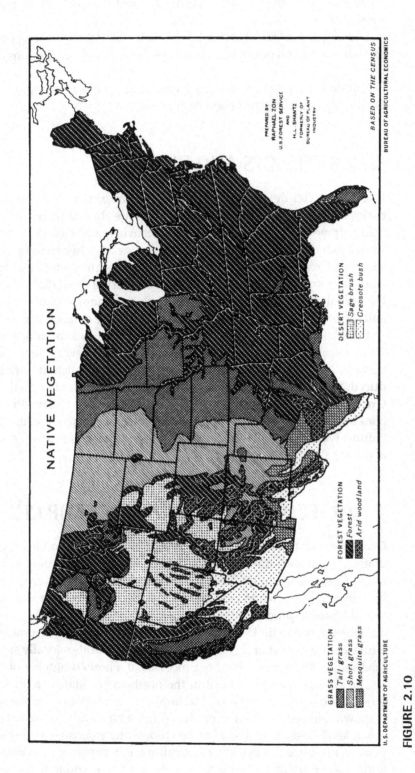

FIGURE 2.10
Native vegetation in the United States.

bunchgrasses and wire grasses.[34] Because such sandy lands have less runoff, they are less subject to crop failure in the drier years than are the short-grass lands in the same locality.

Dense stands of tall sage brush (*Artemisia tridentata*) have long been accepted as an indication of favorable soil conditions for dryland small grains or for irrigated crops.

Salt-desert vegetation, such as greasewood and seepweed, indicates soils that are too high in salinity for good crop growth.

■ 2.6 INSECTS AND DISEASES

Insect pests often determine the success or failure of a crop for a given district. Thus, barley and milo often fail where chinch bugs are abundant. The alfalfa weevil once nearly drove the alfalfa seed crop from certain irrigated valleys of the West. The boll weevil reduced the cotton acreage in the Southeast, while the crop expanded in the southern Great Plains where the weevil was not present. Grasshoppers and the southwestern corn borer have devastated corn in the Great Plains region. Indirectly, this stimulated the growing of sorghum, which seldom is injured. Control of insects and the breeding of resistant varieties may alter the crop situation in the future.

Diseases, likewise, can be responsible for the lack of adaptation of a crop. Thus, sugarbeet production, once discontinued in several localities of the West because of curly top disease, was restored when resistant varieties were produced. Wilt disease caused the flax crop to migrate from the Atlantic Coast to the Great Plains. Flax moved back east to the Mississippi Valley after resistant varieties became available. Losses from virus diseases were so severe that sugarcane culture was almost abandoned in Louisiana before resistant varieties became available about 1926.

■ 2.7 ECONOMIC FACTORS IN CROP CHOICE

Centers of crop production are determined in part by economic forces such as population,[35] demand, transportation, labor problems, and competition with other crops. Cheap lands in the United States are often pastured or grazed, with more intensive crops grown as land values rise. High-priced labor is best employed on land adapted to production of high yields of specialized crops on fertile soils. Special crops may remain localized when a ready market has become established. As a rule, durum wheat or flax must be grown in quantities locally sufficient for carload shipments because they are processed at a limited number of mills.

The average yields of wheat in the northeastern states are much higher than those in the semiarid portions of Kansas and North Dakota where so much wheat is grown. However, wheat is produced more efficiently in the latter areas where broad level fields of rich soil can be cropped by extensive methods with less expense for fertilizers. Also, other agricultural enterprises do not compete seriously with wheat production on the loam soils of the semiarid regions.

Psychological factors as well as considerations of convenience often alter what would appear to be the logical crop choice on the basis of relative economic returns.

For example, farmers prefer growing corn instead of grain sorghum. The field stands of corn are more certain, the crop can be harvested over a longer period, it is easier to store and feed, there is less dust, and there is less damage from birds. With better equipment for seeding and harvesting, including better protection of the operator within a cab, there are fewer reasons for choosing a crop because of ease of handling or irritability of plant chaff such as awns. However, many producers still choose crops based on tradition and personal preferences rather than economics.

In the United States, the potential area for crops—all arable land in pasture and woods that could be put into crop, with drainage and irrigation extended—is greatly in excess of that now used. Large areas of the more level rangeland in the Great Plains and Great Basin could be cropped, provided prices were high enough to justify the risks of submarginal agriculture.

Changes in national economic or trade policies could produce drastic adjustments in the acreage of certain crops. The growing of sugarcane, sugarbeet, flax, and rice are feasible largely because of a protective tariff in some countries.

■ 2.8 CROPPING REGIONS IN THE UNITED STATES

The principal geographic cropping regions in the United States are the Atlantic Coast to the Great Plains, the Great Plains, and the Pacific Coast.

2.8.1 Atlantic Coast to the Great Plains

The Corn and Wheat Belts (Figure 2.11) coincide with the deciduous forest and prairie centers.[3] The production of these crops is greatest between the 60 and 100 percent rainfall-evaporation lines.

The area from Ohio to central Nebraska is remarkably suited to corn production because of the level topography, warm summers, relatively high humidity, and an annual rainfall sufficient for plant growth (Figure 2.12). The geographic center of the Corn Belt is found in central Illinois where the soils are deep and rich in humus, and corn is well adapted. Farther south, corn yields are less because of poorer soils, higher temperatures, intermittent droughts, and insect injury; corn is unable to compete with cotton as a cash crop in this area.

Although the center of wheat production lies west of the best corn lands, wheat is well adapted. In fact, wheat is grown on many farms in rotation with corn throughout the prairie and deciduous forest climaxes. Wheat is higher in protein when grown in a dry climate. Oat is concentrated in the northern portion of the Corn Belt because of the cooler climate.

The region southeast of the 100 percent rainfall-evaporation ratio line is known ecologically as the southeast evergreen center. Cotton is the principal crop in this area because of the high temperatures and the long growing season. In the Cotton Belt, cotton competes with cereals, forage crops, tobacco, soybean, and peanuts, and, on some soils, with sugarcane and rice.

The northern evergreen forest, north and east of the 110 percent rainfall-evaporation ratio line, is devoted agriculturally to tame pastures and hay crops. The need for the bulkier hay and fodder crops, which grow well under the cool temperature conditions, makes cereal production less profitable in this region.

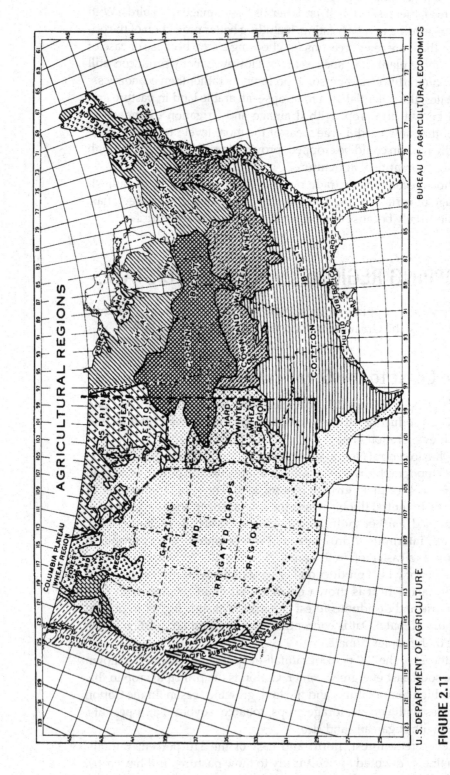

AGRICULTURAL REGIONS

U. S. DEPARTMENT OF AGRICULTURE   BUREAU OF AGRICULTURAL ECONOMICS

FIGURE 2.11

Agricultural regions of the United States. The Great Plains is the west-central area outlined by dashes. The western irrigated cotton-growing area lies within the dotted outline.

48

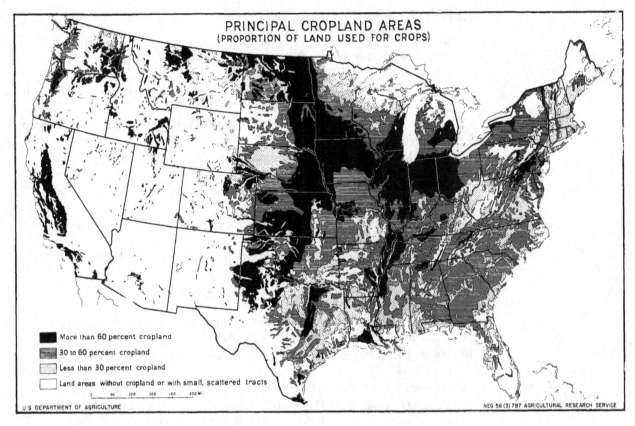

FIGURE 2.12
Deserts, mountains, and infertile soils limit the areas suitable for crops in parts of the United States.

2.8.2 Great Plains

In the western half of the United States, evaporation consistently exceeds precipitation by two or three times. To survive under such droughty conditions, the cultivated plants of the plains must be grown under the best known cultural methods for conservation and utilization of the moisture captured by the soil. Irrigation is practiced where stream flow or underground water supplies can be diverted to supplement the natural rainfall. Wheat is the major crop. Sorghum is grown in some of the more arid portions. Cotton, wheat, and sorghum predominate in the southern areas of the plains. Small grains, corn, flax, sorghum, field bean, and hay are grown in the northern parts.

2.8.3 Pacific Coast

The third distinct crop region, the Pacific Coast, is a winter-rainfall area, while the portion near the ocean has mild winters and cool summers. The northern half of this region is devoted primarily to seed, feed, special purpose, and horticultural crops. Wheat is an important crop in the intermountain region east of the Cascade Mountains in Oregon and Washington. The southern half of the Pacific Coast is devoted to fruits, vegetables, and many different field crops, especially under irrigation.

REFERENCES

1. Roberts, R. H., and B. E. Struckmeyer. "The effect of temperature and other environmental factors upon the photoperiodic responses of some of the higher plants," *J. Agr. Res.* 56(1938):633–678.

2. Klages, K. H. "Geographical distribution of variability in the yields of field crops in the states of the Mississippi Valley," *Ecol.* 11(1930):293–306.

3. Wallace, T. *The Diagnosis of Mineral Deficiencies in Plants by Visual Symptoms.* London: His Majesty's Stationery Off., 1944. pp. 1–116; also Supp., pp. 1–48.

4. Kincer, J. B. "Temperature, sunshine, and wind," in *Atlas of American Agriculture,* USDA, Pt. II, Sec. B, 1928, pp. 1–34.

5. Burr, W. W. "Contributions of agronomy to the development of the Great Plains," *J. Am. Soc. Agron.* 23(1931):949–959.

6. Kincer, J. B. "The relation of climate to the geographic distribution of crops in the United States," *Ecol.* 3(1922):127–133.

7. Kincer, J. B. "Precipitation and humidity," in *Atlas of American Agriculture,* USDA, Pt. II, Sec. A, 1922, pp. 1–48.

8. Hopkins, A. D. "Bioclimatics—a science of life and climate relations," *USDA Misc. Pub.* 280, 1938, pp. 1–188.

9. Robbins, W. W. "Native vegetation and climate of Colorado in their relation to agriculture," *CO. Agr. EXP. Sta. Bull.* 224, 1917.

10. Martin, J. H. "Climate and sorghum," in *Climate and Men,* USDA Yearbook, 1941, pp. 343–347.

11. Waldren, R. P. *Introductory Crop Science,* 5th ed. Minneapolis: Burgess, 2003.

12. Viets, F. G., Jr., and R. H. Hageman. "Factors affecting the accumulation of nitrate in soil, water and plants." *USDA Agr. Handbk.* 413(1971):1–63.

13. Salmon, S. C. "The relation of winter temperature to the distribution of winter and spring grain in the United States," *J. Am. Soc. Agron.* 9(1917):21–24.

14. Sprague, H. B., ed. *Hunger Signs in Crops,* 3rd ed. New York: McKay, 1964. pp, 1–461.

15. Shantz, H. L., and L. N. Piemeisel. "The water requirement of plants at Akron, Colo.," *J. Agr. Rsh.* 34, 12(1927):1093–1190.

16. Pallas, J. E., Jr., and others. "Research in plant transpiration: 1961," *USDA Prod. Rsh. Rpt.* 70, 1962, pp. 1–37. (279 references.)

17. Erie, L. J., and others. "Consumptive use of water by crops in Arizona," *AZ Tech. Bull.* 169, 1965, pp. 1–41.

18. Thomas, B., and D. Vince-Prue. *Photoperiodism in Plants.* San Diego: Academic Press, 1997.

19. Hurd-Karrer, A. M. "Comparative responses of a spring and a winter wheat to day-length and temperature," *J. Agr. Res.* 46, 10(1933):867–888.

20. Frink, C. R. "Frontiers of plant science," CT. *Agr. Exp. Sta.,* New Haven. 1969.

21. Vinall, H. N., and H. R. Reed, "Effect of temperature of other meteorological factors on the growth of sorghums," *J. Agr. Res.* 13, 2(1918):133–147.

22. Borlaug, N. E. "Ecology fever," *Seventh Memorial Lecture at the 16th FAO Conference,* 1971, pp. 21–25.

23. Vavilov, N. I. *World Resources of Cereals, Leguminous Seed Crops, and Flax, and Their Utilization in Plant Breeding.* Washington, D.C.: Offices of Technical Services, U.S. Department of Commerce, 1960.

24. Woodward, J. "Thoughts and experiments on vegetation." *Phil. Trans.* (1699): 382–392.

25. Anon. "Soilless growth or hydroponics," *Can. Dept. Agr. Pub.* 1357(1968):1–7.

26. Gericke, W. F. *The Complete Guide to Soilless Gardening.* Englewood Cliffs, N.J.: Prentice-Hall, 1938, pp. 1–285.

27. Brady, N. C., and R. R. Weil *Elements of the Nature and Properties of Soils, 12th ed.* Englewood Cliffs, NJ: Prentice-Hall, 1999.

28. Sumner, M. E., ed. *Handbook of Soil Science.* Boca Raton, FL: CRC Press, 2000, p. B-241.

29. McMurtrey, J. E., Jr., and W. O. Robinson. "Neglected soil constituents that affect plant and animal development," in *Soils and Men*, USDA Yearbook, 1938, pp. 807–829.

30. Wadleigh, C. H., and C. S. Britt. "Conserving resources and maintaining a quality environment," *J. Soil and Water Cons.* 23, 5(1969):172.

31. Bernstein, L. "Salt tolerance of plants." *USDA Inf. Bull.*, 2838, 1964, pp. 1–23.

32. Shantz, H. L. "Plants as soil indicators," in *Soils and Men*, USDA Yearbook, 1938, pp. 835–860.

33. Aldous, A. E., and H. L. Shantz. "Types of vegetation in the semiarid portion of the United States and their economic significance," *J. Agr. Res.* 28(1924):99–128.

34. Shantz, E. I. L. "Native vegetation as an indicator of the capabilities of land for crop production in the Great Plains area," *USDA Bull.* 201, 1911.

35. Klages, K. H. "Crop ecology and ecological crop geography in the agronomic curriculum," *J. Am. Soc. Agron.* 20(1928):336–353.

Botany of Crop Plants

3.1 GENERAL NATURE OF CROP PLANTS

Most of the important crop plants reproduce by seeds, and all are classified as spermatophytes or seed plants. In the life cycle of these plants the seed germinates and produces a seedling. The vegetative phase is characterized by increases in the number and size of roots, stems, and leaves. Finally the reproductive phase is reached when the plant flowers and produces seeds.

3.2 THE PLANT CELL

Plants are composed of units called cells. A living cell (Figure 3.1), containing 30 to 85 percent water, is composed of two main regions: the cell wall, which is made up of cellulose and is nonliving, and the protoplast, which is the living portion of the cell and is surrounded by a differentially permeable membrane (plasma membrane). The cell wall provides rigidity to the plant's structure while there are vacuoles in the protoplasm that provide a means to remove and store materials, thereby effectively removing them from chemical transformations within the cell.

The protoplasm is the site of the living activity that a plant conducts. The protoplast contains many organelles and structures that are very specific in their contribution to the living cell and plant. The differentially permeable plasma membrane provides a selectivity of materials entering and leaving the cell. The nucleus contains the chromosomes, the carriers of heredity. The chromosomes contain molecules of DNA (deoxyribonucleic acid) in a linear aggregate of four building blocks called nucleotides. The order of these nucleotides determines the genetic message carried by the chromosome. Via this mechanism, plants control the synthesis of enzymes and other proteins.

Mutations occur from chemical changes in the structure of DNA molecules during reproduction of the molecules or as a reaction to radiation or other exposures. The mutations result in changes in the hereditary characters of the plants.

Ribosomes consist of ribonucleic acid (RNA), rich materials on the surface of which amino acids are connected by peptide bonds to form polypeptide

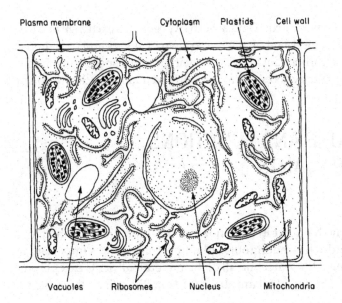

Plasma membrane   Cytoplasm   Plastids   Cell wall

Vacuoles   Ribosomes   Nucleus   Mitochondria

FIGURE 3.1

The plant cell. [As adapted from Samual N. Postlethwait, Henry D. Telinde, and David D. Husband, *A-T Botany, Matter and Mechanics of Cells.* Minneapolis, Minn.: Burgess Publishing Co., 1963]

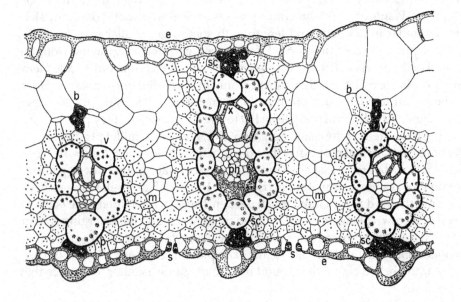

FIGURE 3.2

Cells observable in a cross section of a sugarcane leaf, a typical grass leaf:
(b) bulliform or motor cells that shrink when the leaf wilts and cause the leaf to roll up, and later take up water, expand and flatten out the leaf; *(e)* epidermal cells; *(m)* mesophyll cells; *(p)* palisade-shaped cell of mesophyll; *(ph)* phloem; *(s)* stoma; *(sc)* sclerenchyma (stiffening) cells; *(v)* vascular bundle containing phloem and xylem; *(x)* xylem. [After Artschwager]

units and finally proteins. Ribosomes are essential for protein synthesis; plastids are globular cytoplasmic bodies, the most significant of which is the chloroplast where photosynthesis occurs. Mitochondria are very specialized structures that are involved primarily in energy release. They oxidize organic molecules and make energy available for the life processes of the cell.

Other components of the living cell include microbodies (peroxisomes and glyoxysomes), which are involved in the decomposition of the toxic hydrogen peroxide and play a major role in fat catabolism. Dictyosomes function in the assembly and secretion of products from the cell that lead to cell wall synthesis via polysaccharides produced inside the cell. Spherosomes are involved in the storage and transfer of lipids in plant cells. Cells differ in size and shape in accordance with their special functions (Figure 3.2).

All living cells give off carbon dioxide and absorb oxygen. This exchange of gases, known as respiration, is accomplished by the breakdown of sugars and oxidation of the carbon in the sugar into carbon dioxide. The most important aspect of respiration is the consequent release of energy necessary for growth and maintenance of the plant.

■ 3.3 STRUCTURE AND FUNCTIONS OF CROP PLANTS

3.3.1 Roots

Sometimes as much as one-half, but in certain root crops more than one-half, of a crop plant is underground. The underground plant parts are just as important as the tops because nearly all of the water and all of the mineral nutrients except some nitrogen compounds used by the plant are absorbed from the soil by the roots.[1]

KINDS OF ROOTS Two general types of roots are found in crop plants: fibrous roots and taproots. Plants of cereals and other grasses have fibrous roots (i.e., many slender roots similar in diameter and length with somewhat smaller branches). The seminal or primary roots develop when the seed germinates, growing from the lower end of the embryo. These roots sometimes have been erroneously called temporary roots in the belief that they function only in the earlier growth stages. However, they often remain active until the plant matures.[2, 3, 4] Seminal roots of corn often penetrate to a depth of 5 to 6 feet (150 to 180 cm). In wheat, they provide support to the main stem of the plant. Sorghum, rice, and proso millet produce only a single-branched seminal root. After the young plant unfolds a few leaves, the coronal (crown or nodal) roots that arise from stem nodes underground develop into an elaborate root system. Coronal roots rise about 1 inch (2.5 cm) below the soil surface, even if the seed was planted much deeper. The subcrown internode between the seed and crown elongates until stopped by light striking the emerging coleoptile. At that stage, the crown node is still below the surface. Frequently, roots grow from nodes above the soil surface as, for example, brace (or aerial) roots of corn. They are unbranched above the ground but are the same as other roots once they enter the soil.

Other crop plants, such as sugarbeet, other root crops, and legumes, have a main taproot. The main root develops and pushes straight downward. It may branch throughout its length. The thickened taproot of biennial or perennial plants often serves as an organ for food storage.

The smallest subdivisions of the roots are the root hairs; the single-celled structures found on roots and root branches of both fibrous-rooted and tap rooted plants. Root hairs increase the surface area for absorption of water and nutrients.

EXTENT OF ROOT SYSTEMS Elongation at the rate of ½ inch (1.25 cm) per day is common in roots of many grasses. Seminal roots of winter wheat[1] may grow at that rate for 70 days. The main vertical roots of corn grow 2 to 2½ inches (5 to 6 cm) per day for a period of three or four weeks. Plant roots may extend as far as, or farther, below the ground than the plant does above. The root system of a corn plant, given ample space, may occupy 200 cubic feet (5.66 m³) of soil. A corn plant grown five weeks in a loose soil in Nebraska produced 19 main roots with 1,462 branches of the first order. Upon these were 3,221 branches of the second and third orders. The main roots consisted of 4 to 5 percent of the total root length, the primary branches 47 to 67 percent, and the finer secondary and tertiary laterals 29 to 48 percent of the total. A rather high correlation between root and top growth is found in corn.[5] Given ample space, the roots of a single plant may have a total spread of 4 feet (1.2 m) in wheat, 8 feet (2.4 m) in corn, and 12 feet (3.7 m) in sorghum. Roots of small grains, corn, sorghum, and perennial grasses normally penetrate downward 3 to 6 feet (1 to 2 m), but sometimes they extend 8 or 10 feet (2.4 to 3 m). Alfalfa roots may penetrate more than 30 feet (9 m). The length of the entire root system, which includes all branches of a single large, isolated plant of crested wheatgrass, may be as much as 360 miles (580 km).[6]

Roots grow in moist soil, but most crop plants do not extend roots into water saturated soil. The roots can penetrate short distances into dry soil when moisture is supplied from roots penetrating adjacent moist soil.[7]

ABSORPTION Plants absorb water and also the substances dissolved in it, such as nitrogen and other mineral elements, largely through the root hairs. The root hairs absorb water by osmosis, with their walls acting as differential membranes in the selection of different ions that the plants absorb. When the cell sap is more concentrated than the soil solution, water passes into the root hairs; this occurs except under very abnormal soil conditions. The rate of absorption of water may vary with the osmotic pressure of the cell.[8]

The more water a plant needs, the more vigorously it is absorbed if the water supply remains ample. Water taken up in excess of the ability of the plant to transpire water vapor through the stomata, and especially at night when stomata are closed, is forced out through pores in the leaf called hydathodes. Most of what are called dewdrops is actually water globules from hydathodes. Plasmolysis (or cell collapse) results when the soil solution is more concentrated than the solution in the root hair and other cells. Such a condition may be found in highly saline soils where water is actually withdrawn from the plant.

Nutrients are absorbed independently of the rate of water intake and are absorbed by diffusion as ions. Thus, it is possible for mineral nutrients to be absorbed when the atmosphere is saturated and little water is being taken up and none is being transpired. Minerals in solution tend to pass through the differential cell membrane and reach equilibrium. For instance, potassium nitrate (KNO_3) dissolves and becomes dissociated into K^+ and NO_3^- ions. These ions tend to reach equilibrium regardless of the total concentration in the two solutions, both inside and outside the cell membrane.

3.3.2 Stems and Modified Stems

After a grass seed germinates, the plumule internodes elongate into the first young stem. Stem elongation continues by growth (cell division and cell elongation) at the growth ring (intercalary meristems) above each successive node while the cells are young and active. Growth in the diameter of the stem of cereals and other grasses is due to cell enlargement, not cell division, after the essential stem structures have been formed.

In dicotyledonous (dicot) crop plants, growth is much like that of a tree. New branches arise from buds or adventitious cells, and the branches elongate by cell

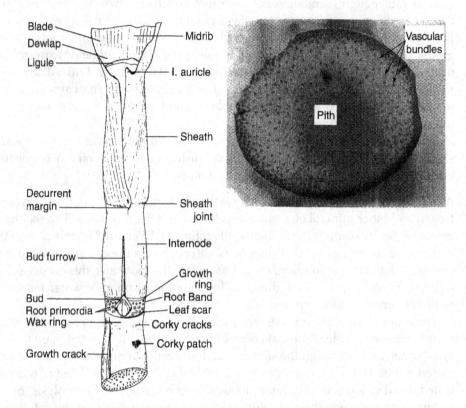

FIGURE 3.3

(Left) A portion of the stem bud at the node. A branch may arise from the corn stalk node. *(Right)* Cross section of sugarcane showing vascular bundles distributed through the pith but more concentrated in the outer rind.

division and cell enlargement near the tips. Growth in diameter comes from cell division in the cambium layer under the bark or periderm, followed by cell enlargement.

GENERAL NATURE OF STEMS The stem is divided into nodes and internodes (joints). The internodes are the stem sections between the nodes (Figure 3.3). In grasses, the internodes when young are solid.[9] In the mature plant, the internodes remain solid in sugarcane, corn, and sorghum (Figure 3.3). Internodes in some grasses, such as wheat and other small grains, are hollow at maturity. These are called culms.

The stems of most grasses die down to the soil surface each year, usually after the seed has matured. Stems that die every year are called herbaceous. In addition to the main stem, branches known as tillers may arise from the lower nodes (Figure 3.4). Large numbers of tillers may arise under favorable environmental conditions, each tiller developing a root system of its own. Small grains and forage grasses produce abundant tillers.

The type of stem may not remain the same during the entire life of the plant.[10] Sometimes the primary axis within the plumule remains so short that the leaves are crowded together, usually in the form of a rosette. Winter annuals (e.g., winter wheat) exhibit this habit until some time in the spring. Sugarbeet and mangel plants have short, contracted internodes during the first year. In the second year, the growing point of the stem, hidden in the center of the rosette, elongates and produces a shoot with long internodes and with the leaves at considerable intervals. Seed is produced on these shoots.

Temperature and day length may affect the growth habit of a plant. Thus, certain varieties of winter oats may show a spreading habit when sown in the fall but

FIGURE 3.4
Tillers and tiller buds on a sorghum plant. The swollen section of the stalk is a node.

A bud is an undeveloped stem, leaf, or flower. Buds may be terminal on the stem, axillary (side), or adventitious (arising out of order). Many buds are dormant.

MODIFIED STEMS Several types of modified stems are recognized.

Rhizomes: Rhizomes are horizontally elongated underground stems with buds at the nodes. Rhizomes develop from nodes by bursting through leaf sheaths of older stems and elongate horizontally below the soil surface. Rhizomes produce new stem shoots from buds at the nodes, which serve to propagate the plant. Some of the most serious weeds, such as quackgrass, johnsongrass, and Canada thistle (Figure 3.5), have rhizomes. Many desirable pasture plants, such as bromegrass and Kentucky bluegrass, have rhizomes and are called sod-forming grasses.

Stolons: Stolons are modified propagative stems produced above ground, as in buffalograss. They produce new plants from buds at nodes similar to rhizomes.

Tubers: A fleshy underground stem is known as a tuber. Examples are the potato and the Jerusalem artichoke. New plants develop from buds or "eyes." Tubers are stems because the buds are equivalent to axillary buds on a normal stem.

LODGING OF STEMS Lodging is the inability of a plant to remain upright. Resistance to lodging, or the capacity of stems to withstand the adverse effects of rain and wind, is an important quality in crop plants. Resistance to lodging often is a varietal character, but its expression is modified by environment. A thick, heavy growth keeps much of the sunlight from reaching the base of the stems. Reduced light or shading causes the cell walls of the stems to be thin and weak, a process

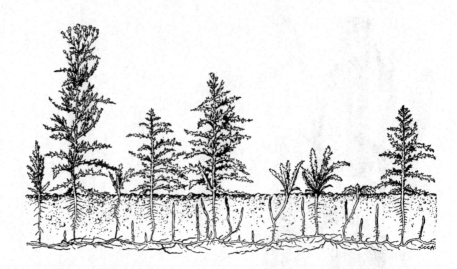

FIGURE 3.5
Lateral roots of Canada thistle.

called etiolation. No single morphological factor can be used as an index of standing power.[11] Abundant soil nitrogen may produce a lush growth that leads to increased lodging. Lodging may result from a low content of dry matter per unit length of stem[12] or a reduced content of lignin or of certain reserve di- and polysaccharides in the stem. Short, thick, heavy stems with thick cell walls are the best insurance against lodging.[13] Culms are inherently weaker and are more likely to lodge than solid stems.

When a grass stem lodges before maturity, it usually bends upward as a result of cell elongation on the lower side of the nodes. When a stem of a dicot plant is lodged, the terminal portion tends to become erect due to the same response. The thickened nodal tissue, which bends up the stem, occurs in both the sheath and culm of some grass plants, such as sorghum and corn, but is found only at the base of the sheath in wheat, oats, barley, and rye. This growth response is the result of a phototropic (light) or an apogeotropic (opposed to gravity) stimulus.

Lodging also may occur by stem breakage, especially after maturity, and some crop plants may fall because of weak root anchorage.[14]

FUNCTIONS OF STEMS The functions of stems are as follows:[15]

1. Conduction of water and mineral solutes from the soil mostly through the vessels of the xylem.
2. Conduction of synthesized food materials through the phloem from the leaves to other parts of the plant.
3. Support and protection of plant organs and display of leaves and flowers to the sunlight to aid pollination.
4. Storage of food materials, as in sugarcane and sweet sorghum.
5. Manufacture of carbohydrates in the young stems that contain chlorophyll.

3.3.3 Leaves

Leaves develop from buds and are side or lateral appendages of the plant stem. Leaves are generally involved in photosynthesis and transpiration.[16]

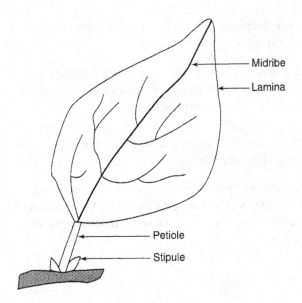

FIGURE 3.6

A simple leaf. The prominent part of the leaf is the lamina. It varies in size, shape, thickness, and other characteristics among species. Some leaves may not have a petiole and are attached directly to the branch.

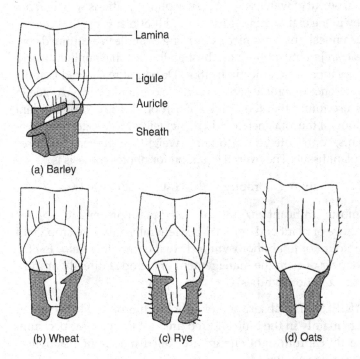

FIGURE 3.7

Grasses have certain unique features at the junction between the lamina and the stem: sheath, collar, and ligule. The sheaths in these small grains are split. The sheaths of rye and oats and the leaf margins of oats have pubescence. They may overlap or be closed in other species. The ligules vary in shape and size. Wheat has a large and rounded ligule. The auricles in barley are long and clasping, while rye has very short auricles. Oats have no auricles.

GENERAL CHARACTERISTICS The principal parts of a dicot leaf are the blade (lamina) and the petiole or leaf stalk (Figure 3.6). Some leaves have stipules, which are two small leaves at the base of the petiole. The leaf is said to be sessile when the petiole is absent. The principle parts of a grass leaf are the blade (lamina) and sheath (Figure 3.7). The sheath usually wraps tightly around the stem. The boundary between the blade and the sheath is called the collar. It may have a ligule and auricles that can be used for identification of the species.

Veins are the vascular bundles (xylem and phloem) in the leaf blade. Ordinary green foliage leaves will have two types of venation:

1. Parallel-vein leaves of the grasses, which contain many veins about equal in size that run parallel and are joined by inconspicuous veinlets. Growth of such leaves takes place by elongation near the base.
2. Net-vein leaves, which have a few prominent veins with a large number of minor veins. Growth occurs along the margin (edge) of the blade. Net-vein leaves are found in all crop plants other than grasses.

Net-veined leaves can have two distinct patterns of venation. Pinnately veined leaves have a prominent vein that runs from the base to the tip of the blade, and other veins branch from it. Palmately veined leaves have several prominent veins that originate at the base of the blade.

3.3.4 Photosynthesis

The seventeenth century physician Van Helmont reported that "164 pounds of wood, bark, and roots arose out of water only," because photosynthesis was then unknown. Joseph Priestley in the eighteenth century found that plants could "restore" air that had been made "undesirable" for mice by burning a candle in a confined container. The Dutch physician Jan Ingenhousz established the fact that sunlight reaching the green parts of a plant restored the air. In 1804, De Saussure showed that the leaves of plants in the presence of light absorb carbon dioxide and evolve oxygen in equal proportions. He also found that the uptake of carbon dioxide through plant stomata could not account for the total increases in plant weight. Hydrogen from water, and mineral nutrients, contributed the additional weight. Hydrogen comprises about 5 percent of the plant tissue. The overall equation for photosynthesis is:

$$6CO_2 + 12H_2O + light + chlorophyll \rightarrow C_6H_{12}O_6 + 6O_2 + 6H_2O$$

Photosynthesis is the most significant enzyme-mediated plant process. It supplies the energy for maintenance of plant and animal life. Generally, ultimate crop yield is determined by the amount of total photosynthesis that occurs in a field. Everything that the crop producer does in the management of a crop is directly or indirectly designed to maximize photosynthesis.

REACTIONS IN PHOTOSYNTHESIS The absorption of light by chlorophyll in the green plant (Figure 3.8) occurs mainly in the blue and red areas of the visible spectrum. The reactions that directly capture light energy are called the "light reactions." They only occur in the presence of light.[17]

The first light reaction involves the displacement of an electron from the chlorophyll molecule. The energy in this excited electron is used to trigger the formation of a high-energy phosphate bond and form an ATP molecule from ADP. This reaction is called photophosphorylation.

The next light reaction is the splitting of water. It is called photolysis or the "Hill reaction." This reaction releases hydrogen to produce NADPH and oxygen from the water molecule (Figure 3.9). The phosphate compounds (ATP and NADPH) participate in the synthesis of sugar. This is followed by the formation of other carbohydrates, as well as proteins and fatty compounds.

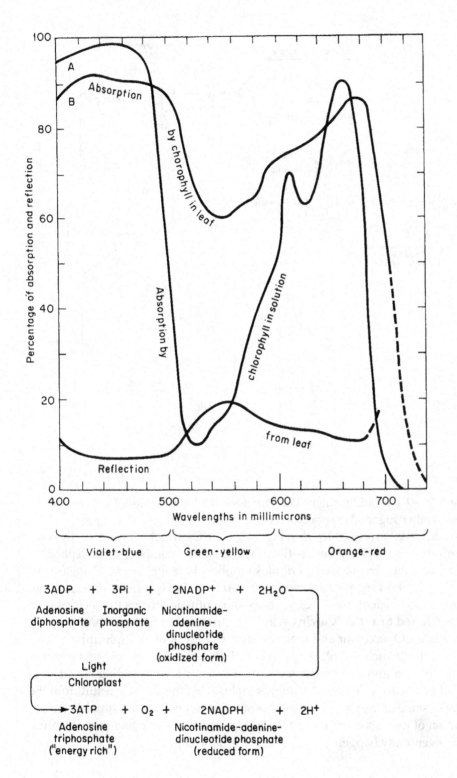

FIGURE 3.8
Absorption and reflection of light by chlorophyll. Curve A: Absorption of light by chlorophyll in solution. Curve B: Absorption by the same concentration of chlorophyll in a leaf. Chlorophyll in the leaf absorbs several times more green light than the same chlorophyll in solution. Curve C: The green color of a leaf is due to a slightly greater reflection of green than of other light. [Adapted from Carl L. Wilson and Walter E. Loomis, *Botany*, Rev. ed. New York: Holt, Rhinehart & Winston, 1957]

FIGURE 3.9
Photosynthetic phosphorylation and photolysis.

The "dark reactions" in photosynthesis (Figure 3.10) occur mostly during the day when the stomata are open and CO_2 is available. This involves the addition of CO_2 to a 5-carbon compound, ribulose diphosphate (*Step 1a*), which is split into two 3-carbon compounds (*Step 2*). These are then changed with the addition of hydrogen and the consumption of NADPH (*Step 3*) to two compounds, which are further converted to

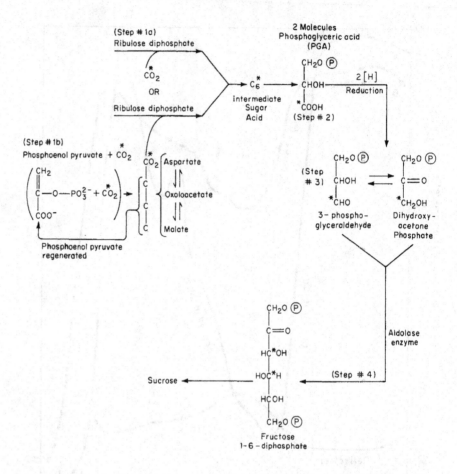

FIGURE 3.10

Simplified scheme showing the path of carbon in photosynthesis. C^* denotes C^{14}-labeled carbon.

fructose (*Step 4*), a 6-carbon sugar. Then fructose, through a series of conversions, is changed to other sugars. This photosynthetic process is called the C_3 system.

In an additional process (*Step 1b* in Figure 3.10), CO_2, which will eventually combine with ribulose diphosphate, is fixed by a 3-carbon compound (phosphoenol pyruvate) and then transferred to ribulose diphosphate for "normal" sugar synthesis. This 3-carbon acceptor process is referred to as the C_4 system. The C_4 system is a much more efficient user of CO_2. The photosynthetic process that begins with *Step 1a* is referred to as the "Calvin cycle." The process that first uses phosphoenol pyruvate as a CO_2 acceptor and then transfers CO_2 to ribulose diphosphate is referred to as the "Hatch and Slack pathway." High dry matter yielding crop plants such as corn, sorghum, and sugarcane are C_4 while the lower yielding wheat, soybean, and alfalfa are C_3 plants. The higher yields from the C_4 cycle result from the trapping of some of the CO_2 that would otherwise escape into the atmosphere. Incorporation of the C_4 system into a C_3 plant type is a goal of many crop scientists that may eventually happen through the use of biotechnology.

3.3.5 Light Energy Utilization

The solar energy reaching 1.5 square miles (3.75 km^2) of the earth's surface in one day is equivalent to a "Hiroshima model" atomic bomb. About a third of the total solar energy is reflected into outer space or absorbed in the atmosphere.

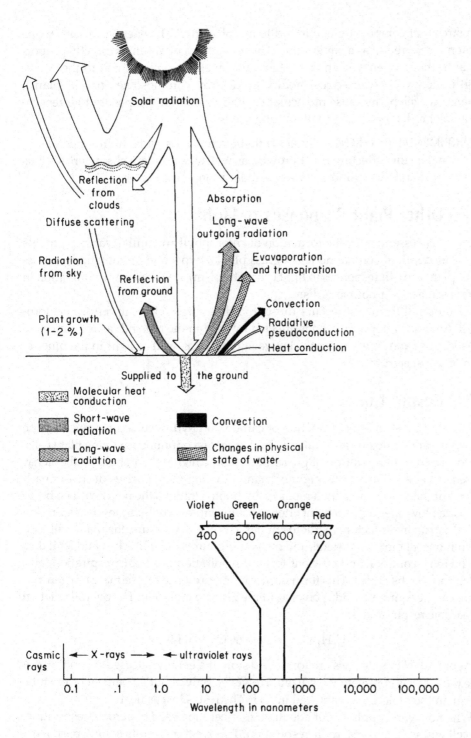

FIGURE 3.11

Energy exchange at noon on a summer day. The arrow width indicates relative amounts of energy transferred. Note that plant growth accounts for a very small part of the total energy budget. [After Geiger, *The Climate Near the Ground.* Cambridge: Harvard Univ. Press. 1950]

FIGURE 3.12

The spectrum of radiant energy plotted on a log scale. [Adapted from Galston, *The Life of the Green Plant.* Englewood Cliffs, N.J.: Prentice Hall, 1964]

Only about 1 to 2 percent of the solar energy received at a crop field during the growth of the plants is stored by photosynthesis (Figure 3.11). Of the total energy falling on the earth's upper atmosphere (3.3 cal/cm²/minute), only 0.146 cal/cm²/minute is available for photosynthesis.

The visible spectrum (300 to 700 nanometers) is light energy largely used by plants in photosynthesis (Figure 3.12). The longer red rays are more effective than the higher energy blue rays in promoting plant growth. Light travels through space as a stream of compact units referred to as "photons." The energy present in one photon is referred to as a "quantum." The energy in a quantum is directly proportional to the frequency of the radiation and inversely proportional to the wavelength. Heavy X-ray dosages modify or destroy chromosomes or kill plants. Infrared wavelengths cause molecular excitation, and long exposure to intensive ultraviolet radiation is fatal to plants and seeds.

ENVIRONMENTAL DIFFERENCES Photosynthetic rate varies with environmental and nutritional factors affecting plant growth, as well as with plant characters such as the leaf area and the position, density, and location of the leaves.

3.3.6 Other Plant Responses to Light

Light is also essential for the formation of chlorophyll and anthocyanin pigments in plants, depth of coronal nodes in grass plants, hypocotyl unhooking in cotyledons, plant growth responses to light (phototropism), dormancy or germination in certain seeds, and photoperiodism.

Chlorophyll formation occurs through a direct light effect upon protochlorophyll. Anthocyanin production is affected by light in the red and far-red ranges of the spectrum and in wavelengths between the two as well as light in the blue region of our spectrum.

3.3.7 Respiration

Respiration consists of the oxidation of carbon and hydrogen accompanied by the release of carbon dioxide and water.[17] This process is continuous in living plant cells. It provides the energy for plant growth during the night when photosynthesis has ceased. As a result, the dry weight of many growing plant species decreases each night, but this loss is more than replaced by photosynthesis the next day. The heavily shaded lower leaves on a plant often consume more carbohydrates by respiration than they can manufacture in photosynthesis. When that occurs, the plant will usually remove all mobile nutrients and other compounds out of the leaf and let it die.

The most common and efficient form of respiration is aerobic respiration, also called the "Krebs cycle." This form occurs in the presence of adequate oxygen and results in the complete oxidation of a simple glucose molecule. The overall reaction for aerobic respiration is:

$$C_6H_{12}O_6 + 6O_2 \rightarrow 6CO_2 + 6H_2O$$

About 40 ATP molecules are formed to store the energy released when the glucose molecule is completely oxidized. However, only about 50 percent of the total energy in the molecule is captured in ATP. The rest is lost as heat.

When oxygen supply is not adequate to meet the needs of aerobic respiration, the cell will switch to anaerobic respiration. This reaction results in only partial oxidation of the glucose and very little ATP formation. The partially oxidized molecule is C_2H_5OH, or ethanol. This compound is toxic to the cell. If anaerobic respiration occurs for very long, the cell will die. This reaction is the basis for fermentation of plant products. Anaerobic respiration can occur in roots when soils are poorly aerated due to flooding or compaction.

A third type of respiration that occurs only in C_3 plants is photorespiration. This occurs in the chloroplasts. It increases with high light intensity, high O_2, and low CO_2

concentrations, the very conditions that would be present in chloroplasts during photosynthesis. Photorespiration is of no value to the plant and reduces net photosynthesis. The presence of this reaction is the primary reason that C_3 plants light saturate (reach maximum rate of photosynthesis) at one-third to one-half full sunlight.

3.3.8 Transpiration

Transpiration is the loss of water in vapor form from a living plant or plant part. Nearly all the water absorbed by plants is transpired, but a plant is unable to grow unless it has sufficient water at its disposal. Transpiration may take place from any part not covered by layer cork or cuticle, but it is usually through the stomata. Stomata evolved to facilitate the entrance of carbon dioxide into the leaf for photosynthesis. The unfortunate side effect of opening up the interior of the leaf to the atmosphere is the loss of water vapor from the leaf. Although the leaf is cooled somewhat by the evaporation of water, and nutrients are transported up from the roots as water flows to the leaves, it is inconsequential compared to the detrimental effects of water loss. Because of transpiration, water is the most limiting factor to yield in most cases.[18]

Transpiration occurs as follows:[19]

1. The cell wall imbibes water from the cell contents. The water in the cell wall is then at a higher concentration than the surrounding gaseous environment of the substomatal cavity.
2. The water evaporates from the cell wall and raises the concentration of water vapor in the cavity.
3. This water vapor diffuses out through the stomata since the air in the substomatal cavity is more highly saturated than the outside air. Guard cells, which surround the stomatal opening, open and close with changes in water balance in the plant and also with light and darkness in many species.

The following conditions influence the rate of transpiration:

1. Capacity of the leaves to supply water
2. Power of the aerial environment to extract water from the leaves (i.e., influence of humidity, light, temperature, and wind)
3. Structure of the transpiration parts (i.e., the number of stomata and the sizes of the openings)
4. Modifications that tend to check excessive evaporation, including reduced plant surfaces, sunken stomata in special epidermal cavities, thickened cuticle, and waxy bloom on the cuticle of leaves and stems. A waxy bloom is conspicuous on sugarcane, sorghum, wheat, and barley.

Crop plants transpire 200 to 1,000 pounds (200 to 1,000 kg) of water for every pound (kg) of dry matter produced, which means that the transpiration ratio usually ranges from 200 to 1,000 for different crops. Thus, each pound (kg) of rainfall stored in the soil and protected from evaporation would produce from 1,000 pounds (kg) down to 200 pounds (kg) of dry matter in the form of crops. For a given crop, the transpiration ratio depends not only upon the previously mentioned factors that influence transpiration rate, but also upon the growth as well as adaptation of the crop to the particular environment. High soil fertility and a large volume of soil in which the roots can feed contribute to a lower transpiration ratio compared with less favorable soil conditions.[20] Cool-season crops such as oat have a comparatively high ratio when the temperatures are too high for optimum growth, while similar conditions would allow warm-weather crops, such as sorghum and cotton, to be relatively more efficient in the use of water.[21] In a cool season, the reverse may be true.

Crops that have the lowest transpiration ratios include millets, sorghum, and corn; the small grains are intermediate, whereas legumes such as alfalfa have relatively high transpiration ratios. The transpiration ratio is not necessarily a measure of the ability of a crop to produce well under drought conditions.[22] It merely measures the water transpired by the plants when the soil is supplied with water for optimum plant growth. For example, field pea and field bean (which have a high ratio) are grown successfully under dryland conditions where fair yields of seed are obtained from relatively small plant growth. The high value of the product compensates for the low production. Corn, sorghum, and certain millets that have a low transpiration ratio also have the efficient C_4 type of photosynthesis for dry matter production.

The adaptation of a crop to drought conditions also depends upon:

1. Drought escapement or evasion (i.e., the ability to complete the growth cycle before soil moisture is exhausted). Early-maturing varieties of small grains, for example, have made this adaptation.
2. Drought tolerance, or the ability of the plant to withstand drying and recover later when moisture becomes available (e.g., sorghum).
3. The ability of the plant to adjust its growth and development to the available water supply. Thus, cotton not only regulates its growth with the moisture supply but also sheds squares (flower buds), flowers, some of its bolls, and finally, many of its leaves when moisture continues to be deficient.

A so-called drought-resistant crop or crop variety will show one or more of these three characteristics.

A summary of the transpiration ratios of several crops and pigweed (*Amaranthus retroflexus*) Russian thistle (*Salsola kali*), and shepherd's purse (*Capsella bursa-pastoris*) at Akron, Colorado,[20] are shown in the following table.

Transpiration Ratios of Crops and Weeds at Akron, Colorado

Crop		Crop	
Alfalfa	858	Proso millet	267
Oats	635	Sudangrass	380
Cowpeas	578	Sugarbeet	377
Cotton	562	Buckwheat	540
Barley	521	Potato	575
Wheat	505	Spring rye	634
Corn	372	Red clover	698
Soybean	646	Hairy vetch	587
Crested wheatgrass	678	Blue grama	343
Millet	287	Sweetclover	731
Sorghum	271	Field pea	747
Flax	783	Pigweed	300
Bromegrass	828	Russian thistle	314
Edible bean	700	Lambsquarter	658

All these figures are based upon total dry matter in the plant above the ground, except for the sugarbeet and potato, in which the roots or tubers also were harvested.

Water balance is the relation between absorption and loss of water by the plant.[21] There is a deficit when absorption is exceeded by water loss. A deficit often occurs at midday when transpiration exceeds absorption (i.e., plants show signs of wilting and curling leaves). Increased and permanent wilting may cause injury and lead to death of the plant. *Permanent wilting* is defined as the point at which a plant cannot recover when placed in a saturated atmosphere. Plant species that endure great water loss without injury (e.g., guayule) are called xerophytes; those that require medium supplies of water (most crop plants) are called mesophytes, while those growing best with abundant water, including aquatics (e.g., rice), are called hydrophytes. Important characteristics of xerophytes are high concentrations of cell sap or high content of colloidal material or coverings that retard evaporation and the ability to recover after drying. The *consumptive* use of water (transpiration plus evaporation from the soil) for growing a crop usually ranges from 8 to 40 inches (200 to 1,000 mm) in total depth but may be up to 96 inches (2,440 mm) for long-season irrigated crops on very permeable soils.

▥ 3.4 REPRODUCTIVE PROCESSES IN CROP PLANTS

Crop plants reproduce either asexually using vegetative parts or sexually by the fertilization of flowers to produce seeds.

3.4.1 Asexual Reproduction

Crop plants that normally reproduce asexually, together with the plant part used in propagation, include:

1. Roots: sweet potato, cassava, and kudzu
2. Tubers: potato and Jerusalem artichoke
3. Stolons: buffalograss, creeping bentgrass, and bermudagrass
4. Rhizomes: bermudagrass and bluegrass
5. Stem cuttings: sugarcane and napiergrass

Asexual reproduction perpetuates uniform progeny except for an occasional bud mutation such as the sudden appearance of a red tuber in a white variety of potatoes. Many varieties that are uniform when propagated asexually are found to be extremely heterozygous (genetically variable) in their hereditary make-up when propagated from seeds.

3.4.2 Sexual Reproduction

The types of flowers involved in sexual reproduction are:

1. Perfect or bisexual flowers containing both stamens and pistils. Most crop plants have perfect flowers although some of them are more or less self-sterile. Some unisexual flowers are found on crop plants that are bisexual (e.g., sorghum and barley).
2. Imperfect or unisexual flowers containing either stamens or pistils, but not both.

When the separate staminate and pistillate flowers are borne on the same plant, the plants are called monoecious. Examples of these are corn, castor, and wild rice. When staminate and pistillate flowers are borne on separate plants, the plants are dioecious. These include hemp, hops, and buffalograss.

An exception to ordinary sexual reproduction is a process called *apomixis*, a form of parthenogenesis that results in the production of seed by a stimulation of the ovary without fertilization. This is a common method of seed formation in Kentucky bluegrass, dallisgrass, species of *Pennisetum* and *Paspalum*, and numerous other plant species.[23] Occasional apomixis occurs in sorghum and many other species. In one type of apomixis, called apospory, somatic cells in the nucleus enlarge and the nuclei divide to form an embryo sac containing an egg and polar nuclei.[24] When Kentucky bluegrass is cross-pollinated, apospory may also occur and increase the number of chromosomes in the cross.

MODE OF POLLINATION Sexually propagated crop plants may be grouped into three general classes on the basis of their normal habit of pollination. These are: (1) naturally self-pollinated, (2) often cross-pollinated, and (3) naturally cross-pollinated.

Naturally Self-Pollinated. Crop plants naturally self-pollinated, including wheat, oat, barley, tobacco, potato, flax, rice, field pea, cowpea, and soybean, usually show less than 4 percent cross-pollination. The percentage of cross-pollination in these crops varies with variety and season or environment.[25, 26, 27, 28, 29] Any environmental or hereditary factor that interferes with normal pollination or fertilization may result in a high proportion of cross-pollination. In naturally self-pollinated plants, the pistil usually is pollinated by pollen from the same flower.

Often Cross-Pollinated. These include cotton, sorghum, foxtail, and proso millet and several cultivated grasses. The pistil may be pollinated by pollen from the same flower or from another flower on the same plant or from another plant. Crossing among cotton flowers is usually due to insects. In the grass crops mentioned previously, pollination is mostly from air-borne pollen. In sorghum, self-pollination occurs in about 94 to 95 percent of the flowers.[30]

Naturally Cross-Pollinated. Cross-pollination is the normal mode of reproduction in many crop plants, including corn, rye, clover, alfalfa, buckwheat, sunflower, some annual grasses, and most perennial cultivated grasses. The extent of self-pollination in corn is less than 5 percent,[31] and sometimes as low as 0.7 percent. Most of the grasses are wind-pollinated, but buckwheat and legumes such as clover and alfalfa are adapted to pollination by insects. Cross-pollination is essential to seed production in many plants of red clover, rye, and common buckwheat because of self-sterility.

DESCRIPTION OF FLOWERS The flower develops from a bud either at the apex of the stem or in the leaf axil. Flowers are usually grouped into an inflorescence on a special shoot or axis. The leaves of the inflorescence are known as bracts, while the axis is usually a rachis or a peduncle. An individual flower stalk is a pedicel. Types of inflorescences are shown in Figure 3.13. The type of inflorescence is valuable in identifying plants.

A flower is composed of the stamens (male organs), pistils (female organs), calyx, and corolla (Figure 3.14). Glumes, lemma, and palea in the grasses replace the last two structures.[32]

The stamens and pistils are the organs of fertilization. The stamen is composed of a filament or stalk that has the anther at its outer end. The interior lobes of the

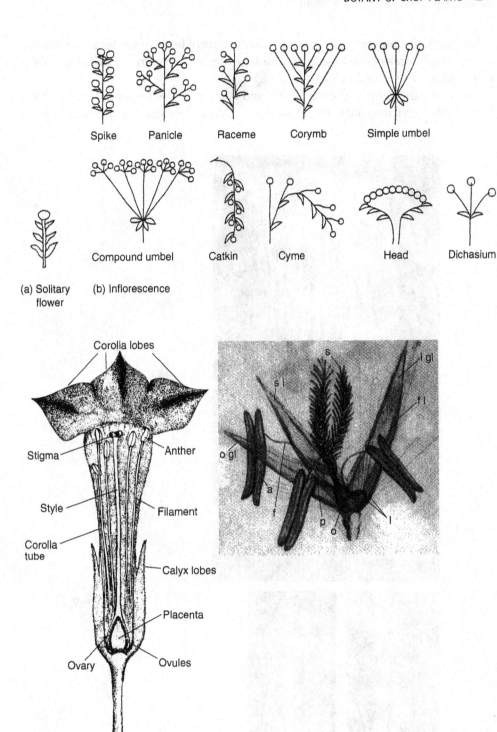

Spike   Panicle   Raceme   Corymb   Simple umbel

Compound umbel   Catkin   Cyme   Head   Dichasium

(a) Solitary flower   (b) Inflorescence

FIGURE 3.13
Types of inflorescence. A simple flower is also solitary, occuring alone on a pedicel. An inflorescence consists of numerous florets arranged in a variety of ways. There are three basic types of inflorescence: head, spike, and umbel.

Corolla lobes

Stigma

Style

Corolla tube

Ovary

Anther

Filament

Calyx lobes

Placenta

Ovules

FIGURE 3.14
(Left) Tobacco flower showing typical floral parts. *(Right)* Parts of an opened grass (sugarcane) flower: *(a)* anther, *(f)* filament, *(fl)* fertile lemma, *(igl)* inner glume, *(l)* lodicules, *(o)* ovary, *(ogl)* outer lemma, *(p)* palea, *(s)* stigmas, and *(sl)* sterile lemma. [After Artschwager]

anthers are hollow spaces or pollen sacs in which the pollen is produced in the form of loose, round pollen grains. Finally, each sac splits open (dehisces) to allow the pollen to escape. The pistil is composed of the stigma, style, and ovary. The ovary is the swollen, hollow base of the pistil (Figures 3.15 and 3.16). The elongated portion is the style, which has the stigma at its apex. The ovary contains ovules, generally egg-shaped, attached to the ovary walls.

In grasses, the flowers are forced open at the time of fertilization by the swelling of two small organs called *lodicules*. The lodicules lie between the ovary and the surrounding lemma and palea.

The perianth, composed of the calyx and corolla, is a nonessential part of the flower, although it may aid in the attraction of insects. The calyx is made up of sepals

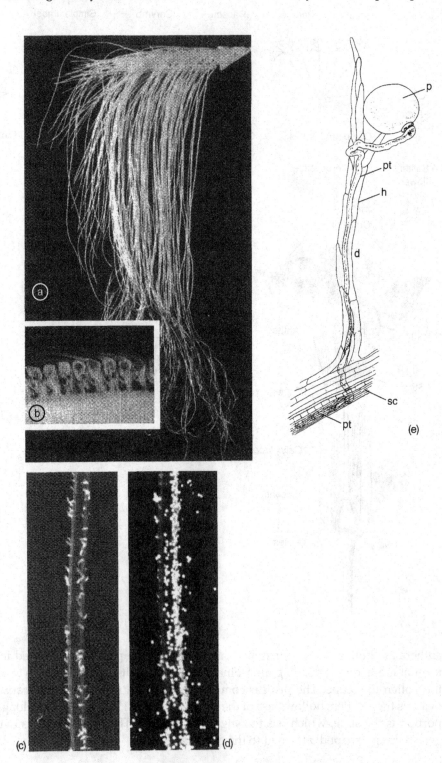

FIGURE 3.15

Pistillate inflorescence of maize: *(a)* unpollinated ear and silks, *(b)* enlarged ovaries showing attachment of silks, *(c)* unpollinated silk, *(d)* pollinated silk, *(e)* germinated maize pollen grain *(p)*, and pollen tube *(pt)* passing down through stigma hair *(h)* into silk *(sc)*.

that enclose the other floral parts in the bud and thus protect the young flower. The corolla, composed of petals, is generally the showy part of the flower (Figure 3.17).

PROCESS OF FERTILIZATION The ovule is composed of a nucellus surrounded by an inner and outer integument. The embryo sac is within the nucellus (Figure 3.16). It contains eight nuclei: two synergids and one egg cell at the micropylar end, three antipodal cells at the opposite end, and two endosperm or polar nuclei near the

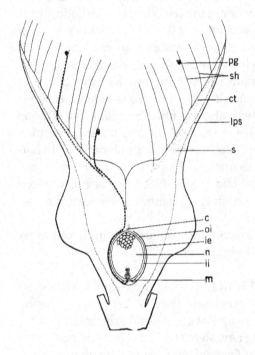

FIGURE 3.16

Barley pistil: *(c)* cone-shaped tip of outer integument, *(ie)* inner epidermis of ovary wall, *(ii)* inner integument, *(lps)* lateral procambial strands, *(m)* micropyle, *(n)* nucellus that contains the embryo sac, *(oi)* outer integument, *(pg)* pollen grain, *(s)* style, and *(sh)* stigma hair. The broken line shows the course of the pollen tube from pollen grain to micropyle *(m)*. Above the micropyle are two synergids and the egg nucleus, and above the latter are the two endosperm nuclei. The three antipodal cells higher up have already divided several times.

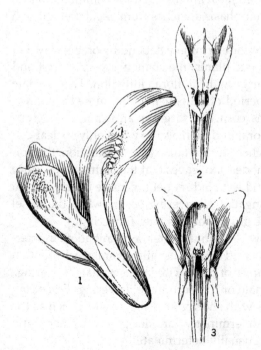

FIGURE 3.17

Alfalfa flower, typical of legumes. *(1)* Opened flower showing position of style and anthers before tripping at left and after tripping at right. The large segment of the corolla at right is the "banner." *(2)* Unopened flower with banner removed. *(3)* Open flower with the banner removed to show the two wings (left and right), the two "petals" of the keel fused at rear, and the style and anthers at front.

center. Pollen grains that fall on the stigma of the pistil germinate in the sticky stigmatic excretion. The germinated pollen grain contains a tube nucleus and two sperm nuclei. The tube nucleus develops a pollen tube that elongates down through the style to the embryo sac. The pollen tube usually grows in the intercellular spaces of the style. One sperm nucleus unites with the egg cell (sexual fertilization) that will develop into the embryo of the seed. The other sperm nucleus fuses with two polar nuclei (triple fusion) that can develop into the endosperm. The process of sexual fertilization and triple fusion is called double fertilization and occurs in all flowering plants whether or not the seed develops an endosperm.

The mechanism of double fertilization accounts for the phenomenon of xenia or the immediate observable effect of foreign pollen on the endosperm. The classical example is the various endosperm colors in maize produced by foreign pollen that are evident as soon as the out-crossed grains ripen. Xenia also has been observed in rye, rice, barley, sorghum, pea, bean, and flax. The process is called metaxenia when the pericarp or other maternal tissue is also modified by double fertilization as a result of adjustment to a change in size or shape of the affected endosperm.

Fertilization in corn takes place within 26 to 28 hours after pollination.[33] In barley, the pollen germinates within 5 minutes after reaching the stigma; the male nuclei enters the egg sac within 45 minutes; and the fertilized egg begins division within 15 hours.[34] Flax pollen germinates and the pollen tube penetrates to the base of the style about 4 hours after pollination.

Crop seeds develop rapidly after pollination, fully maturing within four to six weeks in many crops and up to nine weeks in others.

DETERMINATE AND INDETERMINATE FLOWERING Flowering represents a wide spectrum of physiological and morphological processes. The first step is the changing of the vegetative stem primordia, or growing points, into floral primordia. In other words, the growing points stop producing new stem and leaf tissue and begin producing flower tissue. This change is called *floral initiation*. Floral initiation is a pivotal event in a plant's life cycle as the emphasis switches from vegetative growth to reproductive growth.

Plants are classified as determinate or indeterminate depending on the way that floral initiation progresses. In determinate plants, the change is very abrupt and the plant ceases to initiate vegetative growth at floral initiation. Determinate plants flower only once over a short period of time, usually no more than a few days. All grasses, including small grains, corn, and sorghum are examples of crops with determinate flowering. In corn, floral initiation occurs when only six leaf collars are visible; the plant is about 18 inches (46 cm) tall. At floral initiation, the apical growing point has initiated all the nodes and leaves that the plant will produce and switches to producing the tassel and ears. Subsequent increase in size is from cell division at the nodes and cell elongation. Flowering occurs when the tassel and ears emerge. No further vegetative growth occurs after flowering.

Indeterminate plants continue to grow vegetatively after floral initiation. Indeterminate plants will flower over a longer period of time, usually several weeks. Cotton and some varieties of soybean are examples of crops with indeterminate flowering.

Soybean can exhibit either determinate or indeterminate flowering depending on the variety. Varieties grown in areas with shorter growing seasons, such as the Midwest United States, are usually indeterminate. Varieties grown farther south where the growing season is longer are usually determinate.

3.5 SEEDS AND FRUITS

A representative mature dicot seed consists of the seed coat (testa), nucellus, and embryo. The embryo, or germ, is composed of the plumule (leaves), hypocotyl (stem), radicle (root), and one or two cotyledons. The seed may be defined as a matured ovule. The bean is a good example of a seed. It is attached to the ovary wall (pod) by a short stalk (the funiculus). The hilum is an elongated scar on the bean where the funiculus was attached. Close to one end of the hilum is a small opening called the micropyle, a scar where the pollen tube entered the ovules. A seed coat, or testa, covers the bean, beneath which are two fleshy halves called cotyledons. A typical endosperm is lacking in most dicot seeds because of the cotyledons occupying most of the space within the seed coat.

A fruit is a matured ovary. It contains the seeds or matured ovules. The mature ovary wall is known as the pericarp. The entire bean pod is a fruit (Figure 3.18), but the *beans* are seeds. Such fruits may contain several seeds.

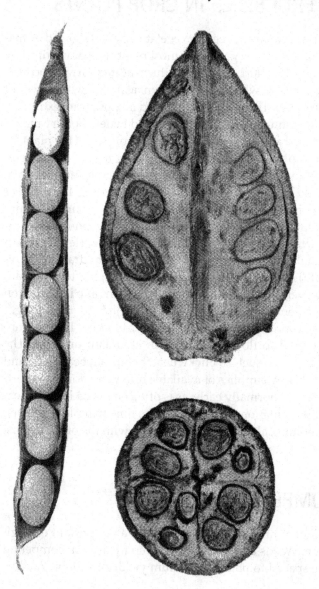

FIGURE 3.18

(Left) Pod (fruit) and seeds (ovules) of the Pinto bean. *(Right)* Half of fruit (boll or capsule), and seeds (ovules) of cotton; the cross section *(below)* shows four locks.

In indehiscent dry fruits, the pericarp is dry, woody, or leathery in texture and does not split or open along any definite lines. The achene is a one-seeded fruit in which the pericarp can be readily separated from the seed as, for example, in buckwheat. The caryopis of grasses is a one-seeded fruit. The seed within it is united with the ovary wall or pericarp. Percival[10] describes the wheat kernel as a kind of nut with a single seed. The wheat kernel (caryopsis) consists of the pericarp, endosperm, and embryo. The outer rim of the embryo is the scutellum, a flattened, somewhat fleshy, shield-shaped structure situated back of the plumule and close to the endosperm. The scutellum is regarded as the single cotyledon of a grass "seed."

In dehiscent dry fruits, the pericarp splits in various ways or opens by pores. The seeds on the interior of the fruit are thus set free. The legume pod or fruit dehisces along two sutures, as in field pea or soybean. In red clover, the capsule dehisces transversally; that is, the upper part of the carpel falls off in the form of a cap or lid.

■ 3.6 GROWTH PROCESSES IN CROP PLANTS

Growing plants pass through a gradual and regular vegetative cycle, followed by maturity and reproduction.[35] The vegetative growth of most plants is characterized by three phases—namely, an initial slow start, a period in which the growth rate becomes gradually faster, and finally, a period when it slows down again. Growth curves of sorghum[36] are shown in Figure 3.19. The processes involved in growth are extremely complex. Growth is usually measured by the amount of solid material or dry matter produced. This is often expressed agriculturally as yield of all or part of the plant.

Vegetative development in plants is interrupted by the reproductive phase in which the flowers and seeds are formed. Different species and crop varieties have a characteristic though highly variable vegetative period. However, the onset of the reproductive period may be altered greatly by changes in environmental conditions. Thus, a soil rich in nitrogen will tend to keep down the carbon-nitrogen ratio, which delays flowering in *nitronegative* crops such as wheat, barley, oat, most pasture grasses, mustard, alfalfa, and clover. The excess of carbohydrates over nitrogen is usually very great (i.e., the carbon-nitrogen ratio is high after the plant commences to flower). However, changes in the carbon-nitrogen ratio in the cotton plant may be merely incidental rather than fundamental to flowering.[37] *Nitropositive* crops, such as corn, sorghum, millet, cotton, tobacco, lupine, sunflower, and pepper, are reported to flower earlier with abundant nitrogen. The flowering time of *nitroneutral* crops, such as buckwheat, hemp, soybean, pea, and bean, is not affected appreciably by supplies of available nitrogen. Moisture shortage may cause plants to flower abnormally early, as is often observed in a dry season. The length of the daily illumination period affects the time that plants flower, called *photoperiodism*. Temperature, alone or in combination with the other factors, also affects the onset of flowering.

■ 3.7 PLANT COMPETITION

Regulation of the kind and amount of plant competition is an important ecological phase of crop production. Weeds are destroyed to keep them from competing with crop plants. Crops are spaced to obtain maximum yields of suitable quality.

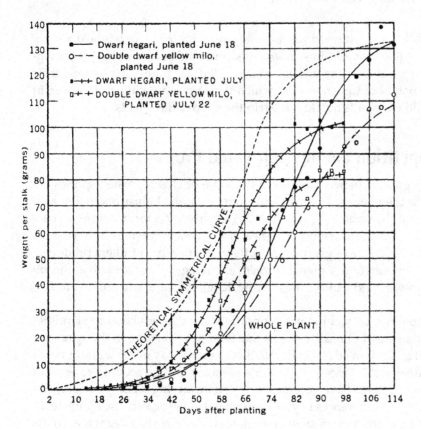

FIGURE 3.19
Most crop plants (e.g., corn and barley) show a growth rate similar to the theoretical symmetrical curve. The sorghums, dwarf hegari and double dwarf yellow milo, have smaller seeds in proportion to the ultimate plant size and therefore show an apparent slower growth rate at the beginning.

A moderate amount of competition between plants is not detrimental on an area basis. A "struggle for existence" results when plants are grouped or occur in communities in such a way that their demands for an essential factor are in excess of the supply. Competition is a powerful natural force that tends to eliminate or wipe out the weak competitors. It is mostly a physical process, in which the amount of water, nutrients, and light that otherwise would be utilized by each individual plant are decreased. Competition increases with the density of the plant population. Excessive populations often reduce the yield of seeds, while stimulating vegetative growth.

Numerous chemicals, called growth regulators, can modify the growth rate, transpiration, height, branching, rooting, flowering, or maturity of plants or plant parts.[38] A number of these are useful in ornamental greenhouses, home gardens, and lawns. In certain field crops, chemicals are used as desiccants, defoliants, or to suppress further branching or flowering.

3.7.1 Weed Competition

In fields, crops are often confronted with serious competition with weeds. Successful competition against weeds depends on the readiness and uniformity of germination under adverse conditions, the ability to develop a large assimilation surface in the early seedling stage, the possession of a large number of stomata, and a root system with a large mass of roots close to the surface and deeply penetrating main roots.[39] Crop plants have been classified in order of decreasing competitive ability against weeds as follows: corn > barley > rye > wheat > oat > flax. Common wild mustard (*Brassica arvensis*) and wild oat (*Avena fatua*) are vigorous competitors among the annual weeds. Wild oat plants may adapt their growth to the height of the grain with which they are competing.[30] However, in a flax field, wild oat plants quickly outstrip the shorter flax plants.

3.7.2 Competition among Cultivated Crops

Competition is greatest between plants of the same or closely related species because they make their demands for moisture, nutrients, and light at about the same time and at about the same level (Figure 3.20). Competition for moisture is especially important in dryland regions. In grass crops, competition for light reduces the number of heads, causes great irregularity in the number of tillers produced, diminishes the amount of dry matter formed, encourages shoot growth at the expense of root growth, and reduces the tillering, length of culm, and number of kernels per head.

The reduction in corn yield is not proportional to the reduction in stand because of the adjustment of nearby plants that benefit by feeding in the space not occupied by plants.[40] Considerable fluctuation in stand may occur without much effect on the final yield. Since it is possible to easily change seeding rates with modern planters, farmers have become interested in varying seeding rates within a field depending on differences in yield potential. However, because plants readily compensate for changes in plant population, there is no advantage unless there are vast differences in yield goal. For example, dryland corners of a center-pivot irrigated field, which would have a much lower yield, would be planted at lower seeding rates. Factors affecting seeding rates are discussed more fully in Chapter 7.

Sugarbeet may respond to additional space from a missing plant, increasing in weight enough to compensate for 96 percent of the loss.[41] Close spacing in flax reduces the growth of plants significantly.[42] As the space between plants increases, there is an increase in number of bolls, yield per plant, weight per plant, and stems per plant.

In sorghum, within certain limits, medium-thick planting makes the plants taller because of greater competition for light. Thick stands make the plants shorter because of inadequate moisture or nutrients to support a good growth in so many plants.

Competition between different crop varieties causes marked changes in population after a few years, even though the original mixture comprised equal numbers of seeds of each variety. The ascendant varieties are not necessarily those that yield best when seeded alone.[43]

3.7.3 Plant Association

Competition between different plant species may not be too severe because one plant has an opportunity to fit in among some others. For example, in a mixture of timothy and clover, the two crops make different demands on the habitat at different times and at different soil depths. The more unlike the plants are, the greater the differences in their respective needs and the less they will compete with each other.

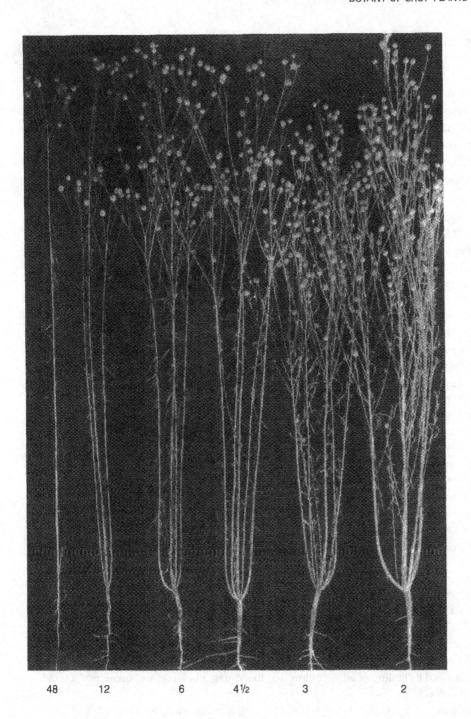

48   12   6   4½   3   2

FIGURE 3.20
Flax plants grown at six different spacings. The single-stemmed plant at the left suffered from competition with other flax plants. The figures indicate the number of plants that occupy a square foot of soil.

Competition played an important role in irrigated pasture mixtures in northern Colorado. The most vigorous seedlings, especially of orchardgrass, soon gained dominance even when seeded in amounts as small as 13 percent of total seed. Kentucky bluegrass, white clover, and timothy suffered greatly in this competition.

Companion crops are annuals planted with new seedlings of alfalfa, red clover, and similar perennial crops in order to secure a return from the land the first year. These companion crops, which compete with the young seedlings, usually are planted at about half the usual rate to reduce competition.

Mixtures of a small grain with a trailing vine legume, such as vetch and field pea, are helpful because the grain stems help support the legume vines. Korean lespedeza is grown with small grains. The planting of corn or sorghum with cowpea, soybean, or velvetbean has been practiced in the Southeast with the object of increasing total yields over the yields produced by the same crops grown alone. Some increase in yield is possible because of the different growing habits of the crops. In crop mixtures in which the cultural requirements for the individual crops are different, there may be sacrifice in quality or yield of one or both crops as well as an increase in production costs.

REFERENCES

1. Weaver, J. E. "Investigations of the root habits of plants," *Am. J. Bull.* 12(1925): 502–509.
2. Krassovsky, I. "Physiological activity of the seminal and nodal roots of crop plants," *Soil Sci.* 21(1926):307–325.
3. Locke, L. F., and J. A. Clark. "Normal development of wheat plants from seminal roots," *J. Am. Soc. Agron.* 16(1924):261–268.
4. Weaver, J. E., and E. Zink. "Extent and longevity of the seminal roots of certain grasses," *Plant Physiol.* 20, 3(1945):359–379.
5. Weihing, R. M. "The comparative root development of regional types of corn," *J. Am. Soc. Agron.* 27(1935):526–537.
6. Pavlychenko, T. "Root systems of certain forage crops in relation to the management of agricultural soils," *Natl. Res. Council Canada and Dominion Dept. Agr. N. C. R. 1088*, Ottawa, 1942. pp. 1–46.
7. Breazeale, J. F. "Maintenance of moisture equilibrium and nutrition of plants at or below the wilting percentage," *AZ Agr. Exp. Sta. Tech. Bull.* 29, 1930.
8. Larson, K. L., and J. D. Eastin, eds. *Drought Injury and Resistance in Crops*, *CSSA Spec. Pub.* 2, 1971, pp. 1–95.
9. Hitchcock, A. S. *A Textbook of Grasses*. New York: Macmillan, Inc., 1914, pp. 1–276.
10. Percival, J. *Agricultural Botany*, 7th ed. London: Duckworth, 1926, pp, 7–24, 34–36, 40–42, 52–60, 113–121, 136–140, 622–646.
11. Anonymous. *Lodging of Cereals, Bibliography*. Cambridge: Imperial Bureau of Plant Genetics, 1930.
12. Welton, F. A., and V. H. Morris. "Lodging in oats and wheat," *OH Agr. Expt. Sta. Bull.* 471, 1931.
13. Atkins, I. M. "Relation of certain plant characters to strength of straw and lodging in winter wheat," *J. Agr. Res.* 56, 2 (1930):99–120.
14. Fellows, H. "Falling of wheat culms due to lodging, buckling, and breaking," *USDA Cir.* 769 (1948).
15. Bobbins, W. W. *Botany of Crop Plants*. Philadelphia: Blakiston, 1931, pp. 22–39.
16. Gausman, H. W., and others. "The leaf mesophylls of twenty crops, their light spectra and optical and geometrical parameters," *USDA Tech. Ball.* 1405,1973, pp. 1–59.
17. Salisbury, F. B., and C. W. Ross. *Plant Physiology*, 4th ed. Belmont, CA: Wadsworth, 1991.
18. Waldren, R. P. *Introductory Crop Science*, 5th ed. Minneapolis: Burgess, 2003.
19. Kiesselbach, T. A. "Transpiration as a factor in crop production," *NE. Agr. Exp. Sta. Res. Bull.* 6, 1916.
20. Kiesselbach, T. A. "Corn investigations," *NE Agr. Exp. Sta. Res. Bull.* 20, 1922.
21. Shantz, H. L., and L. N. Piemeisel. "The water requirement of plants at Akron, Colo.," *J. Agr. Res.* 34, 12(1927):1093–1190.

22. Maximov, N. A. *The Plant in Relation to Water: A Study of the Physiological Basis of Drought Resistance*, London: Allen and Unwin, 1929.

23. Bashaw, A. W., and others. "Apomixis, its evolutionary significance and utilization in plant breeding," *XI International Grassland Congress Proceeding*. 1970.

24. Hanna, W. W., and others. "Apospory in *Sorghum bicolor* (L.) Moench," *Science* 17(1970):338–339.

25. Beachell, H. M., and others. "Extent of natural crossing in rice," *J. Am. Soc. Agron.* 27 (1935):971–973.

26. Dillman, A. C. "Natural crossing in flax," *J. Am. Soc. Agron.* 30(1938):279–286.

27. Garber, R. J., and T. E. Odland. "Natural crossing in soybeans," *J. Am. Soc. Agron.* 18(1926):967.

28. Robertson, D. W., and G. W. Deming. "Natural crossing in barley at Fort Collins, Colo.," *J. Am. Soc. Agron.* 23(1931):402–406.

29. Stanton, T. R., and F. A. Coffman. "Natural crossing in oats at Akron, Colo.," *J. Am. Soc. Agron.* 16(1924):646–659.

30. Sieglinger, J. B. "Cross-pollination in milo in adjoining rows," *J. Am. Soc. Agron.* 13(1921):280–282.

31. Harlan, H. V. "The weediness of wild oats," *J. Hered.* 20, 11(1929):515–518.

32. Bennett, O. D. "Inflorescences of maize, wheat, rye, barley and oats—their initiation and development," *IL Agr. Exp. Sta. Bull.* 721, 1966, pp. 1–105.

33. Miller, E. C. "Development of the distillate spikelet and fertilization in *Zea mays*," *J. Agr. Res.* 18(1919):255–265.

34. Pope, M. N. "The time factor in pollen-tube growth and fertilization in barley," *J. Agr. Res.* 54, 7(1937):525–529.

35. James, W. O. *An Introduction Plant Physiology*, 2nd ed., New York: Oxford A. Press, 1933, pp. 1–263.

36. Bartel, A. T., and J. H. Martin. "The growth curve of sorghum," *J. Agr. Res.* 51, 11 (1938):843–849.

37. Eaton, F. M., and N. E. Rigler. "Effect of light intensity, nitrogen supply and fruiting on carbohydrate utilization by the cotton plant," *Plant Physiol.* 20, 3(1945):380–411.

38. Audus, L. J. *Plant Growth Substances.* London: Leonard Hill, 1959. pp, 1–553.

39. Pavlychenko, T. K., and J. B. Harrington. "Competitive efficiency of weeds and cereal crops," *Can. J. Res.* 10(1934):77–94.

40. Kiesselbach, T. A. "Competition as a source of error in comparative corn yields," *J. Am. Soc. Agron.* 15, (1923):199–215.

41. Brewbaker, H. E., and G. W. Deming. "Effect of variations in stand on yield and quality of sugarbeets grown under irrigation," *J. Agr. Res.* 50, (1935):195–210.

42. Klages, K. H. "Spacing in relation to the development of the flax plant," *J. Am. Soc. Agron.* 24(1932):1–17.

43. Harlan, H. V., and M. I. Martini. "The effect of natural selection in a mixture of barley varieties," *J. Agr. Res.* 57(1938):189–200.

4

Crop Improvement

■ 4.1 POSSIBILITIES IN CROP IMPROVEMENT

The creation of each new superior hybrid or variety (cultivar) is a definite advance in human welfare. The breeding of more productive types is the only stable method of increasing crop yields. Although crop yield and quality can be improved by disease and pest control,[1] fertilizer application, and better cultural methods, such practices must be repeated each season. Improved cultivars of many crops are rapidly replacing inferior local varieties in most countries of the world. The maintenance, production, and appropriate distribution of seed stocks of improved types have resulted in striking increases in crop yields in many countries.[2]

Crop improvement has been in progress since primitive people first exercised a choice in selecting seed from wild plants for growing under cultivation. The greatest advances were made before the dawn of civilization, but material progress also was made thereafter. However, not until the twentieth century, when knowledge of genetics was acquired, did crop breeding become a science with the outcome of breeding methods reasonably predictable.[3] Remarkable increases in yield followed the breeding and wide distribution of semidwarf, photoperiod-insensitive cultivars of rice and wheat that resist lodging under heavy nitrogen fertilization.[4] Considerable progress is still possible by utilizing material from world collections of seeds and plants that are being widely distributed to breeders and from the utilization of genetic engineering.

Striking benefits from crop breeding have been obtained since about 1940.[5] The increased farm income from hybrid corn and hybrid sorghum exceeds the cost of all agricultural research in the United States.

Crop breeders are mainly concerned with more abundant, stable, and economical crop production, but they also are deeply motivated by social and humane wants.[6] Farmers are now producing food products that are higher in protein or certain essential amino acids, as well as sweeter and more tender sweet corn, and lighter bread made from stronger wheat. Crops that are easier for people to harvest, and feed crops that are more palatable or less toxic to livestock, also have been developed. Improved cultivars have helped to alleviate crop failures and their accompanying human and livestock starvation.

■ 4.2 OBJECTIVES IN CROP BREEDING

It is unlikely that a perfect crop variety or hybrid will ever be developed. Every new variety or hybrid has several weaknesses or defects that curtail maximum yields and quality except under rare ideal conditions. The objective of the crop breeder is to correct these defects while developing cultivars with higher yield capabilities. New high-yielding cultivars are being bred in many countries.[7] First-generation hybrids with increased vigor, growth, and reproduction have markedly increased the yields of several crops.

Some of the main objectives in crop breeding are the following improvements in characteristics:

1. Resistance to diseases, insects, drought, cold, heat, lodging, and alkali soils
2. Adaptation to variable photoperiods, shorter seasons, longer seasons, heavy grazing, or frequent cutting
3. Feeding quality, including palatability, leafiness, hull percentage, nutritive value, and texture
4. Market quality, including higher content of textile fiber or of protein, sugar, starch, or other extractives; better processing quality for textiles, foods, beverages, and drugs; and better color
5. Seed quality, including higher or lower seed-setting tendency, greater longevity, higher viability, and larger size[8]
6. Growth habit, including more erect or prostrate stems, more or less tillering or branching, more uniform flowering and maturity, more uniform height, longer life, and better ratio of tops to roots
7. Harvesting quality, including stronger, shorter, or taller stalks; erect stalks and heads; non-shattering qualities; easier processing; and freedom from irritating awns and fuzz
8. Productive capacity, including greater vigor, higher fertility, and faster recovery after cutting.

■ 4.3 METHODS OF CROP BREEDING

Three common methods of crop improvement are introduction, selection, and hybridization. These methods are not wholly distinct because hybridization is usually preceded, followed, or both preceded and followed by some scheme of selection. Irradiation of plant material occasionally produces a new trait that is useful in crop improvement. In addition, genetic engineering allows plant breeders to utilize genes from other species and subsequently incorporate the new traits into existing cultivars through selection and hybridization.

4.3.1 Introduction

Introductions of genetic lines from other countries may be of superior productivity, and they often provide better foundation stocks for breeding. Foreign varieties, that seemingly are worthless, may possess resistance to some disease or insect, or they may have some other useful characteristic that can be transferred to adapted varieties by hybridization.[9] Numerous foreign and domestic wild plant species are being tested for possible cultivation. Collections of exotic and domestic wild and cultivated species and strains of various crop plants are being screened for disease and insect resistance as well as other characteristics that might be useful in breeding.

When some variety, strain, or inbred line is found to be outstanding for some desired trait, such as disease resistance, heterosis capacity, or cytoplasmic sterility, it is likely to be used by most plant breeders. This results in a narrow genetic base, which is disastrous when a new or minor race of a parasite multiplies and almost destroys previously resistant varieties or hybrids that are widely grown. Severe losses have occurred from epidemics of a previously unimportant race of a fungus parasite attacking a large number of varieties or hybrids of wheat, oat, or corn that had a common resistant ancestry.[5, 10] This necessitates a search for plants with genes for resistance to the new race.

Resistance to a specific race, referred to as vertical resistance, provides good protection until other races arise and then multiply rapidly with little competition from any existent race.[11] The alternative is to incorporate genes for horizontal resistance (race nonspecific or generalized resistance) into the crops. Horizontal resistance provides partial protection against, or tolerance for, a number of races of a fungus, but the plants are not wholly free from the disease.[12] Horizontal resistance often is polygenic (involving several genes), but it sometimes is controlled by only a few genes and occasionally is monogenic.

4.3.2 Selection

MASS SELECTION Mass selection is a quick method of purifying or improving mixed or unadapted crop varieties. It involves selecting a large number of plants of the desired type and then increasing the progeny. Seeds from the selected plants are usually bulked together. Mass selection helps eliminate undesirable types. Most of the open-pollinated varieties of corn were products of mass selection.[13] Natural selection, through survival of the more vigorous strains, accomplishes a similar objective. Winter barley and winter rye were improved in winter hardiness by natural selection when grown under cold conditions in the United States. Mass selection accomplishes nothing when selection is confined to vigorous plants that are merely favored by a good environment such as thin stands or more favorable soil or moisture locations in the field.[14]

PURE-LINE OR PEDIGREE SELECTION Pure-line, pedigree, or individual plant selection consists of growing individual progenies of each selected plant so that their performances can be observed, compared, and recorded. Only a few superior strains among the numerous original selections are saved for advanced testing.[15] In crops that are largely self-fertilized, reselection is not necessary except for progenies resulting from occasional natural hybrids or mutations that are still segregating. In crops that are partly or largely cross-fertilized, selection must be repeated until the strains appear to be uniform. Also, cross-fertilized crop plants selected for propagation must be self-pollinated under controlled conditions. This is done by covering the floral parts to exclude foreign pollen, by hand pollinating, or both. In some cases, simply isolating the plants will work. Pure-line selection offers a quick means of segregating desired types from mixed varieties.[16] Numerous varieties were originated by this method. [2, 17, 18] Simple selection fails to introduce new desired traits.

In cross-pollinated crops, or when the original variety and the selections differ by several genetic factors, several (five or more) generations of selection may be required to purify the strains so that each will continue to breed true thereafter while being tested for yield.

Ear-to-row, a form of pedigree selection of corn, involves planting all the seeds from a single ear into a single row. The best ears from the row are selected and the seed is planted again in single rows. The process is continued for several generations. Ear-to-row continued selection and progeny testing was a popular breeding method for the cross-pollinated corn crop from about 1900 to 1920, but the method has been discarded. Ear-to-row testing is effective in isolating some of the better strains of an unadapted variety during the early years of selection, but progressive increases in yield were never obtained.[13] The limitations to the method are the partial inbreeding that reduces yields, as well as the lack of complete control of the pollen parentage of the seed ears. Continued selection in cotton has met with some success because complete purity is never obtained when the selections are grown without protection from insects that can affect intercrossing.

The breeding of cross-pollinated grasses and legumes such as bromegrass, crested wheatgrass, and alfalfa[19, 20] usually have involved continued selection and self-pollination to isolate and purify improved strains. Some of these strains may be combined later into a synthetic variety and allowed to intercross.[20]

PURE-LINE CONCEPT The modern methods of handling, testing, and increasing individual selections of self-fertilized crops had as their basis the pure-line concept proposed in 1903 by W. L. Johannsen who worked with beans. According to this concept, variations in the progeny of a single plant of a self-fertilized species are not heritable but are due to environmental effects. Failure to recognize this principle before, or in some countries as late as 50 years after, its discovery resulted in some futile work in continuously selecting self-fertilized crops such as wheat and barley.

When homozygous self-fertilized plants are selected, the progeny is pure for all traits until outcrossing or mutation occurs.[3] It should be recognized, of course, that varieties that appear to be pure might be pure only for the traits that are observed under a particular environment. A pure-line was originally defined as the descendants of a single, homozygous, self-fertilized individual. The definition commonly used today is that a pure-line comprises the descendants of one or more individuals of like genetic traits that have undergone no significant change in traits.

INBREEDING OF CROSS-FERTILIZED CROPS Before making hybrids in cross-fertilized crops, it may be desirable to select and inbreed the varieties for several generations or until they are reasonably pure for the traits desired. Without such inbreeding (selfing), the hybrids that are obtained from cross-fertilized crops cannot be reproduced. Artificial self-pollination in a normally cross-pollinated crop leads to segregation into pure, uniform (homozygous) lines. There is often a rapid reduction in vigor when self-pollination is practiced. In corn, the reduction in vigor in the first generation of selfing is approximately one-half the differences in vigor between the original corn and that of the homozygous inbred progeny. When appropriate lines of self-pollinated plants are intercrossed, there is usually restoration of vigor, a fact that is applied in hybrid corn production.[21] In addition, certain abnormalities, such as sterility, poor chlorophyll development, dwarf habit, lethal seedlings, and susceptibility to diseases might appear as a result of inbreeding. These types must be discarded. Such abnormalities are mostly recessive traits that, under open-pollinated conditions, are largely suppressed as a result of crossing, usually with normal plants.

4.3.3 Hybridization

Hybridization, the only effective method of combining the desirable traits of two or more crop varieties, offers far greater possibilities. It has been the standard crop improvement method in advanced countries for over 75 years. Most cereal crop varieties grown in the United States by 1970 were selected from hybrid crosses.

The first step in breeding by hybridization is to choose parents that can supply the important trait or traits that a good standard variety lacks. It is important to have definite characteristics in mind, but also to test the material for these characteristics repeatedly. Success has occasionally followed haphazard crossing between productive varieties, but failure to accomplish any improvement by such a procedure is more common.

HYBRIDIZATION TECHNIQUES In crossing self-fertilized crops, the flowers are first emasculated, (i.e., the anthers are removed or killed before they have shed pollen). The flowers are then covered with a small paper, glassine, or cellophane bag to exclude insects and foreign pollen (Figures 4.1, 4.2, 4.3, 4.4, and 4.5). The anthers are usually removed with tweezers, a needle, or a jet of air or water, but rice, sorghum, and certain grass pollens can be killed by immersing the panicles in hot water at about 117°F (47°C) for ten minutes. Pollen from the male parent is applied by brushing or dusting it on the pistil when the stigmatic surface of the latter is receptive. This receptiveness may occur immediately after emasculation, or one or two days later. It is indicated by a sticky exudate covering the stigma.

The seeds that develop from the cross-pollinated flowers produce plants of the first filial (F_1) generation. These plants should show similar traits, with only the dominant and mutual recessive traits expressed. In the second (F_2) generation,

FIGURE 4.1

Cross-pollinating wheat *(right)* after emasculation by removal of the three anthers from flower as shown at left. The white bags protect emasculated flowers from fertilization by undesired foreign pollen.

FIGURE 4.2
Oat floret showing the anthers and stigma within the lemma and palea, with a detached ovary at the right.

FIGURE 4.3
Pollinating a tobacco flower by brushing the stigma with anthers from another species.

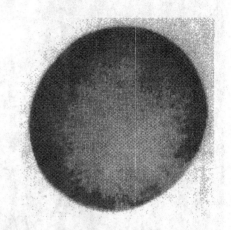

FIGURE 4.4
Barley pollen grain.

FIGURE 4.5
Sorghum florets in bloom.
The exposed stigma often
permit stray foreign pollen
to produce outcrosses
unless the panicles are
covered with bags.

the plants will segregate into all possible combinations of the dominant and recessive traits of the two parents. Plants exhibiting the desired traits are selected;
these and several subsequent generations are handled as previously described
under selection methods. Reselection continues until the desired strains are uniform, usually for three to six generations.

The size of the F_2 population may vary from 200 to 10,000 plants; the proper
number depends upon the number of genetic factor differences involved in the

cross. For example, when closely related varieties are crossed to combine two single-factor traits, one from each variety, an F_2 population of 200 to 500 plants should be ample because approximately one-sixteenth of this number, or about 13 to 31 plants, should breed true for the desired recombination. On the other hand, with a six-factor difference between the two parents (e.g., two factors for smut resistance, two for seed color, and two for plant height) one would expect only 1/4,096, or not more than 3 in 10,000 F_2 plants, to exhibit all the desired traits. However, additional true-breeding plants of the desired recombination would be obtained in later generations from segregating F_3 lines.

The formula $1/4^n$ determines the expected proportion of true-breeding strains of a particular recombination in the F_2 generation in which n is the number of genetic factor differences involved in the cross. For example, with one pair of factors (n = 1), $1/4^n = 1/4$, and for six factors (n = 6), $1/4^6 = 1/4,096$. It is obvious that in wide crosses (i.e., crosses between widely unrelated varieties) the number of factor differences is so large that the chances of finding the desired recombination in the first cross are extremely remote. In such cases, the strain obtained that most closely approaches the desired recombination may be crossed with one of the parents, or with some other variety, and selection then resumed.

The success of these procedures in hybridization and selection is illustrated by the results of crossing two oat varieties, one resistant to smut, and the other resistant to rust. Selections from such a cross are resistant to both smut and rust.[17]

BACKCROSSING Backcrossing is a useful method of breeding when it is desired to add only one or two new traits to an otherwise desirable variety.[3, 10] In this method, the existing variety or elite line (recurrent parent) is crossed with another variety with the new trait (nonrecurrent parent). First-generation plants are backcrossed with the recurrent parent. Then, in each segregating generation a number of plants approaching the desired new recombination are selected and backcrossed with the good commercial (recurrent) parent. Usually, with repeated rigid selection, the improved type is recovered in approximately five backcrosses. Furthermore, little testing is necessary because the recovered variety is practically identical with the original adapted variety, except that it has a new trait such as disease resistance. By omitting several years of testing, the recovered variety may be quickly increased and distributed to farmers. Unfortunately, the procedure fails to introduce new factors for basic higher yields. This results in a phenomenon known as yield lag, where the backcrossed line lags in yield potential compared to other released lines that have been developed during the same period of time. To diminish the effects of yield lag, plant breeders may switch to forward crossing at about the second or third backcross generation. Instead of crossing back to the same elite line, the plant breeder crosses forward to a line very similar to the original elite with greater yield potential.

Backcrossing is also used in the development of genetically engineered hybrids. Often, lines that are transformed with newly designed genes have poor agronomic qualities. In these cases, the elite line serves as the recurrent parent and the transformed line as the nonrecurrent parent. The end product is a high yielding variety containing the transgene.

BULK PROPAGATION OF HYBRID MATERIAL In the bulk method of handling hybrid material, the entire population from a cross is grown and harvested in mass each year. Successive crops are grown for five to ten years to allow natural selection for traits, such as cold resistance or insect resistance, to take place. By this time, the poorly adapted strains are largely eliminated, and the remaining strains are mostly homozygous. Plants are then selected, and the progenies are tested as previously described. Testing of large numbers of unadapted and heterozygous strains is thus avoided. However, desirable types may be lost during the interim because of competition with aggressive but otherwise undesirable plants.

■ 4.4 PRODUCTION OF HYBRID SEED

Nearly all of the corn (Figure 4.6), sorghum, sugarbeet, sunflower, and castor grown in the United States are first-generation hybrids, either single crosses, double crosses, or three-way crosses.

Hybrids of wheat, barley, and other crops have been developed for possible commercial production. The hybrids of wheat, however, have not been commercially successful because the increased cost of the seed was not reliably offset by significant increases in yield and subsequent income.

Commercial hybrid seed of corn can be produced by detasseling the seed-parent rows in a cross-pollination field, or by using cytoplasmic male-sterile lines. Pearl millet hybrids have been produced by planting a mixture of four selected lines. They can also be produced by clipping (topping) the upper part of the heads on the seed-parent plants before they shed pollen, but after the stigmas are exserted, or by pollinating cytoplasmic male-sterile lines. Hybrid seed of crops that are naturally mostly self-pollinated can be produced economically only by the use of male-sterile lines. Plants with cytoplasmic male-sterility have been discovered or developed in several crops. Genetic factors in different varieties or strains of these crops determine whether their progeny is male-sterile or male-fertile when they are used as

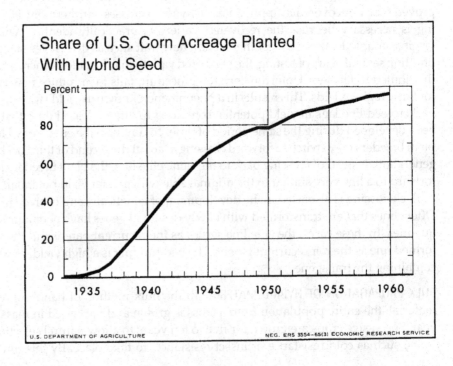

FIGURE 4.6
About 98 percent of the corn acreage of the United States had been planted with hybrid seed by 1969. [Courtesy of USDA]

pollinators on cytoplasmic male-sterile plants. A related (fertile counterpart) strain that carries the recessive sterile-producer gene is used as a pollinator to maintain or to increase the male-sterile line. An unrelated variety or strain that carries the dominant fertility-restorer gene is used as a pollen parent to produce commercial hybrid seed. The pollen parent that is chosen is one that produces a suitable productive hybrid.[8]

▨ 4.5 BREEDING FOR DISEASE RESISTANCE

Some of the most outstanding accomplishments in crop breeding have been in the development of disease-resistant varieties of all major crops grown today. The plant breeding methods are the same as for other objectives except that the progeny must be exposed to the particular diseases either by artificial inoculation or by choosing conditions favorable for natural infection. Plants that resist the disease and are otherwise desirable are selected for further progeny tests.

The breeding of varieties resistant to diseases is sometimes highly complex because of the existence of different physiologic races of the disease organism. Rust-resistant varieties of wheat and oat suffer heavy damage when other races or sub-races of the rust parasite become abundant and are able to attack them. Such races may be new or may have been unimportant because they were unable to multiply on a susceptible host due to competition by races previously present. New races may arise by mutation or by nuclear recombination.

The rust resistance of emmer that had persisted for 50 years in the United States was transferred by hybridization to varieties of common wheat. This failed to protect the wheat crop from the ravages of Race 15B that later became prevalent.[22]

The problem is further complicated by the fact that hereditary factors in both host and parasite affect the disease interaction. The presence of numerous physiologic races often makes breeding for complete resistance rather tedious, because several successive crosses may be necessary to combine all the different essential factors for resistance.[23] Also, some of the genes for resistance may be closely linked with genes for undesired traits. However, crop strains sometimes are obtained from hybrid combinations that, under field conditions, resist all races of the disease organism that are important in a given region. When no current race of the parasite is able to cause appreciable crop damage to a variety, the fact that numerous other races exist is a matter largely of academic interest.

Physiologic races of a disease organism are separated on the basis of their reaction to differential host varieties in greenhouse tests. The existence of such races is undetected until resistant varieties are found. Thus, Race 1 may attack variety A but not variety B; Race 2 may attack variety B but not variety A; and Race 3 may attack both A and B while Race 4 attacks neither. The number of races that can be identified is dependent more upon the number of differential host varieties that are used than upon the number of collections of the disease organism that are available. The reaction to disease is often separated into only two simple categories, resistance and susceptibility, although some intermediate or irregular reactions also are known to occur. Considering merely the two reactions, resistance and susceptibility, the number of possible physiologic races equal 2^n where n is the number of differential hosts. Thus, with 12 differential hosts, 4,096 races might be identified. Some 300 races or sub-races of the fungus that causes stem rust of wheat have been reported. The reaction of wheat seedlings in the greenhouse often differs from the reaction of mature plants in the field.[18] Physiologic races of insect species (e.g., Hessian fly) are also known.

▓ 4.6 BIOTECHNOLOGY AND CROP IMPROVEMENT

The direct manipulation of genetic material as a method of introducing previously unknown traits into plants is a recent phenomenon. The first published report of direct genetic transformation of a crop occurred in 1983 when a fire-fly gene was put into a tobacco cell causing it to glow.[24] Since then, there has been tremendous interest in using biotechnology to develop genetically modified plants. This technique provides tremendous possibilities for improving the food supply for an increasing human population. Potential benefits include the incorporation of specific proteins, amino acids, or enzymes to improve nutrition for humans and livestock. The Rockefeller Foundation funded research at the International Rice Research Institute in the Philippines to use biotechnology to develop new cultivars of rice high in beta carotene, thus providing the means to greatly reduce a common nutritional (vitamin A) deficiency in people that consume rice as the main part of their diet. The production of plants with pharmaceuticals is also possible. Research is currently in progress to develop banana varieties with vaccines for common diseases.

While traditional plant breeding methods are ultimately limited to the theoretically best combination of existing genetic traits, biotechnology can introduce traits from entirely unrelated species. An example is the development of *Bt* cultivars of corn and cotton. *Bacillus thuringiensis (Bt)* is a naturally occurring soil bacterium that infects the larval stage of certain insects. Commercial preparations of cultures of *Bt* have been applied as a natural insecticide to many crops for years. The genes in *Bt* that carry the trait lethal to pest larva were incorporated into corn and cotton plants giving them resistance to pests such as corn borer and cotton bollworm, thus greatly reducing the need for insecticides.

However, the development of genetically modified plants, currently referred to in the public sector as GMO (for genetically modified organism), has met with much controversy in some parts of the world. There is a real concern that extensive use of GMO plants will allow the development of "super pests" with natural resistance to incorporated genetic traits making them very difficult to control in the future. There is also concern, not necessarily scientifically based, that new genetic combinations are "not natural" and could lead to unforeseen problems, such as food safety, in the future.

BIOTECHNOLOGY TECHNIQUES Important developments that have made genetic engineering possible are gene mapping techniques that identify the location of specific genes on the chromosomes. With this information, the biotechnologist can isolate and transfer specific genetic traits between organisms. Gene mapping requires intense research. For example, corn has over 100,000 genes.

Genetic engineering uses recombinant DNA, or genetic material produced in a laboratory by combining parts of chromosomes from different cells. The two most common methods for introducing recombinant DNA into plant cells are the use of a bacterium and a gene gun.

FIGURE 4.7
Gene gun for direct transfer of chromosome segments into a cell nucleus to produce recombinant DNA. [Courtesy University of Nebraska Library of Crop Technology]

The first method uses a soil bacterium, *Agrobacterium tumefaciens*, which has the natural ability to cause tumors by transferring a piece of its genetic material into the chromosomes of plants.[25, 26] This bacterium uses a plasmid to transfer its genes. A plasmid is a genetic element that replicates itself in the cell independently of the chromosomes. Using a special enzyme, scientists remove the genes that cause tumors but keep the genes that enable the bacterium to insert its plasmid into plant cells. Next, scientists insert the new transgene, such as herbicide resistance, into the plasmid and the plasmid into *Agrobacterium*. Small pieces of plant tissue are mixed with the bacteria. *Agrobacterium* is able to enter through the wounds of the cut tissue and transfer the plasmid to a chromosome of the plant cells.

Another method is the use of a gene gun.[27] In this method, microscopic gold or tungsten beads (bullets) are coated with segments of DNA that contain the desired genes. These bullets are then placed in a device that shoots them into the cell where the new genes will combine with the chromosomes already present (Figure 4.7).

In both transformation methods, the altered plant cells are grown in the laboratory until they develop into plants and can be transferred to a growth chamber or greenhouse. Once they flower, they can be used in hybrid crosses like other plants.

▓ 4.7 EXPERIMENTAL METHODS

Progress in crop breeding can be measured only by experimental tests except when breeding for disease resistance, better color, or some other trait that can be readily seen. Even then, the investigator must know how the new strain compares in yield with the variety it is expected to replace. Field-plot experiments are the main activity of most research agronomists (Figure 4.8).

FIGURE 4.8
Cooperative oat-breeding nursery at the State Agricultural Experiment Station, Ames, Iowa. A plant pathologist at the far left assists three famous oat breeders in selection of disease-resistant plants and lines.

The comparative productive value of different varieties is determined by experimental field trials in which all varieties are sown under identical conditions in plots or nursery rows of uniform size, usually ranging from 15 square feet to one-fortieth of an acre (1.4 to 100 m²). An experimental field should be very uniform in topography and soil characteristics. The soil should be representative of the region to which the results are to be applied. Soil preparation, crop sequence, and fertilization should be identical for the entire field area included in a particular experiment. Seeding of all varieties should be done at the same time. It is customary to replicate or repeat the plantings of each variety in three to ten plots or rows in order to equalize differences due to variations in the soil and in mechanical operations that occur in field experimentation. Even the most uniform-appearing field varies considerably in productivity because of differences in soil depth, origin, topography, or treatment. Standing water, runoff, dead furrows, weed clumps, soil compaction, and animal excretions all detract from uniformity. Likewise, inequalities in tillage, soil compaction, seeding depth, cleanness of harvesting and threshing, waste of grain in threshing, weighing, and moisture content of the weighed crop all affect experimental accuracy.

Accepted statistical methods are used to compute probable experimental errors that measure the variations in yield among the different plots of the same variety or treatment due to chance. It gives an estimate of the uniformity or reliability of the experiment, particularly the reliability of differences found between varieties or treatments. When one variety outyields another by a sufficiently consistent margin so that the computed odds for a real difference are 19 to 1 or better (probability ≥ 0.95), it is commonly assumed to be superior. With lower odds, the difference observed between the two varieties is regarded as insignificant, or largely due to chance. Since this arbitrary arrangement may accidentally favor certain varieties, the varieties and treatments are usually arranged in a random or designed order within definite experimental blocks of plots. Special experimental designs, especially the so-called lattice designs,[15] are used as a means of partly adjusting the plot yields to smooth out differences due to soil variation. When the plots of the different varieties or treatments are repeated in the same systematic order with similar or most promising varieties or treatment grouped together, a more accurate observation and comparison is obtained, but the entire experiment shows a higher measurement of error.

The size of plot is determined largely by convenience and by the availability of uniform land for the experiment. Nursery rows give reliable comparisons between varieties when they are repeated several times and protected from competition with adjacent varieties by the planting of border rows. Field plots give a better view of the behavior of the variety under field culture. They also furnish more material for further experimentation or more seed for increase. Field plots occupy more land than do nursery rows, but they are only slightly more reliable in yield tests. Field plots of different varieties of small grain or forage crops are usually 65 to 130 feet (20 to 40 m) long and one drill-width wide. Corn and sorghum are usually tested in 2- to 4-row plots that are 30 to 50 feet (9 to 15 m) in length. Small-grain nurseries are usually planted in rows 12 inches (30 cm) apart with the rows approximately 16 feet (5 m) in length.

REFERENCES

1. Painter, R. G. "Crops that resist insects provide a way to increase world food supply," *KS Agr. Exp. Sta. Bull.* 520, 1968. pp. 1–22.

2. Clark, J. A. "Improvement in wheat," in *USDA Yearbook*, 1936, pp. 207–302.

3. Hayes, H. K. "Green pastures for the plant breeder," *J. Am. Soc. Agron.* 27(1935): 957–962.

4. Borlaug, N. E. *Mankind and civilization at another crossroad.* Seventh Biennial Conference of FAO. Rome, Italy, 1971, pp. 1–48.

5. Salmon, S. C., O. R. Mathews, and R. W. Leukel "A half-century of wheat improvement in the United States," in *Advances in Agronomy*, vol. 5. New York: Academic Press, 1953, pp. 1–151.

6. Martin, J. H. "Breeding sorghum for social objectives," *J. Heredity* 36, 4(1945):99–106.

7. Pal, B. P. "Recent cereal research and future possibilities," *Science and culture* 34 (1968):136–149.

8. Airy, J. M., L. A. Tatum, and J. W. Sorenson, Jr. "Producing seed of hybrid corn and grain sorghum," in *Seeds* USDA Yearbook, 1961, pp. 145–153.

9. Harlan, J. R. "Genetics of disaster," *J. Environ. Quality* 1, 3(1972):212–215.

10. Anonymous. *Genetic Vulnerability of Major Crops.* Natl. Acad. Sci. Washington, D.C. 1972, pp. 1–307.

11. Robinson, R. A. "Vertical resistance," *Rev. Plant Path* 50, 5 (1971): 233–239.

12. Simons, M. D. "Polygenic resistance to plant disease and its use in breeding resistant cultivars." *J. Environ. Quality* 1, 3(1972):232–238.

13. Jenkins, M. T. "Corn improvement," in *USDA Yearbook*, 1936, pp. 455–522.

14. Leighty, C. E. "Theoretical aspects of small grain breeding," *J. Am. Soc. Agron.* 19(1927):690–704.

15. Hayes, H. K., F. R. Immer, and D. C. Smith *Methods of Plant Breeding*, 2nd ed. New York: McGraw-Hill, 1955, pp. 1–551.

16. Noll, C. F. "Mechanical operations of small grain breeding," *J. Am. Soc. Agron.* 19(1927):713–721.

17. Coffman, F. A., H. C. Murphy, and W. H. Chapman. "Oat breeding," in *Oats and Oat Improvement*, New York: Academic Press, 1961, pp. 263–329.

18. Quinby, J. R., and J. H. Martin, "Sorghum improvement," in *Advances in Agronomy*, vol. 6. New York: Academic Press, 1954, pp. 305–359.

19. Hanson, A. A., and H. L. Carnahan. "Breeding perennial forage grasses," *USDA Tech. Bull.* 1145. 1956.

20. Kirk, L. E. "Breeding improved varieties of forage crops," *J. Am. Soc. Agron.* 19(1927):225–238.

21. Jones, D. F. "Crossed corn," *Conn. Agr. Exp. Sta. Bul.* 273, 1926.

22. Martin, J. H., and S. C. Salmon. "The rusts of wheat, oats, barley, rye," in *Plant Diseases*, USDA Yearbook, 1953, pp. 329–343.

23. Aamodt, O. S. "Breeding wheat for resistance to physiologic forms for stem rust," *J. Am. Soc. Agron.* 19(1927):206–218.

24. Herrara-Estrell, Depicker, L., A. M. Van Montegu, and J. Schell. "Expression of chimaeric genes into plant cells using a Ti-plasmid derived vector," *Nature (London)* 303(1983):209–213.

25. Trick, H. N., and J. J. Finer. "SAAT: Sonification assisted *Agrobacterium*-mediated transformation," *Transgenic Research* 6(1997):329–337.

26. Zambriski, P., H. Joos, C. Genetello, J. Leemans, M. Van Montagu, and J. Schell. "Ti plasmid vector for the introduction of DNA into plant cells without alteration of their normal regeneration capacity," *EMBO Journal* 2(1983):2143–2150.

27. Klein, T. M., E. D. Wolf, R. Wu, and J. C. Sanford. "High-velocity microprojectiles for delivering nucleic acids into living cells," *Nature (London)* 327(1987):70–73.

Tillage Practices

5

▓ 5.1 HISTORY OF TILLAGE OPERATIONS

Tillage began before the earliest written records of mankind. The first implements were hand tools to chop or dig the soil, usually made of wood, bone, or stone. They were used to subdue or destroy the native vegetation, make openings in the soil to receive seeds or plants, and reduce competition from native plants and weeds growing among the crops. The next stage of tillage, the use of domestic animals,[1] occurred in parts of the world before the dawn of history. This made possible development of implements that moved forward at a steady pace. Among these were the crooked-stick plow to stir the soil and the brush drag to pulverize the surface. Little further progress was made for many centuries except that eventually some plows were fitted with iron shares despite a common misunderstanding that iron poisoned the soil. The development of steel in 1833 resulted in a plow with sharp edges that cut the soil layer and a curved polished surface that permitted the plow to scour. That straight-line movement of the plow has since been improved with the development of the disk plow, harrow, rotary hoe, and other pulverizing and stirring tools.

Very little was known about the effects of cultural operations in the Middle Ages.[2] In 1733, Jethro Tull published *The Horse-Hoeing Husbandry* in England. He believed that plants took up the minute soil particles. In other words, the more finely the soil was divided, the more particles would be absorbed by the roots.

During the nineteenth century, it became evident that nutrition of plants depended on certain chemical elements from the soil minerals, organic matter, water, and air. The foundation for this concept was proposed by Justus von Liebig and others. The idea became widespread that tillage, by increasing the aeration in the soil, increased the oxidation of chemical compounds in the soil and made them more soluble.

Early American writers believed that tillage allowed the roots to penetrate more deeply or refined the soil to make a greater surface to hold nutrients while others recognized the importance of tillage in weed control. The idea that harmful effects might result from excessive tillage, particularly from greater oxidation of organic matter or from increased erosion, was also pointed out by early writers in this country. Not until after 1890 did experimental evidence begin to show the basic reasons for tillage.

TABLE 5.1 Average Amount of Residue Incorporated by Different Tillage Implements

Tillage	Residue Incorporated (%)
Moldboard plow	100
Oneway disk	40
Tandem or Offset disk	
18–22 inch disks	40
24–26 inch disks	50
Anhydrous applicator	20
Chisel plow	25
Mulch treader	20–25
Sweep plow, 30" or wider blades	10–15
Rod weeder	5–10
Grain drill, double disk openers	20
Planter, double disk openers	10
Slot planter	0
Till planter	20

5.2 PURPOSES OF TILLAGE

The fundamental purposes of tillage[1] are: (1) to prepare a suitable seedbed, (2) to eliminate competition from weed growth, and (3) to improve the physical condition of the soil. This may involve destruction of native vegetation, weeds, or the sod of another crop. Tillage can further involve removal, burial, or incorporation in the soil of manure or crop residue. In other cases, the tillage operation may be solely to loosen, compact, or pulverize the soil. The best system of tillage is, then, the one that accomplishes these objectives with the least expenditure of labor and power.

The incorporation of plant residues and its subsequent effect on soil erosion must also be considered in planning all tillage operations (Table 5.1). The kind of tillage for seedbed preparation on dry lands is governed almost entirely by the effects of tillage upon the conservation of soil moisture and the prevention of surface runoff and wind erosion. It may be necessary to till the soil surface to protect the soil from wind erosion after crops that leave little residue such as cotton and soybean.[3]

Under most conditions, a desirable seedbed is one that allows close contact between the seed and the soil particles. Seeding equipment must cut through any residues present. The seedbed should contain sufficient moisture to germinate the seed when planted and support subsequent growth. In irrigated regions, it is occasionally necessary to plant the crop and then irrigate the field in order to supply sufficient moisture for germination, but this practice is avoided whenever possible because of frequent irregular stands. Irrigation before planting is preferable.

5.3 IMPLEMENTS FOR SEEDBED PREPARATION

Originally, an acre of land was the area tilled by one ox-drawn plow in one day. Tractor-drawn implements cover 0.3 to 0.6 acre per hour per foot of width (0.4 to 0.8 ha/hr/m width).

FIGURE 5.1
A moldboard plow.
[Courtesy Case IH]

5.3.1 Moldboard Plows

The moldboard plow breaks loose or shears off the furrow slice by forcing a triple wedge through the soil. This action inverts the soil and breaks it into lumps. Some pulverization takes place (Figure 5.1). Several different shapes of moldboard plow bottoms are used. At one extreme is the breaker type with a long moldboard adapted to virgin or tough sod for completely inverting the furrow slice without pulverization. At the other extreme is the stubble bottom with a short, abrupt moldboard that pulverizes the furrow slice while turning the soil in order to mix the residue with the soil. The general purpose, or mellow-soil, plow is intermediate between these two types in length, slope, and action. The two-way plow is adapted to steep hillsides because it throws the soil downhill when drawn across the slope in either direction. It is often used on irrigated land to avoid dead furrows. Better incorporation of residue is facilitated on all plows by the use of attachments such as coulters, jointers, rods, and chains. The majority of moldboard plows range in size from 7 to 18 inches (18 to 46 cm) in width of furrow cut.[4] The 14 inch (36 cm) width is common.

5.3.2 Disk Plows

Disk plows (Figure 5.2) are important for use in loose soils but also in those soils too dry and hard for easy penetration of moldboard plows. In some sticky soils, neither type of plow will scour when the soil is wet. In this case, scrapers help keep the disks clean. The disks vary in size from 20 to 30 inches (51 to 76 cm), while the depth of plowing may vary from 4 to 10 inches (10 to 25 cm). The one-way disk plow (Figure 5.3) has been widely used in stubble fields of the wheat belt. The one-way should be avoided for bare summer fallow land, or dry fields

FIGURE 5.2
Disk plow. [Courtesy Case IH]

FIGURE 5.3
The one-way disk plow leaves about one-third of the crop residue on the soil surface. [Courtesy John Deere & Co.]

sparsely covered with stubble or weeds where it tends to pulverize the soil so that the soil blows easily.

5.3.3 Subtillage Implements

Blade or subtillage implements, that leave crop residues on the surface of the soil, are used in semiarid areas.[5] Wide-sweep blades with staggered mountings undercut the stubble and weeds but do not pulverize the soil surface (Figure 5.4). This

FIGURE 5.4
Stubble mulch after the use of sweep blades. [Courtesy Richard Waldren]

so-called stubble mulch reduces soil erosion as well as runoff, but it may not increase soil moisture storage. Special seeding equipment is needed to cut through residue and open seed furrows on land covered with heavy crop residue unless a disc cutter is used to chop up the straw.

5.3.4 Lister

The lister (middlebuster) resembles a double plow with a right-hand and a left-hand bottom mounted back to back. In some of the southern humid sections, it is used to raise beds on which to plant cotton, potato, or other row crops. In the semiarid regions, the lister was widely used to furrow the land. It replaced the moldboard plow in seedbed preparation because it covered the field in half the time. The ridges also checked wind erosion, snow drifting, and runoff when the furrows ran on the contour or across the slope. Row crops are planted in the moist soil at the bottom of the lister furrows. For drilled crops such as small grains, the lister ridges or middles are broken down or leveled with ridge busters. The lister has been largely replaced by sweep blade implements or the one-way plow for the initial seedbed preparation in the semiarid areas.

5.4 SURFACE IMPLEMENTS IN FINAL SEEDBED PREPARATION

Tillage can prepare a seedbed after the land has been plowed by using harrows, field cultivators, or other machines equipped with disks, shovels, teeth, spikes, sweeps, knives, corrugated rolls, or packer wheels. Under most conditions, a smooth, finely pulverized seedbed should not be prepared until just before a crop is to be planted.

5.4.1 Harrows

Harrows, that smooth and pulverize plowed soil, compact it, and destroy weeds, require less power than do plows or listers. The disk harrow cuts, moves, pulverizes the soil, and destroys weeds. The spike-tooth harrow breaks clods, levels the land, and kills small weeds. Both disk and spike-tooth harrows are detrimental when soil is subject to blowing. The spring-tooth harrow consists of flexible spring-steel teeth about 2 inches (5 cm) wide that penetrate sufficiently to tear up deep clods or bring them to the surface while destroying weeds. The corrugated roller is used to crush clods as well as to compact the seedbed.[6] It is used particularly for small seeds to improve the chances of germination.

5.4.2 Cultivators

Several types of cultivators, mainly the duckfoot (field cultivator) and the rod weeder, are used in fallow and seedbed preparation. The field cultivator has sweeps or shovels on the ends of stiff or spring bars. The bars are staggered in two rows, so that the weeds are cut off by the overlapping sweeps. This implement leaves the soil surface rough and cloddy, a condition necessary to prevent wind erosion. The principal feature of the rod weeder is a rotating horizontal square rod at right angles to the direction of travel at a depth of 3 to 6 inches (8 to 15 cm) below the soil surface. This rod pulls out the weeds. The rod is driven by gears or sprocket chains from one of the support wheels, so that it revolves in a direction opposite the direction of travel, which keeps the rod in the ground (Figure 5.5). The rotation of the rod keeps it from clogging, while weeds, straw, and clods are lifted toward the surface, where they check soil erosion.

▦ 5.5 TILLAGE IN SEEDBED PREPARATION

Tillage buries green or dried material, loosens the soil, removes or delays competition with weeds, and roughens the soil surface to reduce runoff of rain water.[7] Other effects of tillage are control of certain insects and diseases and promotion of nitrification.

FIGURE 5.5
Rotary rod weeder
cultivating summer fallow.
[Courtesy John Deere & Co.]

5.5.1 Time for Tillage

Fall-tilled land is warmer in the spring because bare soil, which usually is darker than the plant vegetation, absorbs more solar heat and thus hastens corn emergence and growth.[8]

Land to be summer fallowed is usually not disturbed until the spring following harvest except when it is very weedy. Weeds and stubble are left to hold snow and reduce wind and water erosion. Early spring tillage for fallow is very important[6] to stop weed and volunteer growth and thus conserve moisture and permit accumulation of nitrates. In the more humid regions, early seedbed preparation may be inadvisable because of water erosion.[9, 10]

Land tilled in spring may be higher in nitrates than that left untilled due to rapid mineralization, but the increase is not always evident immediately after tillage.[11] The difference becomes pronounced during the summer season but almost disappears during the winter when nitrates are leached downward.

5.5.2 Depth of Tillage

At the Pennsylvania Experiment Station, deep 12-inch (30-cm) plowing was compared with ordinary 7.5-inch (19-cm) plowing for corn, wheat, oat, barley, alfalfa, clover, and timothy.[12] The average yields failed to show any advantage from the deeper plowing.

The effect of depth of tillage on soil erosion was studied in Missouri.[13, 14] On land where no crop was grown, the annual erosion from tilled soil was greater than from soil that was not stirred and there was no advantage to deeper tillage.

For 100 years, advocates of deeper plowing recommended that the depth of plowing be increased gradually so as not to turn up too much raw subsoil in any one year. The reasoning behind such a practice seems logical, but there is no research that would either prove or disprove the soundness of this recommendation. Varying the plowing depth from year to year may avoid establishing a compact "plow-sole" or "tillage-pan" in heavy soils, but heavy soils commonly have compact subsurface layers. The more friable and muck soils do not form plow soles.

Plows with very large moldboards that penetrate 1 to 4 feet (30 to 122 cm) deep are used for land reclamation operations such as turning under brush or turning up topsoil that has been buried under water-borne sand.

▨ 5.6 SUBSOILING

Tillage below the depth reached by the ordinary plow was once advocated widely. There was a popular belief that plants utilize only the soil moved by tillage, and deeper tillage provides a greater opportunity for root development and moisture storage. Another belief was that to merely loosen, stir, pulverize, or invert the deeper soil layers permits more effective plant growth. Great Plains research that covered ten states showed no general increase in yields from subsoiling or other methods of deep tillage.[15]

The same general results were obtained under other conditions.[12, 16, 17] Increased water infiltration and better crop yields have sometimes followed the deep tillage of land that had a dense hard subsurface layer or a tillage pan that tended to be rather impervious to water. Such benefits may be evident for only one or two years, but the cost of the operation may offset any gain in yield.[18] Deep tillage is ineffective unless the tillage pan is dry enough to be shattered.[19]

The reasons why deep tillage is often ineffective are rather obvious. Heavy soils shrink and leave wide cracks when they dry out, thus providing natural openings for periodical water absorption and aeration. When such soils are wetted, they swell tightly and the cracks close up so that any effect of deep tillage is only temporary. Light soils do not shrink and swell much, and they are always open. Since roots of crop plants ordinarily penetrate several feet below any tilled layer, deep tillage is not essential to deep rooting. The chisel, an implement with a series of points that break up the soil to a depth of 10 to 18 inches (25 to 46 cm) without turning it, is sometimes used on heavy soils. As a substitute for plowing, the chisel opens and roughens the soil so that heavy rains can be absorbed quickly without danger of wind or water erosion. Such deep tillage may favor the growth of sugarbeet because of its large taproot.

▓ 5.7 CONSERVATION TILLAGE

Conservation tillage is a collective term used for any tillage system whose primary goal is maintenance of plant residue on the soil surface.[20] Other more specific terms used include no-tillage, direct-seeding or drilling, and residue farming. With strip-tillage, a narrow strip is tilled over the seeded row. With ridge-till (ridge-plant), ridges are formed and seed is placed in the ridge. Two early proponents of reducing tillage were Faulkner and Scarseth.[21, 22]

Implements with large sweeps or blades leave the residue on the soil surface to form a partial mulch (Figure 5.6). This process must be repeated during the sea-

FIGURE 5.6
Wheat that was planted into stubble mulch fallow.
[Courtesy Richard Waldren]

son to control weeds, because the straw and stubble from a single crop are insufficient to smother all the weeds.[7, 23, 24] However, this partial mulch checks soil blowing, stops runoff, and increases the surface infiltration of rainfall. Soils are cooler under residue mulch in summer. However, heavy residues may keep the soil too cool in the spring for favorable germination of early planted corn, sorghum, or soybean. It also fails to control certain weeds and grasses, especially cheatgrass (*Bromus tectorum*), where herbicides are not fully applicable, unless a machine such as the skew treader or rotary tiller is used to chop through the mulch and cut out the weeds. Damage to the crop from mice, slugs, and birds can increase in heavy mulches. Special equipment such as coulters, chisels, or scalpers to push residue aside are used on planters for mulch tillage fields. Heavy nitrogen applications are necessary to maintain crop yields and hasten the ultimate decay of the residues.[13, 25, 26, 27]

Zero tillage of sod land requires killing the grass plants with chemicals and using a modified planter that cuts through the sod while planting the crop. The dead sod mulch conserves some soil moisture, and the yield of corn may equal that following 8-inch (20 cm) plowing and discing.[28] Zero tillage also reduces water runoff and soil erosion.

The practice of conservation tillage has rapidly increased in recent years (Figures 5.7, 5.8, 5.9). Conservation tillage was used on 41 percent of the corn acreage, 55 percent of the soybean acreage, and 22 percent of the wheat acreage in ten midwest states during the 1995 crop year (USDA NASS). Conservation tillage in northwest Ohio increased from essentially zero in the mid-1980s to about 50 percent of all corn and soybean fields in the mid-1990s and has remained at that level.[29] In 2000, no-tillage was used on more than 51 million acres (21 million ha) in the United States. That is 17.5 percent of total planted acres and is a three-fold increase since 1990.[30] No-tillage is increasing rapidly in other countries as well (Table 5.2).[31]

A minimum of 30 percent residue cover is the common standard used to signify conservation tillage. A 30 percent residue cover reduces erosion by 80 percent.[3] Although there are different methods for estimating residue cover, the line-transect

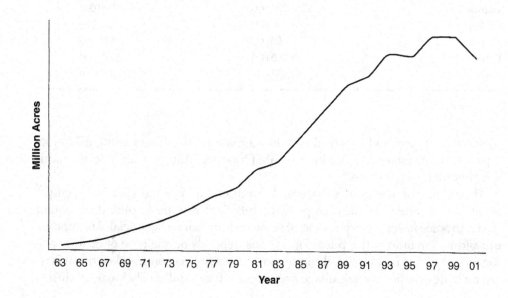

FIGURE 5.7
Conservation tillage trends in the U.S.

FIGURE 5.8
Soybean seedings that were planted without tillage into wheat stubble. [Courtesy USDA NRCS]

TABLE 5.2 Total Area of No-Tillage in Various Countries, 1998–1999

Country	Area	
	Hectares	Acres
United States	19,347,000	47,787,000
Brazil	11,200,000	27,664,000
Argentina	7,270,000	17,000,000
Canada	4,080,000	10,000,000
Australia	1,000,000	2,470,000
Paraguay	790,000	1,950,000
Mexico	500,000	1,235,000
Bolivia	200,000	500,000
Chile	96,000	240,000
Uruguay	50,000	125,000
Others	1,000,000	2,470,000
Total	45,533,000	111,441,000

method is an easy and practical way to estimate residue cover using a 100-foot tape.[32] Residue cover can also be estimated by calculating the amount of residue left after tillage operations.[33]

The amount of residue after harvest depends on the type and yield of the crop.[33] As shown in Table 5.3, most crops leave sufficient residue to provide adequate cover; in some cases, incorporation of some residue can be beneficial. The amount of residue remaining after tillage operations depends on the type of residue and the type of tillage implement. Residue can be classified as non-fragile when it is resistant to destruction during tillage and fragile when it will easily break up during

FIGURE 5.9
No-till row crop planter seeding on the contour.
[Courtesy USDA NRCS]

TABLE 5.3 Crop Residue Classification and Typical Percent Residue Cover After Harvest of Various Crops

Non-Fragile Residue		Fragile Residue	
Crop	% Cover	Crop	% Cover
Alfalfa	85	Canola/rapeseed	70
Corn for grain, 60–120 bu/ac	80	Dry edible beans	15
Barley*	85	Dry peas	20
Corn for grain, 120–200 bu/ac	95	Potatoes	15
Corn for silage	10	Soybeans	70
Forage silage	15	Sugarbeet	15
Grain sorghum	75	Sunflower	40
Hay crops	85	Vegetables	30
Millet	70		
Oat*	80		
Pasture	85		
Popcorn	70		
Rye*	85		
Wheat,* 30–60 bu/ac	50		
Wheat,* 60–100 bu/ac	85		

*For small grains, if a rotary combine or a combine with a straw chopper is used, or if the straw is otherwise cut into small pieces, consider the residue to be fragile.

tillage. Table 5.3 lists types of residues and their fragility. Table 5.4 shows the amount of residue remaining after various tillage operations.

Residue reduces water erosion by forming ponds that reduce runoff velocity and delay runoff and by obstructing and diverting runoff, which increases the length of the down slope flow path (Figure 5.10). Residue also keeps smaller flows

TABLE 5.4 Estimated Percentage of Residue Remaining on the Soil Surface After Specific Implements and Field Operations

Implement	Residue Remaining (%)	
	Non-Fragile Residue	Fragile Residue
Moldboard plow	0–10	0–5
Disk, primary tillage	30–60	20–40
Disk, secondary tillage	40–70	25–40
V-ripper/subsoiler	60–80	40–60
Chisel-subsoiler	50–70	40–50
Disk-subsoiler	30–50	10–20
Chisel-sweep	70–85	50–60
Disk-chisel-sweep	60–70	30–50
Sweep, >30 inches	75–95	60–80
Sweep, <30 inches	70–90	50–75
Sweep-mulch treader	60–90	45–80
One-way disk	40–50	20–40
Spring tooth harrow	60–80	50–70
Rotary rod w/ shovels	70–80	60–70

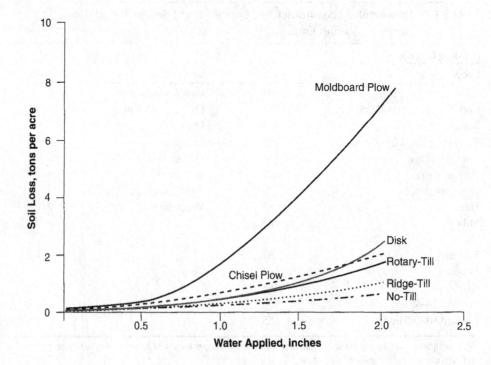

FIGURE 5.10

Soil loss from a silt loam soil on a 10 percent slope with water applied at a rate of 2.5 inches per hour [From Wortman, 2003].

from combining into larger flows that have the capacity to carry more soil particles.[1] Wind erosion is reduced by protecting fragile soil particles and providing microbarriers to the force of the wind. Residues can increase resistance of surface aggregates to the destructive effects of freeze-thaw cycles.[34, 35]

Tillage alters the albedo (reflectivity) of the soil, the surface area exposed to the atmosphere, and the penetration of wind into the soil. Evaporation from a non-tilled soil will be higher initially than a tilled soil, but, as a dry layer develops at the surface of a non-tilled soil, rate of evaporation will be reduced to less than that of a tilled soil. The time and rate depend on soil factors. There are more benefits from non-tillage on fine-textured soils than on coarser-textured soils.[36]

Residues at seeding reduce early season evaporation and conserve water for increased transpiration later in the season.[37, 38] A study of five locations from east-central United States to the Great Plains showed that the water-filled pore space in the top 6 inches (15 cm) was higher in no-tilled soils than in tilled soils.[39] Moisture loss from the surface foot of both irrigated and dryland soils in Utah[40] was retarded by a residue mulch.

Leaving residues on the soil surface affects soil organic matter concentrations and soil microbial populations. Some studies show that soil organic matter is highest near the soil surface, but lower soil horizons are unaffected.[41] In long-term studies of tillage management of winter wheat-fallow systems in Nebraska, leaving residues on the soil surface decreased soil organic matter content in the topsoil. But that decline is partly offset by increased dry matter production due to higher yields and reduced fallow times.[42] Nonsymbiotic nitrogen fixation was two times higher in no-till soils than plowed soils due primarily to higher soil water content which favors microbial development.[43]

When residue is left on soil surface there is increased immobilization of soluble N, especially when residues have a high C/N ratio. Also, mineralization is decreased when residues are not incorporated.[44] Not incorporating residues decreases nitrification and increases denitrification.[45] A depressive effect on nitrate accumulation in soils may occur where straw mulches are heavy enough to suppress weed growth.[10]

There is an increased risk of bacterial and fungal diseases when residues are left on the soil surface that might, in some cases, preclude the use of conservation tillage. There is also an increased risk of phytotoxins when residues are left on the surface. However, these effects are short-lived and should not be overshadowed by the increased erosion protection that the residues provide.[44]

When tillage is reduced or eliminated, weed control is accomplished almost exclusively with herbicides. As shown in Tables 5.5 and 5.6, weed species may change when tillage is reduced.[46] When perennial noxious weeds are present, it may be necessary to combine tillage with herbicides to obtain complete control.

Figure 5.11 shows the short-term and long-term changes that may occur with the adoption of conservation tillage.[34] Earthworm numbers are likely to increase and soil structure will likely improve over the long term. Pest protection, fertilizer requirements, and crop yield will likely remain the same over the long term. Total cost of production and labor will likely decline over the long term. And, although machinery costs will be higher in the short term as new equipment is purchased, machinery costs will be less in the long term as less tillage equipment will be needed.

Conservation tillage has the potential to conserve soil, time, energy, and labor. However, there are restrictions to conservation tillage: (1) soils high in nonexpanding clay minerals, silt, and fine sand may need tillage periodically to reduce soil compaction and improve structure; (2) excess precipitation on poorly drained soils can cause problems when using conservation tillage; and (3) there can be an increase of water-soluble toxins in the residue and toxins from microbial decomposition of residue.[34]

TABLE 5.5 Densities of Annual Weeds in Conservation Tillage Systems Compared to Conventional Tillage Systems (from Moyer, 1994)

Lower Densities	Greater Densities
Velvetleaf (*Abutilon theophrastis*)	Parsley piert (*Alchemilla arvensis*)
Fools parsley (*Aethusa cynapium*)	Blackgrass (*Alopecurus myosuroides*)
Scarlet pimpernel (*Anagallis arvensis*)	*Amaranthus* spp.
Spreading orach (*Atriplex patula*)	Prairie threeawn (*Aristida oligantha*)
Wild oat (*Avena fatua*)	Wild oat (*Avena fatua*)
Wild mustard (*Brassica kaber*)	Browntop millet (*Brachiaria ramose*)
Shepherd's purse (*Capsella bursa-pastoris*)	*Bromus* spp.
Common lambsquarter (*Chenopodium album*)	Shepherd's purse (*Capsella bursa-pastoris*)
Fumitory (*Fumaria officinalis*)	Field sanbbur (*Chenchrus incertus*)
Matricaria spp.	Common lambsquarter (*Chenopodium album*)
Corn poppy (*Papaver rhoeas*)	Horseweed (*Conyza Canadensis*)
Polygonum spp.	Texas croton (*Croton texensis*)
Wild radish (*Raphanus raphanistrum*)	*Descurainia* spp.
Green foxtail (*Seteria viridis*)	*Digitaria* spp.
Common chickweed (*Stellaria media*)	Barnyard grass (*Echinochloa humistrata*)
Field pennycress (*Thalaspi arvense*)	Goose grass (*Eleusine indica*)
Field violet (*Viola arvensis*)	Prostrate spurge (*Euphorbia humistrate*)
Common cocklebur (*Xanthium strumarium*)	Common sunflower (*Helianthus annuus*)
	Kochia (*Kochia scoparia*)
	Prickly lettuce (*Lactuca serriola*)
	Rigid ryegrass (*Lolium rigidum*)
	Matricaria spp.
	Carpetweed (*Mollugo verticillata*)
	Fall panicum (*Panicum dichotomiflorum*)
	Annual bluegrass (*Poa annua*)
	Prostrate knotweed (*Polygonum aviculare*)
	Erect knotweed (*Polygonum erectum*)
	Common purslane (*Portulaca oleracea*)
	Tumblegrass (*Schedonnardus paniculatus*)
	Common groundsel (*Senecio vulgaris*)
	Setaria spp.
	Prickly sida (*Sida spinosa*)
	Cutleaf nightshade (*Solanum triflorum*)
	Shattercane (*Sorghum bicolor*)
	Common chickweed (*Stellaria media*)
	Puncturevine (*Tribulus terrestris*)

▓ 5.8 FALLOW

Land that is uncropped and kept cultivated throughout a growing season is known as fallow. Under humid conditions, however, land that merely lies idle for a year or two is often referred to as fallow. The most important function of fallow is storage of moisture in the soil, but fallowing also promotes nitrification and is a means

TABLE 5.6 Densities of Biennial and Perennial Weeds in Conservation Tillage Systems Compared to Conventional Tillage Systems (from Moyer, 1994)

Biennial Weeds with Greater Densities	Perennial Weeds with Greater Densities
Chervil (*Anthriscus cerefolium*)	Russian knapweed (*Acroptilon repens*)
Biennial wormwood (*Artemisia biennis*)	Woolyleaf bursage (*Ambrosia grayi*)
Plumeless thistle (*Carduus acanthoides*)	Broomsedge (*Andropogon virginicus*)
Musk thistle (*Carduus nutans*)	Hemp dogbane (*Apocynum cannabinum*)
Water hemlock (*Cicuta* spp.)	Common milkweed (*Asclepias syriaca*)
Bull thistle (*Cirsium vulgare*)	Hedge bindweed (*Calystegia sepium*)
Prickly lettuce (*Lactuca serriola*)	Trumpet creeper (*Campsis radicans*)
Sweetclover (*Melilotus* spp.)	Hoary cress (*Cardaria draba*)
Western salsify (*Tragopogon dubius*)	Tumble windmill grass (*Chloris verticillata*)
	Canada thistle (*Cirsium arvense*)
	Field bindweed (*Convolvulus arvensis*)
	Bermuda grass (*Cynodon dactylon*)
	Cyperus spp.
	Quackgrass (*Elytrigia repens*)
	Leafy spurge (*Euphorbia esula*)
	Jerusalem artichoke (*Helianthus tuberosus*)
	Foxtail barley (*Hordeum jubatum*)
	Skeleton weed (*Lygodesmia juncea*)
	Wirestem muhly (*Muhlenbergia frondosa*)
	Dallis grass (*Paspalum dilatatum*)
	Plysalis spp.
	Swamp smartweed (*Polygonum coccineum*)
	Rosa spp.
	Rubus spp.
	Rumex spp.
	Sassafras (*Sassafras albidum*)
	Horsenettle (*Solanum carolinense*)
	Perennial sowthistle (*Sonchus arvensis*)
	Johnsongrass (*Sorghum halepense*)
	Sand dropseed (*Sporobolus cryptandrus*)
	Dandelion (*Taraxacum officinale*)

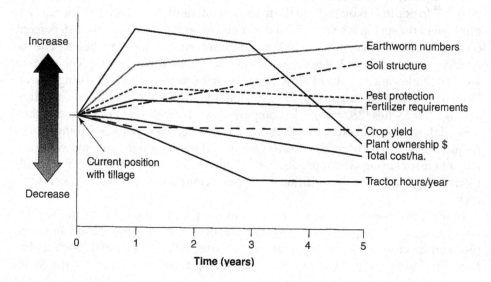

FIGURE 5.11
Short- and long-term trends likely to occur when converting from conventional tillage to conservation tillage. [Copyright CRC Press, Used with permission]

TABLE 5.7 Minimum Amounts of Plant Residue Needed at Wheat Seeding for Erosion Control

Soil Texture	Minimum Residue (pounds per acre)
Sandy to sandy loam	1,700
Very fine sandy loam to silt loam	1,200
Silty clay loam	1,000
Clay loam	750

of controlling weeds. Fallowing is a common practice in semiarid sections where annual precipitation of less than 20 inches (500 mm) is insufficient to produce a satisfactory crop every year. Averages of 15 to 31 percent of the precipitation that fell during the fallow period were stored in the soil at different locations in the dry farming regions.[9, 47] Moisture conservation by fallow is most effective in winter rainfall areas and where summer evaporation is low. Fallow conserves little moisture in coarse sandy soils or in high rainfall areas.

For successful fallow, certain principles must be observed: (1) the surface of the soil must be protected with residues (Table 5.7) or kept sufficiently rough to absorb rains and prevent wind erosion, (2) weed growth must be suppressed to conserve soil moisture, and (3) the operations must be accomplished at a low cost. Fallow tillage should be early in the spring,[5] worked as little as necessary to control weeds, and, when not seeded to winter grains, also ridged in the fall at right angles to the prevailing winds where soil blowing is likely to occur. Sweep plows, field cultivators, or rod weeders are best for fallow tillage because they cut off the weeds and leave the surface rough and cloddy.

In semiarid regions, fallow has been widely used in alternation with small grains (Figure 5.12). In southwestern Kansas, consistently higher yields of winter wheat were obtained on fallow than on land continuously cropped.[48] Fallow was the best preparation for winter wheat at the Central Great Plains Field Station near Akron, Colorado, where the rainfall is about 18 inches (450 mm) per year.[15, 49] In South Dakota, where the mean annual rainfall averaged 16 inches (406 mm), yields of spring wheat increased 74 percent on fallowed land and 46 percent on disked corn land, compared to land continuously cropped to wheat.[50] The respective increases were 109 percent and 48 percent for winter wheat. The relative yields of barley and oat after fallow were much the same as those of spring wheat. Similar results were obtained at Havre, Montana, where the annual precipitation averaged 11 inches (280 mm).[51] In comparison with small grains grown continuously, fallow gives a better distribution of yields between years, with much less frequent failures. At Mandan, North Dakota, fallowing a light soil showed little advantage for spring-wheat production.[52] Fallow may be replaced entirely by cultivated crops under such conditions. The most common is wheat alternated with corn.

In the Great Basin region, where most of the precipitation occurs in the winter months, fallowing has been standard practice for wheat production since the nineteenth century. Yields of winter wheat after fallow are nearly double that from continuous cropping. Continuous cropping is impractical in the driest ar-

FIGURE 5.12
A wheat-summer fallow system in Kansas. [Courtesy Richard Waldren]

eas. Fallow operations seldom conflict with the seeding or harvesting of wheat, which enables a farmer to grow as large an acreage of wheat each year as with continuous cropping. Fallowing often involves only one to three additional rapid tillage operations. The higher yields from fallowing, as compared with continuous cropping, more than compensate for the investment in double the acreage of tillable land.[53]

Stubble mulch fallow is an excellent method for erosion control but excessive amounts of straw can cause seeding problems that can be detrimental to wheat yields. Nitrogen fertilizers applied to the soil avoid yield decreases from medium amounts of straw. The practice of continuous stubble mulch increases infestation of weeds such as cheatgrass (Bromus tectorum). Tillage or herbicides must be used to control it. Special seeding equipment must also be used to cut through heavy residues (Figure 5.13). Stubble much fallow involves tillage with a field cultivator or sweep-blade implement that leaves crop residues on the surface.[9, 23, 24] Subsequent tillage is done with a rod weeder or a blade implement.[5] In the central High Plains, stubble mulch fallow has resulted in better moisture retention and higher yields[54] (Table 5.8).

As better seeding equipment has become available for seeding into residues and more effective herbicides for weed control, the practice of stubble mulch fallow has increased. In some areas, particularly in the Northern Plains, no-tillage practices have increased efficiency of precipitation absorption sufficiently to eliminate fallow altogether. U.S. farmers reduced fallow acreage by 43 percent from 1964 to 1997, with the largest reductions in the Great Plains (USDA-NASS). However, in the Pacific Northwest, adoption of reduced tillage practices in wheat-fallow systems has been slow due to increased economic risks and reduced profit.[55]

FIGURE 5.13
No-till wheat. [Courtesy
Richard Waldren]

TABLE 5.8 Progress in Fallow Systems at Akron, Colorado (adapted from Greb, 1979)

Years	Tillage	Water Storage in Fallow			Wheat Yield	
		(inches)	(mm)	(% of total)	(bu/ac)	(kg/ha)
1916–1930	Maximum tillage (plow, harrow, dust mulch)	4.02	102	19	16	1,070
1931–1945	Conventional tillage (shallow disc, rod weeder)	4.65	118	24	17	1,160
1945–1960	Improved conventional tillage, began stubble mulch in 1957	5.39	137	27	26	1,730
1961–1975	Stubble mulch, began minimum tillage with herbicides in 1969	6.18	157	33	32	2,160
1976–1979	Minimum tillage	7.20	183	40	40	2,690

▓ 5.9 SOIL MULCHES FOR MOISTURE CONSERVATION

Soil mulches are created by stirring the surface soil until it is loose and open. Although research in the early twentieth century advocated shallow tillage to conserve soil water,[56] later studies do not support this theory.[36] Soil mulch reduces evaporation from the soil surface when free water is present only a few feet below the surface of the soil.[57] Although capillary water can move upward 15 to 120 inches (38 to 300 cm) during a season,[58] capillary movement to appreciable heights is very slow,[59] and movement decreases as the soil moisture decreases. Water in sufficient quantities to support crop plants can be raised only a few inches from a moist subsoil.[60] Most of this movement takes place at moisture contents well above the wilting point. Capillarity and gravity combined may move moisture downward for greater distances and with about twice the quantity that occurs in the upward movement against gravity.

Evaporation from a bare soil surface is a three-stage process. The initial stage is energy-limited with a steady rate of evaporation. In the second stage, the surface dries and the rate begins to decline as the water flow to the surface is limited. The third stage consists of a slow, nearly constant rate of evaporation.[38]

In eastern Kansas, plots with soil mulches actually lost more water than did bare, undisturbed plots where the weeds were kept down with a hoe.[7] In experiments in California where the water table was from 18 to 40 feet (5 to 12 m) below the surface, mulching by thorough cultivation failed to save moisture.[61] On upland soils in Pennsylvania where there was no water table in the soil mantle, frequent cultivation did not decrease the evaporation loss materially.[62]

In Nebraska,[63] in soils partly dry and away from a source of free water, capillary movement was so slow that it could barely be detected. In the northern Great Plains, no water moved up appreciable distances to replace that removed by roots. Water was supplied to wheat roots only by the soil they occupied and only that part of the soil suffered exhaustion or reduction of its water content.[64] Once water is in soil deep enough to escape rapid drying at the surface, it mostly remains until it is reached and removed by plant roots.[63]

In semiarid regions, the rainfall is sufficient to wet the soil to a depth of only a few feet.[64] This is usually exhausted each year by growth of native vegetation or crops. The lower layers are dry except in the wettest of seasons. The water table may be more than 20 feet (6 m) below the surface. Formerly, dust mulch was advocated to prevent moisture loss by capillary rise. After some disastrous experiences with soil blowing, the dust mulch theory was abandoned and a clod mulch was advocated. Soon thereafter, the effectiveness of any soil mulch to check moisture loss under dryland field conditions was refuted.

Thus, a soil mulch can reduce surface evaporation where the water table is so shallow that drainage rather than moisture conservation is desirable. In fields, the mulch formed by tillage is merely incidental to weed control.

■ 5.10 CULTIVATION IN RELATION TO SOIL NITRATES

Cultivation, by drying and aerating the soil, which promotes nitrification, is a benefit to a heavy water-logged soil. Controlling weeds by cultivation permits nitrates to accumulate, since nitrates are used up when a crop or weeds occupy the land. Weed control by cultivation also saves soil moisture and thus favors nitrification, which does not operate in dry soil. Cultivation, otherwise, has not produced a regular increase in the accumulation of nitrates, especially in light soils. Greater nitrification occurred in a compacted Kansas soil as compared to an uncompacted soil, up to the point where the moisture content reached two-thirds saturation. In Arkansas, varied depths of cultivation had little effect on accumulation of nitrate nitrogen in a soil of rather open structure.[65] In Pennsylvania research, scraped soil and soil cultivated three to eight times contained almost equal quantities of nitrate nitrogen.[62] Natural processes promote sufficient aeration to admit needed oxygen in the soil. Excessive cultivation reduced the nitrates in Missouri soil where the cultivation kept so much surface soil continually dry that nitrate production apparently was retarded in the upper 7 inches (18 cm).

Other experiments have shown either slight or distinct increases in nitrate accumulation from surface cultivation as compared with soil that was merely scraped.[11, 66, 67, 68, 69] The higher nitrate content in these cultivated soils may have been due to increased aeration.[64] When weed growth was prevented, Kansas research showed that sufficient accumulation of nitrates took place in uncultivated soil to insure an excellent growth of wheat, despite the lower nitrate content.[70] Nitrates leach downward so rapidly after rains that their measurement in the surface foot is of little significance.[67]

▦ 5.11 OTHER EFFECTS OF CULTIVATION

Cultivation can conserve soil moisture by prevention of runoff. Under semiarid conditions, up to 88 percent of the water in a dashing rain can be lost by runoff. A cultivated surface retains more water from a rain than does an uncultivated surface.[63] The faster the rain falls, the greater the difference in the amount of water held between a cultivated and uncultivated surface. On the other hand, excessive pulverization of the surface soil is likely to result in a quickly puddled condition with great runoff losses. Under subhumid conditions, there is little or no relation between the type of tillage treatment for a given crop and amount of soil moisture. Fallow areas merely scraped to control weeds have proved to be slightly less effective in moisture conservation than those that received normal cultivations, probably as a result of increased runoff on the scraped plots.[67]

A loosened soil surface acts as an insulator, so a cultivated soil is slightly cooler than uncultivated soil.[62] Many soils naturally have sufficient aeration for optimum bacterial and chemical activity without cultivation.[2]

▦ 5.12 INTERTILLAGE OR CULTIVATION

The primitive farmer hoed or pulled out the weeds that grew among crops planted at random in small clearings. In ancient and medieval field husbandry, field crops were planted at random or in close rows with the seed dropped in plow furrows. These crops were later weeded by hand or with crude hoes or knives. Jethro Tull introduced intertillage into English agriculture in 1733 when he applied it to crops like turnips that were planted in rows.

5.12.1 Purposes of Intertillage

The primary purpose of intertillage is weed control. Intertillage also breaks a crust that otherwise might retard seedling development, and in some cases roughens the soil sufficiently to increase water infiltration. Some have claimed, without substantial proof, that intertillage brings about aeration of the soil, with the result that plant foods are more readily available because of increased bacterial and chemical action in the soil.

Crops that generally require planting in rows with sufficient space between them to permit cultivation during their growth include corn, cotton, grain sorghum, sugarbeet, sugarcane, tobacco, potato, and field bean. Intertillage controls weeds that grow in the open spaces before the crop can shade the ground.

Crops with relatively slender stems, such as small grains, hay crops, and flax thrive under close plant spacing and cover the land rather uniformly. Without cultivation, these crops tend to suppress weeds by root competition and shading. Yields of these crops are usually low when they are planted in cultivated rows because they do not fully utilize the land.[71] The leaf canopy never closes, resulting in poor utilization of sunlight. Where conditions are so severe that small grain and hay crops succeed only in cultivated rows, they are generally unprofitable.

5.12.2 Intertillage Implements

Many different types of implements are used for intertillage. They come equipped with shovels, disks, teeth, sweeps, or knives.

ORDINARY CULTIVATORS Shovel or sweep cultivators are the types most used for intertillage because they are suited to practically all soil conditions. Cultivator width should correspond with planter width so that it operates over a group of rows that were planted simultaneously (Figure 5.14). Shovels or sweeps are sometimes replaced or supplemented by disks or disk hillers when considerable soil is to be moved or where a considerable amount of residue or roots is to be cut. Shovel cultivators are likely to clog under such conditions.

FIGURE 5.14
Cultivating soybeans.
[Courtesy Case IH]

ROTARY HOE The rotary hoe consists of a series of 18-inch (46 cm) hoe wheels each of which is fitted with teeth shaped like fingers. As the wheels rotate, these teeth penetrate and stir the soil. This hoe is useful for uprooting small weeds by *blind* cultivation before the crop emerges, as well as for early cultivations of young corn and other row crops. Its success depends on its use before weeds have made as much growth as the crop; otherwise, the crop plants would also be uprooted.[4] The rotary hoe is also effective in breaking soil crust that results from hot sunshine after a torrential rain.

OTHER CULTIVATION IMPLEMENTS The ordinary spike-tooth harrow is used with little injury to corn to kill small weeds when the corn plants are small and when the field is comparatively free of residue. Special cultivators for listed crops planted in furrows are necessary for following the furrows and turning the soil along their sides. The lister cultivator is equipped with disks or knives or both and sometimes with disks and shovels. For the first cultivation, the disks are set to cut the weeds on the sides of the furrow with the soil thrown away from the crop row. The shovels can be set to stir the soil near the plants. Hooded shields prevent the soil from rolling in on the young plants. For the second and usually final cultivation, the disks are set to roll the soil into the trench around the plants. This buries the weeds in the row while the rows are being leveled. Knife cultivators (go-devils or knife sleds) cut off the weeds below the surface while slicing the lister ridges.

5.12.3 Cultivation for Weed Control

Today, most weed control is accomplished by the application of herbicides. However, cultivation can be an economical alternative in some cases, especially in organic or sustainable systems when producers want to limit or eliminate the use of chemicals.

In the Corn Belt, about four cultivations are required for surface-planted corn. Experiments at the Illinois Agricultural Experiment Station[72] as early as 1888–1893 showed that corn yielded as well when the weeds were merely cut off with a hoe as when it received four or five deep cultivations.

These results were not taken seriously because of the widespread belief that cultivation was necessary for reasons other than weed control.[73] Later, results were summarized[74] for 125 experiments with corn carried on for 6 years in twenty-eight states, in which regular cultivation was compared with scraping with a hoe to destroy weeds. The average of all the tests showed that the scraped plots produced 95.1 percent as much fodder and 99.1 percent as much grain as did the cultivated plots. Cultivation was not beneficial to the corn except in destruction of weeds. These results were confirmed by additional experiments.[17, 62, 67] Other crops responded in a similar manner.[62] Competition for soil moisture is not the only reason for the low yield of corn in weedy fields, because irrigation of such fields increases the corn yield only slightly.[17]

5.12.4 Cultivation of Drilled Crops

Small grain fields sometimes have been harrowed in the spring, but in general this has been of no benefit to yield, nor has it eliminated all the weeds.[6] Cultivation of alfalfa sod with the toothed renovator or disk harrow to destroy grassy weeds, which formerly was widely advocated, was discontinued after it was found that the resulting injury to the alfalfa crowns permitted entrance of the bacterial wilt organism.

▓ 5.13 ARTIFICIAL MULCHES

Plastic mulches, as well as asphalt-coated paper mulches, increase the mean temperature but also decrease the temperature range of the soil. This may hasten germination and reduce deflocculation of the soil due to freezing and thawing. These mulches reduce soil moisture loss and stop weed growth but may fail to increase yields.[57, 75] The cost of these mulches limits their use for most field crops.

▓ 5.14 TILLAGE IN RELATION TO WATER EROSION

Seventy-five percent of the cultivated land in the United States probably has a slope greater than 2 percent on which the wasteful processes of accelerated soil erosion and runoff occur when the land isn't managed properly.[23, 49, 76] There are three types of water erosion: sheet, rill, and gully. Gully erosion causes deep channels in the field that interfere with cultivation; rill erosion produces small channels; sheet erosion is the nearly uniform removal of the topsoil from the entire slope. Severe sheet erosion on cultivated fields leads to gully formation.

Erosion is greatest in the southeastern states where heavy rainfall, rolling topography, unfrozen soil, and predominance of intertilled crops combine to cause excessive soil loss. Widely spaced crops, such as corn, cotton, tobacco, and soybean, are most generally associated with erosion.[77, 78, 79, 80] The numerous fibrous roots of sorghum and its more complete drying of the soil may account for the lower erosion on sorghum land than is observed on the land devoted to the taprooted cotton plant.[78] The greatest losses are on bare land,[78] being about twice those on corn land.[59] However, on continuously fallowed land the soil is nearly saturated with moisture.[81] Consequently, excess water is lost by runoff. Land in small-grain crops erodes very little except during periods when the soil is bare or nearly so.[23] Pasture and grass crops are very efficient in soil conservation[2] unless they are overgrazed and trampled. Erosion from continuously ungrazed, untrampled sod or meadow is almost negligible. Heavily grazed pasture may have more than three times the runoff of a moderately grazed pasture.[81] A three-year corn, wheat, clover-timothy rotation showed only twice as much runoff as did a moderately grazed pasture.

Natural vegetative cover usually prevents or greatly retards erosion.[82] Davis[83] stated: "A field abandoned because of erosion soon shows these efforts of nature to prevent devastation."

The chief methods of retarding soil erosion[84] are (1) reversion of steep slopes and abandoned cropland to pastures and woodlands, (2) contour tillage and planting, (3) strip cropping, (4) terracing, (5) cover crops, (6) damming of gullies, and (7) minimum tillage as previously described.

Strip cropping has been practiced in Pennsylvania for generations.[80] The alternate strips of thick-growing crops catch the soil water shed from the areas occupied by cultivated row crops. The sloping land is tilled and crops are planted on the contour. A regular rotation can be followed when the strips are of approximately equal width.

FIGURE 5.15
Contour terraces and grass waterways help reduce erosion. [Courtesy USDA NRCS]

The bench terrace, long used in Europe, Asia, and the Philippine Islands, is not feasible in the United States because of its excessive costs. Terracing became a common practice in the Southeast after the Mangum terrace was devised by P. H. Mangum of Wake Forest, N.C., in 1885. Later, many of the terraces were abandoned because of frequent failures of the terraces, high costs of maintenance, and questionable benefits from terracing. Many of these early terraces were poorly built or the land was managed improperly.[85] When a terrace breaks, erosion losses exceed those on unterraced land. Broadbase terraces permit farm machines to pass over them. The Nichols terrace, a broad-channel type, is constructed with less labor. This broadbase terrace is 15 to 25 feet (5 to 8 m) wide at the base and 15 to 24 inches (48 to 61 cm) high. The runoff water is carried away in a gradually sloping broad shallow channel at a low velocity (Figure 15.5). Level terraces are advised where the rainfall is less than 30 inches (760 mm) per year, so all the water is held on the field until it is absorbed.[78] The conservation bench terrace is even more effective because the water above the levee spreads back over a wide bed. The terracing of steep slopes (more than 8 percent) is no longer recommended because of the expense, because all the topsoil may be scraped up to build the terraces,[77] and because steep terrace banks erode badly. As fields and tillage equipment continue to grow larger, many terraces have been abandoned because it is not economical to till the smaller fields that result from terraces. However, if good conservation tillage is practiced, water erosion is less of a problem and may even be eliminated.

Nitrogen, phosphorus, calcium, and sulfur in soil that has been eroded from corn or wheat land can equal or exceed the amounts taken off in the crops.[13] Much of the soil carried by streams comes from stream or gully banks, deserts, and similar areas that have been eroding since the beginning of time. Most important of all, it comes from badlands and breaks. Most of this eroded soil is not topsoil. A combination of computed estimates of the quantity of suspended soil carried into the sea by rivers of the United States[49, 83] indicates that the total is about 1,450 pounds per acre (1,620 kg/ha) per year. Should half of this be topsoil, the equivalent of the total topsoil is washed to the sea every 2,759 years. Thus, only a small part of the erosion losses reported[76, 77] represent soil washed from the continent. Estimates of soil losses can exceed losses actually measured experimentally under the same conditions.[24] Estimates for 1971 suggest that 12 tons of soil per acre (27 MT/ha) were lost by water erosion on farms.[86]

▧ 5.15 TILLAGE IN RELATION TO WIND EROSION

Soil blowing has been going on at times in various parts of the Great Plains for at least 25,000 years. It also occurs elsewhere in the United States when a wind 20 miles per hour (30 km/hr) or more strikes a bare, smooth, loose, deflocculated soil.[87, 88] The organic matter and fine soil particles in the topsoil contain much of the nitrogen.[76, 89] They can be carried great distances and then deposited in some humid area where the additional fertility is needed. When accumulated drifts are stirred and allowed to blow back onto eroded cultivated fields where the soil is caught by vegetation, the aggraded soil that results can be very productive.[90] Uniform sand moves readily with wind where the surface is unprotected. Peat land is subject to serious wind damage. Heavy soils are also subject to movement under certain conditions. The ideal structural condition to prevent erosion is coarse aggregates (clods and crumbs) too large to blow, and yet not so large as to interfere with cultivation and plant growth. There is no advantage in further pulverization.

Summer fallowed land, except when protected by alternate strips of crops, is particularly vulnerable to high winds unless adequate residues are maintained on the soil surface. Land can be ranked as follows for decreasing vulnerability to blowing: (1) bean fields, (2) corn and sorghum stubble, (3) cornstalks, (4) fallow seeded to winter rye or winter wheat, and (5) small-grain or hay stubble.[91] Small grain stubble land rarely blows unless the organic debris is very scanty or has been turned under.

Tillage that furrows the ground or protects the surface soil with small clods and residue is especially desirable in wind erosion control. Furrows should be at right angles to the prevailing wind direction.[87, 92, 93] Among the useful implements are the rod weeder, lister, field cultivator, shovel cultivator, and spring-tooth harrow. The one-way disk plow is satisfactory only on land with heavy stubble because it leaves the residues mixed with and protruding from the surface soil. Cultivation stops wind erosion only temporarily[87] but may delay destructive action until rains start a vegetative cover. Then blowing ceases to be a serious problem.

A permanent vegetative cover is the best protection against soil blowing.[92, 93] The soil is too dry for grass seedlings to become established when soil blowing conditions prevail. When rainfall is ample, weeds cover the land and protect it from erosion without seeding. During severe drought, Russian thistle (*Salsola kali*) and kochia (*Kochia scoparia*) are the predominating weed covers in the Great Plains. These break off and blow away during the winter. In moist seasons, weeds such as wild sunflower (*Helianthus annuus*) and pigweed (*Amaranthus retroflexus*) take root and protect the soil while perennial weeds and grasses become established.

Strips of sorghum planted perpendicular to the prevailing wind direction check soil blowing.[92, 94] Tree belts check soil blowing and accumulate drifted soil. They are, however, of little economic benefit to field crops under semiarid conditions because they preclude crop growth for 4 rods (20 m) on each side and seldom grow more than 20 feet (6 m) high. Tree belts protect farmsteads, gardens, and livestock because they check the wind and catch drifting snow.

REFERENCES

1. Cole, J. S., and O. R. Mathews. "Tillage," in *Soils and Men*, USDA Yearbook, 1938, pp. 321–328.
2. Sewell, M. C. "Tillage: A review of the literature," *J. Am. Soc. Agron.* 11(1919):269–290.
3. Fryrear, D. W., and J. D. Bilbro. "Wind erosion control with residues and related practices," in *Managing Agricultural Residues*, edited by P. W. Unger. Boca Raton, FL: Lewis, 1994, pp. 7–18.
4. Gray, R. B. "Tillage machinery," in *Soils and Men*, USDA Yearbook, 1938, pp. 329–346.
5. Zingg, A. W., and C. J. Whitefield. "A summary of research experience with stubble-mulch farming in the western States," *USDA Tech. Bull.* 1166, 1957, pp. 1–56.
6. Oveson, M. M., and W. E. Hall. "Longtime tillage experiments on Eastern Oregon wheat land," *OR Agr. Exp. Sta. Tech. Bull.* 39, 1957.
7. Call, L. E., and M. C. Sewell. "The soil mulch," *J. Am. Soc. Agron.* 9(1917):49–61.
8. Allmaras, R. R., and others. "Fall versus spring plowing and related heat balance in the western Corn Belt," *MN Agr. Exp. Sta. Tech. Bull.* 283, 1972, pp. 1–22.
9. Duley, F. L. "Yields in different cropping systems and fertilizer tests under stubble mulching and plowing in eastern Nebraska," *NE Agr. Exp. Sta. Res. Bull.* 190, 1960.
10. Evans, C. E., and E. R. Lemon. "Conserving soil moisture," in *Soil*, USDA Yearbook, 1957, pp. 340–359.
11. Albrecht, W. A. "Nitrate accumulation in soil as influenced by tillage and straw mulch," *J. Am. Soc. Agron.* 18(1926):841–853.
12. Noll, C. F. "Deep versus ordinary plowing," *PA Agr. Exp. Sta. Ann. Rpt.* 1912–1913, pp. 39–47.
13. Duley, F. L., and O. E. Hays. "The effect of the degree of slope on runoff and soil erosion," *J. Agr. Res.* 45(1932):349–360.
14. Miller, M. F., and H. H. Krusekopf. "The influence of systems of cropping and methods of culture on surface run-off and soil erosion," *MO Agr. Exp. Sta. Res. Bull.* 177, 1932.
15. Chilcott, E. C., and J. S. Cole. "Subsoiling, deep tilling, and soil dynamiting in the Great Plains," *J. Agr. Res.* 14(1918):481–521.
16. Bracken, A. F., and George Stewart. "A quarter century of dry farming experiments at Nephi, Utah," *J. Am. Soc. Agron.* 23(1931):271–279.
17. Mosier, J. G., and A. F. Gustafson. "Soil moisture and tillage for corn," *IL Agr. Exp. Sta. Bull. 181*, 1915, pp. 563–586.
18. Hobbs, J. A., and others. "Deep tillage effects on soils and crops," *Agron. J.* 53, 5(1961):313–316.
19. Haney, W. A., and A. W. Zingg. "Principles of tillage," in *Soil*, USDA Yearbook, 1957, pp. 277–281.
20. Baker, C. J., K. E. Saxton, and W. R. Ritchie. *No-tillage Seeding: Science and Practice.* Wallingford, UK: CAB International, 1996.
21. Faulkner, E. H. *Plowman's Folly.* Grosset and Dunlap, 1943.
22. Scarseth, G. Ames, IA: Iowa State Univ. Press, *Man and His Earth*, 1962.
23. Duley, F. L., and C. R. Fenster. "Stubble-mulch farming methods for fallow areas," *NE Agr. Ext. Circ.* 54-100 (rev.), 1961, pp. 1–19.
24. Horning, T. R., and M. M. Oveson. "Stubble Mulching in the Northwest," *USDA Inf. Bull.* 253, 1962.

25. Fenster, C. R., and T. M. McCalla. "Tillage practices in western Nebraska with a wheat-sorghum-fallow rotation," *NE Agr. Exp. Sta. SB 515*, 1971, pp. 1–23.

26. Hayes, W. A. "Mulch tillage in modern farming." *USDA Leaflet 554*, 1971, pp. 1–7.

27. Mannering, J. V., and R. E. Burwell. "Tillage methods to reduce runoff and erosion in the corn belt," *USDA Inf. Bull. 330*, 1968, pp. 1–14.

28. Blevins, R. L., and Cook, D. "No-tillage, its influence on soil moisture and soil temperature," *KY Agr. Exp. Sta. Prog. Rpt. 187*, 1970, 1–15.

29. Myers, D. N., K. D. Metzker, and S. Davis. "Status and trends in suspended-sediment discharges, soil erosion, and conservation tillage in the Maumee River Basin-Ohio, Michigan, and Indiana." *U.S. Geol. Surv. Water-Resources Investigations Report 00-4091*, 2000.

30. Köller, K. "Techniques of soil tillage," in *Soil Tillage in Agroecosystems*, edited by Adel El Titi. Boca Raton, FL: CRC Press, 2003, pp. 1–26.

31. Derpsch, R. "Conservation tillage, no-tillage and related technologies," in *Proc. 1st World Congr. Conserv. Agric.* Madrid, Oct. 1–5, 2001. Vols. I and II. 2001.

32. Shelton, D. P., P. J. Jasa, J. A. Smith, and R. Kanable. "Estimating percent residue cover," *NE Coop. Ext. Serv. Nebguide* G95-1132-A. 1995.

33. Shelton, D. P., P. J. Jasa, J. A. Smith, and R. Kanable. "Estimating percent residue cover using the calculation method," *NE Coop. Ext. Serv. Nebguide* G95-1135-A. 1996.

34. Carter, M. R., ed. *Conservation Tillage in Temperate Agroecosystems*. Boca Raton, FL: Lewis, 1994.

35. Slater, C. S. "Winter aspects of soil structure," *J. Soil and Water Conservation 6*, 1(1951):38–41.

36. Jalota, S. K., and S. S. Prihar. *Reducing Soil Water Levels with Tillage and Straw Mulching.* Ames, IA: Iowa Univ. Press, 1998.

37. Duley, F. L., and J. C. Russel. "The use of crop residues for soil and moisture conservation," *J. Am. Soc. Agron. 31*(1939):703.

38. Steiner, J. L. "Crop residue effects on water conservation," in *Managing Agricultural Residues*, edited by P. W. Unger. Boca Raton, FL: Lewis, 1994, pp. 41–76.

39. Mielke, L. N., J. W. Doran, and K. A. Richard. "Physical environment near the surface of plowed and no-till soils," *Soil Till. Res. 7*(1986):355–366.

40. Fenster, C. R., and T. M. McCalla. "Tillage practices in western Nebraska with a wheat-fallow rotation," *NE Agr. Exp. Sta. SB 507*, 1970, pp. 1–20.

41. McCallister, D. L., and W. L. Chien. "Organic carbon quality and forms as influenced by tillage and cropping sequence," *Commun. Soil Sci. Plant Anal. 31*(2000):465–479.

42. Doran, J. W., E. T. Elliott, and K. Paustian. "Soil microbial activity, nitrogen cycling, and long-term changes in organic carbon pools as related to fallow tillage management," *Soil & Till. Res 49*(1998):3–18.

43. Lamb, J. A., J. W. Doran, and G. A. Peterson. "Nonsymbiotic dinitrogen fixation in no-till and conventional wheat-fallow systems," *Soil Sci. Soc. Amer. J 51*(1987):356–361.

44. Doran, J. W., and D. M. Linn. "Microbial ecology of conservation management systems," in *Soil Biology: Effects on Soil Quality*, edited by J. L. Hatfield and B. A. Steward. Lewis, Boca Raton, FL: Lewis, 1994, pp. 1–27.

45. Doran, J. W. "Tillage changes soil," *Crops and Soils Magazine*, Aug-Sep, American Society of Agronomy, Madison, WI. 1982.

46. Lindwall, C. W., F. J. Larney, A. M. Johnston, and J. R. Moyer. "Crop management in conservation tillage systems," in *Managing Agricultural Residues*, edited by P. W. Unger. Boca Raton, FL: Lewis, 1994, pp. 185–210.

47. Thysell, J. C. "Conservation and use of soil moisture at Mandan, N. Dak.," *USDA Tech. Bull. 617*, 1938, pp. 1–40.

48. Von Trebra, R. L., and F. A. Wagner. "Tillage practices for southwestern Kansas," *KS Agr. Exp. Sta. Bull. 262*, 1932, pp. 1–11.

49. Bennett, H. H., and W. C. Lowdermilk. "General aspects of the soil-erosion problem," in *Soils and Men*, USDA Yearbook, 1938, pp. 581–608.

50. Mathews, O. R., and V. I. Clark. "Summer fallow at Ardmore (South Dakota)," *USDA Circ.* 213, 1932.

51. Morgan, G. W. "Experiments with fallow in north-central Montana," *USDA Bull.* 1310, 1925.

52. Sarvis, J. T., and J. C. Thysell. "Crop rotation and tillage experiments of the northern Great Plains Field Station," *USDA Tech. Bull.* 536, 1936, pp. 1–75.

53. Harris, F. S., A. F. Bracken, and I. J. Jensen. "Sixteen years' dry farm experiments in Utah," *UT Agr. Expt. Sta. Bull.* 175, 1920, p. 43.

54. Greb, B. W. "Reducing drought effects on croplands in the west central Great Plains." *USDA Info. Bull.* 420. 1979, pp. 1–31.

55. Juergens, L. A., D. L. Young, H. R. Hinman, and W. F. Schillinger. "An economic comparison of no-till spring wheat and oilseed rotations to conventional winter wheat-fallow in Adams County, WA. *WA State Coop. Ext.* EB 1956E. 2002.

56. King, F. H. *Physics of Agriculture*, 5th ed. Madison, WI: F. H. King, 1910, pp. 158–203.

57. Shaw, C. F. "When the soil mulch conserves soil moisture," *J. Am. Soc. Agron.* 21(1929): 1165–1171.

58. Shaw, C. F., and A. Smith. "Maximum height of capillary rise starting with soil at capillary saturation," *Hilgardia* 2(1927):399–409.

59. McLaughlin, W. W. "Capillary movement of soil moisture," *USDA Bull.* 835, 1920.

60. Lewis, M. R. "Rate of flow of capillary moisture," *USDA Tech. Bull.* 579, 1937.

61. Veihmeyer, F. J. "Some factors affecting the irrigation requirements of deciduous orchards," *Hilgardia* 2(1927):125–284.

62. Merkle, F. G., and C. J. Irwin. "Some effects of inter-tillage on crops and soils," *PN Agr. Exp. Sta. Bull.* 272, 1931.

63. Burr, W. W. "Storage and use of soil moisture," *NE Agr. Exp Sta. Res. Bull.* 5, 1914.

64. Mathews, O. R., and E. C. Chilcott. "Storage of water by spring wheat," *USDA Bull.* 1139, 1923.

65. Sachs, W. H. "Effect of cultivation on moisture and nitrate of field soil," *AR Agr. Exp. Sta. Bull.* 205, 1926.

66. Call, L. E. "The relation of weed growth in nitric nitrogen accumulation in the soil," *J. Am. Soc. Agron.* 10(1918):35–44.

67. Kiesselbach, T. A., A. Anderson, and W. E. Lyness. "Tillage practices in relation to corn production," *NE Agr. Exp. Sta. Bull.* 232, 1928.

68. Lyon, T. L. "Inter-tillage of crops and formation of nitrates in the soil," *J. Am. Soc. Agron.* 14(1922):97–109.

69. Sewell, M. C., and P. L. Gainey. "Nitrate accumulation under various cultural treatments," *J. Am. Soc. Agron.* 24(1932): 283–289.

70. Shaw, C. F. "The effect of a paper mulch on soil temperature," *Hilgardia* 1(1926): 341–364.

71. Sheppard, J. H., and J. A. Jeffrey. "A study of methods of cultivation," *ND Agr. Exp. Sta. Bull.* 29, 1897.

72. Wimer, D. C., and M. B. Harland. "The cultivation of corn," *IL Agr. Exp. Sta. Bull.* 259, 1925.

73. Williams, C. G. "The corn crop," *OH Agr. Exp. Sta. Bull.* 140, 1903.

74. Cates, J. S., and H. R. Cox. "The weed factor in the cultivation of corn," *USDA B.P.I. Bull.* 257, 1912.

75. Magruder, Roy. "Paper mulch for the vegetable garden," *OH Agr. Exp. Sta. Bull.* 447, 1930.

76. Bennett, H. H. "Cultural changes in soils from the standpoint of erosion," *J. Am. Soc. Agron.* 23(1931):434–454.

77. Clark, M., and J. C. Wooley. "Terracing, an important step in erosion control," *MO Agr. Exp. Sta. Bull.* 400, 1938.

78. Dickson, R. E. "Results and significance of the Spur (Texas) run-off and erosion experiments," *J. Am. Soc. Agron.* 21(1929):415–422.

79. Duley, F. L. "Soil erosion of soybean land," *J. Am. Soc. Agron.* 17(1923): 800–803.

80. Miller, M. F. "Cropping systems in relation to erosion control," *MO Agr. Exp Sta. Bull.* 366, 1936.

81. Smith, D. D., and others. "Investigations in erosion control and reclamation of eroded Shelby and related soils at the Conservation Experiment Station, Bethany, Mo., 1930–42," *USDA Tech. Bull.* 883, 1945, pp. 1–175.

82. Jacks, G. V., and R. O. Whyte. "Erosion and soil conservation," in *Imperial Bur. Pastures and Forage Crops Bull. 25*, Aberwystwyth, England, 1938, pp. 1–206.

83. Davis, R. O. E. "Economic waste from soil erosion." *USDA Yearbook*, 1913, pp. 207–220.

84. Brill, G. D., C. S. Slater, and V. D. Broach. "Conservation methods for soils of the northern Coastal Plain," *USDA Inf. Bull.* 271, 1963.

85. Ramser, C. E. "Farm terracing," *USDA Farmers Bull.* 1669, 1931.

86. Hargrove, T. R. "Agricultural research impact on environment," *IA Agr. Exp. Sta. Special Rpt. 69*, 1972, pp. 1–64.

87. Call, L. E. "Cultural methods of controlling wind erosion," *J. Am. Soc. Agron.* 28(1936):193–201.

88. Kellogg, C. E. "Soil blowing and dust storms," *USDA Misc. Pub.* 221, 1935.

89. Daniel, H. A., and W. H. Langham. "The effect of wind erosion and cultivation on the total nitrogen and organic matter content of soils in the southern High Plains," *J. Am. Soc. Agron.* 28(1936):587–596.

90. Whitfield, C. J., and J. A. Perrin. "Sand-dune reclamation in the southern Great Plains," *USDA Farmers Bull.* 1825, 1939, pp. 1–13.

91. Brandon, J. F., and A. Kezer. "Soil blowing and its control in Colorado," *CO Agr. Exp. Sta. Bull.* 419, 1936.

92. Chilcott, E. F. "Preventing soil blowing on the southern Great Plains," *USDA Farmers Bull.* 1771, 1937.

93. Cole, J. S., and G. W. Morgan. "Implements and methods of tillage to control soil blowing on the northern Great Plains," *USDA Farmers Bull.* 1797, 1938.

94. Jardine, W. M. "Management of soils to prevent blowing," *J. Am. Soc. Agron.* 5(1913):213–217.

Fertilizer, Green Manuring, Rotation, and Multiple Cropping Practices

■ 6.1 PURPOSE OF FERTILIZATION

Fertilizers are applied to the soil to promote greater plant growth and better crop quality. Farmers endeavor to apply the kind and amount of each fertilizer constituent that gives the most profitable response.[1] Fertilizer needs are established by field experiments and farm trials for each soil type, environment, crop, and crop variety. Fertilizer needs can be estimated by laboratory tests of soils and plant tissues, by pot cultures, and by observation of deficiency symptoms in plants.[2] Fertilizer recommendations based on suitable tests are usually dependable when previous experience with the soil type and crop behavior in the locality is available. Public or private soil testing facilities are available to all farmers.

Crop residues contribute to soil organic matter and nutrient recycling and reduce the need for fertilizers. Table 6.1 shows the estimated N, P, and K content in residues returned to cropland annually in the United States.[3] The actual amount present in residues in an individual field depends on the type of crop, the crop's yield, and the plant part harvested, all of which determine the total amount of residue left. A residue not utilized by the crop usually remains in the soil for the following crops. Crop residues were discussed in more detail in Chapter 5.

More than a century ago, Justus von Liebig proposed that what is removed in crops should be returned to the soil. This policy is not valid for soils that are especially high in their content of certain elements. It also does not apply to semiarid conditions where only light applications of fertilizer give a yield response, and these can even reduce grain yields in a year of subnormal precipitation.[4, 5]

Soluble compounds of nitrogen, calcium, magnesium, and chlorine are leached downward into the soil, but they contact water supplies only if they reach a drainage outlet or the underlying water table. Phosphorus and potassium compounds become fixed in the upper soil levels and they enter streams or ponds only from fields subjected to water erosion. Very few fertilized fields are as high in fertility as they were in their virgin state. The chemical elements comprising the earth have been entering the seas, streams, and ponds ever since the earth was created.

TABLE 6.1 Estimated N, P, and K Content in Residues Returned to Cropland Annually in the United States

Crop	Area Harvested (1,000 ha)	Harvested Yield* (1,000 Mg)	Harvested Portion	Crop Residue (1,000 Mg)	Nutrients in Residue (Mg)		
					N	P	K
Barley	3,090	6,314	0.4	9,471	65,350	6,630	224,463
Beets	526	25,466	0.9	2,830	76,398	6,225	164,114
Corn	23,571	125,194	0.4	187,791	1,784,015	187,791	2,722,970
Cotton	4,837	3,356	0.4	5,034	45,306	7,551	69,469
Oat	2,239	3,158	0.4	4,737	33,159	2,842	121,741
Peanut	659	1,640	0.4	2,460	41,820	3,690	33,948
Rice	1,174	7,253	0.4	10,880	75,069	8,704	62,013
Rye	241	373	0.4	560	2,686	504	5,427
Sorghum	4,066	15,694	0.4	23,541	195,390	30,603	282,492
Soybean	23,218	42,152	0.3	98,355	819,344	462,267	914,698
Sunflower	777	813	0.3	1,897	15,745	2,846	17,642
Wheat	21,525	49,320	0.4	73,980	429,084	36,990	1,050,516

*Harvested yield is the portion of crop removed at harvest. Harvested yield + residue = total plant biomass at harvest.
Source: USDA, Agricultural Statistics. 1990

Excess phosphates and nitrates in standing or slow-moving waters stimulate the growth of algae, a process called *eutrophication*. Eutrophication will cause a scum of algae to form on the water surface that limits the oxygen supply and thus restricts the fish population. However, many ponds remote from any agricultural area exhibit this scum formation. More than 1 million artificial ponds are distributed over the United States. Many of these have been fertilized to stimulate algae growth in order to produce fish. The algae provide food for minute members of the animal kingdom. They are devoured by a series of larger species, which in turn provide food for the fish.

Soil is the cheapest source of most fertilizer elements. High yields of crops remove more fertilizer elements from the soil than do low crop yields, but this fact does not justify crop neglect. The soil depletion that occurs under good management is not cause for alarm. Columella wrote in the first century AD: "The earth neither grows old nor wears out, if it be dunged." This philosophy is confirmed on fields of Europe and Asia that have been farmed for centuries. Fertilizers can restore depleted soils.

Any practice other than fertilizer application that improves total crop yields, even crop rotation and green manuring, speeds up depletion of the mineral elements of the soil.

The use of improved varieties and other practices that bring profitable crop yields may facilitate soil improvement when they encourage greater use of fertilizers. This has resulted in striking increases in crop yields in the United States since 1940.

The total depth of soil in which annual crop plant roots customarily feed is 3 to 6 feet (1 to 2 m). Phosphorus is usually distributed rather uniformly through the surface, subsurface, and subsoil layers. Potash is distributed uniformly throughout the different soil horizons except in sandy soils and under high rainfall conditions, where the potash content increases with depth. Although nitrates leach downward several feet into the soil, total nitrogen is mostly in the organic residues of the upper horizons, except on soils that have received heavy applications of nitrogen.

Squanto, the first American extension agronomist, advised the Massachusetts colonists to put two dead fish in each hill of corn planted on new land just cleared from the forest. His advice was based upon the known fact that this treatment increased corn yields. Squanto did not know that the fish were rich in nitrogen, phosphorus, and calcium. George Washington found that his land in the coastal plain of northern Virginia would produce two crops before it needed to be fertilized or rested for further cropping to succeed. Forest soils in high rainfall regions on a rolling topography can benefit from fertilizers applied immediately, or soon after they are cleared and farmed, and forever thereafter. On the other hand, the deep black soil of the level, drier prairie of the Red River Valley showed no decrease in productivity in sixty years of continuous cropping to wheat without fertilizer or green manure.[6] Despite this fact, wheat grown there in rotations or with fertilization outyielded continuous wheat grown without fertilizer. Crops grown on newly irrigated desert land usually respond to heavy applications of nitrogen.

6.1.1 Fertilizer Usage

Commercial fertilizer usage in the United States was about 32,000 tons (29,000 MT) in 1860; 1,400,000 tons (1,270,000 MT) in 1890; 2,730,000 tons (2,480,000 MT) in 1900; 8,425,000 tons (7,660,000 MT) in 1930; 12,468,000 tons (11,330,000 MT) in 1944; and 41,000,000 tons (37,300,000 MT) in 1972. The average plant food content was 14.5 percent in 1900, 20.6 percent in 1944, and 41 percent in 1972. The NPK (Nitrogen-Phosphate-Potash) ratio for all mixed or unmixed fertilizers used in the United States is roughly 18.5–12.4–10.3 and 18.5–5.3–8.4 on the elemental basis (Nitrogen-Phosphorus-Potassium).

Between 1972 and 2001, nitrogen fertilizer use increased about 45 percent in the United States, and phosphate and potash fertilizer use remained steady. By the 1970s, the use of fertilizers had become part of accepted crop production practices and further increases were the result of increasing crop yields and further decline in inherent soil fertility levels. In recent years, nitrogen use on corn, the largest user, has changed little while yields have continued to increase. This has resulted in increased efficiency of nitrogen use in that crop.

Global fertilizer use has increased dramatically as less-developed countries improve crop production. Currently, about 150 million tons (140 million MT) are applied globally. Figure 6.1 shows global fertilizer use from 1961 to 2001. The countries applying the most fertilizer to crops are the United States, China, Brazil, India, and France.

Fertilizer is generally used in humid, subhumid, and irrigated areas (Figure 6.2) but is less frequently used in semiarid regions. Eighty percent of the cornfields and practically all of the tobacco, rice, potato, and sugarbeet fields received fertilizers by 1964.[7]

Nitrogen is the main nutrient applied to irrigated cornfields of the West.[8] Currently, about 96 percent of corn receives nitrogen, 79 percent receives phosphate, and 65 percent receives potash fertilizers. About 76 percent of cotton receives nitrogen, 48 percent receives phosphate, and 41 percent receives potash fertilizer.

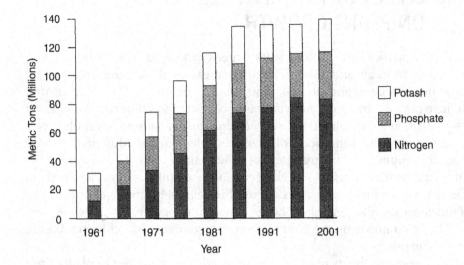

FIGURE 6.1
Fertilizer use in the world
from 1961 to 2001.
[FAOSTAT]

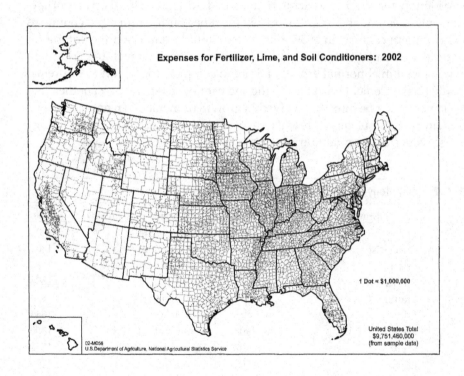

FIGURE 6.2
Fertilizer use in the United
States, 2002. [Source: 2002
Census of U.S. Department of
Agriculture]

Only 11 percent of soybean receives nitrogen, 17 percent receives phosphorus, and 20 percent receives potash fertilizers.

Dairy farms use less commercial nitrogen than do other farms because of the abundance of animal manure. Less potash and phosphate are needed on alluvial soils than on upland soils in the same locality.

Nearly all of the intertilled crops in the irrigated, humid, and subhumid areas receive mineral fertilizers except sometimes when they follow legumes or green manuring.

■ 6.2 EFFECT OF NUTRIENT ELEMENTS ON PLANT GROWTH

Plants require sixteen essential nutrients. Three of these, carbon, hydrogen, and oxygen, come from air and water. The thirteen essential nutrients supplied by the soil are nitrogen, phosphorus, potassium, calcium, sulfur, magnesium, boron, manganese, zinc, iron, molybdenum, copper, and chlorine.[9] At least six others—cobalt, arsenic, selenium, lead, lithium, and vanadium—stimulate certain plants under some conditions. Most of these elements, in addition to iodine, fluorine, chromium, and tin, are essential to human or animal nutrition.[8, 10] Essential plant nutrients used in relatively large quantities are classified as macronutrients, and micronutrients are used in relatively small quantities. The macronutrients are nitrogen, phosphorus, potassium, magnesium, calcium, and sulfur. The micronutrients are copper, iron, manganese, zinc, chlorine, boron, and molybdenum.

Although some nutrients have been recognized as being a part of plants since ancient times, it wasn't until the eighteenth and nineteenth centuries that scientists began to identify many specific elements absorbed by plants. The discovery of specific elements as essential nutrients began in the nineteenth century and continued into the twentieth century. In 1939, Arnon and Stout[11] established the criteria for plant nutrient essentiality that are still used today. These criteria are: (1) a deficiency will result in abnormal growth, failure to complete the life cycle, or premature death; (2) the element must be specific and cannot be replaced by another; and (3) the element must be directly involved in growth or metabolism and not just by antagonizing another element present at toxic levels. Table 6.2 lists the chronology of the discoveries of essential nutrient elements.

TABLE 6.2 Chronology of Discoveries of Essential Nutrient Elements

Element	Discoverer	Year	Discoverer of Essentiality	Year
C	*		DeSaussure	1804
H	Cavendish	1766	DeSaussure	1804
O	Priestley	1774	DeSaussure	1804
N	Rutherford	1772	DeSaussure	1804
P	Brandt	1769	von Liebig	1844
Fe	*		Greiss	1844
Mg	Davy	1808	von Sachs, Knop	1860
K	Davy	1807	von Sachs, Knop	1860
Ca	Davy	1808	von Sachs, Knop	1860
S	*		von Sachs, Knop	1865
Mn	Scheele	1774	McHargue	1922
B	Gay-Lussac and Thenard	1808	Warrington	1923
Zn	*		Sommer and Lipman	1926
Cu	*		Sommer	1931
Mo	Scheele	1778	Arnon and Stout	1939
Cl	Scheele	1774	Stout	1954

*Known since ancient times.

The use of macronutrients nitrogen, phosphorus, potassium, and calcium are described later. Magnesium is supplied in dolomitic limestone applications. Iron and manganese become unavailable in calcareous or alkaline soils, but calcium applications are used to correct manganese toxicity on highly acid soils. Calcium fertilizer improves fruit quality in peanut. The micronutrients are included in mixed fertilizers for certain crops in soils in which these elements are deficient. Fertilizers containing sodium are beneficial for sugarbeet, rape, and mustard, as well as for several other crops when potassium is deficient. Fertilizers containing chlorine are required for flue-cured tobacco. Figure 6.3 shows some effects of certain mineral deficiencies in tobacco. Cobalt, iodine, selenium, or fluorine may be applied to crops to correct a nutritional balance when used as a livestock feed, but they are not essential for plants.[8, 12]

The most common nutrient deficiency symptom in plants is chlorosis (lack of chlorophyll), which gives leaves a yellow color. Other deficiency symptoms include stunting, poorly shaped leaves, purple color, and necrosis (death of tissue). Nutrients that are mobile in the plant will move to younger leaves as deficiencies begin to occur so deficiency symptoms will be visible in the older leaves. Nutrients that are immobile in the plant cannot move from older tissue so deficiency symptoms will be visible in the younger leaves. Table 6.3 lists common deficiency symptoms for nutrients.

Nutrient deficiencies can be detected in growing plants by taking leaf samples. For most crops, the leaf blade is removed, but the petiole should be used for sugarbeet and cotton. A sample of several younger mature leaves are removed from the plant, placed in a paper bag or mailer, and sent to the laboratory. The results are then compared to known sufficiency levels. Sufficiency levels for nutrients have been established in most crops. If the nutrient concentration is below the sufficiency level, fertilizers are added to correct the deficiency. The advantage of leaf sampling is that a deficiency can be detected before symptoms become visible. Table 6.4 lists the sufficiency level for nutrients.

FIGURE 6.3
Effect of mineral deficiencies on tobacco plant growth: plants grown in complete nutrient solution *(A)*, or without nitrogen *(B)*, phosphorus *(C)*, potassium *(D)*, boron *(E)*, calcium *(F)*, or magnesium *(G)*.

TABLE 6.3 Functions and Deficiency Symptoms of Soil Nutrients

Nutrient	Role in Plant	Mobility	Deficiency Symptoms
Nitrogen (N)	Component of proteins, nucleotides, enzymes, alkaloids. 40–50% of protoplasm.	Mobile	Uniform chlorosis of leaves followed by necrosis. Stunting if severe.
Phosphorus (P)	Component of nucleotides, phospholipids, enzymes, ATP. Synthesis of carbohydrates and proteins.	Mobile	Purplish color with purple veins in leaves. Stunted growth. Poor root growth.
Potassium (K)	Functions as enzyme cofactor to regulate photosynthesis, carbohydrate translocation, protein synthesis, etc.	Very mobile	Marginal necrosis in leaves. Leaves may curl.
Calcium (Ca)	Component of cell wall.	Immobile	Flattened tops. Growing tips of both roots and stems may die. Leaves distorted.
Magnesium (Mg)	Component of chlorophyll. Enzyme activator.	Mobile	Interveinal chlorosis, red-purple color on leaves.
Sulfur (S)	Component of all proteins. Plant hormones. Plant flavors such as mustard.	Immobile	Uniform chlorosis of whole plant. Thin stems.
Iron (Fe)	Chlorophyll synthesis. Enzymes for electron transfer.	Very immobile	Interveinal chlorosis.
Zinc (Zn)	Enzyme synthesis. Synthesis of auxin.	Relatively immobile	Small chlorotic leaves with rough margins. Necrotic spots on leaves. Rosette appearance.
Manganese (Mn)	Oxidation-reduction systems. Formation of O_2 in photosynthesis.	Immobile	Interveinal chlorosis.
Copper (Cu)	Catalyst for respiration. Protein formation.	Relatively immobile	Bluish-green, necrotic leaves
Molybdenum (Mo)	Enzyme activator for N fixation and nitrate reduction	Relatively immobile	Uniform chlorosis.
Boron (B)	Translocation and carbohydrate metabolism. Cell wall development.	Immobile	Similar to Ca. Shortening of upper internodes. Brittle growing points.

FIGURE 6.4
Effect of fertilizer on the growth of corn in the Williamette Valley of Oregon. *(Left)* fertilized; *(right)* unfertilized.

TABLE 6.4 Sufficiency Range of Soil Nutrients in Mature Leaves

	Most Crops	Corn
Nitrogen	1.00–6.00%	2.80–3.30%
Phosphorus	0.20–0.50%	0.25–0.35%
Potassium	1.50–4.00%	1.75–2.25%
Calcium	0.50–1.50%	0.21–1.00%
Magnesium	0.15–0.40%	0.20–1.00%
Sulfur	0.15–0.50%	0.21–0.50%
Boron	1–100 ppm	5–25 ppm
Chlorine	50–200 ppm	*
Copper	2–20 ppm	5–20 ppm
Iron	50–75 ppm	20–250 ppm
Manganese	10–200 ppm	20–200 ppm
Molybdenum	0.15–0.30 ppm	*
Zinc	15–50 ppm	25–100 ppm

*Cl and Mo rarely deficient in corn.

6.2.1 Soil–Nitrogen Relations

Except for peat and muck soils, which are high in organic matter, nitrogen fertilizers cause a response in crops grown on most soils of the humid, subhumid, and irrigated areas (Figure 6.4). Nitrogen is absolutely essential to plant growth. Plants grown on soils with sufficient amounts of available nitrogen in the soil make for a thrifty, rapid growth with a healthy deep green color. Ample nitrogen encourages stem and leaf development, while deficiency of this element results in plants of poor color, poor quality, and low production. An oversupply of nitrogen in the soil tends to cause lodging, late maturity, poor seed development in some crops, and greater susceptibility to certain diseases. On semiarid lands, surplus available nitrates cause excessive plant growth that exhausts the soil moisture in dry seasons before grain is produced. In humid areas, abundant nitrogen can prevent firing of the leaves during dry periods.

The nitrogen in the soil is derived originally from the air, which contains about 34,000 tons over each acre of land (38,000 kg/ha). Aside from being fixed artificially, gaseous nitrogen is fixed with other elements through the activities of soil bacteria.[13] It is introduced into the soil as organic nitrogen of plants or plant residues. Upon decomposition of the organic matter, some of the nitrogen passes into the air as elemental nitrogen or ammonia, while the part that remains in the soil is converted into ammonium and nitrites and finally into nitrates.

Ammonia and nitrogen oxides in the air are returned to the soil in rain or snow. Ammonia and nitrates are used by plants, but some are lost in the drainage waters. Some of the nitrogen remains in the soil a long time in the organic matter that is not fully broken down. The nitrogen supply of the soil is maintained by growth of legumes, use of manure, and the addition of nitrogen fertilizers.

Nitrogen is more easily lost from the soil than any other nutrient. The nitrate form, being an anion, cannot attach to the cation exchange sites. As gravitational water moves below the root zone, it will take nitrates with it. Some parts of Nebraska are finding nitrates in aquifers over 100 feet (30 m) due to years of over fertilization and over irrigation. Nitrate-nitrogen can also be lost through denitrification. This occurs

TABLE 6.5 Nitrogen Fixed Annually by Legumes, Based on Solid Stand

Legume	Nitrogen Fixed (pounds/acre/year)
Alfalfa	50–300
Birdsfoot trefoil	40–150
Chickpea	20–100
Cowpea	40–150
Cicer milkvetch	100–200
Crown vetch	90–110
Crimson clover	40–80
Field pea	50–300
Hairy vetch	80–120
Kura clover	20–160
Lentil	120–180
Red clover	60–200
Soybean	20–270
Sub clover	50–170
Sweet clover	80–140
White clover	100–200

in anaerobic conditions when soils are near water saturation. Soil bacteria will use the oxygen in nitrate and convert the nitrogen to nitrous oxides or nitrogen gas, which volatilizes and escapes the soil.

ROLE OF BACTERIA IN NITROGEN FIXATION *Rhizobium* bacteria that multiply in nodules on the roots of legumes fix nitrogen from the air into forms that the plant can utilize (Figure 26.3). This is called symbiotic fixation of nitrogen. An average of 40 to 200 pounds of nitrogen per acre (45 to 224 kg/ha) is added to the soil by *Rhizobium* bacteria annually if the crop is plowed under.[14, 15] Alfalfa and most clovers usually fix more nitrogen than do the large-seeded legumes (Table 6.5).

Rhizobium bacteria, which require molybdenum, cobalt, and iron in order to function, are most effective in soils low in nitrogen. It takes energy for the bacteria to fix nitrogen from the atmosphere. Therefore, if the soil has sufficient levels of nitrogen, the bacteria will simply use the existing "free" nitrogen. *Rhizobium* prefer a nearly neutral soil pH. In addition, the soil should be moist, but not wet, and well aerated.

The relationship of legumes with *Rhizobium* species is a truly symbiotic relationship. The plants provide shelter for the bacteria in the form of nodules and food in the form of sugars from photosynthesis. The bacteria in turn provide the plant with nitrogen.

The species and specific strain of groups of *Rhizobia* are as follows (also referred to as cross-inoculation groups):[16]

Rhizobia meliloti: alfalfa, other *Medicago* species, sweet clover, and fenugreek

R. trifolii: clovers of the genus *Trifolium*, except *T. ambiguum* (kura clover)

R. leguminosarum: peas of genera *Pisum* and *Lathyrus* and vetches of the genus, *Vicia*

R. lupini: lupines (genus *Lupinus*) and serradella (*Arnithopus*)

R. phaseoli: beans of the species *Phaseolus vulgaris* and *P. coccineus*

R. japonicum: soybean

Cowpea group: cowpeas, lespedezas, crotalarias, other species of *Phaseolus*, peanut, velvet bean, kudzu, pigeon pea, guar, Alyce clover, black gram, hairy indigo, and others

Specific strains for each species: birdsfoot trefoil, big trefoil, kura clover (*Trifolium ambiguum*), sainfoin, crown vetch, garbanzo, tropical kudzu, sesbania, and others.

A number of leguminous trees also fix nitrogen by specific *Rhizobium* strains. Some legume varieties of one species require different strains or cultures for effective nitrogen fixation. Strains of bacteria isolated from the same legume also differ in nitrogen-fixing effectiveness. Some strains form nodules but fix no nitrogen. The interior of an active nitrogen-fixing nodule is pink or red.[16]

Effective legume inoculation is provided by thoroughly mixing the seed with the correct, fresh, refrigerated culture of the appropriate bacteria along with a little water, milk, syrup, or other adhesive substance. The inoculated seed should be sown in moist soil that is pressed around the seed, within twenty-four hours if possible. Acid soils should be limed. The culture may not survive in dry soil unless the seed is pelleted. The culture should not be exposed to direct sunlight, either. The *Rhizobium* bacteria enter through the root hairs of the seedling and form the nodules in the cortex of roots. Legume plants obtain their nitrogen from the soil when the bacteria are absent.

Some fifteen other genera of bacteria, including *Azotobacter* and *Clostridium*, are reported to convert lesser amounts of nitrogen into organic combinations by a process called *nonsymbiotic fixation*.[17] A number of genera of blue-green algae growing in ponds or flooded rice fields also fix nitrogen. Root nodulation has been observed on several leguminous and nonleguminous trees and plants.

NITROGEN DISTRIBUTION IN SOILS UNDER NATURAL CONDITIONS Peat soils may contain as much as 200,000 pounds per acre nitrogen (224,000 kg/ha). The nitrogen content is highest in soils of the Prairie and Chernozem types of the central states, where it is estimated to average 16,000 pounds per acre (18,000 kg/ha) to a depth of 40 inches (102 cm) (Figure 6.5). From this region, the amount of nitrogen decreases down to 4,000 pounds per acre (4,500 kg/ha) eastward, westward, and southward. The distribution of nitrogen in soils is closely related to climatic conditions.[13] In the East, the high rainfall leaches out nitrates. In the drier West, sparse plant growth retards accumulation of nitrogen-containing humus. The total nitrogen content of the soil in the United States decreases from north to south as temperature increases. For every 50°F (10°C) rise in mean annual temperature, the average nitrogen content of the soil decreases one-half to two-thirds. The nitrogen in soils ranges from 0.01 to 1.0 percent or more in the United States, while the annual temperature ranges from 32 to 72°F (0 to 22°C). Because of favorable conditions for its preservation, the nitrogen content of soils in the North may be built up. In the South, high temperatures, which favor decomposition, make it practically impossible to increase, except temporarily, the nitrogen content of soils by green-manuring practices while the land is being cropped.

Under natural soil conditions there exists an equilibrium between the formation of organic matter by vegetation and its decomposition by microorganisms, the balance being determined primarily by climatic factors. Cultivation depletes both organic matter and nitrogen. In the wheat production areas, from 20 to 40 percent of the original soil nitrogen was lost in from 20 to 40 years.[18] The end result is a new equilibrium point at a decidedly lower level than the original natural nitrogen content.

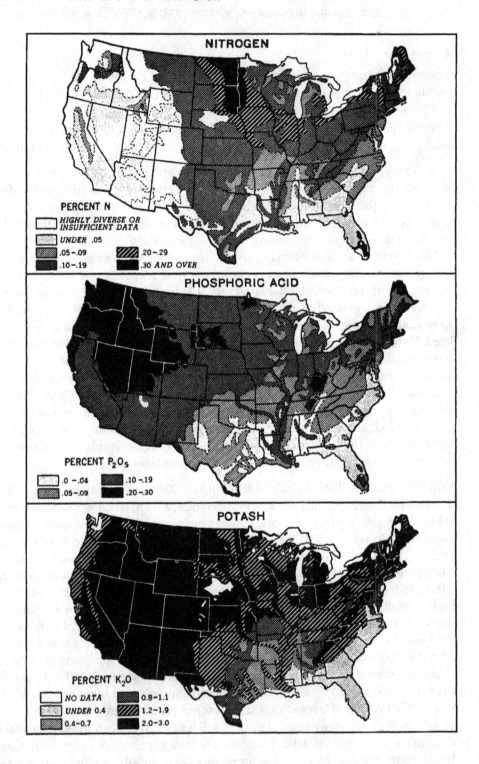

FIGURE 6.5

Distribution of nitrogen, phosphoric acid, and potash in the surface foot of United States soils.

The upper 7 inches (18 cm) of semiarid Kansas soil lost an average of 23 percent of its nitrogen and 36 percent of the organic carbon in 44 years of various cropping methods. Losses from row-cropping and fallowing greatly exceed those from continuous wheat culture.[19]

6.2.2 Soil–Phosphorus Relations

Adequate amounts of phosphorus in soils favor rapid plant growth and early fruiting and maturity, and often improve the quality of vegetation. Liberal supplies of phosphorus enable late crops to mature in the North before they are injured by early freezes. Soils in much of the United States usually give a response to phosphorus applications (Figure 6.4). An exception is parts of the bluegrass region of Kentucky and Tennessee where the soils were formed from rocks high in phosphorus (Figure 6.5). Acid clay soils low in organic matter usually give a greater response to phosphate fertilizers than do other soils of the same region.[20] Crops respond to phosphorus on irrigated lands in the West, particularly after several years of cropping. Calcareous irrigated soils are most likely to respond to phosphorus.[21] Certain western soils are high in both phosphorus and calcium, but the *availability* of phosphorus is low, probably because of the high calcium content.[22]

6.2.3 Soil–Potassium Relations

An adequate supply of potassium in the soil improves the quality of the plant, insures greater efficiency in photosynthesis, increases resistance to certain diseases, helps to balance an oversupply of nitrogen, and aids plants to utilize soil moisture more advantageously. It also insures the development of well-filled kernels and stiff straw in the cereals, encourages growth in leguminous crops, assists in chlorophyll formation, and is particularly helpful in the production of starch- or sugar-forming crops.[12, 23] Potassium is particularly beneficial to tobacco, potato, cotton, sugarbeet, and certain cereals. Potato grown with too little potassium may produce watery tubers low in starch and generally poor in quality. With insufficient potassium, the leaves of tobacco may have poor color and flavor and lack burning qualities. Crops usually respond to potash fertilizers in the Atlantic coastal plain, in regions of high rainfall in the eastern third of the United States (except in certain portions of the Piedmont), and on sandy, muck, and peat soils of the Great Lakes region. Potash is usually abundant in soils of volcanic origin. The ordinary range of potassium expressed as potash (K_2O) in the plow surface of mineral soils is 0.15 percent in sands to 4.0 percent or more in clay soils.[12]

■ 6.3 COMMERCIAL FERTILIZERS

6.3.1 Nitrogen Fertilizers

Nitrate fertilizers are materials with their nitrogen combined in the nitrate form, as in sodium nitrate. Nitrate-nitrogen, readily soluble in water, is rapidly utilized by most crop plants. Sodium nitrate tends to make the soil alkaline.

Ammonium sulfate, ammonium phosphate, anhydrous ammonia, and aqua ammonia carry nitrogen only in the ammonium form. Ammoniacal-nitrogen, applied as anhydrous ammonia, although soluble in water, is less readily leached from the soil than nitrate-nitrogen. It can be used directly by crops, although it is often converted by bacteria to the nitrate form before being utilized by plants. Ammonium sulfate tends to make soils acid because of the presence of the sulfate ion.

The ammoniacal form is better utilized than the nitrate form in wet, saturated soils. Dentrification of the ammoniacal form does not occur under anaerobic conditions, whereas the nitrate form is lost as gaseous nitrogen under the same conditions.

Organic materials, such as tankage or castor bean pomace, contain nitrogen in a complex organic form largely insoluble in water. The nitrogen becomes soluble and available to plants after decomposition of the materials.

Fertilizers that contain nitrogen in amide form include urea and calcium cyanamide. These simple nonprotein compounds dissolve in water, and the nitrogen is usually converted rapidly to the ammoniacal or nitrate forms by soil bacteria. Dissolved nitrogen is available to plants.

The principal nitrogen fertilizer materials in the United States, together with their average nitrogen contents in percentages are anhydrous ammonia (82), nitrogen solutions (20-40), ammonium nitrate (33.5), ammonium sulfate (20.5), ammonium phosphates (11-27), aqua ammonia (20.5), urea (45), and sodium nitrate (16).

All nitrogenous fertilizers, regardless of the form of nitrogen, are eventually converted to the nitrate form by a two-step process called *nitrification*. *Nitrosomonas* bacteria convert ammonium to nitrite, and *Nitrobacter* bacteria convert nitrite to nitrate.

6.3.2 Phosphate Fertilizers

Phosphorus must be dissolved in the soil solution before it can be taken up by plants. The phosphorus in fertilizers is usually in the form of phosphates, mostly of calcium. Phosphoric acid in phosphates of organic origin is more readily available to plants than that in the raw mineral phosphates. The superphosphate industry began in Great Britain in 1843 when it was found that phosphoric acid in rock phosphate could be made available to plants by treatment with sulfuric acid. The principal phosphate fertilizer materials used in the United States are as follows:

1. Concentrated or triple superphosphate with 23 to 48 percent phosphoric acid.
2. Ammonium phosphate, mostly monoammonium phosphate with 11 percent nitrogen and about 48 percent phosphoric acid. Since this has a tendency to increase soil acidity, it is sometimes necessary to add lime to the soil.
3. Superphosphate with 16 to 20 percent phosphoric acid (P_2O_5).
4. Other materials: bone meal, basic slag, and finely ground raw rock phosphate.

When phosphate fertilizer is applied to the soil, it soon reacts to form compounds that are only slightly available. This is known as fixation. Under ordinary conditions, 10 to 20 percent of the phosphorus in broadcast applications is recovered in the first crop. Some residual phosphorus is recovered in subsequent years. Phosphorus seldom leaches from soils. Accumulations may take place from heavy applications year after year, so it is important to analyze soil samples for phosphate to determine the most economical application rate. More efficient use of phosphates has been obtained for row crops when the material is distributed in narrow bands at the side of the seed instead of being broadcast. This banded application is commonly called *starter*. In many cases, it is possible to reduce the rate of phosphate fertilizer by one-half if it is band applied. Concentrating phosphate in a band insures that the plant will obtain an adequate supply of phosphate while still in the seedling stage. Early uptake of phosphate is the most efficient. Broadcast phosphates are fixed rapidly because the particles come in contact with a larger amount of soil.

6.3.3 Potassium Fertilizers

The principal potassium fertilizer materials used commercially are potassium chloride with 47 to 61 percent potash (K_2O), potassium sulfate with 47 to 52 percent, and the manure salts with 19 to 32 percent. Much of the potash is used for cotton, potato, tobacco, and hay crops. The greatest consumption is in the southern Atlantic Coast and Great Lakes states. Potash, like phosphate, is fixed on soil particles and is not easily lost.

6.3.4 Mixed Fertilizers

A mixed fertilizer contains two or more fertilizer elements. About 55 percent of the fertilizer distributed through commercial channels is a mixture. The cost of application for mixed fertilizers is less than when the various components are applied separately. The fertilizer is sold on the basis of its composition as ascertained by chemical analysis. The fertilizer formula indicates the composition of the mixture in terms of the three important elements often designated as NPK. For example, a 10-10-10 fertilizer contains 10 percent each of nitrogen (N), phosphoric acid (P_2O_5), and potash (K_2O). The remainder consists of other elements, usually calcium, sulfates, chlorides, inert materials, and often some micronutrients. It is usually more economical to purchase high-grade mixtures that contain 30 points (30 percent) or more of available plant food constituents.

Mixtures with low nitrogen percentages are usually applied to the soil before or during planting. A top dressing or side dressing with a liquid or granulated nitrogen fertilizer follows later. Winter cereals usually receive a top dressing in the spring. Row crops, such as corn, receive a side dressing after the plants have made some growth. Low-nitrogen mixtures are especially useful for legume crops or legume-grass mixtures that require little or no supplementary nitrogen after the plants are established.

The percentages of phosphorus (P) and potassium (K) in the previous formulas are actually percentages of phosphorus pentoxide (P_2O_5) and potassium oxide (K_2O). This procedure complies with general fertilizer recommendations as well as with the labeling requirements of state fertilizer laws in effect up to 1964 or later. However, some countries label the phosphorus and potassium content of fertilizers on the elemental basis. This procedure was recommended by the American Society of Agronomy in 1963 but was never accepted by the fertilizer industry.

Some companies report the analysis of a fertilizer in both forms. A conversion chart for adjusting the percentages of the two elements is shown in Figure 6.6. Thus, a 10-10-10 fertilizer under the new system would be a 10-4.4-8.3 fertilizer for NPK. About 500,000 tons (455,000 MT) of organic fertilizers are marketed each year.

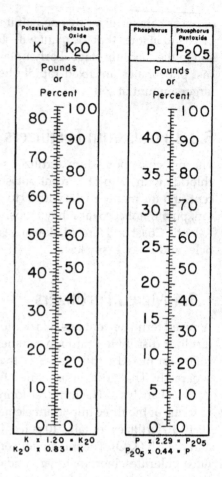

FIGURE 6.6
Fertilizer conversion scales for phosphorus and potassium.

6.3.5 Application of Fertilizers

Commercial fertilizer is applied by such methods as (1) injecting liquid materials into soil with machines, (2) broadcasting or drilling dry materials before or after plowing the soil, (3) placing in bands or hills during planting, (4) side dressing between or beside the crop rows, (5) drilling into slits being opened by deep tillers, or (6) metering liquids into irrigation water that is applied with sprinklers or by gravity flow.

Some crops will utilize nitrogen from dilute solutions of urea sprayed on the leaves, but this method of fertilization for field crops is rarely followed. Zinc, copper, iron, and manganese can also be applied by foliar spraying.

Localized applications of fertilizers near the plant promote early, rapid growth. Localized applications also cause a greater response to small amounts of fertilizer. Advantages of localized placement are less marked in soils of high natural fertility or when heavy applications are made. Application of fertilizer at the sides of, or below, the seed or plant is most efficient. This advantage has been explained by the tendency of fertilizer salts to move up and down in the soil with only a slight horizontal movement. Fertilizer bands vary from ¾ inch to 2 or more inches (2 to 5 cm) in width, with the narrower bands being used for relatively light applications of less than 100 pounds per acre (110 kg/ha). The bands are placed 1 to 2½ inches (2.5 to 6.4 cm) to the side and 2 inches (5 cm) below the seed. The most com-

FIGURE 6.7
Applying anhydrous
ammonia by knifing it into
the soil. [Courtesy USDA
NRCS]

mon spacing is 2 inches by 2 inches (5 cm by 5 cm), commonly called starter application. The distance between the bands of seed and fertilizer ordinarily should be greater for heavy rates of application on coarse soils and for crops more sensitive to fertilizer injury. Plow-sole applications, made by fertilizing before plowing, place the fertilizer at a depth more accessible to the roots of certain crop plants than is obtained from surface applications after the land is plowed. Placement of fertilizers at still greater depths is seldom advantageous.[24]

The most efficient time to apply nitrogen fertilizer is while the crop is growing as side-dress, top-dress, or through a sprinkler irrigation system. Postemergent applications ensure that nitrogen will be taken up by the plant before it is lost by leaching or denitrification. The most efficient time for the application of phosphate and micronutrient fertilizers is at planting as starter. This ensures that the seedling will encounter the nutrients early.

FERTILIZER-DISTRIBUTING MACHINERY Fertilizer-distributing machinery is classified as (1) machines used only for fertilizer application, (2) combination machines for applying fertilizer as well as planting the crop and sometimes applying pesticides in one operation (Figures 7.6 and 36.4), (3) combination machines for applying fertilizer and tilling the soil in one operation, and (4) machines for applying fertilizer in solution with irrigation water. Combination planter-fertilizer distributors for side-band placement of fertilizers are also available. Some transplanters for setting tobacco carry attachments for applying fertilizer at the side of the plant.

High-pressure tanks and equipment are needed for injection of anhydrous ammonia into the soil (Figure 6.7).[10] The liquid exerts a gauge pressure of 211 pounds per square inch (15 kgf/cm^2) at a temperature of 104°F (40°C) and 126 pounds per square inch (9 kgf/cm^2) at 75°F (24°C). The tank must be able to withstand a pressure of 250 pounds per square inch (18 kgf/cm^2). The liquid is injected with a field applicator into the soil at a depth of 4 to 6 inches (10 to 15 cm) in furrows cut by chisels or knives to which the outlet pipes are attached (Figure 36.14). A press wheel or slide covers the furrow with soil to check the escape of ammonia gas. The soil should be friable as well as moist (but not saturated) to retain the ammonia.

Once in the soil, the ammonia will react with soil water to form ammonium. The ammonium ion soon becomes attached to the clay minerals in the soil.

Anhydrous ammonia should be applied one or two weeks before a crop is planted in order to avoid injury to the seedlings. Side-dressings should be applied between the crop rows. Fall applications for a spring crop should be limited to retentive soils in areas of low or medium rainfall with low winter temperatures. The soil temperature should be less than 50°F (10°C) to prevent the nitrification of the ammonium, which attaches to the cation exchange sites, to nitrate, which is easily leached. Heavy losses of nitrogen are thus avoided.

Injection of aqua ammonia, nitrogen solutions, or mixed solutions requires only low-pressure tanks and equipment. The pressures are 25 pounds per square inch (1.8 kgf/cm^2) or less at 104°F (40°C). Low-pressure liquids should be injected at a depth of 2 inches (5 cm) or more, but liquid solutions can be applied to the soil surface. Liquid solutions can contain ammonium nitrate, urea, phosphoric acid, or ammonium phosphates alone or in mixtures.

Liquid fertilizers are applied with less labor than are dry fertilizers. Anhydrous ammonia is usually the cheapest source of fertilizer nitrogen. Much of the liquid fertilizer is applied directly to the field by local dealers on a custom basis. This is usually more economical than for a farmer to buy a tank and applicator.

The use of granular or coated granule fertilizers eliminates the dust that accompanies the application of dry, powdered fertilizers. Coated granules are less subject to caking than are powdered fertilizer materials, while coating also reduces the fire hazard associated with some of the nitrogen compounds.

6.4 LIME AS A SOIL AMENDMENT

Marl, which is high in calcium, was applied to soils by ancient Greeks. About 25 million tons of limestone or equivalent lime materials are applied annually in the United States. Lime is usually regarded as a soil amendment rather than a fertilizer, although calcium, the essential ingredient in lime, may be deficient in rare cases. The principal effect of liming in some cowpea-wheat experiments[25] was to increase the amount of soil nitrogen available for crop use. Soil acidity can be corrected by applications of lime. Soils low in lime are frequently acid, a condition that occurs in humid regions where calcium is leached from the soil or removed by plants.[26] Porous sandy soils leach readily and are more likely to be acid than are heavy soils under the same conditions. In South Carolina, approximately 40 percent of the soils are strongly to extremely acid (pH 5.5 or less).[27] This acidity is too high for satisfactory growth of even the somewhat acid-tolerant crops, tobacco and cotton, without liming. In Missouri, only small yield increases were obtained by liming soils to raise the pH values above 5.0, 5.5, 5.5 to 6.0, and 6.0 for cotton, corn, soybean, and alfalfa, respectively.[28]

6.4.1 Forms of Agricultural Lime

The usual forms of agricultural lime are crushed limestone, burned lime, and hydrated lime. Most lime originally comes from limestone rock where it occurs as calcium carbonate ($CaCO_3$), the active element being calcium. Dolomitic limestone contains considerable quantities of magnesium carbonate. The value of crushed limestone depends upon the purity as well as the degree of fineness of the crushed particles. Limestone becomes more effective as the rock is ground finer. Lime is considered to be sufficiently fine if it all passes through a 10-mesh sieve and 40 percent passes through a 100-mesh sieve.

Burned lime or quicklime (calcium oxide, CaO) is prepared by burning limestone. Hydrated lime [Ca(OH)$_2$] is formed when water is added to burned lime. Since limestone is only slightly soluble, it reacts slowly over a long period of time. Because of its lower cost, however, it is the form used for field application in acid soil regions, except where marl or oyster shells are available locally. The relative neutralization values of liming materials, assuming pure calcium carbonate to be 100 percent, are high-calcium limestone (75 to 95), dolomite limestone (90 to 110), marl (90 to 95), oyster shells (95), hydrated lime (125 to 145), burned lime (150 to 185), and magnesium carbonate (119).

6.4.2 Application of Lime

The limestone requirements to raise the pH of soils are shown in Table 6.6.[29] The annual losses of limestone from leaching or crop removal are 100 to 500 pounds per acre (112 to 560 kg/ha) in the north central states, but as much as 50 percent more in the higher rainfall areas of the Southeast. An application of two tons per acre (4.5 mg/ha) on average soils should last from five to twenty years.

Chemical tests are used to indicate the lime requirement of a soil. Limestone can be applied at any time of the year and to a crop at any stage without injury. Burned lime should not come in contact with the plant foliage. Broadcast applications are generally made in amounts as large as 1,000 pounds per acre (1,120 kg/ha) and applied with a lime spreader (Figure 6.8). It should be thoroughly incorporated into the surface soil. Usually, lime is applied in the spring before any tillage operations.

TABLE 6.6 Amount of Finely Ground Limestone Needed to Raise the pH from 4.5 to 5.5 and from 5.5 to 6.5 in a Seven-inch Layer of Soil

Soil Type	Limestone Requirements per Acre			
	Cool and Temperate Regions[a]		Warm and Tropical Regions[b]	
	pH 4.5–5.5 (tons)	pH 5.5–6.5 (tons)	pH 4.5–5.5 (tons)	pH 5.5–6.5 (tons)
Sand, loam, sand	0.5	0.6	0.3	0.4
Sandy loam	0.8	1.3	0.5	0.7
Loam	1.2	1.7	0.8	1.0
Silt loam	1.5	2.0	1.2	1.4
Clay loam	1.9	2.3	1.5	2.0
Muck	3.8	4.3	3.3	3.8

[a]Ultisol soils.
[b]Alfisol soils.

FIGURE 6.8
Applying lime to a field to correct soil acidity.
[Courtesy USDA NRCS]

■ 6.5 ANIMAL MANURE

6.5.1 Value of Manure

Over a billion tons (900 million MT) of manure are produced annually by livestock on American farms, but only about one-fourth to one-third of its potential value is applied to cropland and pastureland.[30] The remainder is lost by misplacement, drainage, leaching, or fermentation.

Farm manure is a mixture of animal excrements and stable litter. Its value for maintenance of soil productivity has been recognized since very early times. Manure brings about soil improvement because it adds nutrients, certain growth-promoting organic constituents, and humus to the soil.

6.5.2 Composition of Manure

The composition of manures[31] is given in Table 6.7. Manure exposed to rain may lose a large proportion of the nitrogen, phosphorus, and potash by leaching.[32]

Comparisons of equal weights of fresh and rotted manure used as fertilizer invariably favor the rotted product. Manure increases in nutrient concentration as it rots.[33] This increase is obtained by the loss of more than half the organic matter and a considerable loss of nitrogen.[30] The claim that rotted manure does not burn the crop merely reflects the fact that the highly available ammonia nitrogen has been sacrificed in the rotting process.

6.5.3 Application of Manure

Maximum returns from application of manure (Figure 6.9) are obtained when certain facts are recognized.[32]

TABLE 6.7 Average Daily Amount and Composition of Solid and Liquid Excrement of Mature Animals

Animal	Daily Production per Animal		Composition of the Fresh Excrement									
			Dry Matter		Nitrogen		Phosphoric Acid		Potash		Lime	
	Solid (lb)	Liquid (lb)	Solid (%)	Liquid (%)	Solid (%)	Liquid (%)	Solid (%)	Liquid (%)	Solid (%)	Liquid (%)	Solid (%)	Liquid (%)
Horses	35.5	8.0	24.3	9.9	0.50	1.20	0.30	Trace	0.24	1.50	0.15	0.45
Cattle	52.0	20.0	16.2	6.2	0.32	0.95	0.21	0.03	0.16	0.95	0.34	0.01
Sheep	2.5	1.5	34.5	12.8	0.65	1.68	0.46	0.03	0.23	2.10	0.46	0.16
Hogs	6.0	3.5	18.0	3.3	0.60	0.30	0.46	0.12	0.44	1.00	0.09	0.00
Hens	0.1	–	35.0	–	1.00	–	0.80	–	0.40	–	–	–

FIGURE 6.9
Applying liquid manure to a field. [Courtesy USDA NRCS]

1. Manure is relatively deficient in phosphoric acid. A ton (900 kg) of manure is equivalent to 100 pounds (45 kg) of 3-5-10 or 4-5-10 commercial fertilizer. A phosphorus supplement often is profitable.

2. Returns from a given quantity of manure usually are greater with lighter application to a larger area. The common rate of 8 tons per acre (9 kg/ha) is equivalent in value to 1,000 pounds (450 kg) of a 20-unit mixed fertilizer. Residual effects beyond the first year are greater with heavier applications.

3. Because of serious storage losses during the summer months, manure should be applied to spring crops rather than held in storage for fall application.

4. Top dressing may result in improved stands of grass and legume seedlings despite the loss of available nitrogen from the manure.

5. Often, a supplementary application of fertilizer in the hill or row is much more economical than the excessive amount of manure that would be required to meet the early demands of the crop.

6. The greatest profit may be expected from applications to crops of high-acre value and to rotations that include such crops as tobacco or potato. Corn responds well to manure.

7. The returns from a ton of manure applied to land that is low in fertility may be expected to be greater than from an equal quantity applied to land already highly productive.

8. It is sometimes more economical to purchase and apply commercial fertilizers than to hire labor to load, haul, and spread barnyard manure and then later apply the necessary supplemental mineral fertilizers.

9. The application of manure to semiarid land usually results in an increase in average total crop yield. However, in grain yield, manure usually produces increases only in wet years, with decreases in dry years and little or no benefit in average seasons.[34] Thus, in dryland regions, manure should be applied to fields that will be planted to forage crops such as sorghum.

■ 6.6 SEWAGE SLUDGE (BIOSOLIDS)

With greater urbanization there is increasing interest in using sewage as a fertilizer resource instead of simply discarding it after treatment. Treated sewage sludge, commonly referred to as biosolids, is solids, semisolids, or liquids produced during the treatment of municipal wastewater. Wastewater is processed to produce clean water for release back to streams and water bodies. Biosolids are the organic by-products of this process. These biosolids are rich in nutrients but may also contain significant levels of contaminants such as pathogens, pollutants, and synthetic materials discharged into sewers from homes, industries, and businesses. In the past, biosolids were treated as waste products that were often incinerated or dumped in oceans or landfills.

The utilization of biosolids on agricultural lands is regulated by the United States Environmental Protection Agency, which sets limits on the concentration of heavy metals and other pollutants. The spread of pathogens carries little risk if biosolids go through normal sewage treatment and are applied only to crops whose edible parts do not touch the soil.

After normal sewage treatment, biosolids are processed in anaerobic digesters (Figure 6.10) for about three weeks at 100°F (38°C); then they are concentrated into sludge to reduce volume and weight. It is then applied to fields either by spraying it on the surface or injecting it into the soil. Injection reduces odor problems.

When applied to soils, biosolids can supply nutrients and improve soil condition. Biosolids contain the full complement of nutrients required for crop growth and are a good source of nitrogen and phosphorus. Biosolids decompose over several years, gradually releasing nitrogen, sulfur, and micronutrients. In addition to supplying nutrients, biosolids may improve soil organic matter, microbial activity, and soil physical properties. Effects of biosolids on soil physical properties such as increased soil aggregate formation and aggregate stability may be greater than for animal manures due to the stability of organic compounds in biosolids.[35]

FIGURE 6.10
Egg-shaped anaerobic digesters convert sewage into biosolids for use as fertilizer. [Courtesy Richard Waldren]

▓ 6.7 GREEN MANURE AND COVER CROPS

6.7.1 Purpose of Green Manure and Cover Crops

A green manure crop is grown to be turned under for soil improvement while in a succulent condition. A cover crop is one used to cover and protect the soil surface, especially during the winter. Cover crops are commonly turned under for green manure. Green manures are used primarily to increase the yield of subsequent crops as well as to improve the structure of the soil. This is brought about by (1) an increase in the content of organic matter in the soil to counterbalance losses of organic matter through cultivation, (2) prevention of leaching of plant nutrients from the soil during periods between regular crops, (3) an increase in the supply of combined nitrogen in the soil when leguminous plants are turned under, and (4) mobilizing mineral elements. The contribution of organic matter to the soil from a green manure crop is comparable to the addition of 9 to 13 tons per acre (20 to 29 MT/ha) of farmyard manure or 1.8 to 2.2 tons of dry matter per acre (2 to 2.5 MT/ha).[36]

Crops such as buckwheat, rye, lespedeza, and sweetclover are able to grow fairly well on poor soils. When plowed under to decay, they can release nutrients that were unavailable to other crops.

The use of rye for green manure on very sandy irrigated soils adds organic matter that helps hold moisture and retard sand drifting while alfalfa is being established. Heavy green manuring helps cotton as well as sweet potato overcome injury from the *Phymatotrichum* root rot disease.

Turning under a leguminous green manure can add from 50 to 200 pounds per acre (56 to 224 kg/ha) of nitrogen, but neither legumes nor nonlegumes add other mineral elements to the soil. When either type of green manure increases crop yields it also hastens depletion of phosphorus, potash, calcium, magnesium, and sulfur in the soil. A nonlegume speeds up nitrogen depletion also.

No-till planting of corn into a cover crop of barley. [Courtesy USDA NRCS]

The use of green manure was practiced by the Greeks and Chinese before the Christian Era. Lupines, peas, vetch, lentils, and weeds were being turned under as green manure 2,000 years ago.[37] Leguminous green manure crops, as well as buckwheat, oat, and rye, were used by the American colonists.

The most widespread use of cover crops is found along the Atlantic seaboard as well as in the southern states where cover crops are planted mainly to check soil erosion, runoff, and leaching. They are finally turned under for green manure. Their value for soil improvement from the standpoint of yields of subsequent crops appears to be greatest in this general region.[37, 38] Some 3 million acres (1.34 million hectares) of cover crops are planted annually in the United States.

6.7.2 Crops for Green Manure

The most important green manure crops in this country are hairy vetch, rye, crimson clover, lupine, sweetclover, and alfalfa.[37] Buckwheat and other crops are used in limited areas.

Along the Atlantic seaboard, crimson clover, vetch, and burclover are used, especially from Virginia southward to South Carolina. Potato growers in New Jersey use rye as a green manure crop and depend upon commercial fertilizers for their nitrogen supply.

In the Cotton Belt, vetch has been popular because of its winter hardiness. Other crops used in this region include crotalaria, lespedeza, cowpea, mung bean, pigeon pea, soybean, lupine, and velvet beans. White lupin (*Lupinus albus*) is making a comeback in the southern United States due to its high potential in both conventional and sustainable production systems. Lupin can potentially fix 130 to 180 pounds per acre (150 to 200 kg/ha) of nitrogen for the use of a succeeding crop.[39] It has been estimated that if lupin replaced a quarter of the wheat area in the south-

TABLE 6.8 Average Biomass and Nitrogen Yields of Four Legumes Used for Cover Crops or Green Manure (from Sarrantonio, 1994)

Legume	Biomass (tons/acre)	Nitrogen (lbs/acre)
Sweetclover	1.75	120
Berseem clover	1.10	70
Crimson clover	1.40	100
Hairy vetch	1.75	110

eastern United States, 95,000 tons (86,000 MT) of nitrogen fertilizer worth $50 to $60 million per year could be saved.[40] Lespedeza as a cover crop has been effective in reducing erosion in North Carolina.

In the Corn Belt, clover, sweetclover, and alfalfa have been the chief green manure crops. Clover is superior to soybean for the addition of nitrogen.[41] Sweetclover, berseem clover, crimson clover, and hairy vetch are used for green manure in the eastern Corn Belt (Table 6.8). Rye, field peas, sweetclover, and cowpeas have been tried as green manure crops in the semiarid regions of the western states, but their use is seldom beneficial there.[42] Sesbania is a popular green-manure crop on irrigated lands of the Southwest. In the humid or irrigated regions of the Pacific Coast states, vetches, Ladino clover, burclover, sourclover, field peas, rye, and other crops are used for green-manure and winter-cover crops.

On a dry matter basis, green manures of Austrian winter pea, hairy vetch, and alfalfa contain 3 to 4 percent nitrogen, other legumes 2 to 3.5 percent, and cereal plants and ryegrass 1.2 to 1.4 percent.[43] The average availability of nitrogen in harvested green manure material, applied to other lands and turned under, is about 50 to 80 percent.

6.7.3 Decomposition of Green Manure

Green manure must be decomposed before its nutrients become available for plant growth or the organic residues become a part of the soil humus.[44] Young plants and substances high in nitrogen decompose most rapidly. As decomposition progresses, it becomes slower because of the comparatively greater resistance of the residual organic matter to decay. Green manure furnishes energy to microorganisms that bring about decomposition. Water-soluble constituents, largely sugars, organic acids, alcohols, glucosides, starches, and amino acids, are decomposed most rapidly and completely. The free-living, nitrogen-fixing bacteria in the soil use most of these substances as a source of energy. Green manure contains 20 to 40 percent of its total dry matter in water-soluble form. The microorganisms that decompose hemicelluloses and celluloses take up nitrogen from the soil in the process. The proteins decompose quickly, but the nitrogen liberated is immediately assimilated by microorganisms that attack the celluloses and hemicelluloses. The lignins are very resistant to decomposition and usually add to the soil humus.

Postponement of turning under fall cover crops until spring, rather than in late fall, delays decomposition, conserves moisture, and retards leaching, runoff, and erosion.[38] Decomposition of organic matter in soil continues until the carbon:nitrogen ratio drops from perhaps 50:1 or 20:1 down to about 11:1, the humus stage.

The decomposition of organic matter by microorganisms soon brings about the release of gums, slimes, and other products when the organisms die. These or other materials from decaying organic matter leach down into the soil where they increase soil aggregation in clays and loams by cementing soil particles together and improving the soil structure.

Proteinaceous organic materials react with clay minerals, especially montmorillonite and illite, and are retained in the soil. Montmorillonite and illite clay crystals are composed of silica and alumina sheets bonded together in a 2:1 ratio. In montmorillonite, the entire surface of a crystal unit is accessible for surface reactions. In illite, parts of the space between units are accessible. The nonswelling kaolinite clays that predominate in the southeastern states have a low exchange capacity and do not retain much carbon.

Organic matter increases the water—as well as the mineral—holding capacity of sandy soils.

6.7.4 Effects on Subsequent Crop Yields

In research conducted in the Great Plains where moisture is limited, crop yields after fallow were equal to, or better than, yields using green manure.[34, 45] There, the less growth the green manure crop makes before it is turned under, the less moisture it uses up, and the greater the yields of the subsequent crop. Where moisture is the most limiting factor in crop yields, the use of green manure may be expected to remain unprofitable as long as the organic matter in the soil continues at the prevalent general level. However, green manure is beneficial under irrigation. For example, sweetclover turned under has increased the yield of sugarbeet.

Immediate benefits from green manure are usually proportional to its nitrogen content. Turning under a good legume crop has increased subsequent crop yields as much as an application of 30 to 80 pounds per acre (34 to 90 kg/ha) of nitrogen. Crops after rye or other nonleguminous green manure need nearly as much nitrogen fertilizer as crops without the green manure.

In Georgia and Louisiana, increases in cotton yields after legumes were turned under ranged from approximately 22 to 100 percent. Winter legumes plowed under generally increase corn yields from 24 to 78 percent in Georgia, Mississippi, South Carolina, and Virginia. Similar results have been obtained in the Corn Belt when sweetclover used for green manure precedes corn.

Turning under summer legumes also has increased yields. In Alabama, corn planted after velvet bean was plowed under without the addition of phosphate yielded 58 percent more than corn without velvet bean with added phosphate. In Arkansas, the increases in corn yields after turning under the following legumes were: cowpeas, 62 percent; soybean, 7 percent; and velvet beans, 27 percent. When these legumes were cut for hay, with only the stubble turned under, the increased corn yields were 30, 14, and 24 percent, respectively. In Tennessee, wheat yields over a twenty year span averaged 8 bushels per acre (538 kg/ha) more where cowpeas were turned under, compared to yields where the land was similarly fertilized but the cowpeas were removed.[25] Also, in Tennessee, the yield of corn increased 52 to 151 percent by the use of sweetclover green manure, and 27 to 132 percent by lespedeza green manure. Corn after lespedeza in North Carolina gave increased yields of 74 to 310 percent in various trials.

Because of its higher C:N ratio, turning under rye has often resulted in decreased yield of subsequent unfertilized cotton or corn because of the temporary exhaustion of the available soil nitrogen. The depressing effect is increased as the rye approaches maturity before it is turned under.

In a Georgia experiment, rye turned under March 1 increased the yield of the subsequent cotton crop by 39 percent, while the increase after Austrian winter peas was 76 percent. In the same test, rye turned under March 15 depressed the yield of corn grown with nitrogen fertilization by 1 percent, and without the nitrogen by 6 percent. The yields of cotton were depressed by 12 percent in Louisiana when oat was turned under March 20, and by 8 percent by rye similarly treated. Rye 10 to 14 inches high contained 2.5 percent total nitrogen, while mature plants contained only 0.24 percent under the same conditions.

6.7.5 Utilization of Green Manures

A crop is seldom economically feasible for green manure when it requires the entire crop season for its growth, although velvet bean grown on sandy soils in the South may be an exception. The most desirable green manure crop occupies the land for a part of the season without interference with the regular crops in the rotation. Wintercover crops serve effectively as green manures in the South, while catch crops are often used for this purpose in the North. Legumes grown with grain crops can be turned under in the fall.

In the South, when a legume is plowed under in the fall, it should be followed by a wintercover crop to hold the nutrients released by the decomposition of the summer green manure until the crop planted next season can utilize them. On the Norfolk coarse sands of South Carolina, the subsequent yields of both cotton and corn were improved when summer leguminous green manures were followed by winter rye.[15]

When large quantities of green materials are turned under, some time should elapse before a subsequent crop is planted in order to avoid seedling injury from the decomposition products. In the South, a green manure should be plowed under about two weeks before corn is planted and three weeks before cotton.[37]

Heavy applications of nitrogen to continuously grown corn, where all residues are turned under, will produce as high a yield as corn grown without the use of green manures.

▓ 6.8 CROP ROTATION

6.8.1 Principles of Crop Rotation

Crop rotation is defined as a system of growing different kinds of crops in recurrent succession on the same land.[24, 46, 47, 48, 49] A rotation may be good or bad as measured by its effects on soil productivity or on its economic returns. A good rotation that provides for maintenance or improvement of soil productivity usually includes a legume crop to promote fixation of nitrogen, a grass or legume sod crop for maintenance of humus, a cultivated or intertilled crop for weed control, and fertilizers. Perennial legumes and grasses can leave 2 to 3 tons of dry weight per acre (4.5 to 6.7 MT/ha) of root residues in the soil when incorporated by tillage.

Modern crop rotation was established about the year 1730 in England. The famous Norfolk four-year rotation consisted of turnips, barley, clover, and wheat. On a particular field, turnips would be grown the first year, barley the second, clover the third, and wheat the fourth. The wheat was followed by turnips in the fifth year to repeat the rotation. In some rotations, a crop may occupy the land two or more years.

The Rothamsted (England) experiments included rotations that were continued for more than 100 years. In the United States, investigations were conducted in Pennsylvania, Ohio, Illinois, and Missouri for many years.

To maximize success and ease of management, the sequence of annual crops should be the same on all fields in the rotation. If a perennial is included, it is left in a field for several years, after which it is rotated to another field. To keep the rotation system intact, the crop that would have been planted in the field where the perennial was moved is planted in the field where the perennial was. If the perennial was a high nitrogen-producing legume such as alfalfa, it would be important to follow it with a crop, such as corn, which can efficiently utilize the nitrogen left by the perennial.

6.8.2 Factors That Affect Crop Rotations

The choice of a rotation for a particular farm depends upon which crops are adapted to the particular soil, climate, and economic conditions. In addition, weeds, plant diseases, and insect pests may limit the kinds of crops to be grown in a locality.

Rotations, except in the case of an alternate grain-fallow system, provide some diversification of crops. Diversification assures more economical use of irrigation water as well as other facilities. The risk of complete failure due to weather, pests, and low prices is less with several crops than with one. Crops are selected to spread labor throughout the year. Seasonal labor requirements conflict with certain crops such as alfalfa, corn, and winter wheat in Kansas and Nebraska.

6.8.3 Maintenance of Soil Productivity

Loss of organic matter as a result of continuous growth of the same crop has a bad effect on soil structure. Growth of grass, pasture, and deep-rooted legume crops in rotation tends to correct this condition through maintenance of organic matter. Well-arranged systems of crop rotation make practicable the application of manure and fertilizers to the most responsive crops or to those with high cash value. The alternation of deep- and shallow-rooted crops prevents continuous absorption of plant nutrients from the same root zone year after year. Deep-rooted plants like alfalfa improve the physical condition of the subsoil when the underground parts decay.[50]

The nitrogen requirements of nonleguminous crops can be provided by legumes in the rotation, but rotations cannot supply other plant nutrients in which the soil may be deficient. The production of larger crops, made possible by rotation, depletes the soil more rapidly than does continuous cropping. These larger yields cannot continue indefinitely without application of manures and fertilizers. In a corn-wheat-clover rotation in Indiana, the yields of unfertilized plots dropped one-third in ten years.[51]

6.8.4 Legumes in Rotations

For satisfactory growth of legumes, deficient soils may require application of minerals such as lime and phosphorus. Some research in Pennsylvania indicates that crop rotation is unable to maintain yields at a high level unless the soil is fertile enough to also maintain production of clover. When legumes, such as soybean and lespedeza, are harvested from fertile land, more nitrogen can be taken from the soil than is added to it by the legume.

Legumes are more efficient in fixation of nitrogen on soils with low rather than high nitrogen content because they obtain nitrogen from the air only to the extent that the supply in the soil is insufficient. For this reason, a legume will be more effective as a nitrogen gatherer when two or more crops come between applications of barnyard manure. It is usually customary to grow legume crops previous to crops that require large amounts of nitrogen. Increased yields of crops that follow alfalfa may be due mostly to the addition of nitrogen to the soil contributed by the alfalfa crop.[52]

6.8.5 Crop Sequences

Crop yields are greatly influenced by the crop that precedes it. Rhode Island research was designed to study crop sequences in which sixteen different crops were grown for two seasons. Then one of the crops was grown over the entire area the third year.[53] Alsike clover gave the lowest yields after clover and carrot and the highest yields after rye and redtop. The yields probably were influenced by soil acidity.

Growing crops in descending order of their lime requirements sometimes is advisable.[54] For example, for a soil heavy limed for alfalfa, clover, or soybean, the sequence might be: (1) alfalfa, sugarbeet, barley; (2) red clover, tobacco, wheat; (3) alsike clover, corn, oat; or (4) soybean, potato, rye.

The amount of nitrogen left in the soil by a crop can influence the yield of the crop that follows it. In West Virginia, research[55] showed that the yields of wheat, oat, and corn were higher after soybean harvested for hay than the yields after oat harvested for grain.

Corn that follows deep-rooted legumes, such as alfalfa or sweetclover, may yield more from better root penetration as well as from the nitrogen residues. Grasses in a rotation improve the soil structure after it has been damaged by excessive tillage and compaction from growing an intertilled crop like corn. Heavy fertilization often may restore most of the productivity lost by soil depletion.

Crop sequences are very important under dryland conditions because of the differences in residual soil moisture left by different crops, as well as by the length of the fallow period for moisture storage between crops. Thus, small grains yield more after corn than after small grains or sorghum in the Great Plains region because corn leaves more moisture in the soil.[27] Most dryland crops yield poorly after alfalfa because the excessive nitrogen and depleted soil moisture cause the crops to burn except in wet seasons.

6.8.6 Rotations and Pest Control

Crop rotation aids in controlling many plant diseases. Certain parasites that live in the soil tend to increase when a susceptible crop is grown year after year.

Eventually, the disease may become too severe for profitable crop production. Rotation is particularly effective in the control of this group of parasites.

Many insect pests are destructive to only one kind of crop. The life cycle is broken when crops are grown that are unfavorable to the development of the insect pest. Sugarbeet nematode is partly controlled by a rotation in which sugarbeet is grown on the land only once in four or five years. Cotton root knot can be reduced by growing immune crops in the crop sequence. The Delta region of the Mississippi reported cotton lint yield increases ranging from 150 to 400 pounds per acre (170 to 450 kg/ha) the first year following corn.[56]

6.8.7 Rotations and Weed Control

Crop rotation is the most effective practical method for control of many weeds. Some weed species are particularly adapted to cultivated crops, others to small grains, while others thrive in meadows. The continuous planting of one type of crop, such as small grains, on the same land encourages weeds. In Utah, plots continuously cropped to wheat for seven years became so infested with wild oat that yield was seriously reduced.[57] The greatest difficulty affecting continuous growth of wheat on Broadbalk field at Rothamsted, England, has been the weed problem.[58]

There are more than 1,200 species of weeds in the world. Except for some 30 noxious species, most are not able to thrive indefinitely on crop-rotated land.[47] Annual weeds are restricted by a rotation that includes a small-grain crop, a cultivated crop, and a perennial crop. Rotations that include smother crops such as alfalfa, rye, buckwheat, sorghum forage, and sudangrass can also control many weeds.

6.8.8 Cropping Systems and Erosion Control

The potential for erosion is greatest on land that is planted to continuous row crops. Any rotation that uses a sequence of row crops, close-seeded crops, and a perennial will provide adequate erosion protection even on highly erodible land. In Missouri, a corn-small grain-clover rotation was effective in reducing erosion compared with continuous corn.[59] A one-year rotation of lespedeza and winter grain was also effective. Grass-legume mixtures in rotations have been very effective in the reduction of erosion.[46]

6.8.9 Crop Rotation Compared with Continuous Culture

Crop rotation alone may be 75 percent as effective as fertilizers for increasing crop yields as a whole, or 90 percent when only the results for corn, wheat, and oat are averaged.[49] The long-time experiments at the Rothamsted Experimental Station in England indicate the trend in yields when a crop is grown in rotation or in continuous culture. This long-term study (over 60 years) showed that the highest yield occurred where both crop rotation and fertilization were practiced. Similar results were obtained with corn in the Morrow plots in Illinois, operated since 1888.[29, 60]

6.8.10 Rotations in Practice

CORN BELT Crop rotations that supply organic matter and nitrogen are built around corn as the major crop.[18] Continuous corn has resulted in reduced yields where it has been practiced for any length of time unless fertilizer and pest control inputs are increased. A corn-oat-clover rotation is used in the northern part of the region. A four-year rotation of corn, oat, wheat, and alfalfa, clover, or a grass-legume mixture is practiced in the northeastern part of the Corn Belt where corn is a less important crop. A rotation of corn, wheat, and hay is sometimes used in central Indiana. The grass hay crops in the Corn Belt are mostly smooth bromegrass and orchard grass.

Where soybean is grown, a five-year rotation of corn, corn, soybean, wheat, and hay is used.[51] Yields can be maintained when cover crops are seeded in both corn crops and when manure and phosphate are applied to the corn.

In more recent years, a corn-soybean rotation with corn for one to two years followed by one to two years of soybean has become increasingly popular on better lands with less than 3 to 5 percent slope.

COTTON BELT Definite crop sequences are less common in the Cotton Belt than in many other regions.[45] On relatively productive soils, a typical cropping system is one year of corn or an annual hay crop followed by two to four years of cotton. Interplanted legumes are used in this sequence in some instances. Two suggested rotations are:

1. Three-year rotation for land equally adapted to corn and cotton: (1) cotton; (2) summer legumes (cowpea or soybean) for hay or seed, followed by a winter legume; and (3) corn interplanted with cowpea, soybean, or velvet bean.

2. Four-year rotation for areas where half of the cropland is planted to cotton: (1) cotton, followed in part by a winter legume; (2) cotton; (3) summer legumes for hay or seed followed by winter legumes; and (4) corn interplanted with summer legumes.

HAY-PASTURE REGION Hay and pasture predominate in the northeastern states. A three-year rotation of corn for silage, oat, and hay is practiced on dairy farms of the Great Lakes states where considerable land is available for pasture. Potato is sometimes substituted for corn as an intertilled crop. Grass-legume mixtures are sown with spring small grains and kept two years for hay.[61] A large percentage of the cropland is planted to hay on many New England farms.

WHEAT REGIONS In the eastern central region from Missouri to Virginia, various modifications of the corn, wheat, clover, or grass rotation are used. In the northeastern part of the winter-wheat area, a rotation that can be used is corn for two years; oat or barley, one year; and wheat, two years. Some hay crops can be included as a modification of this sequence. Popular hay crops are clover, alfalfa, lespedeza, tall fescue, and orchard grass. In the spring-wheat area of the Red River Valley, a rotation of small grains for two years, alfalfa for two years, and corn or potato or sugarbeet for one year is often used. Definite crop rotations are not prevalent in the major wheat production areas, except for wheat alternating with summer fallow.

In the Pacific Northwest, dryland wheat has been produced under a wheat-summer fallow rotation. Winter or spring wheat is the most important cash crop of the Great Plains and intermountain dryland areas. The rotations in this area are characterized by summer fallow periods and absence of sod crops. In the southern Great Plains, a summer fallow-wheat-grain sorghum rotation may be adapted. Alternate wheat and fallow is a common cropping system in the drier areas that receive 12 to 20 inches (300 to 500 mm) of precipitation per year, while alternate corn and wheat is a practical sequence with slightly more precipitation. Continuous cropping has given satisfactory results in some parts of the semiarid region.

IRRIGATED REGIONS In Utah, a crop rotation that included alfalfa and cultivated crops with rather heavy applications of manure effectively maintained high yields of wheat, potato, sugarbeet, field peas, oat, and alfalfa.[57]

▓ 6.9 ORGANIC FARMING

Organic farming was devised by our aboriginal ancestors. With a few exceptions, such as applying ashes (potassium nitrate) or marl to the soil, it continued in vogue until the nineteenth century. Fertilizers and pesticides came into use following remarkable scientific discoveries. The heaviest use of these materials is in advanced countries with high rates of literacy and longevity and the absence of famines. It is estimated that a complete world reversion to organic farming would solve the population explosion problem because hundreds of millions of people would die from starvation each year.[62, 63]

Plant root hairs absorb ions of mineral nutrient elements from the soil. The root hairs absorb the ions nonpreferentially regardless of whether they come from soil minerals, mineral fertilizers, or decayed organic matter. Soil productivity might be maintained by heavy applications of organic matter, but a much larger area of land would be required to produce the organic material needed by the cropped fields. The "slash and burn" culture system practiced in primitive localities leaves a field idle for three to seven years between croppings, but that practice is rapidly disappearing because increasing populations are putting too much pressure on existing arable land.

All pesticides are purposely toxic to some forms of plant or animal life. Most of them are dangerous to humans when absorbed in excess from misuse or accidents. Several elements essential to human life or health, such as iodine, chlorine, selenium, fluorine, and copper, can be fatal if taken in excess or used in a wrong chemical compound. Vehicles, machines, tools, electricity, and water are fatal when used improperly or when accidents occur. DDT is destructive to bird life but it has saved millions of humans from malaria and typhus diseases. The repression and reoccurrence of malaria in tropical regions is associated with the level of DDT use. The question agronomists must face is how to increase food production without fertilizers and pesticides. Until we can accomplish this challenge, a starving world dictates continued intelligent use of agricultural chemicals. Mercurial pesticides are banned because some ignorant people disregarded labeled directions for proper utilization and used mercury-treated seed grain for food or animal feed. Mercury levels in sea foods are the same as they were 70 years ago. These data are substantiated by analysis of museum specimens collected 70 to 100 years ago.

In the United States, the use of agricultural chemicals expanded rapidly in the last half of the twentieth century. During this period, the death rate decreased, and the younger generations grew larger and matured earlier than did their parents.

Nevertheless, organic farming has gained in popularity in recent years because of increased public concern about pesticide residues in food. Many organic farmers have succeeded by developing niche markets, selling organically produced food at a premium price to people who want it.

To be successful, organic farmers must use crop rotations to aid in maintaining soil fertility and control of pests, diseases, and weeds. They will use more tillage and must be knowledgeable of cultivars with good resistance to pests and diseases. They also have to be more precise with their management practices. Chemicals allow a farmer to correct mistakes. Organic farmers must do it right the first time.

▨ 6.10 MULTIPLE CROPPING

Multiple cropping means growing and harvesting two or more crops from the same land in one year. This is possible in the warm climate areas of all continents. Multiple cropping has been practiced since the dawn of agriculture.[64] There are many types of multiple cropping systems. The most common are intercropping, double cropping, relay intercropping, and strip intercropping. To be economically successful, a multiple cropping system must allow the use of mechanization and minimize labor. Therefore, the only multiple cropping systems used to any extent in modern agriculture are double cropping, relay intercropping, and strip intercropping. Intercropping is too labor intensive because it is difficult to use mechanization and chemicals.

6.10.1 Intercropping

Intercropping is the practice of growing more than one species of crop together at the same time. Since different species have different requirements for light, nutrients, and water, resource utilization is more efficient in an intercropping system.[65] Different species fill different ecological niches. If properly managed, an intercrop will usually outyield the same crops grown separately.

Intercropping is used today in tropical regions and in many gardens, but its application in modern field crop production is limited. Intercropping is used most often in forage crops. Total yield generally was maintained or enhanced when oat was intercropped at a monoculture rate with pea compared to growing oat alone, and protein content was higher when used for forage.[66]

6.10.2 Double Cropping

Double cropping is the practice of growing two crops in one year in sequence. The second crop is seeded soon after the first crop is harvested (Figure 6.11). Winter wheat, because it matures in early to midsummer is the most common choice for the first crop. However, other winter annual crops can be grown as well.

Soybean is the most common choice for the second crop because it can mature earlier when planted later. Over 1 million acres of wheat and soybean are double cropped in Ohio annually. Harvesting wheat when the grain moisture is 18 to 20 percent maintains quality of the wheat and permits soybean planting to be advanced three to five days.[67] Soybean planted after winter wheat will flower about

FIGURE 6.11
Soybean double cropped after winter wheat.
[Courtesy Richard Waldren]

30 days after emergence and does not produce a large plant. Therefore, it is desirable to plant in rows 15 inches (38 cm) or narrower to maximize sunlight interception. A high seeding rate of about 200,000 soybean seeds per acre (495,000 seeds/ha) is necessary for early canopy cover. If planting in 30-inch (76 cm) or wider rows, a higher seeding rate promotes greater plant height and pod set. Select an early maturing variety.[68]

Other crops are common in some regions. Sunflower is used for double cropping after small grains throughout the Corn Belt.[69] In Missouri, amaranth, buckwheat, sunflower, and pearl millet are double cropped after wheat or canola.[70] Winter canola is double cropped after cotton in Georgia.[71]

Two crops of rice per year are frequently harvested from fields in Southeast Asia. In Taiwan, two rice crops with two intervening green vegetable crops have been grown where the rice is transplanted. In a trial in the Philippines, four transplanted crops of early rice varieties were harvested from one field in a year. In southeastern, southwestern, and southern Corn Belt regions of the United States, other crops such as sorghum or soybean can be planted after a small-grain crop is harvested in May or early June and reach maturity in the late autumn.

6.10.3 Relay Intercropping

Relay intercropping is the practice of planting the second crop into the first before it is harvested. Winter wheat is the most common first crop, and soybean is the most common second crop (Figure 6.12).

The biggest advantage of relay intercropping is that it gives the second crop more time to grow and mature, which usually results in higher yields of the second crop. The biggest disadvantage of relay intercropping is the problem the second crop has competing with the first crop for water, light, and nutrients. In the drier

FIGURE 6.12
Soybean relay intercropped
into winter wheat.
[Courtesy Richard Waldren]

areas of the Plains, it is necessary to irrigate shortly before or after interseeding the second crop to insure adequate water for germination and seedling establishment.

Damage to the first crop at interseeding depends on its growth. In Nebraska, the yields of winter wheat relay intercropped with soybean decreased only 9 percent when soybean was interseeded in early May but decreased 52 percent when interseeded in early June because the wheat was taller and was damaged more by the seeding equipment.[72]

Variety selection is important in a relay intercrop system. In a wheat-soybean relay intercrop, a medium to early maturing wheat variety and a full season soybean variety are best. The use of a polymer coating on the soybean seed delays germination and allows earlier seeding.[73]

6.10.4 Strip Intercropping

Strip intercropping is the practice of planting two or more crops in narrow strips (6 to 12 rows) in the field (Figure 6.13). The following year, the crops are alternated in the strips. Strip intercropping allows the use of machinery and chemicals while still maintaining some of the advantages of intercropping and rotations.

One of the biggest advantages of seeding crops in narrow strips is what is called the "border effect." The interaction of two species at the borders of the strips can, in some cases, lead to increased yields in the border rows. For example, border rows of corn can yield more than inner rows because they receive more light. Rarely does total yield in a strip intercrop system fall below the average monoculture performance. In years of adequate rainfall, production of strip crops may outyield monoculture crops by 10 to 20 percent.[74]

In some cases, the rotational advantage in pest control is less effective. Infestations of corn rootworm larvae can be sustained by migration of larvae into corn from soil previously planted to corn in the alternate strips.[75]

FIGURE 6.13
A strip intercrop of six rows of corn and six rows of soybean. [Courtesy Richard Waldren]

REFERENCES

1. Nelson, L. B., and D. B. Ibach. "The economics of fertilizers," in *Soil*, USDA Yearbook, 1957, pp. 267–276.

2. Viets, F. G., and J. J. Hanway. "How to determine nutrient needs," in *Soil*, USDA Yearbook, 1957, pp. 172–184.

3. Schomberg, H. H., P. B. Ford, and W. L. Hargrove. "Influence of crop residues on nutrient cycling and soil chemical properties," in *Managing Agricultural Residues*, P. W. Unger, ed. Boca Raton, FL: Lewis, 1994, pp. 99–122.

4. Brengle, K. G., and B. W. Greb. "The use of commercial fertilizers with dryland crops in Colorado," *CO Agr. Exp. Sta. Bull.* 516-S, 1963.

5. Homer, G. M., and others. "Effect of cropping practices on yield, soil organic matter, and erosion in the Pacific Northwest wheat region," *ID, WA, OR, and USDA Bull.* 1, 1960.

6. Walster, H. L., and T. E. Stoa. "Continuous wheat culture versus rotation wheat culture," *ND Agr. Exp. Sta. Bimonthly Bull.* 5, 1942, pp. 2–8.

7. Ibach, D. B., and J. R. Adams. "Fertilizer use in the United States by crops and areas," *USDA Stat. Bull.* 408, 1967, pp. 1–384.

8. Beeson, K. C. "Soil management and crop quality," in *Soil*, USDA Yearbook, 1957, pp. 258–268.

9. Chapman, H. D., ed. "Diagnostic criteria for plants and soils," *Univ. CA Div. Agr. Sciences*, 1966, pp. 1–793.

10. Adams, J. R., M. S. Anderson, and W. C. Hulburt. "Liquid nitrogen fertilizers," *USDA Handbook* 198, 1961.

11. Arnon, A. I., and P. R. Stout. "The essentiality of certain elements in minute quantity for plants with special reference to copper." *Plant Physiol.* 14(1939):371.

12. Dean, L. A. "Plant nutrition and soil fertility," in *Soil*, USDA Yearbook, 1957, pp. 80–94.

13. Jenny, H. "Relation of climatic factors to the amount of nitrogen in soils," *J. Am. Soc. Agron.* 20(1928):900–912.

14. Dalrymple, D. G. "Survey of multiple cropping in less developed nations," *USDA and USAID FEDR-12*, 1969, pp. 1–108.

15. McKaig, N., W. A. Cams, and A. B. Bowen. "Soil organic matter and nitrogen as influenced by green manure crop management on Norfolk coarse sand," *J. Am. Soc. Agron.* 32 (1940):842–852.

16. Erdman, L. W. "Legume inoculation: What it is; what it does," *USDA Farmers Bull.* 2003, 1953 (revised), pp. 1–20.

17. Stewart, W. D. P. "The nitrogen-fixing plants," *Science* 158(1967):1426–1432.

18. Jenny, H. "Soil fertility losses under Missouri conditions," *MO Agr. Exp. Sta. Bull.* 324, 1933, pp. 1–10.

19. Hobbs, J. A., and P. L. Brown. "Effects of cropping and management on nitrogen and organic carbon contents of a western Kansas soil," *KS Agr. Exp. Sta. Tech. Bull.* 144, 1965, pp. 1–37.

20. Pierre, W. H. "Phosphorus deficiency and soil fertility," in *Soils and Men*, USDA Yearbook, 1938, pp. 377–396.

21. Lyons, E. S., J. C. Russel, and H. F. Rhoades. "Commercial fertilizers for the irrigated sections of western Nebraska," *NE. Agr. Exp. Sta. Bull.* 365, 1944, pp. 1–29.

22. Scott, S. G. "Phosphorus deficiency in forage feeds of range cattle," *J. Agr Res* 38, 2(1929):113–120.

23. Cooper, H. P., O. Schreiner, and B. E. Brown. "Soil potassium in relation to soil fertility," in *Soils and Men*, USDA Yearbook, 1938, pp. 397–405.

24. Johnson, T. C. "Crop rotation in relation to soil productivity," *J. Am. Soc. Agron.* 19(1927):518–527.

25. Mooers, C. A. "Effects of liming and green manuring on crop yields and on soil supplies of nitrogen and humus," *TN Agr. Exp. Sta. Bull.* 135, 1926, pp. 1–64.

26. Truog, E. "Liming-The first step in improving acid soils," *Crops and Soils* 7, 7(1955):12–13.

27. Cooper, H. P. "Fertilizer and liming practices recommended for South Carolina," *SC Agr. Exp. Sta. Cir.* 60, 1939, pp. 1–23.

28. Fisher, T. R. "Crop yields in relation to soil pH as modified by liming soils," *MO Agr. Exp. Sta. Bull.* 947, 1969, pp. 1–25.

29. Allaway, W. H. "Cropping systems and soil," in *Soil*, USDA Yearbook, 1957, pp. 386–395.

30. Salter, R. M., and C. J. Schollenberger. "Farm manure," *OH Agr. Exp. Sta. Bull.* 605, 1939.

31. Salter, R. M., and C. J. Schollenberger. "Farm manure," in *Soils and Men*, USDA Yearbook, 1938, pp. 445–461.

32. Russell, E. J., and E. H. Richards. "The changes taking place during the storage of farmyard manure," *J. Agr. Sci.* 8(1917):495–563.

33. Shutt, F. T. "The preservation of barnyard manure," *Rept. Min. Agr. Canadian Exp. Farm*, 1898, pp. 126–137.

34. Chilcott, E. C. "The relations between crop yields and precipitation in the Great Plains area: Supplement 1; Crop rotations and tillage methods," *USDA Misc. Cir.* 81, 1931, pp. 1–164.

35. Wortman, C. S., and D. L. Binder. "Sewage Sludge Utilization for Crop Production." *NE Coop. Ext. Service NebGuide G02-1454-A.* 2002.

36. Schmid, O., and R. Klay. *Green Manuring: Principles and Practice.* Mt. Vernon, ME: Woods End Agricultural Institute. Translated by W. F. Brinton, Jr., from a publication of the Research Institute for Biological Husbandry. Switzerland, 1984.

37. Pierre, W. H., and R. McKee. "The use of cover and green-manure crops," in *Soils and Men*, USDA Yearbook, 1938, pp. 431–444.

38. Rogers, T. H., and J. E. Giddens. "Green manure and cover crops," in *Soil*, USDA Yearbook, 1957, pp. 252–257.

39. Reeves, D.W., J. T. Touchton, and R. C. Kingery. "The use of lupine in sustainable agriculture systems in the Southern Coastal Plain." In *Abstracts of Technical Papers*, vol. 17, Little Rock, AR: Southern Branch ASA, 1990, p. 9.

40. Bhardwaj, H. L., M. Rangappa, and A. A. Hamama. "Chickpea, faba bean, lupin, mung bean, and pigeon pea: potential new crops for the Mid-Atlantic Region of the United States." pp. 202-. In *Perspectives on new crops and new uses*, ed. J. Janick. Alexandria, VA: ASHS Press, 1999.

41. Roberts, G. "Legumes in cropping systems," *Ky. Agr. Exp. Sta. Bull.* 374, 1937, pp. 119–153.

42. Brown, P. L. "Legumes and grasses in dryland cropping systems in the northern and central Great Plains," *USDA Misc. Pub.* 952, 1964, pp. 1–64.

43. Larson, W. E., and others. "Effect of subsoiling and deep fertilizer placement on yields of corn in Iowa and Illinois," *Agron. J.* 52(1960):185–189.

44. Waksman, S. A. "Chemical and microbiological principles underlying the decomposition of green manures in the soil," *J. Am. Soc. Agron.* 21 (1929):1–18.

45. Funchess, M. J. "Crop rotation in relation to southern agriculture," *J. Am. Soc. Agron.* 19(1927):555–556.

46. Enlow, C. R. "Review and discussion of literature pertinent to crop rotations for erodible soils," *USDA Cir.* 559, 1939, pp. 1–51.

47. Leighty, C. E. "Crop rotation," in *Soils and Men*, USDA Yearbook, 1938, pp. 406–430.

48. Throckmorton, R. I., and F. L. Duley. "Soil fertility," *Ks. Agr. Exp. Sta. Bull.* 260, 1932, pp. 1–60.

49. Weir, W. W. "A study of the value of crop rotation in relation to soil productivity," *USDA Bull.* 1377, 1926, pp. 1–68.

50. Ibach, D. B., J. R. Adams, and E. L. Fox. "Commercial fertilizer used on crops and pasture in the United States," *USDA Stat. Bull.* 348, 1964.

51. Wiancko, A. T. "Crop rotation in relation to the agriculture of the corn belt," *J. Am. Soc. Agron.* 19(1927):545–555.

52. Gardner, R., and D. W. Robertson. "The beneficial effects from alfalfa in a crop rotation," *CO Agr. Exp. Sta. Tech. Bull.* 51, 1954.

53. Hartwell, B. L., and S. C. Damon. "The influence of crop plants on those that follow," *RI Agr. Exp. Sta. Bull.* 175, 1917, pp. 1–32.

54. Bear, F. E. "Some principles involved in crop sequence," *J. Am. Soc. Agron.* 19(1927):527–534.

55. Dodd, D. R., and G. G. Pohlman. "Some factors affecting the influence of soybeans, oats, and other on the succeeding crops," *WV Agr. Exp. Sta. Bull.* 265, 1935.

56. Martin, S. W., F. Cooke, Jr., and D. Parvin. "Economic potential of a cotton-corn rotation," *MS Exp. Sta. Bull.* 1125, 2002.

57. Stewart, G., and D. W. Pittman. "Twenty years of rotation and manuring experiments," *UT Agr. Exp. Sta. Bull.* 228, 1931, pp. 1–32.

58. Russell, E. J., and D. J. Watson. "The Rothamsted field experiments on the growth of wheat," *Imperial Burl Soil Sci. Tech. Comm.* 40, 1940, pp. 1–163.

59. Miller, M. F. "Cropping systems in relation to erosion control," *M. Agr. Exp. Sta. Bull.* 366, 1936, pp. 1–36.

60. De Turk, E. E., F. C. Bauer, and L. H. Smith. "Lessons from the Morrow plots," *IL Agr. Exp. Sta. Bull.* 300, 1927, pp. 105–140.

61. Wiggins, R. C. "Experiments in crop rotation and fertilization," *Cornell Univ. Agr. Exp. Sta. Bull.* 434, 1924, pp. 1–56.

62. McVickar, M. H. *Using Commercial Fertilizers*, 3rd ed. Danville, IL: Interstate, 1970, pp. 1–353.

63. Wiggans, S. C., and B. Williams. "Exploding the myths of organic farming," *Crops and Soils* 24, 4(1972):8–11.

64. Plucknett, D. L., and N. J. H. Smith. "Historical perspectives on multiple cropping," in *Multiple Cropping Systems*, C. A. Francis, ed. New York: Macmillan, 1986.

65. Trenbath, B. R. "Resource use by intercrops." In *Multiple Cropping Systems*, C. A. Francis, ed. New York: Macmillan, 1986.

66. Carr, P. M., E. D. Eriksmoen, G. B. Martin, and N. R. Olson. "Grain yield of oat-pea intercrop. In *Perspectives on New Crops and New Uses*, ed. J. Janick. Alexandria, VA: ASHS Press, 1999, pp. 240–243.

67. Beuerlein, J. "Double-cropping soybeans following wheat." *OH St. U. Ext. Serv. AGF-103-01.* 2001.

68. Moomaw, R., G. Lesoing, and C. Francis. "Two crops in one year: Doublecropping." *NE Coop. Ext. Service NebGuide G91-1025.* 1991.

69. Schmidt, W. H. "Double crop sunflower production," *OH St. U. Ext. Serv. AGF-108-95.* 1995.

70. Pullins, E. E., R. L. Myers, and H. C. Minor. "Alternative crops in double-crop systems for Missouri." *U. MO. Ext. Serv.* G4090, 1999.

71. Buntin, G. D., P. L. Raymer, C. W. Bednarz, D. V. Phillips, and R. E. Baird. "Winter crop, tillage, and planting date effects on double-crop cotton." *Agro. J.* 94(2002):273–280.

72. Moomaw, R., G. Lesoing, and C. Francis. "Two crops in one year: Relay intercropping." *NE Coop. Ext. Service Neb Guide G91-1024-A.* 1991.

73. Beuerlein, J. "Relay cropping wheat and soybeans." *OH St. U. Ext. Serv. AGF-106-01.* 2001.

74. Francis, C., A. Jones, K. Crookston, K. Wittler, and S. Goodman. "Strip cropping corn and grain legumes: A review," *Am. J. Alt. Agric.* 1(1986):159–164.

75. Ellsbury, M. M., D. N. Exner, and R. M. Cruse. "Movement of corn rootworm larvae (*Coleoptera: Chrysomelidae*) between border rows of soybean and corn in a strip intercropping system." *J. Econ. Entomol.* 92, 1(1999):207–214.

Seeds and Seeding

7.1 IMPORTANCE OF GOOD SEEDS

Reasonably good seed is essential to successful crop production, whereas poor seed is a serious hazard for farmers.[1] The variety and the approximate germination and purity of seed should be known before it is planted.[1] Introduction of weeds in the seed often increases the labor for production of the crop, reduces crop yields,[2] and contaminates the current product as well as the seed and soil in future seasons.

7.2 CHEMICAL COMPOSITION OF SEEDS

Table 7.1 shows the chemical composition of some common crop seeds. Ether extract is fat and oil content, ash is mineral content, and nitrogen free extract is carbohydrate content. Seeds store food supply for germination and emergence as fat and oil, or as carbohydrate. Grasses store most of their food supply as carbohydrate while legumes store most of their food supply as fat and oil. Also, seeds that are higher in fat and oil will usually be higher in protein. Oat seed is higher in fiber because the glumes are still attached.

7.3 SEED GERMINATION

7.3.1 Environmental Requirements for Germination

The most important external conditions necessary for germination of matured seeds are: (1) ample supplies of moisture and oxygen, (2) a suitable temperature, and, (3) for some seeds, certain light conditions. A deficiency in any factor can prevent germination.

Good seed shows a germination of 90 to 100 percent in the laboratory. Some sound crop seeds, particularly small grains, show a seedling emergence of as high as 90 percent of the seed when sown under good field conditions. Even corn, which is a rather sensitive seed, often produces stands of 90 percent or more in the field. Sorghum and cotton give a lower percentage of emergence

TABLE 7.1 Chemical Composition of Seeds

	Dry Matter (%)	Crude Protein (%)	Ether Extract (%)	Crude Fiber (%)	Ash (%)	Nitrogen Free Extract (%)
Barley	90.7	14.2	2.1	6.2	3.3	74.2
Corn	89.3	10.9	4.6	2.6	1.5	80.4
Oat	90.3	14.4	4.7	11.8	3.8	65.3
Peanut	—	30.4	47.7	2.5	2.3	11.7
Rice	88.6	9.2	1.4	2.7	1.8	84.9
Rye	—	14.7	1.8	2.5	2.0	79.0
Sorghum	88.7	12.9	3.6	2.5	2.0	79.0
Soybean	—	37.9	18.0	5.0	1.6	24.5
Wheat	88.9	14.2	1.7	2.3	2.0	79.8

Source: *Composition of Cereal Grains and Forages.* National Academy of Sciences, National Research Council, Publ. 505

because they are more susceptible to attack from seed-rotting fungi. Treated sorghum seed may give a field emergence of 75 percent, but 50 percent emergence is all that is normally expected from untreated seed with a 95 percent laboratory germination, even in a good seedbed. However, when the seed germinates only 60 to 70 percent in the laboratory, many of the sprouts will be so weakened that a field emergence of 20 to 25 percent is all that can reasonably be expected. In a poor seedbed, the emergence may be much less. Seeds that germinate slowly may produce weak seedlings. However, the strong seeds in a low-germinating sample may give good yields, provided enough seed is planted.[3]

Small-seeded legumes and grasses are sown at higher rates to compensate for poor germination, low-seedling survival resulting from the necessary shallow seeding, and the failure of hard seeds to germinate when sown.

With aerial seeding, legume seeds are sometimes pelleted to maintain the viability of the *Rhizobia* inoculant on the seed and to repel pests. When sowing in acid soils, the pellet material should be mostly lime or dolomite. Phosphates are helpful for promoting seedling growth. *Rhizobia* to be added to the seed are most frequently carried in peat. These materials, combined with gum arabic or methyl cellulose, are mixed with the seed in a revolving drum to make the pellets.[4]

In seeded grass-pasture mixtures, the species with the most viable seeds often predominates in the immediate stand.[5] Small-seeded grasses are sown at rates in excess of the rates that would be required if all of the seeds were to produce seedlings because the mortality of the seeds and seedlings is likely to be high. Thus, in a bluegrass pasture, a seeding rate of 25 pounds per acre (38 kg/ha) provides more than 1,000 seeds per square foot (11,000/m²), whereas 100 plants per square foot (1,000/m²) would soon provide dense turf.

Commercial seed of Kentucky bluegrass and that of certain other grasses may not germinate over 70 percent. This low germination is due to harvesting when many of the panicles are immature and due to inadequate drying. Any dicotyledonous plant with an indeterminate flowering habit, or a grass that sends up new tillers and panicles over a considerable period, will not mature its seed uniformly. With such crops, immature seed is gathered even though harvesting is delayed until the ripest seeds have already been lost by shattering.

MOISTURE Abundant water is necessary for rapid germination. This is readily supplied by damp blotters or paper towels in a germinator or by soil that contains about 50 to 70 percent of its water holding capacity. Field crop seeds start to germinate when their moisture content (on a dry basis) reaches 26 to 75 percent (e.g., 26 percent in sorghum, millet, and sudangrass; 45 to 50 percent in small grains;[6] and as high as 75 percent for soybean). The minimum moisture for germination of corn is approximately 35 percent in the whole grain and 60 percent in the embryo.[7]

Water usually enters the seed through the micropyle or hilum, or it may penetrate the seedcoat directly. Water enters certain seeds, such as castor and sweetclover, through the strophiole or caruncle, an appendage of the hilum.[8] Water inside the seedcoat is imbibed by the embryo, scutellum, and endosperm. The imbibed water causes the colloidal proteins and starch of the seed to swell. The enormous imbibitional power of certain seeds enables them to draw water from soil that is even below the wilting point, but not in sufficient amounts to complete germination because the adjacent soil particles become dehydrated. Seeds sown in dry soil therefore may fail to germinate, or they may absorb sufficient moisture to swell and partly germinate. Wheat, barley, oat, corn, and pea have been sprouted, allowed to dry, and resprouted three to seven times before germination was fully destroyed.[9] However, germination was lower with each repeated sprouting. Wheat seeds can absorb water from a saturated atmosphere until they reach a moisture content exceeding 30 percent on a wet basis, but this is not high enough to start germination.[10]

OXYGEN Many dry seeds, particularly pea and bean, are impervious to gases, including oxygen. Absorption of moisture may render the seed permeable to oxygen. Seeds planted too deeply or in a saturated soil may be prevented from germinating due to an oxygen deficiency. Rice needs less oxygen than most seeds since it will germinate on the soil surface under 6 inches (15 cm) of water. However, an atmosphere of pure oxygen is as harmful to seeds as it is to humans.

TEMPERATURE The extreme temperature range for the germination of field crop seeds is from 32 to 120°F (0 to 49°C). In general, cool season crops germinate at lower temperatures than warm season crops.

Wheat, oat, barley, and rye may germinate slowly at the temperature of melting ice.[11] Buckwheat, flax, red clover, alfalfa, field pea, soybean, and perennial ryegrass germinate at 41°F (5°C) or less. The minimum temperature for germination of corn is about 50°F (10°C). Tobacco seeds[12] germinate slowly below 57°F (14°C). Of the commonly grown crops, seeds of alfalfa and the clovers will germinate more readily at low temperatures than any others. Since starchy seeds appear to be more easily destroyed by rots, they are less likely to produce sprouts at low temperatures than are oily or corneous seeds of the same species.[13, 14] Smooth, hard, hybrid seed corn gives better stands than do rough, softer types.

A temperature of 59°F (15°C) is about optimum for wheat, with progressively decreasing germination at higher temperatures.[15] Mold increases directly with increased temperatures. Soybean of good quality may germinate equally well at all temperatures from 50 to 86°F (10 to 30°C), but seeds of low vitality germinate best at 77°F (25°C).

The most favorable temperature for germination of tobacco seed is about 88°F (31°C).[12] The optimum laboratory germination for seeds of most cool weather crops is about 68°F (20°C), but certain fescues and other grasses require a somewhat lower temperature. Warm weather crops, particularly the southern legumes and grasses such as crotalaria and bermudagrass, germinate best at 86 to 97°F (30 to 36°C).

Maximum temperatures at which seeds will germinate are approximately 104°F (40°C) or less for the small grains, flax,[16] and tobacco,[12] 111°F (44°C) or less for buckwheat, bean, alfalfa, red clover, crimson clover, and sunflower, and 115 to 122°F (46 to 50°C) for corn, sorghum, and millets. At temperatures too high for germination, the seeds may be killed or be merely forced into secondary dormancy. The killing has been ascribed to destruction of enzymes and coagulation of cell proteins. These reactions as a rule are not observed at temperatures as low as 122°F (50°C), but might occur over the 24 to 28 hours or more necessary to start germination. Secondary dormancy induced by heat may be an oxygen relationship.

LIGHT Light requirement for germination involves the phytochrome system found in most plants. Phytochrome absorbs red light (660 nm) and far red light (730 nm) depending on its configuration. Red light initiates germination, but far red light inhibits germination.[17] Even a flash of light may induce germination in seeds that are wet and swollen.

Light requirement is found mostly in small seeds that need to be close to the soil surface when germinating. If small seeds germinate too deeply in the soil they will exhaust their food supply before reaching the surface. Most weed seeds require light for germination. The absence of light enables such seeds to remain dormant when buried deeply in the soil. Phytochrome is also involved in photoperiodism in plants (Chapter 2).

Most field crop seeds germinate in either light or darkness. Many of the grasses germinate more promptly in the presence of, or after exposure to light, especially when the seeds are fresh. Among these are bentgrass, bermudagrass, Kentucky bluegrass, Canada bluegrass, and slender wheatgrass. Light is necessary for germination of some types of tobacco, except at low temperatures of about 57°F (14°C). Most standard American varieties will germinate in its absence, although the rate and percentage of germination may be considerably retarded. The light requirement in all cases is small.[12]

7.3.2 Process of Germination

When placed under the proper conditions, seeds capable of immediate germination gradually absorb water, until, after approximately three days, their moisture content may be 60 to 100 percent of the dry weight. Meanwhile, the seedcoats have become softened and the seeds swollen. Soluble nutrients, particularly sugars, go into solution. The soluble glucose is transported to the growing sprout by diffusion from cell to cell. It is then synthesized into cellulose, nonreducing sugars, and starch. Proteins are broken down by proteolytic enzymes into amides, such as asparagin, or into amino acids. These are then moved to new tissue and used to rebuild proteins. Fats, which occur mostly in the cotyledons of certain oil-bearing seeds and in the embryos of cereal seeds, are split by enzymes called lipases into fatty acids and glycerol. These, in turn, undergo chemical changes to form sugars, which are used to build up the carbohydrates and fats in the seedlings. Energy for the chemical and biological processes of germination and growth is supplied by respiration or the biological oxidation of carbon and hydrogen into carbon dioxide and water.[17] During germination, respiration proceeds rapidly at a rate hundreds of times faster than in dry seeds.

The energy consumed during germination may amount to one-half the dry weight of the seed.[18] The germination of a bushel of wheat utilizes the equivalent of all the oxygen in 900 cubic feet (25 m³) of air and requires energy equivalent to that expended in plowing an acre of land. Emerging seedlings exposed to light begin photosynthesis early, but even then their dry weight may not equal the dry weight of the seed until seven to fourteen days or more after the seedling appears above the soil surface.

In seeds that germinate promptly, the growing embryo ruptures the seedcoat within one or two days after the seeds are wetted. The radicle, or embryonic root, is the first organ to emerge in nearly all seeds. At this time the seed has absorbed all the available water in its vicinity. The seedling needs additional water to continue growth that is furnished by the radicle. The radicle is soon followed by the plumule or young shoot (Figures 7.1 and 7.2). In many dicotyledonous plants such as the bean and flax, the cotyledons emerge from the soil and function as the first leaves. The plumule emerges in a bent or curved position (Figure 7.3). The arch thus formed serves to protect the cotyledons as they are brought above the surface of the soil by the elongating hypocotyl. This is called *epigeal germination.*

In grasses (monocotyledonous plants), and also in a few legumes such as pea and vetch, the cotyledons remain in the soil. The plumule grows or is pushed upward by the elongation of an epicotyl or a subcrown internode. This is called *hypogeal germination.* The subcrown internode of different grasses has been called a mesocotyl, epicotyl, or hypocotyl, depending upon the seedling node from which it arises.

The coleoptile of grasses emerges from the soil as a pale tube-like structure that encloses the first true leaf. A slit develops at the tip of the coleoptile, and the leaf emerges through it. Then photosynthesis begins and the seedling gradually establishes independent metabolism as the stored food of the seed nears exhaustion. The roots are well developed by that time.

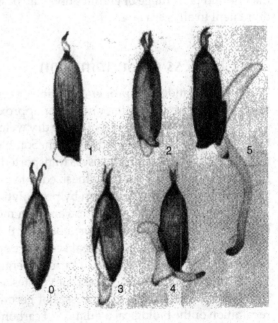

FIGURE 7.1

Six successive stages in the germination of the sugarcane seed.

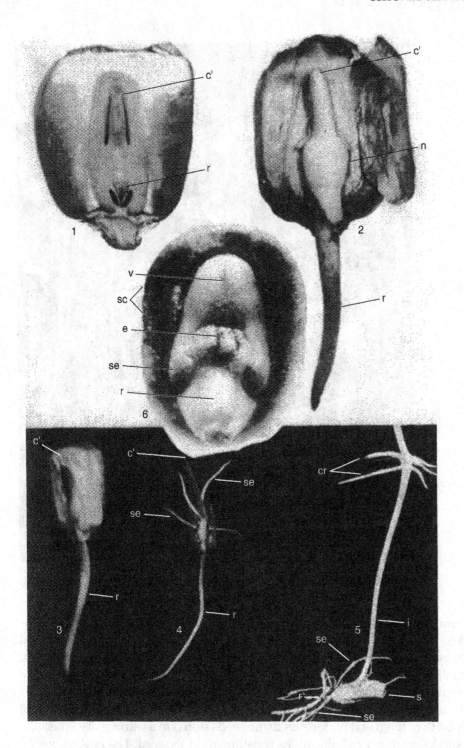

FIGURE 7.2

Stages in corn germination: *(1)* Before germination; *(2)* germinated 36 hours, *(3)* 48 hours, *(4)* 4 days, and *(5)* 8 days. In the two upper views the seedcoat has been removed to expose the embryo. In germinating the radicle or first seminal root *(r)* pushes out quickly; the nodal region *(n)* swells; the coleoptile, which encloses the first leaves and has a vent at the tip *(c')*, grows upward; additional seminal roots *(se)* arise, usually in pairs above the radicle, after three days. Finally the coronal or crown roots *(cr)* develop and the food substance in the seed *(s)* is practically exhausted. At *(6)* a wheat germ enlarged about 25 times, shows the scutellum *(sc)*, vent in coleoptile *(v)* epiblast *(e)*, seminal root swellings *(se)*, and radicle *(r)*, which is enclosed in the coleorhiza. [Courtesy T. A. Kiesselbach]

7.3.3 Qualities in Seeds for Germination

WHOLE VERSUS BROKEN SEEDS A marked decrease in germination of mutilated wheat, corn, and alfalfa seeds occurs when the germ is injured.[19] Broken seeds that contain the embryo germinate less, have a higher seedling mortality, and produce smaller plants than whole seeds.[20, 21] Breaks in the seedcoat of cereals are harmful

FIGURE 7.3

Seedlings of *(1)* bean, *(2)* pea, *(3)* rye, *(4)* sorghum, and *(5)* oat. During germination, the cotyledons *(co)* of the bean are pushed up by the elongating hypocotyl *(hy)*; then the cotyledons separate and the plumule of true leaves *(p)* emerges. In the germinating pea, the epicotyl *(ep)* grows upward from the cotyledons *(co)* and the plumule *(p)* grows out from the tip of the epicotyl. In the cereals, the coleoptile grows or is pushed to the soil surface and then the plumule *(p)* grows out through a slit *(c')* at the tip of the coleoptile. In the rye, as in wheat and barley, the coleoptile base *(c)* arises at the seed *(s)* and the crown node lies somewhere within the coleoptile. If sorghum, as in corn, the coleoptile base *(c)* is at the crown node which is carried upward from the seed *(s)* by the elongating subcrown internode *(i)*. In oats, as in rice, the node at the coleoptile base *(c)* stands just below the crown node and also is carried upward from the seed *(s)* by the elongating subcrown internode. Occasionally adventitious roots *(a)* arise from the subcrown internode *(i)* in oats and other cereals. (The irregular direction of roots resulted from germination between blotters instead of in the soil.)

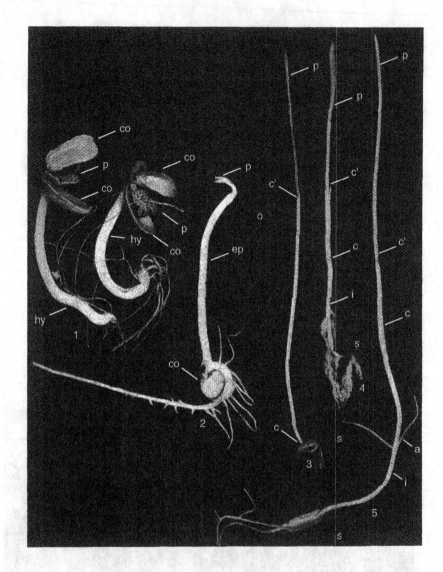

to germination, with injury at the embryo end the most serious.[22] Broken or cracked seeds are more susceptible to mold more than whole seeds.[14] Mechanical injury resulting in broken seedcoats and splitting frequently occurs in field pea. Seeds that consist of the embryo and a single cotyledon, or a part of one, may fail to germinate.[23]

The viability of seeds may be quickly destroyed by mold or heat as a result of the growth of fungi and bacteria on damp seeds stored in a warm place. These organisms break down and absorb the constituents of the seed. The fats are broken down into fatty acids, and germination drops as fatty acids accumulate. After planting, the seeds are exposed to organisms in the soil as well as to those on the seed. The organisms utilize the food materials in the seed, thus starving the young sprout, and certain organisms even invade and kill the young sprouts. Seed-borne and soil-inhabiting organisms often prevent seedling emergence. Sowing healthy seeds at the optimum temperature for germination helps to retard seed rot and seedling blight. Thus, small grains and field pea should be sown when the soil is cool, and planting of corn, sorghum, cotton, peanut, soybean, and millet should be delayed until the soil is warm. The best protection against seed rots and seedling blights is treatment of the seed with approved disinfectants containing a chemical that is toxic to fungi and bacteria.[24] The fungi commonly associated with seed mold or rot and seedling blight are the species of *Pythium, Fusarium, Rhizopus, Penicillium, Aspergillus, Gibberella, Diplodia, Helminthosporium, Cladysporium, Basisporium,* and *Collectotrichum.*[14, 25]

SEED MATURITY Mature seed is preferable to immature seed, but occasionally growers are obliged to plant seeds that have failed to reach full maturity. Prematurely harvested barley kernels have germinated and produced small seedlings when the seeds had attained only one-seventh of their normal weight.[26, 27] Corn seeds grew when gathered as early as twenty days after fertilization of the silks, provided they were carefully dried.[28] Such poorly developed seeds obviously are unsatisfactory for field planting. Table 7.2 shows that corn gathered as early as the denting stage is suitable for seed.[28]

Mature corn produces heavier sprouts than that harvested at immature stages.[29] Immature corn shows more disease infection and yields slightly less than mature corn.[30]

Immature seed,[31] because of its small size, has a low reserve food supply and usually produces poor plants when conditions are adverse at planting time. Immature seeds, high in moisture, are vulnerable to frost injury.

SEED SIZE Small seeds invariably produce small seedlings. The logarithms of seedling and seed weights are directly proportional (Figure 7.4).

In Nebraska research,[32] small seeds of winter wheat, spring wheat, and oat yielded 18 percent less than large seeds when equal *numbers* of seeds were sown per acre at an optimum rate for the large seed, but only 5 percent less when equal *weights* of seed were sown, also at an optimum rate for the large seed. Grain drills sow about equal volumes of large or small seeds of any particular grain, so the latter comparison is the most valid. When unselected seed was used, it yielded 4 percent less than large seed when equal numbers were sown per acre, but only 1 percent less when equal weights of seed were sown.

In comparisons of fanning-mill grades of winter wheat over a seventeen-year period, the heaviest quarter yielded 0.3 percent more, and the lightest quarter yielded 2.0 percent less than unselected seed. Similar results were obtained with oat.

TABLE 7.2 Effect of Maturity of Seed upon the Grain Yield of Dent Corn (Five-Year Average)

Weeks Before Mature	Date Seed Harvested	Days Since Fertilization	Condition of Grain	Field Germination (%)	Yield of Shelled Corn per Acre (bushels)
Mature	September 28	51	Mature	94	55.8
1	September 21	44	Glazing	94	54.4
2	September 14	37	Denting	93	54.9

FIGURE 7.4

The seedling weight (10 to 12 days after planting) shows a direct logarithmic relation to seed weight: (A) relation between weight per seed and weight per stalk 10 days after planting of corn, sorghum, and prove; (B) relation between logarithms of the seed weights and of the stalk weights taken at intervals up to 20 days after planting.

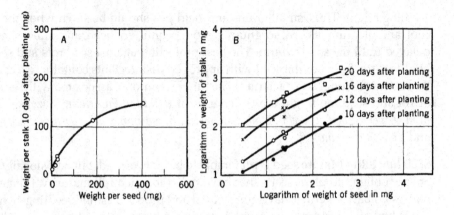

There is no practical gain in grain yield from grading normally developed small-grain seed that is reasonably free of foreign material. Large seeds produce more vigorous seedlings, which better survive adverse conditions, but this advantage within certain limits is largely offset by the greater number of plants obtained from an equal weight of smaller seeds.

The germination and seedling size of shrunken and plump spring wheat, with test weights that ranged from 39.5 to 60.8 pounds per bushel (500 to 780 g/l), was determined.[33] Test weight shows little relation to viability, but shrunken seeds produce such small, weak seedlings that sowing of wheat testing less than 50 pounds per bushel (640 g/l) is not recommended. Kernels of wheat testing 60 pounds per bushel (770 g/l) are about twice as heavy as those testing 50 pounds (640 g/l). A reduction in test weight of one-third (i.e., from 60 down to 40 pounds per bushel [770 to 515 g/l]) reduces the weight of an individual kernel nearly two-thirds,[15] and seedling weights are reduced nearly as much. In general, matured seeds that are less than one-half the normal size are unsuitable for sowing.

Kernel placement on the ear has no effect on corn seed quality. Extensive experiments showed that the average yield of butt seed was 103 percent of that from the middle of the ear, while the yield from tip kernels was 105 percent.[34] In a similar comparison, seeds from the tips, butts, and middles of corn ears yielded comparably.[35] Corn seed will be metered more accurately with plate-type planters if the seed is uniform in size and shape. Modern planters are able to accurately meter mixtures of different sizes and shapes of corn seed.

7.3.4 Seed Dormancy

Seeds of some crop species exhibit dormancy.[36] These seeds fail to grow immediately after maturity, even though external conditions favor germination, until they have passed through an after-ripening period. This is more common in wild plants, but varieties of cultivated plants differ considerably in dormant tendencies. Dormancy in cereal seeds is indicated by the inability to germinate at higher temperatures when they germinate well at 36 to 50°F (2 to 10°C).

CAUSES OF DORMANCY Dormancy may result from seed characteristics or environmental conditions as follows:[31, 37]

1. Thick or hard seedcoats prevent intake of water and probably also of oxygen. The *hard seeds* in many legumes are an example.
2. Seedcoats interfere with the absorption of oxygen. Examples are cocklebur, oat, and barley.
3. In some species, the embryo is still immature and has not yet reached its full development at harvest.
4. The embryos in some seeds appear to be mature but must undergo certain changes before they will germinate. Immature wheat and barley seeds harvested 12 to 24 days after flowering, and dried quickly, may retain their green color but will germinate poorly and produce weak seedlings. Such green seeds also are found in wheat that has been frosted before maturity. Dormant varieties of winter barley apparently become dormant during ripening or drying because seeds of such varieties sprouted in the head before maturity when the seedcoats were kept wet by artificial watering.[38]
5. Germination inhibitors that must undergo natural or applied chemical changes to permit germination.
6. High temperatures during seed maturity may induce dormancy.

HARD SEEDS IN LEGUMES Hard or impermeable seeds prevent penetration of water and cause an apparent enforced dormancy period. Such seeds are common in alfalfa but are also found in most small-seeded legumes.[39] Hard seeds in alfalfa are due to the inability of the palisade cells to absorb water.[40] Apparently, the cuticle does not restrict the intake of water. The percentage of hard seeds varies among different branches of the same plant. Probably as a result of some scarification, machine-threshed seed contains fewer hard seeds than that which is hand-threshed. Plump seed is more likely to be dormant than is shriveled or immature seed.

AFTER-RIPENING Seeds of peanut, alfalfa, clover, and lupin planted soon after maturity under conditions nearly optimum for germination frequently exhibit dormancy ranging up to two years. The rest period appears to be one of after-ripening in peanut.[41]

Seeds of many small-grain varieties require a short period of dry storage after harvest in order to after-ripen and give good germination at temperatures as high as 68°F (20°C). Seeds are usually stored several months before germination tests are made, but in winter cereals it may be necessary to test the seeds soon after threshing to determine their viability for fall planting. Storage of oat at 104°F (40°C) for three months largely eliminated dormancy[42] and also destroyed most of the molds on the seed.

The embryos of cereals are essentially never dormant, the dormancy being imposed by the seedcoat. Artificial dry heating, opening of the coat structures over the embryo, and cutting off the brush ends admit oxygen which induces germination of non-after-ripened or partially after-ripened seeds of wheat, oat, or barley. A temperature considerably below 68°F (20°C) is most satisfactory for germination of freshly harvested seed of these crops.[43]

Immature, poorly cured wheat has a higher percentage of dormancy.[19] Dormancy may decrease to a minimum after four to twelve weeks of storage and can be broken immediately by placing the seeds in cold storage at 40 to 43°F (4 to 6°C) for five days, and then transferring them to alternating temperatures of 68 to 86°F (20 to 30°C) for three days. Some wheat grown under high altitude conditions may be dormant as long as sixty days after harvest.[44] Seedsmen have repeatedly encountered difficulty in getting satisfactory germination in laboratory tests of sound, plump durum wheat, especially in the fall and early winter. After-ripening is completed during warm spring weather. Good stands are obtained when field planting in cool soil even though durum wheat germinates slowly in laboratory tests until spring.[45] After-ripening of mature corn is coincident with loss of moisture.[7] It may be necessary to reduce the moisture content of immature seeds to approximately 25 percent before normal germination occurs. The mechanism that inhibits normal germination of such seeds is believed to occur in the scutellum rather than in the endosperm or pericarp.

Slow germination of freshly harvested seeds is extremely desirable in a wet harvest season when heavy losses occur from grain sprouting in the field.

Dormant varieties do not sprout appreciably.[46, 47] All degrees of prompt, slow, and delayed germination occur in freshly harvested common oat, but cultivated red oat regularly shows slow or delayed germination.[11] Dormancy disappears in most oat varieties after thirty days. Grain sorghums often sprout in the head in the field during rainy periods before harvest.

The usual dormancy in buffalograss seed can be broken by soaking the seed in a 0.5 percent solution of potassium nitrate for twenty-four hours, then chilling it at 41°F (5°C) for six weeks in a cold-storage room and drying it.[48] However, hulling of the seed is fully as effective, and the treatment is more simple and economical.

SECONDARY DORMANCY High temperatures in storage or in the germinator or seedbed may throw seeds of cereals or grasses into a secondary dormancy. Such seeds usually germinate later at normal temperatures after they have been subjected to cold treatment.

Very dry cotton seeds may fail to germinate as promptly or as vigorously as those with a moisture content of approximately 12 percent when planted. A marked increase of hard seeds occurs when they are dried down to 5 or 6 percent moisture.[49] Excessively dry seed gives satisfactory germination when moistened at planting time with about 2 gallons of water per 100 pounds of seed (8 liters per 64 kg). Certain Texas samples entered into a secondary dormancy and failed to produce a crop when the proper seedbed conditions were not present.

7.3.5 Scarification of Hard Seeds

Germination of impermeable seeds in legumes such as alfalfa, sweetclover, and the true clovers may be brought about in several ways.[50]

In a mechanical scarifier, the seed is thrown against a roughened surface to scratch the seedcoats. The scarified seeds imbibe water and germinate in a normal manner. However, mechanical scarification is used less today. In tests in Utah,[51] scarification of alfalfa seed increased germination about 30 percent, but there were more weak and moldy seedlings from the scarified seed. Scarification injured about as much good seed as it made hard seeds germinable. Mechanically scarified seed has been found to deteriorate rapidly in storage. When scarified, seed should be planted immediately.

In New York, fall-sown sweetclover seeds softened and grew the next spring, with 50 to 75 percent of the seeds producing plants. This is about as high a percentage as is ordinarily obtained in field seeding. In fact, 90 to 100 percent of fall-sown hard sweetclover seeds will yield better results under such conditions than scarified seeds sown at any time of the year.

7.3.6 Other Hard-Seed Treatments

Aging brings about slow natural deterioration of the seedcoat in dry storage. In certain experiments, one-third to two-thirds of the hard seeds in red clover were still impermeable after four years, but a majority of the impermeable seeds of alfalfa and hairy vetch became permeable before they were two years old.[52] In another experiment, one-half of the impermeable seeds in alfalfa germinated after 1½ years, while all germinated after eleven years in storage.[44] The percentage of hard seeds in Korean lespedeza is high when the seeds are tested for germination immediately after harvest, but most of them become permeable during the winter.[53] The average percentages of hard seeds in tests made in November, January, and March were 47.25, 12.25, and 11.05, respectively.

Alternate freezing and thawing sometimes stimulates germination of hard seeds of alfalfa and sweetclover but may also destroy some seeds that germinate normally.[54] The breaking of dormancy while in the soil may be due to a period of cold.

The germination of hard seeds of alfalfa or clover also can be achieved by exposure for 1 to 1.5 seconds or less to infrared rays of 1,180 nm wavelength[55] or by exposure for a few seconds to high-frequency electric energy.

7.3.7 Vernalization of Seeds

Temperatures affect the flowering time of many plants. Winter annuals, biennials, and some perennials must be exposed to a period of cool or cold temperatures before they will flower. This process is called *vernalization*. Vernalization assures that flowering will not be induced in the fall resulting in severe damage or death when winter occurs.

Winter wheat sown in the spring fails to produce heads unless the sprouting seeds or growing plants are subjected to cold or cool conditions. Winter wheat generally needs about forty days when the temperature drops below 40°F (4.5°C).

Spring-planted winter varieties of cereal crops will flower normally if the seeds are first soaked for twelve to twenty-four hours, and then stored for four to nine weeks at a temperature of about 36°F (2°C). This process vernalizes the embryos, and the plants will behave like spring varieties. The degree of sprouting during the cold treatment can be restricted by maintaining the moisture content of the grain at approximately 50 percent (on a dry-weight basis). Winter annual legumes and grasses may respond to similar treatments. Vernalization is so laborious and complicated that it is useful only in certain experiments. The sprouted seeds are difficult to store and sow; germination is often damaged by the treatment.[56] Drying and storing the vernalized seed at warm temperatures often causes the seeds to become devernalized and lose much of the effect of the cold treatment.

Vernalized, spring-sown, winter grain yields much less than when sown in the autumn and about the same or somewhat less than adapted spring varieties sown at the same time as the vernalized grain. Certain spring varieties of cereals that have a partial or intermediate winter growth habit usually respond to vernalization treatment when sown late in the spring,[57] but true spring varieties are not affected. Russian workers reported that special vernalization of corn, sorghum, millet, and other warm-temperature crops is effective in hastening flowering. This treatment consists of germinating the seeds in the dark at normal temperature while restricting the sprouting by adding only limited quantities of water, largely in the form of dilute salt solutions. Extensive experiments in other countries failed to substantiate claims for this type of vernalization.[25] The seeds often mold and lose their viability during treatment.

7.3.8 Longevity of Seeds

Most crop seeds are probably dead after twenty-five years or less, even under favorable storage conditions. The alleged germination of seeds after prolonged storage in ancient tombs is known to be a myth, although fresh seeds that have been placed in tombs shortly before the visits of gullible travelers often germinate very well. Authentic seeds from the ancient tombs are highly carbonized and have lost much of their original substance.

SEEDS IN DRY STORAGE The optimum conditions for storing seeds that will endure drying are a 5 to 7 percent moisture content, sealed storage in the absence of oxygen, and a temperature of 23 to 41°F (−5 to 5°C).[58, 59] In moist, hot climates, seeds can be kept viable between seasons only by storing them, well dried, in airtight containers.[60] It has been suggested that seed life is doubled for each drop of 1 percent in moisture content and for each drop of 9°F (5°C) in temperature.[61] Under cool, semiarid conditions in Colorado, wheat, oat, and barley germinated about 10 percent lower when ten years old than when one year old.[62] The germination of soybean decreased about 10 percent in five years, but sorgo sorghum germinated about 97 percent after seventeen years. Yellow Dent corn germinated well for five years, but declined to 32 percent after twenty years. After fifteen years of storage[63] other approximate germination percentages were rye, 8; corn, 36; naked barley, 74; wheat, 80; and unhulled barley, 96. After twenty years, the germination percentages of wheat, barley, and oat were 15, 46, and 50, respectively.[54]

In semiarid eastern Washington, certain varieties of barley, oat, and wheat germinated from 84 to 96 percent after thirty-two years of storage. Corn germinated 70 percent, but rye had lost its viability.[5]

Under Nebraska conditions, corn four years old was satisfactory for seed.[64] Kafir sorghum seed retained its germination well for ten years in western Texas but deteriorated almost completely during the next seven years.[41]

Flaxseed of good quality stored under favorable conditions may be expected to maintain its viability for six to eight years. Seeds nine, twelve, fifteen, and eighteen years old germinated 99, 89, 56, and 58 percent, respectively.[16] Tobacco seed usually retains ability to germinate over many years. One lot of seed germinated 25 percent after twenty years.[23] However, there was a marked retardation in rate of germination in seed more than ten years old.

While many seeds in ordinary storage in drier climates may have a life span of fifteen years, they might live fifty or even one-hundred years in sealed storage at low temperatures in the absence of oxygen and after proper desiccation.[36] Seeds of *Albizzia* germinated after one hundred forty nine years in storage in a British herbarium.

BURIED SEEDS Seeds of some species of wild plants retain their vitality in moist soil for fifty years or more. In research begun in 1902 by the United States Department of Agriculture a total of 107 species of seed were mixed with sterilized soil and buried in pots. None of the cereals or legumes whose seeds are used for food germinated when dug up after twenty years. The seeds of wild plants grew better than those of cultivated plants. Several persistent weeds showed high germination after twenty years in the soil, including dock, lambsquarter, plantain, purslane, jimsonweed, and ragweed. Seeds still alive when finally dug up thirty-nine years after they were buried included Kentucky bluegrass, red clover, tobacco, ramie, and some twenty-five species of weeds.[65] After being buried thirty years, wild morning glory seed germinated within two days after it was dug up.

Another classical experiment with buried seeds was started in 1879 by Dr. W. J. Beal of Michigan State College. Seeds of twenty species were placed in bottles filled with sand and buried under 18 inches (46 cm) of soil. Chess and white clover seeds were dead after five years, while common mallow survived for twenty-five years. Curled dock and mullein germinated seventy years after burial.[65]

Dry arctic lupine seeds found buried in lemming burrows beneath 10 to 20 feet (3 to 6 m) of frozen soil in Yukon Territory, Canada, were able to germinate. Their assumed age was about 14,000 years.[66]

7.4 GERMINATION AND PURITY TESTS

The real value of a seed lot depends upon its purity and the proportion of pure seed in it that will grow. Seeds after threshing usually contain foreign materials such as chaff, dirt, weed seeds, and seeds of other crop plants. These can be removed to a large extent, but not entirely, by cleaning machinery.[24, 61]

The object of laboratory seed testing is to determine the percentages of germination, pure seed, other crop seed, inert matter, the presence and kinds of weed seeds, and, if possible, the kind and variety of the seed sample (Figure 7.5). This serves as an aid in selecting suitable seed, adjusting seeding rates to germination percentages, giving warning of impending weed problems, and reducing dissemination of serious weeds. Seed testing should be done by well-trained seed analysts. Then the label is an accurate guide to the value of the seed.

Germination tests should include 400 seeds counted from the sample indiscriminately and divided into four or more separate tests (Figure 7.6).[67] Purity tests should be made by hand separation of a sample that contains approximately 3,000 seeds. The size of the sample specified for purity tests is indicated in Table 7.3. Most crop seeds are germinated at alternating night and day temperatures of 68 and 86°F (20 to 30°C) in laboratory tests. The alternation of temperatures, which simulates field conditions, favors better germination.

If partly sprouted seeds are present in the sample, the final germination count for legumes with hard seeds is extended for five days. Fresh seeds may require

FIGURE 7.5
A typical seed testing laboratory with equipment to analyze seeds for quality and freedom from weed seeds, insect damage, and seed-borne diseases.
[Courtesy Richard Waldren]

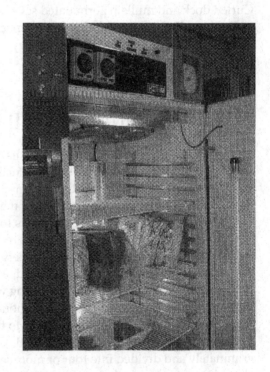

FIGURE 7.6
A seed germinator for determining seed viability and vigor. [Courtesy Richard Waldren]

prechilling, treatment with 0.1 percent or 0.2 percent potassium nitrate, or scarifying. Different seeds are germinated between blotters, on top of blotters, in paper toweling, in rolled towels, in sand, or in soil.

TABLE 7.3 Methods for Testing Typical Seeds

Crop Seed	Seeds per Gram	Minimum Weight for Noxious Weed Examination (grams)	Minimum Weight for Purity Analysis (grams)	Germination Test			Special Treatments
				Temperature (°C)	First Count (days)	Final Count (days)	
Alfalfa	500	50	5	20	3	7–12	—
Alsike clover	1,500	50	2	20	3	7–12	—
Bahiagrass	366	50	10	30–35	3	21	Light; hulling
Barley	30	500	100	20	3	7	—
Bean (field)	4	500	500	20–30	5	8–13	—
Buckwheat	60	300	50	20–30	3	6	—
Crimson clover	330	50	10	20	3	7–12	—
Kentucky bluegrass	4,800	25	1	20–30	7	28	Light
Meadow fescue	500	50	5	20–30	5	14	—

7.4.1 Stress Tests

Stress tests are designed to determine seed vigor under minimum conditions for germination. Usually the seeds are germinated in moist soil at the minimum temperature for germination to simulate field conditions. The seed is placed in soil in containers or wrapped in moist paper towels with soil. Since spring-seeded crops should be planted as early in the season as possible to maximize the growing season, stress tests provide valuable information about how well a particular sample of seeds will perform under these conditions.

The cold test for corn uses soil in containers. The seed is placed in the soil and the soil is moistened to field capacity. The containers are placed in an environment of 50°F (10°C) for seven days. The temperature is then raised to 77°F (25°C) for four days and seedling counts are made.[68]

The cool germination test for cotton is conducted in a different manner. Seeds are placed on moist germination towels as in the standard germination test. The test consists of four replications of fifty seeds each. Rolled towels are placed in a germinator set at a constant 64°F (18°C) for seven days. This temperature is a critical breakpoint for cotton seedling development, and very little variation from this temperature is acceptable. For example, 2°C below (61°F) may result in a greatly reduced number of normal seedlings, while 2°C above (68°F) may result in almost as many normal seedlings as the standard germination test.[69] The cool germination test is also used for sorghum.

7.4.2 Special Tests

In many cases, tests other than germination, purity, and vigor may be needed. These special tests include: the absence or presence of fungicides, diploidy, or tetraploidy; X rays to detect "hollow seeds"; and any other factors that may be of help to a seed buyer or seller.

The viability and vigor of the embryos of many seeds can be quickly evaluated by a staining (tetrazolium) test.[67] A fluorescence test is useful to identify or separate seeds of certain species or varieties that are indistinguishable otherwise.[70]

▓ 7.5 SEED LAWS AND REGULATIONS

7.5.1 Federal Seed Act

For many years, low-grade, foul, and adulterated seeds were sold. The Federal Seed Act was passed in 1912 to protect the American farmer from such seed.[59] The act was amended in 1916 to include a minimum requirement of live pure seed. Another amendment was passed in 1926 to provide for staining imported red clover and alfalfa seed so as to indicate its origin. Investigations have shown that many foreign importations are unadapted to certain regions in this country. Red clover seed from Italy is not adaptable for general use in this country, nor is alfalfa seed from Turkestan or South Africa. The Federal Seed Act of 1939 was much more drastic and inclusive than the original act. The 1939 act required correct labeling as to variety as well as purity and germination, established heavier penalties, and permitted penalties for mislabeling without the government having to prove fraudulent intent. False advertising is prohibited. There have been only minor revisions since 1939. Penalties for violation of the act include fines up to $2,000, seizure of the seed, and orders to "cease and desist" violations of the act.

The act prohibits importation of seed that is adulterated or unfit for seeding purposes. Adulterated seed can contain no more than a 5 percent mixture of other kinds of seed, except for mixtures that are not detrimental. Seed specified as unfit for seeding purposes is that with more than one noxious seed in 10 grams of small-seeded grasses and legumes; one in 25 grams of medium-sized seeds such as sudangrass, sorghum, and buckwheat; and one in 100 grams of grains and other large seeds. The act also prohibits importation of seed that contains more than 2 percent weed seed or (with several exceptions) less than 75 percent pure live seed. Seeds that are allowed to have germination percentages lower than 75 percent are bahiagrass, 50; bluegrass, 65; carrots, 55; chicory, 70; dallisgrass, 35; guineagrass, 10; molassesgrass, 25; and rhodesgrass, 35 percent.

Specified noxious weeds are: whitetop (*Lepidium draba, L. repens,* and *Hymenophysa pubescens*), Canada thistle (*Cirsium arvense*), dodder (*Cuscuta species*), quackgrass (*Agropyron repens*), johnsongrass (*Sorghum halepense*), bindweed (*Convolvulus arvensis*), Russian knapweed (*Centaurea picris*), perennial sowthistle (*Sonchus arvensis*), and leafy spurge (*Euphorbia esula*). There can also be other kinds of seeds or bulblets that, after investigation, the Secretary of Agriculture finds should be included in the list.

The act also prohibits shipment in interstate commerce of agricultural seeds that contain noxious weeds in excess of quantities allowed by the laws of the state to which the seed is shipped, or as established by the Secretary of Agriculture. The seeds must be labeled to include: variety or type, lot number, origin (of certain kinds), percentage of weed seeds including noxious weeds, kinds and rate of occurrence of noxious weeds, percentages of mixtures of other seeds, germination, hard-seed percentage, month and year of germination test, and the name and address of the shipper and consignee. The Federal Seed Act specifies variations in tolerances when determining germination and purity.

It is now possible for an individual or organizations to obtain a federal plant patent on a new, distinct variety or hybrid of any crop they have originated. This protects the plant breeder or employer from unlawful use of unique cultivars. Patents have become increasingly important as biotechnology is used for developing new strains and cultivars.

7.5.2 State Seed Laws

All states have seed laws designed to regulate the quality of agricultural seeds sold within their borders. These laws have done much to reduce the spread of weed seeds, especially noxious weeds. Most of the laws were patterned after the Uniform State Seed Law drawn up by the Association of Official Seed Analysts in 1917, and later were revised to conform more closely with the present Federal Seed Act.[71] There is considerable variation in the details of the laws adopted by the various states, but all require seeds in commerce to be labeled. The specifications embodied in these laws are reflected in the information required on the labels: (1) the commonly accepted name of the agricultural seeds, (2) the approximate total percentage by weight of purity (i.e., the freedom of the seeds from inert matter and from other seeds), (3) the approximate total percentage of weight of weed seeds, (4) the name and approximate number per pound of each of the kinds of noxious weed seeds and bulblets, (5) the approximate percentage of germination of such agricultural seed, together with the month and year that the seed was tested.

The weed seeds regarded as noxious by ten or more states include dodders, Canada thistle, quackgrass, wild mustards, buckhorn plantain, corn cockle, wild oat, wild onion, narrow-leaved plantain, wild carrot, ox-eye daisy, leafy spurge, Russian knapweed, bindweed, perennial sowthistle, and curled dock.

The seed laws are usually enforced by designated state officials. Field inspectors draw samples from seeds offered for sale. These are tested in official laboratories. Penalties are inflicted on dealers who sell seeds in violation of the state law.

▪ 7.6 SEED ASSOCIATIONS

Most states have seed or crop improvement associations of growers who produce quality agricultural seeds under strict regulations. These associations usually cooperate closely with their state agricultural colleges to bring superior crop varieties into widespread use at a reasonable cost.[18]

7.6.1 Registered or Certified Seed

The seed associations, whose rules and regulations differ somewhat among the states, supervise the growing of seeds by their members for certification and registration. The International Crop Improvement Association defines the classes of seed as follows:

1. *Breeder seed* is seed or vegetatively propagated material directly controlled by the originator, or, in certain cases, by the sponsoring plant breeder or

institution that provides the source for the initial and recurring increase of foundation seed.

2. *Foundation seed*, which includes elite seed in Canada, is seed that is handled to most nearly maintain specific genetic identity and purity and that may be designated or distributed by an agricultural experiment station. Production must be carefully supervised or approved by representatives of an agricultural experiment station. Foundation seed shall be the source of all other certified seed classes, either directly or through registered seed.

3. *Registered seed* is the progeny of foundation or registered seed that is so handled as to maintain satisfactory genetic identity and purity and that has been approved and certified by the certifying agency. This class of seed should be of a quality suitable for production of certified seed.

4. *Certified seed* is the progeny of foundation, registered, or certified seed that is so handled as to maintain satisfactory genetic identity and purity and that has been approved and certified by the certifying agency.

7.6.2 Requirements for Registration or Certification

The majority of seed associations require a grower to start with registered seed or foundation seed from an experiment station or equally reliable source.[2] The seed field is inspected before harvest for varietal purity, freedom from disease, and freedom from noxious weeds. An inspector either takes a bin sample after the crop is threshed, or the grower is instructed to send in a representative sample for purity and germination tests. The seed that comes up to the standard set by the association for the particular crop is registered or certified. This seed is sound, plump, and of good color, has high germination, and is free from noxious weeds. It is usually well cleaned and graded. Registered or certified seed is sold under specific tag labels that carry the necessary pedigree as well as other information required under the state seed law.

▓ 7.7 SOURCES OF FARM SEEDS

A farmer who desires a new variety or a fresh seed supply may secure high-quality pure seed of the standard varieties of most field crops from members of a crop improvement association at reasonable prices. These certified seeds are often available in the community. Seed lists issued each year give the name and address of the grower, the crop and variety grown, the amount of seed available, and the price. In general, certified seed is the highest quality of seed available for field crop production. However, the yields obtained from certified seed are not measurably higher than those from uncertified good seed of quality of the same variety with only a small admixture of other varieties. For commercial crop production, an appreciably higher price for certified seed may not be justified when good uncertified seed is available, unless the grower wishes to obtain a new, improved variety or to be assured of seed relatively free from disease.

Good seed of most adapted crops can be grown on the farm with care to prevent admixtures and weed contamination. The seed can be cleaned with an ordinary fanning mill unless it contains weed seeds that can be removed only with the special equipment described in Chapter 9. Consequently, a farmer's home-grown seed often contains many weed seeds.

The farmer usually finds it necessary to purchase hybrid seeds and small seeds such as clover, alfalfa, and forage grasses. Most commercial seed dealers endeavor to sell correctly labeled seeds but use a disclaimer clause for their protection because the crop produced by the farmer is entirely beyond the control of the seed seller. The disclaimer is usually stated as follows:

> The _____ Company gives no warranty, express or implied, as to the productiveness of any seeds or bulbs it sells and will not in any way be responsible for the crop.

This statement on letterheads and seed tags may not exempt dealers from legal redress when the seed is misbranded or fraudulently represented.

■ 7.8 SEEDING CROPS

7.8.1 Implements Used in Seeding

GRAIN DRILLS Combination grain and fertilizer drills are often used. Row spacing is usually from 7 to 14 inches (18 to 36 cm). The single-disk furrow opener is best for penetrating a hard seedbed or cutting through trash (Figure 7.7). Double-disk and shoe openers are best for mellow seedbeds that are firm below. The hoe-type opener is best in loose soil or where soil blowing is likely to occur. The hoe drill turns up clods and trash and does not pulverize the soil to any extent. The surface drill is used in humid regions, while the furrow and semi-furrow drills are widely used for winter wheat, particularly in the Great Plains under dryland conditions.[72] The furrow drill places the seed deep in moist soil. The ridges and furrows hold snow, reduce winter killing, and tend to protect the young plants from wind erosion.

FIGURE 7.7
Grain drill. [Courtesy John Deere]

FIGURE 7.8
No-till row crop planter.
[Courtesy USDA]

Modern grain drills are designed to plant in the heavy residues found in no-till and conservation tillage systems. They usually have a rolling coulter ahead of the opener to cut through residue and either single- or double-disk openers to move residue aside and place the seed in moist soil.

ROW CROP PLANTERS Row planters are widely used for planting intertilled crops. The rows are spaced 20 to 48 inches (50 to 300 cm) apart. By proper selection of drill plates, a plate-type planter can be used to plant corn, sorghum, bean, and other large-seed row crops. The lister planter is a combination tillage machine and planter that has been used under semiarid conditions to plant sorghum, corn, cotton, and other row crops in the bottom of furrows.[72] No-till planting (Figure 7.8) is growing in popularity. Eliminating tillage before planting reduces soil compaction and labor.

Planters can be equipped with rolling coulters that are 18 inches (46 cm) in diameter or larger. Whether the coulters are fluted, rippled, notched, or smooth, they must be sharp, weighted, or spring pressured and set deep enough to cut cleanly through the residue without punching it into the furrow. Residue mixed into the soil above the seed may cause corkscrewing of the seedlings. Smooth coulters give the best performance for stubble cutting. Fluted coulters provide more loose soil for better seed coverage. However, they do not scour properly when the soil is wet, and they do not cut residue adequately if it is damp. Wide fluted coulters throw residue away from the row better than narrow fluted coulters but require more weight for penetration. They also disturb the herbicide barrier more than the other types. By removing straw from the row, soil warms up faster, which is an advantage when soils are cold.[73, 74]

Modern planters have individual seed boxes on each row unit, or they use air to move seed from a central bin to the planter units. Seed metering devices on each row unit allow accurate seeding of a wide variety of crops at faster speeds than were previously possible. Most row crop planters are equipped to apply fertilizer and herbicides in bands while seeding the crop (Figure 7.9).

FIGURE 7.9
Planting row crops and applying fertilizer and herbicides in bands are accomplished in one operation. [Courtesy Case IH]

7.8.2 Method of Seeding

Modern row crop planters and grain drills are designed to place seeds at a uniform depth and space seeds at regular intervals in the row. They are designed to move sufficient soil away from the row to place seeds into moist soil and provide good seed-soil contact. In some cases, planters and drills are equipped with rolling coulters to cut through residues as described previously.

Most crops are planted into level soil with minimum disturbance of the soil. In some cases, crops are planted in the bottom of lister furrows. However, lister planted corn and sorghum start growth slower, and flower and mature later, than level and furrow planted seed.[75, 76] Crop stands are frequently destroyed by heavy rains that wash soil into the bottoms of lister furrows. The soil buries the seedling or covers the ungerminated seeds too deep for emergence. Most of the advantages of lister planting without its disadvantages can be achieved with a surface planter equipped with disk furrow openers.

In high rainfall areas, seeds are planted in raised beds that are formed prior to seeding. Raised beds help prevent flooding of seed rows and help soil warm up faster in the spring. There are some planters that will form the beds and plant the crop in one pass through the field.

Ridge-plant or ridge-till is a commonly used system in the Midwest. During seeding, existing ridges are removed and the seed is placed in the same location as the original ridge (Figure 7.10). During the operation, residues and weed seeds are removed from the seed row and thrown into the old furrow. After the crop has grown to a safe height, soil is thrown over the base of the plant to form the ridges

FIGURE 7.10
Soybean seedlings emerging in the old corn ridges from last year. [Courtesy USDA NRCS]

again.[77] Sometimes shallow tillage of the ridges is performed prior to seeding. Since crops are planted into the same rows every year, machinery following the same tracks can reduce soil compaction except in the track between the rows. Although compaction may be severe in the tracks, crop roots can easily avoid compacted soil since the soil on the other side of the row is not compacted.

7.8.3 Time of Seeding

The time to seed field crops is governed not only by the environmental requirements for the crop but also by the necessity of evading the ravages of diseases and insect pests.

Spring-planted small grains, like other cool season crops, generally are seeded early in the season to permit maximum growth and development toward maturity before the advent of hot weather, drought, and diseases. Nineteen hundred years ago, Columella wrote, "If the conditions of the lands and of the weather will allow it, the sooner we sow, the better it will grow, and the more increase we shall have."

For winter wheat in the western states,[78] the optimum date of seeding occurs when the mean daily temperature lies between 50 and 62°F (10 and 17°C). Higher temperatures prevail in the South and lower temperatures in the North. In the cold semiarid regions, sowing winter wheat early enough in the fall to allow the seedlings to become established before the soil freezes gives maximum protection against cold.

In humid regions, it is essential that wheat plants be well rooted in order to avoid winter injury from heaving. Columella recommended seeding between October 24 and December 7 in temperate regions; for colder regions, he advised seeding October 1, "so the roots of the [wheat grains] grow strong before they be infested with winter showers, frost or hoar frosts." Soil may heave when it freezes. Plants subjected to soil heaving have many of their roots broken and then are killed by desiccation.[79] Losses from heaving are a common occurrence in late-sown small grains in the eastern half of the United States. Seeding late enough to escape severe injury from Hessian fly is important for winter wheat where that pest is prevalent. In the case of cereal crops, practices that lead to maximum average yields are also satisfactory from the standpoint of crop quality.[80]

Early-sown crops mature earlier than those sown later, but they require a longer growing period. Consequently the difference in harvest date is less than the difference in planting date. Planting summer annual crops as early in the spring as possible increases the total amount of photosynthesis during the season, resulting in higher yields. This reaction in corn, which is typical of both long-day and short-day spring-planted crops, is shown in Figure 7.11. Later planted crops are not able to fully utilize the growing season.

Corn is a warm-weather crop that generally utilizes the full season. In Nebraska, early planting of corn in a normal season resulted in earlier maturity, lower grain moisture content, and higher grain viability when the crop was exposed to low temperatures.[80] Under Colorado conditions[81] corn has better yield and quality when planted early, and light frosts do less damage than delayed planting.

In North Carolina, cotton planted after the first week of May declines in yield. Drastic yield declines occurred in cotton planted the last week of May and later.[82] On the Texas High Plains, cotton planted on June 1, June 10, and June 20 yielded 7.6, 23.6, and 48.9 percent less, respectively, than cotton planted on May 15.[83] In California, cotton planted on April 15, April 25, and May 10 yielded 4.0, 8.0, and 17.0%, respectively, less than that planted on April 1.[84] Similar planting date–yield relationships have been established for other production regions in the U.S. Cotton Belt.[85]

In Texas experiments, early-planted cotton had a longer period of development before being attacked by root rot (Phymatotrichum omnivorum), but development of the disease was more rapid in early than in late plantings. The greatest losses were sustained by early plantings.[86]

Perennial legumes may be seeded either in spring or fall. When sown in fall, they should be seeded early enough to permit satisfactory root development before the ground freezes or else so late that the seeds do not germinate until spring.

With many crops, it is a safe practice to seed at higher than normal rates when seeding has been delayed materially beyond the optimum time determined for the

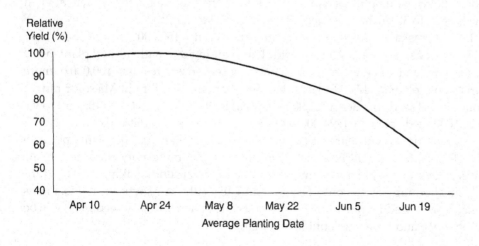

FIGURE 7.11

Effect of planting date on corn grain yield response in Iowa, 1997–2000. [From Farnham, 2001]

region. The plants will use less moisture and the higher seeding rates help maximize light interception in the leaf canopy.

7.8.4 Rate of Seeding

The goal of seeding rate is to match leaf area to expected water supply. Irrigated crops are planted at higher rates than dryland crops. Crops with larger plants are planted at lower rates than smaller plants. Varieties and hybrids that tiller or branch more are planted at lower rates than ones that tiller or branch little. Seeding rate is usually higher in narrow row spacing than in wider.

The objective in spacing crop plants is to obtain the maximum yield on a unit area without sacrificing quality. The rate of seeding is determined by the ultimate stand desired. Small short-season varieties of corn require thicker planting than long-season varieties.[87]

Since not all seeds will germinate and develop into a normal healthy plant that will contribute to yield, seeding rates must be greater than harvest population goals. The percentage of good quality seeds expected to emerge for most crops will be from 85 to 90. An expected field emergence of 85 percent is used for good quality seed with a germination rate of 90 percent or better. To predict expected field emergence of poor quality seed with a germination rate of less than 90 percent, multiply germination rate by 0.85.[29] Adjusted seeding rate is calculated by dividing the desired harvest plant population by the expected field emergence, expressed as a decimal (85% = 0.85).

If plant population is too high, plants will be tall, spindly, and more susceptible to lodging. Yields may decrease because not only does lodging make harvest difficult, resulting in greater harvest losses, but it also disrupts the leaf canopy, often limiting grain development and yield. If plant population is too low, there are fewer plants to contribute to yield and weeds will be a greater risk.

In southeastern Saskatchewan, a planting rate of 18 to 36 pounds per acre (20 to 40 kg/ha) was adequate where wheat yields were less than 20 bushels per acre (1,350 kg/ha).[88] In Alberta, where yield levels range from 38 to 51 bushels per acre (2,550 to 3,425 kg/ha), a planting rate of 90 pounds per acre (100 kg/ha) gives optimum wheat yields.[89] In Utah, where irrigated wheat yields ranged from 60 to 80 bushels per acre (4,000 to 5,475 kg/ha), a rate of 50 to 60 pounds per acre (56 to 67 kg/ha) was adequate except when planted late.[90] In western North Dakota, wheat yields were optimized at a stand rate of one million plants per acre at yield levels of 30 to 35 bushels per acre (2,000 to 2,350 kg/ha).[91]

In Nebraska,[92] seeding soybean at about 150,000 seeds per acre (415,000 seeds/ha) will optimize yield. This planting rate with normal plant losses during emergence and the remaining growing season will result in 100,000 or more harvestable plants (247,000 plants/ha). Soybean seeding rate in Missouri ranges from 130,000 seeds per acre (321,000 seed/ha) in 38- to 40-inch (97 to 102 cm) rows, to 200,000 seeds per acre (494,000 seeds/ha) in 6- to 8-inch (15 to 20 cm) rows.[93]

Soybean fields with harvest population of fewer than 100,000 plants per acre (247,000 plants/ha) will be short, have thick stems, be heavily branched at the lower nodes, and will have many pods close to the ground, making harvest difficult. Furthermore, weed control is more difficult with poor soybean stands. Plants in fields with seeding rates above 150,000 seeds per acre (370,000 seeds/ha) will be tall, spindly, and more susceptible to lodging.

Seeding rates in corn continue to increase as new hybrids are released with excellent stalk strength and increased yield potential. Optimum corn seeding rates in the eastern Corn Belt are about 31,000 to 35,500 seeds per acre (76,600 to 87,700 seeds/ha).[94]

INFLUENCE ON CROP QUALITY Grain quality is only slightly affected by usual variations in seeding rate. Most crop plants have a tremendous ability to alter yield components. Within normal populations, fewer plants will result in more production per plant, and more plants will result in less production per plant. At higher populations, corn will produce fewer kernels per ear, soybean will produce fewer pods and seeds per pod, small grains will tiller less, and cotton will produce fewer bolls. Seed size, however, will be about the same because the main objective of a seed plant is to produce healthy, fully developed seed.

In flax, differences in plant spacing show no consistent influence on the oil content of the seed.[95] Since the fineness of stems adds to the palatability of forage crops, it is desirable to seed them more thickly. Ordinarily, the forage will be finer and leafier without reduction in yield. Corn used for silage will be planted at 2,000 to 4,000 plants per acre (5,000 to 10,000 plants/ha) higher than corn planted for grain.[91] In cotton, higher population may contribute to decreased size of boll.[96]

7.8.5 Depth of Seeding

Seeds will emerge from greater depths in sandy soil than in clay soil and in warm soil than in cold soil. It is customary to plant deep in dry soil in order to place the seeds in contact with moisture. Pea will emerge from a greater depth than will bean when the seeds are the same size because the bean seedling has epigeal emergence and must push the cotyledons up above the soil surface, whereas pea has hypogeal emergence and the cotyledons remain where planted.

Larger seeds have more stored food reserves for germination and emergence. Therefore, in general, the larger the seed, the deeper it can be planted and still emerge from an arable soil (Figure 7.12). Approximately ¼ inch (6 mm) in heavy soil or ½ inch (13 mm) in sandy soil is the most satisfactory depth for seeding small-seeded legumes and grasses under optimum conditions.[97, 98] These include alfalfa,

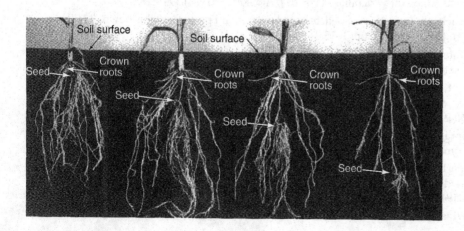

FIGURE 7.12
Corn planted at 2, 4, 6, and 10 inches deep. The crown formed at nearly the same depth regardless of planting depth.

TABLE 7.4 Seeding Depths for Seeds of Different Sizes

Normal Depth of Seeding (inches)	Usual Maximum Depth for Emergence (inches)	Seed Size (no. per lb)	Representative Crops
0.25 to 0.50	1 to 2	300,000 to 5,000,000	Redtop, carpetgrass, timothy, bluegrass, fescues, white clover, alsike clover, and tobacco
0.50 to 0.75	2 to 3	150,000 to 300,000	Alfalfa, red clover, sweetclover, lespedeza, crimson clover, ryegrass, foxtail millet, and turnip
0.75 to 1.50	3 to 4	50,000 to 150,000	Flax, sudangrass, crotalaria, proso, beet (ball of several seeds), broomcorn, and Bromegrass
1.50 to 2	3 to 5	10,000 to 50,000	Wheat, oats, barley, rye, rice, sorghum, buckwheat, hemp, vetch, and mungbean
2 to 3	4 to 8	400 to 10,000	Corn, pea, and cotton
4 to 5		4 to 20 (tubers or pieces)	Potato and Jerusalem artichoke

sweetclover, red clover, alsike clover, white clover, timothy, bromegrass, crested wheatgrass, reed canarygrass, and Kentucky bluegrass. Satisfactory emergence of reed canarygrass was obtained from a 1-inch (25 mm) depth and bromegrass from a 2-inch (50 mm) depth on all soil types. A reduction in the stand of soybean followed seeding deeper than 2 inches (50 mm) in fine sandy loam and 1 inch (25 mm) in a clay soil.[99] However, satisfactory stands were secured at depths up to 4 inches (100 mm) in loam and 2 inches (50 mm) in clay soil. Depth of planting may be an important factor in determining the seedling emergence of many grasses and small-seeded legumes.[98]

The seeding depths for seeds of different sizes under field conditions are shown in Table 7.4. The logarithms of seed size and typical planting depth of most seeds are directly proportional.

REFERENCES

1. Martin, J. H., and S. H. Yarnell. "Problems and rewards in improving shells," in *Seeds*, USDA Yearbook, 1961, pp. 113–118.

2. Parsons, F. G., C. S. Garrison, and K. E. Beeson. "Seed certification in the United States," in *Seeds*, USDA Yearbook, 1961, pp. 394–401.

3. Kiesselbach, T. A., and R. M. Weihing. "Effect of stand irregularities upon the acre yield and plant variability of corn," *J. Agr. Res.* 47(1933):399–416.

4. Plucknett, D. L. "Use of pelleted seed in crop and pasture establishment," *HI Univ. Ext. Circ.* 446, 1971, pp. 1–15.

5. Haferkamp, M. E., E. L. Smith, and R. A. Nilan. "Studies on aged seeds. 1. Relation of age of seed to germination and longevity," *Agron. J.* 45(1953):434–437.

6. McKinney, H. H., and W. J. Sando. "Russian methods for accelerating sexual reproduction in wheat: Further information regarding iarovization," *J. Hered.* 24(1933):165–166.

7. Sprague, G. F. "The relation of moisture content and time of harvest to germination of immature corn," *J. Am. Soc. Agron.* 28(1936):472–478.

8. Martin, J. N., and J. H. Watt. "The strophiole and other seed structures associated with hardness in *Melilotus alba* L. and *M. of officinalis* Willd.," *IA State College Jour Sci.* 18, 4(1944):457–469.

9. Widstoe, J. A. "Dry burning." New York: Macmillan, Inc., 1911, pp. 1–445.

10. Dillman, A. C. "Hygroscopic moisture of flax seed and wheat and its relation to combine harvesting," *J. Am. Soc. Agron.* 22(1930):51–74.

11. Coffman, F. A. "The minimum temperature of germination of seeds," *J. Am. Soc. Agron.* 15(1923):257–270.

12. Johnson, J., H. F. Murwin, and W. B. Ogden. "The germination of tobacco seed," *WI Agr. Exp, Sta. Res. Bull.* 104, 1930.

13. Dungan, G. H. "Some factors affecting the water absorption and germination of seed corn," *J. Am. Soc. Agron.* 16(1924):473–781.

14. Leukel, R. W., and J. H. Martin. "Seed rot and seedling blight of sorghum," *USDA Tech. Bull.* 839, 1943, pp. 1–36.

15. Stoa, T. E., W. E. Brentzel, and E. C. Higgins. "Shriveled lightweight wheat: Is it suitable for cells?" *ND Agr. Exp. Sta. Cir.* 59, 1936, pp. 1–11.

16. Dillman, A. C., and E. H. Toole. "Effect of age, condition, and temperature on the germination of flax seed," *J. Am. Soc. Agron.* 29(1937):23–29.

17. Stanley, R. G., and W. L. Butler. "Life processes of the living seed," in *Seeds,* USDA Yearbook, 1961, pp. 88–94.

18. Palladin, W. *Plant Physiology,* Philadelphia: Blakiston, 1918. pp. 1–320.

19. Whitcomb, W. O. "Dormancy of newly threshed grain," *Proc. Assn. Off. Seed Analysts* 16(1924):28–33.

20. Brown, E. B. "Relative yields from broken and entire kernels of seed corn," *J. Am. Soc. Agron.* (1920):196–197.

21. Sandy, W. J. "Effect of mutilation of wheat seeds on growth and productivity," *J. Am. Sac. Agron.* 31, 6(1939):558–565.

22. Lute, A. M. "Some notes on the behavior of broken seeds of cereals and sorghums," *Proc Assn. Off. Seed Analysts* 17(1925):33–35.

23. Hulbert, H. W., and G. M. Whitney. "Effect of seed injury upon the germination of *Pisum sativum,*" *J. Am. Soc. Agron.* 26(1934):876–884.

24. Gregg, B. R., and others. *Seed Processing.* New Delhi, India: Miss. State Univ. Nat'l Seeds Corp. and USAID, Avion Printers, 1970, pp. 1–396.

25. Anderson, A. M. *Handbook on Seed-borne Diseases,* Assn. Off. Seed corn, *J. Am. Soc. Agron.,* 196–197. 1920.

26. Harlan, H. V., and M. N. Pope. "The germination of barley seeds harvested at different stages of growth," *J. Hered.* 8, 2(1922):72–75.

27. Harlan, H. V., and M. N. Pope. "Development in immature barley kernels removed from the plant," *J. Agr. Res.* 32, 7(1926):669–678.

28. Kiesselbach, T. A. "Corn investigations," *NE Agr. Exp. Sta. Res. Bull.* 20, 1922.

29. Waldren, R. P. *Introductory Crop Science,* 5th ed. Boston: Pearson, 2003, pp. 335.

30. Koehler, B., G. H. Dungan, and W. L. Burlison. "Maturity of seed corn in relation to yielding ability and disease infection," *J. Am. Soc. Agron.* 26(1934):262–274.

31. Bartel, A. T. "Green seeds in immature small grains and their relation to germination," *J. Am. Soc. Agron.* 33, 8(1941):732–738.

32. Kiesselbach, T. A. "Relation of seed size to the yield of small grains," *J. Am. Soc. Agron.,* 16(1924):670–682.

33. Leukel, R. W. "Germination and emergence in spring wheats of the 1935 crop," *USDA B.P.I. Div,. Cereal Crops and Diseases,* Feb. 1936 (processed).

34. Lacy, M. G. "Seed value of maize kernels: Butts, tips, and middles." *J. Am. Soc. Agron.* 7(1915):159–171.

35. Kiesselbach, T. A., and W. E. Lyness. "Furrow vs. surface planting winter wheat," *J. Am. Soc. Agron.* 26(1934):489–493.

36. Crocker, W., and L. V. Barton. *Physiology of Seeds.* Waltham, MA: Chronica Botanica Co., 1953, pp. 1–267.

37. Barton, L. "Seed dormancy," *Encyclopedia Plant Phys.*, 15(1965):699–727.

38. Pope, M. N., and E. Brown. "Induced vivipary in three varieties of barley possessing extreme dormancy," *J. Am. Soc. Agron.* 35, 2(1943):161–163.

39. Stevenson, T. M. "Sweet clover studies on habit of growth, seed pigmentation and permeability of the seed coat," *Sci. Agr.* 17(1937):627–654.

40. Lute, A. M. "Impermeable seed in alfalfa," *CO Agr. Exp. Sta. Bull.* 326, 1928.

41. Lull, F. H. "Inheritance of rest periods of seeds and certain other characters in the peanuts," *FL Agr. Exp. Sta. Bull.* 314, 1937.

42. Brandon, J. F. "The spacing of corn in the West Central Great Plains," *J. Am. Soc. Agron.* 29(1937):584–599.

43. Harrington, G. T. "Forcing the germination of freshly harvested w heat and other cereals," *J. Agr. Res.* 23(1923):79–100.

44. Lute, A. M. "A special form of delayed germination," *Proc Assn. Off. Seed Analysts* 16(1924): 23–29, 1924.

45. Waldron, L. R. "Delayed germination in durum wheat," *J. Am. Soc. Agron.* 1(1908):135–144.

46. Deming, G. W., and D. W. Robertson. "Dormancy in seeds," *CO Agr. Exp. Sta. Tech. Bull.* 5, 1933, pp. 1–12.

47. Harrington, J. B. "The comparative resistance of wheat varieties to sprouting," *Sci. Agr.* 12(1932):635–645.

48. Wenger, L. E. "Buffalo grass," *KS Agr. Exp. Sta. Bull.* 321, 1943, pp. 1–78.

49. Toole, E. H., and P. I. Drummond. "The germination of cottonseed," *J. Agr. Res.*, 28(1924):285–292.

50. Love, H. H., and C. E. Leighty. "Germination of seed as affected by sulfuric acid treatments," *Cornell Agr. Exp. Sta. Bull.* 312, 1912, pp. 294–336.

51. Stewart, G. "Effect of color of seed, of scarification, and of dry heat on the germination of alfalfa seed and some of its imparities," *J. Am. Soc. Agron.* 18(1926):743–760.

52. Harrington, G. T. "Agricultural value of impermeable seeds," *J. Agr. Res.* 6(1916):761–796.

53. Middleton, G. K. "Hard seeds in Korean lespedeza," *J. Am. Soc. Agron.* 25(1933):119–122.

54. Rodriguez, G. "Study of influence of heat and cold on the germination of hard seeds in alfalfa and sweetclover," *Proc. Assn. of. Seed Analysts* 16(1924):75–76.

55. Works, D. W., and L. C. Erickson. "Infrared radiation, an effective treatment of hard seeds in small seeded legumes," *ID Agr. Exp. Sta. Rsh. Bull.* 57, 1963, pp. 1–22.

56. Martin, J. H. "The practical application of iarovization," *J. Am. Soc. Agron.* (note), 26, 3(1934):251.

57. Taylor, J. W., and F. A. Coffman. "Effects of vernalization on certain varieties of oats," *J. Am. Soc. Agron.* 30(1938):1010–1018.

58. Bartel, A. T., and J. H. Martin. "The growth curve of sorghum," *J. Agr. Res.* 57, 11(1938):843–847.

59. Owen, E. B. "The storage of seeds for maintenance of viability," *Commonwealth Bureau Pastures and Field Crops Bull.* 43, 1956, pp. 1–81.

60. Duvel, J. W. T. "The vitality and germination of seeds," *USDA Bur. Plant Industry Bull.* 58, 1904.

61. Harmond, J. E., and others. "Mechanical seed cleaning and handling," *USDA Handbook* 354, 1968, pp. 1–56.

62. Robertson, D. W., and A. M. Lute. "Germination of seed of farm crops in Colorado after storage for various periods of years," *J. Am. Soc. Agron.* 29(1937):822–834.

63. Robertson, D. W., A. M. Lute, and H. Krooger. "Germination of 20-year-old wheat, oats, barley, corn, rye, sorghum, and soybeans," *J. Am. Soc. Agron.* 35(1943):786–795.

64. Kiesselbach, T. A. "Effects of age, size, and source of seed on the corn crop," *NE Agr. Exp. Sta. Bull.* 305, 1937.

65. Quick, C. R. "How long can seed remain alive?" in *Seeds*, USDA Yearbook, 1961, pp. 94–99.

66. Anonymous. "10,000 year old seeds germinate." *Crops and Soils* 23, 8(1971):30–31.

67. Colbry, V. L., and others. "Tests for germination in the laboratory," in *Seeds*, USDA Yearbook, 1961, pp. 433–441.

68. Association of Official Seed Analysts. *Seed Vigor Testing Handbook.* AOSA, 2002.

69. McCarty, W. H., and C. Baskin. "Understanding and using results of cottonseed germination tests." *MS St. U. Ext. Serv. IS* 1364, 2003.

70. Musil, A. F. "Testing seeds for purity and origin," in *Seeds*, USDA Yearbook, 1961, pp. 417–432.

71. Rollin, S. F., and F. A. Johnston. "Our laws that pertain to seeds," in *Seeds*, USDA Yearbook, 1961, pp. 482–492.

72. Hudspeth, E. B., Jr., R. F. Dudley, and H. J. Retzer. "Planting and fertilizing," in *Power to Produce*, USDA Yearbook, 1960, pp. 147–153.

73. Dickey, E. C., and P. Jasa. "Row crop planters: Equipment adjustments and performance in conservation tillage." *NE Coop. Ext. Serv. NebGuide* G83–684, 1989.

74. Wicks, G. A., and N. L. Klocke. "Ecofarming: Spring row crop planting and weed control in winter wheat stubble." *NE Coop. Ext. Serv. NebGuide* G81-551-A, 1989.

75. Jenkins, M. T. "A comparison of the surface, furrow, and listed methods of planting corn," *J. Am. Soc. Agron.* 26(1934):734–737.

76. Salmon, S. C. "Seeding small grain in furrows," *KS Agr. Exp. Stu. Tech. Bull.* 13, 1924, pp. 1–55.

77. Dickey, E. C., and others. "Ridge plant systems: Equipment." *NE Coop. Ext. Serv. NebGuide* G88-876-A, 1988.

78. Martin, J. H. "Factors influencing results from rate and date of seeding experiments with winter wheat in the western United States," *J. Am. Soc. Agron.* 18(1926):193–225.

79. Janssen, G. "Effect of date of seeding of winter wheat on plant development and its relationship to winterhardiness," *J. Am. Soc. Agron.* 21(1929):444–466.

80. Kiesselbach, T. A. "The relation of seeding practices to crop quality," *J. Am. Soc. Agron.* 18(1926):661–684.

81. Robertson, D. W., and G. W. Deming. "Date to plant corn in Colorado," *CO Agr. Exp. Sta. Bull.* 238, 1930.

82. Edmisten, K. "Late planted cotton." *Carolina Cotton Notes.* CCN-97–5a. 1997.

83. Bilbro, J. D., and L. L. Ray. "Differential effect of planting date on performance of cotton varieties on the High Plains of Texas, 1960–65." *TX Agr. Exp. Sta.* MP-934, 1969.

84. Kerby, T. A., S. Johnson, and K. Hake. "When to replant." *California Cotton Review. Univ. of CA Coop. Ext.* 9(1989):7–8.

85. Silvertooth, J. C., J. E. Mulcuit, D. R. Howell, and P. Else. "Effects of date of planting on the lint yield of several cotton varieties planted at four locations in Arizona." *Cotton, A College of Agriculture Report. Series P-77.* Univ. of Arizona. 1989, pp. 69–72.

86. Dana, B. F., H. E. Rea, and H. Dunlavy, "The influence of date of planting cotton on the development of root rot," *J. Am. Soc. Agron.* 24(1932):367–377.

87. Mooers, C. A. "planting rates and spacing for corn under southern conditions, " *J. Am. Soc. Agron.* 12(1920):1–22.

88. Pelton, W. L. "Influence of low seeding on wheat in southwestern Saskatchewan," *Can. J. Plant Sci.* 74(1969):33–36.

89. Guitard, A. A., J. A. Newman, and P. B. Hoyt. "The influence of seeding rate on the yield components on wheat, oats, and barley," *Can. J. Plant Sci.* 41(1961):751–758.

90. Woodard, R. W. "The effect of rate and date of seeding of small grains," *Agron. J.* 48(1956): 160–162.

91. Riveland, N. R., E. W. French, B. K. Hoag, and T. J. Conlon. "The effect of seeding rate on spring wheat yield in western North Dakota-an update," *ND Farm Res. 37*, 2(1979):15–20.

92. Elmore, R. W., and J. E. Specht. "Soybean seeding rates." *NE Coop. Ext. Serv. NebGuide* G99-1395-A, 1999.

93. Helsel, Z. R., and H. C. Minor. "Soybean Production in Missouri." *U. of MO Ag. Pub.G4410.* 1993.

94. Nafziger, E. D. "Corn planting date and plant population." *J. Prod. Ag.* 7(1994):59–62.

95. Klages, K. H. "Spacing in relation to the development of the flax plant," *J. Am. Soc. Agron.* 24(1932):1–17.

96. Tisdale, H. B. "Effect of spacing on yield and size of cotton bolls," *J. Am. Soc. Agron.* 20(1928):298–301.

97. Ahlgren, H. L. "The establishment and early management of sown pastures," *Imperial Bur. Pastures and Forage Crops Bull.* 34, 1945, pp. 139–160.

98. Murphy, R. P., and A. C. Arny. "The emergence of grass and legume seedlings planted at different depths in five soil types," *J. Am. Soc. Agron.* 31(1939):17–28.

99. Stitt, R. E. "The effect of depth of planting on the germination of soybean varieties," *J. Am. Soc. Agron.* 26(1934):1001–1004.

Harvest of Field Crops

■ 8.1 GRAIN AND SEED CROPS

8.1.1 Physiological Maturity

Physiological maturity occurs when plants cease depositing dry matter into the developing seeds. At that point, maximum yield is achieved. However, at physiological maturity most crop seeds are too high in moisture, about 30 to 40 percent, for mechanical threshing and storage. Harvest maturity, when seeds are dry enough for mechanical threshing, occurs when seeds have a moisture content of 20 to 25 percent or less.

Small grains reach physiological maturity when they reach approximately the hard-dough stage or when the moisture content of the grain drops below about 40 percent.[1,2,3] Further ripening consists of continued drying with additional transport of nutrients into the kernel. Ripening is not entirely uniform among different heads or different grains within a head. Consequently, growth may continue until the average moisture content is appreciably below 40 percent. Small grains at the hard-dough stage have a moisture content of 25 to 35 percent, the heads are usually light yellow, and the kernels are too firm to be cut easily with the thumbnail.

Corn has reached physiological maturity when a black layer forms at the base of the seed embryo.[4] It is visible when the seed is cut lengthwise through the embryo (Figure 8.1). This layer is the placenta that regulates movement of dry matter from the plant to the developing embryo. When dry matter stops moving across the placenta, it dies, turns black, and forms an abscission zone. However, black layer formation will occur if dry matter ceases to move across the placenta even if the seed has not reached full dry matter accumulation. This can occur if the plant is damaged from freezing temperature prior to normal maturity. When evaluating damage from an early fall freeze, the seeds need to be checked for black layer formation within hours of the freezing temperatures.[5]

Soybean has reached physiological maturity when about 95 percent of the pods have reached their mature pod color.[6] All the leaves will probably be lost from the plant at this stage. About five to ten days of drying weather will be needed to dry the seeds to harvest maturity of about 15 percent moisture.

Sorghum has reached physiological maturity when a dark spot appears on the opposite side of the kernel from the embryo. This is caused by death of the placenta as described for corn.[7]

KEY TERMS

Harvest maturity
Hay crop
Physiological maturity
Silage
Silo
Stover
Straw

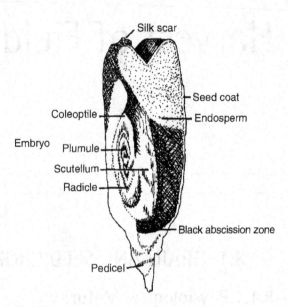

FIGURE 8.1

Mature corn kernel showing black layer (abscission zone). [Courtesy Richard Waldren]

8.1.2 Effects of Premature Harvest

Premature harvest reduces both yield and quality in most grain and seed.[8] Underdeveloped grains are low in test weight, starch content, and market value.[9]

Small grains may be cut with a swather or windrower seven days or more before they are ripe and allowed to dry under cool humid conditions. They can also be cut three to four days before normal harvest under warm dry conditions without appreciable loss in yield or quality.[10] When nearly ripe, wheat kernels sometimes draw material from the straw after it is cut. Considerably more growth has occurred when immature barley grains were left to dry in the head than when they were threshed immediately.[11] However, others[9] have found no significant transfer of material from the straw to the grain during the curing process.

Most forage grasses and legumes ripen irregularly, and the early ripened seeds often shatter. The maximum recovery of marketable seed is obtained by mowing or windrowing when the average seed moisture content is somewhere within the range of 25 to 45 percent, depending upon the species.[12] The effect of moisture content upon field losses at the time of harvest in perennial ryegrass is seen in Figure 8.2.

8.1.3 Effects of Delayed Harvest

When harvested with a combine, small grains must stand in the field for five to ten days after the period of windrow harvest. The additional time is needed to insure the moisture content of the grain has dropped to 14 percent or less, a necessity for safe storage without artificial drying. When wheat grain has a moisture content of 13 percent or less, the rachis of the wheat spike breaks readily, and the straw will burn freely. Losses from delayed harvest are caused by shattering, crinkling, lodging, and leaching.

Cereal stems are likely to break over (lodge) soon after maturity, especially in damp weather.[13, 14, 15] Oat is more susceptible than barley to lodging, while barley is more susceptible than wheat or rye. Most tall-stalked grain sorghums lodge soon after maturity or after a frost, whereas flax stands erect for long periods.

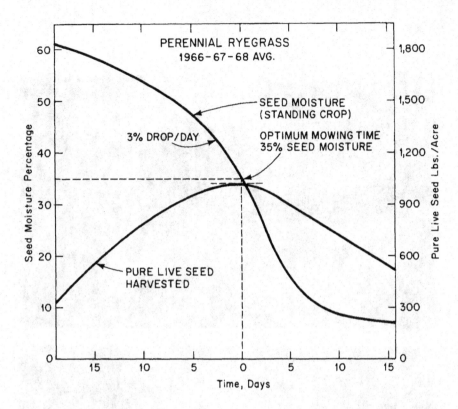

FIGURE 8.2
The influence of seed moisture at mowing time on combine yields of perennial ryegrass.

Weathered or sun-bleached grain is unattractive and often brings a lower grade and price on the market. Ripe grains exposed to wet and dry weathering for long periods in the field are lower in test weight because the grains swell when damp and do not shrink to their original volume after drying. The kernels then are less dense, but the baking quality of the flour from such grains may not be impaired.[16, 17]

Soybean will shatter if not harvested in a timely manner. Newer varieties resist shattering, but it can still be a problem if harvest is delayed during warm, dry weather. Corn will drop its ears or lodge if harvest is delayed, especially if a hard freeze has occurred.

8.1.4 Methods of Harvesting

The reaper replaced the hand sickle, scythe, and cradle for harvesting grains. The binder was put on the market between 1880 and 1890.[18] The header soon introduced still greater savings in the harvest of small grains. Now combines are used in the important grain-growing regions of the world.

Horse-drawn combines (combined harvester-threshers) were used in the Pacific Coast States to harvest wheat and barley before 1880. Their use, usually tractor-drawn, spread across the intermountain states 25 to 35 years later. Combine harvesting began in the Great Plains states during World War I and in other grain-growing states by 1930. By 1971, 96 percent of all combines sold were self-propelled.

Combines (Figure 8.3) are used for the harvest of nearly all small grain, grain sorghum, flax, canola, soybean, cowpea, and safflower in the United States. It has been used since 1960 in harvesting about 95 percent of the dry peas, beans, and seed

FIGURE 8.3
(Top) Harvesting wheat with a slat-type reel. *(Bottom)* Harvesting corn with a row-crop header. [Photos courtesy John Deere & Co.]

FIGURE 8.4
A sectional view of a self-propelled combine. [Courtesy Case IH]

crops of the forage legumes and grasses. About 20 percent of the combine harvesting is done by custom operators. With special attachments for row crops (Figure 8.3), the combine harvests and shells most of the corn.[19, 20, 21] Row crop heads are also used with soybean and other row crops. The grain is collected in a tank from which it is conveyed into a truck or trailer. The straw or stover is usually spread evenly over the ground. When the straw or stover is saved for feed or bedding, it may be deposited in piles or windrows. A self-propelled combine is illustrated in Figure 8.4.

WINDROW PICK-UP MODIFICATION The windrower or swather cuts the crop and places it on the stubble in a windrow (Figure 8.5). After drying, the windrow is gathered by a combine that has a pick-up device attached to the cutter bar (Figure 8.6). The most satisfactory stubble height for windrowed grain is about one-third the total length of the straw.[15] When cutting is too high, the short-stemmed heads will drop to the ground. A fairly heavy windrow may stay in good condition for 30 days, while a light windrow works down into the stubble where the grain may start to sprout.[22] The windrowed grain is usually dry enough to thresh in about five days using a combine with a pick-up attachment.

The windrower is mainly used for flax, peas, field bean, buckwheat, rapeseed and canola, crambe, and small-seeded legumes and grasses.[23] Swathing small grains is advisable in fields that contain many green weeds, are ripening irregularly, or need to be cut early to avoid shattering or crinkling. Green weeds dry in the windrow and thus do not increase the moisture content of the grain or interfere with threshing.

When harvested with the crop, about half the excess moisture in weeds is transferred to the wheat in the first 24 hours of storage and the remainder within 7 days.[15] Damp grain is often dried with forced heated air (Figure 8.7). A fuel shortage or high fuel prices may decrease the amount of artificially dried grain.

OTHER HARVESTING MACHINES Except for seed corn, picker-shellers are seldom used anymore for harvesting corn. Germination of seed corn is improved when the seeds are dried on the cob.

FIGURE 8.5
Self-propelled swather cutting alfalfa. [Courtesy John Deere & Co.]

FIGURE 8.6
Combine with pick-up attachment threshing from window. [Courtesy USDA]

Bean harvesters cut off the roots of beans and similar legumes just below the soil surface and turn the crop from two or more rows into a single windrow. After curing, the beans can be threshed with a combine equipped with a pick-up attachment (Figure 28.4). The first stage of peanut harvest uses machines that dig up the plants

FIGURE 8.7
Continuous grain dryer for reducing grain moisture during storage. [Courtesy Delux Mfg. Co.]

and pods and invert them (Figure 29.6). After drying, the peanut pods are threshed with a combine or harvester.

Field silage cutters and forage harvesters (Figure 8.13) are widely used. They cut the crop and chop it into short lengths in one operation.

Seed of some forage legumes and grasses can be harvested directly with a combine.[24] A suction seed reclaimer, which is attached to the combine, can gather much of the seed that has shattered on the ground (Figure 8.8). Seed of alfalfa, clovers, and certain grasses is combined about three to five days after the crop has been sprayed with a chemical desiccant. This treatment stops growth and dries the leaves and stems. Chemical desiccation before harvest is effective in warm, dry climates. Alfalfa, clover, and certain other seed crops may also be harvested with a swather or a mower with a windrowing attachment.

The seed of Kentucky bluegrass and other grasses used to be harvested with strippers. These machines have revolving spiked cylinders that knock the seed from standing plants into hoppers. The pneumatic type of stripper blows the heads of the plants into the cylinder. The stripped material is then dried.

8.1.5 Straw and Stover

Straw is comprised of the dried stalks or stems and other parts of various crops from which the seed has been threshed in the ripe or nearly ripe stage. Some of the small-grain straw is gathered or saved, usually with a pick-up baler. Additional quantities are recovered by livestock that graze the stubble fields. Most of the harvested straw is used for livestock bedding or for mulching.

Stover is the corn or sorghum plant that remains after the ear or head has been removed. Typical mature corn is 50 to 65 percent stover, while grain sorghum is 45 to 75 percent stover depending on its height.

FIGURE 8.8
Suction seed reclaimer that gathers seed that fall on the ground before or during harvest.

Straw and stover are the least nutritious of all substances used as feed for livestock. Animals cannot live on straw alone unless considerable waste grain has been left in the straw. However, straw is an important supplement in livestock maintenance. Oat and barley straw are usually considered more valuable for feed than either wheat or rye straw. Corn stover is more digestible than the straw of small grains, but waste is significant unless it is shredded. One-third of the total digestible nutrients produced in a corn field remain in the stover after the grain is harvested.

Small-grain straw is bulky and has a low nutrient value. It supplies organic matter when returned to the land. Table 8.1 shows the nutrient content of wheat and barley straw. Approximately a ton (907 kg) of straw is produced for each 20 bushels (545 kg) of yield. Because straw from small grains has a very high C:N ratio, nitrogen may be immobilized as microorganisms use it during decomposition. An application of nitrogen fertilizer may be needed to hasten decomposition and reduce yield depression after a heavy straw residue. In conservation tillage systems, additional nitrogen is usually not needed after the first year as mineralization of existing residues releases nutrients from the straw. Nitrogen immobilization is usually not a problem from the straw or stover of other crops.

In 1924, it was estimated that approximately 15 percent of the straw of small grains was burned, mostly in areas of the United States where the livestock population was small, or where it could not be added advantageously to the land.[25] Western farmers often burned their straw and stubble before plowing because of the difficulty in disposing of a large volume of straw. Burning of straw is objectionable because of air pollution, and is not recommended because repeated burning increases wind and water erosion. Today, most of the straw is scattered over the land by straw spreaders attached to combines. It is then tilled under or left on the surface by conservation tillage implements.

Field burning of residues after the harvest of seed from perennial ryegrass and other grasses has been widely practiced in Oregon and Washington as a means of

TABLE 8.1 Average Nutrient Content per Ton of Wheat, Barley, and Soybean Straw

Crop	N (lb)	P_2O_5 (lb)	K_2O (lb)
Wheat	12	3.7	20
Barley	15	4.1	41
Soybean	18	2.4	18

controlling weeds, blind seed, ergot, nematodes, rust, and other leaf diseases. In California, burning was used to destroy rice straw residues. Burning after harvest also aids in the control of the sod webworm, silvertop, meadow plant bugs, thrips, mites, and other insects.

Burning of residues has been discontinued because of concerns for air pollution. Oregon passed legislation in 1991 mandating a phasing out of residue burning. From 1988 to 1997, Willamette Valley, Oregon, grass seed growers reduced the number of acres burned by more than 70 percent. The reduction in field burning has occurred without a loss in seed yield or quality. Grass seed planting increased from about 333,000 acres (135,000 ha) in 1988 to about 411,000 acres (166,000 ha) in 1997. Sales increased from $190 million in 1988 to more than $300 million in 1997. Baling of seed crop residue has created a grass straw export market. In 1997, approximately 331,000 tons (300,000 MT) were exported and another 30 tons (27 MT) were sold in domestic markets. This new commodity is valued at about $15 million.

Instead of burning residues, producers have resorted to three methods: (1) Much of the straw left after harvest is baled and removed for sale. The remaining residue is chopped and dispersed, allowing the application of chemicals to control weeds and disease organisms. (2) Straw is baled and removed, and a vacuuming machine sucks up leftover straw, seeds, and disease organisms. (3) With the most recent development, producers leave all the straw in the field and chop it several times to spread it out across the field so that the crop is not smothered. The straw mulch helps control some weeds.[26]

Other methods of straw disposal under investigation include its use in feeds, paper, fiber board, fuel, and energy. It is estimated that eastern Washington grain growers produce about 3 million tons (2.7 MT) of straw that is economically feasible to harvest. That much straw has the potential for producing 400 to 425 megawatts of energy each year.[27]

The straw of cowpea, soybean, and other leguminous crops are superior to cereal straw in feed and fertilizer value. They are comparatively high in nitrogen and calcium but similar to cereal straw in other elements (Table 8.1).

▓ 8.2 HAY AND HAY MAKING

8.2.1 Principal Hay Crops

The distribution of forage crops for hay in the United States is shown in Figure 8.9. The principal hay crops are: (1) alfalfa, including alfalfa-grass mixtures, (2) clover and timothy or other grass mixtures, (3) wild hay, (4) grain hay, (5) coastal bermudagrass, and (6) lespedeza. Formerly important hay crops were soybean, cowpea, and peanut vines. Other hay crops include tall fescue, bromegrass, orchardgrass,

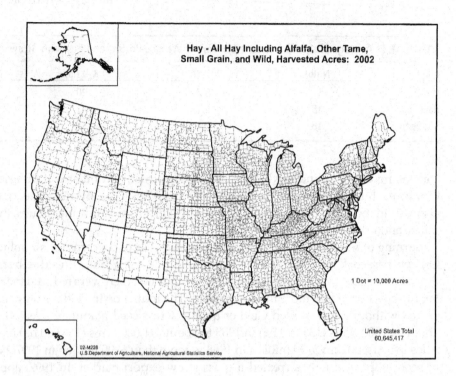

Hay - All Hay Including Alfalfa, Other Tame,
Small Grain, and Wild, Harvested Acres: 2002

1 Dot = 10,000 Acres

United States Total
60,645,417

02-M226
U.S.Department of Agriculture, National Agricultural Statistics Service

FIGURE 8.9
Land from which hay was harvested in 2002. [Source: 2002 U.S. Department of Agriculture]

sudangrass, johnsongrass, crested wheatgrass, sweetclover, vetch, and peas. In 1971, more than 62 million acres (25 million ha) were harvested for hay in the United States. That was 4 million acres (1.6 million ha) less than 60 years earlier, but the production was 55 percent higher. In 2000–2003, U.S. hay production averaged 154 million tons (140 million MT) from 63 million acres (25 million ha). The leading states in hay production are Texas, California, Missouri, South Dakota, Minnesota, and Kansas. Alfalfa production averaged 79 millions tons (72 million MT) from 23 million acres (9 million ha).

Hay is by far the most valuable harvested forage because it does not deteriorate appreciably in storage, it can be shipped and marketed, and it provides a large intake of dry-matter nutrients for livestock.

Hay quality equals nutritive value. The important physical factors[28] that can be practically measured are color, leafiness, maturity when cut, amount of foreign material, condition, and texture. These factors are generally correlated with palatability as well as chemical composition.

8.2.2 Hay Making

Hay is cut with tractor mowers, mower-conditioners, swathers (Figure 8.5), or mower-crusher-windrowers (Figure 8.10). Hay is cured (most of it conditioned by crushing) in the mowed swath or windrow, or in windrows after raking, and occasionally in bunches gathered from the windrow with a stacker, hay rake, or buckrake. Mobile stacking machines load the hay on mechanized trailers that can unload or reload stacks. The windrowing of mowed swaths is mostly done with a side-delivery or finger wheel rake (Figure 8.11) or with a buncher attached to the

FIGURE 8.10
Hay conditioner that hastens drying in the field. The crop is cut, crushed, and windrowed in one operation. [Courtesy John Deere & Co.]

FIGURE 8.11
Wheel rakes. [Courtesy John Deere & Co.]

mower.[18, 29] When the hay is wetted by rain while in the swath or windrow it may be turned over to dry, preferably with a side-delivery rake after partial drying.

Windrowing relatively green alfalfa hay without crushing extends the curing period as compared with swath curing after crushing.[20, 30] Prolonged swath curing may result in a higher loss of leaves and a reduction of almost 10 percent in weight. Mowed alfalfa is best cured by windrowing after it has dried down to 50 percent moisture. The crop should then be raked before it becomes dry enough to shatter the leaves, which sometimes occurs before the hay is dry enough to store. Only a small amount of hay is artificially dried.

8.2.3 Stage of Maturity for Harvesting

The stage of maturity at which hay crops are cut affects color, leafiness, and other factors of quality. It is impossible to produce high-quality hay from late-cut grasses and legumes because lignin (tough fiber) increases as the plants mature.

PLANT STAGE FOR HARVEST The more immature a forage crop is when cut, the smaller the yield and the more palatable and nutritious the product. Because of smaller yields along with increased labor costs from frequent cutting, some sacrifice of quality must be made in the interest of greater yields. Consequently, plants are harvested for hay at an intermediate stage when neither yield nor quality is at its maximum. The best time to harvest grasses and clover for hay is ordinarily between early bloom and full bloom.[28] Alfalfa is best cut at about one-tenth bloom (early bloom).[31] Sweet-clover should be cut in the bud stage to avoid subsequent coarse growth.[32] Crops that are heavy producers of seed, such as cereals, should be cut for hay when the grain is in the soft- to medium dough stage,[33] soybean when the beans are about half grown, and cowpea when the first pods are mature. When a hay of very high quality is required, grasses and alfalfa are sometimes harvested before they bloom. Care should be taken to not cause injury from too frequent cutting or consistently cutting alfalfa before the bud to one-tenth bloom stage. Excessive depletion of food reserves in the roots may cause a reduction in the vigor of the stand.[34, 35]

CHEMICAL COMPOSITION The composition of hay changes as plants approach maturity. The amounts of protein and minerals are higher, and the amounts of the less valuable crude fiber are lower in young than in old plants. Changes in alfalfa[36] and timothy[37] are given in Table 8.2.

Crude fiber (mostly cellulose) of young plants, as in immature pasture herbage, appears to be largely digestible. As the plant matures, a progressively greater proportion of the crude fiber is composed of less digestible lignin, which lowers the net nutritive value of the forage. Increased growth of small grains beyond the medium dough stage is more than offset by shattering of kernels, loss of leaves, leaching, and general deterioration of plant structure. Close associations are evident between the contents of fiber and stems; of total ash and calcium, and leaves; and of phosphorus, crude protein, nitrogen-free extract, and heads.[33]

Carotene content (a measure of vitamin A potency) is higher in early stages of leafy growth because the leaves contain more carotene than the stems, and old leaves lose carotene. In alfalfa, clover, and timothy hays, vitamins B and riboflavin decrease as the plant matures.[38] In general, these vitamins are correlated with the leafiness, greenness, and protein content of the plant.

TABLE 8.2 Effect of Stage of Maturity at Cutting on Chemical Composition of Alfalfa and Timothy Hay

Stage of Maturity	Ash (%)	Crude Protein (%)	Crude Fiber (%)	Nitrogen–Free Extract[a] (%)	Ether Extract (Fat) (%)
Alfalfa					
Prebloom	11.24	21.98	25.13	38.72	2.93
Initial bloom	10.52	20.03	25.75	40.67	3.03
One-tenth bloom	10.27	19.24	27.09	40.38	3.02
One-half bloom	10.69	18.84	28.12	39.45	2.90
Full bloom	9.36	18.13	30.82	38.70	2.99
Seed stage	7.33	14.06	36.61	39.61	2.39
Timothy					
No heads showing	8.41	10.18	26.31	50.49	4.61
Beginning to head	7.61	8.02	31.15	49.14	4.07
Full bloom	6.10	5.90	33.74	51.89	2.38
Seed harmed	5.54	5.27	31.95	54.12	3.13
Seed all in dough	5.38	5.06	30.21	56.48	2.87
Seed fully ripe	5.23	5.12	31.07	55.87	2.72

[a]Mostly carbohydrates; chiefly starches, hemicelluloses, and sugars.

Cutting hay in late afternoon may avoid the depletion of food materials from the leaves and stems by translocation to the roots and by respiration during the night when photosynthesis is suspended. However, it is many times more practical to cut hay in the morning despite these assumed small losses.

8.2.4 Field Curing of Hay

The aim of harvesting hay is to dry the crop to 25 percent moisture or less with as little loss of leaves, green color, and nutrients as possible. Hay crimpers, crushers, or conditioners crush the stems between rolls, which shortens the drying time by one-third to two-thirds. This reduces curing losses. The crushers are towed by a tractor during or after mowing, or a combination mower-crusher is used (Figure 8.10).[18] Loss of leaves is more pronounced with legumes than with grasses, especially when the moisture content is 30 percent or less. The loss of alfalfa leaves begins when the moisture content of the hay drops below 40 percent.

In the first century AD, Columella described the curing of hay in a manner which indicates that hay harvest has changed very little.

It is best to cut down hay before it begins to wither; for you gather a larger quantity of it, and it affords a more agreeable food to cattle. But there is a measure to be observed in drying it, that it be put together neither over-dry, nor yet too green; for, in the first case, it is not a whit better than straw, if it has lost its juice; and, in the other, it rots in the loft, if it retains too much of it; and often after it is grown hot, it breeds fire and sets all in a flame. Some times also, when we have cut down our hay, a shower surprises us. But, if it be thoroughly wet, it is to no purpose to move it while it is wet; and it will be better if we suffer the uppermost part of it to dry with the sun. Then we will afterwards turn it, and, when it is dried on both sides, we will bring it close together into cocks, and so bind it up in bundles; nor will we, upon any account, delay to bring it under a roof.

The leaves of alfalfa apparently do not withdraw water from the cut stems. The moisture loss in alfalfa stems with the leaves attached is about the same as when the leaves are removed.[20, 39] In legumes with large stems, such as sweetclover and soybean, the leaves may aid in the withdrawal of water from the stems.[1]

8.2.5 Losses in Field Curing of Hay

Profound changes in composition occur during the curing of hay in the field.[28, 40] There is an inevitable loss of nutrients when crops are cured for hay. Curing losses are small when drying is rapid. Under adverse conditions, the loss can be 40 percent or more. In some severe conditions, spoilage may be complete. Under normal weather conditions, dry matter loss in field curing of hay may be 10 to 25 percent in different parts of the country (Table 8.3). A loss of 17.6 percent of the dry matter and 21.5 percent of the protein has occurred during the ordinary field curing of alfalfa[40] due to shattering of leaves, pods, seeds, and stems, to leaching and fermentation, and to the respiration that continues in green plants for some time after cutting. Included in these losses was 30 percent of the leaves, with a correspondingly large loss of nutrients. The leaves are the portion of the plant richest in protein, vitamins, phosphorus, and calcium. The leaf loss in alfalfa ranges from 6 to 9 percent of the weight of the total crop. In cereal hay,[33] the leaves constitute one-fifth the weight of the plant. After the milk stage, cereal leaves contain nearly 50 percent of the mineral and 40 percent of the fat of the whole plant.

During curing, carotene starts to decompose immediately by oxidation. Alfalfa hay exposed in the swath and windrow for 30 hours in good curing weather has lost 60 to 65 percent of the carotene, 25 percent of the protein, 15 percent of the leaves, and 10 percent of the total dry matter. The losses in artificially dried hay are less than in sun-cured hay of alfalfa, lespedeza, sorgo, and soybean. Baled alfalfa, timothy, and clover hays stored in a dark, unheated barn may lose 3 percent of their carotene per month in the winter, but the losses are much higher with higher temperatures.[41] The percentage rate of loss of carotene was much more rapid than that of the natural green color. Hay exposed to rain loses a considerable proportion of its vitamin C. Synthesis of vitamin D occurs only when alfalfa is cured in the sun, a condition that results in a loss of vitamin A.

The protein content of alfalfa hay exposed to 1.76 inches (45 mm) of rain distributed over a period of fifteen days was only 11 percent compared with prime hay that

TABLE 8.3 Losses in Harvesting and Preserving Forages

	Losses (%)		
	Field Loss	Storage Loss	Total
Direct cut grass silages	2–3	18–22	20–25
Haylage—65% moisture	11–13	8–12	19–25
Haylage—50% moisture	11–13	3–8	14–21
Baled alfalfa—raked at baling	30–35	2–4	32–39
Baled alfalfa—direct windrowed or raked at 50–60% moisture	12–15	2–4	14–19
Legume-grass—raked at baling	18–23	2–4	20–27
Legume-grass—windrowed or raked at 50–60% moisture	12–18	2–4	14–22

contained 18.7 percent protein.[42] Leached burclover hay, oat hay, and naturally cured range forage showed the greatest percentage loss in minerals.[43] The loss of crude protein ranged from 1 to 18 percent of the total, and the loss of nitrogen-free extract, from 6 to 35 percent of the total according to the nature of the forage. Further losses in nutritive value are probably reflected in impairment of palatability and in loss of color.

8.2.6 Hay Processing

Less than 15 percent of the 1939 hay crop in the United States was baled. By 1959, 82 percent was baled, 11 percent was chopped, and only 7 percent was loose hay stored in outdoor stacks or in barns.[44] Most of the loose hay is wild hay on western ranches where livestock feed on the stacks during the winter.

Baling is done with an automatic field baler, which gathers the hay from the windrow or swath and compresses it (Figure 8.12). Small rectangular or round bales usually weigh 25 to 85 pounds (11 to 39 kg). A throw-type baler ejects the bales into a trailer towed behind the baler, requiring only one person to operate the machine.[18] Other types force the bales up a chute into a trailer or truck or drop the bales on the ground. Many of the dropped bales are later loaded mechanically. Large round bales usually weigh 1,000 pounds (454 kg) or more. Large rectangular bales can weigh from 700 to over 2,000 pounds (318 to over 907 kg) depending on size. These are picked up and loaded onto trucks with special mechanical devices. Bales must be removed from the field and stacked quickly to maintain the quality of the hay. In very dry alfalfa hay (less than 15 percent moisture content),

FIGURE 8.12
Large round baler. [Courtesy John Deere & Co.]

FIGURE 8.13
Forage harvester for gathering green material for silage, soilage, or dehydration. [Courtesy Case IH]

yield losses for large round balers can exceed 25 percent, while a small rectangular baler in the same hay seldom has losses over 5 percent.[45]

Chopping is done with forage harvesters that gather the hay from the swath or windrow, chop it, and blow the material into a trailer (Figure 8.13). When it is still damp, chopped hay can be dried artificially.

Considerable baling and chopping is done by custom operators because this system is more economical than owning the machines unless at least 100 to 200 tons (91 to 181 MT) of hay are harvested each season.

Machines for cubing, pelleting, or wafering hay in the field have been devised for processing hay that contains 12 to 16 percent moisture. The mobile field machines have a lower capacity than field balers or forage harvesters.[18]

Cubed or pelleted alfalfa allows cattle to eat more with less waste than when fed alfalfa hay. Feeding can be done mechanically, and the feed can be stored in less than half of the space required for baled hay.[46, 47] The cubes and pellets are also used in mixed ground feeds. Alfalfa cubes measure $1\frac{1}{4} \times 1\frac{1}{4} \times 2\frac{1}{2}$ inches ($3 \times 3 \times 6$ cm) in size. Pellets are $\frac{1}{4} \times \frac{3}{8}$ inch (0.6×1 cm). Cubing is done in either mobile field machines or in stationary units that also may cube baled hay after the harvest season. When cubing hay of about 10 percent moisture content, it is sprayed with water to raise the moisture content to 14 to 16 percent as it enters the machine. Then the hay is rolled, chopped, and compressed through dies. The emerged cubes are then dried down by heated air to a moisture content of 12 to 14 percent.

8.2.7 Spontaneous Heating of Hay

Some heating is likely to occur when slightly damp hay is stored. Microorganisms, which are able to multiply when the hay is not too dry, consume the material for growth and energy and release heat and moisture. When this moisture is sufficient to be detectable, the hay is said to be going through a *sweat*. Very dry hay and hay in small piles that allow the heat and moisture to be dissipated are not known to sweat. The moisture released by organisms stimulates further microbiological activity and fermentation and contributes to increased heating and sweating. Loose hay that contains less than 25 percent moisture is sufficiently dry at the time of storage to heat or sweat only moderately. Such hay retains its green color and nutritive value. The storage of alfalfa hay with high moisture content causes heavy losses of organic substances, mostly fat, sugar, and hemicellulose. The loss in total weight may be as high as 22 percent.[22]

Stored wet hay is subject to excessive spontaneous heating. Moisture content greater than 25 percent may cause the hay to become brown or black. The green color of hay is destroyed when heating temperatures exceed 122°F (50°C). Clean brown hay suitable for feeding is formed at temperatures above 131°F (55°C) but below 158°F (70°C). Dry matter losses can be very small in brown hay but can be heavy when the hay becomes black. In Kansas research,[48] alfalfa hay stacked with 53 percent moisture sustained a dry matter loss of 39 percent, and a large proportion became black.

Damp hay stored in large masses may develop temperatures sufficiently high to produce ignition.[40, 49] The initial production of heat is due mainly to the action of microorganisms that are capable of raising the temperature of the hay to as high as 158°F (70°C), or slightly higher. Temperatures above this (the death point of microorganisms) are the result of fermentation that produces volatile substances. These gaseous substances can ignite when the temperature rises to 374°F (190°C) or higher.

8.2.8 Dehydrated Hay

Some alfalfa, and occasionally other hays, are dehydrated to produce a high-quality feed product. In dehydration, hay is put through a drier immediately after it is cut and chopped. Forced air in driers conducts heat from the burner to the hay.[10] The rotary drum or high-temperature drier with an inlet temperature, usually between 1,400 and 1,500°F (760 and 815°C), dehydrates the hay in a few minutes. The drum assembly is rotated slowly while the hay is agitated.

The dry meal may be pelleted or mixed with other feeds. In some cases, urea is added to make a high concentrate protein supplement. In addition, separation of the leaves and stems is a frequent operation, with the leaves being used as a protein and xanthophyll source by poultry feed manufacturers.

Baled hay with less than 20 percent moisture usually keeps without spoilage. It dries down until it reaches an equilibrium with the moisture in the air. Hay with more moisture may require artificial drying. Molding and heating occur when the moisture content is 30 percent or more.

8.2.9 Artificial Drying of Hay

High quality hay is usually achieved when the crop is allowed to dry for a day or less down to a 35 to 40 percent moisture content in the windrow. Then it is baled, chopped, or left loose and hauled in for drying down to a moisture content of about 15 percent. Some hay is dried in batches with heated air, either on a drying platform or in specially equipped drying wagons. Most of the drying is done in a barn by use of forced unheated air supplied by a large blower.[50] A system of ducts or slotted panels placed on a tight barn floor is covered with 8 feet (2.5 m) of loose or chopped hay or 10 to 14 layers of baled hay. The air that enters the system is forced upward through the hay. The volume of air for each cubic foot (28 l) of hay should be no less than 2 cubic feet per minute (57 l/min) for loose hay, and 2½ cubic feet per minute (71 l/min) for chopped or baled hay. Drying with heated air is faster than with unheated air, but it involves a fire hazard in barn drying. Prolonged heating of hay at temperatures above 140°F (60°C) lowers the digestibility of the proteins.

Freshly cut alfalfa usually contains 75 percent moisture. This requires the removal of 4,800 pounds (2.2 MT) of water to make one ton (907 kg) of hay that contains 15 percent moisture. When hay is allowed to wilt down to 40 to 50 percent moisture in the field, only 800 to 1,400 pounds (363 to 635 kg) of water needs to be removed by the drier. The temperatures that are permissible range from 250 to 275°F (121 to 137°C) where the air contacts the driest hay.

Nutrients in dehydrated hay are about equal to fresh green plants, except for the unavoidable loss of some carotene and the more volatile nitrogen compounds in the drying process. The dehydrated crop is 2.0 to 2.2 percent higher in protein and also usually higher in carbohydrates and ether extract than the same crop when sun cured.[10] The natural green color, an indicator of carotene content, is retained to a greater extent in artificially dried hay.

Hay dehydration has wider uses where losses from sun curing exceed 25 percent, especially when hay prices are relatively high.[28] The cost of hay dehydration has been high enough to preclude its use by the average farmer.

▨ 8.3 SILAGE

8.3.1 Economy of Silage

Silage is a moist feed that has been preserved by fermentation in the absence of air. The forage, usually green, is commonly chopped into small pieces and stored in a silo. The principal use of the silo is to preserve succulent roughage for winter feedings as well as to save forages that otherwise would be largely wasted, damaged, or lost. While it is impossible to preserve forage crops as silage as cheaply as in the form of hay, properly prepared silage will preserve a greater proportion of the nutritive value of the green plant.[28]

The modern practice of ensiling green forages can be traced directly to the process of making sour hay in Germany in the nineteenth century. Green grasses, clover, and vetches were stored in pits, salted at the rate of 1 pound per 100 pounds (453 g per 45 kg), thoroughly trampled, and covered. The first attempt to ensile green corn occured in Germany in 1861.[51] The first American silo was built in Maryland in 1876. The silo became popular and spread to all parts of the country.

8.3.2 Crops Used for Silage

Corn is the principal silage crop in the United States, with more than 6 million acres (2.4 million ha) producing more than 96 million tons (87 million MT) annually. More than one million acres (400,000 ha), producing more than 11 million tons (10 million MT) of sorghum silage, are harvested in average years. The best silage is made from carbohydrate-rich crops that contain more than two parts of carbohydrates to one part of protein.

CORN Corn is nearly the ideal crop for silage. The yields of silage range from 10 to 30 tons per acre (22 to 67 MT/ha). Corn for silage should be harvested from early dent stage to physiological maturity. Corn harvested before early dent will be too high in moisture, which will reduce silage quality (Figure 8.14).[52, 53, 54] The grain types are usually superior to late maturing "silage" corn, since the nutritive value of corn silage is closely associated with the proportion of grain it contains. Late maturing varieties yield a greater weight of silage per acre but less dry matter than do the grain varieties.[55] The best silage variety is one that utilizes the growing season to the best advantage in production of dry matter but at the same time reaches, at least three years in five, a stage of maturity that may be loosely described as the dough stage. There is less loss of dry matter in storage when corn is ensiled at the more mature stages.

Corn planted for silage will be planted earlier than corn planted for grain, and the seeding rate will be increased 2,000 to 3,000 plants per acre (4,900 to 7,400 plants/ha).[56] A corn crop harvested as silage removes more than twice as much nitrogen, three times as much phosphorus, and ten times as much potassium as when the crop is harvested for grain. The removal of the stalks accounts for the extra nutrient removed from the land. A 20 ton (18 MT) silage crop will remove approximately 175 pounds (79 kg) of nitrogen, 35 pounds (16 kg) of phosphorus, and 175 pounds (79 kg) of potassium.[56]

High-moisture corn grain can be ensiled, providing a high-energy feed to cattle. Storage methods commonly used for high-moisture corn consist of two main types. Ground high-moisture corn is normally stored in bunker or trench silos, whereas whole high-moisture corn is stored in upright sealed silos. Ground or coarse-rolled corn can also be stored in upright structures. With sealed silos, corn can easily be stored at a much lower moisture than is needed for bunker-stored corn.[57]

Corn grain stored in bunker silos should be harvested at moistures above 22 percent. The preferred harvesting moisture is around 24 to 26 percent. Corn stored by this method should be ground or rolled and well packed into the silo. Since proper packing depends on the moisture and particle size, corn that is to be stored in a bunker silo can be coarsely ground (as much as 40 to 50 percent whole corn passing through). However, as the moisture of the corn decreases below 25 percent, finer grinding may be necessary to achieve proper packing. Finer grinds also permit a slower feeding rate once the silo is opened.[57]

SORGHUM Sorghum replaces corn as a silage crop under dry or hot conditions, particularly in the southern Great Plains. Sorghum silage is slightly less palatable than corn silage. Cured sorghum and even damp threshed sorghum grain can be ensiled successfully. Water is added to the grain to raise the moisture content to about 30 to 50 percent, the material is covered, and the usual silage fermentation process

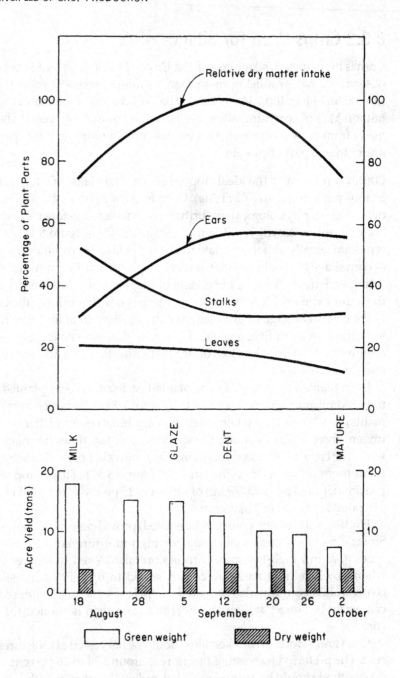

FIGURE 8.14
Corn nutrient yields as related to kernel development.

begins. Sorghum for silage should be cut when the seed is in the stiff-dough stage. The addition of 10 pounds of urea to each ton of sorghum silage increases its nutritive value.[58]

OTHER CROP PLANTS Forage grasses and legumes are widely used as silage crops, especially when weather conditions make it difficult to cure good hay. Grasses can be ensiled alone or with legumes almost as readily as can corn, provided the crop is dried for several hours before it is put in the silo. The material should be finely chopped, packed to exclude air, and possibly weighted. Timothy and other grasses ensiled this way make highly palatable silage with an agreeable odor without excessive losses of dry matter. They are cut for silage when heading begins.

Legumes (alfalfa, soybean, clover, field pea, and vetch) can be ensiled success-fully. Excellent silage can be made from alfalfa, and palatability tests show it to rank next to corn.[59] Alfalfa is cut at a stage of one-tenth to one-fourth bloom, clover at one-half bloom, and the large-seeded legumes when seeds are forming. Sweet-clover, while less palatable as silage than alfalfa, is satisfactory when harvested at a more mature stage than for hay. In one experiment, sweetclover silage contained 16 percent digestible protein and 57 percent total digestible nutrients on a dry ba-sis, as compared with 5 percent digestive protein and 70 percent total digestible nu-trients in corn silage.[60]

Sugarbeet tops can be made into silage. The chief characteristic of this silage is the enormous crude-ash content (ash plus dirt) (Chapter 36) ranging from 20 to 58 percent on a water-free basis. The composition of dirt-free tops on the basis of percentages in water-free material is ash, 15; protein, 15; nitrogen-free extract, 49; fat, 3; and fiber, 18.

Sunflower has been grown as a silage crop where the season was too short or too cool for corn or sorghum.[61] Addition of salt at the time of ensiling apparently re-duces or eliminates resinous odor and flavor and improves the palatability of sun-flower silage.[62] The crude protein content of the silage is highest when the seeds are three-fourths mature. Silage from sunflowers harvested at the bud or full-bloom stage is less palatable than that cut at later stages of maturity. The nutritive value increases as the plants approach maturity.

WEEDS AND OTHER PLANTS Silages made from wild sunflowers and certain other weeds with a rather strong characteristic odor, such as cocklebur and ragweed, are eaten by livestock with reluctance. Livestock refuse to eat silage made from woody plants with a strong undesirable characteristic odor or taste, such as gumweed and wild sage.[59]

Russian thistle (*Salsola kali*) has been used for making silage. It should be har-vested when the spines start to feel prickly but while still fairly soft. Analyses have shown Russian thistle to be equivalent to alfalfa in protein and fat content and su-perior in its carbohydrate:crude fiber ratio. Russian thistle has a high mineral con-tent with over 8 percent potash (K_2O). It is not recommended for silage when reasonably good yields of corn, sorghum, or sunflower can be obtained. In a ma-jority of cases, weeds are only used in an emergency.

8.3.3 Silage Formation

In a tight silo, respiration by microorganisms in the fresh silage uses up the oxygen in a few hours.[2] Then anaerobic bacteria start fermentation. Carbon dioxide increases rapidly for two to three days until it comprises 60 to 70 percent of the gaseous atmo-sphere, with nitrogen as most of the rest. Acid-forming lactic acid bacteria multiply rapidly from 2,000 to 4,000 thousand up to a billion in a gram of good silage.[63]

The temperature of the silage rises for about 15 days after which it subsides and then fluctuates somewhat with the air temperature. Top silage exposed to the air soon spoils. It may reach a temperature of 140°F (60°C). The temperature of the in-terior mass of silage usually does not exceed about 100°F (38°C).

Sugars are the principal food for the bacteria that cause fermentation, but the marked destruction of pentosans and starches indicates that these substances are also used in part. These carbohydrates are converted through the process of fer-mentation to alcohols and then to acids, principally lactic acid with some acetic and succinic acids and a trace of formic acid.[64] The lactic fermentation is the desired one.[28] Crops such as corn and sorghum have a high content of readily fermentable carbohydrates. Because of their content of bases, they soon develop an acidity of pH 4 or below. Forage grasses develop less acidity because of a lower amount of soluble carbohydrates and a higher content of basic elements. The legumes develop still less acid because of a small amount of soluble carbohydrates and a very high calcium content. Good-quality silage has a pH of 4.6 or lower. Poor-quality silage contains less lactic acid, but considerable butyric, acetic, propionic, and succinic acids, as well as ammonia. Butyric acid and ammonia impart bad odors and low palatability to the silage.

The average changes in composition induced in four silage crops by fermentation are shown in Table 8.4. The main difference after fermentation is a reduction in carbohydrates (nitrogen-free extract). This results in a proportionately higher content of the other components. Biological actions release moisture, but the surplus liquid often seeps out of the silo, carrying away dissolved substances. The pH drops from a range of 5.4 to 6.6 in freshly chopped materials, down to 3.6 to 4.6 in good silage. Poor silage fermentation breaks down much of the protein and amino nitrogen into less digestible ammonia forms.[17, 21, 56]

During the ensiling process, some of the green chlorophyll fades, but a fair proportion of the carotene is retained. Considerable carotene is lost in the swath during wilting before the silage is chopped.[28] Good silage has a pleasant aromatic, alcoholic odor and a sour taste. Silage development usually is completed within 12 days.[63]

Losses in dry matter during the fermentation of good silage range from 5 to 20 percent, with an average of 10 percent or more. From 5 to 10 percent of the dry matter is lost as gasses, while seepage losses of similar amounts occur in silages with moisture contents of 72 to 82 percent. Little or no seepage loss occurs in wilted forage. Losses from spoilage depend largely upon the access of air to the silage. This may range from less than 1 to 2 percent in a good silo and up to 40 percent in uncovered silos.[17, 21]

Spoilage is greatly reduced when the silage, except in airtight silos, is covered with a plastic sheet to exclude air (Figure 8.14). The sheet is then weighted down

TABLE 8.4 Typical Composition of Four Crops and Their Silages

Material	Composition of the Dry Matter					
	Moisture (%)	Crude Protein (%)	Ether Extract (%)	Crude Fiber (%)	Ash (%)	N-free Extract (%)
Corn	62.9	8.1	2.9	16.8	5.1	67.0
Corn silage	69.5	8.3	2.8	22.5	7.1	59.4
Alfalfa	75.9	19.6	2.1	26.9	7.9	45.1
Alfalfa silage	76.0	20.1	3.4	30.4	8.4	38.0
Orchardgrass	78.1	16.8	3.4	27.0	8.0	44.8
Orchardgrass silage	78.8	17.3	5.0	30.4	8.4	38.6
Grass-clover	81.7	18.1	2.7	22.0	10.3	47.8
Grass-clover silage	76.9	19.7	4.9	27.7	9.3	40.6

with sawdust, other waste material, or old tires. Any hole or crack in the silo or in the covering causes spoilage for some distance from the opening.[65]

8.3.4 Types of Silage

Direct-cut grass and legume silages usually contain 72 to 82 percent moisture. This involves the hauling and filling of a large volume of water with each ton of dry matter. Such a high-moisture content causes large dry matter losses and often, but not always, results in poor silage of high pH and considerable butyric acid. Good high-moisture silage is obtained by using additives such as ground shelled corn or dried beet pulp at 100 to 150 pounds per ton (50 to 75 kg/MT), ground ear corn at 175 to 200 pounds per ton (87 to 100 kg/MT), liquid blackstrap molasses at 60 to 100 pounds per ton (30 to 50 kg/MT), sodium metabisulfite at 8 to 10 pounds per ton (4 to 5 kg/MT), or calcium formate–sodium nitrate mixture at 5 to 10 pounds per ton (2.5 to 5 kg/MT). Acid additives, such as phosphoric, hydrochloric, and sulfuric acids, lower the pH. They prevent undesirable fermentations but also lower the palatability of the silage. These materials are spread over each load of silage before dumping. When tower silos are being filled, molasses usually is diluted with water to facilitate its movement through the blower, but this increases the moisture content and seepage losses of the silage.[65] The addition of dry feeds, such as corn meal or dried beet pulp, lowers the moisture content and provides carbohydrates that ensure good silage.[17] Fermentation may eliminate up to 20 percent of the nutrients in the added materials. Addition of 0.25 to 0.50 percent of an enzyme, cellulase, increases the digestibility of the cellulose.

Grasses and legumes are cut for silage when the grasses are just starting to head and when alfalfa and clover are in early bloom. Their moisture content does not drop to 70 percent until the seed is ripe and the leaves begin to dry. The nutritive value and palatability are then greatly reduced. Corn can stand in the field until its moisture content is low enough to make good silage by direct cutting. Sorghum silage is often more moist than typical corn silage but its high sugar content permits it to ferment satisfactorily from direct cutting.

Wilting is the preferred method for preparing grass silage. Wilted forage with 62 to 70 percent moisture does not require additives to make good silage. Silage with this moisture content results in a larger intake of dry matter by cattle, resulting in higher milk production. However, the intake and milk production may still be less than that obtained from hay. This system of ensiling usually involves additional labor as well as some loss of nutrients in the field during wilting.[65] Compaction of the silage and the exclusion of air are more difficult than with silages of high moisture content.

Haylage is the name applied to fermented forages with 40 to 60 percent moisture. It is often difficult to preserve haylage that contains less than 50 percent moisture.[66] Alfalfa haylage is stored at 45 to 55 percent moisture.[31] Haylage is best preserved in a sealed silo because of difficulty in compaction and exclusion of air in other types of silos. Decreasing the moisture content of chopped forage from 80 to 50 percent produces silage with increasing lactic acid content and decreasing percentages of total acid, other undesirable acids, and ammonia nitrogen. Barn dried hay often has a higher feeding value for dairy cows than haylage made from the same crop.

The moisture content of chopped green feed can be estimated by a squeeze test. Squeeze a handful into a ball, hold it for 20 to 30 seconds, and then release the grip quickly. The condition of the ball upon release indicates the approximate moisture content as follows:

Moisture Content (%)	Condition of Ball
>75	Holds shape; considerable free juice
70–75	Holds shape; little free juice
60–70	Falls apart; no free juice
<60	Falls apart readily; no free juice

8.3.5 Making Silage

Row crops used for silage, such as corn and sorghum, are harvested with a field silage cutter or a forage harvester (Figure 8.13). Grasses, legumes, and small grains are gathered with a forage harvester. These machines cut and chop the material and blow it into a trailer or truck. Wilted silage and haylage from grasses and legumes are cut with a mower or windrower. Later, the material is gathered and chopped with a forage harvester equipped with a pick-up attachment. The chopped material is hauled to a silo. It is dumped directly into trench and bunker silos. A blower is used to elevate the material into tower and stack silos from a hopper into which the loads are dumped. The material in trench and bunker silos is packed with a tractor that operates continuously during the filling operations. Tower silos may be filled without packing, but tramping or compacting the material in the upper half or two-thirds of the silo helps to exclude air and thus improves the quality of the silage. A revolving distributor placed in the silo also compacts the silage. Stack silage is usually tramped manually.

Well-packed settled silage in a trench or bunker silo weighs about 45 pounds per cubic foot (720 kg/m^3). The density is influenced by depth of silage, the moisture present, the proportion of grain to stalk, and the diameter of the silo. Settled silage in a tower silo weighs 45 pounds per cubic foot (720 kg/m^3) at a depth of 17 to 18 feet (5 to 5.5 m), but the weight varies with the depth in a slightly curvilinear relationship, as seen in Figure 8.15.

8.3.6 Types of Silos

Tower silos are constructed of concrete, brick, vitreous tile, wood staves, or glazed steel.[65] The latter type is airtight with a top seal and has a mechanical unloading device at the bottom (Figure 8.16). Tower silos are circular with vertical walls. They range from 12 to 30 feet (4 to 9 m) in diameter and 30 to 60 feet (9 to 18 m) in height with capacities of 50 to 1000 tons (45 to 900 MT). To avoid spoilage, a 2-inch (5 cm) depth from the top of the silage should be used up each day during cold weather, and 3 inches (8 cm) in summer.

Trench silos are usually 10 to 30 feet (5 to 9 m) wide, 8 to 30 feet (3 to 9 m) deep, and any desired length. A few hold as much as 50,000 tons (45,000 MT) of silage. Many trench silos are dug into the side of an earth bank with one end at ground level. This permits drainage of excess silage juice and also facilitates the removal of silage. Trench silos usually have a concrete floor and the sides are often lined with concrete.

Bunker silos are built on the ground with concrete floors and either concrete or wooden walls (Figure 8.17).[65] The walls are braced with pillars or soil is piled up

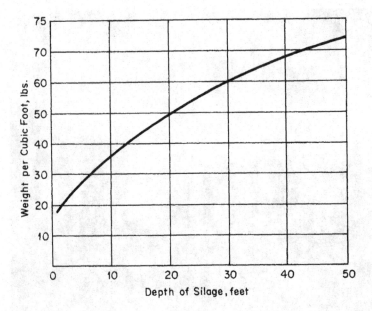

FIGURE 8.15
Weight per cubic foot of silage at different depths.

FIGURE 8.16
Farmstead with upright silos. The two on the left are sealed silos. [Courtesy Richard Waldren]

on the outside. The sides of trench and bunker silos have a slope of 1 or 2 inches (3 to 5 cm) per foot. The sloping sides permit the silage to settle and shrink preventing a gap next to the wall that would start spoilage. The silage in both bunker and stack silos is covered with a heavy impervious sheet.

Silage stored in a trench or bunker can have considerably more dry-matter loss than silage stored in sealed structures. However, bunker losses can be minimized

FIGURE 8.17
Bunker silo with a heavy plastic cover to preserve silage quality. [Courtesy Richard Waldren]

with proper chopping and packing and sealing with a plastic cover. One of the most important factors associated with losses in a trench or bunker silo is the total quantity of silage stored. In general, the larger the pile the lower the percent of dry-matter loss.[31]

Stack silos are temporary circular structures set on the ground. The walls are made of strips of snow fence or of woven wire-fence material spliced at the ends. Two to five overlapping tiers of this fencing give it sufficient height. The walls are lined with sheets of impervious plastic or paper.

About 69 percent of the silos are upright. Most of the newer silos are the bunker type due to the lower cost of construction. Trench silos are popular in the semiarid and arid regions because of their low cost and because there is less interference from rain during filling or emptying the silos in those areas. The trenches are excavated with a bulldozer or power shovel. Much of the trench silage is held over for emergency feeding in drought years when feed production and grazing are limited.

Trench and bunker silos can be adapted to self-feeding by means of movable feeding racks at the end. The animals reach through the rack for feed and move it forward as the feed is consumed. Many others are unloaded with mechanical equipment.

Silage consumption varies with the type and size of the livestock. Dairy or nursing cows will consume the most.

Daily Silage Consumption

Cows (900 to 1,200 pounds)	30–40 pounds
Yearling cattle	15–20 pounds
Fattening cattle	25–35 pounds/1,000 pounds live weight
Sheep	4–5 pounds

REFERENCES

1. Harlan, H. V. "Daily development of kernels of Hannchen barley from flowering to maturity at Aberdeen, Idaho," *J. Agr. Res.* 19, 9(1920):393–429.

2. Olson, G. A. "A study of factors affecting the nitrogen content of wheat and the changes that occur during the development of wheat," *J. Agr. Res.* 24(1923):939–953.

3. Woodman, H. E., and F. L. Engledow. "A study of the development of the wheat grain," *J. Agr. Sci.* 14(1924):563–586.

4. Richie, S. W., J. J. Hanway, and G. O. Benson. "How a corn plant develops." *IA State Univ. Coop. Ext. Serv.* Special Report 48. Revised 1986.

5. Waldren, R. P. *Introductory Crop Science.* Boston: Pearson, 2003, pp. 244–245.

6. Richie, S. W., J. J. Hanway, H. E. Thompson, and G. O. Benson. "How a soybean plant develops." *IA State Univ. Coop. Ext. Serv.* Special Report 53. Revised 1985.

7. Vanderlip, R. L. "How a sorghum plant develops." *KS Agri. Exp. Sta. Publ.* S3, 1993.

8. Reynoldson, L. A., and others. "The combined harvester thresher in the Great Plains," *USDA Tech. Bull.* 70, 1928.

9. Arny, A. C., and C. P. Sun. "Time of cutting wheat and oats in relation to yield and composition," *J. Am. Soc. Agron.* 19(1927):410–439.

10. Arny, A. C. "The influence of time of cutting on the quality of crops," *J. Am. Soc. Agron.* 18(1926):684–703.

11. Harlan, H. V., and M. N. Pope. "Development of immature barley kernels removed from the plant," *J. Agr. Res.* 27(1926):669–678.

12. Klein, L. M., and J. E. Harmond. "Seed moisture—A harvest timing index for maximum yields," *Trans. ASAE* 14, 1(1971): 124–126.

13. Martin, J. H. "The influence of the combine on agronomic practices and research," *J. Am. Soc. Agron.* 21(1929):766–773.

14. Reynoldson, L. A., W. R. Humphries, and J. H. Martin. "Harvesting small grains, soybeans, and clover in the Corn Belt with combines and binders," *USDA Tech. Bull.* 244, 1931.

15. Schwantes, A. J., and others. "The combine harvester in Minnesota," *MN Agr. Exp Sta. Bull.* 256, 1929.

16. Bechdel, S. I., and C. H. Bailey. "Effect of delayed harvesting on quality of wheat," *Cereal Chem.* 5, 2(1928):128–145.

17. Gordon, C. H., and others. "Some experiments in preservation of high moisture hay-crop silages," *J. Dairy Sci.* 11, 7(1957):789–799.

18. Miller, H. F., Jr. "Swift untiring harvest help," in *Power to Produce*, USDA Yearbook, 1960, pp. 164–183.

19. Ferguson, W. L., and others. "Hay harvesting practices and labor used," *Econ. Res. Svc. and Stat. Rpt. Govt. Print. Office.* 1967.

20. Kiesselbach, T. A., and A. Anderson. "Curing alfalfa hay," *J. Am. Soc. Agron.* 19(1927):116–126.

21. Gordon, C. H., and others. "Comparisons of unsealed and plastic sealed silages for preservation efficiency and feeding value," *J. Dairy Sci.* 14, 6(1961):1113–1121.

22. Hoffman, E. J., and M. A. Bradshaw. "Losses of organic substance in the spontaneous heating of alfalfa hay," *J. Agr. Res.* 54(1937):159–184.

23. Friesen, O. H. "Combine operation and adjustment," *Can. Dept. Agr. Publ.* 1464, 1972, pp. 1–30.

24. Harmond, J. E., J. E. Smith, Jr., and J. K. Park. "Harvesting seed of grasses and legumes," in *Seeds*, USDA Yearbook, 1961, pp. 181–188.

25. Leighty, C. E. "The better utilization of straws," *J. Am. Soc. Agron.* 16(1924):213–224.

26. Duncan, D. "Splendor in the mass." *Oregon's Agricultural Progress. OR St. Univ. Ag. Exp. Sta.* Spring/Summer, 1998.

27. Duft, K. D. "Straw to energy? It might be worth a try." *WA St. Univ. Coop. Ext. Serv.* 2002.

28. Woodward, T. E., and others. "The nutritive value of harvested forages," in *Food and Life*, USDA Yearbook, 1939, pp. 956–991.

29. Anonymous. "High quality hay," *USDA Agricultural Research Service Spec. Rept. ARS*, 1960, pp. 22–52.

30. Kiesselbach, T. A., and A. Anderson. "Quality of alfalfa hay in relation to curing practice," *USDA Tech. Bull.* 235, 1931.

31. Mader, T., T. Milton, I. Rush, and B. Anderson. "Feeding value of alfalfa hay and alfalfa silage." *NE Coop. Ext. Serv. NebGuide* G97–1342-A, 1997.

32. Walster, H. L. "Sweet clover as a hay crop," *J. Am. Soc. Agron.* 16(1924):182–186.

33. Sotola, J. "The chemical composition and nutritive value of certain cereal hays as affected by plant maturity," *J. Agr. Res.* 54(1937):399–419.

34. Donaldson, F. T., and J. J. Goering. "Russian thistle silage," *J. Am. Soc. Agron.* 32(1940):190–194.

35. McCalmont, J. R. "Farm silos," *USDA Misc. Bull.* 810, 1367, pp. 1–28.

36. Kiesselbach. T. A., and A. Anderson. "Alfalfa investigations," *NE Agr. Exp. Sta. Res. Bull.* 36, 1926.

37. Trowbridge, P. F., and others. "Studies of the timothy plant, part II," *MO Agr. Exp. Sta. Res. Bid.* 20, 1915.

38. Hunt, C. H., and others. "Effect of the stage of maturity and method of curing upon the vitamin B and vitamin G content of alfalfa, clover, and timothy hays," *J. Agr. Res.* 51(1935):251–258.

39. Westover, L. "Comparative shrinkage in weight of alfalfa cured with leaves attached and removed," *USDA Bid.* 1424, 1926.

40. LeClerc, J. A. "Losses in making hay arid silage," in *Food and Life*, USDA Yearbook, 1939, pp. 992–1016.

41. Kane, E. and others. "The loss of carotene in hays and alfalfa meal during storage," *J. Agr. Res.* 55(1937):837–847.

42. Headden, W. P. "Alfalfa," *CO Exp. Sta. Bull.* 35, 1896.

43. Guilbert, H. R., and others. "The effect of leaching on the nutritive value of forage plants," *Hilgardia* 6(1931):13–26.

44. Van Arsdall, R. N. "Forage mechanization," *Crops and Soils* 15, 6(1962):10–12.

45. Smith, J. A., R. D. Grisso, K. Von Bargen, and B. Anderson. "Management tips for round bale hay harvesting, moving, and storage." *NE Coop. Ext. Serv. NebGuide* G88–874-A, 1988.

46. Dobie, J. B., and R. G. Curley. "Hay cube storage and feeding," *CA Agr. Exp. Sta. Circ.* 550, 1969, pp. 1–16.

47. Forester, E. M., and E. B. Wilson. "The feasibility of operating a stationary hay cubing installation in the Columbia Basin," *WA Ext. E. M.* 3423 (rev.), 1971, pp. 1–24.

48. Swanson, C. O., and others. "Losses of organic matter in making brown and black alfalfa hay," *J. Agr. Res.* 18(1919):299–304.

49. Browne, C. A. "The spontaneous combustion of hay," *USDA Tech. Bull.* 141, 1929, pp. 1–28.

50. Hukill, W. V., R. A. Saul, and D. W. Teare. "Outrunning time; combating weather," in *Power to Produce*, USDA Yearbook, 1960, pp. 183–199.

51. Carrier, L. "The history of the silo," *J. Am. Soc. Agron.* 12(1920):175–182.

52. McCulloch, M. E. "Silage research at the Georgia Station," *GA Agr. Exp. Sta. Rsh. Rpt.* 75, 1970, pp. 1–46.

53. Mueller, J. P., J. T. Green, and W. L. Kjelgaard. "Corn silage harvest techniques," *National Corn Handbook* NCH-49, 1991.

54. Wiggans, R. G. "The influence of stage of growth on the composition of silage," *J. Am. Soc. Agron.* 29(1937):456–467.

55. Nevens, W. B. "Types and varieties of corn for silage," *IL Agr. Exp. Sta. Bull.* 391, 1933.

56. Wheaton, H. N., F. Martz, F. Meinershagen, and H. Sewell. "Corn silage." *U. of MO Coop. Ext. Serv. Agri. Publ.* G4590, 1993.

57. Mader, T., P. Guyer, and R. Stock. "Feeding high moisture corn." *NE Coop. Ext. Serv. NebGuide* G74–100-A, 1983.

58. Becker, R. B., and others. "Silage investigations in Florida," *FL Agr. Exp Sta. Bull.* 734, 1970, pp. 1–31.

59. Westover, H. L. "Silage palatability tests," *J. Am. Soc. Agron.* 26(1934):106–116.

60. Christensen, F. W., and T. H. Hopper. "Digestible nutrients and metabolizable energy in certain silages, hays, and mixed rations," *J. Agr. Res.* 57(1938):477–512.

61. Odland, T. E., and H. O. Henderson. "Cultural experiments with sunflowers and their relative value as a silage crop," *WV Agr. Exp. Sta. Bull.* 204, 1926.

62. Schoth, H. A. "Comparative values of sunflower silages made frolic plants cut at different stages of maturity, and the effect of salt on the palatability of the silage," *J. Am. Soc. Agron.* 15(1923):438–442.

63. Langston, C. W., and others. "Microbiology and chemistry of grass silage," *USDA Tech. Bull.* 1137, 1958.

64. Neidig, R. E. "Chemical changes in silage formation," *IA Agr. Exp. Sta. Res. Bull.* 16, 1914.

65. Anonymous. "Making and feeding hay-crop silage," *USDA Farmers Bull.* 2186, 1962.

66. Gordon, C. H., and others. "Preservation and feeding value of alfalfa stored as hay, haylage, and direct-cut silage," *J. Dairy Sci.* 14, 7(1961):1299–1311.

Handling and Marketing Grain, Seeds, and Hay

■ 9.1 MARKETING GRAIN

Grain from the farm is mostly trucked to elevators or mills built at railroad sidings. An elevator (Figure 9.1) is equipped with a truck scale to measure the load, and the grain is dumped into a pit. The grain is picked up from the pit by buckets attached to an endless belt that carries it up through a *leg*, to the top (cupola) of the elevator, from which it is transferred to bins. The horizontal movement of grain is on an endless belt about 3 to 5 feet (90 to 150 cm) wide (Figure 9.2). Grain is moved by pneumatic systems[1] in some elevators, and some bulked field seeds are moved by vibrating conveyers.[2] Later, grain is transferred into a rail car or semi tractor-trailer from a bin or the cupola.[3, 4] Successful country elevators handle grain equivalent to five or more times their storage capacity during a marketing season. Consequently, they can store only a small proportion of the total crop.

An elevator may be owned by a farmers' cooperative association, an independent operator, or a corporation operating several elevators. The three types are commonly referred to as co-op, independent, and line elevators, respectively. The same prices are nearly always paid by all elevators for grain of a given grade at a particular shipping point. The local price is based upon a card price that is received daily from the terminal market to which the grain is usually shipped. The card price is determined by the terminal price for a particular grade and quality of grain on that day, less the freight to the terminal market, less the fees for handling the grain at the local elevator and the terminal market.

The farm price of grain in any region in the United States is based on prices at an exchange that buys and sells futures contracts on commodities. For example, wheat in the surplus-producing regions brings a lower price as the shipping distance west of Chicago increases. The highest prices are paid in the main deficiency areas, the southern and the Atlantic seaboard states. The lowest prices are in the intermountain states. In the Pacific Coast states, prices are higher than in the intermountain states because of the sea outlet. The most common exchanges are the Chicago Board of Trade for corn, oat, rice, soybean, wheat, and other crops; the Kansas City Board of Trade for wheat; and the Minneapolis Grain Exchange for wheat, corn, and soybean.

FIGURE 9.1
Grain elevator with
concrete silo-type bins.
[Courtesy Richard Waldren]

FIGURE 9.2
The movable automatic
"dump" takes grain from
the belt conveyor and
spouts it into the top of
a bin.

Most grain is shipped to a commission firm, elevator firm, or cooperative agency at a terminal market. The shipper, who uses the bill of lading as security, draws a draft upon the consignee for a substantial portion of the expected selling price of the grain. The draft may be payable on sight or when the grain arrives at the market. The consignee sells the grain. He or she remits any balance due after deducting his or her commission, weighing fees, inspection fees, freight, and other charges.

▧ 9.2 MARKETING SEEDS

Hybrids and varieties of crop seeds are often produced by seed companies, frequently under contract with growers. Large quantities, however, are purchased directly from farmers who produce certified seed or indirectly from local or wholesale seed buyers. Later, individual seed companies may trade with others by buying or selling to provide the quantities required for their expected market. Lesser quantities of seeds are exported to, or imported from, other countries. The regulations concerning seed marketing are described in Chapter 7.

▧ 9.3 HANDLING GRAIN AT A TERMINAL MARKET

Upon arrival at the market, the grain is sampled by licensed inspectors, while other samples taken by authorized representatives are furnished to the consignee. The latter samples, with accompanying grade cards, are placed on tables and offered to buyers on a large cash-grain trading floor at the Grain Exchange.[5]

At the terminal elevator or mill, the grain is unloaded with a power dragline scoop or by a dump that tilts the rail car or tractor trailer and pours the grain into a hopper or pit. The terminal elevators are usually equipped with power windlasses and cables to move rail cars. Inside the terminal elevator, the operations are much the same as in a country elevator. Terminal elevators along a waterfront have marine legs that enclose pneumatic systems or conveyor buckets on an endless belt. These are lowered into the hold of a ship or barge to unload bulk grain. Modern terminal elevators have many large circular bins or silos, each with a capacity of 25,000 to 110,000 bushels (900 to 3,900 m^3). Such an elevator may hold several million bushels of grain. Each bin is equipped with wired thermocouples, about 5 feet (1.5 m) apart from top to bottom. Thus, temperatures in all parts of each bin are measured with a potentiometer located inside the building. Whenever a hot spot, where the grain is 90°F (32°C) or more, is detected, the grain is aerated or moved to another bin so that the moist or heating portion is mixed with the larger bulk of cool sound grain.

At a terminal market, cash grain is sold on the basis of the sample and the official grade. The agreed price is usually a stated premix above the future price for that particular grade of grain. The future price is based upon grain of contract grade (i.e., grain that barely meets the minimum requirements for the particular grade), except for a small allowance for variations. Grain received from a country elevator may grade No. 2 because of the test weight per bushel is below the requirement for No. 1, but it may more than meet the minimum requirements for the No. 1 grade with respect to other grading factors such as foreign material, damage, moisture content, and mixtures of grain of other kinds or classes. Consequently, such grain commands a premium over the price for contract grain of No. 2 grade. These expected premiums are reflected back in the posted card price at a country shipping point. In a typical transaction, a lot of corn of No. 3 grade may be sold at the terminal market for 5 cents a bushel over the July, September, December, or May future price for that grade. The actual sale price is determined soon thereafter when the hedge is closed by the purchase of futures at the prevailing price.

9.4 HEDGING

Speculative risks in buying grain, which result from changes in prices, are avoided by the practice of hedging the purchases for each day.[6] Hedging consists of selling futures whenever cash grain is bought and later closing the hedge by buying futures when the grain is sold. The hedging is done for the country dealer by his or her broker or other representative at the terminal market. For example, a country dealer has just bought 5,000 bushels of wheat from local farmers. He wires his broker to sell 5,000 bushels of wheat futures on the grain exchange. Some days or weeks later the 5,000 bushels of wheat arrive at the terminal market. The broker sells the wheat and buys back the futures for his client. In the meantime, when the price of wheat has advanced 10 cents a bushel, the country dealer makes an extra profit of $500 on the cash grain. But the futures, bought back at the advanced price of wheat, bring a loss of $500. However, when the price of wheat has dropped 10 cents a bushel, the dealer loses $500 on the cash grain. But the futures are bought back for $500 less than the amount for which they are sold, or a profit of $500. Losses or gains due to changes in the market price of the actual (cash) grain are thus balanced by corresponding gains or losses in the transactions with grain futures. A small fee is charged for the futures transactions.[7, 8] Millers and grain dealers at the terminal markets likewise hedge their purchases of grain. This permits a miller to sell flour for delivery several months later without a speculative risk.[9]

9.5 FUTURES TRADING

Dealing in grain futures began in Chicago in 1848. The transactions are conducted by people standing in or around a pit located in the room in which cash grain also is sold (Figure 9.3). The smallest unit for futures trading of grain usually is 5,000 bushels. With two motions of the arm, and accompanied by hand signals (Figure 9.4), a dealer may offer to buy or sell as much as 25,000 bushels of grain at an indicated price. A signal accompanied by a nod indicates acceptance. The shouting around the pit merely helps attract the attention of a buyer or seller. The transactions are recorded as well as conducted under strict rules established by the particular Grain Exchange or Board of Trade. These exchanges, which operate the marketing facilities, are subject to federal regulation under the Commodity Exchange Act. Only members of the particular exchange may conduct future trading operations, but members may act as brokers for others. Speculators as well as grain dealers may trade in futures through their brokers. Prices of grain futures fluctuate constantly with changes in present and prospective supplies, demands for the different cash grains, changes in the volume, and trends in futures trading. The prices of cash grain approximately follow the futures prices. The volume of futures trading is very large in proportion to the

FIGURE 9.3

Trading floor at the Chicago Board of Trade. [Reprinted by permission of the Board of Trade of the City of Chicago, Inc.]

FIGURE 9.4

Hand singals are used to communicate between buyers and sellers at a grain exchange. [Reprinted by permission of the Board of Trade of the City of Chicago, Inc.]

quantity of grain marketed. This is partly due to speculation in grain futures. However, in hedging, the buyer of each cash grain contract sells futures and then later buys them back as a separate transaction. Some grain contracts change hands several times, with hedging transactions each time.

Although the speculative operations may tend to either disrupt or stabilize the market at any given time, they offer additional outlets for, and lend flexibility to, ordinary hedging operations. Without speculators it might be difficult to hedge large sales of grain for export or processing. A speculator is in a *long* position when his or her purchases of futures exceed those that he or she has sold. In that case, he or she expects the price to rise. When he or she expects the price to go down, he or she sells *short*. All sales of grain futures are, in fact, contracts to deliver the grain on or before the end of the month (May, July, September, or December) indicated in the transaction. Only occasionally is delivery called for on a contract. It is much more simple to use a cash settlement. Grain delivered on a futures sale is usually the contract grade stored in a terminal elevator. Contract grain is usually a mediocre quality for processing.

▓ 9.6 GRAIN GRADING

9.6.1 Grain-Grading Factors

The purpose of grading grain is to facilitate its marketing, storage, financing, and future trading.[10, 11] Grain grading standardizes trading practices and reduces the hazards of merchandising. The factors upon which grades are based determine the quality or market value for the purpose for which the particular grain is generally used. Thus plumpness, which usually is measured by the test weight per bushel, indicates a well-developed endosperm. In wheat, higher test weight will result in a high yield of flour. Plump barley gives a higher yield of malt than does thin barley. In all grains, a high test weight indicates a low percentage of hull, bran, or both. This gives a low percentage of crude fiber, resulting in an enhanced feeding value.

The test weight of wheat depends upon its plumpness, shape, density, and moisture content. Plump, rounded grain packs together more closely than does wrinkled, shriveled, or angular grain. Dense vitreous kernels are heavier than starchy kernels. In a given sample of wheat, the test weight varies inversely with the moisture content. Wetting causes the starch to swell. Furthermore, the wheat kernel has a specific gravity of about 1.41 compared with that of water, which is 1.00. When wetted wheat grains dry out they shrink, but not entirely back to their original volume. Each repeated wetting and drying makes the grain lighter in bushel weight. Fully ripe wheat that tests 62 pounds per bushel may test only 55 pounds per bushel when soaked. After several wettings it may test only 59 pounds after it is thoroughly dried, yet it may have deteriorated only slightly in milling and baking quality.

Other grains likewise swell and decrease in test weight when they are wetted. Flaxseed, on the other hand, increases in test weight as its moisture content increases because it contains no starch to absorb water and swell. Furthermore, the 40 percent of oil in the seed has a lower specific gravity than water (about 0.93) and thus the added water increases the weight of the seed.

The test weight per bushel of any grain is increased by cleaning out chaff, shriveled kernels, and other light material. Thereafter, additional cleanings will further increase the test weight of oat and barley by breaking off awn remnants, hairs, and glume tips, which permits closer packing in the test kettle. The test weight of oat is increased merely by transferring the grain from one concrete bin to another because of the abrasive effects just mentioned. The test weight of clean wheat is not altered appreciably by such handling. With flaxseed, however, each successive cleaning or handling reduces the test weight because the resultant roughening of the originally smooth and glossy seed coat causes the seeds to pack less closely.

In addition to plumpness or test weight per bushel, the chief factors that contribute to high quality are soundness, dryness, cleanliness, purity of type, and general condition. Grain that is sour, musty, heated, or moldy is unpalatable, disagreeable in odor, and often unfit for food or feed. Binburnt, severely sprouted, or severely frosted wheat makes flour that has poor baking quality. Dry grain keeps well in storage, whereas wet or damp (tough) grain is likely to spoil. Furthermore, the moisture has no value because the shrinkage in weight when drying occurs is a complete loss to the purchaser. A sample of grain containing 20 percent moisture (80 percent dry matter), when dried down for safe storage to a moisture content of 13 percent (again on a wet basis), shrinks more than 8 percent in total weight. Most of the slightly damp grain that reaches terminal markets is not dried artificially, but is mixed thoroughly with drier grain.

Cleanliness is important except in grain containing easily separated foreign material that has a value in excess of cleaning costs plus cleaning shrinkage. In the grading of wheat, barley, rye, flax, and grain sorghum the foreign material removed by dockage testers or appropriate screening devices is called *dockage*. The weight of the dockage is subtracted from the total weight of these grains in marketing transactions except when the dockage amounts to less than 1 percent of the total weight of the grain. Impurities (foreign material) that cannot be readily separated may detract seriously from the value of the grain. Certain foreign seeds that often remain in wheat after cleaning, such as garlic, rye, and giant ragweed, affect the quality (color, flavor, or texture) of bread made from the wheat. Stones and cinders that are the size of grains are particularly objectionable in grain because they are not separated readily by cleaning machines. As a result, they may damage grinding machinery or contaminate the milled products.

Purity of type is important, except when grain is used for feed. A mixture of grains of different sizes is difficult to clean without heavy shrinkage losses. Different classes of wheat require different tempering and grinding procedures for milling. Admixtures of inferior-quality varieties lower the quality of the entire lot. In toasted wheat products, which are made from white wheats, any red grains present become so dark in color that they appear to be scorched. Mixed barley types do not sprout uniformly during the malting process. Yellow corn mixed with white corn makes an unattractive hominy or meal. Toasted flakes of yellow corn kernels, which are very dark in color, detract front the appearance of corn flakes made from white corn.

Cracked grain is unsuitable for certain grain products. Bleached, weathered, discolored, or off-colored grain is unattractive.

9.6.2 Grain Sampling and Inspection

Grain sampling is the first essential step in grain inspection and grading[12] (Figure 9.5). A carload of bulk grain is sampled by inserting a trier or probe through the grain to the bottom of the car at five or more places and withdrawing it after it fills with grain. The ordinary probe consists of a 63-inch (160-cm) double brass tube with partitioned slots in the core and corresponding slots on one side in the outer shell. Turning the core opens or closes the openings. Sacked grain and seeds are sampled with a small pointed bag trier. A falling stream of bulk grain from an elevator spout is sampled by cutting across the grain stream at intervals with a dipperlike device called a *pelican*. After a brief examination for quality and uniformity,

FIGURE 9.5
Sampling grain with a metal probe that has partitioned slots in the core.

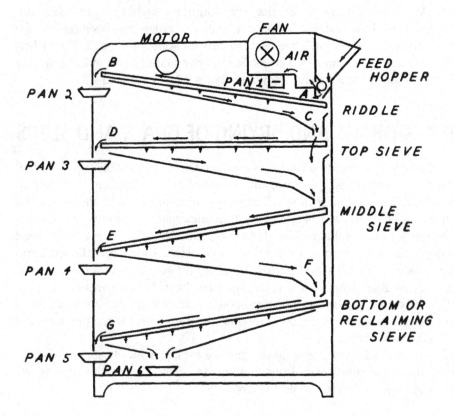

FIGURE 9.6
Screens and a suction fan separate dockage, broken grains, and foreign materials from a sample.

the probe or trier samples are bulked together. Before grading, all samples are mixed and reduced to a suitable size in a Boerner divider.

At the inspection laboratory, the market class, test weight, and dockage (foreign material) are determined on all samples (Figure 9.6). Any other factor that seems likely to determine the grade is also analyzed. For example, the moisture content is determined when the grain feels even slightly damp or tough. Other factors also may be measured, such as damaged grains, foreign material left after the dockage is removed, or grains of other market classes, if they appear to be present in sufficient amount to modify the grade. Grain is graded by inspectors licensed and supervised by the United States Department of Agriculture under the authority of the United States Grain Standards Act passed in 1916 and amended several times since.

The inspection certificate that gives the grade and test weight also designates the car number or elevator bin in which the grain was sampled. Both seller and buyer have access to these records.

The moisture content of the grain is measured almost instantaneously in a moisture meter that depends, for its action, upon a lower electric resistance in grain high in moisture.

9.6.3 Grades of Grain

The United States has established standards for barley, beans, canola, corn, flaxseed, lentils, mixed grain, oat, rice, rye, sorghum, soybean, split peas, sunflower seed, triticale, wheat, and whole dry beans.[13] Grades have been established for these crop and classes have been established for some of them. Table 9.1 lists classes of common grain and seed crops. Summarized tabulations of the grades of wheat, corn, oat, and soybean are shown in Tables 9.2, 9.3, 9.4, and 9.5.

■ 9.7 STORAGE AND DRYING OF GRAIN AND SEEDS

Threshed (shelled) grain and some seeds are stored in comparatively tight bins that allow little moisture to escape. Also, adjustment of grain temperature to the outside temperature takes place slowly. The temperature of the air in the center of a full 1,000-bushel (35 m^3) bin 14 feet (4 m) in diameter lags one to two months behind the temperature of the outside air. Consequently, how well the grain or seed stores depends largely upon its condition when placed in storage. Binned grain that contains more than 13 to 14 percent moisture is likely to deteriorate in warm weather. Grain that deteriorates in storage may heat, become musty, sour, or "sick." Seeds containing excess moisture soon lose their viability. Dry seed stored in woven bags fluctuates in temperature and moisture with the temperature and humidity of the surrounding air.

Corn, peanut, and several other grains and seeds that are moldy sometimes contain dangerous quantities of poisonous aflatoxin. This is caused mostly by the growth of *Aspergillus flavus*, a mold fungus.

TABLE 9.1 U.S. Classes of Grains and Seeds (having more than one class)

Crop	Classes
Wheat	Durum, hard red spring, hard red winter, soft red winter, hard white, soft white, unclassed, mixed
Rice	Long grain rough, medium grain rough, short grain rough, mixed rough
Barley	Barley, malting barley
Corn	Yellow, white, mixed
Sorghum	Tannin, white, mixed
Soybean	Yellow, mixed
Beans	Pea (navy), blackeye, cranberry, yelloweye, pinto, marrow, great northern, small white, flat small white, white kidney, light red kidney, dark red kidney, small red, black, mung, miscellaneous, large lima, baby lima, miscellaneous lima, mixed
Whole, dry peas	Smooth green dry, smooth yellow dry, wrinkled dry, winter dry, miscellaneous dry, mixed dry
Split peas	Green split, yellow split, winter split

TABLE 9.2 Wheat: Grade Requirements for All Classes of Wheat Except Mixed Wheat[a] (Effective May 1993)

Grade	Minimum Test Weight per Bushel		Maximum Limits of Defects						Wheat Other Classes[c]	
	Hard Red Spring Wheat (lb.)	All Other Classes (lb.)	Heat-Damaged Kernels (%)	Damaged Kernels (Total)[b] (%)	Foreign Material (%)	Shrunken and Broken Kernels (%)	Defects (Total) (%)	Total Other Material (Count Limit)[e]	Total Contrasting Classes[d] (%)	Wheat Other Classes (Total) (%)
1	58	60	0.2	2	0.4	3	3	4	1	3
2	57	58	0.2	4	0.7	5	5	4	2	5
3	55	56	0.5	7	1.3	8	8	4	3	10
4	53	54	1.0	10	3.0	12	12	4	10	10
5	50	51	3.0	15	5.0	20	20	4	10	10

[a]*Sample Grade:* Sample grade is wheat that does not meet the requirements for any grades from No. 1 to No. 5 inclusive; or which contains stones; or which is musty, sour, or heating; or which has any commercially objectionable foreign odor except of smut or garlic; or which contains a quantity of smut so great that any one or more of the grade requirements cannot be applied accurately; or which is otherwise of distinctly low quality.
[b]Includes damaged kernels (total), foreign material, shrunken and broken kernels.
[c]Unclassed wheat of any grade may contain not more than 10 percent of wheat of other classes.
[d]Includes contrasting classes.
[e]Included any combination of animal filth, castor beans, crotalaria seeds, glass, stones, or unknown foreign substance.

TABLE 9.3 Corn: Grade Requirements for Corn[a] (Effective September 1996)

Grade	Minimum Test Weight per Bushel (lb)	Maximum Limits Of		
		Broken Corn and Foreign Material (%)	Damaged (Total) (%)	Heat-Damaged Kernels (%)
1	56	2	3	0.1
2	54	3	5	0.2
3	52	4	7	0.5
4	49	5	10	1.0
5	46	7	15	3.0

[a]*Sample Grade:* Sample grade is corn that does not meet the requirements for any of the grades from No. 1 to No. 5 inclusive; or which contains stones in excess of 0.1 percent; or which contains 2 or more pieces of glass or 3 or more crotalaria seeds, 2 or more castor beans, 8 or more cocklebur seeds, or similar seeds singly or in combination; or which contains animal filth in excess of 0.2 percent in 1,000 grams; or which contains 4 or more particles of an unknown foreign substance or a commonly recognized harmful or toxic substance; or which is musty, sour, or heating; or which has any commercially objectionable foreign odor; or which is otherwise of distinctly low quality.

Special Grades:
Flint corn: Corn that consists of 95 percent or more of flint corn.
Flint and dent corn: Corn that consists or a mixture of flint corn and dent corn containing more than 5 percent but less than 95 percent flint corn.
Waxy corn: Corn that consists of 95 percent or more waxy corn.

During heating, grain usually reaches a temperature of 90 to 160°F (32 to 71°C). At higher temperatures, it acquires a browned or mahogany binburnt color and a burned taste. At still higher temperatures that approach spontaneous combustion, grain acquires a charred appearance and taste. It was once believed that the initial

TABLE 9.4 Oats: Grade Requirements for the Classes White Oat, Red Oat, Gray Oat, Black Oat, and Mixed Oat[a] (Effective May 1998)

Grade	Minimum Limits Of		Maximum Limits Of		
	Test Weight per Bushel (lb.)	Sound-Cultivated Oats (%)	Heat-Damaged Kernels (%)	Foreign Material (%)	Wild Oats (%)
1	36	97	0.1	2	2
2	33	94	0.3	3	3
3[b]	30	90	1.0	4	5
4[c]	27	80	3.0	5	10

[a]*Sample Grade:* Sample grade is oats that do not meet the requirements for any of the grades No. 1 to No. 4 inclusive; or which contain 8 or more stones, 2 or more pieces of glass, 3 or more crotalaria seeds, 8 or more cocklebur seeds, 2 or more castor beans, or similar seeds singly or in combination; or which contains 4 or particles of an unknown foreign substance or a commonly recognized harmful or toxic substance; or which contains 10 or more rodent pellets, bird droppings, or equivalent quantity of animal filth per 1-1/8 to 1-1/4 quarts of oats; or which are musty, sour, or heating; or which have any commercially objectionable foreign odor except of smut or garlic; or which are otherwise of distinctly low quality.
[b]Oats that are slightly weathered shall be graded not higher than No. 3.
[c]Oats that are badly stained or materially weathered shall be graded not higher than No. 4.

TABLE 9.5 Soybean: Grade Requirements for All Classes of Soybean and Mixed Soybeans[a] (Effective September 1994)

		Maximum Limits Of							
		Damaged Kernels				Other Materials			
Grade	Maximum Test Weight per Bushel	Total Heat-Damaged Seeds (%)	Foreign Material (%)	Splits (%)	Soybeans of Other Colors (%)[b]	Animal Filth (Count Limits)	Stones (Count Limits)[c]	Glass (Count Limits)	Total (Count Limits)[d]
1	56	2	1	10	1	9	3	0	10
2	54	3	2	20	2	9	3	0	10
3	52	5	3	30	5	9	3	0	10
4	49	8	5	40	10	9	3	0	10

[a]*Sample Grade:* Sample grade is soybeans that do not meet the requirements for any of the grades from No. 1 to No. 4, have a musty, sour, or commercially objectionable foreign odor (except garlic odor), and are heating or otherwise of distinctly low quality.
[b]Disregard for mixed soybeans.
[c]In addition to the maximum count limit, stones must exceed 0.1 percent of the sample weight.
[d]Cannot have more than 1 castor bean seed, 2 crotalaria seeds, or 3 unknown foreign substances. Includes any combination of animal filth, castor beans, crotalaria seeds, glass, stones, and unknown foreign substances. The weight of stones is not applicable for total other material.

heating was caused entirely by the respiration of the live embryo of the grain. Later it was determined that moist grain heats at about the same rate when dead as it does when viable (alive). Thus, most of the heating in grain is caused by the respiration of living and multiplying fungi or mold organisms always present on and in the grain.[14] Damp dead grain does not heat when it is sterile.

The presence of live weevils or other stored crop insects can cause grain to heat. Both molds and insects respire. They release heat, moisture, and carbon dioxide during the respiration process. The released heat and moisture continues to accelerate the heating of the grain. When the temperature increases to about 130°F (54°C), the insects are killed, and bacteria and molds are largely inactivated, but spore-forming organisms survive. Enzymes are inactivated at about 140 to 150°F (60 to 65°C). Heating beyond that point must result from molecular oxidation of organic materials in the grain.

Grain can become musty or moldy when it has enough moisture for the growth of molds but not enough to cause heating. Grain can also become musty when the heat is dissipated rapidly by exposure to air. For example, grain can become musty in the shock, or in the stack before threshing, or in sacks, shallow piles, or bins from which heat escapes readily. Grain of very high moisture content may become sour from fermentation similar to that which occurs in a silo where alcohol and then organic acids are formed. A barrel of grain soaked for hog feed sours in a few days. Wet shelled corn that contains as much as 24 to 30 percent moisture is likely to sour.

Wheat becomes "sick" when the moisture content is about 14 to 16 percent, or slightly above that necessary for good storage quality. It occurs in deep bins, usually in terminal elevators, where the oxygen supply becomes depleted by respiration of the grain and of the fungi and bacteria on the grain. The oxygen content of the air in such bins may drop to 5 or 10 percent while the carbon dioxide content rises from an initial 0.04 percent up to 10 or 12 percent.

The growth of anaerobic fungi or bacteria may be the cause of sick wheat in bins.[15] Anaerobic organisms thrive in the absence of air because they are able to obtain oxygen by breaking down the carbohydrates in the grain. Sick wheat has a dull, lifeless appearance with the germs dead and discolored. Such grain has a high acidity caused by the breaking down of the fats of the embryo into fatty acids. Sick wheat is unsuitable for making bread. Damaged corn that shows similar germ injury is not designated as sick.

Some spoilage can occur in a bin of grain as a result of convection currents that arise with the onset of cold weather. Air around the warm grain at the bottom of the interior portion of the bin moves upward through the grain. It is replaced by cold air that passes downward next to the cold outer walls of the bin. The rising warm air contains moisture, which often condenses on the cool grain near the top of the bin. This results in some molding or caking of the top portion. Such condensation can be avoided by aeration of the grain. A small exhaust fan installed at the top of the bin draws air up through a metal pipe in the center of the bin, which has perforations near the bottom. The fan draws warm air from the bottom. Then cold outside air passes down through the grain from above. Thus, the whole mass of grain is cooled.[16]

▥ 9.8 DRYING GRAIN AND SEEDS

When possible, natural drying before harvest is most economical for the farmer. Damp grain is often sold to buyers who can mix it with dry grain or dry it artificially. Some grain is dried by spreading it out in shallow piles on a barn or bin floor or on the ground outside.

The material used in the construction of a bin has little effect on maintaining the quality or grade of grain while in storage. Metal bins both absorb and radiate heat from the sun more rapidly than do wooden or masonry bins. Insulated and underground bins retard the absorption of summer heat, but they also retain heat that should be dissipated from warm grain in cool weather. Ventilated steel bins with a wind-pressure cowl at the top, and with either perforated sides or bottoms, or perforated pipes or screen-covered flues, will dry grain that is slightly damp or tough. The moisture content of grain may drop only 1 to 3 percent during several months of storage but the grain will remain cool and sound. Grain with a moisture content of 2 percent or more above that essential for maintaining storage quality is likely to lose quality in a ventilated bin. Any type of ventilated bin permits seed to absorb moisture during rains or in damp weather. Many farm driers using either unheated or heated forced air are now in operation[16, 17, 18, 19] (Figure 8–7). Corn should be cleaned before drying to remove trash that obstructs air movement through the grain.[20, 21]

9.8.1 Drying with Unheated Air

Forcing unheated air through a bin of grain or seeds is satisfactory when the grain contains no more than 3 to 4 percent excess moisture. The air fans are only operated on days, or parts of days, when the relative humidity is less than about 70 percent. The humidity should be below 60 percent for final drying of the grain down to a safe moisture content. Grain to be dried with unheated air is stored in a bin with a false-bottom, a perforated floor, or with a connected system of perforated ducts or tunnels set on the bin floor. The forced air enters the grain at the bottom of the bin, passes upward through a 4- to 8-foot (1.2 to 2.5 m) depth of grain, and escapes through ventilators at the top.[18] The air flow ranges from 1 to 6 (usually 2 to 4) cubic feet per minute per bushel (800 to 4,900 l/m^3) of grain. Drying may be completed in one to four weeks. The use of a portable fan permits the drying of several storage bins of grain during and after harvest.

9.8.2 Drying with Heated Air

Drying with heated air is more effective, but also more expensive, than with unheated air. About 2,000 BTUs (500 kcal) are required to evaporate 1 pound (454g) of water.[21] One gallon (4 l) of liquid fuel dries 5 bushels (150 l) of corn.[22] Heated air generally is used for drying grain that contains 18 percent moisture or more. Much of the corn that is harvested with a corn combine or picker-sheller requires heated air. Drying with heated air is usually done in a special bin to which the burner and fan are attached.[17] Drying is completed in a few hours, after which the grain can be moved to a storage bin. Then another batch of grain is dried. The burner uses either natural gas or petroleum fuels. The temperature of the drying grain should not exceed 110°F (43°C) when the grain is to be used for seed, 130°F (54°C) when it is to be processed into starch, or 180°F (82°C) when used for feed. A reduction in the digestibility of protein in the grain occurs sometimes from prolonged heating even at temperatures of 140 to 170°F (60 to 77°C), but quick drying at temperatures above 180°F (82°C) may not be harmful. For most damp field seeds the temperature should not usually exceed 90 to 100°F (32 to 38°C).

Direct heaters utilize the combustion gases mixed with air to dry the grain. Indirect heaters have a heat exchanger so that only hot air enters the bin. This reduces fire hazard but requires about 50 percent more fuel to dry a bin of grain.

Portable driers that use either heated or unheated air can be used to dry ear corn in a crib. The two ends of the crib are covered with impervious panels or sheets; the air is blown into the crib through a canvas funnel that covers one side of the crib.

Farm grain driers are usually large enough to handle the crop from a large farm (Figure 8.7). These are essentially like the commercial driers installed by grain elevators, where drying is done by blowing heated air into chambers between thin layers of grain that are passing down over baffles or through screen-covered flues. Rotary driers similar to hay dehydrators are also used.

Dry seeds can be preserved for long periods in moisture-resistant containers.

9.8.3 Aeration of Grain

In the past, it was customary in elevators to turn grain during cold weather by conveying it from one bin to another. This exposure to cold air may lower the temperature of the grain 20°F (10°C) or more. The moisture content can be lowered as much as 0.25 to 0.50 percent by turning grain when the atmosphere is dry. Often, there is little or no change in moisture content from turning. Cooling the grain retards respiration, heating, and insect damage. The conditioning of grain is now being done more economically by aeration. A large exhaust fan at the bottom of the bin draws cold outside air down through the mass of grain until all portions are about the same temperature. A large silo-type bin of grain can be aerated in about a week when fan draws 0.1 cubic foot (3 l) of air per bushel (35 l) of grain per minute.[23]

Wet grain, especially grain sorghum, for animal feed can be ensiled at a 50 to 60 percent moisture content. Sorghum grain containing 20 to 35 percent moisture can be preserved for feed by adding 0.75 to 1.5 percent (by weight) of propionic acid. Both treatments destroy the germination of the grain and the acid treatment prevents molding.

▨ 9.9 CLEANING GRAIN AND SEEDS

Most grain ultimately requires cleaning before use for any purpose except feed. Market grain rarely is cleaned on the farm because the loss from shrinkage and the cost of cleaning usually exceed the value of the screenings or the dockage deduction. Very little market grain is cleaned at a country elevator except to reduce the dockage content to just below 1 percent so the grain can be sold on a dockage-free basis. Such partial cleaning is done quite often in Canada. Screenings have little sale value in a surplus grain area. Many screenings cannot be shipped without processing because of weed laws. Screenings have more value at a terminal grain market where the edible ingredients can be added to ground mixed feeds that are sold on a guaranteed chemical-analysis basis without specifying the individual ingredients. The value of the dockage is taken into consideration when an offer is made for cash grain at the terminal market. Seed warehouses have elaborate equipment for cleaning various kinds of seeds.[2, 24,]

Terminal elevators usually have equipment for separating some different seeds. In the past, wild oat screened from wheat or barley found a good market for mule feed in the southeastern states where wild oat is not a weed pest. It was sold in carload lots under the designation of mill oat.

During cleaning, grain losses due to shrinkage usually range from 2 to 5 percent, in addition to the foreign material removed, even when the best equipment is used. The invisible loss from dust, moisture evaporation, spilled grain, and grain lodged in the equipment during cleaning in an elevator is likely to be about 0.25 percent. Moving of oat or barley through elevator machinery knocks off portions of the plumes or awns that then become screenings, foreign material, or dockage. Wheat grains lose hairs and bran and suffer cracking during handling.

9.9.1 Types of Cleaners

Separations of different grains, seeds, and foreign material are done mechanically on the basis of differences in length, thickness, shape, or specific gravity.[25] The most common cleaning device found on farms is the fanning mill. It consists of two or more screens to sift out particles or seeds both larger and smaller than the particular seed being cleaned, together with an air blast that blows out the material lighter than the desired clean seed.

Larger machines similar to the fanning mill (Figure 9.7), called receiving separators, are found in many elevators and in all mills and seed warehouses. These machines, which use suction rather than an air blast to remove light or chaffy material, make more accurate separations than a fanning mill. Such machines are excellent for the removal of trash, chaff, certain fine seeds, and cracked or shrunken grain, but they are unsatisfactory for the separation of grains or seeds of similar diameter or specific gravity. For example, oat and wild oat can be only partially separated from wheat, rye, or barley with a screen cleaner because their diameter is much the same, and the differences in specific gravity are too small. Since vetch seed has almost the same diameter and specific gravity as wheat, it cannot be separated using a fanning mill. Mustard seeds will pass through fine screens or sieves that grain cannot penetrate, but many of them roll rapidly down the sloping screen of the cleaner where they are discharged with the cleaned grain. On the other hand, triangular wild buckwheat seeds, which are appreciably larger than mustard seeds, are screened out readily with special sieves because they do not roll.

Indented or cylinder cleaners (Figure 9.7) separate seed of different length. The pockets on revolving disks, or indentations inside a slowly revolving cylinder, pick up shorter particles, seeds, or grains that can be held in shallow indentations while they reject the longer ones. The separated fractions are discharged into different spouts. Such machines have a series of disks or cylinders with different size indentations that permit several excellent separations. Thus, cockle seed (Figure 9–8) can be separated from wheat because of its shorter length. Oat and barley are separated from wheat because of their longer length. A fair separation of mixtures of two varieties of even the same kind of grain can be made when the grains differ appreciably in length. Special indentations separate seeds of different shape.[2] Large cleaners of the screen, disk, or cylinder types can have a capacity of 300 to 600 bushels per hour (10 to 25 m^3/hr).

Ring graders are equipped with rotating, spaced, spring-steel rings to separate grains that do not have the same diameter. With these machines, oat can be separated from barley, and thin barley kernels can be separated from plump grains. Accurate separations are accomplished by different spacings between rings. A needles machine makes separations by allowing the grain to flow over a series of baffles bearing needles (fine spring rods). Slender grains will fall between the

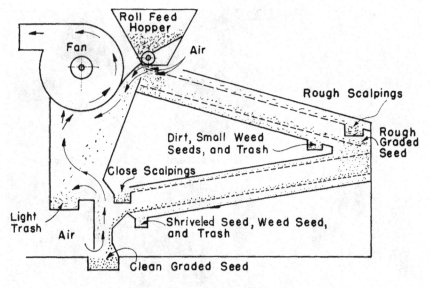

INDENTED CYLINDER SEPARATOR

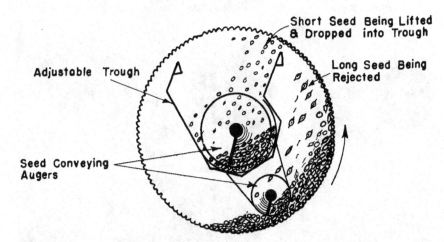

FIGURE 9.7
(Top) Cleaning grain by suction fan and screening. *(Bottom)* Indented cylinder separates seeds or grains of different lengths.

rings. Spiral, horizontal disk, and inclined draper separators remove globular seeds from seeds of other shapes that roll, merely slide, or rest.[19]

Two other types of machines, called gravity and magnetic separators, are used especially for cleaning small-seeded legumes such as alfalfa and clover. The gravity separator has a shaking table with air blowing up through numerous small openings in the table top. This action floats the seeds and material differing in specific gravity over into different compartments or spouts. The gravity separator can remove lighter kernels infested by Angoumois moths from sound corn. Moist sawdust added to clover seed sticks to the mucilaginous seed coats of buckhorn seeds. A gravity separator then removes the sawdust particles from the clover seed, taking with them the otherwise inseparable weed seeds. Dodder seeds have a very rough surface to which small steel or iron filings will cling (Figure 9.9). Filings are mixed with clover seed that contains dodder, after which a magnetic separator is used to draw out the metal-coated dodder seeds. Another type of magnetic separator equipped with powerful electromagnets is placed in grain and seed spouts to

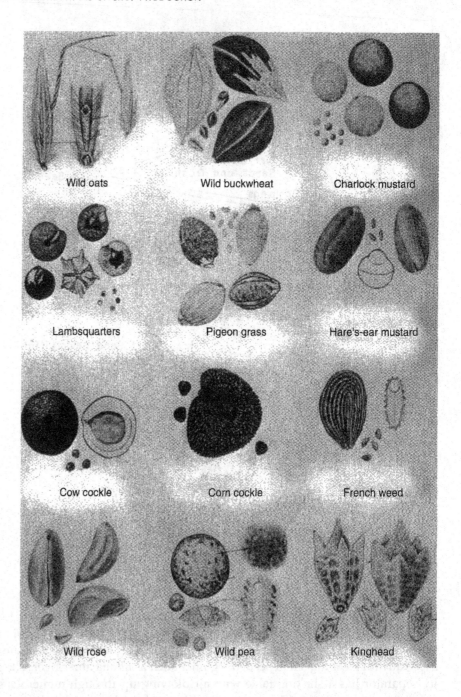

FIGURE 9.8
Weed seeds (enlarged and comparative sizes).

catch any nails, pieces of iron, or steel scrap in a moving stream of grain. This not only prevents damage to grinding or other machinery but also reduces the hazard of fires that might start from accidental sparks from striking metals.

Other cleaning machines include the velvet-roll separator, which removes rough seeds from smooth ones, and electrostatic separators, which operate on the differences in electrical properties of seeds. A machine with a photoelectric cell can separate seeds of different colors. In some admixtures, other varieties or discolored seeds can be removed with this equipment. Awns are removed from some grass seeds with beaters or hammer mills.

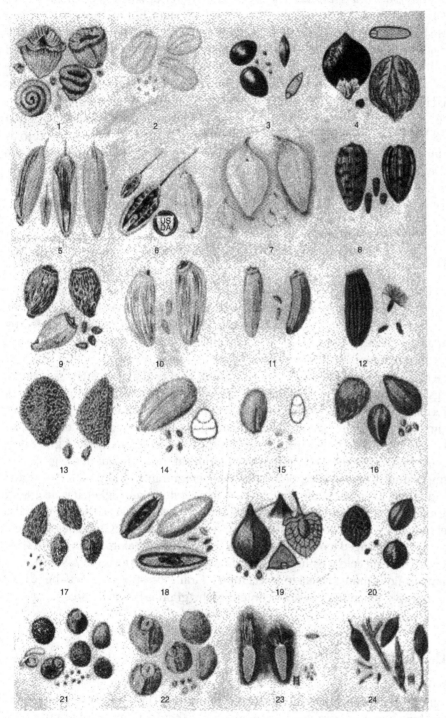

FIGURE 9.9

Seeds of common weeds:
1. Russian thistle
2. Tumbling mustard
3. Rough pigweed or redroot
4. Pennsylvania smartweed
5. Chess or cheat
6. Darnel
7. Wild garlic or onion
8. Wild sunflower
9. Star thistle
10. Bull thistle
11. Canada thistle
12. Perennial sowthistle
13. Field bindweed
14. Peppergrass
15. Whiteweed
16. Poverty weed
17. Broad-leaved plantain
18. Buckhorn or narrow-leaved plantain
19. Curled dock
20. Sheep sorrel
21. Clover dodder
22. Field dodder
23. Horseweed
24. Nutgrass

▧ 9.10 INSECTS IN STORED GRAIN

Grain in storage is subject to damage from the attack of several species of insects.[26, 27] The most common insect pests of stored grain are the rice weevil (*Sitophilus oryzae*), the granary weevil (*Sitophilus granarius*), and the Angoumois grain moth *(Sitotroga cerealella)*. The lesser grain borer or Australian wheat weevil *(Rhizopertha dominica)* is destructive to grain in elevators, but it is not often present in farm bins. The Indian meal moth *(Plodia interpunctella)* damages corn and some-times other grains. The cadelle *(Tenebroides mauritanicus)* not only consumes stored grain but also burrows into the walls of wooden bins. The khapra beetle *(Trogoderma granarium)* is very destructive to stored grain in warm climates. It ap-peared in the United States about 1955, but has since been largely eradicated. Other insects, including beetles, moths, and mealworms, that are found in grain but that subsist largely on broken grains and milled products are popularly referred to as bran bugs. Among the more common bran bugs are the saw-toothed grain beetle *(Oryzaephilus surinamensis)*, the flat grain beetle *(Laemophloeus minutus)*, the con-fused flour beetle *(Tribolium confusum)*, and the red flour beetle *(T. castaneum)*. Species of weevils infest the seeds of beans, peas, and several other legumes.

9.10.1 Means of Insect Infection

The rice weevil, broad-nosed grain weevil, Angoumois grain moth, flour beetles, and certain other insects often fly to the field and attack grain still standing in the field. When harvest or threshing is delayed, the grain may be heavily infested by the time it is placed in storage. The granary weevil has no wings under its wing covers and therefore is unable to fly.

The rice weevil and granary weevil are snout beetles that deposit eggs in small borings in the kernels. The Angoumois grain moth deposits eggs on the surface of the kernels. When the larvae of these insects hatch, they bore inside the kernels, consume the contents, pupate, and then emerge as adults. The grain borers have similar feeding habits. Other insects such as the cadelle, square-necked grain bee-tle *(Cathartus quadricollis)*, and Indian meal moth eat the germ out of the grain.

Temperature is the most important factor in determining the abundance of stored grain insects in a region. In the South, where temperatures are higher, it is necessary to use the greatest precautions to protect grain from storage insects. The rice weevil is dormant at 45°F (7°C) or below, and the granary weevil at 35°F (2°C) and below. These insects die at prolonged temperatures below 35°F (2°C). Therefore, grain is moved to another bin during cold weather to break up hot spots due to insect infestation.

Damp grain is more subject to insect injury. Grain-infesting insects either die or fail to multiply when the moisture content of the grain is below 9 percent.

9.10.2 Control of Insects in Stored Grain

Grain or seed should be cleaned for storage to remove cracked material and live in-sects and then placed in clean bins free from insects. The walls and floors of the cleaned bin should be sprayed with an insecticide before filling with grain.

Stored grain or seed should be fumigated when it becomes infested with insects. Fumigation is the application of a gaseous insecticide to create a poisonous atmo-sphere to control pest insects in grain. Fumigants are dangerous and must be ap-plied by teams of applicators (two people minimum) who are trained, certified, experienced, and who have the proper clothing and safety equipment. Fumigation is not effective unless the storage structure to be treated is well sealed and the grain temperature is well above 50°F (10°C). All instructions must be followed carefully when fumigating, especially those concerning applicator safety.

Currently, only three fumigants are labeled for use in stored grain: methyl bromide, chloropicrin, and phosphine. Each of these products has special limitations and restrictions governing its use.

Methyl bromide is an effective fumigant to control stored-grain insects at all stages of development. It becomes a gas at temperatures above 39°F (4°C) and has no odor or irritating qualities to indicate its presence. The use of methyl bromide, an ozone-depleting chemical, is currently being phased out and was banned in all developed countries in 2005.

Chloropicrin is registered for grain fumigation and also empty bin fumigation. It is especially useful for the control of insects in the sub-floor aeration area of empty bins.

Phosphine-producing materials have become the predominant fumigants used for the treatment of bulk stored grain throughout the world. It is available in solid formulations of aluminum phosphide or magnesium phosphide.[28] Aluminum phosphide is formulated in tablets, pellets, paper sachets, plates, and blister strips. The tablet or pellet formulations are most suitable for farm applications. Solid aluminum phosphide formulations release hydrogen phosphide (phosphine) gas when exposed to moisture and heat. Warm, humid air accelerates the reaction while cool, dry air slows it down. The reaction starts slowly, gradually accelerates, and then tapers off.[6]

Heating grain in a dryer at 130 to 140°F (54 to 60°C) for 30 minutes will kill all insect life.

An atmospheric concentration of 35 to 60 percent carbon dioxide in a tight bin kills insects in seven to four days, respectively. Diatomaceous earths or silica gels applied to grain prevent insect attack.

9.11 MARKETING HAY

The market price of hay depends upon palatability, nutritive value, and appearance. Hay that is a bright green color was cut before it was too ripe, was cured promptly, and is high in carotene. Leafy legume hay that contains 40 to 50 percent or more of its weight in leaves is high in protein. Hay leached by rain loses water-soluble carbohydrates and amino acids, which lowers its nutrient value. Foreign material can be offensive, inedible, less nutritious, or even injurious to livestock. Hay cut when too ripe contains seed that can be rejected or lost by shattering. Seed development withdraws nutrients from the leaves and stems, making the latter less valuable. Stems of overripe hay are usually coarse, woody, and high in crude fiber. Coarse stems are usually rejected, especially by cattle. Horses may eat the coarser stems but refuse shattered seeds and leaves.

9.12 FORAGE QUALITY

There are no established standards, market classes, or grades for forages as there are for grain and seeds. A system was proposed many years ago by the American Forage and Grassland Council (AFGC), but it has not been widely adopted. However, grain and forages that will be fed to livestock should be tested for nutritive content and overall quality to maximize animal performance and minimize feed costs.

Sight, smell, and touch are useful indicators of feed value, but they are of limited value. Stage of maturity at harvest, foreign material, pests, color, and leafiness can be visually detected and provide some limited information on the nutritional value of feed. Musty and foul odors can indicate lower quality due to deterioration in storage. Physical evaluations alone are not sufficient for predicting eventual animal performance. Physical evaluations must be combined with quantitative laboratory tests to adequately evaluate feed quality.

9.12.1 Near Infrared Reflectance (NIR) Spectroscopy

NIR is a fast, low-cost, and reliable method to analyze forage for its nutrient content. NIR uses near infrared light rather than chemicals to identify important compounds and measure their amount in a sample. Forage can be analyzed in less than 15 minutes. Rapid results and lower cost make NIR an attractive method of analysis. Because the NIR method is still relatively new, some feeds and certain nutrients cannot be tested by this method.

9.12.2 Chemical Tests

Nutrient analyses will chemically react with, or extract, important compounds in a laboratory to determine their amount in feed. When representative feed samples are tested chemically, accurate predictions of animal performance can usually be made because nutrient requirements also were determined using chemically tested feeds.

DRY MATTER Dry matter (DM) is the percentage of the forage that is not water. Feed higher in moisture will generally be lower in nutrients because moisture dilutes the concentration of nutrients. Nutrient values should be converted to a dry-matter basis when formulating livestock rations. Feed in a ration should then be converted from the dry basis to an "as fed" basis to obtain the amounts to feed or mix in the ration.

CRUDE PROTEIN Crude protein (CP) measures both true protein and non-protein nitrogen. A CP test will help determine the need for protein supplements. However, crude protein is not a good predictor of the productive energy value of feed.

INSOLUBLE CRUDE PROTEIN All of the following: insoluble crude protein (ICP), acid detergent insoluble nitrogen (ADIN), unavailable nitrogen, and heat-damaged protein, refer to crude protein that has become chemically linked to carbohydrates to form an indigestible compound. This is caused by overheating and is most common in silage stored at less than 65 percent moisture and in hay bales or stacks that contain more than 20 percent moisture. Heat-damaged forage often has an amber, tobacco brown, or charcoal discoloration and a carmelized odor. However, sometimes no discoloration occurs.

Some ICP is usually present in normal forages. When the ICP:CP ratio is above 0.1, excessive heating has probably occurred, causing reduced protein digestibility.

When this happens, crude protein values should be adjusted downward to more accurately balance the rations for protein.

ADJUSTED CRUDE PROTEIN Adjusted crude protein (ACP) is a calculated protein value that corrects for heat damage. Whenever ICP:CP exceeds 0.1, ACP should be used in place of crude protein to balance rations. Most laboratories will compute and report the ACP for forages that have been analyzed for both protein and ICP.

DIGESTIBLE PROTEIN Digestible protein (DP) is reported by some laboratories. Although potentially useful in specialized situations, nearly all common ration formulations, as well as approved nutrient requirements, are based on crude protein and have been adjusted for digestibility. DP values are not used except with the guidance of a reputable nutritionist.

CRUDE FIBER Crude fiber (CF) is an older method to determine fiber content. Newer fiber tests are better measures of nutritional value. Neutral detergent fiber and acid detergent fiber analyses should be used instead of CF to evaluate forages and formulate rations.

NEUTRAL DETERGENT FIBER Neutral detergent fiber (NDF) measures the structural part of the plant, the plant cell wall. NDF provides bulk to the diet, which limits intake. Because NDF can be used to predict intake, it is a very valuable analysis to conduct on forages for dairy rations and can also be useful for beef rations that rely mostly on forages. Low NDF values are desired. As maturity of the plant increases at harvest, cell wall content of the plant increases and NDF increases.

ACID DETERGENT FIBER Acid detergent fiber (ADF) measures cellulose, lignin, silica, insoluble crude protein, and ash, which are the least digestible parts of the plant. Because ADF percentage in forages negatively relates to digestibility, it is used to calculate energy values. ADF is one of the most common analyses made on forages. Forages with low ADF will have higher net energy. As the plant matures, ADF increases.

DIGESTIBLE DRY MATTER Digestible dry matter (DDM) estimates the percentage of forage that is digestible. It is calculated from ADF using the equation

$$DDM \ (\%) = 88.9 - [ADF \ (\%) \times 0.779]$$

NET ENERGY Net energy (NE) is the energy available to an animal in the feed after removing the energy lost as feces, urine, gas, and heat produced during digestion and metabolism. NE is the most useful energy estimate for formulating rations. The net energy value of feed depends on whether the feed is used for maintenance (NE_m), weight gain (NE_g), or milk production (NE_l). These values are obtained from DDM using formulas shown in Tables 9.6 and 9.7.

TOTAL DIGESTIBLE NUTRIENTS Total digestible nutrients (TDN) represents the total of the digestible components of crude fiber, protein, fat ($\times 2.25$), and nitrogen-free extract in the diet. This value is often calculated from ADF. It is less accurate than NE for formulating diets containing both forage and grain. Most rations now are formulated using NE, but TDN is still used to calculate beef cow rations where the diet is primarily forage.

TABLE 9.6 Estimates of Net Energy of Legumes from Percent ADF (from Grant, 1997)

ADF (%)	DDM[1] (%)	TDN[2] (%)	NE$_i$[3]	NE$_m$[4] (Mcal/100 lb)	NE$_g$[5]
20	73	77	81	84	56
22	72	75	78	82	53
24	70	73	76	78	50
26	69	71	73	75	48
28	67	69	71	72	45
30	66	67	69	69	42
32	64	64	66	66	40
34	62	62	64	63	37
36	61	60	62	60	34
38	59	58	59	57	31
40	58	56	57	53	28
42	56	54	54	50	25
44	55	52	52	47	22
46	53	49	50	43	18
48	52	47	47	40	15
50	50	45	45	36	12
52	48	43	43	33	9
54	47	41	40	29	5
56	45	39	38	25	2

[1] $DDM = 88.9 - (.779 \times ADF)$.

[2] $TDN = 4.898 + (NE_i \times .89796)$.

[3] $NE_i = 104.4 - (1.19 \times ADF)$.

[4] $NE_m = (137 \times ME) - (30.42 \times ME^2) + (5.1 \times ME^3) - 50.8$.

[5] $NE_g = (142 \times ME) - (38.36 \times ME^2) + (5.93 \times ME^3) - 74.84$.

$ME = TDN \times .01642$.

TABLE 9.7 Estimates of Net Energy of Corn Silage from Percent ADF (from Grant, 1997)

ADF (%)	DDM[1] (%)	TDN[2] (%)	NE$_i$[3]	NE$_m$[1] (Mcal/100 lb)	NE$_g$[1]
20	73	74	80	78	50
22	72	72	77	77	49
24	70	71	75	76	48
26	69	70	72	75	47
28	67	68	70	73	46
30	66	67	67	72	45
32	64	66	65	71	44
34	62	64	62	70	43
36	61	63	60	68	42
38	59	62	57	67	40
40	58	60	55	66	39

[1] See Table 9.6 footnotes for these measures.

[2] $TDN = 31.4 + (NE_i \times .531)$.

[3] $NE_i = 104.4 - (1.24 \times ADF)$.

DRY MATTER INTAKE Dry matter intake (DMI) estimates the maximum amount of forage dry matter a cow will eat. It is expressed as a percent of body weight and is calculated from NDF using the equation

$$\text{DMI (\% of body weight)} = 120 \: / \: \text{NDF \%}$$

RELATIVE FEED VALUE Relative feed value (RFV) combines digestibility and intake into one number for a quick, easy, effective way to evaluate the quality of forages. It is used primarily with legume or legume/grass forages.

$$\text{RFV (\%)} = [\text{DDM (\%)} \times \text{DMI (\% of BW)}] \: / \: 1.29$$

Relative feed value is most useful for animals on high-forage rations, such as dairy cows and growing animals, because the RFV provides an index to rank a forage according to its digestible energy intake potential. RFV has also been used widely in hay marketing. The relationships among RFV, DDM, DMI, NDF, and ADF are shown in Table 9.8 for alfalfa, along with suggested livestock to best utilize different qualities of forage for greatest return. RFV is only an energy intake index and does not give credit for, or estimate the amount of, protein or minerals in the forage. RFV should not be used alone to formulate or develop rations.

OTHER TESTS When calcium and phosphorus supplements are needed, or when control of their levels is important, an analysis for mineral content may be needed. An example is the control of calcium and phosphorus levels in the dry cow ration of a dairy herd plagued by milk fever. Because of the low cost of supplementation and a cow's ability to tolerate wide variation in ration calcium and phosphorus levels, routine mineral analysis is usually not justified.

Trace minerals are commonly added as insurance against deficiencies. Because of the high cost, trace-mineral analyses are not usually needed. However, trace minerals should be checked periodically or when a specific case suggests the need.

Forages that have been damaged by drought or hail, are stunted, or are harvested before maturity may contain high levels of nitrates. Nitrate levels on these forages should be analyzed to determine whether the forage is potentially toxic.

TABLE 9.8 Relationships among Several Estimators of Alfalfa Quality and Suggested Livestock Uses (from Grant, 1997)

Uses	ADF	NDF	DDM	DMI	RFV
		(% of DM)		(% of Body Weight)	
Prime dairy, fresh and high producers	<31	<40	>65	>3.0	>151
Good dairy, young heifers, excellent backgrounding	31–35	40–46	62–65	2.6–3.0	125–151
Good beef, older heifers, marginal for dairy cows	36–40	47–53	58–61	2.3–3.5	103–124
Maintenance for beef or dry dairy cows	41–42	54–60	56–57	2.0–2.2	87–102
Poor quality[a]	43–45	61–65	53–55	1.8–1.9	75–86

[a]Requires supplementation with higher-quality (energy) feeds as well as possibly other nutrients for most animals. This quality hay could be fed to dry beef cows under some circumstances.

Immature sorghums and sudans and especially the regrowth of previously harvested stunted forage can have excessively high prussic acid levels. However, prussic acid is rarely a problem in harvested forages.

■ 9.13 CROP JUDGING

Exhibition of crop products has considerable educational, cultural, and recreational value. In premium awards for the best products, consideration is given mostly to appearance, uniformity, size, and trueness to type based upon weighted point scores adopted by exhibit officials or judges. Such scores usually reflect market values for the particular crop commodity. However, they have little relation to productivity or to the inherent value of the variety or type being shown. Prizewinning crops usually indicate two things: (1) favorable environment for plant development and (2) meticulous selection and preparation of the exhibit material. The best plants are often found where thin stands have permitted optimum individual plant development at the expense of yield per unit area of land. However, good cultural conditions along with freedom from disease and insect injury are essential to the production of acceptable crop exhibits.

Standards of excellence in crop products are largely arbitrary. In some cases, they are based upon characteristics that actually are harmful to crop yield or quality. For example, deep rough kernels of corn indicate late maturity, slow drying, and susceptibility to kernel rots. Compact heads of sorghum favor worm damage and moldy grain. Tall corn stalks make harvesting difficult. Certain varieties of wheat known to be of inferior quality for making bread regularly take prizes at local, state, and international exhibits. These prizes are won because the grain has a high test weight as well as plump, vitreous, and smooth kernels. The substitution of utilitarian or quality standards does not fully remedy this incongruous situation.

The agronomic merits of a crop or crop variety are determined by comparative experimental tests and by established performance on farms and in markets and processing plants. Experimental facts outweigh the opinions of the best crop judges who examine only a few selected exhibit samples. However, it is incumbent upon crop students to learn to judge and exhibit crops because the present type of agricultural fair will continue in vogue for many years. People thrive on competition. Furthermore, the judging of exhibits develops skill in the appraisal, identification, and grading of crop products.

REFERENCES

1. Brandenburg, N. R., and J. E. Harmond. "Fluidized conveying of seed," *USDA Tech. Bull.* 1315, 1964, pp. 1–41.

2. Harmond, J. E., and others. "Mechanical seed cleaning and handling," *USDA Handbook* 354, 1968, pp. 1–56.

3. Bouland, H. D., and L. L. Smith. "A small country elevator for merchandizing grain," *USDA Marketing Res. Rpt.* 387, 1960.

4. Bruce, W. M., and others. "Planning grain storage elevators," *U. GA Bull.* 51(7d), 1951.

5. Clough, M., and J. W. Browning. "Feed grains," in *Marketing*, USDA Yearbook, 1954, pp. 403–413.

6. Rowe, H. B. "The danger of loss," in *Marketing,* USDA Yearbook, 1954, pp. 309–323.

7. Lutgen, L. H. "Evaluating options vs. futures contracts." *NE Coop. Ext. Serv. NebGuide* G86–771-A, 1986.

8. Lutgen, L. H., and L. A. Todd. "An introduction to grain options on futures contracts." *NE Coop. Ext. Serv. NebGuide* G85–770-A, 1985.

9. Mehl, J. M. "The futures markets," in *Marketing,* USDA Yearbook, 1954, pp. 323–331.

10. Anonymous. "The service of federal grain standards," *USDA Misc. Pub.* 328, 1938, pp. 1–18.

11. Anonymous. "Grain grading primer," *USDA Misc. Pub.* 740, 1957.

12. Carlson, E. *Biennial Rpt. Kans. State Grain Insp. Dept.,* Kansas City, Mo. 1944, pp. 1–29.

13. Anonymous. "Official grain standard of the United States," *USDA SRA-AMS-177* (rev.). January 2002.

14. Christensen, C. M., and H. L. Kaufman. *Grain Storage.* Minneapolis: University of Minnesota Press, 1969.

15. Carter, E. P., and G. Y. Young. "Effect of moisture content, temperature, length of storage on the development of sick wheat in sealed containers," *Cereal Chem.* 22, 5(1945):418–428.

16. Hukill, W. V., R. A. Saul, and D. W. Teare. "Outrunning time; combating weather," in *Power to Produce,* USDA Yearbook, 1960, pp. 183–199.

17. Anonymous. "Drying shelled corn and small grain with heated air," *USDA Leaflet* 331, 1952.

18. Anonymous. "Drying shelled corn and small grain with unheated air," *USDA Leaflet* 332, 1952.

19. Hall, C. W. *Drying Farm Crops.* Reynoldsburg, OH: Agricultural Consulting Associates, Inc., 1957, pp. 1–336.

20. Anonymous. "Drying shelled corn and small grains," *USDA Farmers Bull.* 2214, (rev.), 1971, pp. 1–12.

21. Heard, R. F., and others. "Shelled corn handling systems," *Ont. Dept. Agr. & Food Pub.* 551, 1969, pp. 1–33.

22. Strickler, P. E., and W. C. Hinson. Jr. "Crop drying in the United States, 1966, quantity, equipment and fuel used," *USDA Stat. Bull.* 439, 1969, pp. 1–22.

23. Holman, L. E. "Aeration of grain in commercial storages," *USDA Mktg. Rpt.* 178, 1960.

24. Gregg, B. R., and others. *Seed Processing.* Miss. State Univ., Nat'l Seeds Corp & USAID, New Delhi, India: Avion Printers, 1970, pp. 1–396.

25. Harmond, J. E., L. M. Klein, and N. R. Brandenburg. "Seed cleaning and handling," *USDA Handbook* 179, 1961.

26. Anonymous. "Stored grain pests," *USDA Farmers Bull.* 1260 (rev.), 1958.

27. Anonymous. *Stored-Grain Insects.* USDA Agricultural Research Service, and the Association of Operative Millers, Agriculture Handbook, Number 500. 57 pp. 1986.

28. Harein, P., and B. Subramanyam. "Fumigating stored grain." *U. MN Ext. Serv.* FS-01034-GO, 1990.

Pastureland and Rangeland

■ 10.1 IMPORTANCE OF PASTURES AND RANGE

Range is defined as uncultivated grasslands, shrublands, or forested lands with an herbaceous and/or shrubby understory, used mainly for grazing or browsing by domestic and wild animals.[1] Range includes lands with native vegetation and lands naturally or artificially revegetated with native or adapted forage species. *Pasture* is crop land used temporarily for grazing. Pasture can include irrigated perennial pasture, rotation pasture, annual pasture, and crop aftermath pasture.[2]

About 8,400 million acres (3,400 million ha) of the earth's uncultivated land area is used for grazing. The grazing area in the United States in 1997 was about 491 million acres (199 million ha). Of this, 397 million acres (161 million ha) are grassland pasture, 30 million (12 million ha) are pastured woodland, and 64 million (26 million ha) are pastured crop land. In addition, millions of acres of harvested crop lands are grazed for part of the season to salvage feed from aftermath, stubble, winter grain and hay, and forage seed fields. In 1965, pasture provided 35 percent of the feed for all livestock and 82 percent for sheep and goats.

Development of improved methods of pasture management, together with a better realization of the value of pasturage, has been responsible for an intense interest in the establishment and maintenance of productive pastures since about 1925. In addition, national agricultural policies have favored grassland farming for erosion control.[3] This situation has encouraged renovation of old pastures by tillage, herbicide application, mowing to destroy less desirable plants, fertilization,[4] and reseeding with productive palatable species. On unplowable pastures, the land may be worked with a disk, spring tooth harrow, or other implement. In some sections, a cutaway disk implement, called a *bush-and-bog*, is used for this purpose. Pastures can be improved by replacing Kentucky bluegrass and white clover, which are persistent under heavy grazing, with more productive taller-growing but less permanent crops such as alfalfa, Ladino clover, bromegrass, orchardgrass, tall fescue, and timothy.

More productive forages, with heavy fertilization, can easily double or triple the carrying capacity of pastures in the humid areas. Heavy grazing eventually suppresses the productive species. The pastures then revert to shorter, persistent plants such as Kentucky bluegrass, white clover, common

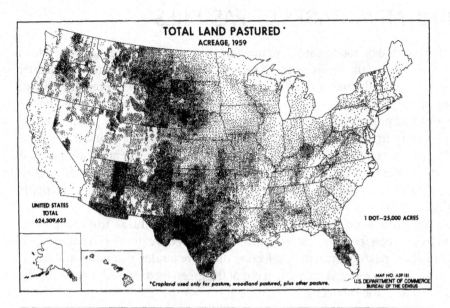

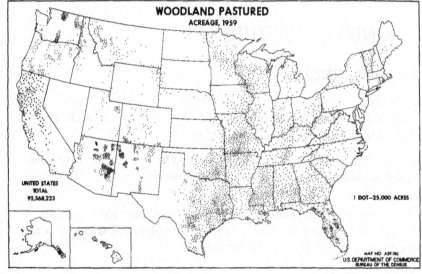

FIGURE 10.1
(Top) Total land pastured in the United States. *(Bottom)* Woodland pastured in 1959. [Courtesy USDA]

bermudagrass, or carpetgrass. Reseeding of improved pastures every two to five years is often necessary.

Millions of acres were plowed for crop production in the semiarid Great Plains and Great Basin regions between 1910 and 1930. The more unfavorable sites were soon abandoned; eventually, they reverted to grassland. The natural process of range restoration on plowed land may require twenty-five to forty years,[5] although good grazing may be available after eight or ten years.[6] Artificial reseeding restores the range quickly, but it is more costly and often fails, except in favorable wet seasons. Improved methods of establishing perennial grasses facilitate range improvement in the drier areas.[7]

The seeding of better grasses, such as crested wheatgrass or wild ryegrass, at suitable sites on western rangeland, may double the forage yield, extend the grazing season, and supplement the native pastures.[8, 9, 10, 11]

▓ 10.2 ADVANTAGES OF PASTURES

The main objective of good pasture maintenance is to provide green succulent feed for livestock during the entire grazing season. Livestock on pastures are cleaner and more comfortable than those kept in dry lots. The animals harvest the crop at a minimum cost. Good pastures produce about two-thirds as much dry matter as the same area in cultivated crops.[12] In Virginia research, the yield of pasturage was 40 to 65 percent that of the same crops allowed to mature and then cut for hay.[13] However, the dry matter of immature grass is higher in protein and vitamins and is more digestible than the dry matter from mature grass harvested for hay. Consequently, pasturing produces about three-fourths as much in digestible nutrients as do the same crops cut for hay[12] (Table 10.1).

Dairy cows produced more milk per unit of dry matter intake from strip grazing than from green feeding or stored feed, due to selective grazing. They rejected 33 percent of the pasture growth in selecting the more tender, palatable, and nutritious herbage. They rejected only 2 percent of the green feed, and 8.5 percent of the coarser portion of the stored feed.[14]

▓ 10.3 KINDS OF PASTURES

10.3.1 Tame Pastures

Tame pastures are lands once cultivated that have been seeded to domesticated pasture plants and are used mostly or entirely for grazing by livestock. The principal kinds of tame pastures include the following:

1. *Permanent pastures* are grazing lands occupied by perennial pasture plants or by self-seeding annuals that remain unplowed for long periods (five years or more).

2. *Rotation pastures* are fields used for grazing that are seeded to perennials or self-seeding annuals but form a unit in the crop rotation plan and are plowed within a five-year or shorter interval.

TABLE 10.1 Digestible Nutrients in Harvested Crops and Pasturage

Crop	Roughage Yield per Acre (tons)	Total Digestible Nutrients Harvested (lb)	Total Digestible Nutrients Grazed (lb)
Alfalfa	2.61	2,793	2,067
Clover: red, alsike, and crimson	1.46	1,504	1,113
Sweetclover	1.82	1,867	1,382
Lespedeza	1.82	1,170	866
Clover and timothy mixed	1.40	1,393	1,030
Timothy	1.26	1,210	895
Grains cut green	1.31	1,292	956
Annual legumes	1.02	1,102	815
Millet, johnsongrass, and sudangrass	1.23	1,225	906

3. *Supplemental pastures* are fields used for grazing when the permanent or rotation pastures do not supply enough feed for the livestock on the farm. Such pastures may be the aftermath of meadows, small-grain stubble, seeded small grains, annuals such as sudangrass, lespedeza, and crimson clover, or biennials such as sweetclover.

4. *Annual pastures* are pastures that are seeded each year to take the place of permanent pasture. Such pastures may include a series of crops such as rye, oat, barley, sudangrass (Figure 10.2), Italian ryegrass, vetch, soybean, and rape.

5. *Renovated pastures* are those restored to former production by tillage, mowing, reseeding, or fertilization.

10.3.2 Natural or Native Pastures

Natural or native pastures are uncultivated lands occupied mostly by native or naturally introduced plants useful for grazing. The main types follow:

1. *Ranges* are very large natural pastures comprised mostly of perennial grasses.

2. *Brush pastures* are areas covered mostly with brush and shrubs. Livestock obtain a large portion of their feed from woody plants.

3. *Woodland pastures* are wooded areas with grass and other edible herbage growing in open spaces and among trees.

4. *Cut-over* or *step pastures* are lands from which trees have been cut, leaving stumps and usually some new tree growth.

▥ 10.4 PERMANENT PASTURES

10.4.1 Pasture Areas

Most of the area in the eastern half of the United States, as well as in the north Pacific coastal region, was originally in forest. As the trees were removed, introduced plants occupied the pastured land. The western area between these general regions includes the prairie lands or tall-grass region, the native short-grass region of the Great Plains, and the native desert grasses and shrubs of the intermountain region.

10.4.2 Environmental Conditions

The 60°F (16°C) annual isotherm marks approximately the northern limit of usefulness of southern pasture plants. The exceptions to this are mostly annuals such as lespedeza and sudangrass. North of this line, southern grasses are subject to winter injury, while to the south bluegrass, orchardgrass, timothy, redtop, and most clovers are unable to thrive during the long period of high temperatures. This temperature effect is particularly important in the humid regions.

Rainfall differences largely determine the flora in the Great Plains and intermountain area. In the intermountain region, the annual precipitation is so low that

FIGURE 10.2
(Top) Sheep range in Colorado. *(Middle)* proved pasture in Mississippi. *(Bottom)* Temporary sudangrass pasture in Nebraska.

desert or semi-desert conditions prevail except in the extreme northern part and at higher altitudes in the mountains. The Pacific Northwest has a fairly abundant winter rainfall and a mild climate due to the Japanese Current.

Although mountain valleys, parks, and clearings furnish considerable pasturage (Figure 10.1), the higher mountain slopes are unimportant from the national pasture standpoint, because in most sections they are almost completely forested or are composed of rock masses with very little productive soil. The most valuable grazing lands are level or rolling areas, particularly those in the Corn Belt. The level grasslands of the Great Plains are also good pasture but are less productive than those in the Corn Belt, due to limited rainfall (Figure 10.3).

Next to climate and topography, soil characteristics have the most influence upon the type of pasture plants that occupy the land. Lespedeza will grow on acid and infertile soils, alsike clover on acid and poorly drained soils, and reed canarygrass on land that is wet. Native wheatgrasses and buffalograss prevail on the heavier soils in the Great Plains, while the gramagrasses are more abundant on the sandy loams.

▨ 10.5 ADAPTED PASTURE SPECIES

In the northeastern region, Kentucky bluegrass *(Poa pratensis)* is probably the most important pasture grass. In mixtures with white clover, it is nearly always present on productive soils in permanent pastures. However, taller hay-type grasses are commonly seeded in pasture mixtures. These include orchardgrass *(Dactylis glomerata)*, smooth bromegrass *(Bromus inermis)*, tall fescue *(Festuca arundinacea)*, timothy *(Phleum pratense)*, and reed canarygrass *(Phalaris arundinacea)*. Other grasses are redtop *(Agrostis alba)* and wild ryegrass *(Elymus* species). The principal pasture legumes are alfalfa, clovers (Ladino, red, and alsike), and birdsfoot trefoil *(Lotus corniculatus)*. Sweetclover is used in the western part of the region, while annual lespedezas are widely grown in the southern portion.

Warm-weather grasses and legumes prevail south of the 60°F (16°C) isotherm, except for some tall fescue and alfalfa in the cooler portions. White clover, vetch, and Austrian pea also are grown there as winter annuals. Tropical grasses are grown on the South Atlantic and Gulf Coastal plains. Native southern grasses are also utilized.[15]

Native short grasses (Figure 10.4) supply much of the pasturage on the Great Plains. The principal grass association in the northern Great Plains consists of blue grama *(Bouteloua gracilis)*, green needlegrass *(Stipa viridula)*, and western needlegrass *(Stipa comata)* with a small admixture of junegrass *(Koeleria cristata)*. Seeded, crested wheatgrass may have twice the grazing value of native grasses.[9, 16] Wild ryegrasses *(Elymus* species) are also productive. The central Great Plains has a very typical association of buffalograss *(Buchloe dactyloides)* and blue grama. Farther south below the Texas Panhandle, black grama *(Bouteloua eripoda)* and curly mesquite *(Hilaria belangerii)* are the most predominant pasture plants. Just east of this region is the tall-grass region with the bluestems *(Andropogon gerardii* and *Schizachyrium scoparium)*, switchgrass *(Panicum virgatum)*, dropseeds *(Sporobolus* species), and sideoats grama *(Bouteloua curtipendula)* among the predominant species. Switchgrass gave much greater steer gains per acre than the seeded native pasture mixture in the central Great Plains.[17] In both the Great Plains and intermountain areas, the

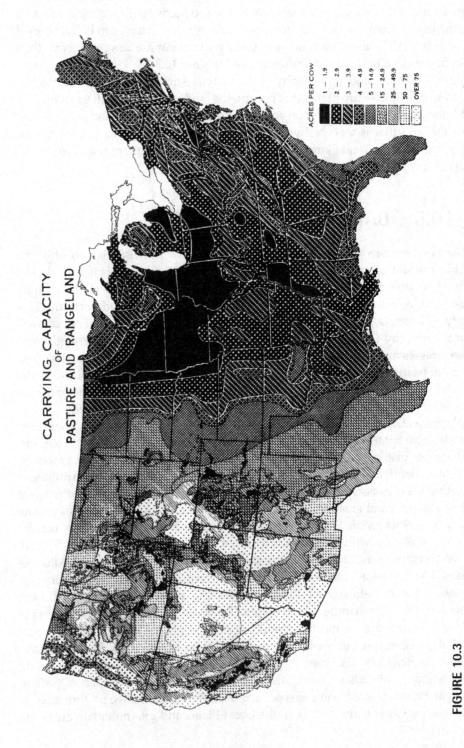

CARRYING CAPACITY
OF
PASTURE AND RANGELAND

ACRES PER COW

1 — 1.9
2 — 2.9
3 — 3.9
4 — 4.9
5 — 14.9
15 — 24.9
25 — 49.9
50 — 75
OVER 75

FIGURE 10.3
Carrying capacity of grazing land in 1947. Pasture and range improvement and better management have increased the carrying capacity of many grazed areas.

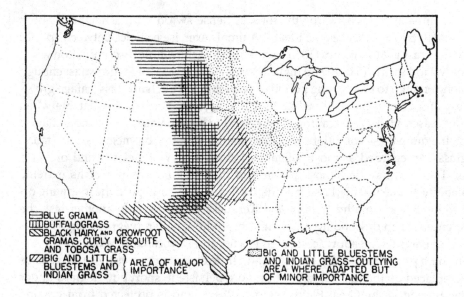

FIGURE 10.4
Leading native pasture grasses of the Great Plains and of prairie regions (dotted area at right). The seeding also of introduced grasses—such as crested wheatgrass, Russian wildrye, and smooth bromegrass—has improved some of the grazing areas.

clovers thrive only in the cooler, irrigated sections. Most of the alfalfa grown there also is irrigated. Sainfoin is sometimes sown for pasture.

The intermountain area has the lowest pasture production per acre because of the low rainfall. In the northern part, thin stands of wheatgrass (*Agropyron spicatum*), little bunchgrass (*Festuca idahoensis*), and Sandberg bluegrass (*Poa sandbergii*) are found.[16] In the central portion of the region (Utah and Nevada) there exists a true desert shrub vegetation characterized by large sagebrush (*Artemisia tridentata*) and shadscale (*Atriplex confertifolia*). During rainy periods, introduced annual bromegrass (*Bromus tectorum*) and similar plants emerge to furnish some grazing. Associated with shadscale is winterfat (*Eurotia lanata*), one of the most valuable grazing plants in the Great Basin. In the southern part of the region, the best grazing lands are found in western New Mexico and eastern Arizona. Drought-resistant grasses that grow among the shrubs include galletagrass (*Hilaria jamesii*), tobosagrass (*Hilaria mutica*), black grama (*Bouteloua eripoda*), and wild ryegrass (*Elymus* species). Disk plowing of the more fertile sagebrush lands, followed by sowing of crested wheatgrass, increased range production several times.[7, 11]

Abundant rainfall and a mild, cool climate along the north Pacific coast permit the establishment of permanent pastures of introduced plants. Grasses that thrive there include tall fescue, ryegrasses, Kentucky bluegrass, orchardgrass, tall oatgrass, timothy, meadow foxtail, reed canarygrass, and bentgrass. Clover and alfalfa also do well in this area.

The rainfall in central and southern California is so low that most of the pastures in the valleys are irrigated. Important pasture plants are alfalfa, Ladino clover, hardinggrass, orchardgrass, subclover, medics, and strawberry clover. Alfileria (*Erodium cicutarium*) and various native grasses provide grazing in the uplands.

■ 10.6 CHARACTERISTICS OF PASTURE PLANTS

Short, young herbage low in fiber content is eaten in preference to old, tall, stemmy, highly fibrous herbage. All grasses are palatable when closely clipped.[18] They become unpalatable when growth stops. A dense sward with a height of about 4 inches (10 cm) approaches the ideal.[19] Animals exercise a choice between species when the grass is 4 to 6 inches (10 to 15 cm) high but show little discrimination when it is only 2 to 4 inches (5 to 10 cm) high.[20] Perennial ryegrass leaves grow fast enough and long enough to have a long period of palatability.[18] Orchardgrass leafage grows rapidly and is very palatable while in active growth, but it is unpalatable when growth ceases. Hairy and scabrous plants tend to be unpalatable. The more mature a leaf, the more these conditions are accentuated. Timothy, bromegrass, and Italian ryegrass are regarded as most palatable, followed by white clover and orchard- grass. Timothy may be almost suppressed in a pasture because of its extreme palatability. In one test in Massachusetts, cattle selected herbage in the following or- der: white clover, timothy, redtop, and Kentucky bluegrass.[20] In another test, the preference was, in descending order, timothy, redtop, Italian ryegrass, English rye- grass, tall oatgrass, meadow fescue, red fescue, and reed canarygrass.

The high palatability of big bluestem and little bluestem causes these tall grasses to be replaced by the less palatable, but more persistent, short buffalograss under heavy grazing in the Great Plains region. Where land is protected from grazing these palatable species are restored and the grass again grows "as high as a horse's belly," as pioneers reported when they observed it for the first time.

■ 10.7 BASIC PRINCIPLES OF PLANT BEHAVIOR

Important principles of pasture plant behavior are:[18]

1. A plant excessively grazed will have its leaf development and root system reduced proportionally. Moreover, the new leaves are reduced in size, resulting in lower functional efficiency. In Nebraska, sod blocks of seven prairie grasses clipped every two weeks produced only 13 to 47 percent as much dry matter as did the unclipped sods.[21] Plants weakened by clipping renew growth slowly.
2. Plants can withstand defoliation in proportion to their ability to develop side shoots and tillers in spite of defoliation. Clipped plants can fail to pro- duce new rhizomes.[22]
3. Plants are damaged in direct proportion to their yielding capacity.
4. Other things being equal, plants are grazed in proportion to their erectness or accessibility.
5. Plants with an erect growth habit are usually less efficient in the produc- tion of side shoots and tillers than prostrate plants.
6. Severe defoliation is more harmful when the plants commence active growth than at other times of the year. Consequently, plants that start active growth exceptionally early in the spring are most subject to damage. The clipping of seedling grass plants[23] decreased the dry weight of forage by 80 to 96 percent.

■ 10.8 PLANTS IN PASTURE MIXTURES

It is seldom desirable to seed land intended for permanent pasture to a single species for the following reasons:[3]

1. Legumes in pastures help to maintain the nitrogen content of the soil.[21]

2. Mixtures result in a more uniform stand and higher production because several soil conditions are often represented in a pasture.[24]

3. Mixtures provide a more uniform seasonal production because the growth and dormancy periods vary among different plants.[25]

4. Mixtures of grasses and legumes provide a better balanced ration since legumes are richer in both proteins and minerals.[26] However, animals often are subject to bloat when the proportion of alfalfa or clover exceeds 50 percent.

Addition of wild white clover to a seeding of Kentucky bluegrass increased the yield of herbage more than 500 percent in New York experiments.[19] The protein content of Kentucky bluegrass grown alone averaged 18 percent while the same grass grown in association with wild white clover averaged 25 percent. A similar gain may be achieved by a heavy application of nitrogen fertilizer.

The number of pounds of the different seeds in a mixture determines only in part the percentages of the different species that emerge or the relative stands that result later, due to differences in the size of seeds, the percentage of live seeds, and the plant competition.[27, 28]

10.8.1 Rates of Seeding

The rates of seeding grasses and legumes in pastures are shown in Table 10.2.

TABLE 10.2 Rates of Seeding Grasses and Legumes in Pasture Mixtures

Plant	Rate (lb/acre)	Plant	Rate (lb/acre)
Cool-Weather Grasses		Boer lovegrass	0.5
Tall fescue	8–10	Weeping lovegrass	0.5
Timothy	3–6	Galletagrass	5
Orchardgrass	5–8	Sand dropseed	0.5–1
Bromegrass (awnless)	4–10	Hardinggrass	2–4
Mounting bromegrass	2–6	Carpetgrass (seeds naturally)	
Reed canarygrass	3–10	Bermudagrass (stolons)	
Tall oatgrass	3–4	Big bluestem	5–6
Meadow fescue	8–10		
Kentucky bluegrass	1–10	Cool-Weather Legumes	
Canada bluegrass	1–3	Alfalfa	2–10
Bulbous bluegrass	1–2	Ladino clover	1–4
Red top	2–10	White clover	1–3
Perennial ryegrass	6–10	Red clover	3–10
Crested wheatgrass	4–11	Alsike clover	2–6
Slender wheatgrass	2–5	Strawberry clover	2–6
Russian wildrye	5	Sweetclover	2–10
		Birdsfoot trefoil	4–6
Warm-Weather Grasses			
Dallisgrass	4–10	Warm-Weather Legumes	
Blue grama	4–8	Lespedeza (Korean and common)	5–15
Sideoat grama	3–4	Low hop clover	1–3
Rothrock graula	3–4	Persian clover	1–3
Buffalograss	1–2	California burclover	2–8
Lehman lovegrass	0.5	Black medic	3

Seeding rates of grasses and legumes are affected by the other species included in the mixture. Seed quality is also a factor. Rates are affected by the number of seeds per pound for a species, pure live seed ratio, amount of foreign material, and other conditions.

■ 10.9 PASTURE MIXTURES FOR DIFFERENT REGIONS

The rates of seeding of pasture mixtures, the adapted pasture plants, and some of the recommended mixtures for different regions and conditions in the United States follow:

Region	Adapted Pasture Plants
Northeast (20–25 pounds per acre)	Orchardgrass, bromegrass, timothy, reed canarygrass, tall fescue, Kentucky bluegrass, redtop, perennial ryegrass, Ladino clover, alfalfa, red clover, alsike clover, birdsfoot trefoil, white clover, Korean lespedeza

Recommended Mixtures:
(1) General use: Orchardgrass; tall fescue, or timothy; and alfalfa, or Ladino clover, or red clover
(2) Poorly drained land: Reed canarygrass or redtop; and alsike clover or Ladino clover
(3) Poorly controlled grazing areas: Kentucky bluegrass and white clover included with Mixture No. 1

North Central (11–20 pounds per acre)	Bromegrass, orchardgrass, timothy, tall fescue, reed canarygrass, redtop, Kentucky bluegrass, alfalfa, Ladino clover, red clover, alsike clover, birdsfoot trefoil, Korean lespedeza, sweetclover

Recommended Mixtures:
(1) Bromegrass; orchardgrass or tall fescue or timothy; and alfalfa or Ladino clover
(2) Timothy, Ladino clover, red clover, alfalfa
(3) For poorly drained soils: timothy, redtop, reed canarygrass, alsike clover, birdsfoot trefoil

Northern Great Plains Eastern portion (12 pounds per acre)	Crested wheatgrass, intermediate wheatgrass, bromegrass, alfalfa, wild rye, feather bunchgrass
Western portion (10 pounds per acre)	Crested wheatgrass, sweetclover, blue grama, sideoats grama, buffalograss, western wheatgrass, wild rye, feather bunchgrass
Northern intermountain irrigated (16–25 pounds per acre)	Bromegrass, orchardgrass, alsike clover, Ladino clover, alfalfa, strawberry clover, alsike clover, and reed canarygrass for wetlands
Northwestern irrigated (16–25 pounds per acre)	Bromegrass; tall fescue; orchardgrass; tall oatgrass; alfalfa; and red, alsike, Ladino, New Zealand white, and strawberry clover; birdsfoot trefoil; big trefoil
Southwestern irrigated (13–30 pounds per acre)	Orchardgrass, perennial ryegrass, tall fescue, Ladino clover, alfalfa, burclover, dallisgrass, rhodesgrass, bromegrass, hardinggrass, yellow sweetclover
Western rangelands (5–13 pounds per acre)	Crested wheatgrass, bluestem wheatgrass, slender wheatgrass, bromegrass, tall oatgrass, sand dropseed, intermediate wheatgrass, wild rye
Southwestern rangelands (8–10 pounds per acre)	Crested wheatgrass, western wheatgrass, bromegrass, galletagrass, yellow sweetclover, Indian ricegrass, domestic ryegrass, weeping lovegrass, subterranean clover, hardinggrass, alfilaria, burnet

Pacific Northwest Coastal wetlands (12–20 pounds per acre)	Tall fescue, reed canarygrass, meadow foxtail, birdsfoot trefoil, big trefoil
Humid uplands (14–20 pounds per acre)	Perennial ryegrass, orchardgrass, tall oatgrass, tall fescue, Chewings fescue, New Zealand clover, subterranean clover, big trefoil, birdsfoot trefoil
Great semiarid (6–11 pounds of per acre)	Crested wheatgrass, Siberian wheatgrass, pubescent wheatgrass, slender wheatgrass, bromegrass, big bluegrass
Southern Great Plains (9–19 pounds of unprocessed seed per acre)	Blue grama, sideoats grama, sand lovegrass, sand bluestem, switchgrass, indiangrass, little bluestem, weeping lovegrass
Southeast (9–32 pounds per acre)	Coastal bermudagrass, tall fescue, hardinggrass, rescuegrass, dallisgrass, bahiagrass, pangolagrass, St. Augustine grass, lespedeza, white clover, hop clover, crimson clover, burclover, kudzu

Recommended Mixtures:

(1) Upper South: Orchardgrass or tall fescue with Ladino clover or alfalfa; Kentucky bluegrass and Ladino clover

(2) Lower South: Coastal bermudagrass and crimson clover; white clover and dallisgrass; carpetgrass establishes itself naturally in Florida and vicinity.

▓ 10.10 ESTABLISHMENT OF PERMANENT PASTURES

Since most of the plant species used for permanent pastures have small seeds, it is necessary to prepare a firm seedbed. The use of good seed of small seeded grasses and legumes is important in order to provide good pasture coverage relatively free from weeds. The seed should be drilled rather than broadcast on a new seedbed. It should be seeded between ¾ and 1¼ inches (2 and 3 cm) deep. Special drills for seeding grasses and legumes have attachments that open furrows, drop fertilizer in bands, and cover the fertilizer with a layer of soil compacted with a press wheel. In the same operation, the seed is drilled into the compacted soil band about 1¼ inches above the fertilizer.[21, 29, 30] Forage crop seeders with two sets of corrugated rollers are very effective on prepared seedbeds. Ladino clover can be established in closely grazed or clipped grass sod either by broadcasting or drilling.

In general, it is advisable to sow cool weather grasses in early fall, but early spring seeding is preferable on extremely heavy soils in the northern states where winter heaving would eliminate many seedlings. Early spring sowing is also advisable in the southern states where the perennial grasses are warm-weather species. Small seeded legumes may be broadcast in late winter or early spring on the fall-sown grass and companion crop fields. Alternate freezing and thawing and rains help to cover the seeds with soil. When seeding with a companion crop, grasses and legumes are seeded with the small grain. Pasture plants are more productive the first year when seeded alone.[3] When grazed properly, they may provide a larger net return than the grain crop, or even offset the grazing value of the companion crop. However, a high-yielding small-grain crop is profitable, and it suppresses weed growth.

When irrigation is used, it is necessary to keep the soil surface moist until the seedings are well started. They may require irrigation every seven to fifteen days. Dryland grass seedings succeed only during favorable moist periods.

Several grasses adapted to Florida are established by vegetative methods, such as root or stem cuttings or sod pieces. These grasses include rapier, Bermuda, para, St. Augustine, pangola, and centipede.

A new pasture should be grazed lightly the first season. The young plants must have time to develop a good root system so as to withstand drought and the strain of grazing.

Pasture maintenance or reseeding, fertilization, and good grazing management can improve productivity.

▓ 10.11 FERTILIZATION OF PASTURES

A ton of dry matter in legume hay contains about 50 to 75 pounds (25 to 38 kg/MT) of nitrogen, 15 to 20 pounds (8 to 10 kg/MT) of P_2O_5, 20 to 25 pounds (10 to 13 kg/MT) of K_2O, and 20 to 25 pounds (10 to 13 kg/MT) of CaO. A ton of grass hay contains about 30 to 40 pounds (15 to 20 kg/MT) nitrogen, 10 to 20 pounds (5 to 10 kg/MT) of P_2O_5, 40 to 50 pounds of (20 to 25 kg/MT) K_2O, and 10 to 15 pounds (5 to 13 kg/MT) of CaO.

Most pasturelands in the humid regions of the country are deficient in calcium, phosphorus, nitrogen, and sometimes potassium.[4] The minerals calcium, phosphorus, and potassium must be applied before much response can be expected from application of commercial nitrogen.[3] Pastures on soils of fair natural fertility, particularly those that have been neglected several years, can be improved by fertilization.[13, 30, 31] An application of 400 to 600 pounds per acre (450 to 675 kg/ha) of a 6-12-6 fertilizer gave the best increases in southern Georgia.[26]

The effect of heavy application of fertilizer is pronounced.[32] In Wisconsin, completely fertilized areas of a bluegrass pasture produced about three times as much herbage as did the unfertilized areas during the first year.[33] The fertilizers used were 100 pounds per acre (110 kg/ha) each of phosphate, potash, and nitrogen. The turf was greatly thickened, the weeds largely disappeared, and white grub injury was practically eliminated. In Ohio, fertilizer treatment doubled the carrying capacity and also lengthened the grazing season by about 15 percent.[34]

Applications of superphosphate alone generally give the greatest response because they encourage the legumes that supply nitrogen to the grasses. However, an excess of clover or alfalfa increases the risk of bloating in cattle.

Applications of nitrogen usually increase the protein content of herbage. Use of nitrogen has increased the crude protein of bluegrass–white clover herbage as much as 12 percent and of orchardgrass from 50 to 100 percent.[35, 36] Nitrogen can promote enough growth so that pastures are ready for grazing as much as three weeks earlier than those without such applications. Nitrogen applications often discourage the growth of legumes in grass-legume mixtures.

Nitrogen is relatively ineffective on Ohio sods with a high clover content.[37] Heavy applications of nitrogen (60 to 200 pounds per acre 167 to 225 Kg/ha) to grass pastures, supplied with ample amounts of other elements, may increase herbage yields from 20 to 100 percent.[31, 38]

When added to a pasture that had been unfertilized for forty years, all fertilizers increased the nitrogen content of the herbage.[39] The averages for each year were about the same for the phosphorus-lime as for the phosphorus-nitrogen treatments. In Ohio, overgrazed bluegrass and white clover pastures have been treated with superphosphate, with lime added where the pH was below 5.5.[34] In the case of lime deficiency, it may be desirable to replace certain pasture plants with others not sensitive to acid soils. For this purpose, lespedezas or birdsfoot trefoil can be used in place of clovers in areas where they are adapted.

Mineral fertilizers, limestone, and barnyard manure can be applied in the fall, winter, or early spring. Commercial nitrogen should be applied at least two weeks before increased growth is desired. Applications of nitrogen are rarely effective except in the presence of adequate soil moisture,[40] but moderate applications are beneficial in the more favorable parts of the semiarid Great Plains.[41]

10.12 RENOVATION AND RESEEDING

Cultivation of pastures to improve grazing is of little value unless accompanied by reseeding or application of fertilizers, or both.[38]

10.12.1 Practices in Humid Regions

Cultivation in conjunction with reseeding and fertilization improved pastures in Vermont and Iowa by eliminating weeds, covering the seed, and mixing the fertilizer in the soil. Grasses and clovers that make a quick growth can be seeded on old pasture sod that has been well disked and fertilized.[42] In Wisconsin,[43, 44] overgrazed bluegrass pastures were improved by the scarification of the sod with a disk or spring-tooth harrow, after which legumes such as alfalfa, sweetclover, and red clover were seeded in the thinned pasture sod. In twenty-seven different pasture renovations, the average total weed population (mainly ragweeds and horseweeds) was reduced by 85 percent after two or three years. Reseeding alone may be desirable in some instances in conjunction with the improvement of old pastures, but it is seldom a complete remedy. Pastures on tillable land may be plowed, tilled, fertilized, and reseeded, either immediately or after an intervening crop. In pastures that are too rough or stony for plowing, broadleaf weeds can be killed with herbicides and then followed by heavy summer grazing. A herbicide, such as glyphosate, can be applied in late summer to kill the grass. The dead sod is disked three to five weeks later and the pasture is then reseeded.

10.12.2 Revegetation of Rangelands

The return of abandoned cultivated land to grass is a difficult problem on the Great Plains as well as in other areas of moisture shortage.[45] From twenty to fifty years are required for buffalograss to become reestablished naturally on abandoned farmland.

Artificial range reseeding in the native sod has often been unsuccessful in the semiarid region unless the seeding was favored by above-average rainfall. However, the seeding of blue grama, buffalograss,[26] and other species sometimes has been successful. Seedings of cultivated grasses (crested wheatgrass, smooth brome, and slender wheatgrass) thrived on areas where the original native vegetation consisted of grama and fescue grasses.[46] The land was disked, the seed was sown broadcast early in the spring, and the field was protected from grazing the first season. Reseeding on depleted mountain meadows, alluvial bottoms, and better sites of mountain slopes has restored range grasses where soil moisture conditions were above average. For cultivated grasses, an annual precipitation of 15 inches (380 mm) or more is essential. In Colorado, range seeding has been unsuccessful on areas that receive less than 10 inches (250 mm) annually.[47]

An overgrazed oak-brush range in Utah was seeded successfully with crested wheatgrass, smooth brome, and mountain brome.[48] The best stands were obtained from seed broadcast on plowed furrows spaced approximately 3 feet (90 cm) apart, and the seed covered with a brush drag. During a period of seven years, this method resulted in an increase of 360 to 900 percent in grazing capacity as compared with open grazed unseeded areas. Some degree of soil preparation was necessary to assure successful reseeding.

Sagebrush lands, as well as other southwestern range areas that have a rainfall of 15 inches (380 mm) or more, were seeded successfully after the land was tilled with a disk plow. Grass production was increased from two- to ten-fold. Adapted grasses were crested wheatgrass, other wheatgrasses, big bluegrass, and smooth bromegrass.[7, 11]

Dropping pelleted grass seed from an airplane on burned or plowed mountain rangeland has been unsuccessful, whereas drilling has produced good stands.[49]

Unless moisture and other conditions are very favorable, two or three years are required to establish a new range. When livestock are kept off for such a period, scattered seedlings have an opportunity to spread. Also, dormant seeds can germinate and produce new seedlings to increase the ground cover.

■ 10.13 GRAZING SYSTEMS

The capacity of native pastures has been increased as much as 50 percent by good grazing management. Controlled grazing is necessary to give palatable species an opportunity to recuperate and produce seed. Persistence of vegetation through a dry summer and a cold winter is directly related to root development. Overgrazing has resulted in poor root growth with very little food accumulation, with the result that the plants are likely to die either from drought or cold.[50] Overgrazing causes large decreases in weight of roots.[18] The total decrease from the early to the late stage of grazing was from 2.17 to 0.95 tons of roots per acre (4.9 to 2.1 MT/ha) in the 0- to 4-inch (0 to 10 cm) depth, and from 0.86 to 0.34 ton per acre (1.9 to 0.76 MT/ha) in the 4- to 12-inch (10 to 30 cm) depth on upland Nebraska soil.

10.13.1 Influence of Grazing on Species

The first signs of an overgrazed range are: the most palatable grasses such as bluegrass, needlegrass, junegrass, wheatgrass, bromegrass, and the fescues become less vigorous, forage production declines, and numbers of palatable species are reduced. The less palatable plants and poisonous plants increase in numbers. As the condition becomes more severe, annual weeds tend to replace perennial weeds and shrubs. In the last stage, bare spots appear and gradually increase, which in turn causes an increase in soil erosion. After the perennial grasses have perished and erosion begins, it will take a long time to restore the range to its original productivity.

TABLE 10.3 Productivity of Successional Types of Range Vegetation

Item	Blue Bunchgrass	Slender Wheatgrass	Porcupine Grass	Rabbit Brush
Density of vegetation (%)	60–80	40–60	30–50	20–40
Grasses (%)	75	65	85	25
Weeds (%)	20	25	10	50
Browse (%)	5	10	5	25
Palatable (%)	62	54	49	25
Surface acres to feed 1 cow 1 month	2	3	4	11

Blue bunchgrass and slender wheatgrass are two valuable range species in Montana. In order to maintain their vigor, these grasses should not be utilized beyond 60 to 70 percent of their foliage production by early summer, or more than 80 to 85 percent at the close of the summer grazing period.[51] The higher the successional stage of the vegetation, the greater the value of the range for grazing. As shown in Table 10.3, the rangeland tends to take on the characteristics of a more arid type as it becomes depleted. The higher successional stages were characterized by greater density of stand, a higher percentage of grass in the stand, and greater grazing capacity.

In the British Isles,[52] grazing management influenced the composition of a mixture of cultivated grasses in a pasture of perennial ryegrass, rough-stalked meadowgrass, and wild white clover. The pasture grazed heavily in March, April, and May had white clover as the most important constituent of the pasture by the middle of the third season. Another pasture resulted in a dominance of grasses when grazing was deferred until April 15. The perennial ryegrass became more vigorous while the white clover was considerably reduced. A third pasture not grazed before April 15, but completely pastured down at each subsequent grazing and then rested for a month, maintained a good balance between perennial ryegrass and white clover.

10.13.2 Deferred Grazing

On the western range, maintaining the important palatable native range plants is a challenge. Grazing too early may injure ranges more than any other practice. Grass grazed too early in the spring may be pulled up by the roots or damaged by trampling in a wet soil. In the spring, when the new grass begins to grow, the water content of the herbage may be as high as 85 percent, with a low feeding value. Excessive early grazing each year may delay satisfactory development of the palatable plants by as much as six weeks. It is desirable to delay grazing until the important forage plants have reached a height of 6 inches (15 cm) or, in the case of the shorter grasses, until the flower heads are in the boot.[51] Normal stand and vigor of bluestem pastures in eastern Kansas were maintained when grazing was deferred until June 15.[22] The deferred system gave an increase in carrying capacity of approximately 25 percent.

10.13.3 Rotation Grazing

Rotation grazing consists of grazing two or more pastures in regular order with intervening rest periods for each pasture.[53] An experimental area on the Jornada Reserve, moderately grazed using the deferred system, was over four times more productive than the outside range heavily grazed all year long.[51] In western ranges, the system of deferred rotation delays grazing until after seed maturity on about one-fourth to one-fifth of the entire area used by the herd. Then that area is grazed. A different area is delayed each year. This results in an increase in carrying capacity of the range, a chance for improvement when the range is depleted, and better growth of animals without losses through nonuse of feed. Continuous grazing is superior to the rotation of range pastures at approximately monthly intervals.

Rotation grazing of Kentucky bluegrass in western Missouri increased beef gains only 5 percent.[32] Larger benefits from rotation grazing occur with taller, heavier-yielding forages. At Beltsville, Maryland, rotation grazing increased the yield of total digestible nutrients by 10.4 percent, heavy fertilization increased the yield by 16.4 percent, and both combined increased the yield by 28.6 percent.[54]

An intensive plan of grassland management, known as the Hohenheim system,[54] was developed in Germany in 1916. This system involves (1) division of the pasture into four to eight paddocks, about equal in size; (2) heavy applications of fertilizers, especially nitrogen; (3) separation of the herd into two groups, producers and non-producers; and (4) frequent rotation of these groups. The cattle are moved progressively to other paddocks at weekly intervals, or whenever the grass reaches a height of 4 or 5 inches (10 to 13 cm). Young stock and dry cows, kept separate from the milk cows, follow into each paddock as the milkers are advanced.

Strip grazing is a modification of rotation grazing. Temporary fences confine the cattle to small areas that they can graze down to the desired height in one or two days. The fences are moved across the field as the herbage is consumed. The animals are returned to the original strips after sufficient new growth has occurred. The strips being grazed are usually bordered with electric fences. Strip grazing largely eliminates selective grazing by the animals but requires more attention than does rotation grazing.

Both rotation and strip systems permit mowing, fertilization, and irrigation of the areas not being grazed. Any herbage in excess of current grazing needs can be cut for hay or silage. The closeness of grazing is under control, and plants are not grazed again until they have made sufficient growth recovery. Milk production from rotation and strip grazing is about equal.[55]

10.13.4 Soiling or Green Feeding

Soiling or green feeding was practically discontinued in the United States by 1918 because of the hand labor required to cut, load, and feed the heavy green forage. The development of forage harvesters and mechanical unloading trucks revived some interest in the practice. Soiling is also called green chop, zero grazing, and mechanical grazing. Soiling eliminates problems of trampling, fencing, and water facilities for pastures. It also eliminates the rejection of the coarser portions of the herbage. The risk from bloating is reduced because the cattle are forced to eat grass and stems along with tender legume herbage. The disadvantages of soiling are the cost, operation, and possible breakdown of machines, as well as the difficulty of getting into the field in extremely wet weather. Animal production per acre from green feeding is superior to that from well-managed rotation or strip grazing, but labor and machinery costs are higher.[13, 55]

■ 10.14 BURNING GRASSLANDS AND BRUSH

Burning grasslands has been practiced for many years, particularly in the Southeast. Authorities differ as to its effectiveness in grassland improvement. Those who advocate burning claim that it: (1) brings about an earlier growth of vegetation; (2) results in more palatable vegetation than that from unburned areas; (3) increases the productivity of the soil through liberation of the lime, phosphoric acid, and potash contained in the ash; (4) improves the character of the herbage by control of weeds and brush; and (5) controls chinch bugs.

Such benefits are more likely to occur in humid regions. Burning is practically necessary in the Gulf Coast region as long as the land is used for both grazing and lumbering, particularly when only the native grasses and legumes are grazed. Burning increases the number of legumes and grasses by destroying the heavy ground cover of pine needles and leaves from other trees.

Burning is often used in western Colorado to destroy relatively dense stands of big sagebrush. Controlled burning is usually done in late summer or early fall when a clean burn can be obtained.[8]

Burning frequently has detrimental effects, particularly in the West. Some of these effects are: (1) removal of the extra vegetative material that would add humus and nitrogen to the soil, (2) destruction of old vegetation in the soil that functions to increase the water holding capacity, and (3) injury to the living vegetation. Burning is detrimental to the vigor of short grasses, the total yield for the season usually being less on burned areas. A single burning may kill shallow rooted grasses, like bluestem and the fescues. The dry forage left protects the young growth from too close grazing.

In Kansas, continued annual burning of native bluestem pastures decreased the total production of grassy yields from burned and unburned areas as shown in Table 10.4.[15, 56] Burning should be just before the native grasses start growth and only in years when an excess amount of dry material is on the pasture from the previous season.[57] Burning has had very little effect on the control of weeds unless done late in the spring. Burning has stimulated early spring growth because of the warmed soil. Periodic burning can be effective in controlling invading woody species, such as eastern red cedar (*Juniperus virginiana*), in the Flint Hills of Kansas and Oklahoma.

In Kansas, buckbrush (*Symphoricarpos* species) was greatly reduced when burned late in the spring.[58] Sagebrush has been eradicated from rangeland if burned in the late fall when it was dry.[46] Removal of the sagebrush permitted grasses such as fescue, wheatgrass, and arid bluegrass to become more productive. Some pastures increased their foliage two to five times from this practice. It took several years for the sagebrush to become reestablished.[59] Burning was equally

TABLE 10.4 Effects of Burning on Bluestem Pastures in Kansas, 1958–1965

	Forage (lb/acre)	Weeds (lb/acre)
Early spring (March 20)	2,612	335
Medium spring (April 10)	3,235	289
Late spring (May 1)	3,529	161
Unburned	3,919	300

damaging to grasses and sand sagebrush in western Oklahoma.[5] Burning as a management practice will likely be decreased, if not eliminated, in the future on the basis of adverse aesthetic effects upon the environment.

10.15 ERADICATION OF WEEDS OR BRUSH

Mowing western Oklahoma pastures in June, in two consecutive years, eliminated much of the sand sagebrush (*Artemisia filifolia*). Resting the pastures for 2½ months after each mowing permitted the grasses to recover. As a result, beef production per acre was more than doubled.[5]

Buckbrush and sumac were eradicated when cut in the flower stage, about May 10 and June 8, respectively, in Kansas.[58] In Connecticut research,[60] July was the best time to mow brush consisting of soft maple, alder, white birch, and blackberry. In northern regions, the critical period for destroying brush is when the roots contain the smallest amount of starch, generally when the plants are flowering. In the southern states, woody shrubs must be grubbed out or killed with herbicides.

Spraying with herbicides is preferable to either mowing or burning for destroying buckbrush, big sagebrush, sand sagebrush, skunkbush, rabbitbrush, and other pasture shrubs.[61] This treatment also kills many broadleaved weeds as well as poisonous plants such as loco, halogeton, deathcamas, orange sneezeweed, woody aster, silvery lupine, water hemlock, princes' plume, and two-grooved milk vetch. Mesquite, oak, and other woody plants, as well as tall larkspur, are controlled with other herbicides. Reseeding the range after weed control is often helpful.

The sprays are most effective when applied in the spring after growth is well established. The herbicides are usually applied with aerial spraying equipment. A repeated spraying after two or more years is usually necessary.

10.16 POISONOUS PLANTS

Losses from poisonous plants are estimated to be 4 percent per year to range livestock. Animals usually eat poisonous plants in harmful quantities only when the more nutritious and palatable plants are inadequate to meet their needs.

Poisonous plants vary greatly with respect to: (1) the condition under which animals are poisoned by them, (2) the portion of the plant that is poisonous, (3) changes in the toxicity of the parts of the plant during growth and drying, (4) the susceptibility of different species of animals to being poisoned by them, and (5) the effects on the poisoned animals.

Some of the important poisonous plants are: arrowgrass (*Triglochin maritima*), deathcamas (*Zygadenus* spp.), horsetail (*Equisetum* spp.), larkspur (*Delphinium* spp.), locoweed (species of *Oxytropis* and *Astragulus*), lupine (*Lupinus* spp.), whorled milkweed (*Asclepias galioides* and *A. mexicana*), poison vetch (*Astragalus* spp.), water hemlock (*Cicuta* spp.), white snakeroot (*Eupatorium urticaefolium*), crazyweed (*Oxytropis* spp.), and sneezeweed (*Helenium* spp.). Halogeton (*Halogeton glomeratus*) is now present in western ranges that cover an area of 11 million acres (4.5 million ha).

Poisonous compounds found in these plants include cyanogenetic glucoside in arrowgrass; an alkaloid, zygadenine, in deathcamas; an alkaloid (equisetin), aconitic acid, and fungi in horsetail; alkaloids (delphinine and others) in larkspur; a toxic base, locoine, in locoweed; alkaloids (lupine and others) in lupine; an alcohol-soluble resin in whorled milkweed; selenium in *Astragalus* species; a resin-like substance, cicutoxin, in water hemlock; oxalic acid in halogeton; and a higher alcohol, tremetol, in white snakeroot.[9, 62]

Grasses with rough awns that injure the mouth, eyes, and noses of grazing animals include foxtail *(Hordeum jubatum)*, cheatgrass *(Bromus tectorum)* and the unpalatable medusahead *(Taeniatherum apserum)*. Control of medusahead involves burning, herbicides, or close grazing followed by disking and seeding of a suitable grass such as crested wheatgrass.[13]

■ 10.17 CROPS FOR TEMPORARY OR ANNUAL PASTURES

The growth of perennial grasses and legumes stops or is greatly reduced each season when the temperature becomes hot or too cold for the particular species, which reduces its carrying capacity. Annual or biennial crops are grown to supplement permanent pastures during unproductive periods. A suitable succession of annual crops provides pasturage for a long season in nearly all parts of the country. Some of the advantages of annual pastures are: greater production per acre, a longer growing season, less trouble from internal parasites, and less danger from noxious weeds. Disadvantages are: greater labor requirement, cost of seed, greater danger of erosion, impracticability of grazing such crops on clay soils in wet weather, and frequent inability to produce good stands.

In many of the humid areas of the North, clover and alfalfa can be temporarily pastured by harvesting one less cutting of hay. Alfalfa hayfields in the irrigated Southwest are pastured during the hot summer months when the growth is too meager for hay production.

Other temporary pasturing schemes include Italian ryegrass in early spring; winter or spring small grains from April to July; rape, field peas, or vetch in early summer; and soybeans or sudangrass from midsummer to fall. Sweetclover may be pastured in the fall in its first year and in early spring of its second year. Some of these crops have two to four times the carrying capacity of comparable permanent pastures in the early spring, midsummer, or late fall.

An example of an effective arrangement of annuals to provide pasturage for dairy herds in southeastern North Carolina is as follows:[3] (1) Abruzzi rye, sown in September and grazed from November 15 to March 15, (2) crimson clover and hairy vetch sown August 15 to September 1 and grazed from March 1 to May 15, (3) sudangrass sown April 1 and grazed from May 15 to November 15, and (4) Biloxi soybean sown March 15 and grazed from June 1 to November 15.

Also in the South, legumes such as lespedeza can be maintained in association with dallisgrass but cannot compete satisfactorily with carpetgrass or bermudagrass. Rye and vetch can be drilled into bermudagrass sod in the fall to provide temporary winter pasture.[26] White clover, a cool-weather perennial legume was once considered only for the North. It provides excellent early pasture in the South where it is maintained mostly as a winter annual. Fall-sown, rust-resistant small grains, alone or with crimson clover or burclover, provide excellent grazing during the fall, winter, or early spring in the South. A mixture of Italian ryegrass and crimson clover provides excellent winter pasture. Pearl millet furnishes abundant summer pasture in the South.

In the southern Corn Belt, Korean lespedeza, grown with winter small grain, provides summer grazing after grain harvest. Then the lespedeza reseeds the field after plowing in preparation for a succeeding small-grain crop. Oat pasturage is more palatable than barley, while barley is more palatable than wheat or rye.

In the Great Plains, grazing is provided for much of the season by (1) native pasture in late spring, early summer, and late fall and (2) by sudangrass from July to September. In eastern Colorado, winter rye provides late fall as well as spring pasture. Native grass pastures or spring-sown small grains provide grazing from late spring to about July 1. Sudangrass can be grazed from early July until frost (Figure 10.2).

■ 10.18 WINTER WHEAT FOR PASTURE

Winter wheat is pastured extensively from October to December and again in the spring in the central and southern Great Plains.[31, 36, 42] From 20 to 65 percent of the winter wheat acreage can be pastured to some extent in favorable years. From 60 to 120 days of grazing are available during the period from November to April. Moderate grazing causes little or no reduction in grain yield in fields of winter wheat well established and well supplied with soil moisture. Yields have been reduced from 5 to 40 percent by heavy grazing or by grazing when the wheat growth was scanty and soil moisture was limited. Research with wheat pasture was conducted in central Oklahoma for three years with variable precipitation. Grain yields were not significantly affected when soil moisture was adequate (Table 10.5).[63] In west Texas, grazing irrigated wheat from autumn to March 20 reduced the wheat yield 20 percent.[36]

Spring grazing may be started when growth is resumed in the spring but should be discontinued when the plants start to grow erect just previous to jointing. This usually occurs about April 10 at the Hays, Kansas, Station.[31] Wheat plants may be injured by grazing at any time after their growing points are above the soil.[64] Permanent injury resulted when the culm tips were grazed, resulting in the loss of the developing heads. An increased seeding rate is recommended where grazing is contemplated.[36] Moderate pasturing is especially beneficial when wheat has been

TABLE 10.5 Effects of Grazing on Yields of Winter Wheat in Central Oklahoma

	1983–84	1984–85	1985–86
	bu/acre	bu/acre	bu/acre
No grazing	52.2 a	38.7 a	28.3 b
Fall, light	43.8 a		35.9 a
Fall, heavy	37.7 b		
Spring, light		40.6 a	20.4 d
Spring, heavy		36.3 a	
Fall and spring, light			26.8 c
Fall and spring, heavy			21.7 d
August–July precipitation, inches	36.8	35.7	32.2

Letter indicates statistical difference between treatments (p<0.05).

sown very early or has made an excessive growth. Pasturing reduces excessive growth and tends to prevent lodging.

Samples of wheat plants taken late in the fall contained 27 to 28 percent protein and 12 to 15 percent ash. The high protein content accounts for the high nutritive value of wheat pasture. However, cattle or sheep grazing in cool weather on small grains or other lush grasses growing on soils that are high in nitrogen and potassium can suffer from grass tetany (hypomagnesaemia). The plants under these conditions have a low content of magnesium in the foliage that causes grass tetany. It is prevented by adding magnesium to any supplementary feed or to the salt blocks and treated by injection of a calcium gluconate solution containing magnesium. Most cases develop after 60 to 150 days of grazing on the lush forage by cows in late pregnancy or when they are nursing calves under 60 days of age.[65] The animals suffer convulsions, with periods of relaxation, and eventually die.

Legumes, but not grasses, may take up enough molybdenum from soils high in molybdenum to be toxic to cattle and sheep.[54]

REFERENCES

1. Vallentine, J. F., and P. L. Sims. *Range science—A guide to information sources.* Detroit: Gale Res. Co., 1980.

2. Vallentine, J. F. "More pasture or just range for rangemen?" *Rangeman's Journal* 5(1978):37–38.

3. Semple, A. T., and others. "A pasture handbook," *USDA Misc. Publ.* 194 (rev.), 1946, pp. 1–88.

4. Mays, D. A., ed., *Forage Fertilization.* Madison, WI: American Society of Agronomy, 1974, pp. 1–621.

5. Savage, D. A. "Grass culture and range improvement in the central and southern Great Plains," *USDA Circ.* 491, 1939, pp. 1–55.

6. Swain, F. G., A. M. Decker, and H. J. Retzer. "Sod-seeding rye, vetch extends Bermuda pasture," *Crops and Soils* 13, 3(1960):20.

7. Cornelius, D. R., and M. W. Talbot. "Rangeland improvement through seeding and weed control on east slope Sierra Nevada and southern Cascade mountains," *USDA Handbook* 88, 1955.

8. Keller, W. "Breeding improved forage plants for western ranges," in *Grasslands*, Am. Assn. Adv. Sci., 1959, pp. 335–344.

9. Lodge, R. W., and others. "Managing crested wheatgrass pastures," *Can. Dept. Agr. Publ.* 1473, 1972, pp. 1–20.

10. McLean, A., and A. H. Bawtree. "Seeding grassland ranges in the interior of British Columbia," *Can. Dept. Agr. Pub.* 1444, 1971, pp. 1–15.

11. Reynolds, H. G., and H. W. Springfield. "Reseeding southwestern range lands with crested wheatgrass," *USDA Farmers Bull.* 2056, 1953.

12. Semple, A. T. *Grassland Improvement.* London: Leonard Hill, 1970, pp. 1–400.

13. Torell, P. J., and L. C. Erickson. "Reseeding medusahead-infested ranges," *ID Agr. Exp. Sta. Bull.* 489, 1967, pp. 1–17.

14. Larsen, H. J., and R. F. Johannes. "Summer forage: Stored feeding, green feeding and strip grazing," *WI Agr. Exp. Sta. Rsh. Bull.* 257, 1965, pp. 1–32.

15. Leithead, H. L., and others. "100 native forage grasses in 11 southern states," *USDA Handbk.* 389, 1971, pp. 1–216.

16. Tisdale, E. W., and others. "The sagebrush region in Idaho, a problem in range resource management," *ID Agr. Exp. Sta. Bull.* 512, 1969, pp. 1–15.

17. Launchbaugh, J. L. "Upland seeded pastures compared for grazing steers at Hays, Kansas," *KS Agr. Exp. Sta. Bull.* 548, 1971, pp. 1–29.

18. Stapledon, R. G. "Four addresses on the improvement of grassland," *U. Col. Wales*, Aberystwyth, England. 1933.

19. Johnstone-Wallace, D. B. "The influence of grazing management and plant associations on the chemical composition of pasture plants," *J. Am. Soc. Agron.* 29(1937):441–455.

20. Beaumont, A. B., and others. "Some factors affecting the palatability of pasture plants," *J. Am. Soc. Agron.* 25(1933):123–128.

21. Brown, E. M. "How to seed new pastures," *MO Agr. Exp. Sta. Bull.* 739, 1959.

22. Aldous, A. E. "Management of Kansas bluestem pastures," *J. Am. Soc. Agron.* 30(1938): 244–253.

23. Robertson, J. H. "Effect of frequent clipping on the development of certain grass seedlings," *Plant Phys.* 8(1933):425–447.

24. Ritchey, G. E., and W. W. Henley. "Pasture value of different grasses alone and in mixture," *FL Agr. Exp. Sta. Bull.* 289, 1936.

25. Cupric, P. O., and V. R. Smith. "Response of seeded ranges to different grazing intensities in the Ponderosa pine zone of Colorado," *U.S. Forest Service Prod. Rsh. Rpt.* 112, 1970, pp. 1–41.

26. Stephens, J. L. "Pastures for the coastal plain of Georgia," *GA Coastal Plain Exp. Sta. Bull.* 27, 1942, pp. 1–57.

27. Hanson, J. C. "Analysis of seeding mixtures and resulting stands on irrigated pastures of northern Colorado," *J. Am. Soc. Agron.* 21(1929):650.

28. Hanson, H. C. "Factors influencing the establishment of irrigated pastures in northern Colorado," *CO Exp. Sta. Bull.* 378, 1931.

29. Anonymous. "Band seeding in the establishment of hay and pasture crops," in *USDA Agricultural Research Service Spec. Rept.* ARS 22–21, 1956.

30. Park, J. K., and others. "Establishing stands of fescue and clovers," *SC Agr. Exp. Sta. Circ.* 129, 1961.

31. Swanson, A. F., and K. Anderson. "Winter wheat for pasture in Kansas," *KS Agr. Exp. Sta. Bull.* 345, 1951.

32. Brown, E. M. "Managing Missouri pastures," *MO Agr. Exp. Sta. Bull.* 750, 1960.

33. Fink, D. S., G. B. Mortimer, and E. Truog. "Three years' results with an intensively managed pasture," *J. Am. Soc. Agron.* 25(1933):441–453.

34. Barnes, E. E. "Ohio's pasture program," *J. Am. Soc. Agron.* 23(1931):216–220.

35. Gordon, C. H., A. M. Decker, and H. G. Wiseman. "Some effects of nitrogen fertilizer, maturity, and light on the composition of orchardgrass," *Agron. J.* 54(1962):376–378.

36. Shipley, J., and C. Regier. "Optimum forage production and the economic alternatives associated with grazing irrigated wheat: Texas High Plains," *TX Agr. Exp. Sta.* MP 1068, 1972, pp. 1–11.

37. Dodd, D. R. "The place of nitrogen fertilizers in a pasture fertilization program," *J. Am. Soc. Agron.* 27(1935):853–862.

38. Brown, E. M. "Improving Missouri pastures," *MO Agr. Exp. Sta. Bull.* 768, 1961.

39. Brown, B. A. "The effects of fertilization on the chemical composition of vegetation in pastures," *J. Am. Soc. Agron.* 24(1932):129–145.

40. Houston, W. R. "Range improvement methods and environmental influences in the northern Great Plains," *USDA Prod. Rsh. Rpt.* 130, 1971, pp. 1–13.

41. Rogler, G. A., and R. J. Lorenz. "Pasture productivity of crested wheatgrass as influenced by nitrogen fertilization and alfalfa," *USDA Tech. Bull.* 1402, 1969, pp. 1–33.

42. Fink, D. S. "Grassland experiments," *MA Agr. Exp. Sta. Bull.* 415, 1943.

43. Fuellman, R. F., and L. F. Grader. "Renovation and its effect on the populations of weeds in pastures," *J. Am. Soc. Agron.* 30(1938):616–623.

44. Graber, L. F. "Evidence and observations on establishing sweetclover in permanent bluegrass pastures," *J. Am. Soc. Agron.* 20(1928):1197–1205.

45. Savage, D. A. "Methods of reestablishing buffalo grass on cultivated land in the Great Plains," *USDA Circ.* 328, 1934.

46. Hanson, H. C. "Improvement of sagebrush range in Colorado," *CO Exp. Sta. Bull.* 356, 1929.

47. Hull, A. C., Jr., and others. "Seeding Colorado rangelands," *CO Agr. Exp. Sta. Bull.* 498-S, 1962.

48. Price, R. "Artificial reseeding on oak-brush range in central Utah," *USDA Circ.* 458, 1938.

49. Hull, A. C., Jr., and others. "Pellet seeding on western rangeland," *USDA Misc. Pub.* 922, 1963.

50. Sims, F. H., and Crookshank, H. R. "Wheat pasture poisoning," *TX Agr. Exp. Sta. Bull.* 842, 1956.

51. Chapline, W. R. "Range research of the U.S. Forest Service," *J. Am. Soc. Agron.* 21(1929):644–649.

52. Jones, A. G. "Grassland management and its influence on the sward; 11, The management of a clovery sward and its effects," *Emp. Jour. Exp. Agr.* 1(1933):122–127.

53. Sarvis, J. T. "Effects of different systems and intensities of grazing upon the native vegetation at the Northern Great Plains Field Station," *USDA Dept. Bull.* 1170, 1923.

54. Turelle, J. W., and W. W. Austin. "Irrigated pastures for forage production and soil conservation in the West," *USDA Farmers Bull.* 2230, 1967, pp. 1–22.

55. Gordon, C. H., and others. "A comparison of the relative efficiency of three pasture utilization systems," *J. Dairy Sci.* 42, 10(1959):1686–1697.

56. Anderson, K. L., E. F. Smith, and C. Owensby. "Burning bluestem range," *J. Range Management* 23(1970):81–92.

57. Towne, G., and C. Owensby. "Long-term effects of annual burning at different dates in ungrazed Kansas tallgrass prairie," *J. Range Management* 37(1984):392–397.

58. Aldous, A. E. "The eradication of brush and weeds from pastures," *J. Am. Soc. Agron.* 21(1929):660–666.

59. Pechanec, J. F. "Sagebrush control on rangelands," *USDA Agr. Handbk.* 277, 1965, pp. 1–40.

60. Brown, B. A. "Effect of time of cutting on the elimination of bushes in pastures," *J. Am. Soc. Agron.* 22(1930):603–605.

61. Sawyer, W. A., and others. "Range robbers—Undesirable range plants," *OR Ext. Bull.* 780, 1963, pp. 1–13.

62. Gilkey, H. M. "Livestock-poisoning weeds of Oregon," *OR Agr. Exp. Sta. Bull.* 564, 1958, pp. 1–74.

63. Christiansen, S., T. Svejcar, and W. A. Phillips. "Spring and fall cattle grazing effects on components and total grain yield of winter wheat," *Agronomy Journal* 81(1989):145–50.

64. Kiesselbach, T. A. "Winter wheat investigations," *NE Agr. Exp. Sta. Res. Bull.* 31, 1925.

65. Holt, E. C., and others. "Production and management of small grains for forage," *TX Agr. Exp. Sta. Bull.* B-1082, 1969.

Weeds and Their Control

11.1 ECONOMIC IMPORTANCE

More than 3,000 species of herbaceous and woody plants of the world are regarded as weeds, but only about 200 are recognized as major problems in world agriculture and cause 90 percent of the crop losses from weeds.[1] In the 1950s, weeds cost American farmers $5 billion annually in crop losses and in the expense of keeping them under control. In 1994, the annual economic impact of weeds on the U.S. economy was estimated to be $20 billion or more with the costs on cropland to be more than $15 billion.[2] Weed control accounts for much of the cost of intertilling row crops, maintaining fallow, preparing the seedbed, and cleaning seed. Another expense is suppression of weeds along highways and railroad rights-of-way, and in irrigation ditches, navigation channels, yards, parks, grounds, and home gardens. Ragweed pollen is a source of annual periodic distress to several million hay fever sufferers. Poison ivy, poison oak, poison sumac, nettles, thistles, sandburs, and puncturevine bring pain to millions of others.[3,4] The barberry bush, which spreads the black stem rust to grains and grasses, can be regarded as a weed. Weeds also serve as hosts for other crop diseases as well as for insect pests.

In some locations, weeds play a beneficial role in reducing soil erosion, supplying organic matter to the soil, and furnishing food and protection to wildlife.

11.2 TYPES OF CROP WEEDS

A weed has been defined as a plant that is useless, undesirable, or detrimental or simply as a "plant out of place."[5,6] Some plants, such as Canada thistle or field bindweed, are always considered weeds, whereas red clover would be considered a nuisance under some circumstances but hardly a weed at any time.

11.2.1 Annual Weeds

Annual weeds live for a single year, produce seed, mature, and die. Most of the common crop weeds belong to this group, among them pigweed, ragweed, velvetleaf, Russian thistle, wild oat, mustard, lambsquarter, kochia, and dodder. Other annual weeds are wild barley, witchgrass, canarygrass, crabgrass, barnyardgrass, darner, smartweed, summer cypress, puncturevine, mallow, false flax, horseweed, wild buckwheat, prickly lettuce, purslane, corn cockle, and sunflower.[7]

11.2.2 Biennial Weeds

Biennial weeds require two seasons to complete their growth. They grow from seeds and store food the first season, usually in short, fleshy roots. The following spring they draw on the stored food to produce vigorous vegetative growth and seeds. Among the biennial weeds are burdock, musk thistle, mullen, wild carrot, and wild parsnip.

11.2.3 Perennial Weeds

Perennial weeds live for more than two years. The majority of simple perennials possess root crowns that produce new plants year after year. They are supported, like the dandelion, by a fleshy taproot or fibrous roots. Except in a few instances, plants of this type depend upon production of seed for survival. Creeping perennials propagate by means of rhizomes or stolons, but also by seeds. Field bindweed, Canada thistle, perennial sowthistle, leafy spurge, quackgrass, and johnsongrass spread by underground rhizomes. Few weeds are stoloniferous. Other perennial weeds are nutgrass, poverty weed, horsetail, bracken fern, Russian knapweed, St. Johnswort, horsenettle, milkweed, curled dock, sheep sorrel, wild rose, blue vervain, chicory, and buckhorn plantain.

11.2.4 Common and Noxious Weeds

Weeds are also classified as common or noxious. Common weeds are annuals, biennials, or simple perennials that are readily controlled by ordinary good-farming practices. Noxious weeds are those that are difficult to control because of an extensive perennial root system or because of other characteristics that make them persistent. Most states have noxious weed laws that designate certain weeds as noxious and place strict limitations on plant populations and seed allowances.

11.3 LOSSES CAUSED BY WEEDS

11.3.1 Decrease Crop Yields

Weeds cause a decrease in yield by removing soil moisture and nutrients and competing for sunlight needed by crop or pasture plants. Most weeds require as much water to produce a pound of dry matter as do cereal crops. In addition, weeds compete with crop and pasture plants for light and soil nutrients.[8] In Kansas research, yields of close-drilled sorghum on bindweed infested land averaged 2.06 tons per acre (4.6 MT/ha), and they averaged 3.92 tons per acre (8.8 MT/ha) on clean land. Barley produced an average yield of 7½ bushels per acre (242 kg/ha) on infested land, but 21½ bushels per acre (1,160 kg/ha) on land free from bindweed.[9]

11.3.2 Reduce Crop Quality

The presence of weed seeds in small grains can lower the quality of the grain. Green weed pieces in threshed grain raise the moisture content so that the grain may not store well. Weeds consumed by dairy cows, such as wild garlic, mustard, fanweed, yarrow, chicory, or ragweed, impart undesirable flavors to dairy products. Black nightshade in soybean stains the seeds. Wild sunflower heads can cause spoilage in harvested wheat.[10] Weed seeds, as regulated contaminants, may prohibit the sale of crops in national and international trade.

11.3.3 Harbor Plant Pests and Diseases

Many weeds act as hosts to organisms that carry plant diseases. Curly top, a serious virus disease of sugarbeet, is carried by the beet leafhopper from such weeds as common mallow, chickweed, and lambsquarters to sugarbeet. The sugarbeet webworm prefers to deposit its eggs on Russian thistle and similar weeds. Weeds of the family *Solanaceae* contribute to the spread of the Colorado potato beetle. Overwintering rhizomes of johnsongrass harbor viruses that cause maize dwarf mosaic and maize chlorotic dwarf virus. During subsequent seasons, these diseases are transmitted to corn by insects.[2]

11.3.4 Increase Irrigation Costs

Weeds on ditch banks or growing in the ditches can seriously impair the efficiency of irrigation channels. Windblown weeds often obstruct headgates and diversion boxes. As a result, ditches must be cleaned every year.

11.3.5 Injure Livestock

Some poisonous weeds, described in Chapter 10, can cause illness or death to livestock. Mature plants of sandbur, three-awned grass, porcupine grass, downy brome (cheatgrass), and squirreltail grass may cause injury to stock that eat them.

11.3.6 Decrease Land Values

The presence of weeds, particularly noxious weeds, reduces the value of land.

■ 11.4 PERSISTENCE OF WEEDS

Weeds can usually survive in competition with crop plants because of a wide range of adaptability as well as effective means of propagation, such as by underground parts. Many weeds produce large amounts of seeds per plant. The average numbers per plant for several species are:[11] field pennycress, 7,040; pigweed, 117,400; Russian thistle, 24,700; lambsquarter, 72,450; green foxtail, 34,000; perennial sowthistle, 9,750; and tumbling mustard, 80,400. Seeds of some weeds escape notice because they are so small.

Many weed seeds remain viable in the soil for many years. Some weed seeds, such as bindweed, wild oat, and cocklebur exhibit dormancy. Their prolonged viability in the soil explains the sudden appearance of certain weeds after years of good cultural practices.

▓ 11.5 DISSEMINATION OF WEEDS

11.5.1 Natural Agencies

Seeds with barbs or hooks, such as cocklebur or devil's claw, become attached to animals and carried long distances. Weed seeds eaten by birds and animals may pass uninjured through the digestive tract. Cactus is spread principally by jackrabbits that eat the fruits that contain indigestible seeds. Seeds equipped with tufts of hair, such as Canada thistle, bull thistle, or sowthistle, may be disseminated by wind for distances up to 15 miles (24 km). Weed seeds are often carried to other locations by rain or streams. Tumbleweeds of various species, such as Russian thistle, tumbling amaranth, and kochia, can roll in the wind for long distances, scattering seeds as they go. Weeds with horizontal rhizomes spread up to several feet each year. Bindweed patches may double their area every five years.

11.5.2 Man-Made Agencies

Weeds are widely spread in impure crop seeds. Some of the most serious weed pests have been introduced from other countries in seeds of wheat, alfalfa, clover, sugarbeet, and other crops. Hay, straw, and other forages have contributed to the spread of weed seeds. Introduced species account for about 65 percent of the total weed flora in the United States; their total impact on the U.S. economy equals or exceeds $13 billion per year.[2]

In irrigated areas, seeds of weeds that grow on the banks of reservoirs, canals, and ditches fall into the water, which carries them to cultivated fields. Several million weed seeds of 81 species floated down a 12 foot (4 m) wide irrigation ditch in Colorado in one day.[12]

Even farm machinery can spread weed seeds. Plows, harrows, and cultivators drag roots or seed-bearing portions of perennial plants to other parts of a field.

Spreading fresh barnyard manure on cultivated fields can disseminate weed seed. In one experiment,[13] 6.7 percent of the weed seeds fed to farm animals were viable when recovered in the fresh manure. After burial for one month in manure, velvetleaf, bindweed, and hoary cress seeds were still viable. Practically all seeds were dead after being buried in manure for three months.

▓ 11.6 CONTROL OF COMMON CROP WEEDS

11.6.1 General Control Measures

Crop seeds must be high in purity and free from noxious weed seeds. An important weed control method is early and frequent cultivation of the land. Row crops should be periodically grown on the land to permit intertillage. Weeds are killed most easily when they are seedlings. The greatest benefit of fallow in the semiarid region is prevention of weed growth. Annual weeds growing in uncultivated areas of a farm can be sprayed with herbicides or mowed to prevent them from going to seed.

Crop rotation is a valuable aid in weed control because many weeds are associated with certain crops. Dodder is troublesome in alfalfa. Wild oat and the mustards become serious pests on land cropped continuously to small grains. Green foxtail and giant foxtail are often troublesome in flax and cornfields. Smother crops make a rapid growth that shades out other plants when grown in thick stands. Common smother crops are alfalfa, foxtail millet, buckwheat, rye, sorghum, and sudangrass.

All types of community farm machinery, particularly separators, combines, and hay balers, should be cleaned free of weed seeds before they are brought to the field.

11.6.2 Clean Cultivation for Noxious Weed Control

A combination of clean cultivation and smother crops can weaken or eliminate perennial weeds. Land clean-cultivated until about July 1, followed by a smother crop of sorghum or sudangrass, can result in almost complete control of bindweed if repeated for several successive seasons. Winter rye and winter wheat have been used as smother crops in some regions, with the land clean-cultivated between harvest and seeding. A smother crop is most effective when it follows a full season of clean cultivation, such as fallow.

Complete eradication of bindweed can be achieved by 20 to 25 cultivations over two years. Repeated cultivation reduces the food materials stored in the roots. Allowing some top growth for six to eight days after the shoots emerge, before repeating cultivation, aids in bindweed eradication.[14] Cultivation every two weeks to a depth of 3 inches (8 cm) throughout the growing season is desirable. The best time for plowing is when the weeds are flowering because root reserves are low at that time.[9, 15, 16] Subsequent cultivation is best with sweep or blade implements.[17] Herbicide use greatly reduces the need for cultivation. The best time to apply herbicide is late summer or early fall when the plants are storing food reserves in roots and rhizomes. The herbicide will be moved below the soil and kill the entire plant.

11.6.3 Flaming

A flaming implement has been used to kill weeds in a cotton field,[18] and in corn and sorghum fields. Flaming of cotton is started a month after planting when cotton plants have a stem diameter of 3⁄16 inch (5 mm) or more. At that time, they are large enough to escape injury from the flame. Weeds are controlled during the first month by cultivation, but thereafter can be controlled satisfactorily by flaming without injury to the cotton. The flame is directed toward the base of the cotton plants and is most effective in killing small weeds. Flaming is not used as much today because of the high cost of propane and other fuels.

11.6.4 Biological Control Methods

Klamath weed (St. Johnswort) is being controlled with parasitic beetles, *Chrysolina* species, imported from Australia. Parasitic insects have also been used to control cactus in Australia, lantana in Hawaii, and yellow toadflax in Canada. Pasturing land with sheep sometimes is an effective method for the control of certain weeds. Sheep are able to suppress field bindweed on land seeded to sudangrass for pasture. They eat the bindweed in preference to the sudangrass, but they make good gains on the latter after eating down the weeds. In certain Texas rangelands, sheep eat many herbaceous weeds that were rejected by grazing cattle; then goats follow to consume woody shrubs. Fish consume algae in flooded rice fields.[19] The development and multiplication of pathogens that attack specific weeds are remote possibilities. During the 1980s, farmers spent about $2.6 billion annually for cultural, ecological, and biological methods of control.[10]

11.6.5 Chemical Control

The most popular method of controlling weeds is the application of herbicides (Figures 7.9, 11.1, and 11.2).[19, 20, 21, 22, 23, 24, 25]

The control of annual, biennial, and perennial weeds through the use of chemical herbicides has expanded remarkably. By 1965, selective herbicides were applied to more than 90 million acres (36 million ha) of crop, grazing, and other land in the United States. During the 1980s, farmers spent over $3 billion annually for chemical weed control.[10] It is important to know the growth stage of the crop at which the herbicide is effective but with the least injury to the crop. Some chemicals are effective in granular form as well as in the usual spray formulas.

About 140 selective herbicides were available for use in 1973, and many more have become available since then. Newer herbicides have a greater selectivity for specific weed and crop plants. The frequent changes in herbicides and in their most appropriate use makes definite recommendations advisable for only limited periods. Crop growers should rely upon the latest local suggestions as to the best herbicide for particular weeds and crops and for the best rate, time, and method of application.

GROUPS OF HERBICIDES Herbicides occur in chemical families with common components. Herbicides can also be grouped according to their use and how they affect the plant. There are two main groups: those applied to foliage and those

FIGURE 11.1
Boxes behind the seed boxes contain a preemergent herbicide that is applied in bands over the planted rows during the planting operation.
[Courtesy Case IH]

FIGURE 11.2
A self-propelled sprayer applies a postemergent herbicide on a field of soybean. [Courtesy John Deere & Co.]

applied to the soil. Herbicides applied to foliage can be further classified as: (1) translocated within the plant and show initial injury on new growth, (2) translocated and show initial injury on older growth, and (3) nontranslocated or contact herbicides showing initial localized injury. The following list includes widely used herbicides but is not intended to be all-inclusive:

Foliage-applied, translocated herbicides that show initial injury on new growth (phloem-mobile):

Auxin-Type Growth Regulators:
    Phenoxy acid: 2, 4-D, 2, 4,5-T, MCPA, 2,4-DB, MCPB
    Carboxylic acid: picloram, clopyralid, triclopyr
    Benzoic acid: dicamba
    Quinoline: quinclorac

Carotenoid Pigment Inhibitors:
    Pyridazinone: norflurazon
    Isoxazole: isoxaflutole
    No chemical family: amitrole, clomazone, fluridone

Branched-Chain Amino Acid (ALS-AHAS) Inhibitors:
    Imidazolinone: imazamethabenz, imazamox, imazapyr, imazethapyr, imazapic
    Sulfonylurea: chlorsulfuron, chlorimuron, primisulfuron, nicosulfuron, triflusurluron, triasulfuron
    Pyrimidinyl oxybenzoate: pyrithiobac
    Triazolopyrimidine sulfonanilide: cloransulam, flumetsulam

Lipid Biosynthesis (ACCase) Inhibitors:
    Cyclohexanedione: clethodim, sethoxydim, tralkoxydim
    Aryloxyphenoxy propionate: diclofop, fluazifop-P, fenoxaprop, haloxyfop

Aromatic Amino Acid (EPSPS) Inhibitor: glyphosate, sulfosate

Organic Arsenicals: DSMA, MSMA

Foliage-applied, Translocated Herbicides that Show Initial Injury on Older Growth (Xylem-mobile):

"Rapidly Acting" Photosynthesis Inhibitors:
Benzothiadiazole: bentazon
Phenylcarbamate: desmedipham, phemedipham
Pyridazinone: pyrazon
Benzonitrile: bromoxynil
Phenlypyridazine: pyridate

"Classical" Photosynthesis Inhibitors:
s-Triazine: ametryn, atrazine, prometon, cyanazine, prometryn, simazine
as-Triazine: metribuzin
Uracil: bromacil, terbacil
Phenylurea: diuron, fluometuron, linuron, tebuthiuron

Foliage-applied, nontranslocated herbicides that show initial localized injury (no movement, contact):

Photosystem I (PS I) Energized Cell Membrane Destroyers:
Bipyridilium: paraquat, diquat

Protoporphyrinogen Oxidase [Protox (PPO)] Inhibitors:
Diphenylether: acifluorfen, fomesafen, lactofen, oxyfluofen
N-phenylphthalimide: flumiclorac
Oxadiazole: oxadiazon, fluthiacet
Triazolinone: carfentrazone, sulfentrazone

Glutamine Synthesis Inhibitor: glufosinate

Soil-Applied Herbicides:
Shoot Inhibitors:
Chloroacetamide: acetochlor, alachlor, flufenacet, metolachlor, dimethenamid, propachlor
Thiocarbamate: butylate, EPTC, cycloate, pebulate
Unclassified: ethafumesate

Root Inhibitors:
Pyridine: dithiopyr
Dinitroaniline: benefin, prodiamine, ethalfluralin, pendimethalin, oryzalin, trifluralin
Amide: pronamid
No chemical family: DCPA

Cell Wall Formation Inhibitors:
Nitrile: dichlobenil
Benzamide: isoxaben

Cell Division Inhibitors:
Phenylurea: siduron
Amide: napropamide
No chemical family: bensulide

New or better herbicides are being discovered every year. Consequently, many current herbicidal recommendations soon become obsolete. Each new herbicide must be tested before it is marketed in order to determine any possible injury to people, animals, or plants that might result from its use. Some of the herbicides used for weed control in various crops are as follows:

Corn—atrazine, flufenacet, metolachlor, cyanazine, pendimethalin
Wheat—2, 4-D, dicamba, bromoxynil, chlorsulfuron, metsulfuron, triasulfuron
Cotton—trifluralin, diuron, MSMA, pendimethalin, norflurazon, prometryn, cyanazine
Soybean—alachlor, trifluralin, propachlor, chlorimuron, ethalfluralin
Canola—trifluralin, ethalfluralin, sethoxydim, quizalofop, clethodim
Peanut—alachlor, benefin, metolachlor, bentazon, acifluorfen
Sorghum—atrazine, metolachlor, dimethenamid, alachlor
Rice—2, 4-D, pendimethalin, acifluorfen, quinclorac
Sugarbeet—TCA, pyrazon, endothall, diallate
Sugarcane—terbacil, MSMA, trifluralin
Dry Bean—EPTC, trifluralin, alachlor, imazethapyr
Pea—MCPB, clomazone, imazethapyr, bentazon
Pasture and Range—2, 4-D, triasulfuron, picloram, metsulfuron

Preplant and preemergent herbicides provide effective weed control for many large seeded crops. These crops are planted deep enough to escape injury from herbicides applied to the surface, except when heavy rains on sandy soils wash the chemicals downward. Preemergent herbicides are applied within two days after planting, or during planting immediately after the crop seed is covered. The applicators are either mounted on the tractor or planter or are trailed behind the planter. A preemergent treatment is often applied in bands 8 to 14 inches (20 to 36 cm) wide directly over the planted row.[26] This reduces the cost of chemicals. Weeds between the rows are controlled by cultivation. Preplant and preemergent applications kill weed seedlings that have sprouted on or near the soil surface. Their effectiveness is greatly impaired by any subsequent cultivation that not only buries some of the chemical but also raises other weed seeds up to the soil surface where they germinate.

Postemergent applications of chemicals are timed to control weeds with the least injury to the crop. Most crop plants are killed or injured in the seedling stage. They are damaged before they reach a height of 5 to 9 inches (13 to 23 cm). Applications in the flowering stage, or shortly before, can cause sterility. Treatments between the different stages are least injurious to the crop. Small weeds are easier to kill than large ones; hence early application is desirable. Grain crops that are near maturity are not harmed by herbicides. Grains can be sprayed then to kill or desiccate green weeds that would otherwise interfere with harvest operations and add moisture to the threshed grain.

Lower rates of herbicide application are more effective on sandy soils than on heavy soils, particularly on organic soils, because less of the chemical is adsorbed on the coarse soil particles. Some of the chemicals are volatile. They have injured cotton plants when hot weather followed their application. The drifting of sprays with their volatile compounds can injure susceptible crops in nearby fields.

Crop injury from herbicides is evident in stunted or distorted growth or dead leaves. Stem bending occurs in sorghum, corn, and other crops that have received a heavy application of 2, 4-D on the growing plants. Such treatment with 2, 4-D also suppresses the development of brace roots and induces stalk brittleness in sorghum and corn, which then causes lodging or stalk breakage.

Most herbicides applied at recommended rates do not have long-term residual effect in the soil.[11] Most postemergent herbicides have little or no residual activity. Preplant and preemergent herbicides need to remain active for several weeks to control weeds in the early growing season. Weed control is essential until the leaf canopy closes. Weeds that emerge after canopy closure have little or no effect on crop yield. Simazine, atrazine, and diuron can remain active for nearly a year and can cause injury to a succeeding crop. Spring oat, sugarbeet, and soybean planted on corn ground that had been sprayed the previous year with simazine or atrazine often show retarded growth.

Most postemergent treatments are applied with field sprayers that can cover a strip from 7 to 30 feet (2 to 9 m) in width. Aerial spraying is common for brushy, woody, or rough pastures and ranges,[11] and for flooded rice fields.[8] It can be used on weedy cornfields that are too wet for cultivation. Aerial spraying of crops that are nearing maturity avoids the damage caused by operation of a tractor-drawn or tractor-mounted field sprayer over the field.

Some weed populations have become herbicide resistant. Herbicide-resistant weeds were first discovered in the early 1970s and have since risen into the hundreds. Herbicide resistance has evolved primarily where producers repeatedly use herbicides or families of herbicides with the same mode of action. Goosegrass, velvetleaf, cocklebur, kochia, water hemp, and johnsongrass are among the many weeds that have become resistant to herbicides. Herbicide-resistant weeds do not usually differ in appearance from individuals of the same species that are susceptible to herbicides. Thus, there is no visual way to identify resistant populations before control becomes ineffective. Rotation of herbicides with different modes of action in a given field can help prevent the buildup of resistant species.

11.6.6 Herbicide Resistance in Genetically Altered Crops

Herbicide-resistant crop varieties and hybrids (cultivars) have been developed to allow the use of herbicides that would damage or kill nonresistance cultivars. These cultivars have been developed by two methods: (1) selecting genetic lines with natural resistance and incorporating the resistance using standard plant breeding techniques, or (2) using genetic engineering to incorporate resistant genes from other plants or organisms. (See Chapter 4 for more information on plant breeding techniques.)

Herbicide-resistant cultivars developed through standard plant breeding techniques include IR (imadazolinone-resistant) and IT (imadazolinone-tolerant) corn, SR (sethoxydim-resistant) corn, and STS (sulfonylurea-tolerant) soybean. IR and IT corn hybrids are commonly referred to as "IMI" hybrids or Clearfield® hybrids; SR corn hybrids are called "Poast Protected" hybrids. Clearfield® hybrids have also been developed for rice, sunflower, canola, and other crops.

Herbicide-resistant cultivars developed using genetic engineering include glyphosate-resistant (Roundup Ready®) corn, soybean, cotton, wheat, and other crops. Gluphosinate-resistant (LibertyLink®) cultivars of corn, soybean, sugarbeet, canola, and other crops have also been developed using genetic engineering.

The development of these genetically altered cultivars allows the use of postemergent herbicides that control a wide variety of weeds with no carryover. However, there is legitimate concern that the widespread use of these cultivars and herbicides will lead to the eventual development of "super weeds" that are resistant to these herbicides and subsequently very difficult to control. There is also concern that resistant cultivars themselves will become weeds. To reduce this possibility, producers are strongly encouraged not to plant all their acres to these cultivars and to use other herbicides on the remaining land.

▓ 11.7 SERIOUS PERENNIAL WEEDS

Some of the most serious perennial noxious weeds in the United States are quackgrass, Canada thistle, bindweed, leafy spurge, hoary cress (whitetop), perennial sowthistle, johnsongrass, Russian knapweed, and nutgrass. These species, serious weed pests in several states, are found in cultivated fields, pastures, and meadows.[6]

Some noxious perennial weeds, such as bindweed, Canada thistle, and sowthistle, are controlled by herbicides such as 2, 4-D, dicamba, glyphosate, and picloram. When combined with fallow, frequent tillage, and smother crops, they are largely suppressed. Quackgrass is suppressed with atrazine, followed by plowing and planting corn. Herbicides for leafy spurge include 2, 4-D, picloram, and imazapic. Hoary cress is controlled with 2, 4-D. Russian knapweed is controlled with dicamba or picloram.

11.7.1 Field Bindweed

Field or European bindweed *(Convolvulus arvensis)* (Figure 11.3) is distributed throughout the United States. It is a serious pest in Iowa, South Dakota, Minnesota, Nebraska, Kansas, Colorado, Utah, Idaho, Washington, and California. This plant, a member of the morningglory family, is a perennial that propagates by seeds and rhizomes. The roots grow 1 inch (2.5 cm) a day and penetrate to a depth of 12 to 20 feet (4 to 6 m) or even more. The rhizomes, which can spread several feet in a year, are mostly found at a depth of 12 to 30 inches (30 to 76 cm). The stems are smooth, slender, slightly angled vines that spread over the ground or twine around and climb any erect crop plant (Figure 11.3). The vines may be 1 to 6 feet (30 to 180 cm) in length.

11.7.2 Hoary Cress

Hoary cress *(Cardaria draba)* is also known as perennial peppergrass, whiteweed, and whitetop. Hoary cress propagates by both seed and rhizomes. The plant is erect, 10 to 18 inches (25 to 46 cm) high, and grayish white in color. Two other species, very similar to *C. draba* and known by the same common names, are *C. repens* (or *C. draba variety repens*) and *Hymenophysa pubescens*. Hoary cress makes vigorous growth on the irrigated alkaline soils of the West. It is a serious pest in the Rocky Mountain region and on the Pacific Coast.

FIGURE 11.3
Field bindweed climbing on barley and oat plants.

11.7.3 Canada Thistle

Canada thistle (*Cirsium arvense*) (Figure 3.5) propagates by seed and rhizomes. The stems are erect, hollow, smooth to slightly hairy, 1 to 4 feet (30 to 120 cm) tall, simple, and branched at the top. The leaves are set close on the stem, slightly clasping, typically green on both sides, sometimes white and hairy beneath, and deeply and irregularly cut into lobes tipped with sharp spines. Canada thistle is widely

FIGURE 11.4
Perennial sowthistle.

distributed in the northern half of the United States. A smooth-leaved type of Canada thistle that is more difficult to control is spreading rapidly in some areas.

11.7.4 Russian Knapweed

Russian knapweed (*Acroptilon repens*) reproduces by seed and black creeping rhizomes. The plant is erect, rather stiff, branched, and 1 to 3 feet (30 to 90 cm) high. The young stems are covered with soft gray hairs. Russian knapweed is generally distributed in the western states.

11.7.5 Perennial Sowthistle

Perennial sowthistle (*Sonchus arvensis*) (Figure 11.4) spreads by seed and rhizomes. The stems are erect, smooth, 2 to 5 feet (60 to 150 cm) high, and unbranched except at the top. The leaves are light green, the lower ones being 6 to 12 inches (15 to 30 cm) long, deeply cut, with the side lobes pointing backwards; the upper leaves are smaller, clasping, with margins slightly toothed, and prickly. This weed is found throughout the northern United States.

11.7.6 Leafy Spurge

Leafy spurge (*Euphorbia esula*) (Figure 11.5) spreads by seed and rhizomes. The plants are pale green, erect, 1 to 3 feet (30 to 90 cm) high, and unbranched except

FIGURE 11.5
Leafy spurge.

for flower clusters. The leaves are long and narrow with smooth margins. The plant is characterized by a milky sap. It is widely scattered throughout the United States.

11.7.7 Quackgrass

Quackgrass *(Agropyron repens)* spreads by seed as well as by long, jointed yellow rhizomes. This grass is erect, 1 to 3 feet (30 to 90 cm) high, with slender stems. The ligule is short and membraneous with large auricles wrapping around the stem. The flat, narrow leaves are rough on top but smooth beneath. The seed is borne in spikes 3 to 7 inches (8 to 18 cm) long. Quackgrass is found throughout the northern half of the United States. Although it has forage value, quackgrass becomes a serious weed because of its persistence. It is related to and resembles some of the native wheatgrasses that are valuable forage and pasture plants.

11.7.8 Johnsongrass

Johnsongrass *(Sorghum halepense)* spreads by seeds and rhizomes. It is an erect plant, 5 to 6 feet (150 to 180 cm) tall. It is closely related to sorghum and sudangrass. Johnsongrass is widespread throughout the southern states. The most economical control method is heavy pasturing or frequent mowing to starve the rhizomes, followed by plowing and planting a clean cultivated crop.

11.7.9 Nutgrass

Nutgrass *(Cyperus rotundus)*, also called cocograss, or nutgedge, is a sedge rather than a grass. It spreads by deepset rhizomes as well as by small tubers (nuts) borne on the rhizomes. The nuts are ¼ to ¾ inch (6 to 19 mm) in diameter. Nutgrass is widely distributed, especially from Virginia to Kansas and southward. It is one of the most serious weed pests in the South.

■ 11.8 SERIOUS ANNUAL WEEDS

The following are considered to be the ten worst annual weeds in the United States: barnyardgrass (*Echinchloa crus-galli*), cocklebur (*Xanthium* species), foxtail (*Setaria* species), lambsquarter (*Chenopodium* species.), morning glory (*Ipomea* species), panicum (*Panicum* species), and pigweed (*Amaranthus* species). In addition, wild oat (*Avena fatua*), wild mustard, and charlock (*Brassica arvensis*) are perhaps the most prevalent weed pests in the spring grain regions. Downy brome or cheatgrass (*Bromus tectorum*) is a serious pest on minimum tillage grain fields and on range-land of the northwestern quarter of the United States. Crabgrass in lawns is very difficult to control.

Lambsquarter, pigweed, Russian thistle, cocklebur, and many other broadleaf annuals are controlled with 2, 4-D, atrazine, metolachlor, dicamba, and numerous other herbicides. Foxtail is controlled with metolachlor, glyphosate, and sethoxy-dim. Barnyardgrass is controlled with glyphosate, fluazifop-P, and metolachlor. Shattercane is controlled with fluazifop-P and sethoxydim. Crabgrass controls include glyphosate, dimethenamid, DCPA, and fluazifop-P.

■ 11.9 PARASITIC PLANTS

Dodder (Figure 11.6), broomrape, and witchweed are flowering plants that parasitize certain crop plants. Alfalfa plants take up a herbicide such as DCPA without injury, but it suppresses the dodder. It must be used only where the alfalfa is to be harvested for seed.

Witchweed (*Striga lutea* or *S. asiatica*) is an annual flowering plant parasite of sorghum, corn, and some other grass species. It is prevalent in Africa and India but also has invaded North Carolina and South Carolina in the United States. Haustoria from germinating seedlings enter roots and slow the growth of the plants be-

FIGURE 11.6
A dodder or "lovevine" (*Cuscuta* species): Seedling comes up and sways around until it encounters a host plant, which it entwines. Haustoria or suckers penetrate the stem of the host and take up nutrients. Then the lower stem of the dodder dies, but the plant proceeds to thrive at the expense of the host.

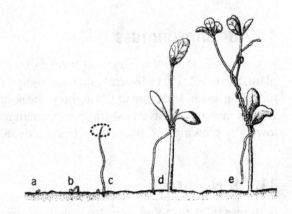

fore the witchweed stems emerge from the soil surface. Spraying with 2, 4-D or other phenoxy compounds kills witchweed, but the damage has already occurred. Planting a trap crop, such as cowpea, which stimulates the germination of witchweed seeds but is not parasitized, is helpful where the practice is feasible. Complete control is difficult because witchweed attacks grasses in uncultivated areas and produces numerous seeds where it is not easily detected.[27]

REFERENCES

1. Holm, L., J. Doll, E. Holm, J. Pancho, and J. Herberger. *World Weeds, Natural Histories and Distribution.* New York: John Wiley and Sons, 1997.

2. Westbrooks, R. *Invasive Plants, Changing the Landscape of America: Fact Book.* Federal Interagency Committee for the Management of Noxious and Exotic Weeds (FICMNEW), Washington, DC 1998.

3. Kingsbury, J. M. *Poisonous Plants of the United States and Canada.* Englewood Cliffs, NJ: Prentice-Hall, 1964.

4. Klingman, D. L., and W. C. Shaw. "Using phenoxy herbicides effectively," *USDA Farmers Bull.* 2183, 1962.

5. Neeley, J. W., and S. G. Brain. "Control of weeds and grasses in cotton by flaming," *MS Agr. Exp. Sta. Circ.* 118, 1944, pp. 1–6.

6. Stevens, O. A. "The number and weight of seeds produced by weeds," *Am. J. Bot.* 19(1932):784–794.

7. McWarter, C. G. "A summary of methods for johnsongrass control in soybeans in Mississippi," *MS Agr. Exp. Sta. Bull.* 788, 1971, pp. 1–8.

8. Thornton, B. J., and H. D. Harrington. "Weeds in Colorado," *CO Agr. Exp. Sta. Bull.* 514-S, 1964.

9. Reed, C. F., and R. O. Hughes. "Selected weeds of the United States," *USDA Handbk.* 333, 1970, pp. 1–463.

10. Ross, M., and C. Lembi. *Applied Weed Science.* Minneapolis: Burgess Publishing Company, 1983.

11. Shaw, W. C., and others. "The fate of herbicides in plants," in *USDA Agricultural Research Service Spec. Rept.* 20–9:119–133, 1960.

12. Egginton, G. E., and W. W. Bobbins. "Irrigation water as a factor in the dissemination of weed seeds," *CO Agr. Exp. Sta. Bull.* 253, 1920.

13. Harmon, G. W., and F. D. Keim. "The percentage and viability of weed seeds recovered in the feces of farm animals and their longevity when buried in manure," *J. Am. Soc. Agron.* 26(1934):762–767.

14. Kiesselbach, T. A., N. F. Petersen, and W. W. Burr. "Bindweeds and their control," *NE Agr. Exp. Sta. Bull.* 287, 1934.

15. Bakke, A. L., and others. "Relation of cultivation to depletion of root reserves in European bindweed at different soil horizons," *J. Agr. Res.* 69, 4(1944):137–147.

16. Barr, C. G., "Preliminary studies on the carbohydrates in the roots of bindweed," *J. Am. Soc. Agron.* 28(1936):787–798.

17. Anonymous. "Chemical control of brush and trees," *USDA Farmers Bull.* 2158, 1961.

18. Muenscher, W. C. L. *Weeds,* rev. ed. New York: Macmillan, Inc., 1961, pp. 1–560.

19. Weldon, L., and others. "Common aquatic weeds," *USDA Handbook.* 352, 1969, pp. 1–43.

20. Anonymous. "Suggested guides for weed control," *USDA Handbook,* 1966.

21. Anonymous. *Principles of Plant and Animal Pest Control.* Vol. 2, *Weed Control,* Washington, DC: National Academy & Science, 1968. pp. 1–471.

22. Ashton, F. M., and W. A. Harvey, "Selective chemical weed control," *CA Agr. Exp. Sta. Circ.* 558, 1971, pp. 1–17.

23. Gerlow, A. R. "The economic impact of canceling the use of 2,4,5-T in rice production," *USDA ERS* 510, 1973, pp. 1–11.

24. Klingman, G. C. *Weed Control as a Science.* New York: Wiley, 1966, pp. 1–421.

25. Klingman, D. L., and W. C. Shaw. "Using phenoxy herbicides effectively," *USDA Farmer's Bull.* 2183, 1971, pp. 1–24.

26. Heddon, O. K., and others. "Equipment for applying soil pesticides," *USDA Handbook.* 287(rev.), 1967, pp. 1–37.

27. Cardenas, J., and others. *Tropical Weeds.* Bogata, Colombia: Inst. Colombiano Agropecuaro, 1972, pp. 1–341.

PART 2 CROPS OF THE GRASS FAMILY

Corn or Maize

Corn (*Zea mays*) is called *maize* in most of the rest of the world outside the United States. The word *corn* merely means *grain* to the British, Germans, and many other nationalities, and often is used to designate the predominant grain in a country or section. Corn usually refers to wheat in England, oat in Scotland, and barley in North Africa. Translations of ancient writings refer to corn that was actually wheat or barley.

■ 12.1 ECONOMIC IMPORTANCE

Corn ranks third, following wheat and rice, in the world production of cereal crops. The world area devoted to corn in 2000–2003 averaged 345 million acres (140 million ha). Corn grain production averaged 70 bushels per acre (4,400 kg/ha) for a total of 24 billion bushels (613 million MT). About 40 percent of the world corn crop is grown in the United States. Other important countries in corn production are China, Brazil, Mexico, and Argentina. Corn is a highly important food plant in Mexico, Central America, and the tropics of South America (Figure 12.1).

In the United States, corn for grain was harvested from about 70 million acres (28 million ha) in 2000–2003. About 6.5 million acres (2.6 million ha) were harvested for silage. The leading states in corn production are Iowa, Illinois, Nebraska, Minnesota, and Ohio. Grain production averaged about 9.6 billion bushels (244 million MT) or 137 bushels per acre (8,600 kg/ha). Silage production was about 103 million tons (94 million MT), or 16 tons per acre (36 MT/ha).

Corn yields in the United States averaged about 26 bushels per acre (1,630 kg/ha) from 1910 to 1914 and have increased steadily since then (Figure 12.2). During this time, commercial corn production became completely mechanized. An average of 135 hours of labor was required to produce 100 bushels (2,540 kg) of corn during the earlier period, compared with less than 7 hours today. Corn is grown in almost every state (Figure 12.3).

Corn supplies three-fourths of the nutrients derived from feed grain and over 80 percent of silage fed in the United States. Corn is used for human food, animal feed, sweetener, biofuel, alcohol, and many industrial products.

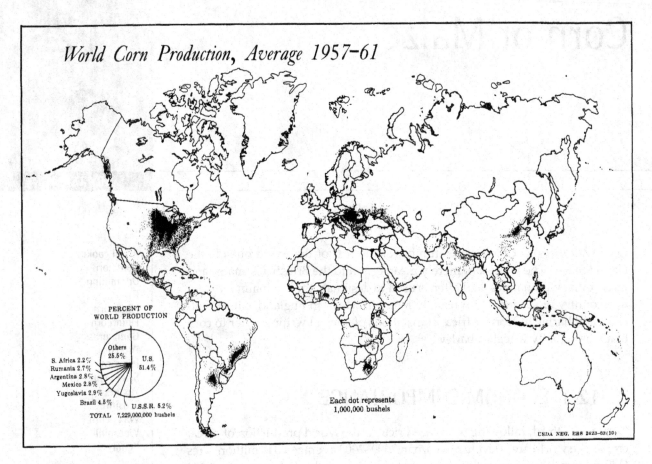

FIGURE 12.1
World corn production. [Courtesy USDA]

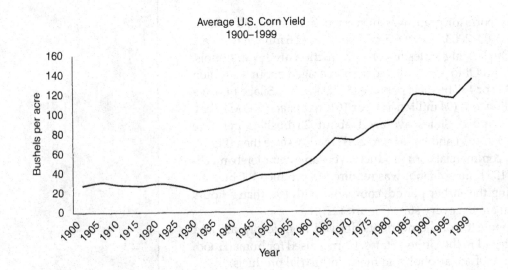

FIGURE 12.2
Increases in corn yield due to hybridization and improved production practices.

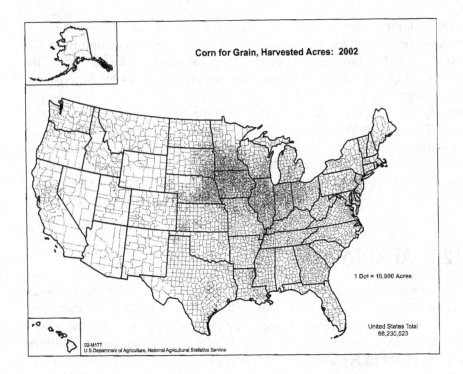

Corn for Grain, Harvested Acres: 2002

1 Dot = 10,000 Acres

United States Total
68,230,523

02-M177
U.S.Department of Agriculture, National Agricultural Statistics Service

FIGURE 12.3
Acreage of corn the United
States. [Source: 2002 Census
U.S. Department of
Agriculture]

U.S. popcorn production is about 800 million pounds (363,000 MT) per year. Practically all of the world's production is in the United States. Nebraska, Indiana, Ohio, Iowa, and Illinois produce about 90 percent of all U.S. popcorn production.

About 700,000 acres (28,000 ha) of sweet corn are grown annually in the United States. About one-third is grown for the fresh market and two-thirds for processing. Average yields are about 13,000 pounds per acre (14,500 kg/ha). The leading states in the production of sweet corn are Florida, California, New York, Georgia, and Pennsylvania.

▪ 12.2 HISTORY OF CORN CULTURE

Corn is perhaps the most completely domesticated of all field crops. Its perpetuation for centuries has depended wholly upon the care of humans. It cannot exist as a wild plant because it cannot efficiently disperse its seed, and the seed has a soft seedcoat that will not survive passage through the gut of an animal.

Corn was seen by Columbus in Cuba on November 5, 1492, on his first voyage to America. The explorations of the sixteenth and seventeenth centuries showed that corn extended from Chile to the Great Lakes. The most likely center of origin of corn is Mexico or Central America, with a possible secondary origin in South America.

The genetic origin of corn has been the subject of much discussion and controversy for many years. Some theories suggested that corn descended from a wild teosinte plant. Others contended that its ancestor was a hypothetical but no longer existent wild corn plant. Still others suggested that it came from a cross between corn and a gama grass (*Tripsacum*) species.[1] With the present ability to map genes, the puzzle is starting to be solved. Genetic evidence points to corn developing from a cross between teosinte and gama grass. The favorable traits resulting from this cross were selected by farmers resulting in the characteristics of modern corn.[2] By 5,500 years ago, kernel size had increased significantly; and by 4,400 years ago, the traits of modern corn were present.

Small ears and grains, similar to those of popcorn but enclosed in floral bracts, were found in deposits in a bat cave in New Mexico. Carbon dating indicates that these specimens are about 5,600 years old. Numerous specimens, which date from 5200 BC to AD 1500 were found in caves in Mexico. They portray the evolution from the wild to the cultivated types of corn. Marked increases in ear size are evident as are reduced glume lengths and departures from the podded types.[3]

▓ 12.3 ADAPTATION

Corn has a remarkable diversity of vegetative types that are adapted to a wide range of environmental conditions. From latitude 58° N in Canada, the culture of corn passes without interruption through the tropical regions and into the Southern Hemisphere (latitude 35 to 40° S). It is grown from below sea level to altitudes of 13,000 feet (4,000 m). Some small, early varieties only 2 feet (0.6 m) tall bear eight to nine leaves and are able to produce mature grain in 50 days. Others with 42 to 44 leaves and growing 20 feet (6 m) tall require as many as 330 days to come to maturity. This shows how corn has been modified to meet the conditions of many environments. The phenomenon has been well described by Jenkins[4] who states: "The greatest plant-breeding job in the world was done by the American Indians. Out of a wild plant, not even known today, they developed types of corn adapted to so wide a range of climates that this plant is now more extensively distributed over the earth than any other cereal crop. Modern breeders are carrying on this work and making important discoveries of their own."

In the United States, corn hybrids (varieties) grown in the northern states are 6 to 8 feet (2 to 2.5 m) tall and mature in 90 to 120 days. In the central and southern Corn Belt, hybrids may grow 8 to 10 feet (2.5 to 3 m) tall and mature in 130 to 150 days. Hybrids in the south Atlantic and Gulf Coast states may grow to a height of 10 to 12 feet (3 to 3.6 m) on fertile soils and require 170 to 190 days to reach maturity.

12.3.1 Climatic Requirement

The region of greatest production of corn in the United States has a mean summer temperature of 70 to 80°F (21 to 27°C), a mean night temperature exceeding 58°F (13°C), and a frost-free season of over 140 days.[4] An average June-July-August temperature of 68 to 72°F (20 to 21°C) is most favorable for maximum yields. Corn flowers and ripens more quickly at 80°F (27°C). Very little corn is grown where the mean summer temperature is less than 66°F (19°C) or the average night temperature is below 55°F (13°C). The minimum temperature for the germination and growth of corn is 50°F (10°C) or slightly below. Corn plants will withstand a light freeze in the seedling stage up to the time when they are 6 inches (15 cm) tall. Thereafter, a light freeze will kill the plants except for those of certain hardy strains. Prolonged low temperatures below 45°F (7°C) but above freezing will kill many strains of corn. Corn is grown extensively in hot climates, but yields are reduced where the mean summer temperatures are above about 80°F (27°C).

The misconception has long existed that corn grows best when nights are hot. Actually accelerated respiration in warm nights reduces the carbohydrates that can accumulate and consequently reduces the dry weight of the plants. Cell division and enlargement may continue in darkness when the temperature exceeds 57°F (14°C).

The annual precipitation where corn is grown ranges from 10 inches (250 mm) in the semiarid plains of Russia to more than 200 inches (5,100 mm) in the tropics of India. In the United States, the best corn regions have an annual precipitation of 24 to 40 inches (600 to 1,000 mm), except where the crop is irrigated. General corn production is limited to regions with an annual precipitation of 15 inches (380 mm) or more and a summer (June to August) precipitation of 8 inches (200 mm) or more. A heavier precipitation is required in southern latitudes and low altitudes. Some corn is grown at high altitudes where the summer precipitation is less than 6 inches (150 mm). Native Americans of New Mexico and Arizona grow corn where the annual precipitation does not exceed 8 inches (200 mm) by planting in widely spaced hills in dry sandy washes that receive extra water whenever a rain is heavy enough to cause runoff.

The corn plant utilizes about half its seasonal intake of water during the five weeks following attainment of its maximum leaf area, which is at about the tasseling stage.[5]

Corn is a short-day plant. Its flowering is hastened and vegetative growth retarded by long nights. Long days increase the leaf number, plant size, and length of the growing period of corn. Northern varieties moved southward to where the days are shorter will mature more quickly with reduced plant growth. Southern varieties moved to the North will make a large vegetative growth but will not reach the silking stage until the short days of autumn. If the altitude is the same, a variety will mature progressively one day earlier or later for about every ten miles it is moved southward or northward, respectively. The varieties that succeed best in any region are those adapted to the particular length of day. Quick-maturing hybrids are grown in the North. Progressively longer-season hybrids are grown as one travels south in the Northern Hemisphere.

In a Nebraska study, adapted types from 43 states showed wide variations in vegetative characteristics. In maturity, the extreme variation was 155 days for Mexican June corn from New Mexico, to 96 days for Northwestern Dent from North Dakota. There was a marked transition from large late types to small early types from south to north across the country. This was associated with a progressively shortened frost-free season and increased maximum length of day from south to north.[6, 7] The number of leaves on the main stalk is the most reliable measure of the length of season required for maturity.[8] The earlier the variety, the fewer the leaves on the main stem. The number of leaves among some 8,000 world varieties and strains ranges from 8 to 48. Canadian varieties vary from 9 to 18 leaves, American from 10 to 27, and Peruvian from 14 to 33. American Corn Belt hybrids generally have from 18 to 21 leaves.

12.3.2 Soil Requirements

Corn requires an abundance of readily available plant nutrients and a soil pH between 5.5 and 8.0 for the best production.[9] Prairie soils seem to be among those most favorable for corn production. Corn makes its best growth on fertile, well-drained, medium-textured soils. On poor soils, vegetative growth may use most of the available nutrients leaving little for grain production later in the season. Under short-season conditions, light sandy soils are preferable for corn because they warm up more quickly in the spring.

12.3.3 Adjustment to the Environment

In most cases, open-pollinated corn, grown for a period of years, becomes adjusted to the local conditions of moisture supply, temperature, and length of frost-free season.[10] This self-adaptation is brought about by natural selection in the mixed or heterozygous population of any open-pollinated corn variety. Adaptation of corn to a region of moisture shortage consists of a reduction in vegetative development with a consequent reduction in the amount of water used by the individual plant.[11] As climatic conditions become more adverse, the stalks in adapted strains are shorter, the leaf area less, the ears shorter, the kernels more shallow, and the yield lower.

Seed of locally adapted strains will ordinarily produce the highest yields. A variety or hybrid from another region can give increased yields, particularly when moved from a less favorable to a more favorable climatic region, provided the length of the growing season permits satisfactory maturity.[12] Ordinarily, corn hybrids adapted within a latitude range of 100 miles are most likely to produce maximum grain yields.

■　12.4　BOTANICAL CHARACTERISTICS

Corn (*Zea mays*) is a coarse annual grass classified in the tribe Maydeae, which has monoecious (separate staminate and pistillate) inflorescences.[13] Among other members of the Maydeae tribe, teosinte (*Zea mays*, spp. *mexicana*), a native of Mexico, is an annual rank-growing grass grown in the southeastern part of the United States as a forage crop. A perennial species of this genus (*Z. perennis*) is also known, as well as other species of teosinte. Gama grass (*Tripsacum dactyloides*) is distributed over the eastern and southern United States and is a useful pasture and meadow grass. Other species of *Tripsacum* occur in Mexico and Central America. Teosinte and gama grass spikes break up into sections upon threshing, revealing a seed enclosed in the thick hulls of each section. Job's tears (*Coix lachryma-jobi*), from southern Asia, is grown in the United States for ornamental purposes. The fruits are used as beads. Thin-hulled strains have been grown in the Philippine Islands and South America as a cereal under the name of *adlay*.

12.4.1 Vegetative Characteristics

STALKS The stalk of corn ranges in length from 2 to 25 feet (0.3 to 7.6 m) and in diameter from ½ to 2 inches (1.3 to 5 cm) in different regional types. The internodes are straight and nearly cylindrical in the upper part of the plant but alternately grooved on the lower part. A bud occurs (usually in the groove) at the base of each internode except the terminal one. When the buds develop, they produce ear shoots. Buds below the soil surface may develop into tillers. Modern field corn hybrids seldom develop tillers in optimum spaced plantings. The outside of the culm is protected by an epidermis that renders it fairly impervious to moisture. Inside the epidermis is the cortex and rind made up of sclerified (thick-walled) parenchyma containing many vascular bundles. Inside the rind, the stem is filled with ground tissue (pith) in which vascular bundles are interspersed (Figure 3.3). Tillers, when present, contribute to grain formation on the main stalk.[14] Removal of tillers may injure the plant and reduce the yield.

Commercial strains and inbred lines of corn vary considerably in lodging. Lodged stalks are those that break over or lean 30 degrees or more from the vertical. Most commercial corn hybrids are resistant to lodging. Increases in percentages of broken stalks occur when they are infected with diseases such as *Diplodia*, *Gibberella*, or *Cephalosporium*, or when starchy seed susceptible to scutellum rot is planted.[15] Increasing plant population can increase lodging. Much of the lodging in corn is due to weak roots. Erect plants have about twice as large a root system as lodged plants.[16] Broken or bent shanks cause the ears to hang down or drop to the ground, which results in greater harvest losses in either case. Growing corn plants increase in height during the night as well as in daytime, although photosynthesis occurs only during daylight.[17]

Corn stalks contain about 8 percent sugar before the grain is formed and as much as 10.5 percent sugar when pollination fails or is prevented.[18]

LEAVES The corn leaf consists of the blade, sheath, and a collar-like ligule. Liguleless lines of corn have more erect leaves, which permit increased light penetration to the lower leaves, while at the same time resulting in greater photosynthetic efficiency. The leaf blade tapers to a point at the tip and also tapers from the central midrib to the thin edges. At the base of the blade, the two edges extend to form two auricles. The sheath clasps the culm.

A typical leaf of a corn plant grown at Ithaca, New York,[19] was about 31 inches (80 cm) long, 3½ to 4 inches (9 to 10 cm) wide, with an average thickness of $\frac{1}{100}$ inch (¼ mm), and contained more than 140 million cells.

The structure of the leaf blade is similar to that of the sugarcane leaf (Figure 3.2). The upper surface has scattered hairs and large stomata. The lower surface is free from hairs and has smaller but more numerous stomata. A cross section of the leaf shows both the upper and lower epidermis that consists of single layers of cells. Between the upper and lower epidermal layers are five or six layers of mesophyll cells, and within the mesophyll are occasional fibrovascular bundles containing both the strengthening and conducting tissues of the leaf. Chloroplasts in the mesophyll tissues surrounding the vascular bundles are a characteristic of plants such as corn and sugarcane (Figure 3.2). They have the C_4 pathway for high photosynthetic efficiency. At intervals along the upper epidermal layer are groups of large bulliform (hygroscopic) cells. Whenever evaporation from the leaf appreciably exceeds the water intake, these bulliform cells shrink and the leaves roll up, thus reducing the surface area exposed to evaporation. When water becomes more abundant, the bulliform cells absorb water, swell, and flatten out the leaf blade again. In the sorghum leaf, in contrast, the bulliform cells are concentrated in large groups near the midrib, so the drying leaf usually folds up like butterfly wings instead of rolling.

The reduction in yield resulting from the loss during storms or other natural hazards becomes progressively less with advances in the development of the plant.[20]

ROOTS The corn plant has three types of roots: (1) seminal; (2) coronal or crown; and (3) brace, buttress, or aerial roots. The seminal roots, usually three or five, originate from within the seed and grow downward at the time of seed germination. Shortly after the plumule has emerged, the coronal roots form at the nodes of the stem, usually about 1 to 2 inches (25 to 50 mm) below the soil surface. The coronal roots, which are ultimately fifteen to twenty times as numerous as the seminal roots, develop at the first seven or eight nodes at the base of the stem. Brace roots arise from the nodes above ground, but their function is similar to that of the coronal roots after they enter the soil.

The usual spread of the main roots can be as much as 3½ feet (1 m) in all directions from the stalk,[21] but most will be within half the distance to neighboring plants. The lateral spread ceases about one or two weeks before the tasseling stage.[22] Later-season growth is mostly downward. Although most of the lateral roots are found in the first and second feet (10 to 20 cm) of soil, some may extend downward as far as 7 to 9 feet (2 to 3 m). The seminal root system will usually be crowded out by the coronal root system. The size of the root system increases with that of the top growth. Compared with a small type, a large type of corn may have a 50 percent greater spread, 10 percent deeper penetration, 65 percent more functional main roots, 92 percent greater combined length of main roots per plant, 311 percent greater root weight, and 23 percent larger diameter of main roots.[23] The rate of root growth as well as top growth is approximately the same for all types until the small varieties tassel. Thereafter, continued vegetative growth causes the medium and large types to surpass small early types. Similarly, as the medium types commence tasseling, their vegetative development is exceeded by that of the large type.

12.4.2 Inflorescence

The corn plant is normally monoecious (i.e., the staminate and pistillate flowers are borne in separate inflorescences on the same plant). The staminate flowers are borne in the tassel at the top of the stalk; the pistillate flowers are located in spikes that terminate lateral branches arising in the axils of lower leaves. The mature pistillate inflorescence is called the ear (Figure 12.4). Occasionally, off-type plants produce seed in the tassel or a tassel on the ear tip.

The central axis of the staminate panicle is the continuation of the main vegetative stem. The branches of the panicle are spirally arranged around the axis. The spikelets are usually arranged in pairs, one sessile and the other pediceled. Groups of three or four may occasionally be found. The spikelet is completely enclosed by two firm, more or less pubescent, ovate plumes. There are two florets per spikelet, the upper being the more advanced in development. Each floret contains three stamens, two lodicules, and a rudimentary pistil. Each anther produces about 2,500 pollen grains or approximately 15,000 for a single spikelet. Thus, a tassel containing 300 spikelets produces 4,500,000 pollen grains to fertilize the 500 to 1,000 silks produced on an ear.

The pistillate inflorescence (ear) is a spike with a thickened axis. The pistillate spikelets are borne in pairs in several longitudinal rows. This paired arrangement explains the customary even number of rows of grains on the ear. An occasional ear with an odd number of rows has lost one member of the pair. The individual spikelet is two-flowered, only one floret ordinarily being fertile. When the second flowers of the spikelet are also fertile, the crowding of the kernels destroys the appearance of rows. The result is an irregular distribution of kernels as seen in the Country Gentleman variety of sweet corn (Figure 12.5).

The two thick and fleshy glumes are too short to enclose the other parts of the spikelet. The lemma and palea, which together constitute the chaff, are thinner and shorter than the glumes. The single ovary in a fertile floret has a long style or silk

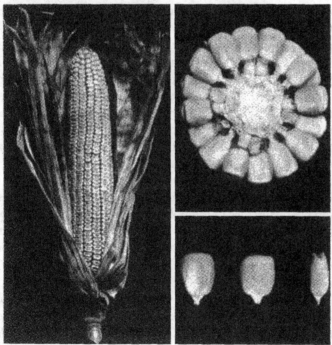

FIGURE 12.4

Top: Pistillate inflorescence—young ear in silking stage. *(Right)* Staminate inflorescence or tassel of the corn plant. *Bottom: (Left)* Mature ear with husks. *(Upper right)* Cross section of ear showing kernel attachment to cob. *(Lower right)* Kernels—the germ side of the kernel faces the tip of the ear except in those that develop from the lower floret of the spikelet.

that is forked at the tip (Figure 3.12). The silk has a sticky stigmatic surface and is receptive to pollen for about two weeks throughout its length. The remnant of the silk is evident on the crown of each mature kernel. The silks ordinarily are 4 to 12 inches (10 to 30 cm) in length. Silks will continue to grow until pollinated. Unpollinated ears may produce silks 20 to 30 inches (50 to 75 cm) long.

FIGURE 12.5
Ear of Country Gentleman sweet corn, which develops kernels in both florets of the spikelet; the resultant crowding prevents the kernels from forming straight rows. The smooth white-capped kernels show xenia that resulted from outcrossing by pollen from field corn.

12.4.3 Pollination

Corn is normally cross-pollinated,[6] being well adapted to wind pollination with the male flowers at the top of the plant and the female flowers about midway up the stalk.

Cold, wet weather retards the shedding of pollen, while hot, dry conditions tend to hasten it. The emergence of the silks is delayed by drought and inadequate nitrogen. Protandry is the shedding of pollen before the appearance of the silks and protogyny is maturation of the stigmas before pollen shedding. Corn is somewhat protandrous, but pollen shedding and the receptivity of the silks overlap sufficiently to permit some self-pollination.

In the fertilization of the corn flower,[24, 25] the pollen grain germinates and establishes a pollen tube within 5 to 10 minutes after it falls on the silk. The pollen tube enters the central core of a hair on the silk and passes down through the silk (Figure 3.12). The two sperm nuclei migrate into the pollen tube and pass down to the embryo sac. Fertilization is accomplished within 15 to 36 hours after pollination, the slower rates occurring at low temperatures and when the silks are long. One of the two sperm nuclei unites with the egg cell to produce the embryo; the other nucleus unites with first one, and then the other, of the two polar nuclei of the embryo sac to give rise to the endosperm. The union of sperm nuclei with both the egg cell and the polar nuclei is called double fertilization. The endosperm often shows certain characteristics of the pollen parent because it develops directly from the endosperm nucleus after union with the male nucleus. The pericarp, on the other hand, does not display characteristics derived from the pollen parent that season. Those characteristics are shown when a new plant develops the next season. The immediate effect of the pollen parent on the characteristics of the endosperm, embryo, or scutellum is called **xenia** (pronounced zee'-nee-ah). Xenia occurs when an ear of sweet corn is pollinated with flint corn and the kernels are smooth and starchy instead of wrinkled and sugary (Figure 12.5). Also, when pollen from a plant with yellow endosperm fertilizes a plant with white endosperm, the kernels have a yellow endosperm because of the Mendelian dominance of yellow over white. When yellow corn is pollinated by white corn, the kernels are yellow but are lighter in color and often capped with white.

Pollen of dent on sweet corn silks may cause a 25 percent increase in average kernel weight because of the formation of a starchy instead of a sugary endosperm.[6] The effect on kernel size is negligible when one commercial variety of dent corn is crossed on another.[26] However, hybrid kernels on pure-line ears have increased as much as 11 percent in weight.

The flowers near the middle of the ear develop the silks early and usually are pollinated first. The flowers in the lower third of the ear develop at about the same time, but, because of their remote position, take more time to extend the silks out of the husk. Those at the tip of the ear are last to develop. Under unfavorable conditions the earlier-pollinated kernels appear to take precedence over the others in food supplies, and the tips and butts of the ears are filled very poorly and produce smaller grains.

12.4.4 Development of the Corn Kernel

In the development of the corn kernel or caryopsis, the ovary wall grows by cell division and cell enlargement to form the pericarp of the kernel.[19, 27] The endosperm and the entire kernel grow rapidly in length and diameter during the period of 10 to 20 days after pollination. This is followed by a correspondingly rapid growth of the embryo during the period from 20 to 40 days after pollination. The embryo grows and differentiates until the various rudimentary structures of a new plant are apparent. The endosperm cells divide rapidly and nearly fill the space within the growing ovary wall within 18 days after pollination. The endosperm cells at first divide in all directions. Later, cell divisions occur largely around the periphery, and rows of cells are forced inward. Final endosperm growth is largely due to cell enlargement. The mature endosperm is 50 times larger than it is in the initial stages of its observable formation. The endosperm cells at first contain simple sugars but later the sugars are converted to starch.

The testa, often called the nucellar layer, arises just inside of the pericarp where the cells of the integuments and nucellus of the embryo sac have degenerated.

The kernel ceases growth about 45 days after pollination. The shape of the individual kernel is determined in part by the degree of crowding on the ear. Isolated grains and many of those at the tip and butt are usually somewhat rounded. Grains in crowded rows on the ear are often angular in shape.

About 1,200 to 1,400 kernels of typical dent corn weigh 1 pound (2,600 to 3,000 kg). The kernels usually range from ¼ to ⅜ inch (6 to 9 mm) wide, ⅛ to ¼ inch (3 to 6 mm) thick, and ⅜ to ⅝ inch (9 to 16 mm) deep. The kernels of Cuzco corn from Peru are about 0.58 inch (16 mm) wide.

The parts of the corn kernel, as shown in Figures 7.2 and 12.6, include a pedicel (tip cap) over the closing layer of the hilar orifice at the end of the kernel where it is attached to the cob.[28] The embryo (germ) lies near one face of the kernel. The outer rim of the embryo is the scutellum. The remainder of the kernel inside the testa is occupied by the endosperm. The outer layer of cells of the endosperm is the aleurone, which comprises about 8 to 12 percent of the kernel and is high in nitrogenous substances. The aleurone may be red or purple in certain varieties. In dent corn the endosperm can be partly corneous (horny) and partly soft or starchy white. The corneous portion may be one-tenth to one-fourth higher in protein. The embryo, comprising about 11 percent of the kernel,

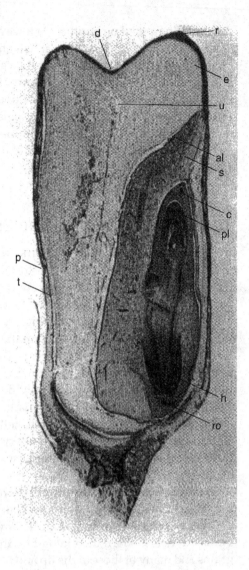

FIGURE 12.6

Half of nearly mature dent corn kernel that shows dent (d), remnant of style (r), endosperm (e), unfilled portion of immature endosperm (u), aleurone (al), scutellum (s), plumule (pl), hypocotyl and radicle (h), testa or true seed coat (t), and pericarp (p). The germ comprises the entire darker area, which includes the scutellum (s), coleoptile (c), and rootcap (ro).

contains five-sixths of the oil, three-fourths of the ash, and one-fifth of the protein in the kernel.

Outside of the aleurone is the testa (true seed coat), which in turn is surrounded by the pericarp. The hull (pericarp and testa), comprising about 6 percent of the kernel, is high in cellulose and hemicelluloses. The hull protects the kernel from mechanical damage and the entrance of disease organisms.

12.4.5 The Ear

A normal ear of corn contains 8 to 28 rows of grains, and each regular row produces 20 to 70 kernels. An occasional freak ear may have only 4 rows when other rows are aborted. Ears with 12, 14, 16, and 18 rows predominate among American dent corns. A large ear of corn may contain about 1,000 kernels.

The average well-dried ear of corn weighs about 8 ounces (227 gm) and contains 500 to 600 kernels. The shelling percentage of dry, well-filled ears averages about 80 to 85 percent. In average dried corn, 70 pounds (32 kg) of ears yield 1 bushel or 56 pounds (25 kg) or 80 percent shelled corn.

◼ 12.5 TYPES OF CORN

Corn is often classified into seven types or groups, based largely upon endosperm and glume characteristics. These are dent, flint, flour, pop, sweet, waxy, and pod corns (Figure 12.7). Earlier, the different types sometimes were regarded as distinct botanical subspecies. However, the use of botanical subspecies names is no longer valid in light of our knowledge of corn heredity. The sweet, waxy, and pod types each differ from dent corn by only a single hereditary factor, and the flint, flour, and popcorn types differ by relatively few hereditary factors.

12.5.1 Dent Corn

Both corneous and soft starches are found in dent corn; the corneous starch is extended on the sides to the top of the kernels (Figure 12.8). When the grain dries, a pronounced wrinkle or dent forms on the top of the kernel, due to the shrinkage of the soft starch. Dent corn is the most widely cultivated type, and is by far the most important one in the United States and almost the only type in the Corn Belt. The kernels have a wide range of colors but are yellow or white in most commercial varieties.

FIGURE 12.7

Ears of *(left to right)* pop, sweet, flour, flint, dent, and pod corn.

FIGURE 12.8
Typical proportions of hard and soft starch in kernels of four corn groups.

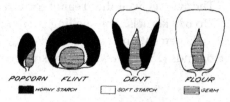

"High lysine" dent corn has a modified opaque endosperm with a high content of lysine, an essential nutritional amino acid for humans, swine, and poultry.

12.5.2 Flint Corn

The endosperm of flint corn is usually soft and starchy in the center but is completely enclosed by a corneous outer layer. The kernels are usually rounded at the tip. Most flint corn, formerly grown in the northern part of the United States, had eight rows of kernels on the ear and were early in maturity. The ears of the tropical flint corn of Louisiana and Florida usually had 12 or 14 rows of kernels. Flint corn is widely grown in Argentina.

12.5.3 Flour Corn

Kernels of flour corn are composed almost entirely of soft starch. They are shaped like flint kernels because they shrink uniformly as they ripen. The kernels exist in all colors, but white, blue, and variegated are the most common. Flour corn is grown, mostly for food, by some of the Native American tribes and by many growers in Central and South America.

12.5.4 Popcorn

This group is characterized by a very hard corneous endosperm and small kernels. There are two subgroups of popcorn: rice, with pointed kernels, and pearl, with rounded kernels. The grains may differ greatly in color and size. Additional ears may be borne on the main stalk or tillers.

12.5.5 Sweet Corn

The kernels of sweet corn are translucent and more or less wrinkled at maturity. Sweet corn, before it is ripe and dry, has a sweeter taste than other types because the endosperm contains sugar as well as starch. It lacks the ability to produce fully developed starch grains. Sweet corn may also produce two or more ears per plant.[29]

12.5.6 Waxy Corn

The kernels of waxy corn have a uniformly dull, rather soft endosperm, showing neither white-starchy nor corneous-translucent layers. The endosperm breaks with a wax-like fracture. The starch in waxy corn consists of amylopectin, which has a branched-chain molecular structure and a high molecular weight (50,000 to 1,000,000). The starch resembles tapioca starch and is used for making special foods or adhesives. The endosperm starch and pollen of waxy corn are stained red by an iodine-potassium iodide solution. Starch of ordinary corn, which stains blue in the iodine test, consists of about 78 percent amylopectin and 22 percent amylose. The latter substance has a straight-chain molecular structure and a molecular weight of 10,000 to 60,000.[30] The iodine test is used to determine the purity of waxy corn before sale or processing.

12.5.7 Pod Corn

In pod corn, each kernel is enclosed in a pod or husk. The ear formed is also enclosed in husks as in other types. Pod corn may be dent, sweet, waxy, pop, flint, or flour corn in endosperm characteristics. It is merely a curiosity and is not grown commercially.

The typical podded ear as described never breeds true. When planted, half the crop is pod corn, a fourth is normal without pods, and the remaining fourth are long-podded ears that contain no grain, although a few seeds may develop in the tassel. Thus, common pod corn is genetically a heterozygous plant, the progeny of which segregates in a typical 1:2:1 Mendelian ratio with the podded character dominant. True-breeding types of pod corn have been recently developed.

▓ 12.6 OPEN–POLLINATED VARIETIES

It has been estimated that over 1,000 varieties of corn were grown in the United States before the spread of corn hybrids (Figure 4.6). They originated through hybridization, segregation, and selection, either natural or controlled. Few of them are grown at the present time.

Varietal names of corn mean less than those of almost any other crop. All are heterozygous. Selection and local adaptation have produced numerous strains and types within varieties. Frequently there are larger differences among strains within varieties than between the varieties.[6] In general, the prolific varieties and hybrids that produce more than one ear per stalk were grown in the southern states.

Native American tribes maintained their own strains of corn, to a large degree, for many decades. Many of those varieties were a colorful mixture of red, purple, yellow, and white kernels, mostly of the flint and flour types.[31] Such mixing helps to avoid the deleterious effects of inbreeding. Ears with mixed kernel colors are now grown and marketed for decorative purposes.

A characteristic of hybrids grown in the South is a long husk on the ear. Long husks partly protect the ears from the corn earworm, weevils, moths, and birds.

▓ 12.7 CORN HYBRIDS

State agricultural experiment stations, the United States Department of Agriculture, and a considerable number of commercial firms have engaged in the development of new corn hybrids. Several hundred different corn hybrids are currently grown on a commercial scale. These are produced from a large number of inbred lines. However, certain outstanding lines enter directly or indirectly into the pedigrees of a large percentage of the hybrids that have gone into production. The original inbred lines were selected largely from some of the most productive open-pollinated varieties. Newer inbred lines are derived mostly from hybrids or synthetic varieties. Many of the varieties that contributed these lines had been carefully selected for many years.

Publicly supported institutions usually release "open-pedigree" hybrids in which the inbred lines and single crosses are revealed. Commercial hybrid seed-corn producers often designate their hybrids by their own firm numbers without disclosing the pedigrees. These are called "closed-pedigree" hybrids.

Most field corn grown today is single-cross hybrids. Double (4-way) cross, three-way cross, and modified-cross hybrids are also grown.

▨ 12.8 POPCORN AND SWEET CORN

Popcorn and sweet corn grown in the United States have shorter, more slender stalks, smaller ears, and mature earlier than the field corn hybrids typical of the Corn Belt. Because of this smaller size and also because of the premature harvesting of sweet corn, it is usually planted at a thicker rate than is field corn.

The leading popcorn hybrids were derived from varieties with large yellow kernels that have a mushroom shape after popping; others have smaller yellow kernels of high popping expansion or pointed white kernels. Some have short, thick ears with 30 to 40 irregular rows of slender pointed white kernels.[32] The most popular hybrids which are three-way crosses, outyield all open-pollinated varieties.

Most of the sweet corn consists of yellow hybrids that have a phenomenal inbred line from the Golden Bantam variety in their ancestry. This line is resistant to the bacterial wilt (Stewart's disease) of corn, has an extremely high quality, and is much more productive than the open-pollinated variety from which it was derived.

Hybrid sweet corn is especially advantageous to growers of corn used for processing because of its more uniform growing period. Since the ears are usually gathered from the field at a single picking, this uniformity of maturity permits a larger portion of the ears to be in prime condition at harvest time.

▨ 12.9 FERTILIZERS

The highest yields of corn are obtained on heavily fertilized land. In 1970, from 95 to 100 percent of the corn acreage was fertilized in the eastern, central, and southern states. In Texas, only 84 percent and in South Dakota only 61 percent of the corn fields received fertilizers.

Barnyard manure is commonly applied to land to be planted to corn when the manure is available and not needed for more intensive crops such as potato or sugarbeet. Much of the corn land in the northeastern states, Great Lakes states, and western irrigated regions is manured.

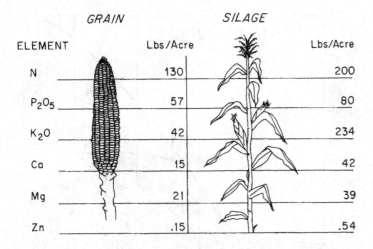

FIGURE 12.9
Corn plant nutrient
removal at yield of
150 bushels per acre.

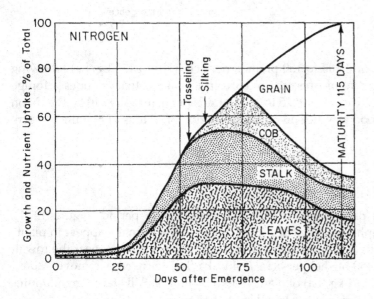

FIGURE 12.10
Nitrogen uptake and
distribution in corn.

Corn producing at the rate of 150 bushels per acre (9,400 kg/ha) will remove from the soil the amounts of fertilizer elements shown in Figure 12.9.

12.9.1 Nitrogen

One bushel of corn contains about 1 pound (18 g/kg) of nitrogen with half that amount present in the associated stover (Figure 12.10). Ample nitrogen should be available throughout the growing season. Nitrogen fertilizer can be applied alone or with other nutrients before planting, during planting, or to the growing crop. The most efficient time to apply nitrogen fertilizer to corn is when the crop is growing, as the roots will quickly absorb it preventing losses from leaching or denitrification. All nitrogen should be applied before tasseling.

High plant populations require heavy fertilization. In Illinois, the most profitable yield responses occur from 160 to 180 pounds per acre (180 to 200 kg/ha) of

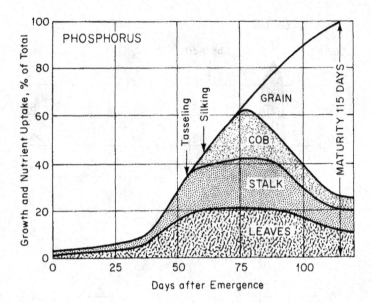

FIGURE 12.11

Phosphorus uptake and distribution in corn.

N for continuous corn; 100 to 150 pounds per acre (110 to 170 kg/ha) following soybean, a small grain, or one year of corn in a rotation that includes a forage legume once in 5 to 6 years; and 75 to 100 pounds per acre (84 to 110 kg/ha) N on corn following a good legume sod or an application of 10 tons of manure per acre (22 MT/ha).

12.9.2 Phosphorus

Corn accumulates phosphorus throughout the growing period (Figure 12.11). Symptoms of phosphorus deficiency (purpling of leaves) usually appear in plants twenty to forty days old. Phosphorus is best applied in a mixture near the row at planting time. A 150-bushel-per-acre (9,400 kg/ha) corn crop removes about 25 pounds per acre (28 kg/ha) of elemental phosphorus. Soils receiving manures or other organic residues require less P in the fertilizer.

Corn seedlings subjected to cool, wet conditions will sometimes exhibit phosphorus deficiency symptoms even though adequate fertilizer was applied. This occurs because the roots cannot absorb adequate phosphate when the soil is cool and wet. Plants with a phosphorus deficiency during these conditions will recover when the soil warms and dries with no loss in yield.

12.9.3 Potassium

Potassium deficiencies (yellow leaf margins) also appear in young plants. The majority of the potassium is taken up by the plant before the tassel stage (Figure 12.12). The total potassium uptake by a corn plant is distributed one third in the grain and two thirds in other plant parts. Therefore, a field of corn harvested for grain will return most of the potassium back to the soil through mineralization of the residue. Corn harvested for silage, however, will remove large amounts of potassium, which will increase the need for fertilizer.

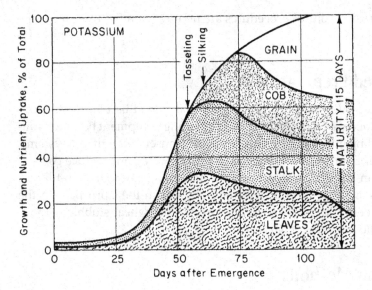

FIGURE 12.12
Potassium uptake and distribution in corn.

12.9.4 Micronutrients

The addition of micronutrients may be necessary on soils that are strongly weathered, coarse textured, alkaline, very acid, or peat and muck. Corn grows best on soils with a pH of 6.0–7.0.

▓ 12.10 ROTATIONS

Rotations in the Corn Belt include a two-year rotation of corn and soybean or corn and small grain; a three-year rotation of corn, small grain, and clover; a four-year rotation of corn, oat, wheat, and clover; and a four-year rotation of corn, corn, small grain, and clover. Sweetclover, alfalfa, or legume-grass mixtures often replace clover in rotations. Soybean often replaces corn or one of the small grains in the rotation. Most rotations in the irrigated regions of the western states include alfalfa and small grain, with corn, sugarbeet, or potato as cultivated crops. Under dryland conditions, alternating corn and wheat, or corn alternating with barley or oat, appears to be most satisfactory. Corn usually shows a relatively small response to the extra moisture provided by fallow.

In much of the South, corn ranks second only to cotton in acreage. Since these two crops do not supplement each other well, it is advisable to grow legumes in a cotton-corn rotation followed by a winter cover crop, such as rye or a legume, to be turned under for green manure as a preparation for corn and cotton.

▓ 12.11 CORN CULTURE

12.11.1 Continuous Corn

Much corn is grown continuously, with success, using heavy fertilization, retaining the crop residues, and controlling insects and weeds with pesticides. The main hazards are a build-up of soil-borne disease or insect organisms and a greater risk from soil erosion. Erosion control procedures are necessary on fields with a slope of 3 to 5 percent or more.

12.11.2 Seedbed Preparation

It used to be an accepted practice to plow fields that would be planted to corn. Plowing has greatly decreased in recent years as planting equipment became available that allows seeding into standing crop residues. Conservation tillage systems, described in Chapter 5, prevent soil erosion, conserve water, and reduce costs of production. Corn can be used with all conservation tillage systems. No-till and ridge plant systems are becoming increasingly popular in the Corn Belt. In the drier western Corn Belt, corn is often seeded directly into wheat stubble without tillage the spring following wheat harvest.[33]

12.11.3 Seeding Methods

Corn is usually planted 2 to 3 inches (5 to 8 cm) deep in rows from 24 to 36 inches (60 to 90 cm) apart. Narrower row spacing allows higher seeding rates and better light interception in the plant canopy. Experiments in six states—Iowa, Kansas, Michigan, Nebraska, Ohio, and Georgia—indicate that yields average about 5 percent more in 30-inch (75-cm) than in 40-inch (102-cm) rows under most conditions. For very early planting in cold soils, 1 to 1½ inches (25 to 38 mm) deep may be advisable, but with a dry surface soil condition, it may be necessary to plant 3 to 4 inches (75 to 100 mm) deep to place the seeds in moist soil. Where soil moisture and fertility are not limited, narrow rows with higher plant populations can lead to yield increases of 10 to 15 percent. Narrow row planting requires special planting, cultivating, and harvesting equipment.

Formerly, most of the corn in the Corn Belt was surface-planted in check-rows planted perpendicular to each other to permit cross cultivation. Now, crop rows run in only one direction and weeds are controlled by herbicides or by more effective tillage. The weeds in the corn row are controlled by preemergent band applications of herbicides.

DATE OF SEEDING Corn planting begins about February 1 in southern Texas and progresses northward through the two eastern thirds of the country at an average rate of 13 miles per day[34] (Figure 12.13). In the Corn Belt, the planting season is at its height about May 15. At a corresponding latitude, planting may be somewhat later in the western third of the country because of cooler weather at the higher elevations. There are two general periods of corn planting in much of the Cotton Belt: one early and one late. The early date corresponds closely to that for planting cotton, while the late period begins in June. In Texas, the highest yields are obtained when corn is planted at the average date of the last spring frost.[35] In western Kansas, planting in June permits the corn to tassel and silk after the extreme midsummer drought and heat are past, and this results in higher yields.

The number of days from planting to maturity decreases as the date of planting is delayed.[36] A variety that required 150 days to mature when planted on April 1 in Tennessee matured in 113 days when planted on June 1.

Corn can be planted earlier on sandy soil than on the heavier clay soils because it will warm up faster. It is futile to plant corn early in a cold, poorly prepared

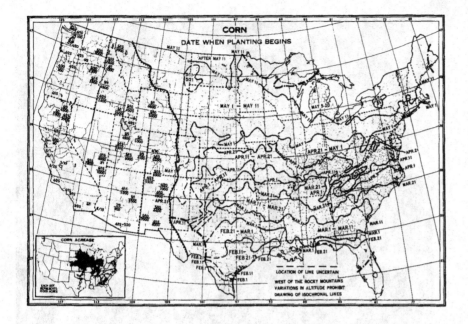

FIGURE 12.13
Dates when corn planting begins.

seedbed. In Colorado corn planted in cold soil on April 20 emerged about the same time as that planted on May 1.[37] The popular idea of Native Americans to delay planting corn "until the oak leaves are as big as squirrels' ears" is a fairly safe rule to follow. However, it is unnecessary to delay planting until all danger of spring frosts is past. Permanent frost damage to corn planted early occurs infrequently because its growing points are still below the soil.

RATE OF SEEDING Optimum spacing of corn ranges from about 3,000 to 4,000 plants per acre (7,000 to 10,000 plants/ha) on poor soils or under extreme semiarid conditions,[10] up to 34,000 per acre (84,000/ha) or more under optimum growing conditions. About 3,500 seeds per acre (8,650 seeds/ha) should be planted for each 20 to 25 bushels (500 to 635 kg) of expected grain yield. About 85 percent of the seeds planted will grow into mature plants.

As corn yields continue to increase, seeding rates continue to increase as well. Modern hybrids are bred with stronger stalks to reduce lodging under higher populations. Irrigated corn (Figure 12.14) will commonly exceed 30,000 plants per acre (74,000 plants/ha). Early hybrids usually have a smaller size and should be planted thicker than late hybrids. Under semiarid conditions, the rate is adjusted to the limited moisture available.

Sweet corn and popcorn are often planted in rows 30 to 40 inches (75 to 100 cm) apart and thick enough in the row that the number of plants per acre ranges from one-fourth more to twice as many as the usual spacing for field corn. Popcorn is usually planted at a rate of 3 to 6 pounds per acre (3.5 to 7 kg/ha) and sweet corn at 8 to 16 pounds per acre (9 to 18 kg/ha).

12.11.4 Weed Control

The principal reason for the cultivation of corn is the control of weeds (Chapter 5), as shown by many comparisons of corn from cultivated plots and corn from plots

FIGURE 12.14
Irrigating corn with a
center pivot irrigation
system. [Courtesy USDA
NRCS]

merely scraped with a hoe to control weeds. Extensive experiments have not shown
any important differences in favor of deep cultivation. Root pruning has frequently
resulted from cultivation deeper than 3 inches (8 cm).[38] Early in the season, weeds
are controlled economically by herbicides or cultivation. Such implements as the
harrow or rotary hoe can be used without damage to the corn plants until they are
approximately 4 inches (10 cm) high. Subsequent cultivation—frequent enough to
control weeds but shallow enough to avoid injury to the corn roots—can be done
satisfactorily with ordinary cultivators. Cultivation of corn in listed furrows saves
subsequent labor by burying the small weeds in the row.

The use of herbicides can eliminate much or possibly all of the need for cultivation of corn. A preemergent application of a suitable herbicide controls most of
the annual broadleaf weeds as well as the germinating seedlings of weedy
grasses. It can be applied in bands with a sprayer immediately behind the planter
to control only the weeds that come up in the row. Late weeds that emerge after
the final cultivation can be suppressed by applications of other herbicides along
the row.

Common preplant and preemergent herbicides used on corn are atrazine, flufenacet, acetochlor, alachlor, and dimethenamid. Common postemergent herbicides
are nicosulfuron, primisulfuron, bromoxynil, mesotrione, and foramsulfuron.

The introduction of glyphosate-resistant or Roundup Ready® (RR) corn hybrids
in 1998 has resulted in a significant shift in weed control. By 2003, 11 percent of all
corn acres in the United States were planted with these genetically engineered hybrids. The simultaneous introduction of other RR crops, especially soybean, has
slowed their use in corn. Many corn producers use soybean in rotation with corn,
and the adoption of RR soybean varieties has increased significantly faster than the
corn hybrids because weed control is easier in corn. Therefore, farmers opt to plant
RR soybean varieties and use conventional herbicides on corn. The use of more

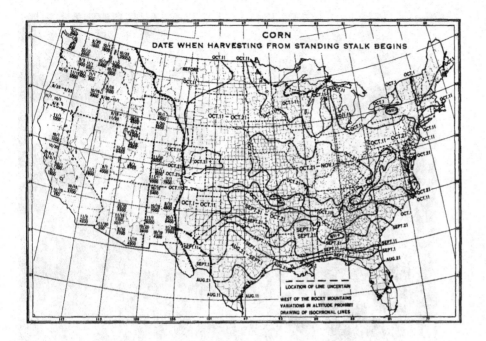

FIGURE 12.15
Dates for harvesting corn for crib storage without artificial drying.

than one RR crop in a rotation is not practical because glyphosate cannot be used to control volunteer plants from a previous crop in a subsequent crop.

12.11.5 Harvesting

Over 90 percent of the corn acreage in the United States is harvested for grain; most of the rest is cut for silage. The harvest of corn for grain begins in late August in the Gulf Coast region and in early October in the northern states (Figure 12.15). A combine is used to harvest nearly all the corn for grain (Figure 12.16). Early harvest followed by artificial drying in a ventilated bin is a popular practice. This method not only saves labor but also avoids greater field losses that result from delayed harvest.

While corn is mature at a moisture content of 30 to 32 percent kernel moisture, harvesting is seldom done at that time. The best time to harvest with minimum yield loss is when the grain moisture content is 26 to 28 percent or less. Early harvesting results in less lodging, less shelling loss in mechanical picking, and there is less delay due to bad weather. A delay from October to December resulted in average losses of 4.6 to 11.8 percent in four Corn Belt states. Most strains of hybrid corn have strong root systems and stand well while awaiting harvest.

Safe long-term storage requires that shelled corn moisture be reduced to at least 13 percent, but it can be kept over winter in cold climates at 15 percent moisture content. Cribbed ear corn on the other hand may be safely stored at 20 percent moisture in the Corn Belt provided the moisture falls to 18.5 percent by late February or early March. Corn keeps ten times as long at 15 percent moisture than at 20 percent moisture, and three times as long at 20 percent than at 25 percent moisture by retarding mold growth. Moldy corn can develop aflatoxin, which is poisonous.[39]

Harvesting corn at a moisture content above that recommended for safe storage requires artificial drying or storage in a high moisture silo. Corn can be dried by

FIGURE 12.16
Harvesting corn with a combine. [Courtesy Case IH]

one of several methods: (1) dry and store in the same structure; (2) dry in one bin as a batch and transfer to another location leaving the final batch in place; (3) portable batch drying, enabling easy movement and transfer of the corn or drier; (4) continuous flow dryer, which dries the corn as a continuous stream of corn moves through the drier. Corn for seed should not be dried at a temperature over 100 to 110°F (38 to 43°C); for processing, this temperature should not exceed 140°F (60°C); while corn for feed can be dried at temperatures of up to 200°F (93°C) with no apparent reduction in feed value.

Corn used for feed that contains 20 to 30 percent moisture can be treated by dry-eration. In this process, the corn is dried down to 16 percent moisture at temperatures of about 200°F (93°C), then held for "tempering" (adjustment of moisture within the kernel). The grain is then allowed to cool, after which it is dried for ten hours down to 14 to 15 percent moisture and again allowed to cool. Grain with 25 percent moisture or less can be dried with a strong blast of unheated air.[40]

Corn can be shelled when it contains 27 percent moisture or less, but the shelled grain cannot be stored safely in bins until it is dried down to less than 13 to 14 percent moisture. Most of the husked ears are stored in cribs until dry enough for safe storage of the shelled grain. Drying is often hastened by ventilating the ears in the crib with forced air, either heated or unheated. Corn cribs have slatted sides and ends to provide free circulation of air. Wet corn that contains more than 30 percent moisture is likely to spoil when ears in cribs or piles are more than 2 feet (60 cm) from the open air. For a crib wider than 4 feet (120 cm), natural drying is aided by inserting an A-shaped or rectangular ventilator in the middle of the crib with the ventilator extended the full length of the crib. In the Great Plains region, much of the husked ear corn is stored in temporary outdoor cribs constructed of woven wire fencing or picket snow fencing.

Damp ear corn, with a grain moisture content exceeding 20 percent or more, should be dried artificially for safekeeping. The screening out of shelled grain and

TABLE 12.1 Weight of Ear Corn Required to Produce 56 Pounds of Shelled Corn at 15 Percent Moisture Content

Moisture in Grain (%)	Ears for 1 Bushel of Shelled Corn at 15 Percent Moisture (%)	Moisture in Grain (%)	Ears for 1 Bushel of Shelled Corn at 15 Percent Moisture (%)
15	70	45	108
20	75	50	120
25	80	55	132
30	85	60	149
35	92	65	170
40	100		

the blowing out of husks and silks improve the circulation of air through the crib and hasten drying. Cleaning, combined with forced-air ventilation, usually takes care of corn that contains 25 to 30 percent moisture.

When corn is killed by freezing before the grain is mature, and high in moisture, the damaged damp grain is referred to as *soft* corn. Even after soft corn is dried, so that it will keep in storage, it is low in test weight per bushel. It is also shrunken or chaffy. However, such corn is nearly equal in feeding value (pound for pound) to plump, sound corn. The lower starch content of soft corn makes it undesirable for processing. The best utilization of soft corn is to feed it promptly to avoid spoilage or the expense and labor of drying it.

The relation between the moisture content of the grain and the number of pounds of ear corn required to produce 56 pounds of shelled corn of 15 percent moisture content, as reported by the Missouri Agricultural Extension Service, is shown in Table 12.1.

Sweet corn is ready to harvest for the market or for canning when the grain is in the late milk stage. It then contains about 70 percent moisture,[41] 5 to 6 percent sugar, and 10 to 11 percent starch. That stage is reached about the time the silks become brown and dry, which is usually about 21 days after silking. The corn remains in prime condition for about two days in hot summer weather and five days in cooler autumn weather. After that, the sugar content decreases, the starch content increases, the appetizing flavor is lost, and the pericarp becomes tough.[42]

Popcorn harvested in the ear should be 18 to 20 percent moisture; for shelled popcorn, 16 to 18 percent. Harvesting equipment should be adjusted accurately and carefully so kernel damage is prevented. Damaged kernels do not pop properly and will cause a discount on the grain market. Use of rotary combines can help prevent kernel damage.[43]

▮ 12.12 CORN STOVER

The leaves and stalks contain about 30 percent of the total nutrients in the corn plant.[44] Stover from corn husked or snapped from the standing stalk is usually utilized only for pasture. Harvested stover is used for feed or bedding. The best utilization of corn stover for bedding is obtained by putting the stalks through a husker-shredder. The coarse parts of the unshredded stalks are not consumed, but they furnish bedding and can be returned to the soil mixed with manure. Even when all the corn stalks are returned to the land, they do not maintain the soil organic matter, except on heavily fertilized fields with high plant populations.[45]

■ 12.13 CORN FODDER

Less than 1 percent of the domestic corn acreage is cut for fodder, grazed, or hogged off. Some corn is cut and fed green. The feeding of corn fodder is most likely in the semiarid as well as in the northern border regions where corn often fails before maturity. Such stalks are more palatable and higher in protein than when they have produced mature ears.[17]

■ 12.14 CORN SILAGE

The most complete utilization of the corn plant is for silage. It is cut when the kernels are in the glazed stage and dented, which is before a serious loss of leaves occurs. The yield may be 20 percent less if corn is ensiled when only half of the kernels are dented. Up to two-thirds of the digestible nutrients of corn silage is in the grain. Corn that fails to produce grain because of drought should be allowed to mature in the field as completely as possible without undue loss of leaves. Nearly all of the corn is harvested for silage with a field silage cutter or a forage harvester (Figure 12.17).

Corn harvested early for silage, because of drought stress, from soils high in nitrates and low in magnesium, may subject cattle to nitrate poisoning or grass tetany, which is sometimes fatal to livestock.[46]

FIGURE 12.17
Chopping corn silage with a self-propelled silage chopper. [Courtesy John Deere & Co.]

▦ 12.15 HARVEST BY LIVESTOCK

A small acreage of corn is hogged or grazed, mostly in the northern Great Plains. Corn may be pastured by cattle, sheep, or hogs, but cattle usually waste a large amount of the crop unless they are followed by hogs. A poor crop or a shortage of labor may make it advantageous to pasture the crop. Cornfields are often pastured after harvest.

▦ 12.16 SEED CORN

Most seed corn is harvested on the ear. Leaving the seeds on the cob during drying increases seed viability because the seeds dry slower due to the high moisture content of the cob.

Harvested seed corn should be placed where it will start to dry the day it is gathered from the field. A free circulation of air is essential to avoid danger of molds and to prevent heating when the moisture content is high. The crop is dried artificially by a forced draft of heated air by most growers of hybrid seed corn. A reduction in the moisture content of the seed to 12 or 13 percent at a temperature range of 105 to 110°F (41 to 43°C) does not reduce the viability.[47]

High-moisture seed corn, exposed to freezing during storage, is likely to be reduced in viability. Death from freezing is directly related to the moisture content of the corn kernel and the duration of the exposure to cold.[48] Corn seed that matures in a natural way becomes progressively more tolerant to freezing as the moisture content diminishes. Corn seed with 10 to 14 percent moisture will stand the most severe winter temperatures without injury to germination.

Well-matured seed corn that was kept dry satisfactorily retained its viability up to four years under Nebraska conditions.[7] Old seed should be tested for germination before it is planted. In Illinois, second-year seed was equal to the first, but the yield from three-year-old seed dropped 7.8 percent, while the decline in yield from seed more than six years old was extremely sharp. The decreased yields from old seed were due to weak plants as well as to low germination.[49]

12.16.1 Seed Selection for Open-Pollinated Varieties

The ideas of the earlier corn growers differed widely as to what constituted the best type of ear to save for seed. This dilemma is eliminated when hybrid seed is used because all ears in a pure hybrid have the same inherent productive capacity if they are sound, viable, and free from disease. When corn shows came into prominence about 1900, judges formulated the corn scorecard as a guide to excellence. The ear was logically regarded as a thing of beauty, but unfortunately, characteristics associated with beauty were erroneously regarded as important from the production standpoint. Physical differences among good seed ears such as number of kernel rows, weight per kernel, ratio of tip to butt circumference, shelling percentage, and filling of tips and butts are valueless as indications of relative productivity.[50, 51] However, yield sometimes is associated with weight of ear and length of ears.[52, 53] In general, ears that are heavy because they are long are more likely to be productive than are heavy ears of a large circumference. Ears with heavier cobs, fewer rows, and heavier kernels with rounded corners and smoother indentation are more productive than the old standard show type.[32] Extremely rough, starchy ears are more susceptible to the rot diseases than the smoother, more flinty types.

Close selection for a particular set of characteristics of a uniform type tends to reduce yields because of a decrease in vigor similar to that obtained by inbreeding.[54]

Field selection of sound seed ears from healthy vigorous plants at about the average date of the first fall frost appears to be the most practical method for yield maintenance or improvement of an open-pollinated variety.

Corn improvement by ear-to-row tests of selections was introduced by C. G. Hopkins at the Illinois Station in 1897 in experiments designed to modify the protein and fat content of the corn kernel. Later, extensive experiments failed to show yield increases by ear-to-ear selection, particularly when practiced continuously.[6]

12.16.2 Hybrid Seed Corn

Inbreeding experiments with corn were started by E. M. East and G. H. Shull before 1905. In 1908, Shull suggested the use of hybrids between inbred lines as a basis for practical corn improvement. The use of double crosses, or crosses between crosses, was advocated by D. F. Jones in 1918. The first commercial production of double cross hybrid corn was in 1921.

About 98 percent of the corn acreage of the United States was planted to hybrid corn by 1965 and almost all of it was in hybrids by 1973. In 1933, only 0.1 percent of the corn acreage was in hybrids. This increased to 14.9 percent by 1938, and the expansion thereafter was rapid. Since 1950, hybrid seed corn has been produced in the Soviet Union, Yugoslavia, Italy, France, Egypt, India, South Africa, Mexico, Argentina, Chile, and many other countries.

First-generation hybrid seed for growing the hybrid crop must be obtained each year. Yields are reduced when seed is saved from the hybrid crop. The yield of the second generation of double crosses is about 84 percent of that of the first generation, or about the same as open-pollinated corn.[55, 56]

The main concern of hybrid corn growers is to obtain seed of the hybrid best suited to conditions in their fields. The published results of state yield tests are a valuable guide in making the proper selection.

The first four hybrids released by the Iowa Agricultural Experiment Station yielded 11 to 26 percent more than the average of open-pollinated varieties in the tests from 1935 to 1940.[57] In Minnesota, the first hybrids yielded 9 to 12 percent more than the check variety.[58]

Corn hybrids were first developed under favorable growing conditions. Hybrids designed especially for adverse conditions later extended the adaptation of hybrid corn. Many superior hybrids are also outstanding for their strength of stalk and lodging resistance.

Production of hybrid seed corn requires the use of more than 200,000 acres (81,000 ha) annually. Seed companies contract with farmers to produce hybrid seed. The company provides the seed and requires very specific planting dates and management practices to maximize seed viability and quality.

Machines are used to detassel the seed or female rows before pollen shedding, but workers (usually teenagers) are still needed to complete the job as not all plants will tassel at the same time. Four to six rows are detasseled. The use of male-sterile lines can eliminate much of the detasseling.

The general methods of developing and producing corn hybrids consist of: (1) development of inbred lines through isolation by self-fertilization and selection of lines that breed more or less true for certain characters, (2) determination of the inbred lines with best combining ability, and (3) utilization of these selfed lines in the production of hybrid seed.

Inbred lines are produced by self-pollination of plants selected from among hybrids or varieties. In self-pollinating (selfing), the pollen from the tassel is placed on the silks of the same plant under controlled conditions. This is done by placing paper bags over the tassels. The ear shoots are covered with small bags before the silks emerge from the husks, and usually the silk is trimmed so that it will grow out as a uniform brush. About one or two days later, the pollen is collected in the tassel bag and then poured on the silks, and the ear shoot is then covered with a large tassel bag. The seed from the better selfed ears is planted in progeny rows. Self-pollinations are then made on several plants in each row, and selection is continued. These selfed lines have less vigor than the parent field variety, but the inbreeding process reveals many undesired abnormalities that can be eliminated. Such abnormalities include sterility, excessive suckers, tassel ears, silkless ears, susceptibility to diseases, and lethal seedlings. After five to seven years of repeated self-pollination and selection, the inbred lines breed relatively true for most genetic traits.

Frequently, after one to three years of inbreeding, some of the seed of each line is planted in a special crossing block with every fourth, third, or alternate row planted to an adapted corn variety or hybrid to be used as a pollen parent. The tassels are removed from all the inbred rows before they shed pollen so that all plants are pollinated by the one-parent variety. The top-crossed seed produced on each row is planted separately to determine which inbred lines have the best general combining ability. In other words, which are likely to produce the highest yields in subsequent crosses. The better lines as shown by the topcross test are retained and inbred until uniform. Meanwhile, most of the inferior inbred lines have been discarded.

Small groups of ten to fifteen high-combining lines are then intercrossed in all possible single-cross combinations. A total of forty-five single crosses [n(n-1)/2] can be obtained with ten lines, and 105 with fifteen lines. These single crosses are tested thoroughly for yield, grain quality, and other characteristics. The performance of all possible double crosses may be predicted from the performance of the single crosses. Six different single crosses are obtainable with any four inbred lines (e.g., A, B, C, and D). These are $(A \times B)$, $(A \times C)$, $(A \times D)$, $(B \times C)$, $(B \times D)$, and $(C \times D)$. The predicted yields of the double cross $(A \times B) \times (C \times D)$ is computed as the average of the yields of the four nonrecurrent single crosses $(A \times C)$, $(A \times D)$, $(B \times C)$, and $(B \times D)$. Seed of only the double crosses predicted to be best is then produced for field testing. The number of possible double crosses is too large to permit field testing of all of them. The number, which is 630 for ten inbred lines and 4,095 for fifteen inbred lines, is computed by the formula [n(n – 1) (n – 2) (n – 3)/8], in which n equals the number of inbred lines. The double crossed hybrids that prove most productive and otherwise desirable are commercially made available for farm use.

A single cross involves a cross of two inbred lines (e.g., $A \times B$). The three-way cross is made from a single cross and an inbred line $(A \times B) \times C$. The double cross is a cross between two single crosses, for example $(A \times B) \times (C \times D)$. There are several types of nonconventional or modified crosses. A common method uses sister lines; inbreds (A_1, A_2) that are closely related. The crossing technique is similar to a three-way cross ($A_1 \times B$) $\times A_2$. Modified crosses are used to enhance a trait from the sister-line parent. The commercial field-corn hybrids in production are double crosses, three-way crosses, modified crosses, or single crosses.

Single-cross hybrids have a high genetic yield potential since the expression of hybrid vigor is greatest in the cross between two inbred lines. Adaptability is lower because there are only traits from two inbred lines in the pedigree. However, uniformity is higher for the same reason. Seed cost is likely to be higher because the seed is produced on inbreds that will have low seed yield. Single-cross hybrids have the potential to perform well in high-yield environments, such as irrigation, and are most popular in developed countries

Double cross hybrids have a lower genetic yield potential since some expression of hybrid vigor is lost in the second cross between two single-cross hybrids. Adaptability will be higher because there are traits from four inbred lines in the pedigree. Uniformity will be lower for the same reason. Seed cost is likely to be less because the seed is produced on single-cross hybrids that will produce more seed. Double cross hybrids have the potential to perform well under less than ideal environments and stress is expected such as dryland production in the western Corn Belt.

Three-way and modified crosses are intermediate in their genetic yield potential, adaptability, and uniformity when compared to single cross and double cross hybrids. Three-way hybrids perform well under conditions where yield potential is less than ideal and are popular in developing countries. Modified crosses are used to address specific soil and environmental stresses. Seed cost will be comparable to double cross seed since the seed is produced on a single cross plant.

For commercial hybrid-seed production,[59] the single crosses or inbred lines are grown in alternate rows or groups of rows in isolated fields. In these crossing fields, all the plants of the seed (female) parent strain are detasseled before they shed any pollen except when male-sterile lines are used. The male or pollen parent rows are not detasseled, and the ears are not gathered for seed. The ratio of male to female rows depends upon the region and the cross to be made, and particularly on the abundance of pollen produced by the male parent. Inbred lines are usually low pollen producers. When they are used to supply the pollen, one row of the male parent is generally planted to every two to four rows of the female parent. When single-cross plants supply the pollen, either one row of male parent to four or six rows of female parent, or a two- and six-row combination (Figure 12.18) is satisfactory under favorable climatic conditions.

■ 12.17 USES OF CORN

About 45 percent of the corn grain produced in the United States from 1970 to 1972 was kept on the farms on which it was produced for use as feed and food. Much of that sold, which includes by-products, is also used for feed. Over 1.3 billion bushels (33 million MT) of corn are used annually for food, alcohol, and other industrial uses.[60]

Corn is used to feed hogs, cattle, poultry, horses, and sheep (in that order). Waxy corn is superior to common field corn in weight gains of steers and lambs. How-

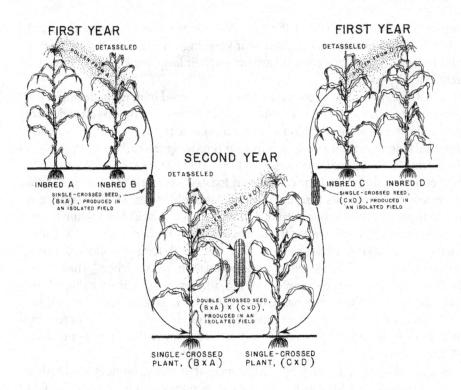

FIGURE 12.18

(Top) Method for producing double cross hybrid seed from four inbred lines of corn. *(Bottom)* Six seed parent (detasseled) rows alternate with pairs of rows of the pollen parent to produce hybrid seed.

ever, reduced yields of waxy corn compared to field corn are not offset by corresponding increases in weight gains.

Exports of corn vary annually depending on market conditions, world production, and existing stocks. In 1998, corn exports were 860 million bushels (21 million MT).

A minor but interesting use of the corn plant is the manufacture of corncob pipes. Much of this industry is localized near Washington, Missouri, where special hybrids of Missouri Cob Pipe corn are grown for their large thick cobs. In this case, the grain is the by-product.

At times, large quantities of corn stalks have been used in manufacturing cellulose products, which include plastic panels. This industry provides a market for the corn stalks but deprives the land of needed organic residues. Pulverized corncobs are used extensively as a mild abrasive for removing the carbon from airplane motors. Corncobs occasionally are used for home fuel or for other purposes.

White corn is usually used for corn-based toasted breakfast foods because the yellow endosperm pigments turn too dark unless slightly heated. Processors pay a substantial premium for white corn of suitable quality. A shift from white to yellow corn occurred about 1920 after the discovery of the higher vitamin-A value of the latter type. A majority of the consumers in the North long ago showed strong preference for corn meal from yellow corn because it "looks richer." This choice was further enhanced by the discovery of the difference in vitamin values. Consumers in the South, on the other hand, use white corn meal almost exclusively because it "looks purer," although they consider yellow corn appropriate for feeding livestock. Per capita consumption of corn flour, meal, and other starchy products in 1997 was 23 pounds (10 kg).

Cornstarch is largely used for food, textile and paper sizing, laundry starch, dextrines, and adhesives. Some cornstarch is used in making cores for foundry molds. Large quantities of starch are hydrolized into glucose (corn sugar), corn syrup, and dextrines by being heated in a dilute solution of hydrochloric acid. The converted solution is then blown into a neutralizing tank that contains sodium carbonate. The glucose is crystallized out of the solution much as cane and beet sugar are crystallized. Using the wet-milling process, a bushel of corn (25 kg) containing 16 percent moisture yields about 1.6 pounds (725 g) of oil and 35 pounds (16 kg) of starch. The starch can be converted into 27.5 pounds (12 kg) of glucose or 40 pounds (18 kg) of syrup. The starch of waxy corn is used for food or for making adhesives such as the gums applied to stamps, envelopes, gummed tape, corrugated paper boxes, and plywood. It can be substituted for tapioca starch for food purposes. The amylose fraction of the starch can be made into films and fibers. Inbred lines that are high in amylose content have been developed.

Corn-steep water has been used in the culture medium for growing *Penicillium notatum*, the organism that produces the antibiotic drug penicillin.

Production of sweetener derived from processed corn is a growing industry. In 1997, per capita consumption of corn-based sweetener was over 86 pounds (39 kg). Over 62 pounds of that was high fructose corn syrup, which is used primarily in soft drinks. From 1972 to 1976, per capita consumption of corn-based sweeteners was only 25 pounds (11 kg), and high fructose corn syrup consumption was less than 4 pounds (2 kg).

Ethanol production is growing rapidly in the United States as a clean-burning, renewable fuel for trucks and automobiles. Most ethanol is mixed with gasoline in a 10 percent ethanol, 90 percent gasoline mixture. Ethanol production consumed 535 million bushels (14 million MT) of corn in 1994, or about 5.3% of the crop. About 667 million bushels (17 million MT) of corn were used for ethanol in 2001. Co-products of ethanol production, such as distillers' grain and corn gluten, are used in food products, animal feed, or exported.

Other industrial uses of corn include biodegradable plastics, starch polymers, and others. The use of these products will likely increase in the future.

Corn meal has never been a generally acceptable food to a majority of northern Europeans even in times of food scarcity.

■ 12.18 MILLING OF CORN

Before milling, corn grain is cleaned by aspiration, sieving, and electromagnetic removal of any foreign metal fragments.

12.18.1 Dry Milling Process

The two types of dry milling are called: new process and old process (or water ground).[61] In milling by the new process, the grain is tempered to a 20 percent moisture content, passed through a degerminator to loosen the germ and hull, and passed through a hominy reel after partial drying. The hull (bran), flour, and germ are removed by sifting and aspirations. The degermed grain can be used for hominy or else ground further into grits or corn meal. Large particles of endosperm are called samp, or pearl or cracked hominy. When coarsely ground, the particles are called hominy grits. The bran is used for feed. The germ is pressed to extract the corn oil and the residue used for feed. Grits are used mostly for food and in brewing. The fine particles of corn flour, which come mostly from the soft part of the endosperm, are used for food or industrial purposes. A bushel (25 kg) of corn yields about 29 pounds (13 kg) of grits and meal, 4 pounds (2 kg) of flour, 1.6 pounds (725 g) of oil, and 21 pounds (10 kg) of feed.

Old-process corn meal is ground between rolls or stone burs usually in small mills, many of which were formerly, or are still, run by water power. The coarser bran particles sometimes are sifted out, but most of the germs remain in the corn meal. Many people relish the flavor of this old-fashioned product, but its keeping quality is impaired by the presence of the oil in the germ, which becomes rancid after prolonged storage. Two bushels (51 kg) of corn yield about 100 pounds (45 kg) of old-process corn meal.

The best corn for dry milling has larger-sized kernels, low kernel size variability, harder kernel texture, and higher protein contents. Harder dent corns or flint corns are best suited for dry milling. Harder corn is desired because it does not break (or crumble) as easily as a soft-textured corn. This results in larger yields of the highest value product: flaking grits.[60]

12.18.2 Alkali Cooking Process

Native Americans invented the alkali cooking process to produce masa flour for making tortillas. They soaked the grain in hot water that contained lye or a bag of wood ashes until the hulls are loosened. The hulled grains were then washed to remove the alkali.[61]

In the alkali cooking process today, the corn is cooked in a boiling lime solution for five to fifty minutes depending on its intended use then steeped for two to twelve hours. The cooked and steeped corn is washed to remove excess alkali and the loose pericarp tissue. The resulting corn product is ground to form masa flour for tortillas and corn-based snack foods.

Snack food and tortilla manufacturers want the highest possible yields of masa. Corn that yields greater amounts of masa has larger-sized uniform kernels that are hard (but not brittle), a lower starch content, and a hull that is difficult to remove. The corn should take up moisture slowly during cooking to reduce overcooking.

Corn used in alkali cooking also must have some additional characteristics. A white cob is preferred. Red cob pieces, which inevitably get into the product, are seen as red specks and considered a defect. Also, corn color is important. Depending upon the geographic region and the particular product being made, white corn or white/yellow corn mixes are used to produce a lighter color product.[60]

12.18.3 Wet Milling Process

In wet milling, the grain is steeped for about two days in warm water that has been treated with sulfur dioxide to prevent fermentation.[40, 45, 62, 63, 64] The softened grain is then passed through degermination mills that crack the kernel and loosen but do not crush the germ. The mass then passes into germ-separation tanks where it is flooded with *starch milk*, and the light, loosened germs float off. The remainder of the kernel is then ground fine and the particles of hull are sifted out. The starch is separated from the protein *(gluten)* by letting the mixture, now suspended in water, flow down long tables. The particles of starch, being heavier, are freed from adhering bits of gluten as they roll over and over, until they finally settle out and come to rest on the bottom of the table. In this process, the gluten passes on over the end. Modern plants separate the starch in a centrifuge. The starch is washed, filtered, and then dried in kilns. The gluten is dried and used mostly for feed. Zein, a purified alcohol-soluble protein of corn, is used in making certain plastics and paints.

Kernel characteristics that help predict wet milling yields are largely unknown. It does appear that larger kernels that are softer or more floury in texture provide higher flour yields. The starch in softer kernels appears to be held less tightly by protein within the kernel, making it easier to separate from the protein and other kernel components. However, kernel size and softness account for only 40 to 50 percent of the observed differences between hybrids when corn is wet milled.[60]

▓ 12.19 MAKING CORN WHISKY

In the manufacture of alcohol or whisky from corn, the grain is cleaned and then either ground or exploded by steam, and the germ is removed. The remaining mass, mixed with water to form a slurry, is heated with steam to gelatinize the starch. The resulting material is cooled to a mashing temperature of about 145°F (63°C) and malt (usually barley malt) is added. The diastase in the malt converts the starch to sugar. The solution (wart) is drawn off, yeast is added, and fermentation is allowed to proceed. The resulting alcohol (whisky) is recovered by successive distillations. Whisky is aged in oaken barrels that are often charred on the inside. During aging, objectionable substances are absorbed by the wood or charcoal, and the whisky absorbs colors and flavors from the wood. Bourbon whisky is made mostly from corn; rye whisky from rye; and Scotch whisky mostly from barley. Other grains or potato can supply part of the starch for any type of whisky. Damp heating grain can be used for making industrial alcohol.

The dried residue, known as distillers' grain, is used for feed. Distillers' slops or distillers' solubles, likewise, may be dried and used for feed.

12.20 POPPING CORN

When popcorn grain that contains 10 to 15 percent moisture is heated sufficiently, this moisture, which is confined in the colloidal protein matrix in which the starch grains are imbedded, is converted into steam. The steam pressure is finally released with explosive force and the grain pops as the starch grains expand. Good popcorn expands 40 volumes or more in popping. Poorer lots may show an expansion as low as 15 volumes.

An important quality measurement for popcorn is the expansion ratio, or the increase in volume that occurs as popcorn pops. It is determined by measuring the volume of popcorn prior to, and after, popping. An expansion ratio of 40 to 44 is usually considered ideal. If the expansion ratio is less than 40, the popped corn can be a little chewy. If the expansion ratio is over 44, the popcorn becomes brittle and tends to break apart into smaller pieces.

Popcorn that contains less than 8 percent or more than 15 percent moisture gives only about half as large a volume on popping as that with 12 to 13 percent moisture.[59] Corneous types of grain sorghum also can be popped. Sweet corn and field corn can be expanded and slightly popped by parching in a kettle with hot edible oils.[13]

12.21 PROCESSING SWEET CORN

About three-fourths of fresh sweet corn is canned and the remainder is frozen. The ears of sweet corn used for processing are snapped closely to leave little or no shank. The ears are usually snapped in the morning and hauled to the cannery as promptly as possible because sweet corn deteriorates rapidly after picking.[65, 66] The kernels can lose half of their sugar as well as much of their flavor within 24 hours. Piles or truckloads of fresh corn will heat unless they are handled soon. At the processing plant, the corn is husked mechanically, sorted, trimmed, and cleaned. The whole kernels may be cut from the cobs by mechanical knives or they may be cut across, the remaining contents scraped out, leaving part of the pericarp on the cob. The latter method produces Maine-style or cream-style canned corn. A ton (900 kg) of ears yields between 600 and 900 pounds (270 and 400 kg) of cut corn. The residues consist of husks, cobs, shanks, and silks, which are usually chopped and made into silage either at the cannery or back on the farm of the grower. Most of the processed crop is grown under contract, with the company supplying the seed and notifying the grower when the corn is ready to be gathered and processed. At the stage for best quality, the dry weight of the kernels is not more than one-half that of mature sweet corn.

Some sweet corn is frozen and marketed as husked ears.

12.22 COMPOSITION OF THE CORN KERNEL

The average composition of a kernel of field corn is as follows:

Substance	Percent
Water	13.5
Protein	10.0
Oil	4.0
Carbohydrates	
Starch	61.0
Sugars	1.4
Pentosans	6.0
Crude fiber	2.3
Ash	1.4
Potassium	0.40
Phosphorus	0.43
Magnesium	0.16
Sulfur	0.14
Other minerals	0.27
Other substances (organic acids, etc.)	0.4
Total	100.0

The germ (embryo) contains about 35 percent oil, 20 percent protein, and 10 percent ash. The comparative vitamin content of yellow corn and wheat is about as follows:

Vitamin	Milligrams per Pound	
	Yellow Corn	Wheat Grain
Vitamin A	1990.00	86.00
Thiamin	2.06	2.25
Riboflavin	0.60	0.51
Niacin	6.40	27.34
Pantothenic acid	3.36	5.83
Vitamin E (as alpha tocopherol)	11.21	16.88

White corn has the same general composition as yellow corn except it is practically devoid of vitamin A. Corn is relatively low in niacin (nicotinic acid), the vitamin that checks pellagra. This disease, which causes dermatitis (inflamed skin), has occurred in the South among people living on diets of large quantities of corn meal and hominy grits. People who consume large quantities of corn may show symptoms of niacin deficiency even with an abundance of supplemental niacin in the diet. This is because corn is lacking the amino acid tryptophan.

Lines of corn that are high in lysine have been developed. Corn is higher in oil and somewhat lower in protein than wheat.

In general, about 450 to 500 pounds (200 to 225 kg) of corn produces 100 pounds (45 kg) of pork when fed to growing pigs. There is a close relationship be-

TABLE 12.2 Nutritive Value of Corn and Other Grains

Test Animal	Percent of Total Digestible Nutrients In								
	Barley	Buck-wheat	Corn	Grain Sorghum	Proso Millet	Oat	Rice (Rough)	Rye	Wheat
Cattle	–	–	78	78	–	72	–	–	–
Sheep	77	62	74	79	69	76	71	79	76
Swine	74	69	80	77	74	74	–	–	73
Poultry	68	65	80	81	72	62	65	60	73

tween the prices of corn and hogs. Under average free-market conditions the price of 100 pounds (45 kg) of hog has been about equal to the cost of 11.4 bushels (289 kg) of corn. When the price of corn rises enough that the hog price per 100 pounds will purchase less than 11 bushels of corn, hog feeding of corn soon decreases.

The relative nutritive value of corn and other grains is indicated in Table 12.2. The true feeding value of corn is not evident from the figures shown, because corn is higher in fat than the other grains. Fat has a caloric value about 2¼ times as high as that of carbohydrates because of a reduced oxygen content. Aside from this difference, the caryopses of other grains are similar to corn in composition and feeding value. Barley, buckwheat, proso millet, oat, and rice grains are enclosed in hulls composed mostly of cellulose, lignin, silica, and other constituents of limited nutritive value. The hull constitutes approximately 13, 22, 22, 28, and 20 percent, respectively, of barley, buckwheat, proso, oat, and rice.

Considerable effort is being expended to increase the protein content of corn, not only for human food but also for animal feed. An international program to develop "quality protein maize" or QPM to improve diets in developing countries has been successful and resulted in the 2000 World Food Prize for two of its leading scientists. In 2003, QPM hybrids were planted on more than 8.75 million acres (3.5 million ha). The protein content of the opaque-2 type corn is no higher than normal corn, but the higher percentage of lysine results in the increased feeding quality.

■ 12.23 DISEASES

The diseases of corn have been described by Dickson,[66] Hoppe,[67] Robert,[68] Ullstrup,[69, 70, 71, 72] and White.[73]

12.23.1 Corn Smut

Corn smut, caused by the fungus *Ustilago maydis,* is the most widely recognized disease of corn. It causes complete barrenness in many plants[74] and reduced grain development in others. The average yield reduction from smut was 30 percent in experiments in Minnesota.[75] Galls on the ear are most destructive while those on

FIGURE 12.19
Galls of common corn smut.

the stalk or leaves above the ear cause more damage than when they attack the lower portion of the plant (Figure 12.19).

The disease appears as galls of various sizes on the aerial parts of the corn plant. The galls are whitish at first but become dark with the development of the black smut spores inside the white membrane. At maturity, the galls break and scatter the powdery spores. Galls may appear at almost any place where there is meristematic tissue, but commonly they are found near the midribs of the leaf or at the nodal buds on the stem and on the ears. Galls are more likely to attack plants that are growing vigorously. Smut is not poisonous to livestock. In fact, young galls cooked and eaten form a palatable substitute for mushrooms, as discovered by Native Americans.[31]

The smut organism lives in the soil. Consequently, seed disinfection is ineffective except for preventing the spread of the disease to new localities. Future infection can be reduced by systematic destruction of smut, by crop rotation, and by refraining from applying infested manure to land that is to be planted to corn. Many hybrids are resistant to the disease.

12.23.2 Head Smut

Head smut is a distinct disease that produces galls on the ear and tassel, entirely destroying these parts. The disease occurs in local areas of the Southwest and Pacific Coast states and rarely elsewhere. The fungus, *Sphacelotheca reiliana*, lives in the soil or on the seed. It enters the young plant and grows into galls in the developing ears or tassels. The galls break and discharge the powdery spores in the wind. The disease can be controlled when infested land is not planted to corn for two years and the seed is treated so that the disease is not carried to new fields. Most field corn hybrids are resistant, but sweet corn hybrids are more susceptible.

12.23.3 Root, Stalk, and Ear Rots

Corn rot can cause reductions in field stand, reductions in vigor of plants that survive, chlorosis, barrenness, general blighting of the plant, and rotting of ears in the fields. Among the organisms that cause seedling blights are the ear-rot fungi *Gibberella zeae*, *Fusarium moniliforme*, and various species of *Aspergillus*, *Penicillium*, and *Pythium*. Root rot is caused by *Pythium arrhenomanes*, *Gibberella zeae*, *G. fujikuroi*, and *Diplodia zeae*. Stalk rot is caused by *Fusarium moniliforme*, *Diplodia zeae*, *Gibberella zeae*, *Macrophomina phaseoli (Sclerotium bataticola)*, and occasionally by *Gibberella fujikuroi*, *Nigrospora sphaerica (Basisporium gallarum)*, and the bacterium *Phytomonas dissolvens*. Ear rots are caused chiefly by *Diplodia zeae*, *D. macrospora*, *Gibberella fujikuroi*, *G. zeae*, *Nigrospora sphaerica*, *Rhizoctonia zeae*, and various species of *Penicillium* and *Aspergillus*.

Although genetic resistance to stalk rot is available, the disease is still a problem under certain conditions. Stalk rot is more prevalent when the plant is stressed. Common stresses include high nitrogen and low potassium fertility, high moisture in mid to late season after a dry early season, moisture stress early in the season and during grain fill, high leaf disease pressure, and insect damage. Selecting hybrids that are tolerant of these stress factors as well as stalk rot tolerance (hybrids with good stand ability) helps reduce the disease. Balanced soil fertility, avoiding high nitrogen and low potassium conditions, is also beneficial.[76]

The ear-rot fungi can continue to grow and damage corn after it is harvested. Mold damage of corn in storage can result from the growth of *Penicillium*, *Aspergillus*, and *Fusarium* species when moisture content of grain exceeds 14 percent. A rot and darkening of the embryo caused by species of *Penicillium* causes the condition called blue eye. Nearly all these organisms are carried in the soil or on decaying corn residues. Most of them are also carried on the seed. The molds are evident on the ears. The dark fruiting bodies of certain stalk and root rots are often seen on either the outside or inside of old corn stalks.

Diplodia develops abundantly on infested kernels; weak, infected plants die following the rotting of roots at the crown. The subcrown internodes of plants grown from *Diplodia* infected seed usually appear dry and brown in contrast to the white tissues in the normal plant. There is little evidence that this fungus advances up the stalk from the rotted roots and rotted crown. Ears may be infected through the shank and may be reduced to a char-like mass. Ears conspicuously rotted with *Fusarium* can be recognized by the characteristic pinkish color of the kernels. *Gibberella* causes both root rot and seedling blight. Charcoal rot (caused by *Macrophomina phaseoli*) destroys the pith in the roots and the lower part of the stalks, leaving the fibrous strands carrying black spores. This disease, which kills the stalks prematurely and causes them to fall down, is most serious under dryland conditions.

The diseases of seedlings and some rots are best controlled by the use of sound, undamaged, disease-free seed and by seed treatment. Corn should be planted after the soil is warm. Some of the stalk rots are checked when the soil nutrient elements are properly balanced by suitable fertilization. The planting of resistant hybrids reduces damage from these diseases.

Chemical seed disinfectants, such as captan, metalaxyl, or fludioxonil, often increase the yield, stand, and vigor through control of the rot diseases when damaged or infested seed must be planted. Favorable results from seed treatment have been obtained on corn planted early, which often rots in cool, waterlogged soil before it can germinate. West of the Missouri River where conditions are drier and rotting organisms less prevalent, seed treatments are usually not as likely to be used.[77, 78, 79]

12.23.4 Leaf Diseases

Northern corn leaf blight *(Exserohilum turcicum)* is characterized by infection of the lower leaves and development of boat-shaped grayish lesions that spread to other leaves in cool, damp weather. Leaf death can occur. Control of the disease is usually economical only in breeding nurseries where the value of the crop justifies treatment with an acceptable fungicide. While this disease is most severe in the East and South, it has caused serious damage as far West as Nebraska. Resistant hybrids help to control the disease.

Southern corn leaf blight *(Bipolaris maydis)* was once considered less of a problem than northern leaf blight. However, in 1970, a severe infestation of this disease occurred throughout the major corn-producing areas of the United States. Especially susceptible to the disease were hybrids in which the female (seed) parent carried the Texas cytoplasmic male sterility gene. Southern corn leaf blight is now established in the Corn Belt. The use of resistant hybrids is recommended. This disease organism needs warm, moist weather to flourish, which is why it is more frequently a problem in the South. Lesions form on the leaf as well as the ear shanks and husks, gradually infecting the ear in severe cases.

Gray leaf spot is caused by the fungus *Cercospora zeae-maydis* and is a significant disease worldwide. It has been present in the United States since 1925 and has been considered a problem in the mid-Atlantic states and the eastern Midwest region of the Corn Belt for decades. The disease prefers cooler temperatures and prolonged periods of overcast days. Increased acreage of reduced tillage and the cultivation of new, high-yielding hybrids have contributed to the increased prevalence and severity of gray leaf spot.[37]

Early lesions from gray leaf spot can be easily confused with lesions caused by other leaf spot and blight pathogens such as eyespot *(Kabatiella)* or anthracnose *(Colletotrichum)*. Mature lesions on leaves are rectangular in shape and restricted by leaf veins. Reverse lighting reveals a yellow halo on most hybrids. Mature lesions can be diagnosed and are easily distinguishable from other diseases. As lesions expand, they coalesce, resulting in a blighting of large portions of the leaf.

Stewart's Wilt of corn is caused by the gram negative bacterium *Pantoea stewartii (Erwinia stewartii)*. It is found throughout the Corn Belt in the United States as well as other countries in South and Central America, Europe, and Asia. Corn flea bee-tles *(Chaetocnema pulicaria)* are the primary vector of the bacterium that causes Stew-art's Wilt. The most common symptoms are long, water-soaked lesions that can extend the length of the leaf of susceptible plants and turn necrotic. Stewart's Wilt lesions on plants after tasseling can be confused with lesions caused by Goss's Wilt.[80]

Goss's Wilt, *Corynebacterium nebraskensis*, causes water-soaked streaks parallel to the leaf veins. Dark, angular, water-soaked spots form next to the leaf veins. The fibrovascular bundles in systematically infected stalks are discolored. Affected plants can be stunted. Plants can be infected, wilt, and die at any stage. The bac-terium overwinters in corn debris near the soil surface and in seed.

Corn rust caused by *Puccinia sorghi* and southern rust caused by *P. polysora* are characterized by small reddish spots on the leaf and aided by humid weather. Re-sistant hybrids are available.

The symptoms of maize dwarf mosaic virus (MDMV) and corn stunt virus (CSV) are the appearance of faint yellowish stripes on plants six to seven weeks old and shortened internodes.[57] Yellowing leaves can eventually turn reddish purple. Excessive tillering with many barren ears may also be observed along with extra long ear shanks.

Maize dwarf mosaic was first identified in the Corn Belt in 1963. It occurs from Pennsylvania southwest to Texas. It also attacks sorghum, johnsongrass, and other grasses. Aphids, particularly the corn leaf aphid *(Rhopalosiphum maydis)*, serve as vectors in transmitting the disease. Resistant lines are available.

The corn stunt virus disease occurs in the southern states from the southeast to California. At least three species of leafhoppers serve as vectors. Wheat streak mo-saic virus (WSMV) also attacks corn in the Pacific Northwest and the Central States and in Ontario, Canada. The vector is the wheat curl mite *(Aceria tulipae)*. Sugar-cane mosaic attacks corn from Louisiana to California. It is transmitted by several aphid vectors. Brown spot, caused by the fungus *Physoderma zeamaydis*, attacks corn in the southern states.

▦ 12.24 INSECT PESTS

The insects that attack corn are reviewed by Dicke.[16]

12.24.1 European Corn Borer

The European corn borer *(Ostrinia nubilalis)* is one of the most serious pests of corn in the United States. The yields of hybrid corn are reduced about 3 percent for each borer per plant.[81] Loss of market sweet corn is about 8 percent per borer per stalk. From 1917, when it was first observed in Massachusetts and New York, to 1946, the European corn borer had migrated as far as Kansas, Nebraska, and the Dakotas. It reached the Rocky Mountains by 1950. The larvae bore within the stalks, tassels, cobs, and ears of corn, eating as they go. The mature larva is about 1 inch (25 mm) long, with a brown head, a grayish to pinkish body with two dark-brown spots on the back of each body segment (Figure 12.20). The insect spends the winter as a full-grown larva in the stalks, stubble, and cobs of corn or in the stalks of other coarse-stemmed crop or weed plants. The larvae begin to pupate from April to July

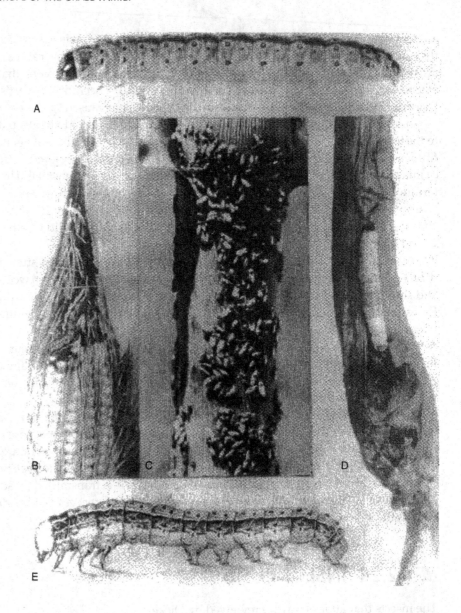

FIGURE 12.20
Insects that attack corn:
(A) European corn borer,
(B) corn earworm,
(C) chinch bugs, *(D)* corn
rootworm, and *(E)* fall
armyworm.

and emerge as moths two or three weeks later. The period of pupation is early in central latitudes and progressively later to the north. However, the multiple-generation strain of the borer pupates earlier than does the single-generation strain. The moths choose the larger plants upon which to deposit the eggs.

Control measures include complete destruction of all waste plant materials around the cornfields and plowing in order to cover all plant residues in the fields. Planting as soon as the soil is warm enough to permit rapid growth improves the vigor of the plants and helps them to survive corn-borer attack. Extremely early or late planting should be avoided. Locally adapted stiff-stalked, nonlodging, disease-resistant hybrids should be planted.

Sweet corn or hybrid seed corn can be protected by insecticides with repeated applications at five-day intervals. Treatment should begin when the first eggs start to hatch. Control with insecticides in corn for grain is advisable when three-fourths of the plants show first-generation larvae feeding in the whorl (Figure 12.21).

FIGURE 12.21
Field sprayer for use in controlling the European corn borer. [Courtesy John Deere & Co.]

12.24.2 Southwestern Corn Borer

The southwestern corn borer *(Diatraea grandiosella)* is a serious pest of corn. The damage it does is similar to that produced by the European born borer. Stalks containing several borers often break over and may fail to produce ears. A native of Mexico, the present range of this insect extends from Arizona to the south central and southeastern states.

The larva of the southwestern corn borer is about $^4\!/_{10}$ inch (1 cm) long and is nearly white, with brown spots scattered over the back. When it occurs on corn roots in the winter, it is creamy without the brown spots.

To avoid loss from this insect, corn can be replaced with grain sorghum, which suffers much less injury. Mid-early planting of corn avoids some damage. The first and second generations of the borer can be controlled by two applications of insecticides ten days apart.[82]

12.24.3 Corn Rootworm

Three species of corn rootworm (Figure 12.20) attack corn in the United States. These are: (1) the southern corn rootworm or 12-spotted cucumber beetle, *Diabrotica undecimpunctata howardi,* which occurs in the South, the East, and the Central States; (2) corn rootworm, *D. longicornis,* of the Great Plains; and (3) the western corn rootworm, *D. virgifera,* of the Mountain States. The larvae bore into bases of seedling plants and ruin the bud or growing point. They also feed on the roots of older corn, causing the plants to fall over. The two last species can be controlled by crop rotation with at least two years intervening between crops of corn on a particular field. Rootworms are controlled by applications of insecticides to the soil before or during planting. Hybrids with vigorous root formation are less likely to lodge when attacked by rootworm.

The development of Bt hybrids has added a new weapon in the arsenal against corn borers and rootworms. Corn hybrids that are genetically engineered with the Bt *(Bacillus thuringiensis)* gene have a natural resistance to corn borers and rootworms. Bt hybrids are engineered to be resistant to corn borers, rootworms, or both. Introduced in 1996, Bt corn hybrids were planted on 25 percent of all corn acres in the United States in 2003.

To retard development of resistance of these pests to the Bt gene, growers using these hybrids are required to use insect resistance management (IRM) plans. Growers may not plant more than 80 percent of their corn acres with Bt hybrids. At least 20 percent of their corn acres must be planted with non-Bt hybrids and treated only as needed with insecticides. Decisions to treat the refuge must be based on IPM economic thresholds. Conventional Bt insecticides must not be used on the non-Bt refuge.

Refuges must be within, adjacent to, or near the Bt cornfields. The refuge must be placed within ½ mile (800 m) of the Bt field, preferably within ¼ mile (400 m), for borer control and adjacent for rootworm control. If a refuge is established as strips within a field, the strips should not be narrower than four rows.[83, 84]

12.24.4 Corn Earworm

The corn earworm *(Heliothis zea)* causes more direct damage to the ear of corn than does any other insect. About 2 percent of the corn crop is destroyed annually by this pest. It occurs throughout the United States wherever corn is grown but is most destructive in the southeastern states. This insect also causes considerable damage to compact-headed grain sorghums. Under the name cotton bollworm, it is a serious enemy of cotton. As the tomato fruitworm, it eats into the nearly ripe fruits of the tomato plant. It also feeds on many other crops.

The number of generations of the corn earworm per year ranges from one in the extreme North and two to four in the Central States to as many as seven in the extreme South.

The corn earworm lives over winter in the pupae stage in burrows 1 to 9 inches deep in the soil. The moth (adult) emerges from these burrows in the spring and the female deposits about a thousand eggs on the leaves of corn or other plants. The larvae hatch in two to eight days and begin feeding. The larvae attains full size in thirteen to twenty-eight days during which time they molt (shed their skins) about five times as they enlarge. The full-grown larva is about ½ inch (38 mm) long, of various colors, and often has conspicuous stripes (Figure 12.20). The eggs are usually deposited on the silks of corn. The young larvae eat their own way down to the ear.

The corn earworm devours developing kernels and fouls the ear so that molds develop. Its tunnels permit weevils to gain access to the ear within the husk. The larvae also feed on leaves, tassels, and silks of corn and often destroy the young growing point or bud of the corn stalk.

The best prevention of corn earworm damage consists of growing hybrids with long, heavy husks. Good husk covering retards or prevents the young larvae from reaching the ear. The best protection is given by husks that extend 4 to 6 inches (10 to 15 cm) beyond the tip of the ear. Early planting partly permits corn to escape damage before the corn earworms become numerous. Early plowing destroys the burrows in which the pupae are hibernating and prevents emergence of many moths. Market sweet corn and hybrid-seed fields, because of their high value, can feasibly be protected by the spraying of insecticides into the leaf whorls and later on the silks.

12.24.5 Chinch Bug

The chinch bug (*Blissus* spp.) (Figure 12.20) is so destructive that in some cases it destroys entire fields. The greatest injury occurs when the corn is planted late, adjacent to a field of barley or other small grain. The bugs, over winter in tall grasses, breed in the small grains and migrate into the corn on foot in June or early July when the small grains ripen and lose their succulence. Chinch-bug losses are greatly reduced by spraying border strips 4 rods (20 m) wide around the cornfield and also along any adjacent field of small grain. Barriers are ineffective when the bugs reach maturity in the small grains and fly into the corn or sorghum fields. Then, the entire field can be sprayed with an insecticide.

Resistant hybrids retard damage from chinch bugs. Inbred lines from eastern and central Kansas and Oklahoma, where chinch bugs usually are abundant, tend to be more resistant than those from the eastern Corn Belt.

12.24.6 Grasshoppers

Several species of grasshopper feed on the foliage of the corn plant. When grasshoppers are abundant, they devour large plants, leaving only the bare stalks or, sometimes, only stubs in the field. Grasshoppers can be controlled with insecticides, preferably applied to the hatching areas when the nymphs are young.

12.24.7 Nematodes

There are over ten species of parasitic nematodes (small, soil-inhabiting roundworms) that can cause damage in corn.[84] They can cause root injury, poor plant color, stunted growth, and reduced grain yields in corn. Symptoms caused by these pests are often confused with root rot diseases, nutritional deficiencies, or climatic stresses. Therefore, special laboratory analyses are necessary to determine if nematodes are the primary cause of reduced corn performance. Although nematodes can be controlled with insecticides, the most economical method is crop rotation.

12.24.8 Weevils

Weevils and other pests of stored corn and other grain were discussed in Chapter 9. The rice or black weevil (*Sitophilus oryza*) and the Angoumois grain moth or fly weevil (*Sitotroga cerealella*) attack corn in the field as well as after it is gathered and

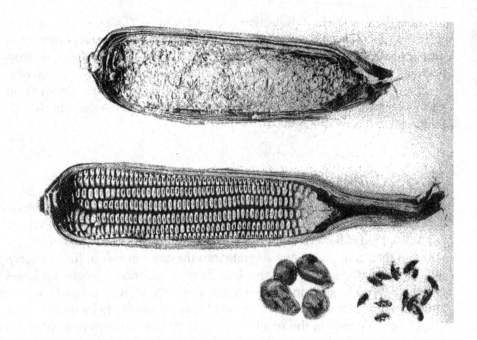

FIGURE 12.22
The grain in the upper ear of corn with the short husks was destroyed by weevils. The long husks on the lower ear protected the grain from the weevils. *(Below left)* Popcorn kernels showing weevil borings. *(Below right)* Adult rice weevils.

stored. These two insects breed in stored grain most of the year, but in the South the adults fly to the field and attack the grain when the corn is in the roasting-ear stage.

The best protection against weevil damage to corn is to grow hybrids with long, heavy husks (Figure 12.22). The Angoumois grain moth cannot penetrate the husk to deposit eggs on the grain. The rice weevil does not usually eat through the husk but attacks ears that have a poor husk covering or those with husks damaged by the corn earworm. In the South, corn is sometimes snapped at harvest and stored in the husk until time to feed the grain. Good undamaged husks offer effective protection against the spread of weevils from infested to sound ears. Repeated fumigation of all grain in storage will keep down the number of insects available to attack corn in the field.

12.24.9 Cutworms

The army cutworm, *Euxoa auxiliaris*, and the pale western cutworm, *Agrotis orthogonia*, are sporadic pests of corn.[85] They damage the plants by cutting off the culm of young plants near the soil surface. The common cutworms are nocturnal feeders. They are controlled by band applications of insecticides applied during planting or foliar applications in growing crops.

12.24.10 Armyworms

The armyworm *(Cirphis unipuncta)* is widely distributed throughout the central latitudes where corn is important. Fall armyworm *(Spodoptera frugiperda)* (Figure 12.20) is found mostly in the South, and the moths frequently migrate far to the North where the larvae damage late-planted corn. Both of these pests devour the leaves of corn; the fall armyworm often attacks the growing point or bud of the plant. They can be controlled with insecticides.

12.24.11 Corn–Root Aphid

The corn-root aphid (*Anuraphis maidiradicis*) is a small, bluish-green insect found on the roots of corn and usually attended by small, brownish ants.[76] The injury is evidenced by weakened plants that turn yellow. The corn-root aphid can be controlled by late fall plowing (which destroys the nests of the attendant ants), by rotation, or by control of weedy hosts in the spring. Where these insects are present, corn should not be grown for more than two years in succession on the same field.

12.24.12 Other Insects

Numerous other insects attacking corn include the corn-leaf aphid (*Rhopalosiphum maidis*), maize billbug (*Sphenophorus maidis*), corn fleabeetle (*Chaetocnema pulicaria*) (which also transmits Stewart's Wilt disease), desert corn fleabeetle (*C. ectypa*), larger corn-stalk borer (*Diatraea zeacolella*), rough-headed corn-stalk beetle (*Euetheola rugiceps*), white grubs (*Phyllophaga* species), corn-root webworm or budworm (*Crambus caliginosellus*), maize billbug (*Calendra maidis*), seed-corn maggot (*Hylemyia platura*), wireworms, webworms, and the Japanese beetle. The Japanese beetle chews off the silks of corn, thus preventing pollination. The corn fleabeetle acts as a carrier (vector) of the bacterial wilt-disease organism. Several corn hybrids are resistant to the cornleaf aphid.[43]

REFERENCES

1. Flannery, K. V. "The origins of agriculture," in *Ann. Rev. of Anthropology*, vol. 2. Palo Alto, CA: Ann. Rev. Inc., 1973, pp. 271–310.

2. Eubanks, M. "The origin of maize: Evidence for Tripsacum ancestry." J. Janick, ed. *Plant Breeding Reviews* 20(2001):15–66.

3. Mangelsdorf, P. C., R. S. MacNeish, and W. C. Galinat. "Domestication of corn," *Science* 143(1964):538–545.

4. Jenkins, M. T. "Influence of climate and weather on the growth of corn," in *Climate and Man*, USDA Yearbook, 1941, pp. 308–320.

5. Kiesselbach, T. A. "Transpiration as a factor in crop production," *NE Agr. Exp. Sta. Res. Bull.* 6, 1916, pp. 1–214.

6. Kiesselbach, T. A. "Corn investigations," *NE Agr. Exp. Sta. Res. Bull.* 20, 1922, pp. 1–151.

7. Kiesselbach, T. A. "Effects of age, size, and source of seed on the corn crop," *NE Agr. Exp. Sta. Bull.* 305, 1937.

8. Kuleshov, N. N. "World's diversity of phenotypes of maize," *J. Am. Soc. Agron.* 25(1933):688–700.

9. Martin, J. H. "Field crops," in *Soil*, USDA Yearbook, 1957, pp. 663–665.

10. Goodding, T. H., and T. A. Kisselbach. "The adaptation of corn to upland and bottomland soils," *J. Am. Soc. Agron.* 23(1931):928–937.

11. Kiesselbach, T. A., and F. D. Keim. "Regional adaptation of corn in Nebraska," *NE Agr. Exp. Sta. Res. Bull.* 19, 1921.

12. Jones, D. F., and E. Huntington. "The adaptation of corn to climate," *J. Am. Soc. Agron.* 27(1935):261–270.

13. Weatherwax, P. *Indian Corn in Old America*. New York: Macmillan, Inc., 1954, pp. 1–253.

14. Dungan, G. H. "Indications that corn tillers nourish the main stalks," *J. Am. Soc. Agron.* 23(1931):662–670.

15. Koehler, B., C. H. Dungan, and J. R. Holbert. "Factors influencing lodging in corn," *IL Agr. Exp. Sta. Bull.* 266, 1925.

16. Hall, D. M. "The relationship between certain morphological characteristics and lodging in corn," *MN Agr. Exp. Sta. Tech. Bull.* 103, 1934.

17. Kiesselbach, T. A. "Progressive development and seasonal variations of the corn crop," *NE Agr. Exp. Sta. Res. Bull.* 166, 1950.

18. Sayre, J. D., V. H. Morris, and F. D. Richey. "The effect of preventing fruiting and of reducing the leaf area on the accumulation of sugars in the corn stem," *J. Am. Soc. Agron.* 23(1931):751–753.

19. Randolph, L. F., E. C. Able, and J. Einset. "Comparison of the shoot apex and leaf development and structure in diploid and tetraploid maize," *J. Agr. Res.* 69, 2(1944):47–76.

20. Dungan, G. H. "Relation of blade injury to yielding ability of corn plants," *J. Am. Soc. Agron.* 22(1930):164–170.

21. Holbert, J. R., and B. Koehler. "Anchorage and extent of corn root systems," *J. Agr. Res.* 27, 2(1924): 71–78.

22. Foth, H. D. "Root and top growth of corn," *Agron. J.* 54(1962):49–52.

23. Weihing, R. M. "The comparative root development of regional types of corn," *J. Am. Soc. Agron.* 27(1935):526–537.

24. Miller, E. C. "Development of the pistillate spikelet and fertilization in *Zea mays*," *J. Agr. Res.* 18(1919):255–265.

25. Randolph, L. F. "Developmental morphology of the caryopsis in maize," *J. Agr. Res.* 53, 12(1936):881–916.

26. Kiesselbach, T. A. "The immediate effect of gametic relationship of the parental types upon the kernel weight of corn," *NE Agr. Exp. Sta. Res. Bull.* 33, 1928, pp. 1–69.

27. Johann, H. "Histology of the caryopsis of yellow dent corn, with reference to resistance and susceptibility to kernel rots," *J. Agr. Res.* 51, 10(1935):855–883.

28. Hopkins, C. G., L. H. Smith, and E. M. East. "The structure of the corn kernel and the composition of the different parts," *IL Agr. Exp. Sta. Bull.* 87, 1903, pp. 79–112.

29. Evans, D. D., and others. "Growth and yield of sweet corn," *OR Agr. Exp. Sta. Tech. Bull.* 53, 1960, pp. 1–36.

30. Hixon, R. M., and G. F. Sprague. "Waxy starch of maize and other cereals," *J. Ind. Eng. Chem.* 34(1942):959–962.

31. Biggar, H. H. "The old and new in corn culture," in *USDA Yearbook*, 1918, pp. 123–136.

32. Brunson, A. M., and J. G. Willier. "Correlations between seed ear and kernel characters and yield of corn," *J. Am. Soc. Agron.* 21(1929):912–922.

33. Wicks, G. A., and N. L. Klocke. "Ecofarming: Spring row crop planting and weed control in winter wheat stubble," *NE Coop. Ext. Serv. NebGuide* G81-551-A, 1997.

34. Neild, R. E. "Effects of weather on corn planting and seedling establishment," *NE Coop. Ext. Serv. NebGuide* G81–552-A, 1981.

35. Mangelsdorf, P. C. "Corn varieties in Texas: Their regional and seasonal adaptation," *TX Agr. Exp. Sta. Bull.* 397, 1929.

36. Mooers, C. A. "Varieties of corn and their adaptability to different soils," *TN Agr. Exp. Sta. Bull.* 126, 1922, pp. 1–39.

37. Robertson, D. W., A. Kezer, and G. W. Deming. "The date to plant corn in Colorado," *CO Agr. Exp. Sta. Bull.* 369, 1930.

38. Nelson, M., and C. K. McClelland. "Cultivation experiments with corn," *AR Agr. Exp. Sta. Bull.* 219, 1927.

39. Anonymous. "Guidelines for mold control in high-moisture corn," *USDA Farmers Bull.* 2238, 1968, pp. 1–16.

40. Anonymous. "Corn in industry," in *Corn Industries Research Foundation*, 5th ed. New York, 1960, pp. 1–63.

41. Doty, D. M., and others. "The effect of storage on the chemical composition of some inbred and hybrid strains of sweet corn," *Purdue U. Agr. Esp. Sta. Bull.* 503, 1945, pp. 1–31.

42. Culpepper, C. W., and C. A. Magoon. "Studies upon the relative merits of sweet corn varieties for canning purposes and the relation of maturity of corn to the quality of the canned product," *J. Agr. Res.* 28, 5(1924):403–443.

43. D'Croz-Mason, N., and R. Waldren. "Popcorn production." *NE Coop. Ext. Serv. NebGuide* G78-426, 1990.

44. Jones, W. J., Jr., H. A. Huston. "Composition of maize at various stages of its growth," *Purdue U. Agr. Exp. Sta. Bull.* 175, 1903, pp. 599–630.

45. Sprague, G. F. "Industrial utilization," in *Corn and Corn Improvement*. New York: Academic Press, 1955, pp. 613–636.

46. Rasby, R., R. Stock, B. Anderson, and N. Schneider. "Nitrates in livestock feeding," *NE Coop. Ext. Serv. NebGuide* G74–170-A, 1988.

47. Kiesselbach, T. A. "Effect of artificial drying upon the germination of seed corn," *J. Am. Soc. Agron.* 31(1939):489–496.

48. Kiesselbach, T. A., and J. A. Ratcliff. "Freezing injury of seed corn," *NE Agr. Exp. Sta. Res. Bull.* 16, 1918.

49. Dungan, G. H., and B. Koehler. "Age of seed corn in relation to seed infection and yielding capacity," *J. Am. Soc. Agron.* 36, 5(1944):436–443.

50. Love, H. H. "Correlations between ear characters and yield in corn," *J. Am. Soc. Agron.* 9(1917):315–322.

51. Olsen, P. J., C. P. Bull, and H. K. Hayes. "Ear type selection and yield in corn," *MN Agr. Exp. Sta. Bull.* 174, 1918.

52. Richey, F. D. "Corn breeding," *USDA Dept. Bull.* 1489, 1927, pp. 1–63.

53. Richey, F. D., and J. G. Willier. "A statistical study of the relation between seed-ear characters and productiveness in corn," *USDA Bull.* 1321, 1925.

54. Garrison, H. S., and F. D. Richey. "Effects of continuous selection for ear type in corn," *USDA Bull.* 1341, 1925.

55. Neal, N. P. "The decrease in yielding capacity in advanced generations of hybrid corn," *J. Am. Soc. Agron.* 27(1935):666–670.

56. Richey, F. D., C. H. Stringfield, and G. F. Sprague. "The loss in yield that may be expected from planting second generation double crossed seed corn," *J. Am. Soc. Agron.* 26(1934):196–199.

57. Sprague, G. F. "Production of hybrid corn," *IA Agr. Exp. Sta. and Ext. Serv. Bull.* 48, 1942, pp. 556–582.

58. Hayes, H. K. *A Professor's Story of Hybrid Corn*. Minneapolis: Burgess, 1963, pp. 1–237.

59. Willier, J. G., and A. M. Brunson. "Factors affecting the popping quality of popcorn," *J. Agr. Res.* 35, 7(1927):615–624.

60. Jackson, D. S. "Corn quality for industrial uses," *NE Coop. Ext. Serv. NebGuide* G92-1115-A, 1992.

61. Anonymous. "Corn and its uses as food," *USDA Farmers Bull.* 1236, 1924, pp. 1–24.

62. Anonymous. "Corn facts and figures," in *Corn Industries Research Foundation*, 5th ed. New York, 1949, pp. 1–48.

63. Majors, K. R. "Cereal grains as food and feed," in *Crops in Peace and War*, USDA Yearbook, 1950–51, pp. 331–340.

64. Zipf, R. L. "Wet milling of cereal grains," in *Crops in Peace and War*, USDA Yearbook, 1950–51, pp. 142–147.

65. Beattie, J. H. "Growing sweet corn for the cannery," *USDA Farmers Bull.* 1634 (rev.), 1945, pp. 1–18.

66. Dickson, J. G. *Diseases of Field Crops*, 2nd ed. New York: McGraw-Hill, 1956, pp. 74–114.

67. Hoppe, P. E. "Infections of corn seedlings," in *Plant Diseases*, USDA Yearbook, 1953, pp. 377–380.

68. Robert, A. L. "Some of the leaf blights of corn," in *Plant Diseases*, USDA Yearbook, 1953, pp. 380–385.

69. Ullstrup, A. J. "Some smuts and musts of corn," in *Plant Diseases*, USDA Yearbook, 1953, pp. 386–389.

70. Ullstrup, A. J. "Several ear rots of corn," in *Plant Diseases*, USDA Yearbook, 1953, pp. 390–392.

71. Ullstrup, A. J. "Diseases of corn," in *Corn and Corn Improvement*. New York: Academic Press, 1955, pp. 465–536.

72. Ullstrup, A. J. "Corn diseases in the United States and their control," *USDA Handbk.* 199 (rev.), 1966, 1–44.

73. White, D. G., ed. *Compendium of Corn Diseases*, 3rd ed. St. Paul, MN: APS Press, 1999.

74. Garber, R. J., M. M. Hoover. "Influence of corn smut and hail damage on yield of certain first generation hybrids between synthetic varieties," *J. Am. Soc. Agron.* 27(1935):38–45.

75. Johnson, L. J., and J. J. Christensen. "Relation between number, size, and location of smut infections to reduction in yield of corn," *Phytopath.* 25(1935):223–233.

76. Salmon, S. C. "Corn production in Kansas," *KS Agr. Exp. Sta. Bull.* 238, 1926.

77. Kiesselbach, T. A. "Field tests with treated seed corn," *J. Agr. Res.* 40(1930):169–189.

78. McClelland, C. K., and V. H. Young. "Seed corn treatments in Arkansas," *J. Am. Soc. Agron.* 26(1934):189–195.

79. Melchers, L. E., and A. M. Brunson. "Effect of chemical treatments of seed corn on stand and yield in Kansas," *J. Am. Soc. Agron.* 26(1934):909–917.

80. Stack, J., J. Chaky, L. Giesler, and R. Wright. "Stewart's wilt of corn," *NE Coop. Ext. Serv. NebFact* NF01-473, 2001.

81. Parker, J. R., and others. "Comparative injury by the European corn borer to open-pollinated and hybrid field corn," *J. Agr. Res.* 63, 6(1941):355–368.

82. Henderson, C. A., and F. M. Davis. "The southwestern corn borer and its control," *MS Agr. Exp. Sta. Bull.* 173, 1969, pp. 1–16.

83. Hunt, T. E., and G. W. Echtenkamp. "Resistance management for European corn borer and Bt transgenic corn: Refuge design and placement," *NE Coop. Ext. Serv. NebFact* NF00-425, 2002.

84. Wysong, D. S., and E. D. Kerr. " Root and soil analyses for nematodes in corn," *NE Coop. Ext. Serv. NebGuide* G84-702-A, 1984.

85. Hein, G. L., J. B. Campbell, S. D. Danielson, and J. A. Kalisch. "Management of the army cutworm and the pale western cutworm." *NE Coop. Ext. Serv. NebGuide* G93-1145-A, 1993.

Sorghum

■ 13.1 ECONOMIC IMPORTANCE

Sorghum (*Sorghum bicolor*) grain was harvested on about 107 million acres (43 million ha) in 2000–2003, with an average production of 2.3 billion bushels (57 million MT) or 21 bushels per acre (1,300 kg/ha). The leading producing countries are the United States, Nigeria, India, Mexico, and Sudan. It is grown to some extent or occasionally grown in all countries of the world except in the cool northwestern part of Europe. Sorghum ranks sixth in acreage and production among all world crops. In 2000–2003, grain sorghum production in the United States averaged about 439 million bushels (11 million MT). It was grown on about 7.8 million acres (3.2 million ha) with an average yield of 56 bushels per acre (3,500 kg/ha). The leading states in grain sorghum production are Kansas, Texas, Nebraska, Missouri, and Arkansas (Figure 13.1). Sorghum ranks fifth in acreage among the crops grown in the United States, exceeded by corn, wheat, soybean, and alfalfa. Since 1948, the yield of grain sorghum has more than doubled and the world production nearly tripled. Sorghum is also an important forage crop in the United States as well as some other countries. An average of about 345,000 acres (140,000 ha) were harvested in 2000–2003 for silage in the United States resulting in about 3.6 million tons (3.2 million MT).

Only a small acreage is harvested for syrup production and broomcorn production. Sudangrass and sorgrass (sorghum-sudan hybrids) are used as silage, pasture, and forage crops, especially in semiarid regions.

■ 13.2 HISTORY OF SORGHUM CULTURE

Sorghum is a native of Africa in the zone south of the Sahara desert, where several closely related wild species are found, and the cultivated types are very diverse. The cultivated type (*Sorghum bicolor*) may have been selected by 3000 BC or several centuries later.[1] It appears to have been grown in India by the beginning of the Christian Era or earlier. It was grown in Assyria before 700 BC and in southern Europe sometime thereafter.[2] Some sorghum seed was brought to the United States from Africa by imported slaves, but noticeable cultivation did not begin until after the introduction of several varieties in 1853 and 1857.

Broomcorn
Durra
Feterita
Johnsongrass
Kafir
Kaoliang
Milo
Prussic acid
Shallu
Sorgo
Sudangrass

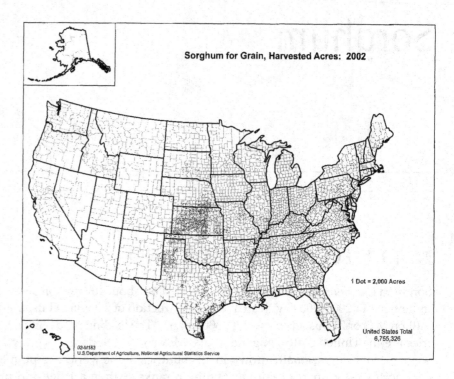

FIGURE 13.1
Sorghum harvested for all purposes in 2002. [Source: 2002 Census of U.S. Department of Agriculture]

Broomcorn was grown in Europe before 1596 but was not cultivated in the United States until late in the eighteenth century after its introduction, presumably by Benjamin Franklin.[3] Sudangrass was introduced from Sudan in 1908.[4]

13.3 ADAPTATION

Sorghum is grown in warm or hot regions that have summer rainfall, as well as in warm, irrigated areas.[5] The most favorable mean temperature for the growth of the plant is about 81°F (27°C). The minimum temperature for growth is 60°F (16°C). Consequently, only part of the frost-free season is available to produce the crop. The sorghum plant seems to withstand extreme heat better than other crops, but extremely high temperatures during the fruiting period reduce the seed yield.[6] Sorghum is a short-day plant.

Sorghum is well adapted to summer rainfall regions where the average annual precipitation is only 17 to 25 inches (430 to 635 mm).[4] The plants remain practically dormant during periods of drought but resume growth as soon as there is sufficient rain to wet the soil. This characteristic accounts in large part for the success of sorghum in a dry season and is the reason why it has been called a *crop camel*. When compared with corn of similar seasonal requirements, sorghum has more secondary roots and smaller leaf area per plant. Sorghum leaves and stalks wilt and dry more slowly than those of corn,[7] enabling sorghum to withstand drought longer. A waxy cuticle apparently retards drying. However, sorghum is also highly productive on irrigated land and in humid areas. About 34 percent of the sorghum acres in the United States were irrigated in 1984.

Sorghum is grown successfully on all types of soil. In moist seasons, the highest yields are obtained on heavy soils. But, in dry seasons, it does best on sandy soil. Sorghum will tolerate considerable soil salinity.

Its resistance to drought and heat as well as to grasshopper, rootworm, and corn borer injury accounts for the growing of sorghum instead of corn in many sections of the United States. The main disadvantages of sorghum compared with corn are: lower yields of succeeding crops, greater uncertainty in getting stands, the necessity of prompt harvesting, greater difficulty in storing the grain, and the greater necessity of grinding the grain before feeding. Although the feeding value of sorghum is comparable to corn, the acreage of sorghum has declined in recent years.

Sorghum acreage in the Great Plains varies annually depending on rainfall. Sorghum acreage increases and corn acreage decreases during drier years. During wetter years, sorghum decreases and corn increases.

■ 13.4 BOTANICAL DESCRIPTION

Sorghum belongs to the family *Poaceae*, tribe *Andropogoneae*. *Sorghum bicolor* includes the annual sorghums with ten pairs of chromosomes; namely, grain sorghum, sorgo, sudangrass, and broomcorn. Sudangrass was once classified as a subspecies *sudanense*, but that designation is no longer valid. Broomcorn was once classified as *S. vulgare*, but now is recognized as *S. bicolor*, var. *technicum*. *S. halepense* (johnsongrass) is perennial in habit, produces rhizomes, and has twenty haploid chromosomes.

Closely related to sudangrass are several species of wild grass sorghums in Africa. One or more of these may be the progenitor of cultivated sorghum. Other related species of sorghum, such as *S. versicolor*, which has only five haploid chromosomes, are found in Africa.[8]

Sorghum is a coarse grass with stalks 2 to 16 feet (0.5 to 5.0 m) in height.[2, 4, 9] The stalks are similar to those of corn, being grooved and nearly oval. The peduncle (top internode) is not grooved. The grooves alternate from one side to the other on each successive internode. Young sorghum plants can be distinguished readily from corn plants because of the sawtooth margins of sorghum leaves. Some varieties have sweet juicy pith in the stalks; others are juicy but not sweet, while still others are deficient in both sweetness and juiciness. A drystalked variety has a white midrib in the leaf. A juicy-stalked variety has a dull or cloudy midrib caused by the presence of juice instead of air spaces in the pithy tissues. A leaf arises at each node, the blades being glabrous with a waxy surface. The surface of the stalks and leaves is glaucous. Buds at the nodes of the stalk can give rise to side (axillary) branches. Crown buds give rise to tillers (Figure 3.4). The total number of leaves on the main stalk, including those formed during the seedling stage, averaged sixteen to twenty-seven per stalk in twenty-one American varieties.[10] The first ten (more or less) of the small leaves arise from the crown nodes underground. Early maturing varieties have few leaves and consequently are limited in plant yield.

The sorghum inflorescence is a loose to dense panicle, having many primary branches borne on a hairy axis. Each branch bears paired ellipsoidal spikelets (Figure 4.5). The sessile spikelet of each pair is perfect and fertile, while the pedicellate spikelet is either sterile or staminate. Two pedicellate spikelets accompany the sessile spikelet at the end of each panicle branch. The two glumes of the fertile spikelet are usually indurate (leathery). There are two florets in the fertile spikelet, the lower sterile and the upper fertile. The lemma and palea are thin and translucent. The lemma may be awned or awnless. Some sorghum has seeds fully covered by the chaff even after being threshed, while others are more than half exposed and thresh completely free from the chaff. The position of the panicle is usually erect but may be recurved. Recurving is the result of heavy, thick panicles being forced out the side of a too-narrow sheath while the peduncle is too flexible to support the panicle in an erect position. Later the peduncle stiffens (becomes lignified) in the recurved condition.[11] Erect varieties have slender panicles during the boot stage. A well-developed panicle may contain as many as 2,000 seeds. Seed weight is generally as follows: grain sorghum, 12,000 to 20,000 per pound (5,500 to 9,100/kg); sorgo, 20,000 to 30,000 per pound (9,100 to 13,600/kg); sudangrass, 50,000 to 60,000 per pound (22,700 to 27,300/kg); and broomcorn, 22,000 to 30,000 per pound (10,000 to 13,600/kg).

Sorghum is about 95 percent self-pollinated in the field,[12] but it will cross naturally with other varieties of sorghum, broomcorn, or sudangrass and occasionally with johnsongrass.

The pigments of colored sorghum seeds are found in the pericarp, the subcoat (testa), or both. When the pericarp only is pigmented, the seeds are yellow or red. When the pericarp is white and a testa is present, the seeds are buff-colored or bluish white in types with a thick starchy mesocarp in the seedcoat. With a colored pericarp and a testa present, the seeds usually are dark brown or reddish brown. Starchy seeds take up water quickly and are subject to invasion by seed-rotting fungi.

Mature seeds of sorghum have a black spot near the base. Immature seeds develop the black spot after drying.

Although considered an annual and usually grown as such, sorghum is a perennial and will continue growing where the temperature is mild and soil moisture is available. It has lived for at least 6 years in fields in southern California and 13 years in a greenhouse. Sorghum planted early in a mild climate may produce a second (ratoon) crop later in the year.[13]

Each new stalk arising from crown buds develops its own root system and a new series of crown buds but remains attached to the old crown.[14] A stalk dies after it has flowered and after all active buds at the stalk and crown nodes also have elongated into stems. Then, its roots die and decay. In the meantime, new stalks have grown alongside the original stalk. This process continues as long as conditions are favorable for vegetative reproduction or until the crowns form too high above the soil surface.

■ 13.5 SORGHUM GROUPS

Grain sorghum originally introduced into the United States could once be classified into rather distinct groups. The sorghum hybrids (Figure 13.2) now grown in the United States and many other countries were derived from crosses between different groups. They represent recombinations of group characters and are a good example of how modern plant breeding can combine the traits from many different plant types into successful, high-yielding hybrids. Although practically all grain and forage sorghum grown in the United States and many other countries are hybrids, the traditional open-pollinated sorghum types are still grown in lesser-developed countries.

Grain sorghum stalks are usually either fairly juicy or comparatively dry at maturity, and the stalk juice is not sweet or is at most only slightly sweet. Grain sorghums, in general, have larger heads and seeds and shorter stalks and produce more seed in proportion to total crop than do the sorgos. The stalks of grain

FIGURE 13.2
A field of hybrid grain sorghum. [Courtesy USDA NRCS]

1   2   3   4

FIGURE 13.3
Panicles of grain sorghum groups: *(1)* shallu, *(2)* milo, *(3)* kafir, and *(4)* feterita. *(Below)* Spikelets *(left)* and kernels *(right)* of kafir.

sorghum hybrids usually range from 30 to 60 inches (75 to 150 cm) in height. The seed threshes free from the glumes.

Among the grain sorghum groups, kafirs have thick and juicy stalks, relatively large, flat, dark-green leaves, and awnless cylindrical heads (Figure 13.3). The

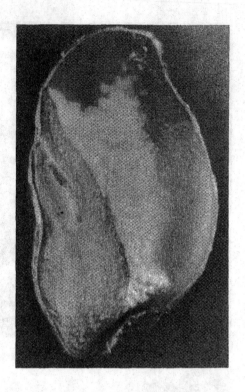

FIGURE 13.4
Longitudinal view of a
sorghum kernel. [Courtesy
Richard Waldren]

seeds are white, pink, or red and of medium size (Figure 13.4). The chaff is either black or straw-colored. Kafirs are grown for both grain and forage. Hegari is similar to kafir in appearance except that the heads are more nearly oval, the seeds are a chalky or starchy white, and the more abundant leaves and sweeter juice make it more prized for forage.

Milo types have somewhat curly light-green leaves and slightly smaller leaves and stalks than kafir and are less juicy. The leaves have a yellow midrib containing carotene. The heads of the true milos are bearded, short, compact, and rather oval in outline, with very dark-brown chaff. The seeds are large and are yellow or white. The plants tiller considerably[15] and in general are earlier and more drought-resistant than those of kafir.

The cytoplasmic male sterility required for hybrid sorghum production was initially obtained from milo but now there are many sources available.

Feterita has few leaves, relatively dry slender stalks, rather oval, compact heads, and very large chalk-white seeds. It matures early.

Durra has dry stalks, flat seeds, very pubescent glumes, and either compact and recurved or loose and erect panicles. Durra is the main type of sorghum in North Africa, the Near East, and India.

Shallu has tall, slender, dry stalks, loose heads, and pearly white seeds. It is relatively late in maturity. The loose heads dry out quickly, do not harbor worms, and make it difficult for birds to roost on the slender branches and eat the grain. Shallu is widely grown in India and also in tropical Africa.

Kaoliang has dry, stiff, slender stalks, open bushy panicles with wiry branches, and rather small brown, or white, seeds. Kaoliang is grown almost exclusively in China, Korea, Japan, and southeastern Siberia.

Sorgo or sweet sorghum (often called *cane*) is characterized by abundant sweet juice in the stalks, which usually range in height from 60 to 120 inches (1.5 to 3 m).

FIGURE 13.5
Panicles of broomcorn.
[Courtesy Colorado Agricultural
Experiment Station]

The seed of some varieties remains enclosed in the glume after threshing. The heads (panicles) may be loose or dense, and the lemmas awned or awnless and black, brown, or red, depending upon the variety. The seeds are small or medium-size and are either white or some shade of brown.

Broomcorn produces heads and fibrous seed branches 12 to 36 inches (30 to 90 cm) long (Figure 13.5) that are used for making brooms and whisk brooms. The stalks range from 3 to 13 feet (1 to 4 m) in height and are dry, not sweet, and of limited value for forage. The lemmas are awned with small brown seeds enclosed in very pubescent tan, red, or dark-brown plumes. The broomcorn industry has been largely replaced by brooms with synthetic fibers.

Sudangrass has slender leaves and stalks and loose heads with small brown seeds (Figure 13.6). Johnsongrass is similar to sudangrass except it is larger, it is a perennial with underground stems, and the seeds are different (Figures 13.6 and 13.7).

▦ 13.6 PRUSSIC ACID POISONING

Young plants, including the roots (Figure 7.3) and especially the leaves of older plants, of sorghum, sudangrass, and johnsongrass contain a glucoside called dhurrin, which upon breaking down releases a poisonous substance known as prussic acid or hydrocyanic acid (HCN). Some losses of cattle, sheep, and goats occur each year from sorghum poisoning when they graze upon the green plants. Silage and well-cured fodder and hay can usually be fed with safety.[16] Silage may contain toxic quantities of prussic acid, but it escapes in gaseous

FIGURE 13.6
Sudangrass panicle.
[Courtesy of Richard Walden]

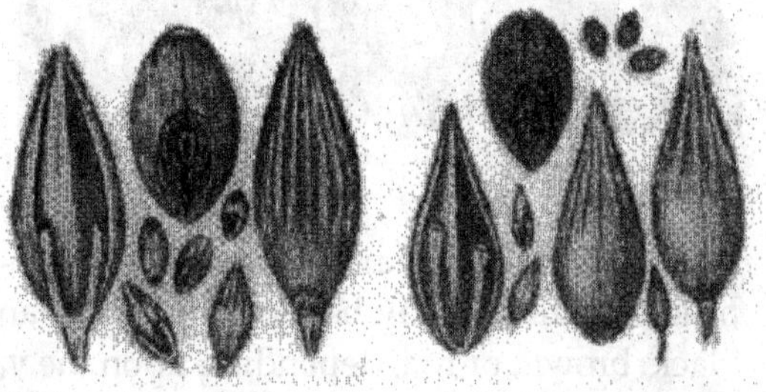

FIGURE 13.7
Seeds, enlarged and natural size of sudangrass *(left half)* and johnsongrass *(right half)*. The swollen tips of the pedicels of the sudangrass seed at extreme left are broken off, but they remain on the corresponding johnsongrass seed.

form while the silage is being moved and fed. The prussic acid content of sorghum hay and fodder decreases during curing so that it is only occasionally dangerous.

The prussic acid content decreases as the plant approaches maturity.[17] Small plants and young branches and tillers are high in prussic acid. The prussic acid content of the leaves is three to twenty-five times greater than that of the corresponding portions of stalks of plants in the boot stage. Heads and sheaths are low in prussic acid. The upper leaves contain more prussic acid than the older lower leaves. The amount of prussic acid varies in different types of sorghum. Sudangrass contains about two-fifths as much prussic acid as many sorghums grown under the same conditions. Sudangrass rarely kills animals unless contaminated with sorghum or sorghum-sudangrass hybrids, except occasionally in the northern states. Even there it is usually safe after the plants are 18 to 24 inches (0.45 to 0.60 m) high.

Freezing does not increase the prussic acid content of sorghum, but it causes the acid to be released quickly from the glucoside form, thus making frosted sorghum very dangerous until it begins to dry out. An abundance of soil nitrates causes sorghum to be high in prussic acid as well as nitrates. Drought-stricken and second-growth plants are dangerous for ruminants because they are small and consist largely of leaves, which are high in prussic acid.

The toxic level of HCN is above 200 parts per million.[18] A mere half-gram of pure prussic acid can kill a cow.

Sorghum is unsafe for pasturing except after the plants are mature and little new growth is present. Individual animals differ in their susceptibility to sorghum poisoning. Sheep seem to be slightly less susceptible to prussic acid than are cattle, while horses and hogs apparently are not injured. However, horses have suffered from a cystitis syndrome when pasturing sudan-sorghum hybrids.[19]

Poisoning is less likely to occur if the animals eat some ground grain before they are turned into the pasture. The growing of varieties of sorgo and sudangrass that are low in dhurrin content reduces losses from poisoning.

The remedy for cyanide poisoning is intravenous injection of a combination of sodium nitrite and sodium thiosulfate. For cattle, 2 to 3 grams of sodium nitrite in water followed by 4 to 6 grams of sodium thiosulfate in water are recommended, and for sheep, up to 1 gram of sodium nitrite and 2 to 3 grams of sodium thiosulfate.

■ 13.7 HYBRIDS

Most forage and grain sorghum produced today in the United States and other developed countries are hybrids. Most grain sorghum hybrids have either a white, floury endosperm with either a red or white pericarp or yellow endosperm with a transparent pericarp. Yellow endosperm sorghum is higher in carotene and xanthophyll, which makes it desirable for poultry feed.

The time of maturity is the most important factor in determining the adaptation of a sorghum hybrid to a particular locality.[20, 21] However, the decision to choose a hybrid for any region is based largely upon its intended use and method of harvesting.[22] Differences in yield, when not too large, are only a secondary consideration provided the variety possesses the other characteristics desired. It is essential that a hybrid reach maturity before it is killed by frost. Consequently, as the average frost-free period becomes shorter, progressively quicker-maturing types are grown.

Sorghums that mature later have large leafy stalks and are suited to regions with short days and a long growing season. They fail to head under the long days in northern latitudes. Quick-maturing hybrids are best adapted to long-day conditions. In the South, they mature very quickly with limited vegetative growth but fail to utilize the full growing season.

Hybrids that involve lines of sudangrass and sorghum-sudangrass hybrids have mostly replaced sudangrass and sorgo varieties. Hybrids have been bred that are low in prussic acid content, resistant to bacterial leaf spots, *Helminthosporium* leaf blight, and other fungus leaf spots as well as those that are juicy and sweet and consequently very palatable pasturage.

The breeding and distribution of new sorghum hybrids have caused marked changes in production since 1928.[8, 23, 24, 25, 26, 27, 28] The development of hybrids or varieties with short, erect stalks have made it possible to harvest the crop successfully with a combine and at the same time eliminate losses from root rot. Thus, grain sorghum production is adapted to large-scale mechanized methods. The development of quick-maturing hybrids extended the grain-sorghum belt into Nebraska, South Dakota, and northeastern Colorado. The breeding of productive hybrids ex-

TABLE 13.1 Approximate Amount of Nutrients in a 100–Bushel Sorghum Crop

	Quantity in Pounds		
Element	Grain	Stover	Total
Nitrogen (N)	79	78	157
Phosphorus (P_2O_5)	44	22	66
Potassium (K_2O)	22	148	170
Sulfur (S)	4	16	20
Magnesium (Mg)	6.7	17	23.7
Calcium (Ca)	3.7	26.7	30.4
Copper (Cu)	0.01	0.03	0.04
Manganese (Mn)	0.04	0.13	0.17
Boron (B)	—	0.04	0.04
Iron (Fe)	—	0.41	0.41
Zinc (Zn)	0.04	0.17	0.21

tended grain-sorghum production eastward into the more favorable subhumid and humid areas. Average yields of grain sorghum in the United States tripled between 1940 and 1970 due to the introduction of hybrids but have only increased about 50 percent since 1970.

13.8 FERTILIZERS

Grain sorghum is an efficient user of soil nutrients because it has an extensive fibrous root system that is about twice the size of corn. A grain sorghum yield of 100 bushels (2,540 kg) will remove about 84 pounds (38 kg) of N, 42 pounds (19 kg) of P_2O_5, and 22 pounds (10 kg) of K_2O.[29] Table 13.1 lists the nutrient content of sorghum grain and stover.

Sorghum responds to applications of barnyard manure of 20 to 40 pounds per acre (22 to 45 kg/ha) of nitrogen in the semiarid Great Plains and 40 to 60 pounds per acre (45 to 65 kg/ha) in the subhumid areas.[30] Little or no benefit from fertilizers is evident in a very dry season. On irrigated land, grain yields usually increase with applications of nitrogen of 90 to 160 pounds per acre (100 to 180 kg/ha). In humid regions, sorghum gives about the same fertilizer response as corn. There, applications of 45 to 135 pounds of nitrogen per acre (40 to 120 kg/ha), plus 35 to 60 pounds per acre (45 to 70 kg/ha) each of P_2O_5 and K_2O are common.[31] Ample nitrogen induces earlier maturity in sorghum unless the application is excessive. Foliar spraying of an iron chelate prevents leaf yellowing in soils of high pH. Zinc chelates may be helpful where topsoil has been heavily graded for irrigation or terracing.

13.9 ROTATIONS

Sorghum follows other crops readily, but caution is recommended in choosing a crop to follow sorghum. Grain sorghum often is grown continuously[32] or is alternated with wheat, sudangrass, soybean, barley, or corn. A three-year rotation of sorghum, fallow, and wheat is popular where fallow is a desirable preparation for winter wheat.[25, 27, 33] In the southern Great Plains, grain sorghum usually produces more after fallow, winter small grains, cowpeas, or cotton than in continuous culture. Sorghum responds well to the additional moisture conserved by fallow. The yields on fallow in western Kansas are up to 75 percent higher than from land continuously in sorghum.[34] Studies in Nebraska showed that sorghum responds well to nitrogen from a previous crop of soybean.[35] In the irrigated areas of southern California and Arizona, the sorghum crop usually follows either wheat or barley in the same season and is followed by some spring-planted crop.

Sorghum is purported to be "hard on the land," the effect being particularly noticeable when fall-sown grains follow immediately. At Hays, Kansas, average wheat yields were 4.1 bushels per acre (257 kg/ha) lower after sorghum than after corn. Winter wheat yields following grain sorghum in Georgia were slightly lower compared to wheat following soybean.[36] Some of the injurious after-effects of sorghum can be explained by the persistence of the sorghum plant—the fact that it keeps growing until killed by frost. Sorghum thus depletes the soil moisture to a greater extent than do other crops.

Sorghum injury to subsequent crops under irrigation, where soil moisture is ample, has been attributed to the high sugar content of sorghum roots and stubble.[37] The sugars in dry roots of different varieties of sorghum at maturity have ranged from 15 to over 55 percent. Corn varieties ranged from less than 1 to about 4.5 percent.[38] These sugars furnish the energy for soil microorganisms that multiply and compete with the crop plants for the available nitrogen in the soil and thus retard the crop growth. This condition lasts for only a few months, or until the sorghum residues have decayed. Ground sorghum roots added to the soil have depressed nitrate accumulation and bacterial development more than have corn roots.[39]

The injurious after-effect of sorghum on irrigated land can be overcome by using nitrogenous fertilizers, barnyard manure, or inoculated legume green manures.[40] Both alfalfa and fenugreek made practically normal growth after sorghum in California. The detrimental influence of sorghum on dry land can be avoided by fallowing the next season, as well as by planting spring crops after sorghum, especially in May or June. By that time, available nitrates will have accumulated, while much of the soil moisture deficiency caused by the sorghums will have been overcome by normal precipitation.

■ 13.10 SORGHUM CULTURE

13.10.1 Seedbed Preparation

Disking is the common practice for seedbed preparation in the humid and irrigated areas. Yields can be increased from 25 to 30 percent by thorough tillage of medium-heavy soils. A warm, mellow seedbed is essential for good seed germination. Weed control before planting is desirable. Stubble-mulch or one-way disk methods till most of the fields in the semiarid region. A long-term study in Kansas on several soil types showed that grain sorghum yields are comparable with no-till, reduced-till, and conventional-till practices.[41]

FIGURE 13.8
No-till planting grain sorghum. [Courtesy USDA NRCS]

13.10.2 Seeding Methods

Sorghum for grain, fodder, or silage is usually planted in cultivated rows 20 to 39 inches (0.5 to 1 m) apart. A sorghum planter, or other type of row planter with special sorghum seed plates, is satisfactory. Planting in shallow furrows is desirable. No-till planting conserves soil moisture (Figure 13.8). Some sorghum is planted with a furrow drill in rows 14 to 20 inches (36 to 54 cm) apart. Such close row spacing is preferable to a 40-inch (1 m) spacing for a suitable rate of planting except under dry conditions (Figure 13.9). Narrow rows provide a more balanced plant distribution for absorbing light rays and better shading to suppress weed growth and soil moisture evaporation. Narrow rows result in higher yields under humid and irrigated conditions.[19]

Planting in narrow rows requires either matching cultivating equipment or the use of herbicides to control weeds after the plants are a few inches in height. Sorghum can be planted in double rows on beds to facilitate irrigation, drainage, and cultivation. The paired rows are 12 to 14 inches (30 to 36 cm) apart, on beds spaced at 40-inch (1 m) intervals.

The amount of seed to plant per acre for a given stand depends upon the condition of the seedbed, seed viability, seed size, and the weather conditions at seeding time. A discrepancy between field and laboratory germination of sorghum seed frequently ranges from 30 to 50 percent when seed of high viability is used.[41] Marked deficiency in field emergence can be expected when the laboratory germination is 85 percent or lower. The stands are improved by seed treatment so that a 75 percent emergence is possible.[30, 42]

Spacing between plants in the row for maximum yields depends upon tillering habits of the hybrid or variety.[13] Grain sorghums that tiller little are planted at higher rates.[43, 44, 45, 46] Producers sometimes use pounds per acre to determine seeding rates, but seeds per acre is much more accurate. Sorghum hybrids can vary from 12,000 to 20,000 seeds per pound (5,500 to 9,100/kg) resulting in wide variation in seeding rates. In regions with less than 20 inches (510 mm) of annual rain-

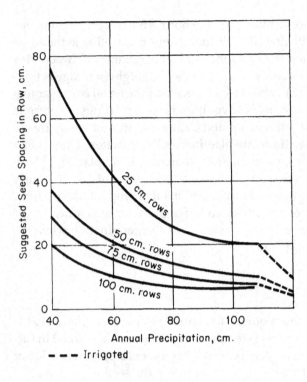

FIGURE 13.9

Row spacing as related to precipitation. [Adapted from *Kans. Ext. Circ. C447*]

fall, recommended populations for Kansas are 24,000 plants per acre (59,300/ha) and 32,000 (79,000/ha) in Texas. Irrigated sorghum should have about 100,000 to 120,000 plants per acre (247,000 to 296,400/ha).[19, ,32, 47, 48]

In order to achieve the highest tonnage of forage from sorghum planted in 40-inch (1 m) rows, the plants should be spaced not more than 4 to 6 inches (10 to 15 cm) apart. Sorgo, grown in rows for forage or silage in semiarid areas, should be planted at a rate of 4 to 5 pounds per acre (4.5 to 5.6 kg/ha), but at 6 to 8 pounds per acre (6.7 to 9 kg/ha) in humid or irrigated areas. The yields of sorghum drill planted in close rows for hay are about the same for all rates of seeding between 15 and 75 pounds per acre (17 to 84 kg/ha). Recommended seeding rates for sorghum forage are 30 pounds per acre (34 kg/ha) west of the 100th meridian in the Great Plains, 45 pounds per acre (50 kg/ha) between the 98th and 100th meridians, and 60 to 75 pounds per acre (67 to 84 kg/ha) east of the 98th meridian.[22]

The seed of sorghum is generally planted 1 to 2 inches (2.5 to 5 cm) deep, insuring good seed-soil contact. The percentage and rapidity of germination is reduced by soil temperatures below 65°F (18°C), and slightly reduced by deep planting (2½ inches; 6.35 cm).[49]

Sorghum yields are usually highest when the crop reaches maturity shortly before the average date of the first killing frost. To achieve this, quick-maturing varieties are planted late while long-season varieties are planted early.[30]

The date of planting of sorghum should occur when the soil is at least 65°F (18°C) so that germination and early growth will take place during the period of moderately high soil temperatures, 70 to 80°F (21 to 26°C), and the blooming and filling occurring at a time that avoids the highest temperatures.[6] In the extreme southern parts of the grain sorghum region, the crop can be planted from late February until August with a good chance to mature. In the northern part of the region, the soil does not become warm until after May 15. In the central as well as in the south central parts of the United States, the best time to plant sorghum is between May 15 and July 1. A safe rule in all localities, except where the sorghum midge is troublesome, is to plant not earlier than about two weeks after the usual corn planting time. The earliest possible safe planting is advisable where chinch bugs are present.

In the irrigated areas of southern Arizona and California, the best yields are obtained from planting in July. Medium-late planting results in better stands, taller stalks, larger heads, and shorter growing periods than does early planting. Planting earlier than necessary for safe maturity usually reduces plant growth. Early planting is sometimes desirable to avoid conflicts with other crops at harvest time, or to clear the land in time for seeding other crops. Sorghum for seed must be ripe before heavy freezes occur because freezing decreases the germination when the seed contains 25 percent moisture or more.[50]

13.10.3 Weed Control

From two to four cultivations are required to control weeds in sorghum that is planted in wide rows. Less cultivation is required when the seed is planted in furrows than when level cultivation is practiced. Surface-planted sorghum in wide or narrow rows can be cultivated with a rotary hoe before the seedlings emerge as well as again shortly thereafter.

There are fewer herbicides for sorghum than for corn because sorghum plants are more sensitive to the chemicals. Common preplant herbicides are atrazine, alachlor, metolachlor, and dimethenamid. Postemergent herbicides include 2, 4-D, dicamba, and bromoxynil. Postemergent applications should be applied when the crop is in early vegetative stages.

13.10.4 Irrigation

From 20 to 25 inches (500 to 650 mm) of water consisting of stored soil moisture, rainfall, and irrigation are required for maximum yields of sorghum.[51, 52] Of this amount, some 8 to 20 inches (200 to 500 mm) of water must be supplied by irrigation in the subhumid, semiarid, and arid regions. Because sorghum has a vigorous root system, preplant irrigation that wets the soil to field capacity to a depth of 6 to 7 feet (2 m) will reduce the number of summer irrigations.[30] Maximum water usage occurs during the boot stage of plant growth when up to 1 inch (25 mm) of water is consumed during three or four days. Water consumption increases in direct proportion to yield, as seen in Figure 13.10.

13.10.5 Harvesting

Grain sorghum is usually harvested with a combine.[44] Flexo-guards on the combine save some heads from being lost during harvesting. Gathering lugs, attached to the cutter bar of row harvester combines, lift lodged stalks. The grain is mature when the seeds are fully colored, have begun to harden, and form a black layer at the base of the embryo.[53] At maturity, the kernel will contain about 18 to 20 percent moisture. For combining, the crop should be allowed to dry until the grain mois-

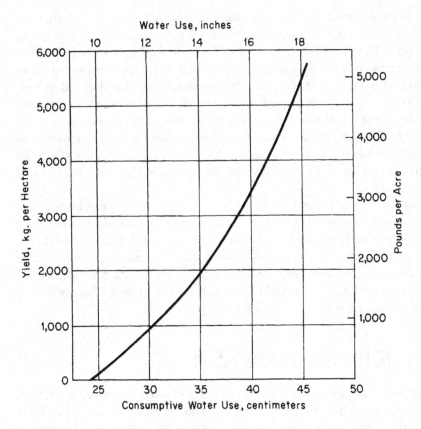

FIGURE 13.10
Water consumption as related to yield. Data obtained at North Platte, Nebraska.

ture is 13 percent or less, unless provision is made for drying the grain. The cheapest method of drying is to dump the threshed grain in long, low tiers or piles on a clean, dry, sodded area. Artificial drying is practiced frequently (Figure 8.6). Desiccant sprays can be used to dry the crop for combining. A few days of dry weather after a severe freeze will usually reduce the moisture content of the grain to the point where it can be combined at a moisture content low enough for safe storage.

The seed of sorghum should be fairly mature before the crop is cut for forage[22] for several reasons: (1) the largest yield of dry matter is obtained from mature plants, (2) the feed is more palatable, (3) the plants contain less prussic acid, and (4) the silage made from mature sorghum is drier and thus has less acid. The total dry weight of the plants may increase about 40 percent between first heading and maturity.

Drilled forage sorghum such as sorghum-sudangrass hybrids are cut for hay with a windrower. Because of the thick juicy stems, it takes a long time to cure forage sorghum sufficiently for stacking or baling unless it is crimped during windrowing. Drilled forage sorghum should be allowed to form seed before it is harvested, except when it has been severely injured by drought and will not head.

▪ 13.11 SUDANGRASS CULTURE

Sudangrass will endure considerable drought. But, like all sorghums, it is sensitive to low temperatures. However, it succeeds in areas where the season is too short for other sorghum, except the very early types.

The crop is grown much the same as other sorghums, being planted after the soil is warm, usually about two weeks after corn is planted. The date of planting varies from April 1 to July 1 from the extreme South to the North.[54] Sudangrass is often drilled in close rows for hay or pasture, but in the southwestern semiarid region it is planted in cultivated rows for all purposes. Since the plant tillers profusely, the final weight of stems per unit area is usually about the same for different rates of seeding. The usual rate for drilling under humid conditions is 20 to 25 pounds per acre (22 to 28 kg/ha), while the rate under semiarid conditions is generally 15 to 20 pounds per acre (17 to 22 kg/ha). When planted in rows 36 to 44 inches (91 to 112 cm) apart, 2 to 4 pounds of seed per acre (2.2 to 4.5 kg/ha) will produce a good stand.

Sudangrass is commonly cut for hay when the first heads appear, being more palatable at this stage than when cut later. Usually two crops are harvested for hay, although three or even four may be cut under favorable conditions. For seed production, the second crop is usually saved for seed, after the first crop was cut rather for hay. Sudangrass grown in rows for seed is usually cut with a windrower. It should be cut after the greatest amount of seed appears to be ripe and before it shatters. The straw is about as valuable for feed as is prairie hay.

13.12 JOHNSONGRASS AS A CROP

Johnsongrass is such a noxious weed pest that its importance as a pasture and forage crop is often overlooked. It is a perennial hay crop of the South, where it also furnishes an appreciable portion of the pasture. Johnsongrass is widely distributed throughout the South and as far north as the 38th parallel from the Atlantic Coast to the Colorado border. It is found even farther north in the Potomac and Ohio Valleys and in California, Oregon, and Washington. Johnsongrass seldom is sown. Much of its spread occurred from crop seeds that contained johnsongrass or by grazing livestock.

Johnsongrass was collected as early as 1696 at Aleppo, a town in Syria. It is still known in Europe and North Africa as Aleppo grass or Sorgho d'Alep. It is regarded as a native of the Mediterranean coastal countries of Europe, Africa, and Asia where it extends eastward into India. It was introduced into South Carolina from Turkey in 1830. Ten years later it was popularized by Colonel William A. Johnson of Selma, Alabama, which accounts for its present name. Despite its weedy habits, johnsongrass was recommended as a crop by certain agricultural authorities as late as about 1890, and again for controlling wind erosion about 1935.

Johnsongrass is most abundant and vigorous on the richer bottomlands. Where such land is reserved for johnsongrass hay meadow, it is often plowed every two or three years to break up and cover the old stubble and clear the land temporarily, which stimulates new growth from the rhizomes. Sometimes winter oat is sown in the fall and harvested in the late spring, and johnsongrass is then allowed to grow for hay. When the land is needed for other crops, johnsongrass is pastured heavily or mowed frequently to deplete the food reserves in the roots and rhizomes and retard development of new stalks or rhizomes. The rhizomes that develop under these conditions are rather shallow. The land is then plowed in the spring and planted to a cultivated crop, or is summer-fallowed. Under these circumstances, johnsongrass is often eradicated by six cultivations at two-week intervals on semiarid land and by ten to fifteen cultivations in humid areas. The rhizomes develop mostly after the plant blooms. Rhizomes and roots live only about a year. Control is aided by repeated applications of suitable herbicides.

Johnsongrass is usually cut for hay before it blooms in order to avoid the development and dispersal of seeds.[55]

▦ 13.13 USES OF SORGHUM

13.13.1 Feed

Most of the sorghum grain grown in Asia and the African tropics is used for human food. Elsewhere, it is generally fed to livestock or poultry.

The composition of grain sorghum (Appendix Table 2) is similar to that of corn except that it is slightly higher in protein and lower in fat.[10] Grain sorghum hybrids and varieties differ widely in digestibility when fed to livestock, and they are generally lower in digestibility than corn. The brown-seeded (bird-resistant) types are high in tannin. Those containing more than 1.6 percent tannin depressed the growth of chicks when they constituted as much as 50 percent of their diet. Sorghum gluten feed is especially high in tannin because it includes the seedcoats that bear the tannin.[56] In other digestion experiments, each additional 1 percent in tannin content lowered digestible dry matter by about 4 percent. Types with high lysine and low tannin contents have now been developed.[57]

Sorghum grain should be ground, steam-rolled or flaked, popped, or dry-heated before feeding to increase its digestibility.[58] In general, 1 pound of sorghum grain is approximately equal to 1 pound of corn in nutritive value for feeding dairy cows or fattening lambs. It is equal to about 95 percent of the feed value of corn on a pound-for-pound basis for fattening beef cattle or hogs.

Forage sorghum fed as roughage requires some supplemental grain, legume hay, or meal to balance the ration. Sorghum silage has about the same composition as corn silage. The comparative feeding value of the two silages varies with the relative moisture content and the proportion of grain they contain. When these two factors are equal, corn silage still has a higher feeding value because the grain is softer and more completely absorbed in the intestinal tract. The silage of sorghum-sudangrass hybrids has about 90 percent of the nutritive value of corn silage.[54]

Sorgo and sudangrass hay are coarser in texture, but similar, in chemical composition to other grass hays, but they are somewhat higher in protein and ash and lower in fat and crude fiber due to the fact that they usually are cut at an earlier stage of growth.

13.13.2 Industrial Uses

Grain sorghum often replaces corn grits in the brewing and distilling industries, and in the manufacture of alcohol. During World War II, the grain of waxy varieties was used for extraction of starch to manufacture a satisfactory substitute for *minute tapioca*. The most important use is in the manufacture of starch, glucose, syrup, oil, gluten feeds, and other products similar to those produced in the wet milling of corn. Grain sorghum flour is used for adhesives.[34, 59] The entire plants of sorgo are sometimes chopped, dehydrated, ground, and pelleted for feed.[41]

13.13.3 Syrup

Sorghum is used for syrup or molasses production in areas of the world that are not suited for sugarcane or sugarbeet production. Mechanical harvesting of sorghum for syrup is similar to that used for sugarcane. Sorgo for syrup is cut when the seed is nearly ripe, or at least in the stiff-dough stage. The juice of the upper internodes and peduncle is much higher in starch and mineral matter than that from the remainder of the stalk. The whole stalks are topped, after which the stalks are chopped mechanically into short sections. Later, fans blow out the leaves. A fair quality of syrup can be made from whole stalks that have been stored as long as about six days before crushing.[60] The juice is expressed from the stripped cane by crushing the stalks between revolving fluted iron rolls in a cane mill equipped with three (or occasionally two) rolls (Figure 13.11). The strained juice that is sometimes allowed to settle for a time is piped to an evaporating pan where it is boiled down to syrup containing about 70 percent sugar. During boiling, the juice is skimmed constantly to remove floating impurities such as chlorophyll, soil, plant fragments, proteins, gums, fats, and waxes. Often, the juice or the partly evaporated semisyrup is treated with lime, clay products, or phosphoric acid to neutralize the acid juice and to precipitate impurities. Treatment with malt diastase, an enzyme product, hydrolyzes the starch into glucose and prevents thickening (jellying) due to the 0.5 to 3 percent of starch usually present in the juice. With such treatment, a concentrated syrup containing as much as 70 percent sugar will pour readily. Sometimes a yeast extract containing another enzyme, invertase, is also added to change part of the sucrose-sugar into dextrose (glucose) and levulose (grape sugar), and thus prevent crystallization (sugaring). The juice of a good variety of sorgo grown under suitable conditions contains 13 to 17 percent sugar, of which 10 to 14 percent is sucrose.

FIGURE 13.11
Crushing sweet sorghum stalks between rolls to extract the juice for boiling down to syrup. [Courtesy Morris Bitzer, University of Kentucky]

A good field should yield 15 tons per acre (34 MT/ha) of fresh sorghum or 10 tons per acre (22 MT/ha) of stripped stalks. A ton of stripped stalks should yield 700 to 1,200 pounds (316 to 544 kg) of juice or 8 to 20 gallons (30 to 76 l) of syrup. However, the average yield of syrup is less than 80 gallons per acre (748 l/ha). Sorgo syrup acquires its flavor or tang from organic acids that are present in the juice. The syrup is rich in iron, particularly when it has been evaporated in an iron pan or kettle. Sorghum syrup is often referred to as molasses, which is a misnomer, since molasses is a by-product of sugar manufacture. The sorghum syrup industry reached its peak in 1920. Today, enterprising farmers who process and market the syrup grow it on only a few acres.

13.13.4 Sugar

Attempts to develop a sorghum sugar industry have not been successful thus far because of certain difficulties and limitations that make the sugar more expensive than that derived from sugarcane or sugarbeet. Sorghum sugar was manufactured on a commercial scale under public subsidy in Kansas and New Jersey about 1890. Extensive research on sorghum sugar manufacture was conducted between 1878 and 1893 and again between 1935 and 1941. The crystallization of the sugar is satisfactory only after the starch and the aconitic acid crystals have been removed by centrifuging. The stalks must be harvested promptly when they reach the proper stage. They may lose sucrose-sugar by inversion to dextrose and levulose when not processed soon. Sorghum often is less uniform in yield and composition than are sugarcane and sugarbeet. The yields of sorghum that are grown during a period of four or five months are obviously less than are obtained from sugarcane that has a growing period of eight to fourteen months.

There has been some interest is growing sweet sorghum for ethanol production as a biomass fuel, but currently the use of grains is more economical.

▓ 13.14 DISEASES

The principal diseases attacking sorghum in the United States[42] are smuts, leaf spots, and root and stalk rots. Other serious diseases attack sorghum in the Eastern Hemisphere.[39] Modern hybrids used today are much more resistant to diseases than the traditional varieties.

13.14.1 Kernel Smut

The kernel smut can reduce seed production materially but probably has less effect on forage yields. All sorgos, kafirs, and broomcorns are susceptible. Each infected ovule becomes a mass of dark-colored spores instead of a sorghum seed (Figure 13.12). Covered-kernel smut *(Sporisorium sorghi)* and loose-kernel smut *(S. cruenta)* can be controlled by seed treatment with fungicides. Some hybrids and varieties of grain sorghum are resistant to certain races of these smut organisms.

FIGURE 13.12
Sorghum smuts. *(Left to right)* Sound head, smut, loose kernel smut, and head smut.

13.14.2 Head Smut

Head smut disease is caused by the fungus *Sporisorium holci-sorghi*. The entire head is replaced by a gall or smut mass (Figure 13.12). The spores are carried in the soil and occasionally on the seed. Seed from fields where the disease is prevalent should be avoided. Some varieties and hybrids are susceptible to head smut. Seed treatment offers no protection against infection carried in the soil. Removal and burning of diseased plants will eliminate reinfestation of the soil.

13.14.3 Sorghum Ergot

Sorghum ergot (*Claviceps africana*), also called sugary disease, has only been a problem in Africa and Asia until recent years when it was discovered in Brazil. It was first observed in the Rio Grande Valley of the United States in 1997. The pathogen infects the flower and prevents pollination. The pathogen eventually replaces the ovary and exudes a sticky liquid that can drip from the head during humid weather. During dry weather, the exudate will dry into crystalline masses. Initial inoculum comes from infected seeds, but spores can spread great distances in wind. Treating seed with a fungicide helps prevent initial infection.

13.14.4 Root and Stalk Rots

Root rot, which attacks the roots and crown of a plant, once caused extensive damage to some varieties, but resistant strains and hybrids replaced the susceptible varieties. A widespread soil-borne fungus, *Periconia circinati*, causes this disease. Charcoal rot attacks the base of the stalk and also the larger roots, rotting away the pith and causing the stalk to fall over. A soil-borne fungus, *Macrophomina phaseoli*, which also attacks several other crop plants, causes the disease. Damage from this disease is limited when the soil is well supplied with moisture and sorghum planting is delayed so that the crop reaches maturity under mild, favorable fall conditions.

13.14.5 Leaf Spots

Bacterial leaf spot diseases include stripe, caused by *Pseudomonas andropogonis*; spot, caused by *Pseudomonas syringae*; and streak, caused by *Xanthomonas campestris* pv. *holcicola*. Fungus leaf spot diseases include sooty stripe, caused by *Ramulispora sorghi*; rough spot, caused by *Ascochyta sorghina*; anthracnose, caused by *Colletotrichum graminicola*; leaf blight, caused by *Setosphaeria turcica*; zonate leaf spot, caused by *Gleocercospora sorghi*; gray leaf spot, caused by *Cercospora sorghi*; and rust, caused by *Puccinia purpurea*.

Sorghums containing a yellow leaf pigment are resistant to bacterial leaf spots. Others are resistant to several of the fungus leaf spot diseases.

13.14.6 Other Diseases

Other diseases attacking sorghum include *Rhizoctonia* stalk rot, caused by *Rhizoctonia solani*; anthracnose stalk rot and red rot, caused by species of *Colletotrichum*; downy mildew, caused by *Sclerospora* species; and crazy top, caused by *Sclerospora sorghi*.

Maize dwarf mosaic virus overwinters in johnsongrass and is carried to sorghum chiefly by aphids.

Weakneck is not a true disease, but decay at the base of the peduncle that occurs after the upper part of the stalk has ceased growth and begun to dry. This decay causes mature heads to fall over and be lost.

▪ 13.15 INSECT PESTS

The most destructive pests on sorghum in the United States are the greenbug aphid, chinch bug, sorghum midge, corn-leaf aphid, corn earworm (Figure 12.20), and sorghum webworm.

13.15.1 Greenbug Aphid

The greenbug aphid, *Schizaphis graminum*, is a light green aphid with a dark green stripe down the middle of the back. The tips of the legs and cornicles and most of the antennae are black. Winged forms of the greenbug are darker than the wingless forms. Greenbugs are most often found feeding on the underside of the lower leaves of sorghum; however, they also may be found in the whorl of seedlings. Greenbugs seldom over winter north of southern Kansas, where they can be a major pest of winter wheat, especially when mild winters permit reproduction and survival. The winged forms are blown north in early spring by southerly winds and infest wheat and other small grains. They cause little damage to small grains. In the early summer, they move into sorghum fields. Research in Nebraska and other states shows that winged greenbugs are more likely to land in fields with bare soil than in fields with higher levels of residue cover.[61]

Unlike some cereal aphids, salivary secretions of the greenbug are toxic to the plant. The toxin kills plant cells resulting in yellow discoloration of the leaves with reddish spotting. The reddish discoloration is especially distinct on susceptible sorghum varieties. Effects of damage are loss of stand, reduced vigor, poorly filled heads, light test weight, and ultimately reduced yields.

Genetic resistance to greenbug damage has been incorporated into sorghum hybrids but biotypes of the aphid that have overcome varietal resistance in sorghum are common. Biological control can sometimes be successful if Integrated Pest Management procedures are followed.

13.15.2 Chinch Bugs

The chinch bug, *Blissus leucopterus*, is a sucking insect that feeds heavily on the leaves and sheaths of sorghum (Figure 12.20). The chinch bug moves to sorghum from nearby small grain fields after the harvest of the latter. In Kansas and northward, they usually migrate on foot from small grains to sorghum, but later in the season they acquire wings and fly to sorghum fields. There they mate and deposit eggs. The eggs hatch a second generation that continues to feed on sorghum plants. A third generation develops in the South in the fall. The most serious injury by chinch bugs is caused in seasons of low rainfall. Many chinch bugs die in cool, wet weather. Spraying border strips with insecticides between the two fields may prevent the migration of chinch bugs on foot from small grain fields to sorghum. South of central Kansas, chinch bugs develop wings by the time small grains are harvested. When they migrate by flight, barriers are ineffective, but injury can be reduced by early planting, resistant varieties, or by spraying with insecticides. The adult chinch bugs hibernate at the base of bunches of grass or in trash.

Common varieties and hybrids of sorghum vary in susceptibility to chinch bug injury.

13.15.3 Sorghum Midge

The sorghum midge (*Stenodiplosis sorghicola*) is abundant in the southern states. It formerly made grain sorghum production unprofitable in that region, but later the damage was less severe, probably because sorghum is more abundant and because more natural enemies of the midge are present. The midge is a small fly with a red body.[22] It lays its eggs inside the glumes at the blossom stage. The egg produces a small white larva that absorbs juices from the sorghum ovary and prevents seed development. The adult midge emerges to start a new generation every 14 to 20 days under favorable conditions. The larva causes injury to the developing seed.

The midge breeds on johnsongrass in the spring and infests sorghum as soon as it heads. All types of sorghum are subject to attack. The midge over winters in the larval stage on sorghum heads, in johnsongrass, and on trash. Planting sorghum very early in the season, so that it blooms before the midge is plentiful, is one means of combating this insect. If an insecticide is used, multiple applications may be necessary.

13.15.4 Other Insects

In some seasons, corn leaf aphid (*Rhopalosiphum maidis*) is a serious pest of sorghum when large numbers occur in the leaf curl. This hinders the exertion of the head from the boot. The honeydew secretions of the aphids form a sticky mass about the plant heads, which encourages the growth of molds and in turn prevents the production of high-quality seed. Aphids winter in the southern states and migrate north.

Sorghum webworm (*Nola sorghiella*) is periodically destructive when it devours developing grains, especially on late planted fields. Banks grass mite (*Oligonychus pratensis*) has acquired tolerance to pesticides and is threatening sorghum in Texas.

Grasshoppers cause little injury to the foliage of sorghum even where corn is completely destroyed. However, they seem to prefer the leaves of certain varieties. Grasshoppers eat developing sorghum grain, but they can be controlled with insecticide sprays. Other pests occur in Africa and Asia.[39]

▨ 13.16 BIOLOGICAL CONTROL OF SORGHUM PESTS

There are several natural enemies of sorghum pests that many times can provide effective control if an Integrated Pest Management (IPM) program is used. For successful use of biological control methods, the use of pesticides must not be used except when other methods have failed.[26, 59]

Native and introduced predators of sorghum pests include several species of lady beetle or ladybug. The lady beetle can provide aphid control. Common species include the twelve-spotted (*Coleomegilla maculata*) and seven-spotted (*Coccinella septempunctata*). Green lacewing (*Chrysoperla carnea*) is a common predator of aphids and other soft-bodied insects. Other parasites of sorghum pests include the larvae of syrphid, or hover fly, and damsel bugs.

A parasitic wasp (*Lysiphlebus testaciepes*) lays eggs on greenbug aphids and its larvae infect them. Chinch bugs can be controlled through the use of a fungal pathogen (*Beauveria bassiana*).

REFERENCES

1. DeWet, J. M. J., and J. R. Harlan. "The origin and domestication of *Sorghum bicolor*," *Econ. Bot.* 25, 2(1971):125–128.

2. Vinall, H. N., J. C. Stephens, and J. H. Martin. "Identification, history, and distribution of common sorghum varieties," *USDA Tech. Bull.* 506, 1936.

3. Martin, J. H. "Broomcorn—the frontiersman's cash crop," *Econ. Bot.* 7(1953):163–181.

4. Martin, J. H. "Sorghum improvement," in *USDA Yearbook,* 1936, pp. 523–560.

5. Martin, J. H. "Climate and sorghum," in *Climate and Man*, USDA Yearbook, 1941, pp. 343–347.

6. Vinall, H. N., and H. R. Reed. "Effect of temperature and other meteorological factors on the growth of sorghums," *J. Agr. Res.* 13(1918):133–147.

7. Martin, J. H. "Comparative drought resistance of sorghums and corn," *J. Am. Soc. Agron.* 22(1930):993–1003.

8. Martin, J. H. "Sorghum and pearl millet," in *Handbuch der Pflanzenzuchtung*, vol. 2. Berlin: Paul Parey, 1959, pp. 565–587.

9. Artschwager, E. "Anatomy and morphology of the vegetative organ of *Sorghum vulgare*," *USDA Tech. Bull.* 957, 1948.

10. Sieglinger, J. B. "Leaf number of sorghum stalks," *J. Am. Soc. Agron.* 28(1936):636–642.

11. Martin, J. H. "Recurving in sorghums," *J. Am. Soc. Agron.* 24(1932):500–503.

12. Sieglinger, J. B. "Cross pollination of mile in adjoining rows," *J. Am. Soc. Agron.* 13(1921):280–282.

13. Nyval, R. F., *Field Crop Diseases*. Ames, IA: Iowa State University Press, 1999.

14. Martin, J. H. "The use of the greenhouse in sorghum breeding," *J. Hered.* 25, 6(1934):251–254.

15. Sieglinger, J. B., and J. H. Martin. "Tillering ability of sorghum varieties," *J. Am. Soc. Agron.* 31(1939):475–488.

16. Vinall, H. N. "A study of the literature concerning poisoning of cattle by the prussic acid in sorghum, Sudan grass, and johnsongrass," *J. Am. Soc. Agron.* 13(1921): 267–280.

17. Martin, J. H., J. F. Couch, and R. R. Briese. "Hydrocyanic acid content of different parts of the sorghum plant," *J. Am. Soc. Agron.* 30(1938):725–734.

18. Gillingham, J. T., and others. "Relative occurrence of toxic concentrations of cyanide and nitrate in varieties of sudangrass and sorghum-sudangrass hybrids," *Agron. J.* 61, 5(1969):727–730.

19. Sumner, D. C., and others. "Sudangrass, hybrid sudangrass and sorghum x sudangrass crosses," *CA Agr. Exp. Sta. Circ.* 547, 1968, pp. 1–16.

20. Brandon, J. F., J. J. Curtis, and D. W. Robertson. "Sorghums in Colorado," *CO Exp. Sta. Bull.* 449, 1938.

21. Swanson, A. F., and H. H. Laude. "Sorghums for Kansas," *KS Agr. Exp. Sta. Bull.* 304, 1942, pp. 1–63.

22. Martin, J. H., and J. C. Stephens. "The culture and the use of sorghums for forage," *USDA Farmers Bull.* 1844, 1940.

23. Martin, J. H. "Breeding sorghum for social objectives," *J. Hered.* 36, 4(1945):99–106.

24. Poehlman, J. M. *Breeding Field Crops*. New York: Dolt, 1959, pp. 1–427.

25. Quinby, J. R. *A Triumph of Research—Sorghum Production in Texas*. College Station, TX: TX A & M Univ. Press, 1971, pp. 1–28.

26. Quinby, J. R. *Sorghum improvement and the genetics of growth*. College Station, Texas: TX A & M Univ. Press, 1974, pp. 1–108.

27. Quinby, J. R., and J. E. Martin. "Sorghum improvement," in *Advances in Agronomy*, vol. 6, New York: Academic Press, 1954, pp. 305–359.

28. Wall, J. S., and W. M. Ross, eds. *Sorghum Production and Utilization*. Westport, CT: 1970, pp. 1–702.

29. Tucker, B. B., and W. F. Bennett. "Fertilizer use on grain sorghum," in *Changing Patterns in Fertilizer*, ed. R. C. Dinauer. Madison WI: Soil Sci. Soc. Amer., 1968, pp. 189–220.

30. Ross, W. M., and O. J. Webster. "Culture and use of grain sorghum," *USDA Handbk.* 385, 1970, pp. 1–30.

31. Musick, J. T., and D. A. Dusek. "Grain sorghum row spacing and planting rates under limited irrigation in the Texas High Plains," *TX Agr. Exp. Sta. MP* 932, 1969, pp. 1–10.

32. Hobbs, J. A. "Yield and protein contents of crops in various rotations," *Agron. J.* 63, 6(1971):832–836.

33. Quinby, J. R., and others. "Grain sorghum production in Texas," *TX Agr. Exp. Sta. Bull.* 912, 1958.

34. Mathews, O. R., and L. A. Brown. "Winter wheat and sorghum production in the southern Great Plains under limited rainfall," *USDA Circ.* 477, 1938, pp. 1–60.

35. Clegg, M. D. "Predictability of grain sorghum and maize yield grown after soybean over a range of environments," *Agric. Systems* 39(1992):25–31.

36. Duncan, R. R. "Agronomic principles and management practices," In R. R. Duncan, ed., *Proceedings Grain Sorghum Short Course. Univ. GA Agri. Exp. Sta. Spec. Pub.* 29, 1985, pp. 5–17.

37. Conrad, J. P. "Some causes of the injurious after-effects of sorghums and suggested remedies," *J. Am. Soc. Agron.* 19(1927):1091–1111.

38. Conrad, J. P. "The carbohydrate composition of corn and sorghum roots," *J. Am. Soc. Agron.* 29(1937):1014–1021.

39. Wilson, B. D., and J. K. Wilson. "Relation of sorghum roots to certain biological processes," *J. Am. Soc. Agron.* 20(1928):747–754.

40. Conrad, J. P. "Fertilizer and legume experiments following sorghums," *J. Am. Soc. Agron.* 20(1928):1211–1234.

41. Swanson, A. F. "Making pellets of forage sorghum," in *Crops in Peace and War*, USDA Yearbook, 1950–51, pp. 353–356.

42. Leukel, R. W., and J. H. Martin. "Four enemies of sorghum crops," in *Plant Diseases*, USDA Yearbook, 1953, pp. 368–377.

43. Bond, J. J., T. J. Army, and O. R. Lehman. "Row spacing, plant populations, and moisture supply as factors in dryland grain sorghum production," *Agron. J.* 56, 1(1964):3–6.

44. Karper, R. E., and others. "Grain sorghum date of planting and spacing experiments," *TX Agr. Exp. Sta. Bull.* 424, 1931.

45. Mann, H. O. "Effects of rates of seeding and row widths on grain sorghum grown under dryland conditions," *Agron. J.* 57(1965):173–176.

46. Martin, J. H., and others. "Spacing and date-of-seeding experiments with grain sorghums," *USDA Tech Bull.* 131, 1929.

47. Amador, J., and others. "Sorghum diseases," *TX Agr. Exp. Sta. Bull.* 1085, 1969, pp. 1–21.

48. Mulkey, J. R., J. Drawee, and E. L. Albach. "Irrigated sorghum response to plant population and row spacing in southwest Texas," *TX A&M Univ. Exp. Sta.* PR-4293, 1985.

49. Martin, J. H., J. W. Taylor, and R. W. Leukel. "Effect of soil temperature and depth of planting on the emergence and development of sorghum seedlings in the greenhouse," *J. Am. Soc. Agron.* 27(1935):660–665.

50. Gangstad, E. O. "DeSoto grass," *TX Res. Foundation Hoblitzelle Ag. Lab. Bull.* 22, 1965.

51. Jensen, M. E., and J. T. Musick. "Irrigating grain sorghums," *USDA Leaflet* 511, 1962.

52. Musick, J. T., J. W. Grimes, and G. M. Harron. "Irrigation water management and nitrogen fertilization of grain sorghum," *Agron. J.* 55, 3(1963):295–298.

53. Vanderlip, R. L. "How a sorghum plant develops," *KS St. Univ. Coop. Ext. Serv.* S-3, 1993.

54. Anonymous. "Sudangrass and sorghum-sudangrass hybrids for forage," *USDA Farmers Bull.* 2241, 1969, pp. 1–12.

55. Vinall, H. N., and M. A. Crosby. "The production of johnsongrass for hay and pasturage," *USDA Farmers Bull.* 1957, 1929, pp. 1–26.

56. Fuller, H. L., and others. "The feeding value of grain sorghums in relation to their tannin content," *CA Agr. Exp. Sta. Bull. N. S.* 176, 1966, pp. 1–14.

57. Singh, R., and J. D. Axtell. "High-lysine mutant gene (hi) that improves protein quality and biological value of grain sorghum," *Crop Sci.* 13, 5(1973):535.

58. Riggs, J. K., and others. "Dry heat processing of sorghum grain for beef cattle," *TX Agr. Exp. Sta. Bull.* 1096, 1970, pp. 1–11.

59. Martin, J. H., and M. M. MacMasters. "Industrial uses for grain sorghum," in *Crops in Peace and War,* USDA Yearbook, 1950–51, pp. 349–352.

60. Coleman, O. H., and I. E. Stokes. "Storage studies of sorgo," *USDA Tech. Bull.* 1307, 1964.

61. Wright, R., S. Danielsen, and Z. B. Mayo. "Management of greenbugs in sorghum," *NE Coop. Ext. Serv. NebGuide* G87-838-A, 1987.

Sugarcane

14.1 ECONOMIC IMPORTANCE

Worldwide, sugarcane was harvested on an average of more than 49 million acres (20 million ha) in 2000–2003. Production averaged 1,428 million tons (1,298 million MT) of cane or 29 tons per acre (65 MT/ha). Leading countries in sugarcane production are Brazil, India, China, Thailand, and Mexico. Production for sugar in the United States in 2000–2003 averaged 34 million tons (30 million MT) of cane. It was harvested from about 936,000 acres (390,000 ha), or about 35 tons per acre (77 MT/ha). The leading states are Florida, Louisiana, Hawaii, and Texas (Figure 14.1). Florida produced about 50 percent of the cane sugar grown in the United States in 2000–2003. The lowest yields in the United States are obtained in Louisiana where the fall-planted cane has a growing period of twelve to fifteen months and the spring-planted cane perhaps nine months. Sugarcane in Florida grows for 15 to 18 months. The ratoon crop makes its growth in one year in both states. In Hawaii, where both planted and stubble sugarcane are allowed to grow for two to three years before harvesting, yields are the highest. The best sugar yields are obtained by harvesting every 20 to 24 months.[1] Total sugarcane production for sugar and seed in the United States in 2003 was about 34 million tons (31 million MT), with about 2 million tons (1.7 million MT) used for seed cane.

Per capita consumption of sugar in the United States in 1973 exceeded 100 pounds annually. However, sugar consumption has continued to decline since then as people have become more diet conscious. In 2002, per capita consumption was only about 63 pounds annually.

14.2 HISTORY OF SUGARCANE CULTURE

Cultivated sugarcanes (*Saccharum* species) probably are derived from indigenous wild species on the islands of Melanesia. It is postulated that they arose by selection from wild canes in New Guinea.[2, 3] They were carried by people as stem cuttings to various centers where they were further modified by natural hybridization with other wild grasses.

Sugarcane of economic usefulness has been transported by people for thousands of years. Sugarcane was cultivated in India as early as 1000 BC, while in

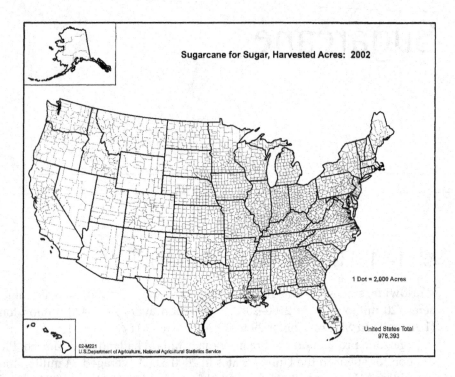

Sugarcane for Sugar, Harvested Acres: 2002

1 Dot = 2,000 Acres

United States Total
978,393

02-M221
U.S.Department of Agriculture, National Agricultural Statistics Service

FIGURE 14.1
Sugarcane harvested for sugar in the United States. [Source: 2002 Census of U.S. Department of Agriculture]

China, crude sugar was being made from sugarcane by 760 BC. The first refined white sugar was processed in Persia (Iran) about AD 600 .[4] Most people still prefer refined sugar.

Sugarcane was taken to Santo Domingo by Columbus on his second voyage, in 1493, but the crop was not established permanently until 1506. Introduction of sugarcane to Louisiana dates from 1751, while the first granulated sugar was produced in that state in 1795.[5]

14.3 ADAPTATION

Sugarcane is primarily a tropical plant that usually requires eight to twenty-four months to reach maturity. The temperatures should be high enough to permit rapid growth for eight months or more. After a period of rapid growth, there is usually, in most cane-producing countries, a period of slow sugarcane growth with increased sugar storage. Sugarcane is considered to be ripe when the sugar content is at its maximum. Sugarcane culture has gradually migrated from the tropics to the subtropical regions in the United States and other countries.

The best soils for sugarcane are those that are well supplied, naturally or artificially, with nutrient minerals and organic matter. In Louisiana, heavy soils are used for sugar production, and lighter soils for cane syrup production. Drainage must be provided for low, wet lands and for areas where the water table is less than 3 feet (90 cm) below the surface. In heavy, poorly drained soils, low oxygen content of the air at a 28-inch (70 cm) soil depth retards root development.[6] In Florida, more than 90 percent of the sugarcane acreage is on muck and peat histosol soils containing 50 to 97 percent organic matter. Soils with lower organic matter contents are preferable. In Hawaii, sugarcane is grown mostly on yellowish-brown or reddish-brown lateritic soils that require heavy fertilization but are often high in manganese. Many of the fields are irrigated. Irrigation is beneficial and often essential for successful sugarcane production, except where a well-distributed rainfall of 45 inches (1,200 mm) or more occurs annually.

▓ 14.4 BOTANICAL DESCRIPTION

Sugarcane is a tall perennial grass[7, 8, 9] with stalks (Figure 3.3) bunched in stools of five to fifty stalks or evenly scattered. The stalks are ¾ to 2 inches (2 to 5 cm) in diameter and may be 16 feet (5 m) or more in height. The internodes are comparatively short, usually 2 to 2¾ inches (5 to 7 cm), often being marked by corky cracks. Mature canes may be green, yellow, purplish, or reddish. A sugarcane plant has a leafy appearance because of the abundant tillers and the numerous nodes, each of which produces a leaf. The stalks are covered with a layer of wax that forms a band at the top of each internode. In addition to the growth ring, there is a root band and a single bud (eye) at each node. The leaf blade is usually erect, ascending, and gently curved, but it may be erect with a drooping tip (Figure 3.2). Fully developed blades measure about 40 inches (1 m) in length. The leaf blade is green, sometimes with a purplish cast. The leaf sheath folds around the stem and serves as a protection to the bud.

When sugarcane is harvested, ratoons (tillers) arise from axillary nodes at the base of the stem. These ratoons will grow and allow several cuttings of sugarcane, called ratoon crops, over a period of several years.

Sugarcane seldom produces seed in the continental United States except in Florida. Consequently, it is harvested before the inflorescence appears. Ordinarily, the panicles do not appear until at least twelve to twenty-four months after the cane is planted. When grown in the tropics most varieties flower after the summer day lengths begin to shorten. A stalk ceases growth when it flowers. The inflorescence, an open, much-branched, plume-like panicle, known as the arrow, is 8 to 24 inches (20 to 60 cm) long (Figure 14.2). The spikelets are arranged in pairs, one spikelet being sessile and the other pedicellate. The spikelets, about ⅛ inch (3 mm) long, are obscured in a tuft of silky hairs two to three times as long as the spikelet. Each flower is subtended by two bracts that form the outer and inner glumes (Figure 3.11). There is also a sterile lemma, but the fertile lemma and palea of the typical grass flower are not present.[9, 10] The fruit is extremely small, about 1/20 inch (1½ mm) long, with a distinct constriction in the region opposite the embryo. The seed, which is enclosed in the glumes and the sterile lemma, has low viability (Figure 7.1).

While the flowers of all sugarcane varieties are perfect, many plants are pollen-sterile or at least self-sterile.[11] There are probably no completely sterile varieties, except under extreme conditions.

▓ 14.5 SUGARCANE SPECIES

The sugarcanes of America are derived from five species and their hybrids. In the good sugar quality *noble* canes, *Saccharum officinarum,* the axis of the inflorescence

FIGURE 14.2
Sugarcane inflorescence or "arrow."

has no long hairs. The stalk diameter is large, the leaves wide, and the fiber content low. The Chinese canes, *S. sinense*, have long hairs on the axis of inflorescence. The northern Indian canes, *S. barber*, differ from the Chinese canes in having narrower blades and more slender stalks.

The wild canes of Asia, *S. spontaneum*, are used in making hybrids with other species. They are a serious weed pest in India. The stalks are very slender, while the leaves are narrow. The wild cane of New Guinea, *S. robustum*,[12] has stalks of medium thickness (1-inch or 2½-cm diameter), very hard, and 25 to 30 feet (7 to 9 m) in height. The nodes are swollen. The leaves are of medium width and rather long. The flowers are very similar to those of *S. officinarum*.

▪ 14.6 VARIETIES

Disease epidemics caused almost total destruction of the commonly cultivated sugarcane varieties in the United States by 1926. Disease-resistant varieties were introduced soon after that, while others were developed in the United States at

Canal Point, Florida. These replaced the noble canes susceptible to mosaic, root rot, and red rot.[3] The resistant varieties of *Saccharum officinarum* were developed by crossing the susceptible, high-sugar noble varieties with other species, especially *S. spontaneum* and *S. barberi*, which are resistant to the most serious sugarcane diseases. These hybrids were then backcrossed to the noble varieties. High-yielding disease-resistant plants that were high in sugar content were selected from the backcross progeny and increased by vegetative propagation. Practically all the improved domestic varieties were derived from all three of the species mentioned here. Current varieties are resistant to some mosaics and partly resistant to the sugarcane borer and red rot.

Sugarcane was crossed with sorghum in India, Taiwan, and the United States,[13] but no improved commercial varieties have yet evolved from this wide intergeneric cross.

▨ 14.7 FERTILIZERS

A 55-ton (50-MT) crop will remove about 65 to 90 pounds (30 to 40 kg) N, 50 to 60 pounds (23 to 27 kg) P_2O_5, and 150 pounds (68 kg) K_2O.[14]

Sugarcane responds mostly to nitrogen in Louisiana and Hawaii,[15, 16] and to some extent on the mineral soils in Florida. Nitrogen is not usually applied to peat and muck soils in Florida.[17] Rates of nitrogen application range from 80 to 140 pounds per acre (90 to 155 kg/ha), with the higher rates for the ratoon crops. Although nitrogen is important in increasing total yield, N applied within six to eight weeks before harvest may increase shoot moisture content and lower sugar yield.[18]

About 40 pounds per acre (45 kg/ha) of P_2O_5 and 80 pounds per acre (90 kg/ha) of K_2O are applied in Louisiana and Hawaii. On Florida mineral soils, about 30 pounds per acre (33 kg/ha) of P_2O_5 and 140 pounds per acre (160 kg/ha) of K_2O are applied. Although the Florida peat and muck soils require no nitrogen, about 18 pounds per acre (20 kg/ha) of P_2O_5 and 140 pounds per acre (160 kg/ha) of K_2O are applied. Higher rates of P_2O_5 and K_2O are applied to the plant cane crop and lesser amounts are applied to ratoon crops. Excessive phosphorus lowers the sugar content of the cane.

An application of 300 to 500 pounds (335 to 560 kg/ha) of sulfur in the row is helpful on Florida peat and muck soils with a pH above 6.6.[17] Sulfur raises the soil pH, which makes micronutrients more available. Lime applications are helpful on Louisiana soils that are below pH 6.0.[19]

Iron and zinc are the most likely micronutrients to be deficient in sugarcane as shown in Table 14.1.[18] Other micronutrients, particularly copper and boron, are required on Florida muck and peat soils, and manganese also is needed on soils of pH 7.5 or higher. These elements may be needed elsewhere, and the need can be determined by plant-tissue tests. In Florida, Hawaii, and some other areas of the world, yield may be improved by the application of silicon even though it is not an essential nutrient.[17, 18]

Dry fertilizers can be placed in the opened furrows during the planting operation. Nitrogen fertilizers can be applied in dry, liquid, or anhydrous forms. In Hawaii, liquid nitrogen solutions may be metered into the irrigation water, or urea may be broadcast on the fields from an airplane.

TABLE 14.1 Probable Nutrient Deficiencies in Sugarcane

Deficiency	Nutrient
Very common	N
Common	P, K, Fe, Zn
Occasional	Mg, S, Mn
Rare	Ca, Cu
Unknown	Mo, Cl, Na, B

14.8 ROTATIONS

Sugarcane is usually grown in a regular rotation with other crops. In Louisiana, after three cane crops have been harvested, the land usually is spring plowed and then planted to a green manure crop or left fallow. The green manure is plowed under in early fall after which the land is again planted to sugarcane. In the Florida everglades, the intervening green manure crop is omitted on the muck and peat soils that already contain excessive quantities of organic matter and nitrogen. In that area, rice is often planted in a rotation with sugarcane.[20] There is very little Hawaiian sugarcane grown in rotation with other crops.

14.9 SUGARCANE CULTURE

14.9.1 Seedbed Preparation

Conventionally, a field that will be planted to sugarcane is deep tilled with a moldboard plow or a chisel plow. In Louisiana, the soil is thrown into beds 1 to 2 feet (30 to 60 cm) wide and 5½ to 6 feet (168 to 183 cm) apart. A water furrow is left between the beds where the land is inclined to be wet. Level culture is practiced in Florida. The Hawaiian crop is usually grown on low beds. Deep tillage before planting is helpful on some Hawaiian soils.

Conservation tillage is satisfactory for sugarcane production. It is more successful in rainfed production where soil moisture conservation is more important than in irrigated fields or areas with high annual rainfall.

14.9.2 Planting

Sugarcane is propagated vegetatively by planting whole canes or 1- to 3-foot sections of cut canes. Cane stalks have an axillary bud (eye) every 2 to 6 inches (5 to 15 cm) at every node. Every bud will sprout when buried in moist soil. Within two to three weeks, shoots will emerge and, under favorable conditions, produce secondary shoots to give a dense stand of cane.

Stalks for seed cane are stripped and cut by hand or by machines. One to three continuous rows of cuttings (canes) are dropped into furrows by people riding a cart or trailer. Even when machine planters with cutters are used, the cut pieces are dropped into the newly opened furrows by people riding the planter. Fertilizers may be applied during this operation.[16] The spacing between rows ranges

FIGURE 14.3
Sugarcane field in Florida.
[Courtesy Richard Waldren]

FIGURE 14.4
Planting sugarcane.
[Courtesy Hawaiian Sugar
Planters' Association, used with
permission]

from 4 to 7 feet (120 to 215 cm) (Figure 14.3). The amount of seed cane planted varies from 1 to 4 tons per acre (2,200 to 4,500 kg/ha), depending upon the stalk diameter as well as upon the number of lines planted in a single furrow (Figure 14.4). The stalks are covered with only enough soil to avoid frost injury, 2 to 8 inches (5 to 20 cm). A light furrow turned from each side covers the seed canes. Higher potential yields from narrowing the row spacing and increasing the plant population are partly offset by higher costs for seed, cultural practices, and harvesting.[18]

In Louisiana, the greater part of the crop is planted for sugar production from August 1 to November 15.[21] In Florida, planting occurs from late August through January. Planting is usually done before sugar harvest to avoid labor conflicts. In the northern portion of the sugarcane-growing area, the crop may be planted in the spring. There, the canes are preserved over the winter in mats or windrows and covered with soil to protect them from the cold. Canes for fall planting are usually cut before the plants are mature enough for sugar harvest. Even when fall-planted, the new crop makes very little growth until spring. In Hawaii and other continuously warm areas, the cane may be planted at any time of the year.

Two to four crops of sugarcane are usually harvested before the field is plowed and replanted. The first crop is known as plant cane, while subsequent crops are known as stubble or ratoon crops. Yields from ratoon crops are usually lower. A common spring operation in Louisiana and Florida is to bar off—that is, throw aside the soil covering either the stubble or nonemerged fall-planted cane—using a plow, harrow, or other implement. This leaves the cane on a narrow bed that warms up quickly to start spring growth. After growth is well started, fertilizer is applied in furrows next to the row and the furrows are leveled. Seasonal cultivation and bed formation move so much soil toward the cane row that it is necessary to bar off each spring.

14.9.3 Weed Control

Weed control is necessary until the plants shade the ground. Herbicides are commonly applied shortly after planting, either banded over the row or broadcast. Herbicides applied at planting or for spring preemergent weed control include atrazine, clomazone, diuron, hexazinone, pendimethalin, and halosulfuron. Postemergent herbicides include 2, 4-D, dicamba, and paraquat. Cultivation can be used between rows.

14.9.4 Harvesting

Sugarcane begins to mature in Louisiana with the advent of cool nights in the fall, usually in October, and is most often harvested between October 15 and January 1. It continues to mature and increase in sugar content during October and November until its growth is checked by frost. Light freezes of 27°F (−3°C) kill leaf tissue and stop sugar accumulation. A freeze of 24°F (−4.5°C) damages the upper part of the stalk, which then must be topped off and discarded. A freeze below 23°F (−5°C) kills the whole stalk, which then deteriorates.

Most harvesting in developed countries is done with a mechanical harvester that cuts the stalk into pieces and conveys it to a waiting truck or trailer (Figure 14.5). The leaves must be removed from the plant prior to harvest, either by burning or stripping the plants by hand. Burning sugarcane before harvest removes about 50 percent of the trash that contributes nothing to the production of sugar. Every 1 percent of trash in harvested sugarcane results in a reduction of about 3 pounds of sugar per gross ton (1,500 g/MT) of sugarcane.[22] Currently, there is no mechanical way to remove the leaves and other trash during or before har-

FIGURE 14.5
Mechanical harvesting of sugarcane. [Courtesy John Wozniak/LSU AgCenter]

FIGURE 14.6
Topping, stripping, and cutting sugarcane by hand.

vest. Careful attention to weather conditions can greatly reduce smoke problems in surrounding areas. Caution must be used to insure that the fire does not spread to other property.

Hand harvesting consists of stripping off the leaves, removing the tops, and cutting off the stalks at the bottom (Figure 14.6). Specially designed knives and stripping implements are used.

The harvested stalks are hauled to a cane mill where the juice is extracted to make sugar or syrup. Practically all the cane except small lots is loaded mechanically onto trucks or trailers even when cut by hand (Figure 14.7). It is hauled to the mill on these vehicles or is reloaded for shipment. Some of it is shipped in cars on narrow-gauge

FIGURE 14.7
Mechanical loading of sugarcane. [Courtesy Hawaii Planters' Association, used with permission]

private railroads or in standard railroad cars or on barges by water. The cane is processed within a few days to avoid sucrose inversion and deterioration.[23, 24]

14.10 SUGAR AND SYRUP MANUFACTURE

In sugar manufacture, the sugarcane stalks are cut, sometimes shredded, and then passed between a series of heavy, grooved iron rollers to press out the juice.[5, 21, 25] Water is added between crushings to dissolve more of the sugar. Milk of lime (hydrated lime in water) is added to the strained juice after which the limed juice is heated with steam. The impurities unite with the lime and appear on the surface as scum or at the bottom of the clear juice as sediment. The clear juice, along with the settlings, is conducted into a series of multiple-effect vacuum evaporators where it is concentrated to semi-syrup that contains about 50 percent sucrose. This syrup is then pumped into other vacuum pans where it is crystallized by further evaporation. The mixture of sugar crystals and molasses, called massecuite, is then separated in a centrifuge chamber with perforated walls. The resulting raw sugar, which contains about 96 percent sucrose, is shipped to refineries.

A large sugar mill can grind 2,000 to 4,500 tons (1,800 to 4,100 MT) of cane in 24 hours and process the crop from 5,000 to 10,000 acres (2,000 to 4,000 ha) in a season. Fresh sugarcane in Louisiana contains an average water content of about 75 percent. The average extraction of juice from sugarcane under normal conditions is also approximately 75 percent for the dry-mill process. In the average juice, 15 to 20 percent is solids, of which some 2 to 3 percent is impurities (invert sugar, nitrogen compounds, salts, fats, wax, fiber, and so on), which leaves 12 to 17 percent sucrose.[25] A ton (907 kg) of average cane yields about 170 pounds (77 kg) of raw (96 percent) sugar in Louisiana, 200 pounds (90 kg) in Florida, 230 pounds (104 kg) in Hawaii, and 240 (109 kg) pounds in Cuba, together with about 5 gallons (19 l) of blackstrap molasses.

The juice of sugar plants (sugarcane, sorghum, or sugar beet) is usually tested with a Baume hydrometer, giving an arbitrary reading in degrees, or with a Brix hydrometer having a scale reading that approximates the percentage of total solids (degrees Brix) based upon the weight of sugar.[26] A similar measure, using only a few drops of juice, can be obtained with a refractometer that measures the refractive index of the liquid. It also has a scale giving a reading of the percentage of total solids. The percentage of sucrose in the juice is measured with a saccharimeter, a special type of polariscope with a sugar percentage scale. The purity of the juice, measured in percent, is then determined by the formula [(sucrose/Brix) × 100].

The primary manufactured products are sugar, syrup, and edible molasses. By-products consist of blackstrap molasses, bagasse or stalk residue, filter-press cake, and bagasse ashes. Blackstrap molasses is largely used for the production of alcohol and for livestock feed. About 500 pounds (227 kg) of bagasse that contains 50 percent or more moisture is obtained from a ton (907 kg) of cane. The fiber in bagasse is used for the manufacture of paper, hardboard, and insulating wallboard.[15] Bagasse is burned for fuel in the operation of some sugar mills, the ashes being returned to the sugarcane fields as fertilizer. Filter-press cake has a fertilizer value equal to barnyard manure.

Most of the sugarcane used for syrup is milled on three-roll mills operated by motor power. These have a lower pressure than the 30 tons per square inch (42 kg/mm^2) used on the rolls of commercial sugar mills, and the juice extraction is less. The juice is then concentrated into syrup in heated evaporating pans.

▪ 14.11 DISEASES

14.11.1 Mosaic

Mosaic is a virus disease caused by the sugarcane mosaic virus (SCMV). It was probably introduced into the United States around 1914.[8, 21, 27] It has been found in every major sugarcane-producing region in the world. Mottling of the leaves is the principal symptom. One type of mosaic appears as longitudinal streaks. Mosaic causes stunting, particularly in the noble canes. The disease is transmitted by several species of aphids. The only practical control for mosaic is the use of resistant varieties.

14.11.2 Red Rot

Red rot caused by the fungus *Glomerella tucumanensis* (*Colletotrichum falcatum*) is one of the major sugarcane diseases.[28] The disease attacks the stalks, stubble, rhizomes, and leaf midribs of the sugarcane plant. The disease is recognized by longitudinal reddening of the normally white internal tissues of the internodes, a discoloration that may extend through many joints of the stalk. Lesions on the leaf midrib are dark, reddish areas. Red rot causes poor stands of both plant and stubble crops, brings about destruction of seed cane in storage beds, and induces inversion of sucrose in mature canes. Among the control measures advocated are seed cane selection, summer planting, and the growing of resistant varieties.[8] Wounds caused by borers and other insects are avenues of entrance for the fungus.

14.11.3 Root Rots

Root rot diseases cause great damage to sugarcane, particularly in Louisiana. The principal cause appears to be *Pythium arrhenomanes.* It can also be caused by *Pachymetra chaunorhiza* or *Marasmius sacchari*, and *M. stenospilus*. Root rot is usually manifested by an unhealthy appearance, less tillering, closing in of the rows due to the lack of tillering, yellowing of the leaves, severe wilting, and even occasional death of young plants.[2, 8, 29] On heavy clay soils, destruction of the roots causes poor stands and failure of the crop. A high water table combined with a fine-textured soil and low winter and spring temperatures are probably indirect causes of severe outbreaks. Cultural practices that promote rapid growth are the most beneficial controls. Root rot damage to susceptible varieties has been markedly reduced by plowing under all cane trash, moderate applications of filter-press cake or barnyard manure, and good drainage.

14.11.4 Leaf Scald

Leaf scald is caused by the bacterium *Xanthomonas albilineans*. It is found in most of the major sugarcane-producing regions of the world. The most common symptom is a white pencil-line streak about 1 to 2 mm wide on the leaf that runs from the midrib to the leaf margin parallel to the veins. A diffused yellowish border of varying widths runs parallel to the pencil-line streak. The pencil-line may have areas of reddish discoloration along part of its length. As the disease progresses, necrosis develops from the tip or margin of the leaf.[30]

14.11.5 Ratoon Stunting Disease

Ratoon stunting disease is caused by the bacterium *Clavibacter xyli* subsp. *xyli*. It may be the most important disease affecting sugarcane production in the world. There are no visible external symptoms other than stunting caused by short internodes. Internally, there is an orange-red discoloration of the vascular bundles. It is easily spread by harvesting machines, knives, and infected canes. To insure that seed cane is not infected, it should be soaked in hot water (120°F, 50°C) for two to three hours before planting.[31]

14.11.6 Pineapple Disease

Pineapple disease occurs in almost all countries where sugarcane is grown. The disease is soil-borne and is caused by the fungus *Ceratocystis paradoxa* that causes seed pieces to decay after planting. The disease derives its name because the rotting seed pieces smell like ripe pineapples. The interior of the stalks will turn black. Prevention includes planting resistant varieties and planting seed cane during optimum growing conditions that result in rapid emergence. Since the disease enters the seed piece from the cut ends, planting pieces with at least three nodes increases the chance that the center buds will be able to develop normally.[32]

14.11.7 Rust

Sugarcane rust is caused by the fungus *Puccinia melanocephela*. This disease is now found almost everywhere sugarcane is grown. The leaves develop small, elongated, yellowish spots that are visible on both leaf surfaces, which increase in length, turn brown to orange-brown or red-brown in color, and develop a slight, but definite, chlorotic halo. The spots range from 2 to 10 mm in length but can reach 30 mm. They are seldom more than 1 to 3 mm in width. The spores are spread by wind. Infection is favored by several hours of wet leaves. Resistant varieties provide some protection from the disease.[33]

14.11.8 Other Diseases

Chlorotic streak, a virus disease, is controlled by planting resistant varieties or by soaking the seed cane in hot water for twenty minutes at 125°F (52°C).[12, 16, 27] Eye spot disease is caused by the fungus *Bipolaris sacchari*. Sugarcane smut disease is caused by the fungus *Ustilago scitaminea*. Pokkah boeng disease is caused by the fungus *Fusarium moniliforme* and/or *Fusarium moniliforme* var. *subglutinans*. It was first discovered in Java. The name is a Javanese term meaning "malformed or distorted top."

More than sixty additional diseases that attack sugarcane in other parts of the world are not very serious or do not occur in the United States.[34]

▓ 14.12 INSECT PESTS

14.12.1 Sugarcane Borer

The sugarcane borer, *Diatraea saccharalis*, may attack the growing point of the plant. In young plants, the inner whorl of leaves dies, resulting in a condition called "dead heart." In mature plants, the tops die or break over. Borings in the stalk reduce sugar yields and contribute to the decay of seed cane. The borers overwinter in sugarcane stalks and stubble and in coarse grasses.[35] Two parasitic wasps, *Agathis stigmatera* and *Cotesia flavipes*, can help control the larvae. It is important to monitor their populations to prevent unnecessary applications of insecticides. There are several insecticides labeled for the control of this pest.[36]

14.12.2 White Grubworms

Several species of grubworms attack sugarcane. The most predominant in Florida is *Ligyrus subtropicus*. It feeds on the roots and underground stems. The leaves will turn yellow, followed by stunting and lodging. There are several bacterial and fungal diseases that help control grubworms. Tillage, flooding, and reducing the number of ratoon crops are beneficial. Pesticides are not usually effective.[13]

14.12.3 Other Pests

Other insects attacking sugarcane include the sugarcane beetle *(Euetheola rugiceps)*, mealy bug *(Pseudococcus longispinus)*, termite, chinch bug *(Blissus leucopterus)*, wireworm, southern cornstalk borer *(Diatraea crambidoides)*, southwestern corn borer *(Diatraea grandiosella)*, and lesser corn stalk borer *(Elasmopalpus lignosellus)*.[23]

Arthropod predators, such as ants, spiders, and beetles, have been shown to help control sugarcane pests in Florida.[12] Careful monitoring of their numbers is necessary to minimize the need for insecticides.

REFERENCES

1. May, K. K., and F. H. Middleton. "Age-of-harvest, a historical review," *HI Planters Rec.* 58, 18(1972):241–263.

2. Artschwager, E., and E. W. Brandes. Sugarcane *(Saccharum officinarum)*, USDA *Agriculture Handbook* 122, 1958, pp. 1–307.

3. Brandes, E. W., and G. B. Sartoris. "Sugarcane: Its origin and improvement," *USDA Yearbook*, 1936, pp. 561–623.

4. Taggart, W. C., and E. C. Simon. "A brief discussion of the history of sugarcane: its culture, breeding, harvesting, manufacturing, and products," in *LA State Dept. Agr. and Immigration*, 4th ed., 1939, pp. 1–20.

5. Martin, L. F. "The production and use of sugarcane," in *Crops in Peace and War*, USDA Yearbook, 1950–51, pp. 293–299.

6. Patrick, W. G., Jr., and others. "Soil oxygen content and root development in sugarcane," *LA Agr. Exp. Sta. Bull.* 641, 1969, pp. 1–20.

7. Artschwager, E. "Morphology of the vegetative organs of sugarcane," *Jour. Agr. Res.* 60(1940):503–550.

8. Edgerton, C. W. *Sugarcane and Its Diseases*. Baton Rouge, LA: LA State U. Press, 1955, pp. 1–290.

9. Van Dillewijn, C. *Botany of Sugarcane*. Waltham, MA: Chronica Botanica Co., 1952, pp. 1–371.

10. Artschwager, E. "Anatomy of the vegetative organs of sugarcane," *Jour. Agr. Res.* 30(1925):197–242.

11. Cowgill, H. B. "Cross pollination of sugarcane," *J. Am. Soc. Agron.* 10:302–306.

12. Brandes, E. W., and C. O. Grassl. "Assembling and evaluating wild forms of sugarcane and closely related plants," *Proc. Int. Soc. Sugar Cane Technologists*, 1939, pp. 128–153.

13. Bourne, B. A. "A comparative study of certain morphological characters of sugarcane x sorgo hybrids," *Jour. Agr. Res.* 50(1935):539–552.

14. Barnes, A. C. *Sugarcane*. London: Leonard Hill Books, 1974.

15. Breaux, R. D., and others. "Culture of sugarcane for sugar in the Mississippi delta," *USDA Handbk.* 417, 1972, pp. 1–43.

16. Stokes, I. E. "The world of sugarcane," *Crops and Soils* 12, 8(1960): 16–17.

17. Rice, R. W., R. A. Gilbert, and R. S. Lentini. "Nutritional requirements for Florida sugarcane," in *Florida Sugarcane Handbook*, R. A. Gilbert, ed. *FL Coop. Ext. Serv.* SS-AGR-239, 2004.

18. Hunsigi, G. *Production of Sugarcane: Theory and Practice.* New York: Springer-Verlag, 1993.

19. Freeman, C. E., and others. "Sugarcane for sugar production guide," *FL Agr. Ext. Circ.* 353, 1971, pp. 1–16.

20. Schueneman, T. "Rice in the crop rotation," in *Florida Sugarcane Handbook*, R. A. Gilbert, ed. *FL Coop. Ext. Serv.* SS-AGR-239, 2004.

21. Hebert, L. P. "Culture of sugarcane for sugar production in Louisiana," *USDA Handbook* 262, 1964, pp. 1–40.

22. Legendre, B. L., F. S. Sanders, and K. A. Gravois. "Sugar production: Best management practices (BMPs)," *LA Coop. Ext. Serv. Publ.* 2833, 2000.

23. Humbert, R. F. *The Growing of Sugar Cane.* New York: Elsevier, 1968.

24. Laurtizen, J. I., and R. T. Balch. "Storage of mill cane," *USDA Tech. Bull.* 449, 1934, pp. 1–30.

25. Spencer, C. L., and C. P. Meade. *Cane Sugar Handbook,* 9th ed. New York: Wiley, 1963.

26. Walton, C. F., E. K. Ventre, M. A. McCalip, and C. A. Fort. "Farm production of sugarcane syrup," *USDA Farmers Bull.* 1874, 1938, pp. 1–40.

27. Abbott, E. V. "Sugarcane and its diseases," in *Plant Diseases*, USDA Yearbook, 1953, pp. 526–539.

28. Abbott, E. V. "Red rot of sugarcane," *USDA Tech. Bull.* 641, 1938, pp. 1–96.

29. Rands, R. D., and E. Dopp. "Pythium root rot of sugarcane," *USDA Tech. Bull.* 666, 1938, pp. 1–96.

30. Comstock, J. C., and R. S. Lentini. "Sugarcane leaf scald disease," in *Florida Sugarcane Handbook*, R. A. Gilbert, ed. *FL Coop. Ext. Serv.* SS-AGR-239, 2004.

31. Comstock, J. C., and R. S. Lentini, "Sugarcane ratoon stunting disease," in *Florida Sugarcane Handbook*, R. A. Gilbert, ed. *FL Coop. Ext. Serv.* SS-AGR-239, 2004.

32. Raid, R. N., and R. S. Lentini. "Pineapple disease of sugarcane," in *Florida Sugarcane Handbook*, R. A. Gilbert, ed. *FL Coop. Ext. Serv.* SS-AGR-239, 2004.

33. Raid, R. N., J. C. Comstock, and R. S. Lentini. "Sugarcane rust disease," in *Florida Sugarcane Handbook*, R. A. Gilbert, ed. *FL Coop. Ext. Serv.* SS-AGR-239, 2004.

34. Martin, J. P., and others. *Sugar Cane Diseases of the World.* New York: Elsevier, 1961.

35. Ingram, J. W., and E. K. Bynum. "The sugarcane borer," *USDA Farmers Bull.* 1884, 1941, pp. 1–17.

36. Cherry, R., and R. Gilbert. "The effect of harvesting and replanting on arthropod ground predators in Florida sugarcane," in *Florida Sugarcane Handbook*, R. A. Gilbert, ed. *FL Coop. Ext. Serv.* SS-AGR-239, 2004.

Wheat

15.1 ECONOMIC IMPORTANCE

Wheat (*Triticum* spp.), the most important crop in the world, is harvested somewhere every month of the year. In 2000–2003, wheat was grown on 527 million acres (213 million ha) with production of 21 billion bushels (577 million MT) or 40 bushels per acre (2,700 kg/ha). The chief producing countries are China, India, the United States, the Russian Federation, and France (Figure 15.1). The leading exporting countries are the United States, Canada, Australia, and Argentina. Wheat ranks third to corn (maize) and soybean in area in the United States. Average production in 2000–2003 was 2 billion bushels (55 million MT) on 50 million acres (20 million ha), or 40 bushels per acre (2,700 kg/ha). The leading states in wheat production are Kansas, North Dakota, Washington, Oklahoma, and Montana (Figure 15.2). All states except Florida, Alaska, Hawaii, and those in New England are commercial wheat producers.

The average yield of wheat in the United States in 1866 was only 11 bushels per acre (740 kg/ha) and had only improved to about 20 bushels per acre (1,344 kg/ha) by 1955. Modern cultivars with increased vigor, better resistance against pests and diseases, and the ability to respond to fertilizer have doubled yields since then. National wheat yields are highest in northwestern European countries having mild winters, cool summers, ample rainfall, and heavy rates of fertilization.

15.2 HISTORY OF WHEAT CULTURE

Wheat played an important role in the development of civilization. It is the preferred grain for food in advanced countries. The origin of cultivated wheat has been well documented by botanists, archaeologists, and geneticists. Evidence indicates that diploid and tetraploid wheat first appeared before 8000 BC in the "Fertile Crescent," an area found within the drainage basins of the Euphrates and Tigris Rivers in present day Syria and Iraq. Some authorities place the domestication of wheat before 10,000 BC.[1] Most wheat grown today is hexaploidal. It originated when emmer, a tetraploid, evolved from a wild type (*Triticum dicoccoides*) that carries the A and B genomes. A genome is a group of chromosomes, seven in each wheat genome. Mutations of emmer resulted in

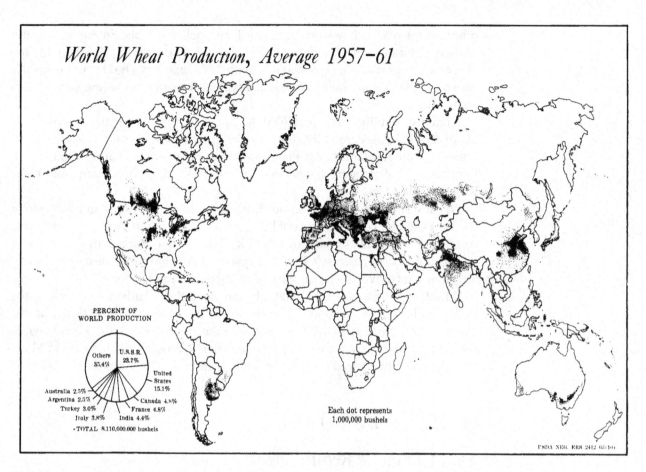

FIGURE 15.1
World wheat production. [Courtesy of USDA]

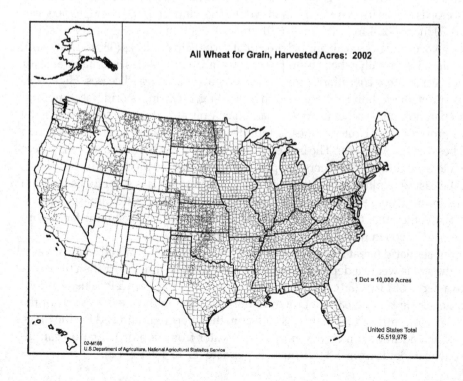

FIGURE 15.2
Distribution of wheat harvested in the United States in 2002. [Source: 2002 Census of U.S. Department of Agriducture]

383

other tetraploids, such as durum, poulard, and polish wheats. Emmer eventually crossed naturally with *Triticum taushii (Aegilops squarrosa)*, a diploid with the D genome, to produce hexaploid wheat with ABD genomes. The D genome greatly expanded the climatic range of wheat, adapting it to areas of severe winters and warm summers.[1]

Emmer was cultivated before 7000 BC. Specimens from Egyptian tombs dated about 3000 BC closely resemble the varieties grown today. Emmer was grown frequently in the northern Great Plains states before about 1930 and is still grown there today as its modern equivalent, durum. It is still an important crop in the Russian Federation.

Wheat was cultivated throughout Europe in prehistoric times and was one of the most valuable cereals of ancient Persia, Greece, and Egypt. Archaeologists have come upon carbonized grains of wheat in Pakistan and Turkey, in the tombs of Egypt, and in storage vessels found in many other countries. Wheat was already an important cultivated crop at the beginning of recorded history.

Wheat culture in the United States began along the Atlantic Coast early in the seventeenth century and moved westward as the country was settled. Captain Gosnold sowed wheat on Elizabeth Island in Buzzards Bay in 1602. It was sown in the Jamestown Colony as early as 1611 and in the Plymouth Colony by 1621.[2] Many varieties were introduced from foreign countries during the Colonial period.

■ 15.3 ADAPTATION

15.3.1 Climatic Requirement

Wheat is poorly adapted to warm or moist climates unless there is a comparatively cool dry season that favors plant growth and retards parasitic diseases. The world wheat crop is grown in temperate regions where the annual rainfall averages between 10 and 70 inches (254 to 1,780 mm), with the exception of drier irrigated areas. It is mostly grown in regions with annual precipitation of 15 to 45 inches (380 to 1,140 mm). High rainfall, especially when accompanied by high temperatures, is unfavorable for wheat because these conditions promote development of wheat diseases. High rainfall also promotes lodging; interferes with harvesting, threshing, storing, seeding, and soil preparation; and leaches nutrients, particularly nitrates, from the soil.[3] Most of the wheat grown in the United States is found where the rainfall averages less than 30 inches (760 mm) per year. The best quality bread wheat is produced in areas where the winters are cold, the summers comparatively hot, and the precipitation moderate.

In the eastern United States, the rainfall in certain seasons is sometimes too high for maximum wheat yields[4] because it promotes certain parasitic diseases such as *Septoria*. On the other hand, insufficient rainfall is the principal climatic factor limiting wheat yields in the major wheat areas of the world.

Under semiarid Great Plains conditions, good yields are obtained in most years when the soil is wet to a depth of 3 feet (90 cm) at seeding time.[5, 6, 7] When the soil is wet to a depth of 1 foot (30 cm) or less at the time of seeding, the yield is usually low.

Large acreages of wheat are grown under irrigation in the western parts of the southern and central Great Plains, as well as in the Intermountain and Pacific Coast states. Each additional inch (25 mm) of soil water from rain or irrigation can increase wheat yields from 1.3 to 6 bushels per acre (87 to 400 kg/ha).

WINTER HARDINESS Many years, low temperatures cause losses nearly as great as those from all wheat diseases combined. An average of one-eighth of the winter wheat seeded is abandoned, much of it due to winterkilling. Winterkilling can have one or more causes that include (1) heaving, (2) smothering, (3) physiological drought, and (4) direct effect of low temperature on plant tissues.[8] Wheat varieties differ widely in winterhardiness.[9]

Winter wheat in a hardened condition has survived temperatures as low as –40°F (–40°C) when protected by snow, and probably as low as –25°F (–32°C) without snow protection. Spring wheat may survive temperatures as low as 15°F (–9°C) in the early stages of growth. A light frost can cause sterility in wheat that has headed or is about to head. A light frost before wheat is fully mature produces blisters on the seed coat and stops further grain development. In general, spring wheat requires a frost-free period of about 100 days or more for safe production. Quick-maturing varieties can be grown where the frost-free period is 90 days or less in Canada and Alaska, where long days hasten flowering.

A cold-resistant plant must have the ability to harden, an inherited characteristic that occurs only after exposure to low temperatures[10] with ample sunlight. Hardening begins at temperatures below about 41°F (5°C); the threshold value above which growth commences. There is a concentration of sugars, amino nitrogen, and amide nitrogen in the tissues of hardy wheat varieties that lowers the freezing point to some extent and decreases protein precipitation.[11] A progressive increase in dry matter and sugar content of winter wheat plants follows low temperature hardening. The hardier the variety, the lower the moisture content of the leaves, and in hardened-off leaves, this moisture is held under a greater force.[12, 13] This condition is believed to be due to the hydrophilic colloid or bound-water content. More hardy wheat plants lose less water after thawing. Hardened plants become more susceptible to cold injury with advanced age and season.

FREEZING PROCESS IN PLANTS Freezing injury occurs when water is withdrawn from the plant cells into the intercellular spaces. This results in concentration of the cell sap, with an accompanying increase in salt concentration and increase in hydrogen ion concentration. These conditions bring about an irreversible precipitation of the cell proteins, as well as other changes in the organization of the cell. Death of the plant or plant tissue occurs when the protoplasm is disorganized beyond recovery. The plant loses water rapidly after thawing. Quick freezing of unhardened tissues ruptures the cells.

More winter wheat plants survive low temperatures in moist soils than in dry soils.[14] Plants in moist soils start to die more quickly, but dying progresses more slowly. An early erect habit of growth often indicates a lack of hardiness, but a recumbent habit of growth does not always indicate hardiness.[15] Injury in wheat plants progresses from the older to the younger leaves, with the crown and meristematic tissues being the most hardy parts.[12] Late, fall-sown wheat with shallow roots heaves more readily than well-rooted wheat before the onset of winter.[16]

15.3.2 Soil Requirements

Wheat is best adapted to fertile, medium- to heavy-textured soils that are well drained. Silt and clay loam, in general, give the highest yields of wheat, but wheat can also be grown successfully on either clay soil or fine, sandy loam. In general, wheat does not yield well on very sandy or poorly drained soils. Wheat tends to lodge on rich bottomlands.

15.3.3 Adjustments to the Environment

FLOWERING Short days or high temperatures stimulate tillering and leaf formation but delay the flowering of wheat plants. Although wheat is a long-day plant, quick-maturing, or photo-insensitive varieties can be grown to maturity in any day length ranging from the twelve-hour days of the equatorial highlands of Ecuador and Kenya to the twenty-hour days of the "land of the midnight sun" at Rampart, Alaska, at latitude 66° N. Wheat plants grown quickly under continuous artificial light are poorly developed and produce small heads and very little grain.

FACTORS AFFECTING WHEAT QUALITY Much of the superior protein content and baking quality of hard wheat is due to the environment in which they are grown. When grown in the soft wheat regions, the grains of hard wheat are starchy or "yellow berry"; they are only slightly if at all higher in protein content and baking quality than are the soft wheats.[17] However, differences in protein quality among varieties may account for considerable differences in the baking qualities of flours of the same protein content. Also, the protein percentage of some varieties may be 1 or 2 percentage points higher than others. In general, high-yielding wheat is expected to be relatively low in protein content except where soil nitrates are abundant.

Most of the plant nutrients derived from the soil are taken up before the plant blooms,[18] and later translocated to the kernel,[18] but nitrogen uptake may continue until the grain is ripe.[19] The ratio of protein to starch in the wheat kernel is largely determined by moisture, temperature conditions, and available nitrates in the soil at the blossom and postfloral periods. The fruiting (postfloral) period tends to be prolonged when the weather is cool and ample soil moisture is available to the plant. A relatively large amount of starch tends to be deposited, resulting in a plump, starchy kernel of low protein content.[18, 20] When sufficient nitrogen has been absorbed before or at the blossom stage, and when moisture conditions are such that the postfloral period is not prolonged, the grain tends to be plump and fairly high in gluten. Starch deposition tends to be affected more than gluten formation in hot, dry weather. This results in small-berried kernels rich in gluten. Dark, hard, and vitreous kernels in bread wheat are usually an indication of comparatively high protein content. When the development of wheat on soil containing ample nitrates is stopped prematurely by drought, the shrunken grain is high in protein and low in starch. Grain shrunken as a result of stem rust disease is low in both starch and protein, because both nitrogenous and carbohydrate compounds are withdrawn from the wheat plant by the growing rust fungus.

In North Dakota research, the protein content of wheat was above 13 percent when the average July temperature was greater than 70°F (21°C).[21] Lower protein content was obtained in years when the July temperature was slightly above 65°F (18°C) and the June temperatures relatively low. Temperatures above 90°F (32°C) lower the baking quality of high-protein wheat.

The influence of the soil on the protein content of wheat is less marked than is that of climate.[20] In many cases, the supply of available soil nitrogen is insufficient, and part, or all, of the endosperm is yellow berry (soft and starchy). Wheat of high protein content is found on soils of high nitrogen content with low moisture content at time of maturity.[20] The addition of irrigation water or prolonged heavy rainfall that leaches out soil nitrates can yield wheat of lower protein content. Applications of nitrogenous fertilizers, particularly inorganic forms, usually result in wheat with higher protein content. Nitrogen applications, particularly foliar sprays of urea solutions, as late as the blossom period, generally will increase the protein in the kernel.

15.4 BOTANICAL DESCRIPTION

Wheat is classified in the grass tribe *Hordeae (Triticeae)* and in the genus *Triticum*. This genus is characterized by two- to five-flowered spikelets placed flat at each rachis point of the spike.

Wheat is a summer annual or winter annual grass (Figures 15.3 and 15.4), with a spike inflorescence consisting of a sessile spikelet placed at each notch of the zigzag rachis. A spikelet is composed of two broad glumes and one to several florets. A floret consists of a lemma, a palea, and, at maturity, a caryopsis. The awn arises dorsally on the tip of the lemma in awned varieties. The caryopsis or grain has a deep furrow (crease) and a hairy tip or brush. The color of the grain may be amber, red, purple, or a creamy white, in different varieties. The purple sorts are not grown in the United States.

The plants normally produce two or three tillers under typical crowded field conditions, but individual plants on fertile soil with ample space may produce as many as 30 to 100 tillers. The average spike (head) of common wheat contains 25 to 30 grains in 14 to 17 spikelets (Figure 15.5). Large spikes may contain 50 to 75 grains. The shape of individual kernels depends in part upon the degree of crowding within a spikelet. The spikelets at the base and tip of the spike usually contain small kernels. Within a spikelet, the second kernel from the base usually is the largest, and the first kernel is next in size but, when present, the third, fourth, and fifth kernels are progressively smaller. A pound of wheat may contain 8,000 to 24,000 kernels (17 to 53/g). The legal weight of a bushel of wheat for the commodity market is 60 pounds (770 g/l), but the average test weight is about 58 pounds (745 g/l). The maximum test weight is about 67 pounds (860 g/l) and the minimum about 40 pounds per bushel (515 g/l).

15.4.1 Pollination

The time required for the wheat flower (Figure 4.1) to open fully and the anthers to assume a pendant position is only about three or four minutes.[22] About 86 percent of the flowers bloom in daylight. Flowers bloom at temperatures ranging from 56 to 78°F (13 to 25°C).

Although wheat is normally a self-pollinated plant, natural cross pollination occurs in 1 to 4 percent of the flowers.[23] Wheat that is partly self-sterile from chromosomal irregularities or adverse environment sometimes becomes crossed extensively.[24] In one season, about six times as much natural crossing occurred in the secondary heads as in the primary heads.

Blooming begins in the spikelets slightly below the middle of the spike and proceeds both upward and downward. Within a spikelet the upper flowers bloom last.

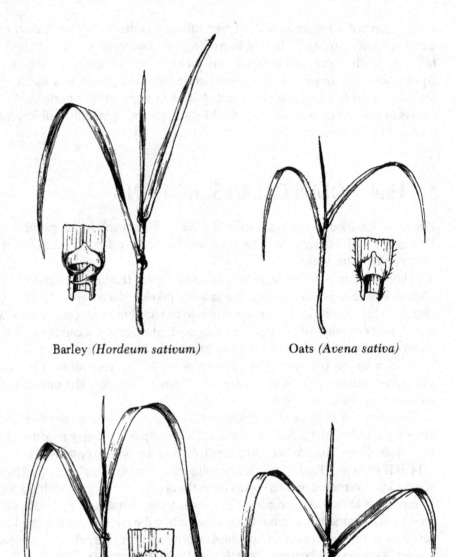

Barley *(Hordeum sativum)*

Oats *(Avena sativa)*

Wheat *(Triticum aestivum)*

Rye *(Secale cereale)*

FIGURE 15.3
Distinguishing characters of the leaves, sheaths, ligules, and auricles of young cereal plants. Barley has long clasping smooth auricles; wheat has shorter hairy auricles; rye has very short auricles; the oat sheath has no auricles. The sheaths of rye and oats and the leaf margins of oats are hairy. The shapes of the shield-like ligules at the base of each leaf blade also are different.

Under ordinary conditions a wheat spike completes its blooming within two or three days after the first anthers appear.

15.4.2 The Wheat Kernel

The wheat kernel or berry is a caryopsis, 0.1 to 0.4 inch (3 to 10 mm) long and 0.1 to 0.2 inch (3 to 5 mm) in diameter. The structure of the kernel is shown in Figure 15.6. The pericarp of the kernel consists of several layers that have developed from the ovary wall. These layers include the cuticle, epidermis, parenchyma, and cross layer.

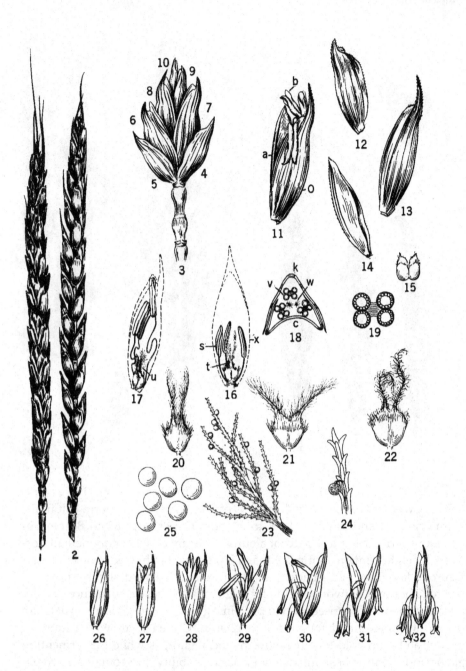

FIGURE 15.4
The wheat inflorescence. *(1)* Spike, dorso-ventral view. *(2)* Spike, lateral view. *(3)* Spikelet, laternal view and subtending rachis. *(4)* Upper glume. *(5)* Lower glume. *(6 to 10)* Florets: 7 is the largest and 6 second largest; 8, 9, and 10 are progressively smaller. *(11)* Floret, lateral view, opened for anthesis. *(12)* Glume, lateral view. *(13)* Lemma, lateral view. *(14)* Palea, lateral view. *(15)* Lodicules, which swell to open the glumes. *(16)* Floret before anthesis, showing position of stamens *(s)* and pistil *(t)*, enclosed in lemma and palea *(x)*. *(17)* Floret at anthesis showing position of pistil *(u)* and elongating filaments of the stamens. *(18)* Cross section of floret: palea *(c)*, lemma *(k)*, stamen *(v)*, and stigma *(w)*. *(19)* Cross section of anther. *(20)* Pistil before anthesis. *(21)* Pistil at anthesis. *(22)* Pistil after fertilization. *(23)* Portion of stigma (greatly enlarged) penetrated by germinating pollen. *(24)* Tip of stigma hair (greatly enlarged) penetrated by germinating pollen grain. *(25)* Pollen grains (enormously enlarged). *(26 to 32)* Florets during successive stages of blooming and anthesis. Time required for stages 26 to 31 is about 2 to 5 minutes, for stages 26 to 32 about 15 to 40 minutes.

The last three layers are sometimes referred to as the epicarp, mesocarp, and endocarp, respectively. The testa (or true seed coat) is often called the perisperm. The thickness of all of these layers may not exceed 3 percent of the thickness of the grain. The testa of red-kerneled varieties contains a brownish pigment, but that of white or amber varieties is practically unpigmented. The aleurone is the single layer of large cells, rectangular in outline, comprising the outer periphery of the endosperm. The rectangular cells of the aleurone are polyhedral when viewed from the top or outside of the grain. Aleurone cells contain no starch; their chief contents are aleurone grains composed of nitrogenous substances, probably proteins, but not gluten. In milling wheat, the bran fraction consists of all of the pericarp layers—the testa, nucellus, and aleurone—and some adhering endosperm.

FIGURE 15.5
Spikes of four varieties of common wheat and one variety of club wheat *(from left)*: fully awnless, *(2nd and 3rd)* awnless (tip awned), fully awned, and club wheat with tip awns.

A plump, well-developed kernel consists of about 2.5 percent germ (plumule, scutellum, radicle, and hypocotyl), 9 to 10 percent pericarp, 85 to 86 percent starchy endosperm, and 3 to 4 percent aleurone. Shrunken wheat of one-half normal kernel weight contains not more than 65 percent starchy endosperm. Under the aleurone layer, the endosperm consists of cells filled with starch grains cemented together by a network of gluten.[25] Gluten is a cohesive substance usually considered as being comprised of two proteins, glutenin and gliadin.[26] There are two types of glutenin and four types of gliadin. Gluten gives wheat flour its strength or ability to hold together, stretch, and retain gas while the fermenting dough expands. Only wheat and rye flour has this ability. Rye contains a protein similar to gluten.

Following fertilization of the egg nucleus, the development of the wheat kernel[27, 28, 29, 30] proceeds by cell division to produce the embryo. The embryo reaches its approximate final size in advance of the full development of the endosperm. The integuments of the ovary become modified and form the layers of the pericarp. Repeated cell division following fertilization of the endosperm nucleus produces endosperm cells. These cells gradually fill with starch grains and proteins, starting from the outer periphery, including the region of the crease, toward the interior of the endosperm (Figure 15.6). Sugars synthesized in the leaves and culms are transported to the endosperm to be deposited finally as starch. The grain increases in percentage of starch until it is fully mature. Amino acids, synthesized in the leaves and transported to the endosperm, change to proteins as the grain begins to dry. A

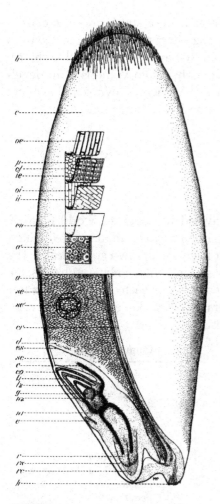

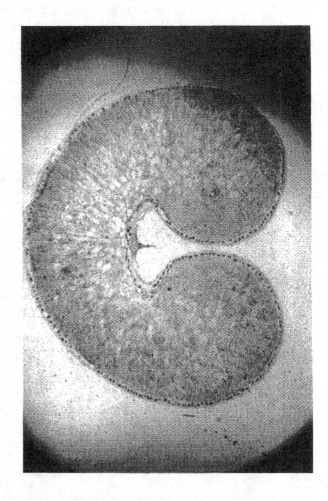

FIGURE 15.6

(Left) Wheat kernel. Brush *(b)*. Pericarp consists of cuticle *(c)*, outer epidermis *(oe)*, parenchyma *(p)*, cross-layer *(cl)*, and inner epidermis *(ie)*. Testa consists of outer integument *(oi)* and inner integument *(ii)*. Nucellar layer—epidermis of nucellus *(en)*. Endosperm consists of the aleurone *(a and a')*, starch, and gluten parenchyma *(se and se')*, and crushed empty cells of endosperm *(d)*. Germ consists of scutellum *(sc)*, epithelium of scutellum *(es)*, vascular bundle of scutellum *(v)*, coleoptile *(co)*, first foliage leaf *(l₁)*, second foliage leaf *(l₂)* growing point *(g)*, second node *(n₂)*, first node *(n₁)*, epiblast *(e)*, primary root *(r)*, root sheath or coleorhiza *(rs)*, and rootcap *(rc)*. Hilum *(h)*. The bran layer is comprised of the pericarp, testa, nucellar layer, and aleurone. [Drawn by M. N. Pope] *(Right)* Cross section of wheat kernel showing the crease and endosperm cells. The dark cells near the periphery constitute the aleurone layer of the endosperm.

part of the nitrogen, phosphorus, potassium, and other mineral constituents absorbed by the plant before the grain is formed are transported to the kernel.

INFLUENCE OF AWNS ON DEVELOPMENT OF WHEAT KERNELS In general, the awn of wheat has been considered a useful structure under certain climatic conditions.[31, 32, 33, 34] Awned varieties often predominate in areas of limited rainfall and warm temperatures, but awnless varieties may yield equally well in cool, humid, or irrigated areas. Many Australian cultivars are awnless. The presence of awns on the florets of wheat can result in the production of heavier kernels[35] as well as higher test weights. The awns serve as a depository of ash. Thus, they can remove substances (possibly silicates) from the transpiration system of the plant at filling time that otherwise might interfere with rapid movement of material into the grain.[32] De-awned heads transpire less than do awned heads.[36] Transpiration through the awns helps to dry out the ripe grains. The tips of awnless spikes are often blasted in hot, dry weather. This suggests that a cooling effect produced by the awns is helpful to the developing grains. Awns contain stomata and considerable chlorophyll. They contribute considerable carbohydrates by photosynthesis to the developing grain, particularly during drought stress.[37] A study of awned and awnless isogenic lines of wheat showed that photosynthesis and transpiration were both higher in the awned line.[38]

◼ 15.5 SPECIES OF WHEAT

The species and groups (subspecies) of cultivated wheat in the world follow.[39] Only common, club, and durum wheat, are currently grown in the United States, but poulard, emmer, spelt, and Polish wheat have been grown sporadically. The groups shown were long regarded as separate species, but members with the same chromosome number intercross readily. They differ from each other by only a few hereditary genes so do not merit rank as separate species. Some of the wild species listed may be ancestors of the cultivated types.

Species	Group	Common Name
Triticum monococcum L.	Diploid (2N = 14 chromosomes) AA genomes	einkorn
T. turgidum L.	Tetraploid (2N = 28 chromosomes) AA BB genomes	
	dicoccon (dicoccum)	emmer
	durum	durum wheat
	turgidum	poulard (branched) wheat
	polonicum	Polish wheat
	carthlicum	Persian wheat
T. timopheevii Zhok.	timopheevii	none
	Zaukovski	none
T. aestivum	Hexaploid (2N = 42 chromosomes) AABBDD genomes	
	spelta	spelt
	vavilovii	none
	aestivum (vulgare)	common bread wheat
	compactum	club wheat
	sphaerococcum	shot wheat

Some Wild Relatives of the Wheats

Triticum aegilopoides—AA genomes
T. speltoides (Aegilops speltoides)—BB genomes
T. taushii (Aegilops squarrosa)—DD genomes
T. cylindricum (Aegilops cylindrica)
T. diccocum (T. dicoccoides)—wild emmer

The spike of common wheat is usually dorsally compressed. The spikelets have from two to five flowers, but usually contain only two or three kernels at maturity. The kernels, which thresh free from the chaff, are either red or white in color and soft or hard in texture. The so-called white kernel is in reality light yellow or am-

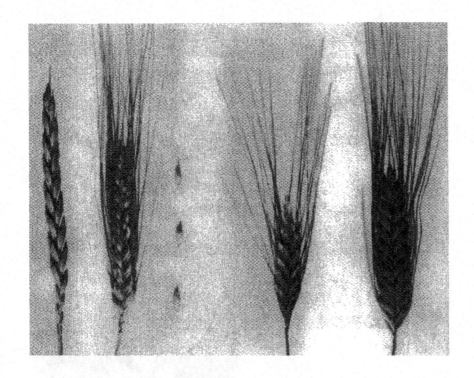

FIGURE 15.7
Left to right: Spikes of spelt, durum wheat, timopheevi, and winter emmer.

ber in color. Common wheat can be a winter or summer annual, awned or awnless, and white, brown, or black in glume color.

The plants of durum wheat are summer annuals. The spikes are compact and laterally compressed (Figure 15.7). The glumes are persistent and sharply keeled and the lemmas awned in the commonly grown varieties. The long, stiff awns, thick compact spikes, and long glumes often cause people to confuse durum wheat with barley. The amber or red kernels are usually long and pointed and the hardest of all known wheat. The kernels of poulard wheat are similar but are shorter, thicker, and not so hard. Some poulard wheat has branched spikes (Figure 15.8).

Club wheat plants may be either winter or summer annual and either tall or short. The spikes, which are usually awnless, may be elliptical, oblong, or clavate in shape. They are short, compact, and laterally compressed. The spikelets usually contain five and sometimes as many as eight fertile florets. The kernels of club wheat, which may be either white or red, are small and laterally compressed.

Most of the kernels of spelt, emmer, and einkorn remain enclosed in the glumes after threshing (Figure 15.9). Emmer is distinguished from spelt by the shorter, denser spikes that are laterally compressed. In the United States, emmer and spelt are used for livestock feed, but they have been grown for food in Europe. Einkorn (Figure 15.10) differs from the other species in that many of the spikelets contain only one fertile floret. Einkorn, or *one-grain*, wheat is grown sparingly in Europe. The wild form *(T. boeticum)* is still abundant in southwestern Turkey. Emmer, spelt, and einkorn have red kernels.

Because of their unusual appearance, Polish and poulard wheat have been fraudulently exploited in the United States (Figures 15.8 and 15.10). Polish wheat, sometimes called corn wheat, has large lax and awned heads that sometimes attain

FIGURE 15.8
Composite or branched
head of poulard wheat.

a length of 6 or 7 inches (15 to 18 cm). The chaff is extremely long, thin, and papery. The kernels, which are sometimes ½ inch (13 mm) long and very hard, frequently have been confused with rye.

▦ 15.6 WHEAT CLASSES

Under the official grain standards of the United States, wheat is separated into seven commercial classes: (1) durum, (2) hard red spring, (3) hard red winter, (4) soft red winter, (5) white, (6) unclassed, and (7) mixed. Red durum is included in unclassed. Emmer, spelt, einkorn, Polish, and poulard wheat are not considered

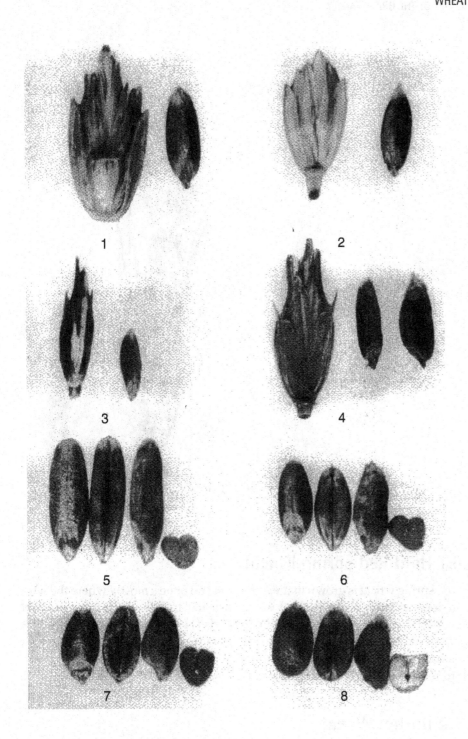

FIGURE 15.9
Spikelet and kernel of: *(1)* spelt, *(2)* spring emmer, *(3)* einkorn, and *(4)* timopheevi wheat. Wheat kernels: *(5)* Polish, *(6)* durum, *(7)* poulard, and *(8)* club.

wheat in the official grades. The distribution of winter and spring wheats are shown in Figure 15.11.

The percentage of the production of the different wheat classes in the United States in 1999 was approximately: hard red winter, 45.2; hard red spring, 20.5; soft red winter, 19.4; white, 10.6; and durum, 4.3.

FIGURE 15.10
Spikes and glumes of Polish wheat *(left)* and einkorn *(right)*.

15.6.1 Hard Red Spring Wheat

Hard red spring wheat is grown in the North Central States; mostly where the winters are too severe for production of winter wheat.[40] In 2000–2003, production of spring wheat averaged about 500 million bushels (14 million MT). Leading states are North Dakota, Minnesota, Montana, Idaho, and South Dakota. Hard red spring is the standard wheat for bread flour.

Hard red spring wheat is also grown in Canada, Russia, and Poland.

15.6.2 Durum Wheat

Durum wheat is produced in the same general region where hard red spring wheat is grown, but the main area lies west of the Red River Valley in North Dakota. Practically all durum wheat production is in North Dakota, Montana, California, Arizona, and South Dakota (Figure 15.12). In 2000–2003, production of durum wheat in the United States averaged about 92 million bushels (2.5 million MT). Durum wheat is used in the making of semolina, from which macaroni, spaghetti, and similar products are manufactured. Durum wheat is grown extensively in North Africa, southern Europe, and Russia.

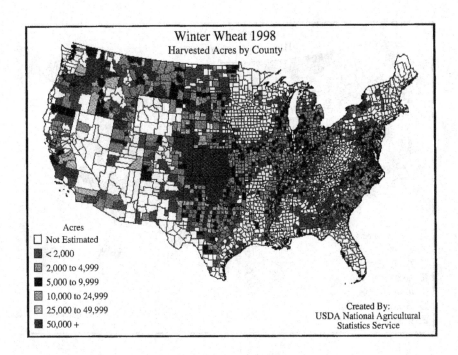

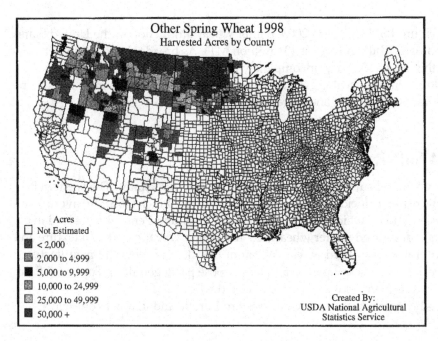

FIGURE 15.11
Top: Winter Wheat harvested in 2002 [Source: 2002 Census of U.S. Department of Agriculture]; *Bottom:* Spring Wheat harvested in the United States in 2002. [Source: Census of U.S. Department of Agriculture]

15.6.3 Hard Red Winter Wheat

Hard red winter wheat is especially adapted to the central and southern Great Plains where the annual rainfall is less than 35 inches (890 mm). Kansas, Oklahoma, Washington, Texas, and Ohio lead in the production of hard red winter

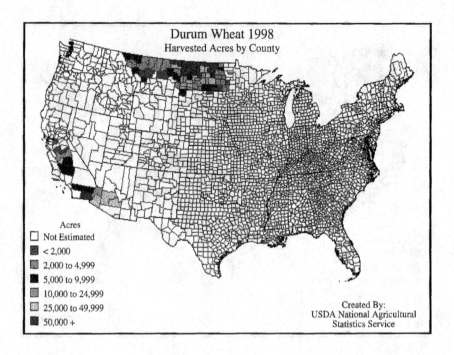

FIGURE 15.12

Distribution of durum wheat in 2002. [Source: 2002 Census of U.S. Department of Agriculture]

wheat (Figure 15.11). In 2000–2003, winter wheat production in the United States averaged over 1 billion bushels (31 million MT). As bread wheat, the better varieties of this class are nearly or completely equal to hard red spring wheat.

Hard red winter wheat is grown extensively in the southern part of Russia, the Danube Valley of Europe, and Argentina.

15.6.4 Soft Red Winter Wheat

Soft red winter wheat is grown principally in the eastern states. The western border of this region coincides roughly with the line of a 30-inch (760 mm) average annual precipitation.[41] Ohio, Missouri, Indiana, Illinois, and Pennsylvania lead in the production of soft red winter wheat. Soft red winter wheat is softer in texture and lower in protein than hard wheat. Wheat of this class is generally manufactured into cake, biscuit, cracker, pastry, and all-purpose packaged flour. For bread flour, it sometimes is blended with flour of hard red wheat.

Soft red winter wheat is grown in Western Europe and in other continents.

15.6.5 White Wheat

White wheat can be either hard white or soft white. Soft white wheat is mostly grown in the far western states and in the northeastern states. Most production is in Washington, Oregon, Idaho, Michigan, New York, and California. Hard white wheat has been grown in the Great Plains in recent years with the development of varieties adapted to that region. About 14 percent of white wheat consists of varieties of club wheat grown in the northwestern states. White wheat is used principally for pastry flour, oriental noodles, bulgur, tortillas, and shredded and puffed breakfast foods. Hard white wheat can also be used for bread and hard rolls.

White wheat is grown in northern, eastern, and southern Europe, and in Australia, South Africa, western South America, and Asia.

■ 15.7 WHEAT RELATIVES

Wheat (genus *Triticum*) is related botanically to certain other grass genera that can be crossed with wheat. Several species of *Aegilops* crossed with wheat include goatgrass *(Aegilops cylindrica)*, a weedy plant. Wheat has been crossed with several species of the genus *Agropyron* (wheatgrass). Crosses of wheat with western wheatgrass *(Agropyron smithii)* and slender wheatgrass *(A. trachycaulum)* have not been successful. Successful crosses with several other species of *Agropyron*, including certain collections of quackgrass (A. *repens*), have been reported. Wheat was crossed with *Agropyron elongatum* in the Russian federation, Canada, and the United States in an endeavor to produce perennial wheat. Thus far, the long-lived perennial types obtained have been decidedly grasslike, whereas the more wheatlike selections have been annuals or short-lived perennials. The desirability of perennial wheat is not fully established since a crop that can grow anytime in the season when conditions are favorable would leave little opportunity for accumulation of moisture or nitrates in the soil.

■ 15.8 HYBRID WHEAT

Extensive effort was devoted to the development of hybrid seed wheat in the 1970s and 1980s, but it was not a commercial success and is not readily available today. The main limitations are the poor sets of seed often obtained in the crossing field and the larger amounts of seed per acre for optimum yields in comparison with those of hybrid corn and sorghum. Both cytoplasmic and genetic male-sterile lines are available. Wheat stigmas are less exposed than those of some other crops and probably intercept less air-borne pollen. Better results may yet be obtained.[42, 43]

Wheat has been crossed with rye, and selected progenies, called triticale, have characters and chromosomes of both parents. (See Chapter 16.) Some wheat varieties carry rye genes, the result of crosses between wheat and triticale. Wheat has also been crossed with the grass genus *Haynaldia*. Reputed crosses of wheat and barley have not been authenticated.

Formerly, there was a popular belief that wheat changed into cheat *(Bromus secalinus)* when subjected to winter injury or when the developing spikes were mutilated in the boot stage. This was refuted by botanists and agricultural authorities more than a century ago. Reports of the reputed change came from areas where cheat (chess) is a common weed in grain fields and waste places and appears in abundance where wheat is thinned out by winterkilling. Cheat formerly was grown extensively as a hay crop in western Oregon. The botanical relationship between wheat and cheat appears to be too remote to permit intercrossing.

■ 15.9 FERTILIZERS

One hundred bushels (2,720 kg) of wheat will require about 120 pounds (54 kg) nitrogen, 56 pounds (25 kg) P_2O_5, 31 pounds (14 kg) K_2O, 13 pounds (6 kg) magnesium, 12 pounds (5.5 kg) sulfur, and 5 pounds (2.3 kg) calcium.[44, 45, 46] About 70 percent of the nitrogen, 75 percent of the phosphorus, and 15 percent of the potassium taken up by wheat plants is deposited in the grain.[47] In humid or irrigated regions, nitrogen applications of 30 to 120 pounds per acre (34 to 134 kg/ha) are applied, except when wheat follows a legume forage crop. The usual response is one bushel (27 kg) of wheat to each 3 to 5 pounds (1.4 to 2.3 kg) of nitrogen. Applications of 10 to 30 pounds (11 to 34 kg) of nitrogen are often beneficial to wheat on sandy soils in semiarid regions, except in the drier seasons. Excessive applications of nitrogen often reduce wheat yields not only by increasing plant lodging, but also by delaying the maturity of the crop so that it is subject to greater damage from rust. Semidwarf wheat with strong culms can utilize more nitrogen without lodging, resulting in higher yields.[48]

Fall applications of nitrogen are especially beneficial in continuous wheat culture where a heavy growth of straw has been turned under. One method to determine if adequate nitrogen is available to a wheat crop is to measure the grain protein. Fields that produced wheat with protein between 11.1 and 12 percent may be deficient in nitrogen.[49] Summer preplant and spring topdressing are the recommended times of nitrogen application for wheat in Kansas.[50] Preplant application allows for deeper placement to avoid poor utilization due to dry soil. Spring applications on the other hand have the advantage of better knowledge of moisture supply and crop condition prior to expenditure of money for nitrogen and a shorter period of capital tie-up prior to harvest.

Fall applications of 30 to 50 pounds per acre (34 to 56 kg/ha) of phosphate (P_2O_5) are helpful on most soils in humid regions as well as occasionally elsewhere. Potassium applications of 40 to 50 pounds per acre (45 to 56 kg/ha) of K_2O are desirable in regions of heavy rainfall.

In humid winter wheat regions, it is customary to apply a mixed fertilizer that contains some nitrogen in the fall during seedbed preparation or seeding. The remainder of the nitrogen is applied as a top dressing in early spring. Early spring applications give the greatest increases in yield.[51] Later applications may give greater increases in the protein content of the grain.

■ 15.10 ROTATIONS

In the eastern states where winter wheat is often seeded on corn stubble, the rotation is likely to contain at least one legume and one or more cultivated crops.

In the Corn Belt, an efficient rotation for much of the area is winter wheat, soybean, and corn. Where lespedeza is an important crop, it may be seeded in the spring on fall-sown wheat. After the wheat harvest, the lespedeza can produce a crop of hay or seed. A common rotation is corn; wheat; lespedeza for pasture, hay, or seed; wheat.

In the Cotton Belt, wheat often cannot be sown immediately after cotton because of the late maturity of the latter crop. A rotation satisfactory for this area is wheat followed by soybeans, cotton, and corn.

In western Kansas, rotations have failed to show any marked gain in wheat yields over continuous cropping, except where wheat follows fallow. A suggested rotation is grain sorghum, fallow, wheat. Continuous winter wheat with no tillage has produced good yields in some areas where weeds and pests are not serious.

Alternate winter wheat and fallow are a practical sequence where the annual precipitation is less than about 22 to 24 inches (560 to 610 mm) in the southern Great Plains, 18 to 20 inches (457 to 508 mm) in the central Great Plains, and 14 to 16 inches (356 to 406 mm) in the northern Great Plains as well as in the inter-mountain region.[52]

In the typical wheat-fallow system, fourteen of twenty-four months are fallow (one crop in two years). With conservation tillage practices that store additional water, a spring crop can replace the fallow period with two crops produced in three years.[53] Several alternatives are possible in the central Great Plains. Sorghum, corn, and sunflower have proven to be profitable crop alternatives. Sorghum is more suited to the semiarid Central Plains, but corn is more productive in northern Kansas and Nebraska. Sunflower is adapted to most regions. Rotations with two crops in three years are the most common, but with sufficient soil water, three crops in four years is possible.

Spring wheat follows an intertilled crop such as corn, potato, or sugar beet most advantageously in the northern spring wheat region. Much of it is sown after wheat or other small grains, or after fallow, because of the relatively small acreage of intertilled crops. Fallow helps in the control of weeds. Land to be fallowed usually is not tilled before early spring of the year after the previous crop harvest. The undisturbed crop residues help to retain snow and rain as well as to reduce wind and water erosion.[52] Green manure crops in preparation for wheat have been unprofitable in semiarid regions.[5]

▦ 15.11 WHEAT CULTURE

15.11.1 Tillage and Seedbed Preparation

Tillage for wheat in the semiarid regions of the United States is used to prevent weeds during fallow and prepare the seedbed. Since soil moisture is often limiting, minimum tillage reduces loss of soil moisture during and following the tillage operation. The first tillage is done with undercutting blade implements, chisels, or one-way disk plow. Subsequent tillage for weed control before seeding is done mostly with the rotary rod weeder or various blade implements.

In a wheat-sorghum-fallow rotation, the amount of soil water was greater at planting time for both wheat and sorghum with minimum tillage, compared to conventional tillage. Wheat yield was increased about 6 bushels per acre (400 kg/ha) under minimum tillage; and grain sorghum yields were 14 bushels per acre (878 kg/ha) greater than under conventional tillage.[53] One of the biggest advantages of minimum tillage is the additional water from snow captured in standing stubble. Two feet (610 mm) of snow can provide 1.6 inches (41 mm) of stored soil water assuming 80 percent efficiency of water capture.[54] Table 15.1 illustrates the effects of tillage on yields of winter wheat at three locations in the Great

TABLE 15.1 Relation of Tillage to Yields of Winter Wheat

Tillage Treatment	Fargo, ND[1]		Sidney, NE[2]	Akron, Co[3]
	1982–85	1983–84	1970–77	1975–87
	Bushels per Acre			
Conventional tillage	7	49	36	42
Reduced-tillagge	33	53	37	44
No-tillage (herbicides)	61	47	38	43

1. *Conventional:* moldboard plow, disk, harrow
   Reduced-till: disked twice.
2. *Conventional:* moldboard plow, field cultivator 2–3 times, rod weeder 1–2 times
   Reduced-till: V-blade 2–4 times, rod weeded 1–2 times.
3. *Conventional:* V-blade twice, rod weeder as needed.
   Reduced-till: V-blade once, rod weeder as needed.

Plains.[55, 56, 57] Use of a moldboard plow significantly reduces yields due to greater loss of soil water from evaporation.

In the eastern states, the seedbed is often prepared with a disk or spring-tooth harrow when wheat follows corn, soybeans, or cotton. When land is to be plowed, as after a green manure crop, three or four weeks should pass between plowing and seeding. Because of potential water erosion in the humid regions, rolling land should not be left bare of vegetation for long periods. Consequently, plowing should not precede seeding by more than four weeks.

When spring wheat is sown after intertilled crops, the land usually is double-disked and harrowed before drilling. Plowing such land should be avoided to reduce soil moisture loss and erosion potential. Spring tillage usually gives higher yields of spring wheat in the drier western region than does fall tillage. The reason is more snow is caught in undisturbed stubble during the winter and more moisture is added to the soil. Fall tillage is usually practiced in the more favorable eastern regions to permit earlier seeding, which is of prime importance.

15.11.2 Seeding Methods

The American wheat crop is sown with a grain drill (Figure 15.13). Drills with disk furrow openers are used most generally in the humid regions. Hoe drills are popular on fallowed land in the semiarid regions of the Pacific Northwest. This type of furrow opener turns up clods and does little pulverizing, which helps to prevent wind erosion. The furrow drill is also widely used in the semiarid western Great Plains where winters are severe. It is a partial insurance against winterkilling and wind erosion, and it places the seed deep enough to reach moist soil. In Montana, average yields of winter wheat from furrow drilling have been higher than from surface drilling.[58]

Most drills are able to plant into soybean and sunflower residue with no tillage. No-till drills are effective for planting into corn and sorghum residues. Sowing no-till fields or those covered with thick stubble mulch require drills with disk or blade cutters to penetrate the residues.

DEPTH OF SEEDING Wheat is usually sown at a depth of 1 to 3 inches (2.5 to 7.6 cm). Wheat seeded after fallow will likely need to be planted deeper because of a lack of moisture near the soil surface. The use of a furrow drill allows the seed to be

FIGURE 15.13
A no-till drill for seeding winter wheat in the semiarid Great Plains.
[Courtesy Case IH]

placed deeper relative to the original soil surface and still keep the seed within 2 inches (5 cm) of the surface.

When wheat germinates, the plumule grows upward from the sprouting seed enclosed in the elongating coleoptile. It can also be pushed upward by the elongation of the subcrown internode that arises below the coleoptile node. Sometimes other internodes also elongate. Coleoptile elongation ceases when light strikes its emerging tip. The coleoptile then opens and allows the first leaves to push out. Then tillering begins. Crown roots start to develop at nodes ¾ inch to 2 inches (2 to 5 cm) below the soil surface. Seeding wheat at a depth of 1 inch (2.5 cm) or less causes the crown to form just above the seed. Unless it is necessary to insure prompt germination, seeding deeper than 1½ inches (4 cm) merely uses up part of the energy in the seed to produce an excessively long sprout and delays and weakens the seedling accordingly. Deep seeding may cause the crown to form somewhat deeper than when seeding is shallow,[59] but the difference in crown depth is much less than the difference in seeding depth. High soil temperatures during seedling emergence cause crowns to be formed at shallow depths. Hardy varieties of winter wheat have deeper crowns than do nonhardy varieties, and spring wheat has the shallowest crowns.[59]

Newer, shorter-strawed varieties cannot be seeded as deep as conventional varieties. The shortened internode length is also expressed in the subcrown internode and the seedling may not reach the soil surface or the crown may be established too deep.

TIME OF SEEDING Factors that determine the best time for seeding winter wheat were discussed in Chapter 7. Medium-season seeding of winter wheat for any locality is usually most favorable. Wheat sown late generally suffers more winter injury, tillers less, and may ripen later the next season. Wheat sown too early may use up soil moisture, suffer from winter injury and foot rots, and become infested with the Hessian fly, where that insect is prevalent.

In the semiarid Great Plains, the optimum date for seeding winter wheat is about September 1 in Montana, and progressively later southward to northwestern Texas, where the best time is about October 15. Planting earlier than the optimum seeding time is a common practice in the Great Plains, either to take advantage of favorable soil moisture conditions when they occur or to provide wheat pasture. The main disadvantage of seeding winter wheat too early is the likelihood of injury from foot rot diseases that develop under warm conditions.[29] Seeding about mid-September is recommended in the Pacific Northwest. November and early December seeding is most favorable in California and southern Arizona. Under the mild climatic conditions in these states, nearly all the varieties have a spring growth habit but are grown as winter wheat. Spring wheat is also grown from fall or winter sowing in China, India, southern Europe, Africa, and parts of Latin America.

In the central and eastern states, the optimum seeding date is about the time of, or a few days earlier than, the Hessian fly-free date,[60] as shown in Figure 15.14. The safe date is the earliest date at which wheat can be seeded to escape damage from this insect pest. Wheat should generally be seeded seven to ten days earlier than the safe date where Hessian fly is not present (Figure 15.15). This gives the plants a better start in the fall.

Winter wheat fails to head when sown in the spring, except when sown very early or under cool conditions (Table 15.2).[61] The critical seeding date for normal heading of winter wheat depends upon the variety (Figure 15.16). This date was February 15 for the one winter variety at Oregon, whereas another variety headed when sown fully a month later.[62] The critical date for spring wheat with a partial winter habit was about April 30, whereas true spring wheat headed when sown about a month later.

Early seeding of spring wheat usually results in the highest yields. Early-sown spring wheat is most likely to escape injury from drought, heat, and diseases, which become more prevalent as the season advances.

The usual procedure is to sow spring wheat as early as the soil can be worked into a good seedbed. This time varies widely between seasons, usually arriving in March in Nebraska and Colorado, April in South Dakota and southern Minnesota, and May along the Canadian border (Figure 15.17). Spring seeding is usually accomplished in March in Washington and Oregon, however, because of a climate milder than that east of the Rockies at the same latitude. When wheat is sown dur-

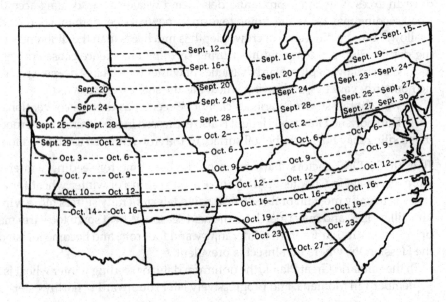

FIGURE 15.14

Fly-free date for sowing winter wheat.

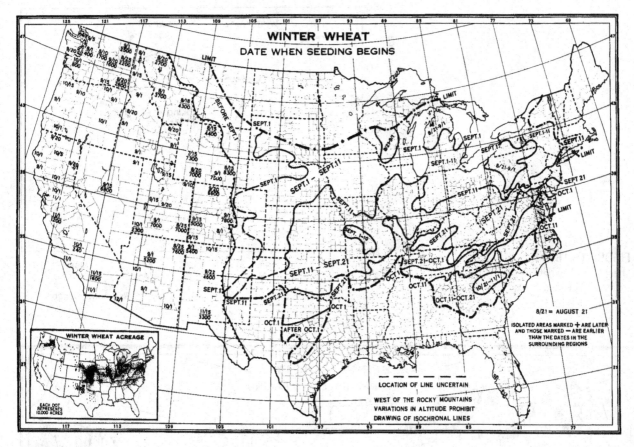

FIGURE 15.15
Date when winter wheat seeding begins.

TABLE 15.2 Wheat Responses to Delayed Planting Dates in Kansas, 1985–1991

Planting Date	Grain Yield (bu/a)	Test Weight (lb/a)	Date Headed	Date Ripe	Mature Height Inches	#Heads per Plant	#Kernels per Plant
Oct. 1	46.1	56.1	5–12	6–17	27.2	3.4	56.4
Nov. 1	35.6	55.1	5–18	6–21	25.6	2.8	54.2
Dec. 1	27.4	53.7	5–22	6–24	25.5	2.9	47.9
Jan. 1	26.3	53.5	5–25	6–35	25.5	2.7	44.1
Feb. 1	18.9	51.5	5–29	6–27	24.4	2.3	31.4
Mar. 1	7.2	22.8	6–6	7–4	21.8	1.4	15.0
Apr. 1	0	—	—	—	—	0	0
LSD (5%)	1.5	0.4	0.1	0.1	0.5	0.3	5.3

ing an extremely early mild period that is followed by a cold snap, the wheat remains in the ground and emerges during subsequent warmer weather usually without evidence of injury. There is no advantage in extremely early seeding except to insure seeding in ample time. Late winter seeding of spring wheat is not advisable except in warm climates.

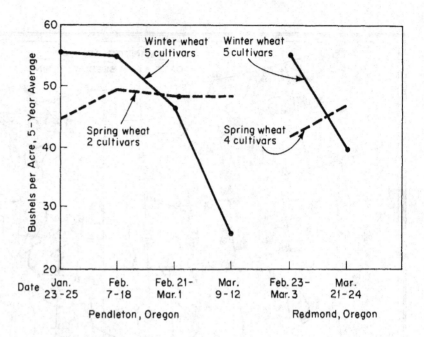

FIGURE 15.16
Effect of seeding date upon
yields of winter and spring
wheat.

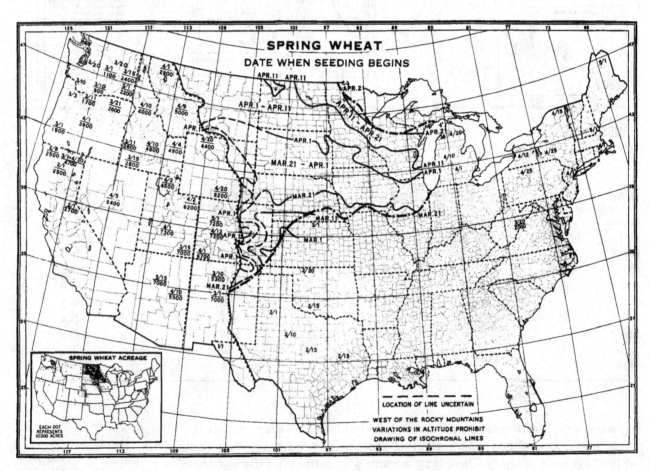

FIGURE 15.17
Dates when spring wheat seeding begins in the United States.

RATE OF SEEDING Wheat has tremendous ability to compensate for soil water supply by varying the number of tillers and the number of kernels per tiller.[63] The rate of seeding of wheat is based on the concept that the greatest number of kernels is produced on the main stem of the plant, and each subsequent tiller produces less than the previous tiller. When water and nutrients are not expected to be limited, higher seeding rates will result in less tillering. The highest percentage of heads will be on the main stem and the primary tiller. As expected water supply diminishes, lower seeding rates will allow the plant to match tiller number to the water supply, thus resulting in a better chance of attaining the maximum yield for each situation.

Winter wheat in the eastern United States is usually seeded at the rate of 90 to 120 pounds per acre (100 to 135 kg/ha). Heavier seeding than usual is advised where seeding is delayed beyond the normal date because there is less opportunity for the plants to tiller. In the Great Plains, seeding rates are usually 40 to 60 pounds per acre (45 to 67 kg/ha) in the semiarid areas and up to 60 to 75 pounds (67 to 84 kg/ha) in areas with more annual precipitation. Irrigated wheat can be planted at rates of 60 to 90 pounds per acre (67 to 100 kg/ha).

15.11.3 Weed Control

Weeds in wheat fields are controlled to a considerable extent by rotation with intertilled crops and by summer fallow. There are no preplant herbicides for wheat. Weedy fields of wheat are often sprayed with postemergent herbicides to control annual broadleaf weeds and wild oat. Several sulfonylurea herbicides, or mixes of these with or without 2, 4-D, can be applied postemergent. Wheat fields also may be sprayed with 2, 4-D or glyphosate a few days before harvest to kill or wilt green weeds that otherwise would add moisture to the wheat grain during harvest. Mature wheat plants are not injured by the herbicides.

Downy brome (cheatgrass, *Bromus tectorum*) is a serious weed in many wheat fields. Stubble-mulch tillage promotes the population of downy brome plants. Sulfosulfuron herbicide applied early postemergent in the fall can provide control. Clearfield® wheat varieties can be sprayed with imazamox. Fallowing provides partial control, but adequate control is obtained by applying herbicides to the fallowed field.

15.11.4 Harvesting

Nearly all of the wheat in the United States has been harvested with a combine since 1960 (Figure 8.3). Wheat harvest begins in May in the southern portion of the United States, and continues through September in the prairie and Great Plains areas of the northern states and the Canadian provinces (Figure 15.18). Wheat is being harvested in some part of the world throughout the year (Figure 15.19).

▓ 15.12 USES OF WHEAT

An estimated 869 million bushels (23 million MT) of wheat was milled for flour in the United States in 1995. A small amount is fed to livestock and poultry, and some is used for seed for the next crop. Feed wheat is cracked or rolled except for poultry.

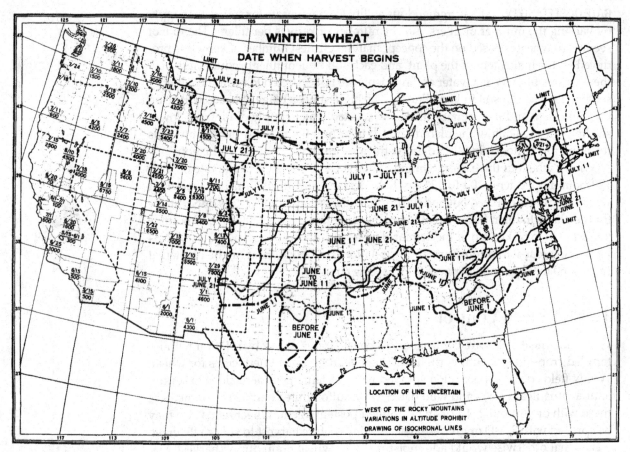

FIGURE 15.18
Dates when winter wheat harvest begins in the United States.

Wheat is a more efficient feed than other grains in weight gains per feed unit, but, when fed as the only grain, the daily intake and daily gain are less. Wheat should thus constitute no more than half of the grain ration mixed with corn or grain sorghum.[62] By-products of wheat milling, which constitute about 30 percent of the uncleaned wheat, are used largely as livestock feeds. The per capita use of wheat products for food in the United States was about 115 pounds (52 kg) in 1970 compared to 330 pounds (150 kg) in 1910.

About 1,250 million bushels (34 million MT) of wheat were exported from the United States in 1996–1997. That is a little over half of the total production. Wheat exported from the United States consists mostly of hard red wheat and Pacific Coast white wheat.

The main food use of wheat flour is for baked products. Other food uses include prepared breakfast foods and pasta made mostly from durum wheat semolina. Varieties of common or durum wheat with large white kernels are often used for puffing. Shredded wheat products are made from soft white wheat varieties. Hard wheat is used mostly for bread flour. Soft wheat, both red and white, is used mostly for cake, cracker, pastry, and all-purpose packaged flour. Asian noodles, flat breads, and tortillas are growing markets for hard and soft wheat. Bulgur, mostly

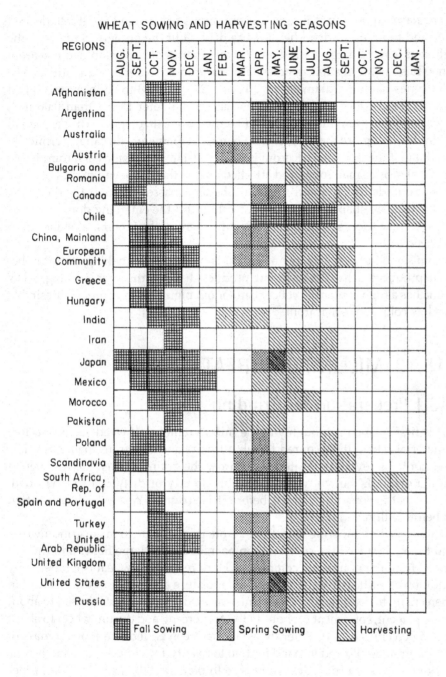

FIGURE 15.19
Normal planting and harvesting season for wheat at selected locations throughout the world. [Sources: [10,12,14,15] and selected reports from U.S. agricultural atlases]

used as a substitute for rice, is prepared from wheat by parboiling, drying, cracking, and removing some of the bran. Since the high bran content of bulgur may induce ulcers or rickets in people who consume large quantities, it is being replaced with a similar product, called world wheat, in which all of the bran is removed.[36]

PASTA In the manufacture of macaroni and other pasta products, semolina or farina is mixed with hot water and kneaded heavily under corrugated rolls until a stiff dough is formed. The dough is then placed in a cylindrical press where the heavy pressures of a plunger force it out through a perforated plate or die at the end. The pasta comes out in a series of continuous strands and is cut into desired lengths and hung up to dry. The pressure drives out the air and compresses the dough, thus leaving a dense product. Various shapes of cut or molded macaroni and noodles, including flat noodles, are made in considerable quantities. The tubular types having a diameter of 0.11 to 0.27 inch (2.8 to 6.9 mm) are called macaroni. Small tubes or solid strands with a diameter larger than 0.06 inch (1.5 mm) and not more than 0.11 inch (2.8 mm) are called spaghetti. Very small solid strands, not more than 0.06 inch (1.5 mm) in diameter, are called vermicelli. The making of the holes in macaroni is relatively simple. Inside each round hole in the die plate is a small round rod attached only at the inside end of the hole. The dough surrounds this rod and forms a tube as it passes through the die plate. When the dies are shaped in the form of letters and the emerging dough strands are cut into very short sections, the basic ingredient of the familiar alphabet soup is produced.

Durum semolina of high protein content is preferred for making macaroni because it produces a hard translucent product that is firm after cooking. Also, semolina has a high content of yellow carotenoid pigments that impart a desirable rich yellow color to the macaroni.

15.13 MILLING OF WHEAT

15.13.1 Preparation for Grinding

Wheat is first cleaned thoroughly at the mill to remove all weed seeds and foreign material. Then, it is scoured to rub off hairs and dirt from the kernel. The scourer contains revolving beaters that throw the grains against the roughened walls of a drum. If the wheat is appreciably smutty or dirty, it also is washed. Smutty wheat is sometimes limed before it is scoured. Washed wheat is often dried before milling.

The reasons for tempering wheat just before grinding are to make grinding easier and to toughen the bran coat so it can be milled off in large flakes and thus be separated from the flour more completely. Hard red winter wheat is tempered by the addition of sufficient water to raise the moisture content from the usual 12 to 13.5 percent up to about 15 to 15.5 percent. Hard red spring wheat is raised to about 15.5 to 16 percent, soft wheat to about 14 to 14.5 percent, and durum wheat to about 16 to 16.5 percent moisture. After the water has been added to a moving mass of wheat, the grain is allowed to stand for two to twenty-four hours, but usually four to six hours, to permit the added moisture to penetrate the kernel to the proper depth for toughening the bran. Very dry wheat is first moistened and allowed to stand for two days or more for the moisture to penetrate the entire kernel. It is then tempered just before milling.

Heat generated during grinding and the aeration of the milled products during their sifting, purifying, and movement through the mill combine to reduce the moisture content in the milled products to about 13 to 13.5 percent. There is very small loss in weight during milling due to evaporation when tempered wheat contains higher moisture than 13.5 percent before tempering. Wheat containing less than 13 percent moisture before tempering usually shows a small gain in milled weight.

15.13.2 The Milling Process

In a large mill, the milled products are divided into numerous streams, each requiring different treatment or disposition.[64] The grain is first cracked or crushed gradually through a series of four to six pairs of chilled iron break rolls. The surface of the break rolls is roughened by sharp, lengthwise corrugations. One member of each pair of break rolls runs about two and a half times as fast as the other, which produces a shearing action on the grain. The grist from each break is sifted and the coarsest particles are conducted to successive break rolls where the process is repeated. Each successive pair of break rolls has finer corrugations and closer distances between the rolls to grind the grain into progressively smaller particles. The finest particles from the breaks are sifted off as flour.

The goal of milling on the break rolls is to get out as large a proportion of middlings as possible. These middlings are granular fractions of the endosperm from which the finer particles (flour), bran, and shorts have been separated. The middlings are reboiled and also aspirated in a middlings purifier to remove small, light bran particles. The purified middlings pass through a series of paired reduction rolls that revolve at speed ratios of about 1:1.5. These rolls, which are smooth, are spaced so that each successive reduction produces finer particles. Flour is sifted out after each reduction. In the milling of straight flour, all the flour streams are combined except for the one called red dog, which is dark in color due to a high content of fine bran particles and much of the aleurone. In usual milling practice, the finer and whiter fractions comprising 70 to 80 percent of the total flour are combined into patent flour. The remaining darker grades, usually called clears or baker's flour, are marketed for special baking uses or are mixed with bran, graham, or rye flours. Low-grade flours (clears and red dog) sometimes are later washed to remove the starch, leaving what is known as gluten flour, which is used mostly by diabetics or by people with digestive ailments.

15.13.3 Bleaching Flour

Most flour is bleached to make it whiter and more attractive. Bleaching is done with small quantities of such chemicals as chlorine, benzoyl peroxide, nitrosyl chloride, chlorine dioxide, or nitrogen peroxide. Bleaching destroys the yellow pigments, mostly xanthophyll, that are naturally present in wheat flour. Chlorine and its compounds improve the quality of the flour to a degree comparable to that attained from natural aging by storage. Extensive research has failed to show any deleterious effect on the wholesomeness of the flour as a result of these bleaching processes.

15.13.4 By-Products of Milling

The by-products of flour milling, often referred to as offal, are usually marketed as bran and shorts and are valuable in dairy feeds and poultry mashes. Some mills make three feed products, namely, bran, shorts (brown shorts), and white middlings. Bran consists mostly of the large particles of pericarp and aleurone with small quantities of starchy endosperm attached.

Shorts (standard middlings) are the final middlings, consisting mostly of coarse particles of endosperm mixed with, or attached to, considerable quantities of fine bran particles. White middlings (not the intermediate milling product called middlings) comprise red dog and other low-grade flours mixed with the finer and whiter particles that otherwise would be included in the shorts. Shorts and middlings are often used as hog feed, with enough water added to make slop. Mill feed from different mills may vary from equal parts to 3:2 proportions of bran and shorts. Bran is a bulky feed, a given volume weighing less than one-third as much as wheat. Most of it is fed to dairy cows.

The germ of wheat usually comes from the break rolls in a stream, called sizings, in which the germ is attached to a large particle of bran. The germ is flattened out and much of the bran detached in the reduction rolls. The germ is usually added to the shorts and thus used for feed. Some of the germ stream, or the extracted oil from the germ, is saved for food purposes because of its high content of vitamins, particularly thiamin (vitamin B) and alpha-tocopherol (vitamin E). This latter vitamin is essential for reproduction in rats and for strengthening the walls of red blood cells in premature newborn human infants. Numerous other reported benefits from increased vitamin E consumption are still debatable. Wheat germ bread is now being baked.[65]

Purified middlings of hard spring wheat are sometimes marketed as *Cream of Wheat®*, and the corresponding product from hard winter wheat, called farina, is also used for a cooked breakfast cereal. Purified middlings from durum wheat, called semolina, are used in the manufacture of macaroni, spaghetti, and other edible pasta products.

15.13.5 Flour Yields from Milling

The milling yield of about 72 percent flour from wheat that consists of nearly 85 percent endosperm is an indication of the difficulty of making a perfect separation of the kernel parts in the milling process. Particles of endosperm, particularly the aleurone, cling to the bran, and these are carried into the feed or by-products streams.

The average straight flour yield from wheat is about 70 to 74 percent. Higher extractions result in darker flour such as the 85 percent dark "victory" flour used in European countries during World War II.

For making about 150 to 160 pounds (68 to 72 kg) of freshly baked bread, 100 pounds (45 kg) of flour are sufficient.

FACTORS AFFECTING FLOUR YIELDS Yields of total flour (including low-grade) range from less than 62 percent for wheat with a test weight of 49 pounds (630 g/l) to more than 79 percent for wheat with a test weight of 64 pounds (825 g/l). With lightweight (shrunken) wheat, which has a low percentage of endosperm, the separation of bran and flour is difficult.

Wetting and drying in the field changes the texture of the wheat kernel, making it softer and starchier in appearance as well as lighter in test weight. This change is the result of additional air space between the starch grains. The low test weight resulting from weathering does not affect the flour yield in milling, and the softening of the endosperm does not impair the baking quality of the flour. In fact, the baking quality may even improve with wetting, as the flour has the characteristics of having been aged. Thus, weathered wheat has a better quality for milling and baking than its test weight and appearance indicate.

Whole wheat (graham) flour is composed of the entire wheat grain ground into flour, and it includes all the bran. Often, graham flour is ground in a bur mill or even between old-fashioned stone burs. Apparently, people of the Stone Age ground their wheat into graham flour. However, the ancient Egyptians learned to bolt (sift) flour through papyrus, and the tendency through the centuries has been to make the whitest flour possible.[66] The ancient Romans made a fairly white flour. George Washington sold bolted flour at his mill. In 1837, Dr. Sylvester Graham published a book extolling the virtues of flour made from the entire wheat kernel. Since then, such flour has been called graham flour. When the germ is retained in the flour as in graham and certain special flours, the fat in the germ tends to become rancid, which makes the flour unpalatable. Graham flour becomes infested with insects more quickly than does white flour. Graham flour usually sells at a higher price because of these storage risks and because of the small quantity marketed.

White flour, until the 1970s, constituted about 97 percent of the wheat flour manufactured in the United States. The remainder was graham and other types of dark flour. White flour is often mixed with dark flours in making bread. Consequently, about 93 percent of the wheat bread baked was made from white flour only, 1 percent graham bread, and the remainder mixed white and dark flours. In recent years the American public has increased its demand for whole wheat foods.

▧ 15.14 CHEMICAL COMPOSITION OF WHEAT AND FLOUR

The approximate chemical composition of the wheat kernel[67, 68] in percentages is starch, 63 to 71; proteins, 10 to 15; water, 8 to 17; cellulose, 2 to 3; fat, 1.5 to 2; sugars, 2 to 3; and mineral matter, 1.5 to 2. The protein content of wheat varies for that grown in different regions. The approximate composition of white and graham flours, and of bran and wheat germ, are shown in Table 15.3. White flour is lower in protein, fat, and ash, and higher in starch than is the original wheat or the graham flour.

TABLE 15.3 Average Composition of Wheat, Flour, Bran, and Germ Containing about 13 Percent Moisture

Composition	Wheat or Graham Flour (%)	White Flour (%)	Bran (%)	Germ (%)
Carbohydrates (nitrogen-free extract)	68	74	50	18
Starch	55	70	10	—
Pentosans	6	3.5	25	6
Dextrins	—	—	4	—
Sugars	2	1.5	1.5	15
Crude fiber	23	0.4	9	2
Fat	2	1	4	11
Crude protein	13	11	17	30
Ash (mineral matter)	2	0.45	7	5

The purest and finest grade of flour, called patent flour, contains roughly 10 to 12 percent of the total thiamin and niacin, 20 percent of the riboflavin and iron, 25 percent of the phosphorus, and 50 percent of the calcium found in the wheat kernel.[69] Enrichment, as practiced for a majority of the flour used in the United States, restores about 60 percent of the niacin and 80 percent of the thiamin and iron and about doubles the concentration of riboflavin in the wheat. Folic acid is also being added.

The germ fraction is high in thiamin, riboflavin, phosphorus, and iron. Bran is high in niacin, phosphorus, and iron. In milling, most of the thiamin is recovered in the red-dog flour and the shorts, and the niacin is found mostly in the bran. The thiamin of the germ is chiefly in the scutellum rather than in the plumule and the embryonic stem and root tissues.

About 75 to 80 percent of the thiamin in wheat is found in the bran and shorts (feed) after milling. When these two products are fed to dairy cows and hogs, respectively, not more than about one-sixth of the thiamin is returned to human food in the milk and edible pork produced. A distinct rise in the incidence of digestive disorders and of rickets in young children has followed mass consumption of high-extraction flours (82 to 100 percent of the grain). Bran contains considerable percentages of indigestible but partly fermentable material that irritates the intestinal tract. Bran also contains phytic acid (inositol hexaphosphoric acid),[70] which inhibits the absorption of calcium from the intestinal contents and induces the development of rickets. Addition of calcium to the branny diet merely reduces phosphorus absorption. If bread were the only food available, graham flour would doubtless be preferable to white flour from the nutritional standpoint. Luckily, no one has to live by bread alone.

15.14.1 Ash Content and Flour Quality

Highly refined patent flour has a low ash content of 0.4 percent or less. The more bran particles a flour contains, the higher the ash content. Therefore, clear and straight flours have higher ash content than does patent flour. Bakers and flour buyers often purchase flour on a guaranteed maximum allowable ash basis. High ash content is not necessarily an indication of high extraction, poor milling, or dirty wheat because wheat varieties differ in ash content. Hard wheat flours usually have a higher ash content than do soft wheat flours. Environment, especially weather, also affects the ash content of wheat and the resultant flour. The main constituents of wheat ash are oxygen, phosphorus, potassium, magnesium, sulfur, and calcium.

15.14.2 Protein Content and Flour Quality

Hard red spring and hard red winter wheat contain an average of about 11 to 15 percent protein when grown in the Great Plains and northern prairie states.[27] Soft wheat contains 8 to 11 percent protein when grown in humid areas. In years when the crop is generally low in protein, premiums may be paid for desirable lots of high-protein hard wheat. Graduated premiums for fractional percentages of protein above 12.5 to 13 percent are usually offered. High protein wheat is blended with other wheat in order to bring up the average protein content of the resulting flour to established standards for the particular flour brand. Often, wheat and flour are sold on the basis of specified protein content. Flours of high and low protein content can be separated from any finely ground flour by air.

Hard wheat flours as a rule are high in bread-making quality, (i.e., the ability to make a large, light, well-piled loaf of bread of good uniform texture and color). Strong wheats that have ample gluten (protein) of good quality are high in water absorption. They usually produce more baked 1-pound loaves of bread from a barrel of flour because of their greater ability to retain moisture as compared with the weaker flours from soft wheat. A light loaf of bread can be made from weak flour, but the pores of such bread are large and the loaf tends to dry out quickly. Hard wheat tends to produce flour of a granular texture regardless of their texture or protein content. The granular flour of hard wheat is not well suited for pastry, but soft wheat produces a fine, soft flour well suited to making cakes, crackers, cookies, and hot breads.

The preferred protein percentages in wheat for various domestic purposes are: macaroni and noodles, 12.5 to 15; white bread, 11 to 14; all purpose flour, 9.5 to 12; crackers, 9.5 to 11; pie crust and donuts, 8 to 10.5; cookies, 8 to 9.5; cake, 8 to 9.5.[71]

The gluten of the wheat kernel contains about 17.6 percent nitrogen. The percentage of nitrogen, determined by analysis, is multiplied by 5.7 to determine the protein content ($100 \div 17.6 = 5.7$). The protein of most feed crops contains about 16 percent nitrogen, and the factor 6.25 is used.

▓ 15.15 DISEASES

15.15.1 Rusts

The three rusts attacking wheat are stem rust or black stem rust (*Puccinia graminis tritici*), leaf rust (*P. recondita*), and stripe rust (*P. striiformis*).[72] These are all fungus diseases, known for many centuries. The Romans attributed the crop damage to the rust god Robigus, who resorted to this means of wreaking his vengeance on a wicked people. Others, noting that rust followed damp weather, believed it to be similar to the rusting of their own iron tools. In fact, some people once believed that their wheat did not rust before the advent of barbed-wire fences around their fields.

STEM RUST Stem rust (*Puccinia graminis tritici*) has been the most destructive disease of wheat. It causes such severe shriveling of the grain of susceptible varieties that often the crop is not worth harvesting. The rust produces masses of pustules on the leaves and stems that contain brick-red spores (Figure 15.20). These spores spread the disease to other plants. Rusted plants transpire water at a greatly accelerated rate. Under the mild climate of the southern United States and in Mexico, the red (uredio) spore stage of the stem rust organism lives throughout the year on seeded and volunteer wheat and on certain grasses. In the spring, the spores multiply and spread to other plants or other spots on the same plant. A new generation of spores is produced every one to three weeks. The spores are caught by wind and air currents and may be carried upward 16,000 feet (5 km) and outward for hundreds of miles. Thus, wheat is infected as the season advances northward. In Texas, an epidemic of stem rust in April can produce a similar epidemic in North Dakota in July when the season favors rust development throughout the central wheat-growing region. A close

FIGURE 15.20
Rusted wheat culms *(left)* and resistant rust-free culms *(right).*

relationship between rust epidemics in Kansas and North Dakota was recognized as early as 1910. Moist, warm weather favors rust development. A lush growth of wheat produced on soils high in nitrogen and moisture is most subject to rust attack.

In the North, the rust organism may pass through additional stages. The red (uredio) spores on the plants are replaced by black (telio) spores as wheat approaches maturity. These spores remain on the straw and stubble over winter, germinate in the spring, and produce basidiospores, which infect the leaves of the common barberry bush *(Berberis vulgaris)* and wild species such as *B. canadensis* and *B. fendleri*. A type of spore (spermagonium or pycnospore) that develops on the barberry constitutes the sexual reproductive cells of the fungus. The union of these cells produces tissues that give rise to cluster cups (aecia) in which aeciospores are borne. The aeciospores then are blown from the barberry bushes to wheat or grass plants, which they infect. Urediospores are produced on the wheat and the life cycle is completed. The teliospores usually fail to survive the long summer weather south of Nebraska, and consequently barberry bushes seldom are infected in the South. On the other hand, the urediospores of stem rust seldom live over winter north of Texas. Where barberry bushes are present in the northern states, they can spread the rust to nearby fields of wheat, and the rust then spreads to other fields of wheat. Wheat near barberry bushes becomes infected with rust fully two weeks earlier than wheat infected from spores from the South.

Eradication of the barberry bushes reduces rust infection in wheat as well as in other grain crops. Since the barberry bush is also a medium on which new races of the rust fungus may arise, its eradication is essential. The eradication of the wild species of barberry bushes in isolated valleys of Colorado, Virginia, and Pennsylvania has greatly reduced local damage from rust. However, in the plains and prairies of the Midwest where rust clouds sweep up from the South, the eradication of barberry bushes only partially reduces total rust damage.

Dusting or spraying fields of wheat with repeated applications of fungicides is possible but is not economically feasible.

Seed treatment has no effect on rust because the disease is not seed-borne. Early-sown wheat and early-maturing varieties partly evade rust injury.

The breeding and distribution of wheat resistant to stem rust has been a spectacular achievement.[52] Genetic resistance to stem rust has been maintained more or less stable, partially due to the lack of new races because of the eradication of barberry. Unfortunately, rust-resistant wheats suffer heavy losses later after new races appear. Additional sources of disease resistance must then be obtained from wheat varieties grown in other countries to use in breeding for resistance. The breeding and testing of wheat for resistance to new and old races of rust is necessarily a continuous process in the wheat-growing countries of the world.

At least 275 distinct physiological races (biotypes) of the stem rust organism have been discovered, but only a few of these are of economic importance at any one time in a region.

LEAF RUST Leaf rust *(Puccinia recondita)* once occurred mostly in the eastern half of the United States but has now spread to the southern Great Plains. The disease is now considered more important than stem rust. Numerous races of the leaf rust organism are known. An early infection of leaf rust can reduce the yield of a susceptible wheat variety 42 to 94 percent.[21] Reduction in yield is due primarily to a reduced number of kernels in the spike. The kernels are also reduced in weight but are not shriveled noticeably. Pustules containing the orange-red spores cover the leaves and part of the stems. The rust fungus lives over winter in wheat plants either in these pustules or as mycelium (vegetative strands) within the leaf tissue. Leaf rust can be controlled by several dustings with a fungicide, but this is too expensive for field-scale operations. The use of resistant varieties is the most practical method of controlling the disease.

STRIPE RUST Stripe rust *(Puccinia striiformis)* in the United States is mostly confined to the western half of the country but is a serious disease in most warm countries of the world. Linear orange-yellow lesions develop on the leaves and, later, on the floral bracts. Damage by stripe rust is similar to that caused by leaf rust, except that stripe rust shrivels the kernels. The urediospores overwinter where the winters are mild or where snow cover protects the foliage. The use of resistant varieties is the most practical control measure in regions where the disease is serious. Many adapted varieties are resistant.

15.15.2 Smuts

BUNT OR STINKING SMUT Bunt (caused by the fungi *Tilletia caries* and *T. foetida*) is found throughout the country but is most severe in the Pacific Northwest where the spores are carried over in the soil. The disease is usually carried over from one crop to the next as black spores on the seed or as smut balls mixed with the seed. The smut spores germinate when bunt-infested seed is sown in moist, cool soils. The fungus infects the young seedlings, grows within them, and produces smut balls, completely filled with black spores instead of kernels in the wheat head (Figure 15.21). The spores have the odor of stale fish. Bunt can be controlled by seed treatment with suitable fungicides.[73]

DWARF BUNT Dwarf bunt *(T. controversa)* causes severe dwarfing of smutted plants.[74] It is more likely to soil borne than common bunt. Seed treatment with fungicides that control the fungus in the soil is effective. Dwarf bunt attacks winter wheat only, chiefly in the northern intermountain states. Resistant varieties are grown where the disease is most prevalent.

LOOSE SMUT Loose smut (caused by the fungus *Ustilago tritici*) may occur wherever wheat is grown, but it is most abundant under humid or irrigated conditions. The

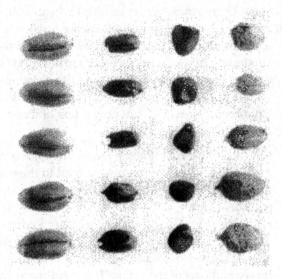

FIGURE 15.21
(Left to right) Sound wheat, nematode galls, cockle seeds, and bunt balls.

floral bracts of the spike are almost completely replaced by black smut masses. The spores are soon disseminated by wind, rain, and other agencies, leaving a naked rachis. The spores infect flowers of sound heads while in bloom. When the spores germinate, an internal infection is produced within the new kernel. Infected kernels are indistinguishable from normal kernels at maturity. When the infected seed is sown, the fungus grows within the new plant as it develops. Surface disinfectants are ineffective against the internal infection. A modified hot-water treatment will control the disease, but it is injurious to the seed and rather expensive. However, it is a practical means of eliminating the disease from a foundation seed supply. The seed wheat is soaked for four or five hours in cold water, then placed in a warming tank at 120°F (49°C) for one minute, put in a hot water tank at 130°F (54°C) for ten minutes, and finally dipped in cold water for one minute. Certain water-soak treatments are less injurious to the seed and about as effective in disease control. During such a treatment, the seed is soaked for six hours, after which it is held in an airtight container for seventy-two hours. The treated seed is spread out to dry and is ready for planting as soon as it is dry enough to feed through a drill. Some varieties are resistant to loose smut.

FLAG SMUT Flag smut (caused by the fungus *Urocystis tritici*) has been reported in Illinois, Missouri, Kansas, and Washington. Losses in yield of 10 to 20 percent may occur in susceptible varieties. Flag smut produces dark stripes on the leaves, sheaths, and culms. Infected plants are more or less dwarfed and often fail to form normal heads. The disease is carried to the next crop both on the seed and in the soil of infested fields. Flag smut can be controlled by the same seed treatments used for bunt if the seed is sown on noninfected soil. Several winter wheat varieties are resistant to the disease.

KARNAL BUNT Karnal bunt (caused by the fungus *Tilletia indica*) is a seed borne disease with limited geographical distribution. It was first discovered in the 1930s in India. It is a problem mostly in warmer climates and has not been detected in the northern wheat-growing regions of North America, Europe, and Asia. It was first detected in the United States in Arizona in 1996. The disease has minimal effect on the yield of wheat, but it is an important disease because there are strict international quarantines in place against wheat infected with the disease. Although it

FIGURE 15.22
Healthy wheat plants and grain *(left)*: scab-diseased plants and scabbed grain *(right)*.

poses no threat to humans or livestock, Karnal bunt can render wheat products unacceptable for human consumption because of odor.

15.15.3 Scab

Wheat scab, caused by *Fusarium graminearum*, is a serious disease in the northern and eastern states, particularly in the Corn Belt. The fungus attacks the heads shortly after they emerge. One or more spikelets can be killed or the development of the kernel prevented. The disease can be identified by the appearance of a pinkish-white fungus growth on or around the dead tissue. The grain in diseased portions is shriveled, almost white, and scabby in appearance (Figure 15.22). The disease is carried from season to season both on the seed and on old crop refuse, such as corn stalks. It is most severe when wheat follows corn in the rotation. Infected seed should be treated with a fungicide before it is sown to prevent seedling blight.

15.15.4 Take-All

Take-all (caused by *Ophiobolus graminis*) is a widely spread disease, but appears to be most serious in Kansas and New York. In some years, it has destroyed 10 to 50 percent of the crop in infested fields in Kansas. Take-all can kill wheat plants in the rosette stage or later after the heads begin to fill. In the latter case, the plants turn almost white. Nearly all plants in certain spots may be killed. Rotting of the roots damages other plants. The bases of infected plants are usually black to a height of 1 or 2 inches (2.5 to 5 cm) above the soil. The take-all fungus lives over in the soil. The only feasible control measure is to keep infested land free from wheat, barley, and rye for at least two to four years.[75] Genetic resistance has helped in the control of this disease.

15.15.5 Mosaics

Wheat mosaics are virus diseases. Wheat soil-borne mosaic virus (WSBMV) produces a severe dwarfing or rosetting of the plants or a mottling of the leaves, and often kills most of the plants in areas scattered over the fields. The virus persists in the soil for several years. The disease occurs in Maryland, North Carolina, and South Carolina, as well as in several states westward to Nebraska, Kansas, and Oklahoma.[76] It can be controlled by sowing resistant varieties.

Wheat streak mosaic virus (WSMV) occurs in the central Plains, some western states, and in Canada. The virus is spread by the wheat curl mite *(Aceria tulipae)*. It is carried over in volunteer wheat plants as well as in several grasses. The symptoms of the disease are yellow stripes and blotches on the leaves and stunted plant growth. Partial control methods are late seeding of winter wheat and destruction of volunteer wheat to control the mite.

Barley stripe mosaic, a seed borne disease of wheat, occurs in several states. Control consists of sowing seed that is free of the virus.

15.15.6 Nematodes

Nematode (eelworm) disease, caused by a small, almost microscopic round worm *(Anguinea tritici)*, occurs in some of the eastern and southeastern states, particularly Virginia, Maryland, North Carolina, and South Carolina. Nematodes living in the soil enter the wheat plants through the roots and move up into the young developing heads before the latter emerge from the sheath. The infested ovary does not develop into a grain but instead forms a hard black gall filled with nematodes. The galls resemble those formed by bunt disease except that they are not easily crushed. The wheat nematode is controlled by rotation with crops other than rye for one or two years and by the planting of clean seed.

15.15.7 Other Diseases

Ergot *(Claviceps purpurea)* sometimes attacks durum wheat rather severely but causes little damage to other wheats. It is controlled by use of clean seed combined with crop rotation.

Powdery mildew *(Erysiphe graminis tritici)*, which occurs under moist conditions, can be controlled by use of resistant varieties.

The foot rot diseases are caused by *Helminthosporium sativum, H. tritici, Pythium arrhenomanes*, other species of *Pythium*, species of *Fusarium*, and other fungi.[77] The *Helminthosporium sativum* fungus also causes the black-point disease of wheat, a discoloration and infection of the germ end, particularly of durum wheat. The western dryland foot rot is associated with a mosaic disease.

Cercosporella foot rot, caused by the fungus *Cercosporella herpotrichoides*, damages wheat mostly in southeastern Washington and adjacent areas. The disease is also called straw breaker or eyespot foot rot. It spreads to basal leaf sheaths of young plants by rain-borne spores from infected stubble in a wheat-fallow crop system. Damage from the disease is reduced by delayed seeding and by spraying with suitable fungicides.

Leaf spot (*Septoria* spp.) attacks wheat over much of the eastern and northwestern parts of the United States when weather conditions favor the fungus. Black chaff and basal glume blotch occur in the Midwest. There is no special control method for these diseases.

Snow mold is caused by the fungi *Calonectria graminicola*, *Typhus incarnata*, and *T. idahoensis*. It frequently damages wheat in the northwestern United States, Europe, and Asia. The disease develops under a deep snow that covers unfrozen ground.[78] Remedies are crop rotation, early seeding, and spring application of a nitrogen fertilizer.

Several other diseases are damaging to wheat.[79]

■ 15.16 INSECT PESTS

More than thirty insect pests are injurious to growing wheat, among them the Hessian fly, wheat jointworm, wheat strawworm, chinch bug, armyworm, and the false wireworm.[80]

15.16.1 Hessian Fly

Hessian fly (*Mayetiola destructor*) is one of the most serious insect enemies of wheat, particularly east of the 100th meridian. Damage occurs both in the fall and the spring. The injury is caused by the maggots (larvae) located between the leaf sheath and stem where they extract the juices of the young stems. Many of the small tillers are killed. The older stems are weakened and break over shortly before harvest. The fall brood may kill many plants outright in cases of serious infestation. The larvae (Figure 15.23) are transformed to a resting stage called a puparium or flaxseed. In the spring, the flaxseed changes into a pupa and then into an adult fly. The female fly deposits eggs on the young wheat plants. Then another generation of larvae hatches out.

The most practical control measure is to sow winter wheat late enough so that the main brood of flies will have emerged and died before the young wheat plants appear above the ground (Figure 15.16). Other methods are crop rotation (except in strip-cropped fields) and the plowing under of infested stubble soon after harvest. Planting resistant varieties is also helpful.[81]

15.16.2 Wheat Jointworm

As a wheat pest, wheat jointworm (*Harmolita tritici*) ranks second only to the Hessian fly in a majority of the wheat states east of the Mississippi River and in parts of Missouri, Utah, and Oregon. It is a small grub that lives in the stem and feeds on

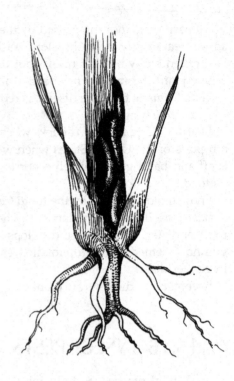

FIGURE 15.23
Larvae or maggots of the
Hessian fly.

the juices. As a result, wart-like swellings on the stem usually occur above the joint or node. The egg from which the grub hatches is laid in the stem by the adult which resembles a small black ant with wings. The damage to wheat caused by the lodging of the infested straw varies from slight injury to total destruction.

Wheat jointworm can be controlled by deeply plowing under wheat stubble after harvest in order to bury it so that the jointworm adults cannot emerge. This may be impractical where grass, clover, or some other legume is seeded with the wheat, except where heavy infestations occur. Rye may be substituted for wheat in the more northerly states where the jointworm is severe.[82]

15.16.3 Wheat Strawworm

Losses caused by the wheat strawworm *(Harmolita grandis)* can be severe throughout the wheat states east of the Mississippi River. It is also an important wheat pest in other wheat-growing states, particularly in western Kansas.[83] Propagation of the worm is favored by infested stubble left on the ground following the use of subsurface tillage implements and the one-way plow.

Two complete generations of the insect occur each year. The first generation, or spring form, kills outright each wheat tiller that it infests. As the larva completes its development, the tiller usually becomes bulb-like at the point of infestation. The spring form is most injurious to winter wheat. The injury caused by the second generation, or summer form, is less severe except in spring wheat.

The wheat strawworm attacks only wheat. Since the spring form is wingless, it can be controlled when wheat is sown 75 yards (70 m) or more from wheat stubble or straw of the previous season. Also, where spring wheat is grown, all volunteer wheat should be destroyed when this pest is abundant, to prevent reinfestation.[84]

15.16.4 Grasshoppers

Wheat is attacked by several species of grasshopper that belong mostly to the genus *Melanoplus*.[84] The grasshopper deposits eggs, enclosed in sacs, about 1 or 2 inches (2.5 to 5 cm) below the surface of cropped, idle, or sod land. Good plowing with a moldboard plow at a depth of 5 inches (13 cm) or more prevents the young grasshoppers from emerging. The young grasshoppers hatch in the spring and soon begin feeding. The grasshoppers eat the leaves and often the stems and heads of the wheat plants.

The main method for the control of grasshoppers is with insecticides. These materials should be applied to the hatching areas when the nymphs are young.

15.16.5 Greenbug

The greenbug or spring grain aphis *(Schizaphis graminum)* occurs in most of the wheat-growing states, and in some years causes severe losses, especially in Texas, Oklahoma, Kansas, and Missouri. It is a sucking insect; adults as well as young bugs feed on wheat plants throughout their lives. They reproduce very rapidly during the summer by vivipary (giving birth to the young) and later by eggs, with or without fertilization. One control measure is destruction of all volunteer wheat, oat, and barley in the summer and fall in each community.[85] Recommended insecticides are helpful.

Several wheat varieties are somewhat resistant to light or medium greenbug attacks.[31] Oats and barley are damaged more severely than is wheat.

15.16.6 Chinch Bug

Chinch bug *(Blissus leucopterus)*, a sucking insect, often is present in damaging numbers from Illinois and Missouri westward to the central Great Plains. It is more destructive to barley, corn, and sorghum than to wheat.[80] No satisfactory methods of controlling infestation in small grains are known. Spraying field borders with insecticides prevents the survival and migration of the chinch bug to fields of corn or sorghum.

15.16.7 Mormon Cricket

The Mormon cricket *(Anabrus simplex)* ranges from the Cascade and Sierra Nevada Mountains to the central Dakotas, devouring range grasses, grains, and other crops. This insect is wingless, but it migrates on foot in dense hordes for considerable distances. Broadcasting poison bait in advance of the horde can be an effective control. Seagulls devoured the crickets that invaded the first crop in Utah.

15.16.8 Wheat Stem Sawfly

The wheat stem sawfly *(Cephus cinctus)* frequently causes severe damage in western Canada and in Montana and other states. The adult female fly splits the wheat stem with a pair of saw-like ovipositor appendages and then deposits an egg inside the stem.[80] The cannibalistic larvae feed downward inside the stem, and the survivor finally chews a groove or ring just above the soil surface, which causes the stems to fall over. The main loss is from the broken stems, which are difficult to save in harvesting. Control measures consist in rotation with other crops, and the turning under of the stubble with a plow that turns the furrow completely. The adults have difficulty emerging through 6 inches (15 cm) of soil. The use of a pick-up reel on the combine can salvage most of the broken stems. Resistant varieties with solid stems are being grown. Two other species of stem sawflies were introduced from Europe.[87] Similar pests, the black grain-stem sawfly *(Cephus tabidus)*, and the European wheat-stem sawfly *(C. pygmaeus)*, are found in the East Central States.

15.16.9 Wireworms

The Great Basin wireworm *(Otenicera pruinina* var. *noxia)* destroys considerable wheat in sections of Washington, Oregon, and Idaho that receive an annual precipitation of less than 15 inches (380 mm). These wireworms, which live in the soil for three to ten years, eat seeded grain and the underground parts of young wheat plants, thus thinning the stands. The adults *(click beetles)* live only a few weeks. Treating the seed with a suitable insecticide aids in controlling the pest. Partial control measures include thick seeding and clean fallowing. Keeping the land free from any growing weed or wheat plant during the fallow season starves the young larvae.

False wireworms *(Eleodes* spp.) have severely damaged wheat in Kansas and Idaho by devouring the seeded grain before it germinates. Control measures include crop rotation, summer fallow, seeding only when the soil is moist enough to insure rapid germination, and seed treatment.

15.16.10 Russian Wheat Aphid

Russian wheat aphid (RWA) *(Diuraphis noxia)* is a recent but important economic pest in wheat and barley. The aphid is small (less than 1/10 inch; 3 mm) and greenish to grayish-green. It is more elongate than other aphids that are tear-drop shaped. The antennae are short and the cornicles are very short. RWA was first detected in Mexico in 1981. The first infestation in the United States occurred in the Texas Panhandle in March 1986. Damaging infestations occurred later that year in New Mexico, Colorado, Oklahoma, and Kansas. The RWA begins feeding at the base of leaves near the top of the plant. The leaf edges will roll inward forming a tube, which makes it difficult to reach them with insecticidal sprays. Salivary toxins injected during feeding cause a purple color in the plant and yellowish and whitish streaks along the leaves. If the aphids attack the flag leaf, the head will not be able to emerge properly.

Some varieties have genetic resistance to RWA. Control of volunteer wheat and barley before seeding winter wheat helps prevent the aphid from moving to the new crop. Later planting is also advisable in infested areas. Many of the beneficial insects that attack other aphids will also attack the RWA including the lady beetle, green lacewing, hover flies, and parasitic wasps. However, leaf curling protects the aphids from natural enemies. Considerable research is currently underway to develop Integrated Pest Management (IPM) guidelines for RWA control.

15.16.11 Other Insects

Fall armyworm (*Laphygma frugiperda*) attacks wheat and numerous other crops in the southern and central states. Control in wheat fields consists of applying insecticides. Delayed seeding of winter wheat, as recommended for Hessian fly control,[88] is helpful.

Armyworm (*Cirphis unipuncta*) has been destructive in many localities in the eastern half of the United States. It is controlled with the insecticides used for the fall armyworm.

Billbugs (*Calendra* species) attack wheat as well as other crops. Crop rotation is the main control method.

Wheat stem maggot (*Meromyza americana*) feeds on the stem just above the top node. It cuts off the peduncle, causing the head to turn white. The percentage of plants injured is usually too small to justify control practices.

Pale western cutworm (*Agrostis orthogonia*) destroys considerable wheat in the dryland sections of Montana, North Dakota, Kansas, and other states. The light gray larvae hatch out in the spring and feed on the underground portions of wheat and grass plants. The moths deposit eggs in the soil in the fall. The best control method is a soil application of an insecticide worked into the soil 6 inches (15 cm) deep before seeding. Summer fallow should be clean to destroy all weeds, grass, and volunteer grain and to starve the larvae.[81]

The insects that attack wheat in storage are described in Chapter 9.

REFERENCES

1. Smith, C. W. *Crop Production: Evolution, History, and Technology.* New York: John Wiley & Sons, 1995.

2. Clark, J. A. "Improvement in wheat," in *USDA Yearbook*, 1936, pp. 207–302.

3. Salmon, S. C. "Climate and small grains," in *Climate and Man*, USDA Yearbook, 1941, pp. 321–342.

4. Welton, F. A., and V. H. Morris. "Wheat yield and rainfall in Ohio," *J. Am. Soc. Agron.* 16(1924):731–749.

5. Army, T. J., and J. C. Hide. "Effects of green manure crops on dryland wheat production in the Great Plains area of Montana," *Agron. J.* 51(1959):196–198.

6. Cole, J. S., and O. R. Mathews. "Relation of the depth to which the soil is wet at seeding time to the yield of spring wheat on the Great Plains," *USDA Circ.* 563, 1940, pp. 1–20.

7. Hallsted, A. L., and O. R. Mathews. "Soil moisture and winter wheat with suggestions on abandonment," *KS Agr. Exp. Sta. Bull.* 273, 1936, pp. 1–46.

8. Salmon, S. C. "Why cereals winterkill," *J. Am. Soc. Agron.* 9(1917):353–380.

9. Quisenberry, K. S. "Survival of wheat varieties in the Great Plains winter-hardiness nursery," *J. Am. Soc. Agron.* 30(1938):399–405.

10. Harvey, R. B. "Physiology of the adaptation of plants to low temperatures," in *World Grain Exhibition and Conference*, vol. 2. Ottawa: Canadian Society of Technical Agriculturists, 1933, pp. 145–151.

11. Zech, A. C., and A. W. Pauli. "Changes in total free amino nitrogen, free amino acids, and amides of winter wheat, crowns during cold hardening and rehardening," *Crop Sci.* 2(1962):421–423.

12. Martin, J. H. "Comparative studies of winter-hardiness in wheat," *J. Agr. Res.* 35(1927):493–535.

13. Newton, R. A. "Comparative study of winter wheat varieties with especial reference to winter-killing," *J. Agr. Sci.* 12(1922):1–19.

14. Klages, K. H. "Relation of soil moisture content to resistance of wheat seedlings to low temperatures," *J. Am. Soc. Agron.* 18 (1926):184–193.

15. Klages, K. H. "Metrical attributes and the physiology of hardy varieties of winter wheat," *J. Am. Soc. Agron.* 18(1926):529–566.

16. Janssen, G. "Effect of date of seeding of winter wheat on plant development and its relationship to winterhardiness," *J. Am. Soc. Agron.* 21(1929):444–466.

17. Fifield, C. C., and others. "Quality characteristics of wheat varieties grown in the western United States," *USDA Tech. Bull.* 887, 1945, pp. 1–35.

18. Alsberg, C. L., and E. P. Griffing. "Environment, heredity, and wheat quality," *Wheat Studies*, Food Res. Inst. Stanford U., 10(1934):229–249.

19. Haunold, A., A. Johnson, and J. W. Schmidt. "Variation in protein content of the grain in four varieties of *Triticum aestivum* L.," *Agron. J.* 54(1962):121–125.

20. Bailey, C. H. *The Chemistry of Wheat Flour*. New York: Chemical Catalogue Co., 1925. pp. 1–324.

21. Johnston, C. O., and E. C. Miller. "Relation of leaf-rust infection to yield, growth, and water economy of two varieties of wheat," *J. Agr. Res.* 49(1934):955–981.

22. Leighty, C. E. and J. W. Taylor, "Studies in natural hybridization of wheat," *J. Agr. Res.* 35:865–887.

23. Garber, R. J., and K. S. Quisenberry. "Natural crossing in winter wheat," *J. Am. Soc. Agron.* 15(1923):508–512.

24. Leighty, C. E., and J. W. Taylor. "Rate and date of seeding and seed-bed preparation for winter wheat at Arlington Farm," *USDA Tech. Bull.* 38, 1927.

25. Bradbury, D., and others. "Structure of the mature kernel of wheat," Pts. I–IV. *Cereal Chem.* 33, 6(1956):329–391.

26. Mecham, D. K., and G. H. Brother. "Wheat proteins, known and unknown," in *Crops in Peace and War*, USDA Yearbook, 1950–51, pp. 621–627.

27. Coleman, D. A., and others. "Milling and baking qualities of world wheats," *USDA Tech. Bull.* 197, 1930, pp. 1–223.

28. Eckerson, S. H. "Microchemical studies in the progressive development of the wheat plant," *WA Agr. Exp. Sta. Bull.* 139, 1917.

29. Robertson, D. W., and others. "Rate and date of seeding Kanred winter wheat and the relation of seeding date to dry-land foot rot at Akron, Colo.," *J. Agr. Res.* 64, 6(1942):339–356.

30. Thatcher, R. W. "The chemical composition of wheat," *WA Agr. Exp. Sta. Bull.* 111, 1913, pp. 1–79.

31. Atkins, I. M., and R. C. Dahms. "Reaction of small-grain varieties to green bug attack," *USDA Tech. Bull.* 901, 1945, pp. 1–30.

32. Lamb, C. A. "The relation of awns to the productivity of Ohio wheats," *J. Am. Soc. Agron.* 29(1937):339–348.

33. Miller, E. C., H. G. Gauch, and G. A. Cries. "A study of the morphological nature and physiological function of the awn of winter wheat," *KS Agr. Exp. Sta. Tech. Bull.* 57, 1944, pp. 1–82.

34. Paterson, F. L., and others. "Effects of awns on yield, test weight, and kernel weight of soft red winter wheats," *Crop Sci.* 2(1962):199–200.

35. Rosenquist, C. E. "The influence of awns upon the development of the kernel of wheat," *J. Am. Soc. Agron.* 28(1936):284–288.

36. Gauch, H. G., and E. C. Miller. "The influence of the awns upon the rate of transpiration from the heads of wheat," *J. Agr. Res.* 61(1940):445–458.

37. McDonough, W. T., and H. G. Gauch. "The contribution of the awns to the development of the kernels of the bearded wheat," *MD Agr. Exp. Sta. Bull. A* 103, 1959, pp. 1–16.

38. Teare, I. D., J. W. Sij, R. P. Waldren, and S. M. Goltz. "Comparative data on the rate of photosynthesis, respiration, and transpiration of different organs in awned and awnless isogenic lines of wheat," *Can. J. Plant Sci.* 52(1972):965–971.

39. Bowden, W. M. "The taxonomy and nomenclature of the wheats, barleys and ryes and their wild relatives," *Can. J. Bot.* 37(1959):657–684.

40. Ausemus, E. R., and R. M. Heerman. "Hard red spring and durum wheats: Culture and varieties," *USDA Farmers Bull.* 2139, 1959.

41. Bayles, B. B., and I. W. Taylor. "Wheat production in the eastern United States," *USDA Farmers Bull.* 2006, 1951.

42. Johnson, V. A., and J. W. anti Schmidt. "Hybrid wheat," *Adv. Agron.* 20(1968):199–233.

43. Roberts, T. H. "The price of hybrid wheat," *Crops and Soils* 22, 1(1969):5–6.

44. Laloux, R., A. Falisse, and J. Peolaert. "Nutrition and fertilization of wheat," in *Wheat*. Ciba-Geigy Ltd. Basle, Switzerland: Technical Monograph, 1980

45. Smith, R. W., and R. D. Hudson. "Wheat production guide," *GA Coop. Ext. Ser. MP* 431, 1990.

46. Whitney, D. A. "Wheat fertilization," *KS St. Univ. Coop. Ext. Ser.* C-529, 1986.

47. Reitz, L. P. "Short wheats stand tall," *USDA Yearbook*, 1968, pp. 236–239.

48. Waldren, R. P., and A. D. Flowerday, "Growth stages and distribution of dry matter, N, P, and K in winter wheat," *Agron. J.* 71(1979):391–397.

49. Goos, R. J. D. G. Westfall, and A. E. Ludwick. "Nitrogen fertilization of dryland winter wheat in eastern Colorado," *CO St. Univ. Coop. Ext. Serv. Service in Action* 554, 1984.

50. Whitney, D. A. "Nutrient management," in *Wheat Production Handbook*, KS St. Univ. Coop. Ext. Serv. C-529. Revised 1997.

51. Doll, E. C. "Effects of fall-applied nitrogen fertilizer and winter rainfall on yield of wheat," *Agron. J.* 54(1962):471–473.

52. Salmon, S. C., O. R. Mathews, and R. W. Leukel. "A half century of wheat improvement in the United States," in *Advances in Agronomy*, vol. 5. New York: Academic Press, 1953, pp. 1–151.

53. Schlegel, A. J., "Effect of cropping system and tillage practices on grain yield and soil water accumulation and use," in *Conservation Tillage Research Report of Progress*. No. 598, KS Agri. Exp. Sta. 1990.

54. Greb, B. W. "Water conservation: Central Great Plains," in *Dryland Agriculture*. Madison, WI: ASA-CSSA-SSSA, 1983.

55. Cox, D. J., J. K. Larsen, and L. J. Brun. "Winter survival response of winter wheat: Tillage and cultivar selection," *Agron J.* 78(1986):795–801.

56. Fenster, C. R., and G. A. Peterson. "Effects of no-tillage fallow as compared to conventional tillage in a wheat-fallow system," *NE Agri. Exp. Sta. Res. Bull.* 289, 1979.

57. Smika, D. E. "Fallow management practices for wheat production in the central Great Plains," *Agron. J.* 82(1990):319–323.

58. May, R. W., and C. McKee. "Furrow drill for sowing winter wheat in central Montana," *MT Agr. Exp. Sta. Bull.* 177, 1925, pp. 1–24.

59. Webb, R. B., and D. E. Stephens. "Crown and root development in wheat varieties," *J. Agr. Res.* 52, 8(1936):569–583.

60. Walton, W. R., and C. M. Packard. "The Hessian fly and how losses from it can be avoided," *USDA Farmers Bull.* 1627, 1936.

61. Witt, M. D. "Delayed planting opportunities with winter wheat in the central Great Plains," *J. Prod. Agric.* 9(1996):74–78.

62. Brethour, J. R. "Feeding wheat to leaf cattle," *KS Agr. Exp. Sta. Bull.* 487, 1966, pp. 1–43.

63. Holliday, R. "Plant population and crop yield," *Field Crop Abstracts* 13(1960): 159–167, 247–263.

64. Dedrick, B. W. *Practical Milling.* Chicago: National Miller, 1924.

65. Pomeranz, K. "Germ bread," *Bakers Digest* 44(1970):30–33.

66. Bailey, C. H. *Constituents of Wheat and Wheat Products.* New York: Reinhold, 1944, pp. 1–332.

67. Kent-Jones, D. W., and A. J. Amos. *Modern Cereal Chemistry,* 5th ed. Liverpool: Northern Publ. Co., 1957, pp. 1–817.

68. Shollenberger, J. H., and others. "The chemical composition of various wheats and factors influencing their composition," *USDA Tech. Bull.* 995, 1949.

69. Andrews, J. S., H. M. Boyd, and W. A. Gortner. "Nicotinic acid content of cereals and cereal products," *J. Ind. Eng. Chem. Anal. Ed.* 14, 8(1942):663–666.

70. Dunlap, F. L. *White versus Brown Flour.* New York: Wallace and Tiernan, 1945, pp. 1–272.

71. Thompson, T. W., and others. "Role of hard red winter wheat in the Pacific Northwest," OR *State Univ. Ext. Circ.* 812, 1972, pp. 1–27.

72. Martin, J. H., and S. C. Salmon. "The rusts of wheat, oats, barley, rye," in *Plant Diseases,* USDA Yearbook, 1953, pp. 329–343.

73. Hanson, E. W., E. D. Hansing, and W. T. Schroeder. "Seed treatments for control of disease," in *Seeds,* USDA Yearbook, 1961, pp. 272–280.

74. Holton, C. S., and V. F. Tapke. "The smuts of wheat, oats, and barley," in *Plant Diseases,* USDA Yearbook, 1953, pp. 360–368.

75. Christensen, J. J. "Root rots of wheat, oats, rye, and barley," in *Plant Diseases,* USDA Yearbook, 1953, pp. 321–328.

76. McKinney, H. H. "Virus diseases of cereal crops," in *Plant Diseases,* USDA Yearbook, 1953, pp. 350–360.

77. Sprague, R. "Rootrots of cereals and grasses in North Dakota," *ND Agr. Exp. Sta. (Tech.) Bull.* 332, 1944, pp. 1–35.

78. McKay, H. C., and J. M. Raeder. "Snow mold damage in Idaho winter wheat," *ID Agr. Exp. Sta. Bull.* 200, 1953.

79. Quisenberry, K. S., and L. P. Reitz, eds. "Wheat and wheat improvement," *Am. Soc. Agron. Mono.* 13(1967):1–566.

80. Packard, C. M. "Cereal and forage insects," in *Insects,* USDA Yearbook, 1952, pp. 581–595.

81. Packard, C. M., and J. H. Martin. "Resistant crops the ideal way," in *Insects,* USDA Yearbook, 1952, pp. 429–436.

82. Phillips, W. J., and F. W. Poos. "The wheat jointworm and its control," *USDA Farmers Bull.* 1006, 1940.

83. Percival, J. *The Wheat Plant: A Monograph.* London: Duckworth, 1921, pp. 1–463.

84. Phillips, W. J., and F. W. Poos. "The wheat strawworm and its control," *USDA Farmers Bull.* 1323, 1937.

85. Parker, J. R. "Grasshoppers," in *Insects,* USDA Yearbook, 1952, pp. 595–605.

86. Walton, W. R. "The green-bug or spring grain aphis," *USDA Farmers Bull.* 1217, 1921, pp. 1–6.

87. Wallace, L. E., and F. H. McNeal. "Stem sawflies of economic importance in grain crops in the United States," *USDA Tech. Bull.* 1350, 1966, pp. 1–50.

88. Walton, W. R., and P. Luginbill. "The fall armyworm, or grassworm, and its control," *USDA Farmers Bull.* 752, 1936, pp. 1–14 (revised).

Rye and Triticale

16.1 ECONOMIC IMPORTANCE

Rye *(Secale cereale)* was harvested from an average of 22 million acres (9 million ha) globally in 2000–2003. Production averaged about 780 million bushels (20 million MT) or 35 bushels per acre (2,200 kg/ha). The leading countries in rye production are the Russian Federation, Germany, Poland, Belarus, and Ukraine. The highest rye yields are obtained in Switzerland and northwestern European countries. Rye production is declining because of the food preference for, and higher yields of, wheat. Average production in the United States from 1970 to 1972 was nearly 38.5 million bushels (1 million MT). In 2000–2003, average production had declined to 7.8 million bushels (197,000 MT) from 287,000 acres (116,000 ha). The average yield in 2000–2003 was 27 bushels per acre (1,700 kg/ha). The leading states in rye production are Oklahoma, Georgia, North Dakota, and South Dakota (Figure 16.1).

In the states listed, rye is grown primarily for grain but occasionally for hay or pasture. In other regions, particularly in the East and Southeast, it is frequently grown for pasture or as a cover and green-manure crop, often in mixtures with vetch or clover.

16.2 HISTORY OF RYE CULTURE

Rye evidently has been under cultivation for a shorter time than has wheat. It was unknown to the ancient Egyptians and Greeks, but is supposed to have come into cultivation in Asia Minor more than 4,000 years ago.[1] Rye is found as a widely distributed weed in wheat and barley fields in southwestern Asia, where it has never been grown as a cultivated crop. Rye appears to be indigenous to that part of Asia. Rye, at first, was not intentionally sown with wheat and barley, but when its value became recognized, rye was sown as a separate crop. Cultivated rye *(Secale cereale)* apparently originated from *S. montanum*, a wild species found in southern Europe and the adjoining part of Asia and possibly grown as a cultivated plant in the Bronze Age. Another wild form of rye, *S. anatolicum*, is found in Syria, Armenia, Persia, Afghanistan, Turkestan, and the Kirghiz Steppe.[2]

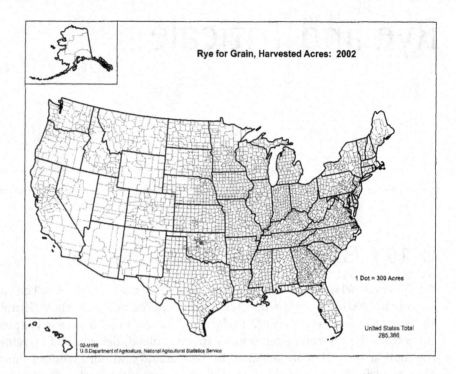

Rye for Grain, Harvested Acres: 2002

1 Dot = 300 Acres

United States Total
285,366

02-M198
U.S. Department of Agriculture, National Agricultural Statistics Service

FIGURE 16.1
Acreage of rye harvested
for grain in 2002. [Source:
Census for U.S. Department of
Agriculture]

▪ 16.3 ADAPTATION

Rye is grown in all states of the United States. An apparent cross between the *cereale* and *montanum* species called Michels grass is an escaped weed in northeastern California.[3] Improved winter hardy rye varieties are the hardiest of all cereals.[4, 5]

The highest yields of rye are usually obtained on rich, well-drained loam soils. Rye is more productive than other grains on infertile, sandy, or acid soils. It is the only small grain crop that succeeds on coarse sandy soils. It is an especially good crop for drained marshlands and cutover areas of the southeastern states when brought under cultivation. Rye usually yields less grain than winter wheat under conditions favorable for the latter crop because of its shorter growing period, heavier straw growth, and lower spikelet fertility. However, rye is usually sown on poorer soils and with poorer seedbed preparation than is wheat. Much higher yields are obtained on fertile or fertilized soils under good cultural conditions.[6, 7] It is a successful hay crop for the high desert soils of eastern Oregon.[8]

Rye volunteers freely because the grain shatters readily, and the seeds and plants thrive under adverse cultural conditions. For this reason, the growing of rye in winter wheat regions is likely to result in rye admixtures in the wheat, with a consequent depreciation in the market value of the wheat.

▪ 16.4 BOTANICAL DESCRIPTION

Rye is a summer annual or winter annual grass classified in the tribe *Hordeae* to which wheat and barley also belong.[9] The stems of rye are larger and longer than those of wheat (Figure 16.2). The leaves of the two plants are similar except that

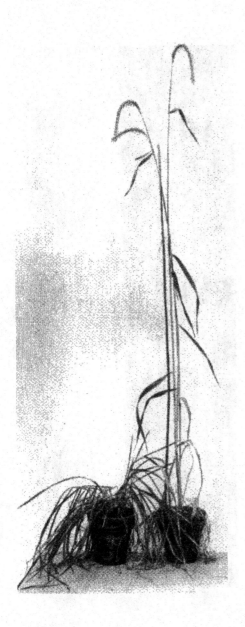

FIGURE 16.2
(Right) Spring rye plant
with spikes in flower. *(Left)*
A winter variety that failed
to head when grown in a
warm greenhouse.

those of rye are coarser and more bluish in color. The roots of rye branch profusely, especially near the soil surface, but some roots can extend to a depth of 5 or 6 feet (1.5 to 1.8 m).[10] This extensive root system is one reason why rye grows better than wheat in certain dry climates and on poor soils.

The inflorescence of rye is a spike with a single spikelet at each rachis joint (Figure 16.3). The spikelet consists of three florets, two fertile and one abortive. The spikelet is subtended by two narrow glumes. The lemma is broad, keeled, terminally awned, and bears barbs on the keel. The palea is thin and two-keeled. The caryopsis (Figure 16.4) is narrower than the wheat kernel. It is usually brownish-olive, greenish-brown, bluish-green, or yellow. The green or blue pigment, when present, is located in the aleurone, but the brownish pigment is in the pericarp. The grain color is determined by combinations of pigmentation in the two tissues. Varieties with kernels characterized by a white wax-like exterior over a blue aleurone are grayish-blue in appearance. Rye is a long-day plant but continuous light can prevent it from jointing.[11]

FIGURE 16.3
Spikes of rye: *(left)* late-blooming stage, and *(right)* mature.

FIGURE 16.4
Kernels (caryopses) of rye.

16.4.1 Pollination

Rye is a naturally cross-pollinated plant. The flower remains open for some time, which facilitates cross-pollination. About 50 percent cross-pollination between adjacent rows of rye was observed in New Jersey.[12] Sterility is frequent in rye. Approximately one-third of the flowers in Wisconsin rye fields failed to set seed.[13] Self-sterility is more common in rye than other small grains.[14] Selfed lines of rye usually show marked reduction in vigor, shriveling of the grain, and some reduction in plant height, as homozygosity is approached.[14] Some inbred lines are fairly productive.

▧ 16.5 VARIETIES

Winter varieties of rye are by far the more important economically.[4] Few varieties of rye maintain distinct characteristics because they comprise a mixture of types produced by cross-pollination. A variety grown in a different environment can quickly adapt itself to the new conditions. Changes in characteristics such as cold resistance, time to maturity, and color are the result of segregation as well as outcrossing with local varieties. Winter varieties adapted to the northern states are totally unsuited to conditions in the South where mild temperatures fail to provide sufficient cold to induce vernalization. The nonhardy varieties grown in the South and Southwest actually have a partial or total summer annual habit of growth. These varieties flower and mature before the advent of unfavorable hot summer weather.

Spring rye is rarely grown from spring sowing in North America, except for grain in cold northern areas where late-spring freezes kill the flowers of the earlier winter rye in the heading stage. Improved winter varieties with resistance to cold, rusts, or mildew are now being grown in many countries.

Productive, disease-resistant varieties grown in the southeastern states provide abundant winter pasture and green manure and good yields of grain.[6]

Tetra-Petkus, a tetraploid variety of winter rye from Germany, was grown for a time in the United States. It has very large bluish kernels.

▧ 16.6 FERTILIZERS

Rye responds to fertilizers, especially to nitrogen up to 120 pounds per acre, when grown for pasture. Mixed fertilizers are applied at seeding time. A spring top-dressing with nitrogen is desirable where the rye is pastured.[4] Heavy nitrogen applications promote lodging in rye grown for grain.[15] Nitrogen should be applied as topdress to established stands. Phosphate and potash are best applied as broadcast and incorporated before planting.[7]

▧ 16.7 ROTATIONS

Rye can replace wheat, oats, or barley in crop rotations that include a small grain. Rye for pasture or green manure is often grown in mixtures with winter legumes such as vetch, Austrian winter peas, or button clover.

▧ 16.8 RYE CULTURE

Cultural requirements for rye are similar to those for wheat.

16.8.1 Seedbed Preparation

Much of the rye grown in the Russian Federation, or in the western half of the United States, is drilled into small grain stubble without previous soil preparation. This practice, which is economical of labor, also leaves the stubble to hold snow and protect the rye plants from winterkilling. It is satisfactory on land that is reasonably free from weeds. Rye can be grown on disked corn land, fall-plowed land, or on summer fallow in the western states, but usually such land is reserved for wheat or other crops.

16.8.2 Seeding Methods

TIME OF SEEDING Winter rye can be seeded at almost any time during the late summer or early fall, but early seeding produces the most fall pasture. Winter rye ordinarily should be sown at about the same time that winter wheat is sown, but the time is less important with rye because it grows better at low temperatures and thus can be sown later. In the eastern states, rye for grain production should be sown from September 1 in the northern part of the country to November 30 in the southern. It should be seeded two to four weeks earlier when grown for pasture, as a cover crop, or for green manure. In the central Cotton Belt, the highest grain yields came from mid-November seeding. Winter rye is sown in mid-August in Alaska and about September 1 to September 15 in southern Canada and the northern part of North Dakota and Minnesota, but at later dates farther south. The optimum date for seeding winter rye in central South Dakota is about September 15.[16] Spring rye, like other spring small grains, should be sown as early as is feasible.

RATE OF SEEDING Rye is generally seeded in the western states at the rate of 55 to 70 pounds per acre (60 to 80 kg/ha).[17] The usual rate in the northeastern states is 100 to 110 pounds per acre (110 to 126 kg/ha), while in the Cotton Belt it varies from 70 to 85 pounds per acre (80 to 95 kg/ha). Heavier rates of seeding prevail where the rye is to be pastured, or for suppressing weeds, but 30 to 60 pounds per acre (35 to 70 kg/ha) may be ample for rye grown for grain in the South. The average rate in the United States is 80 pounds per acre (90 kg/ha).

16.8.3 Weed Control

Herbicides are seldom used in rye fields. Common herbicides, all postemergent, are prosulfuron, bromoxynil, 2, 4-D, MCPA, or combinations of these.

16.9 USES OF RYE

Rye is used as livestock pasture and as green manure in crop rotations. It is an excellent pasture crop before it heads. It can be pastured both autumn and spring, or only during the autumn and then the grain is harvested in the spring. Occasionally, it is grazed in the autumn and then used as a spring cover crop or tilled under as green manure for another crop of higher economic value.[18]

Rye grain is used for livestock feed and in alcohol distilling. Rye flour is used in leavened breads and many other baked products. Rye flour with lower gluten content is considered inferior to wheat flour because the dough lacks elasticity and gas-retention properties essential in production of high volume pan breads. Rye flour is used alone to produce "black" bread, which is popular in Eastern Europe and parts of Asia. Rye is mixed with 25 to 50 percent wheat flour to make "light rye" breads. Many people like the characteristic flavor of rye; small quantities of rye are used in making flat bread, rye crisps, and other baked specialty products.

Rye has a feeding value of about 85 to 90 percent that of corn. It is not highly palatable and is sticky when masticated, so it usually is ground and fed in mixture with other grains. The proportion of rye in mixed feeds is usually less than a third.

Rye grain is also used in production of alcoholic beverages. Distillers prefer a plump light-colored grain when using rye for whisky. Rye is the acknowledged trademark of Canadian whisky.

Rye straw has low digestibility and does not make good livestock fodder, but it is useful as livestock bedding. Small quantities of rye straw are also used in the manufacture of strawboard and paper.[18] Rye straw was formerly used for packing nursery stock, crockery, and other materials, and for stuffing horse collars.

Rye is more winter hardy than wheat and can produce economical yields on poor, sandy soils where no other useful crop can grow. It is grown in many areas that have no alternative crop. Its ability to compete with weeds makes rye a good rotational crop. In some countries, it is used as a pioneer crop to improve the fertility of wasteland and sterile soils. It is an important pasture crop in Argentina and is planted to prevent wind erosion in southern Australia.[18]

16.10 DISEASES

16.10.1 Ergot

Ergot *(Claviceps purpurea)* causes serious losses in rye. The ergot bodies are poisonous to livestock as well as to human beings, and often cause abortion in pregnant animals. Rye that contains 0.50 percent or more of ergot is considered unfit for food or feed. Continued consumption exceeding 0.1 percent can be toxic, inducing lameness and gangrene. Acute ergotism causes convulsions and death.[10] Ergot disease is caused by a fungus that infects the grains at the time of flowering. About seven to fourteen days after infection, the honeydew stage becomes evident. The honeydew contains spores that can infect other grains or grasses in bloom. Soon the fungus threads in the infected grains form ergot bodies (sclerotic), which resemble in shape the grain of the plant on which they develop (Figure 16.5). The purplish-black ergot bodies overwinter in the field or with the seed in storage. They are able, under favorable conditions, to germinate in the spring. Abundant moisture for germination of the ergot sclerotia, as well as warm dry weather for flower infection, are favorable for development of ergot disease in epidemic form.[19] However, rye ergot is a rare occurrence in the United States.[6]

Ergot-infested grain or grasses should be mowed or grazed before the plants flower. Ergot occurrence in grain is as follows: rye > triticale > barley > durum wheat > common wheat > oats. The spores of the sclerotia are disseminated to a great extent by winds.

Ergot can be partly controlled by sowing ergot-free seed on land where rye has not been previously grown for one or two years. The mowing of ergot-infested

FIGURE 16.5
(Left) two spikes of rye infected with ergot. *(Right below)* ergot sclerotia. *(Right above)* rye grains.

grasses adjacent to rye fields is also helpful. Infested rye should be immersed in a 20 percent solution of common salt to separate the ergot bodies. After immersion, the grain is stirred and the ergot bodies will float to the surface where they can be skimmed off. After the salt treatment, the grain should be washed and dried before seeding or feeding. Much of the ergot can be removed by aspiration cleaning equipment available in mills.

Ergot is used extensively in drug preparations. When produced in areas of low summer rainfall, it can be marketed for that purpose. Under peacetime conditions, the supply is largely imported from the Mediterranean countries.

16.10.2 Stem Smut

Stem smut *(Urocystis occulta)* attacks rye wherever it is grown, but particularly in Minnesota and nearby states. Serious losses occur when infection is high.[20] The smut appears on the culms, leaf sheaths, and blades as long narrow stripes. These stripes first appear as lead gray in color but later turn black. Infected plants are more or less dwarfed, and the heads fail to emerge from the sheath. The spores are carried both on the seed and in the soil. Seed treatment, as for stinking smut of wheat, will generally control the disease in the northern states. Crop rotation must be used where the spores are soil-borne.

16.10.3 Other Diseases

Anthracuose *(Colletotrichum graminicola)* is often found on rye in the humid to sub-humid eastern states. Infection of the culm results in premature ripening and death of the tillers. Infected tissues are stained brown on the leaf sheath that surrounds the diseased culm. Head infections can occur later to cause shriveled, light-brown kernels. The lower portion of the culm has a blackened appearance where attacked by the disease. Control measures involve proper crop sequences and plowing under crop residues.

The early maturity of rye usually enables it to escape serious damage from stem rust caused by *Puccinia graminis.* Leaf rust *(Puccinia recondita)* likewise causes little injury. Severe infections, largely confined to the southern range of rye culture, cause a reduction in tillering and decreased yields. The disease overwinters in the leaves of winter rye as dormant mycelium.[21] Destruction of volunteer rye in stubble fields will aid in control of the disease.

▓ 16.11 INSECTS

Rye is attacked by the grasshopper, chinch bug, Hessian fly, jointworm, sawfly, and other common insect pests of small grains. The total losses are not large. However, winter rye sown early furnishes a favorable environment for the depositing of grasshopper eggs, thus promoting grasshopper injury to other crops.

▓ 16.12 TRITICALE (X *TRITICOSECALE*)

Many plant breeders, since 1876, have crossed wheat with rye, sometimes with the hope of transferring the winter hardiness character of rye into wheat. Wheat was the female parent in the successful crosses. Many wheat-like wheat-rye hybrid derivatives merely carried one or more rye chromosomes.[22] These extra rye chromosomes are often lost during meiosis (sexual cell division) in later generations, and the plant reverts to typical wheat when the rye characters are lost. More recently, occasional segments of rye chromosomes have been translocated to a wheat chromosome.

Triticale, which was developed from crosses between wheat and rye, usually carries all of the chromosomes of both parents. The name was coined by combining parts of the parental generic names *Triticum* and *Secale.* Crosses with tetraploid durum wheats have been more successful than those with hexaploid common wheats. The former combination can then be crossed with common wheats and then backcrossed with triticale to retain the rye chromosomes. Triticale kernels are large, and they have a higher content of lysine and of sulfur-containing amino acids than wheat for better nutrition.[23] Most of the triticale strains have shrunken kernels, which accounts for their high protein content. Plump kernels have total protein content similar to that of wheat but higher lysine content. Breeders are attempting to develop plump triticales with short, stiff straw and better disease resistance.

Varieties of triticale bred in Mexico, United States, Canada, and other countries[15, 24, 25, 26] have been grown on a small acreage since 1969,[27] with possibly 200,000 acres (81,000 ha) in the United States in 1971. In 2000–2003, about 7 million acres (2.9 million ha) of triticale were grown worldwide. The leading countries are Poland, France, Germany, China, and Belarus. Little triticale is currently grown in the United States for grain. It is used as a cover crop, grazed, or cut for hay. Protein content and total digestible nutrients of triticale hay are comparable to wheat.

Herbicides are seldom used on triticale. Postemergent herbicides used on wheat can usually be used on triticale.

REFERENCES

1. Khush, G. S. "Cytogenetics and evolutionary studies in *Secale*: III. Cytogenetics of weedy ryes and origin of cultivated rye," *Econ. Bot.* 17(1963):60–71.

2. Khush, G. S., and G. L. Stebbins. "Cytogenetic and evolutionary studies in *Secale*: I. Some new data on the ancestry of *S. cereale*," *Am. J. Bot.* 48(1961):723–730.

3. Suneson, C. A. and others. "A dynamic population of weedy rye," *Crop Sci.* 9.2(1969).

4. Briggle, L. W. "Growing rye," *USDA Farmers Bull.* 2145, 1959.

5. Laude, H. H. "Cold resistance of winter wheat, rye, barley, and oats in transition from dormancy to active growth," *J. Agr. Res.* 54(1937):899–917.

6. Morey, D. D. "Rye, southern style," *Crops and Soils* 24, 7(1972):12–14.

7. Rehm, G., M. Schmitt, J. Lamb, and R. Eliason. "Fertilizer recommendations for agronomic crops in Minnesota." *U. MN Ext. Serv.* BU-06240, 2001.

8. Sneva, F. A., and D. N. Hyder. "Raising dryland rye hay," *OR Agr. Exp Sta. Bull.* 592, 1963.

9. Deodikar, G. B. "Rye, *Secale cereale* L.," *Indian Council Agr. Res.*, New Delhi, pp. 1–152, 1963.

10. Seaman, W. L. "Ergot of grains and grasses," *Can. Dept. Agr. Publ.* 1438, 1971.

11. Wells, D. G., and others. "Jointing and survival of rye plants and clones through successive seasons," *Crop Sci.* 7, 5(1967):473–474.

12. Sprague, H. B. "Breeding rye by continuous selection," *J. Am. Soc. Agron.* 30(1938):287–293.

13. Leith, B. D. "Sterility of rye," *J. Am. Soc. Agron.* 17(1925):129–132.

14. Brewbaker, H. E. "Studies of self-fertilization in rye," *MN Agr. Exp. Sta. Tech. Bull.* 40, 1926.

15. Walker, M. E., and D. D. Morey. "Influence of rates of N. P. and K. on forage and grain production of Gator rye in South Georgia," *GA Agr. Expt. Sta. Cir.* (New Series) 27, 1962.

16. Hume, A. N., E. W. Hardies, and C. Franzke. "The date of seeding winter rye," *SD Agr. Exp. Sta. Bull.* 220, 1926.

17. Robinson, R. R., and others. "Winter rye rate of sowing, row spacing, varietal mixtures and crosses," *MN Agr. Exp. Sta. Misc. Rpt.* 100, 1970, pp. 1–8.

18. Bushuk, W. "Rye production and uses worldwide." In W. Bushuk, ed. *RYE: Production, Chemistry, and Technology*, second ed. St. Paul, MN: AACC, 2001.

19. Brentzel, W. E. "Studies on ergot of grains and grasses," *ND Agr. Exp. Sta. Tech. Bull.* 348, 1947.

20. Leukel, R. W., and V. F. Tapke. "Cereal smuts and their control," *USDA Farmers Bull.* 2069, 1954.

21. Mains, E. B., and H. S. Jackson. "Aecial stages of the leaf rusts of rye, *Puccinia dispersa*, and of barley, *P. anomala*, in the United States," *J. Agr. Res.* 28(1924):1119–1126.

22. Longley, A. E., and W. J. Sando. "Nuclear divisions in the pollen mother cells of *Triticum*, *Aegilops*, and *Secale* and their hybrids," *J. Agr. Res.* 40(1930):683–719.

23. Knipfel, J. E. "Comparative protein quality of triticale, wheat and rye," *Cereal Chem.* 46, 3(1961):313–317.

24. Larter, E. N., and others. "'Rosner,' a hexaploid triticale cultivar," *Can. J. Plant Sci.* 50(1970):122–124.

25. White, G. A. "New crops on the horizon," *Seed World* 110, 4(1972):22–32.

26. Zillinsky, F. J., and N. E. Borlaug. "Progress in developing triticale as an economic crop," *Int. Maize and Wheat Improvement Center (Mexico) Rsh. Bull.* 17, 1971, pp. 1–27.

27. Reitz, L. P., and others. "Distribution of varieties and classes of wheat in the United States in 1969," *USDA Stat. Bull.* 475, 1972, pp. 1–70.

Barley

17.1 ECONOMIC IMPORTANCE

Barley (*Hordeum* spp. *vulgare*) ranks fifth in area among world crops harvested. In 2000–2003, barley was grown on about 136 million acres (55 million ha) with production of 6,400 million bushels (139 million MT) or 46 bushels per acre (2,500 kg/ha). The leading producing countries are the Russian Federation, Germany, Canada, France, and Spain. Production in the United States in 2000–2003 averaged about 267 million bushels (5.8 million MT) on about 4.6 million acres (1.9 million ha), yielding 58 bushels per acre or (2,800 kg/ha). The leading states in barley production are North Dakota, Idaho, Montana, Washington, and Minnesota (Figure 17.1). Most of the production in Canada is in the southern portion of the three Prairie Provinces, with smaller acreages near the southern border of the northeastern Provinces.

17.2 HISTORY OF BARLEY CULTURE

Barley apparently originated by domestication of a wild 2-rowed form, *H. vulgare spontaneum,* in or near the area bordering Syria and Iraq with Iran and Turkey. 6-row barley was being cultivated by about 6000 BC. Barley culture spread to India, Europe, and North Africa during the Stone Age. It was introduced into the Western Hemisphere by Columbus in 1493.[1] It was first grown in the United States in 1602, and additional introductions by English and Dutch settlers occurred during the next three decades.

The English settlers brought mostly 2-row types to the United States but the continental 6-row types were introduced by the Dutch. Spanish pioneers introduced the North African 6-row type of barley from Mexico into Arizona by 1701 and into California by 1771. Winter barley arrived later, possibly from the Balkan-Caucasus region or Korea. Barley production shifted westward, mainly from New York, to the North Central States after 1849.

17.3 ADAPTATION

Barley is grown throughout the more temperate regions of the world. It thrives in a cool climate. It will stand more heat under semiarid conditions than under

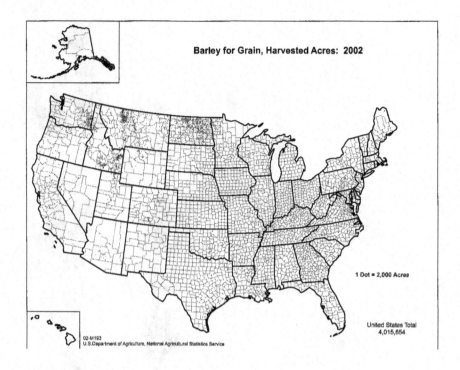

Barley for Grain, Harvested Acres: 2002

1 Dot = 2,000 Acres

United States Total
4,015,654

02-M193
U.S.Department of Agriculture, National Agricultural Statistics Service

FIGURE 17.1
Acreage of barley in the
United States, 2002. [Source:
Census of U.S. Department of
Agriculture]

humid.[2] In the warmer climates, barley is sown in the fall or winter. The best barley soils are well-drained loams. A poor crop, especially in grain quality, is produced on heavy, poorly drained soils in regions with frequent rains. Light, sandy soils are poor for barley because growth often is erratic and the crop is more likely to ripen prematurely by drought. Barley is the most dependable cereal under extreme conditions of salinity, summer frost, or drought. Barley is unsuited to acid soils below pH 6 because of aluminum toxicity that retards root growth. Calcium applications correct the toxicity and promote root growth.

The effect of adaptation on variety survival in mixtures was determined by growing a mixture of eleven varieties at ten stations for several years.[3] Population counts made each year showed a rapid elimination of the types less adapted to the environment and to competitive conditions. The variety that eventually predominated was soon evident at most stations, but a variety that led at one station might be eliminated at another. The varieties that survived best in mixtures were usually, but not necessarily, those that produced the highest yields when grown alone.[4]

17.4 BOTANICAL DESCRIPTION

Barley belongs to the grass tribe Hordeae, in which the spikes have a zigzag rachis. It belongs to the genus *Hordeum*, section Cerealia. In this section are the cultivated species *(Hordeum vulgare)* with a tough rachis and two uncultivated species with a brittle rachis, *H. agriocrithon* and *H. spontaneum*, all with fourteen diploid chromosomes. Other wild species of *Hordeum* include *H. murinum*, *H. bulborsum*, and *H. jubatum* with twenty-eight chromosomes and *H. nodosum* with forty-two chromosomes.

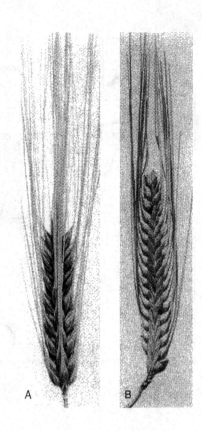

FIGURE 17.2
Spikes of 6-row barley (*A*)
and 2-row barley (*B*).

The vegetative portion of the barley plant is similar to that of the other cereal grasses except that the auricles on the leaf are conspicuous (Figure 15.3). The lateral spread of barley roots varies from 6 to 12 inches (15 to 30 cm), while the depth of penetration varies from 40 to 80 inches (1 to 2 m).

The inflorescence is a spike with three spikelets borne at each rachis node (Figures 17.2 and 17.3). Each spikelet contains a single floret. A spike usually contains ten to thirty nodes. In 6-row forms, all three florets at a node are fertile, while in 2-row barley only the central floret is fertile. Each spikelet is subtended by a pair of glumes, which are normally narrow, lanceolate bracts with short, bristle-like awns (Figure 17.4). The floret is composed of a lemma and a palea, and a caryopsis when fertile. Except in naked (hull-less) varieties, the lemma may terminate in an awn or hood, or it can be merely rounded or pointed. The awns of barley can be rough (barbed) or smooth (Figure 17.5), the latter type usually being smooth at the base and slightly roughened at the tip. Hooded barley has a bifurcate (three-forked) appendage that replaces the awn (Figure 17.3). Awnless varieties are comparatively rare. The rachilla is a small, long- or short-haired structure lying within the crease of the kernel (Figure 17.6).

The grains are about ⅓ to ½ inch (8 to 12 mm) long, ⅛ to ⅙ inch (3 to 4 mm) wide, and 1½ to ⅛ inch (2 to 3 mm) thick. A pound of seed contains about 8,000 to 16,000 kernels (3,600 to 7,300 seeds/kg). About 10 to 15 percent (usually 12 to 13 percent) of the kernel consists of hull, except in the naked varieties, which are free from the hull after threshing. Typical bushel weights for the hulled-type kernels are 48 pounds; and for the naked, 60 pounds.

FIGURE 17.3
Spike of hooded barley.

FIGURE 17.4
Barley spikelets—*(left)*
6-row, *(right)* 2-row:
(a) kernel; *(b)* lateral kernels
on 6-row barley and empty
(sterile) spike 2-row barley;
(c) awn; *(d)* glumes; and
(e) glume awn.

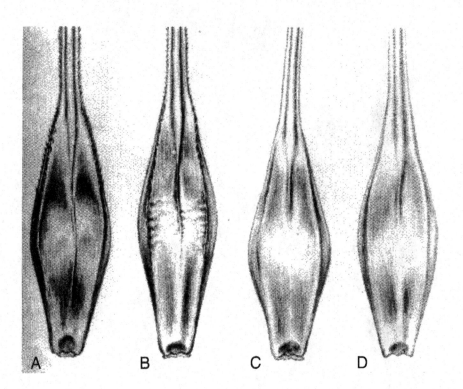

FIGURE 17.5
Rough-awned barleys *(A and B)* and smooth-awned barleys *(C and D)*. The teeth extend down the veins of the glumes of rough-awned kernels. A few teeth occur on *(C)*.

FIGURE 17.6
(Left) Two-row Hannchen barley kernel showing long rachilla hairs and wrinkled palea. *(Right)* The two lateral kernels and the one central kernel of a 6-row barley with short rachilla hairs. The lateral kernels are curved and asymmetrical.

Grain color is determined by the aleurone, a layer of cells just inside the pericarp and testa. Five color conditions are recognized in the barley grain:[5] white, black, red, purple, and blue. The last three colors are due to anthocyanin pigments. When pigments occur in the barley hulls, they are red or purple, but when they occur in the aleurone layer, the grains are blue. The red color usually fades. Naked barley

FIGURE 17.7
Kernels (caryopses) of naked barley.

(Figure 17.7) with red in the pericarp and a blue aleurone appears to be purple. Black comes from melanin-like pigments in the hulls or pericarp. The white and blue barleys are the only ones grown extensively.

17.4.1 Pollination

Barley is generally self-fertilized because pollination occurs while the head is partly in the boot in many varieties. In Minnesota, occasional natural crosses occurred in some varieties when white and black varieties were grown side by side.[6] Similar results were obtained in Colorado, where the commercial varieties showed less than 0.15 percent natural crosses.[7]

During fertilization, pollen can germinate within five minutes after it falls on the stigma. Within six hours after pollination, fertilization of both the egg cell and endosperm nucleus is completed and cell division has begun[8] (Figures 3.13 and 4.4).

■ 17.5 CULTIVATED BARLEY SPECIES

Cultivated barley (*Hordeum* spp.) includes three species based on the fertility of the lateral spikelets: (1) 6-row barley *(H. vulgare)*; (2) 2-row barley, usually listed as *H. distichum*; and (3) 4-row or irregular barley *(H. irregulare* or *H. tetrastichum)*.[1]

The first species, with all florets fertile, includes (1) ordinary 6-row barley with lateral kernels slightly smaller in size than the central one (Figure 17.6) and (2) the intermedium group, in which the lateral kernels are markedly smaller in size than the central one.

The second species with only the central florets fertile is divided into two groups: (1) the common 2-row type with lateral florets consisting of lemma, palea, rachilla, and reduced sexual parts, and (2) a deficiens group with lateral florets reduced and consisting of lemma, rachilla, and, rarely, palea, but with no sexual parts.

The third species, of Ethiopian origin, has sometimes been called Abyssinian intermediate. The central florets are fertile and the lateral florets are reduced to rachillae in some cases, and these are distributed irregularly on the spike. The remainder of the lateral florets can be fertile, sterile, or sexless.

Barley varieties grown on farms are predominately the ordinary 6-row and 2-row types. So-called 4-row barleys are 6-row forms in which 2 rows of lateral florets overlap to form single rows on each side of the spike. The 4-row effect usually occurs only in the upper two-thirds of the spike.

■ 17.6 REGIONAL TYPES AND VARIETIES

Probably more than 4,000 varieties are grown worldwide. In the United States, those grown in the humid spring barley region of the upper Mississippi Valley include mostly 6-row types developed from crosses. Some varieties have smooth awns. Two-row varieties are grown there occasionally.

Both 6-row and 2-row varieties are grown in the semiarid Great Plains as well as in the intermountain and western regions. The 6-row varieties in the Far West are mostly hulled, rough-awned types, largely of North African origin. Some smooth-awned varieties are also grown.

True winter-barley varieties are grown in the area from New York to Colorado and southward, but also occasionally in Utah, Idaho, and the Pacific Coast states. Hooded winter types are rarely grown except in the southern region where smooth-awned varieties are also grown.

Hooded varieties in spring or winter types are grown occasionally, generally for hay. Naked barley, grown rarely, is usually fed to poultry.

■ 17.7 BARLEY IMPROVEMENT

An interesting development in barley improvement was the breeding of smooth-awned varieties (Figure 17.5).[5] This eliminated the severe irritation resulting from handling rough-awned varieties previously grown. Awned varieties are valuable because the awns function in transpiration and photosynthesis and as a depository for mineral matter.[9] When the awns are removed, the ash content of the rachis increases, which may account for the tendency of such spikes to break easily. Smooth-awned barley has no physiological limitations when compared with standard rough-awned varieties.[10] Breeding for resistance to a majority of the barley diseases has been at least partly successful.

Development of winter-hardy varieties extended winter barley northward from Oklahoma into Nebraska and Colorado, and from Maryland and Kentucky northward to New York and the eastern Great Lakes states.

A study of barley crosses in Idaho showed that high-yielding selections from hybrids were obtained more frequently from intercrossing 6-row varieties than from crossing 6-row with 2-row varieties. Hooded segregates were definitely inferior to awned ones, and naked segregates were slightly less productive than covered (hulled) strains. Smooth-awned forms averaged greater floret sterility and were slightly lower in yield than the rough-awned forms, but it seemed likely that some of the smooth strains might prove to be equal to the best rough ones.[11]

■ 17.8 FERTILIZERS

Nitrogen fertilizer is applied at a lower rate for malting barley than for feed barley because malting barley should have lower seed protein. Malting barley will generally receive 10 to 20 pounds per acre (11 to 22 kg/ha) less nitrogen depending on yield goal.[12] Phosphorus and potassium rates are the same for both uses. Total nitrogen requirement is about 70 to 80 pounds per acre (78 to 90 kg/ha) for a 40 bushel per acre (2,200 kg/ha) feed crop, and 150 to 160 pounds (170 to 180 kg/ha) for a yield of 80 bushels per acre (4,300 kg/ha). Heavy applications can induce lodging in some varieties. Barley will likely respond to phosphorus fertilizer, unless the soil level is high, and to potassium in soils with less than 75 to 100 ppm.[12, 13, 14, 15, 16]

■ 17.9 ROTATIONS

Crop rotation helps control many diseases of barley. In general, barley makes its best growth after a cultivated crop such as corn, sugarbeet, or potato. In humid regions, barley is usually rotated with a row crop such as corn or soybean. Sometimes a perennial legume such as alfalfa is included in the rotation and barley is used as a companion crop. On irrigated land in the West, barley is often used as a companion crop for alfalfa. The latter generally is retained for two years or more before being tilled under; it is followed by one or two years of cultivated crops. In the semiarid Great Plains, a common practice is to grow barley after fallow or on disked corn or sorghum land. In the Corn Belt, barley often follows oat, wheat, or soybean instead of corn in order to avoid losses from the scab disease, which is carried on corn. In the Red River Valley, barley will follow sunflower in a multi-crop rotation.[17] In the semiarid portions of eastern Washington and Oregon, barley can alternate with fallow or pea. In the South, winter barley occupies the same place in the rotation as wheat. In Idaho and eastern Washington, rotation of spring barley, fallow, and winter wheat can be used[18] or included in a rotation of pea, barley, winter wheat, clover, lentil, and rapeseed.[19]

■ 17.10 BARLEY CULTURE

The cultural methods for barley are similar in most respects to those for wheat and oat. Weed control measures for barley are about the same as those for wheat.

17.10.1 Seeding Methods

TIME OF SEEDING Maximum yields from spring barley are obtained when the crop is seeded early in the spring. On the northern Great Plains, the most favorable period is from April 1 to 25. Earlier planting increases the likelihood of lower seed protein needed for malting.[12] In southern Minnesota, Iowa, and Wisconsin, where the season is slightly earlier, late seeding is more disastrous than in the northern plains. In New England, the cool summer permits a slightly later seeding. In California and southern Arizona, maximum yields are obtained from late fall or early winter seeding, before December 20. The varieties grown there have a spring growth habit, but they survive the winter under the mild temperature conditions. Cool temperatures and short photoperiods following late fall seeding prolong the vegetative period for about two or three months beyond that required for the same varieties when sown in the spring. The longer growing period favors greater yields. Good results have been obtained over most of the central and southern winter-barley area with September seeding.

RATE OF SEEDING In practice, 90 to 100 pounds per acre (100 to 110 kg/ha) of spring barley are usually seeded in the humid regions. On the northern Great Plains, 45 to 90 pounds (50 to 100 kg/ha) is the usual rate, the lower rate being used in the drier localities. In the very dry localities of the Great Basin, 30 to 40 pounds (34 to 45 kg/ha) are sometimes seeded. In Colorado, about 100 pounds (110 kg/ha) are recommended for irrigated areas.

DEPTH OF SEEDING Barley seed should be sown at a depth at which both moisture and air are available, or about 1½ inches (4 cm) in the humid regions, 2 inches (5 cm) on the northern Great Plains, and 2½ to 3 inches (6.5 to 8 cm) in the Great Basin. The crop is sown with a grain drill.

17.10.2 Weed Control

Weeds that infest wheat and other small grains can also be a problem in barley. However, since barley is usually grown as part of a rotation that includes a wide variety of crops, weeds are less of a problem due to effective control methods with other crops. A few postemergent herbicides are available when needed. Several sulfonylurea herbicides or 2, 4-D can be applied in early vegetative stages before jointing.

17.10.3 Harvesting

The domestic barley crop is threshed with a combine. If the variety is susceptible to shattering, particularly with malting types, the crop can be swathed when the heads have turned a golden yellow, and while the straw is slightly green. It is picked up and threshed after the grain has dried three or four days. About one-third of the acreage is harvested by this method.

Barley is ready for harvest for grain with a windrower when it is physiologically mature, when dry matter ceases to be added to the kernel (Figure 17.8). Maturity is indicated when a thumbnail dent in the kernel remains visible for some time. At this stage, the milky juices have disappeared from the kernel. The ripening process after this time is principally moisture loss.

■ 17.11 USES OF BARLEY

About half of the barley grown in the United States is used for livestock feed; the remainder is used for food, malt, and seed. About 85 percent of the malt is used in making beer, 10 percent for making industrial alcohol and whisky, and the remainder for malt syrup.

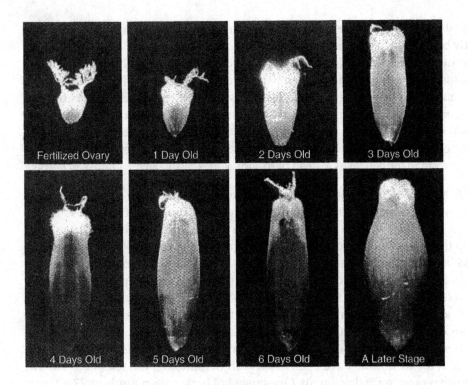

FIGURE 17.8
Barley kernels at different ages of development.

Barley has a feeding value of about 95 percent that of corn. The hulls, comprising about 13 percent of the kernel, detract from the nutritive value. Barley requires grinding or rolling for satisfactory feeding to animals other than sheep. Barley is often steamed to soften the grain before it passes to the rolls for crushing into flakes.

Barley is a better companion crop for legumes and grasses than are most varieties of oat or wheat because it produces less shade and matures earlier. Barley has been cut for hay extensively in California. Winter barley is a good winter-cover crop in the southeastern states. Some of it is pastured.

17.11.1 Malting and Brewing

Maltsters in the United States desire a plump, mellow, small-kerneled barley with tight hulls.[20] Such barley should be all of the same type—starchy, mellow, of high germination capacity, sound, and free from either weather or disease damage. Certain proteins of high solubility are detrimental. Flinty (glassy) kernels, a condition brought about by interrupted growth or other factors, are objectionable because they contain intermediate products undesirable for malting. All of the malting grades specify 70 percent or more of mellow kernels that are not *semisteely*. Barley with a blue aleurone appears somewhat *steely*, but the color itself has no known effect upon malting quality. Broken kernels are objectionable because they will not malt. Skinned or peeled grains fail to germinate or convert properly during the malting process. Musty barley is unacceptable. Damaged barley, of which a maximum of 4 percent is allowed in the malting grades, includes blighted, moldy, heat-damaged, and weathered kernels. In the humid spring region, the 6-row varieties with a white aleurone layer are the most acceptable for malting. Certain varieties with a blue aleurone are acceptable, but these are graded in a special subclass.

Several 2-row varieties are acceptable to certain maltsters. Six-row varieties with a partly blue aleurone also are suitable for malting.

Cultural practices can determine whether the barley crop will meet the requirements for the malting grades. Barley harvested before it is fully ripe in the field can have an objectionable green tinge. The crop should be harvested to provide minimum exposure to unfavorable weather. Skinned or broken kernels usually result from threshing at too high a cylinder speed (600 revolutions per minute), from concave teeth set up too close to the cylinder, or from end play in the cylinder. The use of square teeth in the concaves instead of rounded ones results in fewer broken kernels.

MALTING PROCESS Barley is malted to produce or activate enzymes, mostly alpha-amylase and beta-amylase, which hydrolize starch into sugars. Enzymes also initiate germination, convert proteins into soluble compounds, and develop the aroma and flavor of malt. In the manufacture of malt, the cleaned and sized barley is soaked (steeped), with occasional draining, until the grain contains 44 to 46 percent moisture. It then germinates at a temperature of about 68°F (20°C) in drums or tanks on aprons with frequent but slow stirring with water added to maintain the moisture content. When the sprouts (acrospires) are about 75 to 100 percent of the length of the kernel, kiln-drying the grain down to 4 to 5 percent moisture stops germination. The rootlets (malt sprouts), which in the meantime have been broken off in handling, are screened out. The produce remaining is dried malt. One bushel of barley (48 pounds or 22 kg) produces about 38 pounds (18 kg) dried malt. This quantity of malt with some 17 pounds (8 kg) of adjuncts (corn or rice products, sugar, syrup, and so on) and about 0.7 pound (0.3 kg) of hops and hop extracts is sufficient to make one barrel (31 gallons or 117 l) of beer.

BREWING PROCESS The brewing industry is usually separate from the malting industry. In brewing,[21] the malt is first ground and a portion of the ground malt mixed with unmalted cereals (corn grits, broken rice, and so on). These starchy adjuncts are used to furnish additional fermentable material and to reduce the protein content of the mash. This mixture is then cooked in water to gelatinize and liquefy the starch, after which it is mixed with water and the remaining malt in a mash kettle or tub. The latter has previously been held for half an hour at approximately 118°F (48°C) to favor the breakdown of the proteins. The temperature is raised to 149°F (65°C) and held for a few minutes until the conversion of the starch is complete. Then the temperature is raised to 167°F (75°C), and the mash is mashed off (allowed to settle). The liquid (first wort plus spargings) is drawn off and boiled with hops, after which the wort is then drawn off and cooled. The wort is fermented with yeast for seven to ten days at temperatures ranging from 43 to 59°F (6 to 15°C). Proteins, yeast cells, hop resins, and other insoluble materials settle out during the cooling and the subsequent cool storage, which lasts for three to eight weeks at temperatures that usually range from 32 to 41°F (0 to 5°C). The finished beer is then carbonated and filtered, after which it is ready for the market.

USES OF MALT The spent malt, after brewing, is sold for feed, as dried or wet brewers grains. Malt syrup is used in baking, in candies, and in the textile industries. It is also added to medicines for its laxative effect and as an aid in the control of fevers. Many breakfast-cereal manufacturers use malt syrup. Malt is also used for the production of malted milk, alcohol, vinegar, and yeast.

17.11.2 Other Products of Barley

Pearled barley is made by grinding off the outer portions of the kernel (hull, bran, aleurone, and germ) so as to leave the kernel as a round pellet. These outer parts are removed by the abrasive surface of a whirling, rough, pearling stone. The larger, more spherical types of kernel result in a better pearled product with less loss of outer layers in processing.[6] Plump, large-grained, 2-row, white-grained varieties with a shallow or closed crease are best for pearling. Pearled barley is cooked for human food and is used in baby foods and other products. Partly pearled grains are marketed as pop barley.

Barley flour contains no gluten and it darkens the bread loaf. Large-grained, 2-row varieties with white aleurone layers are best suited to flour milling. Naked barley is grown in parts of Asia for home grinding and cooking. High lysine barley was found in Sweden in 1968.

▦ 17.12 DISEASES

Many of the diseases that attack wheat also attack barley. Some of the more common diseases of barley are listed here. Genetic resistance to many diseases is available for incorporation into modern varieties, but any disease can still be a problem if conditions for its development and spread are ideal.

17.12.1 Covered Smut

Covered smut *(Ustilago hordei)* attacks barley heads and causes formation of hard, dark lumps of smut in place of the kernels. The diseased heads are often borne on shorter stems and appear later than the normal heads. Soon after the smutted heads appear, the membranes that enclose the smut galls begin to split, which releases the spores to be carried to uninfected heads. Infection can occur before the barley is ripe or any time thereafter.[22] Spores that reach the barley kernels frequently germinate and send infection threads beneath the hulls. Covered smut is often found in threshed grain as black, irregular, hard masses.

A fungicidal seed treatment can be effective in control of covered smut. Some varieties are moderately resistant to covered smut.

17.12.2 Brown and Black Loose Smuts

The brown and the black loose smuts attack barley heads, causing the formation of loose, powdery masses of smut in place of the normal spikelets. The dusty spores are blown over the field at the time the normal heads are in blossom. Some spores fall into the open blossoms, resulting eventually in the infection of the next crop. The spore masses of the nude or brown loose smut *(Ustilago nuda)* are olive brown, while those of the nigra or black loose smut *(U. nigra)* are dark brown, almost black. The brown loose smut infests the interior of the seed, while the black loose-smut infestation is superficial.

Use of smut-free seed and a fungicidal seed treatment can provide control. Some varieties are resistant, except when environmental conditions are especially favorable for infection.

17.12.3 Septoria Leaf Blotch

Septoria leaf blotch (*Septoria passerinii*) causes yellowish to light brown elongated spots on the leaves. Spots may merge and almost cover the leaf surface. Margins of leaf may pinch and dry. Black fruiting bodies form in rows in diseased areas.

Some varieties are more resistant than others. Crop rotation and burying infected residue can help prevent initial infection.

17.12.4 Barley Stripe

Barley stripe (*Helminthosporium gramineum*), sometimes called Helminthosporium stripe, is an important disease in the entire United States. The disease causes long white or yellow stripes on the leaves. These stripes later enlarge and turn brown. Many of the stripes may run together so as to discolor the entire plant. The affected plants are stunted, the heads fail to emerge properly or at all, and the grain is discolored and shrunken.

Because the spores are seed-borne, the disease can be controlled by treatment with a fungicide applied to the seed. Several varieties are resistant.

17.12.5 Spot Blotch

Spot blotch (*Cochliobolus sativus*) is widespread in the north, central, and southern portions of the barley area. Dark brown spots coalesce to form blotches on the leaves, stems, and floral bracts. Numerous secondary infections occur up to the time the plant ripens. The disease overwinters on seed and on plant remains in the field.

Both crop rotation and sanitation are important control measures because the organism develops on the crop residues of cereal plants. Seed treatments control only seed infection. Several varieties are resistant to the disease.

17.12.6 Scab or Fusarium Head Blight

Scab (*Fusarium graninearum*), also called fusarium head blight, first appears as brownish spots or lesions at the base of the glumes or on the rachis. The scab organism infects barley heads at the blossom stage. The diseased kernels ripen prematurely and turn pinkish to dark brown in color. The kernels are often shrunken. Infected seed produces seedling blight when it is sown. The scab organism overwinters principally on corn stalks and small grain stubble.

A rotation with broadleaf crops and tillage to bury residue tends to reduce the amount of head blight somewhat, but it is not a complete control. Seed treatment will reduce the amount of seedling blight. Certain varieties have some resistance.

17.12.7 Scald

Scald (*Rhyncosporium secalis*) is a problem in northern regions where cool, wet weather fosters the disease. It produces oval lesions with margins that change from bluish-green to tan or brown rings on the leaves, sheaths, and spikes, causing a shriveling of the grain. The use of resistant varieties, rotation, and burying residue help with control.

17.12.8 Powdery Mildew

Powdery mildew *(Erysiphe graminis hordei)* develops when cool, humid, and cloudy weather persists. White to gray powdery pustules can completely cover the leaves and stems. Several varieties are resistant to several races of the fungus.

17.12.9 Barley Rusts

Stem rust caused by the fungus *Puccinia graminis* results only occasionally in heavy losses in barley. Some productive varieties are resistant.

Leaf rust of barley caused by the fungus *Puccinia hordei* sometimes is damaging in the South. A few varieties are somewhat resistant. Stripe rust caused by *Puccinia striiformis* is controlled with resistant varieties.

17.12.10 Virus Diseases

A seed-borne virus causes barley stripe mosaic virus disease. This serious disease can occur in any barley region of North America. The symptoms are yellow or light-green stripes, or complete yellowing of the leaves, with eventual stunting of the plants. Control measures are sowing disease-free seed and resistant varieties.

Barley yellow dwarf is a virus disease that is transmitted by several species of aphids. The symptoms are a brilliant golden yellowing of the leaves as well as moderate to severe stunting of the plants. Control measures include early seeding, aphid control, and resistant varieties.

17.12.11 Other Diseases

Among other barley diseases of some importance are ergot *(Claviceps purpurea)* (see ergot of rye, Chapter 16), net blotch *(Pyrenophora teres)*, and bacterial blight *(Phytomonas translucens)*. A root and stem disease caused by *Pythium graminicola* damages barley seriously in the Corn Belt states in some years.

▦ 17.13 INSECT PESTS

Among the most important insect pests of barley are the chinch bug, greenbug (spring-grain aphis), and grasshoppers. Barley is very attractive to chinch bugs. Several barley varieties of Asiatic origin, as well as some American varieties, are resistant to greenbugs. Grasshoppers are controlled by insecticides.

The Russian wheat aphid can be a serious pest of barley in areas where it attacks wheat. Chapter 15 gives a complete description and control methods for this pest.

REFERENCES

1. Wiebe, G. A., and others. "Barley: Origin, botany, culture, winter hardiness, genetics, utilization, pests," *USDA Handbk.* 338, 1968, pp. 1–127.

2. Leukel, R. W. "Seed treatment for controlling covered smut of barley," *USDA Tech. Bull.* 207, 1930.

3. Harlan, H. V., and M. L. Martini. "The effect of natural selection in a mixture of barley varieties," *J. Agr. Res.* 57(1938):189–200.

4. Suneson, C. A., and G. A. Wiebe. "Survival of barley and wheat varieties and mixtures," *J. Am. Soc. Agron.* 34, 11(1942):1052–1056.

5. Harlan, H. V., and M. L. Martini. "Problems and results in barley breeding," in *USDA Yearbook*, 1936, pp. 303–346.

6. Stevenson, F. J. "Natural crossing in barley," *J. Am. Soc. Agron.* 20(1928):1193–1196.

7. Robertson, D. W., and G. W. Deming. "Natural crossing in barley at Fort Collins, Colo.," *J. Am. Soc. Agron.* 23(1931):402–406.

8. Pope, M. N. "The time factor in pollen-tube growth and fertilization in barley," *J. Agr. Res.* 54(1937):525–529.

9. Harlan, H. V., and S. Anthony. "Development of barley kernels in normal and clipped spikes and the limitations of awnless and hooded varieties," *J. Agr. Res.* 19(1920):431–472.

10. Hayes, H. K., and A. N. Wilcox. "The physiological value of smooth-awned barleys," *J. Am. Soc. Agron.* 14(1922):113–118.

11. Harlan, H. V., M. L. Martini, and H. Stevens. "A study of methods in barley breeding," *USDA Tech. Bull.* 720, 1940.

12. Danhke, W. C., C. Fanning, and A. Cattanach. "Fertilizing malting and feed barley," *ND St. Univ. Ext. Ser.* SF-723, 1992.

13. Foote, W. H., and F. C. Batchelder. "Effects of different rates and times of application of nitrogen fertilizers on the yield of Hannchen barley," *Agron. J.* 45(1953):532–535.

14. Mahler, R. L., and S. O. Guy. "Spring barley," *Univ. ID Coop. Ext. Ser.* CIS 920, 1998.

15. Mahler, R. L., and S. O. Guy. "Winter barley," *Univ. ID Coop. Ext. Ser.* CIS 954, 1997.

16. Pendleton, J. W., A. I. Lang, and G. H. Dungan. "Responses of spring barley to different fertilizer treatments and seasonal growing conditions," *Agron. J.* 45(1953):529–532.

17. NDSU Extension Service. "Crop rotations for increased productivity," *ND Coop. Ext. Ser.* EB-48, 1998.

18. Hinman, H., and D. Bragg. "Crop enterprise budgets, spring barley-summer fallow-winter wheat rotation," *WA St. Univ. Coop. Ext. Ser.* EB1114, 1990.

19. Painter, K., D. Granatstein, and B. Miler. "Alternative crop rotation enterprise budgets," *WA St. Univ. Coop. Ext. Ser.* EB1725, 1992.

20. Harlan, H. V., and G. A. Wiebe. "Growing barley for malt and feed," *USDA Farmers Bull.* 1732, 1943.

21. Shands, H. L., and J. G. Dickson. "Barley—Botany, production, harvesting, processing, utilization, and economics," *Econ. Bot.* 7(1953):3–26.

22. Tapke, V. F. "Studies on the natural inoculation of seed barley with covered smut (*Ustilago hordei*)," *J. Agr. Res.* 60, 12(1940):787–810.

Oat

▓ 18.1 ECONOMIC IMPORTANCE

Oat *(Avena sativa)* was grown on about 31 million acres (13 million ha) in 2000–2003, with a production of about 29 million tons (26 million MT) or 58 bushels per acre (2,000 kg/ha). The leading countries in oat production are the Russian Federation, United States, Canada, Finland, and Poland (Figure 18.1). Oat was harvested for grain on an average of about 2.1 million acres (860,000 ha) in the United States in 2000–2003. Production was about 132 million bushels (2.9 million MT) or 62 bushels per acre (3,300 kg/ha). The leading five states in oat production, Minnesota, North Dakota, Wisconsin, Iowa, and South Dakota, produce over half of the national crop. The production of oat has declined about 80 percent in the United States in the past 30 years because other feed grains, particularly corn and sorghum, outyield oat. It continues to decline in other countries as well because of other feeds and also because work animals continue to be replaced by motorized equipment.

The lack of yield gains in oat compared to other grains has definitely contributed to its decline. Since 1900, genetic gains in oat yields have only been about 9 to 14 percent[1] compared to gains of 35 to 100 percent or more in other crops.[2] Oat yields in sixty-six varieties only increased 9 percent from 1935 to 1975.[1] One reason for the small increase is because oat germplasm contains very little genetic diversity.

Compared with other grain crops, oat is the easiest and most pleasant to sow, harvest, handle, and feed. It can be fed to horses, sheep, and poultry without grinding. The spring varieties are sown early in the spring before other crops. It also furnishes excellent cereal hay as well as the best straw for feed or bedding. Fall-sown oat is valuable in the South for pasture as well as for retarding soil erosion.

▓ 18.2 HISTORY OF OAT CULTURE

Cultivated oat *(Avena sativa)* was formerly believed to have been derived from two species, common wild oat *(A. fatua)* and wild red oat *(A. sterilis)*.[3] Later evidence[4] indicates that both *A. sativa* and *A. fatua* were probably derived from *A. byzantina* or its progenitor, *A. sterilis*. Apparently, cultivated oat was unknown to the ancient Egyptians, Chinese, Hebrews, and Hindus. Among the earliest

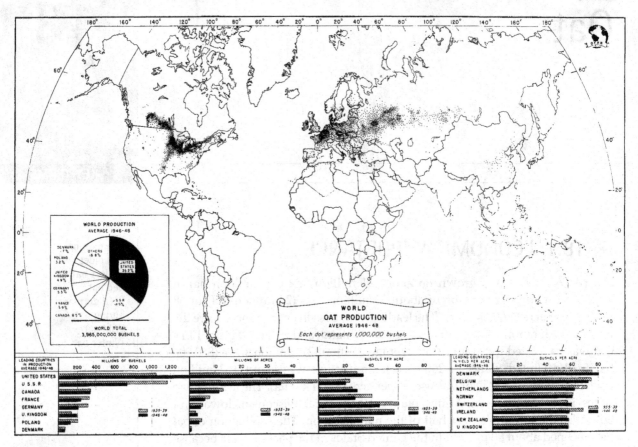

FIGURE 18.1
World oats production, average from 1946 to 1948. Each dot represents 1 million bushels.

indications of the existence of oat are specimens of *A. strigosa* found in the remains of the Swiss lake dwellers of the Bronze Age. Cultivated species of oat were evident from 900 to 500 BC, but classical writers in Roman times mention oat only as a weed that was sometimes used for medicinal purposes.[5] Probably, it was first distributed as a weed mixture in barley and domesticated later. Authentic historical information on cultivated oat appears in the early Christian era. Common oat was reported by writers of this period to be grown by Europeans for grain, while red oat was grown for fodder, particularly in Asia Minor. Common oat, first found growing in Western Europe, spread to other parts of the world. It was believed to have been first cultivated by the ancient Slavonic people who inhabited this region during the Iron and Bronze Ages.

18.3 ADAPTATION

Common oat is best adapted to the cooler, more temperate regions where the annual precipitation is 30 inches (760 mm) or more, or where the land is irrigated (Figure 18.2). Such areas are found in the northeastern quarter of the United States, in the Pacific Northwest, in valleys of the Rocky Mountain region, in northern Europe, and in Canada.[6] High yields of large, plump grains are possible in these areas. The oat crop often fails in the Great Plains because of drought and heat. Hot,

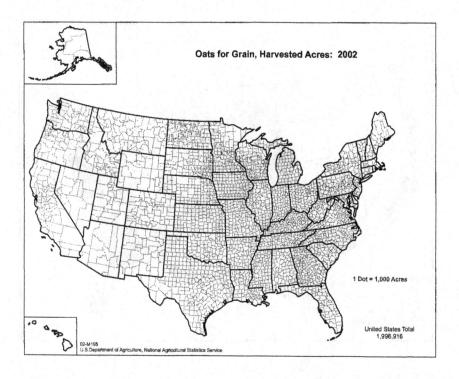

Oats for Grain, Harvested Acres: 2002

1 Dot = 1,000 Acres

United States Total
1,996,916

02-M195
U.S.Department of Agriculture, National Agricultural Statistics Service

FIGURE 18.2
Acreage of oat harvested in the United States in 2002. [Source: 2002 Census of U.S. Department of Agriculture]

dry weather just before heading causes oat to blast (see section 18.13.5), and, during the heading and ripening period, it causes oat to ripen prematurely, with poorly filled grain of light bushel weight. Such damage has been reduced to some extent in the warmer regions by growing early varieties. The greatest heat resistance was shown by varieties adapted to the South that were also resistant to cold. Red oat is grown primarily in warm climates such as in North Africa and Argentina. Heat-tolerant varieties of red oat have been grown in the southern states as well as in the interior valleys of California. Some of these varieties are extremely early. Fall-sown oat is grown successfully where the winters are sufficiently mild, mostly in the South and on the Pacific Coast.[7] The general oat areas are shown in Figure 18.3.

Oat produces a satisfactory crop on a wide range of soil types, provided the soil is well drained and reasonably fertile. In general, loam soil, especially silt and clay loam, is best suited to oat. Heavy, poorly drained clays are likely to cause the crop to lodge or be injured by plant diseases such as rusts and mildew. Excessive available nitrogen and moisture are the main causes of lodging in oat.

18.4 BOTANICAL DESCRIPTION

Oat is an annual grass classified in the genus *Avena.* Under average conditions, the plant[3] produces three to five hollow culms from ⅛ to ¼ inch (3 to 6 mm) in diameter and from 2 to 5 feet (60 to 150 cm) in height (Figure 18.4). Some varieties have hairs on the nodes. The roots are small, numerous, and fibrous and penetrate the soil to a depth of several feet. The average leaves are about 10 inches long and ⅝ inch wide (250 × 16 mm).

The inflorescence is either an equilateral (spreading) or a unilateral (one-sided) panicle (Figure 18.5). The panicle is a many-branched, determinate inflorescence

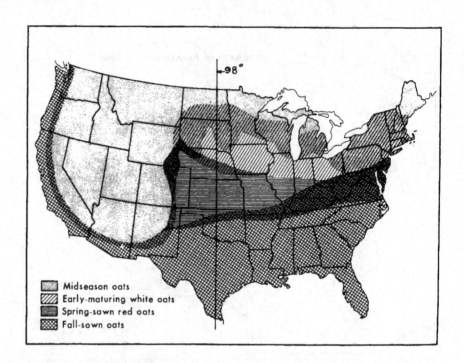

FIGURE 18.3
Areas in which the principal types of oats are grown in the United States. At the northern border of the winter oat area the average minimum winter temperature is about -5°F. Midseason oats are grown in the cool climates of the northern states and the intermountain region. Early white and red oats are grown between the winter-oat and midseason-oat areas.

Legend:
- Midseason oats
- Early-maturing white oats
- Spring-sawn red oats
- Fall-sown oats

FIGURE 18.4
Oat plants in a field.
[Courtesy Richard Waldren]

consisting of a main axis from which arise lateral axillary branches that are grouped on alternate sides of the main axis at the nodes.[8] The main axis and each of the lateral branches terminate in a single apical spikelet. The axis bears four to six whorls of branches.

The oat spikelet is borne on the thickened end of a slender drooping pedicel that terminates the panicle branch. Each spikelet usually contains two or more florets.[8, 9] The lower two florets are usually perfect, while the third, when present, is often sta-

FIGURE 18.5
Oat panicles. *(Left)*
Unilateral, "side," or "horse-mane" with seven whorls of branches. *(Right)*
Equilateral, spreading, or "tree" panicle with five whorls of branches.

minate or imperfect. The two glumes are somewhat unequal, lanceolate, acute, boat-shaped, spreading, glabrous, membranous, and usually persistent. Both usually exceed the lemma in length, except in hulless (naked) oat. The floret is composed of the lemma, palea, and the reproductive organs—namely, the ovary and three stamens (Figure 4.2). The lemma is the lower of the two bracts that enclose the kernel. It ranges in color from white, yellow, gray, and red to black. The base of the lemma may extend into a callous, which often bears basal hairs. Most wild oat species are characterized by hairiness of the callous, lemma, and rachilla (Figure 18.6). The awn of oat is an extension of the midrib of the lemma that usually arises slightly above the middle of the dorsal surface. The awn may vary from almost straight to twisted at the base and bent into a knee (geniculate). Some varieties are awnless. The palea is the inner bract of the floret, thin and parchment-like. In hulled varieties, the lemma and palea firmly enclose the caryopsis. The oat caryopsis is narrowly oblong or spindle-shaped, deeply furrowed, and usually covered with fine hairs, especially at the upper end (Figures 18.7 and 18.8).

The natural separation of the lower kernel from the axis of the spikelet is termed *disarticulation*.[10, 11] In some wild oat species, as well as in their cultivated derivatives, disarticulation leaves a well-defined, deep, oval cavity commonly called sucker mouth. In most cultivated oat varieties, the separation is from a fracture or breaking that results in a roughened tissue with no observable cavity at the base of

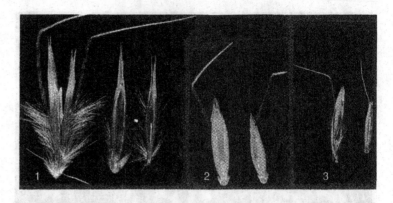

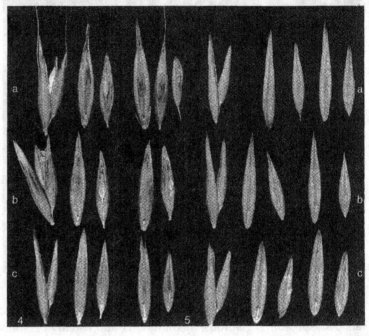

FIGURE 18.6
Spikelets and kernels of wild and cultivated oats. (1) Wild red oat *(Avena sterilis macrocarpa);* (2) fatuoid from *A. sativa;* (3) wild oats *(A fatua);* (4) cultivated red oats *(A. byzantina);* (5) cultivated common oats *(A. sativa).*

the lemma. Most varieties of common white and yellow oat lack the scar commonly found in red oat.

The legal weight per bushel of oat is 32 pounds (410 g/l), but the actual test weight may range from 27 to 45 pounds (350 to 580 g/l). Clipped oat has a higher test weight because the tip of the hull has been removed.

▓ 18.5 SPECIES OF OAT

Oat may be classified into three groups on the basis of chromosome numbers.[12]

Group I: Seven Haploid Chromosomes

A. brevis, short oat

A. wiestii, desert oat

A. strigosa, sand oat

A. nudibrevis, small-seeded naked oat

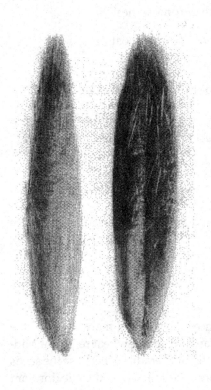

FIGURE 18.7
Caryopses or "groats" of
oats, showing crease in
grain at right.

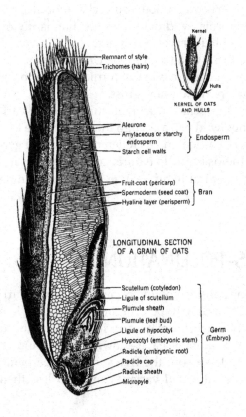

Kernel

Remnant of style
Trichomes (hairs)

Hulls

KERNEL OF OATS
AND HULLS

Aleurone
Amylaceous or starchy } Endosperm
endosperm
Starch cell walls

Fruit-coat (pericarp) }
Spermoderm (seed coat) } Bran
Hyaline layer (perisperm) }

LONGITUDINAL SECTION
OF A GRAIN OF OATS

Scutellum (cotyledon)
Ligule of scutellum
Plumule sheath
Plumule (leaf bud)
Ligule of hypocotyl } Germ
Hypocotyl (embryonic stem) } (Embryo)
Radicle (embryonic root)
Radicle cap
Radicle sheath
Micropyle

FIGURE 18.8
The oat caryopsis. Other
names for its structures are
as follows: starch cell walls
= walls of endosperm cells
that contain starch;
spermoderm = testa;
perisperm = nucellus;
plumule sheath =
coleoptile; ligule of
hypocotyl = epiblast;
hypocotyl = subcrown
internode; radical cap =
rootcap; and radical sheath
= coleorhiza. [Courtesy The
Quaker Oats Company]

Group II: Fourteen Haploid Chromosomes

A. barbata, slender oat   *A. abyssinica,* Abyssinian oat

A. magna[13]

Group III: Twenty-One Haploid Chromosomes

A. fatua, common wild oat   *A. sterilis,* wild red or animated oat

A. sativa, common white or   *A. byzantina,* cultivated red oat
northern oat

A. nuda, large-seeded naked
or hulless oat

Many of the oat varieties grown in the United States were described and also classified by Stanton.[9]

18.5.1 Wild Oat

Common wild oat, *A. fatua* (Figure 18.6), is a noxious weed in many parts of the country, being particularly troublesome in the hard red spring wheat region of Minnesota, the Dakotas, and Montana. Wild oat is difficult to eradicate because the seeds shatter readily and many of the seeds are tilled into the soil, where they lie dormant for many years. The seeds germinate and grow when they are turned up near the surface. However, some collections of wild oat show little or no dormancy.[14] Common wild oat differs from cultivated oat in having taller, more vigorous plants and strongly twisted geniculate awns. The grain has a pronounced sucker mouth at the base and usually a hairy lemma. In California, wild oat grows so abundantly on wastelands and uncropped fields that it is often harvested for hay.

Wild red oat, *A. sterilis,* occurs infrequently as a weed in the United States. Another wild oat, *A. barbata,* is an important range grass in California. In other parts of the world, sand oat, *A. strigosa,* is an important forage grass.

Fatuoids, or false wild oat (Figure 18.6), appear suddenly in cultivated oat varieties and somewhat resemble the common wild oat, although the fatuoids are also similar to the variety in which they occur, particularly as to lemma color and germination habits.[15] The persistence of fatuoids is due to natural crossing and mutation.[16] Many fatuoids originate through natural hybridization between common cultivated oat and wild oat.[17]

■ 18.6 NATURAL CROSS-POLLINATION

Self-pollination is normal in oat but some natural crossing occurs in different varieties and environments ranging from 0 to nearly 10 percent, but it is usually less than 1 percent.[18] Greater numbers of natural hybrids may occur in plants produced from secondary seeds.[19] Open-pollinated fatuoids have shown a maximum of 47 percent of cross-fertilization in a single season, the five-year average being 11.6 percent. Under similar conditions, a cultivated variety contained less than 0.5 percent crosses in any season.[16]

■ 18.7 DORMANCY IN OAT

Freshly harvested seeds of oat often fail to germinate satisfactorily, while others germinate immediately.[6] All degrees of prompt, slow, and delayed germination occur among oat varieties, but all cultivated red oat varieties usually show slow or delayed germination[10, 20] and may cause poor field stands from the use of freshly harvested seed. Winter oat varieties are more likely to show dormancy than are spring varieties.[8] However, there appears to be no association between morphology of the oat kernel and dormancy. Characters associated with wild oat and fatuoids, such as the sucker-mouth base and the presence of basal hairs on the callous, are unrelated to dormancy.[14, 15]

Wild oat, *A. fatua*, seed will lie dormant the winter and germinate in the spring. A cross between common oat and wild oat resulted in a cultivar, "dormoat," that could be seeded in the fall but would not germinate until the spring.[21] The advantage of dormoat was its ability to utilize the favorable growing conditions of early spring, mature earlier, and escape disease. Also, soil conditions are many times more favorable for seeding in the fall than in the early spring. Problems such as volunteer oat plants in subsequent crops and slow, irregular germination in the spring have kept dormoat from gaining wide acceptance.[22]

■ 18.8 OAT TYPES

Seed with white or yellow lemma colors are commonly grown in the North, while red or gray oat is grown mostly in the South. Black oat is rarely grown. Oat is sometimes grouped as early, midseason, and late. A midseason variety may mature 10 to 14 days later than an early variety.

Midseason varieties are well suited to the cooler regions along the northern border and in the irrigated intermountain region. To the south, where hot weather can injure the plants before ripening, the early white and yellow varieties are grown. Further south in the central regions, still earlier spring-sown red varieties are grown,[23] while in the extreme south, fall- or winter-sown red or gray oat is grown, in each case in order that the crop will reach maturity before the advent of severe summer heat or drought.

In the North Central region, early maturing common yellow and white varieties are the main varieties, but midseason varieties are grown in the northern portion, and spring-sown red oat leads in the southern portions of the region.

Hulless (naked) oat can be fed to hogs directly, but they do not often remain sound in storage. Few hulless varieties yield as well as the hulled varieties and are rarely grown.

The spring-sown red oat region lies south of latitude 41° N.[24] The southern border of this area is transitional between the winter and spring oat regions. Medium early as well as midseason varieties are grown in the northeastern states.

Fall-sown oat is grown mostly in Delaware, Maryland, West Virginia, Kentucky, Arkansas, Oklahoma, and southward. The winter oat belt has moved north as winter oat varieties more resistant to cold were developed. Winter oat is now grown in states north of those just mentioned.[25] Midseason varieties prevail in the Rocky Mountain region where oat is mostly grown under irrigation.

Winter gray or white varieties are grown from fall seeding in the western portion of the Pacific Northwest. Varieties of the red oat group are grown in California, mostly from fall seeding. Midseason varieties are the most extensively grown spring oat in Washington and Oregon.[26]

■ 18.9 FERTILIZERS

Nitrogen is the most essential element for oat, but heavy applications are likely to induce lodging. Carry-over nitrogen remaining after intensive corn production may also be at a high enough level to induce extensive lodging. The fertilizer requirements for oat are similar to those for wheat. Oat responds to phosphorus, and often to potassium, in most sections of the humid area. Oat yield is significantly higher when phosphate fertilizer is banded close to the seed compared to broadcast application.[27]

A 70 bushel per acre (2,500 kg/ha) oat crop will remove about 40 pounds per acre (kg/ha) nitrogen, 13 pounds per acre (15 kg/ha) phosphate, and 10 pounds per acre (11 kg/ha) potash. Phosphate fertilizer is not needed on soils high in phosphorus, but oat will respond to potassium fertilizer on all soils except those that contain very high amounts.[27]

■ 18.10 ROTATIONS

Oat fits well into many crop rotations because it is seeded earlier and matures earlier than most other crops. Subsequent crops are easy to seed into oat residue. Oat is used as a companion or nurse crop with red clover or alfalfa in the Corn Belt, and with lespedeza in Missouri and elsewhere. In regions where corn rootworm is a pest, corn should not follow oat or other small grains.[28] If small grain stubble becomes infested with grassy weeds, the adult rootworm beetles will lay eggs in the field resulting in rootworm larvae infestation the following year.

Careful attention to soil nitrogen level is needed when oat follows soybean because the additional nitrogen may cause lodging. In the northeastern states, oat seeded to clover or grass that is left for two years can follow one year of corn. In the northern portion of the spring-sown oat region, a rotation is as follows: corn two years, oat and wheat each for one year, and a grass-legume mixture for two years. In the Great Plains, where a crop can be produced each year, oat usually follows corn, sorghum, or cotton. With irrigation, oat is usually a companion or nurse crop for alfalfa or clover. A typical rotation is alfalfa for three years, sugarbeet or potato or corn for one or two years, and oat seeded with alfalfa as a companion or nurse crop. In the humid Pacific area, oat can occupy the season between row crops and clover.

In southern regions, where crop sequences are rather flexible, fall-sown oat can follow corn or sometimes tobacco. Oat is often harvested in May or early June in time to permit double cropping with a summer crop such as soybean, cowpea, sorghum, or pearl millet.

18.11 CULTURAL PRACTICES

18.11.1 Seedbed Preparation

Where oat follows corn or some other row crop, the seedbed is often prepared by disking and harrowing without plowing.[29] Where abundant weed growth and heavy soils occur, corn-stubble land is usually plowed, disked, and harrowed for seedbed preparation. Oat is sown with a drill (Figure 7.7). If a no-till drill is used, little or no tillage is needed. Oat following soybean needs little or no tillage even with conventional drills.

18.11.2 Time of Seeding

Spring-sown oat should be seeded as early in the spring as a seedbed can be prepared, but after the danger of prolonged cold weather is past. This is especially important in regions subject to heat or drought. Dates of seeding spring oat are shown in Figure 18.9. Oat should be sown before the average temperature reaches 50°F (10°C).[12] Spring seeding is usually later in the more northern regions, but seldom later than May 15, even in New England.

Fall-sown oat can be seeded in the South at any time from October 1 to December 31, but October is often the best month. Earlier seeding is desirable when the oat is to be pastured.[30] Experiments with winter oat sown in Oklahoma and Tennessee showed that seeding around October 1 gave the best yields. North of these states, seeding in September would be desirable. Most oat in California is seeded in November.

18.11.3 Rate of Seeding

The average rate of seeding oat in the United States is 60 to 80 pounds per acre (67 to 88 kg/ha). At an average of 15,000 to 17,000 seeds per pound (6,800 to 7,700

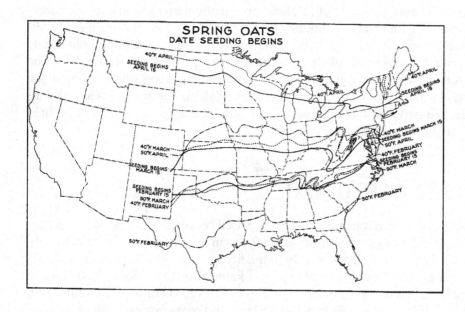

FIGURE 18.9

Seeding of spring oats in the United States begins when the mean daily temperature is between 40 and 50°F.

seeds/kg), that rate results in about 1 million seeds per acre (2.5 million seeds/ha). The yield of most varieties is not affected greatly by planting rate. Oat yields seeded at 48 pounds per acre (54 kg/ha) have been shown to be only slightly less than oat seeded with 80 pounds per acre (90 kg/ha). Therefore, a lower seeding rate should be used when seeding oat as a companion crop with alfalfa or other perennials.[31] Seeding rates for hulless oat should be about 25 percent less because these varieties usually have a higher germination and emergence rate.

18.11.4 Weed Control

Many of the broadleaf weeds in oat fields can be controlled with postemergent herbicides. Common herbicides are the sulfonylureas, bromoxynil, 2, 4-D, and MCPA. Oat is more sensitive than wheat or barley to 2, 4-D. It is more likely to show injury from a residual herbicide when sown on land that had been treated with atrazine the previous season.

18.11.5 Harvesting

Most of the oat acreage sown in the United States is harvested for grain. The remainder is pastured, cut for hay or other forage, or plowed under or abandoned. Nearly 40 percent of the grain oat is windrowed before combining in order to avoid losses from crinkling and lodging of the stems, or from shattering of the grain. Oat is best harvested in the dough stage for hay, but in the milk stage for silage. Fall-sown oat often can be pastured moderately in the autumn and winter until February 1, or sometimes later, and still produce a normal crop of grain.

The growing oat plants, or even oat hay, sometimes contain sufficient nitrates to be poisonous to cattle, sheep, and horses.[32] Oat poisoning occurs occasionally in Colorado, Wyoming, and South Dakota, apparently under soil or climatic conditions in which the uptake of nitrates greatly exceeds the rate of protein synthesis. It possibly results from a relatively slower growth, photosynthesis, and phosphate absorption.[33] Ample phosphate applications may prevent nitrogen excess. The symptoms of nitrate poisoning include rapid breathing along with a bluing of the mucous membranes. Death occurs from asphyxiation. The remedy is an early intravenous injection of 4 milligrams of methylene blue per pound of body weight in a 4 percent solution with distilled water.[34]

■ 18.12 USES OF OAT

Use of oat for human consumption has increased because of purported health benefits. Presently, about 10 percent of total production is milled for human consumption. Oat for milling is a high-quality product that equals or exceeds well-defined specifications of plumpness, kernel weight, freedom from foreign materials, off-flavors, and discoloration.[35] Any cultivar can be used for milling, but the crop should be managed to reduce lodging and control diseases and harvested promptly to reduce weathering and discoloration. Artificial drying may be needed, especially with hulless varieties; direct combining is preferred to swathing if rain is likely.[36]

As horse feed, oat is about 90 percent as efficient as corn. Oat is often fed whole to horses and sheep. Ground or chopped oat is fed to dairy cattle, breeding stock, and young stock. After the hulls are removed, oat is equivalent to corn for swine feed. Oat is fed to poultry, usually without hulling, but sometimes it is sprouted first. Oat flour, in part a by-product of the manufacture of rolled oat, is used in breakfast foods and other foods. The flour possesses a property that retards the development of rancidity in fat products. Paper containers coated with oat flour are used for packaging food products having a high fat content. Oat flour contains soluble phenolic compounds that can be used as an antioxidant. Oat flour is also used as a stabilizer in ice cream and other dairy products.[37]

Oat is one of the best crops for mowing while green to supply dried grass products as well as chlorophyll and carotene extracts for use in food, feed, and medicine. The young leaves are rich in protein and vitamins.

The manufacture of rolled oat consists first of cleaning and sizing the grain.[38] Then the grains are heated in steam-jacketed pans where they dry down to about 6 percent moisture and acquire a slightly roasted flavor. After this drying, the hulls are brittle enough to be readily removed by large hulling stones similar to the old-fashioned bur milling stones. The groats are then separated from the hulls and broken pieces. The whole groats are steamed to be made tougher and are then passed between steel rolls to produce the familiar flakes of rolled oat. So-called quick oats are small thin flakes rolled from steel cut pieces of oat groats. The small size and thinness speed up the cooking process. The hulls, which constitute 27 to 30 percent of the grain, are used in the manufacture of furfural, a solvent widely used in chemical, fat, and petroleum industries.

▦ 18.13 DISEASES

18.13.1 Smuts

Loose smut, *Ustilago avenae*, replaces the oat floret with a loose spore mass.[37] In covered smut, *U. kolleri*, the dark-brown spore mass is more or less enclosed within a grayish membrane, frequently within the lemma and palea of the flower. The two smuts cannot be clearly differentiated in the field. Both can be controlled by seed treatment with suitable fungicides.[38] The percentage of oat plants infected with covered smut may be highest from seed sown when the mean temperatures are relatively high.

Resistant varieties are available. Seed treatment with a fungicide helps control smuts. Products containing carboxin are best for loose smut.

18.13.2 Rusts

Stem rust of oat (*Puccinia graminis avenae*) produces symptoms similar to stem rust of wheat. The alternate host is the common barberry, but in the South the disease overwinters and develops independently of the alternate host and then spreads northward. The principal control method is resistant varieties.

Crown rust or leaf rust (*Puccinia coronata*) is identified by the bright yellow pustules on the leaves. The alternate host is the buckthorn, but in the South the fungus spreads independently of the alternate host. For each unit increase in coefficient of crown-rust infection (percentage of infection × numerical infection type), the yield is decreased from 0.2 to 0.3 bushel per acre (7 to 11 kg/ha).[39, 40, 41] As the eradication of the buckthorn is incomplete, the principal control is resistant varieties.

The breeding of rust-resistant varieties continues in an effort to avoid losses from still newer rust races. Later planted oat is more susceptible.

18.13.3 Bacterial Stripe Blight

Bacterial stripe blight (*Pseudomonas striafaciens*) first appears as sunken, water-soaked dots, which later enlarge to blotches or stripes that become rusty brown to black. The disease survives in crop residue and in seed. The disease develops in seedlings and spreads to other plants by wind and splashing rain. Bacteria enter through stomata, insect punctures, and wounds. It can be more prevalent during wet springs.

Resistant varieties are available. Crop rotations and spring tillage to bury crop residues are also helpful. Seed treatment is not effective.

18.13.4 Oat Red Leaf Virus Disease

Oat red leaf is a virus disease caused by the same pathogen as barley yellow dwarf. It is transmitted by several species of aphids. Affected plants have a bronze to reddish discoloration of leaves and plants are stunted. Yield losses can be severe if infection occurs early in the season. Control measures include aphid control and resistant varieties.

18.13.5 Oat Blast

Oat blast is a physiological disorder that results in white empty spikelets, especially near the base of the head. Oat blast can be caused by disease pathogens, especially barley yellow dwarf and oat blue dwarf viruses. More commonly, it is caused by environmental stress such as high temperatures or drought especially from tillering through pollination. Early planting can help avoid the stress that causes blast.

18.13.6 Other Diseases

There are many other diseases of oat that, while not as important economically, can cause losses in local areas in any given year. The oat cyst nematode (*Heterodera avenae*), a prevalent pest elsewhere, was first observed in the United States in 1974.

18.14 INSECT PESTS

Oat is less subject to insect attack than are barley and wheat. Hessian fly does not attack oat. Chinch bugs, although damaging to oat, greatly prefer barley or wheat.

The bluegrass billbug *(Calendra parvulus)*, certain leafhoppers, armyworm, grain bug *(Chlorocroa sayi)*, grasshopper, and Mormon cricket readily attack oat. The greenbug (or spring-grain aphis, *Toxoptera graminum)* can severely damage oat. Methods of controlling most of these insects are discussed in the chapter on wheat. Other oat pests include wireworms, white grubs, cutworms, leafhoppers, fruit flies, thrips, and winter grain mites.[25]

REFERENCES

1. Langer, I., K. J. Frey, and T. B. Bailey. "Production response and stability characteristics of oat cultivars developed in different eras." *Crop Sci.* 18(1978):938–942.

2. Rodgers, D. M., J. P. Murphy, and K. J. Frey. "Impact of plant breeding on the grain yield and genetic diversity of spring oats." *Crop Sci.* 23(1983):737–740.

3. Stanton, T. R. "Superior germ plasm in oats," *USDA Yearbook,* 1936, pp. 347–413.

4. Coffman, F. A. "Origin of cultivated oats," *J. Am. Soc. Agron.* 38, 11(1946):983–1002.

5. Coffman, F. A. "Origin and history," in *Oats and Oat Improvement,* Am. Soc. Agron. Monographs, vol. 8, 1961, pp. 15–40.

6. Coffman, F. A., and K. J. Frey. "Influence of climate and physiological factors on growth in oats," in *Oats and Oat Improvement,* Am. Soc. Agron. Monographs, vol. 8, 1961, pp. 420–464.

7. Stanton, T. R., and F. A. Coffman. "Winter oats for the South," *USDA Farmers Bull.* 2037, 1951.

8. Brown, E., and others. "Dormancy and the effect of storage on oats, barley, and sorghums," *USDA Tech. Bull.* 953, 1948.

9. Stanton, T. R. "Oat identification and classification," *USDA Tech. Bull.* 1100, 1955.

10. Coffman, F. A., and T. R. Stanton. "Variability in germination of freshly harvested Avena," *J. Agr. Res.* 57(1938):57–72.

11. Stanton, T. R. "Maintaining identity and pure seed of southern oat varieties," *USDA Circ.* 562, 1940.

12. Taylor, J. W., and F. A. Coffman. "Effects of vernalization on certain varieties of oats," *J. Am. Soc. Agron.* 30(1938):1010–1019.

13. Murphy, H. C., and others. "*Avena magna:* An important new tetraploid species of oats," *Science* 159(1968):103–104.

14. Toole, E. H., and F. A. Coffman. "Variations in the dormancy of seeds of the wild oat, *Avena fatua,*" *J. Am. Soc. Agron.* 32(1940):631–638.

15. Coffman, F. A., and T. R. Stanton. "Dormancy in fatuoid and normal kernels," *J. Am. Soc. Agron.* 32(1940):459–466.

16. Coffman, F. A., and J. W. Taylor. "Widespread occurrence and origin of fatuoids in Fulghum oats," *J. Agr. Res.* 52(1936):123–131.

17. Aamodt, O. S., L. P. V. Johnson, and J. M. Manson. "Natural and artificial hybridization of *Arena sativa* with *A. fatua* and its relation to the origin of fatuoids," *Can. J. Res.* 11(1934):701–727.

18. Jensen, N. F. "Genetics and inheritance in oats," in *Oats and Oat Improvement,* Am. Soc. Agron. Monographs, vol. 8, 1961, pp. 125–206.

19. Hoover, M. M., and M. H. Snyder. "Natural crossing in oats at Morgantown, W. Va.," *J. Am. Soc. Agron.* 24(1932):784–786.

20. Deming, C. W., and D. W. Robertson. "Dormancy in small grain seeds," *CO Agr. Exp. Sta. Tech. Bull.* 5, 1933.

21. Burrows, V. D. "Seed dormancy, a possible key to high yields of cereals," *Agricultural Institute Review* 19(1964):33–35.

22. Burrows, V. C. "Yield and disease—Escape potential of fall-sown oats possessing seed dormancy," *Can. J. Plant Sci.* 50(1970):371–377.

23. Coffman, F. A. "Culture and varieties of spring-sown red oats," *USDA Farmers Bull.* 2115, 1958.

24. Etheridge, W. C., and C. A. Helm. "Growing good oat crops in Missouri," *MO Agr. Exp Sta. Bull.* 359, 1936.

25. Coffman, F. A. "Factors affecting survival of winter oats," *USDA Tech. Bull.* 1346, 1965, pp. 1–28.

26. Coffman, F. A. "Oat varieties in the Western States," *USDA Agricultural Handbook* 180, 1960.

27. Dahnke, W. C., C. Fanning, A. Cattanach, and L. J. Swenson. "Fertilizing oat," *ND St. Univ. Ext. Ser.* SF-716, 1992.

28. Wright, R., L. Meinke, and K. Jarvi. "Corn rootworm management," *NE Coop. Ext. Serv.* EC99-1563, 1999.

29. Shands, H. L., and W. H. Chapman. "Culture and production of oats in North America," in *Oats and Oat Improvement*, Am. Soc. Agron. Monographs, vol. 8, 1961, pp. 465–529.

30. Atkins, I. M., and others. "Growing oats in Texas," *TX Agr. Ext. Sta. Bull.* B-1091, 1969, pp. 1–28.

31. Moomaw, R., and C. A. Shapiro. "Oat production in Nebraska," *NE Coop. Ext. Ser.* G79-430-A, Revised 1992.

32. Gul, A., and B. J. Kolp. "Accumulation of nitrates in several oat varieties at various stages of growth," *Agron. J.* 52(1960):504–506.

33. Wright, M. J., and K. L. Davison. "Nitrate accumulation in crops and nitrate poisoning in animals," *Advances in Agronomy*, vol. 16. New York: Academic Press, 1964. pp. 197–247.

34. Binns, W. "Chemical poisoning," in *Animal Diseases*, USDA Yearbook, 1956, pp. 113–117.

35. Stoskopf, N. C. *Cereal Grain Crops*. Reston, VA: Reston Publ. Co, 1985.

36. Clarke, J. M., et al. "Effect of kernel moisture content at harvest and windrow vs. artificial drying on quality and grade of oats," *Can. J. Plant Sci.* 62(1982):845–854.

37. Stanton, T. R. "New products from an old crop," in *Crops in Peace and War*, USDA Yearbook, 1950–51, pp. 341–344.

38. Western, D. E., and W. R. Graham, Jr. "Marketing, processing, and uses of oats," in *Oats and Oat Improvement*, Am. Soc. Agron. Monographs, vol. 8, 1961, pp. 552–578.

39. Simons, M. D., and H. C. Murphy. "Oat diseases," in *Oats and Oat Improvement*, Am. Soc. Agron. Monographs, vol. 8, 1961, pp. 330–390.

40. Hanson, E. W., E. D. Hansing, and W. T. Schroeder. "Seed treatments for control of disease," in *Seeds*, USDA Yearbook, 1961, pp. 272–280.

41. Murphy, A. C., and others. "Relation of crown-rust infection to yield, test weight, and lodging of oats," *Phytopath* 30(1940):808–819.

Rice

▨ 19.1 ECONOMIC IMPORTANCE

Rice (*Oryza sativa*) provides the principal food for over half of the world population. The production of rough rice in 2000–2003 was about 647 million tons (589 million MT) (Figure 19.1.) Tonnage of rice is about the same as wheat. However, rough rice consists of about 20 percent hull so rice ranks second in net kernel production. In 2000–2003, rice was produced on an average of 375 million acres (152 million ha) with an average yield of almost 3,500 pounds per acre (3,900 kg/ha). Leading countries in rice production are China, India, Indonesia, Bangladesh, and Vietnam. Production in the United States in 2000–2003 averaged about 10 million pounds (9 MT) with an average yield of 6,500 pounds per acre (7,300 kg/ha) on 3 million acres (1.3 million ha). The leading states in rice production are Arkansas, California, Louisiana, Mississippi, and Texas (Figure 19.2). Even though the United States is not a large producer of rice, about one-third of its crop is exported annually.

▨ 19.2 HISTORY OF RICE CULTURE

Rice probably originated in India or southeastern Asia where several wild species are found, possibly evolving from a wild form that no longer exists. Another rice species (*O. glaberrima*), grown in western and central Africa, may have originated there as a derivative of the wild *O. breviligulata*.[1] The so-called wild rice of North America, *Zizania aquatica* (or *Zizania palustris*) bears little resemblance to cultivated rice. It belongs to a different grass tribe, *Zizanieae*. Rice culture is believed to have spread into China by 3000 BC and westward into Europe by 700 BC.

The first commercial rice planting in the United States was at Charleston, South Carolina, about 1685. The crop spread to North Carolina and Georgia. Most of the rice was produced largely by hand methods on the delta lands of the South Atlantic states until 1888. Rice growing in this region was adversely affected by the Civil War, after which the acreage increased along the Mississippi River in Louisiana. In 1889, Louisiana became the leading state in rice production. In the Gulf Coast prairie section, it was first grown by animal-powered machine methods. Established rice culture spread from southwestern

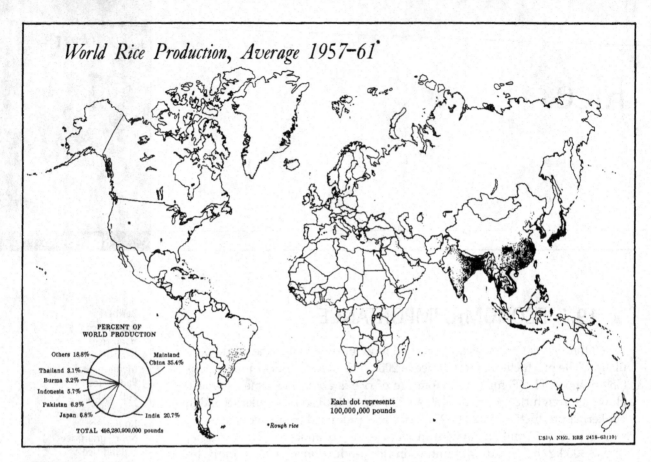

FIGURE 19.1

World rice production. [Courtesy USDA]

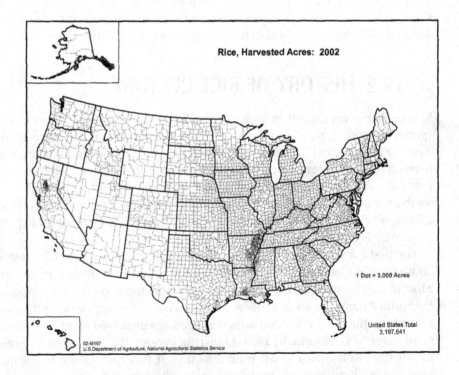

FIGURE 19.2

Rice acreage in the United States. [Source: 2002 Census of U.S. Department of Agriculture]

Louisiana to southeastern Texas about 1900, to eastern Arkansas in 1905, to the Sacramento Valley in California by 1912, and to Mississippi about 1949. Rice has been grown intermittently in Missouri since about 1925.

The yields of rice in the United States in 2000 were over five times those obtained at the beginning of the twentieth century. Rice yields in many Asian countries were greatly improved by growing improved varieties developed by the International Rice Research Institute at Los Blanos in the Philippines. These IRRI varieties have short, stiff culms, which enable them to remain erect on soils that are heavily fertilized with nitrogen. Their yield response to nitrogen fertilization of up to 120 pounds per acre (133 kg/ha) is much superior to that of the tall, weak-stalked varieties formerly grown. The latter types lodge and yield less under even moderate nitrogen fertilization of 30 pounds per acre (34 kg/ha).[2]

■ 19.3 ADAPTATION

Rice is unique among the cereals in being able .o germinate and thrive in water. Other cereal crops are killed if submerged for two or three days in warm weather, because of lack of oxygen for the roots. The rice plant is able to transport oxygen to the submerged roots from the leaves, where oxygen is released during photosynthesis. The water also contains some oxygen. Thus, rice is a hydrophyte that can live in an aquatic environment.[3]

The important factors for rice production are favorable temperatures, a constant supply of fresh water for irrigation, and suitable soils. Rice can be grown most successfully in regions that have a mean temperature of about 70°F (22°C) or above, during the entire growing season of four to six months.[4] Rice yields are higher in warm temperate regions that have a low summer rainfall with a high light intensity[5] than in the humid tropics where rice diseases and soils of low fertility are more prevalent. The countries with the highest average yields are Egypt, Australia, and Greece.

The best soils for rice are slightly acid (pH 5.5 to 6.5). Flooding a rice field increases the pH 0.5 to 1.5 units, and thus releases available phosphorus. Rice is grown on soils that range in pH from 4.5 to 8.5. It is grown most economically on soils of rather heavy texture with an impervious underlying subsoil from 1½ to 5 feet (0.5 to 1.5 m) below the surface. The loss of water by seepage is small through such soils, which is the main reason for selecting heavy soils for rice culture.

Upland rice consists of certain varieties that can be grown without irrigation or submergence in regions of high rainfall where the soil is wet much of the time. Yields are much lower than when rice is grown under field submergence. Upland rice is mostly grown on a small scale by hand methods for home use. It is rarely grown in the United States but is important in parts of Asia, South America, and Africa.

■ 19.4 BOTANICAL DESCRIPTION

Rice belongs to the grass tribe *Oryzeae* characterized by one-flowered spikelets, laterally compressed, and two short glumes. Rice is an annual grass with erect culms 24 to 71 inches (60 to 180 cm) tall (Figure 19.3). The plant tillers freely, four to five culms per plant being common. Despite its annual habit, rice can be propagated

FIGURE 19.3
A mature rice plant.

vegetatively for several years by transplanting rooted tillers. Node branching may occur with early maturity and with ample space for plant development.[6]

The rice inflorescence is a loose terminal panicle of perfect flowers (Figures 19.4 and 19.5). At maturity, the panicle in different varieties can be enclosed in the sheath, partly exserted, or well exserted. Each panicle branch bears a number of spikelets, each with a single floret (Figure 19.6). The flower has six stamens and two long plumose sessile styles, being surrounded by a lemma and a palea at the base of which are two small glumes. The lemma and palea may be straw yellow, red, brown, or black. The lemmas of various varieties can be fully awned, partly awned, tip-awned, or awnless. The rice grain is enclosed by the lemma and palea, these structures being called the hull. The hulled kernels vary from about ⅛ to ⅓ inch (3 to 8 mm) long, 1/15 to ⅛ inch (1.7 to 3 mm) in breadth, and about 1/20 to 1/10 inch (1.3 to 2.3 mm) thick.[7] They are hard, semi-hard, or soft in texture. The color of the unmilled kernel may be white, brown, amber, red, or purple. Commercial American varieties are white, light brown, or amber. An average rice panicle contains 100 to 150 seeds.

19.4.1 Pollination

When blooming, rice flowers open rapidly and the anthers begin pollinating when the flower opens. Sometimes, pollination will begin before the flowers open. The

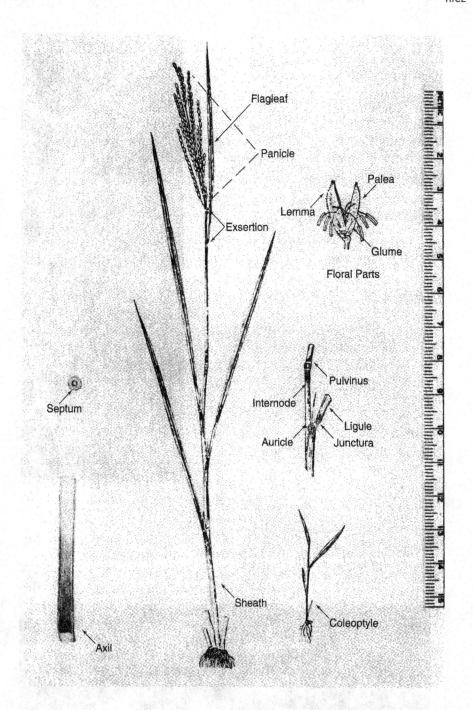

FIGURE 19.4
Parts of the rice plant.

flowers may remain open twenty minutes to three hours.[8, 9] Blooming starts early in the day when temperatures are high but slows down when the sky becomes cloudy causing a drop in temperature.[10]

Rice is normally self-pollinated, but up to 3 or 4 percent natural crossing may occur, although less than 0.5 percent is the average. Varieties differ in the extent of natural crossing.[11, 12] Much more crossing occurs in the South than under the higher temperature and lower humidity conditions prevailing in California.

FIGURE 19.5

Panicle of short grain rice *(right)*. Section of culm, leaf, and sheath *(left)*. Note the long ligule characteristic of the rice plant.

FIGURE 19.6

Spikelets *(top)*, seeds *(middle)*, and kernels *(bottom)* of four types of rice: (1) slender grain, (2) long grain, (3) medium grain, and (4) short grain.

19.5 RICE TYPES

Thousands of varieties of rice are known in Asia but comparatively few are grown in the United States. Varieties grown in the United States include short-, medium-, and long-grain types. The average length of the hulled kernel of the three grain types is: short-grain, ⅕ inch (5.5 mm); medium-grain, ¼ inch (6.6 mm); and long-grain, ⅓ inch (7 to 8 mm). Early-maturing varieties in the southern states require about 120 to 129 days from seeding to maturity, the midseason varieties about 130 to 139 days, and the late varieties 140 days or more. Several varieties of different maturities can be grown advantageously on one farm to distribute harvest labor over a longer period of time.

Short-grain rices, classed as the *Japonica* type, have short, stiff stems that resist lodging. They also respond to heavy applications of nitrogen. The long-grain rices of the *Indica* type usually have taller and weaker stems. The crop often lodges under heavy nitrogen fertilization. First-generation crosses between the *Japonica* and *Indica* types are usually partially sterile. The Ponlai (temperature- and light-insensitive) rices grown in Taiwan were derived from such crosses. The Bulu rices of Indonesia have large kernels and long, but stiff, culms.

Short-grain rices have prevailed in Japan, and improved short-grain IRRI varieties released since 1965 are now widely grown in Asia and other parts of the world. Less than 1 percent of the United States rice crop is short-grain; nearly all of that is grown in California. Long-grain rice accounts for 74 percent of U.S. production. The remaining is medium-grain rice.

The endosperm starch may be nonglutenous (common) or glutenous. Glutenous rices have the amylopectin type of starch suitable for preparing special delicacies. Aromatic (scented) rices exude a distinct mouse-like odor during cooking, which is prized by some Asian consumers. The glutenous and aromatic types are only grown in the United States on small acreages for special consumer groups.

All rices are short-day plants, but varieties differ greatly in their sensitivity to photoperiods. The more sensitive varieties vary considerably in their length of vegetative growth before flowering when seeded in different months or at different latitudes.

19.6 FERTILIZERS

Dwarf types of rice can respond to applications up to 116 pounds per acre (130 kg/ha) or more of nitrogen, preferably in an ammonium form. Tall, weak-stalked varieties can lodge with consequent reduced yields from applications exceeding 30 to 80 pounds per acre (35 to 90 kg/ha) of nitrogen. Ammonium sulfate is the most popular fertilizer, but ammonium phosphate, anhydrous or liquid ammonia, and urea are also applied to rice fields. The rice plant is capable of using both ammonium and nitrate forms of nitrogen, but fertilizers containing nitrate nitrogen are inefficient because nitrate is lost by denitrification in flooded rice soils.[13] Also, nitrates are subject to leaching. Nitrates are often only two-thirds to three-fourths as effective as ammonium forms. Except in excessive amounts, nitrogen fertilizer causes rice to flower earlier. Much of the nitrogen can be applied as a top dressing thirty to seventy days after the rice is seeded.

On many soils, every 4.5 to 9 pounds (2 to 4 kg) of nitrogen increases the yield of rice by 100 pounds (45 kg). Heavy applications of nitrogen are often profitable on California soils unless the rice crop follows a legume green manure. Algae growing in the water on rice fields fix some atmospheric nitrogen. Nitrogen is often applied three times to the rice crop. About 50 to 70 percent of total N is applied preflood, about 15 to 25 percent is applied at internode elongation, and the remainder about two weeks later. Recently, growers have found it more efficient to apply two applications of nitrogen at preflood and internode elongation.[14]

On some Gulf Coast soils, rice responds to 30 to 60 pounds per acre (35 to 65 kg/ha) of P_2O_5 and 60 to 90 pounds per acre (65 to 100 kg/ha) of K_2O with increased yields, especially where large amounts of ammonium nitrogen are applied.[11] Phosphorus fertilizer applied near the time of flooding is more efficient than preemergent applications on alkaline soils.[14]

In Louisiana, rice yields are increased with phosphorus applications on most soils, but also with potassium on some soils. Response to these elements, as well as to zinc or iron,[15] tends to be greater where soil tests show them to be deficient.[16] Phosphorus and potassium fertilizers are usually not required in California.

Continued flooding tends to increase soil pH, and zinc may be needed as soils become alkaline. Zinc can be applied to the soil as a chelated or complex liquid, or directly to the seed before planting.[17] Sulfur is needed on some tropical soils and is usually applied as ammonium sulfate. Silicon, although not an essential element, has shown to benefit rice yield in some Asian soils.[18]

■ 19.7 ROTATIONS

Rice yields become stabilized at low-yield levels when grown continuously on the same land. Weeds also become abundant after two or more rice crops. Heavy, poorly drained rice soils are often unsuited to other crops that otherwise would be grown in rotation with rice. Consequently, in the Gulf Coast region as well as in California, rice is usually grown on the same land for two or three years. The rice land is then sometimes seeded to pasture crops and grazed for two to three years. Grasses, clovers, and lespedeza are often sown in the Gulf Coast area, while Ladino clover, burclover, or strawberry clover may be planted in California. This practice results in higher subsequent rice yields, improved soil structure, higher soil fertility, a lower population of aquatic weeds, and better livestock gains than when the land is pastured without seeding or fertilization.[19]

The aquatic weeds on some California rice lands are partly controlled by an occasional year of clean fallow. A *water fallow* is sometimes practiced by maintaining a depth of water of 2 to 4 feet (60 to 120 cm) in which fish may be grown for two years. Crayfish (*Procambris alleni*) may also be used. This practice may induce excessive accumulation of ammonium nitrogen, in which case corn or sorghum should precede a rice crop.

Crops frequently grown in rotation with rice on the better-drained soils in California include safflower, grain sorghum, wheat, and barley. Purple vetch seed for a green-manure crop can be broadcast on the rice field just before draining the land for harvest. Most of the rice lands in Arkansas, Mississippi, and Missouri are well suited to other crops such as cotton, soybean, corn, oat, lespedeza, and winter legumes. These crops are often grown in two-year to five-year rotations, in which rice often follows lespedeza or a winter legume for green manure. Two successive crops of rice may be grown in the longer rotations.[20] Rice responds well to green manures.[18]

When rotating soybean and rice it is important to maintain soil pH at a level that is compatible with both crops. Rice prefers a pH of 6.5 or less to avoid zinc deficiency, while soybean needs the soil pH to be 5.8 or higher for nitrogen fixation. In some fields, it may be necessary to apply lime to achieve a soil pH within that range.[21]

◼ 19.8 RICE CULTURE

19.8.1 Seedbed Preparation

Land for rice production should be plowed 4 to 6 inches (10 to 15 cm) deep in the fall, winter, or spring. Plowing as deep as 9 inches (23 cm) is sometimes helpful. Seedbed preparation for rice should include destruction of weeds as well as sufficient tillage to produce a mellow, firm surface layer. Clods will slacken after flooding, which buries rice seeds too deep in fields that are drilled or broadcast before flooding. A seedbed with small clods is satisfactory where rice is broadcast in the water.

Levees are laser surveyed and marked on contours either before or after seeding (Figure 19.7). The elevation between levees is seldom more than 2½ inches (6 to 7 cm). The levees may be broad and low enough for machines to pass over them, but some levees are so high and steep that combines or other implements must operate within the levees. Metal or vinyl levee gates or spills are used to maintain a water depth of 2 to 4 inches (5 to 10 cm). The use of plastic levees or asphalt-coated levees permits the field to be harvested as a unit.[22]

19.8.2 Seeding Methods

Seed that contains red rice should be avoided. Red rice, which belongs to the same species as cultivated rice, is grown as a crop in other countries for local use.[23] Wild

FIGURE 19.7
Levees for flooding rice.
[Courtesy Richard Waldren]

red rice is a serious pest in rice fields in the United States because the seeds shatter out early and contaminate the fields. The seed may live in the soil for several years. The presence of red rice spoils the appearance of milled white rice.

Rice is sown with airplane seeders, ground broadcast seeders, or grain drills. The crop is sown from mid-February through April in the southern states,[24] but from April 15 to June 1 in California. Average soil temperature should be at least 60°F (16°C). Seed sown too early may rot because of low temperatures. Rice can be sown over a comparatively long period without serious reduction in yields, but early seeding is desirable.[25] Under very favorable conditions, 80 pounds of seed per acre (90 kg/ha) is sufficient to give good stands. Under ordinary conditions, 90 to 100 pounds (100 to 110 kg/ha) of seed sown with a drill, or 125 to 160 pounds (140 to 180 kg/ha) sown broadcast by airplane, will prove adequate. The goal is a plant population of about 15 to 20 plants per square foot (170 to 220 plants/m^2).[14] Rice should be drilled 1 to 2 inches (2.5 to 5 cm) deep on good seedbeds, with the shallower depth on heavy soils. Seed fungicides are beneficial for early planting, clay soils, reduced tillage (especially no-till), or on fields that usually experience poor seedling emergence and seedling disease.[21]

Most of the rice seed in California and also much of that in the southern states is now broadcast into water by use of an airplane seeder (Figure 19.8). The field is flooded to a depth of 4 to 6 inches (10 to 15 cm) before seeding. Such seeding in wa-

FIGURE 19.8
Soaked rice seed broadcast from the airplane is falling into the water that floods the field.

ter helps to control barnyardgrass as well as other weeds on old rice land, but drilling is preferable on land that has not been cropped to rice for many years. The seed for airplane seeding is soaked for 24 to 36 hours and drained for 18 to 24 hours before sowing. A half-pound (225 g) of sodium hypochlorite (NaOCl) per 100 gallons (380 l) of steep water checks seeding diseases and deactivates germination inhibitors present in the rice hull. The soaked seeds sink down to the soil, whereas many dry seeds will float on the water and drift with the wind before sinking.

Rice is able to germinate with a lower supply of oxygen than is the case with other cereal seeds. Oxygen is released from the seed by fermentation.[26] This permits rice seedlings to emerge through cool water 6 inches (15 cm) or even more in depth. When the submerged rice seed is also covered with 1 inch (2.5 cm) or more of soil, the supply of oxygen is then insufficient for germination, so drilled rice fields may need to be irrigated and then drained, until after the crop is up.[27, 28]

19.8.3 Irrigation

The water requirement of the rice crop varies from 24 to 60 inches (610 to 1,525 mm). About 6 to 20 inches (150 to 510 mm) is supplied by rainfall during the growing season in the southern states. More water, or up to 108 inches (2,745 mm), is required for fields with permeable subsoils.[29]

In the southern states, the soil usually contains sufficient moisture for seed germination as well as for seedling growth on drilled fields. In California, drilled rice is usually irrigated after seeding, then drained, again irrigated, and drained at intervals until about thirty days after the rice seedlings emerge. The land is submerged, or flooded, to a depth of 1 to 2 inches (2.5 to 5 cm) when the young rice plants reach a height of 6 to 8 inches (15 to 20 cm). As the plants grow taller, this depth is gradually increased until it reaches 2 to 4 inches (5 to 10 cm). The water level is maintained at this depth for sixty to ninety days, or until the land is drained prior to harvesting. Levees with overflow outlets are built to maintain the water at the proper level.

After final submergence, especially where the seed was broadcast in the water, the fields are kept flooded until they are drained shortly before harvest. However, occasionally they are drained temporarily for pest control or to facilitate top dressing with nitrogen fertilizers. The final drainage to permit harvesting occurs about ten to fifteen days before the rice is fully mature. At that stage, the plants are fully headed, the panicles turned down, and ripening started in the upper portion.

Some of the water from wells, or from mountain streams, has a temperature of 65°F (18°C) or lower. Germination of the seed of most rice varieties is retarded when the water temperature is below 70°F (22°C). The most favorable water temperatures for rice growth are 77 to 84°F (25 to 29°C). Often, the growth of the rice plant is delayed in the checks at the upper side where the water first enters the field. Shallow basins may be constructed to allow the water to become warmer before it reaches the rice field. Water temperatures that exceed 85°F (29°C) cause poor root development in rice.[20]

Water that contains more than 600 parts per million (ppm) of soluble salts is injurious to the rice plant, especially in the germination, early tillering, and flowering stages.[30, 31] Water from coastal bayous used to irrigate rice may become excessively salty by the invasion of sea water during periods of low rainfall.

19.8.4 Weed Control

The most serious weed pest in rice is red rice. It is the same species as cultivated rice. Red rice is a strong competitor and is difficult to control. Typical red rice is taller than most cultivated rice varieties and is black-hulled and awned. But other strains look similar to domesticated rice. Purchased rice seed must be free of red rice seed. Combines must be thoroughly cleaned after harvesting an infested field. Crop rotation, usually with soybean, allows the use of herbicides to control red rice, and any plants that survive should be rogued from the field. Red rice plants on field borders should also be eliminated.[32]

The release of herbicide-resistant rice varieties is likely to greatly change production practices for red rice control. Currently, research is being conducted to develop rice varieties resistant to glufosinate and imidazolinone. Development of glyphosate-resistant rice is not currently in progress. The first of these varieties, imidazolinone-resistant Clearfield®, was released in 2001.

Other common weeds are barnyardgrass or watergrass (*Echinochloa crusgalli*), Mexican weed (*Caperonia palustris*), tall indigo or coffee weed (*Sesbania exaltata*), redweed (*Melochia corchorifiolia*), and spike rush (*Eleocharia* species). Most of the weeds can be suppressed by suitable recommended herbicides. Seeding rice in water, as described previously, is helpful in the control of barnyardgrass as well as some other weeds. Barnyardgrass (watergrass) is controlled in California by continuous submergence. Barnyardgrass plants are unable to emerge through 8 inches (20 cm) of water, and only a few plants emerge through 6 inches (15 cm) of water, through which rice seedlings will stretch and reach the surface.[28, 33] Crop rotation is also effective.

Common herbicides used in rice include molinate, thiobencarb, propanil, bentazon, 2, 4-D, MCPA, and acifluorfin. Excess scum due to blue-green algae is controlled by adding copper sulfate.

19.8.5 Harvesting

Practically the entire rice crop in the United States is harvested with self-propelled combines equipped with large rubber tires with mud lugs, or with crawler tracks, or halfbacks (Figure 19.9). It is then dried promptly with artificial heat.

In order to produce maximum yields of high milling quality, rice should be harvested when the moisture content of the grain of standing rice has dropped to about 18 to 27 percent.[34, 35] At that moisture, the kernels in the lower portion of the heads are in the hard-dough stage, while those in the upper portions of the head are ripe. Few chalky kernels (a result of immaturity) are found in rice harvested at this stage. Increased shattering and checking of the grains will occur in some varieties when harvested after this stage. During drying, the air temperature should be kept below 110°F (43°C) when drying is to be completed in one operation. Rice is dried in several stages, being held to allow the grain to reach equilibrium between drying operations. After the grain is partly dried, temperatures as high as 130°F (54°C) are not injurious[34] (Figure 19.10).

Other driers include mixing and batch types, farm bins with portable driers, and pothole bag driers for seed rice. Aeration between batch dryings is advisable. Drying with unheated air is sometimes successful, but heating units should be available for high-moisture grain or high humidity conditions.[36]

FIGURE 19.9
Rice harvest with a self-propelled combine. [Courtesy USDA]

FIGURE 19.10
A column-type rice drier: (1) cleaning machinery, (2) columns, (3) dump pit, and (4) hot-air fans.

In lesser-developed countries, rice is often harvested by hand or with a binder. It is then shocked for ten to fourteen days and threshed when the moisture content of the grain is reduced to about 14 percent.

In tropical regions, and along the Gulf Coast of the United States, it is possible to get a second ratoon (stubble) crop in the same season. This is possible when an early-maturing variety is sown early. The crowns of the rice plants will send out additional tillers after harvest. A ratoon crop will usually yield less than half of the main crop,[37] but it can be economical, especially in lesser-developed regions. An application of nitrogen immediately after the first crop is harvested significantly increases the yield. Infestations of rice pests such as the rice water weevil or the rice stink bug may be higher in the ratoon crop.

19.9 RICE CULTURE IN ASIA

A large part of the Asian rice that enters commercial trade is grown by transplanting.[18] The reasons for transplanting include better utilization of land by growing two different crops in a year, the saving of irrigation water, and better weed control. In hand transplanting, the rows are spaced for convenient weeding, harvesting, and fertilizer application, all of which are done by hand. The seed is sown in beds while the fields are still occupied by other crops. When the seedlings are about thirty to fifty days old, preferably thirty days in Peru,[38] they are transplanted in fields (paddies) that have been stirred into a thin, soupy mud into which clumps of three or four seedlings can be pushed easily. Research shows no yield advantage from transplanting compared with ordinary seeding methods when rice is sown in fields or beds at the same time and with the weeds controlled to the same degree.[39] Much of the rice in Asia is harvested by hand and threshed by hand or with small threshers. Considerable quantities of rice for home use are pounded out in a wooden mortar. Rice straw is used for weaving mats, bags, hats, and baskets. It is also used for drinking straws and for making paper and cardboard.

19.10 MILLING OF RICE

The rice kernel, enclosed by the hulls as it leaves the thresher, is known as rough rice or paddy rough. It weighs about 45 pounds per bushel (580 g/l). Rough rice is used for seed but is milled for food. Damaged or low-quality rough rice is used for livestock feed. When milled, rough rice yields about 64 percent whole and broken kernels, 13 percent bran, 3 to 4 percent polish, and 20 percent hulls.

Rough rice is first fanned and screened to prepare it for milling. It is then conveyed to the hulling stones (sheller) where most of the hulls are detached. The mixture of hulled rice, rough rice, and hulls is then aspirated to remove the detached hulls. The remaining portion of the mixture is then passed to paddy machines (separators) where rough rice is separated from the hulled kernels. Rough rice from the paddy machines is returned to the hulling stones for removal of the hulls, which are used for fuel or packing. The hulled kernels (brown rice) (Figure 19.11), are conveyed to hullers where the germ, along with a part of the bran layer, is removed by friction. The rice is then passed to a second set of hullers, and in some mills, also to a pearling cone. From these machines it is passed to the bran reel for separation of the detached bran from the rice. The rice kernels are then conveyed to the brush for polishing. In this process, more of the bran as well as some of the starch cells are rubbed off and screened out. The light-brown powder from the screens, known as rice polish, is used for feed. The polished rice is next conveyed to a revolving cylinder (trumble) where it is steamed. Glucose and talc are also applied when the rice is to be coated. The milled rice is then separated into grades and bagged. A bag containing 100 pounds (45 kg) is called a *pocket*.

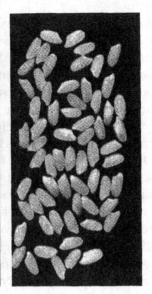

FIGURE 19.11
(Left) Paddy or rough rice.
(Center) Brown rice. *(Right)*
Milled rice.

Parboiling consists of soaking and steaming the rough rice by various procedures until the endosperm starch is partly gelatinized and then drying and milling the grain. The advantages of parboiled (converted) rice are: (1) it is broken less in milling, which results in a higher yield of head rice and helps the kernels remain whole during long cooking; (2) the rice keeps better (probably because of partial sterilization); and (3) more of the water-soluble vitamins are preserved in the kernel. The B vitamins in the bran and hulls are absorbed by the endosperm from the water used in soaking the rice.[40, 41]

Head rice yield is one of the most important criteria for measuring milled rice quality. Head rice is milled rice with length greater or equal to three-quarters of the average length of the whole kernel.[23] Rice that contains mostly half and three-quarter kernels is called second head. Rice composed mostly of halves and thirds of kernels is called screenings. Finely broken rice, called brewers rice, is used as a starchy adjunct in the brewing and distilling industries or for feed. Rice flour is a by-product sifted or ground from the coarser milled particles and used in foods and face powders.

In the milling of long-grain rice, yields of head and second head commonly amount to 50 percent of the weight of rough rice. In short-grain and medium-grain types, the yields of head rice are closer to 60 percent.

The greater breakage of long-grain rice was reported by Thomas Jefferson in 1787. Head rice consists of 11 to 12 percent moisture, about 77 percent starch, 2 percent pentosans, 7 to 8 percent protein, and fractional percentages of fat, minerals, and crude fiber. Like other refined cereal products, it is incomplete from the dietary standpoint. This fact is of little consequence in the United States where the average yearly per capita consumption is less than 30 pounds (14 kg), but it is important in parts of Asia where rice comprises 40 to 80 percent of the food calories, and annual consumption ranges from 125 to more than 400 pounds (57 to 181 kg) per capita. Those who subsist largely on polished rice can develop beriberi, a nerve-damaging nutritional disease. Beriberi is caused by a deficiency of thiamin (vitamin B_1) in the diet when the rice is not properly supplemented with such foods as meat, fish, soybean, and vegetables. Brown rice is richer than polished rice in thiamin, but it tends to become rancid in storage. Also, steady consumption of brown rice by heavy rice eaters often leads to digestive disturbances. However, undermilled rice, eaten by many Asians, contains thiamin. Milled rice is also deficient in beta-carotene, the precursor of vitamin A in the human diet. The use of biotechnology to improve the nutritional quality of rice, as described in Chapter 4, may someday alleviate this problem. Much of the milled rice sold in the United States is enriched by the addition of vitamins and minerals before packaging.

Rice of high milling quality is free from checks (cleavage lines from stresses resulting from rapid drying), free from pecky grains resulting from insect and disease injury, and free from starchy (opaque) spots caused by immaturity or poor environment. These problems increase breakage, thus reducing the yield of head rice in milling. Rice of high cooking quality leaves the kernels flaky, whole, and separated after boiling. Other varieties when cooked are whole but somewhat sticky. However, such rice is highly acceptable to numerous consumers, including those in Europe, the West Indies, Japan, and some other parts of Asia. Rice grains of low cooking quality break down during cooking. Rice, when cooked, absorbs three to five times its weight of water. Long-grain rice has a high water absorption.[41]

For use in canned soups, only parboiled grains of high-quality varieties are acceptable, because other types will disintegrate during the steam-pressure sterilization used in canning. A variety of this type has been grown in the United States only on small acreage.

■ 19.11 DISEASES

The principal diseases of rice are kernel smut, seedling blight, brown leaf spot, narrow brown leaf spot, blast, and stem rot.[16, 36, 42] Resistance to many diseases has been incorporated into newer varieties.

19.11.1 Kernel Smut

Kernel smut is caused by the fungus *Neovossia horrida*. The spores float on the water surface and germinate, releasing secondary spores that infect the flowers and subsequent developing kernels. Infected kernels show small, black pustules or streaks bursting through the glumes. Sometimes the entire kernel is replaced by a powdery, black mass of smut spores. Varieties show varying degrees of resistance. The disease is encouraged by higher rates of nitrogen, especially preflood applications.[35] A fungicide, if needed, is applied at the boot stage.

19.11.2 Seedling Blight

Seedling blight, caused by *Sclerotium rolfsii* and other fungi, attack young seedlings in warm weather. Affected seedlings are slightly discolored. Later, fungus bodies are found on the lower portions of the seedlings. Severely affected seedlings are killed. Rice sown early seems to be more subject to injury by seedling blight than that sown late. Immediate submergence will check the disease. Some fungicides provide considerable protection against seedling blights.

19.11.3 Brown Leaf Spot

Brown leaf spot, caused by *Bipolaris oryzae*, is one of the most serious diseases in Louisiana, Texas, and Arkansas. This fungus attacks rice seedlings, leaves, hulls, and kernels. It may cause seedling blight until the plants attain a height of 4 inches (10 cm). Brownish discolorations first appear on the sheaths between the germinated seed and the soil surface, or on the roots. Small circular or elongated reddish-brown spots appear on the leaves, causing them to dry up on severely affected plants. Spots also appear on the hulls and on the kernels. Resistant varieties may be the only satisfactory control. Seed treatments may protect seedlings.

19.11.4 Narrow Brown Leaf Spot

Narrow brown leaf spot, caused by *Cercospora oryzae*, is one of the most widespread leaf-spot diseases in the southern states. The spots on the leaves are long and narrow. The disease usually appears late in August and in September. The injury to plants is due principally to the reduction of leaf area. The most satisfactory control is the use of resistant or early varieties.[43]

19.11.5 Rice Blast

Rice blast or rotten neck, caused by *Piricularia grisea*, is a disease long known in many countries; it blights the panicles and rots the stems so that they break over. The fungus lives over on crabgrass and rice straw. The only known remedies are the use of strong-stemmed tolerant varieties and avoidance of heavy nitrogen fertilization, which encourages the disease. It is most severe on new rice land and under hot, humid conditions. Certain varieties are resistant to some races of the fungus.

19.11.6 Stem Rot

Stem rot, caused by the fungus *Sclerotium oryzae*, is an important and widespread disease of rice that causes the plants to lodge. Early seeding of quick-maturing tolerant varieties reduces the damage.

19.11.7 Straighthead

Straighthead is a physiological disorder. Heads of affected plants remain erect and fail to set seed. Plants affected with straighthead have dark-green leaves that remain green after the normal plants have matured. Straighthead is most prevalent on long-submerged heavy soils of new rice land or on land uncropped to rice for several years, on which a heavy growth of weeds has been plowed under. The disease can be controlled by draining the land before the rice is in the boot. After the soil has dried on the surface, the land should be submerged again. Several varieties are resistant.

19.11.8 Other Diseases

White tip, which is caused by a seed-borne nematode, *Aphelchoides besseyi*, produces white leaf tips under certain soil conditions. It is controlled by planting nematode-free seed of resistant varieties and by water seeding.

The hoja blanca ("white leaf") disease causes a whitening or striping of the leaves and stunted, unproductive plants that tiller profusely. The disease is caused by a virus that is transmitted by a plant hopper, *Tagosodes orizicolus*. Resistant varieties are available.

Rice is also attacked by sheath spot, sheath rots, kernel spots, leaf smut, and other diseases.[36]

■ 19.12 INSECT PESTS

19.12.1 Rice Stink Bug

The rice stink bug *(Oebalus pugnaz)* sucks the contents from rice kernels during the milk stage, leaving only an empty seed coat when all the milk is withdrawn and causing pecky rice when the milk is only partly consumed. Pecky rice is discolored and is likely to be broken in milling. Control measures include early mowing and winter burning of the coarse grasses in which the insect hibernates.

19.12.2 Stalk Borers

Two species of borers, the sugarcane borer *(Diatraea saccharalis)* and the rice-stalk borer *(Chilo plejadellus)*, damage rice in Louisiana and Texas. They tunnel inside the rice culms, eating the inner parts, which often causes the panicles to turn white and produce no grain, a condition called whitehead. In other culms, grain formation is not completely inhibited, but some of the culms or panicles break off when the culm is badly girdled. The borers prefer large stems. The borers live over winter in the rice stubble. They are killed in their winter quarters by heavy grazing of the stubble, by mashing down of the stubble by dragging, and by flooding the stubble fields by closing up drainage outlets to hold water from winter rains.[36]

19.12.3 Rice Water Weevil

Rice water weevil *(Lissorhoptrus oryzophilus)* larvae reduce yields by feeding on roots, which results in stunting and yield loss from reduction of tillers and panicles or delayed maturity. Adult feeding causes linear slits of varying length on the upper surface of the leaves but usually does not cause economic losses.[44] There is only one generation per year and insecticides must be applied before the adults lay eggs. There are also insecticides for use as seed treatment.

19.12.4 Other Insects

Before the fields are submerged, the rice plants are sometimes damaged by the fall armyworm *(Laphygna frugiperda)*, southern corn rootworm *(Diabrotica duodecim-punctata)*, sugarcane beetle *(Euetheola rugiceps)*, and chinch bug *(Blissus leucopterus)*. Flooding destroys these insects. Threshed rice is attacked by the rice weevil and other insects that attack stored grains.[45]

19.12.5 Other Pests

In California, the tadpole shrimp *(Triops longicaudatus)* digs up and cuts off the leaves of rice seedlings, which prevents their emergence through the water. The shrimp is controlled by applying chemicals such as 10 pounds of granular copper sulfate per acre by airplane about two weeks after the field is submerged.

Additional pests are waterfowl and other birds. Muskrats, nutria, and crayfish burrow through dikes and ditch banks.

▣ 19.13 WILD RICE

Two species of wild rice grow in the United States. Southern wild rice *(Zizania aquatica)* grows in the Atlantic coastal plain. Northern wild rice *(Z. palustris)* ranges from northeast North America through the Great Lakes to south central Canada.[46] Both are annual plants. Wild rice was formerly seldom sown except in marshes where it was desired to attract waterfowl but now is grown on limited acreage for food.

Wild rice grows in lakes as well as in tidal rivers and bays where the water does not contain enough salt to be tasted. A wide range of growth types adapted to local day lengths are found at different latitudes. Wild rice grows mostly in water 2 to 4 feet (60 to 120 cm) deep. On tidal lands, the water may fluctuate through depths from 0 to 3 feet (0 to 90 cm).

The seeds shatter quickly when ripe and fall to the muddy bottom where they remain until spring, when germination occurs. The plants can grow up through the water at depths up to 6 feet (180 cm). The seed fails to germinate when it is allowed to dry. Consequently, it is kept covered with cool water when stored over winter for planting. The plants grow to a height of 3 to 11 feet (90 to 335 cm) and are topped by large open panicles that bear pistillate flowers in the upper half and staminate flowers below. This arrangement favors cross-pollination. The seed (Figure 19.12) is harvested mostly by northern Native American tribes and others who tap the tops with sticks and knock the seed into a boat or canoe. Adequate machines for efficient harvesting are not yet available. It is then partly dried on the ground, parched in an open vessel, and often hulled by pounding or trampling. Because of the labor required to gather and process the grain, wild rice is sold as a high-priced delicacy.

FIGURE 19.12

(1) Wild rice *(Zizania aquantica)* plants growing in water; (2) panicle showing appressed branches of pistillate florets above and spreading branches of staminate florets below; (3) natural-sized unhulled seeds; and (4) hulled seeds.

REFERENCES

1. Yeh, B., and M. T. Henderson. "Cytogenetic relationships between African annual diploid species of Oryza and cultivated rice, *O. sativa L.*," *Crop Sci.* 2(1962):463–467.

2. Chandler, R. F., Jr. "Dwarf rice—a giant in tropical Asia," *USDA Yearbook*, 1968, pp. 252–255.

3. Chapman, A. L., and D. S. Mikkelsen. "Effect of dissolved oxygen supply on seedling establishment of water-sown rice," *Crop Sci.* 2(1963):391–395.

4. Jones, J. W. "Improvement in rice," in *USDA Yearbook*, 1936, pp. 415–454.

5. Tanaka A., and others. "Growth habit of the rice plant in the tropics and its effect on nitrogen response," *IRRI Tech. Bull. 3*, 1964, pp. 1–80.

6. Jones, J. W. "Branching of rice plants," *J. Am. Soc. Agron.* 17(1925):619–623.

7. Ramiah, K. "Rice breeding and genetics," *Indian Council Agr. Res. Science Monograph 19*, 1953.

8. Jones, J. W. "Observations on the time of blooming of rice flowers," *J. Am. Soc. Agron.* 16(1924):665–670.

9. Laude, H. H., and R. H. Stansel. "Time and rate of blooming in rice," *J. Am. Soc. Agron.* 19(1927):781–787.

10. Adair, C. R. "Studies on blooming in rice," *J. Am. Soc. Agron.* 26(1934):965–973.

11. Beacher, R. L. "Rice fertilization: Results of tests from 1946 through 1951," *AR Agr. Exp. Sta. Bull. 552*, 1952.

12. Jodon, N. E. "Occurrence and importance of natural crossing in rice," *Rice J. 62*, 8(1959):8 and 10.

13. Funderburg, E. R., J. K. Saichuk, and P. Bollich. "Fertilization of Louisiana rice." *LA Coop. Ext. Serv. Publ. 2418*, 2000.

14. Snyder, C. S., and N. A. Slaton. "Rice production in the United States: An overview," *Better Crops* 85(2001):3–7.

15. Sedberry, J. E., Jr., and others. "Effects of zinc and other elements on the yield of rice, and nutrient content of rice plants," *LA Agr. Exp. Sta. Bull. 653*, 1971, pp. 1–8.

16. Padwick, G. W. *Manual of Rice Diseases*. Surrey, England: Commonwealth Mycol. Inst., Kew, 1950, pp. 1–198.

17. Slaton, N. A., and others. "Evaluation of zinc seed treatments for rice." *Agron. J.* 93(2001):152–157.

18. De Datta, S. J. *Principles and Practices of Rice Production*. New York: John Wiley & Sons, 1981, pp. 1–618

19. Black, D. E., and R. K. Walker. "The value of pastures in rotation with rice," *LA Agr. Exp. Sta. Bull. 498*, 1955.

20. Sturgis, M. B. "Managing rice soils," in *Soil*, USDA Yearbook, 1957, pp. 658–662.

21. Slaton, N. A., and others. "Rice production handbook," *U. AR Coop. Ext. Serv.* MP-192, 2001.

22. Lewis, D. C., and others. "New rice levee concept," *Rice J.* 65(1962):6–15.

23. Houston, D. F., ed. *Rice, Chemistry and Technology*. St. Paul, MN: Am. Assn. Cereal Chem., 1974, pp. 1–300.

24. Smith, W. D. "Handling rough rice to produce high grades," *USDA Farmers Bull. 1420*, 1940, pp. 1–21.

25. Jodon, N. E., and W. A. McIbrath. "Response of rice to time of seeding in Louisiana," *LA Agr. Exp. Sta. Bull. 649*, 1971, pp. 1–27.

26. Taylor, D. L. "Effects of oxygen on respiration, fermentation and growth in wheat and rice," *Science* 95(1942):116–117.

27. Jones, J. W. "Effect of reduced oxygen pressure on rice germination," *J. Am. Soc. Agron.* 25, 1(1933):69–81.

28. Jones, J. W. "Effect of depth of submergence on control of barnyard grass and yield of rice grown in pots," *J. Am. Soc. Agron.* 25(1933):578–583.

29. Adair, C. R., and K. Engler. "The irrigation and culture of rice," in *Water*, USDA Yearbook, 1955, pp. 389–394.

30. Kapp, L. C. "The effect of common salt on rice production," *AR Agr. Exp. Sta. Bull. 465*, 1947.

31. Pearson, G. A., and L. Bernstein. "Salinity effect at several growth stages in rice," *Agron. J.* 51(1959):654–657.

32. Miller, T., and W. Houston. "Rice–Red rice control." *MS St. Univ. Ext. Serv.* IS930, 2004.

33. Jones, J. W. "Germination of rice seed as affected by temperature, fungicides, and age," *J. Am. Soc. Agron.* 18(1926):576–592.

34. McNeal, X. "When to harvest rice for best milling quality and germination," *AR Agr. Exp. Sta. Bull. 504*, 1950.

35. Smith, W. D., and others. "Effect of date of harvest on yield and milling quality of rice," *USDA Circ. 484*, 1938, pp. 1–20.

36. Adair, C. R., and others. "Rice in the United States: Varieties and production," *USDA ARS Handbk. 289*, (rev.) 1973, pp. 1–154.

37. Craigmiles, J. P., and others. "Rice esearch in Texas, 1971," *Consolidated Progress Rept.,* 1972, pp. 3092–3105.

38. Sanchez, P. A., and N. Larrea. "Influence of seedling age at transplanting on rice performance," *Agron. J.* 64, 6(1972):828–832.

39. Adair, C. R., and others. "Comparative yields of transplanted and direct sown rice," *J. Am. Soc. Agron.* 34, 2(1942):129–237.

40. Jones, J. W., L. Zeleny, and J. W. Taylor. "Effect of parboiling and related treatments on the milling, nutritional and cooking quality of rice," *USDA Circ. 752*, 1946.

41. Rao, R. M. R., and others. "Rice processing effects on milling yields, protein content and cooking qualities," *LA Agr. Exp. Sta. Bull. 663*, 1972, pp. 1–51.

42. Atkins, J. G. "Rice diseases," *USDA Farmers Bull. 2120*, 1958.

43. Adair, C. R. "Inheritance in rice of reaction to *Helminthosporium oryzeae* and *Cercospora oryzeae*," *USDA Tech. Bull. 772*, 1941.

44. Stout, M. J., W. C. Rice, and D. R. Ring. "Integrated management of the rice water weevil," *Louisiana Agriculture* 45, 1(2002):20–22.

45. Anonymous. "Controlling insect pests of stored rice," *USDA ARS Handbk. 129*, 1971.

46. Steeves, T. A. "Wild rice—Indian food and modern delicacy," *Econ. Bot.* 6(1952):107–143.

Millets

KEY TERMS

Browntop millet
Finger millet
Foxtail millet
Japanese barnyard millet
Koda millet
Pearl millet
Proso millet
Teff

■ 20.1 ECONOMIC IMPORTANCE

The millets are minor crops except in parts of Asia, Africa, and Eastern Europe.[1] The total world seed production in 2000–2003 was approximately 30.5 million tons (27.7 million MT) on an average of 89 million acres (36 million ha), or 690 pounds per acre (770 kg/ha). The leading producing countries are India, Nigeria, Niger, China, and the Russian Federation. About 50 percent of the world production consists of pearl millet (*Pennisetum glaucum*), 10 percent finger millet (*Eleusine coracana*), 30 percent foxtail millet (*Setaria italica*) and proso millet (*Panicum miliaceum*), and the remainder of other species (CGIAR). Proso millet is the primary millet in the world import and export market. About 620,000 acres (250,000 ha) of millets were grown in the United States in 2000–2003. Proso millet accounted for 463,000 acres (187,000 ha) of the total. Proso millet is grown in Colorado, Nebraska, and South Dakota.

Other millet species include koda millet (*Paspalum scrobiculatum*), jungle rice or Shama millet (*Echinochloa colonum*), blue panic grass (*Panicum miliare* or *P. antidotale*), funde or fonio (*Digitaria exilis* and *Digitaria iburua*), and teff (*Eragrostis tef*). The genus *Eleusine* belongs to the tribe Chlorideae; the *Eragrostis* genus to the tribe Festuceae; and the other millets belong to the Paniceae tribe.

Thus, the term *millet* is applied to various grass crops whose seeds are harvested for food or feed. Sorghum is called millet in many parts of Asia and Africa. Broomcorn is called broom millet in Australia.

■ 20.2 FOXTAIL MILLET

20.2.1 History

Foxtail millet, also called German or Italian millet, is one of the oldest of cultivated crops. It was grown in China as early as 2700 BC. The crop probably had its origin in southern Asia, but from there its culture spread westward to Europe at an early date. It is grown mostly in eastern Asia.[1] Foxtail millet was rarely grown in the United States during Colonial times, but it became a rather important crop in the Central States after 1849. Its acreage decreased when it was largely replaced by sudangrass as a late-sown hay crop after 1915.

20.2.2 Adaptation

For its best growth, foxtail millet requires warm weather during the growing season. It matures quickly when grown in the hot summer months of the most northern states. The crop is most productive where the rainfall is fairly abundant, but a large part of the acreage is found in the semiarid region. It has a low water requirement.

Foxtail millet lacks the ability to recover after being injured by drought.[2] Because of its shallow root system, it is one of the first crops to show the effects of drought. The millets succeed in localities subject to drought almost entirely through their ability to escape periods of acute drought due to their short growing season. Some early varieties will produce a hay crop in 65 to 70 days. When grown in the dry areas of New Mexico and western Texas, foxtail millet is planted in cultivated rows to permit the crop to succeed. Other crops produce more forage than millet does under irrigation and at high altitudes in Colorado.[3] A fertile loam soil is best for millet, but good drainage is essential.

20.2.3 Botanical Description

Foxtail millet *(Setaria italica)* is an annual grass with slender, erect, leafy stems. The plants vary in height from 1 to 5 feet (30 to 150 cm) under cultivation. The seeds are borne in a spike-like or compressed panicle (Figure 20.1). There is an involucre of one to three bristles at the base of each spikelet. The small, convex seeds are enclosed in the hulls (lemma and palea). The color of the hulls differs with the variety, some of the colors being creamy white, pale yellow, orange, reddish orange, green, dark purple, or mixtures of various colors.[4] The common annual weeds, yellow foxtail *(Setaria lutescens* or *S. glauca)*, green foxtail *(S. viridis)*, and giant foxtail *(S. faberi)* resemble foxtail millet. The foxtail millets are largely self-pollinated, with less than 10 percent natural crossing.[1, 5, 6]

Ten or more distinct varieties of foxtail millet have been grown in the United States.[3, 7, 8] In Nebraska, hay yields of the better varieties were 75 percent that of sorghum.

20.2.4 Rotations

Foxtail millet is usually grown as a catch crop since it matures in seventy-five to ninety days. It can be sown when the date is too late for seeding most other crops. Some crops do not yield well after millet under dryland conditions because there is little soil moisture left after millet matures. Late spring–planted crops (corn or sorghum) follow millet in rotation better than do fall or early spring–sown crops (wheat, oats, or barley). It is usually advisable to grow foxtail millet after small grain or corn.

20.2.5 Culture

Seedbed preparation for foxtail millet is similar to that for spring-sown small grains. A firm seedbed is essential because of the small size of millet seeds. Weeds

FIGURE 20.1
Spikes of foxtail millet varieties. *(Left to right)* Kursk, Hungarian, Common, Siberian, Turkestan, and Golden.

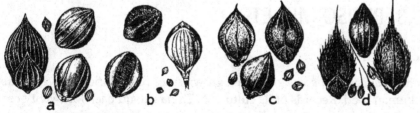

FIGURE 20.2
Millet seeds, enlarged and natural size: *(a)* proso, *(b)* foxtail millet, *(c)* Japanese millet, and *(d)* barnyardgrass.

should be controlled up to the time the crop is planted because the millet seedlings are small and compete poorly with weeds until they have attained some size.

Millet seed (Figure 20.2) should be planted when the soil is warm, or two to three weeks after corn planting time. For late planting, sixty to seventy days before the normal date of fall frost should be allowed. In Colorado, the crop is generally planted between May 15 and July 1 whenever there is sufficient soil moisture to germinate the seed.

Millet is usually seeded with an ordinary grain drill, since close spacing helps the crop to suppress weeds.[3] The plants should be spaced not more than 2 inches (5 cm) apart in the row.[5] The seed should be placed in moist soil even if it is necessary to plant it 1 inch (2.5 cm) deep or slightly deeper.

Foxtail millet is sown at the rate of 25 to 30 pounds per acre (28 to 33 kg/ha). On clean land in semiarid regions 10 to 15 pounds (11 to 17 kg/ha) is sufficient. For cultivated rows, 4 or 5 pounds (4.4 to 5.6 kg/ha) of seeds are ample.[9] The best quality of hay is obtained when it is cut just as the first heads appear, since it is more palatable at that time than when fully mature. For seed, the crop is often windrowed and the seed threshed with a combine with a pick-up attachment. Direct combining allows early-maturing seeds to shatter before the later seeds are ripe.

20.2.6 Utilization

Millet hay is considered inferior to that of timothy. To some extent, this is due to its being less palatable, but the hay contains a glucoside, called setarian, which acts as a diuretic on horses that consume it continuously as the sole roughage.[10] The hay can be fed to cattle or sheep without danger, but the awns may induce sore eyes or lumpy jaws.

In the United States, foxtail millet is seldom utilized as a grain crop, except as bird seed. The seed is slightly less palatable, with only about 83 percent of the feeding value of proso millet.[3] The seed should be ground finely before being fed to livestock. It is used for food or feed in China, Manchuria, Japan, and India. Ripe heads of foxtail millet are sold in the United States for feeding caged birds.

20.2.7 Pests and Diseases

Foxtail millet is subject to several diseases, which include mildew, bacterial blight, and leaf spots. Kernel smut (*Ustilago crameri*) may be controlled by seed treatment.[1] Grasshoppers, armyworms, and chinch bugs can cause economic damage. Foxtail millet can harbor the wheat leaf curl mite, which is a vector of wheat streak mosaic disease in wheat.[11] Therefore, wheat should not be planted after foxtail millet.

■ 20.3 PROSO MILLET

20.3.1 History

Proso millet (*Panicum miliaceum*) has been grown since prehistoric times as a grain crop for human food. Records of its culture in China extend back for twenty centuries or more. It is grown primarily in Africa and Asia, but some is grown in the Americas, Oceania, and the Russian Federation. Proso millet was introduced into the United States from Europe during the eighteenth century. It was grown sparingly along the Atlantic seaboard but began to assume some importance in the North Central States after about 1875 when it was introduced into the Dakotas. It is now grown chiefly in North Dakota, Colorado, Nebraska, and South Dakota.[12] Proso millet is also called broomcorn millet, hog millet, and Hershey millet. A related species (*Panicum sumatrense*) is grown in Southeast Asia.

20.3.2 Adaptation

Proso millet is adapted only to regions where spring-sown small grains are fairly successful. It is a short-season plant that often requires only sixty to sixty-five days from seeding to maturity. Proso millet is readily injured by frost and is not adapted to regions where summer frosts occur. Moderately warm weather is necessary for plant growth.

Proso millet exhibits the efficient C_4-photosynthetic pathway like corn and sorghum, except that it exudes CO_2 during darkness.

While proso millet has the lowest water requirement of any grain crop, it is less resistant to drought than well-adapted varieties of other grains, largely because of its shallow root system.[13] Sometimes it is a complete failure when wheat or barley produces a fair crop. Proso millet yields less than corn under irrigated conditions.[14, 15] Proso millet grows well on all except coarse, sandy soils.[10, 16]

20.3.3 Botanical Description

The proso millet inflorescence is a large open panicle. Proso millet has coarse, woody, hollow stems 12 to 48 inches (31 to 120 cm) high but usually they are about 24 inches (61 cm) high. The stems are round or flattened and generally about as thick at the base as a lead pencil. The stems and leaves are covered with hairs. The stem and outer chaff are green, or sometimes yellowish or reddish green, when the seed is ripe. When threshed, most of the seed remains enclosed in the inner chaff or hull. The seed of proso millet is larger and not as tightly held in the hull as are those of the millets of the foxtail group. The hulls are of various shades and colors, including white, cream, yellow, red, brown, gray, striped, and black. The bran, or seed coat, of all varieties is a creamy white (Figure 20.2).

A large percentage of proso millet flowers are self-fertilized, but considerable cross-fertilization occurs.

20.3.4 Varieties

The varieties of proso millet are divided into three main groups on the basis of the shape of the panicle (Figure 20.3). These shapes are: (1) spreading, (2) loose and one-sided, and (3) compact and erect.

Plump, bright, reddish seeded types are the most attractive seed for use in chick feed and bird seed mixtures.

20.3.5 Rotations

Proso millet can follow any other crop. It is a late-seeded, short-season summer catch crop. It does not respond to additional soil moisture sufficiently to justify its production on fallow land.[17]

Fertilizers are of limited value under semiarid conditions, but 20 to 40 pounds (22 to 44 kg/ha) of N per acre may be helpful when proso millet follows wheat and soil moisture is ample.[14] Phosphate and potash fertilizers can be beneficial in soils deficient in these nutrients.[18]

FIGURE 20.3
Panicles of one-sided proso *(center)* and compact *(right)*. Proso seeds at left.

The average grain yield level of proso millet is about half that of the shortest season sorghum hybrid. Proso millet has a more rapid grain filling period so it matures a week sooner than even the shortest season sorghum. Proso millet matures in such a short season that comparable length maturity sorghums are not available. Thus, proso millet makes an excellent emergency or catch crop.[19]

20.3.6 Culture

Proso millet requires a firm seedbed that has been kept free from weeds up to seeding time. It will not germinate in a cold soil. Since the plants are easily killed by spring frosts, seeding should be delayed until the danger of frost is practically over. Most varieties require sixty to eighty days from seeding to maturity. Generally the crop can be seeded safely from two to four weeks after corn is planted. In eastern Colorado, good yields of proso millet were obtained when it was seeded between June 15 and July 1.[10] Earlier seedings were very weedy, while later seedings often resulted in crop failure.

Proso millet is sown with an ordinary grain drill. A seeding rate of 30 to 35 pounds per acre (34 to 39 kg/ha) has been most satisfactory.[10, 15] Lighter seeding rates are noticeably weedier, but 16 to 20 pounds per acre (18 to 22 kg/ha), or even less, is ample on clean, semiarid soils.[20] Proso millet can be no-till seeded in wheat stubble in semiarid regions. Forty-five to fifty pounds per acre (50 to 56 kg/ha) of nitrogen is applied at planting to reduce nitrogen tie-up during mineralization of the wheat stubble.[21] Herbicides commonly used for weed control are atrazine; 2, 4-D; and prosulfuron.

Proso millet is ready to be harvested when the seeds in the upper half of the heads are ripe. At this stage, the plant is still green. Windrowing is the best method for harvesting the crop followed by a combine with a pick-up attachment a few days later. Proso millet is not adapted to direct combine harvesting because: (1) it shatters soon after it is ripe, (2) it lodges when left standing, and (3) the straw contains too much moisture at harvest time.

20.3.7 Utilization

Proso millet is commonly consumed for food in Asia. The hulled seed is used in soups, and the ground meal is eaten as a cooked cereal. Digestion experiments with bread made from proso millet flour indicate that the carbohydrates are as well utilized as in other cereals, but that only about 40 percent of the protein is digestible.[22]

Proso millet is eaten readily by all kinds of livestock. It should be ground before being fed to livestock. It has about 95 percent of the feeding value of corn or barley for hogs or poultry but only 90 percent or less of the value of corn for fattening cattle or sheep. Proso millet, although occasionally cut green and cured, is a poor hay crop because of the coarse stems and hairy leaves. The chief commercial use of proso millet is as an ingredient of chick feeds and bird seed mixtures.

20.3.8 Diseases

Compared to many other crops, proso millet is relatively free from diseases. A bacterial stripe disease (*Psuedomonas avenae*) has been found on proso millet.[23] Affected plants have brown, water-soaked streaks on the leaves, sheaths, and culms. The long, narrow lesions show numerous thin, white scales of the exudate. The disease is probably seed-borne. Head smut caused by the fungus *Sphacelotheca destruens* attacks proso millet. Another fungal disease is kernel smut (*Ustilago crameri*).

20.3.9 Wild Proso Millet

A wild variety of proso millet is rapidly becoming a problem weed in the Midwest United States. It was first discovered in Wisconsin in the early 1970s. It differs from cultivated proso millet in that the seed is various shades of olive, gray, or black and readily shatters at maturity. The stems have abundant hairs at right angles to the stem. Herbicides to control wild proso millet in corn fields include alachlor, metolachlor, butylate, and nicosulfuron.[24]

20.4 PEARL MILLET

Pearl millet (*Pennisetum glaucum*), also called cattail millet, bullrush millet, candle millet, and penicillaria, is the world's fourth most important food cereal in the tropics with 64 million acres (25 million ha). It is grown mostly in semiarid West Africa and India. In India, it is called bajri, bajra, or cumbo. About 1.5 million acres (600,000 ha) are grown for forage in the United States, mostly in the southern United States. It is an important temporary summer pasture crop on the southern coastal plain. Recently, there is increased interest in growing pearl millet for grain in the Southeast United States.

20.4.1 History

The origin of pearl millet was probably in the African Savannah zone. It has been grown in Africa and Asia since prehistoric times. Pearl millet was introduced in the United States at an early date, but it was seldom grown until after 1875. About 575,000 acres (233,000 ha) were grown in Georgia and other southern states in 1969.

20.4.2 Botanical Description

Pearl millet is a tall, erect, annual grass that grows 6 to 15 feet (180 to 460 cm) in height. The stems are pithy, while the leaves are long-pointed with finely serrated margins. The plant tillers freely. The inflorescence is a dense, spike-like panicle 6 to 14 inches (15 to 36 cm) long, and 1 inch (2.5 cm) or less in diameter (Figure 20.4). The mature panicle is brownish in color. The spikelets are borne in fascicles of two, being surrounded by a cluster of bristles. Each spikelet, subtended by two unequal glumes, has two florets. The lower floret is staminate, being represented in many cases only by a sterile lemma. The upper floret is fertile, with the caryopsis enclosed in the lemma and palea from which it threshes free at maturity. In some cases, spikelets contain two fertile florets. The plant is protogynous; that is, the stigmas appear several days before the anthers protrude. Consequently, pearl millet is mostly cross-pollinated. The spike flowers from the tip downward in four or five days. The late stigmas appear at the base of the spike shortly after the first shedding of pollen starts at the tip of the spike.[25] It is partly self-sterile. Many varieties are known in the Eastern Hemisphere where hybrids are being developed.

20.4.3 Culture

Sandy loam soils are best suited to the crop. Pearl millet is the most productive grain in extremely dry or infertile soils of India and Africa. However, it responds well to heavy fertilization.[26, 27] It should be seeded after all danger of frost is past in warm soils. The date of planting varies from April in the Gulf states to May in the northern states. The seed is generally planted with drills in rows 20 to 42 inches (51 to 107 cm) apart, with the plants 4 to 6 inches (10 to 15 cm) apart in the row. Close planting at a rate of 10 to 20 pounds per acre (11 to 22 kg/ha) is best when the crop is intended for grazing or hay. Heavier seeding rate results in thinner stems, which improves forage quality.[28] From 4 to 6 pounds of seed per acre (4.4 to

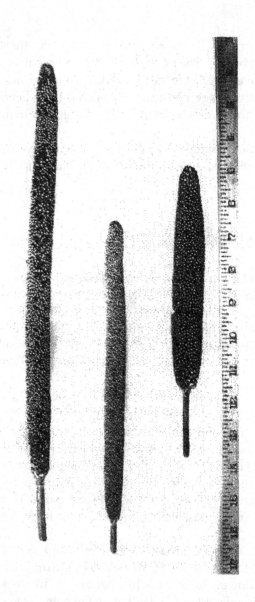

FIGURE 20.4
Panicles of pearl millet.

6.6 kg/ha) are planted for seed production. The seed should be planted at about ½ inch (13 mm) deep. For the best-quality hay, pearl millet is harvested when the heads begin to appear.

Breeding research on pearl millet is being conducted in Georgia, Kansas, and Nebraska. Hybrid pearl millet seed is produced by using a cytoplasmic male-sterile seed.[29]

Common herbicides used on pearl millet are atrazine; 2, 4-D; and prosulfuron. There are few diseases associated with pearl millet. Few insects infest pearl millet. Grasshoppers, armyworms, and chinch bugs can cause economic damage.[11]

20.4.4 Utilization

In the United States, pearl millet is utilized primarily for pasture and occasionally for silage. If harvested before heading, forage quality is excellent with crude protein as high as 15 percent. However, total digestible nutrients are lower than sudangrass or sorghum-sudangrass hybrids.[28] The plant becomes woody as it approaches the flowering stage. The stover is of little value after the seed has matured. Yields as high as 16 tons of dry forage per acre (36 MT/ha) have been recorded under warm favorable conditions. Pearl millet grain is lower quality than corn or wheat when fed to swine.[30]

Pearl millet is utilized as a food grain in India and parts of Sub-Saharan Africa and the remaining stover is fed to livestock, even though it is not high-quality forage.

■ 20.5 JAPANESE BARNYARD MILLET

Japanese barnyard millet or billion-dollar grass (*Echinochloa frumentacea*) is grown as a forage grass in local areas in the United States. It is grown for feed or for its edible seeds in Australia, Japan, Egypt, and other countries. It has been called poor man's millet in Korea, sava millet in India, and Japanese millet. This millet probably originated from the common weed barnyardgrass (*E. crusgalli*), which it closely resembles. A related species (*E. corona*), called shama millet or jungle rice, is grown in India and other countries.

Japanese barnyard millet is an annual that grows 2 to 4 feet (61 to 122 cm) tall. The inflorescence is a panicle made up of from five to fifteen sessile erect branches (Figure 20.5). The spikelets are crowded on one side of the rachis and brownish to purple in color. The spikelet is subtended by two glumes within which are two florets. The lower floret is staminate while the upper one is perfect. Within the glumes is the lemma of the staminate floret followed by the lemma, caryopsis, and palea of the fertile floret. The caryopsis remains enclosed in the lemma and palea. This millet differs from barnyardgrass in its more erect habit, more turgid seeds, and in being awnless.

The cultural methods are similar to those for foxtail millet. The main value of Japanese barnyard millet in the United States is as a pasture or soiling crop, or as food for game birds.[31] The plant is difficult to cure for hay because of its thick stems. Japanese barnyard millet hay is palatable when cut before the plant heads, but much less so as it approaches maturity. High yields are obtained when moisture is ample.

■ 20.6 BROWNTOP MILLET

Browntop millet (*Panicum ramosum*) is grown in the southeastern part of the United States for hay or pasture as well as to provide feed for quail, doves, and other wild game birds. Seed of this plant, a native of India, was introduced into the United States in 1915.

Browntop millet is a quick-growing annual, 2 to 4 feet (61 to 122 cm) tall, with a 2- to 6-inch (5 to 15 cm) open panicle that ranges in color from yellow to brown. Shattering of seed occurs in sufficient amounts to reseed the crop. Consequently, it can become a weed pest on cropped land.

FIGURE 20.5
Plant and detached raceme
of Japanese millet.

Browntop millet is sown from May to July. For hay or pasture, it is drilled at a rate of 10 to 20 pounds per acre (11 to 22 kg/ha). When grown for seed, it is sown in cultivated rows at rates of 4 to 10 pounds per acre (4.4 to 11 kg/ha). It is harvested with a combine.

Browntop millet can be sown in a disked crimson-clover sod after that crop is harvested. This millet will reseed itself thereafter, while a hard-seeded type of crimson clover will also reseed. Natural reseeding of browntop millet also occurs in game preserves.[32]

20.7 OTHER MILLETS

Teff *(Eragrostis tef)* is grown for food in East Africa, mostly in Ethiopia where its production exceeds that of all other grains combined. The iron content of teff is nearly double, and the calcium content is nearly twenty times that of other grains, which favors its nutritional value.[33]

Finger millet *(Eleusine coracana)* is grown extensively in India, East Africa, and elsewhere, with a total production exceeding 3.3 million tons (3 million MT). It is also called ragi and birdsfoot millet. The heads are forked like the feet of a bird.

Paspalum scrobiculatum is grown in India as koda millet and in New Zealand as ditch millet. Little millet or kuthki samai *(Panicum sumatrense)* is grown in India and Pakistan. Fonio *(Digitaria exilis)* is grown in Africa and Asia.

REFERENCES

1. Maim, R., and K. O. Rachie. "The *Setaria* millets—a review of the world literature," *NE Agr. Exp. Sta. Bull.* SB 513, 1971, pp. 1–133.

2. Vinall, H. N. "Foxtail millet: Its culture and utilization in the United States," *USDA Farmers Bull.* 793, 1924.

3. Curtis, J. J., and others. "Foxtail millet in Colorado," *CO Agr. Exp. Sta. Bull.* 461, 1940.

4. Martin, J. H. "Sorghums, broomcorn, sudangrass, johnsongrass, and millets," in *Grassland Seeds*. New York: Van Nostrand, 1957, pp. 1–734.

5. Li, H. W., and C. J. Meng. "Experiments on the planting distance in varietal trials with millet, *Setaria italica*," *J. Am. Soc. Agron.* 29(1937):577–583.

6. McVicar, R. M., and H. R. Parnell. "The inheritance of plant color and the extent of natural crossing in foxtail millet," *Sci. Agr.* 22(1941):80–84.

7. Home, A. N., and M. Champlin. "Trials with millets and sorghums for grain and hay in South Dakota," *SD Agr. Exp. Sta. Bull.* 135, 1912.

8. Leonard, W. H., and J. H. Martin. *Cereal Crops.* New York: Macmillan, Inc., 1963, pp. 740–769.

9. Anderson, E., and J. H. Martin. "World production and consumption of millet and sorghum," *Econ. Bot.* 3(1949):265–288.

10. Curtis, J. J., and others. "Proso or hog millet," *CO Agr. Exp. Sta. Bull.* 438, 1937.

11. Baker, R. D. "Millet production." *NM St. U. Coop. Ext. Serv.* Guide A-414, 2003.

12. Nelson, L. A. "Producing proso in western Nebraska," *NE Agr. Exp. Sta. Bull.* 526, 1973, pp. 1–12.

13. Martin, J. H. "Proso or hog millet," *USDA Farmers Bull.* 1162, 1937.

14. Grabowski, P. H. "Growing proso in Nebraska," *NE Agr. Exp. Sta.* SC 110, 1968.

15. Nelson, C. E., and S. Roberts. "Proso grain millet as a catch crop or second crop under irrigation," *WA Agr. Exp. Sta. Cir.* 376, 1960.

16. McGee, C. R. "Proso—a grain millet," *MI Agr. Ext. Bull.* 231, 1941.

17. Brandon, J. F., and others. "Proso or hog millet in Colorado," *CO Agr. Exp. Sta. Bull.* 383, 1932.

18. Blumenthal, J. M., and D. D. Baltensperger. "Fertilizing proso millet," *NE Coop. Ext. Serv. NebGuide* G89–924-A, 2002.

19. Witt, M. "Proso millet as a crop alternative," *KS Agri. Exp. Sta. Keeping Up with Research* 70, 1983.

20. Hinze, G. "Millets in Colorado," *CO Agr. Exp. Sta. Bull.* 553 S, 1972, pp. 1–12.

21. Ramsel, R. E., L. A. Nelson, and G. A. Wicks. "Ecofarming: No-till ecofallow proso millet in winter wheat stubble," *NE Coop. Ext. Serv. NebGuide* G87-835-A, 1987.

22. Langworthy, C. F., and A. D. Holmes. "Experiments in the determination of the digestibility of millets," *USDA Bull.* 525, 1917.

23. Elliott, C. "Bacterial stripe disease of proso millet," *J. Agr. Res.* 26(1923):151–160.

24. Wilson, R. G. "Wil proso millet," *NE Coop. Ext. Serv. NebGuide* G83-648-A, 1992.

25. Martin, J. H. "Sorghum and pearl millet," in *Handbuch der Pflanzenzuchtung*, II Bd., 2 Aufl. Berlin: Paul Parey, 1959, pp. 565–589.

26. Broyles, K. R., and H. A. Fribourg. "Nitrogen fertilization and cutting management of sudangrasses and millets," *Agron. J.* 51(1959):277–279.

27. Burton, G. W. "The adaptability and breeding of suitable grasses for the southeastern states," in *Advances in Agronomy*, vol. 3. New York: Academic Press, 1951, pp. 197–241.

28. Sedivec, K. K., and B. G. Schatz. "Pearl millet forage production in North Dakota," *ND St. Univ. Ext. Serv.* R-1016, 1991.

29. Andrews, D. J., and others. "Advances in grain pearl millet: Utilization and production research," in J. Janick, ed., *Progress in new crops*. Alexandria, VA: ASHS Press, 1996, pp. 170–177.

30. Adeola, O., D. King, and B. V. Lawrence. "Evaluation of pearl millet for swine and ducks," in J. Janick, ed., *Progress in new crops*. Alexandria, VA: ASHS Press, 1996, pp. 177–182.

31. Schoth, H. A., and J. H. Hampton. "Sudan grass, millets, and sorghums," *OR Agr. Exp. Sta. Bull.* 425, 1945.

32. Craigmiles, J. P., and J. M. Elrod. "Browntop millet in Georgia," *MA Agr. Exp. Sta. Leaflet* 14, 1957.

33. Mengesha, M. H. "Chemical composition of teff *(Eragrostis tef)* compared with that of wheat, barley and grain sorghum," *Econ. Bot.* 20, 3(1966):268–273.

Perennial Forage Grasses

■ 21.1 ECONOMIC IMPORTANCE

About ninety species of perennial pasture and hay grasses have been seeded in the United States. Some forty of these are of considerable economic importance.[1, 2] Most of these are introduced species, but some fifteen native grasses, including little bluestem (Figure 21.1), have been seeded on appreciable acreages.[3] The native species, except for reed canarygrass, have been sown mostly on western ranges or abandoned cropland. Grasses that are fertilized heavily with nitrogen have protein content similar to that of legume forages.[4, 5]

Almost 1 million acres (405,000 ha) of forage and turf grasses are harvested for seed each year in the United States. Much of this seed production is in Oregon and Missouri. The leading grasses are fescue, ryegrass, timothy, bluegrass, orchardgrass, and bromegrass.

The Food Security Act of 1985 initiated the Conservation Reserve Program (CRP), which paid farmers to establish permanent vegetation on highly erodible land. Contracts were for ten to fifteen years. In 1997, there were almost 33 million acres in the CRP. About 7.8 million acres (316,000 ha) of CRP land were seeded to permanent native grasses, and almost 19 million acres were seeded to permanent introduced grasses and legumes. This program created a tremendous demand for grass seed.

■ 21.2 GRASS TYPES

Perennial grasses, with a few exceptions, are largely cross-pollinated, and thus highly variable. Buffalograss plants are dioecious. Kentucky bluegrass and dallisgrass produce much of their seed by apomixis, without pollination. Coastal bermudagrass, gramagrass, guineagrass, paragrass, and napiergrass produce little or no seed, being propagated only vegetatively. Most bermudagrass pastures are established vegetatively.[6]

Perennial grasses can be roughly divided into two classes: bunchgrasses and sod-forming grasses. Bunchgrasses send up tillers from the crown of the plant and form dense tufts or clumps. Sod-forming grasses spread laterally to form a more or less solid sod either by underground stems (rhizomes) or, in a few species, by aboveground stems (stolons).[1] The nodes of rhizomes and

FIGURE 21.1
Little bluestem plant *(left)*, panicle *(center)*, enlarged spikelet *(right)*.

stolons give rise to roots as well as to stems. Tillers, stolons, and rhizomes all arise from buds in the crown of the grass plants below the surface of the soil. When the shoot from the growing bud is on a plant of a species that forms tillers, the growth is erect and a stem (culm) is formed. When in another species the shoot is slender and pliant and is ageotropic (not responsive to geotropic stimuli), a stolon is formed and both roots and culms arise from the nodes. When, as in still other species, the shoot bears scales with a hard, sharp, pointed tip, it is a rhizome that grows along under the surface of the soil and likewise sends up both roots and culms. These distinctions are not absolute, however, because the shoots of certain tillering grasses, such as timothy, may grow somewhat like rhizomes and form roots[7, 8] when buried deeply in soil. Also some grasses, such as bermudagrass, often may bear rhizomes below the soil surface in loose soil and stolons above the surface in compact soil. Nearly all of the sod-forming grasses can be established by transplanting pieces of turf.

Numerous improved varieties of forage grasses are now available as a result of breeding, selection, and introductions.

■ 21.3 ADAPTATION

Some grass species, such as reed canarygrass, can be grown where water stands on the land part of the time, while others are drought resistant but may not withstand prolonged flooding. Many grasses adapted to the South are limited in their northern range because of lack of winter hardiness. Many typical sod-forming northern grasses assume a bunch type of growth and lack aggressiveness when grown in Oklahoma.[9] This probably represents an adaptation to photoperiod and temperature.

■ 21.4 SEEDING

Many species require cool weather for their establishment, while others, such as gramagrass and bermudagrass, make their growth in warm weather. Nearly all perennial grasses have small seeds (Figure 21.2). They are difficult to establish from seeding unless the seedbed is firm enough to avoid covering the seed too deeply and to form a close contact between the seed and soil. Moldboard plowing is often necessary when there is an existing cover of perennial plants.[10] In order to obtain the desired seedbed, tilled land should be harrowed several times or compacted with a corrugated roller before or during seeding (Figure 21.3). When the seed is sown in the spring in a field of grain to be grown as a companion crop, the soil is usually sufficiently compact.

In dryland regions, the sowing of grass is subject to several hazards, mostly caused by drought and blowing soil. Modified no-till drills can be used to seed directly into standing crop residues. Corn, sorghum, sudangrass, and millet residue makes a good cover for winter and spring seedings, especially when rows are at right angles to the prevailing winds or slope.[10] If seeded with conventional drills, stubble should be disked and harrowed to destroy weeds before grass is sown in the spring. Often, a seedbed suitable for spring seeding is found on fallow land that has settled during the winter or on corn or potato land that is disked and harrowed. Seeding into soybean stubble usually requires no tillage. In the northern Great Plains, considerable success has followed the drilling of certain cool-season grasses, such as crested wheatgrass, in weedy fields or untilled grain stubble in fall or early winter, preferably seeding in snow before the ground is frozen. The seed settles into the soil during thawing and germinates in the spring. In the southern Great Plains, the most certain method of establishing grasses has been to grow a

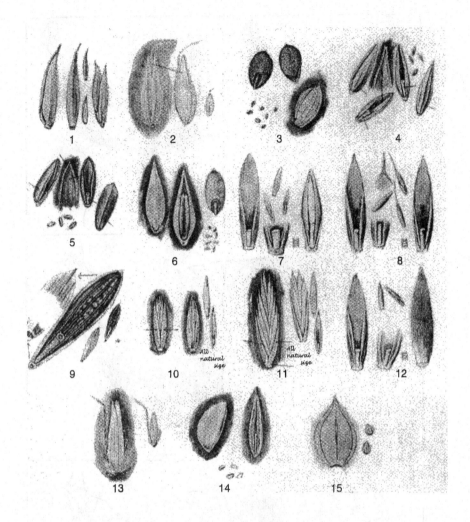

FIGURE 21.2

Seeds of perennial grasses (natural size and enlarged): (1) orchardgrass, (2) meadow foxtail, (3) timothy, (4) Kentucky bluegrass, (5) Canada bluegrass, (6) redtop, (7) perennial ryegrass, (8) Italian ryegrass, (9) smooth bromegrass, (10) slender wheatgrass, (11) western wheatgrass, (12 meadow fescue, (13) tall meadow oatgrass, (14) bermudagrass, (15) dallisgrass.

FIGURE 21.3

Drill for sowing grass seeds. Multiple seed boxes give the flexibility to seed native fluffy seeds, cool-season species, small grains, legumes, and wildflowers. [Courtesy Truax Co.]

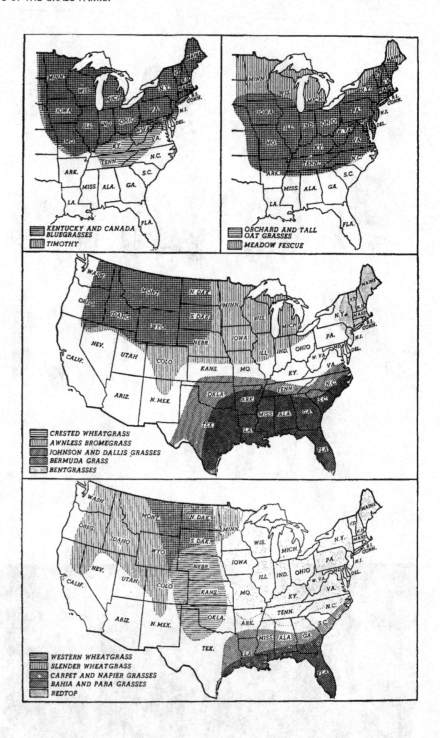

FIGURE 21.4
Sections of the United States in which various grasses are well adapted and of primary importance.

crop of drilled sudangrass or sorghum, mowing often enough to keep it from going to seed, and leaving the swaths on the ground over the winter. The grass seed is drilled into this protective cover the next spring (Figure 21.4).

In the Central Plains and Corn Belt, warm-season grasses are best seeded from mid or early April through mid May to early June.[10, 11, 12] It is possible to seed warm-season grasses in the fall if soils are dry and soil temperatures are too cold for germination (about 45°F). The seeds will lie dormant through the winter and germinate in the spring when the soil warms. Cool-season grasses are best seeded from late August through early September.[10, 12]

Seeding rate is based on pure live seed (PLS). PLS is calculated by multiplying germination percent by the percent purity and dividing by 100. Percent germination and percent purity are listed on the seed tag. Seeding rate should provide about 30 to 40 PLS per square foot (300 to 400 PLS/m²).[10, 11] If broadcast, seeding rates should be increased by 25 percent.

Herbicides can be used to control weeds when seeding perennial forage grasses. Glyphosate can be applied before seeding to control growing weeds. Imazapic can be applied at seeding for preemergence control. After emergence, 2, 4-D; metsulfuron; dicamba; bromoxynil; and imazapic are possible choices. Herbicides can be used on only certain grass species, and only at certain times, so label directions must be followed exactly.

■ 21.5 TIMOTHY

Timothy *(Phleum pratense)* is an important cultivated grass in the United States. It is grown for hay, mostly in mixtures with clover. This mixture yields more than either crop grown alone. It is also higher in feeding value than timothy alone. Timothy is harvested or grazed for two years or longer before it is plowed up for planting other crops in the rotation.[13]

Timothy and clover, which usually are sown with a small-grain companion crop, furnish some pasturage the first year. In the second year, the mixed hay is mostly clover, whereas the third-year crop is largely timothy. When left longer, the crop is nearly all timothy. The states leading in timothy production are New York, Ohio, Illinois, Wisconsin, and Pennsylvania. Most of the timothy seed is produced in Minnesota, Missouri, and Ohio.

Timothy, a native of Europe, was first cultivated in the United States by John Herd, who found it growing wild in New Hampshire in 1711. First called Herd's grass, timothy acquired its present name from a Timothy Hanson, who grew it in Maryland in 1720.

There are several plant breeding programs for timothy in Canada, the United States, Europe, and Japan that continue to release new varieties.

21.5.1 Adaptation

Timothy is best adapted to the cool, humid climate of the northeastern quarter of the United States, but it is grown also to some extent in the cool mountain valleys in the Rocky Mountain region, as well as in the coastal area of the Pacific Northwest. It is also found in southern Canada. It is not adapted to the drier regions, nor does it thrive where the average July temperature exceeds 77°F (25°C). Timothy grows better on clay soils than on the lighter-textured sandy soils.

FIGURE 21.5
Timothy plants.

21.5.2 Botanical Description

Timothy is easily recognized by its erect culms and dense, cylindrical, spike-like inflorescence (Figure 21.5). The spikelets are one-flowered. The plant is largely cross-pollinated under natural conditions. Timothy, a bunchgrass, differs from most grasses in that one of the lower internodes is swollen into a bulb-like base, a haplocorm,[7] often called a corm. Organic food reserves, which accumulate in the haplocorm up to the stage of seed maturity, furnish nutrients for new shoots. Timothy survives for three years or more, but it gradually becomes contaminated with weeds and other grasses. Consequently, it is usually plowed up after the second year.

The leaves of certain selected strains of timothy remain green longer than those of the ordinary timothy and also yield as well or better.[14]

21.5.3 Culture

Timothy is sown with a grain drill equipped with a grass seeding attachment. It is sown either in the fall with winter grain or in early spring with spring oats or barley. Many times, 10 pounds per acre (11 kg/ha) of clover seed are sown in the spring in the same field. When the clover is broadcast in winter wheat or winter rye fields in the spring, timothy may be seeded with it. Seedlings that start in the fall are less likely to be injured by dry weather during late spring and early summer than are those from spring seeding.[2] The usual seeding rates for timothy are

3 to 5 pounds per acre (3.5 to 5.5 kg/ha) in the fall, or 10 pounds per acre (11 kg/ha) in the spring.[13]

Yields of timothy are improved by application of commercial fertilizers, particularly those that contain phosphorus and nitrogen. Complete fertilizer is essential in the Northeast. The protein content of timothy is increased by very heavy application of 33 pounds per acre (37 kg/ha) or more of nitrogen fertilizers. The fertilizer is best applied in the spring.[13]

The largest yield of timothy hay consistent with high quality is obtained by cutting during early blooming or even earlier.[13, 14] The digestibility of timothy declines steadily as the plant develops, beginning as early as when the plants are in full head. Moreover, the more mature plants are usually less palatable because of the increase in crude fiber content. Timothy cut prior to full bloom contains 70 percent or more of the total protein of the hay in the leaf blades, leaf sheaths, and stems of the plant, while that left uncut until 10 percent of the heads are straw-colored contains at least 50 percent of the plant protein in the heads.[15] Much of this protein may be lost in the ripe seeds that shatter during handling and feeding. The total yield increases only slightly after the full-bloom stage.[13] Cutting before blooming reduces the vigor and growth of subsequent cuttings because lower food-reserve storage is present in the haplocorms.

Decreased yields that often occur in crops following timothy are largely due to a temporary deficiency of available nitrogen in the soil until some of the crop residue has decomposed and the nitrogen is released.

Timothy is harvested for seed with a combine after the seed is fully ripe and dry. In threshing, care is required to avoid hulling the seeds. Hulled seeds germinate satisfactorily at first, but they gradually lose their viability after several months.[13]

21.5.4 Utilization

Timothy is a popular hay for feeding horses. Compared with legumes, timothy hay is relatively low in protein and minerals, especially calcium. In mixture with clover, the hay is excellent for cattle and sheep.

Timothy is one of the most palatable pasture grasses, usually grazed in preference to redtop, orchardgrass, or even Kentucky bluegrass. It is often included in mixtures of grasses and legumes sown in permanent pastures. As the pasture becomes older, other grasses gradually replace timothy. Its coarse, bunchy, erect, and semipermanent growth makes it unsuited for lawns.

▨ 21.6 SMOOTH BROMEGRASS

Smooth bromegrass (*Bromus inermis*), introduced into the United States about 1884, soon became a valuable forage plant in the West. The total acreage was small for many years because native grasses, cereals, and legumes furnished most of the forage. Later, bromegrass became a highly important crop in the northern states from the Dakotas and Kansas eastward to Ohio. In 1969, it was grown on nearly 9 million acres, mostly in mixtures with alfalfa. Plant breeding programs in the United States and Canada have released many improved varieties of bromegrass.

21.6.1 Adaptation

Smooth bromegrass is especially adapted to regions of moderate rainfall and cool to moderate summer temperatures, particularly where the native vegetation consists of tall or medium-tall grasses, as in the North Central and Great Lakes states. It is one of the best-introduced grasses in the eastern parts of the Dakotas, Nebraska, and Kansas. Bromegrass is moderately drought-resistant but the more drought-resistant crested wheatgrass outyields and outlasts it in the drier areas.[16]

Bromegrass makes its best growth on moist, well-drained clay to silt loam soils, but it produces satisfactorily on sandy soils when there is sufficient moisture. It is particularly vigorous on fertile soils high in nitrogen. Bromegrass has only a small degree of alkali or salt tolerance.

21.6.2 Botanical Description

Smooth bromegrass is an erect, sod-forming perennial that spreads by rhizomes. The stems vary in height from 2 to 4 feet (60 to 120 cm). The plant produces numerous basal and stem leaves that vary in length from 4 to 10 inches (10 to 25 cm). Frequently, the rough (scabrous) leaves are marked by a transverse wrinkling a short distance below the tip. The inflorescence is a panicle 4 to 8 inches (10 to 20 cm) long that spreads when in flower. The flower head develops a characteristic rich purplish-brown color when mature. There are several florets in a spikelet (Figure 21.6). The seeds are long, flat, and awnless (Figure 21.2).

Solid stands of pure bromegrass are likely to develop a sod-bound condition after two or three years, unless they are heavily fertilized with nitrogen. The plant starts growth early in the spring and continues growth late into the fall.

FIGURE 21.6
Smooth bromegrass:
(1) plant, (2) panicle, and
(3) spikelet.

Northern types of smooth bromegrass are adapted to the northern border states as well as to higher elevations in the intermountain region. Southern types are adapted to the Corn Belt and the eastern Great Plains.[17, 18] A non-creeping strain grown in Canada is finely stemmed and leafier than common bromegrass.

21.6.3 Culture

Late-summer seeding after a small grain crop gives the best results in eastern Nebraska.[19] Smooth bromegrass is seeded there between August 20 and September 15 at the rate of 15 to 20 pounds per acre (17 to 22 kg/ha). It is often included in a mixture that uses 15 pounds per acre (17 kg/ha) of bromegrass and 3 pounds per acre (3.4 kg/ha) of alfalfa. The bromegrass and alfalfa are ready to cut at the same time. Spring seeding, when practiced, should be early. Inclusion of a legume increases both the growth and the protein content of the grass.[20]

Maximum yields of bromegrass are obtained the second and third years after the stand is established. The yield decreases when the plants become sod-bound. The sod-bound condition may be avoided or delayed by applying nitrogen fertilizers or by seeding legumes with the bromegrass.[21]

21.6.4 Utilization

Bromegrass makes hay or silage of excellent quality. Its protein content is high and the crude fiber content relatively low. It is one of the most palatable pasture plants, especially for spring and fall grazing. The best quality of hay is obtained when bromegrass is cut in the bloom stage.[22]

Bromegrass has been used to retard soil erosion because of its heavy mass of roots. It is suitable for steep slopes, buffer strips in strip cropping, small sod dams, drainage outlets, terraces, and waterways through cultivated fields.

When harvesting bromegrass for seed, the stubble should be clipped as soon as possible after harvest to allow light to stimulate tiller initiation at ground level. Since bromegrass has few basal leaves, it should be clipped no less than 4 inches (10 cm) high. Failure to clip the stubble may reduce the next seed crop as much as 30 percent.[23] Seed can be harvested with a combine if done promptly after maturity. Otherwise, the crop should be swathed to prevent seed shattering.

21.6.5 Other Bromegrasses

Mountain brome (*Bromus carinatus*) is a native of the Rocky Mountain region and is adapted to the cooler areas of the West. An improved variety that is taller, later, leafier, and more disease resistant than common mountain brome is grown in Washington and Oregon.[24, 25]

Soft brome (*B. mollis*) is an annual introduced from Europe. It is found in open areas and disturbed sites. It has excellent forage quality when immature.

Two introduced annual weedy grasses with some forage value are cheat (*B. secalinus*) and downy brome, or cheatgrass (*B. tectorum*). The latter provides grazing when the plants are small but later becomes useless because of its irritant awns.

21.7 ORCHARDGRASS

Orchardgrass or cocksfoot *(Dactylis glomerata)*, a native of Europe, is grown to some extent in nearly every state. It grows naturally in many localities. It is grown for pasture, hay, or silage mostly in mixtures with clover or alfalfa. Plant breeding programs in Canada, Northern Europe, and the United States have released many improved varieties of orchardgrass.

21.7.1 Adaptation

Orchardgrass thrives under cool, humid, moist, or irrigated conditions, but it is more tolerant to heat and drought than timothy. In the northern states, orchardgrass provides forage during the spring and summer. In the South, it furnishes grazing during late fall, winter, and early spring. Since it tolerates some shade, orchardgrass is often seeded in woodland or logged-off pastures. It is less resistant to cold than timothy or bromegrass. Consequently, its acreage in the colder northern areas is limited.

21.7.2 Botanical Description

Orchardgrass is distinguished by its large circular bunches, folded leaf blades, compressed sheaths, and, particularly, by the spikelets grouped in dense, one-sided fascicles borne at the ends of the panicle branches (Figure 21.7). A majority

FIGURE 21.7
Orchardgrass during anthesis.

of the orchardgrass acreage in the United States is of the common type, but improved varieties for different areas are now available.[18, 26]

21.7.3 Culture

Orchardgrass is sown at a rate of 3 to 10 pounds per acre (3.4 to 11 kg/ha) in mixtures with alfalfa or ladino clover in the northern and western states, but at 10 to 15 pounds per acre (11 to 17 kg/ha) with lespedeza or clovers in the South. Other grasses or legumes can be included in the mixtures.

Orchardgrass can be sown with a small grain companion crop in the fall in the South or in early spring in the North. In Missouri, 20 to 30 pounds per acre (22 to 34 kg/ha) nitrogen, 50 to 120 pounds per acre (56 to 134 kg/ha) P_2O_5, and 40 to 60 pounds per acre (45 to 67 kg/ha) K_2O should be applied to an orchardgrass-legume mixture at seeding time.[27] Established stands of pure orchardgrass may need 80 to 120 pounds per acre (90 to 134 kg/ha) nitrogen, 40 to 60 pounds per acre (45 to 67 kg/ha) P_2O_5, and 100 to 140 pounds per acre (110 to 157 kg/ha) K_2O annually. Nitrogen fertilizer should not be applied if the stand contains 30 percent or more alfalfa or other legume. Phosphate and potash may be needed, because a ton of bromegrass-legume hay will remove about 10 to 12 pounds (5 to 6 kg/MT) P_2O_5 and 35 to 45 pounds (18 to 23 kg/MT) K_2O.[27]

21.7.4 Utilization

Orchardgrass is primarily a pasture crop because much of the growth is in the lower leaves, but it is also harvested for forage. As a hay crop, it should be cut at first bloom because older stems become coarse and unpalatable. It often reaches this stage in the spring, which allows for high-quality regrowth. Orchardgrass loses its palatability and digestibility much more rapidly than smooth bromegrass.

▓ 21.8 TALL FESCUE

Tall fescue (*Festuca arundinacea*) is grown on over 35 million acres (14 million ha) in the United States, mostly south of the Ohio River, in the southeastern states, in Missouri and Kansas, in the Pacific coastal area, and under irrigation in several western states. Indiana grows over 1 million acres. Plant breeding programs in Europe, Japan, and the United States have released many improved varieties.

21.8.1 Adaptation

Tall fescue, a native of Europe, is a perennial tufted grass suited to various soil and climatic conditions. It is adapted to semi-wet conditions but tolerates temporary drought as well as both acid and alkaline soils.[28] Tall fescue is not suited to semi-arid conditions.

21.8.2 Botanical Description

Tall fescue plants are 2 to 5 feet (60 to 150 cm) tall with dark-green leaves. Since they have a medium-bunch habit, the plants form a fair turf when grazed or mowed regularly. The inflorescence is a long, narrow panicle with several florets per spikelet. The seeds tend to shatter upon ripening.[29]

Tall fescue consists mostly of the Kentucky 31 variety. Other varieties are adapted to the southwestern irrigated areas or for reclaiming saltgrass meadows in Oregon.[28]

21.8.3 Culture

Tall fescue is sown in late summer or early fall at the rate of 2 to 4 pounds of seed per acre (2.2 to 4.5 kg/ha) in mixtures, but up to 15 pounds per acre (17 kg/ha) or more when sown alone. It is usually sown in mixtures with legumes such as (1) ladino clover, alfalfa, or alsike clover in the cooler, humid regions; (2) lespedeza or white clover in the Southeast; and (3) subclover in the western coastal areas. Since it grows best under cool conditions, it is pastured in the South in the winter. Tall fescue, being tolerant of both temporary drought and excessive moisture conditions, has a long growing season.

21.8.4 Utilization

Tall fescue is grown mostly for pasture. It is less palatable than most of the cultivated grasses, but it is usually grazed adequately in mixtures with legumes. High-producing dairy cows may not consume enough for maximum milk yield.[30] Close grazing along with nitrogen fertilization maintains the plants in a succulent, more palatable condition. Grazing animals sometimes develop disorders when confined to tall fescue pastures for long periods. These disorders are fescue foot, bovine fat necrosis, and the accumulation of excessive alkaloids, especially ergovaline.[31] These disorders are caused by a fungal endophyte (Acremonium coenophialum). This fungus lives within the plant's intercellular spaces, so it is not visible upon examination. The fungus has no effect on tall fescue plants. Using seed that is certified to contain less than 5 percent infestation can help prevent the problem.[32] Grazing infected pastures when they are lush and growing rapidly reduces the toxicity.[31]

21.8.5 Other Fescues

Meadow fescue (Festuca elatior) was introduced from Europe many years ago. It was a common ingredient of pasture and meadow mixtures, especially in the southern part of the Corn Belt.[2] It has been largely replaced by tall fescue since 1945, partly because of its susceptibility to crown rust.[33] The plants differ from tall fescue in being about 12 inches (30 cm) shorter, but they also lack fine hairs on the leaf auricle. Moreover, meadow fescue has seven pairs of chromosomes, while tall fescue has fourteen pairs.

Sheep fescue (F. ovina), red fescue (F. rubra), and chewings fescue (F. rubra ssp. commutata) are turf grasses. The last two species, the seed of which is produced mostly in Oregon, are used in shaded lawn areas. Chewings fescue is tufted but does not spread to form a dense turf.

■ 21.9 REED CANARYGRASS

Reed canarygrass (*Phalaris arundinacea*) is native to the northern part of both hemi-spheres. It was cultivated first in Oregon about 1885, and was among the later grasses to assume importance under cultivation.[34] Plant breeding programs in the United States and Canada have released several improved varieties.

21.9.1 Adaptation

Reed canarygrass is grown in the Pacific Coast areas of Oregon, Washington, and northern California, as well as in the North Central states. It makes its best growth in a moist, cool climate but is sensitive to neither heat nor cold. However, it is not very successful where the average mean winter temperature is above 45°F (7°C) or the summer temperature is above 80°F (27°C). This grass is productive on fertile moist soils, being especially suited to swamp or overflow lands. It does well on peat soils and on land that is too wet for other crops.[35, 36] In Iowa, reed canarygrass produced higher hay yields than did timothy, bromegrass, meadow fescue, tall oat-grass, redtop, or orchardgrass.[37]

21.9.2 Botanical Description

Reed canarygrass is a long-lived perennial that spreads by rhizomes. The plants are 2 to 8 feet (60 to 245 cm) tall with leafy stems. They tend to grow in dense tus-socks 2 to 3 feet (60 to 90 cm) in diameter. The leaves are broad, smooth, and light green in color. The inflorescence is a semidense, spike-like panicle, 2 to 8 inches (5 to 20 cm) long (Figure 21.8). The stems become coarse after the panicles begin to appear. The seeds are enclosed in blackish-brown or gray lemmas, while the

FIGURE 21.8
Reed canarygrass: (1) plant, (2) panicle, (3) spikelet, and (4) floret.

paleas are sparsely covered with long hairs. The seeds mature from the top of the panicle downward and shatter readily.

A distinct strain, selected in Oregon, has a more upright growth, greater leafiness, stiffer stems, and better seeding habits than ordinary reed canarygrass.[34] It can grow on uplands that often become dry in spring, summer, or fall. It thrives as well as the ordinary strain in the lowlands.

21.9.3 Culture

Reed canarygrass is grown from fall or spring seeding on the Pacific Coast, but early spring seeding is the most common practice in the North Central states. It is usually seeded at the rate of 5 to 8 pounds per acre (6 to 9 kg/ha) in close drills. It is seldom sown with other grasses, but it can be seeded with small grains on moist, fertile soils. The first crop should be cut for hay as soon as the panicles begin to appear. The later crops usually do not produce panicles.

21.9.4 Utilization

Reed canarygrass furnishes abundant pasturage or silage where soil moisture is adequate because it starts growth early in the spring and continues until late in the fall. It is more palatable than other wetland grasses when grazed closely; otherwise, it becomes coarse. Likewise, the hay is palatable when cut before the stems are too coarse.

21.9.5 Related Species

Hardinggrass (*Phalaris tuberosa*, var. *stenoptera*) is a productive hay crop in California.[6] Canarygrass (*P. canariensis*), a native of the Mediterranean region, is established in many parts of the United States. Canarygrass furnishes the canary seed used so generally for feeding canaries. Most of the seed for this purpose is grown in Europe and Argentina.

▓ 21.10 KENTUCKY BLUEGRASS

Kentucky bluegrass (*Poa pratensis*) is the best known grass in the United States. It is a native of the Old World but has spread naturally or by direct seeding over the humid and subhumid sections in the northern half of the United States to such an extent that it is the dominant species in most of the older pastures. Most of the seed is harvested in the Pacific Northwest, Minnesota, and Kentucky. Plant breeding programs in the United States, Canada, and Europe have released many improved varieties.

21.10.1 Adaptation

Kentucky bluegrass is adapted throughout the northern half of the United States, except where the climate is too dry.[38] It does best under cool, humid conditions on

FIGURE 21.9
Kentucky bluegrass plant
and panicle.

highly fertile limestone soils, but it also thrives on noncalcareous or slightly acid soils with a reaction as low as pH 6. It seems to prefer heavier soils. An ample supply of nitrogen and phosphorus is essential for high production.

21.10.2 Botanical Description

Kentucky bluegrass is a dark-green sod-forming grass and a long-lived perennial. The stems are 1 to 2 feet (30 to 60 cm) in height when allowed to grow uncut. They usually are numerous in a tuft. The plants have narrow leaves 2 to 7 inches (5 to 18 cm) in length.

The inflorescence is a pyramid-shaped panicle about 2 to 8 inches (5 to 20 cm) long (Figure 21.9). The spikelets are composed of three to five florets. Kentucky bluegrass reproduces by rhizomes as well as by seed. New tufts with their roots arise from the nodes along the rhizomes, thus continually occupying the spaces left by the death of the older tufts that fail to survive more than two years. It is classed as a day-neutral plant. Flower heads may be initiated in short, cool, autumn days, but blooming occurs in the longer days of late spring and summer.

21.10.3 Culture

In pasture mixtures, bluegrass is generally seeded with other grasses, and either clovers or alfalfa, at a rate of 1 to 5 pounds per acre (1.1 to 5.6 kg/ha). About 20 to 40 pounds per acre (22 to 45 kg/ha) are sown for pure stands. Usually two to three years are required to produce a good sod from seeding. Annual applications of nitrogenous fertilizers increased the average yield of dry matter 72.1 percent in Wisconsin.[39]

Most bluegrass seed is produced in the far West. It is harvested from the ripening plants with swathers or with mechanical strippers equipped with revolving spiked beaters.[40] It is cured in windrows that are turned frequently to prevent heating. Temperatures above 140°F (60°C) during curing destroy the germination. Partial heating, together with the presence of immature seed and empty chaff, are responsible for much of the commercial bluegrass seed germinating no more than 70 percent. Sound plump seed germinates over 90 percent.

21.10.4 Utilization

Kentucky bluegrass is not a hay crop. It is a leading grass in permanent pastures because of its persistence as well as its ability to withstand close and continuous grazing. It starts growth early in the spring, thus furnishing succulent forage for early grazing. It becomes nearly dormant during the heat of midsummer when daily maximum temperatures approach 90°F (32°C). However, it resumes growth with the advent of cool weather in the fall, when it furnishes additional grazing. It supplies almost continuous summer grazing in the cooler sections of the northeastern states. For rotation pastures, as well as for many seeded pastures in the northeastern and North Central states, Kentucky bluegrass has been partly replaced with orchardgrass, tall oatgrass, and smooth bromegrass. These taller grasses grow more in hot weather.

Because of its dense turf, bluegrass is the most popular lawn grass in America. Lawns are established either by seeding or by transplanting sod. It is sown either in very early fall or very early spring.

21.10.5 Other Bluegrasses

Canada bluegrass (*Poa compressa*), a native of Europe, is grown in Canada. It is distinguished from Kentucky bluegrass by its shorter, compressed stems, which long remain green; its single shoot at the end of each rhizome; and its narrower panicles. Canada bluegrass is found throughout the Kentucky bluegrass region, often on less fertile soils, but it is generally recognized as being decidedly inferior.

Rough-stalked meadow grass or birdgrass (*Poa trivialis*), also a native of Europe, has bright green leaves and thrives in the shade. It is sown in shady locations in lawns.

Among the range grasses native to the United States are big bluegrass (*Poa ampla*), bulbous bluegrass (*P. bulbosa*), Texas bluegrass (*P. arachnifera*), Sandberg bluegrass (*P. secunda*), and mutton or Fendler bluegrass (*P. fendleriana*). Big bluegrass has been seeded on semiarid rangelands of the Pacific Northwest. The Sherman variety, adapted to varied conditions, is a bunchgrass.

Bulbous bluegrass (*Poa bulbosa*) has become well established in certain areas of northern California and southern Oregon. It normally produces bulbs at the base of the stem and bulbils in the inflorescence. It is adapted to a wide range of soil types.[41] The chemical composition of the hay is similar to that of timothy.

Annual bluegrass (*P. annua*) is a serious pest in lawns and seed fields.

■ 21.11 BERMUDAGRASS

Bermudagrass (*Cynodon dactylon*) is a native of Asia but is now distributed throughout the tropical and subtropical parts of the world. Plant breeding programs in several states have released many improved varieties.

21.11.1 Adaptation

Bermudagrass is distributed from Maryland to Kansas, southward to the Gulf of Mexico, and westward from Texas to California. In general, it is adapted to the same area as cotton. The best use of bermudagrass is in pasture mixtures with crimson, white, hop, and Persian clover or with lespedeza. Its growth in the northern states is limited by its lack of winter hardiness, although it is sufficiently hardy to endure the winter conditions in north central Oklahoma.[17] Bermudagrass makes its best growth on fertile lands that are well drained. It shows a marked preference for clay soils, but it grows more or less abundantly on sandy soils. Bermudagrass grows luxuriantly during hot midsummer weather but does not start growth until late spring. It stops growth and becomes bleached with the onset of cold weather in the fall.

The spittlebug (*Prosapia bicincta*) damages bermudagrass in the southeastern states, especially in wet years.

21.11.2 Botanical Description

Bermudagrass is a long-lived perennial with numerous branched leafy stems that vary from 4 to 18 inches (10 to 45 cm) in height. Although the stems of bermudagrass, like those of other grasses, have only one leaf at a node, it may appear to have two to four, due to several contiguous short internodes (Figure 21.10). In ordinary bermudagrass, there are numerous stout rhizomes, which in very hard soil grow above the surface as stolons for 1 to 3 feet. The flowers are borne in slender, spreading spikes arranged in umbels of four to six. Bermudagrass produces very little seed in humid regions.

Common bermudagrass has white rhizomes as large as goose quills, besides the leafy stolons that creep on the soil surface. St. Lucie grass is identical in appearance

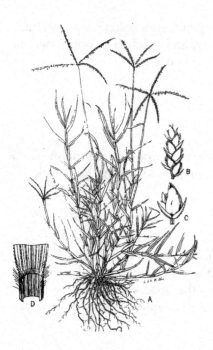

FIGURE 21.10

Bermudagrass: *(A)* plant, *(B)* panicle branch, *(C)* spikelet, and *(D)* ligule. Stolons are shown at the crown region.

but lacks rhizomes and rarely survives the winter north of Florida. Coastal bermudagrass is an improved hybrid strain that produces very little seed. Consequently, it must be propagated entirely by planting stolons.[42] It spreads rapidly, withstands light frosts, and produces well in the southern three-fourths of the Cotton Belt. Its tall growth makes it a productive hay and silage crop as well as a pasture crop. Coastal bermudagrass is an important hay crop in the southern United States. Other varieties are hardier, more drought-resistant, or adapted to soils of low fertility.[43]

Still other varieties are well suited for lawns or for grassed waterways. A heavy bermudagrass turf prevents soil erosion.

21.11.3 Culture

Bermudagrass is established by planting sod, crowns, or pieces of stolons. The most common practice is to break up the sod into small pieces. The pieces are dropped, 2 to 3 feet (60 to 90 cm) apart each way, into furrows on plowed land. These sod pieces can be planted anytime from spring until midsummer, whenever wet weather conditions are likely to prevail. Common and Greenfield bermudagrass are established by seeding about ½ inch (13 mm) deep, at a rate of about 5 pounds per acre (5.6 kg/ha), in the spring after the daily mean temperature reaches 65°F (18°C).[44]

Bermudagrass stands tend to become sod-bound. When this occurs, disking, plowing, or harrowing improves the yields. Bermudagrass often spreads onto cropped land where frequent cultivation or herbicide applications are necessary to keep it under control. Field burning in early April improves the yields of coastal bermudagrass by reducing disease, insect, and weed infestations.

21.11.4 Utilization

Bermudagrass is an excellent summer pasture plant, either alone or in mixtures, particularly when closely grazed. It becomes wiry and tough as it approaches maturity. Sometimes it is planted in mixtures with lespedeza.

■ 21.12 REDTOP

Redtop (*Agrostis alba*) is a cultivated perennial grass introduced during the Colonial period. Improved varieties have been developed in Idaho and New York.

21.12.1 Adaptation

Redtop will grow under a great variety of conditions, as it is one of the best wetland grasses among the cultivated species. It is especially adapted for growth on acid soils that are low in lime where most other grasses fail. Redtop has little tolerance to soil salinity or drought. It is best adapted to cool or moderate temperatures.

FIGURE 21.11
Redtop panicle and plant.

21.12.2 Botanical Description

Redtop spreads by rhizomes, making a coarse, loose turf. The leaves are narrow and the stems slender. The panicle is loose, pyramidal, and reddish in color (Figure 21.11). The spikelets are small and contain one flower. Redtop matures at about the same time as timothy.

21.12.3 Utilization

Redtop is a wetland hay crop used as part of pasture mixtures under humid conditions, as a soil binder, and as an ingredient of hay mixtures.[2] It is seeded either early in the spring or late in the summer at the rate of 4 to 5 pounds per acre (4.5 to 5.6 kg/ha) when used with other grasses for hay. In lawn grass mixtures, redtop does not make a fine, smooth turf but serves as a companion grass while bluegrass is becoming established.

21.12.4 Related Species

Colonial or Rhode Island bentgrass (*Agrostis tenuis*) is found in meadows and pastures in New England, New York, and the Pacific Northwest. Improved varieties are important turf grasses, some of which are injured rather than improved by applying lime to the lawn.

Creeping bentgrass *(A. palustris)* is distinguished by its dense panicle and by its creeping stolons, which may grow as much as 4 feet (120 cm) in a single season. It is common in pastures in many places where the soil is moist and is found in seaside meadows on both the Atlantic and Pacific coasts. Improved varieties and velvet bent *(A. canina)* are sown on golf-course putting greens to provide a fine turf. Other selected strains of creeping bent are established by planting stolons.

■ 21.13 CRESTED WHEATGRASS

Crested wheatgrass is a hardy, drought-resistant perennial bunchgrass native to the cold, dry plains of Russia. It was first introduced to the United States in 1898 but failed to become established from that introduction. After 1915, when research at several locations with new seed from Russia had demonstrated its promise in the northern Great Plains, it was increased for distribution.[2, 45]

21.13.1 Adaptation

Crested wheatgrass is well adapted to the northern Great Plains and the inter-mountain and Great Basin regions where winter temperatures are severe and the moisture supply limited. This grass makes its best growth in cool climates, since forage production is reduced by the higher temperatures in the southern Great Plains. There is no known instance in which the plants were killed either by drought or cold under field conditions.[46] Crested wheatgrass is productive on practically all types of soil, but it is less tolerant of salinity than is western wheatgrass.

21.13.2 Botanical Description

Crested wheatgrass is an extremely long-lived perennial. The fine stems, which vary in height from 1½ to 3 feet (45 to 90 cm), occur in dense tufts. The leaves are flat, somewhat lax, narrow, and sparsely pubescent on their upper surface. The dense spikes are 2 to 4 inches (5 to 10 cm) long, being considerably broader than those of most other species of wheatgrass. The spikelets are closely crowded on the tapered head and tend to stand out from the axis of the spike.

Standard, introduced as *Agropyron desertorum*, is the ordinary commercial strain of crested wheatgrass. It consists of a mixture of many different types or strains that vary in leafiness, stiffness of stems, and size of spike. The short-awned types are considered the most desirable (Figure 21.12). The Nordan variety, developed in North Dakota, produces more and better forage and has better seedling establishment than does the ordinary type.[18, 25]

Fairway, introduced as *Agropyron cristatum*, is a selected strain of crested wheatgrass. It is popular in Canada, but the taller Standard strain is grown more in the United States because of its higher yields and better palatability. Fairway plants are finer-stemmed, leafier, and tiller more than do the Standard plants. The leaves of Fairway plants are covered with fine hairs and are bright green in color, while the leaves of Standard vary from dark green to grayish green. The seeds of Fairway are smaller, bear more awns, and are lighter than those of Standard.

The two strains do not intercross. The Standard type has fourteen pairs of chromosomes, whereas the Fairway and the original *A. cristatum* type have seven pairs. However, a hybrid variety, Hycrest, was developed at Logon, Utah, by crossing Nordan with Fairway. The resulting variety is more robust and larger than the two parental species and is well adapted to drier regions.[47]

FIGURE 21.12
Crested wheatgrass spikes. The two at left are Fairway and the original *A. cristatum*. The three at right are awned and awnless types of the Standard type, *A. desertorum*.

21.13.3 Culture

Crested wheatgrass may be sown either in the fall or spring, in close drills for hay or pasture, or in cultivated rows for seed.[48] Row plantings can be in single or double rows, with the rows 36 to 42 inches (90 to 100 cm) apart in either case. Recommended seeding rates are 5 to 8 pounds per acre (5.6 to 9 kg/ha) for close drills, 1 to 2 pounds per acre (1.1 to 2.3 kg/ha) for single-cultivated rows, and 2 to 3 pounds per acre (2.3 to 3.4 kg/ha) for double-cultivated rows.[45] The seed has been drilled on sagebrush rangeland that has been tilled with a disk plow. It may be drilled directly in small grain stubble where wind erosion is probable. For hay, it should be cut at least by the time it starts to flower.

21.13.4 Utilization

Crested wheatgrass is highly palatable either for hay or pasture. The hay compares favorably with that of western wheatgrass in quality and palatability. The grass becomes harsh as it matures. Crested wheatgrass furnishes pasture earlier in the spring and later in the fall than do other cultivated grasses. In the northern Great Plains it has two to three times the carrying capacity of the native range, especially in early spring.[16] It should be supplemented with other grasses as it tends to become more or less dormant during hot, dry weather. This grass has been utilized effectively for wind or water erosion control in the northern intermountain region as well as in the northern Great Plains because of its persistence and its tough, fibrous root system.

▓ 21.14 WESTERN WHEATGRASS

Western wheatgrass (*Pascopyrum smithii, Agropyron smithii*), also called bluejoint, is a native perennial introduced into cultivation. It is distributed generally through-out the United States, except in the more humid southeastern states, but it is most prevalent in the northern and central Great Plains. Plant breeding programs in the United States and Canada have released several improved varieties.

21.14.1 Adaptation

Western wheatgrass is a cool-season grass adapted to a wide range of soil types. It is extremely drought resistant as well as alkali tolerant. Nearly pure stands of this grass occur in South Dakota on heavy clay soils, in swales or flats where additional moisture from runoff had been received.

21.14.2 Botanical Description

Western wheatgrass has strong creeping rhizomes (Figure 21.13), and the stems are usually are 1 to 2 feet (30 to 60 cm) in height. The leaves usually are 4 to 12 inches (10 to 30 cm) long and less than ¼ inch (6 mm) wide. The upper surfaces of the leaves are scabrous (rough) and prominently ridged lengthwise, while the under-side of the leaf is relatively smooth. The leaves are rather stiff and erect and when dry roll up tightly to give the plant the appearance of having scanty foliage. The entire plant is usually glaucous, which gives it a distinctive bluish-green col-oration. The wheat-like, but more slender, spikes are 2 to 6 inches (5 to 15 cm) long.

21.14.3 Culture

The threshed seed is relatively free of awns or hairs and can be sown with a grain drill without previous treatment. Western wheatgrass is best seeded in the fall or early spring at the rate of 10 to 12 pounds per acre (11 to 13 kg/ha). It is usually in-cluded in a mixture with crested wheatgrass. Germination of the seed is frequently delayed. Consequently, stands of this grass are often slow in becoming established. For this reason, it is most valuable when seeded in a mixture.

▓ 21.15 SLENDER WHEATGRASS

Slender wheatgrass (*Elymus trachycaulus* syn. *Agropyron trachycaulum*) is another native American grass to become established as a cultivated crop. It is adapted to the northern Great Plains and intermountain regions. It is very short-lived and

FIGURE 21.13
Western wheatgrass. Note rhizomes at lower left.

produces well for only two or three years. The inflorescence is a slender, greenish spike on which the closely appressed spikelets are some distance apart. Recommended seeding rates range from 10 pounds to 35 pounds of seed per acre (11 to 39 kg/ha). For hay, it should be cut just before flowering. Slender wheatgrass should be used in pasture or hay mixtures, rather than sown alone. About 2 pounds per acre (2.2 kg/ha) should be included in mixtures sown on dry land, but about 4 pounds per acre (4.5 kg/ha) when sown on irrigated land. Plant breeding programs in Canada and the United States have released several improved varieties.

■ 21.16 OTHER WHEATGRASSES

Intermediate wheatgrass (*Thinopyrum intermedium* syn. *Agropyron intermedium*) is grown for pasture or hay in the northern Great Plains, the intermountain region, and the Pacific Northwest. It was introduced from the former Soviet Union in 1932.[3] The plants are sod forming, while the large seeds facilitate the establishment of stands. It is adapted to parts of the Great Plains and intermountain regions where the rainfall exceeds 15 inches (38 cm), as well as to the Pacific Northwest where it is often grown in mixtures with alfalfa.

Pubescent wheatgrass (*A. trichophorum*), a native of Europe, is grown to some extent in much the same area as intermediate wheatgrass. The glumes are awnless and pubescent.

Tall wheatgrass (*A. elongatum*) was introduced from the former Soviet Union.[49] It is a tall coarse bunchgrass that tolerates soil salinity. It is adapted to sagebrush as well as mountain lands in the central intermountain states. Its long growing period enables it to provide green herbage in midsummer. It has been crossed with wheat in attempts to produce perennial wheat.

Siberian wheatgrass (*A. sibiricum*), introduced from the former Soviet Union, is grown on a limited area in the Pacific Northwest. It is similar to crested wheatgrass but inferior to it except on infertile or arid soils.

Streambank wheatgrass (*A. riparium*) is a native sod-forming species, a selected variety of which is grown on a small acreage in the Pacific Northwest.

Three native bunchgrasses that have been seeded on western rangelands are beardless wheatgrass (*A. inerme*), bluebunch wheatgrass (*A. spicatum*), and thickspike wheatgrass (*A. dasystachyum*). The Whitmar variety of beardless wheatgrass is grown in Washington, Oregon, and Wyoming.

■ 21.17 PERENNIAL RYEGRASS

Perennial ryegrass, or English ryegrass (*Lolium perenne*), was one of the first perennial grasses to be cultivated for forage. It is grown mostly in the Pacific Northwest, intermountain valleys, the Midwest, and the Northeast United States. Plant breeding programs in the United States, New Zealand, and Europe have released many improved varieties.

21.17.1 Adaptation

Perennial ryegrass is less hardy than many other grasses.[50] It is considered a wetland grass in some regions. It can tolerate two to three weeks of flooding when temperatures are below 80°F (27°C).[51] It does best on fertile, well-drained soils; production usually declines as drainage becomes poor. Hot, dry weather adversely affects plant growth.

21.17.2 Botanical Description

Perennial ryegrass is a tufted, short-lived perennial that persists for three to four years. The plants grow 1 to 2 feet (30 to 60 cm) in height. There are numerous long,

FIGURE 21.14
Perennial ryegrass. Plant *(left)*, spike *(right)*.

narrow leaves near the base of the plant. The seed stems are nearly naked. The inflorescence is a spike, with the spikelets set edgewise to the rachis (Figure 21.14). The lemmas of the florets are entirely or nearly awnless.

21.17.3 Culture

Perennial ryegrass can be seeded either in the fall or early spring, but usually in the spring where winters are severe. The crop is sown at the rate of 20 to 25 pounds per acre (22 to 28 kg/ha) where used alone for forage or seed production. Common ryegrass is sometimes seeded with small grain at the rate of 8 to 10 pounds per acre (9 to 11 kg/ha) for annual pastures. For the best quality of hay, it should be cut in the soft-dough stage.

21.17.4 Utilization

Perennial ryegrass is used primarily in permanent pasture mixtures to furnish early grazing while long-lived grasses are becoming established. It is used occasionally for hay or winter cover. It is often used for seeding lawns, particularly in mixtures, and is easily established, but the turf is coarse and not permanent. It usually survives for only one year in the eastern half of the United States.

▨ 21.18 ITALIAN RYEGRASS

Italian ryegrass (*Lolium multiflorum*), together with the so-called domestic or common ryegrass, is grown mostly in Arkansas, California, and the southeastern states. Domestic ryegrass is a mechanical and hybrid mixture of Italian ryegrass with some perennial ryegrass. Plant breeding programs in the United States, Japan, and Europe have released many improved varieties.

21.18.1 Adaptation

Italian ryegrass is a hardy, short-lived annual or winter annual, but some plants live into the second season. It makes rapid growth when seeded in the spring, late summer, or fall.[2]

21.18.2 Botanical Description

Awns are present on the seeds of Italian ryegrass and usually absent in perennial ryegrass (Figure 21.2). The culm of Italian ryegrass is cylindrical, whereas perennial ryegrass culms are slightly flattened. The leaves of Italian ryegrass are rolled in the bud, while those of perennial ryegrass are folded. The plants of Italian ryegrass are yellowish at the base, but those of perennial ryegrass are commonly reddish. Also, the roots of annual ryegrass seedlings exude a substance that emits a bright fluorescent glow when subjected to near-ultraviolet light (300–400 nm). Perennial ryegrass roots lack this characteristic.

21.18.3 Utilization

Italian ryegrass is a very palatable and productive pasture plant. It grows so rapidly that it can be grazed in a short time after seeding. The quick germination and large seeds (and seedlings) account for its prompt establishment.

Italian ryegrass is used as a companion crop for spring-seeded permanent pastures. Sown in combination with winter grains or crimson clover for temporary pasture, it makes a desirable bottom grass and increases the length of the grazing season. It is grown to a considerable extent for hay in western Oregon and Washington. It is an important winter pasture in the Gulf Coast states.[49] Italian ryegrass is seeded in lawns in the South in the fall in order to have a green turf after the warm-season grasses have ceased growth and turned brown.

■ 21.19 OTHER WILDRYE GRASSES

A number of wildrye grasses of *Elymus* species are native to North America and Europe. Canada wildrye *(Elymus canadensis)*, a bunchgrass, is widely distributed over the United States and Canada.[2] It is adapted to the northern Great Plains where it supplies green feed during the summer, except during drought. Mandan, an improved leafy variety is established readily from seeding.

Russian wildrye *(Elymus junceus)* is adapted to the drier areas of the northern Great Plains, as well as somewhat saline irrigated lands.[52] It provides palatable summer grazing. It responds well to nitrogen fertilization on irrigated land. This grass was introduced from the former Soviet Union in 1927. Plant breeding programs in Canada and the United States have released several improved varieties.

Two species native to the western United States are sown occasionally. They are giant wildrye (*E. condensatus*) and blue wildrye (*E. glaucus*).

▦ 21.20 MEADOW FOXTAIL

Meadow foxtail (*Alopecurus pratensis*), a native of Eurasia, is a bunchgrass. It has become established in the humid and subhumid areas of the northern half of the United States. It is best adapted to cool moist climates where it thrives on wetlands.[53] For many years the seed has been included in pasture mixtures, but sometimes it has been cut for hay. New varieties have been released primarily for use in Ontario.

▦ 21.21 TALL OATGRASS

Tall oatgrass (*Arrhenatherum elatius*) is generally grown over the United States, except in the drier areas, but is not important in any locality. It is adapted to well-drained soils, especially those that are sandy. It makes very poor growth in shade. This grass is a hardy, short-lived perennial bunchgrass, growing to a height of 2 to 5 feet (60 to 150 cm). The inflorescence is an open panicle similar to that of cultivated oats, but the seed is much smaller.

Tall oatgrass is often sown in mixtures with red clover, alsike clover, orchardgrass, sweetclover, and occasionally with alfalfa. A stand is difficult to obtain because the seed is of low viability.[2] It is necessary to use 30 to 50 pounds of seed per acre (34 to 56 kg/ha) when this grass is sown alone.

Tall oatgrass is used for pasture and hay. Although succulent, the grass has a peculiar taste to which grazing animals must become accustomed. It is considered palatable and highly nutritious. This grass will furnish an abundance of grazing from early spring to late in the fall. For hay, the crop should be cut at about the bloom stage.

▦ 21.22 BLUE GRAMA

Blue grama (*Bouteloua gracilis*) probably is the most important range grass in the Great Plains. Plant breeding programs in Colorado and New Mexico have developed improved varieties.

21.22.1 Adaptation

Blue grama occurs generally throughout the dry portions of the Great Plains as a component of the short-grass prairie. It predominates in drier areas and on sandier soils than those favored by buffalograss. Southern strains of blue grama tend to make a greater vegetative growth and a lower seed yield when moved northward,

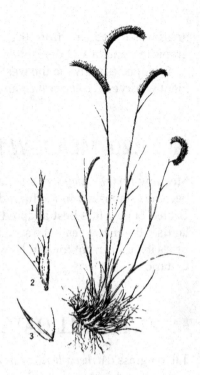

FIGURE 21.15
Blue gramagrass plant and
(1) floret, (2) spikelet, and
(3) glumes.

while northern strains grow too sparingly for southern conditions. Also, southern strains often lack resistance to cold temperatures.

21.22.2 Botanical Description

Blue grama is a low, warm-season, sod-forming bunchgrass perennial with fine, curling basal leaves of a grayish-green color. The leaves are 2 to 5 inches (5 to 13 cm) long and less than ⅛ inch (31 mm) wide, with the ligules sparsely (or occasionally distinctly) hairy (Figure 21.15).

The spikelet consists of an awned fertile floret as well as an awned, densely bearded, rudimentary floret. These appendages, which make the seed light and fluffy, interfere with drill seeding unless they are removed by processing. Blue grama is frequently confused with the staminate plants of buffalograss.

21.22.3 Culture

Stands of blue grama are established by seeding in the spring. When lightly disked and packed before seeding, blue grama has had considerable success when seeded in drilled sudangrass stubble that has left some residue. Seeding may be done with a packer-seeder. Unprocessed seed is sown with a grain drill equipped with cotton planter boxes or is broadcast. The land is packed again after seeding. Processed seed, 60 to 70 percent pure, is sown at the rate of 10 to 12 pounds per acre (11 to 13 kg/ha). It is desirable to include 2 to 4 pounds per acre (2.2 to 4.5 kg/ha) of buffalograss and some side-oats grama with the blue grama seed. The seed should not be covered with more than ½ inch (13 mm) of soil. Blue grama seed is harvested with a bluegrass stripper or a combine.[48]

21.22.4 Utilization

Blue grama is highly palatable and provides choice forage during the summer grazing period. The mature grass cures on the range and retains some of its nutritive value, thus providing excellent fall and winter grazing. The protein content of blue grama decreases toward maturity. At the early-bloom stage, the protein content is about two-thirds that of alfalfa at the same stage.

21.22.5 Related Species

A related species, sideoats grama (*Bouteloua curtipendula*), has been introduced into cultivation. This is a perennial bunchgrass suitable for pasture or hay.[54] Like blue grama, it is adapted to the western Great Plains and parts of the intermountain region. Other grama grasses are black (*B. eriopada*), hairy (*B. hirsute*), and Rothrock (*B. rothrockii*).

■ 21.23 BUFFALOGRASS

Buffalograss (*Buchloe dactyloides*) is a native, long-lived, perennial, sod-forming grass.[55] It is one of the most important range grasses in the Great Plains, probably being second only to blue grama. It has been cultivated only since about 1930.

21.23.1 Adaptation

Buffalograss is found in the Great Plains region from Texas to North Dakota but is most abundant on the heavier soils of the central Great Plains. It is not well adapted to sandy soils.[16] It is highly resistant to drought and heat. Although resistant to cold, it is a warm-season grass that starts growth late in the spring and ceases growth when cold weather arrives in the fall. Buffalograss forms a dense, tough sod under suitable growing conditions and withstands close grazing and tramping better than almost any other grass. Buffalograss furnished abundant forage to buffalo, antelope, and wild ponies in the Great Plains. White settlers used the turf to build their sod houses. With the opening of the region to cattle grazing, buffalograss increased in importance because it soon replaced some of the taller grasses such as the bluestems that could not withstand heavy grazing due to their higher palatability.

21.23.2 Botanical Description

Buffalograss produces fine, grayish-green leaves usually 2 to 4 inches (5 to 10 cm) long. The plant is largely dioecious. The staminate plants send up spikes 4 to

8 inches (10 to 20 cm) in height, while the pistillate plants produce seeds in burs borne down in the turf just above the surface of the soil. It produces rapid-growing long stolons that enable the plants to quickly cover bare spots. Small clumps may cover an area of 4 square feet (3,700 cm^2) or more in a single season.

21.23.3 Culture

Buffalograss seed is difficult to harvest because it is borne among the leaves close to the ground, but harvesters with beater and suction attachments collect much of the seed. Buffalograss seed is low in germination due both to unsound seed and a high percentage of dormancy.[56] The dormancy can be broken by hulling the seed or by special treatments. Seeding of buffalograss is now more feasible than sodding. About 4 or 5 pounds per acre (4.5 to 5.6 kg/ha) of treated seed are sufficient because of the rapid spreading of the plants.

21.23.4 Utilization

Buffalograss is too short to cut for hay, but the yields in the central and southern Great Plains are about as large as those of many of the taller grasses. It furnishes palatable, nutritious pasturage. In seasons of drought, the stand or cover may become very thin, but when moisture conditions again become favorable the few plants that survive soon spread to form a thick turf. Furthermore, dormant seeds in pastures that appear to be ruined germinate and reestablish an adequate ground cover.

There is increasing interest in using buffalograss for low-management turf in lawns and golf courses. New varieties developed in Nebraska have better leaf characteristics and a darker green color. Once established, buffalograss turf requires much less fertilizer and water and experiences fewer pest and disease problems than more conventional turf species.[57] Buffalograss lawns composed of pistillate plants are popular because they require little mowing or watering. The grass withstands drought better than any other grass that produces a dense turf. It is more cold-resistant than bermudagrass and as hardy as Kentucky bluegrass.

■ 21.24 OTHER RANGE GRASSES

21.24.1 Bluestems

Three native grass species of bluestem that are distributed over the Great Plains and western prairie states, along with two introduced species, have been seeded for pasture, hay, and erosion control.[6]

Big bluestem (*Andropogon gerardii*) is a coarse grass commonly found in the eastern half of the six states that extend from North Dakota to Texas and further east. It provides late spring and summer pasture, good ground cover, and good hay when cut before the plants head.

Little bluestem (*Schizachyrium scoparium*) occurs more generally in the semiarid Great Plains area. It is similar to big bluestem except that the plants are smaller.

Sand bluestem *(Andropogon hallii)* is distributed over the western Great Plains, particularly on the lighter soils.

Turkestan or yellow bluestem *(Bothriochloa ischaemum)* was introduced from Asia. It has finer stems than the native bluestems, and is especially adapted to the southwestern region from Texas to Arizona. The King Ranch variety is more productive but less hardy and drought resistant than is the common type. It is adapted to Texas, particularly the southern part.

Caucasian bluestem *(Bothriochloa caucasicus)* is adapted to the same conditions as little bluestem.

21.24.2 Love Grasses

Three species of love grass *(Eragrostis)* have been sown on appreciable acreages.[6]

Weeping lovegrass *(E. curvula)* was introduced from Africa. It is adapted to the area from Oklahoma to Arizona.[2, 6] The grass is palatable in spring and fall, but not in summer unless mowed or closely grazed.

Lehman lovegrass *(E. lehmanniana)*, also introduced from Africa, is adapted to the semi-desert areas of New Mexico, Arizona, and Texas.

Sand lovegrass *(E. trichoides)* is a palatable native bunchgrass adapted to sandy soils in the central and southern Great Plains.

21.24.3 Other Western Grasses

Indian ricegrass or sand bunchgrass *(Achnatherum hymenoides)* is distributed over the western half of the United States, particularly on dry, sandy soils. Its high palatability subjects it to overgrazing. The edible seeds resemble rice grains.

Switchgrass *(Panicum virgatum)* is a sod-forming species grown principally in the central and southern Great Plains. It provides spring grazing as well as summer grazing or hay when soil moisture is ample. Switchgrass is often sown for erosion control. Switchgrass has also been proposed for the production of biomass fuel in the Midwest United States.[58]

Vine mesquite *(Panicum obtusum)* is a vigorous native perennial in the southwestern states. It spreads by means of numerous long stolons. The herbage is palatable when green and succulent.

Blue panicgrass *(Panicum antidotale)* is a tall, coarse, sod-forming grass with thick bulbous stolons. It is adapted to fertile, well-drained soils in Texas, New Mexico, and Arizona. This grass provides palatable pasturage in early spring, especially in mixtures with alfalfa or sweetclover. It is used to hold soil in eroded areas, and the stalks can serve as a windbreak since they may reach a height of 9 feet (275 cm).

Green needlegrass, or feather bunchgrass *(Stipa viridula)*, occurs commonly over the northern and central Great Plains states. It provides leafy grazing or hay.

Indiangrass *(Sorghastrum nutans)* is a tall, coarse grass that occurs in pastures and woodlands over the eastern three-quarters of the United States. It provides a quick ground cover after seeding. It is a productive hay crop.

Dropseeds are grasses of the genus *Sporobolus* that shatter their seeds as soon as they are ripe. They occur over the hot dry areas of the southwestern United States. Two species, sand dropseed *(S. cryptandrus)* and tall dropseed *(S. asper)*, have been used for reseeding rangelands and eroded lands.

■ 21.25 CARPETGRASS

Carpetgrass (*Axonopus affinis*) is a native of the West Indies but is now widespread in the tropics of both hemispheres.

It is grown on the coastal plain soils from southern Virginia to Texas. It is especially adapted to sandy or sandy loam soils, particularly where the moisture is near the soil surface. The plants require abundant heat and moisture. Carpetgrass grows throughout the year except when damaged by severe drought or heavy frost. It tends to become established naturally on pastured land in the South, much as Kentucky bluegrass does in the North, and often replaces more palatable grasses.

Carpetgrass is a perennial creeping grass that forms a dense turf. It is readily distinguished by the compressed two-edged creeping stolons that root at each joint, as well as by the blunt leaf tips (Figure 21.16). The flower stems grow to a height of about 1 foot (30 cm). It is less productive than some of the taller grasses. It is planted only to a limited extent.

Carpetgrass is one of the most common perennial grasses for permanent pastures where it is adapted. It seeds abundantly and forms a dense turf. It can stand heavy continuous grazing.[59] It can be pastured in the South from May to November. Lespedeza and white clover mixed with carpetgrass improve the pasturage but are difficult to maintain because of the dense carpetgrass turf. Carpetgrass lawns are common in the Southeast.

FIGURE 21.16
Carpetgrass: (1) plant,
(2) panicle branch,
(3) glume, (4) ligule,
and (5) floret.

▓ 21.26 NAPIERGRASS

Napiergrass *(Pennisetum purpureum)*, a native of tropical Africa, is grown in the warmer regions of the United States. It is adapted to the southeastern states and southern California. Plant breeding programs in Hawaii and Georgia have developed improved varieties.

This grass is a robust, cane-like, leafy perennial that grows 5 to 7 feet (150 to 215 cm) or more in height. It grows in clumps of 20 to 200 stalks. The inflorescence is a long, narrow, erect, golden spike about 7 inches (18 cm) long. It is propagated by planting root or stem cuttings. It is occasionally grown from seed.

Napiergrass is utilized primarily as a soiling crop in tropical countries. The mature plants are rather woody, but silage made from the mature crop is eaten readily. Napiergrass is palatable and nutritious when grazed rotationally in Florida.[48] The pasture should be stocked so that most of the grass blades are consumed in five to eight days, after which twenty or more days are allowed between grazing for the grass to recover. It does not survive continuous close grazing.

▓ 21.27 DALLISGRASS

Dallisgrass *(Paspalum dilatatum)*, introduced from South America,[2] now occurs abundantly from North Carolina to Florida and west to Texas. It also is grown under irrigation from west Texas to California. Farther north it is too tender for survival. Dallisgrass favors heavy soils that are too wet for bermudagrass.

Dallisgrass is a perennial with a deep root system. It grows in clumps or bunches 2 to 4 feet (60 to 120 cm) in height. The leaves are numerous near the ground, but the stems are practically leafless (Figure 21.17). The slender stems usually droop from the weight of the flower clusters.

Since the seed is very light, the seedbed must be carefully prepared for Dallisgrass. Usually 5 to 10 pounds per acre (5.6 to 11 kg/ha) of hand-picked seeds are sown. In the southern states, the crop is generally sown in October or November. The production of dallisgrass seed is difficult because of heavy ergot attack and a high percentage of sterile florets. Frequent mowing will prevent ergot seeds from maturing. The ergot is poisonous to cattle.

Dallisgrass is primarily a pasture crop because of its tendency to lodge when left for hay. It is a good summer pasture grass for heavy, moist, fertile soils when grown with legumes such as white, hop, or Persian clover. White clover should be sown after the dallisgrass is established. Persistent grazing does not injure dallisgrass, as the leaves are quickly renewed. Permanent pastures of carpetgrass and bermudagrass become more valuable when this grass is included.

▓ 21.28 OTHER SOUTHERN GRASSES

Bahiagrass *(Paspalum notatum)* is a low-growing perennial pasture grass introduced from South America. It differs from dallisgrass not only in being less hardy, but also in producing heavy runners that form a dense sod on sandy soils. Bahiagrass is a tender, warm-season species adapted to the South Atlantic and Gulf

FIGURE 21.17
Dallisgrass plant and culm.

Coast areas. Related species are vaseygrass *(P. urvillei)* and ribbed paspalum *(P. malacophyllum).*[2]

Rhodesgrass *(Chloris gayana)*, introduced from South Africa in 1902, is a leafy, nonhardy grass adapted to the Gulf Coast as well as to southern Arizona and California. It produces abundant seed but also spreads by long stolons. It is a pasture and hay crop that is very tolerant to soil salinity.

Pangolagrass *(Digitaria decumbens)* was introduced from South Africa. It resembles giant crabgrass. It is a nonhardy, rapid-growing, leafy grass grown in central and southern Florida for pasture, silage, and hay. Pangolagrass is very palatable, grows well on sandy soils, and is drought resistant. It must be propagated vegetatively by planting sprigs or stems.

St. Augustinegrass *(Stenotaphrum secundatum)* is a native of the West Indies, Mexico, Africa, and Australia. It is a pasture and lawn grass grown in the Gulf Coast states. This grass forms a dense turf that withstands trampling but requires ample moisture as well as heavy fertilization. It is established by planting rooted stolons.

Paragrass *(Panicum purpurascens)*, a native of Africa, is a tropical perennial grass grown to some extent on wetlands in southern Florida. It is propagated by planting pieces of stolons or stems.

Buffelgrass *(Pennisetum ciliare)* is a warm-season bunchgrass introduced from Africa. It is adapted to the Gulf Coast states and south central Texas. This grass is drought resistant but lacks cold resistance. It provides summer pasture and hay. A variety known as Blue buffelgrass is larger, more erect, more tolerant to cold, and especially adapted to heavy soils.

Smilograss (*Oryzopsis miliacea*), a warm-season bunchgrass from the Mediterranean region, is grown in southern California. Kikuyugrass (*Pennisetum clandestinum*) is a leading pasture grass in Hawaii. It forms a dense sod from spreading stolons and rhizomes.[60]

REFERENCES

1. Hitchcock, A. S. *Manual of the Grasses of the United States*, 2nd ed. (rev. by Agnes Chase), *USDA Misc. Pub.* 200, 1950, pp. 1–1051.

2. Hoover, M. M., and others. "The main grasses for farm and home," in *Grass*, USDA Yearbook, 1948, pp. 639–700.

3. Weintraub, F. C. "Grasses introduced into the United States," *USDA Handbook* 58, 1953.

4. Burton, G. W., and J. E. Jackson. "Effect of rate and frequency of applying six nitrogen sources on coastal bermudagrass," *Agron. J.* 54(1962):40–43.

5. Ramage, C. H. "Yield and chemical composition of grasses fertilized heavily with nitrogen," *Agron. J.* 50(1958):59–62.

6. Anonymous. "Grasses and legumes for forage and conservation," *USDA Agricultural Research Service Spec. Rept.*, ARS 22–43, 1957.

7. Evans, M. W. "The life history of timothy," *USDA Bull.* 1450, 1927, pp. 1–52.

8. Gould, F. W. *Grass Systematics*. New York: McGraw-Hill, 1968, pp. 1–382.

9. Klages, K. H. "Comparative ranges of adaptation of species of cultivated grasses and legumes in Oklahoma," *J. Am. Soc. Agron.* 21(1929):201–223.

10. Anderson, B., "Establishing dryland forage grasses," *NE Coop. Ext. Serv.* G81-543-A, (rev.), 1989.

11. Barnhart, S. K. "Warm-season grasses for hay and pasture," *IA St. Univ. Coop. Ext. Serv.* PM-569, 1994.

12. Robert, C., and J. Gerrish. "Seeding rates, dates and depths for common Missouri forages," *Univ. MO Coop. Ext. Serv.* G4652, 1997.

13. Evans, M. W., F. A. Welton, and R. M. Salter. "Timothy culture," *OH Agr. Exp. Sta. Bull.* 603, 1939, pp. 1–54.

14. Evans, M. W., and L. E. Thatcher. "A comparative study of an early, a medium, and a late strain of timothy harvested to various stages of development," *J. Agr. Res.* 56(1938):347–364.

15. Hosterman, W. H., and W. L. Hall. "Time of cutting timothy: Effect on proportion of leaf blades, leaf sheaths, stems, and heads and on their crude protein, other extract, and crude fiber contents," *J. Am. Soc. Agron.* 30(1938):564–568.

16. Walster, H. L., and others. "Grass," *ND Agr. Exp. Sta. Bull.* 300, 1941.

17. Newell, L. C., and F. D. Keim. "Field performance of bromegrass strains from different regional seed sources," *J. Am. Soc. Agron.* 35(1943):420–434.

18. Hanson, A. A. "Grass varieties in the United States," *USDA Handbook* 170 (rev.), 1972, pp. 1–124.

19. Frolik, A. L., and L. C. Newell. "Bromegrass production in Nebraska," *NE Agr. Exp. Sta. Circ.* 68, 1941.

20. Rather, H. C., and C. M. Harrison. "Alfalfa and smooth bromegrass for pasture and hay," *MI Agr. Exp. Sta. Cir.* 189, 1944.

21. Bourg, C. W., and others. "Nitrogen fertilizer for bromegrass in eastern Nebraska," *Nebr. Agron. Cir.* 97, 1949.

22. Lamond, R. E., J. O. Fritz, and P. D. Ohlenbusch. "Smooth brome production and utilization," *KS St. Univ. Coop. Ext. Serv.* C-402, 1999.

23. Brotemarkle, J., and G. Kilgore. "Seed production management for bromegrass and tall fescue," *KS St. Univ. Coop. Ext. Serv.* MF-924, 1989.

24. Hafenrichter, A. L. "New grasses and legumes for soil and water conservation," in *Advances in Agronomy*, vol. 10. New York: Academic Press, 1958, pp. 349–406.

25. Hanson, A. A., and H. L. Carnahan. "Breeding perennial forage grasses," *USDA Tech. Bull.* 1145, 1956.

26. Hanson, A. A., and F. V. Juska. *Turfgrass Science.* Madison, WI: Am. Soc. Agron. Monograph. 14, 1949, pp. 1–715.

27. Henning, J., and N. Risner. "Orchardgrass," *Univ. MO Coop. Ext. Serv.* G4511, 1993.

28. Rampton, H. H. "Alta fescue production in Oregon," *OR Agr. Exp. Sta. Bull.* 427, 1949.

29. Heath, M. E., D. S. Metcalfe and R. F. Barnes, eds. *Forages.* Ames, IA: Iowa State Univ. Press, 1973.

30. Seath, D. M., and others. "Comparative value of Kentucky bluegrass, Kentucky 31 fescue, orchardgrass, and bromegrass as pasture for milk cows. I. How kind of grass affected persistence of milk production, ton yield, and body weight," *J. Dairy Sci.* 39(1956):574–580.

31. Kilgore, G. L., and F. K. Brazle. "Tall fescue production and utilization," *KS St. Univ. Coop. Ext. Serv.* C-729, 1994.

32. Cherney, J. H., and K. D. Johnson. "Tall fescue for forage production," *Purdue Univ. Coop. Ext. Serv.* AY-98, 1998.

33. Kreitlow, K. W. "Diseases of forage grasses and legumes in the northeastern states," *PA Agr. Exp. Sta. Bull.* 573, 1953.

34. Schoth, H. A. "Reed canary grass," *USDA Farmers Bull.* 1602, 1938.

35. Arny, A. C., and others. "Reed canary grass," *MN Agr. Exp. Sta. Bull.* 252, 1929.

36. Harrison, C. M. "Reed canarygrass," *MI Agr. Exp. Sta. Bull.* 220, 1940.

37. Wilkins, F. S., and H. D. Hughes. "Agronomic trials with reed canary grass," *J. Am. Soc. Agron.* 24(1932):18–28.

38. Evans, M. W. "Kentucky bluegrass," *OH Agr. Exp. Sta. Res. Bull.* 681, 1949.

39. Ahlgren, H. L. "Effect of fertilization, cutting treatments, and irrigation on yield of forage and chemical composition of the rhizomes of Kentucky bluegrass (*Poa pratensis*)," *J. Am. Soc. Agron.* 30(1938):683–691.

40. Spencer, J. T., and others. "Seed production of Kentucky bluegrass as influenced by insects, fertilizers, and sod management," *KY Agr. Exp. Sta. Bull.* 535, 1949.

41. Schoth, H. A., and M. Halperin. "The distribution and adaptation of *Poa bulbosa* in the United States and in foreign countries," *J. Am. Soc. Agron.* 24(1932):786–793.

42. Burton, G. W. "Coastal bermudagrass," *GA Agr. Exp. Sta. Bull.* N52, 1954.

43. Harlan, J. R. "Midland bermudagrass," *OK Agr. Exp. Sta. Bull.* B-416, 1954.

44. Nielsen, E. L. "Establishment of bermuda grass from seed in nurseries," *AR Agr. Exp. Sta. Bull.* 409, 1941.

45. Rogler, G. A. "Growing crested wheatgrass in the western states," *USDA Leaflet* 469, 1960.

46. Westover, H. L., and others. "Crested wheatgrass as compared with bromegrass, slender wheatgrass, and other hay and pasture crops for northern Great Plains," *USDA Tech. Bull.* 307, 1932.

47. Alderson, J., and W. C. Sharp. *Grass Varieties in the United States.* Boca Raton, FL: CRC Press, 1995.

48. Blaser, R. E., W. G. Krik, and W. E. Stokes. "Chemical composition and grazing value of Napier grass, *Pennisetum purpureum*, grown under a grazing management practice," *J. Am. Soc. Agron.* 34(1942):167–174.

49. Weihing, R. M., and N. S. Evatt. "Seed and forage yields of Gulf ryegrass as influenced by nitrogen fertilization and simulated winter grazing," *TN Agr. Exp. Sta. Prog. Rpt.* 2139, 1960.

50. Schoth, H. A., and M. A. Hein. "The ryegrasses," *USDA Leaflet* 196, 1940.

51. Hannaway, D., and others, "Perennial ryegrass," *WA Coop. Ext. Serv.* PNW0503, 1999.

52. Rogler, G. A., and H. M. Schaaf. "Growing Russian wildrye in the western states," *USDA Leaflet* 524, 1963.

53. Schoth, H. A. "Meadow foxtail," *OR Agr. Exp. Sta. Bull.* 433, 1947.

54. Newell, L. C., and others. "Side-oat grama in the central Great Plains," *NE Agr. Exp. Sta. Res. Bull.* 207, 1962.

55. Beetle, A. A. "Buffalograss—a native of the shortgrass," *WY Agr. Exp. Sta. Bull.* 293, 1950.

56. Wenger, L. E. "Buffalo grass," *KS Agr. Exp. Sta. Bull.* 321, 1943, pp. 1–78.

57. Riordan, T. P., F. P. Baxendale, R. E. Gaussoin, and J. E. Watkins. "Buffalograss: An alternative native grass for turf," *NE Coop. Ext. Serv.* G96-1297-A, (rev.), 1998.

58. Teel, A., and S. K. Barnhart. "Switchgrass seeding recommendations for the production of biomass fuel in southern Iowa," *IA St. Univ. Coop. Ext. Serv.* PM1773, 1998.

59. Blaser, R. E. "Carpetgrass and legume pastures in Florida," *FL Agr. Exp. Sta. Bull.* 453, 1948.

60. Mears, P. T. "Kikuyii (*Pennisetum clandestinum*) as a pasture grass. A review." *Trop. Grassl.* 4(1970):134–152.

PART 3 CROPS OF THE LEGUME FAMILY

Alfalfa

22.1 ECONOMIC IMPORTANCE

Alfalfa is the world's leading hay crop. In 2000–2003, alfalfa was grown on an average of 38 million acres (15 million ha) on all continents, but seldom in humid tropical and subtropical areas or on unlimed acid soils or marshland. In the United States in 2000–2003, alfalfa, alone or in mixtures, was harvested for hay on an average of over 23 million acres (9 million ha). It ranked fourth in domestic crop acreages and is also the leading forage legume in Canada. Production in the United States in 2000–2003 averaged over 79 million tons (72 million MT) with a yield exceeding 3 tons per acre (7 MT/ha). The leading states in alfalfa hay production were California, South Dakota, Iowa, Minnesota, and Wisconsin (Figure 22.1). Alfalfa is usually grown alone in the western irrigated areas, but often in grass mixtures in humid areas.

Alfalfa was harvested for seed on about 156,000 acres (63,000 ha) in the United States in 2002 (Figure 22.2). The leading states in seed production are California, Idaho, Washington, Montana, and Oregon.

22.2 HISTORY OF ALFALFA CULTURE

The name *alfalfa*, which comes from the Arabic language, means best fodder. In Europe, it is usually called lucerne. Most authorities believe that alfalfa originated in southwestern Asia, although forms from which it could have arisen are found in China and Siberia.[1] Alfalfa was first cultivated in Iran before 700 BC, and from there it was carried to Arabia, the Mediterranean countries, and finally to the New World. Evidence of the ancient introduction of alfalfa into Arabia is found in the strongly marked characteristics of Arabian varieties,[2] which apparently represented centuries of acclimatization in an arid region.

The first recorded attempt to grow alfalfa in the United States was in Georgia in 1736. Although alfalfa was later grown in the eastern states from time to time, it was not always successful. Introductions into California from Peru in 1841 and from Chile about 1850 started a tremendous expansion.[3] The introduction of a winter-hardy alfalfa into Minnesota by Wendelin Grimm during the mid-1800s further increased the use of this crop.

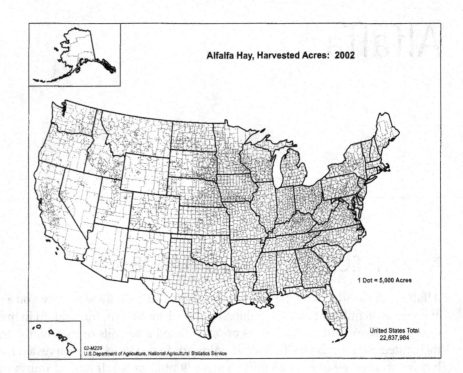

FIGURE 22.1

Alfalfa acreage harvested for hay. [Source: 2002 Census of U.S. Department of Agriculture]

Alfalfa Hay, Harvested Acres: 2002

1 Dot = 5,000 Acres

United States Total
22,637,984

02-M229
U.S.Department of Agriculture, National Agricultural Statistics Service

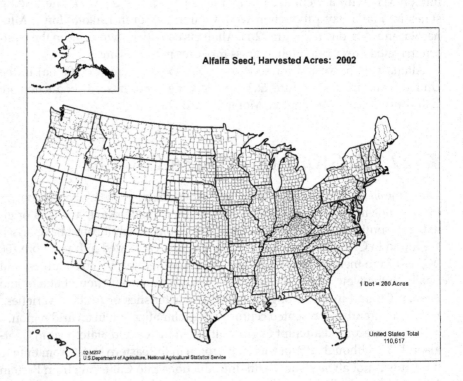

FIGURE 22.2

Alfalfa acreage harvested for seed. [Source: 2002 Census of U.S. Department of Agriculture]

Alfalfa Seed, Harvested Acres: 2002

1 Dot = 200 Acres

United States Total
110,617

02-M237
U.S.Department of Agriculture, National Agricultural Statistics Service

■ 22.3 ADAPTATION

Alfalfa has a remarkable adaptability to various climatic and soil conditions, as is shown by its wide distribution. The alfalfa plant makes its best growth in relatively dry climates where water is available for irrigation. Irrigation requirements may exceed 80 inches (2,000 mm) per year in hot desert areas. It will withstand long periods of drought due to a deep root system but is unproductive under such conditions. Although alfalfa is now grown mostly in the central and eastern United States, it can also succeed in the South.

Alfalfa tolerates extremes of heat and cold. Yellow-flowered alfalfa has survived temperatures as low as –83°F (−64°C). Common alfalfa has been grown in Arabia, as well as in Death Valley, California, where maximum summer temperatures are as high as 120°F (49°C), but the humidity is low. However, alfalfa is relatively dormant during the summer in very hot regions.

Alfalfa is best adapted to deep loam soils with porous subsoils. Good drainage is essential. The plant requires a large amount of calcium for satisfactory growth. It survives on almost all soils in the semiarid region except those high in alkaline salts or those that have a shallow water table.

Alfalfa has spread to the more humid eastern states with the increased knowledge of the requirements for lime, inoculation with nitrogen-fixing bacteria, certain plant nutrients, and also the breeding of hardier, disease-resistant varieties. Except on a few limestone soils, applications of lime are essential for satisfactory growth of alfalfa east of the Mississippi River.

■ 22.4 BOTANICAL DESCRIPTION

Alfalfa is an herbaceous perennial legume that can live fifteen to twenty years or even more in dry climates unless insects or diseases destroy it. The most commonly cultivated species is *Medicago sativa*. Yellow-flowered alfalfa (*M. falcata*) is sometimes regarded as a subspecies (*M. sativa* ssp. *falcata*) of common alfalfa. Yellow-flowered alfalfa is distinguished by its yellow flowers, sickle-shaped seed pods, decumbent growth habit, low-set branching crowns, and a preponderance of branched roots. It is not satisfactory for hay in America because it is somewhat prostrate and usually yields only one cutting a season in the northern states, but it is used for grazing in some dry rangeland areas. It is grown on only limited acres but is of interest primarily for hybridization with common alfalfa to produce hardier varieties.[4]

22.4.1 The Alfalfa Plant

The alfalfa seedling emerges with the two cotyledons, produces one unifoliolate leaf, followed by alternate pinnately trifoliolate leaves thereafter. The plant varies in height from 2 to 3 feet (60 to 90 cm). It has five to twenty or more erect stems that continue to arise from the fleshy crown as the older branches mature and are harvested. Several short branches may grow from each stem (Figure 22.3). The oblong leaflets are sharply toothed on the upper third of the margin, the tip being terminated by the projected midrib (Figure 23.1). About 48 percent of the weight of the plant comes from the leaves.[5]

FIGURE 22.3
Branches and flowers of alfalfa. Leaves at lower right.

The root system consists of an almost straight taproot, which, under favorable conditions, penetrates the soil to a depth of 25 to 30 feet (7 to 9 m) or more. Older varieties had few side branches that extended short distances from the main taproot, but new varieties are much more branched, especially those developed for Midwest climates. The main root normally persists during the entire life of the plant. All varieties of alfalfa develop branch roots in compact soil, while taproots predominate in porous soil.[6]

The flowers of common alfalfa, borne in axillary racemes, are purple except in the variegated types. The fruit is a spirally twisted pod that contains from one to eight small kidney-shaped seeds (Figure 22.4). The seeds are normally olive-green in color (Figure 22.5).

Alfalfa can be propagated vegetatively from stem or crown cuttings.

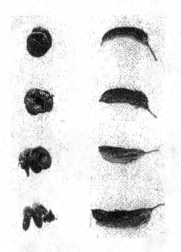

FIGURE 22.4
Coiled seed pods of common alfalfa *(left)* and sickle-shaped pods of yellow-flowered alfalfa *(right).*

22.4.2 Pollination

The plants of commercial varieties of alfalfa are slightly to almost completely self-sterile. Under ordinary climatic conditions, they are nearly incapable of automatic self-pollination.[7] From 7 to as high as 80 percent crossing between closely associated plants of purple-flowered and yellow-flowered alfalfa has been observed.[8, 9] A sizable decrease in both forage and seed yields usually occurs when flowers are self-fertilized.

The external flower structures, the keel, wings, and standard enclose the reproductive tissues and serve to attract insects. Tripping is necessary for pollination and subsequent seed set except in a small percentage of the flowers. Tripping is the release of the sexual column from the keel to the flower.[8, 10] The sexual column includes the style, stigma, and part of the ovary enclosed and surrounded by ten stamens and diadelphous filaments. Tripping, which takes place when the flower is in a turgid condition, is accompanied by an explosive action as though a spring under tension has been released. Tripping naturally, or by mechanical jarring, wind, rain, or sun induces mostly self-pollination and little seed production. Partial self-incompatibility and ovule abortion result in a low percentage of fertilization when alfalfa plants are self-pollinated.[11] Net fertility six days after pollination may be about six times as high in crossed plants as in selfed plants. Insects carry pollen from different plants when tripping the flowers, causing cross-pollination and seed development.

Rupture of the stigmatic surface by tripping is essential to the penetration of the pollen tubes.[12] After tripping occurs, the proper moisture relationship for pollen germination and pollen-tube growth must be maintained to effect fertilization.

Wild bees, mostly leaf-cutter bees (*Megachile* spp.), alkali bees (*Nomia melanderi*), or ground bees (*Andrena* spp.) collect pollen and are the most effective pollinators. Bumble bees (*Bombus* spp.), however, are fairly effective trippers.[13]

About 38 million flowers must be tripped and cross-pollinated to produce 500 pounds (227 kg) of seeds. Two to three good honey bee colonies per acre (5 to 8 per ha) provide the two to three bees per square yard (2 to 3/m²) essential for high seed yields. Only 5 to 20 percent of the bees are pollen collectors. Nectar collectors, however, trip some of the flowers and thus spread pollen. The drones refuse to participate in the pollination.[14] Alfalfa seed growers temporarily rent colonies from bee keepers, placing them in or adjoining the fields, within 800 to 1,400 feet (250 to

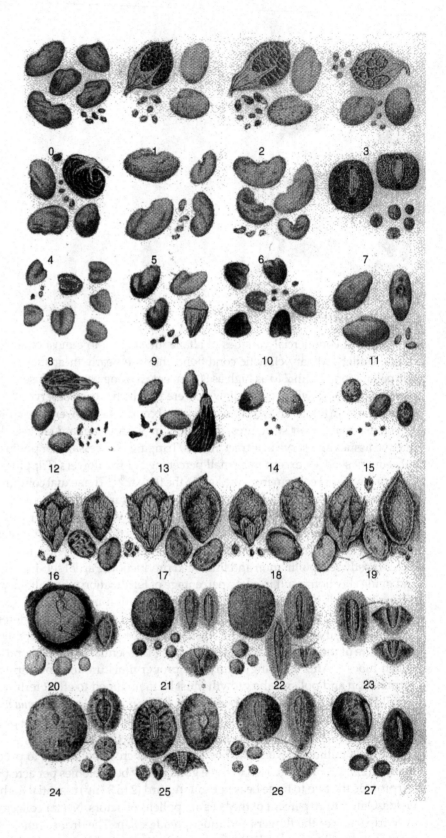

FIGURE 22.5

Seeds of forage legumes (enlarged and natural size) (0) alfalfa, (1) white sweetclover, (2) yellow sweetclover, (3) sourclover (4) black medic, (5) California burclover, (6) spotted burclover, (7) roughpea, (8) white clover, (9) red clover, (10) alsike clover, (11) crimson clover, (12) small hop clover, (13) large hop clover, (14) hop clover, (15) birdsfoot trefoil, (16) common lespedeza, (17) Kob lespedeza, (18) Korean lespedeza, (19) sericea lespedeza, (20) smooth green pea, (21) hairy vetch, (22) common vetch, (23) woolypod vetch, (24) mottled field pea, (25) Hungarian vetch, (26) narrowleaf vetch, and (27) purple vetch.

430 m) from the most distant plants. The wild bee population is ample for pollination in localities that are only partly under cultivation. From a distance, bees select alfalfa flowers by color, with a preference for white, followed by dark-reddish purple, and least by brilliant yellow. When nearby, they are attracted by aroma, and finally by the amount and sweetness of the nectar.[15]

The small alfalfa leaf-cutting bee (Megachile rotundata) is being propagated in Canada and the United States for use in pollination. Some 2,500 to 5,000 female bees are needed for each acre (6,000 to 12,000/ha) of alfalfa. Half of the cocoons produce males, which are ineffective pollinators.[16] Alkali bees are propagated in the United States. Specially treated artificial soil beds are maintained by water and salt control for bee propagation.[17, 18]

Other flowers in bloom may attract bees away from alfalfa and thus reduce the seed set. Lack of tripping by beneficial insects is the most common cause of poor seed yields. Insects rarely collect alfalfa pollen in cool or humid weather. Thin stands of alfalfa, as well as restricted irrigation at flowering time, encourage honey-bee visitations. Thickly planted and lodged plants give poor seed sets. Ample food reserves in alfalfa roots also contribute to good seed sets.

Lygus bugs cause bud damage and flower dropping and often cause a poor seed set in alfalfa.[19, 20]

▨ 22.5 GROUPS

Five somewhat distinct groups of commercial alfalfas are grown in the United States.[1] These are common, Flemish, Turkistan, variegated, and nonhardy.

22.5.1 Common Alfalfa Group

The common group includes the ordinary purple-flowered, smooth alfalfa. Hardy types adapted to the northern states tend to recover more slowly after cutting than those produced farther south. The strains produced in the Southwest recover rapidly after cutting, but they are very susceptible to winter injury except when grown in the southern states. Adaptation, especially to cold resistance, is extremely important in common alfalfa. Seed should be procured from a source where the winters approximate in severity the conditions where it is to be planted.

Most commercial varieties[21] are resistant to bacterial wilt, except for some adapted to regions requiring little or no winter hardiness. These varieties of common alfalfa are adapted to the eastern states, the deep South and Southwest, or California.

22.5.2 Flemish Group

Flemish-type varieties originated in western Europe or were developed elsewhere from imported European varieties. The plants are similar to those of the common group, but they start spring growth earlier, experience medium to late fall dormancy, and are less hardy than the standard varieties. They mature early and recover rapidly after cutting. They are less persistent due to higher susceptibility to bacterial wilt (*Corynebacterium insidiosum*). Flemish types are adapted to the mild humid climates of the eastern and Pacific coastal states.[16] Quick recovery is not advantageous in semiarid areas where only one or two cuttings are obtained. No Flemish varieties are grown in the United States but they are used in crosses when developing new varieties.

22.5.3 Turkistan Group

The Turkistan group is derived from alfalfa that originated in Xinjiang province in China. They are characterized by slow recovery after being cut, early fall dormancy, susceptibility to leaf diseases, winter hardiness, resistance to bacterial wilt, and low seed yields. The growth is somewhat shorter and more spreading, while the leaves are smaller with slightly more pubescence. Its varieties are not important in United States, but many improved varieties have Turkistan in their parentage.

22.5.4 Variegated Group

Introduction, breeding, testing, and distribution of cold-resistant variegated varieties are largely responsible for the successful culture of alfalfa in the northern states.

Variegated alfalfa has likely resulted from natural crossing between the purple-flowered and yellow-flowered species. The predominant flower color is purple, but some brown, green, greenish-yellow, yellow, and smoky to nearly white flowers occur. Seed pods vary in shape from sickle-shaped to coiled. Such variegation is characteristic of many selected plant progenies owing to repeated cross-pollination. Their hardiness is due to the presence of yellow-flowered alfalfa in their ancestry as well as to natural selection under severe climatic conditions. The variegated group has sometimes been considered a separate species, called *Medicago media*. Variegated alfalfa varieties predominate in the North Central states.

22.5.5 Nonhardy Group

Nonhardy varieties introduced from Peru, Africa, or India are adapted to the deep Southwest.[22] The Hairy Peruvian variety has pubescent leaves and stems.

22.5.6 Creeping Alfalfa

Creeping alfalfa[23] multiplies from rhizomes. It often yields less than upright varieties but spreads where stands are thin. It can be grown in western Canada and persists longer under pasturing.[24]

22.5.7 Hybrid Alfalfa

Hybrid alfalfa is produced by selection of clonal lines that combine well, after which they are increased from stem cuttings. A group of plants propagated from a single plant is called a clone. Lines to be crossed are grown in alternate strips from transplanted cuttings. Crosses are sometimes made between purple-flowered and yellow-flowered types. Inbreeding of these lines is unnecessary because they can be reproduced at will by maintaining them as perennials and by propagating them vegetatively. Probably 90 percent of the seed produced is hybrid because the lines used are largely self-sterile.[25]

▦ 22.6 HARD OR IMPERMEABLE SEED

Impermeable seeds, also known as hard seeds, occur commonly in alfalfa, with the average in Colorado-grown seed being 22 percent,[26] and in all areas ranging from 0 to 72 percent. The percentage of hard seed increases with maturity. Alfalfa seed produced in northern areas of the United States has more hard seed than that produced in southern areas, even for the same variety. Hardy varieties exhibit a higher percentage of impermeable seeds than do the less hardy ones. Mature fresh, hard alfalfa seeds are alive and there is no important difference between permeable and hard seeds in the rate of loss of viability when stored the same length of time.[27] Most of the hard seeds of alfalfa that fail to germinate in the laboratory will germinate sooner or later when sown in the field.[28] Alfalfa seed lots with many hard seeds have almost the same agricultural value as those with only a few.

Hard alfalfa seeds are sometimes scarified before seeding to improve germination. Scarification is any process that breaks, scratches, or mechanically alters the seed coat to make it permeable to water and oxygen. Nature provides some scarification. Freezing temperatures or microbial activity change the seed coat throughout winter resulting in increased germination following fall seeding. After scarification, the seeds will have a dull appearance but should not be deeply pitted or cracked, which can damage the embryo. Scarified seeds do not store well and should be planted as soon as possible after treatment.

Scarifying alfalfa seeds is not always practical. Seeding rates can be increased to account for hard seed content if there is more than 25 percent. The hard seeds may improve thin stands with their later germination.

▦ 22.7 WINTER HARDINESS

Resistance of alfalfa to low winter temperatures is important for maintenance of stands in the northern states and Canada. Alfalfa plants have no autonomous rest period but become dormant because of environmental conditions unfavorable for growth.[29] Plants with increased fall dormancy have higher root sugar concentrations and generally overwinter better.[30] Winter hardiness is largely a varietal characteristic.[8] In general, only cold-resistant variegated varieties, or northern strains of common alfalfa, are suitable north of the 40° latitude. The nonhardy varieties are suitable where winter temperatures seldom are lower than 18°F (–8°C).

Hardy alfalfa becomes dormant earlier in the fall and hardens more rapidly than does nonhardy alfalfa.[31] A hardened plant is one that, due to certain environmental changes such as cooler temperatures and shorter days, goes through a hardening process and becomes more capable of withstanding cold and severe weather conditions frequently experienced during winter months. The hardening process in alfalfa is cumulative over the fall and early winter, and then hardening decreases toward spring. Hardy varieties retain their hardening longer than do nonhardy

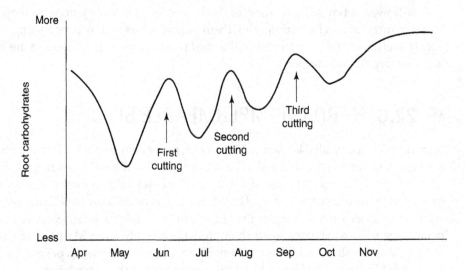

FIGURE 22.6

Root reserves in alfalfa through the growing season.

ones. Plants harden best at 40°F (5°C), but greater hardiness develops under alternating temperatures of 32°F (0°C) for sixteen hours and 68°F (20°C) for eight hours.[10] Hardening progresses with decreasing day length and alternating temperatures in the fall.

Premature or too frequent cutting of alfalfa depletes the organic root reserves, with a subsequent reduction in stand either by winter injury or disease.[32, 33, 34] These reserves are the carbohydrate and nitrogen compounds produced in the leaves, stored mostly in the roots, and later utilized by the plant for maintenance and for future growth. New foliage growth is made partly at the expense of these root reserves (Figure 22.6). The root reserves are increased during the blossoming period.[35] Cool weather in the fall is especially conducive to food storage. The protection afforded by some fall top growth aids in maintenance of stands.[32, 36] In Michigan, the roots of alfalfa plants cut in September were lower in percentage of dry matter than when cut in October.[37]

In areas where winter survival is a problem, alfalfa should not be cut during a two- or three-week period before the final killing frost. Cutting immediately before this period insures adequate buildup of root reserves to insure maximum winter survival. From the standpoint of survival, alfalfa plants should not be cut before the blossom stage, nor permitted to go into the winter in the northern states without some top growth.[25] In Indiana, cutting alfalfa in mid-October reduced root protein and starch concentrations compared with non-harvested plants.[30]

■ 22.8 FERTILIZERS

Alfalfa prefers a soil pH of at least 6.2 for good nodulation. On most soils east of the ninety-fifth meridian, alfalfa responds to lime, to commercial fertilizers, particularly superphosphate, often to potash, and to light applications of boron.

Seed inoculation with effective strains of *Rhizobium meliloti* is required for ample nitrogen fixation in any field that has not produced good alfalfa growth during previous years. Inoculation is especially important in the humid areas. Alfalfa responds to phosphorus applications on many irrigated calcareous soils that have

TABLE 22.1 Soil Nutrient Removal by Alfalfa on a Dry Matter Basis

Nutrient	Lb/T	Kg/Mt
Phosphorus	6–12	3–6
Potassium	48–50	24–25
Calcium	20–30	10–15
Magnesium	4–6	2–3
Sulfur	3–6	1.5–3
Boron	0.05	0.03
Zinc	0.04	0.02
Manganese	0.05–0.12	0.03–0.06
Copper	0.01	0.005
Iron	0.21–0.33	0.1–0.17

been cropped for some time. A low level of available soil phosphorus is one of the causes of poor yields of alfalfa. An application of 300 pounds per acre (335 kg/ha) of P_2O_5 was required on deficient soils to supply the phosphorus needs for a companion crop and three or four years of alfalfa.[38] Ample potash is essential for high yields. On rare occasions, zinc, magnesium, and molybdenum are needed (Table 22.1). Liming and fertilization are required in the humid areas of the Pacific Northwest. Alfalfa will often take up and store in its tissue more potash than is actually needed. This process, known as luxury consumption, will occur when the leaf content of potassium is greater than 2 percent by weight at harvest time. Potassium contents as high as 3.5 percent have been recorded, but the excess has not contributed significantly to yield.

■ 22.9 ROTATIONS

Alfalfa is an important crop in many rotations. In dryland rotations, alfalfa is usually sown with a small grain companion crop. It is cut for hay or seed for two or three years, and often longer when good stands are maintained. The alfalfa is then killed using tillage and/or herbicides. The subsequent crop is planted no-till or in a prepared seedbed. After two to four years of intertilled crops, the land is again sown to alfalfa. When alfalfa sod is turned under, the land can be plowed or disked shallow in the fall to cut off the roots below the crown of the plant, a procedure called crowning. This procedure reduces the need for herbicides.

The roots and residue furnish nitrogen for crops following alfalfa. A good alfalfa crop may provide a successive crop with 100 pounds per acre (110 kg/ha) or more. Fertilizer recommendations based on soil tests usually spread the nitrogen credit over two years due to the fact that not all organic nitrogen will be mineralized the first year. In eastern Nebraska, increased nitrates after alfalfa increased yields in seasons of relatively high rainfall but were detrimental when rainfall was deficient.[8] The increased nitrogen causes excessive vegetative growth early in the season, which increases the crop's need for moisture beyond the supply, and the plants suffer from drought. Where drought is a problem, intervals between alfalfa in rotations should be lengthened. Sorghum or corn is a good crop to follow alfalfa. After alfalfa, the soil may be only a little drier within the rooting zone of annual crops than it is after small grains.

In nonirrigated areas where the annual precipitation is less than 30 inches (760 mm), alfalfa exhausts the subsoil moisture. Yields on land sown to alfalfa for the first time often decline abruptly after four or five years due to subsoil moisture depletion. Subsequent growth is dependent upon current rainfall, which is not sufficient for maximum production under subhumid conditions. In Nebraska and Kansas, alfalfa roots have penetrated 30 to 40 feet (9 to 12 m) in the soil and used the available subsoil moisture at these great depths.[5, 8, 39, 40] In Nebraska research, alfalfa hay yields averaged about 6 tons per acre (13 MT/ha) during the first three years before the deep subsoil moisture was exhausted. During the five years thereafter, the yields averaged only about 2 tons per acre (4 MT/ha), even though good stands were maintained.

When alfalfa is sown again on moisture-depleted land, the yields are far from satisfactory. In Nebraska, only 30 to 53 percent of the moisture removed by the alfalfa was restored during the thirteen to fifteen years of cropping to annual crops. In eastern Kansas, two years of fallow were necessary to restore subsoil moisture on old alfalfa ground to the point where the roots of a newly seeded crop could penetrate through moist soil to a depth of 25 feet (8 m) or more. In semiarid regions, the need for soil moisture explains the success of alfalfa on bottomlands that are subject to overflow or subirrigation in the semiarid regions.

■ 22.10 ALFALFA CULTURE

A firm, moist, well-prepared seedbed is best for obtaining a stand of alfalfa.[1, 41] The land can be fall-plowed for spring seeding or, in the northeastern states, merely disked where alfalfa is to be seeded late in the summer after small grains.

22.10.1 Seeding Methods

Alfalfa is grown alone for seed, market hay, or alfalfa meal in many states. It is also seeded in mixtures with grasses for pasture, silage, or hay, especially in the humid areas.

It is a common practice to seed alfalfa in close rows at a depth of ¼ to 1½ inches (0.6 to 4 cm) on heavy soils and slightly deeper on light soils.[41] Alfalfa is seldom grown in cultivated rows, except strictly for seed production. It is not an economically successful hay crop where the climate is so dry that production in cultivated rows is essential for satisfactory growth.

No-till seeding is possible with the proper seeding equipment. Seeding depth is more shallow, about ¼ to ¾ inch (0.6 to 2 cm).[42] Herbicides such as glyphosate or paraquat are used to kill existing vegetation before seeding. The application of a soil insecticide kills existing pests such as grub worms and wireworms.

22.10.2 Date of Seeding

In the northern half of the country, alfalfa is seeded in the early spring with a companion crop where rainfall is abundant or irrigation water is available. Spring seeding begins when the average night temperature remains above freezing.[43] Sometimes it is sown in August or early September. In the Southeast, the most favorable time ranges from August 15 in the latitude of Washington, DC, to October or November along the Gulf Coast. In the southern Great Plains, as well as in the southern parts in the Corn Belt, late summer seeding or early fall seeding is practiced. October is the best month for seeding in the irrigated areas of the Southwest, although good stands are obtained at other times from August to April 15.[1]

22.10.3 Rate of Seeding

The general rate of seeding east of the Appalachian Mountains has been 15 to 20 pounds per acre (17 to 22 kg/ha) because of frequent difficulties in establishing stands. Most Corn Belt experiment stations advise around 10 pounds per acre (11 kg/ha), which should be ample to produce satisfactory stands on well-prepared seedbeds (Table 22.2). Seeding rates of about 12 pounds per acre (13 kg/ha) with an occasional 20-pound (22 kg) rate,[44] are sown under irrigation in the West. For seed production under irrigation, the usual rate is about ½ to 1½ pounds per acre (0.6 to 1.7 kg/ha) in cultivated rows 24 to 42 inches (61 to 107 cm) apart.[45] Alfalfa is seeded in mixtures with bromegrass, orchardgrass, or intermediate wheatgrass for forage at 6 to 8 pounds per acre (8 to 10 kg/ha).

22.10.4 Weed Control

Few weeds can compete with a dense, rigorous stand of alfalfa that is cut several times during the summer. Therefore, controlling weeds during establishment is most important. As stands age and become thinner, weeds can become a problem, especially winter annuals. Annual broadleaf weeds in alfalfa or alfalfa-grass mixtures can be partly controlled by applying herbicides immediately after harvesting the first crop. Herbicides applied in winter when alfalfa is dormant control many annual weeds, including grasses. Preplant herbicides applied when establishing new stands of alfalfa are benfluralin, trifluralin, and EPTC. Postemergent herbicides on new stands are imazethapyr, imazamox, sethoxydim, clethodim, and bromozynil. On stands one year old or older, some herbicides can be applied when the alfalfa is dormant. They include glyphosate, paraquat, diuron, metribuzin, turbacil, and hexazinone.[46]

TABLE 22.2 Seeding Rates (Pure Live Seed) for Alfalfa Hay, Alone or in Mixtures

Hay Crop	Dryland (lbs per acre)	Irrigated (lbs per acre)
Alfalfa alone	10	12
Alfalfa + oats	10+20	12+20
Alfalfa + orchardgrass + smooth brome	8+2+3	8+4+4
Alfalfa + intermediate wheatgrass	6+8	
Alfalfa + smooth brome	8+4	

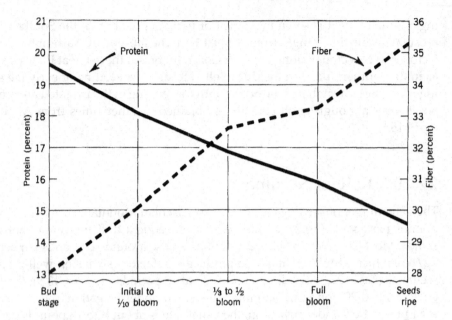

FIGURE 22.7
Alfalfa protein and fiber content at different growth stages.

22.10.5 Harvesting

Alfalfa produces one to two cuttings a season under semiarid conditions, two to four cuttings in the northern states, three to five in the Central and South Central states, and seven to ten cuttings of hay on irrigated land in southern California and Arizona. In some regions, the fields can be pastured during July and August when growth is retarded by hot weather, or in late fall after alfalfa plants go dormant. Cutting too frequently reduces root reserves.

As a crop of alfalfa matures, yield increases until approximately the full-bloom stage, and the percentages of protein and carotene decrease. Protein and carotene reach a maximum at approximately the early-bloom stage.[25] The lignin and crude fiber content increases and the digestible nutrients decrease as the plants approach maturity.[47] Harvesting alfalfa during the period from one-tenth to one-half bloom is a wise practice, taking into account the yields of hay and feed constituents, quality of hay, and permanency of stand.[48] Digestibility is highest at pre-bud stage (67 percent) and declines to 56 percent at maturity.[49] Early to mid-bloom harvest gives the best combination of quality and yield. A modification of this practice to satisfy local conditions can often prove desirable but frequent cutting in more immature stages should be avoided (Figure 22.7).[8] In practice, most alfalfa-grass mixtures for hay may be cut as though they were pure alfalfa.[25]

■ 22.11 ALFALFA QUALITY

Quality of alfalfa hay is related to color, leafiness, fineness of stems, and freedom from foreign material. Protein content is closely correlated with leafiness, while vitamin-A content is closely associated with green color. These factors are definitely involved in the curing process. The leaves, which contain about 70 percent of the total protein in the plant, are easily lost by shattering when the hay is handled improperly. Extensive research[50, 51, 52, 53] indicates that crimping and partial swath curing to hasten the rate of drying, followed by windrowing before the leaves dry sufficiently to cause shattering and prompt baling or storage, appears to be the best farm practice.[44] Alfalfa hay quality is discussed in more detail in Chapter 8.

■ 22.12 DEHYDRATED ALFALFA

Alfalfa dehydration began in the early 1930s. The alfalfa was cut, chopped, and loaded into a dump truck or trailer with a self-propelled or tractor-drawn harvester. The chopped forage was hauled immediately to a nearby central plant where it was dried quickly in a rotary-drum dehydrator that was usually heated with gas or fuel oil. Very little fresh alfalfa is dehydrated anymore due to the cost of drying high-moisture material. Today, most pelleted or cubed alfalfa is made from partly field-dried alfalfa. Drying is completed at the mill. The dried product is ground to a meal in a hammer mill. Much of the meal is pelleted or cubed immediately while the remainder is bagged. The pellets can be stored, shipped, and handled in bulk. Some of the pellets are later ground for mixing with other processed feeds.

Dehydrated alfalfa pellets contain about 17 percent protein and 100,000 International Units (IU) of carotene per pound. The xanthophyll content is reduced by dehydration. The alfalfa is cut in the early-bloom stage when the protein and vitamin contents are still high. The carotene is often preserved by storage with inert gasses in sealed tanks, or by the addition of vegetable oils or antioxidants.[54]

■ 22.13 ALFALFA AS A PASTURE CROP

Despite the hazard from bloating, alfalfa, in mixture with grasses, has become increasingly important for cattle pasture, particularly in the North Central and northeastern states. It is more productive than the usual permanent grass pastures, especially during hot, dry periods. Summer pasturing of alfalfa is a common practice in the hot irrigated area of southern Arizona. In Michigan,[55] alfalfa that is continuously pastured has been as productive as that pastured after a first cutting taken for hay each year. Pasturing the first growth and harvesting the second growth for hay increased the number of annual weeds. Heavy grazing in September proved to be injurious to stands. Alfalfa is an excellent pasture for swine when supplemented with grain.

Alfalfa is less likely to cause bloat when sown in mixtures with adapted grasses.[55] Consequently, alfalfa usually comprises not more than 50 percent of the pastured forage. Bloating in ruminants grazing on alfalfa, clover, and most other forage legumes is caused by the occurrence of a saponin or chloroplastic protein in the plants, particularly in young growth. Exceptions are birdsfoot trefoil, sainfoin, and crownvetch. Bloating is less frequent if the animals graze the legume only briefly at first and are prefed before pasturing, or are given supplementary hay or straw in the pasture. Salt-molasses blocks containing 30 grams of poloxalene (a surfactant) per pound of block placed in the pasture to provide salt usually prevents severe bloat.[56] Fatal bloat is caused by excessive intrarumenal physical gas pressure, which prevents diaphragmatic contractions, resulting in no respiration followed by heart failure.

Alfalfa, clover, and other legumes also contain coumestrol, an estrogen, found particularly in diseased, old, spoiled, or ensiled plants, which may cause infertility in sheep and other animals.[57] Saponins also inhibit growth in chicks that are fed large quantities of alfalfa meal. Certain alfalfa varieties contain less saponin than others.

▨ 22.14 SEED PRODUCTION

About 80 million pounds of alfalfa seed are produced each year in the United States. Alfalfa seed production is most successful where the climate is relatively dry, as in the semiarid and irrigated regions of the West.[41, 45] Yields of alfalfa seed west of the Rocky Mountains average several times those in Midwestern states. California, Idaho, Oregon, Washington, and Montana produce 85 percent of the total. California leads in alfalfa seed production, mostly for shipment to other states. The most consistent seed production occurs when the crop makes a steady, continued growth. Abundant moisture results in increased vegetative growth unfavorable to seed production.[58] The insects important in seed production slow down their activity when the weather is cloudy and stop entirely when it is cold and rainy.[8] As a rule, when seed production is incidental to hay production, growers leave the cutting for seed that matures during the hottest and driest part of the summer.

When alfalfa is grown specifically for seed, the highest average seed yields are obtained where the crop is planted in rows, under either irrigated or dryland conditions. Thin stands increases nectar secretion, bee visitation per flower, and the percentage of cross-pollinated flowers that set seed. A growth regulator sprayed on alfalfa plants increases the number of plant branches and consequent seed yield.[59] Cultivation between the rows suppresses weeds and maintains thin stands. Herbicides provide weed control.

Alfalfa is generally cut for seed when about two-thirds of the pods are brown or black. The crop is harvested with a windrower. Then it is threshed with a combine equipped with a pick-up attachment. Alfalfa seed can be combined directly where the crop has matured uniformly. Often, the field is sprayed with a desiccant to dry the crop just before combining. A desiccant is applied when practically all the pods have ripened.

High-quality alfalfa seed will have a germination rate of at least 85 percent and not more than 15 percent hard seed.

▨ 22.15 DISEASES

Among the most potentially destructive alfalfa diseases in the United States are bacterial wilt, leaf spot, yellow leaf blotch, blackstem disease, crown wart, mildew, rust, sclerotinia stem rot, and fusarium root rots.[26, 60, 61] Other diseases include witches' broom, caused by a virus that is transmitted by leafhoppers; alfalfa dwarf, a virus disease in California transmitted by leafhoppers and spittlebugs;[62] and numerous other virus diseases.[63]

Nematodes also attack alfalfa.[64] Most alfalfa varieties are resistant to several diseases, including bacterial wilt, leaf spot, fusarium root rots, mosaic, downy mildew, alfalfa dwarf, stem rot, stem nematodes, spotted alfalfa aphid, and pea aphids.[21, 40]

22.15.1 Bacterial Wilt

Bacterial wilt (*Clavibacter michiganensis* subsp. *insidiosus*) is the most serious alfalfa disease in the United States being found throughout the country.[60] In Nebraska, it occurs wherever moisture conditions are favorable for a rapid, vigorous growth of the crop.[65] It is less prevalent in dry and hot areas.

The most conspicuous symptom of bacterial wilt, in addition to reduced stands, is a dwarfing of the infected plants. A bunch-growth effect is produced by an excessive number of shortened stems. The leaves are small and yellowish and appear curled. Actual wilting occurs only in the advanced stages of the disease, especially during hot weather. A yellow-to-brown ring is observed under the bark when the roots are cut crosswise.

The disease is spread from infected to healthy plants by mowers, drainage water, and irrigation water. Diseased plants are more susceptible to winter injury than are healthy plants of the same variety.[66] Infection enters the plant through wounds. Longevity of infected plants varies, but most that show conspicuous symptoms of the disease die in the second year after infection. The main control measure is the use of resistant varieties. Nearly all varieties available today have at least some resistance to bacterial wilt.[8, 21, 67]

22.15.2 Anthracnose

Anthracnose, caused by the fungus *Colletotrichum trifolii*, causes diamond-shaped lesions on the stems. Lesions are straw-colored in the center. Subsequent wilting and death of the upper part of the stem gives rise to a curving at the tip called "shepherd's crook." The fungus overwinters on plant debris in the field, which provides new inoculum in the spring. Warm, wet weather favors the disease. It can spread by spores carried in splashing rain or wind and by equipment entering and leaving the field. The primary control is the use of resistant varieties.[54, 68]

22.15.3 Verticillium Wilt

Verticillium wilt is a serious disease of alfalfa. It is caused by the fungus *Verticillium albo-atrum*. The disease was first discovered in Europe in 1918, but it was not known in the United States until 1976. Symptoms are a v-shaped yellowing of leaflet tips and rolling of leaflets along their margins. The disease will progress until all leaves are dead, but the stem is still green. Spores can be carried by wind or flying insects. Harvesting equipment can also spread the disease. The disease is more of a problem on older stands. Resistant varieties are the most important method of control.[69]

22.15.4 Fusarium Wilt

Fusarium wilt, caused by the fungus *Fusarium oxysporum f.* sp. *medicaginis*, causes reddish brown discoloration inside the root and stunting of the plant. It is more prevalent in older stands. It can be more severe in the presence of root knot nematode because the damage by nematodes provides easy entry for the disease. Crowns can show a reddish-brown discoloration. Resistant varieties are available. Resistance to root knot nematode helps break the disease complex.

22.15.5 Phytophthora Root Rot

Phytophthora root rot is caused by the fungus *Phytophthora megasperma f.* sp. *medicaginis*. The disease causes the interior of the roots to rot. Plants will be stunted and die. It is more of a problem in wetter areas of the field. Resistant varieties are available. Cultural practices that improve drainage also help.[70]

22.15.6 Crown and Root Rots

Several fungi, including *Pythium, Phoma, Rhizoctonia,* and *Stagonospora,* can cause rots in crown and roots of alfalfa. Root damage is similar to Phytophthora and Fusarium with a red or reddish discoloration inside the root. Symptoms on the crowns depend on the pathogen. Resistant varieties are the primary method of control.[70, 71]

22.15.7 Leaf Diseases

Leaf spot *(Pseudopeziza medicaginis)* is one of the most widespread alfalfa fungus diseases, being most prevalent in the humid regions in cool, wet weather. It causes small circular dark-brown spots to appear on the leaves, which may cause them to drop. To avoid loss of leaves, the crop should be cut when leaf-spot infection becomes severe. Resistant varieties are available.

Yellow leaf blotch *(Pseudopeziza jonesii),* common in all large alfalfa-growing regions, also attacks the leaves. The disease is characterized by long yellow blotches sprinkled with minute brown dots. Where blotch is serious, the crop should be cut promptly.

22.15.8 Other Diseases

Crown wart *(Physoderma alfalfae)* occurs in greatest abundance in California, being characterized by the appearance of galls on the crown at the base of the stems.[1] Affected plants are seldom killed outright but produce low yields. Blackstem *(Phoma medicaginis)* is most severe in the Southeast and West. It is most serious after winters with little or no snow and during long, wet springs. The disease is characterized by large, irregular, brownish or blackish lesions that occur on the leaves and petioles as well as on the stems, causing serious loss of leaves unless harvested properly. Bacterial stem blight *(Pseudomonas syringae* subsp. *syringae)* causes considerable damage in some sections of the United States.

22.15.9 Stem Nematode

The stem nematode *(Ditylenchus dipsaci)* occurs on alfalfa in North America, South America, and Europe. It has been a serious pest in Nevada and other western states. This colorless eelworm is about 1/20 inch (1.3 mm) long. It infests seedlings at the cotyledonary node, but it also infests older plants in the crown, young buds, stem bases, and sometimes the stems. Infestation results in decay.

▨ 22.16 INSECT PESTS

The spotted alfalfa aphid, alfalfa weevil, potato leafhopper, Lygus bug, grasshopper, alfalfa caterpillar, three-cornered alfalfa hopper, and alfalfa seed chalcid are among the most destructive insect pests of alfalfa.

22.16.1 Spotted Alfalfa Aphid

The spotted alfalfa aphid *(Therioaphis maculata)* was first observed in the United States in New Mexico in 1954. It had spread over much of the central and southern United States by 1962. Parthenogenetic adults overwinter in the South. In the fall, sexual forms deposit eggs that survive the winter in the Central States. Many productive varieties are resistant to the spotted alfalfa aphid.[72] Insecticides also are recommended for control.[27]

22.16.2 Alfalfa Weevil

The alfalfa weevil *(Hypera postica)*, introduced from Europe and Africa, is found throughout the United States. Alfalfa weevil was the first of its kind identified as a pest in the United States; it was discovered in 1904 near Salt Lake City, Utah. By 1960, this pest had spread to New Mexico, Kansas, and all states to the north. In 1951, an eastern strain of the weevil was detected near Baltimore, Maryland, and by 1970 had spread throughout the northeastern United States and was found in Minnesota, Iowa, Kansas, and south to Texas. The Egyptian alfalfa weevil *(Hypera brunneipennis)* was first discovered near Yuma, Arizona, in 1939. The alfalfa weevil is now reported in all 48 of the continental United States. Tolerant alfalfa varieties are available.

Most damage is done by the larvae that feed on plant tips, new leaves as they open, and flower foliage. All but the main veins may be consumed causing damaged fields to appear whitish. Insecticides are available for weevil control in addition to tolerant varieties.

Early first harvest, which is conducive to high yield and quality, increases larval feeding on crown buds and small regrowth.[73] While a light infestation at this stage can be successfully controlled, the danger lies when the first growth is harvested early and the larval feeding is shifted to the regrowth of the second crop. Following late harvest, weevil adults feed on new buds, which delays or even stops new growth and weakens plants, making them more susceptible to winterkill.

22.16.3 Potato Leafhopper

The potato leafhopper (*Empoasca fabae*) causes injury to alfalfa in many locations.[74] It punctures the leaves and causes a yellowing of the plants. Potato leafhopper is usually not a problem with first-cut alfalfa. When the first crop is infested, cutting near the full-bloom stage can help with control. Potato leafhopper eggs attached to the plants of the first crop fail to develop on the cured hay, and the young insects are deprived of succulent food. The pest can be very damaging to new seedlings.

Some genetic resistance is available. Alfalfa can be sprayed with insecticides.

22.16.4 Alfalfa Seed Chalcid

The alfalfa seed chalcid (*Bruchophagus roddi*) causes extensive damage to the alfalfa seed crop. The wasp-like adult chalcid deposits a single egg in each partly developed alfalfa seed. The resulting hatched larva devours the interior of the seed. Several generations may develop in one season. Late-developing larvae hibernate over the winter inside dry seeds that are harvested or in those produced by plants along the ditches and borders of the field. Control measures include destruction of seed-bearing alfalfa and burclover plants along the irrigation-ditch banks and field borders during the winter and harvesting the crop for hay when heavy infestation threatens. Damage is reduced if the same cutting is saved for the seed crop throughout a locality so that the breeding period for the insect does not extend over the entire season. A heavy cleaning of the seed, which removes the lighter infested seeds for prompt destruction, prevents the direct spread of the pest to new fields. Certain strains of alfalfa are resistant to the chalcid.[75]

22.16.5 Other Pests

Three-cornered alfalfa hopper (*Spissistilus festinus*) adults and late-stage (instar) nymphs "girdle" alfalfa stems causing stems to be brittle and plants to break at that site and fall over. Most of the damage caused by the three-cornered alfalfa hopper occurs south of latitude 36° in the United States. Cleaning up hibernation quarters, such as weeds, brush, grass bunches, and rubbish, keeps down the insect population.

Lygus bug (*Lygus* species) causes serious damage to alfalfa seed production.[19, 20] The bugs attack young alfalfa buds, floral parts, and immature seeds. Individual flowers are shed soon after injury by the Lygus bug. The injured seeds turn brown and in many cases shrivel and become papery. The application of insecticides during the early-bud stage protects the seed crop to a considerable extent.

The beautiful yellow butterflies that flit about the alfalfa flowers in nearly all of the alfalfa fields in the western four-fifths of the United States are the adults of the destructive alfalfa caterpillar *(Colias eurytheme)*. They cause the most damage in the southwestern states. The larvae devour the leaves and buds of the alfalfa plants. This pest overwinters in the pupal stage on alfalfa stems, mostly those of alfalfa standing along the ditch banks and field borders. The main control measure is frequent irrigation to maintain a high humidity in the field, which promotes the development of a disease in the caterpillars. Insecticides can be applied as sprays to control the alfalfa caterpillar, spittlebugs, and other insects.

Dodder *(Cuscuta* species) is a parasitic twining weed that attacks alfalfa, lespedeza, and other crops. Dodder seed contamination makes commercial alfalfa seed undesirable or unmarketable. Control methods include flaming, cutting and burning, and tillage of spots of infested alfalfa, and also the use of suitable herbicides.[76]

Grasshoppers can cause extensive damage to alfalfa. They are controlled with insecticides.

REFERENCES

1. Westover, H. L. "Growing alfalfa," *USDA Farmers Bull.* 1722 (rev.), 1941.

2. Hendry, G. W. "Alfalfa in history," *J. Am. Soc. Agron.* 15(1923):171–176.

3. Wheeler, W. A. "Beginnings of hardy alfalfa in North American" Published in *Seed World* in 7 installments in 1950. Reprinted in pamphlet form by Northrup, King, Minneapolis, Minn., 1951, pp. 1–31.

4. Clement, W. M., Jr. "Chromosome numbers and taxonomic relationships in *Medicago,*" *Crop Sci.* 2(1962):25–28.

5. Kiesselbach, T. A., A. Anderson, and J. C. Russell. "Subsoil moisture and crop sequence in relation to alfalfa production," *J. Am. Soc. Agron.* 26(1934):422–442.

6. Carlson, F. A. "The effect of soil structure on the character of alfalfa root systems," *J. Am. Soc. Agron.* 17(1925):336–345.

7. Bohart, G. E. "Insect pollination of forage legumes," *Bee World* 41(1960):5764, 85–97.

8. Tysdal, H. M. "Is tripping necessary for seedsetting in alfalfa?" *J. Am. Soc. Agron.* 32(1940):570–585.

9. Waldron, L. R. "Cross-pollination in alfalfa," *J. Am. Soc. Agron.* 11(1919):259–266.

10. Tysdal, H. M. "Influence of tripping, soil moisture, plant spacing, and lodging on alfalfa seed production," *J. Am. Soc. Agron.* 38(1946):515–535.

11. Cooper, D. C., and R. A. Brink. "Partial self-incompatibility and the collapse of fertile ovules as factors affecting seed formation in alfalfa," *J. Agr. Res.* 60(1940):453–472.

12. Armstrong, J. M., and W. J. White. "Factors influencing seed-setting in alfalfa," *J. Agr. Sci.* 25(1935):161–179.

13. Medler, I. T., and J. F. Lussenhop, "Leafcutter bees of Wisconsin," *WI Agr. Exp. Sta. Rsh. Bull.* 274, 1968, pp. 1–40.

14. McGregor, S. E., and others. "Beekeeping in the United States," *USDA Agr. Handle.* 335, (rev.), 1971, pp. 1–147.

15. Kaufeld, N. M., and E. L. Sorenson. "Interrelations of honeybee preference of alfalfa clones and flower color," *KS Agr. Exp. Sta. Rsh. Publ.* 163, 1971, pp. 1–14.

16. Heinrichs, D. H. "Alfalfa in Canada," *Can. Dept. Agr. Pub.* 1377, 1968, pp. 1–28.

17. Baird, C. R., D. F. Mayer, and R. M. Bitner. "Pollinators," in *Alfalfa Seed Production and Pest Management. Western Regional Ext, Publ.* 12, 1991.

18. Stephen, W. P. "Artificial beds for alkali bee propagation," *OR Agr. Exp. Sta. Bull.* 598, 1965, pp. 1–20.

19. Carlson, J. W. "Lygus bug damage to alfalfa in relation to seed production," *J. Agr. Res.* 61(1940):791–815.

20. Stitt, L. L. "Three species of the genus Lygus and their relation to alfalfa seed production in southern Arizona and California," *USDA Tech. Bull.* 741, 1940.

21. Anonymous. "Varieties of alfalfa," *USDA Farmer's Bull.* 2231, 1968, pp. 1–12.

22. Anonymous. "Varieties and areas of adaptation," *USDA Leaflet* 507, 1962.

23. Graumann, H. O. "Creeping alfalfa," *Crops and Soils* 10, 4(1958).

24. Heinrichs, D. H. "Creeping alfalfa," *Adv. Agron.* 15(1963):317–337.

25. Willard, C. J. "Management of alfalfa meadows after seeding," *Advances in Agronomy,* vol. 3. New York: Academic Press, 1951, pp. 93–112.

26. Froshheiser, F. I. "The worst diseases of crops: Alfalfa," *Crops and Soils* 25, 1(1972):18–21.

27. Howe, W. L., and others. "Studies of the mechanisms and sources of spotted alfalfa aphid resistance in Ranger alfalfa," *NE Agr. Exp. Sta. Res. Bull.* 210, 1963.

28. Weihing, R. M. "Field germination of alfalfa seed submitted for registration in Colorado and varying in hard seed content," *J. Am. Soc. Agron.* 32(1940):944–949.

29. Steinmetz, F. H. "Winterhardiness in alfalfa varieties," *MN Agr. Exp. Sta. Tech. Bull.* 38, 1926.

30. Haagenson, S. M., and others. "Autumn defoliation effects on alfalfa winter survival, root physiology, and gene expression." *Crop Sci.* 43(2003):1340–1348.

31. Peltier, C. L., and H. M. Tysdal. "Hardiness studies with 2-year-old alfalfa plants," *J. Agr. Res.* 43(1931):931–955.

32. Albert, W. B. "Studies on the growth of alfalfa and some perennial grasses," *J. Am. Soc. Agron.* 19(1927):624–654.

33. Leukel, W. A. "Deposition and utilization of reserve foods in alfalfa plants," *J. Am. Soc. Agron.* 19(1921):596–623.

34. Willard, C. J. "Root reserves of alfalfa with special reference to time of cutting and yield," *J. Am. Soc. Agron.* 22(1930):595–602.

35. Graber, L. F., and others. "Organic food reserves in relation to the growth of alfalfa and other perennial herbaceous plants," *WI Agr. Exp. Sta. Res. Bull.* 80, 1927.

36. Grandfield, C. O. "The effect of the time of cutting and of winter protection on the reduction of stands in Kansas Common, Grimm, and Turkistan alfalfas," *J. Am. Soc. Agron.* 26(1934):179–188.

37. Silkett, V. W., C. R. Megee, and H. C. Rather. "The effect of late summer and early fall cutting on crown bud formation and winter-hardiness of alfalfa," *J. Am. Soc. Agron.* 29(1931):53–62.

38. Schmehl, W. R., and S. D. Romsdal. "Materials and method of application of phosphate for alfalfa in Colorado," *CO Agr. Exp. Sta. Tech. Bull.* 74, 1963.

39. Duley, F. L. "The effect of alfalfa on soil moisture," *J. Am. Soc. Agron.* 21(1929):224–231.

40. Kiesselbach, T. A., J. C. Russell, and A. Anderson. "The significance of subsoil moisture in alfalfa production," *J. Am. Soc. Agron.* 21(1929):241–268.

41. Pedersen, M. W., and others. "Growing alfalfa for seed," *UT Agr Exp. Sta. Cir.* 135, 1955.

42. Shroyer, J. P., and others. "No-till alfalfa establishment." *KS St. Univ. Coop. Ext. Serv.* L-875, 2003.

43. Witt, M. D., and C. R. Thompson. "Dormant-season seeding of alfalfa." *KS Agri. Exp. Sta.* SRL 117, 1997.

44. Dennis, R. E., and others. "Alfalfa for forage production in Arizona," *AZ Agr. Exp Sta. Bull.* A-16 (rev.), 1966, pp. 1–40.

45. Pedersen, M. W., and others. "Cultural practices for alfalfa seed production," *UT Agr. Exp. Sta. Bull.* 408, 1959.

46. Wilson, R., and A. Martin. "Weed control in alfalfa." *NE Coop. Ext. Serv. NebGuide* G95–1254-A, 1995.

47. Jensen, E. H., and others. "Environmental effects on growth and quality of alfalfa," *NV Agr. Exp. Sta. Western Regional Rsh. Pub.* T9, 1967, pp. 1–36.

48. Cash, D., and H. F. Bowman. "Alfalfa hay quality testing," *MT St. Univ. Ext. Serv.* MT9302, 1993.

49. Baldridge, D. E., H. F. Bowman, and R. L. Ditterline. "Growing alfalfa for hay." *MT St. Univ. Ext. Serv.* MT8505, 1985.

50. Kiesselbach, T. A., and A. Anderson. "Alfalfa investigations," *NE Agr. Exp. Sta. Res. Bull.* 36, 1926.

51. Kiesselbach, T. A., and A. Anderson. "Curing alfalfa hay," *J. Am. Soc. Agron.* 19(1927):116–126.

52. Kiesselbach, T. A., and A. Anderson. "Quality of alfalfa hay in relation to curing practice," *USDA Tech. Bull.* 235, 1931.

53. Sheperd, J. B., and others. "Experiments in harvesting and preserving alfalfa for dairy cattle feed," *USDA Tech. Bull.* 1079, 1954.

54. Rhodes, L. H., and R. M. Sulc. "Alfalfa anthracnose." *OH St. Univ. Ext.* AC-15-96, 1996.

55. Rather, H. C., and Dorrance, A. B. "Pasturing alfalfa in Michigan," *J. Am. Soc. Agron.* 27(1935):57–65.

56. Foote, L. E., and others. "Controlling bloat in cattle grazing clover," *LA Agr. Exp. Sta. Bull.* 629, 1968, pp. 1–28.

57. Bickoff, E. M., and others. "Studies of the chemical and biological properties of coumestrol and related compounds," *USDA Tech. Bull.* 1408, 1969, pp. 1–95.

58. Tysdal, H. M. "Influence of light, temperature, and soil moisture on the hardening process in alfalfa." *J. Agr. Res.* 46(1933):483–515.

59. Hale, V. Q. "TIBA on alfalfa can help plants produce more seed," *Crops and Soils* 23, 8(1971):7.

60. Jones, F. R., and O. F. Smith. "Sources of healthier alfalfa," in *Plant Diseases*, USDA Yearbook, 1953, pp. 228–237.

61. Leath,. T., and others. "The fusarium root rot complex of selected forage legumes in the northeast," *PA Agr. Exp. Sta. Bull.* 777, 1971, pp. 1–66.

62. Tysdal, H. M., and H. L. Westover. "Alfalfa improvement," in *USDA Yearbook*, 1937, pp. 1122–1153.

63. Crill, P., and others. "Alfalfa mosaic: The disease and its virus incitant; A literature review," *WI Agr. Exp. Sta. Rsh. Bull.* 280, 1970, pp. 1–40.

64. Hunt, O. J., and R. N. Peaden. "Alfalfa nematodes," *Crops and Soils* 24, 6(1972):6–7.

65. Pettier, G. L., and J. H. Jensen. "Alfalfa wilt in Nebraska," *NE Agr. Exp. Sta. Bull.* 240, 1930.

66. Jones, F. R. "Development of the bacteria causing wilt in the alfalfa plant as influenced by growth and winter injury," *J. Agr. Res.* 37(1928):545–569.

67. Elling, L. J., and F. I. Frosheiser. "Reaction of twenty-two alfalfa varieties to bacterial wilt," *Agron. J.* 52(1960):241–242.

68. Watkins, J. E., and B. Anderson. "Alfalfa anthracnose," *NE Coop. Ext. Serv. NebGuide* G89–931-A, 1989.

69. Rhodes, L. H., and R. M. Sulc. "Verticillium wilt of alfalfa." *OH St. Univ. Ext.* AC-43-96, 1996.

70. Watkins, J. E., F. A. Gray, and B. Anderson. "Alfalfa crown and root rots and stand longevity," *NE Coop. Ext. Serv. NebGuide* G89–912-A, 1989.

71. Rhodes, L. H., and R. M. Sulc. "Rhizoctonia root, stem, and crown rot of alfalfa." *OH St. Univ. Ext.* AC-42–96, 1996.

72. Harvey, T. L., and others. "The development and performance of Cody alfalfa, a spotted-alfalfa-aphid resistant variety," *KS Agr. Exp. Sta. Tech. Bull.* 114, 1960.

73. Danielson, S., T. Hunt, and K. Jarvi. "Managing the alfalfa weevil," *NE Coop. Ext. Serv. NebGuide* G94–1208-A, 1994.

74. Danielson, S., and K. Jarvi. "Potato leafhopper management in alfalfa," *NE Coop. Ext. Serv. NebGuide* G93–1136-A, 1993.

75. Hanson, C. H. "Report of the eighteenth alfalfa improvement conference," *USDA* CR-71–62, 1962.

76. Dawson, J. H., and others. "Controlling dodder in alfalfa," *USDA Farmer's Bull.* 2211, 1969, pp. 1–16.

Sweetclover

23.1 HISTORY OF SWEETCLOVER CULTURE

Sweetclover was used as a green manure crop as well as a honey plant in the Mediterranean region 2,000 years ago. Asia Minor appears to be its native habitat. It was first observed in the United States as a wild plant in Virginia about 1739, but its real value was not recognized for many years. As late as 1910, sweetclover was legislated as a weed in some states. Formerly, it was often called Bokhara clover.

Sweetclover is grown mostly in the North Central states for green manure, pasture, seed, silage, hay, and as food for honey bees. Production has declined since 1940 as it is being replaced by alfalfa.

23.2 ADAPTATION

Sweetclover is grown where the annual precipitation, properly distributed, is 17 inches (432 mm) or more. It is grown under irrigation in drier areas. The biennial species tolerate drought once they have been established, but a good supply of moisture and cool temperatures are essential for germination and early seedling growth.[1] Although the plant may persist under dry conditions, little vegetative growth is made. It is more drought resistant than alfalfa or any of the true clovers in the Great Plains region, but it is not a dependable crop under dryland conditions in the central or the southern Great Plains unless moisture and temperature conditions are favorable for some time after seeding. Its consistent survival in Canada, Montana, Michigan, and other northern regions proves that some varieties are very winter hardy.[2,3]

Sweetclover will grow on a wide range of soils provided they are not acid. The soil reaction should be pH 6.5 or higher. It will grow on infertile eroded areas, poorly drained soil, heavy muck and peat soils high in organic matter, or clay soils with almost no organic matter. Most sweetclover failures in humid sections are due to lack of lime in the soil. Sweetclover grows on soils high in salts, even on alkali seepage lands[4] where cereal crops cannot be grown. Sweetcolver is considered a weed in waste places and some cropped fields due to its hard seeds and the tendency of its seeds to shatter. Sweetclover in alfalfa fields

Annual sweetclover
Biennial sweetclover
Coumarin
Dicoumarol
White sweetclover
Yellow sweetclover

harvested for seed contaminates the crop because sweetclover seeds cannot be screened out of alfalfa.

■ 23.3 BOTANICAL DESCRIPTION

Sweetclovers are legumes classified in the genus *Melilotus*. They are tall, erect, annual, or biennial herbs.

The plants grow from 2 to 6 feet (60 to 180 cm) or more in height. The stems are generally well branched, coarse, and succulent. During the first year, only one central stem per plant develops in the biennial sweetclovers, except in certain new strains, which have several. In the second year, numerous branches develop from crown buds formed the previous fall. New growth after cutting arises from buds in leaf axils rather than from crown buds. The leaves are pinnately trifoliolate and possess pointed stipules. The middle leaflet is stalked or petiolated. The leaflets are finely toothed along the margins. Young sweetclover plants can be distinguished from those of alfalfa by the coumarin taste and usually by the presence of serrations (teeth) along the entire margin of the leaflet rather than on the tip only, as in alfalfa (Figure 23.1).

Sweetclover has a fleshy taproot that penetrates to a depth of 5 to 8 feet (150 to 240 cm) in the biennial types.[5] They can develop either a few side roots or many. The taproot of the annual types is relatively short and slender. In the biennial types, the taproot increases in weight slowly until the end of the growing season. The roots may double in weight after September 25, due to food storage.[6,7] During the second year, the roots are depleted rapidly of their stored food for the production of top growth (Table 23.1).

White or yellow sweetclover flowers are borne in short-pedicelled racemes that arise in the axils of the leaves (Figure 23.2). As many as 100 flowers, each with ten stamens, develop on a single large raceme.[8] The inflorescence is indeterminate. Flowering occurs during the first year in the annual forms, but usually not until the second year in biennial types. Some plants of the biennial types flower the first year

FIGURE 23.1

Sweetclover leaf on the left has serrations on the entire leaf margin. Alfalfa leaf on the right has serrations on upper one-third of the leaf margin. [Courtesy Richard Waldren]

TABLE 23.1 Yield and Composition of Tops and Roots of Sweetclover[a] at Successive Dates in the Seeding Year and the Year after Seeding

Date of Sampling	Air-Dry Yield per Acre		Nitrogen In		Nitrogen per Acre			Plants per Square Yard
	Tops (lb)	Roots (lb)	Tops (%)	Roots (%)	Tops (lb)	Roots (lb)	Total (lb)	
Development in the Seeding Year								
July 1	515	—	3.25	—	—	—	—	—
July 15	588	75	2.85	2.30	16	2	18	—
August 1	659	182	2.54	2.53	17	4	21	—
August 15	963	310	2.89	2.90	28	9	37	—
Sept. 1	1,431	577	2.88	3.02	41	18	59	—
Sept. 15	1,544	884	3.09	3.05	48	27	75	—
Oct. 1	1,881	1,273	2.87	3.12	54	40	94	—
Oct. 15	1,714	1,721	2.59	3.25	44	56	100	—
Nov. 1	1,616	2,115	2.38	3.50	39	74	113	—
Nov. 8–10	1,397	2,324	2.32	3.45	32	77	109	—
Development in the Year after Seeding								
April 1	420	2,130	4.35	4.33	19	92	111	125
April 15	690	1,750	4.21	4.15	29	73	102	125
May 1	1,930	1,360	3.90	3.57	75	49	114	117
May 15	3,360	1,280	3.27	2.90	110	37	147	102
June 1	4,940	1,200	2.70	2.38	133	29	162	92
June 15	6,030	1,110	2.35	2.14	142	24	166	82
July 1	7,380	880	1.95	1.94	144	17	161	66
July 15	7,990	790	1.77	1.72	141	14	155	50
August 1	7,290	760	1.54	1.64	112	13	115	40

[a]Sweetclover sown in early oat which was harvested for grain; sweetclover not clipped or pastured after oat harvest.

under favorable growing conditions. Long days hasten flowering and restrict root growth.[9] Biennial types flowered the first year in Palmer, Alaska, where the maximum day length was 19½ hours.

Seeds of sweetclover mature in small, ovoid pods that are indehiscent or finally two-valved. There is usually only one seed per pod, but as many as four can occur in some strains. The pods are dark brown when mature. Sweetclover sets seed readily, but it shatters easily when mature. The mature seeds of white sweetclover are golden yellow in color, but many of those of the yellow varieties are mottled. They are ovoid in shape but not laterally compressed at the ends like alfalfa seeds (Figure 22.5). A large percentage of the seeds are hard or impermeable. Selected strains have a high percentage of permeable seeds.[10]

23.3.1 Pollination

Under natural conditions, pollination of sweetclover is affected principally by honeybees, except for the species, varieties, and individual plants that are spontaneously self-fertilized.[11] Common white sweetclover appears to be highly self-fertile. Plants may set seed when insects are excluded.[12] Self-pollination takes

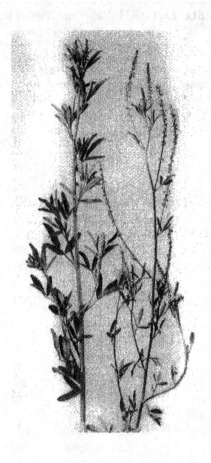

FIGURE 23.2

Stems and flowers of white sweetclover.

place when the pistils and stamens are the same length, but very little occurs when the pistil is longer than the stamens. Some natural crossing takes place under normal field conditions.[13]

Yellow sweetclover is largely, but not completely, self-sterile.[8, 10] It may be unable to set seed unless the flowers are visited by insects.[9]

White sweetclover has been crossed with *M. suaveolens* successfully, but the cross *M. alba* × *M. officinalis* failed to produce any viable seeds.[14] White sweetclover has been crossed with *M. dentata*. The plants from the hybrid seed contain very little chlorophyll. Consequently, they must be propagated by embryo culture, or else grafted on white sweetclover plants to enable them to produce flowers.[13, 15]

■ 23.4 SWEETCLOVER SPECIES

Common biennial white sweetclover (*Melilotus alba*) is an aggregate of various plant types, mostly tall-growing, high-branching, and stemmy with low leaf percentages. The pods are slightly netted. The commonly grown types of biennial yellow sweetclover (*M. officinalis*) are somewhat less upright in growth habit than the white-flowered types, blossom earlier, have finer stems, and tend to be more persistent under pasture conditions. They are usually less productive and mature earlier than the white-blossomed varieties.[10] Yellow varieties are somewhat less winter hardy than white ones in Canada.[12] A yellow-blossom sweetclover (*M. suaveolens*) has growth habits similar to common white. It is usually a biennial, but an annual form is known. A nonbitter sweetclover (*M. dentata*) includes both annual and biennial types that are practically coumarin-free.[16] Sourclover (*M. indica*), a winter annual yellow *Melilotus*, occurs mostly as a volunteer crop in grain fields, where the seed is obtained, in the South and Southwest. It is prevalent in California and Arizona, where it was first observed as an escaped introduction about 1830. The seeds are small, rough, and dark green (Figure 22.5). It is a green manure crop on irrigated lands of the Southwest and along the western Gulf Coast.

▪ 23.5 VARIETIES

The majority of the biennial white-blossom sweetclover is the common unnamed type, but several improved varieties are produced. Some Canadian varieties bloom too early to produce high yields under the shorter summer days of most of the United States[17] but are extremely winter hardy. Improved varieties of biennial yellow sweetclover are grown in the United States and Canada.

Annual white sweetclover is grown as a winter annual in the southern states or as a summer annual in the intermountain states.[17] Annual white sweetclover can be sown in the fall and turned under in the spring in time for planting cotton in central and southern Texas. It can be sown in the late winter or early spring for use as pasture in that region. In the Pacific Northwest, it can be sown in the fall and then turned under in early spring in preparation for potato or sugarbeet. It is sown in the spring in the central and northern states for green manure, pasture, and bee pasture.[18] It grows 2 to 5 feet (60 to 150 cm) high and has numerous strong stems with few leaves. The roots are small and without crown buds. It blooms from about July 15 to September 15. The annual types produce larger yields of tops but lower yields of roots than do the biennials during the first season of growth.[19] Biennial sweetclover produces about 25 percent more total dry matter, and the protein content of the roots is more than double that of the annual variety under Minnesota conditions. The protein content of the different types of biennial sweetclover is very similar.[20]

▪ 23.6 FERTILIZERS

In Indiana,[21] an application of 2 tons per acre (4.5 MT/ha) of ground limestone was recommended for soils of pH 6.0 to 6.2; 3 tons per acre (6.75 MT/ha) for medium-acid soils (pH 5.5 to 5.9); and 5 tons per acre (11 MT/ha) for strongly acid soils (pH 5.0 to 5.5). About 300 pounds per acre (335 k/ha) of superphosphate for thin pastureland was recommended. About 300 pounds per acre (335 k/ha) of 0–20–10 fertilizer has been suggested for soils highly responsive to potash. In other states, much smaller applications of limestone have been satisfactory on acid soils.

▪ 23.7 ROTATIONS

One of the most common crop sequences that includes sweetclover in the northern Corn Belt is corn; small grain (usually oat or barley) seeded to sweetclover; and sweetclover for pasture, hay, or seed. Some grain farms use a shorter rotation of corn and small grain (barley, oat, or wheat) seeded to sweetclover. The sweetclover is plowed under in the spring for corn. This rotation has resulted in increases in corn yields of 30 to 70 percent as compared with the same two-year rotation without the sweetclover in experiments in Ohio. Small increases are also obtained in the small-grain crop.[22]

Where rainfall is adequate, sweetclover may be underseeded with winter wheat in the spring. The next year the sweetclover is grazed or harvested for seed, or both; the residue is plowed under in preparation for fall-sown wheat.

A rotation that is sometimes followed with irrigation is small grain seeded with sweetclover, sweetclover for pasture or silage, sugarbeet, and sugarbeet. Sweetclover is also used as a green manure crop in potato rotations.

Sweetclover can serve as an alternate host for the soybean cyst nematode and should not be included in a rotation designed to control this pest.[23]

▥ 23.8 SWEETCLOVER CULTURE

23.8.1 Seedbed Preparation

The seedbed, inoculation, and lime requirements for sweetclover are similar to those for alfalfa. In Oklahoma research,[24] light applications of finely pulverized limestone, mixed with either rock phosphate or superphosphate, and applied with a grain drill equipped with a fertilizer distributor resulted in excellent yields of sweetclover on land where a very poor growth occurred on unfertilized soil.

23.8.2 Seeding Methods

Most of the sweetclover in the Corn Belt is seeded with small grain as a companion crop. The seeding of sweetclover alone might be advantageous there, except for the growth of annual weeds, which develop freely in the unclipped crop. In Ohio, spring grains are more satisfactory as companion crops than winter grains.[7] Field peas were a better companion crop for sweetclover in the nonirrigated areas of the Pacific Northwest than any of the small grains.[4] Sweetclover should be seeded alone where moisture is a limiting factor, as in many parts of the Great Plains. In the Red River Valley of Minnesota, the maximum total weight of roots, tops, and stubble in October of the first year was obtained when sweetclover was sown alone in April.[18]

Because of the high percentage of hard seed, scarified seed is used most widely, especially for spring seeding. The sowing of unhulled and unscarified seed on winter wheat in January or February has been recommended[25] but is not generally successful.

Sweetclover is usually broadcast or drilled in close rows, except under extremely dry conditions or for seed production. A residue-free seedbed is best.[26] In Oklahoma, sweetclover seeded in 36- to 42-inch (90 to 110 cm) rows and cultivated to control weeds during the first season produced a yield of roots similar to that from seedings in rows 7 inches (18 cm) apart.[24] The wider rows were also advantageous in the western part of the state where the available moisture is relatively

TABLE 23.2 Effect of Seeding Depth of Sweetclover on the Emergence of Seedlings

Depth of Seeding	Days from Seeding to Emergence	Emergence
(inch)	(no.)	(%)
0.5	4	98
1.0	5	88
1.5	6	46
2.0	7	20
2.5	9	2
3.0	None emerged	0

low. However, it is not advisable to grow sweetclover where such methods are necessary. Seeding more than 1 inch (2.5 cm) deep may be responsible for some sweetclover failures[27] (Table 23.2).

Sweetclover is generally sown early in the spring at about the same time as, and usually along with, small grains. Sometimes it is seeded by broadcasting on winter wheat fields in February or March. In Ohio, summer seeding produced lower yields of nitrogen and organic matter for plowing under early in May of the second year than did spring seeding.[7] In Oklahoma and Kansas, sweetclover is frequently seeded in the fall.

When broadcast, the rate of seeding sweetclover ranges from 4 to 25 pounds per acre (4.5 to 28 kg/ha). The most popular broadcast rate is 10 to 15 pounds per acre (11 to 17 kg/ha) of scarified seed. When drilled, only 4 to 10 pounds per acre (4.5 to 11 kg/ha) is needed.[26, 28] For Michigan conditions, 15 pounds per acre (17 kg/ha) of scarified seed, 18 to 20 pounds per acre (20 to 22 kg/ha) of unscarified seed, or 25 pounds per acre (28 kg/ha) of unhulled seed per acre is recommended.[29] For the Pacific Northwest, 10 pounds per acre (11 kg/ha) is recommended for nonirrigated lands, and 20 pounds per acre (22 kg/ha) for irrigated lands.[30] Three to 6 pounds per acre (3.5 to 7 kg/ha) is recommended for 36- to 42-inch rows in Oklahoma,[24] and 10 to 15 pounds per acre (11 to 17 kg/ha) drilled.[31] In Florida, 12 to 15 pounds per acre (13 to 17 kg/ha) broadcast and 8 to 10 pounds per acre (9 to 11 kg/ha) drilled is recommended.[32] In an Ohio experiment, the final stand of sweetclover in the second year was about 30 plants per square yard (36 plants/m²), regardless of the original stand.[7] Little change occurred in the stand during the first year unless the stand was abnormally thick or the season unusually dry.

▨ 23.9 USES OF SWEETCLOVER

23.9.1 Pasture

Sweetclover is a valuable pasture crop, especially in the Great Plains where it is reported to carry more livestock per acre than other plants common to the area. In the Corn Belt, the average period of grazing for the second-year growth was reported as 111 days. The best grazing is from about May 1 to July 20.

When conditions are favorable, sweetclover can be pastured lightly for sixty to seventy-five days during the fall of the seeding year. Some top growth should be left for winter protection in the northern states. The second-year plants are generally grazed early enough in the spring to avoid coarse, woody growth. The danger of bloat is less than on alfalfa, red clover, or alsike clover.[17] When not overgrazed, yellow sweetclover often reseeds itself and thus is more desirable for sowing in permanent grass pastures. Cattle prefer to graze the grasses, which permits some yellow sweetclover plants to produce seed.[33] This seed production, together with the hard seeds that germinate after the first year, often maintains stands for several years.

23.9.2 Hay

Sweetclover hay production has declined because of damage from the sweetclover weevil, and also because of its recognized poor quality of hay. A good first-year growth makes excellent hay when it can be cured in bright fall weather. In the South, the first-year crop is cut for hay, but in the northern Great Plains the first-year crop seldom attains sufficient height. The coarse, juicy stems of the second-year crop are difficult to cure without a heavy loss of leaves, except in dry regions. Crushed stems cure more quickly. Improperly cured sweetclover hay (hay that is allowed to mold) is dangerous to feed because it may cause bleeding in animals. Sweetclover is unable to compete with alfalfa as a forage crop under irrigation.

Sweetclover has about the same feeding value as alfalfa when cut at the proper stage and cured into bright-green hay. When cut too late, the coarse, woody stems are largely rejected by livestock and the feeding value is low.

In West Virginia, cutting biennial sweetclover the first year resulted in a reduced hay yield in the second-year crop, although the total yield was greater when cut both years (Table 23.3).[34] Clipping reduces root development.[35]

Sweetclover should be cut for hay in the bud stage, or at least by the time the first blossoms open. Plants cut the second season at the prebud stage of growth make a more vigorous recovery than when cut later.[34] Delayed cutting until after the plants are in full bloom results in a coarse, woody hay of relatively poor quality.

Sweetclover should be cut at a height of 8 to 12 inches (20 to 30 cm) above the ground where a second crop is desired. Since the new growth arises from buds in

TABLE 23.3 Effect on Yield of Cutting Sweetclover the First Year of Growth

Date Cut	Average Height		Average Yield Dry Hay		
	May 25	June 30	1931	1932	Total
	(in.)	(in.)	(gm)	(gm)	(gm)
Not cut	27.4	49.2	—	1,912	1,912
August 1	23.2	44.6	919	1,571	2,490
August 20	20.8	45.2	1,121	1,542	2,663
September 10	20.4	44.6	1,190	1,385	2,575
September 30	19.4	43.0	1,095	1,370	2,465

the axils of the branches and leaves on the lower stem, the plants are destroyed when cut below these buds. In Utah, no second growth occurred on a 4-inch (10 cm) stubble, a 10 percent crop appeared on an 8-inch (20 cm) stubble, and a full crop on a 12-inch (30 cm) stubble.[36]

23.9.3 Silage

Sweetclover makes a good silage crop if properly processed. It is harvested at the 10 to 20 percent bloom stage with a conditioning windrower and left in the swath for a few hours to wilt. An optimum moisture content when chopping for silage is 65 percent.[26] The silage should be firmly packed in a tight silo and covered with another forage or plastic to prevent spoilage. As in the case of hay, spoiled sweetclover silage is a risky feed. The crop can be made into silage when the weather is too damp for curing hay.

23.9.4 Green Manure

Sweetclover makes an excellent green manure, especially in the humid regions.[35, 37] As is the case with other green manure crops, it has not resulted in any increase in average crop yields in dryland areas. It can, however, be effective in suppressing weeds.[28] Annual sweetclover can be grown where the crop is to be plowed under in the fall of the seeding year because of the difficulty in eradication of the biennial types at that time. The annual type can be grown for soil improvement in areas of Oklahoma where summer drought interferes with the normal development of the biennial varieties.[24] The annual type is also used in mild areas of Texas and the Pacific Northwest irrigated areas where it survives the winter after fall seeding. It can be plowed under in preparation for crops planted in late spring. However, the biennial type produces more total dry matter in the roots and tops as well as more nitrogen under the same conditions.[35]

The best time to plow under biennial sweetclover for soil improvement is in the spring. Under Ohio conditions, plowing between April 20 and May 10 secured 80 percent of the maximum amount of nitrogen accumulated during the season.[7] Rates of 100 to 140 pounds of carryover nitrogen per acre (112 to 157 kg/ha) are possible. This permits the planting of corn on the land.

23.9.5 Seed Production

Seed yields vary from 120 to 600 pounds per acre (134 to 672 kg/ha) and occasionally higher. The seed crop harvested the second season should be cut when three-fourths of the pods have turned brown or black. The seed pods shatter badly when mature, but this can be reduced by cutting the plants when they are damp from rain or dew. The seed crop may be combined directly and then dried artificially, or it may be windrowed and then threshed with a combine equipped with a pick-up attachment. The seed does not ripen uniformly. Consequently, the ripest seeds have shattered while others are still green and damp when the crop is harvested.

■ 23.10 COUMARIN IN SWEETCLOVER

Although newer varieties have been developed that have lower levels of coumarin, many other varieties still have the potential to cause problems. In older varieties, the leaves and flowers, as well as the seeds, stems, and roots, of sweetclover contain considerable quantities of *cis*-O-hydroxycinnamic acid. This converts into a lactone, called coumarin, which causes the bitter taste and low palatability of sweetclover. The maximum coumarin content occurs at the late-bud or early-flower stage, after which it decreases. The content is increased by higher temperatures and light intensity and longer photoperiods.[38] Coumarin is converted to dicoumarol in moldy or decayed sweetclover hay and silage.

Dicoumarol is responsible for the sweetclover disease of cattle.[27, 29] In this disease, blood fails to clot normally when the animals are wounded. Affected animals may bleed to death from minor wounds. The dicoumarol reduces the enzyme thrombin in the blood of the animals that eat the spoiled hay. Thrombin is essential to the clotting of the blood because it converts a soluble plasma protein, fibrinogen, into fibrin. Fibrin consists of an insoluble network of fibers that entrap cells and serum. The use of second-year sweetclover for hay is hazardous because of the possibility of poisoning. Apparently, bright hay free from mold is a safe feed.

The hay of nonbitter sweetclover (*M. dentata*) does not become toxic when spoiled in a similar manner.[39] The low coumarin character of *M. dentata* has been bred into biennial white sweetclover (*M. alba*) by crossing, followed by backcrossing during selection for low coumarin content along with good forage production.[2, 3, 15] Low coumarin content is inherited as a simple recessive character.[40]

Synthetic dicoumarol is used as an anticoagulant for treatment of blood clots in humans, as well as in the making of a rat poison called warfarin.

■ 23.11 DISEASES

Diseases affecting sweetclover are more serious in the humid regions than west of the Iowa-Nebraska boundary.

Blackstem disease caused by the fungus *Mycosphaerella lethalis* is characterized by blackened stems that become evident late in the spring on second-year stems.[41] The leaves, as well as the young stems, may be killed when injury is severe. The disease appears to be most serious when wet weather occurs during the first 6 to 8 inches (15 to 20 cm) of growth.

Other diseases of sweetclover include root rot, caused by *Phytophthora cactorum*, and gooseneck stem blight, caused by *Ascochyta caulicola*.

■ 23.12 INSECT PESTS

The most serious pest of the sweetclover crop is the sweetclover weevil, *Sitona cylindricollis*.[17, 42] The insect, a small, dark-gray snout beetle, feeds upon the leaves of the young seedlings. These insects either kill the plants in a short time or weaken them to such an extent that the plants die from unfavorable summer weather conditions. When the weevil feeds upon the second-year growth, the damage is relatively slight because of the rapid rate of growth of the shoots.

After overwintering, newly emerged adult sweetclover weevils migrate from old stands and wild plants to new seedings. Consequently, sweetclover should not be planted close to an old stand or road ditches and other areas with wild sweetclover when it can be avoided. Insecticides are effective against adult sweetclover weevils.

REFERENCES

1. Aamodt, O. S. "Climate and forage crops," in *Climate and Man*, USDA Yearbook, 1941, pp. 439–458.

2. Goplen, B. P. "Yukon sweetclover," *Can. J. Plant Sci.* 51, 2(1971):178–179.

3. Goplen, B. P. "Polara, a low coumarin cultivar of sweetclover," *Can. J. Plant Sci.* 51, 3(1971):249–251.

4. Hulbert, H. W. "Factors affecting the stand and yield of sweetclover," *J. Am. Soc. Agron.* 15(1923):81–87.

5. Weaver, J. E. "Root development in the grassland formation," *Carnegie Inst. Wash. Pub.* 292, 1920.

6. Smith, D., and L. F. Graber. "The influence of top growth removal on the root and vegetative development of biennial sweetclover," *J. Am. Soc. Agron.* 40(1948):818–831.

7. Willard, C. J. "An experimental study of sweetclover," *OH Agr. Exp. Sta. Bull.* 405, 1927.

8. Smith, W. K., and H. J. Gorz. "Sweetclover improvement," in *Advances in Agronomy*, vol. 17, New York: Academic Press, 1965, pp. 163–231.

9. Kasperbauer, M. J., F. P. Gardner, and I. J. Johnson. "Taproot growth and crown development in biennial sweetclover as related to photoperiod and temperature," *Crop Sci.* 3(1963):4–7.

10. Pieters, A. J., and E. A. Hollowell. "Clover improvement," in *USDA Yearbook*, 1937, pp. 1190–1214.

11. Stevenson, T. M. "Sweetclover studies on habit of growth, seed pigmentation, and permeability of the seed coat," *Sci. Agr.* 17(1937):627–654.

12. Kirk, L. E., and J. G. Davidson. "Potentialities of sweetclover as plant breeding material," *Sci. Agr.* 8(1928):446–455.

13. Fowlds, M. "Seed color studies in biennial white sweetclover, *Melilotus alba*," *J. Am. Soc. Agron.* 31(1939):678–686.

14. Stevenson, T. M., and L. E. Kirk. "Studies in interspecific crossing with *Melilotus* and in intergeneric crossing with *Melilotus, Medicago* and *Trigonella*," *Sci. Agr.* 15(1935):580–589.

15. Hanson, A. A. "Development and use of improved varieties of forage crops in the temperate humid region," in *Grasslands*, Publ. No. 55–53, Am. Assn. Adv. Sci., Washington, DC 1959, pp. 1–406.

16. Harmann, W. "Comparison of heights, yields, and leaf percentages of certain sweetclover varieties," *WA Agr. Exp. Sta. Bull.* 365, 1938.

17. Hollowell, E. A. "Sweetclover," USDA *Leaflet 23* (rev.). 1960.

18. Dunham, R. S. "Effect of method of sowing on the yield and root and top development of sweetclover in the Red River Valley," *J. Agr. Res.* 47(1933):979–995.

19. Arny, A. C., and F. W. McGinnis. "The relative value of the annual white, the biennial white, and the biennial yellow sweetclovers," *J. Am. Soc. Agron.* 16(1924):384–396.

20. Kirk, L. E. "A comparison of sweetclover types with respect to coumarin content, nutritive value, and leaf percentage," *J. Am. Soc. Agron.* 18(1926):385–392.

21. Wiancko, A. T., and R. R. Mulvey. "Sweetclover: Its culture and uses," *Purdue U. Agr. Exp. Sta. Circ.* 261, 1940, pp. 1–14.

22. Cook, E. O., and C. V. Rector. "The effect of fertilizer and sweetclover on oat forage and grain yields, Blackland Experiment Station, 1958–62," *TX Agr. Exp Sta. Prog. Rpt.* 2203, 1964.

23. Giesler, L. J., T. O. Powers, and J. A. Wilson. "Soybean cyst nematode biology and management." *NE Coop. Ext. Serv.* G99–1383-A, 2003.

24. Harper, H. J. "Sweetclover for soil improvement," *OK Agr. Extl. Sta. Circ.* C-94, 1941.

25. Wolfe, T. K., and M. S. Kipps. "Comparative value of scarified and of unhulled seed of biennial sweetclover for hay production," *J. Am. Soc. Agron.* 18(1927):1127–1129.

26. Helm, J., and D. Meyer. "Sweetclover production and management." *ND St. Univ. Ext. Serv.* R-862, 1993.

27. Stevenson, T. M. "Sweetclover in Saskatchewan," *Sask. Agr. Ext. Bull.* 9 (rev.), 1938.

28. Blackshaw, R. E., and others. "Yellow sweetclover, green manure, and its residues effectively suppress weeds during fallow," *Weed Sci.* 49(2001):406–413.

29. Megee, C. R. "Sweetclover," *MI Agr. Exp. Sta. Spec. Bull.* 152, 1926.

30. Hulbert, H. W. "Sweetclover: Growing and handling the crop in Idaho," *ID Agr. Exp. Sta. Bull.* 147, 1927.

31. Redmon, L. A., J. L. Caddel, and J. D. Enis. "Forage legumes for Oklahoma." *OK Coop. Ext. Serv. Fact Sheet* F-2585, 2003.

32. Chambliss, C. G. "Sweetclover production and use in Florida." *FL Coop. Ext. Serv.* SS-AGR-189, 2003.

33. Miles, A. D. "Sweetclover as a range legume," *J. Range Mgt.* 23, 2(1970):220–222.

34. Garber, R. J., M. M. Hoover, and L. S. Bennett. "The effect upon yield of cutting sweetclover *(Melilotus alba)* at different heights," *J. Am. Soc. Agron.* 26(1934):974–977.

35. Stickler, F. C., and I. J. Johnson. "The comparative value of annual and biennial sweetclover varieties for green manure," *Agron. J.* 51(1959):184.

36. Stewart, G. "Height of cutting sweetclover and influence of sweetclover on succeeding oat yields," *J. Am. Soc. Agron.* 26(1934):248–249.

37. Bowren, K. E., and others. "Yield of dry matter and nitrogen from tops and roots of sweetclover, alfalfa and red clover at five stages of growth," *Can. J. Plant Sci.* 49, 1(1969):79–81.

38. Whited, D. A., and others. "Influence of temperature, light intensity and photoperiod on the *o*-hydroxycinnamic acid in plant parts of sweetclover," *Crop Sci.* 6, 1(1966):73–75.

39. Smith, W. K., and R. A. Brink. "Relation of bitterness to the toxic principle in sweetclover," *J. Agr. Res.* 57(1938):145–154.

40. Homer, W. H., and W. J. White. "Investigations concerning the coumarin content of sweetclover. III. The inheritance of the low coumarin character," *Sci. Agr.* 22(1941):85–92.

41. Johnson, E. M., and W. D. Valleau. "Black-stem of alfalfa, red clover, and sweetclover," *KY Agr. Exp. Sta. Bull.* 339, 1933.

42. Munro, J. A., and others. "Biology and control of the sweetclover weevil," *J. Econ. Ent.* 42(1949):318–321.

The True Clovers

The farmer can ameliorate 100 acres with clover more certainly than he can 20 from his scanty dung heap. While his clover is sheltering the ground, perspiring its excrementitious effluvium on it, dropping its putrid leaves, and mellowing the soil with its tap roots it gives full food to the stock of cattle, keeps them in heart, and increases the dunghill.

—J. B. Bradley, *Essays and Notes on Husbandry and Rural Affairs*, Philadelphia, 1801.

24.1 ECONOMIC IMPORTANCE

The introduction of clover into customary crop rotations in the sixteenth century revolutionized agricultural practices throughout the world. The clovers (*Trifolium* species) are grown for hay, pasture, or soil improvement, usually in mixtures with grasses, except when intended for seed production.

24.2 IMPORTANT CLOVER SPECIES

The true clovers are members of the genus *Trifolium* of the plant order *Fabales* and the family *Fabaceae*. About 250 species of *Trifolium* are known world-wide. About thirty of these are native to the United States, mostly in the mountain regions of the West. Zigzag clover (*T. medium*), a perennial species grown in England, has strong creeping rootstocks. It was naturalized in the New England States. Kura clover (*T. ambiguum*), a native of Asia Minor, is being tried as a perennial honey plant in the United States. Twenty species of cultivated clover were described by Hermann.[1] Sweetclover, burclover, sourclover, buttonclover, alyceclover, and lespedeza (formerly called Japan clover) are not true clovers.

Red, white, kura, and crimson clovers are of primary importance in the United States. Clovers important in certain localities include alsike, strawberry, hop, sub, Persian, berseem, ball, lappa, and cluster species. The more important native clovers include seaside (*T. willdenovii*), white-tipped (*T. variegatum*), and long-stalked (*T. longipes*).

▨ 24.3 RED CLOVER

24.3.1 History

Red clover *(Trifolium pratense)*, which includes the type called Mammoth clover, is grown alone or in grass mixtures on about 10 million acres (4 million ha) for hay, pasture, seed, or green manure. It grows wild throughout most of Europe and ranges far into Siberia. The plant was generally cultivated nearly 400 years ago in the Netherlands, whence it was carried to England in 1645. Red clover was grown by the American colonists as early as 1663. The crop has spread to nearly all regions in the United States where there is sufficient rainfall to grow it, but most of it is in the northeastern fourth of the country. Under irrigation in the West, it is inferior to alfalfa for hay, except in some of the cooler intermountain valleys.

24.3.2 Adaptation

In general, red clover is adapted to the humid sections in the northern half of the United States (Figure 24.1). It thrives in a cool, moist climate. In Ohio, yields increased with increases in rainfall for April to June, inclusive.[2] Rainfall probably is the most important climatic factor that determines clover adaptation in the northern states, although stands are often lost by winterkilling. Strains adapted to this area must be able to tolerate a long period of dormancy, which at times may be accompanied by very low temperatures. High summer temperatures apparently cause little injury, provided moisture is abundant.[3]

Red clover makes its best growth on fertile, well-drained soils of pH 6.6–7.6[4] that contain an abundance of lime. Addition of phosphorus is necessary on most soils, while potash is sometimes beneficial. Red clover is not as well adapted to poorly drained or acid soils as is alsike clover. Clover failure, the partial or complete loss

FIGURE 24.1
General adaptation of clovers: (A) red, white, and alsike clovers; (B) white, crimson, and other winter annual clovers (except that crimson clover is seldom grown along the Pacific Coast). Sweetclover is grown mostly in the Great Plains states from Montana and North Dakota southward to Oklahoma. Strawberry clover is grown on wet saline soils in the intermountain region.

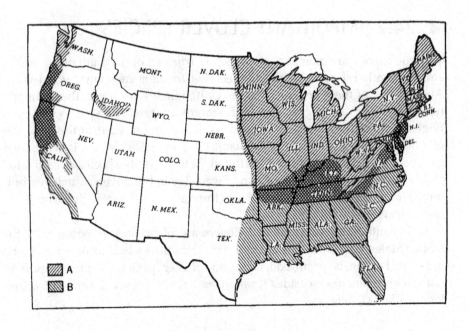

of stands, either in the seeding year or second year, may be caused by an unfavorable soil condition. The addition of lime, phosphorus, and potash, alone or in combination, solves the soil difficulty.[5, 6]

24.3.3 Botanical Description

Wild red clover is extremely variable; most of the plants are short-lived perennials. Types exist that are early, late, smooth, hairy, prostrate, erect, and semierect. One of these types is probably the ancestor of the first clover used in agriculture.[7]

Red clover is an herbaceous plant composed of three to ten or more leafy, hollow, erect stems that arise from a thick crown. The stems generally attain a height of less than 30 inches (75 cm). Each leaf bears three oblong leaflets, usually with a pale spot in the center of each and a prominent stipule at the base. The taproot, which has numerous laterals, may penetrate the soil to a depth of 5 feet (1.5 m). Red clover is a short-lived perennial in the cooler northeastern and high-altitude areas, a biennial in lower-altitude irrigated areas, and a winter annual in the warm southeastern parts of the United States.[4]

The flowers are borne in capitulum inflorescences at the tips of the branches (Figure 24.2). Each inflorescence contains fifty to 100 flowers. The flower consists of a green pubescent calyx with five pointed lobes and an irregular rosy purple corolla of five petals.[1] The flowers, about ½ inch (23 mm) long, are pea-like except that they are more elongated and smaller than those of a pea. The claws of the petals are more or less united to the staminal tube. Nectar, secreted at the bases of the stamens, accumulates in the staminal tube around the base of the ovary. The

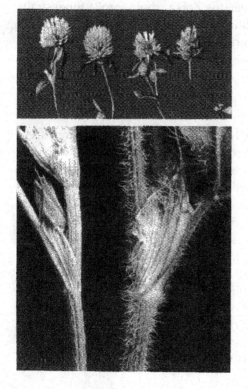

FIGURE 24.2

(Top) Red clover heads, the two at right damaged by clover midge. *(Bottom)* Smooth (or appressed hair) European red clover *(left)*. American domestic red clover with hairy stems *(right)*.

pods are small, short, and break open transversally. Generally, they contain a single kidney-shaped seed, about ⅒ inch (2 mm) long (Figure 22.5). The seeds vary in color from yellow to purple; the color variation is largely due to a complex heredity. Hard seeds are found most frequently among the purple seeds.[8]

POLLINATION In the pollination of red clover, the flowers develop and open from the base toward the top of the inflorescence.[9] The pistil is curved, with the stigma extending beyond the anthers. Flowers have been found with greatly shortened styles. The ovary has two ovules, one of which develops normally while the other aborts. Although the anthers shed their pollen in the bud stage, both stamens and pistil remain in the keel after the floret opens. The pistil is exposed only after the floret has been tripped.

Nearly all red clover plants are self-sterile. Because of the slow growth of the pollen tube in the style, the ovules disintegrate before the generative nucleus from the same plant reaches the egg. Self-fertile progenies have been obtained that continue to set selfed seeds. Other plants produce seeds when selfed, but the progenies may not necessarily be self-fertile. This phenomenon has been termed pseudo-self-fertility. A self-fertile line isolated in Minnesota was much less vigorous than commercial strains.[10] Cross-pollination is necessary for commercial seed production.

Cross-pollination of red clover is affected mainly by bumblebees and honeybees. They visit the flower for pollen or nectar or both. Honeybees principally obtain pollen. The honeybee is the major factor in pollination in western red clover seed fields. In Colorado,[11] plants caged with honeybees produced 61.5 seeds per head as compared with 67.3 from the open field. Bees trip the florets and carry pollen from plant to plant. The nectar of red clover strains with short corolla tubes is more accessible to the honeybee. However, the Zofka (short-corolla tube) strain has less honeybee activity and less seed production than the American or Canadian strain.[12, 13] Maximum seed yields are obtained from about two hives of bees per acre (5 hives/ha) located in or within ¼ mile (400 m) of the clover field.[14]

24.3.4 Regional Strains

Many clover failures were formerly due to the use of seed of unadapted strains, especially when the seed was imported from countries where the climatic conditions are less severe. Foreign clovers, with few exceptions, are unsatisfactory for use in the Northeast.[2, 15] Strains from southern Europe are particularly unproductive. Foreign clovers that overwinter may produce a fair first crop but a poor second cutting. They often die immediately after the first cutting. The plants die because of winter injury, diseases (mostly anthracnose and powdery mildew), or insects (mostly leafhopper). Tests in the Midwest show that seed lots from Europe or South America yielded only 30 to 80 percent as much as did the locally adapted strains. Foreign strains generally are badly winterkilled in Minnesota[16, 17] and Wisconsin. The results shown in Table 24.1 typify the behavior of unadapted clovers from foreign and domestic sources grown in the northern part of the clover region.[17]

Locally adapted seed is preferable to that from other states,[18] unless the seed brought from another area is an adapted variety produced from foundation-seed

TABLE 24.1 Hay Yields of Red-Clover Strains in Minnesota (Two-Year Average)

Source	Hay Yields per Acre			
		15% Moisture Basis		
	Loss in Stand (%)	Ist Year (lb)	2nd year (lb)	Total (lb)
Domestic				
Northern United States	18.1	2862	1122	3983
Central, southern, and western United States	27.3	2409	1264	3673
Foreign				
Western and central Europe	21.3	2228	1016	3244
Southern Europe, England, Wales, and Chile	85.4	488	256	744

stock. Plants in the southern region must be resistant to southern anthracnose (*Colletotrichum trifolii*). Half of the acreage is devoted to improved varieties.

24.3.5 Types, Varieties, and Strains

There are two distinct types of red clover: (1) the early or double-cut that gives two hay crops in a season, and (2) the late or single-cut that produces only one hay crop per season. The double-cut is generally known in this country as medium red clover, while the single-cut is commonly referred to as Mammoth red clover. Mammoth is grown on only 5 percent of the red-clover acreage. Several types of each form are known. The single-cut clovers are taller, bloom ten to fourteen days later, and yield a heavier crop in the first cutting.[15] Except in unimportant particulars, the two types do not differ in leaf shape, habit, root growth, and inflorescence.

Rough pubescent strains typify the red clovers of North America with the hairs attached at right angles to the stems and leaf petioles. On the European types, the hairs are generally appressed or absent; the plants are commonly classified as smooth. The two types can be distinguished in the seedling stage.[19] The American form has developed its characteristic hairiness by the process of natural selection as a defense against leafhoppers, which prefer the smooth types.

Regional strains of red clover, which differ in productivity, winter-hardiness, and disease resistance, have developed as a result of the action of local environment on a highly variable plant. Naturalized varieties originated in Maryland, Louisiana, Pennsylvania, and Mississippi. Improved varieties bred at experiment stations are resistant to anthracnose and powdery mildew.

24.3.6 Cultural Methods

Red clover is well adapted to three-year rotations. A common practice is to seed red clover with small grains the first year, harvest a hay crop the second year, and plow the clover under in preparation for corn the third year. The residue following the hay crop may be pastured before it is plowed under.[20]

The seedbed for red clover should be fine, firm, and moist. Red clover is seeded most frequently either on stands of winter wheat or rye or with small grain in the spring. It is usually broadcast at the rate of 8 to 10 pounds per acre (9 to 11 kg/ha)[5] or is drilled with either a special clover drill or a grain drill with a grass-seeder attachment. Clover is sown at three-fourths of the above rates in mixtures with grasses. The maximum safe depth of planting is about ¼ inch (6 mm) on heavy soils or ½ inch (12 mm) on light soils. The crop is generally seeded early in the spring, although late-summer seedings have been successful as far north as central Indiana. Fall seeding prevails in the South and in the Pacific Coast states.[18] Seed inoculation is beneficial in many cases. Weed control methods are similar to those for alfalfa.

Red clover is cut in the early-bloom stage for the best yield and quality of hay. Many of the leaves are lost when clover is allowed to become too dry in the windrow. The curing practices are similar to those for alfalfa. The use of a hay crusher, together with forced-air drying, greatly facilitates curing red clover into good quality hay. The growth of red clover is improved by mowing the stubble and removing the residue of a companion grain crop immediately after combining. The straw, clover tops, and weeds have some feed value. Mowing the first-year growth in late August provides some forage. It may also improve the yields of the second-year crop.[21]

24.3.7 Utilization

PASTURE Red clover is an excellent pasture plant but close, early grazing is injurious to the stand. Some pasture is ordinarily available in the fall of the seeding year, but some fall top growth should be allowed to remain to prevent winterkilling. There is some danger of bloat from plants that are pastured when young or wet.

SEED PRODUCTION Most of the red clover seed is produced in the Corn Belt, but nearly all of the certified seed is produced in the western states on irrigated land.

The production of red clover seed, especially in the Corn Belt, is generally incidental to hay production. An average set of twenty-five to thirty seeds per head indicates a fair seed crop, while sixty seeds indicate a good seed set. The general conditions that favor seed production are: (1) a vigorous recovery after cutting; (2) clear, warm weather when the second crop is in bloom; (3) an abundance of bees for pollination; (4) absence of injurious insects such as the clover flower midge and the chalcis fly; and (5) good harvesting weather. Wide extremes in soil moisture do not prevent setting of red clover seed.[22]

High seed yields have been obtained in some of the western states.[23] In order to control injurious insects that affect the seed, the first crop should be cut early for hay, or the first growth grazed.[24] Clover for seed should be irrigated sufficiently to produce vigorous growth, but water should be withheld after the seeds are about half mature. Introduction of honeybees in the field when the plants are in full bloom will increase seed formation.

Red clover is ready to cut for seed when the heads have turned brown and the flower stalks are a deep yellow. The crop may stand until the heads are all black, where cutting is done when the atmosphere is damp.[25]

Much of the seed crop is windrowed and then threshed with a combine equipped with a pick-up attachment, but occasionally, under ideal conditions, by direct harvest of the mature crop.

24.3.8 Diseases

Southern anthracnose *(Colletotrichum trifolii)* is one of the most important diseases that cause loss of stands south of the Ohio River, especially in Tennessee. This disease is characterized by a series of elliptical shrunken areas on the stems and leaf petioles that spread until the death of the plant. The disease is particularly virulent in damp, hot weather. Northern anthracnose *(Kabatiella caulivora)* is often severe on the first crop of the second year in the northern Clover Belt, but seldom is it as virulent as the southern form.

Powdery mildew *(Erysiphe polygoni)* is common wherever red clover is grown. It appears as a powdery dust on the leaves of infected plants. The disease is more prevalent when the plants are under stress. Some varieties have partial resistance to the disease. Timely harvest and sanitation of harvesting equipment help prevent losses.

A number of other diseases that attack the leaves cause losses in the quality of the hay. Varieties resistant to the three major diseases are now available. Root rots and viruses also attack red clover.[26, 27, 28]

24.3.9 Insect Pests

Several insects cause damage to red clover, among them the clover root borer *(Hylastinus obscurus)*, the clover seed chalcis fly *(Bruchophagus gibbus)*, the clover seed midge *(Dasyneura leguminicola)*, and the clover root curculio *(Sitonia hispidula)*.

Fields are frequently heavily infested with the clover root borer. The larvae tunnel in the roots, especially in the summer of the second year. One control measure is to plow under the clover after the first hay crop has been removed, preferably between June 15 and August. A three-year rotation is another control method.

The clover seed chalcis fly is one of the worst seed pests in the United States. It is a black, wasp-like insect about the size of a red-clover seed. The larvae develop within the seed. To reduce the damage from this pest, all debris from threshing should be destroyed and the first spring growth should be grazed lightly.

Seed production of red clover can be greatly reduced by the clover seed midge (Figure 24.2). The larvae of this insect attack the flowers, and, as a result, many fail to develop seed. Brown withered petals among the normally pink blossoms are a conspicuous symptom of attack by this pest. Midge may be controlled by cutting the first crop early enough to catch the maggots before they are fully developed. This is usually at the time they begin to change color from creamy white to salmon pink (i.e., early June in most affected areas).

The root curculio gnaws and sometimes girdles the roots, causing wilting or death of the plant.

Other insects that may cause considerable damage are the clover aphids[29] and the potato leafhopper. The former infest clover heads, secreting what is called *honey-dew* at the base of the flowers. They bring about serious reduction in seed yield in the western states as well as causing the seed to be sticky when threshed. The potato leafhopper *(Empoasca fabae)* frequently attacks seedling or second-year plants in the humid regions, often stunting them. The European smooth clovers are particularly susceptible to leafhopper attacks.

Weevils, aphids, leafhoppers, spittlebugs, and grasshoppers that attack clover can be controlled with certain insecticides.

▦ 24.4 ALSIKE CLOVER

24.4.1 History

Alsike clover *(Trifolium hybridum)*, a native of northern Europe, has been cultivated in Sweden since the sixteenth century. It was introduced into England in 1832 and in the United States about 1839. The production of alsike clover in the United States has declined since 1948, but it is still an important export seed crop in western Canada. Today, it is grown primarily in the Pacific Northwest and the Great Lakes.

24.4.2 Adaptation

The geographical distribution of alsike clover is roughly the same as red clover and timothy. It grows well in the intermountain states, especially under irrigation. Alsike clover requires a cool climate with an abundance of moisture. It rarely winterkills, being able to withstand more severe winters than does red clover. Alsike clover makes its best growth on heavy silt or on clay soils that are moist. It is less sensitive to soil acidity or alkalinity and better adapted to low, poorly drained soils than is red clover. Alsike clover sometimes thrives where other clovers fail.

24.4.3 Botanical Description

Alsike clover is a perennial that persists for four to six years in a favorable environment, but usually lasts only two years in the Midwest. The stems may reach a length of 3 feet (90 cm). The stems and leaves are smooth. The flowers are partly pink and partly white in a capitulum inflorescence. Each pod contains two to four seeds. The main axis continues to grow while single flower-bearing branches, each with one or more flower heads, arise successively from each leaf axil.[1, 30] The seeds of alsike clover are various shades of green mixed with yellow and are about one-third as large as those of red clover (Figure 22.5). Alsike clover plants are generally self-sterile and 1,500 bees per acre (3,700 bees/ha) provide pollination for seed production.

24.4.4 Cultural Methods

The cultural methods are similar to those for red clover. The crop is seeded at the rate of 4 to 6 pounds per acre (7 kg/ha) when grown alone. It is more often seeded with timothy or red clover or both. The crop is used for the same purpose as red clover.

Alsike usually yields less hay than does red clover. It produces only one crop in a season. The plant is especially suited for pasture mixtures, particularly in permanent pastures on wet, sour, or occasionally flooded land.

▓ 24.5 KURA CLOVER

24.5.1 History

Kura clover (*Trifolium ambiguum*), also known as Caucasian clover or honey clover, is a perennial that spreads by rhizomes. It is believed to have originated in Caucasian Russia. It was first introduced into the United States in 1911 but was generally unknown until the late 1940s. It was first used for honey production.

24.5.2 Adaptation

Kura clover is cold hardy and does better in cooler climates. It is adapted to a wide diversity of soil conditions, fertility, and acidity. It can tolerate wet soils and goes dormant during drought. Because of this, kura is more persistent in legume-grass mixtures than other clovers.

24.5.3 Botanical Description

Kura clover closely resembles white clover except that it spreads by rhizomes instead of stolons and grows somewhat taller. Each leaf bears three oblong leaflets, usually with a pale spot in the center of each similar to red clover but without stipules. Stems are glabrous. Inflorescence is a capitulum, somewhat larger than white clover (Figure 24.3). Flowers are white to pink. Seeds are yellow-brown and about the same size as red clover. It is more deeply rooted than white clover and thus can better withstand drought. It is closely related to white and alsike clovers and has crossed with these two species naturally. Kura is mostly self-sterile and needs bees and other insects to facilitate cross-pollination. It produces more nectar and has a shorter corolla tube so bees prefer it over other legumes. One U.S. variety, Rhizo, was released in 1990.

24.5.4 Cultural Methods

Kura clover is slow to establish the first year and does not produce seed until the second year. Seedlings are poor competitors with weeds. Because of low seedling vigor, kura clover does not usually do well with a companion crop. Once established, however, it is very persistent because it spreads by rhizomes.

Kura clover is usually seeded in the spring but can be seeded later with irrigation. It can be seeded alone at the rate of 8 to 13 pounds per acre (9 to 15 kg/ha). Rates are reduced one-fourth in mixtures containing grasses.[31] Seeding depth is ¼ to ½ inch (6 to 13 mm). Kura clover seed should be inoculated before planting. Kura requires a different strain of *Rhizobium* (CC283b) than the other clovers. Benefits of inoculation are more pronounced after the initial year of

FIGURE 24.3
Kura clover. [Courtesy
Richard Waldren]

establishment.[32] Kura clover provides adequate soil nitrogen in a mixture with cool-season grasses.[33]

24.5.5 Utilization

Kura clover has a high bloat risk, so it usually is not grazed alone by ruminants. Mixing kura with grasses reduces the risk. Kura clover–grass mixtures provide a high nutritive value for cattle.[34, 35]

A Wisconsin study[36] showed that kura clover can be used as a living mulch with no-till corn.

■ 24.6 WHITE CLOVER

24.6.1 History

White clover *(Trifolium repens)*, which apparently originated in the eastern or Asia Minor region, is now widely distributed in every continent. It is an important crop in New Zealand. It was brought to the United States from Europe by the early settlers. It is a valuable addition to both humid-region pastures and irrigated pastures in America.[37] It is a leading pasture legume grown mostly in mixtures chiefly from Arkansas and Tennessee northward.

24.6.2 Adaptation

The most favorable habitat for white clover is a moist, cool region under which growth is continuous. It will withstand greater temperature extremes than will either red or alsike clover and even survives in Alaska. It grows best between 50 and 85°F (10 and 30°C). In the Southeast,[38] it is grown as a winter annual because most plants die there in early summer after they have produced seed.

White clover is adapted to moist soils, especially to clays or loams abundantly supplied with phosphorus and potash. It does not do well on strongly acid or alkaline soils. Applications of phosphorus, potash, and lime are beneficial to growth on most soil types in humid regions, and often magnesium and sulfur are also helpful.

24.6.3 Botanical Description

White clover is a low, smooth, long-lived perennial plant with solid stems that root at the nodes.[1] The roots are shallow. The capitulum inflorescences arise from the leaf axils on long flower stalks. The flowers are small and white to pink in color. They generally turn brown when mature. The small seed pods usually contain three to five yellow seeds. White clover is extremely variable in leaf size, color, and markings; in size of runners; and in persistence.[7, 39] The spreading habit of the plant is a result of extensive stolon development (frequently referred to as runners). Occasionally plants or branches bear four leaflets that have a V-shaped white mark near the middle of the leaflet.

White clover is practically self-sterile. The florets require cross-pollination for seed formation, usually done by bees.

24.6.4 Types or Varieties

White clovers of agricultural value have been grouped into three types, namely (1) large (*T. repens var. giganteum*), (2) intermediate, and (3) small.

Large varieties include Ladino and others. Ladino clover appears to have been introduced from the vicinity of Lodi in northern Italy and is about two to four times as large as common white in all parts except the seeds.[1] The growth habits are similar to those of common white. It often establishes itself by natural reseeding. Ladino is adapted to a temperate climate where moist soils favor its growth. It will also grow on poorly drained, as well as mildly acid, soils.[40] Ladino is grown in the eastern states from New Hampshire to North Carolina; westward to Iowa, Missouri, and Arkansas; and also to the irrigated regions of the West. It is the leading irrigated pasture legume in California, where most of the Ladino seed is produced.[41]

Intermediate white clover includes common, White Dutch, or other northern types and southern types adapted to warm climates. The cool-season type adapted to the northern states for pasture mixtures or lawns is more persistent than Ladino but produces less herbage. Southern types thrive as winter annuals in the warmer southeastern states, but they lack the hardiness to survive in the North.[29, 42]

Small white clover includes the types commonly found in closely grazed pastures. The small type is represented by what is called English wild white clover.[7] The small or low-growing type is more likely to show the presence of a cyanophoric glucoside, but the quantity of hydrocyanic acid present is so small that it is harmless.[43] The development of varieties or strains has resulted from the action of environment on a variable species.

24.6.5 Cultural Methods

White clover is usually seeded with grasses, either as a new seeding, or on established turf. In new seedings, white clover is mixed with grass seed and the entire mixture seeded at one time, usually early in the spring, although in some places it is sown in early fall.[44, 45] The grass should be clipped short when clover is seeded in turf. The rate of seeding is 2 to 3 pounds of clover seed per acre (2.25 to 3.33 kg/ha).

The yield of top growth was significantly higher in New Jersey experiments where a complete fertilizer of lime, phosphorus, and potash was used.

Cultural methods for Ladino clover are very similar to those for common white.[19] Ladino clover may be sown in mixtures with orchardgrass, timothy, bromegrass, tall fescue, or reed canarygrass. Kentucky bluegrass soon crowds out Ladino clover. When Ladino clover is seeded alone, the usual rate is 1 to 2 pounds per acre (1.1 to 2.25 kg/ha). In a mixture with other legumes, about ¼ pound per acre (280 g/ha) is adequate. When it is seeded in a mixture with grasses, about ½ to 1 pound per acre (0.55 to 1.1 kg/ha) of clover seed is adequate.[46] The best stands are maintained when not mowed until the plants are 6 to 8 inches (15 to 20 cm) high, but after mowing, the stubble should be at least 4 inches (10 cm) above the ground. Close cutting in October is particularly harmful. The crop should be cut for hay when one-tenth of the flower heads have turned brown.[47] Continuous close grazing of Ladino clover is inadvisable. Ladino clover is not suitable for lawns because of its tall growth and inability to withstand close clipping. Ladino clover responds to applications of manure and to phosphate and potash fertilizers in areas where those elements are deficient.

24.6.6 Utilization

The intermediate white clovers are almost always grown in grass mixtures for pasture or lawns. Ladino will produce hay or silage under favorable conditions. Ladino clover pastures carried 30 to 40 percent more stock than an equal acreage of red clover, alsike clover, or ordinary white clover under Idaho conditions.[48]

Bloating of cattle is less likely to occur if the clover is 12 inches (30 cm) tall or if the plants are growing slowly.[49]

■ 24.7 STRAWBERRY CLOVER

24.7.1 History

Strawberry clover (*Trifolium fragiferum*) has assumed considerable importance in the United States. It is a native of the eastern Mediterranean region, but its culture has extended to every continent. It is a profitable pasture crop in the intermountain and Pacific Coast states in places where the subsoil water table is so high that other crop plants are largely or wholly eliminated.

24.7.2 Adaptation

Strawberry clover is adapted to the climate of the Pacific Coast, the intermountain region, and western Nebraska.[19] It is unable to compete with white clover in the eastern states. The primary requirement of strawberry clover is an abundance of water. It is of particular value on wet saline soils. Good growth has been observed where the salt content of the soil was more than 1 percent. Established stands have survived salt concentrations of more than 3 percent for long periods.[50] It tolerates these high salt concentrations when ample water is present. It has survived on soils flooded for one to two months.

24.7.3 Botanical Description

Strawberry clover is a perennial, low-growing plant that spreads vegetatively by creeping stems that root at the nodes. Its leaves, stems, and habit of growth are similar to those of white clover except it is not stoloniferous. It is readily identified by the characteristic strawberry-like capitulum inflorescence. The inflorescence is generally round, with pink to white flowers (Figure 24.4). As the seed matures, the calyx around each seed becomes inflated. The mature head is gray to light brown in color. When ripe, the capsules break from the head. The seed is reddish-brown, or yellow flecked with dark markings. The seeds are much larger than those of white clover but slightly smaller than the seeds of red clover. The flowers are considered to be self-fertile, but abundant insect pollination increases seed yield.[51] An improved variety is adapted to areas of mild winters such as central and southern California.

24.7.4 Cultural Methods

Strawberry clover is ordinarily spring-seeded at the rate of 2 to 5 pounds per acre (2.25 to 5.6 kg/ha). Unhulled seed is generally advised for late winter planting, and

FIGURE 24.4
Strawberry clover.

scarified seed for spring planting in places where the soil is too wet to cultivate. Growth is aided by mowing to reduce the competition of rushes and similar plants, particularly on unprepared seedbeds.

Strawberry clover is almost entirely utilized for pasture, but it has also been used for green manure on saline soils. It offers promise in the reclamation of saline, alkaline soils now considered wastelands in the western states. Strawberry clover forage is as palatable to livestock as is white clover. It survives close grazing, but it is more productive when grazed moderately.

■ 24.8 CRIMSON CLOVER

24.8.1 History

Crimson clover (*Trifolium incarnatum*) is the most important winter annual among the true clovers. It is a native of Europe, being introduced to the United States as early as 1819. It became agriculturally important about 1880. This clover is grown in the central belt of the eastern and southeastern states and on the Pacific Coast.[7,9]

24.8.2 Adaptation

Crimson clover is adapted to cool, humid weather where winter temperatures are moderate. It is sown in late summer or early fall in time to become established before winter. Crimson clover will grow on both sandy and clay soils. While the plant is tolerant of ordinary soil acidity, it is difficult to obtain a stand on very poor soils. This clover is being effectively used as a summer annual in Maine.

24.8.3 Botanical Description

In general, the leaves and stems of crimson clover resemble those of red clover, but the leaflet tips of crimson clover are more rounded. There is a greater covering of hair on both leaves and stems. When sown in the fall, the young plants form a rosette that enlarges under favorable conditions. The inflorescence is an elongated spike-like head (Figure 24.5). The flowers are bright crimson in color, while the seeds (Figure 22.5) are yellow. White-flowered types have been selected.

Crimson clover is a long-day, self-fertile plant that is less variable than either red or white clovers. Insects increase the amount of pollination since the florets are not self-tripping.

24.8.4 Cultural Methods

About half of the crimson clover acreage consists of self-seeding varieties that have hard seeds. The hard seeds do not germinate before autumn when conditions for plant growth are favorable. This provides winter cover and spring pasture where plants are allowed to go to seed. Common crimson clover and some other varieties seldom reseed naturally.

FIGURE 24.5
Flower heads of crimson clover.

Crimson clover is generally seeded in August or September between the rows of cultivated crops, either broadcast and covered with a cultivator or with a drill. Crimson clover is sometimes seeded after a small grain crop or with vetch, Italian ryegrass, and fall-sown grain crops. The seeding rate is 15 to 25 pounds per acre (17 to 28 kg/ha) of hulled seed or 45 to 60 pounds per acre (50 to 67 kg/ha) of unhulled seed. Seed is planted ½ to ¾ inch (13 to 19 mm) deep.[52] Inoculation with suitable cultures is desirable.[44, 53] Crimson clover seed is produced in Oregon and in the Southeast.

24.8.5 Utilization

Crimson clover is used for hay, pasture, winter cover, and green manure. The crop furnishes an abundance of early spring grazing, as well as some fall and winter pasture under favorable growth conditions. It seldom causes bloat. This clover makes excellent hay when cut at the early-bloom stage. As crimson clover reaches maturity, the hairs on the plant become hard and tough. When such hay is fed, hair balls (phytobezoars) may form in the stomachs of horses. These occasionally cause death. As a green manure, crimson clover is generally plowed under two or three weeks before time to plant the next crop. As an orchard cover crop, it is generally allowed to mature, after which it is disked into the soil. It is harvested for seed by combining when the hulls are dark brown or by swathing when light brown.

▓ 24.9 OTHER WINTER ANNUAL CLOVERS

Many other winter annual species of *Trifolium* are found throughout the southern, as well as the south central and Pacific Coast, states. The principal clovers are smaller hop *(Trifolium procumbens)*, small or low hop *(T. debium)*, called suckling

FIGURE 24.6
Plant of small hop clover.

clover in England (Figure 24.6), Persian *(T. resupinatum)*, cluster *(T. glomeratum)*, subterranean *(T. subterraneum)*, hop *(T. agrarium)* (Figure 24.7), ball *(T. nigrescens)*, rose *(T. hirtum)*, striate or knotted *(T. striatum)*, rabbit-foot *(T. arvense)*, Carolina *(T. carolinianum)*, buffalo *(T. reflexum)*, and lappa *(T. lappaceum)*. Hop, large hop, and small hop clovers have yellow flowers. Arrowleaf *(T. vesiculosum)*, as well as bigflower or balsana *(T. michelianum)*, are established in pastures and meadows in the Southeast. Arrowleaf clover is replacing crimson clover in Alabama.[54]

Winter annuals, seeded in the spring in the Cotton Belt or Corn Belt, make only a small growth before the increasing length of day induces them to flower. Many of these winter annuals are perpetuated from year to year by self-seeding in the

FIGURE 24.7
Branch of hop clover.

early summer, from which they volunteer during the fall months.[44] They become agriculturally important only when sufficient germinable seed is present to establish good stands. The seed coats of many of these species are hard.[55] The seeds of the large hop, cluster, subterranean, and Persian clovers germinate when scarified. Even after scarification, germination of the seed of all species is inhibited in varying degrees by temperatures as high as 86 to 95°F (30 to 35°C).

Large hop and small hop clovers are both valuable pasture plants.[56] The association of hop clover with grass appears beneficial to the establishment of the clover, but tall northern grasses may shade it out. The clover is favored when the southern grasses become dormant in the fall. Seeding in September and October promotes nearly complete stands. Hop clover *(T. agrarium)* is becoming established naturally in the northern central states. It is a native of Europe, probably introduced as a mixture in other clovers, but George Washington ordered seed of hop clover from England in 1786.

Subterranean clover is being used in pasture mixtures in the Pacific Coast states, but it also shows promise in the Southeast. It requires cool, mild, moist winters and dry summers together with well-drained soils that are supplied with phosphorus. It provides winter and early spring pasture. Subterranean clover has prostrate stems, 12 inches (30 cm) or more in length, and white flowers with rose stripes. It is best adapted as a pasture crop to nonirrigated, foothill, cut-over lands. It requires phosphorus fertilization on these soils and some response is obtained from small molybdenum supplements. Lime was beneficial on the more acid soils when applied separately from phosphorus. Seed inoculation is essential.[57] The plant buries its maturing seed heads in the soil or in plant residues on the surface. The seed can be gathered with a suction type of seed reclaimer. Subterranean clover is important in Australia as well as in Chile.

Persian clover is a constituent of pasture mixtures in the southeastern states. It is also cut for hay or silage or turned under for green manure. Applications of phosphorus, potash, and, often, lime to the soil are beneficial. The flowers are light purple.[4]

Berseem or Egyptian clover (*T. alexandrinum*) is grown to a slight extent as a winter annual, mostly for pasture, in southern portions of Arizona, California, and Texas. It is the leading winter legume in Egypt. The flowers are yellowish. Temperatures of 20 to 25°F (−4 to −7°C) may kill the plants.

Rose clover is grown in California for pasture or for soil conservation. The flowers are purplish red.[58]

Ball clover, introduced from Turkey, reseeds itself when growing in pasture-grass mixtures in the southeastern states. It withstands late heavy grazing. The flowers are white to yellowish.

Bigflower clover, also from Turkey, is adapted to the southeastern states for hay or pasture. It has roseate flowers.

Striate clover is adapted to heavy soils. It furnishes grazing before the stems become harsh at flowering time. The flowers are reddish to rose.[44]

Rabbit-foot clover grows on sandy soils. It has grayish silky heads.

Carolina clover is a small early species with purplish-red flowers. It is found in pastures as well as on roadsides on poor soils.

Buffalo clover is widely distributed in the South. It has large light brown maturing seed heads, with its flowers reflexed or turned down.

Lappa clover is adapted to the Black Belt of Alabama where it has become naturalized.[59] It is able to reseed itself when used either for pasture or hay. This species has inconspicuous lavender-rose, self-fertile flowers.

REFERENCES

1. Herman, F. J. "A botanical synopsis of the cultivated clovers (Trifolium)," *USDA Monograph* 22, 1953.
2. Welton, F. A., and V. H. Morris. "Climate and the clover crop," *J. Am. Soc. Agron.* 17(1925):790–800.
3. Aamodt, O. S. "Climate and forage crops," in *Climate and Man*, USDA Yearbook, 1941, pp. 439–458.
4. Anonymous. "Growing red clover," *USDA Leaflet* 571(rev.), 1968, pp. 1–8.
5. Fergus, E. N., and E. A. Hollowell. "Red clover," in *Advances in Agronomy*, vol. 12. New York: Academic Press, 1960, pp. 365–436.
6. Pieters, A. J. "Clover problems," *J. Am. Soc. Agron.* 16(1924):178–182.
7. Pieters, A. J., and E. A. Hollowell. "Clover improvement," in *USDA Yearbook*, 1937, pp. 1190–1214.
8. Smith, D. C. "The relations of color to germination and other characters of red, alsike, and white clover seeds," *J. Am. Soc. Agron.* 32(1940):64–71.
9. Anonymous. "Growing crimson clover," *USDA Leaflet* 482 (rev.), 1971, pp. 1–10.
10. Rinke, E. H., and I. J. Johnson. "Self-fertility in red clover in Minnesota," *J. Am. Soc. Agron.* 33(1941):512–521.
11. Richmond, R. C. "Red clover pollination by honeybees in Colorado," *CO Exp. Sta. Bull.* 391, 1932.

12. Armstrong, J. M., and C. A. Jamieson. "Cross-pollination of red clover by honeybees," *Sci. Agr.* 20(1940):574–585.

13. Wilsie, C. P., and N. W. Gilbert. "Preliminary results on seed setting in red clover strains," *J. Am. Soc. Agron.* 32(1940):231–234.

14. Peterson, A. G., and others. "Pollination of red clover in Minnesota," *J. Econ. Ent.* 53, 4(1960):546–550.

15. Pieters, A. J. "The proper binomial or varietal trinomial for American Mammoth red clover," *J. Am. Soc. Agron.* 20(1928):686–702.

16. Arny, A. C. "Winterhardiness of medium red clover strains," *J. Am. Soc. Agron.* 16(1924):268–278.

17. Arny, A. C. "The adaptation of medium red clover strains," *J. Am. Soc. Agron.* 20(1928):557–568.

18. Fergus, E. N. "Adaptability of red clovers from different regions of Kentucky," *KY Agr. Exp. Sta. Bull.* 318, 1931.

19. Hollowell, E. A. "Ladino white clover for the northeastern states," USDA *Farmers Bull.* 1910, 1942, pp. 1–10.

20. Justin, J. R., and others. "Red clover in Minnesota," *MN Ext. Bull.* 343, 1967, pp. 1–15.

21. Torrie, J. H., and E. W. Hanson. "Effects of cutting first year red clover on stand and yield in the second year," *Agron. J.* 47(1955):224–228.

22. Hollowell, E. A. "Influence of atmospheric and soil moisture conditions upon seed setting in red clover," *J. Agr. Res.* 39(1929):229–247.

23. Dade, E., and C. Johansen. "Red clover seed production in central Washington," *WA Agr. Exp. Sta. Cir.* 406, 1962.

24. Hollowell, E. A. "Red clover seed production in the intermountain states," *USDA Leaflet* 93, 1932.

25. McClymonds, A. E., and H. W. Hulbert. "Growing clover seed in Idaho," *ID Agr. Exp. Sta. Bull.* 148, 1927.

26. Fulton, N. D., and E. W. Hanson. "Studies on root rots of red clover in Wisconsin," *Phytopath.* 50, 7(1960):541–550.

27. Hanson, E. W., and D. J. Hagedorn. "Viruses of red clover in Wisconsin," *Agron. J.* 53(1961):63–67.

28. Hanson, E. W., and K. W. Kreitlow. "The many ailments of clover," in *Plant Diseases,* USDA Yearbook, 1953, pp. 217–228.

29. Hollowell, E. A. "White clover," *USDA Leaflet* 119 (rev.), 1947.

30. Pieters, A. J. "Alsike clover," *USDA Farmers Bull.* 1151 (rev.), 1947.

31. Taylor, N. L. "Kura clover in Kentucky," *Univ. KY Coop. Ext. Serv.* AGR-141, 1989.

32. Seguin, P., and others. "Nitrogen fertilization and rhizobial inoculation effects on kura clover growth." *Agron. J.* 93(2001):1262–1268.

33. Zemenchik, R. A., K. A. Albrecht, and M. K. Schulz. "Nitrogen replacement values of kura clover and birdsfoot trefoil in mixtures with cool-season grasses." *Agron. J.* 93(2001):451–458.

34. Mourino, F., K. A. Albrecht, D. M. Schaefer, and P. Berzaghi. "Steer performance on kura clover–grass and red clover–grass mixed pastures." *Agron. J.* 95(2003):652–659.

35. Zemenchik, R. A., K. A. Albrecht, and R. D. Shaver. "Improved nutritive value of kura clover-and birdsfoot trefoil-grass mixtures compared with grass monocultures." *Agron. J.* 94(2002):1131–1138.

36. Zemenchik, R. A., K. A. Albrecht, C. M. Boerboom, and J. G. Lauer. "Corn production with kura clover as a living mulch." *Agron. J.* 92(2000):698–705.

37. Gibson, P. B., and E. A. Hollowell. "White clover," *USDA Agr. Hdb.* 314, 1966, pp. 1–33.

38. Anonymous. "White clover for the South," *USDA Leaflet* 498, 1961.

39. Ahlgren, G. H., and H. B. Sprague. "A survey of variability in white clover (*Trifolium repens*) and its relation to pasture improvement," *NJ Agr. Exp. Sta. Bull* 676, 1940.

40. Ahlgren, G. H., and R. F. Fuelleman. "Ladino clover," in *Advances in Agronomy*, vol. 2. New York: Academic Press, 1950, pp. 207–232.

41. Marble, V. L., and others. "Ladino clover seed production in California," *CA Agr. Exp. Sta. Circ.* 554, 1970, pp. 1–33.

42. Aamodt, O. S., J. H. Torrie, and O. F. Smith. "Strain tests of red and white clovers," *J. Am. Soc. Agron.* 31(1939):1029–1037.

43. Rogers, C. F., and O. C. Frykolm. "Observations on the variations in cyanogenetic power of white clover plants," *J. Agr. Res.* 55(1937):533–537.

44. Benson, R., and E. A. Hollowell. "Winter animal legumes for the South," *USDA Farmers Bull.* 2146, 1960.

45. Morgan, A. "Some common clovers: Their identification," *J. Dept. Agr. Victoria (Australia),* 30(1932):105–112.

46. Henning, J. C., and H. N. Wheaton. "White, ladino and sweet clover." *Univ. MO Ext. Serv.* G4639, 1993.

47. Eby, C. "Ladino clover," *NJ Agr. Exp. Sta. Cir.* 408, 1941.

48. Spangler, R. L. "Ladino clover seed production and the value of Ladino as a pasture crop," *J. Am. Soc. Agron.* 17(1925):84–86.

49. Foote, L. E., and others. "Controlling bloat in cattle grazing clover," *LA Agr. Exp. Sta. Bull.* 629, 1968, pp. 1–28.

50. Hollowell, E. A. "Strawberry clover: A legume for the West," *USDA Leaflet* 464, 1960.

51. Morley, F. H. W. "The mode of pollination in strawberry clover (*Trifolium fragiferum* L.)." *Austral. Jour. Expt. Agr. and Anim. Husb.* 3, 8(1963):5–8.

52. Sattell, R., and others. "Crimson clover," *OR St. Univ. Ext. Serv.* EM 8696, 1998.

53. Anonymous. "Growing crimson clover," *USDA Leaflet* 482, 1961.

54. Hoveland, C. S., and others. "Management effects on forage production and digestibility of Yucki arrowleaf clover (*Trifolium vesiculosum* Savi)," *Agron. J.* 62(1970):115–116.

55. Toole, E. H., and E. A. Hollowell. "Effect of different temperatures on the germination of several winter annual species of *Trifolium*," *J. Am. Soc. Agron.* 31(1939):604–619.

56. Hollowell, E. A. "The establishment of low hop clover, *Trifolium pro-cumbens*, as affected by time of seeding and growth of associated grass," *J. Am. Soc. Agron.* 30(1938):589–598.

57. Jackson, T. L. "Effects of fertilizers and lime on the establishment of subterranean clover," *OR Agr. Exp. Sta. Circ. Information* 634, 1972, pp. 1–15.

58. Knight, W. E., and E. A. Hollowell. "The influence of temperature and photoperiod on growth and flowering of crimson clover (*Trifolium incarnatum* L.)," *Agron. J.* 50(1958):295–298.

59. Sturkie, D. G. "A new clover for the black lands in the South," *J. Am. Soc. Agron.* 30(1938):968.

Lespedeza

KEY TERMS

Chasmogamous flower
Cleistogamous flower
Common lespedeza
Korean lespedeza
Sericea lespedeza
Tannin

▨ 25.1 ECONOMIC IMPORTANCE

Lespedeza is grown mostly in the eastern half of the United States, principally for pasture, either alone or in mixtures, but also for hay, seed, silage, and soil improvement. Its popularity peaked in the 1950s with acreage declining since then. However, annual species of lespedeza are still popular in some areas as there are not many annual legume species that provide nutritious forage for livestock, especially on marginal lands.

Lespedeza provides a legume to help maintain soil productivity in the southeastern states where lime-deficient soils are too acidic for economical production of either alfalfa or red clover. The temperatures there are too high for perennial growth of red clover. It is especially valuable for growing on badly eroded soils in that region. Few crops make as much growth as sericea lespedeza on soils of low fertility.

▨ 25.2 HISTORY OF LESPEDEZA CULTURE

The three cultivated species of lespedeza grown in the United States are of Asiatic origin. The earliest record of common lespedeza or Japanese clover, as it was called, indicates its presence in the southeastern states in 1846. The introduction of Korean and Kobe lespedezas from Korea in 1919 has been the main factor in the widespread use of lespedeza. The agricultural use of sericea lespedeza dates from an introduction from Japan in 1924, although the crop was first tried in North Carolina in 1896.[1] Ten species of native perennial lespedezas have been collected in Kansas.[2]

▨ 25.3 ADAPTATION

The growth of lespedeza is aided by high temperatures, relatively high humidity, and rather high rainfall well distributed throughout the growing season. The annual lespedezas are more drought-resistant than alfalfa or clovers[3] in the early-growth stages, and the perennial sericea will endure extreme drought after it is well established. However, growth is so limited under dry conditions that lespedeza is not suitable for the semiarid Great Plains.

The lespedezas are warm-season plants that begin growth slowly in the spring. Except for Korean, which stops growth with seed maturity, the lespedezas continue to grow until frost.

All the annual lespedezas are sensitive to cold. Seedlings that start growth during warm periods in late winter may be killed by late spring frosts. The perennial sericea lespedeza has survived winter temperatures as low as 36°F (2°C), but it may be killed down to the crowns by heavy spring freezes when actively growing. Lespedeza is most cold-resistant in the cotyledon stage and becomes more tender as the plants develop.[4] In alfalfa and red clover, cold resistance increases with age, a reversal in this respect. A temperature of 23°F (−5°C) for sixteen hours has killed all lespedezas.[4]

Lespedeza fails to set seed under extreme long-day conditions. Photoperiod also affects the adaptation of varieties. Fruiting is prevented in Korean when the day length is more than fourteen hours.

The lespedezas will grow on most soils, especially on soils too acidic for clover without lime applications. However, plant growth is improved by use of lime or fertilizers, or both, on soils deficient in the mineral elements. Lespedeza makes little growth on the sandy soils of the coastal plain unless supplied with phosphorus.[5] In Florida, it was necessary to add calcium, phosphorus, and potassium to the sandy soils of the flat pinelands to produce a satisfactory growth of lespedeza.[6] From 36 to 54 pounds per acre (40 to 60 kg/ha) of both P_2O_5 and K_2O are recommended for soils of the Gulf Coastal Plain.[7] The distribution of the crop is not limited by soil acidity even though lime applications increase growth.

Sericea lespedeza can be grown on most soils without fertilizers or lime[8] but it will respond to additions of lime and fertilizers on poor soils.[1] It fails on poorly drained soils.[3] It appears to thrive on the poor, eroded clays, silts, or silt loams of the Piedmont region and on similar soils elsewhere.

▨ 25.4 BOTANICAL DESCRIPTION

The annual species of lespedeza in the United States are common (*Lespedeza striata*) and Korean (*L. stipulacea*), while sericea (*L. cuneata*) is perennial. The prominent veins or furrowed surface of the leaflets serve to distinguish lespedezas from other legume crop plants. All varieties have a higher leaf percentage than alfalfa does.

25.4.1 Annual Lespedezas

The annual lespedezas are erect or spreading, small, branched, and short-day plants. They attain a height of 24 to 30 inches (60 to 75 cm), but the growth usually ranges from 4 to 12 inches (10 to 30 cm).[9] The leaves are small and trifoliolate. The roots are medium-deep and numerous. The plants produce two kinds of small flowers: petaliferous flowers with purple or bluish petals and the more numerous but inconspicuous apetalous flowers. The apetalous (cleistogamous) flowers actually have very small petals; self-pollination occurs before the calyx opens. The petaliferous (chasmogamous) flowers may be cross-pollinated. Chasmogamous flowers predominate when temperatures are high. The plants bloom from midsummer to early fall. The seeds are about the size of, or slightly larger than, those of red clover, being borne in pods that retain the seeds when threshed (Figure 22.5). The hulls are brown.

Common striate lespedeza is a slender plant, prostrate except in dense stands.[1] It usually grows to a height of 4 to 6 inches (10 to 15 cm), but in southern latitudes it reaches 12 to 15 inches (30 to 38 cm). The flowers are very small, purple, and inconspicuous. The seeds are borne in the axils of the leaves along the entire length of the stem. The seeds are dark purple mottled with white. The hairs on the stem are appressed downward.

Korean lespedeza (Figure 25.1) is coarser and earlier than common and has broader leaflets, larger stipples, and longer petioles. The seed is borne in the leaf axils at the tips of all branches, rather than along the stem as is common. The seeds are a solid dark purple. At maturity, the leaves of Korean turn forward so that the tips of the branches resemble small cones. The leaves of common do not turn forward. The hairs on the stems of Korean are appressed upward.

25.4.2 Sericea Lespedeza

Sericea lespedeza is a short-day perennial that produces coarse, stiff, tough, erect stems. A single stem on each plant, 12 to 18 inches (30 to 45 cm) in height, is produced in the first season. Additional stems, 24 to 66 inches (60 to 150 cm) in height, arise from crown buds in subsequent years. The woody, widely branched roots penetrate the soil to a depth of 40 inches (1 m) or more.[3] Both chasmogamous and cleistogamous flowers occur in sericea, most of the seed being produced from cleistogamous flowers. Pods from chasmogamous flowers are larger and more acute.[7] Sericea flowers are yellow or purple. The species varies in width of leaflets, height,

FIGURE 25.1
Branch of Korean lespedeza.

FIGURE 25.2
Branch of sericea lespedeza.

coarseness, number of stems, and earliness (Figure 25.2). A majority of the chasmogamous flowers are cross-pollinated.

■ 25.5 HARD SEEDS IN LESPEDEZA

Annual lespedezas often contain little or no hard seed, so scarification is unnecessary. In tests made soon after harvest, considerable hard seed was found in Korean lespedeza,[10] but a rapid decrease in hard-seed content occurred from November through January. Most of the hard seeds are small.[11]

Sericea lespedeza has a high percentage of hard seeds. Germination of unscarified seed ranges from 10 to 20 percent, so that scarification is necessary for prompt germination.[3]

■ 25.6 VARIETIES

Varieties of common lespedeza, such as Kobe, are best adapted to the general region from northern Tennessee to the Gulf of Mexico because they require high temperatures.

Korean lespedezas mature earlier than those of common lespedeza and are adapted to more northern latitudes (Figure 25.3). Certain varieties are somewhat resistant to bacterial wilt, or to root-knot nematodes and powdery mildew.[7, 12]

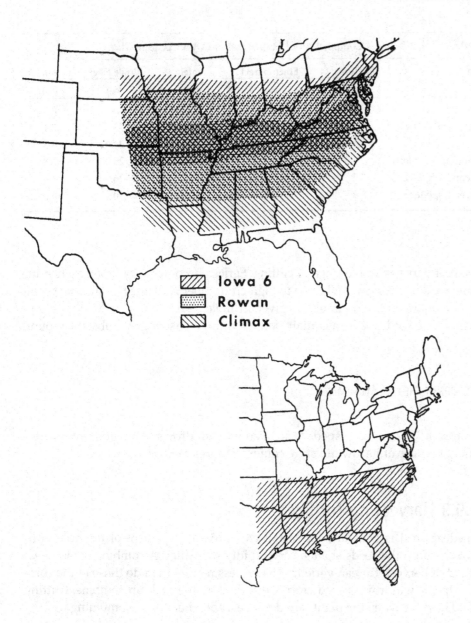

FIGURE 25.3
Areas of the United States where lespedeza varieties are adapted: *(Top)* Korean varieties. *(Bottom)* Common and Kobe varieties.

Iowa 6
Rowan
Climax

Sericea lespedeza is adapted to the zone between the Ohio River and the northern part of Florida where the annual rainfall exceeds 30 to 35 inches (760 to 890 mm).[13] Some species of shrubby perennial lespedezas are being grown for soil conservation, wildlife conservation, and ornamental purposes.

▓ 25.7 CHEMICAL COMPOSITION

At the usual hay-cutting stage, annual lespedezas contain significantly more dry matter than do the other common legumes.[14] Thus, they cure more rapidly. The lespedezas contain 36 to 49 percent dry matter; alfalfa (first three cuttings), 27 to 29 percent; and Laredo soybean, 28 to 30 percent.

Annual lespedezas are nearly equal to other legume hays in feeding value. In general, the digestible protein is slightly lower than alfalfa while the digestible carbohydrate equivalent is higher. The protein content ranges from 9 to 17 percent. In

TABLE 25.1 Chemical Composition of Lespedeza and Other Legume Hays

Crop	Moisture	Ether Extract	Protein	Crude Fiber	Ash
	(%)	(%)	(%)	(%)	(%)
Alfalfa	8.6	2.3	14.9	28.3	8.6
Red clover	12.9	3.1	12.8	25.5	7.1
Soybean	8.6	2.8	16.0	24.9	8.6
Common lespedeza	11.8	2.8	12.1	25.9	5.8
Korean lespedeza	7.2	3.3	16.2	26.0	7.4
Sericea lespedeza	5.9	1.8	12.3	30.1	6.0

sericea, it varies with the stage of cutting. Sericea becomes very woody when the plants are 3 feet (90 cm) high, due to high crude-fiber content. The chemical composition of lespedeza in Illinois[15] is given in Table 25.1.

All species of lespedeza contain some tannin, an astringent substance found

windrower.[19] If mowed, lespedeza is raked into windrows two to four hours after cutting. Because of its low moisture content, it cures rapidly.

25.9.3 Harvesting for Seed

Lespedeza seed is generally harvested when 65 to 75 percent of the hulls turn brown because the seeds shatter very rapidly thereafter. A combine is the usual method of harvest. Korean varieties shatter less readily than do those of the common type. A windrowed seed crop is threshed with a pick-up combine. Cutting should be done when the plants are damp, except when direct combining.

▦ 25.10 CULTURAL METHODS FOR SERICEA LESPEDEZA

Sericea lespedeza, also called Chinese bush clover, is considered to be a serious weed pest is some areas. It has been declared a noxious weed in Kansas and is being considered as a noxious weed in Nebraska, Missouri, and Oklahoma.

Sericea lespedeza is grown for forage mostly in the Southeast United States. It is lower-quality forage than annual lespedezas due to coarser stems and higher tannin concentration. Sericea usually is sown alone, but it may be sown on a field of winter grain. Seeding should be done in early spring. Self-seeding crimson clover can be sown on established stands of sericea. A satisfactory broadcast seeding rate for sericea is 30 to 40 pounds per acre (34 to 45 kg/ha) of hulled scarified seed followed by a cultipacker on noncrusting soils. The soil should be mulched when sericea is seeded on eroded knolls or in gullies. Seedings on established meadows

25.9 CULTURAL METHODS FOR ANNUAL LESPEDEZAS

Like all other crops, lespedeza will respond favorably to good cultural practices. A firm seedbed is essential whether the crop is seeded alone or with small grains. On meadows or pastures seeded with lespedeza, it is a common practice to first loosen the soil with a spring tooth harrow or disk. Inoculation is seldom necessary for most varieties in the South, except possibly on badly eroded soils. It is generally advisable to inoculate lespedeza seed when growing it for the first time in Kentucky and further northward.

The annual lespedezas are seeded either broadcast or ½ inch (13mm) deep (or less with a grain drill). In general, lespedeza is seeded in the spring about two weeks before the last freeze is expected in the locality. In North Carolina, Tennessee, and farther south, it is usually sown between February 15 and March 15. Farther north, seedings are made from March 15 to April 15. Early seedings are subject to late freezes.

Korean or common should be seeded at the rate of 25 to 30 pounds per acre (28 to 34 kg/ha), and Kobe at 30 to 40 pounds (34 to 45 kg/ha).[7, 12] For Missouri conditions, lespedeza should be broadcast at the rate of 15 to 20 pounds per acre (17 to 22 kg/ha), or 10 to 15 pounds (11 to 17 kg/ha) drilled.[20] On pastures, or in pasture mixtures, 5 to 10 pounds per acre (6 to 11 kg/ha) of seed are generally sufficient.

25.9.1 Utilization for Pasture

Annual lespedezas grow actively during the hot summer months when grasses make little growth. They can be maintained in grass-legume mixtures for permanent pastures with grasses that do not form a dense sod. In many places, however, the pasture must be cultivated and lespedeza resown every year or two. In a perennial-grass pasture, close grazing of the grasses in the spring is often practiced to favor the lespedeza. Common lespedeza survives longer in grass mixtures than does improved varieties. Lespedezas persist in pastures because they set their seed near the ground. When grown with small grain, lespedeza is ready for grazing soon after the grain is removed. The principal grazing months are July, August, and September.[7] The fall-sown grain can be pastured in winter, spring, and early summer. The lespedeza is grazed later.

25.9.2 Harvesting for Hay

Annual lespedezas are used as hay crops, particularly where soil conditions are unfavorable for perennial legumes. Hay of excellent quality can be made from lespedeza, especially when cut in early bloom or just before the bloom stage. This usually occurs before August 1 from Missouri to Virginia, or August 1 to 15 in the latitude of North Carolina for the Korean varieties. By this time, a growth of 8 to 10 inches (20 to 25 cm) is attained. Korean may be cut earlier than the other annual varieties because it makes more rapid early growth. Stubble of 3 to 5 inches (8 to 13 cm) is generally left to permit new growth. The crop is usually cut with a windrower.[19] If mowed, lespedeza is raked into windrows two to four hours after cutting. Because of its low moisture content, it cures rapidly.

25.9.3 Harvesting for Seed

Lespedeza seed is generally harvested when 65 to 75 percent of the hulls turn brown because the seeds shatter very rapidly thereafter. A combine is the usual method of harvest. Korean varieties shatter less readily than do those of the common type. A windrowed seed crop is threshed with a pick-up combine. Cutting should be done when the plants are damp, except when direct combining.

■ 25.10 CULTURAL METHODS FOR SERICEA LESPEDEZA

Sericea lespedeza, also called Chinese bush clover, is considered to be a serious weed pest is some areas. It has been declared a noxious weed in Kansas and is being considered as a noxious weed in Nebraska, Missouri, and Oklahoma.

Sericea lespedeza is grown for forage mostly in the Southeast United States. It is lower-quality forage than annual lespedezas due to coarser stems and higher tannin concentration. Sericea usually is sown alone, but it may be sown on a field of winter grain. Seeding should be done in early spring. Self-seeding crimson clover can be sown on established stands of sericea. A satisfactory broadcast seeding rate for sericea is 30 to 40 pounds per acre (34 to 45 kg/ha) of hulled scarified seed followed by a cultipacker on noncrusting soils. The soil should be mulched when sericea is seeded on eroded knolls or in gullies. Seedings on established meadows and pastures have failed.[21]

25.10.1 Harvesting

Sericea is cut for hay when the plants are 10 to 15 inches (25 to 38 cm) high. At this stage, the hay is comparatively high in protein and low in tannin but its feeding value is only about 80 percent of that of alfalfa hay.[13] The plants develop little woodiness up to a height of 12 inches (30 cm), but woody tissue increases rapidly in the later stages. As sericea cures quickly, it should be windrowed within one hour after cutting, or be cut and windrowed in one operation. It should be baled within twenty-four hours after cutting to avoid severe leaf shattering. Two or three cuttings may be harvested under favorable conditions. In Tennessee, yields of 4.14, 3.39, and 3.29 tons of hay per acre (9.3, 7.6, and 7.4 MT/ha) were obtained when sericea was cut two, three, and four times per season, respectively.[21] Four cuttings can seriously reduce the stand. Only one cutting is advisable on very poor soils. When sericea is being cut, stubble of 3 to 5 inches (8 to 13 cm) should be left because the new growth comes from the stems, as in sweetclover.

The first or second growth may be harvested for seed, usually after most of the pods have turned brown. The seed crop is harvested with a combine, a windrower, or a binder.

Sericea is ready to be grazed when it reaches a height of 6 inches (15 cm), being palatable at this stage. The pasture should be grazed or mowed down to a height of 3 inches (8 cm) to avoid coarse growth, which is unpalatable. Grasses suitable for mixed sericea pastures are bahiagrass in the Coastal Plains and tall fescue in northward areas. Sericea competes strongly with warm-season grasses and has been shown to retard the growth of warm-season grasses by producing an allelopathic chemical.[22]

▓ 25.11 ROLE IN SOIL CONSERVATION

Lespedeza is effective in the control of soil erosion due to its dense growth, its ability to establish a cover on poor soil, and the high retention of winter rains by old lespedeza sod. During a twelve-month period, the runoff from Korean lespedeza land was 11.7 percent, with a soil loss of 1.6 tons per acre (3.6 MT/ha).[19] Under comparable conditions, continuous corn showed a loss of 60.8 tons of soil per acre (136 MT/ha) and a 30.3 percent runoff. While the stubble of annual lespedeza tends to prevent erosion during the winter, a winter cover crop after lespedeza will improve the effectiveness.

Sericea, alone or with grasses, is well suited for seeding buffer strips, critical slopes above the flow lines of terraces, small gullies, and depressions for water outlets. Annual lespedezas are used in some of these places, especially in combination with grass.

▓ 25.12 DISEASES

The main diseases of annual lespedezas are bacterial wilt, caused by *Xanthomonas campestris* pv. *lespedezae;* tar spot, caused by *Phyllachora* species; damping off, caused by *Rhizoctonia solani;* southern blight, caused by *Sclerotium rolfsii;* and powdery mildew, caused by *Microsphaera diffusa.* Root-knot nematodes (*Meloidogyne* species) also damage annual lespedeza. Losses from bacterial wilt, powdery mildew, and root-knot nematodes can be reduced by growing resistant or tolerant varieties. Crop rotation reduces damage from tar spot and southern blight.

Sericea lespedeza is relatively free from disease losses, but it is attacked by root-knot nematodes. Because all lespedezas can serve as an alternate host for several important nematode pests, it should not be used in a rotation on infected soils. It is susceptible to cotton root rot, but it is seldom grown where that disease is prevalent.

▓ 25.13 INSECT PESTS

The main insects that attack lespedeza are grasshopper (*Schistocerca americana*), armyworm (*Pseudaletia unipuncta*), fall armyworm (*Laphygma frugiperda*), three-cornered alfalfa hopper (*Spissistilus festinus*), and webworm (*Tetralopha scortealis*). Some of these pests can be controlled by suitable insecticides.

REFERENCES

1. Pieters, A. J. *The Little Book of Lespedeza.* Washington, D.C: Colonial Press, 1934.

2. Anderson, K. L. "Lespedeza in Kansas," *KS Agr. Exp. Sta. Circ.* 251 (rev.), 1956, pp. 1–15.

3. Pieters, A. J. "*Lespedeza sericea* and other perennial lespedezas for forage and soil conservation," *USDA Cir.* 853, 1950.

4. Tysdal, H. M., and A. J. Pieters. "Cold resistance of three species of lespedeza compared to that of alfalfa, red clover, and crown vetch," *J. Am. Soc. Agron.* 26(1934):923–928.

5. Stitt, R. E. "The response of lespedeza to lime and fertilizer," *J. Am. Soc. Agron.* 31(1939):520–527.

6. Blaser, R. E., C. M. Yolk, and W. E. Stokes. "Deficiency symptoms and chemical composition of lespedeza as related to fertilization," *J. Am. Soc. Agron.* 34(1942):222–228.

7. Henson, P. R., and W. A. Cope. "Annual lespedezas—culture and use," *USDA Farmers Bull.* 2113 (rev.), 1964.

8. Bailey, R. Y. "Sericea in conservation farming," *USDA Farmers Bull.* 2033, 1951.

9. Grizzard, A. L., and T. B. Hutcheson. "Experiments with lespedeza," *VA Agr. Exp. Sta. Bull.* 328, 1940.

10. Middleton, G. K. "Hard seed in Korean lespedeza," *J. Am. Soc. Agron.* 25(1933):119–122.

11. Middleton, G. K. "Size of Korean lespedeza seed in relation to germination and hard seed," *J. Am. Soc. Agron.* 25(1933):173–177.

12. Henson, P. R. "The lespedezas," in *Advances in Agronomy*, vol. 9. New York: Academic Press, 1957, pp. 113–157.

13. Guernsey, W. J. "Sericea lespedeza, its use and management," *USDA Farmer's Bull.* 2245, 1970, pp. 1–30.

14. Smith, G. E. "The effect of photo-period on the growth of lespedeza," *J. Am. Soc. Agron.* 33(1941):231–236.

15. Pieper, J. J., O. H. Sears, and F. C. Bauer. "Lespedeza in Illinois," *IL Agr. Exp. Sta. Bull* 416, 1935.

16. Stitt, R. E., and I. D. Clarke. "The relation of tannin content of sericea lespedeza to season," *J. Am. Soc. Agron.* 33(1941):739–742.

17. Wilkins, H. L., and others. "Tannin and palatability in sericea lespedeza," *Agron. J.* 45(1953):335–336.

18. Clarke, I. D., R. W. Frey, and H. L. Hyland. "Seasonal variation in tannin content of *Lespedeza sericea*," *J. Agr. Res.* 58(1939):131–139.

19. Etheridge, W. C., and C. A. Helm. "Korean lespedeza in rotations of crops and pastures," *MO Agr. Exp. Sta. Bull.* 360, 1936.

20. Roberts, C., and J. Gerrish. "Seeding rates, dates and depth for common Missouri forages," *Univ. MO Ext. Serv.* G4652, 2001.

21. Mooers, C. A., and H. P. Ogden. "*Lespedeza sericea*," *TN Agr. Exp. Sta. Bull.* 154, 1935.

22. Ohlenbusch, P. D., and T. Bidwell. "Sericea lespedeza: History, characteristics, and identification," *KS St. Univ. Coop. Ext. Serv.* MF-2408, 2001.

Soybean

26.1 ECONOMIC IMPORTANCE

Soybean is the fourth largest crop in the world grown on an average of 194 million acres (78 million ha) in 2000–2003. World production averaged about 6.5 billion bushels (177 million MT) or 34 bushels per acre (2,250 kg/ha). About one-third of global production is in the United States. Other important soybean-producing countries are Brazil, Argentina, China, and India (Figure 26.1). In 2000–2003, soybean ranked first in area among United States crops with about 73 million acres (29 million ha). Production averaged about 2.7 billion bushels (74 million MT) with an average yield of about 37 bushels per acre (2,300 kg/ha). The leading states in soybean production are Iowa, Illinois, Minnesota, Indiana, and Nebraska.

Soybean is growing in popularity faster than any other crop. U.S. production of soybean rose from less than 5 million bushels (136,000 MT) in 1924, to 1,547 million bushels (42 million MT) in 1973, to 2,418 million bushels (66 million MT) in 2003. World acreage has increased from 58.8 million acres (23.8 million ha) in 1961 to 206 million acres (84 million ha) in 2003.

26.2 HISTORY OF SOYBEAN CULTURE

The soybean is one of the oldest cultivated crops. Its early history is lost in antiquity. The first record of the plant in China dates back to 2838 BC.[1] It was one of the five sacred grains upon which Chinese civilization depended. Cultivated soybean (*Glycine max*) probably was derived in China from a wild type, *Glycine ussuriensis*. It has small seeds and grows in eastern Asia.[1, 2, 3, 4]

Soybean was known in Europe in the seventeenth century, and in the United States in 1804. Little attention was given to soybean as a crop until 1889 when several experiment stations became interested in it. A large number of varieties were imported by the United States Department of Agriculture in 1898. Since that time, there has been a rapid expansion in soybean production, particularly since about 1920.[5] Most of the soybean was grown in the South prior to 1924, when it began to assume importance in the Corn Belt (Figure 26.2).

Determinate plant
Double cropping
Indeterminate plant
Inoculant
Iodine number
Photoperiod
Relay intercropping
Short-day plant
Soybean meal
Soybean oil

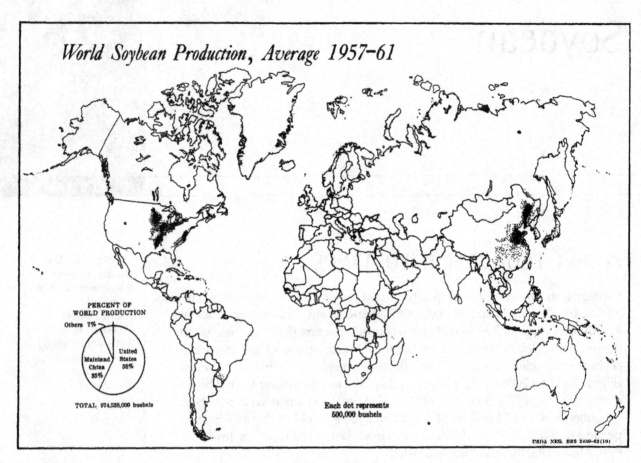

FIGURE 26.1

World soybean production. [Courtesy USDA]

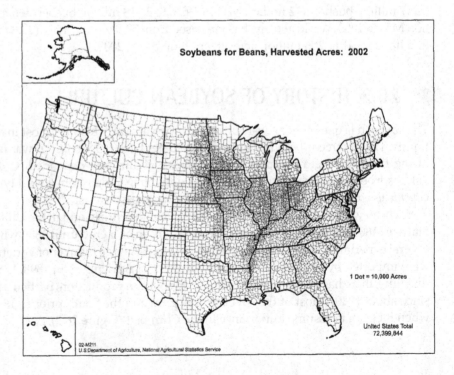

FIGURE 26.2

Acreage of soybean in the
United States in 2002.
[Source: 2002 Census of U.S.
Department of Agriulture]

26.3 ADAPTATION

The climatic requirements for soybean are about the same as those for corn. Soybean will withstand short periods of drought after the plants are well established. Indeed, a well-charged soil profile of 6 to 8 feet (1.8 to 2.4 m) may contain enough moisture to allow the plant to reach bloom stage before additional water is needed after planting. In general, combinations of high temperature and low precipitation are unfavorable. Soybean seed produced under high-temperature conditions tends to be low in oil and oil quality.[6, 7] A wet season does not seriously retard plant growth if the soil is well drained, but soybean is sensitive to overirrigation.[8, 9] Poor soil drainage can reduce yields.[10]

The period of germination is the most critical stage in soybean plant growth, with an excess or deficiency of soil moisture particularly injurious at this time.[11] Soybean is less susceptible to frost injury than corn. Light frosts have little effect on the plants when either young or nearly mature. The minimum temperature for growth is about 50°F (10°C).

The soybean, a short-day plant, is very sensitive to photoperiod.[12] Most varieties need 13.5 hours of day length or less in order to flower.[13] Northern varieties mature quickly with little vegetative growth when grown in the South. Within a variety, variations in time of flowering from year to year with the same day length are closely associated with temperature conditions.

A mean midsummer temperature of 75 to 77°F (24 to 25°C) is optimum for all varieties. Lower temperatures tend to delay flowering.

Soybean grows on nearly all types of soil, but it is especially productive on fertile loams. It is better adapted to low fertility soils than is corn, provided the proper nitrogen-fixing bacteria are present. Soybean will grow on soils that are too acidic for alfalfa and red clover.

26.4 BOTANICAL DESCRIPTION

26.4.1 Seed and Seedling

A soybean seed is composed of two cotyledons, radicle, testa, hypocotyl, and epicotyl. All except the testa are considered to be part of the embryo (Figure 7.3). The cotyledons are the major storage tissue for proteins and oil ranging from 38 to 46 percent protein and 18 to 20 percent oil. The hilum varies from buff to yellow, black, brown, clear, and imperfect black (slate colored with a brown margin). The seeds are prone to mechanical injury during harvest and handling, which reduces germination and/or results in damaged seedlings. The testa is either dull or shiny.

The roots may penetrate 5 to 6 feet (2 m) into the soil, but most of the roots form in the top 8 inches (20 cm) of soil. Nodules (Figure 26.3) begin to form on the roots ten to fourteen days after emergence and continue forming throughout the life cycle of the plant. They serve as the main source of nitrogen for the plant. Active nodules are pink inside.

Upon emergence from the soil, the cotyledons develop chlorophyll and then open to expose the first two true plant leaves. After five to eight days, the plant no longer depends upon its cotyledons as a major food supply. At the base of the

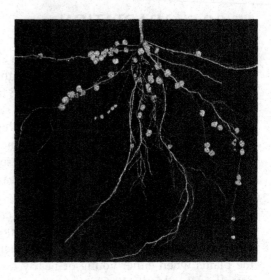

FIGURE 26.3
Characteristic large modules on soybean roots which show that proper nitrogen-fixing bacteria are present.

cotyledons are axillary buds that can develop into main stems in the event that the epicotyl is destroyed by hail or other hazards such as insects or rabbits.

26.4.2 Vegetative Development

The first two leaves are opposite and unifoliate; all subsequent leaves are alternate and trifoliate. The trifoliate will number from one to six or more until floral initiation occurs about six to eight weeks after seedling emergence.

26.4.3 Reproductive Development

Soybean varieties are either indeterminate or determinate in their flowering habit. The small white or purple flowers are borne in axillary racemes on peduncles arising at the nodes. The flowers appear first toward the base of the main stem, then progress upward toward the tip. Frequently, 65 to 75 percent of the flowers abort and fail to produce pods. The pods are small, being straight or slightly curved and covered with long hair (Figure 26.4). They range from very light straw color through numerous shades of gray and brown to nearly black. The pods contain one to four and, occasionally, five seeds. The seeds are round to elliptical. Most varieties have unicolored seeds that are straw yellow, greenish-yellow, green, brown, or black. Bicolored seeds occur in some varieties, the common pattern being green or yellow with a saddle of black or brown.

The seedcoats of many yellow and green seeded varieties sometimes become splashed or blotched with irregular brown or black markings superimposed on the basic color. This mottling, which is sporadic, and does not occur in some areas, is due both to heredity and environment.[14] Rich soils, liberal spacing between plants, and shading are conducive to mottling. Some strains have a high degree of resistance to mottling,[15] and most of the widely used varieties are not subject to mottling.

Soybean is normally self-fertilized because pollination occurs about as soon as the flower opens or a little before. Although natural crossing is much less than 1 percent,[16, 17] it may account for many varietal mixtures. Natural cross-pollination is the work of insects such as thrips and bees.

FIGURE 26.4
Soybean seeds and pods.

26.4.4 General Development

Soybean develops into a rather erect, bushy, leafy plant attaining a height of 2 to 4 feet (60 to 120 cm). Most varieties have a well-defined main stem that branches from the lower nodes when the plants have sufficient space. Determinate varieties cease vegetative growth when the main stem terminates in a cluster of mature pods. Determinate soybean varieties are grown mostly in the southern United States where summer nights are long. Indeterminate varieties develop leaves and flowers simultaneously throughout a portion of their reproductive period, with one to three pods at the terminal apex. These are commonly grown in the northern latitudes of the United States.

Varieties having an indeterminate growth habit begin to flower relatively early and continue to increase in height for several weeks after beginning to flower, while determinate varieties make most of their growth (height) before beginning to flower. At present, many commercially grown varieties of Group IV (Table 26.1) or earlier have an indeterminate growth type, while varieties of Group V maturity or later display determinate growth habits.

As the time of maturity approaches, when the seeds still contain about 20 percent moisture, the leaves begin to turn yellow, dropping off before the pods mature (Figure 26.5). The leaves and stems of nearly all varieties are covered with a fine lining of either grey or tawny-colored pubescence.

▉ 26.5 VARIETIES

As soybean increased in popularity, considerable effort was expended both publicly and privately to develop improved varieties. There may be over one hundred varieties available to soybean growers in any location where the crop is popular.

TABLE 26.1 Maturity Date of the Ten Soybean Variety Groups When Grown in Their Areas of Adaptation

Group	Average Maturity Date for Group	Days from Planting
00	September 10	120
0	September 28	126
I	September 30	126
II	October 3	130
III	October 3	131
IV	October 7	136
V	October 7	139
VI	October 22	148
VII	October 30	156
VIII	November 9	158

Because of different maturities, soybean varieties are limited in their range of best adaptation to latitude zones from 100 to 150 miles (160 to 260 km) wide. When moved north of their zone in the northern hemisphere, the variety may be too late to mature before frost kills the plants. A southward shift makes the plant mature too early for optimum yields.

Temperature affects the time of flowering as well as time of maturity of soybean varieties.[11] The average July temperature decreases to the north about 1.3°F (1°C) for each 100 miles (160 km) of latitude. This is sufficient to delay the flowering and ripening of soybean by about three days in each 100 miles (160 km). Day length and temperature differences together can shift the maturity of a soybean variety from

FIGURE 26.5
Mature soybean plant after the leaves have dropped.

FIGURE 26.6
There are ten maturity classes of soybean varieties. Those varieties adapted for use in southern Canada and the northern most area of the United States are designated 00 and are the earliest maturing. The higher the number, the later the maturity and the farther south the variety is adapted for full-season use. The lines across the map are hypothetical. There are no clearly cut areas where a variety is or is not adapted. [Reprinted with permission of SA Publications. Champaign, IL.]

five to six days with each 100-mile (160 km) difference in latitude. Altitude as well as proximity to large bodies of water also affects the temperature. Consequently, even when zones of variety adaptation, dates of planting, and times of harvest are similar, the results are not exactly parallel.

Representative soybean varieties grown in the United States are categorized in ten maturity classes from 00 to VIII (Table 26.1). Groups 00 and 0 are the earliest-maturing varieties adapted to the northern latitudes of the United States and southern Canada (Figure 26.6). The range in maturity within any one given group varies from ten to twenty-one days due to the acceleration of flowering as a result of decreased day length.

The maturity groups of soybean grown in each state are listed in Table 26.2. All states grow two or more varietal groups. The earlier varieties are usually grown in the northern part of a particular state.[18, 19] They are often used for late planting or are planted on fields to be harvested early before the later varieties are mature. This practice not only extends the harvest season, but also permits the planting of fall-sown crops on the early soybean field. Full-season varieties should be planted early in order to utilize a long growing season and still reach maturity. Partial-season varieties that mature early yield more when planted somewhat later. The longer days of late spring and early summer extend the vegetative period and increase the plant size of the early varieties. When double cropping soybean after wheat, there is no advantage to planting a short-season variety. Photoperiod response will cause the plants to flower at the proper time regardless of seeding date.

26.6 FERTILIZERS

Soybean grows the best on soils of high fertility. Calcium and magnesium are applied to acid soils to raise the pH to 6.0 or 6.5.

TABLE 26.2 Planting Dates for Soybean in Major Production States

States	Maturity Classification of Varieties Grown	Planting Dates for Best Results	Latest Safe Date
Alabama	VI, VII, VIII	May 10 to June 15	July 10
Arkansas	V, VI, VII	May 1 to 20	June 30
Delaware	III, IV, V, VI	May 10 to 30	June 30
Florida	VI, VII, VIII	June 1 to 30	July 15
Georgia	VI, VII, VIII	May 1 to June 15	July 10
Illinois	I, II, III, IV	May 5 to 25	June 30
Indiana	I, II, III, IV	May 5 to 25	June 30
Iowa	I, II, III	May 1 to 30	June 30
Kansas	III, IV, V	May 10 to June 15	June 30
Kentucky	IV, V, VI	May 1 to 30	June 30
Louisiana	VI, VII, VIII	May 10 to June 15	July 10
Maryland	IV, V, VI	May 15 to 30	June 30
Michigan	00, 0, I, II	May 5 to 30	June 30
Minnesota	00, 0, I, II	May 10 to 30	June 30
Mississippi	V, VI, VII	May 1 to June 5	July 5
Missouri	II, III, IV, V, VI	May 5 to 30	June 30
Nebraska	I, II, III	May 15 to June 5	June 30
North Carolina	VI, VII	May 1 to 30	June 30
North Dakota	00, 0	May 15 to 30	June 20
Ohio	I, II, III	May 5 to 25	June 30
Oklahoma	V, VI	May 10 to 30	June 30
South Carolina	VI, VII, VIII	May 1 to 20	June 30
South Dakota	00, 0, I, II	May 10 to 30	June 30
Tennessee	IV, V, VI	May 1 to 25	June 30
Virginia	IV, V, VI	May 15 to June 15	June 30
Wisconsin	0, I, II	May 5 to 30	June 30

About 20 to 40 pounds per acre (22 to 45 kg/ha) of P_2O_5 are advisable on soils very low in available phosphorus, with smaller rates for soils that are shown by soil tests to be only low to medium in phosphorus.

Soybean is a heavy user of potassium. Potassium application is recommended where soil tests indicate less than 150 pounds per acre (168 kg/ha) of K_2O available in the southeastern United States or less than 200 to 250 pounds per acre (220 to 280 kg/ha) of K_2O available per acre in the Midwest. Soybean planted after corn does not need additional fertilizer as there is usually sufficient carryover from the preceding crop.

Sulfur or certain micronutrients are not applied to soybean fields except on strongly weathered, coarse-textured alkaline or organic soils. Iron deficiency in soybean occurs on calcareous soils or soils with a pH above 7.3. Some varieties are less susceptible to iron deficiency chlorosis. Cultivation can increase movement of oxygen into the soil, which helps convert iron to a form that is more readily absorbed by the plants.[20]

▓ 26.7 ROTATIONS

Soybean is often grown in short rotations with corn, cotton, and small grains. As a full-season crop, it can occupy any place in a rotation where corn is used.[21] Soybean usually performs best when following a grass crop such as corn or grain sorghum. Yields of soybean are 5 to 15 percent higher following corn than continuous soybean due primarily to less disease.[22] In many areas, soybean is alternated with corn. A common rotation in the Corn Belt is corn (one or more years), soybean, small grain, and legumes.[4] In the South, soybean is usually grown in rotations with cotton, corn, or rice. Sometimes it is planted after the harvest of early potatoes or winter grain. Winter grain can follow soybean that is harvested before time to seed the grain. Soybean can be substituted for oat in a corn-oat-wheat-clover rotation. It can also be used as a catch crop where new seedings of grass and clover have failed.

On hilly land, soybean cannot take the place of sod legumes such as alfalfa or the clovers because it does not give sufficient protection against soil erosion, particularly when it is planted in cultivated rows. Erosion from soybean land is nearly as high as from land in corn,[23] unless the soybean is drilled in narrow rows.[8]

Soybean can be used in a double-crop rotation with a small grain, usually wheat. Since soybean is very photoperiod sensitive, it will initiate flowering earlier in its vegetative stages when seeded late after wheat harvest. Since the soybean has less vegetative growth, narrow rows and higher seeding rates should be used. Row spacing of 20 inches (50 cm) or less, and seeding rates of 80 to 100 pounds per acre (90 to 110 kg/ha) are recommended for the Corn Belt.[24, 25] Soybean yields will be about 15 to 39 percent lower when planted after wheat harvest because of the shorter growing season.[24]

The decision to plant soybean after wheat should be made immediately after wheat harvest. Soil moisture is the most critical factor. Only rainfall that occurs after the preceding wheat crop is nearing maturity will be available to the subsequent soybean crop. If the soil is dry in the top 2 inches, there will be insufficient water for germination. Preserving soil water by no-till practices will greatly increase the probability of success. At least ninety frost-free days are needed to adequately mature the crop, so the seeding date should be no later than ninety days before expected frost.[25] Double cropping soybean is more common south of the Corn Belt where the growing season is longer. Full-season varieties are used near the Gulf Coast.[26]

26.7.1 Intercropping Soybean

Although soybean is usually grown in a monoculture (one crop in a field), the crop does work well in modified intercropping systems. Strip intercropping, in which soybean and corn are planted in strips four to eight rows wide, has been shown to increase total yields for the field. Soybean gets some protection when the corn acts as a windbreak. The corn has border rows that can take advantage of more light.[27, 28] Inclusion of a third crop such as a small grain has also shown promise.[29]

Seeding soybean into a small grain before the small grain reaches maturity is called relay intercropping. Wheat was the best choice in Nebraska trials.[30] This system increases the growing season for soybean by several additional weeks, which can greatly increase yields. The soybean should be seeded before the wheat reaches boot stage to minimize damage to the growing crop. Taller wheat varieties with stiffer culms work best. Irrigation is necessary, since the germinating seeds and seedlings have very limited root systems, which gives the growing wheat a distinct competitive advantage.

■ 26.8 SOYBEAN CULTURE

Soybean is usually planted in cultivated rows. In the Corn Belt, soybean planted in narrow rows normally yields 10 to 15 percent more than when grown in 36-inch (90 cm) rows. In the South, narrower rows are not beneficial. The primary advantage of narrow rows is more efficient light interception. The canopy closes sooner and plants are spaced farther apart in the row. Yield increases from narrow rows are most likely to occur when other factors, such as water, are not limiting yield.

A grain drill is occasionally used for planting in rows by closing some of the hopper openings. The corn planter, cotton planter, sugarbeet drill, and pea planter can be equipped with soybean plates for planting soybean.

Soybean is usually planted with a drill for hay, soiling, or green manure. A finer quality of hay is obtained by this method. Drilled soybean is weedier than that in cultivated rows, unless it is cultivated two to three times with a rotary hoe, or weeds are controlled with a herbicide. Seed treatment with a fungicide is generally not recommended because of possible damage to the *Rhizobium* inoculant.

26.8.1 Inoculation

Nitrogen-fixing bacteria (*Rhizobium japonicum*) develop nodules on soybean roots (Figure 26.3) that are capable of providing the entire nitrogen needs of the soybean. Inoculation is not required in fields where soybean has been grown and inoculated for many years. But, when in doubt, seed should be inoculated, especially if soybean has not been grown in the past three to five years.[31]

The inoculum is normally mixed with the planting seed. One pint of water to a bushel of dry seed plus the recommended quantity of inoculum applied no sooner than one day before planting is adequate. Treated seed should not be left in direct sunlight since this can destroy the bacteria. Soybean grown for the first time on a field may have higher yields when a soil-applied inoculant is used.[31] Soil-applied inoculants are applied into the seed furrow at planting through the herbicide or insecticide boxes on the planter.

A 40 bushel per acre (2,700 kg/ha) soybean crop contains approximately 240 to 250 pounds (108 to 113 kg) of nitrogen; nearly 150 pounds (68 kg) are in the seed. Well-inoculated soybean usually fails to respond to nitrogen fertilizers, although 10 to 20 pounds per acre (11 to 22 kg/ha) of nitrogen in the starter fertilizer may stimulate early planted soybean before the nodules develop.

Rhizobia do not function efficiently when soils are poorly aerated or when soil pH is below 5.0. Molybdenum is important for nitrogen fixation and may be deficient below pH 6.0. An inoculant supplemented with molybdenum can be used, but the best long-term solution is to apply lime to the soil if soybean will be grown regularly.[31]

26.8.2 Seedbed Preparation

In general, the preparation of the land for soybean should be the same as for corn. A firm seedbed is needed for soybean. Soybean needs more moisture for germination than corn, and good seed-soil contact is essential. Soybean is not a good early weed competitor so a weed-free seedbed is important. Excessive tillage before seeding can result in a drier, less firm seedbed.

26.8.3 Seeding Methods

Soybean is best planted in May or early June in most states, as shown in Table 26.2. Earlier planting hastens flowering and reduces yields. In the North, the soil is usually too cold to plant soybean before late May or early June. Although soybean will germinate when the soil temperature is 50 to 55°F (10 to 12°C), a minimum soil temperature of 60°F (16°C) will result in more uniform germination and emergence.[32]

RATE OF SEEDING Soybean is normally planted at a depth of 1 to 2 inches (2.5 to 5 cm). Seeding rate depends on row spacing and whether the variety is determinate. Seeding rates vary from 100,000 seeds per acre (250,000 per ha) for indeterminate varieties in rows wider than 30 inches to over 200,000 seeds per acre (500,000 per ha) for determinate varieties seeded with a drill in rows less than 10 inches (25 cm).[22, 33, 34, 35]

26.8.4 Weed Control

The relatively slow early growth of soybean necessitates early weed control by cultivation and herbicide applications. Planting in narrow rows restricts weed growth by shading.

Preplant tillage, followed by rotary hoeing when the weeds are in the two- to three-leaf stage after emergence of soybean, is an effective control method. Shallow sweep cultivation controls later weeds.

Chemical weed control can be done on a preplant, preemergence, or postemergence basis. Preplant and preemergent herbicides used with soybean include trifluralin, metribuzin, pendimethalin, clomazone, dimethenamid, and alachlor. Postemergent herbicides include bentazon, chlorimuron, sethoxydim, imazamox, and flumiclorac. Many herbicide combinations are sold. Glufosinate and glyphosate can be applied postemergent on those varieties that have been bred with resistance to those herbicides.

26.8.5 Irrigation

Soybean responds to irrigation when rainfall is limited, but the growth stage when irrigation occurs can affect yield. Abundant water during early vegetative stages can promote excessive vegetative growth and can increase lodging. If water is adequate, irrigation should be delayed on indeterminate varieties until flowering or later.[8] Irrigation before the vegetative apex is finished producing leaves can spur additional vegetative growth at the expense of developing flowers and pods. Irrigation of determinate varieties can begin just before flowering with no yield reduction.[10] Irrigation with all types should be continued until pods are fully developed. If irrigation water is limited to a single irrigation during late-pod development to early seed-filling period will maximize yield.[9]

26.8.6 Harvesting and Storage

HARVESTING SEED Soybean for seed is harvested most efficiently when the moisture content of the seeds drops to 12 percent. Later harvesting increases shattering losses, as well as splitting of the overly dry beans in threshing. Shatter-resistant varieties are available. To minimize split beans, the cylinder speed of the combine should be operated at 300 to 450 revolutions per minute. The seedcoat, cotyledons, and radicle are very sensitive to mechanical damage during harvest. Soybean at 13 percent moisture can be combined directly without windrowing and stored without drying. The height of the cutter bar and speed of the combine are very important in limiting harvesting losses. At faster combine speeds, the cutter bar rides higher and causes increased shattering of lower pods.

SEED STORAGE Soybean should be stored at no more than 13 percent moisture. If the crop will be stored for more than one year, moisture should be 11 percent or less. Soybean can be artificially dried in storage. Drying air temperature should be 130 to 140°F (54 to 60°C). Soybean should not be stirred during drying to avoid cracking the seedcoat. Soybean should be aerated if necessary to maintain seed temperature at 35 to 40°F (2 to 4°C) in winter and 40 to 60°F (4 to 16°C) in summer.[36] Soybean that will be planted for seed should not be stored more than one year because of germination loss during storage.

HAY HARVEST Soybean can be cut for hay anytime from pod formation until the leaves begin to fall. The best quality of hay is obtained when the seeds are about half developed. The weight of leaves increases until the beans are well formed, remains constant for about three weeks, and then decreases rapidly.[37] The hay contains about 60 percent leaves when the beans are well formed, and about 50 percent when the beans appear half grown. In addition to the loss of leaves, the stems become woody when cutting is delayed. Soybean is more difficult to cure than alfalfa or clover because the thicker stems dry out more slowly. Use of a hay crusher facilitates curing. Today, very few soybean fields are cut for hay except after a disaster that prevents the crop from reaching maturity.

▮ 26.9 USES OF SOYBEAN

World production of soybean oil, 25 million tons (23 million MT) in 1999, is over one-half of the total edible vegetable oil production, others being peanut, cottonseed, coconut, sunflower, canola, sesame, and olive oils. The United States produced about 9 million tons (8 million MT) of soybean oil in 1999. A 60-pound (27 kg) bushel of soybean produces about 48 pounds (22 kg) of meal and 11 pounds (5 kg) of oil.

In 1999, about 33 million tons (30 million MT) of soybean, soybean oil, and soybean meal were exported from the United States to other countries. Europe and

Japan are the largest export customers; others include Taiwan, Mexico, Korea, China, and Canada.

Soybean oil is used primarily for shortening, margarine, and salad oil. The lecithin from the oil is used in baked goods, candies, chocolate, cocoa, and margarine.

Soybean cake and soybean meal production in the United States exceeds 30 million tons (27 million MT) annually. Soybean cake and meal are the chief high-protein supplements in mixed feed rations for livestock. The meal is also used in plastics, core binders, glue, and water paints. The oil is utilized in the making of candles, celluloid, biodiesel, disinfectants, electric insulation, enamels, glycerin, insecticides, linoleum, oilcloth, paints, printing ink, rubber substitutes, varnish, and soaps. The casein from vegetable milk produced from the dried bean is used in paints, glue, paper sizing, textile dressing, and for waterproofing.[1, 18, 38]

Soybean flour is a valuable source of vegetable protein. It may be used in mixture with wheat flour in various baked products such as bread, cake, cookies, and crackers. Soybean flour is used in ice cream, artificial creamers, ice cream cones, candies, puddings, and salad dressing. Because of its low starch content, it has found a place as a diabetic food. Soybean flour makes a good bread when mixed with wheat flour in any combination containing up to 20 percent soybean flour.[39]

In oriental countries, soybean is used in the production of soybean milk and curd, various soy sauces, fermented products, bean sprouts, and numerous other foods. Soybean is also used in confections, soups, potted meats, food drinks, breakfast foods (puffed beans, flakes, and prepared meal), and as salted roasted beans, soybean butter, and as a substitute for coffee.[40] Most of the products, except the fermented ones, are now readily available in the United States. Soybean is used as green shelled beans and dry beans.[41] Vegetable varieties of soybean cook up more readily than do the field varieties. These vegetable varieties, which are used both as shelled green beans and as dry beans, have large seeds.

26.10 CHEMICAL COMPOSITION

The chemical composition of soybean and soybean products is given in Table 26.3.

The percentage of nitrogen in the leaves of soybean hay is nearly twice as high as that of the stems.[42] The percentage of nitrogen in the total of the tops decreases

TABLE 26.3 Analyses of Soybean Forage, Seeds, and Meal

Item	Moisture	Ash	Crude Protein	Carbohydrates		Oil or Fat
				Crude Fiber (%)	N-Free Extract (%)	
	(%)	(%)	(%)			(%)
Green forage	75.1	2.6	4.0	6.7	10.6	1.0
Hay	8.4	8.9	15.8	24.3	38.8	3.8
Seeds	6.4	4.8	39.1	5.2	25.8	18.7
Oil meal (hydraulic or expelled process)	8.3	5.7	44.3	5.6	30.3	5.7
Oil meal (solvent process)	8.4	6.0	46.4	5.9	31.7	1.6

during the period of rapid growth and increases as the seed matures. More than half of the nitrogen in the total of the tops is stored in the seed at maturity.

The oil content of the seeds ranges from 14 to 24 percent or more,[18] while the protein ranges from 30 to 50 percent. The breeding of varieties with a high oil content has since raised the average oil content to more than 20 percent. In general, soybean seeds with a low fat content are high in protein and vice versa. Soybean protein contains all the essential amino acids for animal feed and human food. Soybean contains two to three times as much ash as wheat, and is a valuable source of calcium and phosphorus. Like other edible legume seeds, soybean is high in thiamin (vitamin B_1).

■ 26.11 SOYBEAN–OIL EXTRACTION

Soybean oil is first extracted using an expeller. A solvent-extraction process is then used to extract the remaining oil. In the solvent-extraction process, the oil is extracted from the beans by a chemical solvent such as commercial hexane. The solvent, recovered from the oil by distillation, is used again. Oil obtained by this process has superior bleaching properties. The meal, since it contains less oil, is less likely to become rancid. Oil extraction by the solvent method is 95 percent complete, only 0.5 to 1.5 percent oil being left in the meal. Before extraction, the beans are cleaned, dried to 10 to 11 percent moisture, crushed into grits, tempered for fifteen to twenty-five minutes at 130 to 170°F (54 to 77°C), and rolled into flakes about 0.007 inch (178 μ) thick.[43]

26.11.1 Quality of Soybean Oil

One of the important qualities in soybean oil is the drying property as measured by the iodine number. A high iodine number indicates good drying quality for paint purposes. The iodine number of soybean oil ranges from 118 to 141 in different varieties, while that of linseed oil is about 180.[1, 44] However, soybean oil can be fractionated to produce oils of both high and low iodine number. In the hydrogenation process to produce solid fats for shortening or margarine, an oil with a low iodine number is desirable.

■ 26.12 DISEASES

Soybean, while once considered a disease-free crop in the United States, has been grown long enough that many diseases have become a potential problem. A few of the many diseases that attack soybean are described.

26.12.1 Bacterial Blight

Bacterial blight, caused by *Pseudomonas glycinea* and reported from various regions, occurs on the leaves of the soybean plant as small angular spots. These spots at first are yellow to light brown, but later they become dark brown to almost black. Diseased tissues of the leaves may finally become dry and drop out. The spots sometimes spread to the stems and pods. Under favorable conditions, bacterial blight spreads rapidly in the field. The bacteria are seed-borne, but they may overwinter on dead leaves.[45, 46] Crop rotation, and the turning under of soybean residues, aids in reducing the disease. Some varieties are partly resistant.

26.12.2 Bacterial Pustule

Bacterial pustule disease, caused by *Xanthomonas phaseoli* var. *sojense*, has been reported in several states. The disease is confined largely to the foliage.[45, 46] In the early stages, the spots are light green in color. Later, the disease is characterized by angular reddish-brown spots on the leaves that may become large, irregular brown areas. Portions of the larger spots frequently drop out. The organism overwinters in diseased leaf material as well as on the seed of diseased plants. Some control of the disease has resulted from crop rotation and from disposal of dead, diseased leaf material by plowing after harvest. Resistant varieties are available. Bacterial pustule is a warm-weather disease that occurs later than bacterial blight.

26.12.3 Wildfire

Wildfire, caused by the bacteria *Pseudomonas tabaci*, is a serious disease in the South. Black or brown spots on the leaves are surrounded by yellow halos and occur only where bacterial pustules are found. Some resistant varieties are available.

26.12.4 Sclerotinia Stem Rot

Sclerotinia stem rot is caused by the fungus *Sclerotinia sclerotiorum*. This fungus has caused white mold disease in dry, edible beans for years, but it has recently infected soybean. It overwinters in plant residue and soil. It requires cool, wet summers for infection, so it will not be a problem every year. It is mostly confined to the Corn Belt. The spores infect blossoms, and the disease is most visible during pod development. Leaves will wilt and turn gray-green before turning brown and dying.[20] There is no effective control for this disease.

26.12.5 Phytophthora Root and Stem Rot

Phytophthora root and stem rot, caused by the fungus *Phytophthora sojae*, is one of the most serious diseases of soybean in the United States. It occurs when the crop is exposed to cool, wet conditions, particularly when the soil is not well drained. It can kill seeds and seedlings during germination and emergence. Seedling symptoms are similar to other "damping-off" fungal diseases; the seedlings wilt and appear water-soaked. When the disease attacks older plants, the first symptoms are yellowing and wilting of leaves with dark brown discoloration of the stem from the soil line upward. Secondary infection by other fungi, particularly *Fusarium* spp., may also occur, producing a white fungal growth in the phytophthora-infected areas. Below ground, the taproot is dark brown and the root systems rotted.[3] The best control is resistant varieties. Application of a fungicide seed treatment can help control the disease at germination.[47]

26.12.6 Brown Stem Rot

Brown stem rot, caused by a soil-inhabiting fungus *Cephalosporium gregatum*, decays the interior of the stem. It attacks the plant early in the season, but symptoms are not usually seen until later in the year. The center of the stem in diseased plants is reddish brown. Crop rotation, which does not have soybean on a field for two to three years, is an effective cultural control.

26.12.7 Stem Canker

Stem canker is caused by a seed-borne fungus *Diaporthe phaseolorum* var. *batatatis* that girdles the stem and kills the plant. A brown lesion that girdles the stem is usually found at the base of a branch or leaf petiole. Resistant varieties are available.

26.12.8 Pod and Stem Blight

Pod and stem blight, caused by the fungus *Diaporthe phaseolorum* var. *sojae*, is found wherever soybean is grown. The stems and pods are heavily dotted with black, spore-filled sacs *(pycnidia)*. The fungus may girdle the stem, kill the plant, and prevent seed development. Sometimes the fungus penetrates the seed and destroys subsequent germination. The fungus is carried over winter on diseased stems and infected seeds. Recommended control measures for the disease are crop rotation and the planting of disease-free seed.

26.12.9 Frog Eye Leaf Spot

Frog eye, a leaf spot disease caused by the fungus *Cercospora sojina*, attacks the leaves, pods, and stems of the soybean plant. It may go through the pod and enter the seed. The lesions are reddish in color when young, but change to brown and then to smoky gray with age. The fungus overwinters on refuse of diseased plants left in the field after harvest. Frog eye leaf spot is introduced into new fields by diseased seed. Seed disinfectants have failed to give satisfactory control of the disease, but several varieties are resistant.

26.12.10 Brown Spot

Brown spot, caused by the fungus *Septoria glycines*, attacks leaves, stems, and pods and causes early defoliation. The small, reddish-brown spots on the leaves are angular. The fungus overwinters on the crop residues, which should be plowed under. Crop rotation reduces the disease losses.

26.12.11 Target Spot

Target spot is caused by the fungus *Cornespora casucola*, which produces reddish-brown spots that are encircled. Resistant varieties are available.

629

26.12.12 Downy Mildew

Downy mildew is caused by the fungus *Peronospora manchurica*. The disease is seed-borne. When infected seeds are planted, the first leaves to unfold are frequently covered with the characteristic mildew growth. The lack of resistant varieties requires crop rotation, plowing under residue, and planting disease-free seed to control the disease.

26.12.13 Mosaic

Mosaic is one of the most common diseases. Infected leaves may show characteristic puckering or wrinkles. A reduction in seed numbers in the pods of affected plants is also noticeable. At temperatures above 80°F (27°C), soybean outgrows the symptoms of the disease. The virus that causes the disease is transmitted by aphids. There is no practical control for the disease.

26.12.14 Purple Stain

This disease is caused by several species of the *Cercospora* genus. The symptom most frequently observed is a discoloration of the seed varying from pink to dark purple. *Cercospora kikuchii* is the most common causal organism. Excessive "purple stain" may lead to reduction of seed value in the market place. Seedlings that develop from diseased seed may die. Control measures include treating the seed and using high-quality seed to avoid the infestation.

26.12.15 Other Diseases

Sclerotial blight, caused by the fungus *Sclerotium rolfsii*, is a rotting of the base of the stem. It occurs mostly in sandy soil areas in the South. The disease attacks many other crop plants.

The charcoal rot disease, caused by the fungus *Macrophomina phaseoli (Sclerotium bataticola)*, also rots the base of the stems. After the plant is dead, the stem and roots bear numerous black sclerotia (spore bodies). The fungus lives in the soil. No remedy for the disease is known.

Other diseases of soybean include (1) root rots caused by the soil-inhabiting fungi *Pythium debaryanum, Rhizoctonia solani, Phymatotrichum omnivorium*; (2) anthracnose, caused by *Glomerella glycens*; and (3) fusarium blight caused by *Fusarium oxysporum f. tracheiphilum*. The last disease occurs on sandy soils in the South; several varieties are resistant.

▪ 26.13 NEMATODES

Nematodes are a major problem in soybean production primarily in the southern United States. Root-knot nematode causes knot-like swellings or galls on the roots. The resulting stunted plants wilt during dry weather and show marginal firing of the leaves. Rotation with nonsusceptible crops such as small grains and sorghum may be the best control.

Plants severely attacked by the soybean cyst nematode *Heterodera glycines* are usually stunted and the foliage is prematurely yellowed. The female, lemon-shaped nematode is found on the roots of infected plants. Using resistant varieties is the major control recommended. Rotation and preventing movement of infected soil are also important control measures.

▦ 26.14 INSECT PESTS

Integrated pest management (IPM) strategies are developed for all common soybean pests. Careful monitoring of beneficial insects as well as pests and knowledge of treatment thresholds can provide satisfactory results with minimum cost.

One insecticide application is usually sufficient to control insect attacks. Leaf-feeding insects, including cabbage loopers, green clover worms, velvet bean caterpillars, fall armyworms, beetles, and grasshoppers, are the major problems in soybean. As high as 35 percent foliage loss during blooming, and no greater than 20 percent during pod filling, may be tolerated before spraying is normally recommended.

26.14.1 Leaf Feeding

Two pests that can cause sufficient leaf loss to affect yields are the two-spotted spider mite and bean leaf beetle. The two-spotted spider mite feeds on the undersides of leaves by sucking plant juices. Damage shows first on the top side of leaves as a yellow or whitish spotting of leaf tissues. Webbing will be evident on the under side. Severely infested leaves will die.[48]

Bean leaf beetle adults feed on leaves, causing defoliation, and on pods, causing scarring.[49] The larvae feed on the roots and root nodules below ground. This pest also spreads the mottling virus.

26.14.2 Pod and Flower Feeding

Pod and flower feeding insects cause a direct yield reduction, especially if infestation occurs late in the blooming period. Bollworms and stink bugs, Mexican bean beetle, grasshoppers, green clover worm, and velvet bean caterpillars attack pods or flowers in soybean.

26.14.3 Stem Feeding

Major stem-feeding insects are the three-cornered alfalfa hopper and weed borer. Control measures are rarely needed to prevent yield reductions from these pests.

26.14.4 Seed and Seedling Feeding

Seed corn maggot, seed corn beetle, wireworm, grape calapsis, white grubs, thrips, southern corn rootworm, bean leaf beetle, and garden symphylans all feed on the seed, seedling, plant roots, or leaves. All of these pests, except garden symphylans, can be controlled by seed treatment or use of a soil insecticide.

REFERENCES

1. McClelland, C. K., and J. L. Cartter. "Improvement in soybeans," *USDA Yearbook*, 1937, pp. 1154–1189.

2. Caldwell, B. E., and others. eds. *Soybeans: "Improvement, production and uses."* ASA Monograph 16. Madison, WI: Am. Soc. Agron., 1975.

3. Johnson, H. W., and R. L. Bernard. "Soybean genetics and breeding," in *Advances in Agronomy*, vol. 14. New York: Academic Press, 1962, pp. 149–221.

4. Weiss, M. C. "Soybeans," in *Advances in Agronomy*, vol. 1. New York: Academic Press, 1949, pp. 77–157.

5. Pendleton, J. W., R. L. Bernard, and H. H. Hadley. "For best yields grow soybeans in narrow rows," *Illinois Research* (IL Agr. Exp. Sta.). Winter, 1960.

6. Cartter, J. L., and T. H. Hopper. "Influence of variety, environment, and fertility level on the chemical composition of soybean seed," *USDA Tech. Bull.* 787, 1942, pp. 1–66.

7. Howell, R. W., and F. L. Collins. "Factors affecting linoleic acid content of soybean oil," *Agron. J.* 49(1957):593–597.

8. Benham, B. L., J. P. Schneekloth, R. W. Elmore, D. E. Eisenhauer, and J. E. Specht. "Irrigating soybeans," *NE Coop. Ext. Serv.* G98-1367-A, 1998.

9. Helsel, D. G., and Z. R. Helsel. "Irrigating soybeans," *Univ. MO Coop. Ext. Serv.* G4420, 1993.

10. Thomas, G. T., and A. Blaine. "Soybean irrigation," *MS St. Univ.Ext. Serv.* 2185, 1999.

11. Howell, R. W. "Physiology of the soybean," in *Advances in Agronomy*, vol. 12. New York: Academic Press, 1960, pp. 265–310.

12. Johnson, H. W., H. A. Borthwick, and R. C. Leffel. "Effects of photoperiod and time of planting on rates of development of the life cycle," *Bot. Gaz.* 122, 2(1960):77–95.

13. Thomas, B., and D. Vince-Prue, *Photoperiodism in Plants*, 2nd ed., San Diego: Academic Press, 1997.

14. Owen, F. V. "Hereditary and environmental factors that produce mottling in soybeans," *J. Agr. Res.* 34(1927):559–587.

15. Dimmock, F. "Seed mottling in soybeans," *Sci. Agr.* 17(1936):42–49.

16. Etheridge, W. C., C. A. Helm, and B. M. King. "A classification of soybeans," *MO Agr. Exp. Sta. Res. Bull.* 131, 1929.

17. Woodworth, C. M. "The extent of natural cross-pollination in soybeans," *J. Am. Soc. Agron.* 14(1922):278–283.

18. Beeson, K. E., and A. H. Probst. "Soybeans in Indiana," *Purdue U. Agr. Ext. Bull.* 231 (rev.), 1961.

19. Bernard, R. L., C. R. Mumaw, and D. R. Browning. "Soybean varieties for Illinois," *IL Agr. Exp. Sta. Circ.* 794, 1958.

20. Steadman, J. R., S. Rutledge, D. Merrill, and D. S. Wysong. "Sclerotinia stem rot in soybeans," *NE Coop. Ext. Serv.* G95-1270-A, (Rev.), 1995.

21. Johnson, H. W., J. L. Cartter, and E. E. Harturg. "Growing soybeans," *USDA Farmers Bull.* 2129, 1959.

22. Helsel, D. G., and Z. R. Helsel. "Soybean production in Missouri," *Univ. MO Coop. Ext. Serv.* G4410, 1993.

23. Duley, F. L. "Soil erosion on soybean land," *J. Am. Soc. Agron.* 17(1925):800–803.

24. Anonymous. "Double cropping winter wheat and soybeans in Indiana," *Purdue Univ. Coop. Ext. Serv.* ID-96, 1996.

25. Minor, H. C., and W. Wiebold "Wheat-soybean double-crop management in Missouri," *Univ. MO Coop. Ext. Serv.* G4983, (Rev.), 1998.

26. Blaine, A. "Soybeans: double-cropping soybeans and wheat in Mississippi," *MS St. Univ. Ext. Serv.* 1380, 1998.

27. Francis, C. A., A. Jones, K. Crookston, K. Wittler, and S. Doodman. "Strip cropping corn and grain legumes: A review," *Am. J. Altern. Agric.* 1(1986):159–164.

28. Ghaffarzadeh, M. "Strip intercropping," *IA St. Univ. Coop. Ext. Serv.* PM1763, 1999.

29. Ghaffarzadeh, M., F. Garcia, and R. M. Cruse. "Grain yield response of corn, soybean, and oats grown in a strip intercropping system," *Am. J. Altern. Agric.* 9(1994):171–177.

30. Moomaw, R., G. Lesoing, and C. Francis. "Two crops in one year: relay intercropping," *NE Coop. Ext. Serv.* G91-1024-A, 1991.

31. Elmore, R. W."Soybean inoculation—when is it necessary?" *NE Coop. Ext. Serv.* G84-737-A, 1985.

32. Elmore, R. W., and A. D. Flowerday. "Soybean planting date, when and why," *NE Coop. Ext. Serv.* G84-687-A, 1984.

33. Blaine, A. "Soybeans: plant populations and seeding rates," *MS St. Univ. Ext. Serv.* 1194, 1994.

34. Christmas, E. P. "Plant populations and seeding rates for soybeans," *Purdue Univ. Coop. Ext. Serv.* AY-217, 1993.

35. Elmore, R. W., R. S. Moomaw, and R. Selley. "Narrow-row soybeans," *NE Coop. Ext. Serv.* G90-963, 1990.

36. Hurburgh, C. R., Jr., "Soybean drying and storage," *IA St. Univ. Coop. Ext. Serv.* PM-1636, 1995.

37. Willard, C. J. "The time of harvesting soybeans fin hay and seed," *J. Am. Soc. Agron.* 17(1925):157–168.

38. McClelland, C. K. "Soybean utilization," *USDA Farmers Bull.* 1617, 1930.

39. Bailey, L. H., R. G. Capen, and J. A. LeClerc. "The composition of soybeans, soybean flour, and soybean bread," *Cereal Chem.* 12(1935):441–472.

40. Woodruff, S., and H. Klaas. "A study of soybean varieties with reference to their use as food," *IL Agr. Exp. Sta. Bull.* 443, 1938, pp. 425–467.

41. Drown, M. J. "Soybeans and soybean products," *USDA Misc. Pub.* 534, 1943, pp. 1–14.

42. Cartter, J. L., and E. E. Hartwig. "The management of soybeans," in *Advances in Agronomy*, vol. 14. New York: Academic Press, 1962, pp. 360–412.

43. D'Aquin, E. L., E. A. Gastrock, and O. L. Brekke. "Recovering oil and meal," in *Crops in Peace and War*, USDA Yearbook, 1950–51, pp. 504–512.

44. Lloyd, J. W., and W. L. Burlison. "Eighteen varieties of edible soybeans," *IL Agr. Exp. Sta. Bull.* 453, 1939, pp. 385–438.

45. Johnson, H. W., and D. W. Chamberlain. "Bacteria, fungi, and viruses on soybeans," in *Plant Diseases*, USDA Yearbook, 1953, pp. 238–247.

46. Johnson, H. W., D. W. Chamberlain, and S. G. Lehman. "Soybean diseases," *USDA Farmers Bull.* 2077, 1955.

47. Wrather, J. A., and L. E. Sweets. "Soybean disease management," *Univ. MO Ext. Serv.* G4452, (Rev.), 1998.

48. Wright, R., R. Seymour, L. Higley, and J. Campbell. "Spider miter management in corn and soybeans," *NE Coop. Ext. Serv.* G93-1167-A, 1993.

49. Hunt, T., J. F. Witkowski, R. Wright, and K. Jarvi. "The bean leaf beetle in soybeans," *NE Coop. Ext. Serv.* G90-974, (rev.) 1994.

Cowpea

27.1 ECONOMIC IMPORTANCE

Cowpea is an important edible legume crop in many parts of the world, particularly Central and West Africa. In the United States, it is commonly called blackeye bean or blackeye pea. Global production of cowpea seed in 2000–2003 averaged nearly 3.9 million tons (3.5 million MT), grown on more than 23 million acres (9 million ha), and yielding about 335 pounds per acre (375 kg/ha). Over half of the world production is in Nigeria. Other countries leading in cowpea production are Burkina Faso, Niger, Myanmar, and Mali. In the United States, cowpea seed was harvested on an average of 32,000 acres (13,000 ha) in 2000–2003. Average yields were about 1,800 pounds per acre (2,000 kg/ha). Total production averaged about 28,000 tons (26,000 MT). Practically all of the cowpea is grown in California and Texas.

For years, cowpea was the leading legume in the southeastern United States. However, since 1941, it has been gradually replaced by soybean, clovers, and other legumes. Low yields of hay and seed, together with high production costs, make cowpea less desirable than other crops. About 50,000 tons (45,000 MT) of cowpea hay was harvested in the United States in 1969, but later statistics on the declining crop are unavailable.

27.2 HISTORY OF COWPEA CULTURE

Cowpea is native to central Africa where wild forms are found at the present time. Hybrids of the wild plant and the cultivated cowpea are readily obtained.[1] Cowpea has been cultivated for human food since ancient times in Africa, Asia, and Europe. Cowpea is probably the *Phaseolus* mentioned by the Roman writers.[2] It was introduced into the West Indies by early Spanish settlers. The crop was grown in North Carolina in 1714 and spread throughout the southern states later.

27.3 ADAPTATION

Cowpea is a short-day, warm-season crop grown primarily under humid conditions. It has been grown to some extent as far north as southern Illinois,

Indiana, Ohio, and New Jersey.[3] Cowpea is similar to corn in its climatic adaptation except that it has a greater heat requirement. It is sensitive to frost both in the fall and spring. Severe drought generally prevents formation of seed in most varieties.[4] It can be grown in rows for hay or seed in the southern Great Plains where the average precipitation is as low as 17 inches (432 mm), but the yields obtained scarcely justify the labor and expense involved in that method of production.

Cowpea is adapted to a wide range of soils. It grows as well on sandy soils as on clays, but Fusarium wilt and root knot are more prevalent in sandy soils than in soils that contain more clay.[5] The plant thrives better than clover on either infertile or acid soils, but saline and alkaline soils are unsuitable.[6] There is a general tendency for it to produce a heavy vine growth with few pods on very fertile soils or when planted too early. The primary soil requirements are good drainage and the presence of, or inoculation with, the proper nitrogen-fixation bacteria cultures.

■ 27.4 BOTANICAL DESCRIPTION

Cowpea (*Vigna unguiculata* syn. *V. sinensis*) is more closely related to beans (*Phaseolus* spp.) than to pea (*Pisum* spp.) It differs from the common bean in that the keel of the corolla is only slightly curved instead of twisted or slightly coiled.[1] Two related subspecies are the asparagus bean (*Vigna unguiculata* subspp. *sesquipedalis*) and the catjang cowpea (*V. unguiculata* subspp. *cylindrica*).

Cowpea is an annual herbaceous legume. The plants are viny or semi-viny and fairly leafy with trifoliolate leaves (Figure 27.1). The leaflets are relatively smooth and shiny. The growth habit of the cowpea is indeterminate. The plant continues to blossom and produce seed until checked by adverse environmental conditions.

The white or purple flowers are borne in pairs in short racemes. The pods are smooth, 8 to 12 inches (20 to 30 cm) long, cylindrical, and somewhat curved. They are usually yellow, but brown or purple pods are found in some varieties. Seeds are smooth or wrinkled and generally bean-shaped but usually short in proportion to their width. The multicolored seeds may be variously spotted, speckled, or marbled. The seeds weigh about 60 pounds per bushel (770 g/l). A pound (0.454 kg) contains 1,600 to 4,400 seeds.

The cowpea is largely self-pollinated. Natural crosses rarely occur in the field in most regions.

Catjang cowpea is an erect, semibushy plant with small, oblong seeds borne in small pods 3 to 5 inches (8 to 13 cm) long. The pods are erect or ascending when green and usually remain so when dry. Varieties of this group are very late and not very prolific in the United States, so they are seldom grown there.

The asparagus bean or yard-long bean plants are viny with pendant pods 12 to 36 inches (30 to 90 cm) long that become more or less inflated, flabby, and pale before ripening. The seeds are elongated and kidney-shaped. None of the varieties produces as much seed or forage as do the better cowpea varieties. Some people grow the asparagus bean as a novelty vegetable; it has pods so large that one suffices for a meal. It is not grown extensively because other beans are more palatable.

FIGURE 27.1
Plant of Victor cowpeas.

27.5 VARIETIES

There are over fifty varieties of cowpea that can be classified into several types (Table 27.1). The crowder varieties produce seeds more or less flattened on the ends as a result of being crowded in the pods. The seeds are either uniformly colored or multicolored. The more common solid colors are buff, clay, white, maroon, purplish, or nearly black, with a second color usually concentrated about the hilum[1] (Figure 27.2).

Good forage cowpea varieties are vigorous, erect, prolific, and disease-resistant, bear the pods well above the ground, and retain their leaves late in the season.[4, 7] Early maturity is important in the northern part of the cowpea region. The viny habit is considered desirable when the crop is planted in mixtures with corn or sorghum. White-seeded varieties and the crowder and blackeye types are commonly grown for table use or canning as green beans.[5]

TABLE 27.1 Classes or Types of Cowpea

Type	Seed Description
Blackeye or purple eye	White seeds with a black eye around the hilum that darkens to a dark purple. Some varieties have pink, purple, or red eyes. Seeds are oblong or kidney-shaped.
Brown eye	Immature seeds are medium to dark brown in color. Seeds more round than kidney in shape.
Crowder	Black, speckled, or brown seeds with a brown eye around the hilum. Seeds are round in shape.
Cream	Cream-colored seeds with an inconspicuous hilum. Seeds oblong or kidney-shaped.
Clay	Medium to dark brown seeds. Kidney-shaped.
White acre	Tan seeds. Kidney-shaped with blunt ends.

Viny varieties that mature late are preferred in the Mississippi Delta and nearby areas and for forage in Florida.[8] In the northern part of the cowpea region, earlier varieties resisting wilt and nematodes are desired. Chinese Red, an early erect dwarf variety with small seeds, is preferred in the semiarid western part of Oklahoma. Improved strains of California blackeye are grown for food and seed in California.[6]

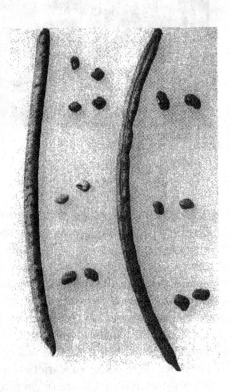

FIGURE 27.2
Pod and seeds of Brabham cowpea *(left)* and Groit cowpea *(right)*.

■ 27.6 ROTATIONS

Cowpea is grown in corn and cotton rotations in the South to supply nitrogen and organic matter to the soil. Cowpea often precedes or follows winter oats or winter wheat in rotations.

At the Arkansas Experiment Station, cowpea was interplanted with corn to determine its effect on the yields of subsequent crops. It was planted: (1) in the same row with corn, (2) between 44-inch (110 cm) corn rows at cultivation time, and (3) before cultivation in wide 58-inch (147 cm) rows. The nine-year average yields of oats following the three methods were 15.5, 9.5, and 5.9 percent, respectively, higher than for oats following corn alone.[9] Cotton showed the greatest increase when it followed cowpea planted in the same row and at the same time as corn.

In Africa, cowpea is usually intercropped or double cropped with pearl millet or other grain crops to provide protein in the diet.

■ 27.7 COWPEA CULTURE

Cowpea succeeds on a seedbed prepared for corn. Sometimes it is seeded as a catch crop after small grains, with disking as the only soil preparation, especially where the land is free from weeds. Fertilizer requirements are similar to those for soybeans. Inoculation is generally unnecessary, except where the crop is being grown for the first time. Seed treatment may reduce damage from seed rots and seedling blight.[5]

27.7.1 Seeding Methods

For best results, cowpea must be planted in warm soil after all danger of frost is past. For hay or seed production, the crop is usually planted at about the same time as corn or up to two weeks later. The crop can be planted in the South for green manure, pasture, or hay as late as August 1, or up to ninety days before the first frost. In northern Virginia, cowpea should be planted in May or early June for seed and in late June for best yields of hay. In California, the best time to plant is between May 1 and June 15.[6]

Cowpea is planted with either a grain drill or a corn planter. For seed production, the crop is commonly planted in rows 24 to 40 inches (60 to 100 cm) apart with the seeds 2 to 3 inches (5 to 8 cm) apart in the row. The rate of seeding varies from 20 to 45 pounds per acre (22 to 50 kg/ha). For forage or green manure, cowpea is generally drilled in close rows at the rate of 75 to 120 pounds per acre (84 to 134 kg/ha). Seed is planted 1 to 1½ inches (2.5 to 3.8 cm) deep.

27.7.2 Weed Control

Adequate control of weeds is necessary for satisfactory yields of cowpea. Mechanical cultivation using a rotary hoe or row cultivator should be done early in the growing season.[10] Herbicides can be used, but the crops listed on the label are blackeye peas, blackeye beans, or southern peas instead of cowpea.

27.7.3 Harvesting

HAY. Cowpea is harvested for hay when the pods begin to turn yellow. The best quality of hay is obtained when the pods are fully grown and a considerable number of them are mature. The vines are difficult to cure when harvested earlier. Delay beyond this stage results in tough woody stems as well as an excessive loss of leaves. A hay crusher facilitates curing of the coarse stems. Hay yields range from 1 to 2 tons per acre (2 to 4 MT/ha) under good growing conditions, with a single crop harvested each season.

SEED Cowpea should be harvested for seed when one-half to two-thirds of the pods have matured. Large fields are usually harvested with a combine after most of the seed is ripe. The crop can also be harvested with a windrower, allowed to dry, and picked up with a combine that is equipped with a pick-up attachment. Cowpea grown for home food is usually picked by hand.

27.7.4 Cowpea in Mixtures

Cowpea can be grown in mixtures with corn for silage. It is also grown for hay in combination with such crops as sorghum, Sudangrass, and johnsongrass. When grown for silage with corn, cowpea and corn are usually planted by the corn planter in one operation. Cowpea, grown in combination with other crops for hay, produces a larger yield of more readily curable hay than cowpea grown alone.

■ 27.8 USES OF COWPEA

A portion of the threshed cowpea is used for planting. Surplus seed, or that damaged or otherwise unfit for planting or human consumption, is fed to livestock. Although high in feed value, cowpea seed is generally not an economical livestock feed. The seed of certain varieties is a popular food in the South, being used in the pod, shelled green, or shelled dry, in preference to other types of beans.[2, 3, 5] The protein content of cowpea seed is about 25 percent.

Cowpea hay that has been well cured is considered equal to red clover hay in nutritive value. For a silage crop, cowpea grown in mixture with corn or sorghum is cut when the first cowpea pods begin to turn yellow. Cowpea, although high in protein and low in carbohydrates, will make good silage alone when wilted to the proper moisture content of 60 to 68 percent before being put into the silo.[11] Cowpea is utilized for soil improvement in the southern states. It makes a good growth on soils too poor for soybeans. Often, cowpea is used for pasture or some seed is picked before being plowed under. On very poor soils, it is advisable to plow under the entire crop in the green state. About 85 percent of the fertilizing value is in the hay and about 15 percent in the roots and stubble.[2] Thus, the soil benefits even when the crop is cut for hay.

Because of its nutritional value and ease of growing, cowpea has been researched as a possible crop for use in controlled ecological life support systems for future exploration of outer space.[12]

27.9 DISEASES

Fusarium wilt, caused by the fungus *Fusarium oxysporum* var. *tracheiphilum*, causes the leaves to yellow and fall prematurely[4] and finally results in the death of the plant. The stems turn yellow, the plants become stunted, and seed setting generally fails. Diseased stems are brown to black inside. Wilt is generally observed about midseason, being spread by cultivation implements, drainage water, and other agencies. The most satisfactory control measure is the use of resistant varieties.

Cowpea root knot, caused by a nematode (*Meloidogyne* species), is identified by galls over the entire root system. The roots soon turn brown, decay, and often die. The most practical measure is the use of resistant varieties in combination with the planting of other immune crops in rotations. Suitable rotation crops that are immune include winter grains, velvetbean, corn, sorghum, and some soybeans.

Several varieties are resistant to bacterial blight or canker caused by *Xanthomonas axonopodis* pv. *vignicola*. Charcoal rot caused by *Macrophomina phaseoli* (*Sclerotium bataticola*) damages cowpea severely under certain soil conditions.

Other fungus diseases of the cowpea include zonate leaf spot, caused by *Aristastoma oeconomicum*; red leaf spot, caused by *Cercospora cruenta*; and mildew, caused by *Erysiphe polygoni*. Most American varieties are resistant to the bean rust caused by the fungus, *Uromyces phaseoli*, that attacks introduced varieties. Red stem canker, caused by *Phytophthora cactorum*, is of minor importance.[11]

Virus diseases include cucumber mosaic, southern bean mosaic, and curly top.[7]

Diseases and viruses of cowpea can be controlled by using resistant varieties, treating seed or soil with a fungicide, avoiding throwing soil against the plants during cultivation, using a four- or five-year rotation with other crops, controlling weeds, and adopting cultural practices that reduce stress.[10]

27.10 INSECT PESTS

The main insect pests of cowpea are two species of weevil that damage the seed, namely, cowpea weevil or cowpea bruchid (*Callosobruchus maculatus*), and the southern cowpea weevil or four-spotted bean weevil (*Mylabris quadrimaculatus*). These weevils lay their eggs on the pods or in the seeds in the field and later in the threshed seeds in storage. The larvae bore into the seeds and complete their life cycle there. New generations of the weevil continue to develop unless the temperature falls too low. Control of the weevils consists of insecticide application, fumigation, or heat treatment of the stored seeds.[4] The cowpea curculio (*Chalcodermus aeneus*) infests developing seeds. It is controlled by the application of insecticides to the field when blossoming begins.

Other pests include lygus bugs, corn earworm, lima bean pod borers, mites, cowpea aphids, bean thrips, yellow-striped and beet armyworms, and root-knot nematodes.

REFERENCES

1. Piper, C. V. "Agricultural varieties of the cowpea and immediately-related species," *USDA Bur. Plant Industry Bull.* 229, 1912.

2. Morse, W. J. "Cowpeas: Utilization," *USDA Farmers Bull.* 1153, 1920.

3. Ligon, L. L. "Characteristics of cowpea varieties," *OK Agr. Exp. Sta. Bull.* B-518, 1958.

4. Morse, W. J. "Cowpeas: Culture and varieties," *USDA Farmers Bull.* 1148 (rev.), 1947.

5. Lorz, A. P. "Production of southern peas (cowpeas) in Florida. 1. Cultural practices and varieties." *FL Agr. Exp. Sta. Bull.* 557, 1955.

6. Sallee, W. R., and F. L. Smith. "Commercial black-eye bean production in California," *CA Agr. Exp. Sta. Circ.* 549, 1969, pp. 1–15.

7. Wright, P. A., and R. H. Shaw. "A study of ensiling a mixture of Sudan grass with a legume," *J. Agr. Res.* 28(1924):255–259.

8. McKee, R., and A. J. Pieters. "Miscellaneous forage and cover crop legumes," *USDA Yearbook*, 1937, pp. 999–1031.

9. McClelland, C. K. "Variety and inter-cultural experiments with cowpeas," *AR Agr. Exp. Sta. Bull.* 343, 1937.

10. Davis, D. W., and others. "Cowpea," in *Alternative Field Crops Manual.* WI Coop. Ext. Serv. and MN Ext. Serv. 1991.

11. Weimer, J. L. "Red stem canker of cowpea, caused by *Phytophthora cactorum*," *J. Agr. Res.* 78(1949):65–73.

12. Bubenheim, D. L., C. A. Mitchell, and S. S. Nielsen. "Utility of cowpea foliage in a crop production system for space," in J. Janick and J. E. Simon, eds., *Advances in New Crops.* Portland, OR: Timber Press, 1990.

Field Beans

KEY TERMS

Bush type
Day–neutral plant
Determinate plant
Hardshell
Indeterminate plant
Long-day plant
Short–day plant
Vining type

28.1 ECONOMIC IMPORTANCE

The field bean, or dry, edible bean, is an important human food of high protein content. In 2000–2003, it was grown on an average of 64 million acres (26 million ha), with a production of about 20 million tons (18 million MT), or 620 pounds per acre. The leading producing countries are Brazil, India, China, Mexico, Myanmar, and the United States. Production in the United States in 2000–2003 averaged about 1.24 million tons (1.1 million MT) or 1,660 pounds per acre (1,860 kg/ha) on about 1.5 million acres (600,000 ha). The leading states in bean production are North Dakota, Nebraska, Michigan, Minnesota, and California (Figure 28.1).

28.2 HISTORY OF BEAN CULTURE

The common bean *(Phaseolus vulgaris)* was probably domesticated from a wild form having a long slender vine that is found in Mexico and Central America.[1] Kidney beans were being cultivated there some time between 5000 and 3000 BC. The tepary bean *(P. acutifolius)* was grown before 3400 BC. The big and little types of lima bean *(P. lunatus)* are assumed to have arisen in Peru and Mexico, respectively. The common bean was collected by early explorers and was being grown in Europe by 1542. Lima bean seed was brought to the United States from Peru in 1824, and a bush lima bean was found growing along a Virginia roadside in 1875.

28.3 ADAPTATION

The bean plant is a warm-season annual adapted to a wide variety of soils. Its optimum mean temperature is 65 to 75°F (18 to 24°C). High temperatures interfere with seed setting, while low temperatures are unfavorable for growth.[2] Field beans are produced most successfully in areas where the rainfall is light during the latter part of the season. This reduces weather damage to the mature beans. Under dryland conditions, the bean plant adjusts its growth, flowering,

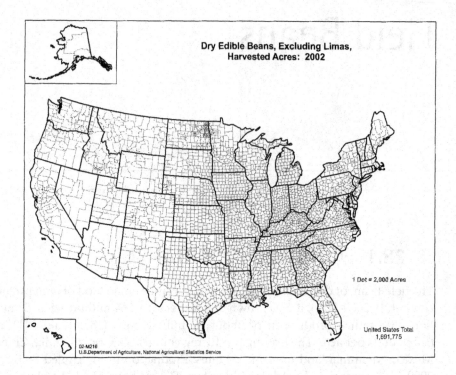

FIGURE 28.1
Acreage of dry, edible beans, excluding lima beans, in the United States in 2002. [Source: 2002 Census of U.S. Department of Agriculture]

and pod setting to the soil moisture supply to a marked degree. Beans require a minimum frost-free season of about 120 to 130 days in order to mature seed.[3]

Lima bean is grown mostly in California, navy bean in Michigan and North Dakota, and the Great Northern bean in Nebraska. Pinto bean is grown mostly in Colorado and Nebraska, kidney bean in Michigan and Minnesota, and small red bean in Idaho and Michigan. Cranberry bean and black bean are grown mostly in Michigan, blackeye bean in California, and garbanzo bean in California and Idaho.

Hardshell, a form of dormancy caused when seedcoats are impermeable to water, is prevalent in a large number of bean varieties.[4] The condition is accentuated by hot, dry winds at harvest time or by storage in a heated room.

Most types of field beans are short-day plants, but the Boston yellow eye and cranberry types are day-neutral and will flower in twenty-six to thirty-nine days after planting at all day lengths from ten to eighteen hours,[5] and the red kidney shows only a partial response to different day lengths. Lima bean *(P. lunatus)* and chickpea *(Cicer arietinum)* are day-neutral; tepary bean, mung bean, velvetbean, and cowpea are short-day plants; and the scarlet runner bean and yellow lupine are long-day plants.

28.4 BOTANICAL DESCRIPTION

Dry, edible beans, including the common bean as well as lima, tepary, runner, and mung bean, belong to the genus *Phaseolus*.[6] All are more or less flat-seeded.

Plants of the common bean *(Phaseolus vulgaris)* are either determinate or indeterminate in growth habit, and either bush or vining type. Common beans can be classified into four plant growth habits that are useful for identification and classification of new cultivars:[7, 8]

Type I: Determinate, bush
Type II: Indeterminate, upright short vine
Type III: Indeterminate, prostrate vine
Type IV: Indeterminate with strong climbing tendencies

Field beans bear racemes in the leaf axils. The leaves are pinnately trifoliolate. Leaves are either glabrous or pubescent. The flowers are white, yellow, or bluish purple. The pods are straight or distinctly curved, 4 to 8 inches (10 to 20 cm) long, and end in a distinct spur (Figure 3.16). The seeds may be white, buff, brown, pink, red, blue-black, or speckled in color (Figure 28.2). The immature pods of snap beans are either yellow or green, but those of field varieties are green. Their fibrous pods distinguish the field varieties of the common bean from the snap beans, which have very little or no fiber in the pods.[9] A satisfactory dry bean bears its pods above the ground, ripens uniformly, and does not shatter appreciably at maturity. Beans are normally self-pollinated with less than 1 percent natural crossing.[10]

Dry beans, soybean, and some peanut varieties contain considerable low molecular-weight carbohydrates, especially raffinose and stachyase. Anaerobic bacteria in the human intestine release carbon dioxide, methane, ammonia and hydrogen sulfide gas from the seeds, causing flatulence.

The commonly recognized types of dry beans grown in the United States are listed in Table 28.1. They average about 1,000 to 2,000 seeds per pound.

◼ 28.5 BEAN TYPES

28.5.1 Field Beans

Navy (white pea) bean matures in 110 to 120 days. Navy bean is a small, semi-trailing plant with white flowers and small white seeds (Figure 28.2). Most varieties are Type I or II. U.S. production in 2000–2003 averaged about 187,000 tons (170,000 MT). Leading states are North Dakota, Minnesota, Michigan, Idaho, and South Dakota.

Red kidney bean requires a frost-free season of at least 140 days to mature. Because of the red color of its seed it is less susceptible to staining and therefore is better adapted to areas of high rainfall than are the white beans.[11] The red kidney bean plant is a bush type (Type I). The small red varieties are mostly Type III. They have bright red seeds. The flowers are lilac in color. The seeds are large, flattened, and pink when newly harvested and mostly dark red when old. The matured pods are often splashed with purple. U.S. production of light red kidney bean in 2000–2003 averaged about 56,000 tons (51,000 MT). Leading states are Nebraska, Michigan, New York, Colorado, and Minnesota. Production of dark red kidney bean averaged about 47,000 tons (42,000 MT). Over half of the dark red kidney bean is grown in Minnesota. Other leading states are Wisconsin, Michigan, North Dakota, and California.

Great Northern bean is Type III or IV, the flowers white, and the seeds white, large, and flattened. It is medium-late in maturity. U.S. production in 2000–2003 averaged about 115,000 tons (95,000 MT). Almost 80 percent of the Great Northern bean is grown in Nebraska. Other leading states are North Dakota, Idaho, Wyoming, and Washington.

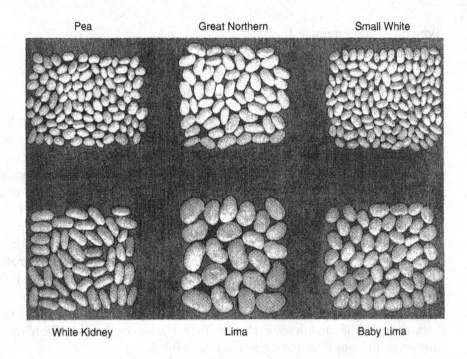

Pea Great Northern Small White

White Kidney Lima Baby Lima

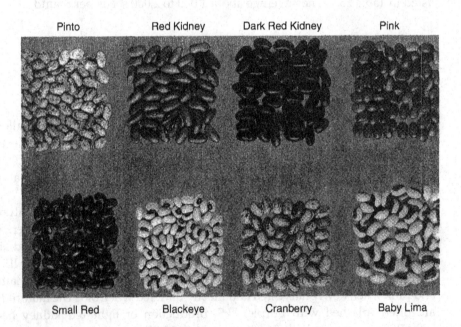

Pinto Red Kidney Dark Red Kidney Pink

Small Red Blackeye Cranberry Baby Lima

FIGURE 28.2
Principal types of field beans, showing relative sizes (about two-fifths natural size).

Pinto bean is mostly a Type III or IV plant with white flowers. The seeds are medium to large, somewhat flattened, buff-colored, and speckled with tan to brown spots and splashes. The Pinto possesses a distinctive flavor not found in other dry beans. U.S. production in 2000–2003 averaged about 539,000 tons (490,000 MT). About half is grown in North Dakota. Other leading states are Colorado, Nebraska, Idaho, and Wyoming.

TABLE 28.1 Average Annual Production of Bean Classes in the United States, 1997–1999

Class	Production (1,000 Cwt)
Field bean	
Black	3,019
Pinto	12,090
Great northern	2,298
Light red kidney	1,377
Dark red kidney	959
Small white	118
Pink	812
Small red	524
Cranberry	540
Navy	5,568
Other	879
Lima bean	
Large lima	395
Baby lima	594
California blackeye (culinary cowpea)	929
Chickpea	
Garbanzo (*Cicer arietinum*)	517

The pink bean is a Type III with white flowers and medium-size pink seeds. It is very heat resistant. Cranberry bean, Type I and III, has deep red splashes on a buff seed coat. Black bean is mostly a Type II plant.

28.5.2 Tepary Bean

Tepary bean (*P. acutifolius* var. *latifolius)* has been grown by the Native Americans of Arizona and New Mexico since early times but was not tested in field experiments before 1910.[12] It has small leaflets and white seeds somewhat smaller than those of the navy bean. The beans are considered to be harder to cook and less palatable than the common edible beans and therefore have only a limited market. They do best in arid climates. They are grown only occasionally for home use in the United States.

28.5.3 Mung bean

Mung bean, also called green gram or golden gram (*Vigna radiate* syn. *Phaseolus aureus)*, is grown mainly for human food in the southern half of Asia and nearby islands and also in Africa. It is the source of the canned bean sprouts found on the market and used in chop suey and similar foods. Sprouted seeds of mung bean are popular additions to salads. Mung bean has been known in the United States since 1835 under such names as Chickasaw pea, Oregon pea, Neuman pea, Jerusalem pea, or chop suey bean. Prior to World War II, nearly all of the mung bean used in the United States was imported. Production was begun in 1945, mostly in Oklahoma.

FIGURE 28.3
Pods and leaflets of mung bean.

Practically all of the sprouting mung bean grown in the United States is grown in Oklahoma.[13] The beans are suitable for feed, and the plants make good silage and fair hay. The plants have glabrous leaves and are very similar to those of the cowpea except for being smaller and more bushy (Figure 28.3). The pale yellow flowers are borne in racemes or clusters of ten to twenty-five. The ripe black or brownish pods are 3 to 5 inches (8 to 13 cm) long and contain ten to fifteen small globose or oblong seeds. Most varieties have green, yellow, golden-brown, or marbled seeds.[14]

Mung bean is seeded in rows 20 to 30 inches (50 to 75 cm) wide in a warm soil. Seeding rates are 10 to 15 pounds per acre (11 to 17 kg/ha). Mung bean plants do not mature uniformly and are usually swathed and allowed to dry before threshing with a combine.[13]

The blackgram, urd, or mungo bean (*Vigna mungo*), grown mostly in India for human food and for hay, is somewhat similar to the mung bean.

28.5.4 Lima Bean

Large lima bean (*Phaseolus limensis*) is a perennial that is grown as an annual. The small (baby) lima (*P. lunatus*) is an annual. These are now consumed mostly as green limas, either frozen, canned, or fresh. U.S. production in 2000–2003 of large lima bean averaged about 18,000 tons (17,000 MT). Small lima bean production averaged about 20,000 tons (18,000 MT). The largest domestic production is found in California.

When allowed to mature, lima bean is handled and used in a manner similar to other dry, edible field beans. Temperature and moisture extremes may limit the fertilization of the lima bean and induce pod shedding.[2]

28.5.5 Broad bean

Broad bean (horse bean, faba bean, or Windsor bean), *Vicia faba*, is related botanically to the vetches but differs from them in having coarse, erect stems, large leaflets, large pods, and large flattened seeds. It is mostly cross-pollinated and was grown by ancient Egyptians. This is an important food crop in North Africa. China is the chief producer, and the world production exceeds 3.3 million tons (3 million MT).

28.5.6 Chickpea

Chickpea (*Cicer arietinum*), usually called garbanzo or gram, is grown on a commercial scale in the United States almost entirely in California. Global production in 2000–2003 averaged about 8.3 million tons (7.6 million MT). Over half of the global production is in India. Other leading countries in chickpea production are Turkey, Mexico, Iran, and Australia. Chickpea is a native of Europe. It was taken to California from Mexico during the Spanish mission period.[15] It is best adapted to warm, semiarid conditions, where most of it is grown. Average U.S. production in 2000–2003 was about 53,000 tons (48,000 MT). Leading states are California, Idaho, Washington, North Dakota, and Montana.

The plant is a low, bushy annual with hairy stems. Each leaf is comprised of several pairs of small, rounded or oblong leaflets. The flowers are white or reddish, small, and borne singly at the tip of axillary branches. The seeds are roughly globular, flattened on the sides, and somewhat wrinkled. The chickpea is grown and handled about like the field bean. The threshed seeds are prepared for food in much the same manner as is dried lima bean. Roasted seeds are used as a confection or snack, and sometimes as a coffee substitute. The herbage is low in yield and is toxic to animals.

28.5.7 Other Beans

The culture of cowpeas, including the California blackeye variety, is discussed in Chapter 27. The adzuki bean *(Vigna angularis)* is an important crop in Japan and China. In the United States, it is grown only occasionally for home use.

28.6 LENTIL

Lentil *(Lens culinaris)* has been an important food item in the Mediterranean region and Asia Minor since the Bronze Age. Global production in 2000–2003 was about 3.4 million tons (3 million MT). The leading countries in lentil production are India, Canada, Turkey, Australia, and China. Lentil was introduced into the United States by 1898[16] but was grown only to a limited extent, mostly in home gardens. The commercial culture of lentil in the United States began in the state of Washington in 1937.[17] Average U.S. production in 2000–2003 was about 137,000 tons (124,000 MT). Leading states are Washington, Idaho, North Dakota, and Montana.

The lentil plant is a branched, weakly upright or semi-viny (Type II) annual 18 to 22 inches (45 to 55 cm) tall, with pinnately compound leaves. It has a general resemblance to vetch. The flowers are white, lilac, or pale blue. It is highly self-pollinated.[16] The pods contain two or three seeds, which are thin and lens-shaped, usually smaller than pea seeds, and of various cotyledon colors, including yellow, orange, or both.

Most lentil is grown in rotations preceding fall-sown small grains. It competes poorly with weeds, especially wild oat. The plants require molybdenum, sulfur, and sometimes phosphorus fertilization for good growth.[18] For best results, inoculated seed is sown in April at rates of 60 to 80 pounds per acre (67 to 90 kg/ha), depending upon seed size. The crop is swathed when the pods are a golden color, and then threshed with a pick-up combine about ten days later. Swathing when the pods are dewy and tough reduces seed shattering.

Lentil is susceptible to many viruses that attack pea, clover, or alfalfa, and the fields should be isolated from those of other legumes. The chief insect pests are cowpea and black bean aphids.

28.7 FERTILIZERS

Field beans are moderate users of phosphorus. Phosphate fertilizer is usually not needed if the soils have medium levels of available phosphate unless the yield goal is over 2,000 pounds per acre (2,240 kg/ha).[7, 19] Application rates can be reduced by one-third to one-half if the phosphate is band applied at seeding. Field beans will usually not require potash fertilizer if the soil has medium or higher levels of available potash.[7, 19] No fertilizer should be placed with the seed. At least a 1-inch separation between seed and fertilizer is required.[7, 19]

Since field beans are legumes, inoculating the seed with *Rhizobium phaseoli* bacteria can reduce the need for nitrogen fertilizer. However, if the seed is treated with a chemical to control bacterial blight, the inoculant may not be effective. Newer strains or formulations of inoculant appear to resist seed treatment.[7] The relationship between *Rhizobia* and field beans is not as strong as with soybean, so field beans may respond to additional nitrogen fertilizer. Nitrogen fertilizer is usually necessary when yield goals are over 2,000 pounds per acre (2,240 kg/ha) or when soils are low in organic matter.[19]

Field beans respond to zinc fertilizer when the soil test level is below 0.8 ppm using the DTPA procedure.[7, 19] Band application of zinc at seeding reduces the amount needed by 80 percent.

▦ 28.8 ROTATIONS

It is advisable to grow beans in long rotations with other crops in the humid and irrigated regions. An interval of three to four years between bean crops reduces the risk from soil-borne disease infection, especially if white mold is present.[7] Beans succeed well after green-manure crops,[20] legume-grass hay crops, small grain, corn, or potato. Typical rotations in New York and Michigan bean areas include such crops as winter wheat, clover and timothy, corn, and potatoes. A rotation of alfalfa, potato, bean, sugarbeet, and grain is popular in the irrigated intermountain regions.

In semiarid regions, beans can replace summer fallow in the alternate wheat-fallow cropping system. The yields from wheat that has followed beans are almost as good as those from wheat that has followed summer fallow. The beans are usually harvested in time for the fall seeding of wheat. In California, beans may be grown continuously or in rotations following alfalfa or an intertilled crop.[3] The continuous culture of beans may result in severe soil erosion. On the drylands of New Mexico, Colorado, and Kansas, beans can follow small grains, sorghum, or corn.[21] Strip cropping of beans is advisable in order to reduce soil blowing after bean harvest. Thus, strips of four to twenty-four rows of beans alternate with strips of sorghum or corn of equal width.[22, 23]

▦ 28.9 BEAN CULTURE

A firm seedbed, free from clods and coarse debris, is desirable for field beans. Conservation tillage methods are advised, if needed, to prevent soil erosion. Corn, wheat, or other residue-producing crops must precede field bean when using conservation tillage.[8] Field beans are sensitive to soil compaction and reducing tillage minimizes compaction.

28.9.1 Seeding Methods

Beans should be planted in warm soil, preferably above 65°F (18°C), after all danger of frost is past. In the Northwest, the time varies from about May 20 to June 10, in Colorado from May 20 to June 15, in New Mexico from May 15 to July 1, and in California from April 10 to July 10. Cold, wet soil is likely to result in low germination when beans are planted in western New York before June 1.[3, 8]

Beans are generally planted in drilled rows 20 to 30 inches (50 to 75 cm) apart. Yields are increased when row spacing is below 30 inches (75 cm).[8] Seed is planted with a two-row to eight-row bean planter, or with sugarbeet or corn planters equipped with bean plates. The seeds are planted at a depth of 2 to 3 inches (5 to 8 cm) in semiarid areas or where irrigated surface soils may dry out before germination is complete. A planting depth of 1½ to 2 inches (4 to 5 cm) is desirable where soil moisture in the seedbed is ample.[3] The desired stand under favorable irrigated or humid conditions is 60 to 90 pounds per acre (67 to 100 kg/ha). Dryland seeding rates are 18 to 30 pounds per acre (20 to 34 kg/ha).[8] Large-seeded types such as red kidney, pinto, and Great Northern are planted at a rate of 60 to 80 pounds per acre (67 to 90 kg/ha). For the smaller-seeded red Mexican and navy types, the desired rates are about 60 and 40 pounds per acre (67 and 45 kg/ha), respectively. The planting rate for large lima beans is 80 to 100 pounds per acre (90 to 110 kg/ha), whereas that for baby limas is 40 pounds per acre (45 kg/ha). As the plant spacing increases, the number of pods per plant increases. Close spacing of plants does not increase the percentage of immature pods.

Dryland beans in southwestern Colorado and northern New Mexico are planted in 30 to 42 inch (76 to 107 cm) rows at a rate of about 30 pounds per acre (34 kg/ha) or less. One plant every 10 inches (25 cm) in the row is ample.

Seeding rates should be increased by 10 to 15 percent if the soil is cool and wet, as those conditions result in poor germination.[7]

28.9.2 Weed Control

Beans can be cultivated with a rotary hoe before the germinating seedlings emerge, or after the plants are 2 to 4 inches (5 to 10 cm) tall. Later cultivations are usually made with sweep implements.

A variety of herbicides are approved for use with field beans. Preplant herbicides include EPTC, metolachlor, dimethenamid, trifluralin, and alachlor. The herbicides metolachlor and dimethenamid can also be applied preemergent. Postemergent herbicides include bentazon, imazethapyr, sethoxydim, and clethodim. Paraquat and sodium chlorate can be used as a desiccant after 80 percent of the leaves are yellowing.

28.9.3 Irrigation

Field beans require 12 to 18 inches (300 to 460 mm) of soil moisture during the growing season. Generally the larger, bushier bean types will use more water than the shorter, narrow types.[7] Field beans do not root as deeply as many other crops. About 90 percent of the roots will be found in the top 2 feet (60 cm) of the soil, which is considered the effective depth for irrigation. This rooting depth is usually reached about thirty to thirty-five days after emergence.

At planting time, the soil in an irrigated bean field should be wet nearly up to its field carrying capacity to a depth of 2 to 3 feet (60 to 90 cm). This may necessitate irrigating just before final seedbed preparation. The seedbed is then prepared and the beans planted as soon as the soil is dry enough to work. The field should not be irrigated after planting until the seedlings have emerged. Additional irrigations should be applied within three to five days after the bean foliage turns a dark bluish green. This change in plant color develops as the soil moisture becomes more deficient. Usually, three or four irrigations, but sometimes more, are necessary after the beans have emerged. Irrigation for the season should cease when one-fourth of the bean pods have turned yellow.

White mold infection can be greatly reduced by keeping the soil surface below the plant canopy as dry as possible during pod filling and maturity.[24] Irrigation water should be applied only as necessary during this time.

28.9.4 Harvesting

Field beans are generally harvested when most of the pods have turned yellow but before they are dry enough to shatter from the pods. The vines are cut below the surface with blades attached to a tractor that combines two rows of beans into one windrow.[25, 26] It is a common practice to throw two to four windrows together with a side-delivery rake during or immediately after cutting. Tillage that produces 3 to 4 inch (8 to 10 cm) high ridges under the plant row is necessary to efficiently cut the plants.[8]

Beans can be cut above the ground with a windrower. Cutting and raking, when the pods are damp with dew, reduces shattering losses. The beans are threshed with a pick-up combine after curing in the windrow for five to fifteen days (Figure 28.4). Cutting beans later to reduce the time in windrows is becoming more popular, particularly when harvesting later in the season. Frequently, beans are combined directly. The soil surface should be level and free of large clods when cutting above the soil.[8]

Special bean combines are equipped with two or three rubber-faced cylinders. Beans should be threshed at low cylinder speeds of 200 to 400 revolutions per minute and with ample concave clearance to avoid cracking and splitting. This is especially important in threshing beans for seed.[27] Choice beans for the retail trade may be handpicked after cleaning. Discolored beans can be removed with an electric-eye sorter.

FIGURE 28.4
Threshing beans after curing in windrows.
[Courtesy Shari Rosso, Neb. Panhandle Res. & Ext. Center]

When threshed beans are being dried with heated air, the temperature should not exceed 100°F (38°C) and not more than 3 percent moisture should be removed in one drying stage.

28.10 DISEASES

The most serious losses from bean diseases occur in the humid and irrigated regions. More than sixty parasitic diseases attack the crop.[28]

28.10.1 Bacterial Blights

Blights are among the most serious diseases in the important bean areas. Three types of bacterial blight occur in field beans: (1) common or fuscous blight (*Xanthomonas campestris* pv. *phaseoli*), (2) halo blight (*Pseudomonas syringae* pv. *phaseolicola*), and (3) brown spot (*Pseudomonas syringae* pv. *syringae*). The most striking symptoms are spots that may enlarge rapidly and produce dead areas on the leaf. On the stem, a characteristic lesion known as stem girdle appears. This lesion may weaken the stem so that it breaks at the diseased node. Pod lesions, usually reddish, water-soaked, and very irregular in outline, may prevent the proper filling of the pods and cause shriveling of the beans. Infected white bean seed may turn a butter-yellow color. The principal control measure is to plant disease-free seed. Crop rotations of three to four years reduce inoculum from residues. Some resistant varieties are available.

28.10.2 Anthracnose

Bean anthracnose (*Colletotrichum lindemuthianum*) is a serious disease in the humid East. The disease is usually first noted on the leaves where dark-colored areas appear. The fungus may destroy the veins, while the blade shows numerous cracks or holes with shriveled blackened margins. Large, round, dark-colored lesions finally appear on the pods. Large oval cankers may also be observed on the stem, which may so weaken it that it is easily broken when cultivating or by a strong wind. The disease is most serious in wet seasons. It overwinters in the seed. The most satisfactory control measures are: (1) use of disease-free seed grown in the semiarid regions, (2) resistant varieties, (3) crop rotation, (4) keeping workers out of the field when the plants are wet, and (5) spraying with suitable fungicides.

28.10.3 Bean Common Mosaic Virus (BCMV)

BCMV is a seed-borne virus disease that is also spread in the field by several species of aphids. The mottling of the leaves may form various patterns of dark green and light green areas.[29] The dark-green areas often occur along the midvein. The leaves of infected plants may be curled downward. Diseased plants are usually paler green and more dwarfed than healthy ones. Early infections may cause a complete failure to form pods. The use of resistant varieties is the only practical control. The yellow mosaic disease spreads to beans from red clover, crimson clover, and sweetclover.

28.10.4 White Mold

White mold is a serious fungal disease *(Sclerotinia sclerotiorum)*. It is difficult to control because fungal sclerotia can survive more than ten years in the soil. It germinate to produce fruiting structures that release air-borne spores throughout the growing season. The disease can spread rapidly after spore release; an entire field can be lost only days after initial symptoms are detected.[30] The disease can be introduced into a field through infected seed, flood or irrigation water, through windblown soil or contaminated equipment.

Initial symptoms are wet, soft spots or lesions that enlarge into a watery, rotten mass of tissue covered by a white moldy growth. Wet, cool weather favors disease development. Because the sclerotia can survive for years, crop rotations will not control the disease but will reduce the infection level. Planting disease-free seed, using plant populations and row spacings that reduce the likelihood of dense growth, and careful water management can help reduce the severity of the disease. Applications of fungicides are effective in some cases. Some varieties have genetic tolerance to the disease.

28.10.5 Bean Rust

Bean rust *(Uromyces appendiculatus)* appears as small brown pustules on the leaves and, to a lesser extent, on the stems and pods. The pustules turn black for the winter stage. Control measures include: (1) dusting the plants with a fungicide, (2) crop rotation, and (3) plowing to bury residue.[31]

28.10.6 Other Diseases

Curly top is a virus disease of bean, as well as sugarbeet. It causes losses to beans in the area west of the continental divide. It is carried by the beet leafhopper (or white fly) *(Circulifer tenellus)* from plants infected with the virus to other healthy plants. Bean plants are more subject to injury while in the younger stages of growth. The growing point is killed by the virus and then drops off. The plant will then die. On larger plants the first symptom of curly top is the downward curling of the first trifoliolate leaf. The curly-top virus is not carried in the seed. Several improved varieties are resistant to the disease.

Root rot *(Fusarium* spp.) attacks the roots and causes reddish-colored lesions that later turn dark brown. If roots are heavily infected, upper plant parts are yellowed and stunted and often wilted.[30] Long rotations can reduce the amount of fungus in the soil.

Aphids spread bean yellow mosaic virus (BYMV). It causes stunting and contrasting areas of dark green and yellow tissue. Some varieties are resistant.

28.11 INSECT PESTS

Western bean cutworm can be a serious pest. The adults lay eggs on the underside of bean leaves and the larvae begin feeding on leaves and blossoms during the night. As the larvae grow, they will move to the pods and developing seeds. Larger larvae may move to the soil during the day. If they have not completed development, the larvae may continue to feed on pods and seeds after the crop has been windrowed.[32] The pest can also infect cornfields so rotation with corn is not effective. Although some varieties are somewhat resistant, insecticides are the best method of control.

Beans are often seriously damaged in storage and in the field by the bean weevil (*Acanthoscelides obtectus*). It can be controlled by planting weevil-free seed, by fumigation of infected seeds as soon as possible after harvest, and by field sanitation.

The Mexican bean beetle (*Ephilachna varivestris*) (Figure 28.5) is a serious pest in many bean-growing areas. The larvae feed on the undersides of the bean leaves. They remove the epidermis but seldom cut through to the upper surface. Spraying or dusting with insecticides can control this insect.

FIGURE 28.5
Mexican bean beetle feeding on the underside of a bean leaf. Note the spotted adults, hairy larvae, and egg mass.

Other insects that attack beans include potato leafhopper, seedcorn maggot, Pacific Coast wireworm, white-fringed beetle, lygus bugs, leafhoppers, bean aphid, armyworm, and cutworm.[33, 34]

REFERENCES

1. Gentry, H. S. "Origin of the common bean (*Phaseolus vulgaris*)," *Econ. Bot.* 23(1969):55–69.

2. Howe, O. W., and H. F. Rhoades. "Irrigation of Great Northern field beans in western Nebraska," *NE Agr. Exp. Sta. Bull.* 459, 1961.

3. Mimms, O. L., and W. J. Zaumeyer. "Growing beans in the western States," *USDA Farmers Bull.* 1996, 1947, pp. 1–42.

4. Gloyer, W. O. "Percentage of hardshell in pea and bean varieties," *NY Agr. Exp. Sta. (Geneva) Tech. Bull.* 195, 1932.

5. Allard, H. A., and W. J. Zaumeyer. "Responses of beans (*Phaseolus*) and other legumes to length of day," *USDA Tech. Bull.* 867, 1944, pp. 1–24.

6. Hedrick, V. P., and others. *The Vegetables of New York*, vol. I, Part 1, *Beans of New York*, Albany, New York N.Y. Agr. Exp. Sta., 1931, pp. 1–110.

7. Berglund, D., and others. "Dry bean production guide," *ND St. Univ. Ext. Serv.* A-1133, 1997.

8. Schwartz, H. F., M. A. Brick, D. S. Nuland, and G. D. Franc. "Dry bean production and pest management," *CO St. Univ. Coop. Ext. Serv. Regional Bull.* 562A, 1996.

9. Steinmetz, F. H., and A. C. Arny. "A classification of the varieties of field beans, *Phaseolus vulgaris*," *J. Agr. Res.* 45(1932):1–50.

10. Mackie, W. W., and F. L. Smith. "Evidence of field hybridization in beans," *J. Am. Soc. Agron.* 27(1935):903–909.

11. Hardenburg, E. V. "Experiments with field beans," *Cornell U. Agr. Exp. Sta. Bull.* 776, 1942.

12. Freeman, G. F. "Southwestern beans and teparies," *AZ Agr. Exp. Sta. Bull.* 68, 1912.

13. Sholar, R., and L. Edwards. "Mungbean production in Oklahoma," *OK St. Univ. Coop. Ext. Serv.* F-2050, 1986.

14. Ligon, L. L. "Mungbeans—a legume for seed and forage production," *OK Agr. Exp. Sta. Bull.* 284, 1945, pp. 1–12.

15. Allard, R. W., and F. L. Smith. "Dry edible bean production in California," *CA Agr. Exp. Sta. Cir.* 436, 1954.

16. Wilson, V. E., and A. G. Law. "Natural crossing in *Lens esculenta* Moench," *J. Am. Soc. Hort. Sci.* 97, 1(1972):142–143.

17. Youngman, V. E. "Lentils, a pulse of the Palouse," *Econ. Bot.* 22, 2(1968):135–139.

18. Entermann, F. M., and others. "Growing lentils in Washington," *WA Ext. Bull.* 590, 1968, pp. 1–6.

19. Rehm, G., M. Schmitt, and R. Eliason. "Fertilizer recommendations for edible beans in Minnesota," *Univ. MN Ext. Serv.* FO-6572-GO, 1995.

20. Rather, H. C., and H. R. Pettigrove. "Culture of field beans in Michigan," *MI Agr. Exp. Sta. Spec. Bull.* 329, 1944, pp. 1–38.

21. Brandon, J. F., and others. "Field bean production without irrigation in Colorado," *CO Agr. Exp. Sta. Bull.* 482, 1943, pp. 1–22.

22. Greig, J. K., and R. E. Gwin. "Dry bean production in Kansas," *KS Agr. Exp. Sta. Bull.* 486, 1966, pp. 1–19.

23. Paur, S. "Growing pinto beans in New Mexico," *NM Agr. Exp. Sta. Bull.* 378, 1953.

24. Steadman, J. R., H. F. Schwartz, and E. D. Kerr. "White mold of dry beans," *NE Coop. Ext. Serv.* G9-21103-A, 1992.

25. Schrumpf, W. E., and W. E. Pullen. "Growing dry beans in central Maine, 1956," *MA Agr. Exp. Sta. Bull.* 577, 1958.

26. Harrigan, T. M., S. S. Poindexter, and J. P. LeCureux. "Harvesting Michigan navy beans," *Michigan St. Univ. Ext. Bull.* 24089201, 1999.

27. Toole, E. H., and others. "Injury to seed beans during threshing and processing," *USDA Cir.* 874, 1951.

28. Hall, R., ed. *Compendium of Bean Diseases.* St. Paul, MN: APS Press, 1991.

29. Zaumeyer, W. J., and H. R. Thomas. "Bean diseases—how to control them," *USDA Handbook* 225, 1962.

30. Venette, J. R., and H. A. Lamey. "Dry edible bean diseases," *ND St. Univ. Ext. Serv.* PP-576, (rev.), 1998.

31. Steadman, J. R., H. F. Schwartz, and D. T. Lindgren. "Rust of dry bean," *NE Coop. Ext. Serv.* G95-1250-A, 1995.

32. Seymour, R. C., G. L. Hein, R. J. Wright, and J. B. Campbell. "Western bean cutworm in corn and dry beans," *NE Coop. Ext. Serv.* G98-1359-A, 1998.

33. Anderson, A. L. "Dry bean production in the eastern States," *USDA Farmers Bull.* 2083, 1955.

34. LeBaron, M., and others. "Bean production in Idaho," *ID Agr. Exp. Sta. Bull.* 282, 1958.

Peanut

■ 29.1 ECONOMIC IMPORTANCE

Peanut *(Arachis hypogaea)*, also called goober, pindar, groundnut, or earthnut, was grown on an average of 61 million acres (25 million ha) in 2000–2003. The leading countries growing peanut are China, India, Nigeria, United States, and Indonesia (Figure 29.1). Production in 2000–2003 averaged about 38.5 million tons (35 million MT) or about 1,300 pounds per acre (1,400 kg/ha). About 1.9 million tons (1.7 million MT) were produced in the United States in 2000–2003 on about 1.3 million acres (540,000 ha). Yield averaged about 2,800 pounds per acre (3,100 kg/ha). The leading states in peanut production are Georgia, Texas, Alabama, North Carolina, and Florida (Figure 29.2). Peanut is grown for oil in much of the world, but all but about 15 percent of the U.S. crop is used for food. About 6 pounds per capita are consumed as peanuts and peanut products in the United States.

■ 29.2 HISTORY OF PEANUT CULTURE

The peanut plant is a native of South America where closely related wild species are found. It may have originated in the mountainous parts of northern Argentina,[1] in the state of Matto Grosso, Brazil,[1,2] or in Peru where it was being cultivated by 1000 BC.[3] Peanuts were grown in Mexico by the beginning of the Christian era. Early slave ships carried peanuts to Africa and years later introduced them to the colonial United States from Africa. Commercial development of a peanut industry began about 1876. A rapid increase in production occurred after 1900 when the boll weevil caused serious damage to the cotton crop.

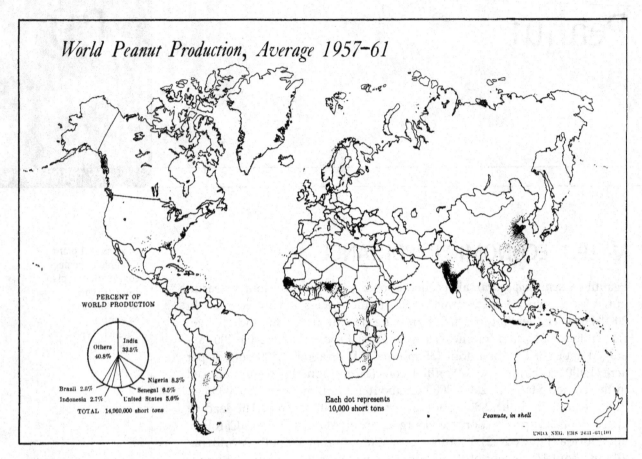

FIGURE 29.1
World peanut production. [Courtesy of USDA]

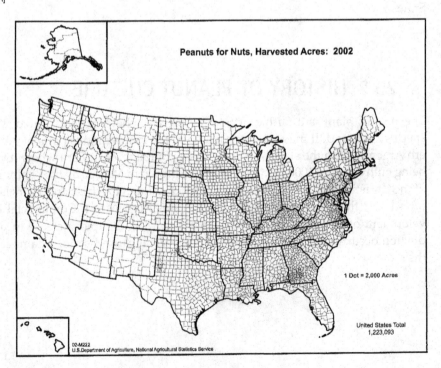

FIGURE 29.2
Peanut acreage in the
United States in 2002.
[Source: 2002 Census of U.S.
Department of Agriculture]

29.3 ADAPTATION

The most favorable climatic conditions for peanuts are moderate rainfall during the growing season, an abundance of sunshine, and relatively high temperatures.[4] The plants need ample soil moisture from the beginning of blooming tip to two weeks before harvest. The best crops are obtained where the annual rainfall is between 42 and 54 inches (1,070 to 1,370 mm), but peanuts for home use are grown where the precipitation is less than 19 inches (480 mm). They are grown under irrigation in eastern New Mexico. The peanut region has an average frost-free period of 200 days or more. The growing periods for different varieties range from 110 to 150 days.

The largest yields of the best quality of market peanuts are secured on well-drained, light, sandy loam soils. Dark-colored soils stain the hulls, which causes them to be rejected for many commercial uses. For forage, the crop can be grown on almost any type of soil except heavy clay soils low in organic matter. Peanuts produce the highest yields on soils with a pH of 5.8 to 6.2.[5,6] Poorly drained soils have generally been unsatisfactory. Light, sandy soils offer less resistance to (1) the penetration of the pegs that must enter the ground in order for the pods to develop, and (2) the recovery of the nuts from the soil when digging. The soil should be free from small stones.

The three major regions of peanut production are: (1) the southeastern states (Alabama, Florida, and Georgia) that grow primarily runner types, (2) the southeastern Virginia–northeastern North Carolina area that grows primarily Virginia types, and (3) the southwestern area (Texas and Oklahoma) that grows primarily Spanish types. Valencia types are grown primarily in New Mexico.

29.4 BOTANICAL DESCRIPTION

Cultivated peanut (Arachis hypogaea) is a member of the legume family. It is a pea rather than a nut. It is unknown in the wild state, but about fifteen species that bear some resemblance to the cultivated type are found in South America.

Peanut has been considered a short-day plant, but it is mostly photoperiod insensitive (day neutral) because floral initials may be present in ungerminated seeds.[7]

The peanut plant is a low-growing annual with a central upright stem. The numerous branches vary from prostrate to nearly erect. The pinnately compound leaf consists of two pairs of leaflets, but occasionally a fifth leaflet is borne on a slender petiole. Peanut varieties are readily separable into bunch and runner types. The nuts are closely clustered about the base of the plant of the erect or bunch type (Figure 29.3). The runner varieties have nuts scattered along their prostrate branches from base to tip. The peanut has a well-developed taproot with numerous lateral roots that extend several inches into the ground. Most roots have nodules but bear very few root hairs (Figure 29.3).

Peanut is an indeterminate plant that flowers while still growing vegetatively. The flowers are borne in the leaf axils, above or below ground, singly or in clusters of about three. Under field conditions it is not uncommon to find the blossoms with their yellow petals 3 inches below the soil surface. The calyx consists of a long, slender tube crowned with five calyx tips (Figure 29.4). The corolla is borne at the end of the calyx tube. After pollination takes place, the section immediately behind the

FIGURE 29.3
A prolific peanut plant of the bunch type, showing leaves above and nodule bearing roots below.

ovary (the "gynophore" or peg) elongates and pushes the ovary into the soil, where the pod develops (Figure 29.5). Peanut flowers are perfect and self-pollination is the general rule, but natural crossing between varieties sometimes occurs.[8] Flowering and pod set can occur two or more times during the season. This is caused by periods of favorable weather that occur several weeks apart, separated by a period of dry weather.[9] The plants can produce a matured early-set followed by a later-set limb crop that later causes harvest difficulties.

The peanut fruit is an indehiscent pod containing one to six (usually one to three) seeds. The pods form only underground (Figure 29.5). The seed is a straight

FIGURE 29.4
Portion of peanut plant showing blossom *(b)*, calyx tips *(c)*, gynophores or "pegs" *(g)*, leaflets *(l)*, and developing pod *(p)*.

embryo covered with a thin, papery seedcoat. The outer layer varies in color, but generally it is brick red, russet, or light tan, and occasionally black, purple, flesh-colored, or white. The weight per bushel of unshelled peanuts is about 22 pounds (280 g/l) for the Virginia type, 28 pounds (360 g/l) for the Southeastern Runner types, and 30 pounds (390 g/l) for the Spanish type.

Seed dormancy is characteristic of the runner types.[10] Dormant peanuts planted soon after maturity frequently require rest periods ranging up to two years before germination occurs.[11] The average time for emergence of fresh seed of different strains from the Spanish and Valencia groups ranges from nine to fifty days, while in a more dormant group that includes runner peanuts, it ranges from 110 to 210 days. The rest period is broken after several weeks or months in dry storage. The Spanish type shows no dormancy.

Peanut seed retains its viability for three to six years under proper storage conditions.

29.5 TYPES OF PEANUT

American peanuts are classified into four types: runner, Virginia, Spanish, and Valencia. The runner type leads in production. Virginia varieties were traditionally grown on the most acres, but newer, higher-yielding runner varieties have displaced Virginia types as the most popular varieties.

Runner varieties[12] mature over a range of 130 to 150 days. A Virginia bunch variety, which has 700 seeds per pound (1,540/kg), is classed commercially as a runner. Runner peanuts are grown mostly in Georgia, Alabama, Florida, Texas, and Oklahoma and account for about 75 percent of total U.S. production. Runners are used mostly in peanut butter and candy but can be used in all products.

Virginia varieties have dark green foliage, large pods, large seeds (approximately 500 per pound) (1,100/kg), and russet seedcoats. Occasionally, pods will

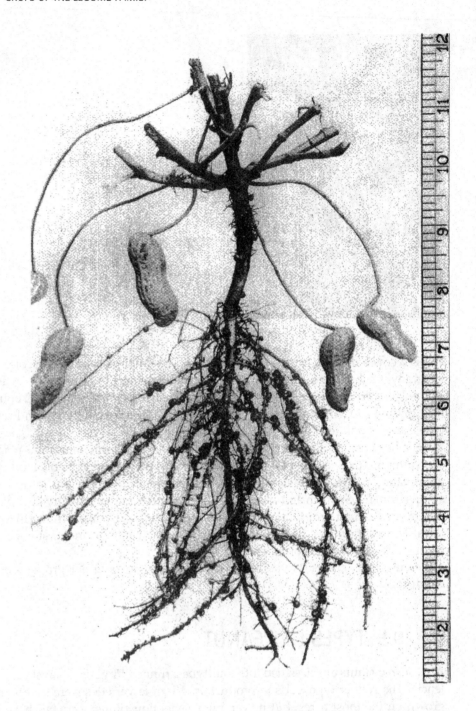

FIGURE 29.5

Plant of the runner type showing attachment of pods.

have three or four seeds. Virginia varieties mature in 135 to 140 days after planting. Virginia varieties are grown in Virginia, in North Carolina, in South Carolina, and to some extent in other southeastern states. Some Virginia peanuts are grown in north Texas under irrigation.[13] Virginias account for much of the peanuts roasted and eaten in the shell and shelled salted peanuts. Virginia peanuts account for about 21 percent of total U.S. production.

The Spanish group consists of erect types with light green foliage, rarely more than two seeds in a pod, short seeds, and tan seedcoats. Both pods and seeds are small. Spanish peanuts are grown in Texas, Oklahoma, and to a large extent in Georgia and Alabama. The varieties have 1,000 to 1,400 seeds per pound (2,200 to 3,100/kg). Spanish-type peanuts account for about 4 percent of U.S. production. Spanish peanuts are mostly used in salted nuts, candy, and peanut butter.

The Valencia type is characterized by many pods with three or four seeds, as well as by very sparse branching habits. The varieties are erect. The foliage is dark green, the seeds long or short, while the seed coats may be purple, red, russet, or tan. Valencia peanuts are now relatively unimportant, except in New Mexico where the acreage is limited. Valencia peanuts account for less than 1 percent of the U.S. market. They have a very sweet taste and are usually roasted and sold in the shell or boiled for fresh use.

Well-developed peanuts have a shelling percentage of 70 to 80 percent, but the seeds usually comprise 60 to 70 percent of the unshelled product, or average about 67 percent.

Another groundnut, called bambarra (*Boanzeia subterranea*), is a native African legume grown for food in tropical African countries. The pods and seeds are formed underground as in peanuts.[14]

29.6 FERTILIZERS

Peanuts are usually grown on leached sandy soils of relatively low fertility. Peanuts respond better to residual soil fertility than to direct fertilizer application. The previous crop should be fertilized for maximum yields and to increase residual soil fertility.[5] Grass crops, such as corn or grain sorghum, respond well to high rates of fertilizer. Peanuts have a deep root system and can utilize nutrients that leach below the rooting zone of grass crops.[13] The pegs as well as the roots can absorb mineral elements from the soil.[15] Soils testing medium or higher for potassium and phosphorus will not need additional fertilizer.[5, 13] High concentrations of potassium, magnesium, and sodium may interfere with calcium absorption. Peanut seldom responds to additional nitrogen fertilizer.

Peanut is one of the few crops that respond to calcium fertilizer. Applications of calcium increase the quality of the seeds. Calcium is directly absorbed from the soil by the peg and developing pods; it cannot be translocated from the roots.[5] Therefore, calcium must be readily available in the upper 3 inches (8 cm) of the soil solution where the pods are developing. Gypsum (calcium sulfate) is usually applied. Calcium supplied through gypsum is relatively water-soluble compared to other sources of calcium. Gypsum is more necessary in soils with a pH greater than 7.0 because other forms of calcium may be less available.[13] Gypsum may also control or reduce the pod rot disease complex.[5] The exchangeable calcium in the soil should be at least 600 pounds per acre (670 kg/ha) of calcium carbonate equivalent.[9] A calcium deficiency causes a darkened plumule[16] in the seed, or "pops" (unfilled shells and reduced yields),[15, 17] especially in dry weather. The calcium is best supplied by band applications over the tops of the plants when blooming begins.[9, 13, 18, 19] Calcium fertilizer is more important with large-seeded Virginia types, but smaller-seeded runner types are less susceptible.

Manganese deficiency may occur when soils are overlimed, soil pH is above 7.0,[5] or soil pH is very low.[9] A foliar application of a compound containing manganese sulfate may be needed. Boron can be deficient on deep, sandy soils. Boron deficiency causes peanut seeds to have cotyledons with depressed, darkened centers called "hollow heart." Foliar application of boron is recommended when needed. Iron, copper, and zinc may be deficient in high pH soils.[9, 13] Foliar applications of iron and copper, and foliar or soil applications of zinc, may be needed.

▓ 29.7 ROTATIONS

Peanut needs a long rotation program to help control pests. A three-year rotation with two years of grass crops helps reduce nematode infestations, soil-borne diseases, and many weeds.[5] A four-year rotation can provide more control.[13] Grass crops, such as corn or small grains, respond to heavy fertilization but leave adequate nutrients in the residue for the subsequent peanut crop. Peanuts following tobacco or other legumes have higher disease losses. Cotton can precede peanut if the cotton taproots are ripped up in the fall to allow decomposition before spring seedbed preparation. A small-grain cover crop planted in the fall between cotton and peanut helps to reduce disease, nematodes, and erosion.[5] In Texas, cotton is a good alternative crop.[9, 13]

In the southern Great Plains, peanuts should be planted in strips alternating with strips of crops such as sorghum or corn to reduce soil erosion because the removal of the peanut crop leaves the soil loose and bare.

In Georgia, growing peanuts in rotations reduces attacks by crown rot, Southern blight (white mold), nematodes, and leaf spot. Best yields are obtained following grass pasture. Other alternatives are rye or finally other grains and grasses. Peanuts should not follow soybeans or lupines because of disease hazards. Southern blight is often serious if peanuts follow cotton or if any crop litter is left uncovered in the field.[6]

▓ 29.8 PEANUT CULTURE

Either shelled or unshelled peanuts can be planted, but today practically all seed is shelled. Shelled seed gives quicker and often better stands. Care must be used when shelling seed peanuts to minimize damage to the seed and seedcoat. Breaking or splitting the seeds makes them useless for seeding. Seeds shelled at a moisture content of 7 to 8 percent have less than half the damage of seeds shelled at 4 to 5 percent.[20] Shelled seed should be stored in a moderately cool, dry place with relative humidity about 60 percent the temperature between 32 and 70°F (0 and 22°C). Seed must be handled gently to minimize damage.

Peanuts are graded to screen out the small, immature, poorly developed or wrinkled seeds that produce small, weak seedlings when planted.[21] More medium-sized seeds per acre should be planted than are optimum for large seeds.

Treatment of shelled seed with dust fungicides increases the assurance of satisfactory stands.[9, 22] A light coating of a pine tar–kerosene mixture repels crows and squirrels from the planted peanuts.

29.8.1 Seed Inoculation

Inoculation of peanuts with the proper culture of *Rhizobium* bacteria is advisable unless inoculated peanuts, cowpeas, or velvetbeans have been grown on the soil previously. Most peanut seed is treated with a fungicide to control diseases during germination and emergence. Inoculant applied to the soil instead of directly to the seed can help reduce injury to the bacteria culture.[20]

29.8.2 Seedbed Preparation

Thorough seedbed preparation for peanuts is essential. A well-tilled soil aids the penetration of the pegs. Fall or winter plowing is practiced except where erosion is serious. Plowing 8 to 9 inches (20 to 22 cm) deep, while completely covering the plant residues, greatly reduces losses from the stem and peg root disease caused by *Sclerotium rolfsii*.[23] Burying the residue from the previous crop hastens decomposition, making nutrients more available to the peanuts. Although low beds are preferable, growers often bed up peanut fields where there is a tendency for the soil to be wet.

When soils are highly erodible, no-tillage, and strip tillage is practiced. Strip tillage equipment will usually include an in-row subsoiler that is run 10 to 16 inches (25 to 40 cm) deep in the seed row. Strip tillage can be performed before seeding or during seeding.[24] Strip tillage before seeding allows incorporation of herbicides, fertilizer, fumigant, and insecticides in the seed row if appropriate. No-tillage may increase incidence of some diseases as well as reliance on herbicides that do not require incorporation. Burndown herbicides will have to be used to kill existing weeds before seeding. Fertilizer is best applied to the previous crop when using no-till. In 1998, only about 10 percent of peanut farmers in North Carolina used reduced tillage.[24]

29.8.3 Seeding Methods

Peanuts are generally planted after the danger of late spring frosts is past and when soil temperature, measured at noon at a 4-inch (10 cm) depth, is 65°F (18°C) or higher for three days.[5] In the commercial peanut region, the main crop is planted between April 10 and May 10. Planting about March 15 is recommended for southern Georgia. Spanish peanuts will mature in the Gulf Coast region when planted as late as July 1, but higher yields result from earlier plantings. In Texas, peanuts are planted about the same time as, or a little later than, cotton. The actual dates of planting range from about March 1 in southern Texas to May 15 or later in the northern part of the state. Peanuts should not be planted in dry soil.

Shelled peanut planting is done with either special peanut planters, corn or cotton planters equipped with peanut plates, or air planters. Peanut seed is easily damaged and must be handled gently. Planters with two inclined seed plates per row have a lower speed that reduces seed damage.[25] Seeding rates for maximum yields range from 35 pounds per acre (39 kg/ha) of shelled, small-seeded Spanish types in 36-inch (90 cm) rows, to 100 pounds per acre (112 kg/ha) of large-seeded types in 24-inch (60 cm) rows.[20] Under favorable growing conditions, better yields are obtained from thicker spacing of 3 to 8 inches (8 to 20 cm) apart in 30-inch (76 cm) rows for the larger types, and 3 inches (8 cm) apart in 24-inch (60 cm) rows for the small types. The average seeding rate in the United States is about 67 pounds per acre (75 kg/ha) on an unshelled basis, or about 45 pounds per acre (50 kg/ha) of shelled kernels.

29.8.4 Weed Control

The cultivation of peanuts is similar to that of soybeans grown in rows. One or two early cultivations with a rotary hoe or a flexible shank weeder are helpful. Later shallow cultivation is done with row equipment using flat blades that cut few peanut roots or pegs. Level cultivation is essential. Bedding up (moving soil up) to the peanut plants during cultivation causes stem and peg rots to develop.[23]

Annual weeds in peanuts are controlled by herbicides. Preplant herbicides for peanut include pendimethalin, ethalfluralin, vernolate, s-metolachlor, and imazethapyr. Preemergent herbicides include norflurazon, alachlor, and imazethapyr. Postemergent herbicides include imazapic, chlorimuron, bentazon, acifluorfen, and sethozydim. Herbicides can also be applied at the "cracking stage" (just before plants emerge) to control existing weeds. "Cracking stage" herbicides include paraquat + bentazon, imazethapyr, and acifluorfen.

29.8.5 Irrigation and Water Use

Peanut requires about 20 to 28 inches (500 to 700 mm) of water for satisfactory yields. Daily water use increases during vegetative growth, peaks during the reproductive period, and declines as pods begin to mature. During blooming, water stress can delay formation of flowers and inhibit flowering under extreme conditions. After bloom, peg penetration into the soil requires adequate moisture, so the pegging zone should be kept moist, even if adequate moisture is present in the soil profile. Also, a moist pegging zone facilitates the uptake of calcium by the pods. Low relative humidity and high soil temperatures can reduce the ability of pegs to penetrate soil and develop pods. Since peanut is usually grown on sandy soils that hold less water, the crop may benefit from irrigation, even if subsoil moisture is adequate.[9] After the pods are set, irrigation should be reduced to prevent the possibility of saturated soil from rainfall that can increase susceptibility to pod rot organisms.[9, 13]

Irrigation water quality is very important. The addition of salts from irrigation water can interfere with the uptake of calcium by the pegs and pods.[26]

29.8.6 Harvesting

Digging at the proper time is one of the most crucial decisions for a grower. Peanuts can gain from 300 to 500 pounds per acre (336 to 560 kg/ha) in yield and 1 to 2 percent in grade in the period before the optimum harvest date.[9] Since peanut has an indeterminate growth habit, the pods will not mature uniformly. Immature peanuts have poor flavor and deteriorate quickly in storage. They are more difficult to cure and more likely to be affected by undesirable mold growth, including aflatoxin. Optimum yield and quality is attained when pods of runner and Virginia types are 70 to 80 percent mature and Spanish types are 80 percent mature.[9] Mature pods will have dark veins on the inside, and the exposed, scraped hull will also be dark brown to black in color. The most accurate method for determining peanut maturity is the hull-scrape method. A representative sample of pods from several plants is scraped, and their color is compared to a chart. By using this method, the grower can determine percent maturity.

Peanuts in commercial fields are removed from the soil with special diggers equipped with blades or half-sweeps that cut off the roots below the nut zone and lift the plants from the ground. The vines are then windrowed with a shaker-windrower or with a side-delivery rake. Most common digging losses occur when digger blades run too shallow and peanuts are left in the soil, and when peanuts separate from the vines as they are lifted from the soil. Additional losses can occur when plants are shaken to remove dirt and placed in windrows. Proper synchronization of ground speed and shaker speed is necessary to minimize losses.[25]

Shaking the peanuts free from dirt is accomplished most readily in light, sandy soils free from clods. Small-podded varieties are separated from the soil with less loss than are the large-podded Virginia types. Most growers use a digger-shaker (Figure 29.6) and a windrower or a combination digger-shaker-windrower. They often reshake the windrow without digging blades attached to the shaker in order to remove more of the dirt from the plant roots. Peanuts are threshed with a combine when the kernels have about 18 to 24 percent moisture content.[9] At that stage the kernels rattle slightly in the shell. Combining at 25 to 35 percent moisture can reduce hull damage but will increase drying cost.[25] It is safe to wait for lower moisture if weather permits. Gentle handling of the crop is essential to minimize threshing losses. A well-adjusted combine will produce minimum vine breakage during threshing.

Peanuts are immediately dried to a moisture content of about 11 to 11.5 percent using forced air. Air temperature should never rise more than a 15°F (8°C) during drying, nor should the temperature exceed 95°F (35°C).[9, 25] After drying, peanuts will continue to lose moisture to about 9 to 10 percent moisture. Stored peanuts should not be allowed to dry below 8.5 percent moisture.[9]

Rapid curing by direct sunlight reduces seed viability, increases seed breakage as well as skinning during shelling, and impairs the edible quality of the peanuts. Large, loose windrows limit the number of pods that are exposed to sunlight yet permit the pods at the bottom and interior of the windrow to cure without molding.

▦ 29.9 PROCESSING PEANUTS

Before shelling, peanuts are cleaned to remove rocks, sticks, and stems as well as the shells. The nuts are passed through various screens and then through a machine sheller. Blanching gives a whiter and more homogeneous appearance to the nuts. This treatment involves removal of the seedcoat by a combination of drying, heating, rubbing between soft surfaces, and blowing an air current through the nuts. Shelled nuts may become rancid after storage for two months or more when exposed to the air. Rancidity is retarded when the nuts contain 5 percent moisture or less, and the relative humidity of the storage room is about 60 percent. Mechanical sorters equipped with photoelectric cells remove discolored kernels. More than half of the shelled edible peanuts consumed in the United States are

FIGURE 29.6
Inverters dig the peanut plants, shake them to remove soil from the peanuts, and invert them to dry and await harvest.
[Courtesy Kelley Manufacturing Co. (KMC)]

used to make peanut butter by grinding dry roasted salted nuts.[20] The remainder are mostly dry roasted until the nuts develop a brown color or are used in making peanut candy. Sometimes the nuts are roasted in oil, especially when they are to be salted.

For oil extraction, peanuts are shelled, cleaned, hulled, and crushed to open the oil cells as much as possible. From the rollers, the pulp goes to a cooker where the material is heated at about 235°F (113°C) in a humid atmosphere for 90 minutes. The oil is then extracted by the hydraulic press-plate method or by the expeller (cold-press) method. The crude peanut oil is collected in storage tanks, while the residual cake is ground into meal. The crude oil may be filtered to remove the foots (particles of meal), or the residue may be allowed to settle to the bottom of the tank and then removed.

An average ton (900 kg) of cleaned, unshelled peanuts yields about 530 pounds (240 kg) of oil, 820 pounds (372 kg) of meal, and 650 pounds (295 kg) of shells.

29.10 USES OF PEANUT

Peanuts that are not harvested are used for livestock feed (hogged off). The equivalent of more than 100 million pounds (45 million kg) of unshelled peanuts, or 6 percent of the threshed crop, is used for planting. An average of more than 500 million pounds (225 million kg) of shelled peanuts is crushed, producing more than 200 million pounds (90 million kg) of oil. About 6.5 pounds (3 kg) per capita are consumed as peanuts and peanut products, excluding crushing. Annual world production of peanut oil is about 5.5 million tons (5 million MT). It is an important edible vegetable oil in Europe, Asia, and Africa where peanut butter is almost unknown.

Green, leafy peanut hay is as nutritious as good-quality alfalfa hay.[27] Nutrient content of field-cured hay may be reduced 25 percent or more through loss of leaves, weathering, excessive stems, mustiness, and dirt content.

29.11 CHEMICAL COMPOSITION

Peanuts are high in both protein and oil. In general, the vines with the nuts contain 10 to 12 percent protein, or, without the nuts, about 7 percent protein. Unshelled peanuts contain about 25 percent protein and 33 percent fat. The nuts contain 40 to 48 percent oil, and from 25 to 30 percent protein. All of the embryo cells contain oil.[28] The nuts contain the amino acids, including cystine, that are essential for animal growth.[1] They are rich in phosphorus. Like other large legume seeds, peanuts are an excellent source of the vitamins thiamin and riboflavin as well as a good source of niacin. An average thiamin content of 9.6 micrograms per gram has been reported.[29] In thiamin content, the skins are highest, followed by the cotyledons, and then the germ. The average niacin content is about 17.2 milligrams per 100 grams.

29.12 DISEASES

Several diseases cause problems in peanut. Rotation with a grass crop and deep plowing at least 6 inches (15 cm) help to control most diseases. Other control methods are listed for each disease.

29.12.1 Peanut Leafspot

Peanuts suffer some losses from Cercospora leafspots. Early leafspot is caused by *Cercospora arachidicola* and late leafspot by *Cercosporidium personatum*.[22]

The leafspot diseases appear as small, brown-black spots on the leaves. They usually become severe in wet weather when the leaves may turn yellow and fall off. Applications of fungicides can help control the disease.

29.12.2 Southern Stem Rot

The fungus *Sclerotium rolfsii* causes Southern stem rot, also called Southern blight and white mold. It infects stems, pegs, rods, and nuts. Applications of fungicides can help control the disease. Some varieties have slight resistance.[22]

29.12.3 Seedling Blight

Several types of fungi cause seedling blight. The disease causes pre- or postemergent death of seedlings. Symptoms are elongated, sunken, dark areas in young plant parts. Seed treatment with a fungicide helps in control. Vigorous, high-germinating seed is more resistant.

29.12.4 Fusarium Root Rot or Wilt

Caused by *Fusarium* sp. fungi, the disease causes leaf yellowing and wilting of a branch or entire plant. Symptoms are more prevalent during hot periods of the day with plants recovering during cooler and wetter periods. Avoiding deep planting, if possible, and reducing drought stress help reduce incidence of this disease.

29.12.5 Crown Rot

Crown rot, also called black mold (*Aspergillus niger*), causes weak and wilted plants. Seedlings show water-soaked tissue on crown and roots, followed by a powder mass of black spores at the crown. More mature plants show lesions on the stem from the crown to above the soil. Seed treatment and high-germinating seeds help in control.

29.12.6 Sclerotinia Blight

Sclerotinia blight, *Sclerotinia minor*, is a serious disease in some areas.[30] It first appears after limbs have lapped the row middles and adequate moisture has occurred. Straw-colored lesions develop in the central portion of the stems. Fluffy white fungus growth on the margins of infected tissue is seen on moist mornings. As the disease progresses, stems will shred and darken. Black, elongated fungus structures (sclerotia) will develop. Sclerotia will overwinter on residue. Rotation with cotton, corn, or wheat helps to control the disease. Some varieties are resistant.

29.12.7 Cylindrocladium Black Rot (CBR)

The fungus *Cylindrocladium parasiticum* causes CBR. Symptoms are yellowing and wilting of leaves followed by browning of leaf margins. The taproot and other roots may turn black. Brown to black lesions occur on pegs and pods. To control this disease, peanut or soybean should not be planted in infected fields for at least five years. Avoid planting in cool soils. Thorough cleaning of tractors and other equipment before entering fields helps prevent spread. Some fungicides help reduce damage after infection.

29.12.8 Peg and Pod Rots

Peg and pod rots, caused by several species of fungi, can be a problem especially during wet periods. Avoiding excessive irrigation and controlling soil insects and nematodes help to control these diseases. Adequate calcium supply makes plants more resistant, but excess potassium makes plants more susceptible.

29.12.9 Other Diseases

Other diseases of peanut include root rot (*Rhizoctonia* spp. and *Fusarium* spp.), aerial blight and limb rot (*Rhizoctonia solani*), peanut rust (*Puccinia arachidis*), and alternaria leafspot (*Alternaria* spp.). Yellow mold (*Aspergillus flavus*) is a source of aflatoxin. *Aspergillus parasiticus* can also cause aflatoxin. Virus diseases include peanut mottle, peanut stripe, and tomato spotted wilt.

▓ 29.13 INSECT AND MITE PESTS

Most insects and mites damage peanut plants by feeding on foliage or feeding on roots, pegs, or pods. Using Integrated Pest Management (IPM) practices can effectively identify pests and determine the best control measures.[31] Since peanut is a food crop, careful use of insecticides is necessary to avoid contamination of the peanut fruits.

Foliage pests include corn earworms, fall armyworms, leafhoppers, spider mites, whiteflies, thrips, velvetbean caterpillars, and whitefringed beetles. Root-peg-pod pests include cutworms, lesser cornstalk borers, Southern corn rootworms, and white grubs. Rednecked peanut worms feed in the buds of plants.

REFERENCES

1. Leppik, E. E. "Assumed gene centers of peanuts and soybeans," *Econ. Bot.* 25, 2(1971):188–194.

2. Loden, H. D., and E. M. Hildebrand. "Peanuts—especially their diseases," *Econ. Bot.* 4, 4(1950):354–379.

3. Smith, C. E., Jr. "The new world centers of origin of cultivated plants and the archeological evidence," *Econ. Bot.* 22, 3(1968):253–266.

4. Beattie, J. H., and others. "Growing peanuts," *USDA Farmers Bull.* 2063, 1954.

5. Jordan, D. L. "Peanut production practices," in *North Carolina Peanut Production Guide,* NC Coop. Ext. Serv. rev, 1999.

6. McGill, J. F., and L. E. Samples. "Peanuts in Georgia," *GA Ext. Bull.* 640, 1969, pp. 1–39.

7. Wynne, J. C., and others. "Photoperiodic responses of peanuts," *Crop Sci.* 13, 5(1973):511–514.

8. Reed, E. L. "Anatomy, embryology, and ecology of *Arachis hypogea*," *Bot. Gaz.* 78(1924):289–310.

9. Lemon, R. G., and others. "Texas peanut production guide," *TX Agri. Ext. Serv.* 3M-0–96, 1996.

10. Stansel, R. H. "Peanut growing in the Gulf Coast Prairie of Texas," *TX Agr. Exp. Sta. Bull.* 503, 1935.

11. Hull, F. H. "Inheritance of rest period of seeds and certain other characters in the peanut," *FL Agr. Exp. Sta. Tech. Bull.* 314, 1937.

12. Higgins, B. B., and W. K. Bailey. "New varieties and selected strains of peanuts," *GA Agr. Exp. Sta. Bull.* New Series 11, 1955.

13. Lemon, R. G., and T. A. Lee. "Production of Virginia peanuts in the Rolling Plains and southern High Plains of Texas," *TX Agri. Ext. Serv.* L-5140, 1995.

14. Doku, E. V., and S. K. Karikari. "Bambarra groundnut," *Econ. Bot.* 25, 3(1971):255–262.

15. Brady, N. C. "The effect of calcium supply and mobility of calcium in the plant on peanut fruit filling," *Soil Sci. Soc. Am. Proc.* 12(1948):336–341.

16. Harris, H. C. "Calcium and boron effects on Florida peanuts," *FL Agr. Exp. Sta. Tech. Bull.* 723, 1968, pp. 1–17.

17. Scarsbrook, C. E., and J. T. Cope, Jr. "Fertility requirements of runner peanuts in southeastern Alabama," *AL Agr. Exp. Sta. Bull.* 302, 1956.

18. Bailey, W. K. "Virginia type peanuts in Georgia," *GA Agr. Exp. Sta. Bull.* 267, 1951.

19. Futral, J. G. "Peanut fertilizers and amendments for Georgia," *GA Agr. Exp. Sta. Bull.* 275, 1952.

20. Woodroof, J. G. *Peanuts: Production, Processing, Products,* 3rd ed. Westport, CT: AVI Pub. Co., 1983, pp. 1–291.

21. Mixon, A. C. "Effect of seed size and vigor and yield of runner peanuts," *Auburn U. Agr. Exp. Sta. Bull.* 346, 1963.

22. Kucharek, T. "Disease management in peanuts," *FL Coop. Ext. Serv.* PMDG-V1–10, 1997.

23. Mixon, A. C. "Effects of deep turning and non-dining cultivation on bunch and runner peanuts," *Auburn U. Agr. Exp. Sta. Bull.* 344, 1963.

24. Jordan, D. L., and others, "Agronomic and pest management considerations in reduced tillage peanut production," in *North Carolina Peanut Production Guide*, NC Coop. Ext. Serv. rev, 1999.

25. Roberson, G. T. "Planting, harvesting, and curing peanuts," in *North Carolina Peanut Production Guide*, NC Coop. Ext. Serv. rev, 1999.

26. McFarland, M. L., R. G. Lemon, and C. R. Stichler. "Irrigation water quality: Critical salt levels for peanuts, cotton, corn, and grain sorghum," *TX Agri. Ext. Serv.* SCS-1998-02, 1998.

27. Ronning, M., and others. "Feeding tests with threshed peanut hay for dairy cattle," *OK Agr. Exp. Sta. Bull.* B-400, 1953.

28. Stokes, W. E., and F. H. Hull. "Peanut breeding," *J. Am. Soc. Agron.* 22(1930):1004–1019.

29. Higgins, B. B., and others. "I. Peanut breeding and characteristics of some new strains. II. Thiamin chloride and nicotinic acid content of peanuts and peanut products. III. Peptization of peanut proteins," *GA Agr. Exp. Sta. Bull.* 213, 1942.

30. Bailey, J., and B. Shew. "Sclerotinia blight control," in *North Carolina Peanut Production Guide*, NC Coop. Ext. Serv. rev, 1999.

31. Johnson, F., and R. Sprenkel. "Insect management in peanuts," *Univ. FL Coop. Ext. Serv.* ENY-403, 1999.

Miscellaneous Legumes

Legumes that are grown for seed, forage, green manure, or as cover crops include field pea, vetch, velvetbean, burclover, black medic, buttonclover, kudzu, crotalaria, trefoil, sesbania, lupine, guar, Florida beggarweed, roughpea, hairy indigo, alyceclover, crownvetch, pigeon pea, fenugreek, sainfoin, and seradella. They may not occupy a large area in the country as a whole but are important where they are well adapted. Except for field pea, crownvetch, sainfoin, and trefoil, these crops are grown almost entirely in the southern and Pacific Coast states.

◼ 30.1 FIELD PEA

30.1.1 Economic Importance

In 2000–2003, global, dry, edible pea was grown on an average of about 15 million acres (6.3 million ha) annually. Annual production averages about 11 million tons (10 million MT) or 1,480 pounds per acre (1,650 kg/ha). The countries leading in pea production are Canada, France, the Russian Federation, China, and India. Production in the United States in 2000–2003 averaged about 215,000 tons (195,000 MT) on about 246,000 acres (99,000 ha). Yield was about 1,800 pounds per acre (2,000 kg/ha). The leading states in dry pea production are North Dakota, Washington, Idaho, Montana, and Oregon. It is also grown in southern Canada.

Sixty percent of the domestic seed pea production consists of Alaska and other smooth, green, seeded types used as dry peas or as planting seed by growers of fresh garden, canning, or frozen peas. The smooth yellow types, also produced in Canada, are largely consumed as dry split peas, while the white (Canada) types produced mostly in Minnesota and North Dakota serve as feed or for planting for hay or silage.[1] A considerable portion of the dry pea production is exported, mostly to Europe. In the Pacific Coast states, winter pea seed production has been largely replaced by blue lupine because of destruction by diseases. The chief disease is black stem caused by the fungi *Ascochyta pinodella* and *Mycosphaerella pinodes*.

30.1.2 Adaptation

A cool growing season is necessary for successful production of field pea. High temperatures are more injurious to the crop than are light frosts, which are injurious only when the plants are in blossom. Climatic requirements limit field-pea production to the northern states as a summer crop and to the southeastern states and the mild coastal sections of the Pacific Northwest as a winter crop.

Field pea is most productive where rainfall is fairly abundant, but it succeeds in cool, semiarid regions. Well-drained clay loam soils of limestone origin are best suited to field pea. Inoculation is essential or beneficial, except in fields where nodulated pea had been grown in recent years, leaving sufficient inoculum in the soil.[2]

30.1.3 Botanical Description

Field pea is classified botanically as *Pisum arvense*. Garden and canning peas are usually classified as *P. sativum*. However, several canning and garden varieties are also grown as dry field pea.[3] Garden pea tends to be sweeter and more wrinkled than field pea, but, since this does not hold for all varieties, there can be no sharp distinction between the two species. The pea is an annual herbaceous plant with slender, succulent stems 2 to 4 feet (60 to 120 cm) long. The foliage is pale green with a whitish bloom on the surface. The leaf consists of one to three pairs of leaflets and terminal branched tendrils. In most varieties, the blossoms are either reddish-purple or white. The pods are about 3 inches (8 cm) long and contain four to nine seeds. These seeds may be round, angular, or wrinkled. The seed cotyledons are mostly yellow or green (Figure 22.5).

Field pea is generally self-fertilized, but some cross-pollination is evident when varieties with white and colored blossoms are grown in alternate rows.[4]

The green types have green seedcoats and cotyledons. The yellow and white types have yellow or white seedcoats, but both have yellow or orange cotyledons. The name Canada or Canadian has been applied to white- and yellow-seeded varieties. Varieties with gray, brown, or mottled seeds, including the winter types, are grown for soil improvement or feed.[5, 6, 7]

30.1.4 Field-Pea Culture

In the Palouse area of Washington,[8] field pea can replace summer fallow in wheat rotations, as well as in the more semiarid sections of the Pacific Northwest. However, the harvesting of the two crops conflicts with each other. Pea fits into hay, small-grain, and corn rotations in most northern states.

Fall plowing is generally practiced for seedbed preparation in northern latitudes to facilitate early spring planting. Where grown as a summer crop, field pea is commonly sown with a grain drill as early in the spring as possible, usually in April. In the South and in the coastal section of the Pacific Coast states, the usual practice is to seed winter pea from September 15 to October 15. Annual weeds, including wild oats, in pea fields can be controlled with suitable herbicides.

Phosphorus fertilization for pea fields is advisable where soil tests show phosphorus availability to be deficient.[2, 9] The application of sulfur and molybdenum can sometimes be beneficial.

Seeding rates with a drill range from 45 to 150 pounds per acre (50 to 170 kg/ha) for small-seeded types to 80 to 180 pounds per acre (90 to 200 kg/ha) for large-seeded types. The lower rates apply to semiarid conditions, while the heaviest rates are suitable for well-fertilized soil in irrigated or humid areas. The desired stand in Idaho and Washington is six to nine plants per square foot (65 to 100 plants/m^2). In the Cotton Belt, winter pea is seeded at 30 to 50 pounds per acre (34 to 56 kg/ha).[1,2, 10,11]

For hay or silage, field pea is often sown in mixtures with oats or barley. The grain stems help to support the pea vines, reduce lodging, and make a better-balanced feed. Field pea is harvested for hay when most of the pods are well formed. Since pea often lodges badly, it is usually cut with a mower equipped with special lifting guards and a windrow attachment. Field pea may be harvested for seed in much the same manner when the pods are mature. The pick-up combine is often used to thresh the crop from the windrow or bunch. Most of the seed crop is combined directly. Defective and weevil-infested seeds are separated from sound seed by floating them off in brine.

30.1.5 Utilization

In the Cotton Belt, winter pea has been planted as a fall-sown cover and green-manure crop. Sufficient growth is generally made by March for the crop to be turned under. In northern states, field pea may be used for hay, pasture, and silage. For best results, pea should be allowed to mature before being grazed by hogs or sheep.

Whole dry peas are occasionally soaked and then cooked as a substitute for canned peas. Certain yellow and green varieties listed as dry, edible peas are marketed largely as split peas for soups. Split peas are the separated cotyledons of the pea, with the seed coats and most of the embryos removed. The commercial process of splitting peas is a trade secret. Pea can be hulled and split in a burr mill, with the burrs set far enough apart to avoid serious cracking of the cotyledons. The hulls, which have been removed by aspiration, and the embryos and seed fragments, which have been screened out, are useful in mixed feeds. Field-pea seed can be ground, mixed with grains, and fed to livestock. Pea and its by-products are popular pigeon feeds.

30.1.6 Diseases

Ascochyta blight of leaves and stems is caused by three different fungi: *Ascochyta pisi*, *Mycosphaerella pinodes*, and *Ascochyta pinodella*. It occurs in all states east of the Mississippi River, and it may cause heavy losses in seasons of abundant rainfall.[12] This disease is characterized in part by the formation of black to purplish streaks on the stem. The lesions are more conspicuous at the nodes. They enlarge into brown or purplish areas scattered from the roots to 10 or more inches (25 cm) up the stem. Spots on the leaves may be small to large, purplish, and irregular or circular. The entire leaf may shrivel and dry up. Similar spots on the pods are shrunken. The fungi that cause ascochyta blight are seed-borne, but they can also live from one season to the next on field residue.[13] To control the disease, only clean seed should be planted, such as western-grown seed. Crop residues should be deeply plowed under as soon as the pea crop is harvested. A three-year to five-year rotation should be practiced.[12] Resistant varieties are now available.[9]

Bacterial blight, caused by *Pseudomonas pisi*, produces olive-green to olive-brown water-soaked areas on the stems, leaves, and pods. The infected plants may be killed. The bacteria that cause the disease live over winter in the seed. Control measures include rotations and the use of clean seed.

Fusarium wilt of pea is caused by a soil-borne fungus, *Fusarium oxysporum* f. sp. *pisi*. Early plant symptoms are yellowing lower leaves, stunted plant growth, and downward curling of the margins of the leaves. Infected small plants may die without the production of seeds, and infected older plants may develop poorly filled pods. The disease can only be controlled by resistant varieties that are now available.[13]

Other diseases that attack field pea and the fungi that cause them include Septoria blight or blotch *(Septoria pisi)*, powdery mildew *(Erysiphe polygoni)*, downy mildew *(Peronspora pisi)*, anthracnose *(Colletotrichum pisi)*, various root rots, and several virus diseases.

Seed is usually treated with a protectant to reduce damage from seed rots and seedling blights. Graphite can be added to overcome the increased friction caused by the fungicidal dusts, which facilitates the flow of seed through the drill. Insecticides are often applied along with the fungicides.

30.1.7 Insect Pests

Pea weevil *(Bruchus pisorum)* is one of the most serious insect pests of the field pea. The adult weevil deposits its eggs on the pods while the seed is in the immature stage. The egg hatches and produces a larva that finally bores into the young seed, where it feeds on the embryo. The principal injury is the destruction of the seed, impairing it as food. In regions where the pea weevil is prevalent, the seed should be treated with an insecticidal fumigant, dust, or spray as soon as it has been threshed. Crop rotation and the burning of pea residue are also helpful. Application of insecticides will control the weevil when applied to the foliage during the early-bloom period before the eggs are laid.

Pea aphid *(Acyrthosiphon pisum)* attacks all kinds of pea. It sucks the sap from the leaves, stems, blossoms, and pods of the plants on which it feeds.[14] It can be checked in Austrian winter pea in western Oregon by delayed sowing until after October 16 to 20, and it can be controlled with insecticides.

Alfalfa looper and celery looper are controlled with insecticides. Wireworms are partly controlled by broadcasting an insecticide on the field, which is then disked under before pea is planted.

Other enemies of the pea include the pea moth *(Cydia nigricana)* and the root-knot nematode.

▓ 30.2 VETCH

30.2.1 Economic Importance

In 2000–2003, vetch *(Vicia* spp.) was harvested globally for seed on about 2.3 million acres (930,000 ha). Average global production was about 1 million tons (940,000 MT), or 3,200 pounds per acre (1,000 kg/ha). The leading countries in vetch seed production are the Russian Federation, Turkey, Spain, Ukraine, and Ethiopia. In the United States, vetch seed was grown on limited acres and production statistics are no longer available. Vetch is grown primarily for green manure or a winter cover crop, but also for pasture, hay, or silage.

30.2.2 Adaptation

Vetch makes its best development under cool-temperature conditions. The vetches vary in winter hardiness, but all are less hardy than alfalfa or red clover. Hairy vetch is the most winter hardy. Hungarian vetch and smooth vetch survive temperatures as low as 0°F (−18°C) in regions where temperature fluctuations are slight or where snow cover exists. Common vetch will not withstand zero temperatures and was killed out completely at Stillwater, Oklahoma.[15] Hungarian and woolypod vetches are almost as hardy as smooth vetch. Monantha, bard, and purple vetches are nonhardy. They are grown as winter crops only in mild climates. Vetches are grown in the winter in the Southeast as well as in the South Central and Pacific Coast states. Hairy and smooth vetches are winter crops in colder northern states. Common vetch is a spring crop in the Northeast and in the Central States. Vetch is unadapted to semiarid conditions.

All vetches grow well on fertile loam soils. Hairy, smooth, and monantha vetches are productive on poor sandy soils, while Hungarian vetch is productive on heavy wet soils. As a group, the vetches are only moderately sensitive to soil acidity.

When grown for seed, some of the seeds shatter and volunteer. The volunteer vetch contaminates wheat grown on the land as long as the vetch continues to re-seed itself. Wheat and vetch seed cannot be separated in the thresher or in a fanning mill, but only in special disk, cylinder, and spiral incline machines.

30.2.3 Botanical Description

Some thirty-six species of native or naturalized vetches are growing in North America.[16] Most of the commonly grown vetches are annuals. Hairy vetch is either an annual or a biennial. The common agricultural species are viny.[17] Stems may grow 2 to 5 feet (60 to 150 cm) or more in length. All the cultivated species have leaves with many leaflets. The leaves are terminated with tendrils in most of the species. The flowers are generally borne in racemes. The seeds of the vetches are more or less round, while the pods are elongated and compressed. Common vetch (Figure 30.1) is self-pollinated. The flowers of most vetches are violet or purple, the purple vetch having reddish-purple flowers. Hungarian vetch has white flowers.[18]

The most widely grown vetches are smooth, hairy, purple, common, woolypod, monantha, and bard. Narrowleaf vetch (*Vicia angustifolia*), often called wild vetch or wild pea, occurs as an introduced weed in the North Central states, where the seed contaminates wheat. It is sometimes used as an orchard cover crop in the Southeast. The seed is usually salvaged from wheat screenings. Bitter vetch (*V. ervilia*) is grown in southern Europe and in the Near East, but not in the United States. It could be grown along the Pacific Coast but not in the Southeast.[17]

Hairy or sand vetch (*V. villosa*) was once the most widely grown type in the South, but it has been replaced to a large extent by smooth vetch, which was selected from hairy vetch in Oregon. The stems of hairy vetch are viny and ascend

FIGURE 30.1
Branch of common vetch.

with support. It is characterized by pubescence on the stems and leaves as well as by the tufted growth at the ends of the stems. Smooth vetch is similar to hairy vetch except that it has less pubescence and lacks the tufted growth at the ends of the stems. Although winter hardy in the southern states, smooth vetch will not survive as far north as hairy vetch does.[17, 19]

Common vetch (*V. sativa*), sometimes called spring vetch or tares, is a semiviny plant with slightly larger leaves and stems than those of hairy vetch. One variety has a yellowish seed.

Hungarian vetch (*V. pannonica*) is less viny than either hairy vetch or common vetch. The plants tend to be erect when the growth is short or when they have support. They have a grayish color because of the pubescence on the stems and leaves.

Woolypod vetch (*V. dasycarpa*) has slightly smaller flowers and a more oval seed, but otherwise it is very similar to Hungarian vetch. It is adapted to the southeastern and Pacific Coast states, where it makes more early spring growth than hairy vetch.[17]

Purple vetch *(V. bengalensis)* is similar in growth habit to hairy vetch. The pods of purple vetch are hairy, while those of hairy vetch are smooth. It is the best adapted vetch for the coastal sections of California.[17]

One-flower vetch *(V. articulata)* has finer stems and leaves than hairy vetch. Single flowers are borne on long stems. It is grown safely as a winter legume only in Florida and other states bordering the Gulf, and in the Pacific Coast states where it is grown for seed.

Monantha, or bard, vetch *(V. monantha)* is similar to one-flower vetch except that the seeds are more rounded, while the flowers (two in a cluster) are purple instead of light lavender. Bard or monantha vetch is adapted only to the lower Colorado River Basin.

30.2.4 Vetch Culture

Hairy vetch is planted in the fall wherever it is grown. North of latitude 40°, all other vetches are planted in early spring, except on the Pacific Coast. In the Cotton Belt, the crop is seeded in early fall, usually in September or October. Vetch is planted from about September 15 to October 15 in Washington and Oregon. The rate of planting depends largely upon the size of seed, but planting twelve seeds per square foot (130/m^2) will give a satisfactory stand in California. Seeding rates of several species range from 20 to 50 pounds per acre (22 to 56 kg/ha) in the southern states to 30 to 80 pounds (34 to 90 kg/ha) in the North and West. The quantity of vetch seed should be reduced about one-fourth and that of small grain about one-half when they are grown together in mixtures. A common practice is to plant vetch with small grain when it is grown for forage. The grain stems support the vetch vines, which avoids damage from contact with the ground and also makes cutting easier.

Vetch is generally cut for hay when the first pods are well developed. It can be cut with a windrower, or a mower equipped with lifter guards and a windrow attachment. If grown as a seed crop, common, hairy, and smooth vetches are harvested when the lower pods are ripe to avoid shattering. Purple and Hungarian vetches are usually cut after 75 to 90 percent of the pods are ripe. The seed crop can be windrowed, dried, and then threshed with a pick-up combine.

30.2.5 Utilization

Vetches are most widely grown in the southern states as cover and green-manure crops, although occasionally they are cut for hay. In Alabama, vetch or Austrian winter pea grown in a two-year rotation (cotton–winter legume–corn) increased the corn yield 18 bushels per acre (1,100 kg/ha).[20] The second-year residue from these legumes increased the cotton yield by 213 pounds per acre (240 kg/ha). Large increases in the yields of corn and cotton were obtained when vetch was turned under in Alabama as early as March 15. Vetch, or vetch and rye, can be drilled into bermudagrass sod in the fall and then grazed during the winter.

In the Pacific Northwest, vetches are used for hay either alone or in mixtures with oats. Common and Hungarian are grown most generally for hay. The vetches have been used for soiling. They can also be grazed during the winter, spring, and early summer.

30.2.6 Diseases

Anthracnose, caused by the fungus *Colletotrichum acutatum*, attacks the leaves and stems of smooth and Willamette vetch. Other common vetches and Hungarian are resistant. False anthracnose, caused by *Kabatiella nigricans*, girdles the stem and sometimes infects the vetch seeds. Oregon common, Hungarian, and monantha vetches are resistant. Downy mildew, caused by *Peronospora viciae*, damages vetch in the Pacific Northwest as well as in the South. Blackstem, caused by *Ascochyta pisi*, discolors the stems. It may occur wherever vetch is grown. The disease is especially serious in thick stands of smooth and hairy vetch. Root rot, which is caused by several different fungi, damages the plants and often kills vetch seedlings in all areas. Smooth vetch is partly resistant.

Other diseases and the causal organisms include rust (*Uromyces fabae*), leaf and stem spot, and *Septoria* scald. Damage from root-knot nematodes, where present, is reduced by late seeding.

30.2.7 Insect Pests

Many insect pests of other forage legumes also attack vetch. Some of these are aphids, corn earworms, grasshoppers, cutworms, armyworms, various weevils, and leafhoppers. These and lygus bugs (*Lygus* species) can be at least partly controlled with suitable insecticides.

A specific pest, the vetch bruchid or vetch weevil (*Bruchus brachialis*), can cause heavy reduction in the seed yield of hairy, smooth, and woolypod vetch. It does not infest the seed of common or Hungarian vetch. It is found throughout the vetch-growing area of the Middle Atlantic states, but it is also considered a serious pest in some of the western states. Dusting the fields with insecticides when the first pods appear is an effective control method.

Heavy infestations of the pea aphid (*Macrosiphum pisi*) destroy the seed crop and seriously damage the quality of the hay. It can be controlled by spraying the field with an insecticide.

▪ 30.3 VELVETBEAN

Velvetbean has been grown in the United States as a field crop only since about 1890. Its popularity declined after 1941.

30.3.1 Adaptation

Velvetbean (*Mucuna* spp.) is a warm-weather crop that requires a long frost-free season to mature the seed. The crop was confined to Florida and other Gulf Coast states until the introduction of early varieties extended its range to the northern limits of the Cotton Belt. Velvetbean is well adapted to the sandier, less fertile soils of the southeastern states, especially those of the coastal plain.[21] It makes poor growth on cold, wet soils. It has been used extensively as a green-manure crop on newly cleared woodlands.

30.3.2 Botanical Description

The velvetbean is a vigorous summer annual. With the exception of the bush varieties, the vines attain a length of 10 to 25 feet (3 to 8 m). The leaves are trifoliate, with large, ovate, membranous leaflets shorter than the petiole. The flowers are borne singly or in twos and threes in long pendant clusters. The numerous hairs on the pods of velvetbean sting like nettles and make the crop unpleasant to handle. The pubescence usually sheds soon after maturity. The pods range from 2 to 6 inches (5 to 15 cm) in length. The seeds may be grayish, marbled with brown (Figure 30.2), white, brown, or black. Velvetbean has numerous fleshy surface roots that often reach 20 to 30 feet (6 to 9 m) in length.

Florida velvetbean (*Mucuna deeringianum*) was the only species grown in this country until 1906. It required 240 to 270 days to mature. Much earlier varieties are now available.

The purple flowers are borne in clusters 3 to 8 inches (8 to 20 cm) long.[22] The pods are 2 to 3 inches (5 to 8 cm) long, nearly straight, and covered with black velvety pubescence. The seeds are nearly spherical, grayish, and marbled with brown.

The Lyon velvetbean (*M. pruriens*) is a long-season type that matures about 10 days earlier than the Florida. It has white flowers borne in racemes sometimes 2 to 3 feet (60 to 90 cm) long. The pods are 5 to 6 inches (13 to 15 cm) long, covered with gray hairs, and have a tendency to shatter when mature. The seeds are ash-colored.

Earlier varieties from a cross between the Florida and Lyon species are heavy seed producers.

FIGURE 30.2
Leaves, pod, and seeds of the velvetbean.

The Yokohama velvetbean *(M. hassjo)* produces a smaller vine growth than the other species. It matures within 110 to 120 days. The flowers are purple. The pods are 4 to 6 inches (10 to 15 cm) long, flat, and covered with gray pubescence. This variety is seldom grown.

30.3.3 Velvetbean Culture

Since velvetbean does not tolerate cold, it should be planted in warm soils after all danger of frost is past. In the northern part of the Cotton Belt the planting time is about the same as for corn. Farther south, late varieties are planted as soon as the soil is in good condition, but planting of early varieties may extend over a period of forty to sixty days.[6] Velvetbean should be planted in rows 3 to 4 feet (90 to 120 cm) apart at the rate of 30 to 35 pounds per acre (34 to 39 kg/ha) of seed when grown for grazing, green manure, or as a smother crop. When grown for seed, it is interplanted with corn at a rate of about 6 pounds per acre (7 kg/ha) or more. Fertilization with phosphorus and, often, potash may be beneficial.

Velvetbean plants produce little seed unless they are grown with an upright crop like corn so that the vines can climb and bear their flowers where there is air circulation. This also prevents pod decay. The pods are picked by hand and threshed later, after drying. No machine is suitable for harvesting the beans. The corn is also harvested by hand because the long, tangled velvetbean vines clog the mechanical harvesters. The heavy labor requirements discourage velvetbean seed production.

30.3.4 Utilization

Velvetbean is now mainly grown for fall and winter grazing after the crop has matured. Cattle can chew the beans after they have been softened by rains. As summer pasture, velvetbean is much inferior to soybean or pearl millet in palatability and in animal gains.

Velvetbean is a good green-manure crop. In experiments in Alabama, the yield of seed cotton after cotton was 918 pounds per acre (1,030 kg/ha); after velvetbean cut for hay, 1,126 pounds (1,260 kg/ha); and after velvetbean was plowed under, 1,578 pounds (1,770 kg/ha). In southern Georgia, corn that was grown after velvetbean was plowed under in the winter yielded about twice as much as when following corn.

Velvetbean is an unsatisfactory hay crop because of difficulties in mowing and curing the mass of tangled vines. Moreover, the cured vines and leaves are black and unattractive.

Ground or soaked velvetbean is sometimes fed to livestock, but it is an uneconomical feed. It is much less nutritious than cottonseed meal. Ground beans soon become rancid.

30.3.5 Pests

Velvetbean is remarkably free from disease or insect pests.[22] The velvetbean caterpillar *(Anticarsia gemmatilis)* may cause serious injury to the crop south of central

Georgia. The caterpillar seldom attacks the crop until the plants begin to bloom. The damage is caused by defoliation, especially on late varieties. The caterpillars are controlled by dusting with insecticides.

▩ 30.4 ANNUAL MEDICS

Burclover, black medic, and buttonclover are naturally reseeding winter annual legumes.[14, 23, 24] They are members of the genus *Medicago*, to which alfalfa belongs. All burclovers are native to the Mediterranean region.

30.4.1 Adaptation

The burclovers are primarily adapted to the mild, moist winters of the Cotton Belt and the Pacific Coast. Few legumes make more growth in the Gulf Coast area in cool weather than does burclover. The crop will grow on moist soils, but it is most productive on well-drained loam soils. In California, burclover grows vigorously in adobe soils, which are often poorly drained. The plant thrives on soils rich in lime but succeeds also on somewhat acid soils.

Most soils on the Pacific Coast are inoculated with the proper legume bacteria, but inoculation is essential in many southern states. In Alabama tests, nodule formation was greater when inoculation bacteria were applied to the unshelled seed than to the shelled seed.[25]

30.4.2 Botanical Description

Burclover plants branch at the crown and have from ten to twenty decumbent branches that are from 6 to 30 inches (15 to 75 cm) long. The roots do not extend very deeply into the soil. Burclover has small yellow flowers borne in clusters of five to ten. The coiled pods are covered with spines that form the so-called bur. A vigorous plant may produce as many as 1,000 pods. The seeds resemble those of alfalfa.

Spotted or southern burclover *(Medicago arabica)*, widely grown in the Cotton Belt, has a purple spot in the center of each leaflet. The pods contain from two to eight seeds. California or toothed burclover *(M. hispida)* is the most common species on the Pacific Coast. It lacks the spots on the leaflet. The pods have three and sometimes five seeds. California burclover is less winter hardy than spotted burclover. Tifton burclover *(M. rigidula)*, a more recent introduction grown on the coastal plain of Georgia, is more resistant to extremes of temperature, is more resistant to disease, and produces more seed than the other common burclovers.[23] Cogwheel burclover *(M. tuberculata)* has spineless pods. This makes it preferable for sheep pasture because the burs do not become entangled in the wool. The burs look like cog-wheels.

Black medic *(M. lupulina)* has slender, finely pubescent procumbent stems, 1 to 2 feet (30 to 60 cm) long. The leaves are pinnately trifoliolate with long petioles. The leaflets are finely pubescent, obovate, rounded, and slightly toothed at the tips, and are ½ inch (13 mm) or less in length. The small, bright-yellow flowers are borne in dense heads about ½ inch (13 mm) long. The seeds (Figure 22.5), which closely resemble those of alfalfa, formerly were imported from Europe as an adulterant of alfalfa seed. Black medic (often called yellow trefoil) is distributed naturally in pastures, waste places, and meadows over much of the humid area of the United States. In certain sections of the Southeast, it furnishes valuable pasturage in late winter or early spring when other pasture plants are largely dormant.[26]

Buttonclover *(M. orbicularis)* has spineless, flat, coiled, button-shaped seed pods. It is adapted to soils relatively high in lime where it is a valuable pasture plant.

30.4.3 Medic Culture

Burclover, black medic, and buttonclover are seeded in the fall from September to December. Seeding must be early enough for the plants to become established before winter.

Seed in the bur is generally broadcast at the rate of 3 to 6 bushels per acre (260 to 520 l/ha), while the shelled seed is drilled at the rate of 12 to 20 pounds per acre (13 to 22 kg/ha).[27] Good stands are obtained thereafter without additional seeding, provided the land is plowed after the burs are ripe. The plants reseed indefinitely on established pasture lands. Lighter seeding rates than those mentioned previously are advisable when sowing with grasses or in established pastures.

Seed of the annual medics is sometimes harvested with a combine equipped with a tined reel and special lifter guards. In other cases, the vines are raked and removed after the burs have fallen to the ground. The ripe burs are then gathered with a combine equipped with brush and suction attachments. Formerly, the burs were swept up and gathered by hand.

30.4.4 Utilization

Burclover is primarily used as a pasture crop. It is valuable on California rangelands where animals graze the pods, particularly after the pods are softened by rains, or where the burclover cures and dries after the rainy season is over.[28] Weathering of the cured forage reduces its digestibility. A combination of burclover and bermudagrass is often used as a permanent pasture in the South.

Burclover alone has been widely used in California orchards or rice fields as a green-manure crop. It has proved to be a valuable cover and green-manure crop in the South. It is often seeded in cotton and corn to control soil erosion. Burclover is seldom used for hay.

Black medic sown with locally adapted (Florida or Alabama) seed is recommended for Florida pastures.[26] Imported (European) seed produces slow-growing, prostrate plants that fail to produce seed under high-temperature, short-day conditions. It is adapted to well-drained, open, limed, or calcareous soils. It is sown at a rate of 7 to 12 pounds per acre (8 to 13 kg/ha).

Buttonclover is a good hay crop, especially in mixtures with grass or small grains.[29]

30.4.5 Pests

Cercospora leaf spot attacks spotted burclover, while Pseudopeziza leaf spot and rust attack black medic. Rust also attacks California burclover. Most of the annual medics suffer from virus diseases. Insects that attack these crops are the alfalfa weevil and the spotted alfalfa aphid.

30.5 KUDZU

Kudzu *(Pueraria montana* var. *lobata)* is a native of China and Japan. It was introduced into the United States in 1876. It became popular in the southern states because of its rapid growth, which provided exceptional erosion control in highly eroded fields and road banks. By 1945, there were about 500,000 acres (200,000 ha) of kudzu in the southeastern states.[30] It is estimated that kudzu now covers over 7 million acres (3 million ha) in the South.

30.5.1 Adaptation

Kudzu is a warm-season plant that is somewhat drought resistant. The leaves and stems are sensitive to frost. Kudzu makes its seasonal growth from the time the soil becomes warm in the spring until the first frost in the fall. The plant will grow on many soil types, being able to thrive on soils too acidic for clover. It served as a cover crop for rough, cultivated lands. Kudzu is not adapted to soils with a high water table or to the lime soils of the Black Belt in Alabama.

30.5.2 Botanical Description

Kudzu is a perennial leguminous vine. The plant has many stems or runners that may grow as much as 70 feet (21 m) in one season. The stems have nodes that send out roots wherever they come in contact with the soil. The internodes die and the rooted nodes become separate plants called crowns. The leaves resemble those of the velvetbean except that they are hairy. The deep-purple flowers are borne in clusters (Figure 30.3). Seed production is usually very poor. More seed is set on old than on young vines,[31] and more on climbing than on prostrate vines. The seeds are

FIGURE 30.3

Leaves and flowers of kudzu.

FIGURE 30.4
Kudzu vines have completely covered a forest landscape in Mississippi.
[Courtesy Richard Waldren]

mottled and about one-fifth as large as those of pea. Tropical kudzu (*P. phaseoloides*) grows in the West Indies as well as in other hot climates.

30.5.3 Utilization

Because of its rapid growth, ability to flourish on marginal soils, and lack of natural predators and diseases in the United States, kudzu has become a serious weed pest where it grows (Figure 30.4). In the 1950s, kudzu was removed from the federal government's list of recommended cover crops. In the 1960s, the focus of research shifted from how to utilize the crop to how to eradicate it. In 1970, kudzu was officially declared a weed by the USDA, and in 1997 it was upgraded to a noxious weed. All states where kudzu survives have added it to their noxious weed list. Today, kudzu is often used as an example of the dangers of importing plants into a new region.

Kudzu can be controlled by heavy grazing, frequent mowing, or spraying with picloram, dicamba, or glyphosate. A bioherbicide developed from the sicklepod fungus (*Myrothecium verrucaria*) has recently been released that shows great promise in combating kudzu infestations.

Kudzu can be used for hay or pasture. Its feeding value compares favorably with that of other legumes. Kudzu furnishes grazing during the summer drought period when many other pastures are unproductive.

■ 30.6 CROTALARIA

Crotalaria is an annual or perennial that has been grown in the southeastern states as a summer annual cover crop. The five species in the United States are *Crotalaria*

FIGURE 30.5
Showy crotalaria.

intermedia, C. striata, C. spectabilis, C. lanceolata, and *C. juncea.*[32] *Crotalaria juncea* is called Sunn hemp in India where it is grown frequently as a fiber crop. Sowing of crotalaria is now banned or discouraged even as a cover crop because all crotalaria seeds, and also the plants of the *spectabilis* and *juncea* species, are poisonous to livestock.[21] It is being replaced by pigeon pea and other legumes.

30.6.1 Adaptation

Crotalaria is native to tropical regions of heavy rainfall. It requires a warm season for vigorous plant growth. The plant is sensitive to frost, being killed or seriously injured at 28 to 29 °F (−2°C).[33] Showy crotalaria *(C. spectabilis)*(Figure 30.5) will mature seed as far north as North Carolina.[32] Crotalaria is adapted to light sandy soils, especially to those of the Coastal Plains area of the southeastern states. Well-drained soils are essential to growth. The soils in this region are naturally inoculated with the proper legume bacteria. Crotalaria will grow on poor acidic soils. It is also resistant to the root-knot nematode.[5]

30.6.2 Botanical Description

The *Crotalaria* species, *C. spectabilis* and *C. striata*, are moderately branched, upright annuals that attain a height of 3 to 6 feet (90 to 180 cm). The leaves are large and numerous. *C. striata* has numerous yellow flowers borne on long terminal racemes.[33] From forty to fifty seeds are borne in pods similar to those of pea. The seeds are kidney-shaped and vary in color from olive-green to brown. The leaves are trifoliate in this species, but monofoliate in *C. spectabilis*. Both species have a bitter taste in the green state.

30.6.3 Crotalaria Culture

When used as a cover crop, crotalaria is generally planted from March 15 to April 15 in Florida, or from April 1 to 30 in other Gulf Coast states.[32] As a hay crop, it is often planted in June so as to be ready for harvest by October 1. Crotalaria can be sown by broadcasting, or in close drills at the rate of 15 to 30 pounds of seed per acre (17 to 34 kg/ha). The higher rate is generally followed for unscarified seed or when seeding conditions are poor. Crotalaria can be seeded in fields of winter oat in the lower South,[32] where it makes a rapid growth after the oat is harvested. Where crotalaria is grown for seed, it is commonly planted at a rate of 2 to 6 pounds per acre (2 to 7 kg/ha) in cultivated rows.

Crotalaria should be cut for hay at least by the bloom stage and preferably in the bud stage or earlier.[34] To obtain a satisfactory second growth, it is necessary to leave stubble of at least 8 to 10 inches (20 to 25 cm). Seed may be harvested with a combine or left to fall on the ground to reseed the crop. Much of the seed requires scarification before it will germinate promptly.

30.6.4 Utilization

Crotalaria was once used as a cover and green-manure crop.[33, 35] Showy crotalaria (*C. spectabilis*) was often planted in pecan and tuna groves, but striped crotalaria (*C. striata*) harbors injurious insects in citrus orchards.

Crotalaria is rarely used as a forage crop because it is not very palatable. The coarse, fibrous stalks also make a poor quality of hay. A fair quality of hay and silage can be made from *C. intermedia* when the plants are cut early.[34] The green forage, hay, and silage of *C. spectabilis* and *C. juncea* can cause death when fed to livestock. They contain an alkaloid called monocrotaline. The immature plants of *Crotalaria intermedia*, *C. lanceolata*, and *C. striata* are not poisonous.

▨ 30.7 TREFOIL

30.7.1 Economic Importance

Trefoils are perennial long-day legumes that persist for many years in pastures. They were naturalized, or grown after seeding, on more than 1.7 million acres (690,000 ha) of pasture and meadowland in the United States in 1969, mostly in mixtures with grasses and other legumes. Most of the acreage was in California, Ohio, Iowa, New York, Pennsylvania, and Vermont. It is also of some importance in Illinois, Michigan, Oregon, and Minnesota. Most of the trefoil is the broadleaf birdsfoot trefoil, *Lotus corniculatus*. Narrowleaf trefoil, *L. tenuis*, is grown in New York, but also under irrigation in California, Oregon, Idaho, and Nevada. Big trefoil, *L. uglinosus* (or *L. major*), is grown in Washington and Oregon.[36]

30.7.2 Adaptation

Trefoils are adapted to the cooler climates of United States north of latitude 37°. It is suited to the acidic and less-fertile soils and to poorly drained soils.[17] Birdsfoot trefoil has a place on fertile, well-drained soils in permanent pastures, but present varieties are less productive for hay than is alfalfa. It is poorly adapted to the southeastern states, except in the cooler Appalachian Mountains.

Narrowleaf trefoil is well suited to heavy, poorly drained clay soils in the same general region.[17] It is an important species in pastures in New York as well as in California. This species grows well in soils that contain large quantities of soluble salts. It is less hardy than birdsfoot trefoil.

Big trefoil is well adapted to the acidic coastal soils of the Pacific Northwest where the winters are mild. It is not tolerant to drought owing to its shallow rooting habit, but it grows well on low-lying soils that are frequently flooded during the winter months.[17] Big trefoil is productive on soils of pH 4.5 to 5.5. It has promise as a pasture legume in the acidic flatwood soils of the southeastern coastal regions when adequately fertilized.[37]

30.7.3 Botanical Description

Birdsfoot trefoil is broadleaved, with a well-developed branching taproot. It has few to many ascending stems that develop from each crown,[17] and it is similar to alfalfa in growth habit. The plants usually are erect or ascending, reaching a height of 12 to 30 inches (30 to 75 cm). The leaves are borne alternately on opposite sides of the stem. They are composed of five leaflets, of which one is terminal, two apical, and two basal, at the base of each petiole. The last two resemble stipules. Two to six flowers are borne in umbels at the extremity of a long peduncle that arises from the leaf axil. The pea-like flowers are yellow with faint red or orange stripes that are usually present in young flowers. Pods form at right angles at the end of the peduncle in the shape of a bird's foot. The seeds are oval to spherical and (Figure 22.5) vary in color from light to dark brown. Improved varieties of birdsfoot trefoil are replacing the common type.[11, 38]

Narrowleaf trefoil has narrow, linear lanceolate leaflets on slender, weak stems with comparatively long internodes. Flowers are slightly smaller, are fewer in number, and usually change from yellow to orange-red at maturity. Otherwise, it is similar to the birdsfoot species.

Big trefoil is similar to birdsfoot trefoil in appearance but differs in that it has rhizomes.[17] It also has more flowers per umbel than either of the other two species. The seeds of big trefoil are almost spherical. They vary in color from yellowish to olive-green without any speckling. Big trefoil seeds are much smaller than those of the other two species.

30.7.4 Trefoil Culture

Trefoils may be grown alone or in mixtures with grasses or other legumes. Competition with companion crops, grasses, and weeds often causes poor stands. Shading from a companion crop depressed seedling growth and nodulation more in birdsfoot trefoil than it did in alfalfa or red clover.[39]

Shallow sowing of trefoil on a well-prepared compact seedbed is essential. In California, birdsfoot trefoil is sometimes broadcast from an airplane on flooded, irrigated fields. The sprouting seeds settle in the mud when the fields are drained forty-eight hours after sowing. Early spring seeding is advisable in the northern or central states, while fall or late-winter seeding occurs in California and Florida. When grown alone, birdsfoot or narrowleaf trefoil may be sown at 4 to 6 pounds of seed per acre (4.5 to 7 kg/ha). These rates are based on good-quality seed that is scarified as well as inoculated with a specific trefoil bacterial inoculum. Simple mixtures of a grass, with 4 to 6 pounds of trefoil seed per acre (4.5 to 7 kg/ha), usually have been successful.[17]

Trefoil seedlings are small and develop slowly the first year as they are very sensitive to shading. Seeding above a band of fertilizer helps to establish trefoil stands. Weeds or weedy grasses often restrict their growth unless the weeds are mowed or treated with herbicides.

Trefoil hay yields are comparable to those of red clover under the same conditions. The hay and silage have a composition and quality similar to those of clover and alfalfa. Its value for pasture lies mostly in its succulent growth in midsummer and late summer when other herbage has nearly ceased growing. Under good management, the stands are maintained when the plants are occasionally allowed to reseed naturally.

Trefoil seed is threshed with a combine after windrowing or after the application of a desiccant to the crop. Clean seed yields of approximately 100 pounds per acre (110 kg/ha) are typical because of uneven flowering and the shattering of seed soon after the pods are ripe.

30.7.5 Pests

Trefoils may be attacked by leaf blight, crown rot, root rots, leaf spot, stem canker, and nematodes. Special control methods are unavailable. Insect pests include spittlebugs, leafhoppers, lygus bugs, stink bugs, grasshoppers, cutworms, and seed chalcids. They are controlled largely by insecticides. Trefoils are not damaged by the alfalfa weevil.

▦ 30.8 CROWNVETCH

Crownvetch (*Coronilla varia*), a native of the eastern hemisphere, was introduced into the United States and grown as an ornamental. Its agricultural value was recognized many years later.[40]

30.8.1 Adaptation

Crownvetch is most successful in Pennsylvania but is well adapted to the area from northeastern Oklahoma to northern Iowa and eastward to the Atlantic Ocean. It is also grown in Minnesota where the less hardy varieties may winterkill. It can be grown in the humid areas of the Pacific Northwest. Crownvetch requires good drainage but ample soil moisture. It grows on infertile soils but responds well to applications of phosphorus, potassium, calcium, and magnesium. A soil pH of 6.5 to 7.0 is preferable.

30.8.2 Botanical Description

Crownvetch is a hardy, long-lived perennial that spreads by creeping underground rootstalks. The leafy stems are hollow and decumbent unless supported by other vegetation. They may reach a length of 2 to 4 feet (60 to 120 cm). The leaves have a superficial resemblance to the true vetches (*Vicia* species). The variegated flowers are white to rose or violet.[40, 41] Insect pollination is essential for good seed yields.[1] Fully mature seeds are mostly hard and require scarification, sulfuric acid soaking, or dipping in liquid nitrogen in order to germinate promptly.

30.8.3 Crownvetch Culture

Mature, scarified, inoculated, and treated seed should be sown ¼ to ½ inch (6 to 13 mm) deep in a well-prepared seedbed.[42] Young plants grow slowly and compete poorly with weeds or with companion crops unless they are sown in alternate drill rows. A period of one to four years is required to establish a full productive stand.

30.8.4 Utilization

The main use of crownvetch is erosion control on steep slopes and banks.[41] It provides fair pasturage but recovers slowly after mowing or close grazing. The hay is similar to other forage legumes in protein and fiber content. The plants have high tannin content and are not highly palatable, but livestock become accustomed to it. The development of improved varieties has greatly expanded crownvetch as a forage crop.

▦ 30.9 CICER MILKVETCH

Cicer milkvetch (*Astragalus cicer*) is a native of Europe. It was introduced into the United States in the 1920s. Improved varieties have been released from Wyoming, Montana, and Colorado.

30.9.1 Adaptation

Cicer milkvetch is adapted to a wide range of environments, from Alaska to New Mexico, and from the Central Plains to the Cascade-Sierra ranges. However, its principal range of adaptation is the Rocky Mountain region. It is adapted to all soil textures, but it does best on moderately coarse-textured soils. It tolerates weak soil acidity and is very tolerant of soil alkalinity. It has been grown on wet soils with a pH as high as 9.8.

Cicer milkvetch requires an annual precipitation of 15 inches (380 mm) or more and does well in regions with 18 to 35 inches (460 to 900 mm). It is tolerant of a high water table and thrives in sub-irrigated meadows. It does well in higher elevations and has been grown in Montana at 10,000 feet (3,000 m). It is more frost tolerant than alfalfa.

30.9.2 Botanical Description

Cicer milkvetch is a long-lived perennial with vigorous, creeping rhizomes. Stems are hollow and succulent. Stem growth is upright when young, but stems become decumbent to trailing as growth continues. Plant height is usually about 3 feet (90 cm) but may reach 4 feet (120 cm) or more when flowering. Leaves are 4 to 8 inches (10 to 20 cm) long and pinnately compound with 21 to 27 leaflets. The lower sides of leaflets are sparsely hairy. The inflorescence of fifteen to sixty flowers is a raceme that originates at the base of the leaves. Flowers are pale yellow to white.

30.9.3 Cicer Milkvetch Culture

Cicer milkvetch requires a well-prepared, firm seedbed. Seed is drilled ½ to ¾ inch (13 to 19 mm) deep at a rate of 6 to 12 pounds per acre (7 to 14 kg/ha) PLS. When seeding mixtures with grasses, the seeding rate is 5 pound per acre (5.6 kg/ha). For spring seeding, the seed should be scarified. Scarification is not necessary for fall seeding. Seed should be inoculated with the appropriate strain of *Rhizobium* bacteria.

30.9.4 Utilization

Because the plants are slow to recover, only two cuttings of cicer milkvetch are obtained in areas that yield three cuttings of alfalfa. The protein content is comparable to other legumes. The plants should be crimped during windrowing, and turning the windrows will hasten drying.

Cicer milkvetch is well suited as pasture. It is palatable to all kinds of livestock and has a low incidence of bloat. Its rhizomes help the plant resist overgrazing. Grasses compatible with cicer milkvetch are creeping foxtail, meadow bromegrass, orchardgrass, and tall fescue.

It generally takes two years to produce an economical seed crop under irrigation, but it may take three years under dryland conditions. Higher soil moisture levels later in the season promote vegetative production at the expense of seed production. The crop is windrowed when most of the seed pods are mature; cicer milkvetch does shatter as readily as other legumes. After drying, the crop is harvested with a combine with a pick-up attachment. Pods thresh more easily when dry. High humidity or dew can make the pods leathery and reduce threshing efficiency.

30.9.5 Pests

Cicer milkvetch is susceptible to rots caused by *Sclerotinia trifoliorum*, a disease common to most forage legumes. Symptoms are wilting and death of aerial parts of the plant. Black sclerotial bodies are found on, or inside, the stem. New growth from rhizomes helps the plants recover when infestations are light. Insect damage is rarely significant.

■ 30.10 SAINFOIN

Sainfoin (*Onobrychis viciaefolia*) is a forage legume native to the Mediterranean region. It has been grown for more than 400 years in Europe and Asia, now mostly in the Russian Federation and Turkey. George Washington introduced sainfoin into the United States in 1786, but it was grown rarely until improved hardy varieties were developed and released in Montana and western Canada in the 1960s.[43] About 25,000 acres (10,000 ha) were grown in Montana in 1970. It has also been called Saint Foin, meaning holy grass, because it is reputed to have filled the manger in which Christ was born.

The sainfoin plant is a deep-rooted perennial with numerous coarse, hollow stems that arise from the crown and may reach 33 inches (1 m) in height. The leaves bear twelve to fifteen leaflets. The rosy-pink flowers are borne on spike-like heads. The pod contains one smooth, kidney-shaped, olive-brown to dark brown seed about ⅛ inch (3 mm) long. Sainfoin forage does not induce bloating.[29] The plants are immune to damage from the alfalfa weevil, but they suffer somewhat from seed weevil and sweetclover weevil attacks. Sainfoin is adapted to poor and calcareous soils but cannot tolerate saline, acid or wet soils, or a high water table. It is a suitable dryland hay or pasture crop in western Canada where the annual precipitation is sufficient. Its response to irrigation is less than that of alfalfa. Sainfoin starts spring growth about two weeks before alfalfa and matures four to six weeks earlier, but it yields only 80 to 90 percent as much hay. Its slow recovery after mowing permits only one cutting of hay each year.

Podded inoculated seeds are sown at a rate of 30 to 35 pounds per acre (34 to 40 kg/ha) for hay on irrigated land, at 12 to 17 pounds per acre (13 to 19 kg/ha) in 12 to 18 inch (30 to 46 cm) rows on dryland for hay or pasture, and at 6 to 9 pounds per acre (7 to 10 kg/ha) in 24 to 36 inch (60 to 90 cm) intertilled rows for seed production. It is often sown in mixture with grasses for dryland pastures.

■ 30.11 SESBANIA

Sesbania (*Sesbania macrocarpa*) is a coarse, upright annual legume, native to North America. It is found in various localities in the southern part of the United States from California to Georgia, extending as far north as Arkansas. It is abundant on the overflow lands of the lower Colorado River in Arizona and California where it is referred to as wild hemp or Indian hemp. There, the Native Americans used the fiber for making fish lines and other twines. It also occurs as a weed in the rice fields of southern Texas and Louisiana where it is known as tall indigo.

In hot weather, with ample rainfall or irrigation and good soil, sesbania grows to a height of 6 to 8 feet (180 to 245 cm) in a few weeks. The yellow flowers are borne in racemes that arise from the leaf axils. Long, slender seed pods are produced later. Sesbania is not relished by livestock. This legume is grown for green manure in the irrigated valleys of southern California, in southern Arizona, and in the lower Rio Grande Valley of Texas. Seeded in June, it provides a heavy growth of green manure to be turned under in late August, in preparation for fall-planted vegetable crops.[21] It should be sown at the rate of about 20 pounds per acre (22 kg/ha). The seed is usually gathered from wild stands.

Sesbania roots are attacked by nematodes and the seed by a small weevil. Soil disinfection controls the nematode and seed fumigation controls the weevil. Seed inoculation is unnecessary.

▓ 30.12 LUPINE

Lupine has been known for 2,000 years. More than 3.9 million acres (1.6 million ha) of seed were harvested in 2003. About 90 percent of total world production is grown in Australia. Other leading countries in lupine production are France, the Russian Federation, Chile, and Morocco. Although cultivated in Europe for 200 years, lupine did not become well established in the United States before 1943. Extensive acreages have been grown since that time. The culture of lupine for winter cover, green manure, or grazing is limited to the South Atlantic and Gulf Coast states.[44] The lupine seed crop is largely produced in Georgia, South Carolina, and Florida.

Many perennial and annual species of lupine are native to North America. Among them is the bluebonnet, *Lupinus subcarnosus*, the state flower of Texas. The three species now grown commercially in the Gulf Coast area, white lupine (*L. albus*), yellow lupine (*L. lutens*), and blue lupine (*L. angustifolius*), were introduced from Europe. The value of these lupines lies in the fact that the seed can be produced in the South, where they also make a heavy winter growth for plowing under. The plants are upright, long-day winter annuals with coarse stems, medium-size digitate (fingered) leaves, and large attractive flowers of the colors indicated by the names (Figure 30.6). The bitter blue and white lupines were used only for green manure because they contain alkaloids, particularly in the seeds and pods, which are poisonous to livestock. Later, sweet (nonpoisonous) varieties of blue and white lupines were acquired and grown for grazing as well as for green manure. Sweet yellow lupines were acquired from Germany.

Inoculated lupine seed is drilled 1 to 2 inches (2.5 to 5 cm) deep into the soil in the fall in the southeastern states, where temperatures do not go below 15°F (−9°C). The seeding rate is 90 to 120 pounds per acre (100 to 135 kg/ha) of white lupine, 60 to 90 pounds per acre (67 to 100 kg/ha) of blue lupine, or 50 to 80 pounds per acre (56 to 90 kg/ha) of yellow lupine.[23] The seed can be harvested with a combine. Lupine diseases include brown spot, powdery mildew, Ascochyta stem canker, root rots, seedling blights, Southern blight, and virus attacks. Disease-free seed, seed treatment, and crop rotations reduce some of the damage. Insect enemies of lupines include the root weevil, lupine maggot, thrip, and white-fringed beetle.[44]

FIGURE 30.6
White lupine: plant *(left)*,
flower raceme *(right)*.

▨ 30.13 GUAR

Guar (*Cyamopsis psoroloides* or *C. tetragonolobus*), sometimes called clusterbean, is a native of India, where it is grown for fodder, for pods used for food or feed, and for beans for export. Guar is also grown in Africa, but its use as an edible bean in Nigeria is limited due to its high prussic acid content. Guar was introduced into the United States in 1903. It is used as a leguminous green-manure crop in southern California, Arizona, and Texas,[17] particularly on soils infested with the cotton root rot organism (*Phymatotrichum omnivorum*). It is the most resistant to this disease of any legume tested. Guar is also partly resistant to root-knot nematode.

The guar plant is a coarse summer annual. It is an upright, drought-resistant herb 2 to 4 feet (60 to 120 cm) in height with angular, toothed trifoliate leaves (Figure 30.7), small purplish flowers borne in racemes, and long leathery pods.

Guar is grown for seed mostly in northern Texas[45] and southern Oklahoma.[46] It is sown in late spring at the rate of 40 to 60 pounds per acre (45 to 65 kg/ha) drilled, or 10 to 30 pounds per acre (11 to 34 kg/ha) in rows. Yields average 300 to 700

FIGURE 30.7
Guar plant.

pounds per acre (335 to 785 kg/ha). The seed contains a gum or mucillage, manno-galactan, which is useful in paper manufacture as a substitute for carob gum. It is also used in textile sizing, as a stabilizer or stiffener in foods and various other products,[47] and in drilling mud and ore flotations.[12] The beans have considerable value for livestock feed.

Sheep will graze green guar, but other animals reject it because of the bristle-like hairs on the plants. Cured hay is eaten readily. The seed crop is harvested with a combine.[39, 45, 46]

Purple stain is caused by a fungus (*Cercospora kikuchii*) that attacks the stems and seeds of guar.[48]

■ 30.14 FLORIDA BEGGARWEED

Florida beggarweed (*Desmodium tortuosum*) has been cultivated on a small acreage in the South for more than sixty years. It is a native of tropical America, but is established in the southern United States, mostly in cultivated fields where it volunteers freely. Florida beggarweed is adapted to the southern coastal plain.[21] When sown in northern states, it makes a fair vegetative growth but will not produce seed. It tolerates acid soils and is resistant to the root-knot nematode.

The Florida beggarweed plant is an upright, herbaceous, short-lived perennial legume that attains a height of 4 to 7 feet (120 to 215 cm). It usually lives as an annual in most of the United States. The pubescent main stem is sparsely branched. The leaves are trifoliate with large ovate pubescent leaflets. Racemes of inconspicuous flowers terminate the main stem and lateral branches. The seeds are borne in jointed pods, segments of which adhere to cotton lint and clothing.

The seed is sown in late spring or early summer following a cultivated crop or is planted at the last cultivation of early planted corn. About 10 pounds per acre (11 kg/ha) of hulled or scarified seed, or 30 to 40 pounds per acre (34 to 45 kg/ha) of unhulled seed, are sown on a compact seedbed.

The main uses of Florida beggarweed are pasture, hay, green manure, and production of seeds for quail feed. It cannot be maintained in permanent pastures.

■ 30.15 ROUGHPEA

Roughpea (*Lathyrus hirsutus*), also known as wild winter pea, caley pea, and singletary pea, is a reseeding winter annual. It is grown in North Carolina, Tennessee, Arkansas, eastern Oklahoma, and states to the south. Roughpea is a native of Europe that escaped to pastures as well as to cultivated land in southeastern United States many years ago.

The roughpea plant resembles the sweetpea (*L. odoratus*) as well as the flatpea or wagnerpea (*L. sylvestris*). Seeds of the roughpea, like those of wagnerpea, are poisonous to livestock. Roughpea has weak stems that are decumbent except in thick stands. Growth is slow until late winter or early spring when rapid growth begins. The plant matures in late spring or early summer. The crop thrives on lime soils but will grow well on slightly acidic soils, heavy clay soils, and poorly drained land.

Roughpea is sown in the fall, alone or in mixtures with grasses or small grains. When roughpea is sown alone, the seeding rate is 20 to 25 pounds per acre (22 to 28 kg/ha) of scarified seed or 40 to 60 pounds per acre (45 to 67 kg/ha) of unscarified seed. The crop responds to phosphorus fertilizer and sometimes also to potash and lime.

Roughpea can be safely grazed or cut for hay only before pods are formed. The seeds contain a poison that causes lameness in cattle that eat the pods. However, animals recover when roughpea is replaced by other feed.[23] Delayed growth limits the usefulness of roughpea for green manure, except when it precedes a crop that is planted late.

■ 30.16 HAIRY INDIGO

Hairy indigo (*Indigofera hirsuta*) is a native of tropical Asia, Australia, and Africa. It was introduced into cultivation in Florida in 1945 after experiments had shown its value as a hay and green-manure crop on sandy soils low in calcium. It is adapted to the coastal-plain area from Florida to Texas.

The hairy indigo plant is a summer-annual legume that attains a height of 4 to 7 feet (120 to 215 cm). The somewhat coarse stems become woody with age. The pinnately compound leaves resemble those of vetch. A late strain of hairy indigo that matures in November is adapted to southern Florida, but a smaller earlier strain is adapted northward to central Georgia. Hairy indigo reseeds naturally when it is allowed to mature. The plant is resistant to root-knot nematode.[21]

The seed of hairy indigo is drilled in the spring at a rate of 3 to 4 pounds per acre (3.4 to 4.5 kg/ha). Artificial inoculation is unnecessary. Many of the seeds are hard. In cornfields, the hairy indigo seeds dropped the previous winter will germinate and establish a stand after the last cultivation of corn. The application of 30 to 50 pounds per acre (34 to 56 kg/ha) of both phosphorus and potash in a fertilizer mixture is advisable.

Hairy indigo is useful for green manure. It also makes good hay when cut before the plants exceed 3 feet in height.

■ 30.17 ALYCECLOVER

Alyceclover (*Alysicarpus vaginalis*) is a summer annual native to tropical Asia. It is grown in the Gulf region for hay, as an orchard cover crop, or occasionally for pasture. It is not adapted to wet or very infertile soils, or to lands infested with root-knot nematode. The plants are 3 feet (90 cm) or less in height, with coarse leafy stems, broadly oval unifoliate leaves, and a spreading habit of growth except in thick stands. Uninoculated seed is sown in May at the rate of 15 to 20 pounds per acre. The seed shatters quickly after it is ripe. Phosphorus and potash fertilizers improve the yields.[21]

■ 30.18 PIGEON PEA

Pigeon pea, or red gram (*Cajanus cajan*), is grown on nearly 7 million acres (3 million ha), mostly in India where it probably originated. Other leading countries in pigeon pea production are Uganda, Myanmar, Malawi, and Kenya. Practically all of the U.S. seed production is in Hawaii at elevations below 3,300 feet (1,000 m). It grows best with an annual rainfall of 20 to 60 inches (500 to1,500 mm) a year.

It is grown mostly for the seed, which is a popular food. The tall stalks provide abundant fodder for livestock, or an abundance of vegetation for green manure. The pigeon pea is a short-lived perennial, but it must be grown as an annual where frost occurs. Only the earlier varieties can mature seed north of southern Florida. The plants are usually 5 to 7 feet (150 to 215 cm) tall, but old perennials may grow to 20 feet (6 m) in the tropics. It is grown as a forage or green-manure crop in the southeastern United States to replace crotalaria that bear poisonous seeds. The coarse stalks make undesirable hay. Seed yields in Hawaii range from 600 to 1,000 pounds per acre (670 to 1,100 kg/ha). Pigeon pea is planted in cultivated rows at the rate of 10 to 15 pounds per acre (11 to 17 kg/ha), or in drills at 25 pounds per acre (28 kg/ha).

■ 30.19 OTHER FORAGE LEGUMES

Fenugreek (*Trigonella foenum-graecum*) is grown in India, Egypt, and the Near East for forage as well as for the seeds, which are used as a condiment and in horse-conditioning powders. It has been grown occasionally in the milder parts of California as a winter-annual cover crop in orchards.

Partridge pea *(Cassia fasiculata)*, an annual legume native in the eastern half of United States, was grown as a cover or forage crop from about 1800 to 1870. Improved strains collected in Missouri and Illinois outyielded annual sweetclover, Korean lespedeza, and berseem clover in tests in Illinois but yielded less than soybeans. It is adapted to eroded low-fertility soils and might be suitable for erosion control, green manure, wildlife cover, or forage in the area where Korean lespedeza is now grown.[49]

Siratro *(Macroptillium atropurpureum)* is a long-lived perennial legume with large taproots and trailing stems grown in Florida in mixtures with pangolagrass or bahiagrass. It was developed in Australia by crossing two native Mexican strains. This summer-growing crop is suitable also for grazing, green chop, hay, or wildlife.[50, 51]

Koa Haole *(Leucaena leucocephala)* is a shrubby legume grown on a small acreage in Hawaii for pasture. Many new superior strains tested are capable of high yields.[52]

Townsville lucerne *(Stylasanthes humilis)*, a self-fertile annual legume indigenous to Mexico and Central America, which was naturalized and grown in Australia after 1900, is now grown in Florida. It is used for pasture, green chop, silage, or hay in mixtures with pangolagrass.[53]

Serradella *(Ornithopus* spp.) is a native of Europe, where it is commonly cultivated. In the United States, it has produced good yields in only a few trials, but it has not been established as a crop. It is an annual legume with a vetch-like appearance with long, often procumbent stems. It grows as a winter annual in warm climates.

Other forage or soil-improvement legumes that have been grown on farms in the United States include Sulla or French honeysuckle *(Hedysarum cornarium)* and Tangier pea *(Lathyrus tingitanus)*.

REFERENCES

1. Robinson, R. G. "The shortcomings of crownvetch," *Crops and Soils* 21, 9(1972):18–19.

2. Reisenauer, H. M., and others. "Dry pea production," *WA Agr. Exten. Bull.* 582, 1965.

3. Hedrick, U. P., and others. *The Vegetables of New York*, vol. 1, part 1, *The Peas of New York*, N.Y. Albany, NY: Agr. Exp. Sta., 1928, pp. 1–132.

4. Koonce, D. "Field peas in Colorado," *CO Agr. Exp. Sta. Bull.* 416, 1935.

5. Anonymous. "Grasses and legumes for forage and conservation," *USDA Agricultural Research Service Spec. Rept.* ARS 22–43, 1957.

6. McKee, R., and A. J. Pieters. "Culture and pests of field peas," *USDA Farmers Bull.* 1803, 1938.

7. Wallace, A. T., and G. B. Killinger. "Big trefoil—a new pasture legume for Florida," *FL Agr. Exp. Sta. Cir.* S-49, 1952.

8. Seancy, R. R., and P. R. Henson. "Birdsfoot trefoil," *Adv. Agron.* 22(1970):119–157.

9. Ali-Khan, S. T., and R. C. Zimmer. "Growing field peas," *Can. Dept. Agr. Pub.* 1433, 1972, pp. 1–8.

10. Hulbert, H. W. "Uniform stands essential in field pea variety tests," *J. Am. Soc. Agron.* 19(1927):461–465.

11. Pierre, J. J., and J. A. Jackobs. "Growing birdsfoot trefoil in Illinois," *IL Ext. Cir.* 725, 1954.

12. Zaumeyer, W. J. "Pea diseases," *USDA Handbook* 228, 1962, pp. 1–30.

13. White, G. A. "New crop on the horizon," *Seed World* 110, 4(1972):20–22.

14. Dudley, J. E., Jr., and T. E. Bronson. "The pea aphid on peas and methods for its control," *USDA Farmers Bull.* 1945 (rev.), 1952.

15. Klages, K. H. "Comparative winterhardiness of species and varieties of vetches and peas in relation to their yielding ability," *J. Am. Soc. Agron.* 20(1928):982–987.

16. Gunn, C. R. "Seeds of native and naturalized vetches of North America," *USDA Handbk. 392,* 1971, pp. 1–42.

17. Henson, P. R., and H. A. Schoth. "The trefoils—adaptation and culture," *USDA Handbook* 223, 1962.

18. Hermann, F. J. "Vetches in the United States—native, naturalized and cultivated," *USDA Handbook* 168, 1960.

19. Goodding, T. H. "Hairy vetch for Nebraska," *NE Agr. Exp. Sta. Cir.* 89, 1951.

20. Varney, K. E. "Birdsfoot trefoil," *VT Agr. Exp. Sta. Bull.* 608, 1958.

21. Anonymous. "Growing summer cover crops," *USDA Farmers Bull.* 2182, 1962.

22. Piper, C. V., and W. J. Morse. "The velvetbean," *USDA Farmers Bull.* 1276 (rev.), 1938.

23. Henson, P. R., and E. A. Hollowell. "Winter annual legumes for the South," *USDA Farmers Bull.* 2146, 1960.

24. McKee, R. "Bur clover cultivation and utilization," *USDA Farmers Bull.* 1741 (rev.), 1949.

25. Duggar, J. F. "Root nodule formation as affected by planting of shelled or unshelled seeds of bar clover, black medic, hubam, and crimson and subterranean clovers," *J. Am. Soc. Agron.* 26(1934):919–923.

26. Blaser, R. E., and W. E. Stokes. "Ecological and morphological characteristics of black medic strains," *J. Am. Soc. Agron.* 38, 4(1946):325–331.

27. Hughes, H. D., and J. M. Scholl. "Birdsfoot still looks good," *Crops and Soils,* 11(1959):3.

28. Guilbert, H. R., and S. W. Mead. "The digestibility of bur clover as affected by exposure to sunlight and rain," *Hilgardia* 6(1931):1–12.

29. Davis, J. H., E. O. Gangstad, and H. L. Hackerott. "Button clover," *Hoblitzelle Agr. Lab. Bull.* 6, 1957.

30. McKee, R., and J. L. Stephens. "Kudzu as a farm crop," *USDA Farmers Bull.* 1923 (rev.) 1948.

31. Taylor, T. H., and others. "Management effects on persistence and productivity of birdsfoot trefoil (*Lotus corniculatus* L.)," *Agron. J.* 65, 4(1973):646–648.

32. McKee, R., and others. "Crotalaria culture and utilization," *USDA Farmers Bull.* 1980, 1946, pp. 1–17.

33. Tabor, P. "Observations of kudzu (*Pueraria thunbergiana*) seedlings," *J. Am. Soc. Agron.* 34(1942):500–501.

34. Ritchey, G. E., and others. "Crotalaria for forage," *FL Agr. Exp. Sta. Bull.* 361, 1941, pp. 1–72.

35. Leukel, W. A., and others. "Composition and nitrification studies on *Crotalaria striata,*" *Soil Sci.* 28(1929):347–371.

36. Anonymous. "Trefoil production for pasture and hay," *USDA Farmers Bull.* 2191, 1963.

37. Weimer, J. L. "Austrian winter field pea diseases and their control in the South," *USDA Cir.* 565, 1940.

38. Rachie, K. O., and A. R. Schmidt. "Winterhardiness of birdsfoot trefoil strains and varieties," *Agron. J.* 47, 4(1955):155–157.

39. McKee, G. W. "Effects of shading and plant competition on seedling growth and nodulation in birdsfoot trefoil," *PA Agr. Exp. Sta. Bull.* 689, 1962, pp. 1–35.

40. Henson, P. R. "Crownvetch," *USDA Agricultural Research Service Spec. Rept.* ARS 34–53, 1963.

41. Musser, H. B., W. L. Hottenstein, and J. P. Stanford. "Penngift crownvetch for slope control on Pennsylvania highways," *PA Agr. Exp. Sta. Bull.* 576, 1954, pp. 1–21.

42. McKee, G. W., and others. "Seeding crownvetch? Watch seed maturity," *Crops and Soils* 22, 3(1972):8.

43. Hanna, M. R., and others. "Sainfoin for western Canada," *Can. Dept. Agr. Pub.* 1470, 1972, pp. 1–18.

44. Henson, P. R., and J. L. Stephens. "Lupines," *USDA Farmers Bull.* 2114, 1958.

45. Brooks, L. E., and C. Harvey. "Experiments with guar in Texas," *TX Agr. Exp. Sta. Cir.* 126, 1950.

46. Matlock, R. S., D. C. Aepli, and R. B. Street. "Growth and diseases of guar," *AZ Agr. Exp. Sta. Bull.* 216, 1948.

47. Hymowitz, T., and R. S. Matlock. "Guar in the United States," *OK Agr. Exp. Sta. Bull.* B-611, 1963.

48. Johnson, H. W., and J. P. Jones. "Purple stain of guar," *Phytopath.* 52, 3(1962):269–272.

49. Foote, L. E., and J. A. Jacobs. "Partridge pea management and yield comparisons with other annual forage legumes," *Agron. J.* 57, 6(1966):573–575.

50. Kretschmer, A. E., Jr. "Siratro (*Phaseolus atropurpureus* D.C.) a summer-growing perennial pasture legume for central and south Florida," *FL Agr. Exp. Sta. Circ.* S-214, 1972.

51. Stokes, W. E. "Crotalaria as a soil-building crop," *J. Am. Soc. Agron.* 19(1927):944–948.

52. Brewbaker, J. L., and others. "Varietal variation and yield trials of *Leucaena leucocephala* (koa haole) in Hawaii." *HI Agric. Exp Sta. Bull.* 166, 1972.

53. Kretschmer, A. E., Jr. "*Stylosanthes humilis*, a summer-growing self-generating annual legume for use in Florida pastures," *FL Agr. Exp. Sta. Circ.* S-184, 1968.

PART 4 CROPS OF OTHER PLANT FAMILIES

Buckwheat

KEY TERMS

Common buckwheat
Japanese buckwheat
Silverhull buckwheat
Tartary buckwheat
Winged buckwheat

■ 31.1 ECONOMIC IMPORTANCE

Buckwheat was harvested globally on about 7 million acres (2.8 million ha) in 2000–2003 with an average production of about 2.8 million tons (2.5 million MT) or 760 pounds per acre (880 kg/ha). Leading countries producing buckwheat are China, the Russian Federation, Ukraine, France, and Poland. In 2000–2003, buckwheat was harvested in the United States on about 161,000 acres (65,000 ha), yielding some 72,000 tons (65,000 MT) or 900 pounds per acre (1,000 kg/ha). The leading states in buckwheat production are North Dakota, Washington, Minnesota, New York, and Pennsylvania.

Production was about 22 million bushels (480,000 MT) in 1866 and more than a million acres (405,000 ha) were harvested in the United States in 1918, but the acreage has declined steadily since then. Recent exploitation of buckwheat as a "natural" food may increase domestic consumption. Buckwheat yields less than the cereal grains because of its short growing season, poor response to nitrogen fertilization, and failure to develop any improved variety until very recently. Also, buckwheat has a lower feeding value for livestock than other grains, and the human consumption of buckwheat cakes is now very limited.[1] About 150,000 acres (60,700 ha) were harvested in Canada in the late 1970s and early 1980s. Production has since declined to 30,000 to 40,000 acres (12,000 to 16,000 ha) annually. Manitoba is the major producer of buckwheat in Canada with an average of 70 percent of the acreage. More than half of the Canadian production is exported, mostly to Japan.[2]

■ 31.2 ADAPTATION

Buckwheat makes its best growth in a cool, moist climate. It is grown in the North as well as in the higher-altitude areas of the East. It is sensitive to cold, being killed quickly when the temperature falls much below freezing. Buckwheat requires only a short growing season of ten to twelve weeks. The crop is sensitive to high temperatures and dry weather when the plants are in blossom. For this reason, seeding generally is delayed to allow plant growth to take place in warm weather and seed to form in the cooler weather of late summer.

When the climate is favorable, buckwheat will produce a better crop than other grain crops on infertile, poorly tilled land. Its response to fertilizer applications is less than that of other crops.[3] Buckwheat is able to extract more nutrients from raw rock phosphate than are other grain crop plants.[4] It is well suited to light, well-drained soils such as sandy loams or silt loams, and it grows satisfactorily on soils too acidic for other grain crops. Buckwheat usually produces a poor crop on heavy wet soils. The crop is likely to lodge badly on rich soils high in nitrogen.[5] Buckwheat is often sown as a catch crop with little regard for the best conditions for its growth. This accounts for the low average yield in the United States.

■ 31.3 HISTORY OF BUCKWHEAT CULTURE

Common buckwheat appears to have been cultivated in China for at least 5,000 to 6,000 years.[6] It probably originated in the mountainous regions of that country. During the Middle Ages, the crop was introduced into Europe and from there it was brought to the United States. It was grown by Dutch colonists along the Hudson River before 1625.

Buckwheat is not a cereal, nor is it really a grain because it does not belong to the grass family. Because the fruits or seeds are used like grain, buckwheat is commonly regarded and handled as a grain crop. The name buckwheat was coined by the Scotch from two Anglo-Saxon words, *boc* (beech), because of the resemblance of the achene or fruit to the beechnut, and *whoet* (wheat), because of its being used as wheat. In Germany, the name is *Buchweizen*, likewise meaning beech wheat.[3]

■ 31.4 BOTANICAL DESCRIPTION

Buckwheat belongs to the *Polygonaceae*, or buckwheat, family. The principal species of common buckwheat, *Fagopyrum esculentum (F. sagittatum)*, appears to have been derived from *F. cymosum*, a wild perennial species of Asia. Another species, *F. tataricum*, or tartary buckwheat, is also grown. Winged buckwheat is another genus of *Polygonaceae, Eriogonum alatum*.

The buckwheat plant is an annual, 2 to 5 or more feet (60 to 150 cm) in height, with a single stem and usually several branches.[7, 8] The stems are strongly grooved, succulent, and smooth, except at the nodes (Figure 31.1). The stems range in color from green to red but turn brown with age. More reddening is evident when the plants have a poor seed set. The leaf blades, 2 to 4 inches (5 to 10 cm) long, are triangular and heart-shaped. Leaves are petiolate at the bottom of the plant and become sessile near the top.

The plant has a taproot with numerous short laterals that may extend 3 to 4 feet (90 to 120 cm) or less in the soil. The root system comprises about 3 percent of the weight of the plant, compared with 6 to 14 percent in the cereal grains.[3]

The inflorescence of common buckwheat consists of axillary or terminal racemes or cymes with more or less densely clustered flowers (Figure 31.2). The inflorescence is partly determinate. The flowers, composed of petal-like sepals, are white or tinged with pink. They have no petals. There are eight stamens and a triangular, one-celled ovary that contains a single ovule, with a three-parted style with knobbed stigmas.

FIGURE 31.1
Plant of Japanese buckwheat.

Buckwheat is indeterminate and will flower for several weeks. The buckwheat plant begins to bloom four to six weeks after seeding. The flowers are dimorphic; that is, some plants bear "pin" flowers with stamens shorter than the styles, while in other "thrum" flowers, the stamens are longer.[9] The few flowers that have styles and stamens of the same length are usually sterile, but self-fertile lines have been isolated.[10] Sterility occurs following either self-pollination or cross-pollination between flowers of the same type because pollen tube growth is inhibited. Cross-pollinations between the two types usually induce fertility. Thus, common buckwheat is mostly naturally cross-pollinated and self-sterile.[5] Bees and other insects distribute the pollen, but cross-fertilization can occur where bees are excluded.[10] Tartary buckwheat is self-fertile. Experiments in Europe indicate that boron sprays increase seed setting in buckwheat.

FIGURE 31.2
Branch bearing flowers,
immature seeds, and leaves
of Japanese buckwheat.

The fruit is an achene that is brown, gray-brown, or black in color. The point of the seed is the stigmatic end, while the persistent calyx lobes remain attached at the base. The hull is the pericarp. The seed or matured ovule inside the hull has a pale brown testa, which is triangular, as is the fruit.

The endosperm is white, opaque, and starchier than cereal grain endosperms. The embryo, embedded in the center of the endosperm,[11] possesses two cotyledons that are folded in the form of the letter S.

31.5 TYPES

Japanese buckwheat seed is brown and usually large, ¼ inch (7 mm) long and ⅕ inch (5 mm) wide, resulting in about 15,000 seeds per pound (33,000/kg). The seed is nearly triangular in cross section. The plants are tall with large leaves and coarse stems.

Silverhull buckwheat has smaller, glossy, silver-gray seeds, or 20,000 per pound (44,000/kg). The sides of the seed are rounded between the angles, which makes them appear less triangular. The stems and leaves are smaller than those of Japanese, while the stems are more reddened at maturity. Common gray is like Silverhull[5] but may have smaller seeds. Mechanical and hybrid mixtures of Japanese and Silverhull can be designated as common buckwheat.[3] A new tetraploid-seeded variety is now grown in the United States.

Winged or notch-seeded buckwheat occurs only as a mixture. It is merely a type of common Japanese buckwheat in which the angles of the hulls are extended to form wide margins or wings, making the seeds look large.

Tartary or mountain buckwheat (*F. tataricum*) is grown occasionally for feed in the mountains of North Carolina and Maine and in certain other areas. It has escaped from cultivation to become a weed, particularly in the prairie provinces of Canada. It is sometimes called India wheat, duck wheat, rye buckwheat, Mountain Siberian, wild goose, or Calcutta.

Tartary buckwheat is distinguished from the common species by its more indeterminate habit, simple racemes, and smaller seeds (26,000 per pound; 57,000/kg) that are nearly round in cross section and usually pointed. The seed color ranges from dull gray to black, while the pericarp (hull) varies from smooth to decidedly rough and spiny. The plants are somewhat viny. The flowers are very small and inconspicuous with greenish-white sepals.[3, 5]

Common buckwheat completes terminal growth six to eight weeks after planting, and tartary in ten weeks in Pennsylvania.[3] The percentages of hull in the seed ranges from 18 to 22 percent. Hybrid buckwheat seed is being produced in the Russian Federation by interplanting two strains and setting beehives in the field.

Wild buckwheat (*Polygonum convovulvus*) is a common weed.

31.6 FERTILIZERS

Little or no fertilizer is needed in fields with moderate fertility or better. Application of nitrogen is usually less than 50 pounds per acre (56 kg/ha) as excess N increases lodging.[12, 13] About 20 to 40 pounds per acre (22 to 45 kg/ha) of phosphate fertilizer is applied to soils low in available phosphate. Need for potash fertilizer is less than 70 pounds per acre (78 kg/ha) even with high yield goals on potassium-deficient soils.[14] Moderate applications of lime have shown some benefit, but heavy liming is detrimental.

31.7 ROTATIONS

Injurious effects of buckwheat on subsequent crops on unfertilized land are attributed to excessive removal of mineral nutrients by the rapidly growing, shallow-feeding buckwheat plants.[3] A winter cover crop should follow buckwheat not only to increase organic matter, but also to reduce erosion.[3] Buckwheat stubble land is more subject to erosion than is small-grain land due to the loose, friable condition of the soil.

31.8 BUCKWHEAT CULTURE

31.8.1 Seedbed Preparation

Early plowing is advantageous. The soil preparation as well as the seeding depth is about the same as it is for small grains.[3, 5, 15] The seedbed should be free of weeds as there are no herbicides registered for use on buckwheat. Buckwheat can be used in rotation with other crops. In Missouri, it can be double cropped after a small grain. Buckwheat is sensitive to broadleaf herbicides such as trifluralin, triazine, and sulfonylurea. Care in the use of these herbicides on the preceding crop is necessary.[6]

31.8.2 Seeding Methods

Buckwheat is generally seeded in the Northeast between June 15 and July 1, and June 1 to June 20 in Manitoba, Canada. It is seldom advisable to seed after July 15. Seeding time in a locality can be calculated fairly accurately by allowing a period of twelve weeks for growth before the average date of first fall frost. Seeding in late May is preferable in northern Europe where the summer climate is cooler.[16] Buckwheat germinates best when the soil temperature is about 80°F (27°C), but it will germinate at any temperature between 45 and 105°F (7 and 41°C). The rate of seeding varies from 40 to 60 pounds per acre (45 to 67 kg/ha) for common buckwheat, while 24 pounds per acre (27 kg/ha) is sufficient for the smaller-seeded tartary buckwheat. Seeding depth is 1 to 2 inches (2.3 to 5 cm). The crop may be either drilled or broadcast.

31.8.3 Harvesting

Buckwheat is often harvested with a swather when 75 to 80 percent of the seeds have turned brown or black. The seed is then threshed with a combine equipped with a pick-up attachment. If directly combined, 90 to 95 percent of the seeds should be brown or black. Combine cylinder speed should be set at 600 to 800 revolutions per minute and the concave setting at ½ inch (13 mm).[6] It is usually necessary to dry seed harvested by direct combining before it can be stored safely, because the plants do not bloom or ripen uniformly. Seed should be dried to a moisture of 12 to 13 percent. Air temperature should not exceed 110°F (43°C).

■ 31.9 USES OF BUCKWHEAT

Buckwheat has been grown primarily for human food; but since only the seed of better quality buckwheat is bought by the millers, most of it is fed to livestock or poultry. About 6 to 7 percent is used for seed. Buckwheat flour is consumed mostly in the form of buckwheat cakes or in mixed pancake flours. A continued heavy diet of buckwheat cakes results in development of a skin rash in those who are allergic to buckwheat protein. White-haired animals that are fed a buckwheat ration also develop a rash if they are exposed to light.[5]

Buckwheat is one of the best temporary honey crops, since it produces blossoms for thirty days or more. It will supply enough nectar on 1 acre for 100 to 150 pounds (275 to 415 kg/ha) of honey.[15] Buckwheat honey is dark, with a distinctive flavor that some people do not relish. Bees ordinarily do not collect nectar from tartary buckwheat.

The leaves and flowers of buckwheat were formerly the commercial source of rutin, a glucoside used medicinally to check capillary hemorrhages, help reduce high blood pressure, prevent frostbite gangrene, and act as a protection against the aftereffects of atomic radiation.[17] Synthetic rutin has replaced that from buckwheat.

Whole seeds are fed to poultry, but the seed often is hulled for other livestock. Buckwheat is a valuable green manure summer cover crop, smother crop, or catch crop.

Buckwheat shorts or middlings are a valuable constituent in mixed feeds for livestock. Buckwheat hulls have very little feeding value. Most mills burn the hulls, but some are sold for packing.

■ 31.10 CHEMICAL COMPOSITION

The percent chemical composition of the whole buckwheat seed is about water 10, ash 1.7, fat 2.35, protein 11, fiber 12, and nitrogen-free extract 60–64. The lysine percentage of the crude protein exceeds that of cereal grains.

■ 31.11 MILLING

Buckwheat is milled for either flour or groats (hulled seeds).[18] Tartary buckwheat is not milled because the flour has a dark color and a bitter taste.

In flour milling, buckwheat is cleaned and dried to 12 percent moisture, then scoured and aspirated to remove dust, fuzz, and the calyx that adheres to the fruit. The seed is then passed through break rolls where the hulls are cracked and loosened. Since some moisture may be taken up during this process, the material is again dried to 12 percent moisture to aid in the separation of the hulls from the seeds. After the hulls have been sifted out, the broken seeds are further ground and sifted. Some buckwheat flour is milled so fine that it is as white as wheat flour. However, mills generally use coarse bolting cloths, through which small particles of hulls pass and remain in the flour. These particles give the flour a characteristic dark color. The dark, coarse-particled residue from flour extraction is called middlings or shorts.

About 100 pounds (45 kg) of clean buckwheat seed can yield 60 to 75 pounds (27 to 34 kg) of flour, 4 to 18 pounds (2 to 8 kg) of middlings, and 18 to 26 pounds (8 to 12 kg) of hulls,[5] but only about 52 pounds (24 kg) of pure white flour are obtained.

In milling for groats, separated, uniform, medium-sized kernels are passed between two millstones, which are adjusted to crack the hulls without a grinding action. The purified whole groats are used in porridge, soups, and breakfast food, mostly by Europeans and immigrants from Europe. Broken groats are eaten as roasted broken kernels and farina.

■ 31.12 PESTS

Buckwheat suffers relatively little damage from either diseases or insects. The diseases most frequently reported are a leaf spot caused by the fungus *Ramularia* species and a root rot caused by *Rhizoctonia*. Wireworms, aphids, birds, deer, and rodents attack buckwheat occasionally.

REFERENCES

1. Marshall, H. G. "Description and culture of buckwheat," *PA Agr. Exp. Sta. Bull.* 754, 1969, pp. 1–26.

2. Ali-Khan, S. T. "Growing buckwheat," *Can. Dept. Agr. Publ.* 1468, 1972, pp. 1–5.

3. White, J. W., F. J. Holben, and A. C. Richer. "Experiments with buckwheat," *PA Agr. Exp. Sta. Bull.* 403, 1941.

4. Truog, E. "The utilization of phosphates by agricultural crops, including a new theory regarding the feeding power of plants," *WI Agr. Exp. Sta. Res. Bull.* 41, 1916.

5. Sando, W. J. "Buckwheat culture," *USDA Farmers Bull.* 2095, 1956.

6. Myers, R. L., and L. J. Meinke. "Buckwheat: A multi-purpose, short-season alternative," *Univ. MO Ext. Serv.* G4306, 1999.

7. Percival, J. *Agricultural Botany.* 7th ed. London: Duckworth, 1926, pp. 350–355.

8. Robbins, W. W. *Botany of Crop Plants*, 3rd ed. Philadelphia: Blakiston, 1931, pp. 276–286.

9. Stevens, N. E. "Observations on heterostylous plants," *Bot. Gaz.* 53(1912):277–308.

10. Marshall, H. G. "Isolation of self-fertile homomorphic forms in buckwheat, *Fagopyrum sagittatum* Gilib," *Crop Sci.* 9, 5(1969):651–653.

11. Stevens, N. E. "The morphology of the seed of buckwheat," *Bot. Gaz.* 53(1912):59–66.

12. Oplinger, E. S., and others. "Buckwheat," in *Alternative Field Crops Manual. WI Coop. Ext. Serv.* and *MN Ext. Serv.* 1989.

13. Taylor, R. W. "Buckwheat." *Univ. DE Ext. Serv.* AF-02, 1998.

14. Rehm, G., M. Schmitt, J. Lamb, and R. Eliason. "Fertilizer recommendations for agronomic crops in Minnesota." *Univ. MN Ext. Serv.* BU-06240, (Rev.). 2001.

15. Cormancy, C. E. "Buckwheat in Michigan," *MI Agr. Exp. Sta. Spec. Bull.* 151, 1926.

16. Lewicki, S., M. Ruszkowski, and Z. Kaszlej. "Studies on buckwheat. Pt. 9. Grain and green forage yields as depending on 10 different seeding times," *Off. Tech. Serv. U.S. Dept. Commerce.* 1963.

17. Naghski, J., and others. "Effects of agronomic factors on the rutin content of buckwheat," *USDA Tech. Bull.* 1132, 1955.

18. Coe, M. R. "Buckwheat milling and its by-products," *USDA Cir.* 190, 1931.

Flax

CHAPTER

32

KEY TERMS

32.1 ECONOMIC IMPORTANCE

Flax *(Linum usitatissimum)* is a multipurpose crop that can be harvested for seed to produce linseed oil and for the fiber in the stem that can be used for linen cloth and paper. Flax was harvested on an average of 7.5 million acres (3 million ha) from 2000 through 2003. Seed production was about 2.3 million tons (2 million MT), or about 700 pounds per acre (800 kg/ha) from 6.3 million acres (2.6 million ha). Canada is the leading country for seed-flax production followed by China, the United States, India, and the Russian Federation. About 1.2 million acres (470,000 ha) of flax were harvested for fiber. The fiber yield was about 740,000 tons (670,000 MT), or 1,300 pounds per acre (1,400 kg/ha). China is the leader in fiber-flax production followed by France, the Russian Federation, and the Netherlands. In the United States, flax seed was harvested on an average of 595,000 acres (241,000 ha), yielding 316,000 tons (287,000 MT) or 1,100 pounds per acre (1,200 kg/ha) in 2000–2003. North Dakota produces about 95 percent of the U.S. production, followed by Minnesota, Montana, and South Dakota (Figure 32.1) About 15,000 tons (14,000 MT) of flax are planted for seed annually in the United States.

The production of flax for fiber in the United States was discontinued in 1956, due to a lack of demand and cheaper imports of linen. But a small amount is decorticated from seed-flax straw for making cigarette paper. Recently, there has been interest in growing fiber-flax in the Southeast United States as an alternative fall-planted crop.[1]

Flaxseed production has declined because other products substitute for linseed oil in paints. However, the health benefits of flaxseed consumption have helped strengthen markets for the crop. Flaxseed and flaxseed oil are rich in alpha-linolenic acid (ALA), an essential fatty acid that appears to be beneficial for heart disease, inflammatory bowel disease, arthritis, and a variety of other health problems. ALA belongs to a group of substances called omega-3 fatty acids that help reduce inflammation.

Fiber flax
Linen yarn
Lodine number
linseed oil
Retting
Scutching
Seed flax
Tow

713

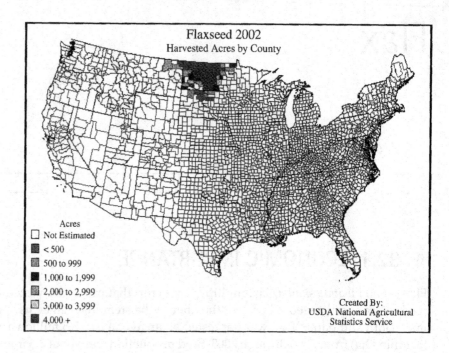

FIGURE 32.1

Flaxseed acreage in the United States in 2002. [Source: 2002 Census of U.S. Department of Agriculture]

32.2 HISTORY OF FLAX CULTURE

Flax was cultivated long before the earliest historical records. Remnants of flax plants were found in Stone Age dwellings in Switzerland. The art of making fine linen from flax fibers was practiced by ancient Egyptians. Primitive people probably used the seeds of wild flax for food.[2]

Flax is an Old World crop that was probably cultivated first in southern Asia as well as in the Mediterranean region. It may have originated from wild flax *(Linum angustifolium)*[2] native to the Mediterranean region, the only species which crosses readily with cultivated flax.

Cultivation of fiber-flax began in America during the Colonial period when it was used to supply linen for hand spinning in the home. The commercial extraction of linseed oil began about 1805. Flaxseed production moved westward with the settlement of new lands not only because it was a good first crop on newly broken sod, but also because of the damage from flax wilt on old, cultivated fields. The development of wilt-resistant varieties has made flax a dependable crop on old, cultivated lands.[3]

32.3 ADAPTATION

Flax requires moderate to cool temperatures during the growing season. Seed-flax is generally grown where the average annual precipitation ranges from 16 to 30 inches (400 to 750 mm), but it is also grown under irrigation in dry climates. Drought and high temperatures, about 90°F (32°C), during and after flowering, reduce the yield, size, and oil content of the seed as well as the quality of the oil.[4] The fiber-flax plant requires adequate moisture as well as a cool temperature during the growing season, but after maturity, dry weather facilitates harvesting, curing, and drying after retting. Cool weather from March to June, followed by warm, dry weather in July, affords excellent conditions for fiber-flax production.[5]

Flax may be damaged or killed at temperatures of 18 to 26°F (–7 to –4°C) in the seedling stage, while a light freeze of 30°F (–1°C) may cause injury in the blossom or green boll stage. Between these stages, the plants may survive temperatures of 15°F (–10°C) or even lower.[6] Frost injury may occur in the northern states, either when flax is sown too early in the spring, or when sown so late that it is frosted in an immature stage in the fall. Flax sown in early October in California or southern Texas may be damaged by a sudden freeze in February when it is in the blossom stage. That sown in late November or in December may be damaged in the seedling stage. Varieties released in Texas are a distinctly winter type resistant to cold.

Flax makes its best growth in well-drained, medium-heavy soils, especially silt loams, clay loams, and silty clays. Light soils are unsuited to seed-flax, particularly in regions of deficient rainfall.[7] Since flax has a relatively short root system, it is dependent on moisture largely in the top 24 inches (60 cm) of soil. Fiber-flax grown on heavier soils has consistently outyielded that grown on lighter soils.[8] Weedy land should be avoided.

32.4 BOTANICAL DESCRIPTION

Flax is an annual herbaceous plant that may be grown as a winter annual in warm climates. The plant grows to a height of 12 to 48 inches (30 to 120 cm) (Figure 32.2). It has a distinct main stem and a short taproot. Two or more basal branches may arise from the main stem just above the soil surface unless the

FIGURE 32.2

Flax plants shortly after flowering. [Courtesy Richard Waldren]

FIGURE 32.3
Individual flax plant
showing branching and
flower development at top
of branches. [Courtesy
Richard Waldren]

stand is thick (Figure 32.3). The main stem and basal branches give rise to the primary, secondary, and tertiary branches that bear the leaves, flowers, and bolls (Figure 3.17).

Three principal tissue areas are recognized in the stems: pith, wood, and bark. The bark contains the comparatively long bast or flax-fiber cells that constitute the linen fibers.

The flax flower has five petals and a five-celled boll or capsule that contains ten seeds when each of the cells bears the complete set of two seeds each (Figure 32.4).

A typical boll may contain eight seeds or less at maturity because some of the ovules fail to develop or cease to develop under stress conditions. Small-seeded varieties produce more seeds per boll than large-seed varieties in order to equalize seed yields.[9]

The flowers open at sunrise on clear, warm days, the petals falling before noon. Flowering is indeterminate and continues until growth is stopped. The petals are blue, pale blue, white, or pale pink, depending on the variety. The bolls are semi-dehiscent in most varieties grown in the North Central states.[10] The bolls of this type rarely dehisce so far as to allow the seeds to fall out. Most varieties of Indian and Argentine origin are indehiscent.

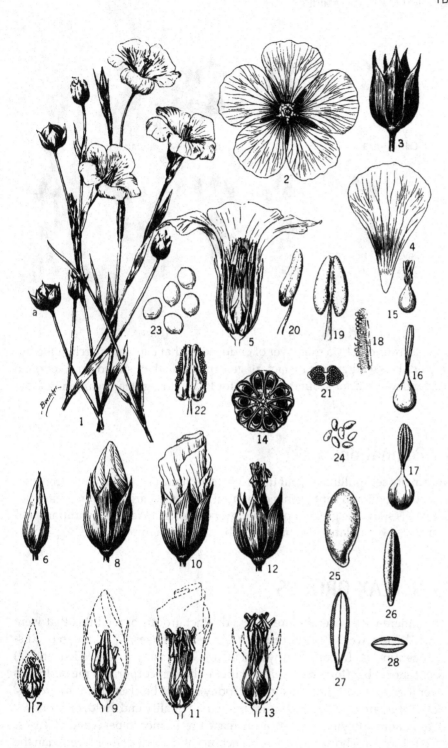

FIGURE 32.4

Flax inflorescence: (1) upper stems showing leaves, flowers, and boll *(a)* about natural size; (2) expanded flower; (3) calyx after anthesis and shedding petals; (4) upper surface of a petal; (5) section of flower showing calyx, corolla, the five stamens, and five stigmas; (6 to 13) four stages of flower opening and anthesis—(6 and 7) two days before anthesis, (8 and 9) late afternoon of day before anthesis, (10 and 11) anthesis occurring at sunrise, (12 and 13) three to six hours after anthesis; (14) cross section of boll, showing the ten ovules (seeds) developed in the five carpels; (15 to 17) pistil before and during anthesis; (18) portion of stigma, greatly magnified, showing adhering pollen grains; (19) dorso-ventral view of anther with a portion of the filament; (20) lateral view of anther; (21) cross section of anther; (22) dehisced anther; (23) pollen grains, greatly magnified; (24) seeds, natural size; (25 and 26) seed, magnified; (27) seed dorso-ventral longitudinal (sagittal) section, showing cotyledons and surrounding endosperm; (28) cross section of seed showing cotyledons and surrounding endosperm.

The seeds of flax vary from ⅛ to ⅕ inch (3.2 to 5 mm) in length. A thousand seeds weigh 3.8 to 7.0 grams. There are 65,000 to 120,000 seeds in a pound (14,000 to 54,000 per kg). The seeds are usually light brown in color, although in certain varieties they are yellow, mottled, greenish-yellow, or nearly black. The seeds have a smooth, shiny surface that results from a mucilaginous covering (Figure 32.5). The

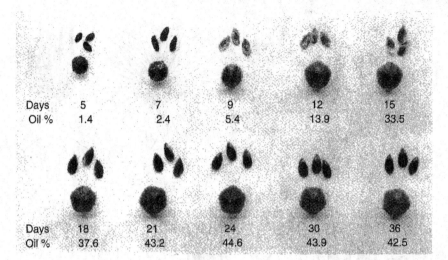

| Days | 5 | 7 | 9 | 12 | 15 |
| Oil % | 1.4 | 2.4 | 5.4 | 13.9 | 33.5 |

| Days | 18 | 21 | 24 | 30 | 36 |
| Oil % | 37.6 | 43.2 | 44.6 | 43.9 | 42.5 |

FIGURE 32.5

Ten stages of seed and boll development, and oil content of flaxseeds from five to thirty-six days after pollination.

embryo is surrounded by a thin layer of endosperm that contains starch in the immature seed. Flaxseed gives satisfactory germination when stored for five to ten years under dry conditions, but that stored for fifteen to eighteen years shows low viability.[11]

32.4.1 Pollination

Flax is normally self-pollinated, but up to 2 percent natural crossing may occur.[12, 13] Insects seem to be important agents of natural crossing. Large-flowered varieties with flat petals show the greatest percentage of crossing.[12] Weather conditions and the distance between plants also influence the amount of out-crossing.[13]

■ 32.5 FLAX GROUPS

In the past, there were several more or less distinct groups of seed-flax that were recognized.[14] These were: (1) wilt-resistant, short-fiber flaxes, (2) common or Russian, (3) Argentine, (4) Indian, (5) Abyssinian, (6) Golden or yellow-seeded, and (7) indehiscent short fiber (Indian). Modern varieties have incorporated the best traits from several groups and most flax grown today cannot be classified into groups. Modern varieties are bred for maximum linseed oil quality and yield, or for maximum fiber quality and yield, as well as increased resistance to pests and diseases.

The seed-flaxes are shorter, are more branching, and produce more seed than the fiber-flaxes. Seed flaxes range in height from 15 to 30 inches (38 to 76 cm), while fiber-flaxes range from 37 to 48 inches (76 to 112 cm) (Figure 32.6). Seeds of seed-flax may be large, medium, or small, but those of all fiber varieties are small. The fiber in seed-flax is short and sometimes harsh. Consequently, it is not used for production of fine linen yarn. The small stems and the absence of basal branches of fiber-flax are merely a result of thick seeding.[2]

FIGURE 32.6

Flax: *(A)* ripe bundles of (left) fiber type and (right) seed type; *(B)* field in bloom; *(C)* young plants of five varieties—(3) Roman, winter type; (4) Punjab, (5) Cirrus, fiber type, (6) Bison, and (7) Rio. Cotyledons are still attached at the bases of the branches in (4) and (7).

▇ 32.6 ROTATIONS

Flax is most productive on clean land because it is a poor competitor with weeds. Flax is a good crop to follow newly broken pastures or meadows. It is best grown after either a clean-cultivated row crop or a legume. It does not do well after canola or mustard because those crops may leave a toxin in the residue that inhibits flax seedlings.

In the North Central states, flax grows well when it follows corn. A satisfactory sequence of crops includes a small grain, a legume, corn, and flax. In southeastern Kansas,[15] flax after soybean yielded much better than flax following corn, sorghum, or oat. Where it is adapted, flax has proven to be a good companion crop for alfalfa, clover, or grass because it is less competitive for light than small-grain crops. Flax rarely does well after small grains due to the abundance of weeds, unless the grain stubble is plowed early and then worked to stimulate germination of weed seeds. The weeds then are killed by winter freezes, and the flax is sown in the spring after a shallow disking. Flax sown after potato or sugarbeet is likely to be weedy because the digging of these crops brings weed seeds to the surface.

Fiber flax may follow sod or corn.[5] Because of disease problems, flax should not be planted in a field more often than once every three to six years.

■ 32.7 SEED-FLAX CULTURE

Flax requires a firm, weed free seedbed. Clean-cultivated land for row crops is usually disked instead of plowed in preparation for flax. Except in the drier areas, fall plowing is generally practiced where plowing is necessary.

32.7.1 Fertilizers

In the North Central states and Canada, little fertilizer is applied after corn, although nitrogen may be needed if the soil level is low. Manure and nitrogen fertilizers appreciably increase flax yields in southeastern Kansas, while phosphate and lime are also beneficial. Nitrogen fertilizer is often used on irrigated lands in California.

Flax is sensitive to deficiencies of iron and zinc, especially on calcareous soils. A temporary iron deficiency may occur in the spring when soils are cold and wet, but this problem will disappear when growing conditions improve. It has no effect on yield.

32.7.2 Seeding Methods

Flax is generally sown in the spring as early as possible after the seeding of spring small grains is completed. The highest yields are obtained when the crop makes its growth during comparatively cool weather. In the northern Great Plains, flax produces well when sown early in May. Flax is sown in April in Iowa, Idaho, and Washington. Fall seeding between November 15 and December 15 is recommended for southern Texas and California. Flax sown in the Imperial Valley of California on September 20, when the soil temperature was above 100°F (38°C), gave poor stands.[11]

The general practice is to sow flax with a grain drill at a depth of 1 inch (2.5 cm) or less. It is sown at the rate of 42 pounds per acre (47 kg/ha), or slightly more, under humid conditions in Minnesota,[3] Iowa, Kansas, and elsewhere,[16] or under irrigation in the western states. Large-seeded varieties are sometimes sown at a rate of 56 pounds per acre (63 kg/ha).[17] With drier conditions in Montana and the Dakotas, 25 to 45 pounds per acre (28 to 50 kg/ha) is a common rate of seeding. Seed treatment is often helpful. Higher rates help the crop compete with weeds.[18]

32.7.3 Weed Control

Flax does not compete well with weeds. A heavy weed infestation in the field may reduce flax yields by as much as 40 to 70 percent. Sowing clean seed on clean soil is essential to good flax yields. Wild oat, foxtail, mustard, lambsquarters, canarygrass, and barnyard grass commonly occur in flax fields. These can be partly controlled with herbicides such as bromoxynil, clopyralid, MCPA, sulfentrazone, or trifluralin.

32.7.4 Harvesting Seed-Flax

Seed flax is harvested after a majority of the bolls are ripe. The bolls of most varieties are semi-dehiscent when dry enough to thresh readily with a combine. Partial dehiscence occurs when the seeds contain 9 to 11 percent moisture.[10]

Seed-flax may be harvested directly with a combine if the crop is thoroughly dry and free from weeds. Many fields ripen unevenly or contain green weeds. Such fields are harvested with a windrower and pick-up combine, or the field may be sprayed with a desiccant before combining. A relative humidity well below 75 percent appears to be necessary for effective drying.[19] Air-dried flaxseed usually contains about 6 to 10 percent moisture compared with 10 to 14 percent in wheat under the same conditions. The lower water absorption of flaxseed, as compared with wheat and other starchy seeds, is characteristic of oleaginous (oil-bearing) seeds. Flaxseed that contains more than 9 to 11 percent moisture is likely to deteriorate in storage.

▨ 32.8 LINSEED OIL EXTRACTION

Flaxseed contains 32 to 44 percent oil, based on dry weight. Large seeds are highest in oil content. A gallon of linseed oil weighs about 7½ pounds (900 g/l). In commercial crushing using the solvent process, about 2.67 gallons (10 l), or 20 pounds (9 kg), of oil are obtained from a bushel (56 pounds, 25 kg) of cleaned flaxseed. The oil content of flax continues to increase until eighteen to twenty-five days after flowering, but total oil per seed and total dry weight continue to increase until the seeds are mature[6, 20] (Figure 32.5). The oil content is usually low when drought occurs at the filling stage, or within a period of about thirty days after flax flowers. Shriveled seeds are low in oil.

The iodine number is a chemical test for the drying quality of the oil. It is a measure of the quantity of oxygen the unsaturated chemical bonds in the oil will absorb in drying to form the characteristic paint film. Iodine number is defined as the number of grams of iodine that 100 grams of oil will absorb. The iodine number of linseed oil usually ranges from 160 to 195. Linseed oil must have an iodine number of not less than 177 to meet standard specifications. The unsaturated fatty acids in linseed oil are oleic with one double bond, linoleic with two, and linolenic with three double bonds. A comparison of the principal vegetable oils is shown in Table 32.1.

The specific gravity of the previously mentioned oils ranges from 0.91 to 0.93, with the exception of castor oil, 0.96; tung oil, 0.94; and oiticica, 0.97. The plants that produce tung, oiticica, olive, palm, and coconut oils are trees. The other oils are from field crops discussed elsewhere in this book. The oils from seeds of soybean, cottonseed, peanut, sunflower, sesame, safflower, corn, mustard, coconut, and most of the rape are regarded as edible oils. Industrial oils are linseed, castor, oiticica, hemp seed, tung, perilla, palm kernel, babassu, and olive residue.

Linseed oil is expressed from flaxseed with a hydraulic press, a continuous-screw press, or, more recently, by a combination of pressing and solvent extraction. In the third method, cleaned linseed is ground to a meal, heated with steam to a temperature of 190 to 200°F (88 to 93°C), and then forced through a tapered steel bore by continuous screw action. About two-thirds of the oil in the meal is squeezed out through the fine grooves and perforations of the bore. Most of the oil that remains in the press cake is then extracted with a solvent as described for soybean oil extraction in Chapter 26. The residue left after the oil is expressed is known as linseed cake, or, when ground, as linseed meal. The cake contains from 1 to 6 percent oil, depending on the method of extraction. The protein content of the oil meal is about 35 percent.

TABLE 32.1 Characteristics of the More Common Oil Seeds and Vegetable Oils

Crop	Botanical Name	Oil Content (%)	Iodine Number	Average World Oil Production 2000–2003 (1,000 metric tons)
Drying oil				
Perilla	*Perilla frutescens*	40–58	182–206	—
Flax (linseed)	*Linum usitatissamum*	35–45	170–195	642
Tung (China wood)	*Aleurites fordii*	40–58	160–170	77
Hemp seed	*Cannabis sativa*	32–35	145–155	1
Safflower	*Carthamus tinctorius*	24–36	140–150	159
Oiticica	*Licania rigida*	60–75	140–148	—
Drying or semi-drying oil				
Soybean	*Glycine max*	17–18	115–140	28,417
Semi-drying oil				
Sunflower	*Helianthus annuus*	29–35	120–135	8,743
Corn (germ)	*Zea mays*	50–57	115–130	1,977
Cottonseed	*Gossypium hirsutum*	15–25	100–116	3,830
Rapeseed/canola	*Brassica napus*	33–45	96–106	12,563
Nondrying oil				
Sesame	*Sesamum indicum*	52–57	104–118	798
Peanut	*Arachis hypogeae*	47–50	92–100	5,298
Castor	*Ricinus communis*	35–55	82–90	505
Coconut	*Cocos nucifera*	67–70	8–12	3,401
Olive	*Olea europaea*	—	86–90	2,630
Olive residue	*Olea europaea*	—	10–31	238
Palm	*Elaeis guineensis*	—	49–59	25,139
Palm kernel	*Elaeis guineensis*	—	204–207	3,112
Crambe	*Crambe abyssinica*	30–50	95–108	—

■ 32.9 FIBER-FLAX CULTURE

The seedbed and fertilizer requirements for fiber-flax are essentially the same as for seed-flax. With some exceptions, phosphorus increases the fiber content, while nitrogen decreases it.[21] A 4–16–8 fertilizer mixture is usually recommended for fiber-flax where little is known regarding the particular soil conditions.

32.9.1 Seeding Methods

Fiber-flax is sown at the rate of 75 to 85 pounds per acre (84 to 95 kg/ha). Such thick seeding produces tall, nonbranching plants that favor fiber production instead of seed production. Lodging frequently occurs when the seeding rate is appreciably heavier.

32.9.2 Weed Control

Weed control methods for fiber-flax are the same as for seed-flax. A fiber-flax field is better able to compete due to the higher plant populations.

32.9.3 Harvesting

Fiber-flax is usually harvested when one-third to one-half of the seed bolls are brown or yellow with fully developed brown seeds. At this stage, the stems have usually turned yellow, while the leaves have fallen from the lower two-thirds of the stems.[5] The fibers of flax plants harvested too early tend to be fine and silky but lacking in strength; fibers of flax harvested late are coarse, harsh, and brittle, with poor spinning qualities.

Fiber-flax plants are pulled from the ground with special pulling machines, windrowed, and baled. The plants can be mowed instead of pulled but that reduces fiber yield. Yields of 1¾ tons per acre (4,000 kg/ha) of cured pulled flax are typical. Such a quantity yields about 400 pounds per acre (450 kg/ha) of scutched, or processed, fiber (including tow) and about 300 to 500 pounds per acre (340 to 560 kg/ha) of seed.

▨ 32.10 PROCESSING FIBER–FLAX

In the mill, fiber-flax is first threshed or deseeded. The next steps are retting, breaking, and scutching.[8, 22]

Retting (partial rotting) dissolves gums that bind the fibers to the wood and destroys the thin-walled tissues that surround the fibers. Retting is caused by common soil-inhabiting bacteria that are present on the straw when it is harvested.

Two common methods of retting are practiced: dew retting and water retting. Water retting is more satisfactory because of better uniformity.

32.10.1 Dew Retting

In dew retting, the flax is spread on the ground where it is grown. It is retted by molds and bacterial action promoted by frequent rains and dews. Retting is usually completed in fourteen to twenty-one days. The straw must be thinly and evenly spread on the ground for uniform retting.

32.10.2 Water Retting

In water retting, the straw is placed in tanks of water[22] or in ponds or sluggish streams. The retting is accomplished in six to eight days when the water is free of impurities and kept at 80°F (27°C). The water in the tank should be circulated for uniform retting. Cooler water lengthens the retting period. The bundles of retted straw are taken to a field where they are placed in shocks to dry.

32.10.3 Scutching

The breaking process consists of breaking the woody portions of straw into fine pieces (shives) by passing the dry retted straw between fluted rollers. At the same time, the shives are broken or loosened from the fiber. The shives are then beaten off by a cylinder or wheel in the next process: scutching. The long fibers are strong and flexible enough to resist breaking during these processes. Good fiber averages 20 inches (50 cm) in length. Single cells average 25 millimeters in length and 0.23 millimeter in diameter.

■ 32.11 USES OF FLAX

About 81 percent of linseed oil is used in paints and varnishes, 11 percent in the manufacture of linoleum and oilcloth, 3 percent in printers' inks, and the remainder in soaps, patent leather, and other products. Linseed cake or meal is used as feed for livestock. Animals that are fed it have glossy coats.

Seed-flax straw has had a limited market for the manufacture of upholstery tow, insulating material, and rugs. Fiber from flax straw is used in the making of cigarette paper, bible pages, currency, and other high-grade papers. Such paper was formerly made entirely from linen rags. To make this paper, straw from fields with a good, dense, uniform growth of flax that is relatively free from weeds and grasses is gathered for processing. Such fields yield about 1,000 to 1,500 pounds of straw per acre (1,100 to 1,700 kg/ha), with a fiber content of about 10 percent. The straw from combined fields is raked into windrows, sometimes after the stubble has been mowed. It is then baled and hauled to a processing plant. There, the straw is crushed and hackled to remove the shives. Then the fiber is baled and shipped to a paper mill.

The feeding value of flax straw is comparable to wheat or oat straw. Ground flaxseed has long been used in medicine as a conditioner or for making a poultice. An ancient practice was to place a flaxseed in the eye to remove a cinder. Processed edible linseed oil was shipped to the Soviet Union during World War II.

Flax fiber is spun into linen yarns, which are used in threads and twines of various kinds. The yarn is also woven into toweling, clothing fabrics, table linen, handkerchiefs, and other textiles. The short-tangled fibers, called tow, usually a by-product, are used for upholstering, paper manufacture, and packing.

■ 32.12 DISEASES

32.12.1 Flax Wilt

Flax wilt is a fungal disease (*Fusarium oxysporum*) that generally causes infected plants to wilt and die. It infects the live plant as well as the dead plant material in the soil. The fungus may remain in the soil for as long as twenty-eight years. The use of resistant varieties is the only satisfactory control on wilt-infected soil. Varieties grown in the United States are resistant to wilt.

32.12.2 Flax Rust

Flax rust, caused by the fungus *Melampsora lini*, frequently damages flax in wet seasons. Bright orange pustules, the uredineal stage on the leaves, are followed by black shiny areas on the stems (the telial stage) late in the season. The fungus is carried over winter on infected stubble and straw.[23] Crop rotation aids in reducing damage from the disease, but resistant varieties are the only effective means of control.

32.12.3 Pasmo

Pasmo, caused by the fungus *Septoria linicola*, appears on the foliage of young plants as yellow-brown circular lesions. As the plant reaches maturity, brown to black blotches are observed on the stems.[23] The disease is seed-borne. Diseased plants produce smaller seeds and sometimes fewer seeds per boll.[24] Some control measures are crop rotation, seed treatment, and use of resistant varieties.

32.12.4 Other Diseases

Anthracnose or canker, caused by the organism *Colletotrichum lini*, has caused some loss as a seedling disease. Seed treatment stops the spread of the disease.

Heat canker is caused by high temperatures at the soil surface when the plants are small. The young stems, girdled at the soil line, finally break over and die. The damage is prevented by early seeding.

Seed treatments improve stands and yields and reduce damage from seedling blight, particularly when the flaxseed has been damaged during threshing, as is usually the case.

Two virus diseases of flax, aster yellows and curly top, cause losses, but adequate control measures are unavailable. Resistant varieties have been developed. Aster yellows occurs mostly in the northern states where it also attacks several other crops and many weeds. The virus is transmitted by the six-spotted leafhopper, *Macrosteles fascifrons*. Curly top also attacks sugarbeet, beans, and several weeds. The disease is prevalent in the intermountain and Pacific Coast states. The virus is transmitted by the beet leafhopper, *Circulifer tenellus*. Some varieties show tolerance to the disease.

■ 32.13 INSECT PESTS

The most common insect injury to flax is damage from various species of grasshoppers and crickets that chew off the pedicels and allow the bolls to drop to the ground. These pests are controlled with poison baits or sprays. Other insects that sometimes damage flax include cutworms, particularly the pale western cutworm *Agrotis orthogonia*, in Montana and the western Dakotas; armyworms, especially the Bertha armyworm in North Dakota and the beet armyworm in California; false chinch bugs in California; stink bugs in Texas; and the flax worm, *Cnephasia longana*, in Oregon. The last pest is partly controlled by crop rotation. Corn earworm damages flax in Texas. Insecticides control most of these pests.

REFERENCES

1. Foulk, J. A., D. E. Akin, R. B. Dodd, and D. D. McAlister III, "Flax fiber: Potential for a new crop in the Southeast," in J. Janick and A. Whipkey eds., *Trends in New Crops and New Uses*. Alexandria, VA: ASHA Press, 2002, pp. 361–370.

2. Dillman, A. C. "Improvement in flax," in *USDA Yearbook*, 1936, pp. 745–784.

3. Culbertson, J. O., and others. "Growing seed flax in the North Central states," *USDA Farmers Bull.* 2122, 1958.

4. Dillman, A. C., and T. H. Hopper, "Effect of climate on the yield and oil content of flaxseed and on the iodine number of linseed oil," *USDA Tech. Bull.* 844, 1943, pp. 1–69.

5. Robinson, B. B. "Flax-fiber production," *USDA Farmers Bull.* 1728, 1940.

6. Dillman, A. C. "Cold tolerance in flax," *J. Am. Soc. Agron.* 33(1941):787–799.

7. Klages, K. H. W. "Flax production in Idaho," *ID Agr. Exp. Sta. Bull.* 224, 1938.

8. Robinson, B. B., and R. L. Cook, "The effect of soil types and fertilizers on yield and quality of fiber flax," *J. Am. Soc. Agron.* 23(1931):497–510.

9. Ford, J. H. "Relation between seed weight and seeds per plant," *Crop Sci.* 5, 5(1965):475–476.

10. Dillman, A. C. "Dehiscence of the flax boll," *J. Am. Soc. Agron.* 21(1929):832–833.

11. Dillman, A. C., and E. H. Toole, "Effect of age, condition and temperature on the germination of flaxseed," *J. Am. Soc. Agron.* 29 (1937):23–29.

12. Dillman, A. C. "Natural crossing in flax," *J. Am. Soc. Agron.* 30(1938):279–286.

13. Henry, A. W., and C. Tu, "Natural crossing in flax," *J. Am. Soc. Agron.* 20(1928):1183–1192.

14. Dillman, A. C. "Classification of flax varieties, 1946," *USDA Tech. Bull.* 1064, 1953.

15. Davidson, F. E., and H. H. Laude, "Flax production in Kansas," *KS Agr. Exp. Sta. Bull.* 191, 1938, pp. 1–14.

16. Hill, D. D. "Seed-flax production in Oregon," *OR Agr. Exp. Sta. Cir.* 133, 1939.

17. Reddy, C. S., and L. C. Burnett, "Flax as an Iowa crop," *IA Agr. Exp. Sta. Bull.* 344, 1936.

18. Stevenson, F. C., and A. T. Wright, "Seeding rate and row spacing affect flax yields and weed interference." *Can. J. Plant Sci.* 76(1996):537–544.

19. Dillman, A. C. "Hygroscopic moisture of flax seed and wheat and its relation to combine harvesting," *J. Am. Soc. Agron.* 22(1930):51–74.

20. Dillman, A. C. "Daily growth and oil content of flaxseeds," *J. Agr. Res.* 37, 6 (1928):357–377.

21. Robinson, B. B. "Some physiological factors influencing the production of flax fiber cells," *J. Am. Soc. Agron.* 25(1933):312–328.

22. Harmond, J. E. "Processing fiber flax in Oregon," in *Crops in Peace and War*, USDA Yearbook, 1950–51, pp. 484–488.

23. Flor, H. H. "Wilt, rust and pasmo of flax," in *Plant Diseases*, USDA Yearbook, 1953, pp. 869–873.

24. Frederiksen, R. A., and J. O. Culbertson, "Effect of pasmo on the yield of certain flax varieties," *Crop Sci.* 2(1962):434–437.

Cotton

33.1 ECONOMIC IMPORTANCE

Despite the use of synthetic fibers, cotton is the major textile fiber in the world. In 2000–2003, average area harvested was about 80 million acres (32 million ha), chiefly in China, the United States, Pakistan, India, and Uzbekistan (Figure 33.1). Cotton lint production averaged about 21 million tons (19 million MT), or 310 pounds per acre (350 kg/ha). Cottonseed production exceeded 61 million tons (56 million MT), or 1,500 pounds per acre (1,700 kg/ha). In 2000–2003, cotton was grown in the United States on about 13 million acres (5 million ha) (Figure 33.2). Lint production averaged about 18 million bales of 480 pounds each (4 million MT) or 680 pounds per acre (770 kg/ha). Cottonseed production was about 6.7 million tons (6.1 million MT). The average ratio of lint to seed production in the United States is about 36 to 64. The leading states in cotton production are Texas, California, Mississippi, Georgia, and Arkansas. Over 95 percent of cotton grown in the United States is upland cotton.

33.2 HISTORY OF COTTON CULTURE

There are probably several centers of origin of the cotton plant: Indochina and tropical Africa in the Old World, and South and Central America in the New World.[1] Separate origins are indicated by the fact that consistently fertile hybrids have never been obtained from crosses between the 26-chromosome American cottons and the 13-chromosome Asiatic cottons.[2]

Cotton has been grown in India for making clothing for more than 2,000 years, and in certain other countries for several hundred years. Early European travelers returned from southern Asia with weird tales of seeing wool growing on trees. Early herbalists sometimes illustrated the cotton plant by drawings of sheep hanging from the branches of a tree. Apparently, tree types of cotton were grown to a considerable extent at that time. A cotton plant growing as a perennial in the tropics can attain the size of a small tree. Even today, Germans call cotton Baumwolle (tree wool).

Columbus found cotton growing in the West Indies. Cotton fabrics about 800 years old have been found in Native American ruins in Arizona. Cotton was grown in the Virginia Colony in 1607.[3] Its culture soon spread

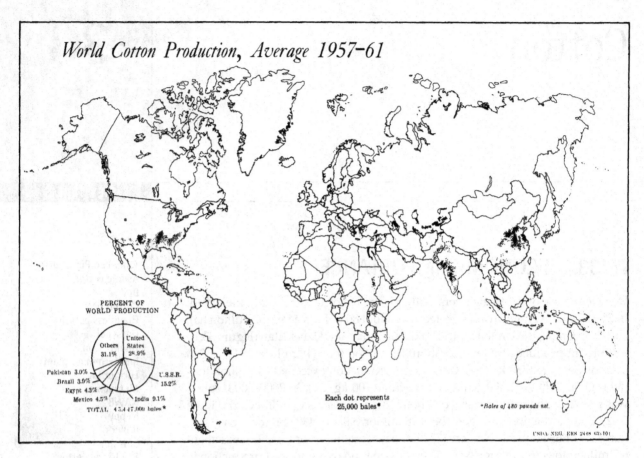

FIGURE 33.1

World production of cotton. [Courtesy USDA]

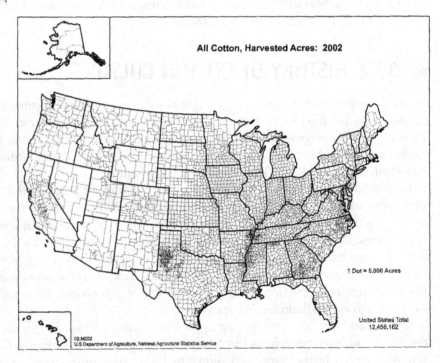

FIGURE 33.2

Acreage of cotton harvested in the United States in 2002. [Source: 2002 Census of U.S. Department of Agriculture]

throughout the South, but large-scale production began after the invention of the cotton gin in 1794.

The three distinct types of cotton grown in the United States—upland, Sea Island, and American Pima—are of American origin. Upland cotton was probably cultivated in Mexico about 3500 BC, and Sea Island and American Pima cottons were grown in Peru about 2500 BC.[4] Upland cotton is assumed to have descended from Mexican cotton or from natural crosses of Mexican and South American species.

▥ 33.3 ADAPTATION

Climatic conditions are favorable for cotton where the mean temperature of the summer months is not less than 77°F (25°C). The zone of cotton production lies between 37° N and 32° S latitude. An exception is Ukraine, where cotton is grown up to 47° N latitude. Three climatic essentials are freedom from frost for a minimum growing and ripening season, an adequate supply of moisture, and abundant sunshine.[5]

General requirements for growing cotton are:

1. A mean annual temperature of over 60°F (16°C). Where the distribution of rainfall, sunshine, and temperature is favorable, a mean temperature of over 50°F (10°C) would probably be sufficient.
2. A frostless season of 180 to 200 days.
3. Minimum rainfall of 20 inches (500 mm) a year with suitable seasonal distribution. A maximum of 60 to 75 inches (1,525 to 1,829 mm) would not be excessive if distribution were favorable.
4. Open, sunny weather. Areas with consistent "half cloudiness" have too little sunshine to be safe, and areas that are over three-fifths cloudy are unsuitable for cotton.[5]

The Sea Island and American Pima types require about six months to reach maturity, while the period for upland varieties is about five months. In the United States, cotton is limited to areas that have annual rainfall of 16 inches (400 mm) or more without irrigation; but, in parts of the irrigated southwestern cotton region, the average annual rainfall is less than 6 inches (150 mm). Cotton is grown almost exclusively on irrigated land in California and Arizona. Most of the crop in New Mexico, western Texas, and southwestern Oklahoma is also irrigated, and some irrigation is applied in other states.[6] All the cotton in Egypt and the Sudan is irrigated, as well as much of it in the Near East and Middle East.

The growing conditions most favorable for cotton are a mild spring with light, frequent showers; a warm, moderately moist summer; and a dry, cool, prolonged autumn. Rainy weather when the bolls begin to open retards maturity, interrupts picking, and damages the exposed fiber. Irregular growth in cotton when irrigation is delayed causes lack of uniformity and lack of strength in the fiber.[7] Early killing frosts in the fall and high evaporation during the flowering period limit the yields of American Pima cotton in Arizona.[8] American upland cotton, with its indeterminate growth habit and lack of photoperiod sensitivity,[9] will produce flowers throughout the year under warm conditions.

The minimum, optimum, and maximum temperatures for the germination and early growth of cotton are about 60, 93, and 102°F (16, 34, and 39°C), respectively. The most rapid growth and flowering of the plants occur at temperatures of 91 to 97°F (33 to 36°C).[10] The highest cotton yields in the United States are in areas where the mean July temperature averages 81 to 83°F (27 to 28°C). When calculating heat units or growing degree days (GDD) for cotton, 60°F is used as the base temperature (GDD=(°F Max + °F Min Temp)/2-60).

Cotton grows well on moderately fertile soils. The soils in the cotton regions range from sands to very heavy clays with ranges in acidity from pH 5.2 to pH 8+. The best cotton lands are mixtures of clay and sandy loam that contain a fair amount of organic matter and a moderate amount of available nitrogen, phosphorus, and potash.[3] Heavier soils promote later maturity, larger vegetative growth, and greater boll weevil damage. The best cotton regions, from the standpoint of both yield and quality, are perhaps the Mississippi Delta and the irrigated valleys of the Southwest. Cotton tolerates some salinity but thrives best on nonsaline soils. A salinity of –5 bars of root osmotic potential may reduce cotton yields 50 percent.

Environmental conditions that promote early maturity in Mississippi are important because they help avoid excessive shedding due to midseason and late-season stress conditions, avoid boll weevil infestation, and have the crop ready to pick before the onset of fall rains and cold weather.[11] Potash-deficient soils, low soil moisture, and other environmental contributions to early maturity usually result in decreased yields.[12] Thick spacing, which rarely reduces yields, was once the most practical way to obtain earliness.[13] Today, plant growth regulators are used to control vegetative growth and control maturity.

■ 33.4 BOTANICAL DESCRIPTION

Cotton belongs to the family *Malvaceae*, or mallow family. Upland cotton (*Gossypium hirsutum*) fibers range from ¾ to 1¼ inches (19 to 32 mm) or more in length and are of medium coarseness. The flowers are creamy white when they first open, but they soon turn pink or red. The lint fibers adhere strongly to the seed. The bolls usually contain four or five locks (Figure 3.15). A lock, or loc, is a division of the ovary called a locule that contains several developing seeds. When the boll opens the number of locks will be evident (Figure 33.3). *G. hirsutum* is photoperiod insensitive.[4]

Sea Island and American Pima cottons (*G. barbadense*) have extra-long, fine fibers, 1½ to 2 inches (38 to 51 mm) long, or even longer in some cases. The lint is readily detached from the seed. The petals are yellow with a purple spot at the base or claw. The bolls usually contain three locks. The term *American Egyptian* was changed to *American Pima* in 1970. *G. barbadense* is considered to be a short-day photoperiod sensitive plant.[4]

The Asiatic cottons are classified as *G. arboreum* and *G. herbaceum*. The fibers are coarse and short, with a length from ½ to ⅞ inch (13 to 22 mm). Cotton developed by Native Americans, such as Hopi, also has short fibers. Upland and Asiatic cottons have shorter boll periods than do the Pima and Sea Island cottons.[14]

The cotton plant is usually considered an annual, although it is a long-lived perennial in the tropics where the mean temperature of the coldest months does not fall below 65°F (18°C). The plant is herbaceous, with a long taproot. It can attain a height of 2 to 5 feet (60 to 150 cm) or more and has a main stem from which

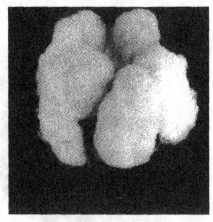

FIGURE 33.3
Open cotton bolls.

many branches arise. The taproot grows downward into moist soil at a rate of about 1 inch (2.5 cm) per day up to five weeks and averages nearly ⅜ inch (1 cm) daily for the entire 175-day growing period.[15] The leaves arise on the main stem in a regular spiral arrangement.

At the base of each cotton leaf petiole are two buds, the true axillary bud, which continues to make vegetative growth, and an extra-axillary bud, which produces the fruiting branch.[16] Leaves are petiolate and stipulate, with palmate venation and three, five, or seven lobes arising at each node above the cotyledons (Figure 33.4). The leaves and stems are usually covered with fine hairs. The leaves are green except in a few red leaf varieties.[17] Red leafed varieties have been grown occasionally in the United States. Varieties with colored lint—green, brown, yellow, blue or red—are grown in the Russian Federation and other countries.

The flowers are arranged on alternate sides of the fruiting branch. Additional flowers appear on a branch at about six-day intervals. There are three relatively large leaf-like bracts at the base of the flower above which is a true calyx that consists of five unequally lobed sepals. The corolla consists of five petals that range in color from white to yellow to purple in different types. The staminal column bears ten more or less double rows carrying ninety to one hundred stamens, while the pistil consists of three to five carpels. The stigma tip extends up above the anthers. The fruit (Figure 3.15) is the enlarged ovary that develops into a three- to five-loculed capsule or boll. The bolls are 1½ to 2 inches (4 to 5 cm) long among the common varieties. About sixty to eighty bolls are required to produce a pound (450 g) of seed cotton.[18] Other types have bolls not over 1 inch (2.5 cm) long. The earliest and latest bolls on a plant are usually smaller, with shorter fiber, than the intermediate bolls. The boll dehisces or splits open at maturity (Figure 33.3). Late, unopened bolls are called bollies. The seeds are covered with lint hairs (the fibers)[19] and usually with short fuzz.

33.4.1 Reproductive Growth

Flower buds, or squares, will appear about thirty to thirty-five days after emergence. The first square will usually form on a fruiting branch arising from the axil of the fifth to seventh node. Lower plant populations can lower the node of the

FIGURE 33.4
Cotton leaves, flowers, square *(upper right)*, and unopened boll *(right)*.

first fruiting branch. High plant populations, night temperatures below 60°F (15°C) or above 80°F (27°C), or thrip damage can raise the node of the first fruiting branch.[20] Flowers will open about twenty-one days after the square is visible. Cotton that begins fruiting higher on the plant is more likely to grow too tall, especially if early squares are shed.[20] Early shedding of squares can be caused by cool, cloudy weather with night temperatures below 55°F (13°C) due to decreased photosynthesis. Water-saturated soils and pest damage can also cause early square shedding.[20]

Cotton is indeterminate and will normally bloom for seven or eight weeks. Stress from drought, nematodes, and fertility can shorten the bloom period. Poor fruit retention or excess nitrogen can lengthen the flowering period. Counting nodes above white bloom (NAWB) is used to assess crop condition during flowering. NAWB involves counting the number of nodes above the uppermost first-position white bloom on the stem. Each node above the first-position white bloom is counted if the main stem leaf attached to the node is larger than a quarter.[20]

NAWB should be eight to ten at first bloom depending on variety and growing conditions. NAWB will be closer to eight for short season varieties and closer to ten for full-season varieties. Environmental stress can lower NAWB. Poor square retention or excess nitrogen can cause higher NAWB. NAWB will decrease as the growing season progresses and terminal growth will cease when it reaches five. If NAWB decreases too rapidly, the producer should identify the stress or stresses involved and attempt to reduce them.[20, 21]

Many of the buds, flowers, or young bolls drop off. As a result, only 35 to 45 percent of the buds produce mature bolls normally. Most boll shedding occurs from three to ten days after pollination.[10, 22] The period between flowering and the opening of the mature boll is about six to eight weeks, becoming longer as the season extends into cool autumn weather. The cotton plant shows a remarkable adjustment to its environment. Under severe drought conditions, the plant may be 6 inches (15 cm) high and bear one boll. A plant of the same variety grown under irrigation may be 5 feet (150 cm) high and produce forty bolls. The number of bolls that develop is kept in balance with the growth of the plant.[10, 23]

33.4.2 Pollination

Cotton is readily self-pollinated. Out-crossing is less than 5 percent in some parts of Texas but over 50 percent in some states to the east. The extent of cross-pollination varies with the population of bees and other insects.[24]

Metaxenia, or the immediate effect of foreign pollen on the fertilized ovules, has reduced the lint length of Pima (long-staple) cotton when pollinated with Hopi (short-staple).[25]

33.4.3 Lint or Fiber

Cotton fibers are slender, single-cell hairs that grow out from certain epidermal cells of the cottonseed.[14] Fiber growth starts on the ovules about the time the flower opens. The fibers lengthen rapidly and attain full length in about fifteen to twenty-five days when the seed has attained full length. For an additional twenty-five to forty days, the cell walls of the fiber continue to thicken. Thickening occurs by the growth of two additional spiral rings each day on the inner surface of the cell wall. The fiber is cylindrical before the boll opens but collapses and becomes more or less flattened and twisted (with convolutions) with the opening of the boll (Figure 33.5). The fiber is thin-walled, weak, and poorly developed when unfavorable conditions prevail during the time that the fiber is thickening. A pound (450 g) of lint contains 100 million or more fibers. The fibers range in length from ¼ inch (6 mm) to over 2 inches (50 mm) (Figure 33.6), and the thickness will vary from 0.015 to 0.020 mm. Varieties with long fibers tend to have a low lint percentage in seed cotton, whereas those with short fibers have the highest percentage of lint.

The fiber length is enhanced by ample rainfall while the bolls are developing.[26] Thin-walled or immature fibers produce yarn that is unduly neppy (full of knots and snarls) that cannot all be removed by the spinner. Dark-colored, plump, well-developed seeds yield a high percentage of mature fibers.[27]

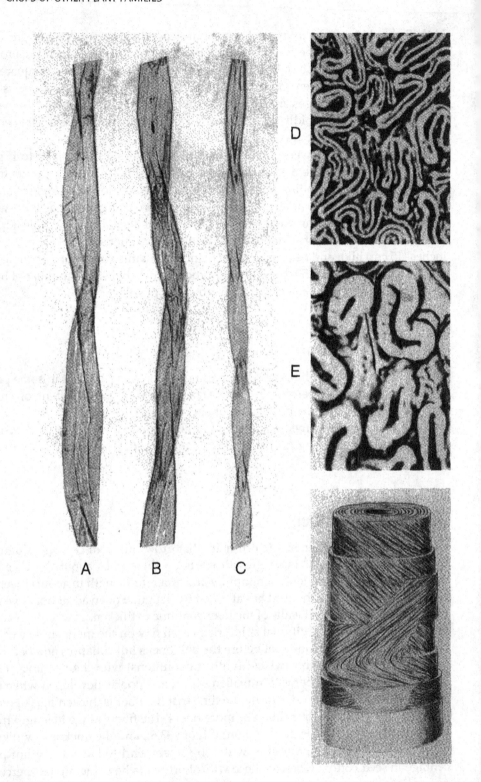

FIGURE 33.5
(Left) Short portions of cotton fibers magnified 400 diameters: *(a)* coarse fiber; *(b)* medium fiber of an American upland variety; and *(c)* fine fiber of Sea Island cotton. *(Upper right)* Cross sections of dry cotton fibers, *(d)* four weeks after pollination, and *(e)* at maturity showing thickened walls. Highly magnified. *(Lower right)* Diagram of a mature cotton fiber showing the slope of the strands that comprise the daily growth rings within the cell wall.

Yarn strength is determined by the length, strength, and fineness of the fiber by an approximate respective ratio of 47:37:16.[14] Small-diameter fibers with thin walls contribute to yarn strength but tend to cause a neppy or poor appearance in the yarn unless a combing process is used in spinning. Other things being equal, long

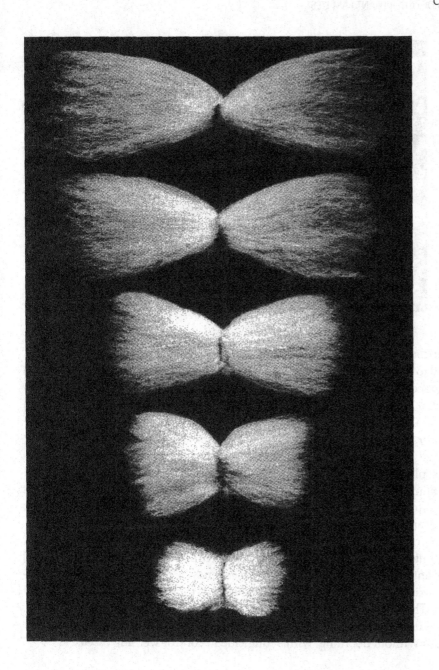

FIGURE 33.6
Combed fibers attached to the seeds of five types of cotton: (from top to bottom) Sea Island, American Pima, American upland long staple, American upland short staple, and Asiatic. (All natural size.)

fibers give a smoother and stronger yarn than do short fibers. Consequently, long staple cotton is used in making the better grades of yarn. Short staple varieties are grown in certain localities because they are more profitable.

33.4.4 Seed

Normally, there should be nine seeds in each lock, or twenty-seven to forty-five per boll. However, a lock usually contains one or two undeveloped (aborted) seeds called motes. The seeds are usually ovoid in shape (Figure 33.7). Most of the upland cotton varieties have dark brown seeds covered with fuzz, while Sea Island

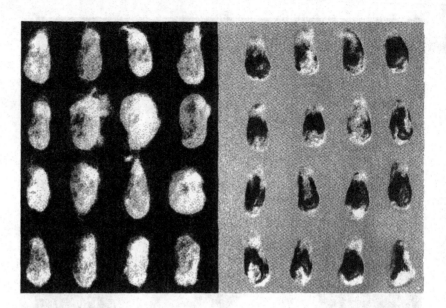

FIGURE 33.7
Cottonseed: *(left)* gin run; *(right)* delinted.

and American Pima seeds are black and practically free from fuzz. A pound contains 3,000 to 5,000 seeds (6,600 to 11,000 seeds/kg). Large-bolled varieties tend to have large seeds. The seedcoat is a tough, leathery hull that constitutes about 25 to 30 percent of the weight of the seed. The oil content of the hulled kernel (or meat) varies from 32 to 37 percent.[28]

Seed, stored in tight containers for seven years at moisture below 8 percent, has retained its viability with only a slight decrease.[29] Seed, especially that intended for planting, may require artificial drying after ginning in order to avoid deterioration.[30]

Seed is often delinted before planting to allow it to pass through the planter box more freely and germinate more rapidly. Seed is delinted mechanically with delinting saw gins or by treatment with concentrated sulfuric and hydrochloric acids. Seed may be delinted very closely with less than 1 percent saw-cut injury. From 100 to 150 pounds of linters per ton (90 to 75 kg/MT) of seed may be removed with safety.[30] Linters are a mixture of soft or flaccid fibers and fuzz that escaped removal from the seed in ginning. The sale value of linters helps to defray the cost of mechanical delinting. Mechanical delinting is frequently done at oil mills. Often, the seed is treated after delinting, or without delinting, with a fungicide to control seedling diseases.[30]

■ 33.5 VARIETIES

More than 500 varieties of cotton were grown in the United States in the early years of the twentieth century.[31] By 1970, strains of only six varieties occupied 67 percent of the cotton acreage. Mediocre varieties have been eliminated due to the release of improved varieties, followed by regional testing and the establishment of numerous one-variety communities. Continued breeding efforts have moved from one-variety communities to the multiple varieties now available. Some cotton varieties are widely adapted and are grown in several states in the humid cotton region. The average staple length of the American cotton crop increased 10 percent from 1928 to 1962 by breeding for better staple quality.

Cotton varieties are classified into three groups: (1) short-season, more determinate plants; (2) medium-season plants; and (3) full-season, more indeterminate plants.[4] The difference in maturity between short-season and full-season varieties is only about three weeks.

Varieties differ in adaptation, staple length, fiber quality, boll size, disease resistance, and retention of the cotton in the locks. The extra long staple cottons grown today, with fibers 1¼ to 1⁹⁄₁₆ inches (32 to 40 mm) long, are the American Pima varieties. Sea Island cotton with even longer fiber length was once grown in the South Atlantic states, but its production had practically ceased by 1944. Long staple upland cotton varieties, with fibers 1¹⁄₁₆ to 1³⁄₁₆ inches (27 to 30 mm) long, include selected strains of a Mexican variety, Acala, grown in California, Arizona, New Mexico, western Texas, and Oklahoma. Long staple upland varieties grown in the eastern and central states of the Cotton Belt[32] have a staple length ranging from 1¹⁄₁₆ to 1³⁄₁₆ inches (27 to 30 mm). Medium length staple is 1 to 1¹⁄₁₆ inches (25 to 27 mm). Short staple cottons are mostly storm-proof types grown in northwest Texas and adjacent areas, with staple lengths of ⅞ to 1 inch (22 to 25 mm).[33]

American Pima varieties have small bolls, 100 to 135 bolls per pound (220 to 300 bolls/kg) of seed cotton. Most upland varieties have 70 to 85 bolls per pound (155 to 190 bolls/kg), but large-boll varieties have 55 to 70 per pound (120 to 155 bolls/kg). About 85 percent of total U.S. production of American Pima long staple cotton is grown under irrigation in California. Other states are Arizona, Texas, and New Mexico.

Recent efforts in plant breeding have resulted in the development of genetically engineered varieties of cotton that are resistant to certain pests and herbicides. Transgenic cotton varieties that contain genes from *Bacillus thuringiensis* (Bt cotton) are resistant to the tobacco budworm *(Heliothis virescens)* and the pink bollworm *(Pectinophora gossypiella)* and show some resistance to the bollworm *(Helicoverpa zea)*.[34] Buctril-tolerant (BXN) varieties allow the postemergent application of bromoxynil herbicide on cotton. Roundup-Ready® (RR) cotton varieties allow the use of glyphosate postemergent herbicide. Research to develop other transgenic varieties will undoubtedly result in subsequent release of more bioengineered varieties.

▓ 33.6 FERTILIZERS

Nitrogen is the most critical nutrient for cotton production. Nitrogen deficiency will cause yields to drop sharply. Too much nitrogen, or nitrogen application at the wrong time, will produce plants that are too tall, slow to flower, more attractive to pests, late to mature, quick to develop boll rot, more difficult to defoliate, and produce lower quality lint.[35, 36] Cotton needs about 60 to 70 pounds of nitrogen per acre (67 to 78 kg/ha) to produce one bale of cotton. Careful testing of the soil to determine soil nitrogen level is needed to determine the need for additional nitrogen fertilizer. Most nitrogen is taken up after squares begin to set, so large preplant applications are usually not needed unless the soil is low in nitrogen. Nitrogen uptake will peak about two weeks after first bloom and continue at a high rate for several weeks during reproductive growth.[36] Most nitrogen fertilizer should be applied in split applications during reproductive growth to reduce nitrogen losses from the soil.

An accurate method for determining nitrogen supply to the plant during the growing season is to analyze leaf petioles for nitrogen. Sampling should begin one week before first bloom. About twenty leaves with petioles should be sampled, and all leaves should be the same physiological age. The petiole from the first fully expanded leaf on the main stem from the top of the plant is selected, usually the third leaf below the quarter-sized main stem leaf in the terminal.[37] Sampled petioles are sent to a laboratory for analysis. Complete information on crop condition and growth stage should be included with the sample.

Phosphorus and potassium are much less likely to be deficient in cotton. The nutrients are usually applied only when soil test levels are less than 45 ppm P_2O_5 using the sodium bicarbonate test, and below 240 ppm K_2O using the Mehlich-3 test.[4, 38] Sulfur fertilizer may be needed if indicated by soil tests. Micronutrients that may be needed for cotton are boron, copper, zinc, and manganese. A starter application at seeding is the most effective method of applying nutrients other than nitrogen unless soils are very sandy.

The optimum soil pH range for cotton is 6.2 to 6.5.[36] Many soils in the cotton region will benefit from the application of lime. Low soil pH reduces nutrient availability and increases the chances of aluminum toxicity. Saline and sodic soils may occur in arid regions where cotton is grown. Cotton is considered a relatively salt-tolerant crop and may succeed where other crops would not.[38]

Seed cotton equivalent to a bale of lint contains about 35 pounds (16 kg) of nitrogen, 14 pounds (6.3 kg) of phosphoric acid, 14 pounds (6.3 kg) of potash, 3 pounds (1.4 kg) of lime, and 4 pounds (1.8 kg) of magnesium. The lint portion contains about 1 pound (0.45 kg) of nitrogen and 6 pounds (2.7 kg) of mineral nutrients.

■ 33.7 ROTATIONS

Rotation is essential in some areas in order to reduce some cotton diseases such as root rot. Corn, small-grain, and grain sorghum rotations will reduce reniform nematode populations.[35] Any of these crops may be followed by a winter legume that is plowed under before cotton is planted. However, recent research shows that legume cover crops are hard to kill, make it more difficult to plant the crop, and may delay cotton maturity. Small-grain cover crops, mostly wheat and rye, are better preceding cotton.[35, 39, 40] Double-crop cotton following winter wheat has been successful in southern Georgia, but late-planted cotton is more at risk to early frost.[35] In irrigated areas, cotton may follow alfalfa, sorghum, or an oilseed crop. Cotton is often grown continuously on semiarid lands because it does not fit well in sequence with winter wheat or sorghum, the other important crops. Continuous cotton is also possible on clean land free from wilt infestation in the Yazoo-Mississippi Delta.[41]

■ 33.8 COTTON CULTURE

33.8.1 Seedbed Preparation

An increasing amount of cotton acreage is being grown using conservation tillage, including minimum tillage, strip tillage, and no-tillage.[39, 40, 42] Strip tillage is conducted over the seed row, usually 4 to 12 inches (10 to 30 cm) wide. In-row subsoiling at a depth of 10 to 16 inches (25 to 40 cm) may or not be done at the same time.[40] Some growers use combination tool bars that strip till and plant in one operation. If ripping is done at planting, seed rows should be offset 2 to 3 inches (5 to 8 cm).[35]

Cotton is planted on ridges or beds in most of the humid Cotton Belt, while level or furrow plantings are more common in the western half of the Cotton Belt, except under irrigation. Low beds are preferable from the standpoint of weed control as well as for moisture conservation. Higher beds may be desirable in wet areas. If bedding is used with conservation tillage, either fall bedding or re-using the bed from the previous growing season is usually required.[40] Fall bedding may be difficult in areas where cotton harvest is not complete until late November or December. Research in North Carolina showed that fall ripping is as effective as spring ripping when a cover crop was seeded after ripping and strip tillage was conducted before planting.[40] If strip tillage is not done, then a cover crop should be killed at least two to three weeks before planting.[42] Cotton planted without beds is offset up to 6 inches (15 cm) from previous rows to avoid stalks from the previous crop.[40]

33.8.2 Seeding Methods

Cotton is best planted when the soil at a depth of 8 inches (20 cm) has warmed up to at least 60°F (15°C)[29] and there is a high probability of an air temperature of 68°F (20°C) for ten days following planting. Most of the crop is planted in March, April, or May, the later dates being applicable to the northern part of the Cotton Belt (Figure 33.8). Cotton seeding in North Carolina begins after April 15 and should be completed before May 5. Yields decline about 12 pounds per acre for every day planting is delayed after that date. Cotton planted after May 20 yields significantly less.[43]

Planting on low W-type beds is helpful not only for obtaining good stands, but also for subsequent weed control, drainage, and irrigation. The W-type beds are alternate ridges and furrows. In irrigated areas, the field is often watered before planting, but after the beds are prepared. Attachments on the planter scrape off dry soil from the top of the beds and smooth them for planting. Rubber-tired wheels at the rear of the planter press the seeds into moist, compacted soil. Knife or disk furrow openers are used.

Most varieties range from 4,000 to 5,000 seeds per pound (8,800 to 11,000 seeds/kg). Seed is planted 1 to 2 inches (25 to 50 mm) deep. The most common row spacing is 38 inches (97 cm). Seeding rate should be about 3 or 4 seeds per foot (10 to 13 seeds/m) of row in the humid regions resulting in a final stand of 2 or 3 plants per foot (7 to 10 plants/m). Irrigated cotton is seeded at 4 to 6 seeds per foot (13 to 20 seeds/m). Plant populations greater than 3 plants per foot (10 plants/m) increase the percentage of the crop set at the first position of fruiting branches and reduces the total number of fruiting branches, resulting in higher yield potential.[43] However, more than 5 plants per foot (16 plants/m) increases the node number at which fruiting begins, resulting in lower yield potential.[35]

Although the cotton plant is adaptable to produce satisfactory yields over a wide range of populations, moderately close spacing has given the best yields. From

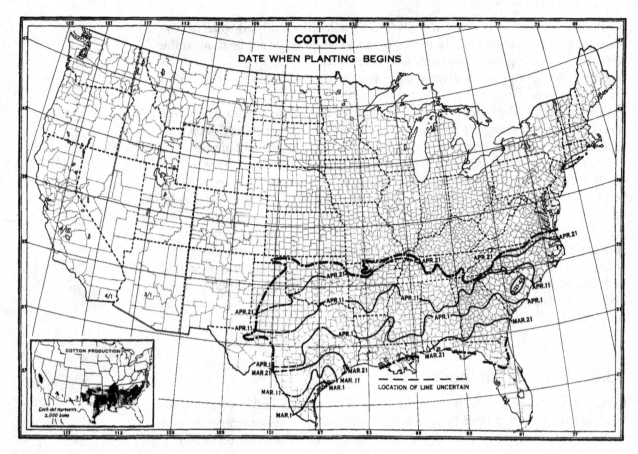

FIGURE 33.8
Dates of planting cotton in the United States.

25,000 to 50,000 plants per acre (62,000 to 125,000 plants/ha) are ample for maximum yields under humid or irrigated conditions. [32, 44] These stands are approximately 1 plant every 3 to 8 inches (8 to 20 cm) in the typical 38- to 40-inch (97 to 102 cm) rows.

Narrow-row cotton with a row spacing of 7 to 10 inches (18 to 25 cm) is receiving increased interest with growers since the introduction of broad spectrum postemergent herbicides and genetically engineered herbicide-tolerant varieties. A seeding rate of 2 or 3 seeds per foot (7 to 10 seeds/m) results in a plant population approaching 150,000 plants per acre (370,000 plants/ha). Each plant will produce only a few bolls, but acceptable yields can be produced in a short period of time.[35] The availability of a suitable harvester is essential before attempting narrow-row cotton. Use of growth regulators, such as mepiquat chloride, to keep plant height no greater than 30 inches (76 cm), may also be necessary.

Soil crusting can greatly reduce emergence of cotton seedlings. Easily crusted soils can be planted at higher seeding rates to increase seedling pressure on the crust. The increased stress of emerging through a crusted soil will reduce seedling survival resulting in normal plant population. However, if crusting does not occur, thinning may be needed. Soil crust can also be broken by the use of a rotary hoe or similar implement.

33.8.3 Weed Control

Most grass seedlings and some annual broadleaf weeds in cotton fields can be controlled by spraying herbicides on the field and then incorporating the material into the surface soil before planting. Common herbicides used to kill existing weeds before planting are paraquat and glyphosate. Preplant herbicides include pendimethalin and trifluralin that are incorporated during strip tillage.

Preemergent herbicides control early-season, small-seeded annual weeds when applied during planting or shortly thereafter. Band application over the planted rows during planting reduces chemical costs, but cultivation or postemergent herbicides must then be applied to control the weeds between the rows. Common preemergent herbicides are fluometuron, norflurazon, clomazone, prometryn, and metolachlor.

Some postemergent herbicides used in cotton fields are applied as basally directed sprays after the cotton plants are 3 to 6 inches (8 to 15 cm) tall or taller. Some common postemergent-directed herbicides are MSMA/DSMA, fluometuron, cyanazine, oxyfluorfen, and lactofen. Postemergent herbicides applied over the top of cotton plants include clethodim, pyrithiobac, quizalofop, sethoxydim, and fluazifop. Glyphosate or bromoxynil can be applied to varieties that have been genetically engineered to be resistant to them.

Early cultivations are often made with a rotary hoe. Later, shallow cultivations with sweep or blade implements cut off the weeds with little injury to the cotton roots.

33.8.4 Irrigation

All the cotton in California and Arizona is irrigated. The crop is also grown under irrigation in nearly all of New Mexico, in much of Texas, and on appreciable acreages in Oklahoma, Arkansas, Louisiana, and other states. Cotton is more tolerant of dissolved salts in irrigation water than corn, grain sorghum, and peanut.[45]

When water is available, daily consumption by transpiration and soil evaporation in the cotton field averages about ⅕ inch (5 mm) per day. Maximum daily usage ranges from ¼ to ⅖ inch (6 to 10 mm) at the peak of growth in midsummer. Thus, for high yields, from 24 to 42 inches (740 to 1,070 mm) of water in the soil from rainfall or irrigation is used in the growing season. The soil should contain available moisture throughout the root zone of 4 to 6 feet (120 to 180 cm) when the cotton is planted. Irrigation before planting time makes this possible.

Frequent irrigation is desirable to keep the soil moisture from dropping much below 50 percent of its field capacity. In hot, dry climates, this requires irrigation every eight to sixteen days. The soil moisture content can be determined with a tensiometer, or it may be estimated by an examination of a soil core taken with a probe or auger. The cotton plant needs water when the leaves turn to a dark or dull bluish-green color followed by wilting during the hot part of the day.[44] A need for water is evident when vegetative growth decreases to the point where flowers are opening near the top of the plants. Adequate, frequent irrigation keeps 4 to 6 inches (10 to 15 cm) of growth above the upper flowers on the plant. Furrow or basin irrigation often produces higher yields than does sprinkler irrigation.

FIGURE 33.9
Mechanical cotton picker.
[Courtesy Case IH]

33.8.5 Harvesting

Cotton in the United States is harvested using mechanical pickers or strippers. Mechanical pickers (Figure 33.9) are equipped with rotating steel spindles attached to revolving drums or moving bars. The spindles are fluted rods, smooth rods, or barbed cones.[46] Moistened spindles pull the lint and seeds from the opened bolls, after which they are pushed from the spindle with a doffer (revolving cylinder, or vibrating bar, with teeth) and then conveyed to a basket mounted on the picker. The harvested cotton contains fewer leaf fragments when smooth leaf varieties are grown.

Strippers (Figure 33.10) snap the open and unopened bolls from the plants as the machine moves across the field. Strippers are equipped with inclined rollers that straddle the plant rows. The long-brush type of stripping roll is very effective.[47] Some strippers are equipped with a stick and bur remover so that less trash is hauled to the gin.

A finger-type stripper is used to harvest narrow row or broadcast cotton. The strippers are operated after frost, defoliants, or desiccants have removed most of the leaves from the plants. The burs on stripped cotton are removed at the gin, which also is equipped with boll breakers and extractors to remove the seed cotton from unopened bolls.

A cotton combine is a picker or scrapper that cuts off the plants, strips out the seed cotton with spindles, and then chops the stalks and branches and returns them to the soil (Figure 33.11). Some fields receive two pickings to avoid losses from weathering of the early cotton and save much that might drop to the ground before the late harvest.

Lint and seed of cotton picked when the plants are wet from rain or dew will be damaged unless they are dried promptly.[22] Tramping or packing machine-picked

FIGURE 33.10
Mechanical cotton stripper.
[Courtesy John Deere & Co.]

FIGURE 33.11
Unloading picked cotton in
the field. [Courtesy Case IH]

cotton in the trailer or module builder likewise contributes to deterioration. Modern gins are equipped with dryers as well as cleaning equipment. Picking should be postponed until the relative humidity drops below 50 percent each day,[32] unless the cotton is to be dried at the gin (Figure 33.12).

During the period from 1910 to 1914, about 276 hours of labor were expended in producing, harvesting, and handling a bale of cotton. This was accomplished in only twenty-six hours in 1970–71, following mechanized operations and higher yields.

■ 33.9 DEFOLIATION, DESICCATION, AND TOPPING

Cotton that is mechanically harvested is often chemically treated to remove or dry the leaves.[48] These treatments eliminate much dampness, fiber staining, and trash from seed cotton. They also reduce difficulties caused by clogging of picker spindles with trash or leaf juices. Leaf removal provides better ventilation to reduce boll rotting on the lower branches of the plants. It also reduces lodging of the plants.

Defoliation induces abscission (separation) of the cells where the leaves are attached to the branches.[22] Most of the leaves drop off within a few days after proper treatment. Desiccation kills the leaves so that they dry and break off. Either defoliation or desiccation hastens drying so that picking can be started about one hour earlier in the morning. Defoliants are most effective at temperatures between 90 and 100°F (32 and 38°C), but they are relatively ineffective below 55°F (13°C) or above 100°F (38°C).

At least 50 to 60 percent of the bolls should be open, and the youngest boll to harvest mature, when defoliants are applied, because boll and fiber development is stopped when the leaves are removed or killed.[49, 50] When mature, a boll is difficult to cut in cross section, and seed will be fully filled, with no jelly in the center. Another method for determining time for defoliation is when the first-position cracked boll is within four nodes of the last boll ready for harvest.[49] Bolls need forty to sixty days from flowering to maturity.[50]

The best time for defoliation of irrigated cotton in semiarid and arid regions is when the plants are under mild to moderate water stress. Water stress promotes crop senescence, but the plant needs sufficient water to allow leaves to form the abscission layer and maintain enough leaf weight to cause the leaf and petiole to break off. Defoliation when available soil water is about 70 percent depleted gives the best results.[38]

There are two types of defoliants. Those with herbicidal activity include Def, Folex, and Harvade. Defoliants with hormonal activity contain ethephon or thidiazuron, usually with other chemicals.

Desiccants are more drastic than defoliants in their effect on plant development. Consequently, they should be applied five to seven days later than defoliants. Desiccants are not generally used when cotton will be harvested with spindle-type pickers. Dessicants kill the entire plant and burn immature bolls, so 85 to 90 percent of the bolls should be open before applications.[50] If desiccants are applied to kill weeds or regrowth vegetation, a defoliant is applied first and a dessicant is applied after leaf drop. Dessicants contain paraquat or sodium chlorate.

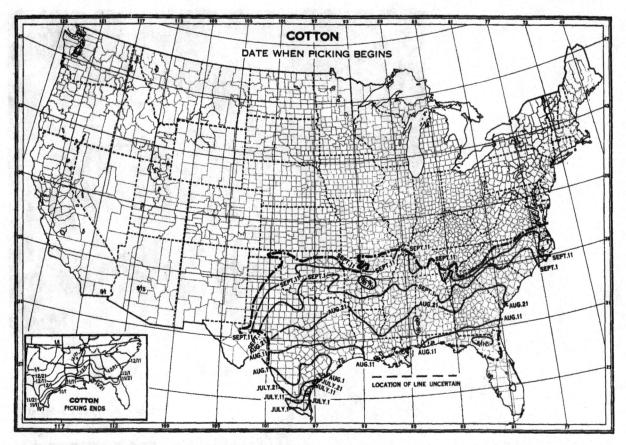

FIGURE 33.12
Dates when cotton picking begin.

Cotton plants that have ceased growing after exhaustion of the soil moisture or mineral nutrients, or for other causes, can drop their leaves without the need of defoliants.

Mechanical topping of cotton plants with high clearance cutters is practiced in California to reduce lodging of the plants as well as to facilitate defoliation and the operation of spindle pickers. The plants are topped at a height of about 4 feet (120 cm), or by cutting 8 to 12 inches (20 to 30 cm) off the top of shorter plants.[51, 52]

33.10 COTTON GINNING

After it is harvested, cotton is hauled in bulk to the gin. Cotton ginned within a day after the harvest has more long fibers of improved spinning quality.[53]

Cotton ginning involves separation of the fibers from the seed after removal of dirt, hulls, and other trash. Damp or wet cotton is put through a drier before ginning.[45] Dry lint contains 8 percent moisture or less.

Seed Cotton Inlet

Front

Section through a Gin Stand
in the U.S.A.

Seed
Roll

Lint to
Condenser

Brush

Ginning
Ribs

Saws

Seed
Discharge

Mote Board

Motes

FIGURE 33.13

(Top) A cotton gin plant.
(Bottom) Section of a
stand. [Courtesy Steve Brown,
U of GA]

The mechanism that separates the seed from the lint is called a gin stand. There
are two types of gins: saw gin and knife gin. The essential part of a saw gin is a set
of circular saws that revolve rapidly with a portion projecting through a narrow slit
between parallel ribs or bars of iron.[3] The teeth of the saws catch the lint, draw it between the ribs, and pull it from the seeds (Figure 33.13). Since the seeds are too large

to pass through, they are dropped into a conveyer or suction pipe. An air blast or a brush removes the lint from the saws. Suction conveyers carry the lint to the press.

American Pima cotton is ginned on a roller gin. Roller gins use knives to jerk the seed from the fiber. The fiber is then pulled away between a roller and a fixed knife. Roller ginning is more gentle on the fibers, which helps prevent damage to the longer fibers of American Pima.

At the press or baler, the fluffy lint is pressed into bales of about 500 pounds (230 kg). The bale is covered with heavy jute or plastic and held by six asphalt-coated flat steel ties. The standard bale is 54 to 55 inches (137 to 140 cm) long, 20 to 21 inches (51 to 53 cm) wide, and 33 inches (84 cm) or less thick. The volume is 17 cubic feet (0.48 m^3) with a density of 28 pounds per cubic foot (472 kg/m^3). Each bale contains 480 pounds (218 kg) of lint and 20 pounds (9 kg) of bagging and ties. For more economical export, transportation, and storage, the bales are often further compressed.

Seed cotton from the field will yield about 43 percent cottonseed products, 29 percent trash, and 27 percent lint, with the rest being motes or short fibers.[4]

▦ 33.11 USES OF COTTON

Cotton lint is spun into thread to be woven into various fabrics (Figure 33.14). The finer threads are made from American Pima cotton, while the coarser ones are made from the upland varieties. About 73 percent of domestic cotton is used for clothing and household goods. The remainder is used in industry, mostly for awnings, bags, belts, hose, and twine. Short lint is utilized in carpets, batting, wadding, and low-grade yarns, as well as for stuffing material for pads and cushions.[52] Linters or fuzz is used mostly for stuffing and for making rayon and other cellulose products.

About 7 percent of the cottonseed produced is planted each year, while small quantities are used for feed. The remainder is crushed for oil, and the meal used as a high-protein food or feed. A ton of cottonseed yields about 320 pounds (160 kg/MT) or 43 gallons (180 l/MT) of oil, 900 pounds (450 kg/MT) of meal, 135 pounds (67 kg/MT) of linters, and 500 pounds (250 kg/MT) of hulls. The remainder of 145 pounds (72 kg/MT) is trash, waste, and invisible losses. The production of linters averages more than 1½ million bales annually. The seed is delinted before it is crushed for oil extraction.

From 40 to 70 percent of the cotton produced each year, as well as some cotton fabrics, are exported. About one-third of the cottonseed oil, and some cottonseed meal, is exported. About 20 percent of the total oil production is made into shortening, and 7 to 8 percent into margarine by the process of hydrogenation.

Gossypol is a natural toxin that helps protect cotton plants from insect damage. It is secreted by glands on the leaves and in the seeds. Cottonseed meal is a high-protein feed, but the presence of gossypol makes it toxic to livestock if fed in large quantities. This limits the proportion of cottonseed meal that can safely be added to poultry or livestock rations. Cotton with few glands on the leaves and in the seeds is low in gossypol. Glandless varieties of cotton are available but fiber yield is less than normal cotton and the cost of pest control is higher. Use of glandless cotton, however, reduces the cost of oil processing and increases the protein meal quality of the seed meal. Some seed processors

FIGURE 33.14
A cotton mill converting
yarn to thread.

are paying a premium for the seed to offset the reduced fiber yield and insure a
supply of glandless cotton.

▨ 33.12 MARKET QUALITY OF COTTON FIBER

Cotton is classed according to fiber length and uniformity. Fiber length is the average length of the longer one-half of the fibers (upper half mean length). It is
reported in both 100ths and 32nds of an inch. It is measured by passing a
"beard" of parallel fibers through a sensing point. The beard is formed when
fibers from a sample of cotton are grasped by a clamp, then combed and brushed
to straighten and parallel the fibers. In any sample, an array of different fiber
lengths can be found, ranging from very short to somewhat longer than the typical length. Upland cotton, having fibers 36/32 (1⅛ inches; 29 mm) or longer, is
classed as long staple. Length uniformity is the ratio between the mean length
and the upper half mean length of the fibers and is expressed as a percentage. If
all fibers were the same length, the uniformity index would be 100. Samples

with very high uniformity have a uniformity index of 85 or greater. Other grades are high (83–85), intermediate (80–82), low (77–79), and very low (below 77).

Fiber fineness is often recorded as micronaire values,[54] which approximate the weight of an inch of fiber in micrograms. It actually measures the airflow at a standard pressure through a comparable weight and volume of fiber.

Quality factors that determine grade include color, leaf and other foreign matter, ginning preparation, and maturity. These values are estimated by visual comparison with standard samples. Character, including strength, fineness, pliability, and uniformity of length, is determined by sight and touch. Instruments for determining length, strength, fineness, and structure are used, mostly in research studies of the fiber. Recognized color classes are extra white, white, blue-stained, gray, spotted, yellow-tinged, light-stained, and yellow-stained. Within these classes are thirty-two grades for quality. For American upland white cotton, the numerical grades are as follows:

No. 1 or Middling Fair

No. 2 or Strict Good Middling

No. 3 or Good Middling

No. 4 or Strict Middling

No. 5 or Middling

No. 6 or Strict Low Middling

No. 7 or Low Middling

No. 8 or Strict Good Ordinary

No. 9 or Good Ordinary

Grades 1 to 7 are deliverable on future contracts.[14, 55] Grade 5, Middling, is the basic grade for commercial market. Premiums are allowed for higher grades, while discounts are applied for grades lower than Middling. Grades are also established for American Pima and Sea Island cotton and for linters.

When growers take cotton to the gin, they may sell it directly. However, they usually have it custom-ginned and baled, then stored at a warehouse, or hauled home or to a buyer. The seed may be sold to the gin, hauled home, or hauled to an oil mill. The identity of the cotton of each grower is maintained at the gin and at the warehouse where the bales are stored. Prices offered the grower are based upon the current future price for Middling cotton of 1 inch staple at a terminal market, plus any premium for quality above that base (or less any discount for lower quality), less costs of handling, shipping, and the margin of the buyer.[56] A typical cotton warehouse is a large, one-story, open-sided building with the bales set on end in a single layer.

■ 33.13 DISEASES

33.13.1 Seed and Seedling Diseases

Seedling blights, and damping-off diseases, are caused by species of *Rhizoctonia*, *Fusarium*, and *Pythium*. They are partly controlled by treating seeds with a fungicide. Fungicides should be mixed with the seed before planting, or they can be applied to the seed in the seed hoppers of the planter. A spray or granule fungicide may also be mixed with the soil that covers the row during the planting operation.

33.13.2 Root Rot

Root rot, caused by the fungus *Phymatotrichum omnivorum*, is a destructive cotton disease, especially on calcareous soils of the Southwest as well as on the black waxy soils of Texas. Affected leaves and roots are yellow or bronze, and the plants wilt and die. The fungus survives in the soil on dead roots of cotton or other taprooted plants, or in a dormant sclerotial stage. Diseased plants occur in irregular areas in the field. The fungus attacks most dicotyledonous trees, shrubs, and herbs but does not attack plants of the grass family.

The severity of root rot disease is reduced by rotation with grain or grass crops for two or three years, combined with deep tillage after these crops are harvested. Effective control was obtained in Arizona by turning under large amounts of organic matter in rows and then planting over the row.[57, 58]

33.13.3 Fusarium Wilt

Fusarium wilt, caused by the fungus *Fusarium oxysporium* f. sp. *vasinfectum*, is a serious problem on light sandy soils from the Gulf Coast plain to New Mexico. Infected plants are stunted early in the season, and the leaves turn yellow along the margins and between the veins. The cut stem of a wilted plant shows brown or black vascular tissues inside. The disease can be controlled by the use of resistant varieties and adequate amounts of potash fertilizer. Several varieties are resistant to the fusarium wilt as well as to the nematode wilt complex (Figure 33.15).

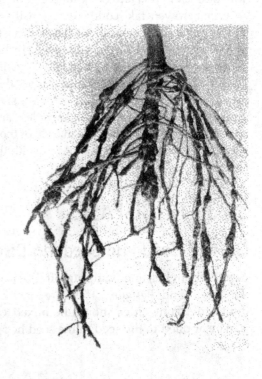

FIGURE 33.15
Root knot of the cotton plant.

33.13.4 Verticillium Wilt

Verticillium wilt is a disease caused by a soil-inhabiting fungus *(Verticillium alboatrum)* that attacks the nuts of the cotton plant. It causes wilting, mottling, shedding of the leaves, and vascular discoloration of the roots and stems. Some varieties are resistant to the disease, while others are somewhat tolerant.

33.13.5 Bronze Wilt

Bronze wilt is a relatively new disease of cotton. It first appeared in Mississippi and Louisiana during the hot summer of 1995. Since then, it has spread through most of the cotton growing region of the United States. The cause of the disease is not certain, but there is evidence to show that a unique strain of *Agrobacterium tumefaciens* may be involved. Symptoms are a bronze tint in leaves, together with wilting. Symptoms usually occur during fruit development and become more severe as bolls mature. The disease is more prevalent when temperatures are warmer than normal during the growing season, with accompanying water stress. Some varieties are resistant to the disease. Early planting, irrigation to prevent water stress, moderate use of nitrogen fertilizer, and prevention of other nutrient deficiencies reduce damage to susceptible varieties.[59]

33.13.6 Nematodes

The root-knot, reniform, stubby-root, sting, lance, and meadow nematode species are very damaging to cotton. Resistant varieties are available only for root-knot nematodes. Crop rotation is the best method to prevent the buildup of nematodes in the soil. Cover crops of rye or wheat can suppress reniform and Columbia lance nematodes but can be hosts for root-knot, sting, and stubby-root nematodes. Plowing out roots of cotton and other crops soon after harvest reduces nematode carryover. Nematodes may be partly controlled by application of insecticide to the soil during planting.[60]

33.13.7 Other Diseases

Other diseases of cotton and their causal organisms are leaf blights, *Ascochyta gossypii,* and bacterial blight, or angular leaf spot or black arm, *Xanthomonas malvacearum* (Figure 33.16). These organisms are seed-borne, but they also live over winter on cotton residues. Crop rotation, trash disposal, and seed treatment reduce these and some other leaf diseases. Empire cottons are somewhat tolerant to *Ascochyta* blight, and some varieties are resistant to certain races of the bacterial blight organism. The fungus, *Glomerella gossypii,* causes anthracnose or sore shin.

Common soil-borne fungi that include species of *Aspergillus, Fusarium, Nigrospora,* and *Rhizopus* cause cotton boll rots. Topping and defoliation keep the plants drier and reduce boll rotting. Types with "okra" leaves or "frego" boll bracts also have less boll rots. Bolls that are attacked by bollworms are often infected with boll rots.

Several different fungi cause fiber deterioration in damp cotton. A rust fungus, *Puccinia stakmanii,* also attacks the crop.

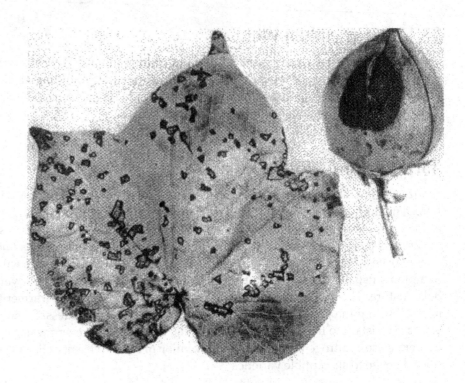

FIGURE 33.16
Angular leaf spot on leaf
and boll.

Nonparasitic disorders of cotton include crazy top, caused by irregular irrigation; a potash deficiency sometimes called rust; and crinkle leaf, which is a manganese toxicity that occurs on acid soils deficient in calcium.

■ 33.14 INSECT PESTS

The most serious insect pests of cotton are the boll weevil (*Anthonomus grandis*),
tobacco budworm (*Heliothis virescens*), pink bollworm (*Pectinophora gossypiella*),
cotton leafworm (*Alabama argillaceae*), bollworm (*Heliocoverpa zea*), cotton fleahopper (*Psallus seriatus*), cotton aphid (*Aphis gossypii*), red spider mite
(*Tetranychus bimaculatus*), garden webworm (*Loxostege similalis*), lygus bug or tarnished plant bug (*Lygus pratensis*), cotton stainer (*Dysdercus suturellus*), and thrip
(*Frankliniella spp.*).[32, 61] Other insects that attack the cotton plant include the cotton leaf perforator (*Bucculatrix thurberiella*), cutworm, grasshopper beet armyworm (*Spodoptera exigna*), and fall armyworm (*Spodoptera frugiperda*). At least
fifteen other insects are common pests of cotton in the United States. Cotton
pests of the Eastern Hemisphere but not of the United States include the jassid
(*Empoasca* species), cluster caterpillar (*Prodenia litura*), and spiky bollworm
(*Earias insulana*).

Boll weevil and some of the other insects are controlled by destruction of old
cotton stalks and weeds in and near the cotton fields that harbor the pests.
Plowing cotton stalks more than 6 inches (15 cm) deep immediately after harvest is recommended as a control measure for the pink bollworm and boll weevil. The pink bollworm overwintering in the cottonseeds can be killed with
heat.[62] The spread of the pink bollworm is being held in check by severe quarantine regulations.

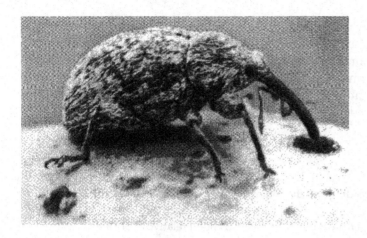

FIGURE 33.17
Boll weevil puncturing cotton boll. (Much enlarged.)

All the important cotton pests can be kept under control by one or more insecticides. Repeated sprayings are necessary to control the pink bollworm. It is the most serious pest in the southwestern states.

33.14.1 Boll Weevil

The boll weevil is probably the most destructive insect, but its depredations are confined to sections that have an annual rainfall of more than 25 inches (740 mm). The adult female weevil punctures the squares or bolls, in which it deposits its eggs (Figure 33.17). This destroys the squares, and the larvae that develop within the boll destroy the lint in all or some of the locks. The bollworm and the pink bollworm also destroy the contents of the boll. The bollworm, also called the corn earworm, attacks corn, sorghum, and other crops. Genetically engineered cotton varieties that contain genes from *Bacillus thuringiensis* (Bt cotton) are resistant to the tobacco budworm and the pink bollworm and show some resistance to the bollworm.[40] Although still a problem in other countries, a program to eradicate the boll weevil has been successful in the United States, where it has been a major pest.

33.14.2 Thrips

Thrips are tiny insects that migrate into cotton from wild hosts as cotton seedlings are just emerging. They have rasping mouthparts with which they rupture cells and imbibe the released contents. Symptoms are ragged, split leaves that may also be cupped upward. Thrips are most damaging during cool weather when plants are not actively growing. Application of a systemic insecticide at planting is recommended, but additional applications of insecticide may be needed with heavy infestations.[63]

33.14.3 Other Pests

The leaf worm and webworm devour the leaves and thus retard boll development. Aphids, red spiders, and fleahoppers, all sucking insects, check the development of the plants, bolls, and lint. Aphids cause curling and shedding of leaves, and they also secrete honey dew, which may contaminate the lint of opened bolls. Red spiders, which may have seventeen generations in a year, feed on the under side of the leaves, causing them to redden and drop from the plant. The fleahopper feeds on the juices of the terminal buds, small squares, and other young, tender plant parts. The injured squares may shed while they are too small to be observed. The leaves become deformed.

REFERENCES

1. Ware, J. O. "Plant breeding and the cotton industry," in *USDA Yearbook*, 1936, pp. 657–744.
2. Webber, J. M. "Cytogenetic notes on cotton and cotton relatives," *Science* 80(1934):268–269.
3. Bennett, C. A. "Ginning cotton," *USDA Farmers Bull.* 1748 (rev.), 1956.
4. Smith, C. W., and J. T. Cothren. *Cotton: Origin, History, Technology, and Production.* New York: John Wiley and Sons, 1999.
5. Doyle, C. B. "Climate and cotton," in *Climate and Man*, USDA Yearbook, 1941, pp. 348–363.
6. Woolf, W. E., and others. "An economic analysis of irrigation, fertilization and seeding rates for cotton in the Macon Ridge area of Louisiana," *LA Agr. Exp. Sta. Bull.* 620, 1967, pp. 1–32.
7. King, C. J. "Effects of stress conditions on the cotton plant in Arizona," *USDA Tech. Bull.* 392, 1933, pp. 1–35.
8. Fulton, H. J. "Weather in relation to yield of American-Egyptian cotton in Arizona," *J. Am. Soc. Agron.* 31(1939):737–740.
9. Eaton, F. M., and N. E. Rigler. "Effect of light intensity, nitrogen supply, and fruiting on carbohydrate utilization of the cotton plant," *Pl. Physiol.* 20, 3(1945):380–411.
10. Tharp, W. H. "The cotton plant—how it grows and why its growth varies," *USDA Handbook* 178, 1960.
11. Neely, J. W. "The effect of genetical factors, seasonal differences, and soil variations upon certain characteristics of upland cotton in the Yazoo-Mississippi Delta," *MS Agr. Exp. Sta. Tech. Bull.* 28, 1940.
12. Ludwig, C. A. "Some factors concerning earliness in cotton," *J. Agr. Res.* 43(1931):637–657.
13. Eaton, F. M., and D. R. Ergle. "Mineral nutrition of the cotton plant," *Pl. Physiol.* 32, 3(1957):169–175.
14. Anonymous. "Better cottons," USDA Progress Report. 1947.
15. Dennis, R. E., and R. E. Briggs. "Growth and development of the cotton plant in Arizona," *AZ. Agr. Exp. Sta. Bull.* A-64, 1969, pp. 1–21.
16. Hancock, N. I. "Relative growth rate of the main stem of the cotton plant and its relationship to yield," *J. Am. Soc. Agron.* 33(1941):590–602.
17. Silvertooth, J. C., and E. R. Norton. "Cotton monitoring and management system," *Univ. AZ Coop. Ext. Serv.* AZ1049, (rev.), 1999.
18. McNamara, H. C., D. R. Hooton, and D. D. Porter. "Differential growth rates in cotton varieties and their response to seasonal conditions at Greenville, Tex.," *USDA Tech. Bull.* 710, 1940, pp. 1–44.
19. Gore, U. R. "Morphogenetic studies on the influence of cotton," *Bot. Gaz.* 97(1935):118–138.
20. Edminsten, K. "The cotton plant," in *North Carolina Cotton Production Guide. NC State Univ. Coop. Ext. Serv.* AG-417, (rev.), 2000.

21. Monks, C. D., C. H. Burmester, M. G. Patterson, and D. Delaney. "Monitoring cotton growth and development in Alabama," *AL Coop. Ext. Syst.* ANR-995, (rev.), 1996.

22. Hall, W. C. "Physiology and biochemistry of abscission in the cotton plant," *TX Agr. Exp. Sta.* MP-285, 1958.

23. Eaton, F. M. "Physiology of the cotton plant," *Ann. Rev. Pl. Physiol.* 6(1955):299–328.

24. Stephens, S. G., and M. D. Finkner. "Natural crossing in cotton," *Econ. Bot. 7*, 3(1953):257–269.

25. Harrison, G. J. "Metaxenia in cotton," *J. Agr. Res.* 42(1931):521–544.

26. Rhyne, C. L., F. H. Smith, and P. A. Miller. "The glandless leaf phenotype in cotton and its association with low gossypol content in the seed," *Agron. J.* 51(1959):148–152.

27. Hawkins, R. S. "Field experiments with cotton," *Az. Agr. Exp. Sta. Bull.* 135, 1930.

28. Bailey, A. E. *Cottonseed and Cottonseed Products.* New York: Interscience, 1948, pp. 1–960.

29. Smith, C. E., Jr. "The new world centers of origin of cultivated plants and the archaeological evidence," *Econ. Bot.* 22, 3(1968):253–266.

30. Franks, G. N., and J. C. Oglesbee, Jr. "Handling planting-seed at cotton gins," *USDA Production Res. Rpt.* 7, 1957.

31. Tharp, W. H. "Cotton growing in the U.S.A.," *World Crops*, April 1957:139–143.

32. Shaw, C. S., and C. N. Franks. "Cottonseed drying and storage at cotton gills," *USDA Tech. Bull.* 1262, 1962.

33. Thurmond, R. V., J. Box, and F. C. Elliott. "Texas guide for growing irrigated cotton," *TX Agr. Exp. Sta. Bull.* B-896.

34. Moore, G. C., J. H. Benedict, T. W. Fuchs, R. D. Friesen, and C. Payne. "Bt technology in Texas: A practical view," *TX Agri. Ext. Serv.* L-5169, 1999.

35. Brown, S. M., and others. "Cotton production guide," *Univ. GA Coop. Ext. Serv.* Rev. 2000.

36. Hodges, S. C. "Fertilization," in *North Carolina Cotton Production Guide. NC State Univ. Coop. Ext. Serv.* AG-417, (rev.), 2000.

37. Hickey, M. G., C. R. Stichler, and S. D. Livingston. "Using petiole analysis for nitrogen management in cotton," *TX Agr. Ext. Serv.* L-5156, 1996.

38. Simpson, D. M. "Factors affecting the longevity of cottonseed," *J. Agr. Res.* 64(1942):407–419.

39. Monks, C. D., and M. G. Patterson. "Conservation tillage cotton production guide," *AL Coop. Ext. Syst.* ANR-952, (rev.), 1996.

40. Naderman, G. C., K. L. Edmisten, A. C. York, and J. S. Bacheler. "Cotton production with conservation tillage," in *North Carolina Cotton Production Guide. NC State Univ. Coop. Ext. Serv.* AG-417, (rev.), 2000.

41. Surgeon, W. I., and F. T. Cooke, Jr. "Cost reduction research for cotton production systems in the Yazoo-Mississippi delta," *MS Agr. Exp. Sta. Bull.* 783, 1971, pp. 1–12.

42. McCarty, W. H., A. Blaine, and J. D. Byrd, Jr. "Cotton no-till production," *MS St. Univ. Ext. Serv.* 1695, 1998.

43. Edminsten, K. "Planting decisions," in *North Carolina Cotton Production Guide. NC State Univ. Coop. Ext. Serv.* AG-417, (rev.), 2000.

44. Peterson, G. D., Jr., R. L. Cowan, and P. H. Van Schaik. "Cotton production in the lower desert valleys of California," *CA Agr. Exp. Sta. Cir.* 508, 1962.

45. Montgomery, R. A., and O. B. Wooten. "Lint quality and moisture relationships in cotton through harvesting and ginning," *USDA Agricultural Research Service Spec. Rept.* ARS 42-14, 1958.

46. Tavernetti, J. R., and H. F. Miller, Jr. "Studies on mechanization of cotton fanning in California," *CA Agr. Exp. Sta. Bull.* 747, 1954.

47. Humphries, R. T., J. M. Green, and E. S. Oswalt. "Mechanizing cotton for low cost production," *OK Agr. Exp. Sta. Bull.* B-382, 1952.

48. Tharp, W. H., and others. "Effect of cotton defoliation on yield and quality," *USDA Prod. Res. Rpt.* 46, 1961.

49. McCarty, W. H., C. Snipes, and D. B. Reynolds. "Cotton defoliation," *MS St. Univ. Ext. Serv.* IS529, 1998.

50. Patterson, M. G., and C. D. Monks. "Cotton defoliation," *AL Coop. Ext. Syst.* ANR-715, (rev.), 1996.

51. Carter, L. M., R. F. Colwick, and J. R. Tavernetti. "Topping cotton to prevent lodging and mechanical harvesting," *USDA Agric. Res. Service Spec. Rept.* ARS 42-67, 1962.

52. Duggan, I. W., and P. W. Chapman. "Round the world with cotton," *USDA Adjustment Adm. Publ.*, 1941, pp. 1–148.

53. Ward, J. M., and J. W. Graves. "Effects of ginning of cotton on fiber quality as reflected in certain fiber properties," *TX Agr. Exp. Sta. Bull.* 1039, 1965, pp. 1–8.

54. Anonymous. "Cotton fiber and spinning tests," Report of National Cotton Council, Memphis, Tenn. 1956.

55. Anonymous. "The classification of cotton," *USDA Misc. Pub.* 310, 1938, pp. 1–54.

56. Anonymous. *Agricultural Statistics, 1962, USDA Ann. Publ.*, 1963, pp. 1–741.

57. King, C. J. "A method for the control of cotton root rot in the irrigated Southwest," *USDA Cir.* 425, 1937, pp. 1–9.

58. Presley, J. T. "Cotton diseases and methods of control," *USDA Farmers Bull.* 1745, 1954.

59. Baumann, P. A. "Suggestions for weed control in cotton," *TX Agri. Ext. Serv.* B5039, 1998.

60. Koenning, S. "Disease management in cotton," in *North Carolina Cotton Production Guide. NC State Univ. Coop. Ext. Serv.* AG-417, (rev.), 2000.

61. Reynolds, E. B., and D. T. Killough. "The effect of fertilizers and rainfall on the length of cotton fibers," *J. Am. Soc. Agron.* 25(1933):756–764.

62. Anonymous. "Controlling the pink bollworm on cotton," *USDA Farmers Bull.* 2207, 1965, pp. 1–12.

63. Bacheler, J. S. "Managing insects on cotton," in *North Carolina Cotton Production Guide. NC State Univ. Coop. Ext. Serv.* AG-417, (rev.), 2000.

Tobacco

"By Hercules! I do hold it and will affirm it, before any prince in Europe, to be the most sovereign and precious weed that ever the earth tendered to the use of man."

"By Gad's me!" rejoins Cob, "I mar'l what pleasure a felicity they lease in taking this roguish tobacco. It is good for nothing but to choke a man and fill him full of smoke and embers."

—Ben Johnson, *Every Man in His Humour*, 1593.

Air-cured
Burley
Cased leaves
Cigar binder
Cigar filler
Cigar wrapper
Fire-cured
Flue-cured
Perique
Priming
Stalk cutting
Topping

34.1 ECONOMIC IMPORTANCE

In 2000–2003, tobacco leaf production averaged about 7 million tons (6.3 million MT). It was grown on an average of 9.9 million acres (3.9 million ha) with an average yield of 1,400 pounds per acre (1,600 kg/ha). The countries leading in tobacco production are China, Brazil, India, the United States, and Turkey (Figure 34.1). Production in the United States (Figure 34.2) in 2000–2003 averaged about 465,000 tons (423,000 MT) with an average acre yield of 2,100 pounds (2,400 kg/ha) on 435,000 acres (176,000 ha). The leading states in tobacco production are North Carolina, Kentucky, Tennessee, South Carolina, and Georgia. U.S. tobacco production has declined since 1975 due to increased concerns about the effects of tobacco on public health.

34.2 HISTORY OF TOBACCO CULTURE

Tobacco *(Nicotiana tobacum)* is a native of the Western Hemisphere from Mexico southward. Two sailors saw a Native American smoking a cigarette with a corn husk or palm leaf wrapping on San Salvador on November 6, 1492. The Native Americans of Yucatan were cultivating tobacco when the Spaniards first arrived there in 1519.[1,2] It was grown in Spain and France by 1561 and in England in 1565, probably twenty years before Sir Walter Raleigh popularized it.[3]

Tobacco culture was begun by John Rolfe at Jamestown, Virginia, in 1612, and fifteen years later 500,000 pounds (230,000 kg) of tobacco were shipped to England. Tobacco culture began in southern Maryland in 1635, and tobacco continues to be the chief cash crop of that region.

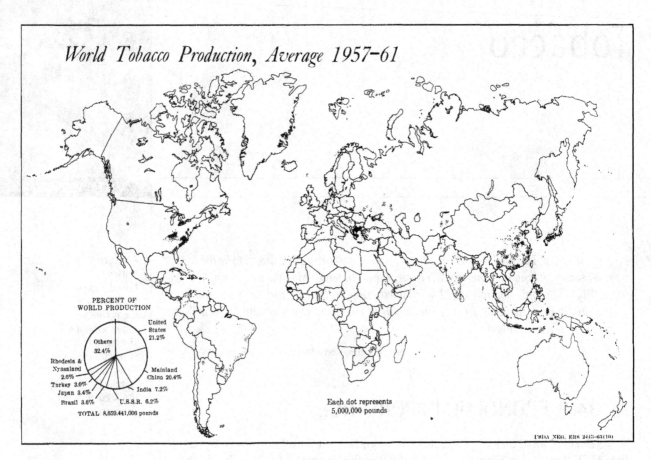

FIGURE 34.1

World tobacco production. [Courtesy USDA]

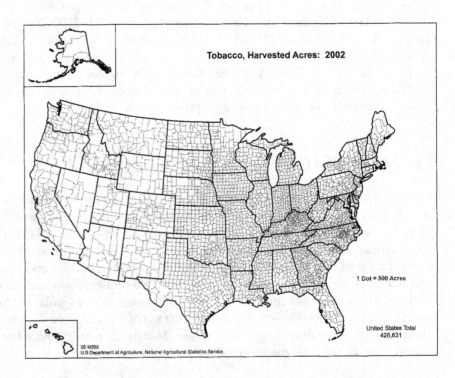

FIGURE 34.2

Acreage of tobacco in the United States in 2002. [Source: 2002 Census of U.S. Department of Agriculture]

In parts of Mexico, and in the United States from the Mississippi River eastward and northward to Canada, Native Americans were growing another species, *Nicotiana rustica*. The word *tobacco* comes from a Native American word for pipe. Three other tobacco species formerly grown by Native Americans in the United States[1] are *N. attenuata*, grown in the Southwest, in most of the plains area, and along the northern Pacific coast; *N. quadrivalvis*, grown in the Northwest; and *N. trigonophylla*, grown by the Yuma tribe of the Southwest.

▩ 34.3 ADAPTATION

Tobacco can be grown under a wide range of climatic and soil conditions, but the commercial value of the product depends largely on the environment under which it is produced. In the United States the culture of each important commercial type of tobacco is highly localized, due primarily to the influence of the climate and soil on the quality of the leaf.

Tobacco is grown from central Sweden at latitude 60° N southwest to Australia and New Zealand at latitude 40° S. The optimum temperature for germination of tobacco seed is about 88°F (31°C).[4] During the six to ten weeks that tobacco seedlings are in a cold frame or hotbed, the optimum temperature for growth is about 75 to 80°F (24 to 27°C), the minimum about 50°F (10°C), and the maximum 95°F (35°C) or higher. A very heavy growth of tobacco is undesirable because of its consequent poor quality. Tobacco in the field grows most rapidly at about 80°F (27°C). It will mature at this temperature in about seventy to eighty days. In southern Wisconsin and the Connecticut Valley tobacco regions, where the mean summer temperature is about 70°F (21°C), a frost-free period of about 100 to 120 days from transplanting to maturity is required to mature the crop. In the central tobacco regions, the average temperature during the growing season is about 75°F (24°C), while in northern Florida it is about 77°F (25°C).[5] Flue-cured tobacco thrives best where the daytime temperature is 70 to 90°F (21 to 32°C).[6]

A low rainfall results in tobacco high in nitrogen, nicotine, ether extract, acids, and calcium, but low in potash and soluble carbohydrates.[7] Most of the tobacco crop in America is grown where the annual precipitation is 40 to 45 inches (1,000 to 1,150 mm). An autumn humidity that permits the harvested leaves to cure at the proper rate is important, yet the humidity should be sufficiently high at times during the year to make the leaves pliable enough to be handled without breakage. High winds are damaging to tobacco leaves in the field. Tobacco is unsuited to semiarid regions not only for this reason, but also because of the low humidity, low summer rainfall, and alkaline soils high in nitrogen.

Tobacco does the best on a well-drained soil. Each type of tobacco has its special soil requirements. Most tobacco production is on sandy loam or silt loam soils, but soils are sandy in Georgia[8] and Florida, clay in Virginia, and clay loam in the Miami Valley of Ohio. Fire-cured tobaccos are grown on heavy loam soils not adapted to other types. In some tobacco areas, such as Lancaster County in Pennsylvania and the Miami Valley of Ohio, the soils are highly productive loams[9] of limestone origin, but in most of the tobacco region of the Piedmont and coastal plain of the South Atlantic states, the soils are low in both organic matter and mineral nutrients. Heavy soils and fertile soils tend to produce tobacco that is high in nitrogen, nicotine, and calcium but low in potash and carbohydrates. Light sandy soils have an opposite effect.[10]

Since the high returns per acre from good tobacco justify good, heavy fertilization, natural fertility is usually a secondary consideration. The proper relative proportions of calcium, magnesium, and potassium, which are essential to good burning qualities in tobacco, especially cigar tobacco, are usually attained by fertilizer application. A somewhat acidic soil (pH 5.8 to 6.5) is best for tobacco.[11] Excessive acidity (below pH 4.5 to 5) is often detrimental to quality or yield because it retards absorption of calcium, magnesium, and phosphorus and leads to excessive absorption of manganese and aluminum. If lime is needed to correct soil acidity, it should be applied two or three years ahead of the crop.[12] A soil near the neutral point favors development of the black root rot disease. Tobacco is unsuited to soils in which the water table is less than 3 feet (90 cm) from the surface.

■ 34.4 BOTANICAL DESCRIPTION

Tobacco belongs to the family *Solanaceae*, the nightshade family. *Nicotiana tobacum*, the common cultivated tobacco, is an amphidiploid, assumed to have originated from a natural cross between *N. sylvestris* and some other native species. The plants are 4 to 6 feet (120 to 180 cm) in height. They are terminated by a raceme of up to 150 or more funnel-shaped flowers (Figure 34.3). The terminal flowers are first to open. The five petals are pink, but they may be white or red in some varieties. The tobacco fruit is a capsule that splits into two or four parts at maturity. It may contain 4,000 to 8,000 seeds. A single tobacco plant sometimes produces 1 million seeds. The seeds are extremely small, about 5 million per pound (1,100/g). The plants bear 12 to 25 leaves that range up to 2 feet (60 cm) in length and are covered with sticky hairs.

Flowers of cultivated tobacco are normally self-fertilized, although occasional out-crosses occur due to wind-borne or insect-borne pollen (Figure 4.3).

FIGURE 34.3
Tobacco flowers. [Courtesy
Richard Waldren]

Of the two species of tobacco cultivated for their leaves, *Nicotiana tobacum*, which is native to Mexico and Central America, is grown in the United States as well as in most other tobacco-producing countries. The species native to the United States, *N. rustica*, is grown in India and certain other foreign countries. However, in the United States it is grown only sparingly or experimentally for making insecticides because of its high content of nicotine.[13] *N. rustica* has pale yellowish flowers and a short, thick corolla tube with rounded lobes. It is characterized further by thick, broadly ovate leaves covered with sticky hairs and by a distinct naked petiole or leaf stalk. It grows to a height of only 2 to 4 feet (60 to 120 cm).

More than 50 species of *Nicotiana* exist, mostly in the Western Hemisphere and Australia. Of these species, *alata*, *glauca*, *sylvestris*, and *sanderae* are grown as ornamentals, while *rustica*, *longiflora*, *debneyi*, *glutinosa*, and *megalosiphon* have been used in breeding varieties of tobacco that resist diseases and nematodes.[14]

The leaves of *N. tobacum* are arranged in a spiral on the stalk.[15] A leaf of flue-cured tobacco averages about 21 inches (53 cm) in length and 11 inches (28 cm) in width. The lamina (blade between veins) of the leaf is about 0.01 inch (0.3 mm) thick. The average weight of a green leaf is about 1½ ounces (42 g), while that of a cured leaf is about ⅕ ounce (6 g); making 72 leaves per pound (160/kg). Cell division ceases when a leaf reaches about one-sixth its normal size. All subsequent growth thereafter is due to cell enlargement and thickening of cell walls together with increases in dry weight of the cell contents. During curing a leaf shrinks about 9 percent in length, 15 percent in width, and 24 percent in area.[16, 17]

The upper leaves on the plant tend to be high in nitrogen, nicotine, and acid but low in ash, calcium, and magnesium. The potash content shows a less definite decrease from top to bottom.[18, 19, 20]

◼ 34.5 CLASSIFICATION OF TOBACCO

The type and quality of tobacco produced depend on the location where it is grown. Also, various tobacco products require leaves of different characteristics. Therefore, a standard system of classification by the U.S. Department of Agriculture (USDA) designates six major classes of U.S. tobacco (Table 34.1). The locations where most types are grown are shown in Figure 34.4.

The first three classes are named on the basis of the method used in curing. The last three classes are all cigar leaf and designate their traditional use in cigars. Most cigar tobaccos are air-cured. Some grades from any type may go into loose-leaf chewing.

TABLE 34.1 Tobacco Classes in the United States

Type	Class No.
Flue-cured	1
Fire-cured	2
Air-cured	3
Cigar filler	4
Cigar binder	5
Cigar wrapper	6

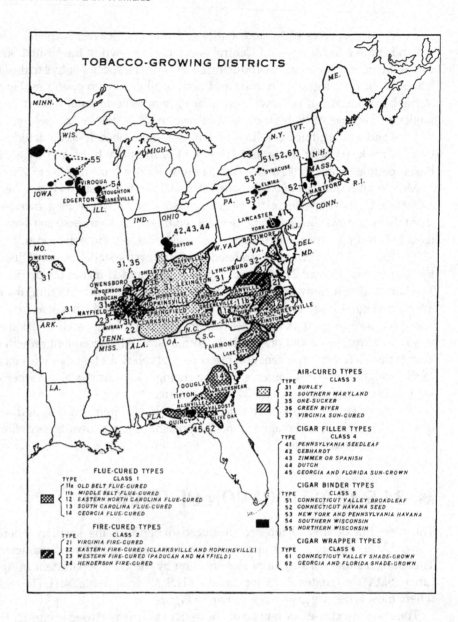

FIGURE 34.4

Types of tobacco, with the districts in which they are grown.

Each class comprises two or more different types. Individual types of flue-cured tobacco are no longer easily identified, and the type designation usually refers only to a marketing area. Each class is further divided into grades. These grades are related to stalk position, quality, color, and other characteristics of the leaf.

34.5.1 Flue-Cured (Class 1)

About 51 percent of all tobacco produced in the United States in 1999 was flue-cured. The name originates from the flues (metal, brick, or stone) of the heaters originally used in curing barns. The tobacco is yellow to reddish-orange and has a thin to medium body and a mild flavor. About 95 percent of flue-cured tobacco is used in cigarettes. Sometimes it is used in smoking and chewing tobacco. Flue-cured tobacco is grown primarily in North Carolina, South Carolina, Virginia, Florida, and Alabama.

34.5.2 Fire-Cured (Class 2)

The tobacco is light to dark brown and has a medium to heavy body and a strong flavor. The name originates from the smoky flavor and aroma received from drying it over open fires in curing barns. Fire-cured tobacco is used for roll and plug chewing tobacco, strong cigars, and heavy smoking tobacco. Kentucky, Tennessee, and Virginia grow fire-cured tobacco.

34.5.3 Air-Cured (Class 3)

Air-cured tobacco is further classified as light or dark. About 97 percent of light air-cured production is burley, but Maryland types are also included. Burley tobacco accounted for 43 percent of all tobacco produced in the United States in 1999. Burley is normally cured without supplementary heat. It is the most widely grown tobacco across the United States but production is second behind flue-cured. Over 90 percent of burley is used in cigarettes. Its flavor and aroma make it popular for cigarette blends. It is also used for pipe tobacco and plug and twist chewing. Most burley is produced in Kentucky and Tennessee, but many other states also produce it. Maryland tobacco is usually considered to have ideal burning qualities for use in cigarette blends. Almost all of it is used in cigarettes.

Dark air-cured tobacco (class 3) is light to medium brown and has a medium to heavy body. It is used mostly in chewing tobacco and snuff, and, to some extent, in smoking tobacco and cigars. Kentucky, Tennessee, and Virginia grow dark air-cured tobacco.

34.5.4 Cigar Filler (Class 4)

This tobacco has a medium to heavy body and is mainly used in the core of cigars. The main factors are flavor, aroma, and burning quality.

34.5.5 Cigar Binder (Class 5)

This tobacco was originally used for binding bunched filler tobacco into the form and shape of the cigar. However, practically all cigars now use a reconstituted tobacco sheet for the inner binder. Natural leaf binders must have good burning quality, aroma, and elasticity. Today, most cigar binder tobacco is used in loose-leaf chewing tobacco.

34.5.6 Cigar Wrapper (Class 6)

This tobacco is used primarily for the outside cover on cigars. Leaves must be elastic, free of injury, and uniform in color and must have good burning qualities. They should also be very thin, smooth, and of fine quality. To produce such leaves, producers must protect the leaves against the sun and weather extremes. Wrapper tobacco may also be used as binder and filler in cigars.

■ 34.6 FERTILIZERS

Fertilizers are applied to all tobacco fields in the United States. A ton of flue-cured tobacco leaves contains about 32 pounds (16 kg/MT) of nitrogen, 8 pounds (4 kg/MT) of P_2O_5, 40 pounds (20 kg/MT) of K_2O, 50 pounds (25 kg/MT) of calcium, 52 pounds (26 kg/MT) of magnesium, and 12 pounds (6 kg/MT) of sulfur. A liberal amount of potash in tobacco leaves is essential to good burning quality. Nitrogen is necessary for good growth of tobacco, but an overabundance of nitrogen results in a low-quality[2] leaf. Excessive phosphorus shortens the leaf burn duration.[21] Magnesium is essential to prevent sand drown, a nutritional disorder caused by a deficiency of that element in some soils. Magnesium and calcium can be supplied in dolomitic limestone.[22] A 2,000-pound crop of tobacco leaf requires 24 to 35 pounds per acre (12 to 18 kg/MT) of water-soluble magnesium, with five times as much available calcium in the soil.

The amount of fertilizer needed depends somewhat on the type of tobacco being produced. Flue-cured tobacco needs the least amount of fertilizer and burley the highest. Exact amounts depend on the fertility of the soil.

An application of 8 to 10 tons per acre (18 to 22 MT/ha) of well-rotted manure is recommended for burley tobacco in Ohio.[12]

Some tobacco-transplanting machines are equipped to apply fertilizer in double bands 3 or 4 inches (76 to 100 mm) to each side of the row and below the level of the roots of the tobacco plants as they are being planted. When the rate of application exceeds 750 pounds of fertilizer per acre, the remainder should be applied as a side dressing or the entire amount should be broadcast before planting.

■ 34.7 ROTATIONS

The chief consideration in determining a crop rotation system for growing tobacco is the effect of the previous crop on the quality and yield of tobacco and on the incidence of tobacco diseases.[9, 23, 24] Most adapted crops succeed as well or better after tobacco than they do following other intertilled crops such as corn. Consequently, conditions that favor the high-priced tobacco crop obviously should govern the rotation. In Colonial times, the best tobacco crops were often obtained on cleared forest land or on idle land that had gone back to weeds, grass, and brush. In southern Maryland, tobacco following a natural two-year weed fallow gave better returns than tobacco following either bare fallow or legume crops in rotations.[25] In southern Maryland experiments, tobacco yielded well after red clover in a tobacco–wheat–red clover three-year rotation; it also yielded well when continuously cropped, with hairy vetch as a winter cover crop, but the tobacco quality was poorer than that of tobacco following ragweed and horseweed.

Flue-cured tobacco that follows corn, cotton, and small grain has excellent leaf quality.[26] When grown after such legumes as cowpeas, crotalaria, lespedeza,

peanuts, soybean, and velvetbean, the tobacco ranges from poor to good. For tobacco following legumes, an appreciable increase in quality is obtained when nitrogen fertilization is reduced or heavy applications of potash are made to balance
the deleterious effect of the excess nitrogen left over from legumes.

In Pennsylvania, cigar-binder tobacco is grown mostly by continuous cropping, whereas cigar-filler tobacco follows a legume in three-year or four-year rotations that include wheat or wheat and corn. A five-year rotation of tobacco,
wheat, two years of grass, and corn is also practiced. In Kentucky, Ohio, and Virginia, tobacco usually follows a legume or grass.[6, 27, 28] In Connecticut, much of
the tobacco is grown in alternation with fall-sown grain, usually rye, as a cover
crop, or after a grass meadow crop, because tobacco following a legume has a
dark leaf.[11]

■ 34.8 TOBACCO CULTURE

34.8.1 Seeding Methods

Transplanted tobacco seedlings are used to establish most tobacco fields. Tobacco
seed is so tiny (5 million per pound or 1,100/g) that conventional seeding equipment cannot accurately meter it and establish uniform stands. Tobacco seedlings
are transplanted to the field when they are about 6 to 10 inches (15 to 20 cm) tall.
There are several methods commonly used to produce tobacco seedlings.

FLOAT SYSTEM The float system is becoming increasingly popular in countries
where land and labor costs are high. It is now the most common method for
seedling production in the United States. Over 70 percent of seedlings in North
Carolina were produced using the float system in 1996.[29]

Tobacco seedlings are grown in polystyrene trays filled with a soilless growing
medium, usually a mixture of vermiculite, peat, and perlite. There are about 200 to
400 cells per tray. The trays are placed in heated greenhouses or covered outdoor
water beds. The trays float on a water solution with plant nutrients added. Urea-
based nitrogen should be limited to reduce the risk of nitrite toxicity to the
seedlings. A big advantage to the float system is that the seedlings are transplanted
with their root systems intact. Seedlings tend to be more uniform and the plants
mature more uniformly.[29]

BEDS OR COLD FRAMES Tobacco seed is planted in beds or cold frames 3 to 9 feet (90
to 275 cm) wide and as long as needed to provide sufficient plants for planting the
field.[30] The minimum area of bed in square yards necessary to insure ample plants
for an acre of tobacco land is about as follows: for cigar types, 33 (68 m^2/ha); for
fire-cured, dark air-cured, and perique, 50 (103 m^2/ha); for flue-cured, 67
(138 m^2/ha); for Maryland and aromatic, 100 (220 m^2/ha); and for burley, 100
(220 m^2/ha).

Clean land for the beds is plowed or disked and then sterilized with wood
fires, steam, or chemical disinfectants. Since the use of methyl bromide is being
phased out because it destroys the ozone layer of the atmosphere, alternatives
are being sought for disinfecting transplanting beds. A mixture of metam sodium
and 1,3-dichloropropene has become a good alternative to methyl bromide.[29]

Insecticides also may be applied to the bed. Combinations of these soil treatments can kill soil-borne fungi, nematodes, weed seeds, and insects.

Beds for flue-cured tobacco should be fertilized at the following rates per 100 square feet: 3.5 pounds nitrogen, 1.75 pounds phosphate, 1.75 pounds potash, and 1.33 pounds MgO (1.6 kg N/10m^2, 0.8 kg P$_2$O$_5$/10m^2, 0.8 kg K$_2$O/10m^2, and 0.6 kg MgO/10m^2). Rates of nitrogen, phosphate, and potash for burley tobacco should not exceed 5 pounds per 100 square feet (2.3 kg/10m^2) each.[31] Most or all of the nitrogen should be in the nitrate form.[32]

Tobacco seed is usually mixed with sand, sifted wood ashes, land plaster, or other inert material to permit better distribution of the seed in the bed. Seed is scattered over the bed three times to ensure more even distribution. The bed is then rolled to compact the seedbed and press the seeds into the soil. Then the bed is covered over with plastic, cloth, or glass.[30] A cloth is adequate in warm climates. Sometimes cloth is also laid on the soil until the seed germinates. The bed may require occasional watering. A seeding rate that results in one seedling every 2 × 2 inches (10 × 10 cm) of seedbed will result in vigorous seedlings that are easy to transplant.[29]

34.8.2 Transplanting into the Field

Land for tobacco is usually plowed in the spring or fall, at which time the cover crop or weed growth is turned under. The land is worked down into a good tilth before the tobacco is transplanted. Where rainfall is heavy the land may be thrown up into beds. Tobacco is transplanted six to ten weeks after the beds are sown, when the plants have six to eight leaves and are 6 to 8 inches (15 to 20 cm) in height.[33] Transplanting is done in June in the northern districts, in May in the central districts as well as in the mountainous southern sections, in April in South Carolina, and from March 20 to April 15 in Florida and southern Georgia. Transplanting is done by machine or by hand. The plants are watered when they are being set in the field unless the soil is wet.

Distances between rows vary from 32 to 54 inches (80 to 135 cm), depending on local customs. Wider rows permit the use of beds and water furrows.

Plant population depends on the type of tobacco being grown. Dark air-cured and fire-cured tobacco is the lowest at about 3,600 plants per acre (8,900 plants/ha). Flue-cured and burley tobacco will be about 6,100 plants per acre (15,000 plants/ha).[34] Optimum plant population is determined more by leaf quality than by yield.

Tobacco is cultivated at frequent intervals, beginning a week to ten days after transplanting. The high value of the crop justifies intensive cultivation. Herbicides can also be used for weed control. Common preemergent or preplant herbicides used on tobacco fields are pendimethalin, clomazone, pebulate, napropamide, and sulfentrazone. Tobacco plants can effectively compete with weeds six weeks after transplanting.

Cigar-wrapper tobacco is grown under artificial shade (Figure 34.5). The object of shading is to produce smaller, thinner, and smoother leaves with smaller veins more suitable for fine cigar wrappers.[11] Shading increases humidity and reduces air currents, thereby retarding evaporation from the plants and the soil.[35] The shading material, a loosely woven cloth (eight to ten threads to the inch), is spread over and tied to wires that are supported by stout posts extending 8 feet (2.4 m) above

FIGURE 34.5
Shade tobacco in Florida.
Three primings of the lower
level have been harvested.

the ground. In the Connecticut Valley, the posts are placed 33 feet (10 m) apart. Most of the cloth is used for only one year.

Tobacco plants seem to grow very slowly at first because the seeds and consequently the seedlings are extremely small. At transplanting, the dry weight of the seedling is less than 0.05 percent of the final dry weight of the plant.[36] Even twenty-six days after transplanting, less than 2 percent of the total growth has been produced. Growth proceeds rapidly after that stage.

34.8.3 Topping Tobacco Plants

Topping consists of removing the top or crown of the plant at about the third branch below the flower head (Figure 34.6). Tobacco plants are topped to keep them from producing seed. This forces the synthesized carbohydrate and nitrogen materials to remain in the leaves for further growth and enrichment. The top leaves that are removed are highest in nicotine content.[19] Topping results in larger, thicker, and darker leaves that mature earlier and more uniformly than do those on un-topped plants (Figure 34.7).

The height of the plant when topped depends on the type of tobacco being grown. Dark air-cured and fire-cured tobacco plants are topped at eight to fifteen leaves. Early topping prolongs the vegetative period resulting in larger, heavier, darker leaves with higher nicotine content. Flue-cured tobacco plants are topped at about eighteen leaves, which will result in moderate thickness and nicotine content. Burley tobacco plants are topped at about twenty-four leaves, producing lighter leaves and medium to high nicotine content.[34]

After topping, it is necessary to remove suckers every seven to ten days. Suckers are branches that develop in the leaf axils after the tops are removed. Sucker growth can be suppressed by the application of chemicals such as maleic hydrazide, vegetable oils, or fatty acids to the plants immediately after topping.

FIGURE 34.6
Mechanically topping tobacco plants. [Courtesy USDA NRCS]

FIGURE 34.7
Flue-cured tobacco that has been topped.

Maleic hydrazide treatment increases yield,[33] leaf weight, sugar content, and equilibrium moisture content, especially in flue-cured tobacco. It also reduces the specific volume or the filling power of the tobacco in the cigarette.[37] This reduced specific volume impairs the drawing quality of the cigarette, which is reflected in the value of the tobacco. Chemicals, such as maleic hydrazide, butralin, and flumetralin, suppress sucker growth without a loss in quality.[38]

34.8.4 Harvesting

The leaves of tobacco are ripe and ready to harvest when they turn a lighter shade of green[39] and have thickened so that when a section of the leaf is folded, it creases or cracks on the line of folding. This stage is reached in the bottom leaves when the seed heads form, but in the middle leaves it is reached about two or three weeks later. Flue-cured and fire-cured types are harvested when fully ripe with the leaves showing light yellow patches or flecks. Stalk cutting is done when the middle leaves are ripe.

Two general methods of harvesting are practiced: priming and stalk cutting. Priming is practiced on nearly all flue-cured tobacco and most of the cigar-wrapper and cigar-binder tobacco. The other types are stalk cut (Figure 34.8). Priming is picking the leaves when they are in prime condition, beginning with the lower three or four leaves of commercial size, followed by four or five successive pickings of two to four leaves each at intervals of about five to ten days. The successive harvestings of the lower leaves cause gains in weight of the remaining upper leaves. This weight gain, together with the avoidance of translocation of materials from the leaf to the stalk during curing, results in primed leaves that weigh 20 to 25 percent more than the leaves on cut stalks. A single priming of the lower leaves of burley tobacco increases the total yield, but the quality of the primed leaves is not typical of stalk-cured burley. After priming, the leaves are placed in baskets, in mounted bins, or on burlap sheets and conveyed to the curing barn. The leaves are strung on sticks about 4½ feet (140 cm) long. A string is attached to one end of the stick, after which the leaves are strung either by a needle pushed through the base of the leaf midrib or by looping the string around the base of two to five leaves. When the string is full, it is tied to the other end of the stick (Figure 34.9). A stick holds sixty to eighty or more leaves. This system has been largely replaced by loose-leaf handling and bulk curing. Several men riding on a moving machine strip off the leaves and place them on a metal rack.

In stalk cutting, the stalks are cut off near the base and left on the ground until the leaves are wilted. The stalk is then hung upon a lath. This is done by piercing

FIGURE 34.8
Spearing tobacco stalks and hanging them on the stick.

FIGURE 34.9
Tobacco leaves strung and
hung on sticks.

the stalk near the base with a removable metal spearhead placed on the end of the lath and sliding the stalks along the lath (Figure 34.8). A lath holds six to eight stalks. Sometimes the stalks are hung on the lath with hooks or nails. Fire-cured and burley stalks are split down toward the base; after cutting, they are placed astride the laths. Splitting cut stalks hastens drying of the stalks and quickly stops translocation. The laths, which bear about six plants, are placed upon a rack and hauled to the curing barn.

The use of mechanical harvesters reduces the intensive labor involved in manual harvest, whether by priming or stalk cutting. However, yields and quality are lower when compared to manual harvest.[40]

◼ 34.9 CHANGES IN TOBACCO PRODUCTION

There is a lot of labor involved in producing a tobacco crop. About half of the labor is involved in harvesting, curing, and marketing the crop. Such high labor requirements limit the acreage of tobacco that a farm family can operate.

Motorized equipment for tillage, transplanting, topping, and spraying has reduced labor somewhat. Chemical sprays to suppress sucker growth can replace hand picking of suckers. A crew riding on a mobile machine can strip and load tobacco leaves more rapidly than when walking. Much of the tobacco is now cured and marketed as loose-leaf without tying it. Untied tobacco reduces labor by more than 20 percent.[41] Machines that strip the leaves mechanically are now in use. Tobacco plants can be topped with mechanical cutters, but with some loss of leaves. Heating the curing barns with fuel oil or gas requires less labor than the former practice of stoking a wood fire.[42]

■ 34.10 CURING TOBACCO LEAVES

Curing involves the processes of drying, decomposition of chlorophyll until the green color disappears from the leaf, changes in nitrogen compounds including release of ammonia, hydrolysis of starch into sugars, and respiration or fermentation of the sugars.[43, 44, 45] Mineral salts also crystallize out, thus producing the grain of the leaf. Practically all the sugars are used up when tobacco is air-cured, but considerable quantities remain in the leaves of fire-cured and flue-cured types. Losses of nicotine during curing range from 10 to 33 percent.

Too-rapid drying, like bruising the leaf, kills the tissue prematurely. This prevents complete decomposition of chlorophyll, which results in an undesirable green color. It also stops some of the other desirable chemical reactions. During curing there is a decrease in total weight of 84 to 88 percent,[46] which includes a loss of dry matter of 12 to 20 percent in primed leaves, with an additional 10 to 12 percent of material translocated to the stalks when the leaves are not primed. Harvested tobacco contains about 75 to 80 percent moisture. After the leaf is cured, it is allowed to regain moisture (the regained moisture is known as order or case) until the leaves contain about 24 to 32 percent moisture and are pliable enough to handle without breaking. Thus, the final net loss in total weight of the leaf during curing is about 75 percent. For example, 2 tons of fresh leaves will yield 1,000 pounds of cured and cased product (250 kg/MT). Eight tons of cut stalks will yield about 1,800 pounds of cured leaf (112 kg/MT).

34.10.1 Tobacco Barns

Barns for curing tobacco usually are high enough (16 to 20 feet (5 to 6 m) to the plate) to hold three to five tiers of suspended leaves or stalks. The barns are built in trusses with poles spaced at the proper distances for having sticks or laths hung over them. For primed tobacco on sticks, the poles are spaced 50 inches (130 cm) apart horizontally and 22 to 30 inches (56 to 76 cm) apart vertically. For stalk curing on laths, the pole centers are 46 inches (117 cm) apart horizontally and 2½ to 5 feet (75 to 150 cm) apart vertically. Since the tobacco should hang at least 3 feet (90 cm) above the ground, the lower tier of poles is 6 to 9 feet (1.80 to 2.75 m) high. A space of 1½ to 2½ cubic feet (0.04 to 0.07 m^3) per plant is required for curing tobacco. A barn 17 feet (5 m) high, 32 to 36 feet (10 to 11 m) wide, and 140 to 160 feet (43 to 49 m) long may hold the crop from 3 or 4 acres (1.2 to 1.6 ha). Barns for loose-leaf curing on metal racks are much lower.

34.10.2 Air–Curing

Tobacco is air-cured in huge, tight barns sometimes 300 feet (90 m) long, equipped with numerous ventilating doors on the walls that can be opened or closed to regulate the humidity and temperature (Figure 34.10). The relative humidity at the beginning should be about 85 percent, but after the leaves begin to turn brown a lower humidity that permits rapid drying is advisable.[37] A condition called pole sweat or house burn occurs when the humidity exceeds 90 percent for 24 to 48 hours with the temperature above 60°F (16°C). This condition causes injured,

FIGURE 34.10
Barn for air curing stalk-cut cigar-binder tobacco in Connecticut valley. Note the center driveways.

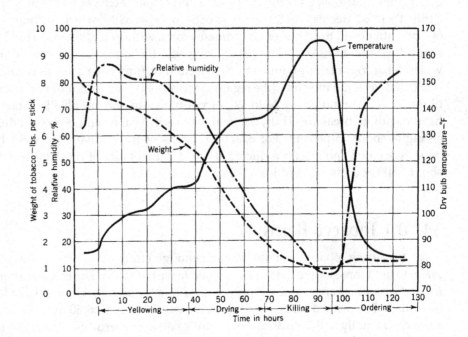

FIGURE 34.11
Temperature and humidity requirements for flue-cured tobacco. [After Moss and Teeter, N. Car. Agr. Exp. Sta.]

partly cured, and even fully cured leaves to begin to soften and decay. This damage is prevented by the use of artificial heat to lower the relative humidity and promote drying. This may be provided by LP gas burners, coke stoves, or several small charcoal fires built on the earth floor inside the barn. These fuels do not impart undesirable odors to the tobacco leaf. Artificial heat is also required when the temperature drops below 50°F (10°C) so that curing processes other than drying may continue. Air-curing may require from four to eight weeks.[47]

34.10.3 Flue-Curing

The objective in flue-curing is to hasten the early curing stages so that drying will be completed while the leaves are still a light yellow color. Flue-cured or bright tobacco is riper and lighter green than other types when curing begins. The typical temperature and humidity requirements for the yellowing, drying, killing, and ordering stages of flue-curing are illustrated in Figure 34.11. The yellow color of the leaf is fixed during the later period of the yellowing stage. The temperature is allowed to rise to 130°F (54°C) in about fifty hours, with a range in relative humidity of 65 to 85 percent in the yellowing stage. Drying is completed in about forty or more hours as the temperature is raised to about 165°F (74°C). The leaves are killed during this stage, while the relative humidity drops to less than 10 percent. The heating is stopped after four or five days and the tobacco is allowed to take up moisture. When drying is too rapid, the humidity in the barn can be increased by wetting the floor and walls, and when the humidity is too high the roof ventilators are opened. Improper curing may result in leaf discolorations such as black splotching, green veins, green leaf, dark sponging, scalding, and black stem.

For flue-curing, a large sheet-iron pipe connected to the top of each furnace is usually extended to the far side of the barn, placed along the wall for a short distance, and then turned back and out through the wall on the furnace side.

Ventilation is provided for incoming air by openings near the bottom of the barn and for the departing moist air by ventilating doors in the roof. The cold incoming air can be passed over hot flues inside the barn.[48] Most of the furnaces fired with wood or coal[49] have been replaced with types that burn propane gas, natural gas, stove oil, or furnace oil. The combustion gases pass through the tobacco by forced air ventilation or exhaust fans.[50] The temperatures for forced-air drying are 95 to 105°F (35 to 40°C) during yellowing, 105 to 135°F (40 to 57°C) during color fixing, and 135 to 165°F (57 to 74°C) during final drying.

34.10.4 Fire-Curing

The usual system of fire-curing is to allow the tobacco to yellow and wilt in the barn without fires for three to five days. Slow fires are then started that maintain the temperature at 90 to 95°F (32 to 35°C) until yellowing is completed. Finally, the temperature is raised to 125 to 130°F (52 to 54°C) until the leaves are dry. The fires are kept up for three to five days or more. The smoke from the open, hardwood fires in the barn imparts the characteristic odor and taste to the tobacco desired for chewing plug and snuff.

34.10.5 Perique Tobacco

Perique tobacco is air cured for eight to fourteen days with the whole stalks hung on wire. Then the dry brown stripped leaves are moistened, stemmed, formed into twists, and packed in casks under heavy pressure with occasional loosening for a month until the leaves are black. Thereafter they are allowed to ferment in their own juice for nine months. Perique is used in pipe tobacco blends and the only production in the United States is in St. James Parish in Louisiana.

▓ 34.11 HANDLING AND MARKETING

After tobacco on the stalk is cured and cased enough to be pliable, it is ready to be stripped from the stalks by hand. In that condition, the leaf contains 24 to 32 percent moisture. Often, the tobacco is bulked in piles without being removed from the sticks

FIGURE 34.12
Loose-leaf tobacco is stuffed into bales for storage after drying.
[Courtesy USDA]

when it is first taken down after curing. The bulking not only prevents further drying but helps to decompose any remaining chlorophyll. Both stripped and primed leaves are sorted into grades and sometimes tied into hands of fifteen to thirty leaves, with a leaf used as a binder. Most of the flue-cured tobacco is now being sold as loose leaf without tying. A pound of cured wrapper tobacco contains sixty to ninety leaves.[51]

The hands of cigar tobacco are packed in cases or in bundles (bales) that then are covered with paper. Other types, handled as loose-leaf tobacco, are delivered in baskets or in bundles wrapped temporarily in burlap or other cloth (Figure 34.12). Most tobacco is sold by auction[49] after it is graded and labeled. Cigar tobacco is often sold at the farm soon after curing. The marketing season is July to December for flue-cured types, and December to February or March for burley fire-cured and air-cured types. However, Maryland tobacco is sold from May to July in the year after its production.[49]

Marketed tobacco is moist enough to be pliable. This pliability prevents severe losses from breakage, but usually such tobacco is too moist for storage or export.[52] It is redried on racks in heated chambers or containers, or by a high-frequency current diathermic process. The latter process has been used for drying and heat fumigation of tobacco in large casks or barrels destined for export. Before redrying, tobacco is sometimes sweated or fermented in bulk piles to complete processes begun during curing. Export tobacco contains about 11 percent moisture. Fermentation consists of allowing a pile or box of slightly moist tobacco to heat at temperatures of 85 to 120°F (29 to 49°C), the temperature depending upon the type of tobacco and method of bulking. Fermentation causes losses of 5 to 12 percent in dry matter. Decreases in starch, sugar, nicotine, amino nitrogen, citric acid, and malic acid occur during fermentation.

Fermented tobacco is aged for one to three years before being used in manufacture. In aging, tobacco undergoes a decrease in nitrogen, nicotine, sugars, total acids, and pH, as well as in irritating and pungent properties. The aroma is improved by aging.[53]

The leaves with the midrib (stem) removed are known as strips. Stemming may be done just before the leaf is used in manufacturing or two years earlier when it is shipped and stored. The stem comprises about 25 percent of the weight of the leaf but varies from 17 to 30 percent. Tobacco used for snuff is not stemmed. Stems, along with tobacco waste, are used in manufacturing nicotine sulfate for insecticides, although the stems contain only about 0.7 percent nicotine as compared with the leaf, which usually contains 1 to 5 percent. Some of the stems, which contain about 6 percent potash, are used as fertilizer or mulch. *Nicotiana rustica* leaves are very high (5 to 10 percent) in nicotine content. When leaves of this species are grown, they are used for insecticide materials.[13] Most of the nicotine can be extracted by forcing the leaves through a series of rollers. Some nicotine is extracted from waste tobacco.

The shrinkage of tobacco purchased from the grower ranges from 30 to 40 percent as a result of stemming, drying, sweating, and handling. A pound (454 g) of tobacco that includes this waste will make about 40 cigars, 350 cigarettes, 1.06 pounds (480 g) of snuff, or 500 modern filter-type cigarettes. Tobacco stems, licorice, sugar, and other materials are added to unstemmed tobacco for making snuff. Chewing plug consists of about 60 percent tobacco; the remainder is comprised of sweetening and flavoring materials. A considerable quantity of cigar-filler tobacco is produced in Puerto Rico, but some is imported from other countries. Large quantities of aromatic tobacco for cigarette and pipe tobacco blends are received from Turkey, Greece, and countries near them.

▨ 34.12 TOBACCO QUALITY

Tobacco varies widely in quality, which depends upon the conditions encountered in growing, curing, sorting, and handling the crop.[54] The production of a good-quality leaf is essential to success, since low-grade tobacco brings less than its cost of production. Color, texture, and aroma are important characteristics. Dark fire-cured tobacco is expected to be strong, dark, and gummy. Flue-cured cigarette tobacco is usually a light lemon-yellow color, while air-cured cigarette tobaccos are light brown. Cigar-wrapper leaf should be thin, uniform, and free from blemishes. The color of the wrapper, whether dark or light, is not a true measure of the mildness of a cigar, as is popularly assumed. A heavy sweating process that darkens the leaf also reduces the nicotine content. Furthermore, the color of the thin wrapper leaf does not reveal the color or mildness of the binder and filler that make up the greater part of a cigar. Most domestic cigars are now being made with a "reconstituted sheet" as a binder. The sheet is rolled out of ground cigar binder leaf, stems, and scraps and mixed with a cohesive.[49] The highest prices are paid for shade-grown wrapper, followed by flue-cured and light air-cured types. The composition of flue-cured and burley tobacco is shown Table 34.2. Dark air-cured and fire-cured tobacco contains 4 to 4.5 percent nicotine.[30]

A high-quality tobacco is usually high in soluble carbohydrates and potash but relatively low in crude fiber, nicotine, nitrogen, calcium, ash, and acids. An extremely low content (1 to 1.3 percent) of nitrogen is not desirable. A pH not lower

TABLE 34.2 Typical Composition of Tobacco in Percents (Except for pH)

	Flue-Cured	Burley
Total nitrogen	2.0	3.4
Nicotine	2.3	3.0
Petroleum-ether extract	5.0	5.5
Total sugars	18.75	0.5
Acids	14.0	27.0
pH	5.1	6.0
Ash	12.0	19.0
Potash (K$_2$O)	3.25	
Calcium oxide (CaO)	3.25	
Magnesia (MgO)	7.0	
Chlorine	0.70	
Sulfur	0.60	

than 5.3 is preferred. The popular brands of cigarettes contain about 1.75 percent nicotine. Cigars of all classes contain about 1.5 percent nicotine. The nicotine contents of granulated pipe mixtures, scrap chewing tobacco, and chewing plug are about 2 percent, 1 percent, and 2 percent, respectively.

Low nicotine content is desired in tobacco grown for domestic use, despite the fact that nicotine is the active stimulant. Nicotine content varies greatly with climate, season, soil, maturity, and general cultural practices.[55] Low-nicotine varieties of tobacco are available if a market for them is ever developed.

Nicotine is an alkaloid, is an oily liquid soluble in water, and has considerable volatility at ordinary temperatures. It boils at 480°F (250°C) and is extremely poisonous. The chemical formula is $C_{10}H_{14}N_2$. When oxidized, it yields nicotinic acid, the amide of which is the familiar vitamin called niacin.

Cultivated tobacco contains traces of several other alkaloids. The nicotine content is higher in the leaf lamina (blade) than in the veins, but it is higher in the veins than in the midribs or stalks.[56] The alkaloid increases as the plant grows, with the result that the greatest amount is probably present when the plant is fairly mature. It is believed to decrease after this time.

Tobacco smoke contains carbon, carbon dioxide, carbon monoxide, oxygen, hydrogen sulfide, hydrocyanic acid, ammonia, nicotine, pyridine, methyl alcohol, resins, phenols, ethereal oils, nicotinic and glutamic acids, and other compounds.[5] An extensive campaign to stop cigarette smoking and other tobacco products because of their tendency to increase the incidence of emphysema and lung cancer [57] has reduced the per capita consumption of tobacco products in the more-developed countries, but consumption is increasing in developing countries. Pipe and cigar smokers are less likely to inhale the fumes into the lungs, so their cancers are usually restricted to the mouth and throat.

■ 34.13 DISEASES

The important tobacco diseases are black root rot, blackfire, wildfire, mosaic, blue mold, and frenching.[30, 43, 58]

34.13.1 Black Root Rot

Root rot, caused by *Thielaviopsis basicola*, is a fungus disease that produces black lesions and eventual decay of the roots. Diseased plants are small and stunted. Tile roots are small and begin decay at the tips. Frequently, the plants become yellowish. The plants usually outgrow the disease in the field when the temperatures exceed 80°F (27°C). The fungus lives over in the soil. The most effective control measures are crop rotation, seedbed disinfection, and resistant varieties.[59]

34.13.2 Mosaic

The mosaic disease (also called calico or walloon) is the most destructive virus disease of tobacco. It is present wherever tobacco is grown.[43] Plants infected with mosaic at transplanting time may have a reduction in yield of 30 to 35 percent, and a 50 to 60 percent reduction in value.[60] The disease is characterized by a mosaic pattern of light and dark green or yellow areas. The different strains of the virus give various symptoms. Some strains cause mild mottling with no distortion of the leaves, while others cause prominent mottling or distortion of new leaves. The virus that causes the common mosaic of tobacco is a crystalline nucleoprotein of high (40,000,000) molecular weight. The crystals are rod-shaped, about 270 millimicrons in length, and 15 millimicrons thick as revealed in an electron microscope photograph. Those virus crystals possess many characteristics of living organisms such as reproduction and mutation. The leaves and leaf juices of plants containing mosaic virus are very infectious. The virus is spread to other plants by direct or indirect contact.

The tobacco mosaic virus remains infectious for long periods—as long as 30 years—in dead plant material.[61] It will overwinter in dry, compact, or water-logged soils. Perennial wild hosts and aphids that feed on them as well as on the tobacco plants are important factors in the dissemination of the disease. The disease is also spread by workers who use tobacco.

The disease can be partly controlled by (1) sterilizing the tobacco bed and tobacco cloths, (2) destroying near-by Solanaceous weeds, (3) removing diseased plants, (4) destroying all tobacco refuse around the bed, (5) washing hands with a trisodium phosphate solution after handling diseased plants, and (6) completely abstaining from use of natural leaf tobacco products during cultural operations.[62] Resistant varieties are grown.

34.13.3 Blackfire or Angular Leaf Spot

Blackfire is a destructive bacterial leaf-spot disease in Virginia and Kentucky caused by *Pseudomonas syringae*. Distinct dark, angular spots between the veins occur on leaves. Rotating plant beds and fields with pasture crops, burial of infected plant residues, and avoidance of high rates of nitrogen and lime help prevent the disease. Care should be used to assure that transplants are not infected.

34.13.4 Wildfire or Bacterial Leaf Spot

Wildfire, caused by the bacterium *Pseudomonas tabaci*, spreads rapidly through the field. The symptoms are lemon-yellow spots on the leaves with small dead spots in the center of the yellow spots. The disease is best controlled by spraying the plants in the bed with fungicides shortly after they come up and at weekly intervals thereafter until transplanted.

34.13.5 Bacterial (Granville) Wilt

The organism *Ralstonia solanacearum*, which causes bacterial wilt or Granville wilt, enters the roots, where it multiplies and clogs the water-conducting vessels, causing the plants to wilt. Control methods[63] are rotations that involve immune crops such as corn, cotton, sweet potato, grass, clover, and small grains, in which tobacco is grown not more often than once in four or five years. Since the wilt organism attacks other plants (tomatoes, peppers, peanuts, and various weeds), these plants must be excluded from tobacco land. Resistant varieties are available.

34.13.6 Blue Mold

The blue mold (downy mildew) disease is caused by the fungus *Peronospora tabacina*, which attacks and destroys the leaves in the seedbed. It often kills large numbers of plants. Symptoms are a yellowing and cupping of leaves. The disease occurs in a mild form from Florida to Maryland every year. Serious damage occurs sporadically, depending upon weather conditions. Spore production occurs when the temperature is 46 to 86°F (8 to 30°C).[64] The best remedy is regular spraying or dusting with fungicides.

34.13.7 Root Diseases

At least six root diseases,[43, 58] which include the Granville wilt already described, damage the tobacco plant. Southern stem rot, caused by *Sclerotium rolfsii*, is controlled by crop rotation and soil fumigation. Fusarium wilt, caused by *Fusarium oxysporum* var. *nicotianae*, can be controlled by the growing of resistant varieties, by soil fumigation to control nematodes, and by crop rotation. Black shank, caused by *Phytophthora parasitica* var. *nicotianae*, can be controlled by growing resistant varieties as well as by crop rotation that avoids Solanaceous plants.

Root knot is caused by a nematode or eelworm (*Meloidogyne* species) that produces galls on the roots. This disease as well as nematode root rot, caused by the *Pratylenchus* species, is controlled by soil fumigation and crop rotation. Several varieties are resistant to the root-knot nematode.[42]

34.13.8 Other Diseases

Other diseases include two leaf spots, frog eye (or Wisconsin leaf spot), caused by the fungus *Cercospora nicotiane*, (Figure 34.13) and brown root rot. Several varieties are somewhat resistant to some leaf spots. Two virus diseases, etch and vein banding, are checked by the control of aphids.[45]

Broom rape, a parasitic flowering plant, lives upon the tobacco plant in much the same manner as dodder lives on clover.

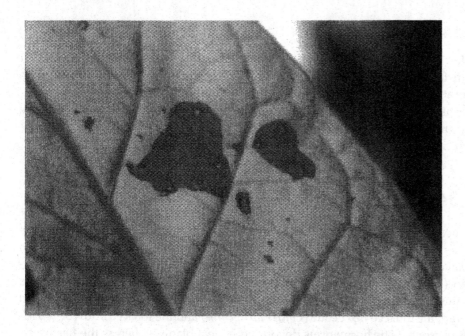

FIGURE 34.13
Frog eye lesion on a
tobacco leaf. [Courtesy USDA
ARS]

▓ 34.14 NUTRITIONAL OR PHYSIOLOGICAL DISORDERS

Tobacco serves as an excellent test plant for determination of mineral deficiencies in soils or culture solutions, because the deficiencies produce symptoms that can be readily recognized.[22, 65, 66]

34.14.1 Frenching

Although considered to be a physiological disease, frenching is caused by a toxin produced by a nonpathogenic soil bacterium, *Bacillus cereus*. Yellowing begins along the edge of young leaves and can progress to include all the area between the veins that remain green. Leaves are narrowed and drawn, the tips sometimes bending sharply downward so as to form a cup of the underside of the leaf. The disease is more prevalent in seasons of abundant rainfall, especially on poorly drained soils with higher pH. Usually only scattered spots or plants in the field are affected. No control method is recommended. Frenching[65] has been confused with mosaic.

34.14.2 Sand Drown

Sand drown (magnesium hunger) is caused by deficiency of available magnesium together with low available sulfur. It occurs most frequently after heavy rains on sandy soils subject to the leaching of soluble minerals, hence its name. The leaves turn nearly white at the tips and along the margins, especially on the lower leaves. Sand drown is prevented by the application of a potassium-magnesium sulfate or dolomitic limestone that is high in magnesium. Manure and other organic fertilizers also tend to prevent the disease.

34.14.3 Fleck

Fleck causes small gray or white spots on the upper surface of tobacco leaves.[67] It occurs following high concentrations, four or more parts per 100 million, of ozone in the atmosphere. Ozone apparently is formed by photochemical reactions that involve nitrogen dioxide, certain hydrocarbons, and other air-borne chemicals. Fleck occurs most frequently in fields that are in the proximity of metropolitan and manufacturing areas where industrial and automotive fumes are abundant.

■ 34.15 INSECT PESTS

The most serious insect pests of tobacco in the United States[45, 68] are the hornworms or greenworms, of which there are two important species: the tomato hornworm *(Manduca quinquemaculata)* and tobacco hornworm *(M. sexta)* (Figure 34.14). Caterpillars (larvae) of the hornworms can reach 4 inches (10 cm) in length. They have a prominent horn on the rear, which gives them their name. Other serious insect pests are the tobacco flea beetle *(Epitrix hirtipennis)* and the tobacco budworm *(Heliothis virescens)*. The hornworms devour entire leaf blades, whereas the flea beetle riddles the leaves with small holes. The flea beetle causes the most damage on young plants. The budworm feeds in the top of the plants and cuts holes in the young leaves. By chewing into the bud it may puncture several unopened leaves at once. This insect is prevalent in the southern tobacco growing areas. Cutworms attack the plants in the beds as well as after they are set in the field. All of these pests as well as aphids can be controlled with insecticides.[45] Destruction of tobacco-plant residues, unused plants in the beds, and weeds near the beds and fields help control insect populations.

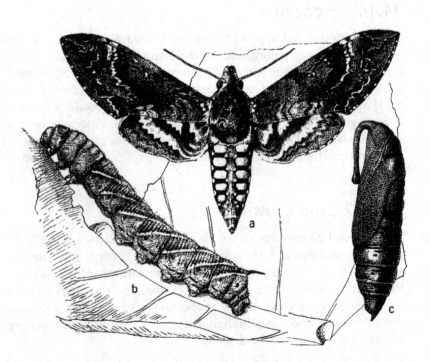

FIGURE 34.14
Southern tobacco hornworm: (*a*) adult, (*b*) larva, and (*c*) pupa.

The tobacco or cigarette beetle *(Lasioderma serricorne)* damages stored and manufactured tobacco. The tobacco moth *(Ephestia elutella)* attacks flue-cured and imported Turkish cigarette tobacco in storage.[69] The larvae devour the entire leaves except for the larger veins. They also foul other leaves with webs and excrete. They infest tobacco stored in cases, hogsheads, or piles in warehouses or storage sheds. This pest is kept under control by periodic fumigation with hydrocyanic-acid gas in closed storages. In open storages, pyrethrum sprays and dust are effective.

REFERENCES

1. Garner, W. W., H. A. Allard, and E. E. Clayton. "Superior germ plasm in tobacco," *USDA Yearbook*, 1936, pp. 785–830.

2. Garner, W. W., and others. "The nitrogen nutrition of tobacco," *USDA Tech. Bull.* 414, 1934.

3. Hahn, P. M. *Sold American.* Louisville, Ky: The American Tobacco Company, 1954.

4. Johnson, J., H. F. Murvin, and W. B. Ogden. "The germination of tobacco seed," *WI Agr. EXP Sta. Res. Bull.* 104, 1930.

5. Garner, W. W. "Climate and tobacco," in *Climate and Man,* USDA Yearbook, 1941, pp. 364–372.

6. Matthews, E. M., and T. B. Hutcheson. "Experiments on flue-cured tobacco," *VA Agr. Exp. Sta. Bull.* 329, 1941.

7. Darkis, F. R., and others. "Flue-cured tobaccos: Correlation between chemical composition and stalk position of tobaccos produced under varying weather conditions," *J. Ind. Eng. Chem.* 28 (1936):1214–1223.

8. Carr, J. M. "Bright tobacco in Georgia," *GA Coastal Plain Exp. Sta. Bull.* 10, (1928), pp. 1–31.

9. Colwell, W. E. "Tobacco," in *Soil,* USDA Yearbook, 1957, pp. 655–658.

10. Darkis, F. R., L. F. Dixon, and P. M. Gross. "Flue-cured tobacco: Factors determining type and seasonal differences," *J. Ind. Eng. Chem.* 27(1935):1152–1157.

11. Anderson, P. J. "Growing tobacco in Connecticut," *CT Agr. Exp. Sta. Bull.* 654, 1953.

12. Anonymous. "Burley tobacco production," in *Ohio Agronomy Guide. OH St. U. Ext. Bull.* 472, 1996.

13. McMurtrey, J. E., Jr., C. W. Bacon, and D. Ready. "Growing tobacco as a source of nicotine," *USDA Tech. Bull.* 820, 1942.

14. Burk, S. G., and H. E. Heggestad. "The genus *Nicotiana:* A source of resistance to disease of cultivated tobacco," *Econ. Bot.* 20(1966):76–88.

15. Allard, H. A. "Some aspects of the phyllotaxy of tobacco," *J. Agr. Res.* 64(1942):49–55.

16. Avery, G. S., Jr. "Structure and development of the tobacco leaf," *Am. J. Bot.* 20(1933):565–592.

17. Wolf, F. A., and P. M. Gross. "Flue-cured tobacco: A comparative study of structural responses induced by topping and suckering," *Bull. Torrey Bot. Club,* 64 (1937):117–131.

18. Bowman, D. R., B. C. Nichols, and R. N. Jeffrey. "Time of harvest—its effect on the chemical composition of individual leaves of burley tobacco," *TN Agr. Exp. Sta. Bull.* 291, 1958.

19. Hamner, H. R., O. E. Street, and P. J. Anderson. "Variation in chemical composition of cured tobacco leaves according to their position on the stalk," *CT Agr. Exp. Sta. Bull.* 433, 1940, pp. 177–186.

20. Swanback, T. R. "Variation in chemical composition of leaves according to position on the stalk," *CT Agr. Exp. Sta. Bull.* 422, 1939.

21. Bowling, J. D. "Phosphorus, potassium, calcium and magnesium requirements of Maryland tobacco grown on Monmouth soil," *MD Agr. Exp. Sta. Bull.* A-151, 1967, pp. 1–48.

22. McMurtrey, J. E., Jr. "Relation of calcium and magnesium to the growth and quality of tobacco," *J. Am. Soc. Agron.* 24 (1932):707–716.

23. Gaines, J. G., and F. A. Todd. "Crop rotations and tobacco," in *Plant Diseases*, USDA Yearbook, 1953, pp. 553–561.

24. Jones, J. P. "Influence of cropping systems on root-rots of tobacco," *J. Am. Soc. Agron.* 20 (1928):679–685.

25. Brown, D. E., and J. E. McMurtrey, Jr. "Value of natural weed fallow in the cropping system for tobacco," *MD Agr. Exp. Sta. Bull.* 363, 1934.

26. Clayton, E. E. "The genes that mean better tobacco," in *Plant Diseases*, USDA Yearbook, 1953, pp. 548–553.

27. Bachtell, M. A., R. M. Salter, and H. L. Wachter. "Tobacco cultural and fertility tests," *OH Agr. Exp. Sta. Bull.* 590, 1938.

28. Roberts, G., E. J. Kinney, and J. F. Freeman. "Soil management and fertilization for tobacco," *KY Agr. Exp. Sta. Bull.* 379, 1938.

29. Smith, W. D. "Seedling production," in D. L. Davis and M. T. Nielsen, eds., *Tobacco Production, Chemistry, and Technology*. Oxford: Blackwell Science, 1999, pp. 70–76.

30. Miles, J. D. "Producing flue-cured tobacco plants under plastic cover," *GA Agr. Res.* 3, 2 (1961):8–10.

31. Palmer, G., B. Maksymowicz, and J. Calvert. "Transplant production," in *Tobacco in Kentucky. KY Coop. Ext. Serv. Bull.* ID 73, 1993.

32. Dean, C. E., H. Seltmann, and W. G. Woltz. "Some factors affecting root development by transplanted tobacco transplants." *Tobacco Sci.* 4 (1960):19–25.

33. Smiley, J. H. "Kentucky farm tobacco," *Rsh. Rpt.* 195, 1971, pp. 1–24.

34. Flower, K. C. "Field practices," in: D. L. Davis, and M. T. Nielsen, eds., *Tobacco Production, Chemistry, and Technology*. Oxford: Blackwell Science, 1999, pp. 76–104.

35. Hasselbring, H. "The effect of shading on the transpiration and assimilation of the tobacco plant in Cuba," *Bot. Gaz.* 57 (1914):257–286.

36. Vickery, H. B. "Chemical investigations of the tobacco plant. V. Chemical changes that occur during growth," *CT Agr. Exp. Sta. Bull.* 375, 1935, pp. 557–619.

37. Jeffrey, R. N., and E. L. Cox. "Effects of maleic hydrazide on the suitability of tobacco for cigarette manufacture," *USDA Agricultural Research Service Spec. Rept.* ARS 34–35, 1962.

38. Tso, T. C. *Physiology and Biochemistry of Tobacco Plants*. Strousburg, PA: Dowden, Hytchinson and Ross, 1972, pp. 1–393.

39. Moseley, J. M., and others. "The relationship of maturity of the leaf at harvest and certain properties of the cured leaf of flue-cured tobacco," *Tobacco Sci.* 7 (1963):67–75.

40. Peedin, G. F. "Flue-cured tobacco," in: D. L. Davis and M. T. Nielsen, eds., *Tobacco Production, Chemistry, and Technology*. Oxford: Blackwell Science, 1999, pp. 104–143.

41. Hayert, J. H. "Feasibility of Maryland tobacco leaf untied," *MD Agr. Exp. Sta. M.P.* 788, 1971.

42. Elliot, J. M., and C. F. Marks, "Control of nematodes in flue-cured tobacco in Ontario," *Can. Dept. Agr. Pub.* 1465, 1972.

43. Clayton, E. E., and J. E. McMurtrey, Jr. "Tobacco diseases and their control," *USDA Farmers Bull.* 2033 (rev.). 1958.

44. Cooper, A. H., and others. "Drying and curing bright leaf tobacco with conditioned air," *J. Ind. Eng. Chem.* 32 (1940):94–99.

45. McMurtrey, J. E., Jr. "Tobacco production," *USDA Info. Bull.* 245, 1961.

46. Moss, E. G., and N. C. Teter. "Bright leaf tobacco curing," *NC Agr. Exp. Sta. Bull.* 346, 1944, pp. 1–25.

47. Donaldson, R. W. "Quality tobacco in Massachusetts," *J. Am. Soc. Agron.* 23 (1931):234–241.

48. Moss, E. G., and others. "Fertilizer tests with flue-cured tobacco," *USDA Tech. Bull.* 12, 1927.

49. Doub, Albert, Jr., and A. Wolfe. "Tobacco in the United States," *USDA Misc. Pub.* 367, 1961.

50. Walker, E. K., and L. S. Vickery. "Curing flue-cured tobacco," *Can. Dept. Agr. Publ.* 1312, 1969, pp. 1–27.

51. Jenkins, E. H. "Studies on the tobacco crop of Connecticut," *CT Agr. Exp. Sta. Bull.* 180, 1914, pp. 1–65.

52. Cooper, A. H. "The curing and drying of tobacco," *Refrig. Engr.* 38 (1939):2079, 244–245.

53. Dixon, L. F., and others. "The natural aging of flue-cured cigarette tobacco," *J. Ind. Eng. Chem.* 28 (1936):180–189.

54. Ward, G. M. "Physiological studies with the tobacco plant," *Dominion Dept. Agr. Tech. Bull.* 37, Ottawa, 1942.

55. Thatcher, R. W., L. R. Streeter, and R. C. Collison. "Factors which influence the nicotine content of tobacco grown for use as an insecticide," *J. Am. Soc. Agron.* 16(1924):459–466.

56. Bacon, C. W. "Some factors affecting the nicotine content of tobacco," *J. Am. Soc. Agron.* 21(1929):159–167.

57. Wynder, E. L., and D. Hoffman. *Tobacco and Tobacco Smoke: Studies in Experimental Carcinogenesis.* New York: Academic Press, 1967, pp. 1–730.

58. Clayton, E. E., and others. "Control of flue-cured tobacco root diseases by crop rotation," *USDA Farmers Bull.* 1952, 1944, pp. 1–12.

59. Hadden, C. H., S. Bost, and D. Hensley. "Plant diseases: Black root rot of tobacco," *TN Agric. Ext. Serv.* SP 277–8, 1999.

60. McMurtrey, J. E, Jr. "Effect of mosaic disease on yield and quality of tobacco," *J. Agr. Res.* 38, 5 (1929):257–267.

61. Johnson, J., and W. B. Ogden, "The overwintering of the tobacco mosaic virus," *WI Agr. Exp. Sta. Res. Bull.* 95, 1929.

62. Lehman, S. G. "Practices relating to control of tobacco mosaic," *NC Agr. Exp. St. Bull.* 297, 1934.

63. Smith, T. E. "Control of bacterial wilt (*Bacterium solanacearum*) of tobacco as influenced by crop rotation and chemical treatment of the soil," *USDA Circ.* 692, 1944, pp. 1–16.

64. Kucharek, T. "Common leaf diseases of flue cured tobacco." *FL Coop. Ext. Serv.* PP-15, (rev.), 2001.

65. Karraker, P. E., and C. E. Bortner. "Studies of trenching of tobacco," *KY Agr. Exp. Sta. Bull.* 349, 1934.

66. McMurtrey, J. E., Jr. "Nutrient deficiencies in tobacco," in *Hunger Signs in Crops*, 3rd ed., New York: McKay, 1964, pp. 99–141.

67. Heggestad, H. E., and J. F. Middleton. "Ozone in high concentrations as cause of tobacco leaf injury," *Science* 129, 3334 (1959):208–210.

68. Lacroix, D. S. "Insect pests of growing tobacco in Connecticut," *CT Agr. Exp. Sta. Bull.* 379, 1935, pp. 84–130.

69. Anonymous. "Stored tobacco insects, biology and control," *USDA Handbk.* 233, (rev.), 1971, pp. 1–43.

Sugarbeet

35.1 ECONOMIC IMPORTANCE

Sugarbeet was grown on an average of about 17.8 million acres (6 million ha) in 2000–2003. Production averaged about 266 million tons (242 million MT) or 18 tons per acre (40 MT/ha). The leading countries in sugarbeet production are France, the United States, Germany, Turkey, and Ukraine. Average production in the United States in 2000–2003 was about 29 million tons (26 million MT) on 1.3 million acres (540,000 ha). Average yield was 22 tons per acre (49 MT/ha). The leading states in sugarbeet production were Minnesota, Idaho, North Dakota, Michigan, and California (Figure 35.1). Mechanized planting, thinning, harvesting, and cultivation and the application of selective herbicides, together with some semiautomatic sprinkler irrigation, have eliminated most of the hand labor formerly required for sugarbeet production in the United States.

Sugarbeet has been one of the most uniformly profitable cash crops in many irrigated valleys in the western United States. Sugarbeet not only allows diversification, but also provides an intertilled crop in rotations with hay and grain crops. An extensive sheep and cattle feeding industry has formed around each sugarbeet factory due to the use of the by-products of beet production and manufacture as livestock feed.

35.2 HISTORY OF SUGARBEET CULTURE

Sugarbeet dates back only about 250 years, which makes it a relatively new crop.[1] In 1747, a German chemist named Andrew S. Marggraf found that the kind of sugar in two cultivated species of beets was identical to that in cane. The first factory for the extraction of sugar from sugarbeet, built in 1802 in Silesia, was a failure. The percentage of sucrose in sugarbeet then was very low. Louis Vilmorin in France selected beets by progeny test methods and raised the sugar content from 7.5 percent to 16 or 17 percent. By 1880, sugarbeet had practically as high a sugar percentage as the varieties of today. The first successful commercial factory in America was erected at Alvarado, California, in 1870. General success of the industry in the United States dates from 1890.

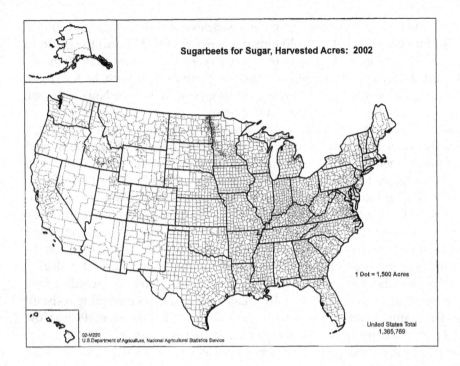

Sugarbeets for Sugar, Harvested Acres: 2002

1 Dot = 1,500 Acres

United States Total
1,365,769

02-M220
U.S.Department of Agriculture, National Agricultural Statistics Service

FIGURE 35.1
Acreage of sugarbeet in the
United States in 2002.
[Source: 2002 Census of U.S.
Department of Agriculture]

About one-fourth of the world sugar supply is furnished by sugarbeet. A popular prejudice against beet sugar, now almost forgotten, continued for many years despite the fact that beet sugar could not be distinguished from cane sugar. The pure sugar (sucrose) from the beet and the pure sugar from the cane are identical.

35.3 ADAPTATION

Sugarbeet is grown in favorable localized areas within feasible shipping distances from sugarbeet factories. A factory represents an investment of several million dollars and usually has a capacity of 1,000 to 5,500 tons (900 to 5,000 MT) of beets daily. Therefore, it is essential that adequate area, usually 10,000 acres (4,000 ha) or more of beets be grown in order to keep the factory in operation during the season of 65 to 200 days after harvest begins.

Successful sugarbeet production is found only on fertile soils. It thrives best in soils of pH 6 to 7.5. A good percentage of soil organic matter supplied naturally, by application of manure or by legume residues, favors beet yields. Loam or sandy loam soils predominate in most sugarbeet areas. Heavy clay soils are used to some extent but are usually less favorable for high yields. Heavy soil also increases the difficulties of lifting the beets and freeing them from adhering soil. Sugar content of beets is highest on soils that also produce the best yields.[2] However, nitrogen fertilizer applied in excess, or late in the season, lowers the sugar content. Sugarbeet can endure large quantities (1 to 1.5 percent) of saline salts (alkali) after it becomes established.[3]

Irrigation water is necessary for successful commercial sugarbeet production where annual rainfall is less than about 18 inches (460 mm), even in the cool northern boundary states of the United States. Seasonal water usage (evapotranspiration) for beet yields of 20 to 30 tons per acre (45 to 67 MT/ha) ranges from 21 inches (530 mm) in the cool area in southern Canada to more than 40 inches (1,000 mm) in Arizona. Commercial sugarbeet production west of the 100th meridian is on irrigated land. Most sugarbeet grown in eastern North Dakota, Minnesota, and states to the east is not irrigated.

Sugarbeet seed will germinate when the soil temperature is about 60°F (16°C), but germination is most rapid at 82°F (28°C). The sugarbeet root has the highest sugar content when the summer temperature averages about 67 to 72°F (19 to 22°C). The optimum temperature for plant growth is about 75°F (24°C), but 63 to 68°F (17 to 20°C) is the best temperature for root growth.[4] The plant is not injured by cool nights. Cool autumn weather (59°F, 15°C) favors sugar storage in the roots. Temperatures of 86°F (30°C) or above retard sugar accumulation. Sugarbeet makes good vegetative growth in the South, but the roots are low in sugar content.[5] On the other hand, sugarbeet planted in the winter in California and harvested during very hot July weather has a high sugar content. The newly emerged seedling may be killed by a temperature of 25°F (−4°C). Later, the plants become more resistant to cold. The mature plant is able to withstand fall frosts, but a temperature as low as 26°F (−3°C) usually causes foliage injury. After such injury, a period of warm weather may bring about growth of new leaves as well as a sharp decline in the sugar content of the roots.[6]

Sunlight has a sanitary effect in disease control. The intensity of light is associated more closely with the utilization of nutrient elements than with their absorption. The relative sugar content is not influenced by light until growth is inhibited[2] by high or low intensities.

Initiation of seed stalks and flowers of sugarbeet (a long-day plant) is brought about mainly by the cumulative effect of exposure to low temperatures (vernalization) followed or accompanied by the effects of long photoperiods. This combined effect of light and temperature has been termed *photothermal induction*.[7] Seed stalks are produced most abundantly and seed yields are highest when the maximum temperature is less than 70°F (21°C), the weather is wet and cloudy (with less than 10.6 hours sunshine per day) for a period of about six weeks, followed by two weeks of cool, dry weather to stimulate seed production.[8] Early planted beets are able to flower at cooler temperatures than are those planted late.[9]

▓ 35.4 BOTANICAL DESCRIPTION

Sugarbeet (*Beta vulgaris*) is an herbaceous biennial dicotyledon, a member of the family *Chenopodiaceae*, characterized by small, greenish, bracteolate flowers. The flowers are perfect, regular, and without petals. A large fleshy root develops. The species includes four groups: (1) sugarbeet, (2) mangel-wurzel or mangel, (3) garden beet, and (4) leaf beets such as chard and ornamental beets. Sugarbeet is sometimes classed as a separate species, *B. saccharifera*, but there appears to be little justification for such a separation.

It is believed that a wild type of *Beta vulgaris*, often referred to as *Beta maritima*, was the progenitor of sugarbeet. Several wild species of *Beta* native to Europe[1] have been crossed with *B. vulgaris*.

35.4.1 Vegetative Development

Sugarbeet, a biennial, normally completes its vegetative cycle in two years. During the first year, it develops a large succulent root, in which a large quantity of reserve food is stored. During the second year, it produces flowers and seeds. Prolonged cool periods cause a seed stalk to develop the first year, a behavior known as bolting. Some strains of beet bolt more readily or at higher temperatures than do others.

The mature beet (Figure 35.2) is an elongated, pear-shaped body composed of three regions: the crown, the neck, and the root.[10] The crown is the broadened, somewhat cone-shaped apex that bears a tuft of large succulent leaves and leaf bases. Just below it is the neck, a smooth thin zone that is the broadest part of the beet. The root is cone-shaped and terminates in a slender taproot. The root is flattened on two sides and is often more or less grooved. The two depressions, which extend vertically downward, or form a shallow spiral, contain the lateral rootlets indistinctly arranged in two double rows. The surface of the beet is covered by a thin cork layer that is yellowish white except on the aerial parts at places of injury.[11]

In cross section, the beet is made up of a number of rings or zones of growth. The center of the beet has a more or less star-shaped core. Most cultivated beets have ivory-white flesh, but in poor strains the flesh is a watery-green or yellowish hue.

The beet root is richest in sugar slightly above the middle, with sugar quantity decreasing toward both ends. The tip of the root is lower in sugar than any other part except the center of the crown. There is little consistent relationship between the internal structure, size, or shape of the root and the sugar content.[12]

The beets grown in one season may have a higher average sugar percentage than those grown in another. Individual beets vary widely in sugar content but usually range from 10 to 20 percent.

The leaves are arranged on the crown in a close spiral.[10] Leaf growth after unfolding occurs by cell enlargement, but root growth results from cell division.[4]

FIGURE 35.2
Sugarbeet top and root showing characteristic groove.

35.4.2 Flower and Seed Development

Sugarbeet normally sets seed as a result of wind or insect pollination. Either the bagging of branches or plant isolation is necessary for the production of pure strains.[13, 14]

During the second year, beets first produce a rosette of leaves like those of the first year. After about six weeks of growth, the newly formed leaves become progressively smaller and then the flower stalk develops.[15] The flower stalk grows rapidly and produces many branches. The mature inflorescence, or *seed bush*, is composed of large, paniculate, more or less open spikes that bear the flowers and later the seeds (Figure 35.3). The calyx and stigma adhere to the mature fruit. The flower stalk is 5 to 8 feet (150 to 250 cm) tall.

FIGURE 35.3
Sugarbeet plant producing seed. A branch bearing monogerm seed balls is shown at lower left.

Sugarbeet fruit is an aggregate, formed by the cohesion of two or more flowers that have grown together at their bases. They form a hard and irregular dry body, the so-called seed ball, which usually contains two to five seeds (multigerm). The mature seed is a shiny, lentil-like structure about 3 millimeters long and 1½ millimeters thick. The mature reddish-brown outer seed coat is very brittle and separates easily from the seed.[15]

A plant with monogerm seed borne in separated flowers was found in western Oregon in 1948.[16] The progeny of this plant was crossed with several multigerm varieties. Selection, backcrossing, and reselection transferred the monogerm character into improved disease-resistant varieties as well as into lines used for producing hybrid sugarbeet seed (Figure 35.4). Sugarbeet plants with cytoplasmic male-sterile flowers were reported in 1945.[17] This character was also bred into improved strains of sugarbeet, which made hybrid sugarbeet seed production a commercial success. The hybrid beets in the United States are diploid plants but anisoploids are widely used in Europe.[18]

▦ 35.5 HYBRIDS OR VARIETIES

Since 1967, most of the sugarbeet grown in the United States are improved hybrids bred for disease resistance, monogerm seed, slow bolting, regional adaptation, high production, or other desired characters.[16, 19] Sugarbeet breeding for disease resistance was urgent due to the heavy losses from curly top, leaf spot, downy mildew, and blackroot diseases.

Multigerm

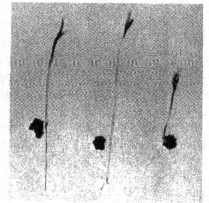

Monogerm

Hand Thinned

Mechanized

FIGURE 35.4

The use of monogerm seed permits thinning, where necessary, by a down-the-row thinner. Unprocessed multigerm seed requires hand thinning and either hand or mechanical blocking.

Beet-sugar companies in the United States have developed varieties or hybrids adapted to their particular areas. Many of these are resistant to curly top or leaf spot. Practically all sugarbeet is grown under contract with a sugar company that furnishes technical assistance and usually produces or procures the seed for the growers.

Genetically modified cultivars have been developed to be resistant to glyphosate herbicide.

▓ 35.6 FERTILIZERS

Nitrogen is the most limiting nutrient for sugarbeet growth, and its management during the growing season is critical.[20] Sugarbeet requires about 15 pounds of available or applied nitrogen per ton (8 kg/MT) of roots. Some nitrogen fertilizer should be applied before seeding and the remainder sidedressed before mid-season. Nitrogen applied late in the season may decrease sugar production and increase impurities in the beet root.[20] It is important to carefully determine the residual soil nitrogen levels by soil analysis in order to calculate the proper amount of fertilizer needed.[20, 21] Suggested nitrogen applications for sugarbeet are 0 to 60 pounds per acre (0 to 67 kg/ha) on highly fertile soils, 60 to 120 pounds per acre (67 to 134 kg/ha) on medium soils, and 120 to 180 pounds per acre (134 to 200 kg/ha) on soils of low fertility or where a heavy growth of residues is turned under.

High phosphorus fertility is needed to enhance sugar production in sugarbeet, so phosphate fertilizer is applied unless phosphate levels are very high.[20, 21] Up to 100 pounds per acre (110 kg/ha) of phosphate may be needed depending on soil analysis. Application of phosphate fertilizer is determined primarily by soil phosphate level and somewhat by yield goal.[20, 21] Potassium requirement is 200 to 400 pounds per acre (225 to 450 kg/ha). Most sugarbeet is grown on soils well supplied with potash. However, muck soils often require the application of potash. Phosphate fertilizer is best applied in a band application close to the seed at planting. Potassium fertilizer should be broadcast applied and incorporated during seedbed preparation unless only small amounts are needed.

Small applications of borax, not exceeding 15 pounds per acre (17 kg/ha), are beneficial on soils deficient in boron. On some muck soils, beet responds to applications of 25 to 50 pounds per acre (28 to 56 kg/ha) of copper sulfate. Zinc or iron applications may be needed on heavily graded irrigated spots. Sodium is an essential element for sugarbeet, but the soil supplies ample amounts.

▓ 35.7 ROTATIONS

Sugarbeet is grown almost exclusively in rotations that involve legumes (alfalfa, red clover, and sweetclover), small grains, and, frequently, potato, corn, or bean.[22] The growing of sugarbeet in continuous culture soon results in depressed yields[23] and often encourages infestation of the land by the sugarbeet nematode or by disease-producing fungi and bacteria. If sugarbeet nematodes are present, canola should not be included in a rotation with sugarbeet, as the nematode will infest that crop as well.[24]

Planting sugarbeet immediately after a legume is usually not advisable. The early spring preparation essential to early planting of sugarbeet destroys the legume crop before it can make enough spring growth to provide much green manure. Consequently, such land is best left for later plowing in preparation for corn or potato. Furthermore, the frequent failure to kill all alfalfa and sweetclover plants by tillage is detrimental to sugarbeet yields.

Growing an intertilled crop, such as potato or corn after a legume crop, facilitates disease and weed control and is also favorable to beet stands and yields. Certain legume crops and weeds promote the occurrence of soil-borne organisms that cause damping off and black root of sugarbeet. Better stands of beets are obtained following corn, potato, small grain, or soybean than after alfalfa, sweetclover, or red clover. Thus, sugarbeet usually follows the crop that follows the legume. A small grain, usually wheat or rye, is sometimes used as a cover crop preceding sugarbeet. The sugarbeet crop can be planted no-till into wheat or rye with a nonselective postemergence herbicide such as glyphosate used to kill the small grain before the sugarbeet seedlings emerge.

■ 35.8 SUGARBEET CULTURE

35.8.1 Seedbed Preparation

Fall plowing in preparation for sugarbeet is usually advisable except for very friable soils. Deep plowing 8 to 12 inches (20 to 30 cm) is usually recommended,[6, 25] although research shows that plowing friable soils deeper than 8 inches (20 cm) is not justified. Chiseling or deep plowing annually or occasionally is practiced on very hard soils. Sugarbeet roots penetrate below any plowing depth with the feeding roots going down 5 or 6 feet (150 to 180 cm).[22] Tillage subsequent to plowing should provide a mellow seedbed for the small beet seedlings.

35.8.2 Seeding Methods

Planting is usually done with special beet planters (Figure 35.5) with rows 18 to 22 inches (45 to 55 cm) apart, or in 12- to 14-inch (30 to 36 cm) double rows spaced 36 to 40 inches (90 to 100 cm) between centers. Planting is done in April or early May in northern areas, in February and March or April in milder and cooler areas of the United States, and in the winter or autumn in parts of California and Arizona. Unprocessed monogerm seed or processed multigerm seed is planted at rates of 5 to 8 pounds per acre (6 to 9 kg/ha), or at least 10 to 12 seeds per foot (33 to 39 seeds/m) of row. Processed monogerm seed, polished to remove some of the surrounding cork, is planted at the rate of 1.5 to 4 pounds per acre (1.7 to 4.5 kg/ha). One pound of seed per acre (1.1 kg/ha), or 4 to 6 seeds per foot (13 to 20 seeds/m) of row, is sufficient to provide a stand of beets under favorable conditions. Such a rate with good seed placement may eliminate the need for thinning.

FIGURE 35.5
Sugarbeet planter with fertilizer attachments. [Courtesy American Crystal Sugar Co.]

35.8.3 Thinning

The development of monogerm seed and precision seeding equipment has eliminated the need for thinning in most sugarbeet fields. The use of pelleted seed to improve uniformity has also increased seeding precision. Thinning, when needed, is usually done with electronic-eye thinning machines in weed-free fields. The desired plant spacing is 8 to 12 inches (20 to 30 cm) apart in the row, or 21,000 to 36,000 plants per acre (52,000 to 89,000 plants/ha).

When unprocessed multigerm seed was planted, the rows were blocked out by hand hoeing or by machine. Each clump that remained was thinned to a single plant by hand. A multigerm seed ball produces two or more plants in one place, but two adjacent plants do not develop normal beets. Hand blocking and thinning require 20 to 40 man-hours per acre (50 to 100/ha). Thinning is often done when the plants have six to eight leaves, but it usually begins about three weeks after planting when the seedlings have about four leaves. Thinning should be completed by the time the plants have eight to ten leaves.

35.8.4 Weed Control

Weed control is more essential in sugarbeet than in most other crops because the plants do not form a sufficiently dense leaf canopy for effective weed control. Before the development of effective herbicides for sugarbeet, weeds were destroyed by hand labor and machine cultivation. Beet cultivators equipped with both knife-type and small sweep-type shovels were most satisfactory. Small weeds in the row are suppressed by cultivation with a harrow or flexible-shank weeder before the beets are up and shortly thereafter. Some hand hoeing prevents scattered weeds from going to seed.

Herbicides are applied to most of the sugarbeet acreage in the United States and Europe. Common herbicides applied preplant or preemergent with sugarbeet include ethofumesate, cycloate, EPTC, and trifluralin. Common postemergent herbicides include quizalofop-D, sethoxydim, clopyralid, and triflusulfuron methyl.

Glyphosate can be applied to sugarbeet varieties that have been genetically modified to be resistant to that herbicide.

35.8.5 Irrigation

In the western United States, beets are irrigated about every ten to fourteen days, when water is available, or whenever the plants show the need for water by their dark green color and the continuation of leaf wilting after sunset. Better yields result when the field is irrigated before the plants show signs of water need. The highest yields are obtained when the soil moisture at a depth of 1 foot (30 cm) is maintained at more than 50 percent of the total available water holding capacity of the soil. The usual irrigation is 2 to 6 inches (50 to 150 mm) at each application, or 12 to 36 inches (300 to 900 mm) for the season. The final irrigation should provide sufficient moisture for the crop to complete its growth. In addition, it should leave the soil moist enough for the roots to be dug. In some areas, it is necessary to irrigate in the spring before or after planting, or both, in order to germinate the seed.

Sugarbeet is relatively drought resistant, and most studies involving amounts and timing of irrigation have proved inconclusive.[18] Stressed plants may have higher sugar concentrations in the roots but also may have increased impurities. Excessive irrigation may leach nitrogen and other nutrients below the active root uptake zone in the soil.

35.8.6 Harvesting

Sugarbeet should be left in the field until it reaches maximum sucrose content. Maturity is indicated by a browning in the lower leaves and a yellowing of the remaining foliage.[26] The contracting sugar company usually makes sugar analyses and instructs the growers as to when harvesting should begin.

The harvesting of sugarbeet in the United States was fully mechanized by 1958. The former hand operations of pulling the beets, knocking off the dirt, and piling, topping, and loading beets are now accomplished by machines.[27] The beets are topped by the cutting off of the crown at the base of the lowest leaf scar (Figure 35.6). The crown contains little sugar, but it is high in mineral salts that can interfere with sugar extraction. The beets are then dug up by a beet harvester that shakes excess soil from the beets and elevates them to a hopper (Figure 35.7). Some machines cut off the tops of the standing plants and then lift, shake, and elevate the roots to a hopper in one operation. Other harvesters lift the beets from the soil and convey them to rotating disk blades for topping in one operation.

The topped beets are hauled in trucks to a railroad siding or directly to the sugar factory. At the factory, the beets are either processed immediately or are piled for processing at a later time.

▪ 35.9 SUGAR MANUFACTURE

In the manufacture of sugar, the topped beets are first washed in a flume of rapidly flowing water.[28] They are then sliced mechanically with V-shaped knives into thin, angular strips called cosettes, which are about the diameter of a lead pencil. The

FIGURE 35.6
Beet tops are removed
using a field chopper.
[Courtesy Shari Rosso, Neb.
Panhandle Res. & Ext. Center]

FIGURE 35.7
A beet harvester digs the
beets and transfers them to
a hopper. [Courtesy Shari
Rosso, Neb. Panhandle Res. &
Ext. Center]

sugar is extracted from the slices by the diffusion process in large drums that contain warm juice followed by warm water at a temperature of 176 to 183°F (80 to
84°C). After separation of the juice from the pulp, milk of lime is added to the juice
in large tanks to precipitate impurities and to neutralize oxalic and other organic

acids. The acids combine with calcium to form less-soluble salts that settle out of the solution. Excess calcium is precipitated as calcium carbonate by carbonation of the limed juice. The juice is filtered, further clarified, decolorized with sulfur dioxide, and again filtered. The juice is then concentrated to syrup by being boiled under reduced pressure in steam-heated vacuum pans or evaporators called effects. The syrup is treated with sulfur dioxide, again filtered, and evaporation is continued until the sugar crystallizes. This mixture of sugar crystals and molasses is separated in centrifuges with perforated inner walls. The washed sugar crystals are then separated and dried in a granulator. The molasses or mother liquor is reworked several times to recover additional sugar, leaving a residue of final-discard molasses.

The average yield of refined sugar obtained in the United States is about 258 pounds per ton (129 kg/MT) of beets. The average sugar content of the beet exceeds 15 percent, but 16 percent of the sugar is left in the molasses, pulp, or residues.

By-products of beet-sugar manufacture are pulp, molasses, and lime cake or waste lime from the filter presses. Beet pulp is the wet fibrous material left after the sugar is extracted from the sliced beets. The yield of wet pulp that contains 90 to 95 percent water is about 1,600 pounds per ton (800 kg/MT) of beets. The final yield of wet pulp, after removal of excess water by pressing or partial drying, is 400 to 600 pounds per ton (200 to 300 kg/MT) of sliced roots. The pulp is used for fresh, ensiled, or dried stock feed. Beet pulp contains about 17 to 22 percent crude fiber that is highly digestible and about 8 to 10 percent crude protein. It has at least 85 percent of the energy value of corn and 95 percent of the energy value of barley for livestock.[29]

Usually 2 to 3 gallons or more of final-discard molasses are obtained from each ton (7 to 10 l/MT) of beets. This molasses contains about 20 percent water, 60 percent carbohydrates (mostly sucrose, arabinose, and raffinose), and about 10 percent ash or mineral matter in which potassium compounds predominate. The molasses is often added to the pulp and used as dried-molasses feed. Considerable quantities are fed in mixture with other feeds. The remainder is used mostly in the manufacture of alcohol. Molasses also contains glutamic acid,[28] some of which is recovered for making monosodium glutamate. When added to foods, monosodium glutamate stimulates human taste buds to a greater appreciation of food flavors.

The dry matter in the lime cake contains the equivalent of more than 80 percent calcium carbonate; 10 percent organic matter; and traces of potash, phosphorus, and nitrogen. It is suitable for liming soils, but most soils in the western sugarbeet areas are not in need of lime. For each ton of sliced beets, about 100 pounds (500 kg/MT) of burned limestone are used, with water, to form milk of lime.[30]

By weight, beet tops consist of about one-third crown and two-thirds leaves. The green weight of tops ranges from 75 to 80 percent of the weight of topped beets. Beet tops are fed mostly to sheep and cattle, either ensiled, fresh, or in dry form as cured in small piles in the field. Often, livestock are turned into the field to eat the piled tops, but this is a wasteful practice. Beet tops are palatable and nutritious, but they are dangerous when fed in large quantities to horses and pigs because of their abundance of cathartic salts and oxalic acid. Ruminants are able to utilize large quantities without injury. Beet tops contain about two-thirds of the digestible nutrients found in corn silage.

■ 35.10 SEED PRODUCTION

It was discovered about 1925 that beets could overwinter in the field in the mild climates of the Southwest and Pacific Coast. By 1932, the American sugarbeet seed industry had started in those areas.[1, 16] Today, all sugarbeet seed in North America is produced in the Willamette Valley of Oregon. The area has mild winters, low disease pressure, and dry weather at harvest, which favor sugarbeet seed production. Crop rotations of five to eight years are used to minimize pests and diseases. Beet seed is planted in rows in late August or early September. Multiple applications of herbicides and insecticides are applied in the fall before and after planting. Strips of four pollinator (male) rows are planted adjacent to eight to sixteen female (male-sterile) rows. The hybrid seed will then be harvested from the female rows. Figure 35.8 shows male-fertile and male-sterile sugarbeet flowers. The male rows are usually destroyed after pollination the next year.

The sugarbeet seed crop is harvested about a year after planting. The crop is harvested with a swather equipped with both horizontal and vertical cutting bars to minimize dragging of plants that can increase seed shattering and loss. Swathers

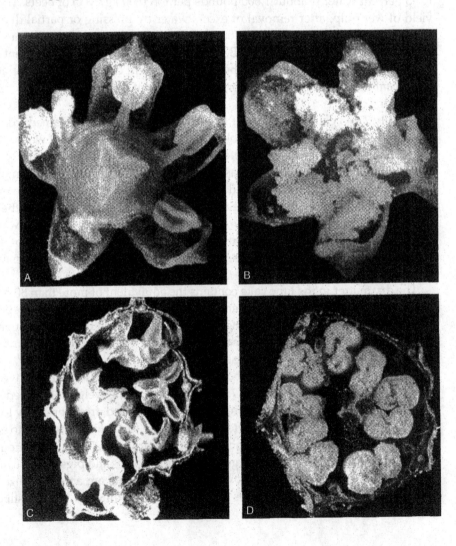

FIGURE 35.8
Sugarbeet flowers:
(A) male-sterile, *(B)* male-fertile, *(C)* clipped empty male-sterile anthers, and *(D)* clipped male-fertile anthers full of pollen.

are typically equipped with draper belts. The best time for swathing is during periods of high humidity, which reduces shattering. A combine with a pick-up attachment threshes the crop.

Seed is run across round-hole screens to remove sticks and other undesirable material, as well as any large, multigerm, self-pollinated seed balls from the smaller monogerm seed. It is then run across a draper (machine with cloth belts) that will carry additional trash up the inclined belt while the seeds roll down the belt. All seed-cleaning equipment must be cleaned between hybrids.

The biennial sugarbeet plant will become a winter-annual seed producer when grown in environments suitable for photothermal[7] induction of flowering. Thermal induction occurs at mean temperatures of about 45 to 55°F (7 to 10°C) and lasts for 90 to 110 days. The lengthening days of spring that follow cause a photoperiodic response, after which flowers and seeds are produced.

Earlier, production of sugarbeet seed was largely a European enterprise. Nearly all of the seed planted in America was imported because of the labor involved in beet-seed production. By the old methods, the beets for seed were planted in the usual manner but were left closer together in the row at thinning time in order to obtain more but smaller beets, usually called stecklings. These were dug in the fall and stored in a pit or pile covered with straw or earth or both. In the spring, the stecklings were set out in the field for seed production.

▓ 35.11 DISEASES

35.11.1 Cercospora Leaf Spot

Cercospora leaf spot, caused by the fungus *Cercospora beticola*, is one of the most prevalent of sugarbeet diseases.[31, 32] Circular sunken spots scattered over the leaf are about 1 to 2 millimeters in diameter. The center of a spot is ashy gray, but frequently the spot has a reddish-purple margin. When the spots are numerous, they coalesce, causing the leaf blades to become brown and dry. The outer leaves of the plant often show the blighting, which is a result of infection that took place when they were unfolding. Under severe attack, the affected leaves die, becoming brown or black, and the field looks brown or scorched.

The principal control measures for leaf spot are resistant varieties or hybrids and fungicidal sprays.[33]

35.11.2 Curly Top

Curly top is a virus disease that once caused heavy losses to sugarbeet growers in the intermountain region, and in California. Production declined so much that several beet-sugar factories were abandoned. The insect vector of curly top is the beet leafhopper (*Circulifer tenellus*), often called the white fly.[31] This leafhopper breeds and feeds upon about forty plant species, principally weeds of the goosefoot and mustard families.[34] Russian thistle and other weeds that spread on overgrazed range and abandoned fields caused the beet leafhopper and curly-top disease to increase. The leafhoppers move to sugarbeet fields after the range plants dry up in the spring. A portion of the leafhoppers carries the curly-top virus after feeding upon weeds that are subject to curly top. The beet leafhoppers introduce and spread the virus throughout a sugarbeet field.

The typical curling is upward. It is usually accompanied, more or less, by a roughening and distortion of the leaf veins. These symptoms are accompanied by a shortening of the petiole as well as a general retardation in the growth of the entire plant. The stunted beets commonly develop large numbers of tiny rooflets. The most satisfactory control is the planting of resistant hybrids. Most of those grown west of the Rocky Mountains are at least fairly resistant to curly top.

35.11.3 Rhizomania

The beet necrotic yellow vein virus causes Rhizomania disease. A soil fungus vector *(Polymyxa betae)* transmits the virus. The disease reduces taproot development and causes the production of a mass of fibrous roots to develop. In some cases, it will also cause chlorosis followed by necrosis of leaf blades and petioles. The fungal resting spores can survive in fields for fifteen to twenty years, so crop rotations provide little control. Careful sanitation of seeding, cultivating, and harvest equipment is necessary to prevent the spread of the disease from infected fields.[35] Water runoff from infected fields can also spread the disease. Seed potatoes from infected fields can spread the disease to sugarbeet fields when potato is included in a crop rotation.[18] The use of fungicides has had only limited success. Soil fumigants such as methyl bromide have been shown to reduce infection, but these are very expensive and highly regulated. The development of resistant varieties and hybrids offers the best solution to controlling this disease.

35.11.4 Seedling Diseases

A general complex of damping-off diseases leads to the death of the sugarbeet plant at the time that sprouting takes place or as the seedling is emerging from the soil. This complex also includes later phases of attack which occur on plants that have partly recovered from damping off. This disease complex has been called black root by growers.[31] Death of the plant may occur at the time of sprouting or when the plants emerge from the soil. Some plants persist in spite of fungus attack, but these may remain stunted or eventually die.

A number of fungus pathogens are responsible for sugarbeet black root. Soil-inhabiting organisms such as species of *Pythium* and *Rhizoctonia* cause the death of sugarbeet plants. A seed-borne fungus, *Phoma betae*, is also serious. The most serious loss, however, is apparently caused by *Aphanomyces cochlioides*, a water mold. This fungus does not kill the plant outright but dwarfs it because of a persistent attack on the lateral or feeding roots.

The control of seedling diseases or black root involves: (1) long rotations and crop sequences that keep soil infestation at a minimum, (2) proper drainage and fertilization so that the plants make vigorous growth, (3) seed treatment with fungicides, and (4) the growing of resistant hybrids.

35.11.5 Root or Crown Rots

Sugarbeet roots may rot in midseason because of the attack of a number of fungus pathogens, including *Pythium* and *Phymatotrichum*. *Rhizoctonia* crown rot is probably the most serious.

In many cases, the rotting of the mature root is a carry over from an attack of the fungus in the seedling stage. This is followed by partial recovery and then a renewed spread of the fungus on the half-grown plants. Such plants usually show a cleft top, but sometimes the entire crown breaks away. Since the fungus tends to spread down the row, it is common to find a number of contiguous plants affected with crown rot. Crop rotation, good culture, and the control of seedling diseases serve to check the losses from rots to a considerable extent.

35.11.6 Other Diseases

Downy mildew, caused by the fungus *Peronospora schachtii*, occurs in the coastal areas of California, but resistant sugarbeet hybrids are being grown there.[19] Virus yellows, a common disease of sugarbeet in Europe, is now present in several producing areas of the United States. Resistant varieties and hybrids are available.

35.11.7 Storage Diseases

Sugarbeet must be harvested before severe freezing weather occurs. Consequently, the rapid harvest greatly exceeds the processing capacity of the beet-sugar factory, and the roots are piled at the factory where they may be stored for several months. The roots keep well as long as they are alive and reasonably cool. Frozen roots, wounded roots, roots topped excessively low, and those roots whose tails die because of excessive drying are subject to attack by fungi. These fungi include not only such parasites as *Phoma betae*, *Rhizoctonia*, and *Phycomycetes*, but also saprophytes such as *Fusarium*, *Penicillium*, *Aspergillus*, and the *Mucors*. Prevention of storage losses requires careful handling of the beets and avoidance of excessive drying.[36] Blowing cold night air into the beet piles lowers the temperature, reduces respiration, and checks storage rots. A controlled atmosphere of 6 percent CO_2 and 5 percent oxygen at 36°F (2°C) gives the best sugar retention.

▩ 35.12 DEFICIENCY DISORDERS

Sugarbeet shows typical deficiency symptoms when the supply of any of the essential mineral elements is inadequate. Nitrogen deficiency produces typical yellowing effects, as does sulfur deficiency. Potash deficiency manifests itself by a reddish coloration or bronzing. When phosphorus is deficient, the plants show stunting and unbalanced proportions of roots to tops as well as characteristic necrotic blotches on the blades of the older leaves. In severe cases, the leaves dry and shrivel, with the leaf rolling in on the midrib from the tip to give a fiddle-neck effect. The severe aspects of phosphate deficiency have been called black heart,[37] a name that indicates the symptoms very poorly. Plants with severe phosphorus deficiency have dull-colored leaves, with occasional slight bronzing and purple spotting near margins

Boron deficiency causes blackening followed by death of the root heart, leaves, blackening or black markings on the inner faces of the petioles, and cankers on the roots. The flesh beneath the dried necrotic spot or canker is brown or lead-colored. Sugarbeet requires a relatively large amount of boron to avoid the deficiency symptoms. Applications of about 10 pounds of boron per acre (11 kg/ha) are adequate where this element has merely been leached out. In soils in which boron is bound as a result of high soil salinity, much heavier applications are necessary.[38] Boron deficiencies may occur in organic soils with a pH above 5.5 or in mineral soils with a pH above 6.6.

■ 35.13 NEMATODES

The sugarbeet nematode (Heterodera schachtii), a minute whitish eelworm, attacks the roots of sugarbeet and kills or stunts the plants.[39] Stunted roots may be covered with short, hairy rootlets. Infestation is spread through soil (dump dirt) that is transferred to the beet dumps on the roots and then returned to the fields. The pest is controlled by long rotations with immune crops such as alfalfa, bean, potato, small grains, and corn, with sugarbeet grown at intervals not less than every four or five years. Fumigation of the soil with chemicals is restoring sugarbeet culture to fields once abandoned because of nematode infestation. Such fumigation is expensive and highly regulated. The land must also be kept free from susceptible weeds such as mustard, lambsquarters, knotweed, ladysthumb, purslane, curly dock, and black nightshade, as well as from susceptible crops such as mangels and turnips. Early seeding when the soil temperature is below 60°F (16°C) allows sugarbeet to develop its root system before the nematodes become active.[40]

■ 35.14 INSECT PESTS

The larvae of the sugarbeet root maggot (Tenanops myopaeformis) feed on the sugarbeet root by scraping the root surface with rasping mouthparts. Damage is more severe during the seedling stage and early growth. Fully developed larvae overwinter and pupate the next spring, and the adult flies lay eggs in sugarbeet fields. Early planting gives sugarbeet plants a head start before feeding begins. Sugarbeet plants that have ten to fourteen true leaves at peak fly flight are better able to tolerate maggot feeding.[41] Deep fall plowing brings larvae to the soil surface increasing winterkill.[42] The application of a soil insecticide at planting may be necessary if infestations are high.

The beet leafhopper (Circulifer tenellus)[26, 43, 44] causes some damage by feeding on the leaf sap, but the main injury from this insect is the spread of curly-top virus. Losses are minimized by growing hybrids resistant to the virus. The beet webworm (Loxastege sticticalis) and the beet armyworm (Spodoptera exigua) devour the leaves. The latter insect may also attack the crown and roots of the plant. Flea beetles puncture the leaves. Other insects that attack sugarbeet include grasshoppers, leaf miners, wireworms, white grubs, sugarbeet root aphid *(Pemphigus betae)*, darkling beetles, salt marsh caterpillars, and spider mites.[45] Damage from these pests is reduced by applying suitable insecticides.

REFERENCES

1. Coons, G. H. "Improvement of the sugar beet," *USDA Yearbook,* 1936, pp. 625–655.

2. Tyson, J. "Influence of soil conditions, fertilizer treatments, and light intensity on growth, chemical composition, and enzymatic activities of sugar beets," *MI Agr. Exp. Sta. Tech. Bull.* 108, 1930.

3. Headden, W. P. "A soil study: The crop grown—sugar beets," *CO Agr. Exp. Sta. Bull.* 46, Part 1, 1898.

4. Terry, N. "Developmental physiology of the sugar-beet," *J. Exp. Bot.* 21, 67(1970):477–496.

5. Brandes, E. W., and G. H. Coons. "Climatic relations of sugarcane and sugar beet," in *Climate and Man,* USDA Yearbook, 1941, pp. 421–438.

6. Nuckols, S. B. "Sugar beet culture under irrigation in the northern Great Plains," *USDA Farmers Bull.* 1867, 1941, pp. 1–52.

7. Stewart, D. "New ways with seeds of sugar beets," in *Seeds,* USDA Yearbook, 1961, pp. 199–205.

8. Kohls, H. L. "The influence of some climatological factors on seedstalk development and seed yields of space-isolated mother beets," *J. Am. Soc. Agron.* 29(1937):280–285.

9. Palmer, T. B. *Sugar Beet Seed.* New York: Wiley, 1918, pp. 1–120.

10. Artschwager, E. "Anatomy of the vegetative organs of the sugar beet," *J. Agr. Res.* 33(1926):143–176.

11. Artschwager, E., and R. C. Starrett. "Suberization and wound-cork formation in the sugar beet as affected by temperature and relative humidity," *J. Agr. Res.* 47(1933):669–674.

12. Artschwager, E. "A study of the structure of sugar beets in relation to sugar content and type," *J. Agr. Res.* 40(1930):867–915.

13. Brewbaker, H. E. "Self-fertilization in sugar beets as influenced by type of isolator and other factors," *J. Agr. Res.* 48(1934):323–337.

14. Down, E. E., and C. A. Lavis. "Studies on methods for control of pollination in sugar beets," *J. Am. Soc. Agron.* 22(1930):1–9.

15. Artschwager, E. "Development of flowers and seed on the sugar beet," *J. Agr. Res.* 34(1927):1–25.

16. Coons, G. H., F. V. Owen, and D. Stewart. "Improvement of the sugar beet in the United States," *Adv. Agron.* 7(1955):89–139.

17. Owen, F. V. "Cytoplasmically inherited male-sterility in sugar beets," *J. of Agri. Res.* 71(1945):423–440.

18. Cooke, D. A., and R. K. Scott. *The Sugar Beet Crop.* London: Chapman & Hall, 1993.

19. McFarlane, J. S., F. V. Owen, and A. M. Murphy. "New hybrid sugar beet varieties for California," *J. Am. Soc. Sugar Beet Tech.* 11, 6(1961):500–506.

20. Mortvedt, J. J., D. G. Westfall, and R. L. Croissant. "Fertilizing sugar beets," *CO Coop. Ext. Serv.* 0.542, (rev.), 1999.

21. Cattanach, A., W. C. Dahnke, and C. Fanning. "Fertilizing sugarbeet," *ND Ext. Serv.* SF-714, (rev.), 1993.

22. Tolman, B., and A. Murphy. "Sugar beet culture in the intermountain area with curly top resistant varieties," *USDA Farmers Bull.* 1903, 1942, pp. 1–52.

23. Nuckols, S. B. "Yield and quality of sugar beets from various rotations at the Scotts Bluff (Nebr.) Field Station, 1930–35," *USDA Circ.* 444, 1937, pp. 1–14.

24. McMullen, M. P., and H. A. Lamey. "Crop rotations for managing plant disease," *ND Ext. Serv.* PP-705, (rev.), 1999.

25. Lill, J. G. "Sugar beet culture in the North Central States," *USDA Farmers Bull.* 2060, 1964.

26. Harris, F. S. *The Sugar Beet in America.* New York: Macmillan, Inc., 1926, pp. 1–342.

27. Barmington, R. D., and S. W. McBirney. "Mechanizing of sugar beets," *CO Agr. Exp. Sta. Bull,* 420-A, 1952.

28. Stout, M., S. W. McBirney, and C. A. Fort. "Developments in handling sugar beets," in *Crops in Peace and War,* USDA Yearbook, 1950–51, pp. 300–307.

29. Schroeder, J. W. "By-products and regionally available alternative feedstuffs for dairy cattle," *ND Ext. Serv.* AS-1180, 1999.

30. Skuderna, A. W., and E. W. Sheets. "Important sugar beet by-products and their utilization," *USDA Farmers Bull.* 1718, 1934, pp. 1–29.

31. Coons, G. H. "Some problems in growing sugar beets," in *Plant Diseases,* USDA Yearbook, 1953, pp. 509–524.

32. LeClerg, E. L. "Control of leaf spot of sugar beets," *MN Agr. Exten. Circ.* 46, 1934.

33. Kerr, E. D., and A. Weiss. "Cercospora leaf spot of sugar beet," *NE Coop. Ext. Serv. NebGuide* G98-1348-A, 1998.

34. Cook. W. C. "Life history, host plants and migrations of the beet leafhopper in the western United States," *USDA Tech. Bull.* 1365, 1967, pp. 1–122.

35. Smith, J. A., and E. D. Kerr. "Practices that reduce risk of spreading rhizomania," *NE Coop. Ext. Serv. NebGuide* NF93-121, 1993.

36. Tompkins, C. M., and S. B. Nuckols. "The relation of type of topping to storage losses in sugar beets," *Phytopath.* 20(1930):621–635.

37. Maxson, A. C. "Manure and phosphate," *Through the Leaves,* 18, 1(1930):33–37.

38. Cox, T. R. "Relation of boron to heart rot in the sugar beet," *J. Am. Soc. Agron.* 32, 5(1940):354–370.

39. Maxson, A. C. *Principal Insect Enemies of the Sugar Beet in the Territories Served by the Great Western Sugar Company.* Denver: Great Western Sugar Company, 1920, pp. 1–157.

40. Kerr, E. D., and F. A. Gray. "Sugar beet nematode," *NE Coop. Ext. Serv.NebGuide* G98-1348-A, 1998.

41. Glogoza, P. "Sugar beet insects," in *Field Crop Insect Management Recommendations, ND Ext. Serv.* E1143, (rev.), 2000.

42. Bechinski, E. J., R. L. Stoltz, and J. J. Gallian. "Integrated pest management guide for sugarbeet root maggot," *U. ID Coop. Ext. Serv.* CIS 999, 1993.

43. Cook, W. C. "The beet leafhopper," *USDA Farmers Bull.* 1886, 1941, pp. 1–21.

44. Peay, W. E. "Sugarbeet insects—how to control them," *USDA Farmer's Bull.* 2219, 1968 (rev.).

45. Dawson, J. H. "Weed control in sugarbeets with cycloate," *USDA Tech. Bull.* 1436, 1971, pp. 1–24.

Potato

KEY TERMS

Red potato
Russet potato
Seed potato
Solanin
Suberization
Tuber
White potato

36.1 ECONOMIC IMPORTANCE

Potato ranks as the third crop in worldwide tonnage and provides more calories than rice. Except for cereals, it is the world's most important food crop. World potato production averaged about 350 million tons (318 million MT) in 2000–2003. Potato was grown on about 48 million acres (19 million ha) with an average yield of about 7.5 tons per acre (17 MT/ha). The leading countries in potato production are China, the Russian Federation, India, the United States, and Ukraine (Figure 36.1). In the United States, potato was harvested on about 1.3 million acres (505,000 ha) in 2000–2003. Production was about 23 million tons (21 million MT) with an average yield of 18 tons per acre (40 MT/ha). The leading states in potato production are Idaho, Washington, Wisconsin, Colorado, and North Dakota (Figure 36.2).

36.2 HISTORY OF POTATO CULTURE

Potato originated in the Andes highlands, probably in Peru, Chile, or Bolivia. Hundreds of wild species of *Solanum* are found there, several of which have been cultivated. It has been proposed that two diploid species intercrossed naturally to produce a tetraploid cultivated type. This was collected by European explorers and later reselected to evolve the current typical cultivated species, *Solanum tuberosum*.[1] Potato apparently was cultivated well before 400 BC.

Spanish explorers found potato under cultivation in western South America in 1537, but not in Mexico. They had introduced it to Europe by 1565 or 1580. Potato reached England and Ireland somewhat later, but it did not become widely grown in any European country until after 1750. The potato was taken from England to Bermuda, and from there it was taken to Virginia in 1621. It was not generally grown in the United States until after it was introduced at Londonderry, New Hampshire, in 1719. The planting stock was obtained from Ireland by Scotch-Irish immigrants, hence the name Irish potato.[2, 3]

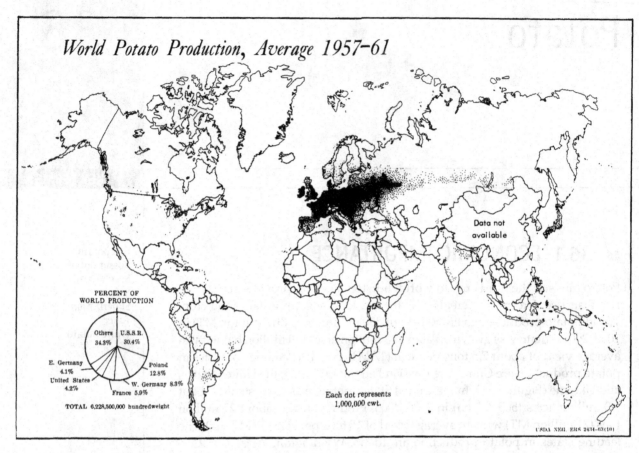

FIGURE 36.1
World potato production [Courtesy USDA]

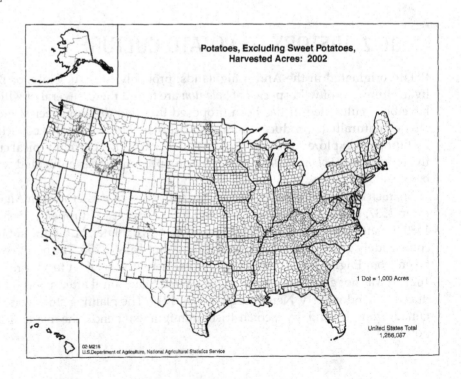

FIGURE 36.2
Potato growth in the United States. [Source: 2002 Census of U.S. Department of Agriculture]

■ 36.3 ADAPTATION

Potato is grown in every state in the United States, but most of the commercial crop is highly localized (Figure 36.3). The concentrated areas are either especially suited to potato production or are favorably situated to supply certain seasonal or regional markets. Winter potato is produced in California and Florida, which have mild winters, with summers too hot for growing fall potato. Spring potato is produced mostly in California, North Carolina, Texas, and Arizona. Colorado and Texas are leading states in summer potato. Other states with significant summer potato production are California, Missouri, Nebraska, Illinois, and Virginia. Idaho and Washington are the leading states in fall potato production. Other states with significant fall potato production are Colorado, North Dakota, Wisconsin, Oregon, and Minnesota.

Potato is a cool-weather plant that makes its best growth where the mean July temperature is about 70°F (21°C) or lower.[4] Major potato production is found mostly in cool climates such as those found in northern Europe and the northern United States. Young sprouts develop best at soil temperatures of about 75°F (24°C), but the best temperature for tuber growth is 60 to 65°F (16 to 18°C). Tuber production is retarded at soil temperatures above 68°F (20°C). Second

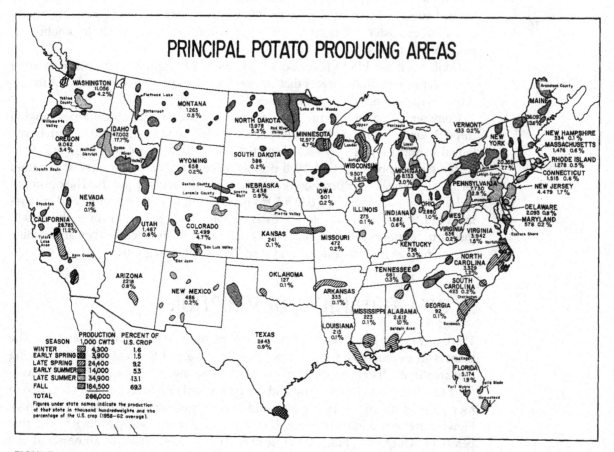

FIGURE 36.3

Areas where potato is produced for different seasonal markets.

growth may occur at 80°F (27°C), and growth is completely inhibited at 84°F (29°C), above which point the carbohydrates consumed by respiration exceed those produced by photosynthesis.[5] Potato grown in the South is planted in early spring or in the fall or winter so that growth takes place while the weather is cool. Where temperatures of 77°F (25°C) or higher prevail, the symptoms of mosaic are indistinct.[6] The potato plant withstands light frosts but frequently is injured by freezing in the fall, winter, or early spring. The freezing point of the tubers is 28 to 30°F (−2 to −1°C). When completely frozen, the tissues disintegrate soon after thawing.

Long days, high temperatures, and high amounts of nitrogen favor a heavy growth of stems and leaves[6] and delay tuber production. Short days, cool temperatures, or a deficiency of nitrogen favors early tuber production. Days of intermediate length, cool temperatures, and ample nitrogen favor maximum tuber production. Although flower primordia (rudimentary flowers) can be formed in either long or short days and even in darkness,[7] flowering and seed formation are favored by long days and cool temperatures. Thus, potato plants commonly produce seeds in the more northern states but not in the middle latitudes or in the South. However, varieties differ in their fertility.[8] Lack of fertility may be caused by abnormal chromosome behavior at the time of pollen formation. Considerable shedding of buds may occur even before pollen formation.

A uniform supply of soil moisture is essential to the production of good, well-formed tubers. Interrupted growth followed by later favorable growing conditions, or excessive nitrogen or high temperatures, may result in knobby or pear-shaped tubers.

Cool, moist conditions favor development of late blight disease. Warm weather favors reproduction of insects that transmit the mosaic viruses, which in turn accelerates the spread of mosaic diseases.

Commercial potato production is found on a wide range of soil types ranging through sandy loams, silt loams, loams, and peat.[9] Good yields of potato can be obtained in some fertile clay soils, but sticky soil adhering to the tubers interferes with digging and marketing. Well-drained porous soils are most desirable. Either fertile soil or heavy fertilization is essential to high yields of potato. The optimum soil pH for potato is about 5.0 to 5.5 from the standpoint of both yield and scab retardation. Scab disease develops mostly on soils above pH 5. However, the soils in the north central potato areas are nearly neutral in reaction, while they are alkaline in the irrigated intermountain and Great Plains potato regions.

■ 36.4 BOTANICAL DESCRIPTION

The cultivated potato (*Solanum tuberosum*) belongs to the family *Solanaceae*. The plant is an annual having stout, erect, branched stems 1 to 2 feet (30 to 60 cm) long that are slightly hairy and distinctly winged at the angles.[2] The slightly hairy leaves are 1 to 2 feet (30 to 60 cm) long and are comprised of one terminal leaflet, two to four pairs of oblong acute leaflets, and two or more short leaflets (Figure 36.4). Flowers are borne in compound, terminal cymes with long peduncles. The five petals are white, rose, lilac, or purple in color. The flower has five anthers and one pistil with a long style.[2, 10] The potato ball or fruit is a smooth, globose, green, or brown berry (Figure 36.5) less than an inch (25 mm) in diameter. The rhizomes,

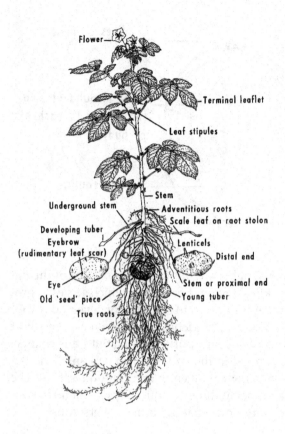

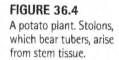

FIGURE 36.4
A potato plant. Stolons, which bear tubers, arise from stem tissue.

FIGURE 36.5
Potato "balls" that contain seeds.

which are 2 to 4 inches (5 to 10 cm) long, enlarge at the outer end to form a tuber. The tuber is a modified stem with lateral branches forming what are known as potato eyes.[11] Each eye contains at least three buds protected by scales. The eyes are arranged around the tuber in the form of a spiral.

The interior of the tuber shows a pithy central core with branches leading to each of the eyes. Surrounding the pith is the parenchyma, in which most of the starch is deposited. Toward the outer part of the tuber is the vascular ring that contains the cambium, as well as an outer cortex that contains the pink, red, or purple pigment of colored-skinned varieties (Figure 36.6). The potato skin (periderm) is a layer six to ten cells deep, composed largely of cork (or suberin) having a basic composition similar to that of fatty substances. Scales form on the outer surface of the periderm. Openings in the periderm, called lenticels, become enlarged under moist conditions.

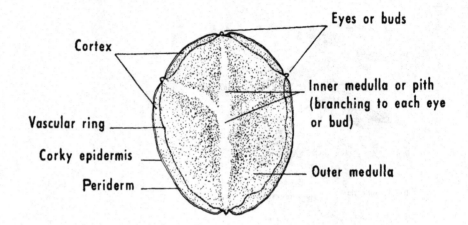

FIGURE 36.6

Cross section of a potato tuber. In red varieties, the pigment occurs in the periderm or outer cortex. The pith has a water-soaked appearance.

When a potato is cut and left in a suitable environment, the surface of the cut *suberizes* (corks over) to form a new skin—a wound periderm, which serves to protect the tuber from decay. When a tuber is exposed to sunlight for several days, either before or after digging, the skin of most varieties turns green as a result of development of chlorophyll. Some varieties turn purple. Along with this change, increased quantities of solanin are formed in the cortex. Solanin is an alkaloidal glucoside, bitter in taste, and poisonous when taken in sufficient quantities. The poisonous alkaloid in solution is called solanidine. The quantity of solanin in sunburned tubers may be more than twenty times that in normal tubers. Most of the solanin is discarded with the peelings.

▓ 36.5 VARIETIES

The genetic diversity of cultivated and wild potato is greater than that of any other major world food crop. In addition to *S. tuberosum*, there are seven other cultivated species and over 200 wild species of potato.[12] Most cultivated potato is tetraploid with forty-eight chromosomes. Cultivated potato is very heterozygous, but asexual propagation maintains varietal characteristics. Due to heterozygosity, the probability of selecting offspring that are superior to the parents is very low.

Present varieties grown today belong to three basic types:

1. *White:* white or light skinned with oblong to round shape. Varieties include Superior, Kennebec, Norchip, Atlantic, Shepody, and Sebago. White potato is used mostly for boiling and chipping.
2. *Red:* Red skinned with round shape. Varieties include the Norlands, Red Pontiac, and Chieftain. Red potato is produced primarily for the fresh market.
3. *Russet:* Russeted skin with long cylindrical or slightly flattened shape. Varieties include Russet Burbank, Russet Norkotah, and Centennial Russet. The russet varieties are used primarily for baking and French fries.[13]

Figure 36.7 shows four representative varieties of potato. Table 36.1 lists characteristics of each type. Ideally, tubers should have shallow eyes to facilitate peeling.

Consumers of the United States rejected two productive yellow-fleshed varieties distributed in the 1940s. This yellow-flesh characteristic is popular in several

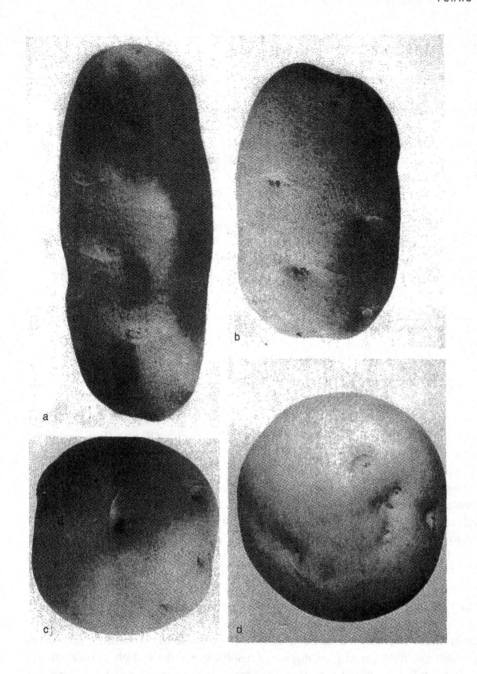

FIGURE 36.7
Tubers of four types of potato: *(a)* russet, *(b)* oval round, *(c)* round red, *(d)* round white.

European and Latin American countries because of its assumed high vitamin A value, as well as its rich appearance. Some potato grown in South America has purple flesh.

Development of new, improved varieties and elimination of inferior or un-adapted varieties have been important factors in the improvement of potato yields. Along with this has been the use of better planting stocks, increased use of fertilizer, better control of insects and diseases, better cultural methods, and concentration of the crop in the more favorable localities. All of these factors together are responsible for the consistent increases in potato yields from an average of 4,800 pounds per acre (5,400 kg/ha) around 1890 to about 35,000 pounds per acre (39,000 kg/ha) at the end of the twentieth century.

TABLE 36.1 Potato Types and Their Characteristics

	Red	Russet	White
Skin color	Red	Brown	Tan
Skin texture	Smooth	Rough	Smooth
Shape	Round/oblong	Oblong/long	Round/oval
Key appearance	Redness	Longness	Roundness
Flesh color	White	White	White
Specific gravity	1.055–1.075	1.070–1.085	1.075–1.105
Dry matter	15–19% (low)	17–21% (medium)	18–24% (high)
Starch	10–13% (low)	12–15% (medium)	13–18% (high)
Cooking texture	Waxy/pasty	Flakey/meally	Grainy/dry
Glucose	>0.035% (high)	>0.020% (medium)	>0.030% (high)
Frying color*	Brown	Tan/off-white	White
Taste	Slightly sweet	Bland/soft	Bland/dry
Main use	Salad/boiled	Fries/baked	Chips/frying/baking

*At harvest, after good growing conditions.
Source: Pavlista, 1997

▨ 36.6 FERTILIZERS

Potato responds well to applied fertilizers. A 20-ton (18 MT) crop of potato will absorb about 171 pounds (78 kg) of nitrogen, 22 pounds (10 kg) of phosphorus, and 195 pounds (88 kg) of potassium.[14]

Most nitrogen uptake by the entire potato plant occurs early in the season, and uptake slows as the plant nears maturity. Uptake by the tubers, however, occurs at a constant rate from initiation to maturity, and considerable nitrogen is translocated from the vines to the tubers.[15] The amount of nitrogen fertilizer is directly related to soil nitrogen level and yield goal. Application of one-third to one-half of the total nitrogen fertilizer before planting is advisable unless the soil is very high in residual nitrogen. Additional application of nitrogen the first half of the growing season minimizes nitrogen losses.[14, 16] Late application of nitrogen can delay maturity. If the potato crop is sprinkler irrigated, injecting nitrogen through the irrigation system is an efficient method of application.

Phosphorus fertilizer will be needed on most potato fields. A banded application of phosphorus fertilizer at planting may be sufficient if the soil phosphate level is high. If the soil is lower in phosphate, a broadcast application of phosphorus fertilizer before planting during seedbed preparation may be needed in addition to a banded application at planting.[14, 16] Banded application of phosphorus assures that potato plants have access to a sufficient supply of phosphate early in the growing season.

Potassium fertilizer will usually be needed unless the soil potassium level is very high. If only small amounts of potassium fertilizer are needed, it can be band applied near the seed at planting. High amounts of potassium fertilizer should not be banded near or with the seed as it may interfere with sprouting and early growth of the potato plants.[17] On soils with less residual potassium, a broadcast application of potassium during seedbed preparation may be needed.[14, 16]

Calcium and magnesium fertilizers may be needed on acid soils that have not been limed. There is a reluctance to lime potato fields as the incidence of potato scab increases when soil pH is above 5.2.[14] Calcium and magnesium fertilizer should be broadcast applied before planting. A banded application of sulfur may be needed on sandy soils.

Micronutrients that may be deficient in potato are boron, manganese, iron, and zinc. A banded application of zinc fertilizer may be needed when soil pH is greater than 7. Iron deficiency may occur on calcareous soils. Iron deficiencies are difficult to correct, since the problem is unavailability of iron to the plant, rather than a true soil deficiency. Some potato varieties are less susceptible to iron deficiency. Boron is most likely to be deficient on very acid soils or calcareous soils.

■ 36.7 ROTATIONS

Rotations in which potato is grown three or more years apart are essential to help control soil-borne diseases. Potato usually succeeds best after a green-manure crop or after a legume crop such as clover, alfalfa, sweetclover, vetch, or pea, with the residues turned under in the fall. In the northeastern states, the rotation is usually potato, small grain, and legume. In the Corn Belt, the rotation may be corn, potato, small grain, and soybean. The potato field leaves a good seedbed that requires little tillage for small grains. Sometimes two consecutive crops of potato are grown. In the irrigated regions, potato often follows alfalfa that has been *crowned* (crowns cut) in the fall and replowed in the spring. Sugarbeet follows the potato crop advantageously, and the beet field makes a good seedbed for sowing small grain and alfalfa.

In the South, potato planted in early spring often follows an early fall-sown green-manure crop such as crimson clover, vetch, or winter pea. Potato planted in the fall or winter can follow a summer legume such as lespedeza, cowpea, or soybean.

Potato responds well to green manure[18] except on peat and muck soils. Rotations in which potato is grown three or more years apart reduce losses from diseases caused by soil-borne organisms. Such diseases include scab, verticillium wilt, Fusarium wilt, and blackleg.

In the dryland regions, potato is usually planted on fallowed land or after corn or beans. The potato crop is likely to fail on dry lands unless the soil is moist to a depth of 3 feet (90 cm) or more at planting time. A spring grain crop usually follows potato.

■ 36.8 SEED POTATO

The planting of certified, disease-free seed is essential to reduce infection by viruses causing mosaic, leaf roll, spindle tuber, and other diseases. Some seed tuber production is done in carefully managed fields that are regularly inspected for diseases. A relatively recent trend is production of seed tubers under sterile laboratory conditions designed to prevent diseases and produce a uniform-sized seed tuber. In these cases, seed tubers are produced from *in vitro* cuttings to increase seed tuber production, assure freedom from pathogens, and produce a very uniform product in only forty to fifty days.

Certified seed tubers are produced for up to six generations in most states before it is necessary to return to foundation or "nuclear" seed stock.[13, 19] If seed tuber production is tightly controlled, there is no reduction in yield when seed tubers are produced from later generations of seed stock.

Potato growers can purchase small seed tubers that are planted whole, or they can purchase larger tubers that are divided shortly before planting. Regardless of the type of seed tuber used, careful attention to storage before and after purchase helps to maintain seed tuber viability.

Certified seed production in the United States is ample for the total commercial domestic planting. Certified potato seed is exported in considerable quantities, but some is imported from Canada. The leading states in certified seed production are Idaho, North Dakota, Maine, Colorado, and Minnesota. Cool seasonal conditions in these states are very suitable for producing certified potato seed because mosaic-disease symptoms come to full expression in the field, and the diseased plants can be rogued or removed easily. Also, the cool temperature conditions in the North are suited to the growing of well-developed tubers, and the late harvest and cold winters facilitate the storage of seed supplies. Both field and bin inspections are required for certification.

36.8.1 Storage

Tubers left in the air or in dry soil will produce sprouts but no roots. In moist soil or some other wet medium, both sprouts and roots are formed. Large vigorous sprouts are an indication of quick emergence and relatively early maturity. In darkness or subdued light, the sprouts are long and lack green color. In sunlight, the sprouts are short, and soon turn green, and the tuber also becomes green. Occasionally, tubers are prepared for planting by being exposed to light, a process called greening.

Since each potato eye usually contains several buds, sprouts will be produced even after some of the sprouts are removed or damaged. The removal of one crop of sprouts is only slightly detrimental to the tuber, although growth and removal of sprouts reduce the vigor of subsequent sprouts.[13, 20] However, all eyes do not sprout at once unless the tubers are cut up because of apical dominance. The eyes nearest the bud (outer) end of the tuber sprout first. Long storage destroys apical dominance.[20]

Sprouting of potato tubers in storage occurs at temperatures of 40°F (4°C) or higher, but only after completion of a rest period, which is caused by restricted oxidation.[21] The rest period may range from four to sixteen weeks when the potato is stored at 70°F (21°C) or longer at cooler storage temperatures. Immature tubers may have a longer rest period than do those that are fully mature.

Seed tubers should be stored in clean, dry containers that have no traces of fertilizer or pesticide residues, or the sprout inhibiting chemical CIPC. It is best to disinfect storage containers with bleach, iodine, or another approved chemical and then thoroughly rinse and allow to dry.[22]

36.8.2 Size of Seed Piece

The planting of certified, nearly disease-free seed and the prevalent practice of closer planting have altered the former situation in which the planting of small tubers was hazardous because they were often diseased.[2] The best size range for planting whole tubers is 1¾ to 2½ ounces (50 to 70 g).[13] Specialized seed potato growers often plant at higher populations so the tubers will not be too large. Larger tubers that are cut with mechanical seed cutters are cut into 1.5- to 3-ounce (40 to 85 g) pieces.[13] Blocky seed pieces are best. Uniformity of seed tubers is important to allow mechanical potato planters to uniformly space seed tubers in the field.

36.8.3 Preparation of Seed Tubers

Where conditions permit, potato pieces are cut immediately before planting. The cut seed is sometimes dusted with fir bark dust to keep the pieces from sticking together. Freshly cut pieces will suberize or heal after planting.[13] Often, labor is more readily available when fields are too wet for tillage operations, which makes it advisable to cut the seed before planting time. In the latter case, it is usually recommended that tubers be suberized before planting. This consists of handling the tubers so as to promote development of cork tissue on the cut surface. The tubers are held in crates or sacks for a day or two after cutting and then emptied out to separate any pieces that have stuck together. Throughout the ten-day period of suberization, the temperature should be kept at about 60°F (16°C), with a high humidity (about 85 percent) and good ventilation. Humidity can be maintained by wetting the storage room or keeping wet sacks over the pile of cut seed.[23] Suberized potato pieces can usually be stored up to thirty days after cutting without a reduction in stand or yield in the resultant crop. Shrinkage of the suberized tubers should not exceed 2 or 3 percent in thirty days under good storage conditions.[24] Cutting may be done by special machines or by hand.[25]

Seed potato pieces may be treated with disinfectants for the control of fusarium and rhizoctonia when disease-free certified seed is not used.[26]

Clean potato pieces need no treatment. Clean pieces, or even treated pieces, planted in scabby soil or in soil heavily infested with the rhizoctonia organism will produce diseased tubers. A fungicide such as mancozeb, thiophanate, or fludioxonil may be applied to the cut seed to prevent decay and control seed diseases.

▓ 36.9 POTATO CULTURE

36.9.1 Seedbed Preparation

Plowing for potato is often done in the fall to turn under legume residues, but spring plowing is satisfactory when it can be completed two to four weeks in advance of planting, so that any vegetation has time to decompose. Deep plowing for potato is usually recommended. It is essential that the seedbed be prepared at least 2 inches (5 cm) deeper than the 3- to 5-inch (8 to 13 cm) depth at which potato is planted. Most potato roots are found within 1 foot (30 cm) of the surface of the soil, but since they can penetrate to a depth of 5½ inches (170 cm), there is no advantage in plowing deeply to facilitate root growth. Subsoiling in preparation for potato is of no advantage even in heavy clay underlain by a clay subsoil.[27] However, heavy soils in California frequently are tilled with a chisel at a depth of 12 to 16 inches (30 to 40 cm) in preparation for potato. A very compact and fine seedbed, which is desirable for small-seeded crops, is not essential for potato. In the South, the land is often bedded up with a lister so that the potato rows on the beds will be well drained.

36.9.2 Planting Methods

Four general types of potato planters are used: cup type, assisted feed, tuber unit, and automatic picker.[13] The most popular is the automatic picker planter that selects the seed piece by jabbing it with a pointed spike, which may spread diseases to healthy seed pieces. Cup type planters that require one or two operators are most suitable for planting whole tubers. The assisted feed and tuber unit planters require one or two persons to help distribute the seed pieces in the compartments so that one piece is dropped each time. The last two give the most uniform stands but operate slower than the other two. Planting is usually 4 to 6 inches (10 to 15 cm) deep. It is deeper in sandy soil than in heavy soil, in dry soil than in wet soil, in late planting than in early planting, and in warm soil than in cold soil.

The amount of tuber seed planted depends on seed piece size, row spacing, and variety characteristics. The rate of planting potato in the United States is about 1,700 to 2,500 pounds per acre (1,900 to 2,800 kg/ha), but the rates can range from about 1,100 to 4,300 pounds per acre (1,200 to 4,800 kg/ha).[13, 28] Lower rates are used in certain dryland areas. Under irrigated or humid conditions where the soil is fertile or heavily fertilized, close planting results in better yields and fewer oversized tubers. About 2,200 pounds per acre (2,500 kg/ha) are required for planting 2- to 2½-ounce (55 to 70 g) seed pieces 1 inch (25 cm) apart in 36-inch (90 cm) rows. Varieties that produce few tubers per hill require thick planting, often 6 inches (15 cm) apart in the row, so the tubers will not be too large for the best market. In general, the best yield and quality of tuber are obtained from relatively close spacing.[29] Thick planting and large seed pieces reduce the incidence of oversized tubers, hollow heart, and growth cracked tubers.

In the northern part of the United States, the planting of potato often begins as early as the soil can be tilled after the land has thawed and has become sufficiently dry for seedbed tillage. The minimum temperature for any sprout growth in the potato is 40°F (4°C), but 50°F (10°C) is the recommended minimum soil temperature to reduce the risk of failure. Planting in the late winter and spring is done when the mean air temperature has risen to about 50 to 55°F (10 to 13°C). In the spring, the soil temperature at a 4-inch (10 cm) depth, where potato is usually planted, is about the same as the mean air temperature. In the northern portion of the country, this temperature is reached about ten to fourteen days before the average date of the last killing frost. In the central latitudes of the United States, planting is generally about four weeks before, but in the southern latitudes about six weeks before, the average last killing frost in the spring. The milder cold spring periods in the South do not freeze the ground as deep as the potato is planted. This permits relatively earlier planting with little fear of spring frosts. Winter potato is planted in October or November in the southern parts of California and Florida (Figure 36.8). Spring planting begins about January 20 to February 1 in southern Georgia at latitude 31° N and about May 10 to 20 in Aroostook County, Maine, at about latitude 46° N. Therefore, in the low altitude Atlantic coastal region, the planting date differs by one day for about every 8 miles (13 km) north or south, or

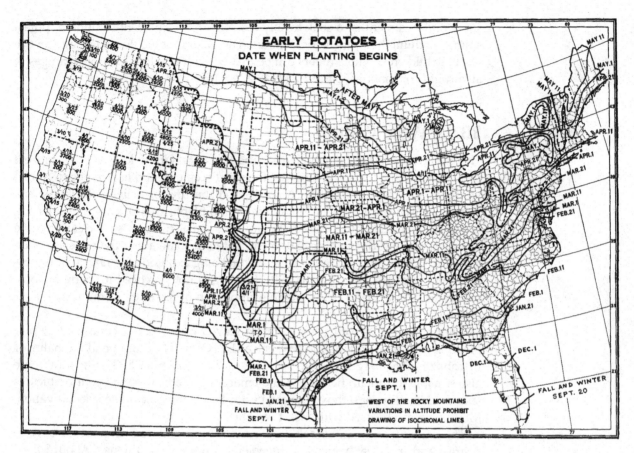

FIGURE 36.8
Dates when potato planting begins.

one day for every ⅛ degree difference in latitude. For higher elevations at a given latitude, planting is delayed about one day for each additional 100 feet (30 m) in altitude.

The period between planting and digging is about 3½ to 4 months for the early crop in all regions and for the main crop in the northern border states, and 3 to 3½ months for the late or fall crop where early growth is not retarded by cold weather. The late crop is usually planted sufficiently early for the tubers to be fully mature at or before the date of the average first killing frost. Thus, quick-maturing varieties can be planted two to four weeks later than late (long season) varieties. A late variety tends to produce the highest yield when it is planted early, and the tubers are able to reach maturity before frost. An early variety, however, usually gives the highest yields when it is planted late.[30] Most of the potato crop consists of fall potato grown in the northern half of the United States.

Two crops of potato can be grown per year, sometimes in the same field, in the area from Maryland and Kentucky southward. The fall or late crop is planted from July to September in time for the tubers to mature before freezing weather. The seed for the late crop is obtained either from stored tubers, usually from the North, or from the early southern crop. Certain varieties of these early crop tubers do not sprout promptly because the rest period has not been broken. Freshly harvested tubers of certain varieties will sprout promptly after chemical treatment. A fairly successful treatment is to soak freshly cut seed in a 1½ percent solution of potassium or sodium thiocyanate for 1½ hours. The tubers should be cut through an eye or at the bud end. Whole tubers can be treated with ethylene chlorohydrin gas in an airtight container at 75 to 80°F (24 to 27°C) for five days.[2]

36.9.3 Weed Control

There is no benefit[31] to more tillage beyond that necessary to kill weeds, because it merely injures the roots and dries out the soil. Preemergent tillage to destroy small weeds is an accepted practice. A spike tooth harrow or a light, flexible shank weeder is used once or twice before the potato plants are up. This is usually followed by deep cultivation close to the row as soon as the plants are clearly visible to loosen the soil for tuber development. Later, in some areas, the rows are often ridged to prevent the shallower tubers from being exposed to sunburn, or to protect them from freezing (Figure 36.9). Ridging is not practiced in many regions, particularly on dryland fields where evaporation of moisture from the soil is increased by ridging.

Annual grasses as well as many broad-leaved weeds can be controlled by pre-planting treatment with herbicides such as EPTC, metolachlor+pendimethalin, or metribuzin. Preemergent herbicides such as rimsulfuron or EPTC+trifluralin are also suitable for potato fields. For late-emerging weeds, postemergent herbicides such as clethodim, or sethoxydim can be applied at the time of the last cultivation. Metribuzin can be used with russet varieties.

The growing of potato under straw mulch lowers the soil temperature several degrees and appreciably increases the yields under hot conditions.[32] About 8 to 10 tons per acre (18 to 22 MT/ha) of straw will leave a mulch 4 inches (10 cm) deep after settling. In the cooler northern border states, mulching has lowered the yields.

FIGURE 36.9
A potato field that was ridged to better bury developing tubers. [Courtesy USDA NRCS]

Mulches should be applied after the soil begins to warm. Weeds are able to penetrate a straw mulch when the rate of application is less than 8 tons per acre (18 MT/ha).

36.9.4 Irrigation

Under irrigated conditions, water should be applied often enough to keep the crop well supplied with soil moisture (above 65 percent of field capacity), but water should always be applied before the plants show the need for moisture by a dark-green color in the leaves. Keeping soil moisture relatively uniform prevents the development of knobby tubers, growth cracks, and decreases common scab infection. Potato is usually irrigated in furrows, but sprinkler irrigation can also be used. The field should not be flooded. A seasonal total of 18 to 24 inches (460 to 600 mm) of water in several applications is usually ample for potato in the intermountain and Great Plains regions. On sandy soils in warmer regions, heavier irrigation may be necessary. The largest rate of water use is after the crop blooms and the tubers are making rapid growth. Potato roots may absorb 57 percent of the water used by the crop from the surface foot (30 cm) of soil, 23 percent from the second foot (30 to 60 cm), 13 percent from the third foot (60 to 90 cm), and 7 percent from the fourth foot (90 to 120 cm). Roots feed below the second foot only later in the season.

36.9.5 Vine Killing

Killing the vines before harvest stops growth hastens maturity, stops the formation of oversized and hollow heart tubers, and reduces the spread of blight infection from the vines to the tubers.[13, 33] Killing the vines may permit digging before heavy freezing occurs and limits the tuber size of seed potato. Tubers ripen quickly, skins thicken, and digging is easier after the vines are killed. Various chemicals, including sulfuric acid, diquat, and paraquat have been used for vine killing.[34, 35] Vines are also killed with mechanical beaters and rotary mowers and by flaming. Rollers are sometimes used to crush stems and mash down vines and improve coverage of vine killing sprays, but they can spread disease throughout the field.[13] Vine destruction facilitates mechanical digging.

36.9.6 Harvesting

Potato is harvested in all months of the year from January to December in some part of the United States. Early potato is ordinarily dug as soon as the tubers are large enough for the market. These new potatoes, in addition to being small, have thin, poorly suberized skins that are easily damaged or rubbed off in handling. They are watery and somewhat low in starch content. High seasonal prices justify their premature harvest. In potato not harvested early, the number of tubers per hill and the percentage of starch, protein, and ash continue to increase until most of the leaves are dead. The total weight of tubers may increase fivefold between blossoming and the stage at which all the leaves are dead[20] (Figure 36.10). In this same period, the sugar content may decrease from 1.0 down to 0.2 percent, and the moisture content drops from 84 down to 80 percent, but the crude-fiber content remains almost constant. The starch content of tubers may increase from 11.5 percent to

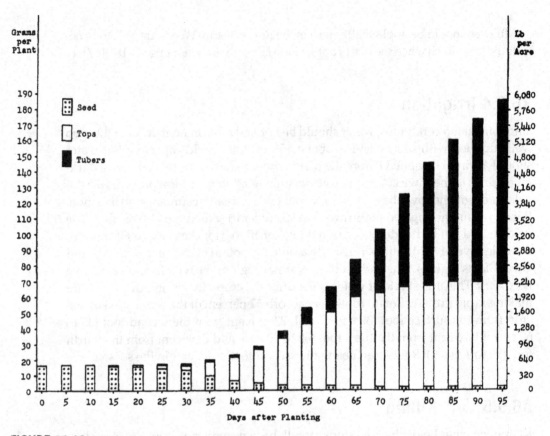

FIGURE 36.10
Dry weight of potato tops, tubers, and seed pieces at various stages of growth. [Courtesy Maine Agricultural Experiment Station]

17 percent during the three weeks between the time when all the blooms have fallen and when 80 percent of the leaves are dead. The total weight of tubers may double in the last twenty days of the growing period.[36]

Commercial potato is dug with a mechanical digger. The small two-row and four-row diggers shake and screen out the soil through round bars and deposit the tubers directly into bulk trucks. In some areas two, four, or more rows are harvested with a potato windrower, and the tubers are placed between two undug rows. Harvesters then dig the remaining two rows and deposit the tubers into waiting trucks. The harvest is passed over sorting machines to screen out small tubers, either at a warehouse or shed, or sorted by machine in the field before they are hauled. Windrowing increases the acres that can be harvested with a machine during harvest season, reduces machine traffic over the soil, and increases the volume of potato within the harvester.[13] Careful operation of the windrower and harvester is essential in order to keep mechanical bruising of the tubers at a minimum.

Careful handling in all operations will avoid bruises or other damage to the tubers. Harvesting when soil moisture is between 60 and 80 percent of field capacity for medium and coarse-textured soils reduces bruising.[13] Tuber temperature equals soil temperature and bruising increases as soil temperature drops from 60 to 50°F (16 to 10°C). Tubers are especially susceptible to bruising when soil temperature is less than 50°F (10°C).[13] Delaying harvest up to three weeks after vine kill toughens the skin on tubers and reduces bruising. In hot, dry weather, the tubers must be picked up within a few hours after digging in order to avoid sun scald. Most of the tubers are washed with jets of water before they are sorted and graded.

36.10 STORAGE

Potato storage occurs in three phases—curing, holding, and warming. For best keeping quality, potato going into storage should be cured at about 50 to 60°F (10 to 16°C) with a humidity of 85 percent for two weeks to permit injured tubers to heal. Lower temperatures are used for potato that will be processed into frozen French fries, dehydrated products, or sold on the fresh market.[13] During the holding phase, the temperature should be lowered to 40 to 50°F (4 to 10°C),[13] when it is desired to keep them from sprouting for an indefinite period. Chip potatoes are stored at a minimum of 50°F (10°C) to prevent increase in sugar content. Potato for the fresh market or processing should be stored at 38 to 40°F (3 to 4°C), for chips at 50°F (10°C), and for French fries at 45°F (7°C).[13] Seed potato is stored at 36 to 40°F (2 to 4°C).

At 40°F (4°C), potato will remain dormant for three to five months after harvest. Potato stored at 32°F (0°C) does not freeze, but it does not keep as well as when stored at 40°F (4°C). Potato is stored occasionally in refrigerated chambers, but usually, in cold climates, they are kept in pits, cellars, or warehouses, the latter that are mostly or partly underground. At low temperatures, potato can be stored in large bins without danger of heating. Ventilation of the storage place, often by forced air, is necessary to prevent spoilage, because the tubers are living organisms that continue respiration with evolution of heat, water, and carbon dioxide. The ventilating airflow should be 0.5–0.7 cubic feet per minute (14 to 20 l/min) for every 100 pounds (45 kg) of potato.

Warm air admitted to cold potato tubers causes condensation of moisture, which in turn encourages decay. Consequently, ventilators should be closed when the outside temperature is higher than that in the cellar. The relative humidity should be kept at 95 percent to prevent shrinkage of the tubers.

Conversion of starch into sugar occurs more rapidly when tubers are stored below 50°F (10°C). At 36°F (2°C), 25 to 30 percent of the starch is converted into sugar within six weeks,[2] and the tubers are sweet and soggy instead of mealy, and thus of poor cooking quality. When slices of such potato are French fried or cooked into chips, they have an unattractive dark brown color. Starch conversion is much less at 40°F (4°C) and scarcely perceptible at 50°F (10°C). The sugar reverts largely to starch when the storage temperature is raised to 50 to 70°F (10 to 21°C)[35] for a few weeks or sometimes for only a few days. For immediate or early use, potato should be stored at 50 to 70°F (10 to 21°C). At 60 to 70° F (16 to 21°C), nearly all the initial sugar present is eventually used up in respiration, and the tubers become mealier.

For home storage at moderate temperatures, sprouting can be reduced by storing apples in the same compartment. The apples emit ethylene that stops sprouting. When it is necessary to hold potato tubers for some time at higher storage temperatures, or to ship them southward, sprouting of tubers can be retarded by dusting, spraying, or fume circulation with chemicals such as CIPC in the storage area.[37]

Irradiation with gamma rays after seventy days of normal storage kept tubers from sprouting for 502 days, and they were firm and salable 300 days after treatment.[38]

▦ 36.11 USES OF POTATO

In the United States, the annual per capita consumption was about 198 pounds (90 kg) in 1910 and declined to 103 pounds (47 kg) in 1952, but then it increased to over 130 pounds (59 kg) by the 1990s. About 65 percent of potato consumed is processed as chips, dehydrated, or frozen (Figure 36.11).[13] The Incas of South America practiced freeze drying of potato.[3] Dehydrated potato is also used in whisky manufacture. Some 8 to 12 million pounds (3,600 to 5,400 MT) are processed into starch and flour. In Germany, much potato is processed into starch and alcohol or is fed to livestock. Russians make vodka from potato.

The average whole potato consists of about 79 percent water, 1 percent ash, 2 percent protein, 0.1 percent fat, 0.6 percent crude fiber, and 17 to 18 percent nitrogen-free extract. About 20 percent of a large sound tuber consists of the peel (periderm or skin) plus some of the layers of the cortex.[39] The remainder is composed of about 78 percent water, 2.2 percent protein, 1 percent ash, 0.1 percent fat, 0.4 percent crude fiber, and 18 percent starch, sugar, and other carbohydrates. Dry matter consists of about 70 percent starch, 15 percent other carbohydrates, 10 percent protein, 4.4 percent ash, and 0.3 percent fat. High starch content, high amount of solids, and high specific gravity are associated with mealiness and good cooking quality for chips, baking, boiling, mashing, or French frying. Potato solids and starch content can be measured by determination of the specific gravity of the whole tuber.

The mealiest baking potato has a specific gravity of 1.085 or higher, which indicates solids content above 20 percent as well as starch content of about 15 percent (Table 36.1). Tubers most suitable for boiling, mashing, or French frying have specific gravities of 1.080 or higher, with at least 19.8 percent solids and 14 percent starch. Potato can be sorted mechanically for these differences by immersion in

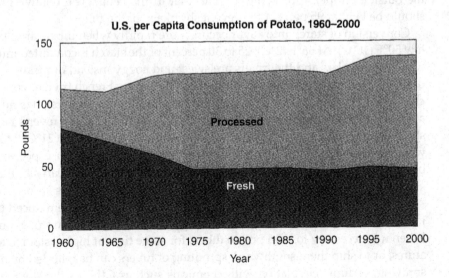

FIGURE 36.11
Trends in potato consumption patterns. [Courtesy USDA]

tanks that contain salt brine or other solutions of the desired specific gravity. Sorted potato is later washed and rinsed to remove the salts, which would otherwise damage the tubers.[2, 40]

Potato of high specific gravity may be produced by: (1) early planting and delayed harvest to permit full maturity, (2) avoiding excessive applications of potassium and nitrogen, (3) providing adequate but not excessive amounts of irrigation water or fertilizers, (4) using slow-action chemicals for vine killing, and (5) when possible, growing the crop where sunny weather with temperatures below 70°F (21°C) prevails late in the season.[2, 41, 42]

▓ 36.12 DISEASES

36.12.1 Diseases Caused by Viruses

Virus diseases include leaf roll, mild mosaic, rugose mosaic, latent mosaic, and calico. Tubers from infected plants carry the virus, and, when planted, the plants become diseased. The affected plants should be removed from fields producing seed stock, but it is not necessary to remove them from plantings for table use. The virus is spread considerably by insects, especially aphids. Use of certified seed reduces losses from virus diseases.[26, 43, 44]

Leaf roll is caused by the potato leaf roll virus (PLRV). The symptom is leathery leaflets that roll upward at the edge. These rolled leaves become yellow-green instead of dark green, and the plants are considerably dwarfed. Affected leaves are brittle to the touch. Aphids spread the disease. The yield of affected plants may be reduced by one-third to one-half of normal. Only certified seed stock, known to have come from plantings that were free from leaf roll, and resistant varieties should be used for planting. Controlling aphid vectors and destroying diseased plants when they are detected help minimize the disease.

Mild mosaic, caused by potato virus A, is characterized by definite mottling of yellowish or light-colored areas, alternating with similar areas of normal green tissue in the leaf (Figure 36.12). A slight crinkling is usually present. Under conditions favorable for the disease, the margins of the leaves may be wavy or ruffled. Aphids help spread this disease. Usually, affected plants will not produce more than three-fourths of the normal yield. Virus-free seed should be planted. Controlling aphid vectors and destroying diseased plants when they are detected help minimize the disease. Several varieties are resistant to this disease.

Rugose mosaic, caused by potato viruses Y + X, is more destructive than mild mosaic. Plants from diseased tubers are dwarfed and the leaves are mottled. The mottled areas are smaller and more numerous than in mild mosaic. A distinct crinkling is always present. The underside of lower leaves generally show blackening and death of veins. Aphids help spread this disease. Tubers from infected plants from the current season will produce plants with rugose mosaic the following season. Virus-free seed should be planted. Controlling aphid vectors and destroying diseased plants when they are detected help minimize the disease. Resistant varieties are available.

Latent mosaic, caused by potato virus X, shows only a slight mottling of the leaflets. Virus-free seed should be planted. Resistant varieties are available.

FIGURE 36.12
Potato leaflet showing mild
mosaic.

36.12.2 Common Scab

Common scab[43, 45, 46] is caused by a soil-inhabiting bacterium (*Streptomyces scabies*) that is also carried on the tubers. The disease is characterized by round, corky pits on the skin of the potato (Figure 36.13).

Development of common scab is favored by an alkaline soil pH, but it also occurs in slightly acidic soils. An increase in soil acidity will stop development of the fungus. Addition of lime or barnyard manure to any soil, except a very acidic one, tends to increase scab. The fungus develops when soil moisture is slightly below the optimum for development of the potato plant. In loose, well-aerated soils, it may develop readily even under very wet conditions. Application of sulfur or sulfur-containing fertilizer on lighter soils to make the soil more acidic may reduce the severity of the attack somewhat. Disease-free seed or resistant varieties should be used.

36.12.3 Rhizoctonia Canker (Black Scurf)

Rhizoctonia canker (on the stem), or black scurf (on the tuber), is caused by a widespread soil-inhabiting fungus (*Rhizoctonia solani*) that sometimes causes considerable reduction in stand and yield.

The most common symptom of the disease, irregular small black crusts on the skin, does little harm except to the appearance of the mature tubers (Figure 36.13). When infested tubers are planted, the fungus may attack and destroy the young sprouts. In this stage, brown decayed areas on the white underground stems often girdle the stems or stolons. Plants attacked in this manner may turn yellowish, the leaves may roll, and small greenish tubers may appear on the stems above the ground. The fungus thrives under moist, low-temperature conditions. It is most likely to attack potato planted in the cooler part of the year.

Rhizoctonia canker can be controlled in clean fields by treatment of infected seed stock with a fungicide. Planting on a ridge tends to reduce the seriousness of

FIGURE 36.13
Tuber diseases: *(a)* scab, *(b)* russet scab, *(c)* black scurf, and *(d)* rhizoctonia.

this disease, since the soil on a ridge warms up more rapidly than on flat places, thus making conditions unfavorable for the fungus.

36.12.4 Spindle Tuber

Spindle tuber is caused by a viroid. In most varieties, spindle tuber is characterized by a pronounced elongation of the tubers. The tubers are small and may become pointed at one or both ends. The plants are more erect than normal and somewhat spindly in growth. Also, there is a narrowness of shoots, dwarfing, and a decidedly darker green color of the foliage. The disease may cause a marked reduction in yield. Affected tubers are often so poorly shaped that they are of low commercial grade. Certain insects, such as grasshoppers, fleabeetles, tarnished plant bug, and

Colorado potato beetle, spread the disease. Cutting knives and picker planters also spread it. It can be controlled by use of good certified seed.

36.12.5 Purple Top Wilt

Purple top wilt is caused by the aster-yellows phytoplasma. Phytoplasmas are prokaryotes lacking cell walls that are currently classified in the class *Mollicutes*. The disease is transmitted by leafhoppers. It attacks both vines and tubers. The leaflets are discolored and rolled at the base. Control measures include control of the leafhoppers, destruction of weeds that carry the virus, and the planting of certified seed.

36.12.6 Bacterial Ring Rot

Bacterial ring rot, caused by *Corynebacterium sepedonicum*, is a very destructive seed-borne disease. The first symptom of the disease in the foliage is a wilting of the tips of leaves and branches. Later, the leaves become slightly rolled and mottled and fade to pale-green, followed by a pale-yellow color. Dead areas develop on the leaves, and the affected plants gradually die.[43]

When diseased hills are dug into, they usually contain tubers ranging from those that are apparently sound to those that are completely decayed. Decay of the tuber begins in a region immediately below the skin, thus causing a ring-rot appearance.[47]

Infected seed should not be planted. Only dependable certified seed should be used to avoid possible loss from ring rot. Whole-seed tubers are less likely to be infected. Some varieties will show fewer symptoms.

36.12.7 Late Blight

Late blight, caused by *Phytopthora infestans*, is usually seen first on the margins of the lower leaves. It works inward until entire leaves are affected and killed. Irregular water-soaked spots develop at the margins of the leaves. Under conditions of high moisture, with warm day temperatures and cool nights, the spots enlarge rapidly and the infection spreads to other leaves. All the plants in a field may be killed in a few days. A noticeable odor is given off from the dead foliage in fields that are severely attacked. Late blight was the chief cause of the potato famine in Ireland in 1845–1846.

The late-blight organism may also attack the tubers and produce slightly sunken brownish or purplish spots that enlarge until the entire tuber is affected. When such tubers are cut, the interior shows granular brick-red blotches, a distinctive characteristic of late blight. Infected tubers may rot in the field or in storage.[47]

A fungicide can be applied when the disease first appears, but it may not provide complete control. Use of disease-free seed and elimination of cull piles help prevent infection. Killing the vines ten days to two weeks before harvest helps minimize tuber losses. Few resistant varieties are available.

36.12.8 Early Blight

Early blight, or target spot, caused by the fungus *Alternaria solani*, is a very common disease of potato that causes noticeable damage in some years. It usually becomes serious several weeks before harvest. Small, scattered, dark, circular spots are first produced on the lower leaves, which often become yellow. These spots enlarge, and the affected tissue dies. The enlarged spots develop a series of concentric rings that produce a target-board effect. Fungicides help in the control of early blight.

36.12.9 Physiological Tuber Disorders

Brown Center, or Hollow Heart, may occur when tubers grow rapidly after cool temperatures and water stress. The tuber will have a brown center that will develop into a hollow heart. Jelly End, or Glassy End, occurs when water stress is fostered by low soil temperature, high air temperature, dry winds, and excessive top growth. The stem end of the tuber will be pointed and flaccid. The interior of the tuber stem end will look opaque or glassy. Internal Heat Necrosis may occur with high temperatures at harvest together with acid soil and low calcium. The interior of the tuber will show light-tan or reddish-brown flecks or specks. Vascular discoloration may occur with low soil moisture and rapid vine death from frost or other causes. The vascular tissue in the interior of the tuber will be discolored. Blackheart is caused by an oxygen deficit during harvest, storage, or transport. The center of the tuber will turn brown or black.

▓ 36.13 INSECT PESTS

36.13.1 Colorado Potato Beetle

About 120 different insects attack potato in the United States.[48, 49] The Colorado potato beetle *(Leptinotarsa decemlineata)* is one of the most widespread, and formerly was one of the most destructive, insect pests of potato in the United States. It spread into Europe also. The adult, which appears in the spring, is a hardshelled beetle about ⅜ inch (10 mm) long, stout and roundish, light yellowish color, with ten black stripes down its back (Figure 36.14). The female beetles lay orange-red eggs on the undersides of the potato leaves. The eggs hatch into small larvae, which range in color from lemon to reddish brown marked with two rows of black spots on each side. The heads and legs are black. The larvae feed greedily and grow rapidly for a period of approximately two weeks, during which they devour large quantities of potato foliage. There may be one to two generations of

FIGURE 36.14
Colorado potato beetle
(Much enlarged.)

this insect each year, depending upon climatic conditions. Colorado potato beetle can be controlled with insecticides, but the beetles can develop resistance.

36.13.2 Potato Fleabeetle

The adult potato fleabeetle *(Epitrix tuberis)* feeds on the foliage, but that usually has little effect on tuber yield. Feeding by the larvae on the tubers is more likely to cause damage. The adult beetle is about 1/16 inch (1 to 2 mm) long, black in general appearance, with yellow legs. When disturbed, it jumps quickly and can readily disappear from sight. It eats holes in the leaves, which causes some leaves to dry and fall. The slender, white, worm-like larvae, approximately 1/5 inch (5 mm) long, feed on the roots and tubers, producing tiny tunnels near the surface of the tubers and pimple-like scars on the surface. Potato fleabeetle on the foliage can be controlled with the same insecticides used to control the Colorado potato beetle.

36.13.3 Potato Psyllid

The potato psyllid *(Paratrioza cockerelli)* greatly reduces the yields of potato in Colorado, Nebraska, Wyoming, Texas, and New Mexico. The adult psyllids often migrate long distances from their winter hibernation quarters along the Rio Grande Valley. The young psyllids or nymphs, which are flat, scale-like, and light yellow or green, feed on the underside of the leaves. The feeding causes curling and yellowing of the leaves, known as psyllid yellows, as well as the production of many small tubers in the hill. Several generations of the psyllid develop within a year. They are controlled by dusting or spraying or by band applications of granulated insecticides at planting time.

36.13.4 Potato Tuber Worm

The potato tuber worm *(Phthorimaea operculella)* is the larva of a small gray moth that deposits its eggs on the foliage of potato and related plants. In stored potato, the moth lays eggs near the eyes of the tubers or in other depressions. The eggs soon hatch into small, white tuber worms that develop a pinkish cast along the back as they mature. In the field, the young tuber worms are leaf or stem miners. When the infested foliage dries, they find their way to the tubers through cracks in the soil. Tuber worms tunnel throughout the tubers and render them unmarketable.

Tuber worm is controlled in the field with suitable insecticides. All culls and infested tubers should be destroyed or fed to stock immediately after harvest. All volunteer potato plants should be destroyed. Tuber worms in stored potato can be killed by dusting the tubers with insecticides.

36.13.5 Potato Aphid

Potato aphid *(Macrosiphum euphorbiae)* attacks potato foliage and transmits several virus diseases, including leaf roll, mosaic, and others. The pest is found throughout North America. A wide variety of crops and weeds can serve as alternate hosts. The pest can overwinter in the egg stage, or female nymphs will overwinter in sheltered sites and lay eggs in the spring. There can be several generations during the growing season. When abundant, these aphids reduce the yield of tubers by sucking juices from the foliage. Predaceous insects can help provide some control. High temperatures, damp weather, fungus diseases of aphids, and hard rains may also help. Potato aphid is controlled with dust or spray insecticides.

36.13.6 Seed Corn Maggot

Seed corn maggot (*Hylemya cilicrura*) is the immature, or maggot, stage of a small fly that lays its eggs on soil and decaying vegetable matter. Its feeding on potato seed pieces in the soil is accompanied by decay. The infested young potato plants are either killed or become so weakened that their growth and yield are reduced. The best control for seed corn maggot is to allow the potato seed pieces to heal (suberize) before they are planted or to mix suitable insecticides in the surface soil before planting.

36.13.7 Potato Leafhopper

Potato leafhopper (*Empoasca fabae*) causes a very destructive disease-like condition known as hopper burn. This condition begins with a yellowing of the leaf around the margin and tip, followed by a curling upward and rolling inward. The leaf changes in appearance from yellow to brown and then becomes dry and brittle. When the leafhopper infestation is heavy, the entire plant may die prematurely. Potato leafhopper is a small, green wedge-shaped insect, about ⅛ inch (3 mm) in length. Both young and adults feed on the underside of the leaves and suck juices from the plants.

Leafhoppers and the resultant hopper burn can be controlled by spraying or dusting with suitable insecticides.

36.13.8 Wireworms

Potato tubers are often rendered unmarketable by small holes caused by wireworms. Wireworms bore clean tunnels that are usually perpendicular to the surface of the tuber and are lined with a new growth of plant tissue. Lands known to be infested with wireworms should be avoided for potato culture. Partial control is obtained by soil treatments with insecticides.

36.13.9 Nematodes

Root-knot nematodes (*Meloidogyne* species) that attack potato roots and tubers can be partly controlled by application of a nematicide before planting. With high levels of infestation it is necessary to rotate potato with small grains, rapeseed, or vegetables.[50]

The golden nematode (*Heterodera rostochiensis*), a serious pest of potato in Europe, occurs in parts of Long Island in New York, with occasional small outbreaks elsewhere in the United States. The infested areas are under quarantine to restrict further spread of the pest.

Crop rotations of four to six years that include cereals and other grasses help to control this pest. Careful sanitation of cultivation and handling equipment and removal of infested cull piles is necessary. Application of a nematicide to the soil helps reduce infection.

REFERENCES

1. Urgent, D. "The potato. What is the botanical origin of this important crop plant, and how did it first become domesticated?" *Science* 170, 3963(1970):1161–1166.

2. Kehr, A. C., R. V. Akeley, and G. V. C. Houghland. "Commercial potato production," *USDA Handbook* 267, 1964.

3. Stevenson, F. J. "The potato—its origin, cytogenetic relationships, production, uses, and food value," *Econ. Bot.* 5(1951):153–171.

4. Smith, J. W. "The effect of weather on the yield of potatoes," *U.S. Monthly Weather Rev.* 43(1915):222–236.

5. Bushnell, J. "The relation of temperature to the growth and respiration of the potato plant," *MN Agr. Exp. Sta. Tech. Bull.* 34, 1925, pp. 1–29.

6. Boswell, V. R., and H. A. Jones. "Climate and vegetable crops," in *Climate and Man*, USDA Yearbook, 1941, pp. 373–399.

7. Jones, H. A., and H. A. Borthwick. "Influence of photoperiod and other factors on the formation of flower primordia in the potato," *Am. Potato J.* 15, 12(1938):331–336.

8. Longley, A. C., and C. F. Clark. "Chromosome behavior and pollen production in the potato," *J. Agr. Res.* 41, 12(1930):867–888.

9. Boswell, V. R. "Vegetables," in *Soil*, USDA Yearbook, 1957, pp. 692–698.

10. Clark, C. F., and P. M. Lombard. "Descriptions of and key to American potato varieties," *USDA Circ.* 741, (rev.), 1951.

11. Artschwager, E. "Studies on the potato tuber," *J. Agr. Res.* 27, 11(1924):809–835.

12. Horton, D. *Potatoes: Production, Marketing, and Programs for Developing Countries.* Boulder, CO: Westview Press, 1987.

13. Sieczka, J. B., and R. E. Thornton. *Commercial Potato Production in North America.* Potato Assn. of Amer. 1993.

14. Rosen, C. J. "Potato fertilization in irrigated soils," *Univ. MN Ext. Serv.* FO-3425-GO, (rev.), 1991.

15. Sullivan, D. M., J. M. Hart, and N. W. Christensen. "Nitrogen uptake and utilization by Pacific Northwest crops," *U. ID Coop. Ext. Syst.* PNW 512, 1999.

16. Mortvedt, J. J., R. T. Soltanpure, R. T. Zink, and R. D. Davidson. "Fertilizing potatoes," *CO St. Coop. Ext. Serv.* 0.541, 1999.

17. Dahnke, W. C., C. Fanning, and C. Cattanach. "Fertilizing potato," *ND St. Univ. Ext. Serv.* SF-715, (rev.), 1992.

18. Terman, G. L. "Effect of rate and source of potash on yield and starch content of potatoes," *ME Agr. Exp. Sta. Bull.* 481, 1950.

19. Love, S. L., P. Nolte, and M. K. Thornton. "Potato seed myths," *U. ID Coop. Ext. Syst.* CIS 1028, 1995.

20. Appleman, C. O. "Potato sprouts as an index of seed value," *MD Agr. Exp. Sta. Bull.* 265, 1924, pp. 239–258.

21. Appleman, C. O. "Study of rest period in potato tubers," *MD Agr. Exp. Sta. Bull.* 183, 1914, pp. 181–216.

22. Secor, G. A., N. C. Gudmestad, D. A. Preston, and H. A. Lamey. "Guidelines for seed potato selection, handling, and planting," *ND St. Univ. Ext. Serv.* SF-715, (rev.), 1997.

23. Wright, R. C., and W. M. Peacock. "Influence of storage temperatures on the rest period and dormancy of potatoes," *USDA Tech. Bull.* 424, 1934, pp. 1–22.

24. Lombard, P. M. "Suberization of potato sets in its relation to stand and yield," *Am. Potato J.* 14, 10(1937):311–318.

25. Leach, S. S., and others. "Mechanical potato cutters can produce high quality seed pieces," *Rsh. in Life Sci.* 19, 16(1972):1–8.

26. Folsom, D., G. W. Simpson, and R. Bonde. "Maine potato diseases, insects, and injuries," *ME Agr. Exp. Sta. Bull.* 469, (rev.), 1955.

27. Shepperd, J. H., and O. O. Churchill. "Potato culture," *ND Agr. Exp. Sta. Bull.* 90, 0.103, 1911, pp. 81–126.

28. Croissant, R. L., and J. W. Echols. "Planting guide for field crops," *CO St. Univ. Coop. Ext. Serv.* 1999.

29. Edmundson, W. C. "Distance of planting Rural New Yorker No. 2 and Triumph potatoes as affecting yield, hollow heart, growth cracks and second-growth tubers," *USDA Circ.* 338, 1935, pp. 1–18.

30. Edmundson, W. C. "Time of planting as affecting yields of Rural New Yorker and Triumph potatoes in the Greeley, Colo., District," *USDA Circ.* 191, 1931, pp. 1–7.

31. Lombard, P. M. "Comparative influence of different tillage practices on the yield of the Katahdin potato in Maine," *Am. Potato J.* 13, 9(1936):252–255.

32. Bushnell, J., and F. A. Welton. "Some effects of straw mulch on yield of potatoes," *J. Agr. Res.* 43, 9(1931):837–845.

33. Kunkel, R., W. C. Edmundson, and A. M. Binkley. "Results with potato vine killers in Colorado," *CO Agr. Exp. Sta. Tech. Bull.* 46, 1952.

34. Nelson, D. C., and R. E. Nylund. "Effect of chemicals on vine kill, yield and quality of potatoes in the Red River Valley," *Am. Pot. J.* 46, 9(1969):315–322.

35. Parent, R. C., and others. "Potato growing in the Atlantic Provinces," *Can. Dept. Agr. Pub.* 1281, 1967, pp. 1–28.

36. Carpenter, P. N. "Mineral accumulation in potato plants," *ME Agr. Exp. Sta. Bull.* 562, 1957.

37. Lewis, M. D., M. K. Thornton, and G. E. Kleinkopf. "Commercial application of CIPC sprout inhibitor to storage potatoes," *U. ID Coop. Ext. Syst.* CIS 1059, 1997.

38. Sparks, W. C., and W. M. Iritani. "The effect of gamma rays from fusion product wastes on storage losses of Russet Burbank potatoes," *ID Agr. Exp. Sta. Rsh. Bull.* 60, 1964, pp. 1–26.

39. Caldwell, J. S., C. W. Culpepper, and F. J. Stevenson. "Variety, place of production, and stage of maturity as factors in determining suitability for dehydration in white potatoes II," *The Canner*, June 3 and 10, 1944.

40. Kunkel, R., and others. "The mechanical separation of potatoes into specific gravity groups," *CO Agr. Exp. Sta. Bull.* 422-A, 1952.

41. Cunningham, C. E., and others. "Yields, specific gravity and maturity of potatoes," *ME Agr. Exp. Sta. Bull.* 579, 1959.

42. Murphy, H. J., and M. J. Goven. "Factors affecting the specific gravity of the white potato in Maine," *ME Agr. Exp. Sta. Bull.* 583, 1959.

43. Harris, M. R. "Diseases and pests of potatoes," *WA Agr. Ext. Bull.* 553, 1962.

44. Rost, C. O., H. W. Kramer, and T. M. McCall. "Fertilizers for potatoes in the Red River Valley," *MN Agr. Exp. Sta. Bull.* 385, 1945, pp. 1–16.

45. Afanasiev, M. M. "Potato diseases in Montana and their control," *MT Ext. Bull.* 329, 1970, pp. 1–46.

46. Schultz, E. S. "Control of diseases of potatoes," in *Plant Diseases*, USDA Yearbook, 1953, pp. 435–443.

47. Rowberry, R. G., and others. "Potato production in Ontario," *Ont. Dept. Agr. & Food Publ.* 534, 1969, pp. 1–50.

48. Shands, W. A., B. J. Landis, and W. J. Reid. Jr. "Controlling potato insects," *USDA Farmers Bull.* 2168, (rev.), 1965.

49. Shands, W. A., and B. L. Landis. "Controlling potato insects," *USDA F.B.* (rev.), 2168, 1970, pp. 1–15.

50. Ingham, R., R. Dick, and R. Sattell. "Columbia root-knot nematode control in potato," *OR St. Univ. Ext. Serv.* EM-8740, 1999.

Sweet Potato and Yam

■ 37.1 ECONOMIC IMPORTANCE

Sweet potato (*Ipomoea batatas*) is grown in the warm parts of all continents. Over 95 percent of sweet potato is grown in developing countries where it is the fifth most important crop. In 2000–2003, world production of sweet potato was about 145 million tons (132 million MT). China produces over 80 percent of the crop. Other leading countries producing sweet potato are Uganda, Nigeria, Indonesia, and Vietnam. Sweet potato was grown on an average of 22 million acres (9 million ha) in 2000–2003, with an average yield of 13,000 pounds per acre (14,500 kg/ha). The average production of sweet potato in the United States from 2000–2003 was about 713,000 tons (645,000 MT) on 91,000 acres (37,000 ha), or about 15,700 pounds per acre (17,600 kg/ha). The states leading in sweet potato production were North Carolina, Louisiana, California, and Mississippi.

Production of sweet potato declined about 50 percent from 1944 to 1964 but has been maintained since that time by higher yields despite a continued decline in acreage. The per capita consumption of sweet potato in the United States declined from 31 pounds in 1920 to about 4.6 pounds in 2000. The production of sweet potato in the United States declined partly because of heavy labor requirements for producing the crop.

■ 37.2 HISTORY OF SWEET POTATO CULTURE

Sweet potato apparently originated in tropical America where a wild species, *Ipomoea trifida*, occurs.[1, 2] The plant appears to have been introduced into the Pacific Islands in early times. Its culture may also have spread to China and other Asiatic countries before 1492. Columbus found sweet potato growing in the West Indies. The crop was grown in Virginia as early as 1648.

■ 37.3 ADAPTATION

Sweet potato thrives only in moderately warm climates. It requires a growing period of at least four months of warm weather. A slight frost kills the foliage, while prolonged exposure of the plant or roots to temperatures much below 50°F (10°C) causes damage.[3] However, no part of the United States is too hot or too moist for sweet potato to grow. Maximum yields of roots are obtained where the mean temperature for July and August exceeds 80°F (27°C). Where sweet potato is grown commercially, the mean temperature for the summer months (June, July, and August) is above 70°F (21°C). It is higher than 75°F (24°C) in all regions except Maryland, Delaware, and New Jersey, as well as in the less important growing areas of Iowa, southern Illinois, and southern Indiana.

Differences in day lengths at different latitudes in the United States do not greatly modify root production. Long days promote heavy vegetative growth. The plants make their growth during the long days of summer. Relatively short days, high temperatures, and a long growing season are required for sweet potato to produce an abundance of seed or flowers. Consequently, the plants seldom bloom in the continental United States when grown in the field, but most varieties flower readily when grown in a warm greenhouse in the winter. An eleven-hour day is favorable for flowering.[4] Sweet potato readily produces seed in the West Indies and in other tropical regions.

Nearly all the nonirrigated commercial sweet potato crop is grown where the average rainfall exceeds 35 inches, but irrigation increased yields in Louisiana.[5]

Sweet potato grows best on sandy loam or loamy fine sand soils with a pH of 5.6 to 6.5.[6] Neutral soils induce the pox and scarf diseases.[7] A sandy soil with a clay subsoil is desirable, but good yields are obtained in very sandy soils that are heavily fertilized.[8] Clay soils are undesirable because of harvest difficulties. The roots often have poor shapes when grown in clay or muck soils. Very fertile soils favor a heavy growth of vines and less root growth. However, the crop succeeds on fertile soils in the Southwest. A well-drained soil is desirable, but bedding or ridging of the field helps to overcome the effects of inadequate surface drainage. Sweet potato thrives under subirrigated conditions in California, where the water table is only 4 or 5 feet (120 to 150 cm) below the surface.

37.4 BOTANICAL DESCRIPTION

Sweet potato is a member of the *Convolvulaceae*, or morningglory, family. Sweet potato (*Ipomoea batatas*) has funnel-shaped flowers ¾ to 1½ inches (19 to 38 mm) wide and 1 to 1½ inches (25 to 38 mm) long. The flowers usually have rose-violet or bluish petals, with a darker color in the throat. Each flower has five stamens and an undivided style (Figure 37.1). One to four small, black, flattened seeds are borne in each fertile capsule. Nearly all plants are self-sterile.[9]

The sweet potato plant is a perennial, but in the United States it is killed by cold weather each winter. Therefore, it is grown in this country only as an annual. The vines are usually 4 to 16 feet (120 to 490 cm) long, green or red to purple or both, and often somewhat hairy, especially at the nodes. The leaves are tinged with purple in some varieties. The characteristic leaves in many varieties are somewhat heart-shaped in outline, with the margins entire or toothed in the more popular varieties and deeply lobed in others. The roots have white, yellow, salmon, red, or purplish-red skin, while the flesh is white or various shades of yellow, orange, or salmon.[2] Roots with purple or partly purple flesh are encountered occasionally.

FIGURE 37.1
Flowers and leaves of the sweet potato.

Since it is not a tuber, the sweet potato root lacks the eyes or buds found in potato, but shoots arise from adventitious buds.[10] Starch is stored in the parenchyma cells of both the central core and the cortex.

■ 37.5 VARIETIES

Sweet potato is often considered as falling into two culinary classes: the firm (dry-fleshed) or Jersey type, and the soft (moist-fleshed) type or "yam." The industrial and feed varieties represent still other classes. The Jersey type was formerly favored in the markets of the North, while the soft-fleshed varieties were popular in the South. The difference between the two types lies in the structure and appearance of the flesh, because the Jersey type is higher in moisture content.[11]

The name "yam," as applied either to types or varieties of sweet potato, is a misnomer. To avoid confusion, the USDA requires that such produce be labeled as "sweet potato yam." True yam, a tropical plant that produces tubers instead of fleshy roots, belongs to the *Dioscoria* genus and *Dioscoreaceae* family. True yam is a monocot, while sweet potato is a dicot. Table 37.1 lists characteristics of sweet potato and yam. True yam is discussed later in this chapter.

Most of the older sweet potato varieties originated from root selections. The newer varieties are selected from controlled crosses or from bud mutations.[12,13] The crossing is done in the greenhouse.

New varieties released to growers since 1940 have largely replaced the older varieties.[13] New varieties similar to the Jersey type have orange-colored flesh, rather than the firm yellow flesh typical of the older Jersey types.

Varieties with a salmon or orange flesh have a higher carotene content than do the white or yellow varieties. Sweet potato with white flesh, or with purple streaks in the flesh, is popular in some other countries but not in the food markets of the

TABLE 37.1 Comparison of Sweet Potato and Yam

	Sweet Potato	Yam
Botanical class	*Dicotyledonae*	*Monocotyledonae*
Plant family	*Convolvulaceae* (**morning glory**)	*Dioscoreaceae* (yam)
Origin	Tropical America	West Africa, Asia
Edible plant part	Root	Tuber
Propagation	Roots, vine cuttings	Tuber pieces
Plant appearance	Viny stem, heart-shaped leaves	Viny stem, heart-shaped leaves
Growing season	90–150 days	180–360 days
Appearance	Smooth, thin skin	Rough, scaly skin
Taste	Sweet	Starchy

United States. Soft-fleshed varieties, in which more of the starch is converted to sugars and dextrin, are most widely grown.

Varieties with a cream-colored flesh are suitable for stock feed or industrial use, but they are also preferred for human food in many countries.

37.6 FERTILIZERS

General fertilizer recommendations for high yields of sweet potato are 50 to 100 pounds per acre (55 to 110 kg/ha) nitrogen, 100 to 160 pounds per acre (110 to 180 kg/ha) phosphate, and 60 to 150 pounds per acre (65 to 170 kg/ha) potash.[8, 14] Exact fertilizer needs must be determined by soil analysis. Too much nitrogen can result in excessive vine production at the expense of root production. Half the nitrogen should be applied before planting, the rest when the vines begin to spread.

The optimum pH range for sweet potato is 5.8 to 6.2. If lime is needed, it should be applied and incorporated several months before planting to allow adequate adjustment of the soil pH. Dolomitic lime should be used if the soil's magnesium level is low.[14] Moderate liming may be helpful where the soil pH is below 5.0,[8] but a slightly acidic soil prevents injury from scarf. Deep placement of fertilizer and lime may tend to develop long roots.

37.7 ROTATIONS

Rotations of five years or longer with other crops such as corn, cotton, potato, small grains, and legume hay are helpful in sweet potato production. Such rotations reduce the infestation by soil-borne organisms that cause sweet potato diseases such as black rot and scurf.[15] Sweet potato advantageously follows winter cover crops because there is usually time to plow under the cover crop before the sweet potato plants are set in the field.[8] Legume crops sown in the fall may save about one-half to two-thirds of the quantity of nitrogen fertilizer that would otherwise need to be applied. Excessive nitrogen merely increases vine growth.

■ 37.8 SWEET POTATO CULTURE

37.8.1 Seedbed Preparation

Seedbed preparation for sweet potato is about the same as that for other row crops in a particular region. Throwing up ridges or beds about 10 inches (25 cm) high in the field is believed to encourage growth of long roots.

37.8.2 Seeding Methods

Sound small to medium-size sweet potato roots are first treated with a recommended fungicide. The roots are usually pre-sprouted in the spring by exposure to temperatures of 68 to 86°F (20 to 30°C) for two to six weeks. The sprouted roots are then placed 2 to 3 inches (5 to 8 cm) deep in beds for further sprouting. In the North, the roots are bedded in a greenhouse or an electrically heated bed. Sometimes the beds are mulched to retain the heat until the sprouts emerge. In the South, where the season is longer, outdoor beds are covered with plastic to increase the temperature until the threat of frost is over.

Sprouts are pulled from the beds by hand when they are about 1 foot (25 cm) in length with six to twelve leaves[7, 16] (Figure 37.2). About 150 to 240 square feet (14 to 22 m²) of bed planted with 10 to 14 bushels (350 to 498 l) of roots provide sufficient slips for planting an acre (about 1 m³/ha). Less will be needed if two or three pullings are made. A new crop of sprouts develops in two or three weeks after the first sprouts are pulled. The lower yields usually obtained from the second and third pullings are merely a consequence of later planting. Prompt planting of the

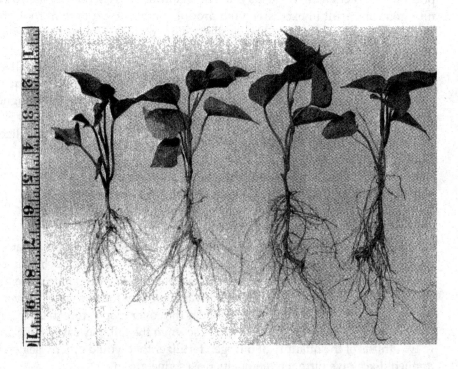

FIGURE 37.2
Sweet potato sprouts or slips ready for planting.

pulled sprouts is essential to obtain the highest yields.[16, 17] Four- to eight-row mechanical planters are used to transplant the sprouts in the field.[7, 18, 19]

Sweet potato for seed is usually grown from vine cuttings. These cuttings are also used for the late plantings,[8] especially in the South. Cuttings are taken from plants as soon as the vines start to run; 8- to 12-inch (20 to 30 cm) sections of vine are pushed into the soil at desired intervals. Vine cuttings are cheaper than sprouts and less likely to spread diseases, but their use delays planting. One acre of plants will supply cuttings for eight acres of crop.

Sweet potato is usually planted 10 to 12 inches (25 to 30 cm) apart in 40- to 50-inch (100 to 120 cm) rows. This will result in a plant population of about 16,000 plants per acre (40,000 plants/ha). Plant spacing within the row can range from 10 to 24 inches (25 to 60 cm) without much effect upon total yield, yield of marketable roots, or starch content of the roots.[20] On highly productive soils, the plant population can be higher. Thinner spacing tends to produce fewer small roots but more large (Jumbo) roots.[21] Uniform medium-size roots are desired for the food market, but small roots are desired for canning.

Sweet potato should be transplanted about one month after the average date of the last spring frost, or when the leaves of most oak trees have reached full size. Any delay beyond the optimum date not only results in marked decreases in yield, but also produces slender roots of low starch and carotene content.[17, 20]

37.8.3 Weed Control

The field can be cultivated to control weeds until the vines spread between the rows. Early cultivation may be done with a modified rotary hoe.[8] For other cultivations, the prevailing local implements, such as sweep cultivators, are satisfactory. A few herbicides are available. Metolachlor and clomazone can be applied immediately after transplanting. EPTC can be applied preplant.

37.8.4 Irrigation

Sweet potato requires about 1 inch (25 mm) of water per week. Although sweet potato is drought tolerant, it still requires a sufficient and uniform moisture supply to produce quality roots. Fluctuations in the water supply can cause growth cracks in the roots, which render them unmarketable. Excessive moisture can cause excessive vine growth and undesired elongation of roots. Although not always needed, the ability to provide irrigation can mean the difference between producing high yields of quality roots and poor yields of poorly shaped, unmarketable roots, especially in dry summers.

37.8.5 Harvesting

Sweet potato harvest begins in August in the South, September in New Jersey, and October in California. The maximum yield of sweet potato is reached at the end of the growing season. The roots should be dug before, or immediately after, the first light frost. The number and, often, size of roots continue to increase for a month after the maximum starch content is reached.[22] Since the sweet potato vines interfere with the operation of digging machines, the vines are usually cut either during or before digging. The vines can be cut with a mower, shallow-running vertical knives, or, preferably, with shielded 8-inch colters (20 cm) attached to the digger.[19] Cutting vines prematurely significantly decreases yields. Even pruning vines to obtain cuttings for planting decidedly reduces yields when more than one-half the runners are removed.[17]

As far as food use is concerned, the most important factor in digging is to avoid cutting or bruising the roots, which would promote rapid decay.[23] A wide-bottom (16-inch, 40 cm) plow run at the side of the row is preferable for digging sweet potato that is to be stored. Usually the roots (Figure 37.3) are allowed to dry for a few hours before they are picked up by hand. A two-row sweet potato harvester digs, tops, sorts, and loads the roots into containers. It saves labor but may injure more roots than hand picking.

■ 37.9 CURING

Uncured sweet potato roots decay quickly at temperatures of 50°F (10°C) or lower.[24] About 20 to 40 percent of the harvested sweet potato crop may spoil before the roots are used. The storage room, as well as the containers, should be fumigated before the roots are harvested. The proper conditions for curing are a temperature

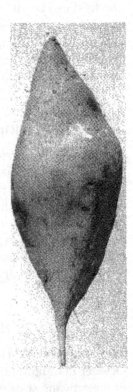

FIGURE 37.3
Sweet potato root.

of 85 to 90°F (29 to 32°C) and a relative humidity of 85 to 90 percent maintained for four to seven days. Such conditions promote the healing of wounds in the roots, which check future fungus infection.[10, 23]

▓ 37.10 STORAGE

After curing, sweet potato roots should be stored at 55 to 60°F (13 to 16°C), with a relative humidity of 85 to 90 percent. Storage of cured roots at 40°F (4°C) for two weeks or more tends to promote decay. Ventilation is occasionally necessary, not only to regulate the temperature, but also to prevent condensation of moisture in the storage room or on the roots. The storage room should provide for artificial heat to keep the temperature from falling below 55°F (13°C).[24] Storage in baskets or crates causes less bruising and consequent shrinkage and decay than occurs when the roots are stored in slatted bins.

Curing causes shrinkage of about 10 percent due to moisture evaporation and loss of dry matter that result from respiratory oxidation of sugars. Subsequent shrinkage occurs during a five-month storage period,[25] but total shrinkage and spoilage under good conditions should not exceed 15 percent. In storage, some of the starch and dextrin is converted to sugar. Under proper conditions the flesh softens somewhat, while the cooking quality, color, and flavor improve.[7, 23]

▓ 37.11 USES

Sweet potato is used primarily for food, prepared mostly from the fresh state. However, large quantities are frozen, canned, chipped, or dehydrated. Culls are used mostly for livestock feed. Sweet potato vines make a satisfactory roughage or silage, but the best practice is to leave them on the field for soil improvement.[26]

Extraction of starch from sweet potato[26] was carried on in Mississippi and Florida from 1934 to 1946. The starch was used mostly for textile sizing or in laundering. The average yield was 10 to 11 pounds of starch per bushel (130 to 140 g/l) of roots. Roots of high starch content (23 percent) should yield 20 percent starch or 12 pounds per bushel (60 pounds) with efficient operation.

Consumers prefer roots that are about 4 to 6 inches (10 to 15 cm) long, are 1¾ to 3½ inches (4.5 to 9 cm) in diameter, and weigh not more than 1 to 1½ pounds (450 to 680 g).[27] Thick roots longer than 10 inches (25 cm), called Jumbos, are usually not shipped. A standard bushel of sweet potato varies in weight among different states from 50 to 60 pounds.

Peeled sweet potatoes contain about 69 percent water, 1.1 percent ash, 1.8 percent protein, 0.7 percent fat, 1 percent crude fiber, 28 percent nitrogen-free extract (of which about 3 to 4.5 percent is sugars and the remainder mostly starch), 0.2 percent calcium, and 0.05 percent phosphorus. The sweet potato contains about 100 micrograms of carotene per gram as well as fair quantities of ascorbic acid and B vitamins. Compared with the Irish potato tuber, the sweet potato root is higher in dry matter, starch, sugar, crude fiber, and fat but lower in protein.

■ 37.12 DISEASES

There are two main types of sweet potato diseases:[7, 15, 28] (1) those that attack the plants or roots in the field and (2) storage diseases that attack the mature harvested roots. Some organisms attack the roots both in the field and in storage.

37.12.1 Stem Rot or Wilt

The stem rot or wilt disease is caused by two species of fungi, *Fusarium batatatis* and *F. hyperoxysporium*. The fungus is carried over in infected roots as well as in the soil, where it remains for several years. It causes a wilt in the sprouts in the bed and in the vines in the field, as well as a rotting of the vines. It forms a blackened ring inside the root. Control measures consist of bedding disease-free roots that have been disinfected for five to ten minutes in 1:1,000 bichloride of mercury, in clean or sterilized beds. The roots are then transplanted to noninfected fields. A good method for getting clean seed is selection of roots to be bedded from hills in which the stems, when split at harvest time, are somewhat resistant.

37.12.2 Foot Rot or Die-Off

Foot rot or die-off, caused by the fungus *Plenodomus destruens*, occurs in Virginia, Ohio, Iowa, Missouri, and other states. It stunts or kills the sprouts, rots or girdles the base of the stems, and spreads to the roots, where it develops brown spots. The black fruiting bodies of the fungus are visible on diseased stems. Infection usually spreads in the plant bed. Sanitary measures recommended for stem rot aid is control of foot rot.

37.12.3 Texas Root Rot

The Texas root-rot disease is caused by the fungus *Phymatotrichum omnivorum*. It occurs in irregular areas in many fields as well as in uncultivated areas, mostly in Texas, Oklahoma, New Mexico, Arizona, and California. On the sweet potato it causes a brown rot of the roots and a browning of the stems. Control for sweet potato consists of avoiding areas in the field known to be heavily infested. Rotation with cereal or grass crops or green manuring with these crops is helpful.

37.12.4 Scurf

Scurf (also called soil-stain, rust, and Jersey mark) is caused by the fungus *Monilochaetes infuscans*. It affects the appearance of the roots by producing brown or blackish spots on the root surfaces. It is not a decaying organism, but it causes excessive shrinkage in storage. Control consists of crop rotation along with the use of clean planting stock. Seed treatment with bichloride of mercury (1:1,000) plus 3.7 ounces of wettable sulfur per gallon of solution is helpful. Roots produced from vine cuttings are more likely to be free from infection. Crop rotation is helpful but does not offer complete protection where the disease has occurred previously.[15] Several varieties are resistant or partly resistant.

37.12.5 Black Rot

Black rot, caused by the fungus *Ceratosomella fimbriata*, causes widespread heavy losses in the field and in storage.[5] It often destroys sprouts and yellows, wilts, and dries the foliage, after which black cankers develop in the underground parts of the stem. The infected roots show dark brown or black spots that enlarge until most of the root is affected. Control measures are similar to those for stem rot. Precautionary sanitary measures in and around the plant beds help prevent the spread of black rot. Resistant varieties or partly resistant varieties are available.

37.12.6 Soil Rot or Pox

The soil rot or pox disease of sweet potato is caused by the bacteria-like fungus *Actinomyces ipomoea*, which infests the soil. It is a serious widespread disease in certain important sweet potato areas. Affected plants are stunted, have decayed roots, and often succumb to the disease. The surface of diseased mature roots shows pits or cavities with irregular jagged or roughened margins. These pits are filled with dead tissues at first, but this later sloughs off. Soil rot does not develop on soils in Louisiana that have a pH below 5.2.[15] The disease was largely checked there by applying sufficient sulfur to acidify the soil down to pH 5.2 or less. Several varieties are resistant to disease.

37.12.7 Virus Diseases

Yellow dwarf and ring spot are virus diseases that are evident on sweet potato leaves. The planting of relatively virus-free seed stocks is desirable as a control measure.

37.12.8 Storage Rots

Soft rot, caused by the fungus *Rhizopus stolonifer*, is perhaps the most destructive storage disease of the sweet potato. The mold enters the root through wounds or at the attachment end. It spreads quickly over the root, which causes complete decay. Control consists of good storage and proper curing of the roots to heal wounds.

Some varieties are resistant to the disease.

Dry rot, caused by the fungus *Diaporthe batatatis*, produces a firm brown rot. Java black rot, caused by *Riplodia tubericola*, makes the roots dry and hard. Charcoal rot is caused by *Macrophomina phaseoli (Sclerotium bataticola)*. It breaks down the interior of the root where it leaves black spore bodies.

Internal cork is a virus disease carried in the roots and spread in the field by aphids. Virus-free seed stocks should be planted.

■ 37.13 INSECT PESTS

37.13.1 Sweet Potato Weevil

The sweet potato weevil (*Cylas formicarius* var. *elegantulus*)[29] deposits its eggs in small punctures in the stems or roots. The newly hatched larvae bore through the stems and roots. They leave a residue that is so unpalatable that even hogs will not eat badly damaged roots. A complete weevil generation develops every few weeks throughout the year. It causes the most damage in the Gulf Coast region. The weevil is controlled by applying powdered insecticides in 7-inch (18 cm) bands along the row as well as on the plant bed. Lighter field applications may be desirable later. Other helpful control measures involve sanitation practices such as feeding or destruction of all plant residues, including small roots in the field and plant beds; destruction of related wild hosts that belong to the genus *Ipomoea*; and use of weevil-free planting roots, vine cuttings, or slips. Keeping sweet potato off the field for a year is effective, provided all volunteer plants and weed hosts in the immediate vicinity are destroyed. The seed in storage may be dusted with insecticides.

37.13.2 Other Sweet Potato Insects

The adults of the sweet potato leaf beetle (*Typophorus viridicyaneus*), a pest of much less importance than the weevil, emerge in late May to feed on sweet potato leaves. The larvae feed on leaves and roots, enter the soil, and pupate in the fall. Some of the larvae, still in the roots at digging time, are transported to other fields.

Other insects that attack sweet potato include grasshoppers, flea beetles, termites, wireworms, and white-fringed beetles.[30]

Insects harmful to sweet potato can be controlled with suitable insecticides.

■ 37.14 YAM

In 2000–2003, global yam production averaged 41 million tons (39 million MT). Over 65 percent is produced in Nigeria. Other leading countries are Ghana, Ivory Coast, Benin, and Togo. Yam was grown on an average of 10 million acres (4 million ha) in 2000–2003, with an average yield of 8,000 pounds per acre (9,000 kg/ha). No yam is grown commercially in the United States. The principal species are *Dioscorea alta, D. rotunda,* and *D. esculenta,* but several other species are grown occasionally. The plants are vines that climb on poles. The tubers are similar to potato in food value but of poorer keeping quality than sweet potato. The plants require a warm growing season of six to ten months, ample soil moisture, good drainage, and fairly productive soils. The crop is propagated by planting cut, or small whole, tubers.[27, 31]

REFERENCES

1. Nishiyama, I. "The origin of the sweet potato plant," *Contribution No. 291 from the Laboratory of Genetics.* Kyoto, Japan: Faculty of Agriculture, Kyoto University, 1961.
2. Smith, C. E., Jr. "The new world centers of origin of cultivated plants and the archaeological evidence," *Econ. Bot.* 22, 3(1968):253–266.

3. Boswell, V. R., and H. A. Jones. "Climate and vegetable crops," in *Climate and Man*, USDA Yearbook, 1941, pp. 373–399.

4. McClelland, T. B. "Studies of the photoperiodism of some economic plants," *J. Agr. Res.* 37(1928):603–628.

5. Hernandez, T. P., and others. "The value of irrigation in Louisiana," *LA Agr. Exp. Sta. Bull.* 607 1965, pp. 1–15.

6. Boswell, V. R. "Vegetables," in *Soil*, USDA Yearbook, 1957, pp. 692–698.

7. Steinbauer, C. E, and L. J. Kushman. "Sweet potato culture and diseases," *USDA Handbk.* 388 1971, pp. 1–74.

8. Boswell, V. R. "Commercial growing and harvesting of sweet potatoes," *USDA Farmers Bull.* 1950.

9. Poole, C. F. "Sweet potato genetic studies," *HI Agr. Exp. Sta. Tech. Bull. 27*, 1955.

10. Artschwager, E. "On the anatomy of the sweet potato with notes on internal breakdown," *J. Agr. Res.* 27, 3(1924):151–166.

11. Caldwell, J. S., and C. A. Magoon. "The relation of storage to the quality of sweet potatoes for canning purposes," *J. Agr. Res.* 33, 7(1926):627–643.

12. Harter, L. L. "Bud sports in sweet potatoes," *J. Agr. Res.* 33, 6(1926):523–525.

13. Steinbauer, C. E. "Principle sweet potato varieties developed in the United States, 1940 to 1960," *Horticultural News* 43, 1(1962).

14. Sanders, F, and M. Cannon. "Sweet potato production BMPs," *LA Coop. Ext. Serv. Publ.* 2832, 2000.

15. Hildebrand, E. M., and H. T. Cook. "Sweet potato diseases," *USDA Farmers Bull.* 1059, (rev.), 1959.

16. Beattie, J. H., V. R. Boswell, and J. D. McCown. "Sweet potato propagation and transplanting studies," *USDA Circ.* 502, 1938, pp. 1–15.

17. Clark, C. F. "Sweet potato varieties and cultural practices," *MS Agr. Exp. Sta. Bull.* 313, 1938, pp. 1–27.

18. Greig, J. H. "Sweet potato production in Kansas," *KS Agr. Exp. Sta. Bull.* 498, 1967, pp. 1–27.

19. Park, J. K., M. R. Powers, and O. B. Garrison. "Machinery for growing and harvesting sweet potatoes," *SC Agr, Exp. Sta. Bull.* 404, 1953.

20. Anderson, W. S., and others. "Regional studies of time of planting and hill spacing of sweet potatoes," *USDA Circ.* 725, 1945, pp. 1–20.

21. Beattie. J. H., V. R. Boswell, and E. E. Hall. "Influence of spacing and time of planting on the yield and size of the Puerto Rico sweet potato," *USDA Circ.* 327, 1934, pp. 1–10.

22. Kimbrough, W. D. "Studies of the production of sweet potatoes for starch or feed purposes," *LA Agr. Exp. Sta. Bull.* 348, 1942, pp. 1–13.

23. Kushman, L. J., and F. S. Wright. "Sweet potato storage," *USDA Handbk.* 358, 1969, pp. 1–35.

24. Lutz, J. M., and J. W. Simons. "Storage of sweet potatoes," *USDA Farmers Bull.* 1442, (rev.), 1958.

25. Poole, W. D. "Harvesting sweet potatoes in Louisiana," *LA Agr. Exp. Sta. Bull.* 568, 1963, pp. 1–24, 1963.

26. Thurber, F. H., E. A. Gastrock, and W. F. Guilbeau. "Production of sweet potato starch," in *Crops in Peace and War*, USDA Yearbook, 1950–51, pp. 163–167.

27. Coursey, D. G. *Yams, an Account of the Nature, Origins, Cultivation and Utilization of the Useful Members of the Dioscoreaceae.* London: Longmans, 1967, pp. 1–230.

28. Elmer, O. H. "Diseases of sweet potatoes," *KS Agr. Exp. Sta. Bull.* 495, 1967, pp. 1–43.

29. Anonymous. "The sweet potato weevil and how to control it," *USDA Leaflet* 431, 1958.

30. Cuthbert, F. P., Jr. "Insects affecting sweet potatoes," *USDA Agric. Handbook* 1967, pp. 1–28.

31. Martin, F. W. "Yam production methods," *USDA Prod. Rsh. Rpt.* 1471-1-17, 1972.

Rapeseed/Canola

Argentine type
Double low standard
Erucic acid
Glucosinolate
HEAR
LEAR
Polish type

38.1 RAPESEED AND CANOLA

Canola, a low erucic acid variety of rapeseed, is a relatively new crop. Canola was first developed in Canada in 1956–1957, and continuing research in Canada during the 1950s and 1960s further established the crop. Canola acreage grew rapidly until it is now third in production of edible oils behind soybean and palm.

"Canola" is a trade name registered by the Western Canadian Oilseed Crushers Association. The name was derived from "CANadian-Oil-Low-Acid." However, "canola" has become a generic term for edible rapeseed oil and is no longer considered an industry trademark. Canola must have less than 2 percent erucic acid in its oil, and less than 20 micromoles of total glucosinolates per gram in its seed (see Section 38.9). Varieties that meet that standard are called "double low." In 1985, the U.S. Food and Drug Administration gave canola oil a "Generally Recognized as Safe" designation, which opened up markets in the United States and other countries.

Rapeseed oil and meal are often referred to as HEAR: High Erucic Acid Rapeseed. Canola is referred to as LEAR: Low Erucic Acid Rapeseed.

In this chapter, information on rapeseed includes canola varieties unless otherwise stated.

38.2 ECONOMIC IMPORTANCE

Global production of rapeseed, most of which is canola, averaged about 40 million tons (36 million MT) in 2000–2003. It was grown on 58 million acres (23 million ha) with an average yield of 1,400 pounds per acre (1,550 kg/ha). Production of rapeseed oil averaged about 13 million tons (12 million MT) in 2000–2003. The leading countries in rapeseed production are China, Canada, India, Germany, and France. U.S. canola production in 2000–2003 averaged 886,000 tons (802,000 MT) on 1.32 million acres (536,000 ha). The average yield was 1,340 pounds per acre (1,500 kg/ha). Practically all of the canola is produced in North Dakota and Minnesota. Rapeseed production in the United States averaged about 1,900 tons (1,700 MT) in 2000–2003.

■ 38.3 HISTORY OF RAPESEED CULTURE

The "rape" in rapeseed comes from the Latin word *rapum*, which means turnip. Rapeseed and other *Brassica* species are believed by some to have originated in the Himalayan region of Asia.[1] However, the greatest genetic diversity of wild types is found in the Mediterranean Region, which is usually the accepted criterion for establishing origin.[2] In India, rapeseed was found in archaeological sites as old as 2300 BC, and Sanskrit writings show that rapeseed was used as early as 1500 BC. Rapeseed was grown in China as early as 5000 BC, and the Chinese word for rapeseed appears in writings about 2,500 years ago. Two thousand years ago, rapeseed was grown from China and Korea to northern Europe. Cultivation of rapeseed in Europe began in the thirteenth century.

Rapeseed was introduced into Canada in 1936 when a Polish immigrant brought seed from his homeland. The rapeseed brought by him was *Brassica rapa*, and it was called Polish rapeseed. The first significant production in the United States occurred in 1943 to replace supplies cut off during World War II. The seed, obtained from Argentina, was *B. napus* and was called Argentine rapeseed. The names "Polish" and "Argentine" are still used today to refer to these two types of rapeseed.

Industrial rapeseed has been grown in the Pacific Northwest since about 1960.

■ 38.4 ADAPTATION

There are both winter annual and summer annual varieties of rapeseed grown in the United States. The adaptation and growing season of winter and spring rapeseed varieties in a region will coincide somewhat with the adaptation and growing season of winter and spring wheat. Only spring varieties are grown in Canada. There is a biennial variety grown in northern Europe that could be adapted to the Pacific Northwest, Central Plains, and Northeast United States, but it is not currently grown to any extent.

Fertile, well-drained soils are the best for rapeseed. It prefers a soil pH between 5.5 and 8.3. Plants are moderately tolerant of soil salinity. They cannot tolerate poorly drained or flooded soils during establishment.

Ideal temperatures for growth range from 50 to 86°F (10 to 30°C), with the optimum temperature about 68°F (20°C). Plants are very sensitive to high temperatures during flowering. Consistent daytime temperatures over 86°F (30°C) can cause poor pollination and fertilization, resulting in significant yield loss in both irrigated and dryland fields. During pod filling, plants are less affected by high temperatures. Temperatures of 50 to 60°F (10 to 15°C) will increase seed oil content as the crop matures.

Sensitivity to cold temperatures depends on the degree to which the plants have become hardened. Temperatures below 50°F (10°C) will begin the hardening process. Hardened winter varieties can survive temperatures of –5 to 5°F (–15 to –20°C). Dehydration during sunny and windy days can increase vulnerability to winter kill. In the fall, winter rapeseed should have well-developed rosettes with six to ten leaves before winter dormancy to minimize winter kill. Winter rapeseed requires about three weeks of near freezing temperatures for vernalization. Research using controlled environments showed that eight weeks at 40°F (4°C) will fully vernalize plants.[2] Spring varieties do not require vernalization.

38.5 BOTANICAL DESCRIPTION

Rapeseed belongs to the mustard family, *Cruciferae*. There are two species of cultivated rapeseed, *Brassica napus* and *B. rapa* (formerly *B. campestris*). Most of the canola varieties are *B. napus*. Other cultivated species of *Brassica* are *B. oleracea* (cabbage, Brussels sprouts, and broccoli) and *B. juncea* (leaf mustard or mustard greens).

Plants have a taproot and grow 2 to 5 feet (60 to 150 cm) high with a much branched central stem. Leaves are sessile and deeply lobed with wavy margins (Figure 38.1). Flowers are bright yellow with four petals (Figure 38.2). Pods 1 to 1½ inches (25 to 40 mm) long and ⅛ inch (3 mm) wide develop after fertilization. Rapeseed is indeterminate in its flowering habit and will flower for five to six weeks. Seeds (Figure 38.3) are brown to black in color, round, and very small, averaging about 6,000 per ounce (200/g).

38.6 FERTILIZERS

Rapeseed is a heavy user of nitrogen. In research in Canada and Illinois, the crop used about 150 pounds (70 kg) of nitrogen to produce 2,000 pounds (900 kg) of seed.[3] All the required nitrogen can be applied in the early spring, or 35 to 50 pounds per acre (40 to 55 kg/ha) can be applied in the early spring and the remainder during the rosette stage. Adequate sulfur increases utilization of the nitrogen, but excessive sulfur can increase the glucosinolate content in rapeseed

FIGURE 38.1

Rapeseed plant. *(Left)* Upper stem with pods and flowers. *(Right)* Lower stem with sessile leaves. [Courtesy Richard Waldren]

FIGURE 38.2
Rapeseed flowers. [Courtesy Canola Council of Canada]

FIGURE 38.3
Rapeseeds. [Courtesy Canola Council of Canada]

meal. A highly available form of sulfur, such as ammonium sulfate or ammonium thiosulfate, is better than elemental sulfur.[3] Rapeseed will use about twice as much sulfur as will wheat.[4]

Phosphorus and potassium requirements for rapeseed are similar to the needs of high-yielding wheat crops. Preplant application is best. Boron fertilizer may be needed in sandy and high pH soils.

38.7 ROTATIONS

Canola should not be grown in fields that were recently seeded to rapeseed, canola, or other Brassica crops. Canola varieties should never be planted next to a field of rapeseed varieties as the plants may cross-pollinate and raise the erucic acid content of the canola seeds, making the seeds unfit for the canola market. A four-year rotation will help decrease damage from many diseases.

Rapeseed is very sensitive to several common broadleaf herbicides, including triazine, imadazolinone, picloram, and dicamba. Care should be exercised when using these herbicides on crops preceding rapeseed.

Spring varieties can be used as a double crop after winter wheat.

38.8 RAPESEED CULTURE

38.8.1 Seedbed Preparation

Because the seed is so small, the seedbed must be firm and free of debris. The usual sequence is disking followed by harrowing or rolling. Excessive tillage can result in loss of soil moisture and surface crusting. The final tillage operation should be less than one week before planting to kill weeds. Conventional tillage is more effective for rapeseed than no-till.[3]

38.8.2 Seeding Methods

The minimum soil temperature for seeding spring varieties is 45°F (7°C). For maximum germination, winter varieties should be seeded when the soil temperature is less than 86°F (30°C). Rapeseed can be seeded using an ordinary grain drill, but mixing the seed with filler such as cracked grains (three parts filler to one part seed) increases accuracy and reduces seed damage.[5] More precise seeding can be done with an alfalfa or grass drill. Row spacing of 6 to 14 inches (10 to 36 cm) allows the crop to quickly cover the soil surface and reduce weed competition. Seeding depth is ½ to 1 inch (13 to 25 mm) deep. Seed can also be broadcast. Following broadcast seeding with a harrow or roller helps to incorporate the seed.

Winter varieties are seeded two to four weeks earlier than winter wheat in an area to ensure sufficient plant development before winter dormancy.[5] Plants with six to ten leaves, a taproot 5 to 6 inches (13 to 15 cm) long, and a crown ⅜ inch (1 cm) in diameter will have good winter survival. Spring varieties are planted in March and early April in the Central Plains, and April to early May in the North.[4, 5] In Canada, research has shown that dormant seeding of spring varieties in the fall, after the soil temperature has dropped below freezing, can increase yields.[6] The seed will not germinate until spring, but the additional time before traditional early spring seeding gives plants time for further growth and avoids higher temperatures before maturity.

Seeding rates of 4 to 8 pounds per acre (4.5 to 9 kg/ha) are common. Rates of 8 to 10 pounds per acre (9 to 11 kg/ha) are used if the crop will be irrigated. If broadcast, the rate is increased about 20 percent.

38.8.3 Weed Control

Since rapeseed forms a dense leaf canopy, it is able to compete with most weeds. However, it is important to control weeds in the field before seeding. The occurrence of wild lettuce, wild mustard, and other *Brassica* weeds can complicate the use of herbicides. Wild mustard seed is a serious contaminant in canola and its presence can result in price dockage or even rejection at the elevator.

Common preplant herbicides used on canola are trifluralin and ethalfluralin. Sethoxydim, quizalofop, and clethodim are common postemergent herbicides. Three herbicide-resistant lines of canola have been released. Liberty-Link® varieties are resistant to glufosinate, Roundup-Ready® varieties are resistant to glyphosate, and Clearfield® varieties are resistant to imidazolinone.

38.8.4 Harvesting and Storage

Rapeseed is ripe when plants drop their leaves (Figure 38.4), turn a straw color, and seeds are dark brown. Rapeseed can be directly combined, but the crop usually matures nonuniformly, and the pods easily shatter when mature. Therefore, it is common to windrow the crop when the bottom pods are brown, and the pods at the top of the plant are green but well filled. Moisture content at this stage is about 35 percent.[4] Reel speed should match ground speed to reduce shattering. Windrows of canola can be fluffy, so the crop should be cut just below the lowest pods to leave as much stubble as possible to prevent the windows from blowing away. A roller behind the swather firms the windrow and further reduces movement (Figure 38.5).

When the seed has dried to 15 percent moisture or less, the crop is threshed using a combine with a pick-up attachment (Figure 38.6). The moisture content of

FIGURE 38.4
Mature rapeseed plants ready for windrowing.
[Courtesy Richard Waldren]

FIGURE 38.5
Rapeseed is windrowed to dry and await threshing. Notice the swather in the background is pulling a roller to firm the windrow and prevent blowing. [Courtesy Canola Council of Canada]

FIGURE 38.6
A combine picks up swathed rapeseed for threshing. [Courtesy Canola Council of Canada]

rapeseed going into storage should not be more than 10 percent, so producers may postpone harvest to further dry the seed if the weather cooperates. Since rape seed is very small, proper adjustment of the combine is critical to minimize threshing losses. Combine cylinder speed should be one-half to three-fourths that used for wheat.[7] A slower ground speed and less air volume also reduce threshing losses.

Handling rapeseed is similar to handling flaxseed, as the seeds are small, round, and heavy, and they run freely. Trucks and storage bins must be very tight. Seed in long-term storage should be 9 percent moisture or less. Heated or natural air drying is used to reduce moisture content of the seed. If using heated air, the temperature should be no more than 110°F (43°C) to maintain seed quality.[4]

Newly harvested seed may increase in moisture content (sweat), so air movement should be provided in all canola storage bins. Canola seed is small enough to pass through many bin aeration floors. Therefore, nylon window screen, or something similar, may be needed on the bin floor and other ventilation channels.

▓ 38.9 USES OF RAPESEED AND CANOLA

Nonedible uses of rapeseed include the production of rubber products, lubricants, continuous casting of steel, brake fluids, printing inks, plasticizers, oiled fabrics, and others. In developing countries, it is used as lamp fuel and for making soap. Edible uses of canola include vegetable oil, shortening, salad oil, mayonnaise, bread, cake mixes, and others.

LEAR meal can be fed to livestock without problems. HEAR meal can be used for livestock feed, but amounts should not exceed 10 percent of the ration for mature poultry or cattle, 20 percent for sheep, or 4 percent for young pigs. HEAR meal is unsuitable for young poultry and pregnant or nursing sows. If necessary, the toxins can be extracted from the meal with water.[8]

▓ 38.10 CHEMICAL COMPOSITION

Erucic acid is a mono-unsaturated fatty acid that is toxic when ingested in sufficient quantities. It has been shown to affect muscle tissue, including buildup of fat in heart muscle and heart lesions in laboratory rats. The erucic acid content of rapeseed (HEAR) oil ranges from 0.3 to 50.6 percent.

Glucosinolates are sulfur-containing organic anions that release cyanate compounds when metabolized. These compounds have been shown to decrease fertility and interfere with the uptake of iodine in monogastric animals such as poultry, swine, and humans. This reduces thyroid activity, which can cause goiters.

Solvent-processed canola meal is similar to soybean meal in protein and amino acid content but lacks the vitamins and minerals (Table 38.1).

TABLE 38.1 Chemical Composition of Alfalfa Cubes, Canola Meal, and Soybean Meal

	Dry Matter	Energy	Crude Protein	Ash	Ca	P	Mg	Na	S	Cu	Mn	Zn
	(%)	(MJ/kg)	(% DM)	(% DM)	(% DM)	(% DM)	(% DM)	(% DM)	(% DM)	(mg/kg DM)	(mg/kg DM)	(mg/kg DM)
Alfalfa cubes	91.0	9.9	17.9	10.9	1.38	0.27	0.24	0.01	0.28	4.4	22.7	17.0
Canola meal	89.8	11.4	40.3	7.8	0.75	1.25	0.57	0.03	0.28	—	—	68.0
Soybean meal	88.8	12.8	49.7	6.9	0.40	0.69	0.31	0.08	0.45	46.4	40.0	64.0

■ 38.11 OIL EXTRACTION

Rapeseed contains about 42 percent oil. Most seed is processed using a pre-press solvent extraction system. The seed is first rolled or flaked to fracture the seedcoat and rupture the oil cells. Flakes are then cooked to rupture any intact cells that remained after the flaking process. The flaked and cooked seeds are then subjected to a mild pressing process, which removes about two-thirds of the oil and compresses the fine flakes into large cake fragments. These fragments are then extracted with a solvent to remove most of the remaining oil.

Solvent is removed from the oil by a solvent recovery system that ensures solvent-free oil. This oil is combined with the pre-pressed oil to form crude oil, which is passed through another process to remove undesirable gummy impurities. The degummed oil is then refined to further remove impurities.

The major type of refining in Canada is physical refining with steam, although alkali refining is also used. The oil can be high in chlorophyll pigments that cause the oil to become rancid quickly if they are not removed by bleaching with bleaching clays. This bleaching also removes other minor impurities from the oil. Additional processing is completed according to product requirements. Deodorization is usually the final refining step and removes compounds that impart undesirable odor and taste. Different treatments are used for salad oils, margarine oils, and shortenings.

After oil extraction, the remaining solvent in cake fragments is removed by injection of live steam into the meal. The final stripping and drying of the cake is accomplished in kettles, and the cake that emerges, free of solvent, contains 1.5 percent residual oil and has a moisture content of 10 to 12 percent. After cooling, the cake (Figure 38.7) is often granulated to a uniform consistency and is either pelletized or sent directly to storage for marketing as a high protein feed supplement for livestock and poultry.

FIGURE 38.7
Rapeseed cake after extraction of oil. The cake will be ground into meal or pelleted. [Courtesy Canola Council of Canada]

▓ 38.12 DISEASES

Several diseases attack rapeseed, but proper management can help keep them from significantly reducing yields. The use of crop rotations in which rapeseed is not planted sooner than every four years is necessary to control most diseases. Destroying infected residues in fields and planting certified disease-free seed also help to control many diseases. Some varieties are resistant to certain diseases, but cultural control is necessary in heavily infected fields.

38.12.1 Downy Mildew

Downy mildew, caused by the fungus *Perenospora parasitica*, is the most common disease of rapeseed. It causes yellow discoloration of the upper leaf surface and white fungal growth on the lower surface. Although it is most severe in the autumn on winter annual varieties, the disease can also occur in cool, wet springs on all varieties.

Crop rotation and burial of infected residue help control the disease.

38.12.2 White Mold and Stem Rot

White mold and stem rot, caused by the fungus *Sclerotinia sclerotiorum*, can infect stems, branches, leaves, and pods. Symptoms begin with the development of water-soaked spots and subsequent development of white cottony mycelia (Figure 38.8).

Excessive nitrogen fertilizer, high plant populations, and high humidity favor the disease. A rotation to at least three years of nonhost crops, deep plowing to bury infected residues, and certified disease-free seed help to control the disease. Other crops susceptible to this disease are sunflower, tobacco, field bean, alfalfa, and several vegetables.

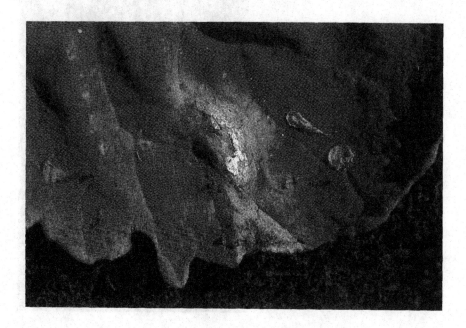

FIGURE 38.8
White mold mycelia.
[Courtesy Canola Council of Canada]

Fungicides are available to help control white mold. Timing of application is critical to maximize fungicide effectiveness. Most fungicides should be applied at early bloom. Fungicides should only be used when: (1) yield potential is above average, (2) weather leading to early bloom has been wet, (3) more rain and high humidity are expected, and (4) the disease has been a problem in recent years. A fungicide is more likely to be needed if rotations of three years or less are used.[4]

38.12.3 Blackleg

Blackleg is caused by the fungus *Leptosphaeria maculans*. There are two strains of the disease, a mild strain and a virulent strain. The mild strain is common in rapeseed-producing regions, but it infects plants late in the season and rarely causes yield loss. The virulent or aggressive strain can infect plants early, causing leaf spots any time from the seedling stage to crop maturity. Leaf spots are tan or buff color and round to irregular in shape. Stem lesions are gray and dark bordered, later turning dark gray to black (Figure 38.9). Early infection causes premature dying of the plant or lodging. Pods and seeds also may be infected. Infected pods split open, resulting in seed loss. Seed produced in infected pods may be shriveled and gray in color.

Control of blackleg includes crop rotations of at least three years of other crops, deep plowing to bury infected residues, control of weed hosts, and planting certified disease-free seed.[9] Many varieties have at least moderate resistance to blackleg.

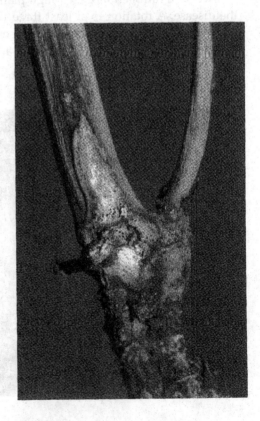

FIGURE 38.9
Blackleg lesion on a rapeseed stem. [Courtesy Canola Council of Canada]

38.12.4 Black Spot

Black spot, caused by the fungi *Alternaria brassicae* and *A. raphani*, overwinters on plant residue and seed. The disease can attack any above-ground part of the plant at any growth stage from emergence to maturity. The disease begins as small, yellow lesions that enlarge into dark spots with concentric rings. Leaf spots can grow up to 2 inches (5 cm) in diameter.

A rotation of at least three years of other crops, disease-free seed, and controling weedy mustards and volunteer rapeseed plants help to control the disease.

38.12.5 Stem Canker

Stem canker, caused by the fungus *Phoma lingum*, is especially a problem in Canada. An early symptom of the disease is the formation of a wedge or V-shaped yellow spot that extends from the leaf margin inward along the midrib of the leaf. As the disease progresses, the yellow regions become tan, dry, and papery or brittle. The leaf will eventually fall from the plant.

Deep plowing to bury infected residues and a rotation of at least three years of other crops can help the control the disease. The use of certified disease-free seed is necessary. Treating certified seed with benomyl assures that the disease will not be introduced with the seed.

38.12.6 Other Diseases

Seed rots, seedling blights, and root rots caused by *Fusarium* and *Pythium* species of fungi can reduce stands in rapeseed fields. Light leaf spot (*Pyrenopeziza brassicae*) attacks all parts of rapeseed plants.

▓ 38.13 INSECT PESTS

Several pests are problems in areas where rapeseed has been grown for years. The most common are the flea beetle, cabbage seedpod weevil, armyworm, and aphid.

38.13.1 Flea Beetles

Flea beetles are small, jumping beetles. There are several species that infect rapeseed and other plants. The crucifer flea beetle, *Phyllotreta cruciferae*, is the most common species infesting rapeseed. The adult is a small, oval-shaped, blackish beetle with a bright blue sheen on its forewings (Figure 38.10). They are about 1/32 to 1/8 inch (2 to 3 mm) long. Flea beetles will jump quickly when disturbed. The adults overwinter in plant residue and emerge to feed on seedlings. Early seeding helps plants grow enough before infestation to minimize damage. Seed treatment with a systemic fungicide helps control the pest. Occasionally, a later generation of flea beetle will feed on developing pods.

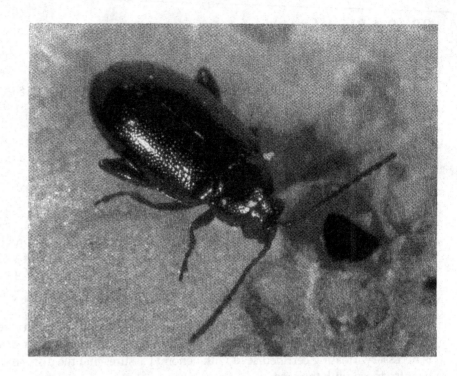

FIGURE 38.10
Flea beetle adult on a rapeseed leaf. [Courtesy Canola Council of Canada]

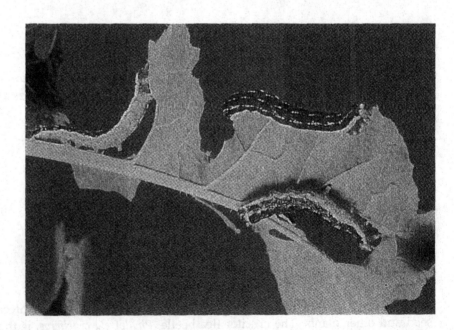

FIGURE 38.11
Bertha armyworm larvae feeding on rapeseed leaves. [Courtesy Canola Council of Canada]

38.13.2 Armyworms

Several species of armyworms can cause damage to rapeseed. The Bertha armyworm *(Mamestra configurata)* is the most important species and is a serious pest in Canada. Damage is done by the larvae that feed on leaves (Figure 38.11). Later, the larvae will feed on stems, causing flowers and pods to drop on the ground. When rapeseed plants are nearing maturity and dropping their leaves, the larvae will

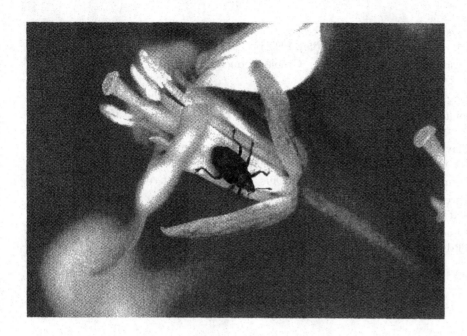

FIGURE 38.12
A cabbage seedpod weevil feeding on a rapeseed flower. [Courtesy Canola Council of Canada]

feed on the pods. Infestations are more serious when adequate snow cover over the winter protects overwintering pupae from cold temperatures. Insecticide application is necessary before infestations reach the level of economic damage.

38.13.3 Cabbage Seedpod Weevil

The cabbage seedpod weevil (*Ceutorhynchus obstrictus*) is a problem mostly in winter rapeseed, which flowers earlier in the season than spring varieties. Adults overwinter in the soil and emerge in the spring. They feed on rapeseed and other plants in the spring to early summer. When rapeseed begins to flower, the weevils feed on flower buds. Later, the females lay eggs on the pods and the larvae will burrow into the pods. Infested pods will look distorted. Adult weevils are ash-grey and about ⅙ inch (3 to 4 mm) long (Figure 38.12). They have a prominent, curved snout that is typical of most weevils. There are only limited insecticides available. Controlling alternate hosts, such as weeds of the mustard family and volunteer rapeseed, helps to control the pest.

38.13.4 Other Pests

Other pests that may cause problems in rapeseed are grasshoppers, aphids, harlequin bugs, cutworms, and others.

REFERENCES

1. Downey, R. K. "The origin and description of the *Brassica* oilseed crops," in J. K. G. Kramer, F. D. Sauer, and W. J. Pidgen, eds., *High and Low Erucic Acid Rapeseed Oils*. New York: Academic Press, 1983, pp. 1–20.

2. Sovero, M. "Rapeseed, a new crop for the United States," in J. Janick, and J. E. Simon, eds., *New Crops*. New York: Wiley, 1993, pp. 302–307.

3. Grombacher, A. and L. Nelson. "Canola production," *NE Coop. Ext. Serv. NebGuide* G92-1076-A, 1992.

4. Berglund, D. R., and K. McKay. "Canola production," *ND St. Univ. Ext. Serv.* A-686, 2002.

5. Johnson, D. L., and R. L. Croissant. "Rapeseed/canola production," *CO St. Univ. Ext. Serv.* 0.110, 1992.

6. Clayton, G., K. Turkington, N. Harker, J. O'Donovan, and A. Johnston. "High yielding canola production." *Better Crops* 84, 1(2000):26–27.

7. Schmidt, W. "Canola production in Ohio," *OH St. Univ. Ext.* AGF-109-95, 1995.

8. Owen, D. F., and others. "A process for producing nontoxic rapeseed protein isolate and an acceptable feed by-product," *Cer. Chem.* 48, 2(1971):91–96.

9. Lamey, A. "Blackleg of canola: Biology and management." *ND St. Univ. Ext. Serv.* PP-1024, (rev.), 1996.

Miscellaneous Industrial Crops

Several crops are grown for industrial oils, such as castorbean. Some are grown for use as food additives or flavorings, such as mint, hops, and mustard. Some are grown for their edible oils, such as sunflower and safflower. Others are grown for fiber, such as hemp and ramie. Although global production is not extensive, these crops are important locally and regionally and may find increasing use in the future.

◼ 39.1 HOPS

39.1.1 Economic Importance

In 2000–2003, hops were harvested on an average of 163,000 acres (66,000 ha), with an average production of about 111,000 tons (101,000 MT) or 1,370 pounds per acre (1,500 kg/ha). The leading countries in hop production are Germany, the United States, China, the Czech Republic, and the United Kingdom. From 2000–2003, the United States produced an average of nearly 31,000 tons (28,000 MT) on about 32,500 acres (13,100 ha) with an average yield of 1,900 pounds per acre (2,100 kg/ha). Practically all of the hops are grown in Washington, Oregon, and Idaho.

39.1.2 History of Hop Culture

Hop is thought to be indigenous to the British Isles,[1] but cultivated hops were grown in Germany in AD 768, and in England late in the fifteenth century.[2] Hops were taken to America as early as 1629. The first commercial hop field, commonly referred to as a hop yard, was established in New York in 1808. Cultivation of hops later extended into Wisconsin, but before 1869 they had become established in the Pacific Coast States.

39.1.3 Adaptation

Hops are grown under a wide range of climatic conditions, but mostly they are cultivated in mild regions with abundant early rainfall or irrigation, followed by dry, warm weather. Such conditions are found in the Pacific Northwest, the valleys of the northern half of California accessible to cool ocean breezes, western New York, and northern and central Europe. The preferable soils are deep sandy or gravelly loams.[3]

39.1.4 Botanical Description

Hop belongs to the subfamily *Cannabineae* of the family *Urticaceae*, the nettle family. The hop plant *(Humulus lupulus)* is a long-lived perennial.

VEGETATIVE GROWTH Hop stems are herbaceous, roughened, angular, and hollow. Their color among different varieties is purplish-red, pale green, or green streaked with red. The stems above ground bear thin opposite lateral branches. The slender trailing or twining vines may grow to a length of 25 to 30 feet (7 to 9 m). The stems, as well as the leaf petioles, have several lines of strong hooked hairs that help them cling to their supports. After being cut back each year, new stems arise each spring from buds on the underground rhizomes.

INFLORESCENCES The hop is a dioecious plant with sterile flowers in racemes or panicles and fertile (pistillate) flowers in clusters or catkins. Occasional plants are monoecious. The commercial hops borne on the female plants are the spike-like pistillate inflorescences[4] that somewhat resemble fir cones in shape (Figure 39.1). Botanically, the cones are called strobiles, but colloquially they are referred to as beer blossoms. The strobiles, borne on lateral branches of the main stem, are 1 to 2½ inches (25 to 64 mm) in length.

The staminate flowers are about ¼ inch (6 mm) in diameter. Since the male vines produce no hops, large numbers of them are undesirable. In a crop, 1 or 2 percent male plants uniformly distributed throughout the hop yard will provide sufficient pollen for fertilization of the distillate plants.[3]

FRUIT AND SEED The fruits of the hop are purple, oval, and about the size of a mustard seed. The hop seed possesses a curved embryo and a very small amount of endosperm. The crop is seldom reproduced from seed because of the great variability

FIGURE 39.1
The hops plant. [Courtesy USDA-NRCS PLANTS Database/Britton, N.L., and A. Brown. 1913. Illustrated flora of the northern states and Canada. Vol. 1:633]

in the progeny. Seedless (unfertilized) hops weigh about 30 percent less than those that develop seed. Their quality is better than that of seeded hops, but the yield is smaller. Hop seeds contain a bitter, unpalatable oil.[5] Because of their higher market value, seedless hops are now being grown to a considerable extent by excluding all male plants.

The outer surface of the bracteoles, perianth, and bases of the bract-like stipules are covered with yellow pollen-like resinous grains. These grains are the lupulin or hop meal. The commercial value of well-cured hops is based upon the amount and quality of the lupulin.

The principal active ingredients in the lupulin glands are:

1. Hop oil. A volatile essential oil with a characteristic aroma; hop oil constitutes about 0.2 to 0.8 percent of the active lupulin-gland ingredients.
2. Nonresinous bitter compounds.
3. Resins. Of these, the hard resin is practically tasteless and is devoid of brewing value. The soft alpha and beta resins are bitter. They impart this taste to the beer wort and have a preservative effect. Hops contain 10 to 18 percent or more total resins.
4. Tannin, which is also bitter, constitutes 4 to 5 percent of the hop.

The alpha-acids in the resins, which are the most important bitter compounds, constitute from 4½ to 7 percent of the weight of the dried hops in the important domestic varieties and 8 to 12 percent in some English varieties.

39.1.5 Hop Culture

Hops are propagated by cuttings from numerous runners (rhizomes) sent out by the plant just below the surface of the soil. The excess shoots are pruned once a year, usually in the spring. These are cut into pieces 6 to 8 inches (15 to 20 cm) long that bear at least two sets of buds that are promptly planted.[3]

The hops are set in rows 6½ to 8 feet (200 to 240 cm) apart. Cuttings are spaced at least 3 feet (90 cm) apart in the row. Very few strobiles are produced the first year. In fact, the vines do not reach full production until the second or third year.

Hop yards, once established, are maintained for many years. Missing hills will be replaced at pruning time each spring. Winter cover crops such as vetch or small grain are grown between the hop rows to be plowed under in the spring. Fertilizers that are applied usually consist of 80 to 160 pounds of nitrogen per acre[6] and 50 to 100 pounds of P_2O_5. Most soils in the present domestic hop-growing areas contain ample potash, but sulfur, zinc, or boron may be needed.[3] These elements are supplied by applications of calcium sulfate, zinc sulfate, and borax.

Hop plants are trained on high wire trellises, usually about 18 feet (5.5 m) above the ground (Figure 39.2). Heavy wires attached to upright poles set four hills apart are stretched across the field. Lighter trellis or "stringing" wires are strung across the heavy wires over each row or over the middle between two rows. The trellis wires are hung under the cross wires with hooks where hops are picked by hand. Large end poles at the borders of the field are slanted outward and well anchored.

FIGURE 39.2
Hops growing on trellises.
[Courtesy Richard Waldren]

Coir twine tied to the trellis wires and pegged to the ground at or near each hill provides strings for the vines to climb. The vines wind clockwise around the strings.[7]

HARVEST Ripe hops have the best quality. The strobiles or cones change from a bright green to a bright yellowish green. They become more resilient, sticky, crisp, and papery, while the pericarp on the hop fruits takes on a dark purple color. The characteristic lupulin odor becomes very marked as hops approach ripeness.

By 1974, nearly all domestic hop vines were cut and loaded into trucks by mobile cutting machines that move down the rows, with men on elevated platforms cutting the vine tops free from the trellises (Figure 39.3). The cut vines are hauled to a stationary picker adjoining a drier. The picking machines strip the hops and most of the leaves from the vines with fingers attached to rotating drums. These then pass through a forced air stream to remove most of the leaves and stems and are conveyed onto a series of inclined "dribble belts" that carry the trash upward while the cleaned hops roll off the belt (Figure 39.4). The hops are then conveyed to the drier.

39.1.6 Curing

Artificial drying reduces the 65 to 80 percent moisture content of hops down to 8 to 10 percent by blowing air heated to 140 to 150°F (60 to 66°C) through layers of hops on a moving belt or in stationary chambers. Dried hops are then moved to a cooling chamber where they remain for five to seven days or longer.[7] Hops are then pressed into bales.

Most of the hops are bleached and conditioned by the burning of 1 to 4 ounces of sulfur per 100 pounds (630 to 2,500 mg/kg) of green hops, to produce the sulfur dioxide fumes that are blown up through the hops during drying.

FIGURE 39.3
Hops being harvested with a mobile machine. [Courtesy Oregon State University, used with permission]

39.1.7 Uses of Hops

The principal use of hops is in the manufacture of beer. The sweet beer wort is boiled with the hops. The flavor of the wort is improved by the extraction of the active ingredients of the hops. An appreciable portion of the crop is now sold as the extract from the dried hops. The essential oil of the lupulin glands imparts an aroma to the beer. The tannin probably serves to precipitate albuminous substances. The malic and critic acids in the hops tend to increase the acidity of the wort. About 8 to 13 ounces (230 to 370 g) of hops are used for each barrel of beer.[8]

39.1.8 Diseases

Downy mildew is the most serious disease of hops in the world. This disease is caused by a fungus *(Pseudoperonospora humuli).* The disease attacks the shoots,[9] stems, and flower parts. It is spread in damp weather by wind and from cuttings. Downy mildew may be controlled by destroying the cut vines, escaped plants, and affected shoots, as well as by spraying or dusting. Stripping all leaves from the stems from the bottom up to within 4 feet (120 cm) of the trellis wire reduces the

FIGURE 39.4
Hand-cut hop vines
entering a stationary picker.
[Courtesy Oregon State
University, used with
permission]

spread of downy mildew as well as of the red spider and the hop aphid. Sprays
should be applied: (1) when the vines are first trained, (2) when the vines get to the
cross wires, (3) just before the female flower buds open, and (4) at other times when
weather conditions are favorable to the spread of the disease. Streptomycin sprays
are effective against downy mildew.[10]

Sooty mold *(Fumago vagans)* blackens the cones. The infection is carried into the
cones by aphids. Control of aphids largely prevents the trouble. Powdery mildew
(Sphaerotheca humuli) is controlled by sulfur dusting. Plants affected by virus diseases, crown gall, and rots should be removed from the field and replaced by
healthy cuttings.

39.1.9 Insect Pests

The chief insect that attacks hops is the hop aphid *(Phorodon humuli).* It is controlled
by stripping the lower leaves from the vines and by insecticidal sprays and dusts.
Other insects that attack hops include the cucumber beetle, thrip, leaf tier, hop-
plant borer, hop butterfly, hop fleabeetle, and webworm.

The red spider is controlled by stripping the lower leaves or by dusting with
malathion or other insecticides.

▦ 39.2 MINT

39.2.1 Economic Importance

Global production of peppermint in 2000–2003 averaged 62,000 tons (56,000 MT). Most of the world's production of peppermint is in Morocco. Peppermint production in the United States in 2000–2003 averaged nearly 3,400 tons (3,100 MT). Leading states in peppermint production are Oregon, Washington, Idaho, Indiana, and Wisconsin. U.S. spearmint production in 2000–2003 averaged about 2 million pounds (900 MT). Over 70 percent of the U.S. production is in Washington. Other leading states in spearmint production are Wisconsin, Indiana, and Oregon.

39.2.2 History of Mint Culture

The distillation of mint oil for use in medicines was practiced in Egypt as early as AD 410. The Japanese were distilling mint oil for eyewash as early as AD 984. Production began in England about 1696 and in the United States, at Ashfield, Massachusetts, in 1812. Production extended to Michigan in 1835, to Indiana about 1855, and to the Pacific Coast States over a half-century later.

39.2.3 Adaptation

Mint is mostly grown in humid areas where summer temperatures are moderate and the long summer days promote a high oil content. Soils best suited to mint production are deep, rich, and well drained, especially muck or other soils that are high in organic matter. Muck soils are usually somewhat acidic. However, mint is grown in the alkaline, irrigated sandy loam soils of the Yakima Valley of Washington. The preferred pH range is 6.0 to 7.5.

39.2.4 Botanical Description

Mint belongs to the mint family *Lamiaceae*. Plants of this family have a four-sided or square stem. The three perennial species grown commercially are peppermint (*Mentha piperita*), spearmint (*M. spicata* or *M. viridis*), and Scotch spearmint (*M. gentiles*). Japanese mint (*M. arvensis* var. *piperascens*) is grown to a limited extent. The inconspicuous flowers, which are borne in whorls on a spike about 2 inches (5 cm) long, produce very few seeds. The plants produce numerous rhizomes that are used for propagation or for field planting. Peppermint plants (Figure 39.5) are usually 3 feet (90 cm) or less in height.

Three types of peppermint are: (1) black peppermint (black mint, English peppermint, or Mitcham mint), which has dark green to purple stems and deep-green, broadly lanced, slightly toothed leaves, and, usually, light purple flowers; (2) American peppermint (American mint or State mint), which has green stems and lighter green leaves, and is otherwise similar to black mint but not very commercially satisfactory; and (3) white peppermint, which is grown in England.[11]

FIGURE 39.5

Flowers and leaves of peppermint *(left)*; flowers, leaves, and rhizomes of spearmint *(right)*.

Spearmint has light green, pointed, spear-shaped leaves and pointed flower spikes that are longer than those of peppermint. Spearmint is the common plant grown in home garden for its leaves, which give the mint flavor as well as the green decoration to drinks and foods (Figure 39.5).

Native species of mint, which include *M. canadensis*, contain oil of an inferior quality. They should be eradicated where they occur as weeds in commercial mint fields. Catmint, or catnip *(Nepeta cataria)*, is another member of the mint family. This native of Europe, which escaped to the United States, is attractive to cats.

39.2.5 Mint Culture

Lands planted to mint should be comparatively free from weeds,[12] because many species of weeds affect the color or odor of the distilled oil. Ragweed is the most objectionable weed because it contains a light, volatile oil that distills over and contaminates mint oils with a disagreeable flavor. A mixture of 1 percent ragweed in spearmint discolors the oil and renders chewing gum unfit for sale.[13]

Other weeds, such as smartweed, pigweed, horseweed, nettle, hemp, purslane, lambsquarter, buttermold, and Canada thistle impart disagreeable odors. Since their oils are heavier than water, they separate from mint oils after distillation. Clean land is best attained by planting the mint after fallow or after an intertilled crop such as potato or onion, which also are adapted to muck soils. Preemergent and postemergent herbicides, such as terbacil, bentazon, and sethoxydim, are used on mint crops.

Fertilizer application on muck soils usually consists of 120 to 200 pounds per acre (135 to 225 kg/ha) nitrogen, 50 pounds per acre (56 kg/ha) phosphate, and 100 to 150 pounds per acre (110 to 170 kg/ha) potash.[14] Irrigated sandy loams in eastern Washington and eastern Oregon receive about 120 pounds per acre (135 kg/ha) nitrogen. Potassium, and sometimes phosphorus, is also beneficial.

Mint fields should be prepared in time for early spring planting of the stolons. Sometimes strips of rye or other tall crops or permanent willow windbreaks are planted at intervals across the field to prevent drifting sand particles from damaging the young mint plants.

The fields that supply the rhizomes should have been cut early the previous year so the runners have time to build up adequate plant food reserves. The rhizomes are plowed out, forked free from soil, and thrown into piles, and the piles are covered lightly to prevent excessive drying. One acre produces enough rhizomes to plant 10 acres or more.

The runners are planted with special machines or dropped into the furrows by hand, lengthwise, end to end. Mint is often propagated by pulling up young rooted plants that arise from the rhizomes in established mint fields in the spring. These are set out with vegetable or tobacco transplanting machines.

A rotary hoe, light weeder, or harrow is used for cultivation until the mint plants are 5 to 6 inches (13 to 15 cm) high. Rubber-toothed harrows are often used to cultivate mint fields. After the first year, row cultivation is not needed because the plants spread between the rows. Spring tooth or other harrows or the rotary hoe may be used early in the spring. After harvest, the fields are left until late fall or early spring when they are plowed. Fall plowing covers the rhizomes enough to protect them from winter cold. The plants come up from rhizomes each year. Fall plowing is not recommended in the coastal districts where the land is subject to overflow during the winter or early spring.

HARVEST The maximum yield of oil for both peppermint and spearmint occurs when the plants are in bloom. In Indiana, the oil and menthol yield continues to increase up to the stage of 50 percent bloom,[13] but a better quality of oil is sometimes obtained by cutting before blooming begins.[12] The crop must be cut before leaf shedding occurs because the oil is contained in small glands on the lower leaf surface. The crop is usually harvested after the plants are well in bloom in Oregon and Washington, except where cutting in early bloom permits the cutting of two crops in a season. Cutting is done in July, August, and September. The mint is cut with a mower equipped with lifter guards. It is raked a day or two later after it is partly dried. The crop is dry enough for distilling when the moisture content is down to 40 to 50 percent, but it must be gathered and hauled to the still before the leaves are dry enough to shatter. The partly dried mint is gathered and chopped with a forage harvester and blown into a trailer or into portable distilling tubs,[11] or gathered from the swath with a hay loader and unloaded into the distilling tub with slings (Figure 39.6).

39.2.6 Distillation of Mint Oil

Steam is used to distill mint oil from the leaves (Figure 39.7). Distillation is completed in forty-five to sixty minutes. The vapors are condensed in water-cooled coils and passed into a tank receiver. About 6 volumes of water boil over with each volume of oil. The oil, which is in the top layer, is drawn off into metal cans or drums.

39.2.7 Uses of Mint

Peppermint oil should consist of 50 percent or more menthol and 4 to 9 percent esters. Peppermint oil is used in ointments, salves, and medicines as a flavoring and

FIGURE 39.6
Gathering mint from a windrow and loading into a distilling tub.

FIGURE 39.7
Unloading mint into a distillation vat.

as a remedy for digestive disturbances. It has an important use in confections, toothpastes, tooth powders, and perfumes. Spearmint oil contains at least 50 percent carvone or carvol. The chief use of spearmint oil is in chewing gum, mint sauces, and mint flavors. A substantial portion of mint oil is evaporated for use as a dry ingredient.

39.2.8 Diseases and Pests

The chief fungal diseases of mints, and their causal organisms, are Verticillium wilt (*Verticillium albo-atrum*), rust (*Puccinia menthae*), and anthracnose or leopard spot (*Sphaceloma menthae*). Wilt is controlled with resistant varieties and by postharvest flaming that kills the spores without harming the rhizomes. Rust and anthracnose are reduced by plowing under all mint residues or by applying fungicides. Rusted, early emerging mint shoots may be killed by herbicides or by flaming to reduce the amount of later infection. Powdery mildew, leaf spots, root rots, and nematodes also attack mints.[15]

The gray looper moth (*Rachiplusia ou*), mint bud mite (*Flordotarsonemus* sp.), and mint fleabeetle (*Longitarsus waterhousei*) are the main insect pests of mint.[12] Other loopers, caterpillars, and cutworms can cause problems in local areas.

39.3 SUNFLOWER

39.3.1 Economic Importance

Sunflower seed was harvested globally on an average of about 50 million acres (20 million ha) in 2000–2003. Global production averaged about 27 million tons (20 million MT) or 1,100 pounds per acre (1,200 kg/ha). The leading producing countries are Argentina, the Russian Federation, Ukraine, China, and France. Global production of sunflower seed oil is about 9.6 million tons (8.7 million MT) annually.[16] Average U.S. production in 2000–2003 was 3 million tons (2.7 million MT). About 2.4 million acres (970 million ha) were harvested annually in the United States in 2000–2003, with an average yield of 1,260 pounds per acre (1,400 kg/ha). Leading states in sunflower production are North Dakota, South Dakota, Kansas, Colorado, and Minnesota. About 80 percent of U.S. production is for sunflower seed oil. Sunflower can be grown from Canada to Texas.[17]

39.3.2 History of Sunflower Culture

The cultivated sunflower is a native of America. It was taken to Spain from Central America before the middle of the sixteenth century. It was being grown by Indians for food at Roanoke Island in 1586, and in New England for hair oil in 1615.[18] Established culture of the sunflower in the United States followed introduction of improved varieties that had been developed in Europe before 1600.

Wild sunflower, the state flower of Kansas, flourishes in overgrazed pastures, in fields, and in waste places over the western United States, particularly in the Great Plains States, strictly as a weed. When the rains came in 1940 and 1941 after the southwestern dust bowl, wild sunflower dramatically took possession of the wind-eroded fields and pastures where the efforts of people had failed. Its rank-growing branched stems protected the lands from further wind injury while perennial vegetation became re-established.

39.3.3 Adaptation

The sunflower is adapted for seed production where corn is successful in the northern two-thirds of the United States. It was formerly grown for silage in cool northern and high altitude regions where corn does not thrive.[19] The young plants will withstand considerable freezing until they reach the four- to six-leaf stage. The ripening seeds likewise suffer little damage from frost. Between those stages the plants are more sensitive to frost. The plants show relatively little photoperiodic response.

FIGURE 39.8
Sunflower head. Seeds are produced in the disk flowers. The ray flowers around the rim are sterile.
[Courtesy Richard Waldren]

39.3.4 Botanical Description

Sunflower *(Helianthus annuus)* belongs to the family *(Asteraceae Compositae)*. Plants of this family bear heads in which the fertile flowers are aggregated and are bordered by rays, the corollas of sterile flowers.

Cultivated sunflower is a stout erect annual, 3 to 6 feet (90 to 180 cm) in height.[18] Some varieties can reach 20 feet (6 m) in height. Plants have a rough hairy stem 1 to 3 inches (2.5 to 8 cm) in diameter, which terminates in a head or disk 3 to 24 inches (8 to 60 cm) in diameter (Figure 39.8). The top of the disk is brown to nearly black. The rim of the disk is surrounded by pointed scales and 40 to 80 yellow rays. The seed is an elongated rhomboid achene. Seeds of the oil-type varieties are smaller and dark brown to black in color. The confectionary varieties have larger seeds with light tan stripes alternating with dark brown to black. The stalk may produce several branch heads, but most varieties produce only one head. The flowers are almost entirely cross-pollinated. Cytoplasmic male sterile types are now used for producing hybrid seed.[20]

The stalk of wild sunflower (also called *H. annuus*) is usually 1 inch (2.5 cm) or less in diameter, has many branches, and bears numerous heads. Wild sunflower seeds are too small and chaffy for economic utilization.

Sunflower gets both its common and botanical names from the noticeable characteristic of the heads that face toward the sun throughout the day (Figure 39.9). Their heliotropic movement results from a bending of the stem, a process called nutation, which tilts the head to the west in the afternoon.[21] After sunset, the stem

FIGURE 39.9
A field of sunflowers.
[Courtesy Richard Waldren]

gradually straightens until it becomes erect about midnight. Thereafter, the stem gradually bends in the opposite direction up to as much as 90 degrees, so that the head faces east by sunrise. Soon afterward, the stem starts to straighten until the head is erect again at noon. The leaves likewise face east in the morning, west in the evening, and upward at noon and midnight. However, at 10 P.M., the leaves are drooping downward. Stripping the leaves from the stalk stops all bending of the stem. Nutation ceases when anthesis (pollen shedding) begins, or shortly thereafter. Fully 90 percent of the heads are facing east or northeast as they hang at maturity except where strong winds occur. Growers of tall varieties have taken advantage of this eastward nodding habit by planting the rows north and south. At harvest time, they drive along the east side of each row and cut off the overhanging heads.[22]

39.3.5 Varieties

Most sunflowers planted today are hybrids. These types grow to a height of 3 to 5 feet (90 to 130 cm). Tall late hybrids or varieties are grown in warm climates and in home gardens.[23]

39.3.6 Sunflower Culture

The seedbed for sunflowers should be prepared the same as for corn. No-till planting can be used. Planting is done with a row crop planter or with a grain drill with some of the feed cups closed off. Sunflower is planted when the minimum soil temperature is 60°F (16°C). Planting is usually in May except in the central latitudes of the United States where April planting may be more favorable. Seeding rate should be adjusted to provide about 14,000 to 21,000 plants per acre (35,000 to 52,000 plants/ha) dryland and 20,000 to 25,000 plants per acre (49,000 to 62,000 plants/ha) irrigated.[24] Row spacing can vary from 12 inches (30 cm) for smaller hybrids to 30 inches (74 cm) for taller hybrids.[25]

HARVEST Sunflowers are mature when the backs of the heads are yellow and the outer bracts are beginning to turn brown. The seeds are ready to thresh and store when they contain not more than 8 to 9 percent moisture. Harvest at higher moisture content is possible, followed by artificial drying in storage. This is sometimes necessary when lodging, birds, or head rot is causing losses. Most hybrids are harvested with a combine. Extra slats or screening on the reel arms of combines are necessary to avoid catching and throwing out the crook-necked heads.

39.3.7 Uses of Sunflower

The head of the mature sunflower contains about 50 percent of the dry matter of the whole plant.[18] Nearly one-half the weight of dried heads is seed. About 35 to 50 percent of the seed consists of hull. The whole seed contains 24 to 45 percent oil, whereas the hulled kernel contains 45 to 55 percent or more. The expressed oil yield from whole seed ranges from 20 to 35 percent. The residue is oil cake, which contains about 35 percent protein when made from whole seed.

Sunflowers are now grown primarily for the production of oil. This began in the United States in 1967 using high-oil varieties developed in the Soviet Union. Hybrids or varieties with large, striped seeds are eaten as a confection, either roasted, raw, or in candies. The seeds are also fed to poultry, caged birds, or wild birds. The meal is fed to livestock but would be suitable for human feed in the form of chips or breads.[17] The meal is free from toxins. Discoloration of the white defatted meal into shades of beige, green, or brown is mostly due to oxidation of chlorgenic acid at pH above 7.0.

Sunflower oil is mostly poly-unsaturated. It is used primarily in shortening, cooking, salad oils, and margarine. It smokes at about 302°F (150°C), which is too low for deep-fat frying at the usual temperature of about 390°F (200°C). Sunflower oil is also used for making soaps and paints. It is a semidrying oil with an iodine number of about 130. The oil cake is mostly fed to livestock.

Growing sunflowers for silage in the colder parts of the United States ceased when forage harvesters became available. These machines facilitated the ensiling of adapted grasses and legumes that are more palatable and nutritious than sunflowers.

39.3.8 Diseases

Sunflowers are attacked by rust, Sclerotinia stalk and head rot, Verticillium wilt, charcoal rot, downy mildew, powdery mildew, and black leaf.

Sclerotinia stalk and head rot, caused by the fungus *Sclerotinia sclerotiorum*, is a problem in more humid regions. This fungus also causes white mold in dry bean so sunflower should not be in a rotation with that crop. Infected plants develop a canker that girdles the stem resulting in wilting. If the disease occurs after flowering, it will infect the head.

Sunflower rust, caused by the fungus *Puccinia helianthi-mollis*, is the most common disease. It develops numerous brown (and later black) pustules on the leaves, which cause them to dry up. Destruction of so much leaf tissue stops the development of the stalk and head. Resistant hybrids are available. European varieties, the ornamental sunflower *(Helianthus argophyllus)*,[19] and the North Texas wild sunflower are also resistant.

Charcoal rots cause wilting and lodging following the decay of the pith in the stalk. Control consists of late planting, an ample soil moisture supply, and avoidance of severely infested soils.

39.3.9 Insect Pests

The chief insect enemies of the sunflower are sunflower head moths *(Homoeosoma ellectellum)*, cutworms, wireworms, white grubs, grasshoppers, aphids, thrips, sunflower beetles, seed weevils, carrot beetles, sunflower budworms, and webworms. Head moths are numerous in warm areas, but they can be controlled with insecticides. At the present time, no known method, cultural or chemical, is available for control of the carrot beetle, which feeds on the root system of the plant. Some sunflower varieties or hybrids are resistant to the sunflower beetle.

▧ 39.4 MUSTARD

39.4.1 Economic Importance

In 2000–2003, mustard was grown globally on a average of 1.7 million acres (700,000 ha). Average production was 540,000 tons (490,000 MT) for a yield of 620 pounds per acre (690 kg/ha). Leading countries producing mustard are Canada, Nepal, the Russian Federation, Myanmar, and the Czech Republic. An average of 92,000 acres (37,000 ha) of mustard was grown in 2000–2003 in the United States. Average production was about 3,400 tons (3,100 MT) or 800 pounds per acre (890 kg/ha).

39.4.2 Adaptation

Mustard seed is best adapted to regions with relatively cool summer temperatures, fair supplies of soil moisture during the growing season, and dry harvest periods. Such conditions are found in northern Montana and in the low rainfall but foggy coastal valley sections of southern California. In Montana, the soils where mustard is grown chiefly are medium loams that are slightly alkaline. In California, the brown mustard is grown mostly on sandy loam, while the yellow mustard is produced on either a heavy type of sandy loam or a light adobe soil.

39.4.3 Botanical Description

Mustard belongs to *Cruciferae*, the mustard family, and the genus *Brassica*, which also includes rape, turnip, rutabaga, and kale. The plants of this family are biennial or annual. They bear flowers with four petals arranged in the form of a cross. The genus *Brassica* has yellow flowers, six stamens, and globose seeds borne in two-celled pods. The species grown for their seeds are all annual herbaceous plants

FIGURE 39.10
Seeds of black mustard
(left), yellow mustard
(center), and wild mustard
or charlock *(right)*.

about 2 to 3 feet (60 to 90 cm) in height. They include: (1) yellow or white mustard (*Brassica alba*) (Figure 39.10), (2) brown mustard (*B. juncea*), and (3) black mustard (*B. nigra*). The mustards that occur as weeds, the seeds of which are sometimes processed, include white mustard and black mustard, as well as charlock or wild mustard, *B. arvensis;* ball mustard, *Neslia paniculata;* and, sometimes, other species.

39.4.4 Mustard Culture

The cropping system, seedbed preparation, and seeding date for mustard crops are all very similar to those for small grains in any particular region. Thus, in Montana, most of the mustard is sown in the spring on fallow land or as a catch crop where winter wheat has winterkilled. In California, mustard is sown from January to March on fall plowed land. It is sown as a companion crop with clover in Minnesota. The seed is sown with a grain drill about 1 inch (2.5 cm) deep at a rate of 3 to 4 pounds per acre (3.3 to 4.5 kg/ha).

Mustard is harvested just before the pods open in order to avoid heavy losses from shattering of seed. When the field is free of weeds and the crop has matured uniformly, mustard can be harvested with a combine. When necessary, it is cut with a swather and then threshed with a combine equipped with a pick-up attachment.

39.4.5 Uses of Mustard

As a condiment, the main use of mustard is the ground seed or paste spread on cooked meats and other foods. The counter-irritant properties of the nondrying mustard oil are the basis for the familiar mustard plaster as well as other medicinal applications. Brown mustard is best for medicinal uses. Whole seeds of yellow mustard are used for seasoning pickles.

■ 39.5 SESAME

39.5.1 Economic Importance

Sesame seed (benne) was grown on about 17 million acres (7 million ha) in 2000–2003, mostly in China, India, Myanmar, Sudan, and Uganda. Average production in 2000–2003 was about 3.3 million tons (3 million MT) or about 380 pounds per acre (430 kg/ha). World production of sesame oil in 2000–2003 was about 725,000 tons (656,000 MT). Production in the United States in recent years has been very limited. Sesame has been grown in Texas, Arizona, New Mexico, Oklahoma, and Kansas.

Sesame may have originated in East Africa where several wild species are found. It was cultivated in the Near East by about 3000 BC and in Egypt by 1300 BC.[26] It reached the United States in the seventeenth century. The crop was grown sporadically, but regular commercial production began about 1950. It usually is grown under contract.

39.5.2 Adaptation

Sesame requires a warm climate with a frost-free period of 150 days or more. Medium-textured soils are most favorable.[27]

39.5.3 Botanical Description

Sesame (*Sesamum indicum*) belongs to the plant family *Pedaliaceae*, characterized by bell-shaped flowers and opposite leaves. It is an erect annual plant that reaches a height of 3 to 5 feet (90 to 150 cm). The tubular, two-lipped flower is about ¾ inch (2 cm) long, with a pink or yellow corolla. The lower flowers begin blooming two or three months after seeding, but blooming continues for some time until the upper pods open.

The upright pods split open at the top at maturity, which gave rise to the expression "open sesame" (Figure 39.11). The seeds drop out when the plant is inverted, except with varieties with indehiscent pods. The seeds of different varieties are creamy white, dark red, brown, tan, or black.[27] The seeds somewhat resemble flaxseed in size, shape, and sometimes in color. The yields and oil contents of non-shattering varieties are usually below those of the shattering varieties.

39.5.4 Sesame Culture

Sesame should be planted in late spring after the soil temperature is 70°F (21°C) or higher. It is planted in rows, usually with row crop planters equipped with vegetable seed planting plates. The rows are usually about 36 to 40 inches (90 to 100 cm) apart, although narrower rows of 18 to 30 inches (46 to 76 cm) often give higher yields. Seeding rate is 290,000 to 350,000 seeds per acre (325,000 to 390,000 seeds/ha).

Shattering varieties are harvested with a row binder or grain binder when the early pods are about to open. The bundles are cured in shocks, after which they are threshed with a combine drawn up to each shock. Special precautions are taken to avoid losses from shattered seeds when feeding the bundles into the combine. Nonshattering varieties can be harvested with a combine.

39.5.5 Uses of Sesame

Sesame is mostly used for the fixed (nonvolatile) oil in the seed that is utilized for edible purposes. The oil content may range from 50 to 56 percent. It is suitable for a salad oil when combined with equal or larger proportions of other palatable oils. The oil has been used for lighting purposes in Asia. Decorticated seeds of sesame

FIGURE 39.11
Spike and capsules of sesame.

are sprinkled on the surface of certain types of bread, rolls, cookies, and cakes just before baking. This use is most familiar to Americans. Whole or ground sesame seeds are included in many food preparations.

39.5.6 Diseases and Pests

Some improved varieties of sesame are resistant to leaf-spot diseases, but the older varieties are susceptible. Wilts, blights, and charcoal rot also attack sesame.

Aphids, thrips, and other insect pests can be controlled with insecticides.

39.6 SAFFLOWER

39.6.1 Economic Importance

World production of safflower oil in 2000–2003 averaged about 675,000 tons (615,000 MT). Leading countries are India, the United States, Mexico, Ethiopia, and China. Average annual production in the United States in 2000–2003 was 133,000 tons (121,000 MT). Annual global production of safflower oil is about 170,000 tons (154,000 MT.)

FIGURE 39.12
Safflower seeds. [Richard Waldren]

Safflower probably originated in the Near East. It has been grown there and in India for centuries. It has been grown in Europe for more than 200 years, mainly for its reddish dye (carthamin).[28] Safflower was introduced into the United States over 100 years ago, but its culture was not established until after 1940. U.S. production peaked in 1964, when over 276,000 tons (251,000 MT) were produced.

39.6.2 Adaptation

Safflower is adapted to semiarid and irrigated regions of the western half of the United States where there is a frost-free period of 110 days or more.[29] The plants (Figure 39.12) are subject to diseases under humid conditions.[12] Young plants withstand temperatures of 20 to 25°F (–7 to –4°C) or lower, but buds, flowers, and developing seeds are damaged by frost. The moisture requirements for safflower are similar to those for successful small-grain production. Warm weather and low humidity during the flowering period are helpful.

The plant tolerates high temperatures, considerable soil salinity, and rather high levels of sodium salts.[30]

39.6.3 Botanical Description

Safflower *(Carthamus tinctorius)* is a member of the family *Compositeae*. It is an annual, erect, glabrous, deep-rooting herb, 1 to 4 feet (20 to 120 cm) high, and branched at the top, with white or yellowish smooth pithy stems and branches. The flower heads are globular, ½ to 1½ inches (13 to 38 mm) in diameter, with white, yellow, orange, or red florets. It is largely self-pollinated except when many insects are present. The seed is smooth, obovoid, four-angled, and white or cream-colored, and it resembles a small sunflower seed. The leaves and outer floral bracts of the best adapted varieties bear short spines, which make the plant look like a thistle (Figure 39.13). Varieties without spines are being tested. The plant is an annual, but it is grown as a winter annual in the southern parts of California and Arizona. The growing season is 110 to 150 days from spring sowing, but 200 or more days from fall sowing.

39.6.4 Safflower Culture

Safflower usually replaces small-grain crops in rotations. It may follow an intertilled crop that has been kept free from weeds. It is often sown on fallowed land in

FIGURE 39.13
Safflower head. [Courtesy Richard Waldren]

dryland areas and is irrigated in Arizona and other arid areas. In northern California, safflower can follow rice. The rice lands usually contain ample moisture after the winter rains, along with that from flooding of the rice fields the previous year. Safflower fields in the semiarid regions are often unfertilized, but irrigated lands may receive from 50 to 120 pounds per acre (56 to 134 kg/ha) nitrogen and sometimes phosphate.

Safflower seed is sown from February 15 to April 20 in California, but from April 10 to May 20 in other areas.[29, 31] It is sown as a winter crop from early November to mid-January in the warm southern irrigated areas. It is usually sown with a grain drill at a rate of 20 to 30 pounds per acre (22 to 25 kg/ha) in dryland areas, but at somewhat higher rates under irrigation.

Safflower is a poor weed competitor, especially early in the growing season. Herbicides, such as EPTC, trifluralin, and metolachlor, can be used with safflower.

HARVEST Safflower is ready to harvest when most of the leaves are brown and little green is visible on the bracts of the latest head. It is harvested with a combine when the stems are dry, because the crop does not lodge, the seeds seldom shatter, and the seeds are not eaten by birds.

39.6.5 Uses of Safflower

The whole seed of improved varieties contains 32 to 40 percent oil, while the meal contains from 50 to 55 percent. The meal from unhulled seed contains 18 to 24 percent protein, but hulled seed has from 28 to 50 percent protein.

Safflower oil is a drying oil with an iodine number of about 140 to 155. The oil is used for edible purposes as well as in soaps, paints, varnishes, and enamels. Although of lower drying value, it is superior to linseed oil for white inside paint because the painted surface remains white instead of turning yellow as white linseed-oil paint does. This is due to the absence of linolenic glycerides. The increased popularity of safflower oil for edible uses is due to its content of unsaturated fatty acids (75 percent linoleic acid).

The flower heads of safflower yield a red dye, carthamin, formerly extracted commercially in India.

39.6.6 Diseases and Pests

Safflower is comparatively free from diseases in dry seasons, but it may be seriously damaged by leaf spot, rust, wilt, and root rot under moist conditions.[32] Rust, caused by *Puccinia carthami*, and leaf spot, caused by *Alternaria carthami*, infest the seeds. Partial control is achieved by planting treated disease-free seed. Pseudomonus bacterial blight, *Pseudomonus syringae*, appears after heavy rains. Resistant varieties and disease-free seed help control this disease. Root rot, caused by *Phytophthora drechsleri*, is a severe disease on irrigated land. Some varieties are resistant. Verticillium wilt and Botrytis head rot attack safflower under humid or foggy conditions where the crop probably should not be grown.

The chief insects that attack safflower are lygus bugs, wireworms, aphids, leafhoppers, green peach aphids, and thrips.

39.7 CASTORBEAN

39.7.1 Economic Importance

In 2000–2003, castorbean was grown on about 3.3 million acres (1.3 million ha) annually with a production of 1.3 million tons (1.2 million MT). The leading producing countries are India, China, Brazil, Ethiopia, and Paraguay. World production of castor oil in 2000–2003 averaged about 560,000 tons (505,000 MT). There is little commercial production of castorbean in the United States. Castorbean has been imported for oil extraction in crushing plants along the Atlantic and Pacific coasts.[33]

39.7.2 History of Castorbean Culture

Castorbean is generally believed to be a native of Africa. However, it may have originated in India because wild forms occur in that country. Castorbean was grown in Kentucky and New York as early as 1803. By 1850, some twenty-three oil extraction mills were operating in the East and South and in the Mississippi Valley. After 1865, the chief area of production shifted from Missouri to Kansas and then to Oklahoma. The domestic crop declined to negligible quantities after 1900.

Production was revived by government supports during World War I, World War II, and the Korean conflict, but the growing of castorbeans practically ceased when peace was restored. Commercial mechanized production was achieved after 1957 when efficient harvesting machines and improved varieties became available.[34]

39.7.3 Adaptation

Castorbean is adapted to areas that have a frost-free period of 160 to 180 days, an average warm season (April to September), precipitation between 15 and 25 inches (380 to 635 mm), an average July temperature above 76°F (24°C), and an average relative humidity at noon during July of less than 60 percent.[35] The crop is well adapted to irrigated regions.

Gray mold disease makes the castorbean a risky crop in the humid high-rainfall areas that comprise the Atlantic and Gulf coastal plains regions.

The soils for castorbeans should be fertile, be well drained, and contain enough sand so that it warms up early in the spring. The crop should be planted on sites that are not subject to water erosion. Castorbean is grown successfully in Egypt on irrigated coarse sandy soils of pH 8.7.[36]

39.7.4 Botanical Description

Castorbean belongs to the family *Euphorbiaceae*, the spurge family. Castorbean (*Ricinus communis*) is a short-lived perennial in the tropics where it grows into a tree 30 to 40 feet (9 to 12 m) in height. In the United States, it is an annual, 3 to 12 feet (90 to 365 cm) in height, except in Florida and parts of Texas and California, where it can survive the winters (Figure 39.14).

FIGURE 39.14
Castorbean plant showing spike *(left)*; castorbean seeds *(right)*. The protuberance at the end of the seed is a strophiole or carnucle.

The leaves are large, usually 4 to 12 inches (10 to 30 cm) wide or wider, alternate, and palmately divided into five to eleven lobes. The leaves of different varieties, which include numerous ornamental forms, may be green, purple, or red. The stems may also be green or red. Greenish-yellow flowers without petals are borne in racemes. The plant has an indeterminate blooming habit. It is also monoecious, with the pistillate flowers in the upper part of the raceme and the staminate flowers below. Consequently, it is usually cross-pollinated. The fruit is a spiny or smooth, dehiscent or indehiscent, three-celled capsule. Each cell contains a hard-shelled seed that is usually mottled.

The pods of many varieties forcibly eject the seeds at maturity. Improved varieties for seed production retain most of the seeds for a time after maturity.

The general appearance of the seed is somewhat similar to that of a mottled bean, except that it is obovoid instead of kidney-shaped. It has a prominent hilum or caruncle at the end instead of a depressed hilum on the edge of the seed as in a bean.

Since castorbean is not a legume, the seed is not truly a bean. Furthermore, the oil is not used to lubricate castors. The name *castor* apparently was coined by English traders who confused the oil with that from *Vitex agnus-castus*, which the Spanish and Portuguese in Jamaica called *agno-casto*. Castorbean is also called the castor-oil plant, Palma Christi, and mole bean.

Hybrids have been developed using pistillate plants that are pollinated by normal monoecious strains. Dwarf varieties and hybrids, 3 to 5 feet (90 to 150 cm) tall, have been developed.[37, 38]

39.7.5 Castorbean Culture

Soil organic matter affects the need for nitrogen fertilizer. Nitrogen is applied at rates from 40 pounds per acre (45 kg/ha) on high organic matter soils to 120 pounds per acre (135 kg/ha) on low organic matter soils. From 20 to 30 pounds per acre (22 to 34 kg/ha) phosphate and 40 to 70 pounds per acre (45 to 78 kg/ha) potash are applied to soils deficient in those nutrients. Seedbed preparation for castorbean is about the same as for cotton or corn.[38, 39, 40] Castorbean grown under irrigation is usually planted on beds, either in shallow furrows or in moist soil that has been exposed by scraping off the ridge by a blade on the planter.

Treated seed is planted in 40-inch (1 m) rows, at a depth of ½ to 3 inches (4 to 8 cm) in moist soil on dryland, and in narrower rows on irrigated land.[7] Spring planting occurs after the soil has warmed up to a temperature of 60°F (16°C). Desirable stands are 10,000 to 13,000 plants per acre (25,000 to 32,000 plants/ha) on dryland, and 25,000 to 30,000 plants per acre (62,000 to 74,000 plants/ha) under irrigation. The seed size is 1,000 per pound.

Castorbean is a poor early weed competitor. Early cultivation is done with a rotary hoe, while later cultivation is performed with sweep or shovel cultivators. Trifluralin can be applied for weed control where the label is registered for castorbean.

HARVEST Castorbean is harvested when all the capsules are dry and the leaves have fallen from the plants. Delay in harvest after the crop is ready may result in "shattering" losses when the seeds pop out of the capsules. Indehiscent (nonshattering) hybrids are grown so that harvest can be deferred until the crop is ripe and dry. Much of the crop is harvested after the plants have been killed by freezes.[38, 39] A killing frost ten to fourteen days before harvest helps dry the plants. Defoliants are applied when the crop is ripe before frost occurs. However, defoliants may reduce yields.

Most castorbean today is harvested with a combine. In some areas, special self-propelled harvesters are used (Figure 39.15). The machines have beaters that shake the capsules from the spikes. Augers and elevators convey the capsules to the

FIGURE 39.15
Two-row castorbean harvester.

huller, while the hulled and cleaned beans are carried to a hopper bin on the machine.[38, 40, 41]

From 5 to 10 percent of the seeds may drop on the ground during or before harvest. These seeds produce volunteer plants the next spring, which must be eradicated before the planting of another crop. Stray castorbean plants or seeds in a forage or food crop could be fatal when consumed. Grain sorghum is planted late enough for castorbean to be destroyed by cultivation. Stray or surviving plants can be killed with herbicides.

39.7.6 Uses of Castorbean

Hulled castorbean seed is 35 to 55 percent oil. It will average about 50 percent in well-developed seed. The oil is extracted by pressure without grinding or decorticating, and usually without heating the beans. Castor oil is a nondrying oil with an iodine number that ranges from 82 to 90. A process that dehydrates castor oil will give it drying oil properties, and it is now widely used in paints and varnishes. The oil is also used in producing sebacic acid, which is used in making synthetic lubricants for jet aircraft and as a solvent or plasticizer in the manufacture of certain types of nitrocellulose plastics and fabrics. Castor oil is also used in the manufacture of linoleum, oil cloth, printer's ink, artificial leather, and dyeing textiles.[42] Its use as a laxative for ailing youths is all too well known to former generations. Now, only 1 percent of the oil is used as a laxative. Castor oil is used frequently in hydraulic cylinders, lubricants, cosmetics, soaps, and other products.[38] Considerable castor oil is imported into the United States, chiefly from Brazil, but castorbean seed is no longer imported.

Castorbean seed, as well as the press cake after oil extraction, is poisonous to humans, livestock, and poultry. The toxic principles are ricin, which is an albumen, and ricinine, which is an alkaloid. The press cake is largely used in mixed fertilizers under the name *castor pomace*. The oil is nonpoisonous, but the leaves and stems of the plant contain a sufficient quantity of ricinine to be toxic. Eating one castorbean seed can produce nausea, and ingestion of appreciable numbers can be fatal.[39]

39.7.7 Diseases and Pests

Alternaria leaf spot, caused by *Alternaria ricini*, can defoliate castorbean, while the same fungus can also cause the young capsule to mold.[39] Bacterial leaf spot, caused by *Xanthomonas ricinicola*, has been destructive in Texas.[38] Some varieties are resistant to these leaf spot diseases.

Gray mold, caused by *Sclerotinia ricini*, attacks and destroys the racemes. It occurs in warm humid regions.[43] Castorbean is highly susceptible to the Texas root-rot disease caused by the fungus *Phymatotrichum omnivorum*. Consequently, the crop should not be planted in fields infested with that organism.

Insects that attack the castorbean include the false chinch bug, armyworm, corn earworm, stink bug, leafhopper, leaf miner, caterpillar, grasshopper, and Japanese beetle, but usually the damage is not great. Nematodes also attack castorbean to some extent.

■ 39.8 GUAYULE

39.8.1 Economic Importance

Guayule (pronounced gway-oo-lay or wy-oo-lay) is the Native American name for a North American rubber crop. More than 33,000 acres (13,400 ha) of guayule were planted in California, Arizona, and Texas from 1942 to 1944. During World War II, other rubber sources were not available. Guayule production was discontinued in the United States by 1950 because synthetic rubber supplies were abundant.

39.8.2 History of Guayule Culture

Guayule is a native of northern Mexico or southwestern Texas. The plant was first collected for identification in 1852.[44] Native Americans learned to extract the rubber by chewing the wood. The first experiments on the commercial extraction of rubber from guayule were begun about 1888. Several factories were established in Mexico in 1904 and shortly thereafter, and one at Marathon, Texas, in 1909. Attempts to cultivate guayule were begun in Mexico about 1907. A total of 8,000 acres (3,200 ha) were planted in California, Arizona, and elsewhere in the United States after 1912, but most of this had been plowed up by 1942.

39.8.3 Adaptation

Guayule is distributed naturally over an area of 130,000 square miles (337,000 km^2), which extends southward from the vicinity of Fort Stockton, Texas, into Mexico.[44] This area has an annual rainfall of 10 to 15 inches (255 to 380 mm) with altitudes of 3,000 to 7,000 feet (915 to 2,130 m). In the United States, guayule could be grown in a belt about 150 miles (340 km) wide north of the Mexican border, extending from the mouth of the Rio Grande to southern California and northward to central California. In a dormant condition, the plants can withstand temperatures of 5 to 10°F (−15 to −12°C), but while actively growing they are injured by temperatures of 20°F (−7°C). Guayule thrives under hot conditions. It can be grown either on irrigated or dryland fields.

39.8.4 Botanical Description

Guayule (*Parthenium argentatum*) is a slow growing, widely branching, woody, long-lived perennial shrub that belongs to the family *Compositae*. Plants in the wild may live forty to fifty years. It is seldom more than 30 inches (76 cm) tall at maturity. The plant develops a strong taproot, as well as a number of long, shallow lateral roots from which new shoots may arise. The leaves are silver gray, due to a covering of thick, short white hairs. Leaves are 1 to 2 inches (2.5 to 5 cm) in length. The fruit is an achene with ray and disk flowers attached after maturity. A pound (454 g) of fruit consists of about 600,000 achenes. About 90 percent of the achenes contain no seeds due to sterility that results from irregular chromosome numbers and other factors. Many seeds are produced without fertilization by apomixis. Fresh seed of guayule germinates very poorly, but after-ripening is completed in about a year, so that year-old seed can be planted without treatment. Calcium hypochlorite is used to break dormancy in guayule seed.

39.8.5 Guayule Culture

Cultural methods[45] for guayule include planting treated, pre-sprouted seed in beds with the seeds covered by ⅒ inch (3 mm) of sand by the planter. After eight months, the seedlings are topped and then undercut using special machines. Seedlings are then transplanted into 40-inch (1 m) rows about 12 inches (30 cm) apart in the row. Weeds are controlled by cultivation and oil sprays.

The plants are normally suitable for harvest after four years of growth on irrigated land or five years on dryland. Plants are harvested either by digging plants or by clipping plants about 4 inches (10 cm) above the soil. Clipping, instead of digging the plants, reduces rubber yield 25 to 35 percent, but it eliminates the expense of establishing the crop. After clipping and regrowth, the crop is ready for harvest in two or three years.

A special rotary cylinder machine with vacuum suction equipment is used to gather seed from mature plants in the field.

39.8.6 Rubber Extraction

Extraction of rubber from guayule involves crushing the material in corrugated rolls, after which it is macerated in water in a series of pebble mills to release the rubber. The particles of rubber aggregate into what are called worms that float and can be skimmed off. Worms are separated from adhering bark, scrubbed, dried, and pressed into blocks.

39.8.7 Uses of Guayule

Good varieties of guayule yield about 20 percent rubber or about 1,600 pounds per acre (1,800 kg/ha) of rubber.[46]

Harvested guayule rubber ordinarily contains 15 to 20 percent resins. Removal of these resins (deresinating) leaves a product very similar to Hevea rubber, except that it does not make a satisfactory rubber cement. However, for many purposes, such as the fabrication of tires, the presence of resin is beneficial in reducing friction. Some use may be found for the resins in the future. Guayule plants make a hot fire when burned. Formerly, they were used for smelting in Mexico. The native shrub is browsed by burros and goats.

39.8.8 Diseases and Pests

The principal diseases that attack guayule are Verticillium wilt, caused by the fungus *Verticillium albo-atrum*, and Texas root rot, caused by *Phymatotrichum omnivorum*. There is no control for the former disease. The latter, which is not serious, can be evaded only by avoiding infected fields when planting. Several other root-rot diseases attack guayule.

Insects that attack guayule include grasshopper, lygus bug, lacebug, harvester ant, leaf-cutting ant, and termite. Dodder weed is sometimes a parasite on guayule plants.

39.8.9 Other Sources of Natural Rubber

Hevea rubber trees *(Hevea brasiliensis)* are unable to survive in the continental United States, except in the southern tip of Florida. The roots of Russian dandelion or kok-saghyz *(Taraxacum koksaghyz)* yield an excellent quality of rubber, but yields of present varieties cannot be expected to exceed 50 to 60 pounds per acre (56 to 67 kg/ha) of rubber per year under average conditions. Goldenrod *(Solidago leaven-worthii)* gives low yields of low-grade rubber by double solvent extraction of the leaves that contain 4 to 7 percent rubber. A vine-like rubber-bearing shrub, *Cryptostegia grandiflora*, grown in India and Madagascar, may yield 100 pounds per acre (110 kg/ha) per year by excessive expenditure for hand labor.

Rabbitbrush *(Chrysothamus nauseosus)*, a native shrub distributed over the drier regions of western United States, yields a good quality of rubber, but the percentage is too small for commercial success.

■ 39.9 HEMP

39.9.1 Economic Importance

Hemp was harvested for fiber on an average of about 160,000 acres (64,000 ha) in 2000–2003. Leading countries are China, Korea, Spain, Chile, and Romania. Annual production in 2000–2003 was about 74,000 tons (68,000 MT). The average fiber yield is about 940 pounds per acre (1,050 kg/ha). Also, about 32,000 tons (29,000 MT) of hemp seed is harvested, mostly in China and France. Hemp was grown in the United States from early Colonial days but declined to about 200 acres by 1933. Production was chiefly in Kentucky but also through the North Central States. Production resumed from 1940 to 1946 to replace imported abaca (Manila hemp), which was then unavailable during World War II. The fiber was needed for marine cordage. No hemp is commercially grown in the United States today.

39.9.2 History of Hemp Culture

Hemp is probably native to central Asia. It has been grown in China for many centuries. It was grown by the ancient Greeks.

Hemp escaped from cultivation in the United States many years ago. It now occupies large areas of uncultivated land, chiefly in the bottomlands along the lower Missouri River and its lower tributaries. Harvesting the wild growth has not been feasible because of scattered and irregular stands and growth and because of mixtures with other types of vegetation. It is now being eradicated,[47] with herbicides, flaming, mowing, and tilling.

39.9.3 Adaptation

Hemp is adapted to areas that produce a good crop of corn. It requires a frost-free season of about four months for fiber production or five months for seed production.[48] It will withstand light frosts. For good uniform growth of hemp, rainfall or soil moisture should be ample throughout the season. A higher quality of fiber is usually obtained in the more humid climate east of longitude 95° due to better conditions for dew retting than commonly occur west of the meridian.

Well-drained, deep, fertile, medium-heavy loams have produced the best hemp. Barnyard manure, commercial fertilizers, or lime should be applied to soils where other crops respond to these supplements.

39.9.4 Botanical Description

Hemp (*Cannabis sativa*) belongs to the family *Urticaceae*, tribe *Cannabinaceae*. It is a stout, erect annual that grows to a height of 5 to 15 feet (150 to 460 cm). It is dioecious; the male plants, which comprise one-half of the crop, produce no seed.

Hemp is of considerable historical interest because Camerarius, in 1694, used hemp to discover sexual reproduction in plants. It has continued to serve as the subject for determining the nature of sex in plants. Under normal field conditions, hemp is strictly dimorphic (has two forms) as to flowers and vegetation parts. The pistillate and staminate plants show almost no sex reversal. However, individual bisexual plants occur under unfavorable conditions,[49] such as reaction to old pollen,[37] mutilation,[50] and reduced light.[41]

The main stem of hemp is hollow and produces a few branches near the top. The best fibers are not as fine as those of flax, even when the plants are grown close together. The staminate inflorescences are in axillary, narrow, loose panicles.[10] The pistillate flowers are in erect leafy spikes (Figure 39.16).

The hemp ovary matures into a hard ovoid achene.[51] The seeds mature on the lower part of the spikes first and on the upper part last.

39.9.5 Types or Varieties

The common fiber hemp has larger stalks than the drug plant type, which is grown in the hot climates of India, Syria, and elsewhere, often being classified as a separate species, *Cannabis indica*. The dried leaves and flowers of this species are called hashish. The type grown in the United States, known as Kentucky or domestic hemp, is of Chinese origin.

39.9.6 Hemp Culture

Hemp should be planted on fall-plowed land just before the time for planting corn.[48] Hemp responds to fertilizers on less fertile soils, but heavy nitrogen fertilization reduces the strength of the fibers.[52]

For fiber production, hemp should be sown with a grain drill ½ to 1 inch (13 to 25 mm) deep at the rate of 55 pounds per acre (62 kg/ha) to produce fine uniform stems suitable for machine processing. Hemp is harvested for fiber when the male

FIGURE 39.16
Pistillate (seed-bearing)
hemp plant *(left)*;
staminate plant *(right)*.

plants are in full flower and shedding pollen.[53] Earlier harvesting yields weak fiber. Harvesting is done with a modified rice binder or a self-rake reaper that delivers the stalks in an even swath on the ground. The stalks are left in this position for dew retting to separate the fibers. After retting is complete, the stalks are gathered up and bound by a pickup binder. The bundles are shocked and later hauled to the processing plant any time after drying.

39.9.7 Preparation of Hemp for Market

In dew retting, the stalks lying on the ground are exposed to cool moist weather or, sometimes, to alternate freezing and thawing. The process is complete when the bark separates easily from the woody portion of the stem. The stalks are turned once or twice to provide uniform exposure. Water retting is practiced in Europe.

In the breaking process, the inner cylinder of brittle woody tissue is broken into short pieces called hurd, while the long, flexible fibers are left largely intact. This permits the wood to be separated from the fibers.[54] Scutching is the beating or scraping off of the broken pieces of wood.

The rough fiber is combed cut by drawing it over coarse hackles. Hemp tow consists of short, more or less tangled strands. A ton of dry retted hemp stalks processed by modern machine methods yields about 340 to 360 pounds (170 to

180 kg/MT) of fiber, of which slightly more than half is line (long) fiber and the remainder is tow.

Hemp is used in commercial twine, small cordage, thread, hemp carpet twines, oakum, and marline.[54] It ranks next to Manila hemp for marine ropes.

39.9.8 Hemp Seed and Leaves

Hemp seed can be fed to caged birds, but the embryo of the seed must be killed in accordance with federal regulations (in order to prevent its use for planting). The seed contains 20 to 25 percent of an oil that sometimes is extracted for use in making soft soaps or is used as a substitute for linseed or other oils. The glandular hairs secrete both a volatile oil and the strong narcotic resin (cannabin). The drug, called marijuana in North America and hashish or bhang in Asia, is obtained from the flowers and leaves of both *Cannabis sativa* and *C. indica*. However, *C. sativa* yields little drug. Hemp is the source of "reefer" cigarettes that are sold illegally. Hemp can be grown only under a license issued by the Bureau of Narcotics. The principal hallucinogen of marijuana is 9-tetrahydrocannabinol (THC).

39.9.9 Hemp-Like Plants

In addition to the true hemp *(Cannabis sativa)*, several fiber plants sometimes called hemp are grown in other countries.[55] These include abaca or Manila hemp *(Musa textilis)*, a relative of the banana plant grown in the Philippines and other Pacific Islands; Sunn hemp *(Crotalaria juncea)*, a legume cover crop also harvested for fiber in India and other parts of Asia; Mauritius hemp, green aloe, or piteira *(Furcraea gigantea)* of Africa; New Zealand hemp *(Phormium tenax)*;[56] jute or India hemp *(Corchorus capsularis)*; and sisal hemp. The last type includes sisal *(Agave sisalana)* grown in Africa and the West Indies, and henequen or Mexican or Cuban sisal *(Agave fourcroydes)*. Henequen, sisal, and abaca are the chief fibers used in binder twine. Jute is used in making burlap. All the fibers named are used to make rope or twine. Bowstring hemp *(Sansevieria guineensis)* has been tested in Florida as a possible substitute for Manila hemp.

◾ 39.10 RAMIE

39.10.1 Economic Importance

Global production of ramie in 2000–2003 averaged about 237,000 tons (215,000 MT). Over 98 percent of the ramie is grown in China. Other countries that grow ramie are Brazil, Laos, and the Philippines.

39.10.2 History of Ramie Culture

Ramie, also called Chinese grass, grasscloth, and rhea, has been grown in the Orient since the earliest records of agriculture. It was used for Chinese burial shrouds over 2,000 years ago, long before cotton was introduced in the Far East. Ramie was introduced into the United States in 1855. It has been tested in many states, and small acreages were grown occasionally. Several hundred acres were planted in the Florida Everglades region in 1945 and 1946, but production ceased after 1961. Interest in the crop is stimulated not only by its fiber, which is the strongest of vegetable fibers, but also by the lustrous appearance of ramie cloth.

39.10.3 Adaptation

Ramie is a semitropical plant adapted to areas of the Southeast where the annual rainfall exceeds 40 inches (1,000 mm). It is also adapted to interior irrigated valleys of California, where frost does not penetrate into the soil more than an average of 3 inches (8 cm). The capacity of the plants to make a heavy vegetative growth necessitates a rich alluvial soil for a satisfactory crop. Excellent results have been obtained on the muck soils of the Florida Everglades.[57] Sandy soils have often proven unsatisfactory.

39.10.4 Botanical Description

Ramie (*Boehmeria nivea*) belongs to the family *Urticacae*. The plant is a perennial that sends up new stalks from the rhizomes. Plants often reach a height of 8 feet (245 cm). The stems are ½ inch (13 mm) or less in diameter and covered with inconspicuous hairs (Figure 39.17). The plant is monoecious. Flowers are small with the staminate flowers in the lower part of the cluster. Established plantings live ten or more years.[58]

39.10.5 Ramie Culture

Ramie will usually need a complete fertilizer, including manganese, copper, and zinc, on peat and muck soils.[57] Ramie is propagated by dropping pieces of rhizomes about 6 inches (15 cm) long in an upright or slanting position in furrows

FIGURE 39.17
Leaves and inflorescence of ramie.

with a transplanting machine. The furrows are then filled so that the upper ends are covered with about 2 inches (5 cm) of soil. The rhizome pieces are spaced 12 to 18 inches (30 to 45 cm) apart in rows 3½ to 4 feet (107 to 122 cm) apart. Fields are cultivated not only to control weeds, but also to prevent the stand from becoming too thick. Young plants require six to eight cultivations before the plants are large enough to shade the ground.

HARVEST Ramie is ready to harvest when the growth of the stem has ceased and the staminate flowers are beginning to open. It is advisable to defoliate the plants by spraying a chemical defoliant.[59] Ramie is gathered with a harvester-ribboner that cuts the stalks and separates the wood and bark from the ribbons of fiber. These ribbons contain cortical tissue and gums along with the fiber. They are hauled directly to a decorticator, which removes the cortical tissue. The ribbons are then squeezed through rollers to remove part of the water, brushed to untangle the fibers,[60] and then dried with artificial heat down to a moisture content of 10 to 12 percent. Gum is later removed with chemical solvents.

39.10.6 Uses of Ramie

Ramie fiber is similar to linen in appearance and use. The fabrics are called Swatow grass cloth, grass linen, Chinese linen, and Canton linen. Commercial fiber strands are 3 to 5 feet (90 to 150 cm) long and 0.002 to 0.003 inch (50 to 75 μm) in diameter. The best fiber cells average 6 to 8 inches (15 to 20 cm) long.

▨ 39.11 KENAF

39.11.1 Economic Importance

Kenaf *(Hibiscus cannabinus)* is grown for its soft fibers. It has been grown in its native habitat of Thailand, India, and Pakistan since ancient times. Kenaf was grown in Florida on a small acreage starting in 1943, but it did not result in a firmly established fiber industry. In 1968, about 5,000 acres (2,000 ha) of kenaf stalks were cut in Florida for use as bean poles. The plant has been tested extensively as a possible source of pulp for paper making,[61] but other sources are more economical.

39.11.2 Adaptation

Kenaf is mostly grown in tropical regions. It is adapted to the warm southeastern states of the United States where there is 20 to 25 inches (500 to 635 mm) of rainfall or reserve moisture during the four- to five-month growing season. A well-drained soil with ample organic matter is desirable.

39.11.3 Botanical Description

Kenaf, sometimes called bimli-jute, is a member of the plant family *Malvaceae*.[62] The plants grow to a height of up to 12 feet (366 cm) or more, with deeply lobed leaves.

The yellow flowers with red centers open for six to eight hours and then close. The petals drop off later. The flowers are largely self-pollinated. Kenaf is a short-day annual plant,[63] and most varieties bloom and produce seed only when the day length is less than 12½ hours. Some varieties are photoinsensitive (day neutral).

39.11.4 Kenaf Culture

For pulp, kenaf is planted at 5 to 8 pounds per acre (6 to 9 kg/ha), in rows 20 to 36 inches (50 to 90 cm) apart. This will produce a population of 75,000 to 100,000 plants per acre (185,000 to 250,000 plants/ha). The best time to sow kenaf seeds is in the spring (March or April in South Texas), when the soil temperature reaches 55°F (13°C) or higher.[34] For fiber, the seed is drilled at 30 to 35 pounds per acre (34 to 39 kg/ha) in 7 to 8 inch (18 to 20 cm) drill rows.[64, 65] The crop responds to heavy fertilizer applications on sandy soils.

39.11.5 Uses of Kenaf

Kenaf is harvested for fiber when the first flowers open, with a harvester-ribboner equipped with crushing rolls. The ribbons of long bast fiber are retted in tanks where bacterial action frees the fiber from adhering substances. The ribbons are then cleaned in a burnishing machine and dried.[60] Fiber yields as high as 2,000 pounds per acre (2,240 kg/ha) have been obtained under good growing conditions.

For pulp, the plants should be harvested when they cease growth or are killed by frost.[61, 66] The bast or outer bark fiber and the inner thick core of short, woody fibers provide the pulp.[61, 67]

For seed production, kenaf is sown in southern latitudes in late summer so that the shortening days will induce flowering.[62, 63] The seed crop is harvested with a binder equipped with a high reel.

39.11.6 Diseases

Kenaf is susceptible to a fungus, *Colletotrichum hibisci*. Some selected strains of kenaf are resistant to certain races of this fungus. A rot caused by a species of *Botrytis* occurs under high humidity. It can be controlled by spraying with fungicides.

■ 39.12 PYRETHRUM

39.12.1 Economic Importance

Global pyrethrum production in 2000–2003 averaged about 13,800 tons (12,500 MT). About two-thirds of the world's production is in Kenya. Pyrethrum is grown for its flowers, which contain two active compounds, pyrethrin I and pyrethrin II, that have insecticidal properties. It has been grown at times in California, Colorado, Pennsylvania, and other states, but there is no commercial production today in the United States. Synthetic pyrethrins are now available.

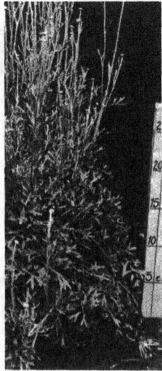

FIGURE 39.18
Pyrethrum plant in bloom
(left); leaves and pods of
pyrethrum *(right).*

39.12.2 Botanical Description

Pyrethrum *(Chrysanthemum cinerariifolium)* is a herbaceous perennial that belongs to the family *Compositae.* The flower is a typical white daisy in appearance. The plant forms a clump 6 to 12 inches (15 to 30 cm) in diameter at the base from which the stems flare upward and outward (Figure 39.18). The height ranges from 15 to 40 inches (38 to 100 cm). The plants do not flower during their first year, but thereafter they bloom each year. When grown in the southern states, pyrethrum often lives only as an annual or biennial.[68]

39.12.3 Pyrethrum Culture

The methods of establishing pyrethrum fields are very similar to the early phases of tobacco culture (Chapter 34). About 9 ounces (250 g) of seed planted in 300 square feet (28 m²) of bed space should produce 15,000 to 30,000 plants. The plants from spring sowing reach a height of 3 to 5 inches (8 to 13 cm) about three months after sowing. They are then ready for transplanting. Pyrethrum can also be planted in late summer or in the fall. The plants are then set out the next spring. Methods of transplanting are the same as for tobacco. The plants are set in rows at a rate of about 10,000 plants per acre (25,000 plants/ha). Fertilizer is applied between the rows and worked into the soil either in early spring or after the crop is harvested.[68]

HARVEST Pyrethrum flowers are ready for gathering when one-half to three-fourths of the florets on most of the disks are opened. A higher content of pyrethrins as well as higher total yields are obtained when harvesting is early compared to delaying harvest until the plants are three-fourths in bloom.[68, 69] An efficient hand worker can gather only about 100 pounds (45 kg) of fresh flowers in a day, equivalent to 25 pounds (11 kg) of dried flowers. Strippers operated by hand or by machine gather too much trash along with the flowers.

Soon after picking, the flowers must be dried on trays in the sun, on floors of large drying sheds, or in artificial driers.

39.12.4 Uses of Pyrethrum

Yields of dried pyrethrum flowers average about 700 to 800 pounds per acre (785 to 900 kg/ha) where the crop is adapted. No crop is obtained in the year the plants are set out. Fair yields are obtained in the second year, but better yields are produced in the third and fourth years. Thereafter, yields start to decline. Dried flowers contain about 1 percent total pyrethrins, the stems about 0.1 percent, and the leaves practically none. Pyrethrin content is affected only slightly by soil type and fertilization. The powdered dried flowers of pyrethrum constitute the well-known insect powder used to control fleas, lice, bedbugs, and other household pests. Pyrethrins dissolved in kerosene or other petroleum derivatives, when supplemented by other chemicals, were once used in common fly sprays. An important use of pyrethrum is dusting vegetables nearly ready for harvest, when other insecticides would leave a residue of poisons on edible portions of the plants. Synthetic pyrethrins have replaced much of the pyrethrum.

39.12.5 Diseases and Pests

Several diseases attack pyrethrum, causing the death of plants, which then require annual replacements. These diseases are caused by the fungi *Septoria chrysanthemella*, *Diplodia chrysanthemella*, *Sclerotium rolfsii*, *S. delphinii*, *Rhizoctonia*, and various species of *Fusarium*, *Sclerotinia*, *Alternaria*, and *Phytophthora*.

Insects have not been especially damaging to pyrethrum. However, the fact that the plant contains an insecticide is no assurance that certain insects might not attack pyrethrum severely.

■ 39.13 OTHER INDUSTRIAL CROPS

39.13.1 Dill

Dill (*Anethum graveolens*) belongs to the family *Umbelliferae*, the parsnip family. The dill plant is a perennial that develops flowers and seeds the year of seeding when the time of sowing is early enough to satisfy the cold temperature requirements for vernalization. The plants reach a height of 3 to 4 feet (90 to 120 cm).

Although a native of the Mediterranean region, dill is adapted to the relatively cool climate of the northern United States, where it grows well on fertile soils. Dill is sown in cultivated rows 1½ to 3 feet (45 to 90 cm) apart either in early spring or so late in the fall that the plants do not emerge until early spring. The plants are later thinned so they are 6 to 15 inches (15 to 38 cm) apart in the row.

The seed crop is cut with a windrower or mower when the first seeds are ripe, and while the plants are damp enough to avoid excessive losses from shattering. The crop is partly cured, after which it is threshed, all possible precautions being used to prevent shattering of the seed. The seed is then spread out in shallow piles, with occasional stirring, until dry. The yield of seed under good cultural conditions is reported to average about 500 to 700 pounds per acre (560 to 785 kg/ha).

When harvested for oil production, dill is cut when the most nearly ripe seed is turning brown. After partial field curing, the crop is hauled to the still where it is handled similar to the process described for the distillation of mint oil. Oil yields are approximately 20 pounds per acre (22 kg/ha).

Dill is grown commercially mostly for the fruits or seeds that are used to flavor foods, mainly pickles, or for the volatile oil distilled from the harvested plants that is used for similar purposes and in perfumes and medicines. The leaves and seed heads, often grown in herb gardens, are also used to season prepared foods.[70]

39.13.2 Wormseed

Wormseed (*Chenopodium ambrosioides* var. *anthelminticum*) belongs to the *Chenopodiacea* or goosefoot family. The genus *Chenopodium* includes such plants as the lambsquarter weed, *C. album*, and *C. quinoa*.

Wormseed is also known as Chenopodium, American wormseed, and Jerusalem oak. The plant, a native of Europe, has become naturalized in waste places from New England to Florida and westward to California. It is a many-branched annual herb, 2 to 4 feet in height (Figure 39.19). The entire plant, including the fruit and seed, contains a volatile oil of disagreeable odor that is used to

FIGURE 39.19
Wormseed inflorescence.
[Photo: Robert H. Mohlenbrock
@ USDA-NRCS PLANTS
Database / USDA NRCS. 1992.
Western wetland flora: Field
office guide to plant species.
West Region, Sacramento, CA]

treat infections by parasitic intestinal worms. The seeds are also used in worm remedies for humans and other animals.[71] The oil contains more than 65 percent ascaridole, the active principle that kills several internal parasites.

Wormseed is sown in beds in the spring and transplanted to the field when the plants are about 4 inches (10 cm) tall. The plants are spaced about 18 inches (46 cm) apart in rows approximately 3 feet (90 cm) apart. The crop is harvested about the middle of September, usually with a modified binder. The rather loose bundles are placed in shocks and then hauled to the still when partly dried. The oil is distilled into a long trough type of condenser. Distillation is much like that described for mint. Yields of oil range from 30 to 90 pounds per acre (34 to 100 kg/ha).[69] A federal permit must be obtained to operate any still.

39.13.3 Quinoa

Quinoa (*Chenopodium quinoa*) is grown in the Andes region of Peru, Bolivia, and Equador. In 2000–2003, average production was about 56,000 tons (51,000 MT) on over 160,000 acres (66,000 ha). There it is called *quinua* (keen-wha). The seeds are consumed as a food substitute for cereals. Attempts to produce quinoa in the Rocky Mountain region of the United States have been disappointing.

39.13.4 Wormwood

Wormwood, absinth, or madderwort *(Artemisia absinthium)* was once produced in Michigan and Oregon. The plant is a hardy, long-lived perennial, 2 to 4 feet (60 to 120 cm) in height. It is related botanically to sagebrush. The hairy, silvery shoots bear grayish-green leaflets and small yellow flower clusters. The seed is planted in a bed or directly in the field. The plants can also be propagated by division. Wormwood is harvested once a year with a mower or windrower.[71]

The oil distilled from the leaves and tops was used in tonics and external proprietary medicines. It is no longer used as a worm remedy because of its toxic properties when taken internally.

39.13.5 Perilla

Perilla is grown for its seeds, which contain a drying oil of excellent quality that in the United States has been used in paints, varnishes, and linoleum manufacture. It has not been grown in this country except in preliminary field trials. It is grown in China, Korea, Japan, and India.

Perilla requires a warm growing season of five to six months with an ample supply of moisture. It can be grown successfully in the United States only in the humid Southeast and in the irrigated Southwest. The plants can withstand light frosts. Sandy loam soils of average fertility are suitable for perilla.

Perilla (*Perilla frutescens*, or *P. ocymoides*) belongs to the *Lamiaceae*, or mint, family. A subspecies, *P. frutescens nankinensis*, is also grown. Perilla is a coarse annual plant indigenous to India, Japan, and Mainland China, where it is cultivated. The plants are 3 to 6 feet (90 to 180 cm) in height and much branched unless crowded. They bear coarse-toothed leaves and have small, inconspicuous white flowers borne in racemes 3 to 8 inches (8 to 20 cm) long. The seeds are round and brownish and resemble mustard seeds.

The oil content of perilla seed ranges from 35 to 45 percent, with an average of about 38 percent. The extracted oil is yellow or greenish. It is similar to linseed oil in odor and taste, but it is higher in drying quality (iodine number about 200). Therefore, it is suitable for quick-drying paints and varnishes. In Japan, the oil is used in the manufacture of paper umbrellas, oil papers, artificial leather, printer's ink, paint, varnish, and lacquer. Processed oils have largely replaced perilla oil in the United States.[27]

39.13.6 Crambe

Crambe, or Abyssinian mustard *(Crambe abyssinica)*, originated in the Mediterranean region. Crambe is prevalent across Asia and Western Europe. The former Soviet Union started agronomic testing of crambe in 1932. Crambe was introduced into the United States in 1950. Performance trials in the late 1950s and 1960s identified crambe as a promising new alternate crop for the United States. In 1992, North Dakota produced 20,000 acres (9,000 ha) of crambe under contract.

Crambe is a member of the family *Cruciferae*. It is an erect annual with large pinnately lobed leaves, white flowers, and spherical seeds. Crambe is a cool-season crop adapted to North Central and northwestern states and adjacent areas of Canada. Seed yields under semiarid conditions range from 500 to 1,000 pounds per acre (560 to 1,100 kg/ha) but may exceed 2,000 pounds per acre (2,200 kg/ha) under irrigation.[61, 72]

Crambe is sown with a grain drill at rates of 20 pounds per acre (22 kg/ha) in 6- to 7-inch (15 to 18 cm) rows, and 8 to 15 pounds per acre (9 to 17 kg/ha) in 12- to 14-inch (30 to 36 cm) rows.[73] Seeding depth is ½ to ¾ inch (13 to 19 mm). The best time for sowing lies between the seeding of spring grain and the planting of corn. Fertilizer requirements are similar to those for small grains. Crambe does not compete well with many weeds. Trifluralin herbicide can be applied for weed control. The crop is mature when the seed pods are yellow or brown. It is usually harvested by swathing and then combining. The threshed seeds are retained in the hulls, which comprise 20 to 30 percent of the weight.

The unhulled seeds usually weigh 5 to 7 grams per 1,000 seeds. Good quality seed contains about 32 percent oil and 20 to 30 percent protein. The hulled seed may contain 44 percent oil. The erucic acid content of the oil ranges from 50 to 60 percent. Potential uses for the oil include plasticizers, rubber additives, waxes, and a lubricant for molds in continuous casting of steel. The seed meal contains sulfur compounds (thioglucosides) that are unpalatable to livestock.

39.13.7 Sansevieria

Sansevieria (bowstring hemp or snake plant) is grown for cordage fiber in tropical and subtropical countries. More than 50 species of the genus *Sansevieria* are native to Africa and Asia. Sansevieria is a popular ornamental garden plant in warm climates. It is a common potted house plant in the United States. Several species of *Sansevieria* have been grown experimentally in Florida as potential sources of marine cordage.[60, 64]

The plants are perennials and members of the lily family, *Liliaceae*. The leaves are harvested for fiber extraction every three to five years. The leaves are 2 to 3 feet (60 to 90 cm) long. Sansevieria is propagated by planting leaf cuttings 6 to 8 inches (15 to 20 cm) long with the lower end of the cutting set in the soil in rows. Machines have been devised for making cuttings, for harvesting the leaves, and for separating the fiber. The cuttings are set with a transplanting machine.[45]

39.13.8 Taro

Global production of taro or cocoyam *(Colocasia esculenta)* in 2000–2003 was about 100 million tons (91 million MT) on 3.8 million acres (1.5 million ha). Leading countries producing taro are Nigeria, Ghana, China, the Ivory Coast, and Papua New Guinea. Only about 6,900 acres (2,800 ha) are grown in the United States, primarily in Hawaii. Dasheen, a taro variety of Chinese origin, is grown in Florida for the vegetable market. Taro, a native of Asia, spread to Africa and the Pacific Islands many centuries ago. It is a popular food crop in tropical countries. It is grown for its starchy, edible corms and cormels.

Taro is a member of the plant family *Aracea*. It greatly resembles the ornamental elephant's ear *C. antiquorum*. The plants with their enormous leaves are usually 3 to 5 feet (90 to 150 cm) tall.

Taro culture is similar to that for growing potatoes, except that the plants need more space. The corms are often planted in Hawaii in mud flats.[74] Corms are planted in the spring. The crop is dug after the tops are dead, about eight months later.

Taro corms are processed into flour or bread. Fresh corms are peeled, mashed, and fermented to produce the dish called *poi*. Taro is also baked or boiled like potatoes. Most varieties of raw taro are inedible because of numerous irritating needle-like oxylate crystals in the flesh.

39.13.9 Teasel

Teasel, a native of southern Europe, belongs to the *Dipsaceae* family. Teasel was introduced into the United States from Europe about 1840. It escaped from cultivation in England, New York, Oregon, and other states.

Fuller's teasel plant *(Dipsacus fullonum)* is an erect herbaceous, branched, monoecious biennial. It grows 2 to 6 feet (40 to 180 cm) in height, with opposite or whorled leaves. The upper pairs of leaves are united to form a cup that holds water around the stem. The blue or lilac flowers are in heads or whorls surrounded by a many-leaved involucre. A low, leafy plant is produced in the first season, but flower stalks shoot up the second year.

The teasel crop is propagated by seeds planted in cultivated rows in the spring. It is ready for harvest in the summer of the second year. In harvesting, the stems are cut a few inches below the heads with a knife as soon as the blossoms have fallen. The cut heads are then hauled to a drying yard.

The commercial product of teasel is the dried, dense, ovoid flower head that bears numerous pointed bracts that end in hooked spines. These heads are used to raise the nap on woolen cloth (Figure 39.20). The heads of the fuller's teasel or clothier's teasel were formerly attached to revolving cylinders that engaged fibers of a passing bolt of cloth and raised the nap to make the goods feel softer and

FIGURE 39.20
Teasel head.

warmer.[75] Hooks made of plastic have replaced teasels, and there culture in the United States has ceased.

39.13.10 Chicory

Global chicory production in 2000–2003 was about 990,000 tons (900,000 MT) on 60,000 acres (24,000 ha). Chicory is grown in Belgium, France, and other European countries as well as in South Africa. The chicory plant, a native of Europe, was introduced into the United States for production many years ago. Chicory culture began in Michigan in 1890 and continued until about 1960. It is now naturalized over a large part of the North, where it is a common weed pest in meadows and waste places.

Chicory thrives best where the average temperature during the growing season does not exceed 70°F (21°C).[76] It will tolerate considerable drought. The soil requirements are similar to those for sugarbeet.

Chicory (*Cichorum intybus*) is a member of the family *Compositae*. It is also called succory, blue sailors, blue daisy, coffee weed, and bunk. Chicory is a perennial, erect herb 1½ to 6 feet (46 to 180 cm) in height, with long, slender, fleshy roots that resemble parsnips. The ray flowers are bright blue and fertile.[77]

Sugarbeet machinery was used for planting and cultivating chicory in Michigan. The roots were dug and topped in the same way as sugarbeets. Yields of fresh chicory roots range from 5 to 10 tons per acre (11 to 22 MT/ha).

Chicory roots were shipped to a drying plant where they are cut up into 1-inch (25 mm) cubes and then dried. The dried product is shipped to factories where it is roasted and ground as needed.

Dried chicory root is mixed with coffee or used as a coffee substitute. Chicory imparts a special flavor and bitterness to the beverage brewed from the coffee mixture that is familiar to those who have tasted the French drip coffee of the deep

South. The leaves of seedlings, as well as shoots that sprout from roots planted in a dark cellar, are eaten as fresh vegetables.

39.13.11 Belladonna

Belladonna was grown as an emergency crop on 275 acres in 1918, and again on about 700 acres in 1942 when imports of this essential drug plant were cut off by war. Belladonna contains alkaloids that can be used to dilate the pupils of the eyes and relieve muscle spasms. Another compound, atropine, is an antidote for organic phosphate poisoning. Belladonna is best adapted to the northeastern, North Central, and Pacific Coast states.[78]

The name *belladonna*[69] means beautiful lady. The belladonna plant (*Atropa belladonna*) is a poisonous perennial herb of the *Solanaceae* (nightshade) family. It grows 2 to 6 feet (60 to 180 cm) in height, with ovate, entire leaves and purple bell-shaped flowers.

The sowing, transplanting, culture, harvest, and curing of belladonna are very similar to the methods used for tobacco. The 1942 crop was produced by tobacco growers in Kentucky, Tennessee, Virginia, Ohio, Pennsylvania, and Wisconsin. It can be cut two to four or more times in a season.

The leaves, stems, and roots of belladonna contain the alkaloid atropine used to dilate the pupils of the eyes. Extracts of the plant are used internally or externally as a sedative.

Belladonna plants are attacked not only by the Rhizoctonia disease, but also by the same insects that attack potatoes and other solanaceous crops. These insects include the Colorado potato beetle, cutworm, fleabeetle, and tobacco worm.[71]

39.13.12 Henbane

The leaves, flowers, and stems of henbane (*Hyoscyamus niger*) contain several poisonous alkaloids that are used in the preparation of medicines. Henbane was grown on a small acreage in Montana, Michigan, and other states during World War II. Only about 200 acres will supply our ordinary domestic requirements. Supplies are usually imported. Henbane was introduced from Europe, but it escaped from cultivation.

Henbane is adapted to the relatively cool, long-day, summer conditions of the northern border and Pacific Coast states. The plant is an annual or biennial that belongs to the family *Solanaceae*. The biennial type, which is the type usually grown, produces a crop the first year, but it may not live over the winter. It can be harvested the second year when it survives. The plant has a nauseating odor. It is covered with hairs. The flowers have a five-toothed calyx and an irregular funnel-shaped corolla (Figure 39.21). The seeds are borne in a capsule.[71]

Henbane is usually sown in pots under glass in midwinter with the seedlings transplanted to 3-inch (8 cm) pots and later transplanted to the field in May. The plants are set about 15 inches (38 cm) apart in rows 30 inches (76 cm) or more apart. The leaves and stems are harvested by hand when the plants are in full bloom, which is usually in August. The drying process and facilities required are about the same as for tobacco.

FIGURE 39.21
Henbane plant. [Courtesy
USDA-NRCS PLANTS Database /
Britton, N.L, and A. Brown.
1913. Illustrated flora of the
northern states and Canada.
Vol. 3: 169]

Extracts of henbane are used as sedatives for the relief of coughs, spasmodic asthma, spasms, and other ailments.

The henbane plant is extremely susceptible to tobacco-mosaic virus. It often is destroyed by the disease.

39.13.13 Ginseng

Ginseng (*Panax quinquefolium*), also called American ginseng, sang, red berry, and five-fingers, occurs sparingly in shady sloping locations in rich, moist soil in the hardwood forests from Maine to Minnesota and southward to the mountains of northern Georgia and Arkansas. The gathering of wild ginseng for export began in the eighteenth century, but natural supplies eventually were greatly depleted. Ginseng culture was started after 1870 by Abraham Whisman at Boones Path, Virginia.[40]

Ginseng might truly be regarded as a glamour crop. In the United States, it has been the cause of visions of great wealth because of its high price per pound, although the profits hoped for by its promoters seldom have been realized. In eastern Asia, it has been regarded as a medicine that possesses extraordinary virtues as a cure-all, an aphrodisiac, a heart tonic, and a remedy for exhaustion of body and mind. Older, branched roots that resemble the human form are especially prized.[71] The main market for ginseng is in the Orient, although it is sold in food stores in the United States.

Ginseng can be produced by dividing the roots of old plants, but this entails a risk of transmitting diseases.[79] Ginseng is usually propagated by seeds that are stratified in moist sand or sawdust. They are kept in a cool, damp place until they are ready to germinate, which is usually in the second spring after they are gathered. The seeds are then planted under dense natural or artificial shade in a bed of deep leaf mold or compost. The plants are spaced about 6 or 8 inches (15 to 20 cm) apart. The roots are ready to dig and dry for the market not earlier than the sixth year after seeding when they have reached a length of about 4 inches (10 cm).

39.13.14 Goldenseal

Goldenseal *(Hydrastis canadensis)* was formerly grown in the United States. Goldenseal is native to the United States, being found in patches in high, open woods, and usually on hillsides or bluffs, from western New England to Minnesota and south to Georgia and Missouri.

The goldenseal plant is an erect perennial with a hairy stem, about 1 foot (30 cm) in height, with two branches at the top. One branch bears a leaf, while the other bears a leaf and a flower (Figure 39.22). The greenish-white flower is followed by a fleshy, berry-like head that contains ten to twenty small, hard, black, shiny seeds. The rhizome is about 2 inches (5 cm) long, ¾ inch (2 cm) thick, and bright yellow when fresh, with yellow flesh. It bears numerous fibrous rootlets. Goldenseal, which belongs to the *Ranunculaceae* or crowfoot family, is also called yellowroot, yellow puccoon, orange-root, yellow Indian paint, turmeric root, Indian turmeric, Ohio curcuma, ground raspberry, eyeroot, eyebalm, yelloweye, jaundice root, and Indian dye.

Goldenseal was used by Native Americans as well as by the early settlers of eastern North America as an external remedy for sore mouth and inflamed eyes. It was also used internally as a bitter tonic for stomach and liver troubles.[80] Commercial demand began about 1860. The collection of goldenseal from wild areas for domestic and export use had so largely exhausted the supply by 1904 that its cultivation was begun by ginseng growers.

Goldenseal is propagated from seeds, rhizome division, or buds or sprouts on the fibrous roots. Rhizome division is the most popular method. The seeds are

FIGURE 39.22
Goldenseal plant. [Courtesy Thomas G. Barnes]

partly separated from the pulp of the fresh fruit, stratified in moist sand until October, and then planted in beds. They are transplanted a year or two later.

The soil and shade conditions for goldenseal are about the same as for ginseng, except that the need for potash fertilizer is greater. The rhizomes are ready for digging three or four years from the time of rhizome or bud planting, or five years after seed planting. They are dug in the autumn after the tops are dead, and then sorted, washed, and dried in trays. Yields of 2,000 pounds per acre (2,240 kg/ha) or less of dried roots can be obtained.

Goldenseal is subject to certain diseases that preclude its culture in the southeastern Piedmont states.

39.13.15 Poppy

Seed of the opium poppy (*Papaver somniferum*) that has been rendered unviable is used in the bakery trade. Global production in 2000–2003 averaged about 57,000 tons (52,000 MT). The leading producing countries are the Czech Republic, Turkey, France, Germany, and Hungary. The seeds are sprinkled on certain types of bread and rolls before baking. The seeds produce edible oil that has an iodine number of about 135. White-seeded types are fed to birds under the name of maw seed. Considerable quantities of opium for medicinal purposes are imported from Asia Minor, India, and Mainland China. Federal narcotic laws prohibit the growing of the opium poppy in the United States except under permit. There is no commercial production here because of the risk of facilitating illicit drug traffic.

The opium poppy is adapted to regions where certain other species of the Oriental poppy are known to be adapted as ornamentals. *P. somniferum* is an annual plant that belongs to the family *Papaveraceae*. It is a native of Greece and the Orient. The plants grow to a height of 4 feet (120 cm) and have large white, purple, reddish-purple, or red flowers.

Opium is the dried milky juice obtained by lancing the fertile but unripe seed capsules. Opium contains about nineteen distinct alkaloids, including morphine and codeine. Paregoric and laudanum are manufactured from opium. Morphine can be extracted from the dried plant.[78, 81]

REFERENCES

1. Polhamus, L. G. "Rubber from guayule," *Agriculture in Americas* 5, 2(1945):1–4.

2. Sievers, A. F., M. S. Lowman, and W. M. Hurst. "Harvesting pyrethrum," *USDA Circ.* 581, 1941, pp. 1–18.

3. Brooks, S. N., C. E. Homer, and S. T. Likens. "Hop production," *USDA Inf. Bull.* 240, 1961.

4. Kadans, J. M. *Modern Encyclopedia of Herbs.* West Nyack, NY: Parker Publ. Co., 1970, pp. 1–256.

5. Kuhlman, G. W., and R. E. Fore. "Cost and efficiency in producing hops in Oregon," *OR Agr. Exp. Sta. Bull.* 364, 1938, pp. 1–57.

6. Brooks, S. N., and K. R. Keller. "Effect of time of applying fertilizer on yield of hops," *Agron. J.* 52(1960):516–518.

7. Brigham, R. D. "Performance of dwarf internode castor (*Ricinus communis* L.) under two row spacings," *TX Agr. Exp. Sta.* MP-1070, 1972, pp. 1–14.

8. Smith, D. C. "Varietal improvement in hops," in *USDA Yearbook*, 1937, pp. 1215–1241.

9. Hoerner, G. R. "Downy mildew of hops," *OR Ext. Bull.* 440, 1932, pp. 1–11.

10. Homer, C. E. "Chemotherapeutic effects of streptomycin on establishment and progression of systemic downy mildew infection in hops," *Phytopath.* 53, 4(1963):472–474.

11. Green, R. J., Jr. "Mint farming," *U.S. Dept. Agr. Info. Bull.* 212, (rev.), 1963.

12. Dunlap, A. A. "Septoria leaf spot on safflower in north Texas," *Plant Disease Reporter* 25, 14(1941):389.

13. Ellis, N. K., and others. "A study of some factors affecting the yield and market value of peppermint oil," *Purdue U. Agr. Exp. Sta. Bull.* 461, 1941, pp. 127.

14. Weller, S., and others. "Mint production and pest management in Indiana," *Purdue Univ. Coop. Ext. Serv.* PPP-103, 2000.

15. Maloy, O. C., and C. B. Skotland. "Diseases of mint," *WA Ext. Circ.* 357, 1969, pp. 1–4.

16. Trotter, W. K. "Potential for oilseed sunflowers in the United States," *USDA* AER 237, 1973, pp. 1–55.

17. Talley, L. J., and E. E. Burns. "Sunflower, a survey of the literature, June 1967 to January, 1971," *TX Agr. Exp. Sta.* MP 992, 1972, pp. 1–21.

18. Wiley, H. W. "The sunflower plant: Its cultivation, composition, and uses," *USDA Div. Chem. Bull.* 60, 1901.

19. Vinall, H. N. "The sunflower as a silage crop," *USDA Bull.* 1045, 1922.

20. Fick, G. N., and C. M. Swallers. "Higher yields and greater uniformity with hybrid sunflowers," *ND Farm. Rsh.* 29, 6(1972):7–9.

21. Schaffner, J. H. "The notation of Helianthus," *Bot. Gaz.* 29 (1900):197–200.

22. Hensley, H. C. "Production of sunflower seed in Missouri," *MO Coll. Agr. Ext. Cir.* 140, 1924.

23. Robinson, R. G., and others. "The sunflower crop in Minnesota," *MN Ext. Bull.* 299, 1967, pp. 1–30.

24. Schild, J., and others. "Sunflower production in Nebraska," *NE Coop. Ext. Serv. NebGuide* G91-1096-A, 1991.

25. Kittock, D. L., and J. H. Williams. "Effects of plant population on castorbean yield," *Agron. J.* 62, 4(1970):527–529.

26. Nayar, N. M., and K. L. Mehra. "Sesame: Its uses, botany, cytogenetics and origin," *Econ. Bot.* 24, 1(1970):20–31.

27. Anonymous. "Sesame production," *USDA Farmers Bull.* 2119, 1958.

28. Peterson, W. F. "Safflower culture in the west-central plains," *USDA Inf. Bull.* 300, 1965, pp. 1–22.

29. Anonymous. "Growing safflower—an oilseed crop," *USDA Farmers Bull.* 2133, (rev.), 1961.

30. Dennis, R. E., and D. D. Rubis. "Safflower production in Arizona," *AZ Agr. Exp. Sta. Bull.* A-47, 1966, pp. 1–24.

31. Knowles, P. F., and M. D. Miller. "Safflower in California," *CA Agr. Exp. Sta. Manual* 27, 1960.

32. Thomas, C. A., J. Klisewicz, and D. Dimmer. "Safflower diseases," *USDA Agricultural Research Service Spec. Rept.* ARS 34-52, 1963.

33. Shrader, J. H. "The castor-oil industry," *USDA Bull.* 867, 1920, pp. 1–40.

34. Whiteley, E. L. "Influence of date of harvest on the yield of kenaf (*Hibiscus cannabinus* L.)," *Agron. J.* 63, 3(1971):509–510.

35. Classen, C. E., and T. A. Kiesselbach. "Experiments with safflower in western Nebraska," *NE Agr. Exp. Sta. Bull.* 376, 1945, pp. 1–28.

36. El-Hamidi, A., and others. "Effects of nitrogen and spacing on castorbean in sandy soils in Egypt," *Expl. Agric.* 4(1968):61–64.

37. Bessey, E. A. "Effect of age of pollen upon the sex of hemp," *Am. J. Bot.* 15(1928):405–411.

38. Brigham, R. D., and B. R. Spears. "Castorbeans in Texas," *TX Agr. Exp. Sta. Bull.* B-954, 1961.

39. Anonymous. "Castorbean production," *USDA Farmers Bull.* 2041, (rev.), 1960.

40. Williams, L. O. "Ginseng," *Econ. Bot.* 11, 2(1957):344–348.

41. Schaffner, J. H. "The fluctuation curve of sex reversal in staminate hemp plants induced by photoperiodicity," *Am. J. Bot.* 18(1931):424–430.

42. Schoenleber, L. G. "Mechanization of castor bean harvesting," *OK Agr. Exp. Sta. Bull.* B-591, 1961.

43. Godfrey, G. H. "Gray mold of castor bean," *J. Agr. Res.* 23(1923):679–715.

44. Lloyd, F. E. *Cuayale, a Rubber Plant of the Chihuahuan Desert.* Washington, D.C., Carnegie Inst. Wash., 1911, pp. 1–213.

45. Hellwig, R. E., and M. El. Byrom. "Sansevieria planting machinery," *USDA Agr. Res. Serv.* ARS 42-34, 1959.

46. Polhamus, L. G. *Rubber—Botany, Production and Utilization.* New York: Interscience, 1962, pp. 1–449.

47. Eaton, B. J. "Identifying and controlling wild hemp," *KS Agr. Exp. Sta. Bull.* 555, 1972, pp. 1–12.

48. Robinson, B. B. "Hemp," *USDA Farmers Bull.* 1935, 1943, pp. 1–16.

49. McPhee, H. C. "The genetics of sex in hemp," *J. Agr. Res.* 31(1925):935–943.

50. Pritchard, F. J. "Change of sex in hemp," *J. Hered.* 7(1916):325–329.

51. Percival, J. *Agricultural Botany,* 8th ed. London: Duckworth, 1936, pp. 1–839.

52. Jordan, H. V., A. L. Lang, and G. H. Enfield. "Effects of fertilizers on yields and breaking strengths of American hemp, *Cannabis sativa,*" *J. Am. Soc. Agron.* 38, 6(1946):551–562.

53. Robbins, W. W. *Botany of Crop Plants.* Philadelphia, PA: Blakiston, 1931, pp. 1–639.

54. Humphrey, J. R. "Marketing hemp," *KY Agr. Exp. Sta. Bull.* 221, 1919, pp. 1–43.

55. Dewey, L. H. "Hemp," *USDA Yearbook,* 1913, pp. 283–346.

56. Puri, Y. P., and others. "Anatomical and agronomic studies of phormium in western Oregon," *USDA Prod. Rsh. Rpt.* 93, 1966, pp. 1–43.

57. Neller, 1. F. "Culture, fertilizer requirements and fiber yields of ramie in the Florida Everglades," *FL Agr. Exp. Sta. Bull.* 412, 1945, pp. 1–40.

58. Robinson, B. B. "Ramie fiber production," *USDA Circ.* 585, 1940, pp. 1–15.

59. Byrom, M. H. "Ramie production machinery," *USDA Info. Bull.* I56, 1956.

60. Byrom, M. H., and H. D. Whittemore. "Long fiber burnishing, ribboning, and cleaning machine," *USDA Agricultural Research Service Spec. Rept.* ARS 42–49, 1961.

61. White, G. A., and others. "Culture and harvesting methods for kenaf—an annual crop source of pulp in the Southeast," *USDA Prod. Rsh. Rpt.* 113, 1970, pp. 1–38.

62. Haarer, A. E. *Jute Substitute Fibers.* London: Wheatland Journals, Ltd., 1952, pp. 1–210.

63. Walker, J. E., and M. Sierra. "Some cultural experiments with kenaf in Cuba," *USDA Cir.* 854, 1950.

64. Byrom, M. H. "Progress with long vegetable fibers," in *Crops in Peace and War,* USDA Yearbook, 1950–51, pp. 472–476.

65. Wilson, F. D., and J. F. Joyner. "Effects of age, plant spacing and other variables on growth yield and fiber quality of kenaf (*Hibiscus cannabinus* L.)," *USDA Tech. Bull.* 1404, 1969, pp. 1–19.

66. Whiteley, E. L. "Influence of date of planting on the yield of kenaf (*Hibiscus cannabinus* L.)," *Agron. J.* 63, 1(1971):135–136.

67. White, G. A. "New crops on the horizon," *Seed World* 110, 4(1972):20–22.

68. Culbertson, R. E. "Pyrethrum—a new crop," in *Proceedings 2nd Dearborn Conference of Agriculture, Industry, and Science.* Dearborn, MI: Farm Chemurgic Council and the Chemical Foundation, Inc., 1936, pp. 292–297.

69. Sievers, A. F. "Belladonna in the United States in 1942," *Drug and Cosmetic Industry,* May 1943, p. 7.

70. Williams, L. O. "Drug and condiment plants," *USDA Handbk.* 172, 1960, pp. 1–37.

71. Sievers, A. F. "Production of drug and condiment plants," *USDA Farmers Bull.* 1999, 1948, pp. 1–99.

72. White, G. A., and J. J. Higgins. "Culture of crambe—a new industrial oilseed crop," *USDA Prod. Rsh. Rpt.* 95, 1966, pp. 1–20.

73. Grombacher, A., L. Nelson, and D. Baltensberger. "Crambe production," *NE Coop. Ext. Serv. NebGuide* G93–1126-A, 1993.

74. Greenwell, A. H. B. "Taro—with special reference to its culture and uses in Hawaii," *Econ. Bot.* 1, 3(1947):276–289.

75. Bailey, L. H. *Cyclopedia of American Agriculture, vol. II, Crops.* New York: Macmillan, Inc., 1907.

76. Cormany, C. E. "Chicory growing in Michigan," *MI Agr. Exp. Sta. Spec. Bull.* 167, 1927, pp. 1–11.

77. Kains, M. G. "Chicory growing," *USDA Div. Bot. Bull.* 19, 1898, pp. 1–52.

78. McCollum, W. B. "The cultivation of guayule," *India Rubber World* 105, 1(1941):33–36.

79. Whetzel, H. H., and others. "Ginseng diseases and their control," *USDA Farmers Bull.* 736, 1916, pp. 1–23.

80. Van Fleet, W. "Goldenseal under cultivation," *USDA Farmers Bull.* 613, 1914, pp. 1–15.

81. Fulton, C. C. *The Opium Poppy and Other Poppies.* U.S. Treas., Dept. Bureau of Narcotics, 1944, pp. 1–85.

Miscellaneous Food and Forage Crops

There are several crops of botanical families other than grasses and legumes that have local or regional importance. Some, like Jerusalem artichoke, beet, and cassava, are used for both livestock feed and human food. Others, like chufa, comfrey, and burnet, are used almost exclusively for livestock feed.

40.1 JERUSALEM ARTICHOKE

40.1.1 Economic Importance

Jerusalem artichoke, or sunchoke, has been grown only to a limited extent in the United States, although it has been advocated[1] as having great possibilities for food, feed, sugar, or alcohol. It was grown in the Pacific Coast states for many years and later in Nebraska.

The Jerusalem artichoke, native to America, was used for food by Native Americans who gathered the wild tubers or grew them in their clearings. It was found by the explorer Samuel de Champlain on Cape Cod in Massachusetts, who took it France in 1605. He thought it tasted like artichokes, a name that he carried back to France. The Native Americans called them sun roots and introduced these perennial tubers to the pilgrims who adopted them as a staple food. Small quantities of the tubers are produced and sold for food in the United States.

40.1.2 Adaptation

The highest yields of Jerusalem artichoke tubers in the United States have been obtained in cool, mild, humid sections of the Pacific Northwest. The tops are killed by the first frost in the fall. The crop responds to a good supply of moisture.[2] The plants require a growing season of 125 days, but they do not develop flowers or tubers until the approach of shorter days in August. Jerusalem artichoke is a short-day plant.[3] The crop is well adapted to productive light or medium loams. Heavy soils make digging the numerous small tubers rather difficult.

40.1.3 Botanical Description

The name *artichoke* was early applied to the Jerusalem artichoke because its taste resembled that of the edible bracts of the true artichoke: the globe, French, or bur artichoke *(Cynara scolymus)*. The Jerusalem part of the name has no reference to the Holy Land but arose from a corruption of the word *girasol*, the Italian name for sunflower.

Jerusalem artichoke *(Helianthus tuberosus)* is a member of the sunflower family, *Asteraceae (Composilae)*. A native of North America, it grew wild along the eastern seaboard from Georgia to Nova Scotia.

Jerusalem artichoke grows to a height of 6 feet (180 cm). It is similar to the wild sunflower except that the center of the head (disk) is light yellow instead of dark brown. The flower heads usually are about 2 to 3 inches (5 to 8 cm) across. The tubers (Figure 40.1), which resemble hand grenades, are ovoid but irregular due to knobs and rings or sections of different diameter. The skin is much thinner than that of potato and not corky. Well-developed tubers weigh 1 to 5 ounces (20 to 140 g). Sugar and other carbohydrates stored in the stem stimulate growth in the tubers after the leaves cease photosynthesis.[4] Since the tubers remain alive in the ground over winter, they maintain the crop as a perennial. Jerusalem artichoke becomes easily established as a weed because it is extremely difficult to gather all of the tubers.

Many varieties of Jerusalem artichoke grown in Europe have tubers with white, yellow, or red skin.[5]

FIGURE 40.1
Tubers of Jerusalem artichoke.

40.1.4 Jerusalem Artichoke Culture

Cultural, fertilizer, and rotation practices for Jerusalem artichoke are similar to those for potato.[6] Liming increases the yield of sugar per acre.[7] Either whole or cut tubers, preferably the former, may be used for planting. Recommended planting rates for 2 to 3 ounce (55 to 85 g) tubers or cut pieces range from 300 to 1,300 pounds per acre (340 to 1,460 kg/ha) in rows 3½ to 5 feet (106 to 150 cm) apart.[2, 5, 8, 9] Large seed pieces produce more stems but do not increase the size of tubers.

Tubers should be planted 4 inches (10 cm) deep as early in the spring as the land can be prepared. When some of the tubers are left during digging, a crop will develop the next year. The best method for eradication of Jerusalem artichoke is to plow the land in late spring or early summer when the tops are about 1½ feet (45 cm) high and then immediately plant a smother crop.

40.1.5 Harvest

The digging of tubers starts after frost. It is done with a modified elevating potato digger, which also breaks down the stalks and permits people to ride on the machine while sorting and bagging the tubers.[10]

When hogs gather the tubers, they leave enough seed tubers in the ground for a stand the following year. Tubers for planting are not often dug until spring because they keep better in the ground than in storage.[6]

Fresh green tops are suitable for feed, but when mature they are harsh, woody, unpalatable, low in digestible nutrients, and therefore seldom harvested.

Average yields of tubers in the Midwest and East are about 5 to 6 tons per acre (11 to 13 MT/ha), but yields of 10 tons per acre (22 MT/ha) or more are obtained under favorable conditions.[6]

The best temperature for storing tubers is 31 to 32°F (−.5 to 0°C), with a relative humidity of 90 to 95 percent.[11]

40.1.6 Uses of Jerusalem Artichoke

The Jerusalem artichoke tuber is used mostly for animal feed or human food. It is very low in digestibility and watery when cooked. In gross chemical composition, the Jerusalem artichoke tuber is similar to potato, but it contains no starch. The carbohydrates are mostly several polysaccharides, chiefly synanthrine,[12] which, when hydrolyzed by acids or enzymes, produce a simple very sweet sugar called levulose, fructose, or fruit sugar. Levulose often is prescribed for diabetics. Hydrolyzed corn starch, or potato starch, produces dextrose (glucose, or grape sugar). When cane sugar is hydrolyzed, or partly digested, it produces the sweet invert sugar that consists of equal quantities of dextrose and levulose. Honey is composed mostly of invert sugar. Levulose is 3 to 73 percent sweeter than cane sugar, but this depends upon who does the tasting.[3] Dextrose is 20 to 50 percent less sweet than cane sugar. Equal mixtures of levulose and dextrose therefore should be sweet as honey. This is the basis for the theory that, by producing and mixing levulose from artichokes and dextrose from grain or potato, the domestic sugar requirements of the United States could be supplied at home.[4]

■ 40.2 ROOT AND LEAF CROPS

The main root crops grown for forage include mangel (or mangel-wurzel, marigold, or stock beet, *Beta vulgaris*), turnip (*Brassica rapa*), rutabaga or swede (*Brassica napus*), and carrot (*Daucus carota*).[13] The chief leaf crops for forage are biennial rape (*Brassica napus*) and thousand-headed kale (*Brassica oleracea* var. *acephala*). These root and leaf crops, except rape, were grown mostly in cool sections of the northern border states to supply succulent feed during the fall, winter, and spring, mostly on farms too small to afford a silo. In northern Europe and parts of Canada, where corn does not succeed, root crops have been important sources of winter feed.

The chief handicaps to growing root crops are the labor involved in growing, harvesting, and slicing them, as well as the limited period of storage as compared with corn or grass silage.[14] Under favorable conditions, mangel, rutabaga, and turnip yield 20 to 30 tons per acre (45 to 67 MT/ha). Carrot yields about half as much, and sugarbeet yields two-thirds as much. Corn silage is higher in digestible nutrients.[15]

40.2.1 Adaptation

Root crops thrive where the mean summer temperature is 60 to 65°F (15 to 18°C). They do not succeed as summer crops where the mean temperature is much above 70°F (21°C). Mangel, turnip, rutabaga, kale, and rape will withstand light freezes. A moist climate together with deep loam soils and soils high in organic matter are favorable for most root crops. Carrot thrives in light, friable soils.

40.2.2 Origins

Beets, of which the mangel is a representative, like the turnip, rutabaga, and carrot, are native to Eurasia. Varieties grown in the United States are mostly of European origin or were developed from European varieties. Kale is native to Europe. Turnip and rutabaga were grown for food and feed in the Roman Empire in the first century AD.

40.2.3 Botanical Description

Mangel differs from sugarbeet (Chapter 35) chiefly in its larger roots, larger acre yields, and smaller sugar content, which is only about one-third or one-half as much. Mangel roots usually extend up above the surface of the soil for about two-fifths their length.

Turnip and rutabaga belong to the mustard family, *Cruciferae*, and the genus *Brassica*, which also includes kale. The plants of this family are biennial or annual pungent-tasting herbs, which bear flowers with four petals arranged in the form of a cross. The genus *Brassica* has yellow flowers, six stamens, and round seeds borne in two-celled pods. Rutabaga can be distinguished from turnip because it forms a neck on the top of the root to which the leaves are attached. Also, the older leaves of rutabaga are smoother and have a lighter bluish color. The roots of most rutabaga varieties have yellow flesh in contrast with the white flesh of most turnip varieties.

Biennial rape plants resemble cabbage when young, but later they grow to a height of 18 to 30 inches (45 to 75 cm). They produce only leaves and branches without forming a head. Kale plants, likewise, are branched and leafy but are larger than those of rape.

Varieties of mangel[16] are either red-fleshed, yellow-fleshed, or white-fleshed. Types with long roots are not well suited to shallow soils. Carrot varieties for feed include white and yellow varieties.

Thousand-headed kale is a single variety. A similar crop, but with thick fleshy stems, is called marrow kale or marrow cabbage.[17]

40.2.4 Fertilizers

Root crops respond to applications of manure.[15, 18] Phosphorus may be needed unless the soil is well supplied. Acidic soils may require lime.

40.2.5 Root Crop Culture

The seedbed for root crops should be prepared as described for sugarbeet in Chapter 35. On wet lowlands, planting the crops on ridges or beds is recommended.[19] A fine, compact seedbed is essential to good stands.[20]

Rows can be spaced at the desired distance of 18 to 24 inches (45 to 60 cm) apart with special planters, sugarbeet drills, or grain drills. Seeds of most root crops are planted in a seedbed of good tilth, about ⅓ to ¾ inch (8 to 20 mm) deep. Carrot is planted ¼ to ½ inch (6 to 13 mm) deep, the shallower depths applying to heavy soils. The usual rates of planting[18, 21] are 6 to 8 pounds per acre (7 to 9 kg/ha) for mangel, 1½ to 2 pounds per acre (1.7 to 2.2 kg/ha) for carrot, and 1 pound per acre (1.1 kg/ha) for rutabaga and turnip. When labor is available for thinning, the final desired spacing within the row is: mangel, 12 inches (30 cm); rutabaga, 10 to 12 inches (25 to 30 cm); turnip, 10 inches (25 cm); and carrot, 6 to 8 inches (15 to 20 cm). Broadcast turnip is planted at a rate of 4 to 5 pounds per acre (4.5 to 5.6 kg/ha).

Thousand-headed kale is sown in a bed or cold frame. Later, the plants are set out about 18 to 36 inches (45 to 90 cm) apart in rows about 42 inches (107 cm) apart. A pound (450 g) of seed is sufficient for 4 or 5 acres (1.5 to 2 ha).[22] When sown directly in the field to be thinned later, 8 to 12 ounces of seed per acre (225 to 340 g/ha) are usually sufficient.

Mangel and carrot should be planted in the spring as soon as the danger of heavy freezes is past, or a little before corn planting time. Turnip and rutabaga may be planted in early spring for a summer crop, in the summer for a fall crop in cool climates only, and in early fall in the milder sections of the South and Pacific Coast when a winter crop is desired. Rape, likewise, may be sown in the spring, summer, or fall, depending upon the region as well as on the period in which the crop is to be pastured or fed. Early spring planting is recommended in Missouri.[23] In cool sections, rape is often sown in the summer as a catch crop or is sown with grain in the spring and then pastured in the fall or late summer. Thousand-headed kale seed is planted in the spring. The plants are usually set out in June and are cut during the fall and winter as needed for feeding in a fresh condition.

40.2.6 Harvest and Storage

Root crops are usually harvested before a heavy freeze, or as soon as the leaves wither and turn yellow, an indication that growth has ceased. Harvest equipment is similar to that used with sugarbeet. Before digging, the tops are removed with a mower or topped with a sharp hoe. If hand dug, the tops can be twisted or cut off after pulling. The crown is not cut off as with sugarbeet. The harvested roots are usually stored in root cellars or outdoor pits. The best storage conditions are those described for potato, namely, a temperature of 36 to 40°F (2 to 4°C), ample ventilation, and humidity high enough to avoid excessive shrinkage. Turnip does not keep as long as mangel, rutabaga, or carrot. Consequently, it should be used first for feed. When kept for three months in a root cellar in West Virginia without any special control of temperature, ventilation, or humidity, mangel lost considerable water and protein, while rutabaga lost water, carbohydrates, and some protein.[24]

40.2.7 Uses

Root crops replace silage as a succulent winter feed. They may be fed in quantities of 1 to 2 pounds daily per 100 pounds (10 to 20 g/kg) of live weight of the animal.[18] Large roots may be fed whole to poultry but should be sliced or chopped for livestock. Mangel, rutabaga, turnip, and carrot contain roughly 90 percent water. In mixed rations for hogs, 440 to 780 pounds (200 to 350 kg) of roots can replace 100 pounds (45 kg) of grain.

40.2.8 Diseases and Insect Pests

Root crops suffer relatively little damage from diseases in the regions where they are grown. The main damage is rotting in storage.

While root crops usually suffer relatively little damage from insects in the main production areas, they occasionally are attacked by cutworms, grasshoppers, aphids, flea beetles, cabbage worms, and root maggots. These can be controlled with suitable insecticides.

■ 40.3 PUMPKIN AND SQUASH

40.3.1 Economic Importance

Pumpkin and squash were once important crops for fall and winter feeding of livestock in the United States. By 1959, these crops were rarely grown for feed. The former practice of interplanting pumpkins with corn interferes with mechanized field operations.

40.3.2 Adaptation

Field or stock varieties of pumpkin and squash are adapted to humid conditions.[25, 26] They also thrive under a wide range of temperatures but lack tolerance to frost or prolonged exposure near freezing. The large varieties require 110 to 120 frost-free days to reach maturity. They thrive in rich, well-drained, light soils. They prefer full sunlight but develop satisfactorily in a cornfield.

40.3.3 Botanical Description

Pumpkin and squash belong to the genus *Cucurbita*, of the family *Cucurbitacea* or the gourd family, which bears a fleshy fruit called a pepo. The large showy yellow flowers are monoecious.

The common or field pumpkin belongs to the species *Cucurbita pepo*; Cushaw (or crookneck) pumpkin is *C. mixta*, Kentucky Field pumpkin is *C. moschata*, and the large yellow varieties of field pumpkins (or squash) are *C. maxima*. The *C. pepo* species includes some small table varieties of pumpkins and summer squash such as zucchini. The *moschata* species include the Winter Crookneck and Butternut varieties of squash. The *maxima* species includes the Hubbard and Banana winter squashes. Thus, the distinction between a pumpkin and a squash is rather confusing.

The peduncle of *C. pepo* is hard, angular, and ridged; the *C. moschata* peduncle is hard, smoothly angular, and flared; the *C. mixta* peduncle is hard, basically angular, and enlarged by hard cork; the *C. maxima* peduncle is soft, basically round, and enlarged by soft cork.[25, 27, 28, 29]

These species are normally cross-pollinated but do not show a loss of vigor when inbred. They are usually pollinated by insects. The different species do not readily intercross except with *C. moschata*. They do not cross with other cucurbits.

Both pumpkin and squash originated in the Western Hemisphere and were cultivated by the Native Americans.[29]

40.3.4 Pumpkin and Squash Culture

Field varieties of pumpkin and squash are planted in hills 10 to 12 feet (3 to 3.5 m) apart, 3 to 6 seeds per hill, and later thinned to 1 or 2 plants. On soils of low or moderate fertility, it is helpful to stir in a forkful of manure at each hill, leaving a cover of about 2 inches (5 cm) of unmixed soil in which the seed is planted. Planting is done after the danger from frost is past. The crop requires clean cultivation until the trailing vines start to run and the ground is partly shaded by the large leaves.

40.3.5 Harvest

For best storage quality, field pumpkins and squash should be harvested upon the approach of the first frost. Light frost kills the vines but does not appreciably damage the fruits. Field pumpkins and squash store well when a short section of the stem is left on the fruit when it is cut from the vine. Careful handling at harvest is essential to prevent decay in storage. The best storage conditions for pumpkins and squash are temperatures of 50 to 55°F (10 to 13°C) with a relative humidity of 70 to 75 percent. Such conditions are often available in a livestock barn or a house cellar. Preliminary curing for about two weeks at 80 to 85°F (27 to 20°C) will ripen immature fruits and aid in healing mechanical injuries. Field pumpkins and squash do not often keep more than two or three months.

40.3.6 Uses

Sliced or chopped pumpkins and squash are fed as succulents to hogs during late fall and early winter. They are also used in pies, casseroles, and ornamental arrangements.

■ 40.4 OTHER CROPS

Several crops that are neither grasses nor legumes have been grown on limited acreages in the United States. They are grown on only limited acreage because of their low value or high production costs.

40.4.1 Cassava

Cassava *(Manihot esculenta)*, also called mandioca or yuca, is a perennial root crop native to Brazil, Venezuela, and Mexico.[30] Global cassava production in 2000–2003 averaged 200 million tons (185 million MT). The leading countries in cassava production are Nigeria, Brazil, Thailand, Indonesia, and Congo. There is no significant commercial production of cassava in the United States. It has been grown for forage in Florida and adjacent states. Cassava is an important food plant in some developing countries.

The plant is a shrubby perennial sometimes more than 15 feet (4 m) tall with various colored stems. It is propagated from stem cuttings and from sprouts. The flowers are monoecious and cross-pollinated mostly by insects. Cassava roots (Figure 40.2) are the source of tapioca starch, which is used for food, adhesives, and sizing.

The cortical layers of cassava roots contain varying quantities of linamarin, a cyanogenetic glucoside, and an enzyme which releases the poisonous hydrocyanic acid. The roots are often prepared for food by grating and pressing to remove the juice that contains most of the glucoside. Boiling the roots destroys the enzyme, and the linamarin is dissolved in the cooking water, which is discarded.

40.4.2 Cactus

Spineless cactus was seldom planted for forage because of its recognized slow growth and limited feed value. Nearly spineless forms of prickly pear cactus *(Opuntia* species), native to Mexico, have long been observed. In emergencies, wild spiny prickly pear has been gathered, the spines singed off, and the pear fed to

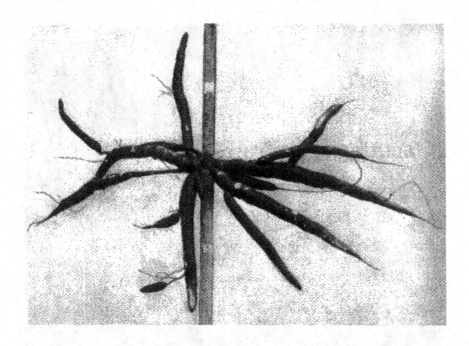

FIGURE 40.2
Cassava roots.

livestock, mostly in southern Texas. Prickly pear is watery, as well as high in crude fiber, but it is satisfactory for feed.[20]

40.4.3 Chufa

Chufa *(Cyperus esculentus)* is occasionally grown in the United States, mostly in Florida. Chufa is a sedge-like plant with creeping rhizomes that produce small tubers or nuts. In many places it is a serious weed called yellow nutgrass or nutsedge. Although chufa is a monocot, it is not a grass. It belongs to the sedge family, *Cyperaceae*. The stems are solid and triangular in cross section. Most leaves are found near the base. The stem has a distinct set of three leaves just below the inflorescence (Figure 40.3). The inflorescence is an open panicle with spikelets grouped in clusters. The fruit is an achene.

Chufa is planted from late spring to midsummer by dropping tubers 6 to 12 inches (15 to 30 cm) apart in rows spaced 30 to 36 inches (75 to 90 cm) apart. The planting rate is about 15 to 40 pounds per acre (17 to 45 kg/ha). The crop is dug with a plow or potato digger, the plants are allowed to dry, and then the tubers are knocked off with a flail. Chufa has been used to pasture hogs, which will feed on the tubers.[31] The tops can be used for hay. Chufa suffers little loss from diseases, but the tubers can be infested by the Negro bug.

40.4.4 Comfrey

Comfrey has been grown in the United States as forage since about 1952, but the acreage is limited by high production costs and its unsuitability for mechanical harvesting. It is soon destroyed by grazing, and the thick leaves and stems with a moisture content of nearly 90 percent make it difficult to preserve as hay or silage.

FIGURE 40.3
Chufa, also known as yellow nutgrass, is a serious weed in many areas. [Photo: Robert H. Mohlenbrock @ USDA-NRCS PLANTS Database / USDA SCS. 1989. Midwest wetland flora: Field office illustrated guide to plant species. Midwest National Technical Center, Lincoln, NE]

Comfrey, a member of the *Boragineae* plant family, comprises three species, *Symphytum officinale, S. asperrium,* and *S. peregrinum.* It is a perennial that attains a height of 2 to 4 feet (60 to 120 cm), with large hairy leaves and blue, white, or pink flowers. It is propagated by root divisions or root cuttings. Comfrey has a high protein content, but it is low in digestibility. Economically, it cannot compete with alfalfa or clover as forage.

Medical folklore credits comfrey with curative powers for more human ailments than are claimed from any other plant.[32]

40.4.5 Other Plants

Burnet *(Sanguisorba minor)*, a deep-rooted perennial native to Europe, is utilized occasionally in pastures in California and Oregon. Sacaline *(Polygonum sachalinese)*, a member of the *Polygonaceae* (buckwheat) family grown as an ornament, has been occasionally exploited as a forage crop, often under some other name such as Eureka clover.

REFERENCES

1. Boswell, V. R. "Growing the Jerusalem artichoke," *USDA Leaflet* 116, 1936, pp. 1–8.

2. Kiesselbach, T. A., and A. Anderson. "Cultural tests with the Jerusalem artichoke," *Jour. Amer. Soc. Agron.* 21, 10(1929):1001–1006.

3. Shoemaker, D. N. "The Jerusalem artichoke as a crop plant," *USDA Tech. Bull.* 33, 1927, pp. 1–2.

4. Incoll, L. D., and T. F. Neales. "The stem as a temporary sink before tuberization in *Helianthus tuberosus* L.," *J. Exp. Bot.* 21, 67(1970):469–476.

5. Boswell, V. R., and others. "Studies of the culture and certain varieties of the Jerusalem artichoke," *USDA Tech. Bull.* 514, 1936, pp. 1–70.

6. Burlison, W. L. "Growing artichokes in America," in *Proceedings 2nd Dearborn Conference of Agriculture, Industry, and Science.* Dearborn, MI: Farm Chemurgic Council and the Chemical Foundation, Inc., 1936, pp. 111–120.

7. Sprague, H. B., N. F. Farris, and W. G. Colby. "The effect of soil conditions and treatment on yields of tubers and sugar from the American artichoke," *J. Am. Soc. Agron.* 27, 5(1935):392.

8. Johnson, H. W. "Storage rots of the Jerusalem artichoke," *J. Agr. Res.* 43, 4(1931):337–352.

9. Kiesselbach, T. A. "Should more Jerusalem artichokes be grown?" *U. NE Agr. Ext. Circ.* 108, 1937, pp. 1–4 (mimeographed).

10. Lehmann, E. W., and R. I. Shawl. "New method is devised for harvesting artichokes," *47th Ann. Rpt. IL Agr. Exp. Sta.* 1934, p. 200.

11. Jackson, R. F., C. G. Silsbee, and M. J. Proffitt. "A method for the manufacture of levulose," *J. Indus. and Eng. Chem.* 16(1924):1250–1251.

12. Bates, F. J. "Discussion," in *Proceedings 2nd Dearborn Conference of Agriculture, Industry, and Science.* Dearborn, MI: Farm Chemurgic Council and the Chemical Foundation, Inc., 1936, p. 159.

13. Westover, H. L., and H. A. Schoth. "Experiments in growing roots as feed crops," *USDA Tech. Bull.* 416, 1934, pp. 1–15.

14. Fraser, S., J. W. Gilmore, and C. F. Clark. "Root crops for stock feeding," *Cornell U. Agr. Exp. Sta. Bull.* 243, 1906, pp. 45–76.

15. Delwiche, E. J. "Profitable root crops," *WI Agr. Exp. Sta. Bull.* 330, 1921, pp. 1–22.

16. Piper, C. V. *Forage Plants and Their Culture*, rev. ed. New York: Macmillan, 1924, pp. 1–671.

17. Chapin, L. J. "Thousand-headed kale and marrow cabbage," *WA Agr. Exp. Sta. Spec. Bull.* 6, 1912.

18. Westover, H. L., H. A. Schoth, and A. T. Semple. "Growing root crops for livestock," *USDA Farmers Bull.* 1699, 1933, pp. 1–12.

19. Carrier, L. *The Beginnings of Agriculture in America.* New York: McGraw-Hill, 1923, pp. 1–323.

20. Griffiths, D. "Prickly pear as stock feed," *USDA Farmers Bull.* 1072, 1920, pp. 1–24.

21. Schoth, H. A. "Root crop production for livestock," *USDA Leaflet* 410, 1957.

22. Stookey, E. B. "Kale and root crops," *WA Agr. Exp. Sta. Monthly Bull* 8, 1(1920):8–10.

23. Hutcheson, C. B. "Growing rape for forage," *MO Agr. Ext. Serv. Circ.* 3, 1915, pp. 1–4.

24. Morrow, K. S., R. B. Dustman, and H. O. Henderson. "Changes in the chemical composition of mangels and rutabagas during storage," *J. Agr. Res.* 43, 10(1931):919–930.

25. Tapley, W. T., W. D. Enzie, and G. P. Van Eseltine. *The Vegetables of New York,* vol. 1, Part IV, *The Cucurbits.* Albany, NY: N.Y. Agr. Exp. Sta. 1937, pp. 1–131.

26. Thompson, R. C. "Production of pumpkins and squashes," *USDA Leaflet* 141, 1937, pp. 1–8.

27. Castetter, E. F., and A. T. Erwin. "A systematic study of squashes and pumpkins," *IA Agr. Exp. Sta. Bull.* 244, 1927, pp. 107–135.

28. Russell, P. "Identification of the commonly cultivated species of *Cucurbita* by means of seed characters," *WA Acad. Sci.* 14(1924):265–269.

29. Whitaker, T. W., and G. W. Bohn. "The taxonomy, genetics, production, and uses of the cultivated species of *Cucurbita,*" *Econ. Bot.* 4, 1(1950):52–81.

30. Rogers, D. J. "Some botanical and ethnological considerations of *Manihot esculenta,*" *Econ. Bot.* 19, 4(1965):369.

31. Killinger, G. B., and W. E. Stokes. "Chufas in Florida," *FL Agr. Exp. Sta. Bull.* 419, 1946, pp. 1–16.

32. Kadans, J. M. *Modern Encyclopedia of Herbs.* West Nyack, NY: Parker Publ. Co., 1970, pp. 1–256.

Appendix

TABLE A.1 Seed and Plant Characteristics

Temperature type: C = cool-weather growth; W = warm-weather growth
Growth habit: A = summer annual; W = winter annual; B = biennial; P = perennial; P(A) = perennial but grown as annual
Photoperiodic response: L = long day; S = short day; N = day neutral or indeterminate; I = intermediate

Crop	Botanical Name	Seeds per Pound (1,000)	Gram (no.)	Weight per Bushel (lb)	Germination Time (Days)	Temperature Type	Growth Habit	Photoperiodic Response
Alfalfa	*Medicago sativa*	220	500	60	7	C	P	L
Alyceclover	*Alysicarpus vaginalis*	275	660	60	21	W	A	
Bahiagrass	*Paspalum notatum*	150	336		21	W	P	
Barley	*Hordeum vulgare*	13	30	48	7	C	A; W	L
Bean								
Adzuki	*Phaseolus angularis*	5	11	60	10	W	A	
Field	*Phaseolus vulgaris*	1–2	4	60	8	W	A	S; N
Lima	*Phaseolus limensis*	0.4	2	56	9	W	P(A)	S
Lima (baby)	*Phaseolus lunatus*	0.5				W	A	S; N
Mung	*Phaseolus aureus*	11	24		7	W	A	S
Tepary	*Phaseolus acutifolius latifolius*	2	5			W	A	S
Beet (see mangel and sugarbeet)								
Beggarweed (Florida) unhulled	*Desmodium purpureum*	194	400	60	28	W	P(A)	S
Bentgrass								
Brown	*Agrostis canina*	11,000	24,000		21	C	P	
Common	*Agrostis tenuis*	5,500	12,000		28	C	P	
Creeping	*Agrostis palustris*	7,700	17,000		28	C	P	L
Belladonna	*Atropa belladonna*	480	1,000				P	
Bermudagrass	*Cynodon dactylon*	1,800	3,900	40(14)	21	W	P	L
(unhulled)		1,300	2,900					
Berseem (see Clover, Egyptian)								
Big trefoil	*Lotus uliginosus*	1,000	1,900	60	7	C	P	L
Birdsfoot trefoil	*Lotus corniculatus*	375	800	60	7	C	P	

(Continued)

Crop	Botanical Name	Seeds per Pound (1,000)	Gram (no.)	Weight per Bushel (lb)	Germination Time (Days)	Temperature Type	Growth Habit	Photoperiodic Response
Black medic	*Medicago lupulina*	300	600	60 (hulled)	7	C	A	L(?)
Bluegrass								
Annual	*Poa annua*	1,200	2,600		21	C	A	N
Bulbous	*Poa bulbosa*	450	1,000		35	C	P	L
Canada	*Poa compressa*	2,500	5,500		28	C	P	L
Kentucky	*Poa pratensis*	2,200	4,800	14	28	C	P	N
Nevada	*Poa nevadensis*	1,000	2,300		21	C	P	
Rough	*Poa trivialis*	2,500	5,600		21	C	P	
Texas	*Poa arachnifera*	1,100	2,500		28	C	P	
Wood	*Poa nemoralis*	3,200	7,000		28	C	P	
Bluestem								
Big	*Andropogon gerardii*	150	340		28	W	P	S
Little	*Schizachyrium scoparius*	260	560		28	W	P	L
Sand	*Andropogon hallii*	100	230		28	W	P	
Bonavista	*Dolichos lablab*	1	2	60		W	P(A)	N
Bromegrass								
Mountain	*Bromus marginatus*	64	140		14	C	P	
Smooth	*Bromus inermis*	137	300	14	14	C	P	L
Broomcorn	*Sorghum vulgare*	25	60	(44–50)	10	W	A	S
Buckwheat								
Common	*Fagopyrum esculentum*	20	45	48	6	W	A	L; N
Tatar	*Fagopyrum tataricum*	26	60	48			A	
Buffalograss	*Buchloe dactyloides*							
	(burs)	50	110		28	W	P	
	(caryopses)	330	738					
Buffelgrass	*Cenchrus ciliaris*					W	P	S
Burclover								
California	*Medicago polymorpha*							
Spotted	(out of bur)	209	303	50	14	C	W	
	Medicago arabica (in bur)	22	49	8–12	14	C	W	
Buttonclover	*Medicago orbicularis*	150	330			C	A	
Burnet	*Sanguisorbe minor*					W	P	
Canarygrass	*Phalaris canariensis*	68	150	50			A	L
Carpetgrass	*Axonopus compressus*	1,350	2,500	18–36	21	W	A	N
Carrot (stock)	*Daucus carota*	405	800	50	28	C	B	L
Cassava (roots)	*Manihot esculenta*					W	P	N
Castorbean	*Ricinus communis*	1	2	50		W	P(A)	L
Cheat	*Bromus secalinus*	71	160			C	W	
Chicory	*Cichorium intybus*	420	940		14	C	P	L
Chickpea	*Cicer arietinum*	1	2		7	C	A	N;L
Chufa	*Cyperus esculentus*			44		W	P	

Crop	Botanical Name	Seeds per Pound (1,000)	Seeds per Gram (no.)	Weight per Bushel (lb)	Germination Time (Days)	Temperature Type	Growth Habit	Photoperiodic Response
Clover								
Alsike	*Trifolium hybridum*	680	1,500	60	7	C	P	L
Alyce (see alyceclover)								
Bur (see burclover)								
Cluster	*Trifolium glomeratum*	1,300	2,900	60	10	C	W	L
Crimson	*Trifolium incarnatum*	150	330	60	7	C	W	L
Egyptian (berseem)	*Trifolium alexandrinum*	210	460	60	7	C	W	L
Hop	*Trifolium agrarium*	830	1,800	60			W	L
Ladino	*Trifolium repens*	860	1,900	60	10	C	P	L
Lappa	*Trifolium lappaceum*	680	1,500	60	7	C	W	L
Large hop	*Trifolium procumben*	2,500	5,400	60	14	C	W	L
Low hop	*Trifolium dubium*	860	1,900	60	14	C	W	L
Persian	*Trifolium resupinatum*	640	1,400	60	7	C	W	L
Red	*Trifolium pratense*	260	600	60	7	C	P	L
Rose	*Trifolium hirtum*	160	360					
Sour (see sourclover)								
Strawberry	*Trifolium fragiferum*	290	640	60	7	C	P	L
Sub		55	120	60	14	C	W	L
Sweet (see sweetclover)								
White	*Trifolium repens*	700	1,500	60	10	C	P	L
Zigzag	*Trifolium medium*						P	L
Cloudgrass	*Agrostis nebulosa*							
Corn								
Field—for grain	*Zea mays*	1.2	3	56	7	W	A	S
Pop	*Zea mays*	3	7	56	7	W	A	S
Sweet	*Zea mays*	2	4	50	7	W	A	S
Cotton						32		
Upland	*Gossypium hirsutum*	4	8	28–33	12	W	P(A)	N
Amer. Pima	*Gossypium barbadense*	4	9		12	W	P(A)	N
Cowpea	*Vigna sinensis*	2–6	8	60	8	W	A	S
Crambe	*Crambe abyssinica*	86	190	27		C	A	
Crested dogtail	*Cynosurus cristatus*	860	1.900		21	C	P	L
Crested wheatgrass (see wheatgrass)								
Crotalaria								
Lanceleaf	*Crotalaria lanceolata*	170	375	60	10	W	A	S
Showy	*Crotalaria spectabilis*	30	80	60	10	W	A	S
Slenderleaf	*Crotalaria intermedia*	96	210	60	10	W	A	S
Striped	*Crotalaria mucronata*	75	215	60	10	W	A	S
Sunn hemp	*Crotalaria juncea*	16	35	60	10	W	A	S

(Continued)

Crop	Botanical Name	Seeds per Pound (1,000)	Seeds per Gram (no.)	Weight per Bushel (lb)	Germination Time (Days)	Temperature Type	Growth Habit	Photoperiodic Response
Crownvetch	*Coronilla varia*	140	300				P	
Dallisgrass	*Paspalum dilatatum*	340	485	12–15	21	W	P	S
Digitalis (foxglove)	*Digitalis purpurea*						B	L
Dill	*Anethum gravolens*	410	900		21		A; P	L
Fenugreek	*Trigonella foenum-graecum*			60			A	L
Dropseed, sand	*Sporobolus cryptandurs*	5,450	12,000		42	W	P	
Fescue								
Chewings	*Festuca rubra* var. *commutata*	615	1,400	14–30	21	C	P	
Hair	*Festuca capillata*	1,500	3,200		28	C	P	
Meadow	*Festuca elatio*	230	500	14–24	14	C	P	
Red	*Festuca rubra*	400	900		21	C	P	
Sheeps	*Festuca ovina*	530	1,167	10–30	21	C	P	
Tall	*Festuca arundinacea*	227	500			C	P	
Feterita (see sorghum)								
Field pea	*Pisum arvense*	4	8	60	8	C	A	L
Austrian winter	*Pisum arvenu*	5	11		8	C	W	
Flax	*Linum usitatissimum*	82	180	56	7	C	A	L
Gamagrass	*Tripsacum dactyloides*	7	15			W	P	S
Giant panic grass	*Panicum antidotale*	610	1,450		28		A	
Goldenseal	*Hydrastis canadensis*						P	
Grama								
Black	*Bouteloua eri poda*	560	1,200			W	P	
Blue	*Bouteloua gracilis*	900	1,980		28	W	P	N; S
Hairy	*Bouteloua hirsuta*	980	2,200					
Side oats	*Bouteloua*	200	442		28	W	P	
(caryopses)	*curtipendula*	730	1,600					
Grasspea	*Lathyrus sativus*				60		A	
Guar	*Cyamopsis psoralides*	5	11	60		W	A	
Guayule	*Parthenium argentatum*	600	1,300				P	
Guineagrass	*Panicum maximum*	1,000	2,200		28	W	P	
Hairy indigo	*Indigofera hirsuta*	200	450	55		W	A	S
Hardinggrass	*Phalaris tuberose* var. *stemoptera*	340	750		28		P	
Hegari (see sorghum)								
Hemp	*Cannabis sativa*	27	45	44	7	C	A	S
Henbane	*Hyoscyamus niger*						A; B	L
Hop	*Humulus lupulus*						P	
Horsebean	*Vicia faba*	1–2	3	47	10	C	A	L
Hungariangrass (see millet-foxtail)								

Crop	Botanical Name	Seeds per Pound (1,000)	Gram (no.)	Weight per Bushel (lb)	Germination Time (Days)	Temperature Type	Growth Habit	Photoperiodic Response
Indiangrass	*Sorghastrum nutans*	170	365		21	W	P	I
Jerusalem artichoke	*Helianthus tuberosus*					C	P	S; N
Job's tears	*Coix lachyryma-jobi*						A	N
Johnsongrass	*Sorghum hale pense*	130	290	28	35	W	P	S
Kafir (see sorghum)								
Kale (thousand-headed)	*Brassica oleraceae*	140	315		10	C	B	N; L
Kenaf	*Hibiscus cannabinus*	14	31		7	W	A	S
Kidney vetch	*Anthyllis vulneraria*	150	320	60				
Kudzu	*Pueraria thunbergiana*	37	81	54	14	W	P	
Lentil	*Lens esculenta*	5–8	9	20	60	W	A	L
Lespedeza								
Chinese (Sericea)	*Lespedeza cuneata*	372	820	35	28	W	P	S
(scarified)		335	820	60				
Common	*Lespedeza striata*	343	750	25	14	W	A	S
Kobe	*Lespedeza striata*	185	750	30	14	W	A	S
Korean	*Lespedeza stipulaceae*	240	525	45	14	W	A	S
Siberian	*Lespedeza juncea*	370	820	60	21	W		
Lovegrass,								
Weeping	*Eragrostis curvula*	1,500	3,300		14	W	P	
Lehmann	*Eragrostis lehmanniana*	4,245	9,400			W	P	
Sand	*Eragrostis trichoides*	1,800	4,000			W	P	
Lupine								
Narrowleaf	*Lupinus angustifulius*	3	7	60	10	C	A	N; S
White	*Lupinus albus*	3	7	60	7	C	A	N
Yellow	*Lupinus luteus*	4	9	60	21	C	A	N; L
Mangel	*Beta vulgaris*	22	48			C	B	L
Meadow foxtail	*Alopecurus pratensis*	540	1,200	6–12	14	C	P	L
Millet								
Browntop	*Panicum ramosum*	140	300		14	W	A	
Foxtail	*Setaria italica*	220	470	50	10	W	A	L
Japanese	*Echinochloa frumentacea*	155	320	35	10	W	A	
Pearl (cattail)	*Pennisetum glaucum*	85	190		7	W	A	S
Proso	*Panicum miliaceum*	80	180	56	7	W	A	S
Ragi (finger)	*Eleusine coracana*					W	A	S
Milo (see sorghum)								
Mint								
Peppermint	*Mentha piperita*						P	
Spearmint	*Mentha spicata*						P	
Molasses grass	*Melinis minutiflora*	6,800	15,000		21	W	P	

(Continued)

Crop	Botanical Name	Seeds per Pound (1,000)	Seeds per Gram (no.)	Weight per Bushel (lb)	Germination Time (Days)	Temperature Type	Growth Habit	Photoperiodic Response
Mustard								
Black	*Brassica nigra*	570	1,250		7	C	A	L
Brown	*Brassica juncea*	280	624		7	C	A	
White	*Brassica alba*	73	160		5	C	A	
Napiergrass	*Pennisetum purpureum*	1,402	3,100		10	C	P	
Narrowleaf trefoil	*Lotus tenuis*	375	800	60	7	C	P	L
Oat								
Common	*Avena sativa*	14	30	32	10	C	A; W	L
Red	*Avena byzantina*	14	30	32	10	C	A; W	L
Oatgrass, tall	*Arrhenatherum elatius*	150	330	11–14	14	C	P	
Orchardgrass	*Dactylis glomerata*	590	1,440	14	18	C	P	N; L
Paragrass	*Panicum purpurescens*					W	P	
Patridgepea	*Cassia fasiculata*							
Pea (see field pea)								
Peanut	*Arachis hypogeae*	1	1–3	20–30	10	W	A	S
Pepper	*Capsicum frutescens*	75	165		14	W	A	N; S
Perilla	*Perilla frutescens*			37			A	S
Pigeonpea	*Cajanus cajun*	8	18	60			P	
Popcorn (see corn)								
Poppy (opium)	*Papaver somniferum*			46			A	L
Potato (tubers)	*Solanum tuberosum*			60		C	P(A)	L; N
Pumpkin	*Cucurbita pepo*	2	4		7	W	A	
Cushaw	*Cucurbita moshata*	2	4				A	
Pyrethrum	*Chrysanthemum cinerariifolium*						P	L
Ramie	*Boehmeria nivea*						P	
Rape							50	
Oilseed	*Brassica napus*	160	345		7	C	A; W	L
Biennial	*Brassica rapa*	240	535		7	C	B	L
Redtop	*Agrostis alba*	5,100	11,000	14	10	C	P	
Reed canarygrass	*Phalaris arundinacea*	550	1,200	44–48	21	C	P	L
Rescuegrass	*Bromus catharticus*	70	145	8–12	35	W		
Rhodes grass	*Chloris gayana*	1700	4,700	8–12	14	W	P	
Rice	*Oryza sativa*	15	65	45	14	W	A	S
Ricegrass, Indian	*Oryzopsis hymenoides*	140	310		42	C	P	
Roughpea	*Lathyrus hirsutus*	14	40	55	14	C	W	
Roughstalked meadow grass (see bluegrass, rough)								
Rutabaga	*Brassica napobrasica*	200	430	60	14	C	B	L
Rye	*Secale cereale*	18	40	56	7	C	A; W	L
Ryegrass								
Italian	*Lolium multiflorum*	227	500	24	14	C	A; W	L
Perennial	*Lolium perenne*	330	500	24	14	C	P	L

Crop	Botanical Name	Seeds per Pound (1,000)	Gram (no.)	Weight per Bushel (lb)	Germination Time (Days)	Temperature Type	Growth Habit	Photoperiodic Response
Safflower	*Carthamus tinctorius*	22	45			A		L; N
Sainfoin	*Onobrychis viciaefolia*	23	50	55	10		P	
Sand dropseed	*Sporobolus cryptandrus*	3–5	5,000			W	P	
Sesame	*Sesamum indicum*	100	220	46	10	W	A	S
Sesbania	*Sesbania macrocarpa*	40	105	60	7	W	A	
Smilograss	*Oryzopsis miliacea*	990	2,000		42		P	
Sorghum	*Sorghum bicolor*					W	A	
Feterita		13	33	56	10	W		
Hegari		20	44	56	10	W		
Kafir		20	55	56	10	W		
Milo		15	33	56	10	W		
Sorgo		28	50	50	10	W		
Sorgo (sumac)		40	88	50	10	W		
Sourclover	*Melilotus indica*	300	660	60	14	C	W	
Soybean	*Glycine max*							
Small-seed		8	18	60	8	W	A	S; N
Medium-seed		2–3	6–13	60				
Large-seed		1	2	60				
Spelt (see wheat)								
Squash	*Cucurbita maxima*	6	14		7	W	A	
Sudangrass	*Sorghum bicolor*	55	120	40	10	W	A	S
Sugarbeet	*Beta vulgaris*	22	48	15		C	B	L
Sugarcane	*Saccharum officinarum*	Veg.						
Sulla	*Hedysarium coronarium*	100	220	60			P(A)	S
Sunflower	*Helianthus annuus*	3–9	13	24	7		A	
Sweetclover								
White	*Melilotus alba*	250	570	60	7	C	B; A	L
Yellow	*Melilotus officinalis*	250	570	60	7	C	B	L
Sweet potato	*Ipomoca batatas*	Veg.		55–60		W	P(A)	S
Sweet vernal grass	*Anthoxanthum odoratum*	730	1,600		14	C	P	
Switchgrass	*Panicum virgatum*	370	815		28	W	P	
Tall meadow oatgrass (see oatgrass)								
Teasel	*Dipsacus fullonum*						B	
Teosinte	*Zea* spp.	7	15				A; P	
Timothy	*Phleum pratense*	1,230	2,500	45	10	C	P	L
Tobacco	*Nicotiana tobacum*	5,000	11,000		14	W	A	N
Trefoil (see big trefoil, narrowleaf trefoil, and birdsfoot trefoil)								
Turnip	*Brassica rapa*	154	535	55	7	C	B	L

(Continued)

Crop	Botanical Name	Seeds per Pound (1,000)	Gram (no.)	Weight per Bushel (lb)	Germination Time (Days)	Temperature Type	Growth Habit	Photoperiodic Response
Vasey's grass	*Paspalum urvillei*	440	970		21	W	P	S
Velvetbean	*Mucuna* spp.	1	2	60	14	W	A	
Velvetgrass	*Holcus lanatus*	1,200	2,800		14	C	P	
Vetch								
Common	*Vicia sativa*	7	19	60	10	C	A; W	L
Hairy	*Vicia villosa*	21	36	60	14	C	W; B	
Hungarian	*Vicia pannonica*	11	24	60	10	C	A; W	L
Monantha	*Vicia monantha*	10	23	60	10	C	A; W	L
Narrowleaf	*Vicia angustifolia*	27	60	60	14	C	A	
Purple	*Vicia* spp.	9	20	60	10	C	W; A	
Woolypod	*Vicia villosa* ssp. dasycarpa	11	25	60	14	C		
Wheat								
Club	*Triticum compactum*	20–24	48	60	7	C	A; W	L
Common	*Triticum aestivum*	12–20	35	60	7	C	A; W	L
Durum	*Triticum durum*	8–16	26	60	10	C	A	L
Emmer	*Triticum dicoccoides*			40	7	C	A; W	L
Spelt	*Triticum spelta*			40	7	C	A; W	L
Wheatgrass								
Crested	*Agropyron cristatum*	320	714		14	C	P	
Slender	*Agropyron trachycaulum*	150	320		14	C	P	
Western	*Agropyron smithii*	110	235		35	C	P	L
Wild rice	*Zizania aquatica*						A	S
Wild rye, Canada	*Elymus canadensis*	120	261		21	C	P	
Wormseed	*Chenopodium ambrosioides* var. *anthelminticum*						A	
Wormwood	*Artemisia absinthium*						P	

Temperature type: C = cool-weather growth, W = warm-weather growth
Growth habit: A = annual; W = winter annual; B = biennial; P = perennial; P(A) = perennial but grown as annual
Photoperiodic response: L = long day; S = short day; N = day neutral or indetermminate; I = intermmediate

TABLE A.2 Composition of Crop Products

(Ether Extract = Fat, Nitrogen—Free Extract = Carbohydrates except fiber)

Product	Moisture (%)	Ash (%)	Crude Protein (%)	Ether Extract (%)	Crude Fiber (%)	Nitrogen-Free Extract (%)	Calcium (%)	Phosphorus (%)
Grains, Seeds, and Mill Concentrates								
Barley	9.6	2.9	12.8	2.3	5.5	66.9	0.07	0.32
Barley feed	7.9	4.9	15.0	4.0	13.7	54.5	0.03	0.41
Bread, kiln-dried	10.5	2.1	12.5	1.6	0.4	72.9	0.03	0.12
Brewers' dried grains:								
18–23 percent protein	7.9	4.1	20.7	7.2	17.6	42.5	0.16	0.47
23–28 percent protein	7.7	4.3	25.4	6.3	16.0	40.3	0.16	0.47
Brewers' rice	11.6	0.7	7.0	0.8	0.6	79.3	0.03	0.25
Buckwheat	12.6	2.0	10.0	2.2	8.7	64.5		
Buckwheat middlings	12.4	4.6	28.0	6.6	5.3	43.1		
Corn, shelled	12.9	1.3	9.3	4.3	1.9	70.3	0.01	0.26
Corn bran	10.0	2.1	10.0	6.6	8.8	62.5	0.03	0.14
Corn chop	11.3	1.4	9.8	4.1	2.1	71.3	0.01	0.26
Com (ear) chop	10.7	2.0	8.2	3.4	9.2	66.5		
Corn-feed meal	10.8	1.9	10.5	5.3	2.9	68.6	0.04	0.38
Corn-germ meal	7.0	3.8	20.8	9.6	7.3	51.5	0.05	0.59
Corn-gluten feed	9.5	6.0	27.6	3.0	7.5	46.4	0.11	0.78
Corn-gluten meal	8.0	2.2	43.0	2.7	3.7	40.4	0.10	0.47
Corn-oil meal	8.7	2.2	22.1	6.8	10.8	49.4	0.06	0.62
Cottonseed, whole pressed	6.5	4.3	29.6	5.8	25.1	28.7		
Cottonseed cake	7.5	5.9	44.1	6.4	10.3	25.8		
Cottonseed feed, 32 percent protein	8.3	4.8	32.1	6.4	15.3	33.1	0.20	0.73
Cottonseed hulls	8.7	2.6	3.5	1.0	46.2	38.0		
Cottonseed meal:								
33–38 percent protein	7.4	5.2	36.6	5.6	15.3	29.9	0.28	1.30
38–43 percent protein	7.3	6.1	41.0	6.5	11.9	27.2	0.19	1.11
> 43 percent protein	7.2	5.8	43.7	6.5	11.1	25.7	0.18	1.15
Distillers' (corn) grain	7.0	2.4	28.3	9.4	14.6	38.3	0.04	0.29
Distillers' (rye) dried grain	6.1	2.4	17.9	6.3	15.9	51.4	0.13	0.43
Feterita	9.1	1.7	14.2	2.9	1.4	70.7		
Hemp cake	10.8	18.0	30.8	10.2	22.6	7.6		
Hempseed, European	8.8	18.8	21.5	30.4	15.9	4.6		
Hominy feed	9.5	2.9	11.2	8.3	6.3	61.8	0.03	0.44
Kafir	11.9	1.7	11.1	3.0	2.3	70.0	0.01	0.25
Kafir-head chops	10.4	3.9	10.9	2.5	6.0	66.3	0.09	0.20
Linseed meal:								
33–38 percent protein	8.5	5.6	35.3	5.4	8.3	36.9	0.36	0.84
38–43 percent protein	8.5	5.3	40.4	5.8	7.5	32.5	0.33	0.74
Malt	7.7	2.9	12.4	2.1	6.0	68.9		
Malt sprouts	7.3	6.1	28.1	1.8	13.3	43.4	0.26	0.68
Mesquite beans and pods	6.6	4.5	13.0	2.7	22.8	50.4		
Millet, foxtail	10.1	3.3	12.6	4.3	8.4	61.3		
Millet, proso or hog millet	9.8	3.4	12.0	3.4	7.9	63.5		
Milo	9.3	1.6	12.5	3.2	1.5	71.9		

(Continued)

Product	Moisture (%)	Ash (%)	Crude Protein (%)	Ether Extract (%)	Crude Fiber (%)	Nitrogen-Free Extract (%)	Calcium (%)	Phosphorus (%)
Grains, Seeds, and Mill Concentrates (*continued*)								
Milo-head chops	10.4	4.3	10.7	2.6	7.i	64.9		
Molasses, cane	24.0	6.8	3.1			66.1	0.35	0.06
Oat, grain	7.7	3.5	12.5	4.4	11.2	60.7	0.10	0.40
Oat chops	8.9	3.9	12.8	5.0	11.8	57.6	0.10	0.36
Oat clips	9.0	9.3	11.8	4.5	22.7	42.7		
Oat groats, ground rolled	10.4	2.6	17.3	6.6	1.8	61.3	0.08	0.43
Oat hulls	5.8	6.5	4.3	1.9	30.8	50.7	0.09	0.12
Oatmeal	8.9	2.3	16.5	4.8	3.6	63.9	0.08	0.43
Oat millfeed	6.9	6.0	6.3	2.2	27.9	50.7	0.20	0.22
Peanuts, kernels	5.5	2.3	30.2	47.6	2.8	11.6	0.06	0.38
Peanuts, shells on	6.0	2.8	24.7	33.1	18.0	15.4		
Peanut meal:								
38–43 percent protein	6.4	4.4	41.6	7.2	16.0	24.4	0.10	0.50
43–48 percent protein	6.7	4.6	45.1	7.2	14.2	22.2	0.17	0.55
> 48 percent protein	7.0	5.0	51.4	4.8	9.2	22.6		
Rapeseed, brown Indian	5.7	6.4	21.0	41.2	12.5	13.2		
Rapeseed, common	7.3	4.2	19.5	45.0	6.0	18.0		
Rice, rough	9.7	5.4	7.3	2.0	8.6	67.0	0.10	0.10
Rice, bran	8.8	12.2	12.8	13.8	12.2	40.2	0.10	1.84
Rice hulls	6.5	21.9	2.1	0.4	44.8	24.3	0.08	0.06
Rice polish	10.0	7.6	12.4	13.2	2.8	54.0	0.03	1.52
Rice-stone bran	8.4	11.9	12.5	13.0	11.1	43.1		
Rye	9.5	1.9	11.1	1.7	2.1	73.7	0.04	0.37
Rye feed	10.2	4.0	15.6	3.2	4.3	62.7		0.59
Rye middlings	9.5	4.4	16.7	3.7	5.5	60.2		
Sesame seed	5.5	6.5	20.3	45.6	7.1	15.0		
Sesame-seed cake	9.8	10.7	37.5	14.0	6.3	21.7		
Sorgo	12.8	2.1	9.1	3.6	2.6	69.8		
Soybean	8.0	4.8	38.9	18.0	4.8	25.5	0.22	0.67
Soybean meal:								
38–43 percent protein	7.8	5.8	41.7	5.8	6.2	32.7	0.29	0.67
43–48 percent protein	8.2	6.0	44.7	4.6	5.8	30.7	0.34	0.71
Sunflower seed	6.9	3.2	15.2	28.8	28.5	17.4		
Sunflower hulls	10.5	2.6	4.4	3.4	57.0	22.1		
Sunflower kernels	6.9	4.2	29.4	43.9	2.6	13.0		
Velvetbean	9.8	3.1	26.2	4.8	6.0	50.1		
Wheat	10.6	1.8	12.0	2.0	2.0	71.6	0.05	0.38
Wheat bran	9.4	6.4	16.4	4.4	9.9	53.5	0.10	1.14
Wheat, brown shorts	10.8	4.0	17.8	4.8	5.8	56.8		
Wheat-flour middlings	10.4	3.3	18.8	4.0	4.2	59.3	0.09	0.80
Wheat, gray shorts	11.0	4.1	17.5	4.4	5.4	57.6	0.08	0.86
Wheat, mixed-feed	9.9	4.4	18.2	4.4	6.9	56.1	0.11	0.96
Wheat, red-dog	11.1	2.2	18.3	3.4	2.3	62.7	0.12	0.83
Wheat, standard middlings	10.4	3.9	17.0	4.3	5.4	59.0	0.09	0.90
Wheat, white shorts	10.9	2.2	15.6	3.7	2.4	65.2		
Wheat waste, shredded	8.0	1.6	12.4	1.6	2.6	73.8		

Product	Moisture (%)	Ash (%)	Crude Protein (%)	Ether Extract (%)	Crude Fiber (%)	Nitrogen-Free Extract (%)	Calcium (%)	Phosphorus (%)
Dried Forages								
Alfalfa hay	7.2	8.0	15.4	1.6	30.3	37.5	1.51	0.21
Alfalfa-leaf meal	8.5	14.4	20.9	2.6	15.7	37.9	1.42	0.25
Alfalfa meal	8.2	10.0	15.2	2.2	27.5	36.9	1.56	0.22
Alfalfa meal, dehydrated	6.6	10.0	16.9	2.6	25.4	38.5		
Alfalfa-stem meal	9.1	7.7	11.4	1.3	36.1	34.4		
Alsike clover hay	10.5	8.8	14.4	2.5	24.7	39.1	0.78	0.20
Australian saltbush hay	6.7	16.9	16.1	1.8	21.5	37.0		
Barley hay	15.0	6.4	6.7	1.6	21.4	48.9	0.17	0.25
Barley straw	14.2	5.7	3.5	1.5	36.0	39.1		
Bermudagrass hay	8.9	7.9	7.2	1.7	24.9	49.4	0.60	0.16
Black grama hay	5.5	7.0	4.3	1.3	31.4	50.5	0.22	0.09
Blue grama hay	10.9	8.5	6.7	1.8	27.9	44.2		
Bluegrass hay, immature	7.3	7.9	15.2	3.0	23.7	42.9	0.45	0.35
Bluegrass hay, bloom	11.9	7.0	9.3	3.4	27.9	40.5	0.30	0.21
Bluejoint grass hay	7.5	6.9	6.7	3.0	34.2	41.7		
Bromegrass hay	14.0	9.7	9.3	1.8	26.6	38.6		
Buckwheat straw	9.9	5.5	5.2	1.3	43.0	35.1		
Buffalograss hay	6.2	10.8	5.6	1.7	26.1	49.6		
Burclover hay	8.7	12.3	15.7	3.0	25.5	34.8	1.11	0.15
Corncobs	10.7	1.4	2.4	0.5	30.1	54.9		
Corn fodder	11.8	5.8	7.4	2.4	23.0	49.6		
Corn husks	9.8	2.9	2.9	0.7	30.7	53.0		
Corn leaves	11.8	8.5	8.1	2.2	24.4	45.0		
Corn stalks	11.7	4.6	4.8	1.8	32.7	44.4		
Corn stover	10.7	6.1	5.7	1.5	30.3	45.7	0.45	0.10
Cowpea hay	9.7	12.9	17.5	2.8	20.5	36.6	1.84	0.25
Cowpea straw	9.7	5.3	7.4	1.3	41.5	34.8		
Crabgrass hay	9.0	7.9	6.5	2.2	32.1	42.3	0.33	0.17
Crimson clover hay	9.6	8.6	15.2	2.8	27.2	36.6	1.18	0.13
Feterita fodder	13.3	6.4	8.7	1.9	21.5	48.2	0.27	0.19
Field-pea hay	10.6	8.3	16.1	2.7	24.8	37.5		
Flax straw	6.2	3.8	7.8	2.1	46.9	33.2		
Hegari fodder	13.5	8.2	6.2	1.7	16.7	53.7	0.17	0.18
Hegari stover	15.1	9.7	4.5	1.9	26.6	42.2	0.38	0.09
Johnsongrass hay	7.2	7.2	8.1	2.8	30.4	44.3	0.55	0.40
Kafir fodder	9.1	7.8	6.6	2.1	28.4	46.0	0.31	0.05
Kafir stover	12.6	9.0	5.8	1.7	27.5	43.4		
Lespedeza hay	7.9	6.2	11.9	2.8	28.5	42.7	0.80	0.25
Little bluestem hay	8.6	4.9	4.0	1.6	35.4	45.5		
Meadow fescue hay	11.6	7.0	6.6	2.0	31.6	41.2		
Millet hay, foxtail	7.0	8.2	9.2	2.8	28.0	44.8		
Millet hay, pearl or cattail	10.1	9.7	9.0	1.8	32.3	37.1		
Natalgrass hay	7.5	4.8	3.7	1.4	39.5	43.1	0.49	0.32
Oatgrass, tall, hay	8.1	6.4	9.4	2.7	29.8	43.6		
Oat hay	11.8	5.7	6.1	2.4	27.1	46.9	0.27	0.22
Oat straw	8.1	7.6	4.4	2.5	36.2	41.2	0.23	0.20

(Continued)

Product	Moisture (%)	Ash (%)	Crude Protein (%)	Ether Extract (%)	Crude Fiber (%)	Nitrogen-Free Extract (%)	Calcium (%)	Phosphorus (%)
Dried Forages (*continued*)								
Orchardgrass hay, immature	9.9	6.0	8.1	2.6	32.4	41.0	0.31	0.18
Orchardgrass hay, mature	9.9	7.0	6.9	3.0	32.7	40.5		
Prairie hay (Colorado, Wyoming)	5.5	7.2	7.0	2.4	31.3	46.6		
Prairie hay (Kansas, Oklahoma)	9.5	7.5	4.4	2.3	30.4	45.9	0.55	0.07
Prairie hay (Minnesota, S. Dakota)	11.6	7.2	6.0	2.4	30.3	42.5	0.44	0.11
Red-clover hay	7.0	10.0	16.1	2.6	23.6	40.7	1.01	0.14
Red clover, mammoth, hay	12.2	7.5	12.8	3.3	27.1	37.1		
Redtop hay	8.9	5.2	7.9	1.9	28.6	47.5	0.35	0.18
Rhodesgrass hay	8.6	8.4	5.3	1.2	33.4	43.1		
Rice straw	8.9	13.5	4.5	1.6	34.0	37.5	0.18	0.05
Rye hay	6.4	4.7	5.9	2.0	37.4	43.6	0.27	0.22
Rye straw	7.1	3.2	3.0	1.2	38.9	46.6		
Ryegrass perennial, hay	10.2	8.6	8.6	4.1	24.5	44.0	0.17	0.11
Ryegrass, Italian, hay	8.5	6.9	7.5	1.7	30.5	44.9		
Ryegrass hay	8.3	8.5	6.3	2.0	33.0	41.9		
Sedge, western species	5.4	6.7	11.6	2.4	27.4	46.5		
Slender wheatgrass	7.5	6.6	7.8	2.1	30.8	45.2		
Sorgo fodder	11.6	6.0	5.3	2.4	26.0	48.7	0.27	0.15
Sorgo hay	5.8	9.5	9.5	1.9	26.8	46.5	0.31	0.09
Soybean hay	8.4	8.9	15.8	3.8	24.3	38.8	1.26	0.22
Soybean straw	8.7	7.4	5.7	2.5	34.6	41.1		
Sudangrass hay	5.3	8.1	9.7	1.7	27.9	47.3	0.47	0.24
Sweetclover hay	8.1	7.5	16.2	2.8	25.9	39.5	0.74	0.08
Sweetclover straw	5.1	3.4	6.7	1.2	49.6	34.0		
Timothy hay	7.1	5.8	7.5	2.9	30.2	46.5	0.31	0.13
Vetch, hairy, hay	13.1	8.4	20.9	2.7	24.2	30.7	0.25	0.30
Western needlegrass hay	9.9	6.2	5.5	2.7	33.2	42.5		
Western wheatgrass hay	8.6	8.7	8.4	2.3	31.9	40.1		
Wheat hay	9.6	4.2	3.4	1.3	38.1	43.4	0.14	0.15
Wheat straw	6.8	5.4	4.3	3.4	36.8	43.3		
White clover hay	7.2	9.4	15.6	2.2	22.7	42.9	1.31	0.28
Wiregrass hay	8.5	7.3	6.6	1.3	34.6	41.7		
Green Forages								
Alfalfa, immature	79.4	2.9	5.2	0.7	3.8	8.0	0.28	0.09
Alfalfa, in bloom	77.2	1.8	3.2	0.6	7.8	9.4	0.39	0.07
Alsike clover, immature	81.2	2.4	4.9	0.6	3.1	7.8	0.26	0.09
Alsike clover, in bloom	74.8	2.0	3.9	0.9	7.4	11.0	0.21	0.06
Barley, immature	83.4	1.5	2.8	0.7	3.6	8.0	0.06	0.07
Barley, mature	77.1	1.6	2.2	0.5	6.4	12.2	0.05	0.07
Bluegrass, Kentucky, immature	70.5	2.5	5.0	1.2	7.5	13.3	0.15	0.13
Bromegrass, immature	77.5	2.9	4.3	0.9	5.2	9.2	0.14	0.10
Cabbage	90.5	0.9	2.4	0.3	1.2	4.7	0.06	0.02
Canada bluegrass, immature	74.1	2.5	4.3	1.3	6.8	11.0	0.11	0.12

Product	Moisture (%)	Ash (%)	Crude Protein (%)	Ether Extract (%)	Crude Fiber (%)	Nitrogen-Free Extract (%)	Calcium (%)	Phosphorus (%)
Green Forages (*continued*)								
Corn fodder:								
Dent, immature	79.0	1.2	1.7	0.5	5.6	12.0		
Dent, mature	73.4	1.5	2.0	0.9	6.7	15.5		
Flint, immature	79.8	1.1	2.0	0.7	4.3	12.1		
Flint, mature	77.1	1.1	2.1	0.8	4.3	14.6		
Cowpea	82.5	2.5	3.4	0.5	4.0	7.1	0.18	0.05
Crimson clover	80.9	1.7	3.1	0.7	5.2	8.4	0.28	0.04
Kafir	73.0	2.0	2.3	0.7	6.9	15.1		
Lespedeza, Korean, immature	74.1	2.4	4.6	0.8	5.8	12.3	0.34	0.11
Meadow fescue, immature	78.8	2.6	4.0	0.9	4.7	9.0	0.15	0.11
Meadow foxtail, immature	73.9	2.8	4.5	1.2	5.6	12.0	0.15	0.12
Millet, foxtail	71.1	1.7	3.1	0.7	9.2	14.2	0.09	0.05
Oatgrass, tall, immature	78.4	3.0	4.3	1.0	4.6	8.7	0.11	0.13
Oats, immature	82.6	1.7	2.9	0.7	3.3	8.8	0.07	0.07
Oats, mature	72.0	2.1	2.7	0.9	7.4	14.9	0.08	0.08
Orchardgrass, immature	78.3	2.8	3.4	1.0	5.3	9.2	0.14	0.13
Orchardgrass, in bloom	73.0	2.0	2.6	0.9	8.2	13.3		
Pricklypear	78.9	4.3	.7	0.4	2.6	13.1		
Rape	85.7	2.0	2.4	0.6	2.2	7.1		
Red clover, immature	81.2	2.7	5.0	0.8	3.0	7.3	0.27	0.10
Red clover, in bloom	70.8	2.1	4.4	1.1	8.1	13.5	0.44	0.07
Red fescue, immature	70.5	2.8	4.1	0.9	8.2	13.5	0.16	0.13
Redtop, immature	76.8	2.8	4.1	0.9	5.4	10.0	0.15	0.10
Reed canarygrass, immature	80.7	2.4	3.5	0.7	4.3	8.4	0.13	0.10
Rye, immature	80.8	2.3	4.5	1.1	3.4	7.9	0.10	0.10
Rye, mature	76.6	1.8	2.6	0.6	11.6	6.8	0.08	0.06
Ryegrass, Italian, immature	77.3	2.5	3.5	1.0	5.2	10.5	0.13	0.12
Ryegrass, perennial, immature	75.9	3.0	3.8	0.9	5.4	11.0	0.15	0.12
Sorgo	77.3	1.3	1.5	1.0	6.2	12.7		
Soybean	73.9	2.9	4.0	1.1	7.6	10.5	0.28	0.05
Sweetclover, immature	75.3	2.2	5.3	0.7	6.7	9.8	0.26	0.07
Sweet corn	79.1	1.3	1.9	0.5	4.4	12.8		
Timothy, immature	74.9	2.3	4.1	0.9	5.4	12.4	0.12	0.11
Timothy, in bloom	61.6	2.1	3.1	1.2	11.8	20.2	0.13	0.05
Wheat, immature	82.3	2.1	3.8	0.9	3.0	7.9	0.07	0.10
Wheat, mature	68.7	2.6	2.4	0.7	8.6	17.0	0.06	0.08
White clover, immature	82.0	2.1	4.9	0.6	3.1	7.3	0.23	0.09
White clover, wild, immature	81.2	2.2	5.2	0.6	2.9	7.9	0.25	0.10
Silages, Roots, Tubers, and By-Products								
Alfalfa silage	68.9	2.7	5.7	1.0	8.8	12.9		
Alfalfa-molasses silage	68.6	3.4	5.8	1.0	8.4	12.8		
Beet pulp, dried	9.2	3.2	9.3	0.8	20.0	57.5	0.66	0.06

(Continued)

Product	Moisture (%)	Ash (%)	Crude Protein (%)	Ether Extract (%)	Crude Fiber (%)	Nitrogen-Free Extract (%)	Calcium (%)	Phosphorus (%)
Silages, Roots, Tubers, and By-Products (*continued*)								
Beet pulp, molasses, dried	8.0	5.2	11.6	0.7	16.4	58.1	0.59	0.09
Carrot	88.6	1.0	1.1	0.4	1.3	7.6		
Cassava	63.8	1.4	1.0	0.3	0.8	32.7		
Corn silage	73.8	1.7	2.1	0.8	6.3	15.3	0.08	0.08
Corn silage, immature	79.1	1.4	1.1	0.8	6.0	11.0		
Corn silage, mature	70.9	1.4	2.4	0.9	6.9	17.5		
Corn stover silage	80.7	1.8	1.8	0.6	5.6	9.5		
Cowpea silage	77.8	2.1	3.2	0.9	6.5	9.5		
Hegari silage	66.3	3.4	2.3	0.8	6.7	20.5		
Jerusalem artichoke	78.7	1.1	2.5	0.2	0.8	16.7		
Mangel-wurzel	90.8	1.0	1.4	0.2	0.9	5.7	0.02	0.02
Napiergrass silage	67.5	1.8	1.2	0.7	14.4	14.4	0.10	0.10
Parsnip	80.0	1.3	2.2	0.4	1.3	14.8		
Pea-vine silage	75.1	1.7	3.0	0.9	8.1	11.2		
Potato	78.9	1.0	2.1	0.1	0.6	17.3	0.01	0.06
Red-clover silage	72.0	2.6	4.2	1.2	8.4	11.6		
Rutabaga	88.6	1.2	1.2	0.2	1.3	7.5	0.05	0.04
Sorgo silage	74.7	1.4	1.6	1.0	6.9	14.4	0.09	0.04
Soybean silage	75.6	2.6	2.4	0.8	9.6	9.0	0.29	0.10
Sugarbeet	78.0	1.0	1.5	0.1	2.9	16.5	0.05	0.06
Sugarbeet pulp	90.5	0.4	.9	0.2	2.2	5.8		
Sunflower silage	77.9	2.1	1.8	1.6	6.5	10.1		
Sweetclover silage	70.2	2.9	6.1	1.0	9.7	10.1		
Sweet potato	71.1	1.0	1.5	0.4	1.3	24.7	0.02	0.05
Turnips	90.6	0.8	1.3	0.2	1.2	5.9	0.05	0.05

Source: From *USDA Yearbook*, 1939. For further information on the composition and digestibility of crops, crop products, and range plants, see *Atlas of nutritional data on United States and Canadian feeds.* Washington, DC: Natl. Acad. Sci. 1971 pp. 1–772

TABLE A.3 Conversion Tables

Mass

1 microgram (ug) = 0.000001 g
1 milligram (mg) = 0.001 g
1 centigram (cg) = 10 mg = 0.01 g
1 decigram (dg) = 100 mg = 0.1 g
1 gram (g) = 1,000 mg = 15.432356 grains
1 dekagram (dkg) = 10 g
1 hectogram (hg) = 100 g
1 kilogram (kg) = 1,000 g = 2.204622341 lb
1 metric quintal = 100 kg = 220.46 lb
1 metric ton = 1,000 kg = 2,204.62 lb

Avoirdupois Weights

1 grain (gr) = 64.798918 mgm
1 ounce (oz) = 437.5 gr = 28.349527 g
1 pound (lb) = 16 oz = 7,000 gr = 453.5924 g
1 short hundredweight = 100 lb
1 long hundredweight = 112 lb
1 short ton = 2,000 lb
1 long ton = 2,240 lb

Length

1 kilometer (km) = 1,000 m = 0.62137 mile
1 hectometer (hm) = 100 m = 328 ft 1 in
1 dekameter (dkm) = 10 m = 393.7 in
1 meter (m) = 1 m = 39.37 in = 3.28 ft
1 decimeter (dm) = 1/10 m = 3.937 in
1 centimeter (cm) = 1/100 m = 0.3937 in
1 millimeter (mm) = 1/1000 m = 0.03937 in
1 micron (u) = 1/1000 mm
1 millimicron (mu) = 1/1000μ
1 Angstrom (Å) = 1/10 mμ
1 mile = 5,280 ft = 1,760 yd = 320 rd = 80 chains
1 chain = 66 ft = 22 yd = 4 rd = 100 links
1 rod = 16½ ft = 5½ yd = 25 links
1 yard = 3 ft = 36 in = 0.9144 m
1 foot = 12 in = 30.48006 cm
1 link = 7.92 in
1 inch = 2.540005 cm = 25.4 mm

Area

1 centare (ca) = 1 square meter
1 are (a) = 100 sq m
1 hectare (ha) = 10,000 sq m = 2.47104 ac
1 labor (lah-bore') = 177⅓ ac
1 square league = 25 labors = 4,409 ac
1 Old Spanish league = 2.63 miles (Parts of Texas are surveyed in leagues and labors)
1 sq in = 6.4511626 sq cm
1 sq ft = 144 sq in
1 sq yd = 9 sq ft
1 sq rd = 1 perch = 30¼ sq yd = 272¼ sq ft
1 acre = 160 sq rd = 4,839 sq yd = 43,560 sq ft = 0.404687 ha
1 sq mi = 1 section = 640 ac
1 township = 36 sections = 23,040 ac

Volume

1 cubic centimeter (cc) = 0.06102338 cu in
1 cubic decimeter = 1,000 cc = 61.02338 cu in
1 cubic meter = 1,000,000 cc = 35.31445 cu ft
1 cu in = 16.387162 cc
1 cu ft = 1,728 cu in = 28,317.016 cc
1 cu yd = 27 cu ft = 46,656 cu in

Capacity (Liquid Measure)

1 gill = 118.292 ml
1 pt = 4 gi = 473.167 ml = 0.473167 1
1 qt = 2 pt = 8 gi = 0.946333 1
1 gal = 4 qt = 8 pt = 3.785332 1 = 231 cu in
1 liter (1) = 1,000 milliliters (ml or cc) = 2.11342 pt = 1.05671 qt
1 teaspoon = 4.93 ml
3 teaspoons = 1 tablespoon = 14.79 ml
2 tablespoons = 1 fluid oz = 29.578 ml
16 tablespoons = 8 fluid oz = 1 cup = 236.583 ml
2 cups = 16 fluid oz = 1 pt = 473.167 ml

Capacity (Dry Measure)

1 liter = 61.025 cu in = 0.908102 dry qt
1 dekaliter (dkl) = 10.1 = 0.28378 bu
1 hectoliter (hl) = 100 1 = 2.8378 bu
1 U.S. bu per ac = 0.8708 hl per ha
1 hl per ha = 1.1483 bu per ac
1 dry pt = 33.6003125 cu in = 0.550599 1
1 dry at = 2 dry pt = 1.101198 1
1 peck = 8 dry qt = 8.80958 1
1 bushel = 4 pk = 32 dry qt = 35.2383 1
1 Winchester (U.S.) bu = 1.244 cu ft = 2,150.42 cu in
1 Imperial (British) bu = 2,219.36 cu in = 1.0305 Winchester bu

Conversion Centigrade to Fahrenheit Temperature

$1°C = 1.8°F$ $1°F = 5/9°C$
$°C = (°F - 32) \times 5/9$ $°F = 9/5°C + 32$

°C	°F	°C	°F	°C	°F
100	212	50	122	0	32
90	194	40	104	−10	14
80	176	30	86	−20	−4
70	158	20	68	−30	−22
60	140	10	50	−40	−40

(Continued)

Water

1 gal = 8.355 lb

1 cu ft = 7.48 gal = 62.42 lb

1 acre-inch = 113 tons (approximately)

1 acre-foot = 43,560 cu ft = 323,136 gal

1 second foot = 1 cu ft per sec flow past a given point

1 second foot = 7½ gal per sec = 450 gal per min

1 second foot in 24 hours = 1.983 acre-feet

1 second foot in 1 hour = 1 acre-inch (approximately)

1 second foot = 40 miner's inches (in some states by statute)

1 second foot = 50 miner's inches (in other states)

Dry Soil

1 cu ft muck = 25 to 30 lb

1 cu ft clay and silt = 68 to 80 lb

1 cu ft sand = 100 to 110 lb

1 cu ft loam = 80 to 95 lb

1 cu ft average soil = 80 to 90 lb

1 acre foot (43,560 cu ft) = 3,500,000 to 4,000,000 lb (2,000 tons)

The soil surface plow depth (6⅔ in) is usually calculated as 2 million lb (1,000 tons) per acre.

The volume of compact soil increases about 20 percent when it is excavated or tilled.

Grain Storage

To compute the approximate capacity of a grain bin in bushels from the measurements of the bin in feet:

Rectangular bins—Length × width × height × 0.8 = bu

Round bins—Diameter squared × height × 5/8 = bu

Ear corn in crib:

Length × average width × average depth in feet × 0.4 = bu

Weights, Measures, and Conversion Factors Used by the U.S. Department of Agriculture

Barley

Flour 1 bbl (196 lb) ≈ 9 bu barley

Malt 1.1 bu (34 lb) ≈ 1 bu (48 lb) barley

Beans, dry 1 sack = 100 lb

Beer 1 bbl = 31 gal

Broomcorn 1 bale = 333 lb; 6 bales = 1 ton

Buckwheat flour 1 bbl (196 lb) ≈ 7 bu buckwheat

Corn

Meal (degermed) 1 bbl (196 lb) ≈ 6 bu corn

Meal (nondegermed) 1 bbl (196 lb) ≈ 4 bu corn

Ear corn (dry) 2 level bu or 3 heaping half-bushels weighing 70 lb ≈ 1 bu (56 lb) shelled corn

Cotton 1 lb ginned ≈ 2.86 lb unginned

Lint 1 bale (gross) = 500 lb

1 bale (net) = 480 lb

Flour 1 bbl = 196 lb 1 sack = 100 lb

Hops 1 gross bale = 200 lb

Jerusalem artichoke 1 bu = 50 lb

Linseed oil 2½ gal (19 lb) ≈ 1 bu flaxseed

Oatmeal 1 bbl (196 lb) ≈ 10 8/9 bu oats

Oil (corn, cottonseed, soybean, peanut, linseed, etc.) 1 gal = 7.7 lb (in trade), castor oil 1 gal = 8 lb

Potatoes 1 bu = 60 lb; 1 bbl = 165 lb; 1 sack = 100 lb

Rice

Rough 1 bag = 100 lb; 1 bbl = 162 lb

Milled 1 pocket (100 lb) ≈ 154 lb (3.42 bu) rough (undulled)

Rye flour 1 bbl (196 lb) ≈ 6 bu rye

Sorgo sirup 1 gal = 11.55 lb

Sugar 1 ton domestic 96° raw sugar ≈ 0.9346 tons refined sugar

Sugarcane sirup 1 gal = 11.35 lb

Sweet potatoes 1 bu = 55 lb

Tobacco 1 lb stemmed ≈ 1.33 lb unstemmed

Maryland (hogshead) = 600–800 lb

Flue-cured (hogshead) = 900–1,100 lb

Burley (hogshead) = 1,000–1,200 lb

Dark air-cured (hogshead) = 1,000–1,250 lb

Virginia fire-cured (hogshead) = 1,050–1,350 lb

Kentucky and Tennessee fire-cured (hogshead) = 1,350–1,650 lb

Cigar leaf (case) = 250–365 lb

(bale) = 150–175 lb

Turnips (without tops) 1 bu = 54 lb

Vegetables and other dry commodities 1 bbl = 7,056 cu in = 105 dry qt

Wheat flour 1 bbl (196 lb) ≈ 4.7 bu wheat

100 lb ≈ 2.33 bu wheat

(Bushel weights of grains and seeds are shown in Appendix Table A.1.)

Glossary

A horizon The surface and subsurface soil that contains most of the organic matter and is subject to leaching.

Abortive Imperfectly developed.

Abscission The natural separation of leaves, flowers, and fruits or buds from the stems or other plant parts by the formation of a special layer of thin-walled cells.

Achene (akene) A dry, hard, one-celled, one-seeded, indehiscent fruit to which the testa and pericarp are not firmly attached.

Acid soil A soil with a pH reaction of less than 7.0 (usually less than 6.6). An acid soil has a preponderance of hydrogen ions over hydroxyl ions. Litmus paper turns red in contact with moist acid soil.

Acuminate Gradually tapering to a sharp point, as in a leaf of grass.

Acute Sharp-pointed but less tapering than acuminate.

Adventitious Arising from an unusual position on a stem or at the crown of a grass plant (e.g., roots and buds).

Aerial roots Roots that arise from the stem above the ground.

Aftermath (rowen) The second or shorter growth of meadow plants in the same season after a hay or seed crop has been cut.

Aggregate A mass or cluster of soil particles or other small objects.

Ageotropic Lacking a geotropic response, as in stolons, rhizomes, and lateral roots that grow neither erect nor downward.

Agrobiology A phase of the study of agronomy dealing with the relation of yield to the quantity of an added or available fertilizer element.

Agroecology The study of the relation of crop varietal characteristics to their adaptation to environmental conditions.

Agrology The study of applied phases of soil science and soil management.

Agronomy The science of crop production and soil management. The name is derived from the Greek words *agros* (field) and *nomos* (to manage).

Aleurone The outer layer of the cells of the endosperm of the seed. It sometimes contains pigment.

Alfisol soil formed under native deciduous forests or savanna.

Alkali soil A soil, usually above pH 8.5, containing alkali salts in quantities that usually are deleterious to crop production.

Alkaline soil A soil with a pH above 7.0, usually above 7.3.

Alternate Placed singly rather than opposite (e.g., one leaf at a node), or flower parts of one whorl opposite to intervals of the next.

Ammonification Formation of ammonia or ammonium compounds, in soils.

Glossary terms and definitions adapted from (1) reports of committees on terminology of the American Society of Agronomy; (2) glossaries in USDA yearbooks for 1936, 1957, and 1961; and (3) various botanical glossaries.

Amylopectin   The branched-chain (waxy or glutinous) starch.

Amylose   The straight-chain fraction of normal starch.

Andisol   Soil formed from volcanic ash or cinders.

Angiosperms   The higher seed plants.

Annual   A plant that completes its life cycle from seed in one year.

Annular   Forming a ring; circular.

Anther   The part of the stamen that contain the pollen.

Anthesis   The period during which the flower is open and, in grasses, the period when the anthers are extended from the glumes.

Anthocyanin   A water-soluble plant pigment that produces many of the red, blue, and purple colors in plants.

Apetalous   Without petals.

Apiculate   Having a minute, pointed tip.

Apogeotropic   Turning upward in response to a stimulus opposed to the force of gravity (e.g., seedling stems growing erect).

Apomixis   A type of asexual production of seeds, as in Kentucky bluegrass.

Appressed   Pressed close to the stem.

Aquatic plant   A plant that lives in water.

Arid climate   A dry climate with an annual precipitation usually less than 10 inches (250 ml) and not suitable for crop production without irrigation.

Aridisol   Soil formed in arid regions.

Aristate   Awned.

Articulate   Jointed.

Arviculture   Crop science.

Asexual reproduction   Reproduction without involving the germ or sexual cells.

Ash   The nonvolatile residue resulting from complete burning of organic matter.

Association   A biologically balanced group of plants.

Attenuate   Gradually narrowing to a slender apex or base.

Auricles   Ear-shaped appendages, as at the base of barley and wheat leaves.

Auxins   Organic substances that cause stem elongation.

Awn   The beard or bristle extending from the tip or back of the lemma of a grass flower.

Axil   The angle between the leaf and stem or between a branch or pedicel and its main stem.

Axillary   In the axil.

Axis   The main stem of a flower or panicle.

B horizon   The subsoil layer in which certain leached substances (e.g., iron) are deposited.

Backcross   The cross of a hybrid with one of the parental types.

Banner   The upper or posterior, usually broad, petal of a papilionaceous legume flower; the standard.

Barbed   Furnished with rigid points or hooks.

Beak   A point or projection, as on the glume of the wheat flower.

Beard   The awn of grasses.

Bed   (1) A narrow, flat-topped ridge on which crops are grown with a furrow on each side for drainage of excess water. (2) An area in which seedlings or sprouts are grown before transplanting.

Bed up   To build up beds or ridges with a tillage implement.

Biennial   Of two years' duration; a plant germinating in one growing season and producing seed the next.

Binder   A machine for cutting a crop and tying it into bundles with twine.

Bine   A twining vine stem, as in the hop.

Binomial   The two Latin names for a plant species.

Biometry   The application of statistical methods to biological problems.

Biotechnology   Biological techniques developed through basic research that now applied to research and product development; mostly in the context of the use of recombinant DNA, cell fusion, and other bioprocessing techniques by industry.

Blade   The part of the leaf above the sheath.

Blind cultivation   Cultivating with a harrow, weeder, rotary hoe, or other implement to kill weeds before a seeded or planted crop has come up.

Boll   The subspherical or ovoid fruit of flax or cotton.

Bolt   The formation of an elongated stem or seed stalk as in the sugarbeet.

Boot (1) The upper leaf sheath of a grass. (2) The stage at which the inflorescence expands the boot.

Brace root An aerial root that functions to brace the plant, as in corn.

Bract A modified leaf subtending a flower or flower branch.

Bran The coat of a caryopsis of a cereal removed in milling consisting of the pericarp, testa, and usually the aleurone layer.

Branch A lateral stem.

Bristle A short stiff hair.

Broadcast To sow or scatter seed on the surface of the land by hand or by machinery.

Browse (1) The leaves and twigs of woody plants eaten by animals. (2) To bite off and eat leaves and twigs of woody plants.

Bud An unexpanded flower or a rudimentary leaf stem or branch.

Bulb A leaf bud with fleshy scales, usually subterranean.

Bur A prickly or spiny fruit envelope.

Bush-and-bog A heavy cut-away disk-tillage implement used on rough or brushy pasture land.

C horizon The layer of weathered parent rock material below the B horizon of the soil but above the unweathered rock.

Calcareous soil An alkaline soil containing sufficient calcium and magnesium carbonate to cause visible effervescence when treated with hydrochloric acid.

Caliche A more or less cemented deposit of calcium carbonate often mixed with magnesium carbonate at various depths characteristic of many of the semiarid and arid soils of the Southwest.

Callus A hard protuberance at the base of the lemma in certain grasses.

Calyx The outer part of the perianth composed of sepals.

Cambium The growing layer of the stem.

Capillary (1) Hair-like. (2) A very small tube.

Capitate With a globose head.

Capsule (pod) Any dry dehiscent fruit composed of more than one carper.

Carbohydrates The chief constituents of plants including sugars, starches, and cellulose in which the ratio of hydrogen molecules to oxygen molecules is 2:1.

Carbonate accumulation zone A visible zone of accumulated calcium and magnesium carbonate at the depth to which moisture usually penetrated in semiarid soils before they were brought under cultivation.

Carinate Keeled.

Carotene A yellow pigment in green leaves and other plant parts (the precursor of vitamin A). Beta carotene has the formula $C_{40}H_{56}$.

Carpel A simple pistil, or one element of a compound pistil.

Caruncle A strophiole.

Caryopsis The grain or fruit of grasses.

Catkin A spike with scale-like bracts.

Cell The unit of structure in plants. A living cell contains protoplasm, which includes a nucleus and cytoplasm within the cell walls.

Cellulose (1) Primary cell-wall substance. (2) A carbohydrate having the general formula $(C_6H_{10}O_5)n$.

Cereal A grass cultivated for its edible seeds or grains.

Chaff (1) A thin scale bract or glume. (2) The glumes and light plant-tissue fragments broken in threshing.

Chalaza That part of the ovule where all parts grow together.

Chartaceous Having the texture of paper.

Check row A row of seeds or plants spaced equally in both directions.

Chisel A tillage implement with points about a foot apart that stir the soil to a depth of 10 to 18 inches.

Chlorophyll The green coloring matter of plants that takes part in the process of photosynthesis. It occurs in the chloroplasts of the plant cell.

Chlorosis The yellowing or blanching of leaves and other chlorophyll-bearing plant parts caused by a lack of chlorophyll.

Chromosome That which carries genes (or factors), the units of heredity. Chromosomes are dark-staining bodies visible under the microscope in the nucleus of the cell at the time of cell division.

Ciliate Fringed with hairs on the margin.

Clavate Club-shaped.

Clay   Small mineral soil particles less than 0.002 millimeter in diameter (formerly less than 0.005 millimeter).

Cleistogamous   Fertilized without opening of the flowers.

Climate   The total long-time characteristic weather of any region.

Clone   A group of organisms composed of individuals propagated vegetatively from a single original individual.

Coleoptile   The sheath covering the first leaf of a grass seedling as it emerges from the soil.

Coleorhiza   A sheath covering the tip of the first root from a seed.

Collar   The area on the outer side of the leaf at the junction of the sheath and blade.

Colloid   A fine particle usually 10–6 to 10–4 millimeters in diameter, which carries an electric charge. A wet mass of colloidal particles is glue-like in consistency.

Columella   The axis of a capsule which persists after dehiscence.

Combine   A machine for harvesting and threshing in one operation.

Companion crop   A crop sown with another crop, as a small grain with forage crops. Formerly called a nurse crop.

Conservation tillage   A collective term used to describe tillage operations that are designed to leave plant residues on the soil surface.

Consumptive use   The use of water in growing a crop, including water used in transpiration and evaporation.

Continental climate   Climate typical of the interior of large land masses having wide extremes of temperature.

Contour furrows   Furrows plowed at right angles to the slope, at the same level or grade, to intercept and retain runoff water.

Convolute   Rolled longitudinally, with one part inside and one outside.

Cordate   Heart-shaped with point upward.

Coriaceous   Leathery.

Corm   The hard, swollen base of the stem.

Corolla   The inner part of the perianth, composed of petals.

Correlation   A relation between two variable quantities such that an increase or decrease of one is associated (in general) with an increase or decrease in the other.

Correlation coefficient (r)   The degree of correlation, which ranges from +1 to −1. A correlation coefficient of zero means that the two variables are not interrelated. An r value of +1 or −1 indicates complete association. A positive (plus) correlation means that high values of one variable are associated with high values of the other. A negative (minus) correlation means that as one variable increases the other tends to decrease.

Corymb   A flat-topped or convex flower cluster in which the pedicels are unequal and the outer flowers blossom earliest.

Cotyledons   (1) The first leaves of a plant as found in the embryo. (2) The major portion of the two halves of a pea or bean (legume) seed.

Cover crop   A crop grown between orchard trees or on the field between cropping seasons to protect the land from leaching and erosion.

Crenate   Dentate (toothed) with rounded teeth.

Cross   (1) A hybrid. (2) To produce a hybrid.

Cross drill   To drill seed in two directions, usually at right angles to each other.

Cross-fertilization or cross-pollination   Fertilization secured by pollen from another plant.

Crown   The base of the stem where roots and/or fillers arise.

Culm   The jointed, hollow stem of grasses.

Cultivar   See *variety*.

Cure   To prepare for preservation by drying or other processes.

Cuticle   The outer corky or waxy covering of the plant.

Cutin   A complex fatty or waxy substance in icell walls particularly in the epidermal layers.

Cutting   A part of a plant to be rooted for vegetative propagation.

Cyme   A somewhat flat-topped determinate inflorescence.

Cytology   The study of the structure function and life history of the cell.

Cytoplasm   The contents of a cell outside the nucleus.

Cytoplasmic male sterility Male sterility caused by the cytoplasm rather than by nuclear genes; transmitted only through the female parent.

Deciduous Plants or trees that shed leaves or awns at a particular season or stage.

Decumbent Curved upward from a horizontal or inclined position.

Deflocculate To separate or break down soil aggregates of clay into their individual particles.

Dehiscence The opening of valves or anthers or separation of parts of plants.

Denitrification The reduction of nitrates to nitrites ammonia and free nitrogen in the soil.

Dentate Toothed.

Detassel To remove the tassel.

Determinate inflorescence Flowers that arise from the terminal bud and check the growth of the axis.

Determinate plant A plant that completes vegetative growth before flowering.

Diadelphous (stamens) Collected in two sets.

Dicotyledonous plants Plants producing two cotyledons in each embryo.

Differentiation The process whereby cells and tissues become structurally unlike during the process of growth and development.

Digitate Fingered; compound with parts radiating from the apex of support.

Dioecious Having stamens and pistils in separate flowers upon different plants.

Diploid Having two sets of chromosomes. Body tissues of plants are ordinarily diploid.

Disarticulating Separating at maturity.

DNA Deoxyribonucleic acid. The material inside the cell nucleus that carries information concerning genetic traits.

Dominant Possessing a character that is manifested in the hybrid to the apparent exclusion of the contrasted character from the other (the recessive) parent.

Dormancy An internal condition of a seed or bud that prevents its prompt germination or sprouting under normal growth conditions.

Dorsal Relating to the back or outer surface.

Double cross The result of mating two single crosses, each of which had been produced by crossing two distinct inbred lines.

Drill (1) A machine for sowing seeds in furrows. (2) To sow in furrows.

Drill row A row of seeds or plants sown with a drill.

Duck-foot cultivator A field cultivator equipped with small sweep shovels.

Echinate Armed with prickles.

Ecology The study of the mutual relations between organisms and their environment.

Ecotype A variety or strain adapted to a particular environment.

Edaphology Soil science, particularly the influence of soil upon vegetation.

Egg The female reproductive cell.

Elluviation Removal of material by solution or suspension.

Emarginate Notched at the end.

Embryo The rudimentary plantlet within a seed; the germ.

Embryo sac The sac in the embryo containing the egg cell.

Emergence Coming out of a place, as a seedling from the soil or a flower from a bud.

Endemic Indigenous or native to a restricted locality.

Endocarp Inner layer of pericarp.

Endosperm The starchy interior of a grain.

Ensilage Silage.

Ensile To make into silage.

Entisol Soil with little or no profile development. Very recently deposited.

Epicarp (exocarp) The outer layer of pericarp formed from the ovary wall.

Epicotyl The stem of the embryo or young seedling above the cotyledons.

Epidermis External layer of cells.

Epigynous flower Flower with a corolla that seems to rise from the top of the ovary.

Erosion The wearing away of the land surface by water, wind, or other forces.

Ether extracts Fats, oils, waxes, and similar products extracted with warm ether in chemical analysis.

Exotic plant An introduced plant not fully naturalized or acclimated.

Exserted Protruding beyond a covering.

F_1 The first filial generation; the first-generation offspring of a given mating.

F_2 The second filial generation; the first hybrid generation in which segregation occurs.

Factor (genetic or hereditary) The gene or unit of heredity.

Fallow Cropland left idle, usually for one growing season, while the soil is being cultivated to control weeds and conserve moisture.

Fasicle A compact bundle or cluster of flowers or leaves.

Fertile plant A plant capable of producing fruit.

Fertility (plant) The ability to reproduce sexually.

Fertility (soil) The ability to provide the proper compounds in the proper amounts and in the proper balance for the growth of specified plants under the suitable environment.

Fertilization (plant) The union of the male (pollen) nucleus with the female (egg) cell.

Fertilization (soil) The application to the soil of elements or compounds that aid in the nutrition of plants.

Fibrous root A slender thread-like root, as in grasses.

Filament The stalk of the stamen which bears the anther.

Filiform Thread-shaped.

Fleshy root A thickened root containing abundant food reserves (e.g., carrot, sweet potato, and sugarbeet).

Float (1) A land leveller. (2) A plank clod masher.

Flocculate To aggregate individual particles into small groups or granules.

Floret Lemma and palea with included flower (stamens, pistil, and lodicules).

Fodder Maize, sorghum, or other coarse grasses harvested whole and cured in an erect position.

Foliate Leaved.

Forage Vegetable matter, fresh or preserved, gathered and fed to animals.

Friable Easily crumbled in the fingers; nonplastic.

Fruit The ripened pistil.

Fungicide A chemical substance used as a spray, dust, or disinfectant to kill fungi infesting plants or seeds.

Fungus A group of the lower plants that causes most plant diseases. The group includes the molds and belongs to the division *Thallophyta*. Fungi reproduce by spores instead of seeds, contain no chlorophyll, and thus live on dead or living organic matter.

Fusiform Feather-shaped, enlarged in the middle, and tapering at both ends.

Gellisol Soil with permafrost or churning from permafrost.

Gene The unit of inheritance, which is transmitted in the germ cells.

Genetic engineering The process of directly modifying, removing, or adding genes to a DNA molecule in order to change the information it contains.

Genetics The science of heredity, variation, sex determination, and related phenomena.

Geniculate Bent abruptly or knee-like, as the awn of wild oats.

Genotype The hereditary make up of characteristics of a plant or a pure line or variety.

Geotropic Turning downward in response to a stimulus caused by the force of gravity (e.g., the roots of a seedling growing downward).

Germ cell A cell capable of reproduction or of sharing in reproduction.

Glabrous Smooth.

Glandular Containing or bearing glands or gland cells.

Glaucous Covered with a whitish, waxy bloom, as on the sorghum stem.

Glumes The pair of bracts at the base of a spikelet.

Gluten The protein in wheat flour that enables the dough to rise.

Glutinous Glue-like.

Gopher (1) A crop-damaging burrowing rodent. (2) A type of plow.

Grain (1) A caryopsis. (2) A collective term for the cereals. (3) Cereal seeds in bulk.

Grass A plant of the family *Poaceae*.

Green manure Any crop or plant grown and plowed under to improve the soil, especially by addition of organic matter.

Gully erosion Erosion that produces channels in the soil.

Gynophore A pedicel bearing the ovary (e.g., the peg which penetrates the soil and bears the peanut pod).

Habit Aspect or manner of growth.

Halophyte A plant tolerant of soils high in soluble salts.

Haploid Single; containing a reduced number of chromosomes in the mature germ cells of bisexual organisms.

Hardpan   A hardened or cemented soil horizon.

Haulm   A stem.

Hay   The herbage of grasses or comparatively fine-stemmed plants cut and cured for forage.

Head   A dense, roundish cluster of sessile or nearly sessile flowers on a very short axis or receptable, as in red clover and sunflowers.

Heliotropic   Turning towards the sun.

Herb   A plant that contains little wood and that dies down to the ground each year.

Herbaceous   Having the characteristics of an herb. No woody growth.

Herbage   Herbs, collectively, especially the aerial portion.

Herbicide   A weed killer or any chemical substance used to kill herbaceous plants.

Hermaphrodite   Perfect, having both stamens and pistils in the same flower.

Heterosis   Increased vigor, growth, or fruitfulness that results from crosses between genetically different plants, lines, or varieties.

Heterozygous   Containing two unlike genes of an allelomorphic pair in the corresponding loci (positions) of a pair of chromosomes. The progeny of a heterozygous plant does not breed true.

Hexaploid   A plant or crop that carries three genomes (sets of paired haploid chromosomes).

Hilum   The scar of the seed. Its place of attachment.

Hirsute   Hairy.

Histosol   Soil with high organic soil material extending over a foot deep.

Homogamy   The maturing of anthers and stigmas at the same time.

Homonym   The same name for two different plants.

Homozygous   Containing like germ cells. Homozygous plants produce like progeny for the character under observation.

Horizon, soil   A layer of soil approximately parallel to the land surface with more or less well-defined characteristics. See also *A horizon, B horizon,* and *C horizon.*

Hormone   A chemical growth-regulating substance that can be or is produced by a living organism.

Hull   (1) A glume, lemma, palea, pod, or other organ enclosing a seed or fruit. (2) To remove hulls from a seed.

Humid climate   A climate with sufficient precipitation to usually support forest vegetation. In the agricultural sections of the United States the precipitation in humid regions usually exceeds 30 to 40 inches (750 to 1,000 ml) annually.

Humus   The well-decomposed, more or less stable part of the organic matter of the soil.

Husk   (1) The coarse outer envelope of a fruit, like the glumes of an ear of maize. (2) To remove the husks.

Hyaline   Thin and translucent or transparent.

Hybrid   The offspring of two parents unlike in one or more heritable characters.

Hybridization   The process of crossing organisms of unlike heredity.

Hydrophyte   A plant adapted to wet or submerged conditions.

Hydroponics   The growing of plants in aqueous chemical solutions.

Hypocotyl   The stem of the embryo or young seedling below the cotyledons.

Hypogynous   With parts under the pistil.

Imperfect flower   Flower lacking either stamens or pistils.

Inbred   Resulting from successive self-fertilization.

Indehiscent   Not opening by valves or slits.

Indeterminate inflorescence   An inflorescence in which the flowers arise laterally and successively as the floral axis elongates.

Indeterminate plant   A plant that continues to grow vegetatively after flowering begins.

Indurate   Hard.

Inflorescence   The flowering part of the plant.

Insecticide   A chemical used to kill insects.

Inseptisol   Soil with the beginning of profile development.

Integuments   Coats or walls of an ovule.

Internode   The part of the stem or branch between two nodes.

Intertilled crop   A crop planted in rows, followed by cultivation between the rows.

Involucre   A circle of bracts below a flower or flower cluster.

Ion   An electrically charged element, group of elements, or particle.

Joint   (1) A node. (2) The internode of an articulate rachis. (3) To develop distinct nodes and internodes in a grass culm.

Keel (1) A ridge on a plant part resembling the keel of a boat, as the glume of durum wheat. (2) The pair of united petals in a legume flower.

Kernel The matured body of an ovule.

Lamina The blade of a leaf, particularly those portions between the veins.

Laminated In layers or plates.

Lanceolate Lance-shaped, as a grass leaf.

Land plane A large, wheeled machine used to level the land.

Laterally compressed Flattened from the sides, as the spike of emmer.

Lax Loose.

Leach To remove materials by solution.

Leaf The lateral organ of a stem.

Legume (1) Any plant of the family *Fabaceae*. (2) The pod of a leguminous plant.

Lemma The outer (lower) bract of a grass spikelet enclosing the caryopsis.

Lethal substance A substance or hereditary factor causing death.

Ley A pasture used for a few seasons and then plowed for other crops.

Lignin A constituent of the woody portion of the fibrovascular bundles in plant tissues, made up of modified phenyl propane units.

Ligule (I) A membranous projection on the inner side of a leaf at the top of the sheath of wheat and many other grasses. (2) A strap-shaped corolla as in the ray flower of the sunflower.

Lime Calcium oxide or quicklime (CaO); often, also, calcium carbonate ($CaCO_3$), or calcium hydroxide, hydrated or slaked lime ($Ca[OH]_2$).

Limestone Rock composed essentially of calcium carbonate.

Linear Long and narrow, with nearly parallel margins.

Lister An implement for furrowing land, often having a planting attachment.

Loam A soil composed of a mixture of clay, silt, and less than 52 percent sand.

Lobe A rounded portion or segment of any organ.

Lodicules The pair of organs at the base of the ovary of a grass floral that swell and force open the lemma and palea during anthesis.

Loess Geological deposit of relatively uniform fine material, mostly silt, presumably transported by wind.

Longevity Length of life, usually of seeds or plants of longer than average life.

Marl A crumbling deposit of calcium carbonate mixed with clay or other impurities.

Maslin (meslin) Grains grown in mixture or the milled product thereof.

Meadow An area covered with fine-stemmed forage plants, wholly or mainly perennial, and used to produce hay.

Meiosis The division of the sexual cells in which the number of chromosomes is halved.

Mellow soil A soil that is easily worked or penetrated.

Membrane A thin, soft tissue.

Mesocarp Middle layer of pericarp.

Mesocotyl The subcrown internode of a grass seedling.

Mesophyll Tissues between two epidermal layers, as in the interior of a leaf.

Mesophyte A plant that thrives under medium conditions of moisture and salt content of the soil.

Metabolism The life processes of plants.

Metaxenia Direct effect of the pollen on the parts of a seed or fruit other than the embryo and endosperm.

Microclimate (1) A local climatic condition that differs from surrounding areas because of differences in relief, exposure, or cover. (2) Experimental, controlled plant-growth chambers.

Micronutrient A mineral nutrient element that plants need only in trace or minute amounts.

Micropyle Opening through which the pollen tube passes.

Middlebuster A double shovel plow or lister.

Minimum tillage A collective term used to describe tillage operations that are designed to leave plant residues on the soil surface.

Mitosis Cell division involving the formation and longitudinal splitting of the chromosomes.

Moisture tension The force at which water is held by the soil. One atmosphere equivalent tension equals a unit column of water of 1,000 centimeters.

Mollisol Soil formed under prairie grasses.

Monadelphous Uniting stamens into one set.

Monocotyledon A plant having one cotyledon, as in the grasses.

Monoecious Having stamens and pistils in separate flowers on the same plant.

Morphological Referring to structure or texture.

Mow (mo) Cut with a mower or scythe.

Mow (mou) (1) A place for indoor hay storage. (2) To place hay in a mow.

Muck Fairly well decomposed organic soil material relatively high in mineral content, dark in color, and accumulated under conditions of imperfect drainage.

Mulch A layer of plant residues or other materials on the surface of the soil. Formerly, a loose layer of cultivated surface soil.

Mutation A sudden variation that is later passed on through inheritance.

Native pasture A pasture covered with native plants or naturalized exotic plants.

Naturalized plants Introduced species that have become established in a region.

Necrosis Discoloration, dehydration, and death of plant parts.

Nectary Any gland or organ that secretes sugar.

Nematocide A substance that can be used to kill nematodes.

Nerve A vein on a leaf, glume, or lemma.

Neutral soil A soil neither acid nor alkaline, with a pH of about 7.0 or between 6.6 and 7.3.

Nitrification Formation of nitrates from ammonia.

Nitrogen fixation The conversion of atmospheric (free) nitrogen to nitrogen compounds brought about chemically, by soil organisms, by organisms living in the roots of legumes.

Nitrogen-free extract The unanalyzed substance of a plant (consisting largely of carbohydrates) remaining after the protein, ash, crude fiber, ether extract, and moisture have been determined.

No tillage Seeding directly into crop residue without previous tillage. Zero tillage.

Node The joint of a culm where a leaf is attached.

Nodule A tubercle formed on legume roots by symbiotic nitrogen-fixing bacteria of the genus *Rhizobium*.

Nucellus The ovule tissue within the integuments.

Nucleolus A small spherical body within a cell nucleus.

Nucleus A body of specialized protoplasm containing the chromosomes within a cell.

Nurse crop See *companion crop*.

Nutrient A chemical element taken into a plant that is essential to the growth, development, or reproduction of the plant.

Obovate Egg-shaped with larger end above.

Obovoid Inversely ovoid; roughly egg-shaped, with narrow end downwards.

Obtuse Rounded at the apex.

Oceanic climate A climate modified by the tempering effect of ocean water.

One-way A tillage implement having a gang of disks that throw the soil in one direction.

Organic farming Growing crops without applying pesticides and mineral fertilizers in an inorganic form.

Osmosis Diffusion of substances through a cell wall or other membrane.

Outcross A cross to an individual not closely related.

Ovary Seed case of the pistil.

Ovoid Egg-shaped.

Ovules Unripe seeds in the ovary.

Oxidation A chemical change involving the addition of oxygen or its chemical equivalent or involving an increase in positive or decrease in negative valence. Burning is oxidation.

Oxisol Soil that is highly weathered. Found in tropical regions.

Palea (palet) Inner (upper) bract of a floret lying next to the caryopsis in grasses. It usually is thin and papery.

Palisade cells Elongated cells perpendicular to the epidermis on the upper side of a leaf.

Palmate Radiately lobed or divided.

Panicle An inflorescence with a main axis and subdivided branches, as in oats and sorghum.

Papilionaceous Butterfly-shaped, like the flower of legumes.

Pappus The teeth, bristles, awns, and so on that surmount the achene of the sunflower and other *Compositae*.

Parasitic Living in or on another living organism.

Parenchyma Soft cellular tissue.

Parthenogenesis The development of a new individual from a germ cell without fertilization.

Pasture An area of land covered with grass or other herbaceous forage plants, used for grazing animals.

Pasture renovation   Improvement of a pasture by tillage, seeding, fertilization, and, sometimes, liming.

Pasture succession   A series of crops for grazing in succession.

Peat   Slightly decomposed organic matter accumulated under conditions of excessive moisture.

Pedicel   (1) A branch of an inflorescence supporting one or more flowers. (2) The stalk of a spikelet.

Pedology   Soil science.

Peduncle   The top section of the stalk that supports a head or panicle.

Pepo   The fruit of cucurbitaceous plants, as the pumpkin, squash, and gourd.

Perennial   Living more than one year but, in some cases, producing seed the first year.

Perfect flower   A flower having both pistil and stamens.

Perianth   The floral envelope including the calyx or calyx and corolla.

Pericarp   The modified and matured ovary wall, like the bran layers of a grain.

Perigynous   Located around the pistil.

Perisperm   Nucellus.

Permanent pasture   A pasture of perennial or self-seeding animal plants kept for grazing indefinitely.

Petal   A division of the corolla.

Petiole   The stalk of a leaf.

pH   The designation for degree of acidity or hydrogen-ion activity. pH value = logarithm 1/CH, in which CH is the concentration of active hydrogen ions expressed in gram ions per liter of a solution.

Phenotype   The organism as exemplified by its expressed characters but not necessarily as all its progeny will appear.

Phloem   Portion of a vascular bundle containing the sieve Pubes through which are transported the food materials manufactured in the plant leaves.

Photoperiodism   The response of plants to different day lengths or light periods.

Photothermal induction   The initiation of flowering in a plant resulting from the combined effect of light and temperature.

Phototropism   The growing or turning toward the light.

Phyllotaxy   The arrangement of leaves upon the stem.

Phytobezoar   A ball in the intestinal tract of an animal, formed from plant hairs.

Phytotoxic   Injurious to plant life or life processes.

Pick-up   An attachment for a combine or other implement to gather cut crops from a windrow and convey them to the machine.

Pilose   Covered with soft slender hairs.

Pinnate leaf   A compound leaf with leaflets arranged on each side of a common petiole, as in the pea leaf.

Pistil   The seed-bearing organ of a flower consisting of the ovary, style, and stigma.

Pistillate   Provided with pistils but without stamens.

Placenta   The part of the ovary to which the ovules are attached, as in the pea pod.

Plant   (1) Any organism belonging to the plant or vegetable kingdom. (2) To set plants or sow seeds.

Planter   A machine for opening the soil and dropping tubers, cuttings, seedlings, or seeds at intervals.

Plastid   A protoplasmic body in a cell that may or may not contain chlorophyll.

Plumose   Feather-like.

Pod   Any dry dehiscent fruit.

Pollen   The male germ cells produced in the anthers.

Pollination   The transfer of pollen from the anther to the stigma.

Polyadelphous (stamens)   Separate, or in more than two groups.

Polycross   A selection, clone, or line naturally out-crossed through random pollination by all other strains in the same isolated block.

Polyploid   A plant having three or more basic sets of chromosomes.

Primary root   (1) A seminal root. (2) A main root.

Procambium   Fibrovascular tissue formed before the differentiation into xylem and phloem.

Productivity (of soil)   The capability of a soil to produce a specified plant or sequence of plants under a specified system of management.

Profile (of soil)   A vertical section of the soil through all its horizons and extending into the parental material.

Protandrous   Having anthers that shed their pollen before the stigmas are receptive.

Protein   Nitrogenous compounds formed in plants, composed of 50 to 55 percent carbon, 19 to 24 percent oxygen, 6.5 to 7.3 percent hydrogen, 15 to 17.6 percent nitrogen, usually 0.5 to 2.2 percent sulfur, and 0.4 to 0.9 percent phosphorus. Carbohydrates, the chief constituents of carbonaceous feeds, contain only 40 to 44 percent carbon.

Protogynous   Having pistils ready for fertilization before anthers are matured.

Protoplasm   The contents of a living cell.

Puberulent   Minutely pubescent.

Pubescent   Covered with fine, soft, short hairs.

Pulse   Leguminous plants or their seeds, chiefly those plants with large seeds used for food.

Pulvinus   The swelling at the base of the branches of some panicles, which causes them to spread.

Punctate   Dotted.

Pure line   A strain of organisms that is genetically pure (homozygous) because of continued inbreeding, self-fertilization, or other means.

Race   A group of individuals having certain characteristics in common because of common ancestry—generally a subdivision of a species.

Raceme   An inflorescence in which the pediceled flowers are arranged on a rachis or axis.

Rachilla (little rachis)   Tile axis of a spikelet in grasses.

Rachis   The axis of a spike or raceme.

Radicle   That part of tile seed that, upon vegetating, becomes the root.

Range   An extensive area of natural pastureland. If unfenced, it is an open range.

Ray   The branch of an umber; marginal ligulate flowers of a composite head.

Reaction (of soil)   Tile degree of acidity or alkalinity of the soil expressed as pH.

Receptacle   The swollen summit of flower stalk.

Recessive   Possessing a transmissible character that does not appear in the first-generation hybrid because it is masked due to the presence of the dominant character coming from the other parent.

Reciprocal cross   A cross between the same two strains but with the pollen and distillate parents reversed.

Recombinant DNA   DNA that has been altered by joining genetic material from two different sources, usually from two different species.

Replication   Multiple repetition of an experiment.

Respiration   The process of absorption of oxygen and giving out of carbon dioxide.

Reticulate   In a network.

Retting   A process for separating fibers in plant stems, usually by soaking in water.

Rhizome   A subterranean stem, usually rooting at the nodes and rising at the apex; a rootstock.

Rill erosion   Erosion producing small channels that can be obliterated by tillage.

Rogue   (1) A variation from the type of a variety or standard, usually inferior. (2) To eliminate such inferior individuals.

Root   The part of the plant (usually subterranean) that lacks nodes.

Root cap   A mass of cells protecting the tip of a root.

Root hair   A single-celled protrusion of an epidermal cell of a young root.

Rootlet   A small root.

Rootstock   A rhizome.

Rosette   A cluster of spreading or radiating basal leaves.

Rotation pasture (ley)   A pasture used for a few seasons and then plowed for other crops.

Rudimentary   Underdeveloped.

Rugose   Wrinkled; rough with wrinkles.

Runner   A creeping branch or stolon.

Rush   A plant of the family *Juncaceae*.

Sac   A pouch.

Sagittate   Shaped like an arrowhead.

Saline soil   A soil containing an excess of soluble salts, more than approximately 0.2 percent, but with a pH of less than 8.5.

Salt   The product, other than water, of the reaction of a base with an acid.

Sand   Small rock or mineral fragments having diameters ranging from 0.05 to 2.0 millimeters.

Saprophytic   Living on dead organic material.

Scabrous   Having a rough surface.

Scale   Any thin appendage; morphologically, a modified degenerate leaf.

Sclerenchyma   Lignified tissue in referring to thick-walled fibers.

Scraper   (1) A machine for beating the spikelets from broomcorn fibers. (2) A blade for removing soil from revolving disks or wheels on field implements.

Scutellum   A shield-shaped organ surrounding the embryo of a grass seed, morphologically a cotyledon.

Second bottom   The first terrace level of a stream valley above the flood plain.

Secondary root   A branch or division of a main root.

Sedge   A plant of the family *Cyperaceae*, especially of the genus *Carex*.

Seed   (1) The ripened ovule enclosing a rudimentary plant and the food necessary for its germination. (2) To produce seed. (3) To sow.

Seedling   (1) The juvenile stage of a plant grown from seed. (2) A plant derived from seed (in plant breeding).

Segregation   Separation of hybrid progenies into the different hereditary types representing the combination of characters of the two parents.

Selection   The choosing of plants having certain characteristics for propagation. Natural selection occurs under competitive conditions or adverse environment.

Selfed (self-pollinated)   Pollinated by pollen from the same plant.

Semiarid climate   A climate which, in the United States, usually has an annual precipitation of between 10 and 20 inches or more. Prevailing plants in a semiarid region are short grass, bunchgrass, or shrubs.

Seminal   Belonging to the seed.

Seminal root   A root arising from the base of the hypocotyl.

Sepal   A division of the calyx.

Septum   A partition or dividing wall.

Serrate   Having sharp teeth pointing forward.

Sessile   Without a pedicel or stalk.

Setaceous   Bristle-like.

Sheath (boot)   The lower part of the leaf that encloses the stem.

Sheet erosion   Erosion by removal of a more or less uniform layer of material from the land surface.

Shock   (1) An assemblage or pile of crop sheaves or cut stalks set together in the field to dry. (2) To set into shocks.

Shoot   (1) A stem with its attached members. (2) To produce shoots. (3) To put forth.

Siblings (sibs)   Offspring of the same parental plants.

Silage   Forage preserved in a succulent condition by partial fermentation in a tight container.

Silo   A tight-walled structure for making and preserving silage.

Silt   Small mineral soil particles of a diameter of 0.002 to 0.05 millimeter.

Simple   (1) In botany—without subdivisions, as opposed to compound (e.g., a simple leaf). (2) In heredity—inheritance of a character controlled by not more than three factor-pair differences.

Single cross   The first-generation hybrid between two inbred lines.

Sinuous   Wavy.

Slip   A cutting, shoot, or leaf to be rooted for vegetative propagation.

Sod   (1) Turf. (2) Plowed meadow or pasture.

Soil   The natural medium for the growth of land pleats on the surface of the earth, composed of organic and mineral materials.

Solum   The upper part of the soil profile; the A and B horizons.

Sow   To place seeds in a position for growing.

Spatulate   Shaped like the tip of a spatula.

Specific combining ability   The response in vigor, growth, or fruitfulness from crossing with other particular inbred lines.

Spicate   Arranged in a spike.

Spike   An unbranched inflorescence in which the spikelets are sessile on the rachis, as in wheat and barley.

Spikelet   The unit of inflorescence in grasses, consisting of two glumes and one or more florets.

Spodosol   Soil formed from coarse-textured, acid parent material.

Spore   Single-celled reproductive bodies produced by fungi.

Sport   Abrupt deviation from type of a plant.

Sprout   (1) A young shoot. (2) To produce sprouts. (3) To put forth sprouts from seeds. (4) To remove sprouts, as from potato tubers.

Square   An unopened flower bud of cotton with its subtending involucre bracts.

Stalk   A stem.

Stamen   The pollen-bearing organ of a flower.

Staminate   Having stamens but no pistils.

Stand   The density of plant population per unit area.

Standard   See *banner.*

Stellate   Star-shaped.

Sterile   Incapable of sexual reproduction.

Stigma   The part of the pistil that receives the pollen.

Stipule   One of two leaf-like appendages arranged in a pair at the base of the petiole.

Stock   A supply of seed of a crop or variety.

Stolon   A modified propagating, creeping stem above ground that produces roots.

Stoloniferous   Bearing stolons.

Stoma   An opening in the epidermis for the passage of gases and water vapor.

Stomata   Plural of stoma.

Stool   The aggregate of a stem and its attached tillers (i.e., a clump of young stems arising from a single plant).

Strain   A group of plants derived from a variety.

Straw   The dried remnants of fine-stemmed plants from which the seed has been removed.

Striate   Marked with parallel lines or ridges.

Strip cropping   The growing of dense crops alternating with intertilled crops or fallow in long, narrow strips across a slope approximately on a line of contour or across the direction of prevailing winds.

Strobile   An inflorescence or fruit with conspicuous imbricated bracts, as in the hop.

Strophiole (caruncle)   A swollen appendage near the micropyle of a seed.

Structure   The morphological aggregates in which the individual soil particles are arranged.

Stubble   The basal portion of the stems of plants left standing after cutting.

Style   The slender part of the pistil supporting the stigma.

Suberize   To form a protective corky layer on a cut surface of plant tissue.

Subhumid climate   A climate with sufficient precipitation to support a moderate to dense growth of tall and short grasses but usually unable to support a dense deciduous forest. In the United States, this usually means a precipitation of 20 to 30 inches (500 to 750 ml) or more.

Subsoil   That part of the solum below plow depth or below the A horizon.

Subspecies   A taxonomic rank immediately below that of a species.

Succulent   Juicy, fleshy.

Sucker   (1) A tiller. A shoot produced from a crown or rhizome, or, in tobacco, from axillary buds. (2) To produce suckers. (3) To remove suckers.

Surface soil   The upper 5 to 8 inches (13 to 20 cm) of the soil, or, in arable soils, the depth commonly stirred by the plow.

Suture   The line of junction between contiguous parts.

Sward   Fine grassy surface of a pasture.

Swath   A strip of cut herbage lying on the stubble.

Sweat   To emit moisture as does damp hay or grain, usually with some heating taking place at the same time.

Sweep   A double-bladed, V-shaped knife on a cultivating implement.

Symbiotic nitrogen fixation   The fixation of nitrogen by bacteria infesting the roots of legumes while benefiting the legume crop.

Synonym   A different name for the same species or variety.

Tame pasture   A pasture covered with cultivated plants and used for grazing.

Taproot   A single central root.

Tassel   (1) The staminate inflorescence of maize composed of panicled spikes. (2) To produce tassels.

Taxonomy   The science of classification.

Tedder   An implement for stirring hay in the swath or windrow.

Temporary pasture   A pasture grazed during a short period only—that is, not more than one crop season.

Tendril   A leaflet or stem modified for climbing or anchorage, as in the pea.

Terete   Cylindrical and slender, as in grass culms.

Terminal   Situated at or forming the extremity or upper bud, flower, or leaf.

Terrace   A channel or embankment across a slope approximately on a contour to intercept runoff water.

Testa　The seed coat.

Tetraploid　Having four times the primary chromosome number.

Till　To plow or cultivate soil.

Tiller　(1) An erect shoot arising from the crown of a grass. (2) To produce tillers.

Tilth　The physical condition of the soil with respect to its fitness for the planting or growth of a crop.

Topsoil　The surface soil, usually the plow depth or the A horizon.

Tow　The coarse and broken fibers left after separation from whole fibers.

Transpiration　The evaporation of moisture through the leaves.

Trichome　A plant hair.

Truncate　Ending abruptly as if cut off across the top.

Tuber　A short, thickened subterranean branch.

Turf　The upper stratum of soil filled with the roots and stems of low-growing living plants, especially grasses.

Turgid　Distended with water or swollen.

Ultisol　Soil with a well-developed clay horizon.

Umbel　An indeterminate type of inflorescence in which the peduncles or pedicles of a cluster seem to rise from the same point, as in a carrot-flower cluster.

Undulate　Having a wavy surface.

Unisexual　Containing either stamens or pistils, but not both.

Unit character　A hereditary trait that is transmitted by a single gene.

Valve　One of the parts of a dehiscent fruit or a piece into which a capsule splits.

Variation　The occurrence of differences among individuals of a species or variety.

Variety (cultivar)　A group of individuals within a species that differs from the rest of the species.

Vein　A bundle of threads of fibrovascular tissue in a leaf or other organ.

Venation　The arrangement of veins.

Ventral　On the lower or front side.

Verrucose　Warty.

Verticillate　Whorled.

Vertisol　Soil with high concentrations of swelling and cracking clays.

Villous　Bearing long, soft, straight hairs.

Virus　Ultramicroscopic protein bodies, the presence of certain types of which causes mosaic and other diseases in plant tissues. Viruses reproduce in plant tissues.

Viscid　Sticky.

Weed　A plant that, in its location, is more harmful than beneficial.

Whorl　An arrangement of organs in a circle around the stem.

Windrow　(1) Curing herbage dropped or raked into a row. (2) To cut or rake into windrows.

Wing　The lateral petal of the papilionaceous flower of a legume.

Winter annual　A plant that germinates in the fall and blooms in the following spring or summer.

Xerophyte　A plant adapted to arid conditions.

Xylem　The woody part of a fibrovascular bundle containing vessels, the water-conducting tissue.

Zero tillage　Seeding directly into crop residue without previous tillage. No tillage.

Zygote　Product of united gametes.

Index